Introduction to Fungi

John Webster
University of Exeter

and

Roland Weber
University of Kaiserslautern

Third Edition

CAMBRIDGE
UNIVERSITY PRESS

CAMBRIDGE UNIVERSITY PRESS

Cambridge, New York, Melbourne, Madrid, Cape Town, Singapore, São Paulo

Cambridge University Press
The Edinburgh Building, Cambridge CB2 2RU, UK

Published in the United States of America by Cambridge University Press,
New York

www.cambridge.org
Information on this title: www.cambridge.org/9780521807395

© J. Webster and R. W. S. Weber 2007

First published 2007

Printed in the United Kingdom at the University Press, Cambridge

A catalogue record for this publication is available from the British Library

Library of Congress Cataloguing-in-Publication data

Webster, John, 1925-
Introduction to fungi / John Webster and Roland Weber. − 3rd ed.
    p. cm.
Includes bibliographical references and index.
ISBN 0-521-80739-5 (hardback) − ISBN 0-521-01483-2 (pbk.)
1. Fungi. 2. Fungi−Classification. I. Weber, Roland, 1968- II. Title.
QK603.W4 2006
579.5−dc22

                                                    2006036496

ISBN-13   978-0-521-80739-5   hardback
ISBN-10   0-521-80739-5   hardback

ISBN-13   978-0-521-01483-0   paperback
ISBN-10   0-521-01483-2   paperback

To Philip M. Booth

Contents

Colour plate section appears between pages 412 and 413

Preface to the first edition

There are several available good textbooks of mycology, and some justification is needed for publishing another. I have long been convinced that the best way to teach mycology, and indeed all biology, is to make use, wherever possible, of living material. Fortunately with fungi, provided one chooses the right time of the year, a wealth of material is readily available. Also by use of cultures and by infecting material of plant pathogens in the glasshouse or by maintaining pathological plots in the garden, it is possible to produce material at almost any time. I have therefore tried to write an introduction to fungi which are easily available in the living state, and have tried to give some indication of where they can be obtained. In this way I hope to encourage students to go into the field and look for fungi themselves. The best way to begin is to go with an expert, or to attend a Fungus Foray such as those organized in the spring and autumn by mycological and biological societies. I owe much of my own mycological education to such friendly gatherings. A second aim has been to produce original illustrations of the kind that a student could make for himself from simple preparations of living material, and to illustrate things which he can verify for himself. For this reason I have chosen not to use electron micrographs, but to make drawings based on them.

The problem of what to include has been decided on the criterion of ready availability. Where an uncommon fungus has been included this is because it has been used to establish some important fact or principle. A criticism which I must accept is that no attempt has been made to deal with Fungi Imperfecti as a group. This is not because they are not common or important but that to have included them would have made the book much longer. To mitigate this shortcoming I have described the conidial states of some Ascomycotina rather fully, to include reference to some of the form-genera which have been linked with them. A more difficult problem has been to know which system of classification to adopt. I have finally chosen the 'General Purpose Classification' proposed by Ainsworth, which is adequate for the purpose of providing a framework of reference. I recognize that some might wish to classify fungi differently, but see no great merit in burdening the student with the arguments in favour of this or that system.

Because the evidence for the evolutionary origins of fungi is so meagre I have made only scant reference to the speculations which have been made on this topic. There are so many observations which can be verified, and for this reason I have preferred to leave aside those which never will.

The literature on fungi is enormous, and expanding rapidly. Many undergraduates do not have much time to check original publications. However, since the book is intended as an introduction I have tried to give references to some of the more recent literature, and at the same time to quote the origins of some of the statements made.

Exeter, 27 April 1970   J.W.

Preface to the second edition

In revising the first edition, which was first published about ten years ago, I have taken the opportunity to give a more complete account of the Myxomycota, and to give a more general introduction to the Eumycota. An account has also been given of some conidial fungi, as exemplified by aquatic Fungi Imperfecti, nematophagous fungi and seed-borne fungi. The taxonomic framework has been based on Volumes IVA and IVB of Ainsworth, Sparrow and Sussman's *The Fungi: An Advanced Treatise* (Academic Press, 1973).

Exeter, January 1979   J.W.

Preface to the third edition

Major advances, especially in DNA-based technology, have catalysed a sheer explosion of mycological knowledge since the second edition of *Introduction to Fungi* was published some 25 years ago. As judged by numbers of publications, the field of molecular phylogeny, i.e. the computer-aided comparison of homologous DNA or protein sequences, must be at the epicentre of these developments. As a result, information is now available to facilitate the establishment of taxonomic relationships between organisms or groups of organisms on a firmer basis than that previously assumed from morphological resemblance. This has in turn led to revised systems of classification and provided evidence on which to base opinions on the possible evolutionary origin of fungal groups. We have attempted to reflect some of these advances in this edition. In general we have followed the outline system of classification set out in *The Mycota* Volume VII (Springer-Verlag) and the *Dictionary of the Fungi* (ninth edition, CABI Publishing). However, the main emphasis of our book remains that of presenting the fungi in a sensible biological context which can be understood by students, and therefore some fungi have been treated along with taxonomically separate groups if these share fundamental biological principles. Examples include *Microbotryum*, which is treated together with smut fungi rather than the rusts to which it belongs taxonomically, or *Haptoglossa*, which we discuss alongside *Plasmodiophora* rather than with the Oomycota.

Molecular phylogeny has been instrumental in clarifying the relationships of anamorphic fungi (fungi imperfecti), presenting an opportunity to integrate their treatment with sexually reproducing relatives. There are only a few groups such as nematophagous fungi and the aquatic and aero-aquatic hyphomycetes which we continue to treat as ecological entities rather than scattered among ascomycetes and basidiomycetes. Similarly, the gasteromycetes, clearly an unnatural assemblage, are described together because of their unifying biological features.

However, in all these cases taxonomic affinities are indicated where known. We have also included several groups now placed well outside the Fungi, such as the Oomycota (Straminipila) and Myxomycota and Plasmodiophoromycota (Protozoa). This is because of their biological and economic importance and because they have been and continue to be studied by mycologists.

There have been major advances in other areas of research, notably the molecular cell biology of the two yeasts *Saccharomyces* and *Schizosaccharomyces*, 'model organisms' which have a bearing far beyond mycology. Further, much exciting progress is being made in elucidating the molecular aspects of the infection biology of human and plant pathogens, and in developing fungi for biotechnology. These trends are represented in the current edition. Nevertheless, the fundamental concept of *Introduction to Fungi* remains that of the previous two editions: to place an organism in its taxonomic context while discussing as many relevant aspects of its biology as possible in a holistic manner. Many of the illustrations are based on original line drawings because we believe that these can readily portray an understanding of structure and that drawing as a record of interpretation is a good discipline. However, we have also extended the use of photographs, and we now provide illustrated life cycles because these are more easily understood. As before, our choice of illustrated species has been influenced by the ready availability of material, enabling students and their teachers to examine living fungi, which is a cornerstone of good teaching. At their first introduction most technical terms have been printed in bold, their meanings explained and their derivations given. The page numbers where these definitions are given have been highlighted in the index.

The discipline of mycology has evolved and diversified so enormously in recent decades that it is now a daunting task for individual authors to give a balanced, integrated account of the fungi. Of course, there will be omissions or misrepresentations in a work of this scale, and we offer our apologies to those who feel that their work or that of others has not been adequately covered. At the same time, it has been a fascinating experience for us to write this book, and we have thoroughly enjoyed the immense diversity of approaches and ideas which make mycology such a vibrant discipline at present. We hope to have conveyed some of its fascination to the reader in the text and by referring to as many original publications as possible.

Exeter and Kaiserslautern, 1 March 2006

J.W. and R.W.S.W.

Acknowledgements

We are indebted to many people who have helped us in our extensive revisions to *Introduction to Fungi*. This edition is dedicated to Mr Philip M. Booth in profound gratitude for his financial support and his encouragement over many years. We have acknowledged in the figure legends the many friends and colleagues who have responded so enthusiastically to our call for help by providing us with illustrations, sometimes previously unpublished, and we thank numerous publishing houses for permission to include published figures. We thank Caroline Huxtable and Rob Ford (Exeter University Library) and Jennifer Mergel and Petra Tremmel (Kaiserslautern University Library) for help beyond the call of duty in obtaining inter-library loans. Dr Wolf-Rüdiger Arendholz and Dr Roger T.A. Cook have read the entire manuscript or parts of it, and their feedback and corrections have been most valuable to us. We are immensely grateful to Professors Heidrun and Timm Anke (Kaiserslautern) for their support of this project, their encouragement and for providing such a stimulating environment for research and teaching of fungal biology.

By far the heaviest toll has been paid by our families and friends who have had only cursory sightings of us during the past six years. We owe a debt of gratitude to them for their patient forbearance and unwavering support.

Introduction

1.1 | What are fungi?

About 80 000 to 120 000 species of fungi have been described to date, although the total number of species is estimated at around 1.5 million (Hawksworth, 2001; Kirk *et al.*, 2001). This would render fungi one of the least-explored biodiversity resources of our planet. It is notoriously difficult to delimit fungi as a group against other eukaryotes, and debates over the inclusion or exclusion of certain groups have been going on for well over a century. In recent years, the main arguments have been between taxonomists striving towards a phylogenetic definition based especially on the similarity of relevant DNA sequences, and others who take a biological approach to the subject and regard fungi as organisms sharing all or many key ecological or physiological characteristics – the 'union of fungi' (Barr, 1992). Being interested mainly in the way fungi function in nature and in the laboratory, we take the latter approach and include several groups in this book which are now known to have arisen independently of the monophyletic 'true fungi' (**Eumycota**) and have been placed outside them in recent classification schemes (see Fig. 1.25). The most important of these 'pseudofungi' are the Oomycota (see Chapter 5). Based on their lifestyle, fungi may be circumscribed by the following set of characteristics (modified from Ainsworth, 1973):

1. *Nutrition*. Heterotrophic (lacking photosynthesis), feeding by absorption rather than ingestion.

2. *Vegetative state*. On or in the substratum, typically as a non-motile mycelium of hyphae showing internal protoplasmic streaming. Motile reproductive states may occur.

3. *Cell wall*. Typically present, usually based on glucans and chitin, rarely on glucans and cellulose (Oomycota).

4. *Nuclear status*. Eukaryotic, uni- or multinucleate, the thallus being homo- or heterokaryotic, haploid, dikaryotic or diploid, the latter usually of short duration (but exceptions are known from several taxonomic groups).

5. *Life cycle*. Simple or, more usually, complex.

6. *Reproduction*. The following reproductive events may occur: sexual (i.e. nuclear fusion and meiosis) and/or parasexual (i.e. involving nuclear fusion followed by gradual de-diploidization) and/or asexual (i.e. purely mitotic nuclear division).

7. *Propagules*. These are typically microscopically small spores produced in high numbers. Motile spores are confined to certain groups.

8. *Sporocarps*. Microscopic or macroscopic and showing characteristic shapes but only limited tissue differentiation.

9. *Habitat*. Ubiquitous in terrestrial and freshwater habitats, less so in the marine environment.

10. *Ecology*. Important ecological roles as saprotrophs, mutualistic symbionts, parasites, or hyperparasites.

11. *Distribution*. Cosmopolitan.

With photosynthetic pigments being absent, fungi have a heterotrophic mode of nutrition. In contrast to animals which typically feed by ingestion, fungi obtain their nutrients by extra-cellular digestion due to the activity of secreted enzymes, followed by absorption of the solubilized breakdown products. The combination of extracellular digestion and absorption can be seen as the ultimate determinant of the fungal lifestyle. In the course of evolution, fungi have conquered an astonishingly wide range of habitats, fulfilling important roles in diverse ecosystems (Dix & Webster, 1995). The conquest of new, often patchy resources is greatly facilitated by the production of numerous small spores rather than a few large propagules, whereas the colonization of a food source, once reached, is achieved most efficiently by growth as a system of branching tubes, the **hyphae** (Figs. 1.1a,b), which together make up the **mycelium**.

Hyphae are generally quite uniform in different taxonomic groups of fungi. One of the few features of distinction that they do offer is the presence or absence of cross-walls or **septa**. The Oomycota and Zygomycota generally have aseptate hyphae in which the nuclei lie in a common mass of cytoplasm (Fig. 1.1a). Such a condition is described as **coenocytic** (Gr. *koinos* = shared, in common; *kytos* = a hollow vessel, here meaning cell). In contrast, Asco- and Basidiomycota and their associated asexual states generally have septate hyphae (Fig. 1.1b) in which each segment contains one, two or more nuclei. If the nuclei are genetically identical, as in a mycelium derived from a single uninucleate spore, the mycelium is said to be **homokaryotic**, but where

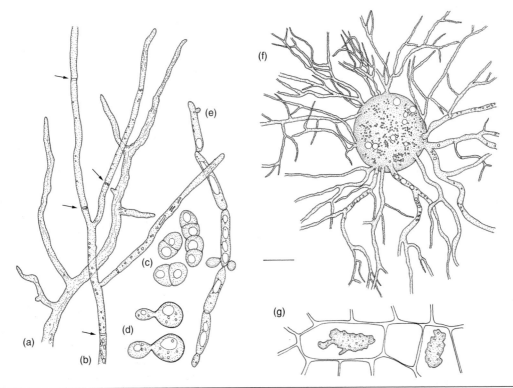

Fig 1.1 Various growth forms of fungi. (a) Aseptate hypha of *Mucor mucedo* (Zygomycota). The hypha branches to form a mycelium. (b) Septate branched hypha of *Trichoderma viride* (Ascomycota). Septa are indicated by arrows. (c) Yeast cells of *Schizosaccharomyces pombe* (Ascomycota) dividing by binary fission. (d) Yeast cells of *Dioszegia takashimae* (Basidiomycota) dividing by budding. (e) Pseudohypha of *Candida parapsilosis* (Ascomycota), which is regarded as an intermediate stage between yeast cells and true hyphae. (f) Thallus of *Rhizophlyctis rosea* (Chytridiomycota) from which a system of branching rhizoids extends into the substrate. (g) Plasmodia of *Plasmodiophora brassicae* (Plasmodiophoromycota) inside cabbage root cells. Scale bar = 20 μm (a,b,f,g) or 10 μm (c–e).

a cell or mycelium contains nuclei of different genotype, e.g. as a result of fusion (**anastomosis**) of genetically different hyphae, it is said to be **heterokaryotic**. A special condition is found in the mycelium of many Basidiomycota in which each cell contains two genetically distinct nuclei. This condition is **dikaryotic**, to distinguish it from mycelia which are **monokaryotic**. It should be noted that septa, where present, are usually perforated and allow for the exchange of cytoplasm or organelles.

Not all fungi grow as hyphae. Some grow as discrete **yeast** cells which divide by fission (Fig. 1.1c) or, more frequently, budding (Fig. 1.1d). Yeasts are common, especially in situations where efficient penetration of the substratum is not required, e.g. on plant surfaces or in the digestive tracts of animals (Carlile, 1995). A few species, including certain pathogens of humans and animals, are **dimorphic**, i.e. capable of switching between hyphal and yeast-like growth forms (Gow, 1995). Intermediate stages between yeast cells and true hyphae also occur and are termed **pseudohyphae** (Fig. 1.1e). Some lower fungi grow as a **thallus**, i.e. a walled structure in which the protoplasm is concentrated in one or more centres from which root-like branches (**rhizoids**) ramify (Fig. 1.1f). Certain obligately plant-pathogenic fungi and fungus-like organisms grow as a naked **plasmodium** (Fig. 1.1g), a uni- or multinucleate mass of protoplasm not surrounded by a cell wall of its own, or as a **pseudoplasmodium** of amoeboid cells which retain their individual plasma membranes. However, by far the most important device which accounts for the typical biological features of fungi is the hypha (Bartnicki-Garcia, 1996), which therefore seems an appropriate starting point for an exploration of these organisms.

1.2 | Physiology of the growing hypha

1.2.1 Polarity of the hypha

By placing microscopic markers such as small glass beads beside a growing hypha, Reinhardt (1892) was able to show that cell wall extension,

measured as an increase in the distance between two adjacent markers, occurred only at the extreme apex. Four years earlier, H. M. Ward (1888), in an equally simple experiment, had collected liquid droplets from the apex of hyphae of *Botrytis cinerea* and found that these 'ferment-drops' were capable of degrading plant cell walls. Thus, the two fundamental properties of the vegetative fungal hypha – the polarity of both growth and secretion of degradative enzymes – have been known for over a century. Numerous studies have subsequently confirmed that 'the key to the fungal hypha lies in the apex' (Robertson, 1965), although the detailed mechanisms determining hyphal polarity are still obscure.

Ultrastructural studies have shown that many organelles within the growing hyphal tip are distributed in steep gradients, as would be expected of a cell growing in a polarized mode (Girbardt, 1969; Howard, 1981). This is visible even with the light microscope by careful observation of an unstained hypha using phase-contrast optics (Reynaga-Peña *et al.*, 1997), and more so with the aid of simple staining techniques (Figs. 1.2a–d). The cytoplasm of the extreme apex is occupied almost exclusively by secretory vesicles and microvesicles (Figs. 1.2a, 1.3). In the higher fungi (Asco- and Basidiomycota), the former are arranged as a spherical shell around the latter, and the entire formation is called the **Spitzenkörper** or 'apical body' (Fig. 1.4c; Bartnicki-Garcia, 1996). The Spitzenkörper may be seen in growing hyphae even with the light microscope. Hyphae of the Oomycota and some lower Eumycota (notably the Zygomycota) do not contain a recognizable Spitzenkörper, and the vesicles are instead distributed more loosely in the apical dome (Fig. 1.4a,b). Hyphal growth can be simulated by means of computer models based on the assumption that the emission of secretory vesicles is coordinated by a 'vesicle supply centre', regarded as the mathematical equivalent of the Spitzenkörper in higher fungi. By modifying certain parameters, it is even possible to generate the somewhat more pointed apex often found in hyphae of Oomycota and Zygomycota (Figs. 1.4a,b; Diéguez-Uribeondo *et al.*, 2004).

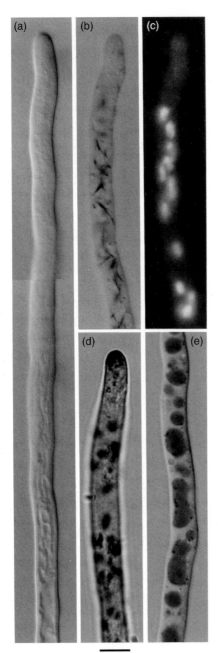

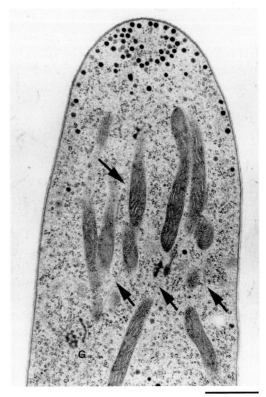

1 μm

Fig 1.3 Transmission electron microscopy of a hyphal tip of *Fusarium acuminatum* preserved by the freeze-substitution method to reveal ultrastructural details. The vesicles of the Spitzenkörper as well as mitochondria (dark elongated organelles), a Golgi-like element (G) and microtubules (arrows) are visible. Microtubules are closely associated with mitochondria. Reproduced from Howard and Aist (1980), by copyright permission of The Rockefeller University Press.

5 μm

Fig 1.2 The organization of vegetative hyphae as seen by light microscopy. (a) Growing hypha of *Galactomyces candidus* showing the transition from dense apical to vacuolate basal cytoplasm. Tubular vacuolar continuities are also visible. (b–e) Histochemistry in *Botrytis cinerea*. (b) Tetrazolium staining for mitochondrial succinate dehydrogenase. The mitochondria appear as dark filamentous structures in subapical and maturing regions. (c) Staining of the same hypha for nuclei with the fluorescent DNA-binding dye DAPI. The apical cell contains numerous nuclei. (d) Staining of acid phosphatase activity using the Gomori lead-salt method with a fixed hypha. Enzyme activity is localized both in the secretory vesicles forming the Spitzenkörper, and in vacuoles. (e) Uptake of Neutral Red into vacuoles in a mature hyphal segment. All images to same scale.

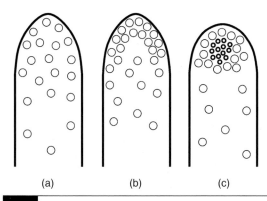

(a) (b) (c)

Fig 1.4 Schematic drawings of the arrangement of vesicles in growing hyphal tips. Secretory vesicles are visible in all hyphal tips, but the smaller microvesicles (chitosomes) are prominent only in Asco- and Basidiomycota and contribute to the Spitzenkörper morphology of the vesicle cluster. (a) Oomycota. (b) Zygomycota. (c) Ascomycota and Basidiomycota.

A little behind the apical dome, a region of intense biosynthetic activity and energy generation is indicated by parallel sheets of endoplasmic reticulum and an abundance of mitochondria (Figs. 1.2b, 1.3). The first nuclei usually appear just behind the biosynthetic zone (Fig. 1.2c), followed ultimately by a system of ever-enlarging vacuoles (Fig. 1.2d). These may fill almost the entire volume of mature hyphal regions, making them appear empty when viewed with the light microscope.

1.2.2 Architecture of the fungal cell wall

Although the chemical composition of cell walls can vary considerably between and within different groups of fungi (Table 1.1), the basic design seems to be universal. It consists of a structural scaffold of fibres which are cross-linked, and a matrix of gel-like or crystalline material (Hunsley & Burnett, 1970; Ruiz-Herrera, 1992; Sentandreu et al., 1994). The degree of cross-linking will determine the plasticity (extensibility) of the wall, whereas the pore size (permeability) is a property of the wall matrix. The scaffold forms the inner layer of the wall and the matrix is found predominantly in the outer layer (de Nobel et al., 2001).

In the Ascomycota and Basidiomycota, the fibres are **chitin** microfibrils, i.e. bundles of linear β-(1,4)-linked N-acetylglucosamine chains (Fig. 1.5), which are synthesized at the plasma membrane and extruded into the growing ('nascent') cell wall around the apical dome. The cell wall becomes rigid only after the microfibrils have been fixed in place by cross-linking. These cross-links consist of highly branched **glucans** (glucose polymers), especially those in which the glucose moieties are linked by β-(1,3)- and β-(1,6)-bonds (Suarit et al., 1988; Wessels et al., 1990; Sietsma & Wessels, 1994). Such β-glucans are typically insoluble in alkaline solutions (1 M KOH). In contrast, the alkali-soluble glucan fraction contains mainly α-(1,3)- and/or α-(1,4)-linked branched or unbranched chains (Wessels et al., 1972; Bobbitt & Nordin, 1982) and does not perform a structural role but instead contributes significantly to the cell wall matrix (Sietsma & Wessels, 1994). Proteins represent the third important chemical

Table 1.1. The chemical composition of cell walls of selected groups of fungi (dry weight of total cell wall fraction, in per cent). Data adapted from Ruiz-Herrera (1992) and Griffin (1994).

Group	Example	Chitin	Cellulose	Glucans	Protein	Lipid
Oomycota	*Phytophthora*	0	25	65	4	2
Chytridiomycota	*Allomyces*	58	0	16	10	?
Zygomycota	*Mucor*	9*	0	44	6	8
Ascomycota	*Saccharomyces*	1	0	60	13	8
	Fusarium	39	0	29	7	6
Basidiomycota	*Schizophyllum*	5	0	81	2	?
	Coprinus	33	0	50	10	?

*Mainly chitosan.

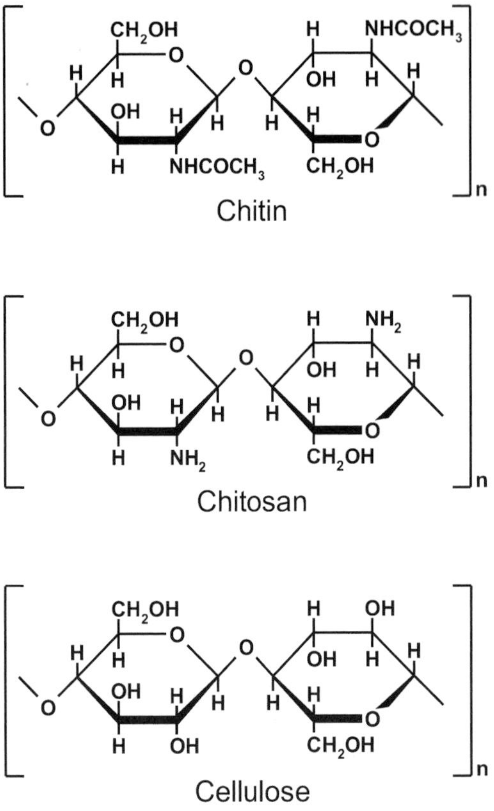

Fig 1.5 Structural formulae of the principal fibrous components of fungal cell walls.

constituent of fungal cell walls. In addition to enzymes involved in cell wall synthesis or lysis, or in extracellular digestion, there are also structural proteins. Many cell wall proteins are modified by glycosylation, i.e. the attachment of oligosaccharide chains to the polypeptide. The degree of glycosylation can be very high, especially in the yeast *Saccharomyces cerevisiae*, where up to 90% of the molecular weight of an extracellular protein may be contributed by its glycosylation chains (van Rinsum *et al.*, 1991). Since mannose is the main component, such proteins are often called **mannoproteins** or mannans. In *S. cerevisiae*, the pore size of the cell wall is determined not by matrix glucans but by mannoproteins located close to the external wall surface (Zlotnik *et al.*, 1984). Proteins exposed at the cell wall surface can also determine surface properties such as adhesion and recognition (Cormack *et al.*, 1999). Structural

proteins often contain a glycosylphosphatidyl-inositol anchor by which they are attached to the lumen of the rough endoplasmic reticulum (ER) and later to the external plasma membrane surface, or a modified anchor which covalently binds them to the β-(1,6)-glucan fraction of the cell wall (Kollár *et al.*, 1997; de Nobel *et al.*, 2001).

In the Zygomycota, the chitin fibres are modified after their synthesis by partial or complete deacetylation to produce poly-β-(1,4)-glucosamine, which is called **chitosan** (Fig. 1.5) (Calvo-Mendez & Ruiz-Herrera, 1987). Chitosan fibres are cross-linked by polysaccharides containing glucuronic acid and various neutral sugars (Datema *et al.*, 1977). The cell wall matrix comprises glucans and proteins, as it does in members of the other fungal groups.

One traditional feature to distinguish the Oomycota from the 'true fungi' (Eumycota) has been the absence of chitin from their cell walls (Wessels & Sietsma, 1981), even though chitin is now known to be produced by certain species of Oomycota under certain conditions (Gay *et al.*, 1993). By and large, however, in Oomycota, the structural role of chitin is filled by **cellulose**, an aggregate of linear β-(1,4)-glucan chains (Fig. 1.5). As in many other fungi, the fibres thus produced are cross-linked by an alkali-insoluble glucan containing β-(1,3)- and β-(1,6)-linkages. In addition to proteins, the main matrix component appears to be an alkali-soluble β-(1,3)-glucan (Wessels & Sietsma, 1981).

1.2.3 Synthesis of the cell wall

The synthesis of chitin is mediated by specialized organelles termed **chitosomes** (Bartnicki-Garcia *et al.*, 1979; Sentandreu *et al.*, 1994) in which inactive chitin synthases are delivered to the apical plasma membrane and become activated upon contact with the lipid bilayer (Montgomery & Gooday, 1985). Microvesicles, visible especially in the core region of the Spitzenkörper, are likely to be the ultrastructural manifestation of chitosomes (Fig. 1.6). In contrast, structural proteins and enzymes travel together in the larger secretory vesicles and are discharged into the environment when the vesicles fuse with the plasma membrane

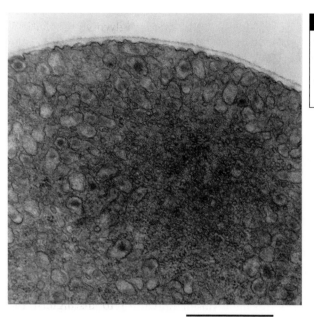

Fig 1.6 The Spitzenkörper of *Botrytis cinerea* which is differentiated into an electron-dense core consisting of microvesicles (chitosomes) and an outer region made up of larger secretory vesicles, some of which are located close to the plasma membrane. Reprinted from Weber and Pitt (2001), with permission from Elsevier.

1 μm

(Fig. 1.6). Whereas most proteins are fully functional by the time they traverse the plasma membrane (see p. 10), the glucans are secreted by secretory vesicles as partly formed precursors (Wessels, 1993a) and undergo further polymerization in the nascent cell wall, or they are synthesized entirely at the plasma membrane (Sentandreu *et al.*, 1994; de Nobel *et al.*, 2001). Cross-linking of glucans with other components of the cell wall takes place after extrusion into the cell wall (Kollár *et al.*, 1997; de Nobel *et al.*, 2001).

Wessels *et al.* (1990) have provided experimental evidence to support a model for cell wall synthesis in *Schizophyllum commune* (Basidiomycota). The individual linear β-(1,4)-*N*-acetylglucosamine chains extruded from the plasma membrane are capable of undergoing self-assembly into chitin microfibrils, but this is subject to a certain delay during which cross-linking with glucans must occur. The glucans, in turn, become alkali-insoluble only after they have become covalently linked to chitin. Once the structural scaffold is in place, the wall matrix can be assembled. Wessels (1997) suggested that hyphal growth occurs as the result of a continuously replenished supply of soft wall material at the apex, but there is good evidence that the

softness of the apical cell wall is also influenced by the activity of wall-lytic enzymes such as chitinases or glucanases (Fontaine *et al.*, 1997; Horsch *et al.*, 1997). Further, when certain Oomycota grow under conditions of hyperosmotic stress, their cell wall is measurably softer due to the secretion of an *endo*-β-(1,4)-glucanase, thus permitting continued growth when the turgor pressure is reduced or even absent (Money, 1994; Money & Hill, 1997). Since, in higher Eumycota, both cell wall material and synthetic as well as lytic enzymes are secreted together by the vesicles of the Spitzenkörper, the appearance, position and movement of this structure should influence the direction and speed of apical growth directly. This has indeed been shown to be the case (López-Franco *et al.*, 1995; Bartnicki-Garcia, 1996; Riquelme *et al.*, 1998).

Of course, cell wall-lytic enzymes are also necessary for the formation of hyphal branches, which usually arise by a localized weakening of the mature, fully polymerized cell wall. An *endo*-β-(1,4)-glucanase has also been shown to be involved in softening the mature regions of hyphae in the growing stipes of *Coprinus* fruit bodies, thus permitting intercalary hyphal extension (Kamada, 1994). Indeed, the expansion of mushroom-type fruit bodies in general seems to be based mainly on non-apical extension of existing hyphae (see p. 22), which is a rare exception to the rule of apical growth in fungi.

The properties of the cell wall depend in many ways on the environment in which the hypha grows. Thus, when *Schizophyllum commune* is grown in liquid submerged culture, a significant part of the β-glucan fraction may diffuse into the liquid medium before it is captured by the cell wall, giving rise to mucilage (Sietsma *et al.*, 1977). In addition to causing problems when growing fungi in liquid culture for experimental purposes, mucilage may cause economic losses when released by *Botrytis cinerea* in grapes to be used for wine production (Dubourdieu *et al.*, 1978a). On the other hand, secreted polysaccharides, especially of Basidiomycota, may have interesting medicinal properties and are being promoted as anti-tumour medication both in conventional and in alternative medicine (Wasser, 2002).

Another difference between submerged and aerial hyphae is caused by the **hydrophobins**, which are structural cell wall proteins with specialized functions in physiology, morphogenesis and pathology (Wessels, 2000). Some hydrophobins are constitutively secreted by the hyphal apex. In submerged culture, they diffuse into the medium as monomers, whereas they polymerize by hydrophobic interactions on the surface of hyphae exposed to air, thereby effectively impregnating them and rendering them hydrophobic (Wessels, 1997, 2000). When freeze-fractured hydrophobic surfaces of hyphae or spores are viewed with the transmission electron microscope, polymerized hydrophobins may be visible as patches of rodlets running in parallel to each other. Other hydrophobins are produced only at particular developmental stages and are involved in inducing morphogenetic changes of the hypha, leading, for example, to the formation of spores or infection structures, or aggregation of hyphae into fruit bodies (Stringer *et al.*, 1991; Wessels, 1997).

Some fungi are wall-less during the assimilative stage of their life cycle. This is true especially of certain plant pathogens such as the Plasmodiophoromycota (Chapter 3), insect pathogens (Entomophthorales; p. 202) and some members of the Chytridiomycota (Chapter 6). Since their protoplasts are in direct contact with the host cytoplasm, they are buffered against osmotic fluctuations. The motile spores (zoospores) of certain groups of fungi swim freely in water, and bursting due to osmotic inward movement of water is prevented by the constant activity of water-expulsion vacuoles.

1.2.4 The cytoskeleton

In contrast to the hyphae of certain Oomycota, which seem to grow even in the absence of measurable turgor pressure (Money & Hill, 1997), the hyphae of most fungi extend only when a threshold turgor pressure is exceeded. This can be generated even at a reduced external water potential by the accumulation of compatible solutes such as glycerol, mannitol or trehalose inside the hypha (Jennings, 1995). The correlation between turgor pressure and hyphal growth might be interpreted such that the former drives the latter, but this crude mechanism would lead to uncontrolled tip extension or even tip bursting. Further, when hyphal tips are made to burst by experimental manipulation, they often do so not at the extreme apex, but a little further behind (Sietsma & Wessels, 1994). It seems, therefore, that the soft wall at the apex is protected internally, and there is now good evidence that this is mediated by the cytoskeleton.

Both main elements of the cytoskeleton, i.e. microtubules (Figs. 1.7a,b) and actin filaments (Fig. 1.7c), are abundant in filamentous fungi and yeasts (Heath, 1994, 1995a). Intermediate filaments, which fulfil skeletal roles in animal cells, are probably of lesser significance in fungi. Microtubules are typically orientated longitudinally relative to the hypha (Fig. 1.7a) and are involved in long-distance transport of organelles such as secretory vesicles (Fig. 1.7b; Seiler *et al.*, 1997) or nuclei (Steinberg, 1998), and in the positioning of mitochondria, nuclei or vacuoles (Howard & Aist, 1977; Steinberg *et al.*, 1998). They therefore maintain the polarized distribution of many organelles in the hyphal tip.

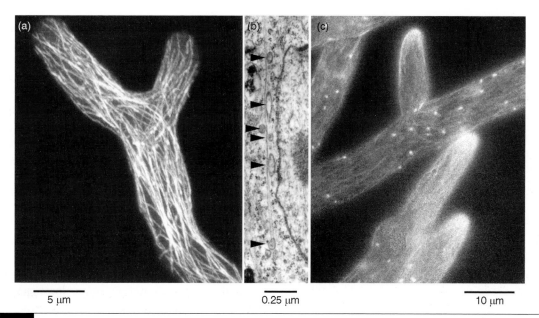

5 μm   0.25 μm   10 μm

Fig 1.7 The cytoskeleton in fungi. (a) Microtubules in *Rhizoctonia solani* (Basidiomycota) stained with an α-tubulin antibody. (b) Secretory vesicles (arrowheads) associated with a microtubule in *Botrytis cinerea* (Ascomycota). (c) The actin system of *Saprolegnia ferax* (Oomycota) stained with phalloidin—rhodamine. Note the dense actin cap in growing hyphal tips. (a) reproduced from Bourett *et al.* (1998), with permission from Elsevier; original print kindly provided by R. J. Howard. (b) reproduced from Weber and Pitt (2001), with permission from Elsevier. (c) reproduced from I. B. Heath (1987), by copyright permission of Wissenschaftliche Verlagsgesellschaft mbH, Stuttgart; original print kindly provided by I. B. Heath.

Actin filaments are found in the centre of the Spitzenkörper, as discrete subapical patches, and as a cap lining the inside of the extreme hyphal apex (Heath, 1995a; Czymmek *et al.*, 1996; Srinivasan *et al.*, 1996). The apical actin cap is particularly pronounced in Oomycota such as *Saprolegnia* (Fig. 1.7c), and it now seems that the soft wall at the hyphal apex is actually being assembled on an internal scaffold consisting of actin and other structural proteins, such as spectrin (Heath, 1995b; Degousée *et al.*, 2000). The rate of hyphal extension might be controlled, and bursting prevented, by the actin/spectrin cap being anchored to the rigid, subapical wall via rivet-like integrin attachments which traverse the membrane and might bind to wall matrix proteins (Fig. 1.8; Kaminskyj & Heath, 1996; Heath, 2001). Indeed, in *Saprolegnia* the cytoskeleton is probably responsible for pushing the hyphal tip forward, at least in the absence of turgor (Money, 1997), although it probably has a restraining function under normal physiological conditions. Heath (1995b)

has proposed an ingenious if speculative model to explain how the actin cap might regulate the rate of hyphal tip extension in the Oomycota. Stretch-activated channels selective for Ca^{2+} ions are known to be concentrated in the apical plasma membrane of *Saprolegnia* (Garrill *et al.*, 1993), and the fact that Ca^{2+} ions cause contractions of actin filaments is also well known. A stretched plasma membrane will admit Ca^{2+} ions into the apical cytoplasm where they cause localized contractions of the actin cap, thereby reducing the rate of apical growth which leads to closure of the stretch-activated Ca^{2+} channels. Sequestration of Ca^{2+} by various subapical organelles such as the ER or vacuoles lowers the concentration of free cytoplasmic Ca^{2+}, leading to a relaxation of the actin cap and of its restrictive effect on hyphal growth.

In the Eumycota, there is only indirect evidence for a similar role of actin, integrin and other structural proteins in protecting the apex and restraining its extension (Degousée *et al.*, 2000; Heath, 2001), and the details of

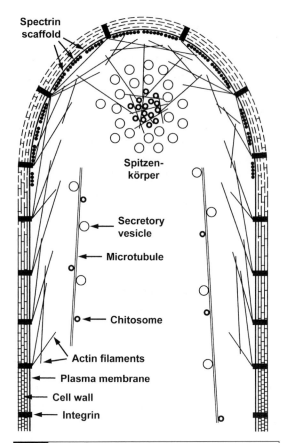

Spectrin scaffold

Spitzen-körper

Secretory vesicle

Microtubule

Chitosome

Actin filaments

Plasma membrane

Cell wall

Integrin

Fig I.8 Diagrammatic representation of the internal scaffold model of tip growth in fungi proposed by Heath (1995b). Secretory vesicles and chitosomes are transported along microtubules from their subapical sites of synthesis to the growing apex. The Spitzenkörper forms around a cluster of actin filaments. An actin scaffold inside the extreme apex is linked to rivet-like integrin molecules which are anchored in the rigid subapical cell wall. The apex is further stabilized by spectrin molecules lining the cytoplasmic surface of the plasma membrane. Redrawn and modified from Weber and Pitt (2001).

regulation are likely to be different. Whereas a tip-high Ca^{2+} gradient is present and is required for growth, stretch-activated Ca^{2+} channels are not, and the apical Ca^{2+} seems to be of endogenous origin. Silverman-Gavrila and Lew (2001, 2002) have proposed that the signal molecule inositol-(1,4,5)-trisphosphate (IP_3), released by the action of a stretch-activated phospholipase C in the apical plasma membrane, acts on Ca^{2+}-rich secretory vesicles in the

Spitzenkörper region. These would release Ca^{2+} from their lumen, leading to a contraction of the apical scaffold. As in the Oomycota, sequestration of Ca^{2+} occurs subapically by the ER from which secretory vesicles are formed. These therefore act as Ca^{2+} shuttles in the Eumycota (Torralba et al., 2001). Although hyphal tip growth appears to be a straightforward affair, none of the conflicting models accounts for all aspects of it. A good essay in hyphal tip diplomacy has been written by Bartnicki-Garcia (2002).

Numerous inhibitor studies have hinted at a role of the cytoskeleton in the transport of vesicles to the apex. Depolymerization of microtubules results in a disappearance of the Spitzenkörper, termination or at least severe reduction of apical growth and enzyme secretion, and an even redistribution of secretory vesicles and other organelles throughout the hypha (Howard & Aist, 1977; Rupeš et al., 1995; Horio & Oakley, 2005). In contrast, actin depolymerization leads to uncontrolled tip extension to form giant spheres (Srinivasan et al., 1996). Long-distance transport of secretory vesicles therefore seems to be brought about by microtubules, whereas the fine-tuning of vesicle fusion with the plasma membrane is controlled by actin (Fig. 1.8; Torralba et al., 1998). The integrity of the Spitzenkörper is maintained by an interplay between actin and tubulin. Not surprisingly, the yeast S. cerevisiae, which has a very short vesicle transport distance between the mother cell and the extending bud, reacts more sensitively to disruptions of the actin component than the microtubule component of its cytoskeleton; continued growth in the absence of the latter can be explained by Brownian motion of secretory vesicles (Govindan et al., 1995; Steinberg, 1998).

1.2.5 Secretion and membrane traffic

One of the most important ecological roles of fungi, that of decomposing dead plant matter, requires the secretion of large quantities of hydrolytic and oxidative enzymes into the environment. In liquid culture under optimized experimental conditions, certain fungi are capable of secreting more than 20 g of a single enzyme or enzyme group per litre culture broth within a few days' growth (Sprey, 1988; Peberdy, 1994). Clearly, this aspect of fungal physiology holds considerable potential for biotechnological or pharmaceutical applications. However, for reasons not yet entirely understood, fungi often fail to secrete the heterologous proteins of introduced genes of commercial interest to the same high level as their own proteins (Gwynne, 1992). There are still great deficits in our understanding of the fundamental mechanisms of the secretory route in filamentous fungi, although much is known in the yeast *S. cerevisiae*. An overview is given in Fig. 1.10.

As in other eukaryotes, the secretory route in fungi begins in the ER. Ribosomes loaded with a suitable messenger RNA dock onto the ER membrane and translate the polypeptide product which enters the ER lumen during its synthesis unless specific internal signal sequences cause it to be retained in the ER membrane. As soon as the protein is in contact with the ER lumen, oligosaccharide chains may be added onto selected amino acids. These glycosylation chains are subject to successive modification steps as the protein traverses the secretory route, whereby the chains in *S. cerevisiae* become considerably larger than those in most filamentous fungi (Maras *et al.*, 1997; Gemmill & Trimble, 1999). Paradoxically, even though filamentous fungi possess such powerful secretory systems, morphologically recognizable Golgi stacks have not generally been observed except for the Oomycota, Plasmodiophoromycota and related groups (Grove *et al.*, 1968; Beakes & Glockling, 1998). In all other fungi, the Golgi apparatus seems to be much reduced to single cisternae (Howard, 1981; see Fig. 1.3), with images of fully fledged Golgi stacks only published occasionally (see e.g. Fig. 10.1). In *S. cerevisiae* and probably also in filamentous fungi, the transport of proteins from the ER to the Golgi system occurs via vesicular carriers (Schekman, 1992), although continuous membrane flow is also possible (see p. 272). Membrane lipids seem to be recycled to the ER by a different mechanism relying on tubular continuities (Rupeš *et al.*, 1995; Akashi *et al.*, 1997).

In the Golgi system, proteins are subjected to stepwise further modifications (Graham & Emr, 1991), and proteins destined for the vacuolar system are separated from those bound for secretion (Seeger & Payne, 1992). Both destinations are probably reached by vesicular carriers, the secretory vesicles moving along microtubules to reach the growing hyphal apex (Fig. 1.7b), which is the site for secretion of extracellular enzymes as well as new cell wall material (Peberdy, 1994). Collinge and Trinci (1974) estimated that 38 000 secretory vesicles per minute fuse with the plasma membrane of a single growing hypha of *Neurospora crassa*. Microvesicles (chitosomes) probably arise from a discrete population of Golgi cisternae (Howard, 1981).

There is mounting evidence that fungi, like most eukaryotes, are capable of performing endocytosis by the inward budding of the plasma membrane at subapical locations. Endocytosis may be necessary to retrieve membrane material in excess of that which is required for extension at the growing apex, i.e. endocytosis and exocytosis may be coupled (Steinberg & Fuchs, 2004). The prime destination of endocytosed membrane material or vital stains is the vacuole (Vida & Emr, 1995; Fischer-Parton *et al.*, 2000; Weber, 2002). In fungi, large vacuoles (Figs. 1.2e, 1.9) represent the main element of the lytic system and are the sink not only for endocytosed material but also for autophagocytosis, i.e. the sequestration and degradation of organelles or cytoplasm. Autophagocytosis is especially prominent under starvation conditions (Takeshige *et al.*, 1992). Careful ultrastructural studies have revealed that adjacent vacuoles may be linked by thin membranous tubes, thereby providing a potential means of transport (Rees *et al.*, 1994). These tubes can extend even through the septal pores and show peristaltic movement, possibly explaining why especially mycorrhizal fungi are capable of rapid translocation of solutes over long hyphal distances (Fig. 1.9; Cole *et al.*, 1998; Ashford *et al.*, 2001).

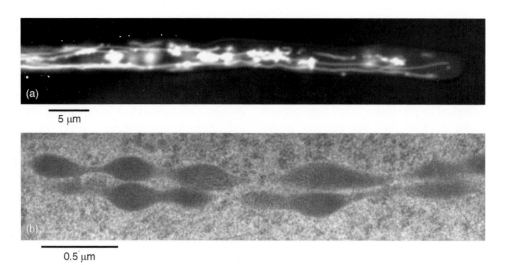

5 μm

0.5 μm

Fig 1.9 Tubular continuities linking adjacent vacuoles of *Pisolithus tinctorius*. (a) Light micrograph of the vacuolar system of *Pisolithus tinctorius* stained with a fluorescent dye. (b) TEM image of a freeze-substituted hypha. Reproduced from Ashford *et al.* (2001), with kind permission of Springer Science and Business Media. Original images kindly provided by A. E. Ashford.

1.2.6 Nutrient uptake

One of the hallmarks of fungi is their ability to take up organic or inorganic solutes from extremely dilute solutions in the environment, accumulating them 1000-fold or more against their concentration gradient (Griffin, 1994). The main barrier to the movement of water-soluble substances into the cell is the lipid bilayer of the plasma membrane. Uptake is mediated by proteinaceous pores in the plasma membrane which are always selective for particular solutes. The pores are termed **channels** (system I) if they facilitate the diffusion of a solute following its concentration gradient whilst they are called **porters** (system II) if they use metabolic energy to accumulate the solute across the plasma membrane against its gradient (Harold, 1994). Fungi often possess one channel and one porter for a given solute. The high-affinity porter system is repressed at high external solute concentrations such as those found in most laboratory media (Scarborough, 1970; Sanders, 1988).

In nature, however, the concentration of nutrients is often so low that the porter systems are active. Porters do not directly convert metabolic energy (ATP) into the uptake of solutes; rather, ATP is hydrolysed by ATPases which pump protons (H^+) to the outside of the plasma membrane, thus establishing a transmembrane pH gradient (acid outside). It has been estimated that one-third of the total cellular ATP is used for the establishment of the transmembrane H^+ gradient (Gradmann *et al.*, 1978). The inward movement of H^+ following its electrochemical gradient is harnessed by the porters for solute uptake by means of solute–porter–H^+ complexes (Slayman & Slayman, 1974; Slayman, 1987; Garrill, 1995). Different types of porter exist, depending on the charge of the desired solute. Uniport and symport carriers couple the inward movement of H^+ with the uptake of uncharged or negatively charged solutes, respectively, whereas antiports harness the outward diffusion of cations such as K^+ for the uptake of other positively charged solutes. Charge imbalances can be rectified by the selective opening of K^+ channels. Porters have been described for NH_4^+, NO_3^-, amino acids, hexoses, orthophosphate and other solutes (Garrill, 1995; Jennings, 1995).

The ATPases fuelling active uptake mechanisms are located in subapical or mature regions of the plasma membrane, whereas the porter systems are typically situated in the apical membrane (Harold, 1994), closest to the site where the solutes may be released by the activity of extracellular enzymes. Thus, mature hyphal segments make a substantial direct contribution

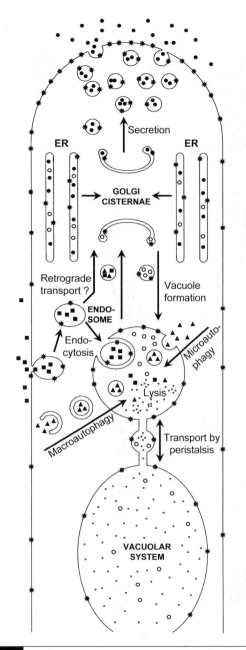

Fig 1.10 Schematic summary of the pathways of membrane flow in a growing hypha. Secretory proteins (●), vacuolar luminal proteins (○), membrane-bound proteins (✳), endocytosed (■) and autophagocytosed (▲) material is indicated, as are vacuolar degradation products (•). Redrawn and modified from Weber (2002).

(Fig. 1.11), which was at one time thought to be a causal factor of hyphal tip polarity but is now regarded as a consequence of it (Harold, 1994).

Proton pumps fuelled by ATP are prominent also in the vacuolar membrane, the tonoplast (Fig. 1.11), and their activity acidifies the vacuolar lumen (Klionsky et al., 1990). The principle of proton-coupled solute transport is utilized by the vacuole to fulfil its role as a system for the storage of nutrients, for example phosphate (Cramer & Davis, 1984) or amino acids such as arginine (Keenan & Weiss, 1997), or for the removal of toxic compounds from the cytoplasm, e.g. Ca^{2+} or heavy metal ions (Cornelius & Nakashima, 1987).

1.2.7 Hyphal branching

Assimilative hyphae of most fungi grow monopodially by a main axis (**leading hypha**) capable of potentially unlimited apical growth. Branches arise at some distance behind the apex, suggesting some form of apical dominance, i.e. the presence of a growing apex inhibits the development of lateral branches close to it. Dichotomous branching is rare, but does occur in *Allomyces* (see Fig. 6.20d) and *Galactomyces geotrichum*. In septate fungi, branches are often located immediately behind a septum. Branches usually arise singly in vegetative hyphae, although whorls of branches (i.e. branches arising near a common point) occur in reproductive structures. Branching may thus be under genetic or external control (Burnett, 1976). An even spacing between vegetative hyphae results from a combination of chemotropic growth towards a source of diffusible nutrients, and growth away from staling products secreted by other hyphae which have colonized a substratum. The circular appearance of fungal colonies in Petri dish cultures arises because certain lateral branches grow out and fill the space between the leading radial branches, keeping pace with their rate of growth. This **invasive growth** is the most efficient way to spread throughout a substratum. In nature, it may be obvious even to the naked eye, for example, in the shape of fairy rings (see Figs. 19.18a,b).

to the growth of the hypha at its tip. The spatial separation of H^+ expulsion and re-entry generates an external electric field carried by protons

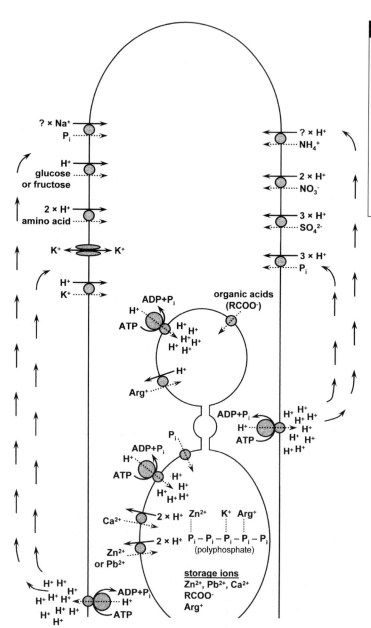

Fig 1.11 Ion fluxes in a growing hypha. The proton (H⁺) gradient across the plasma membrane is generated by subapical ATP-driven expulsion of protons. It is used for the active uptake of nutrients by porters. Channels also exist for most of the nutrients but are not shown here, except for the K^+ channel which operates to compensate for charge imbalances. Dotted arrows indicate movement of a solute against its concentration gradient; solid arrows indicate movement from concentrated to dilute. For details, see Garrill (1995).

1.3 Hyphal aggregates

Whereas plants and animals form genuine tissues by their ability to perform cell divisions in all directions, fungi are limited by their growth as one-dimensional hyphae. None the less, fungi are capable of producing complex and characteristic multicellular structures which resemble the tissues of other eukaryotes. This must be controlled by the positioning, growth rate and growth direction of individual hyphal branches (Moore, 1994). Further, instead of spacing themselves apart as during invasive growth, hyphae must be made to aggregate. Very little is known about the signalling events leading to the synchronized growth of groups of hyphae. However, it may be speculated that the diffusion of signalling molecules takes place between adjacent hyphae, i.e. that a given hypha is able to influence the gene expression of adjacent hyphae by secreting chemical messengers. This may be facilitated by an extrahyphal glucan matrix within which aggregating hyphae are typically embedded (Moore, 1994). Such matrices have been found in rhizomorphs (Rayner *et al.*, 1985), sclerotia (Fig. 1.16c; Willetts & Bullock, 1992) and fruit bodies (Williams *et al.*, 1985). The composition of proteins on the surface of hyphal walls may also play an important role in recognition and adhesion phenomena (de Nobel *et al.*, 2001).

1.3.1 Mycelial strands

The formation of aggregates of parallel, relatively undifferentiated hyphae is quite common in the Basidiomycota and in some Ascomycota. For instance, mycelial strands form the familiar 'spawn' of the cultivated mushroom *Agaricus bisporus*. Strands arise most readily from a well-developed mycelium extending from an exhausted food base into nutrient-poor surroundings (Fig. 1.12a). When a strand encounters a source of nutrients exceeding its internal supply, coherence is lost and a spreading assimilative mycelium regrows (Moore, 1994). Alternatively, mycelial strands may be employed by fungi which produce their fructifications some distance away from the food base, as in the stinkhorn, *Phallus impudicus*. Here the mycelial strand is more tightly aggregated and is referred to as a **mycelial cord**. The tip of the mycelial cord, which arises from a buried tree stump, differentiates into an egg-like basidiocarp initially upon reaching the soil surface (Fig. 1.12b).

The development of *A. bisporus* strands has been described by Mathew (1961). Robust leading hyphae extend from the food base and branch at fairly wide intervals to form finer laterals, most of which grow away from the parent hypha. A few branch hyphae, however, form at an acute angle to the parent hypha and tend to grow parallel to it. Hyphae of many fungi occasionally

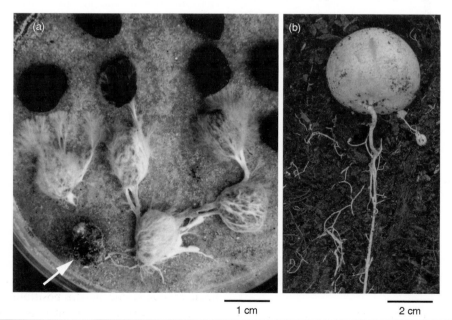

1 cm   2 cm

Fig 1.12 Mycelial strands. (a) Strands of *Podosordaria tulasnei* (Ascomycota) extending from a previously colonized rabbit pellet (arrow) over sand. Note the dissolution of the strand upon reaching a new nutrient source, in this case fresh sterile rabbit pellets. (b) Excavated mycelial cords of the stinkhorn *Phallus impudicus*, which can be traced back from the egg-like basidiocarp primordium to the base of an old tree stump (below the bottom of the picture, not shown).

grow alongside each other or another physical obstacle which they chance to encounter. A later and specific stage in strand development is characterized by the formation of numerous fine, aseptate 'tendril hyphae' as branches from the older regions of the main hyphae. The tendril hyphae, which may extend forwards or back- wards, become appressed to the main hypha and branch frequently to form even finer tendrils which grow round the main hyphae and ensheath them. Major strands are consolidated by anastomoses between their hyphae, and they increase in thickness by the assimilation of minor strands. A similar development has been noted in the strands of *Serpula lacrymans*, the dry- rot fungus (Fig. 1.13), which are capable of extending for several metres across brickwork and other surfaces from a food base in decaying wood (Jennings & Watkinson, 1982; Nuss *et al.*, 1991).

By recovering the nutrients from obsolete strands and forming new strands, colonies can move about and explore their vicinity in the search for new food bases (Cooke & Rayner, 1984; Boddy, 1993). Mycelial strands are capable of translocating nutrients and water in both direc- tions (Boddy, 1993; Jennings, 1995). This property is important not only for decomposer fungi, but also for species forming mycorrhizal symbioses with the roots of plants, many of which produce hyphal strands (Read, 1991).

1.3.2   Rhizomorphs

In contrast to mycelial strands or cords which consist of relatively undifferentiated aggrega- tions of hyphae and are produced by a great variety of fungi, rhizomorphs are found in only relatively few species and contain highly differentiated tissues. Well-known examples of rhizomorph-forming fungi are provided by *Armillaria* spp. (Figs. 1.14 and 18.13b), which are serious parasites of trees and shrubs. In *Armillaria*, a central core of larger, thin-walled, elongated cells embedded in mucilage is surrounded by a rind of small, thicker-walled cells which are darkly pigmented due to melanin deposition in their walls. These root-like aggrega- tions are a means for *Armillaria* to spread underground from one tree root system to another. In nature, two kinds are found — a dark, cylindrical type and a paler, flatter type. The latter is particularly common beneath the bark of infected trees (see p. 546). Rhizomorphs on dead trees measure up to 4 mm in diameter. It has been estimated that a rhizomorph only 1 mm in diameter must contain over 1000 hyphae aggregated together. The development of rhizomorphs in agar culture has been described by Garrett (1953, 1970) and Snider (1959). Initiation of rhizomorphs can first be

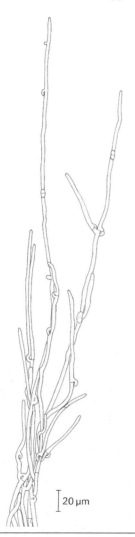

$20\,\mu m$

Fig 1.13   The tip of a hyphal strand of *Serpula lacrymans* (Basidiomycota). Note the formation of lateral branches which grow parallel to the direction of the main hyphae. The buckle-shaped structures at the septa are clamp connections.

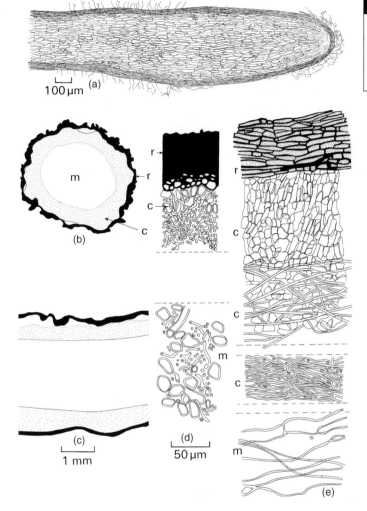

Fig 1.14 Rhizomorph structure of *Armillaria mellea* (Basidiomycota). (a) Longitudinal section. (b) Transverse section, diagrammatic. (c) L.S. diagrammatic. (d) T.S. showing details of cells in the rind (r), cortex (c) and medulla (m). (e) L.S. showing details of cells.

observed after about 7 days' mycelial growth on the agar surface as a compact mass of darkly pigmented hypertrophied cells. These pigmented structures have been termed **microsclerotia**. From white, non-pigmented points on their surface, the rhizomorphs develop. The growth of rhizomorphs can be several times faster than that of unorganized hyphae (Rishbeth, 1968). The most striking feature of the development of rhizomorphs is their compact growing point at the apex, which consists of small isodiametric cells protected by an apical cap of intertwined hyphae immersed in mucilage which they produce. Because of its striking similarity with a growing plant root, the rhizomorph tip was initially interpreted as a meristematic zone (Motta, 1967), but its hyphal nature can be demonstrated by careful ultrastructural observations (Powell & Rayner, 1983; Rayner *et al.*, 1985). Behind the apex there is a zone of elongation. The centre of the rhizomorph may be hollow or solid. Surrounding the central lumen or making up the central medulla is a zone of enlarged hyphae 4–5 times wider than the vegetative hyphae (Fig. 1.14e). Possibly these **vessel hyphae** serve in translocation (Cairney, 1992; Jennings, 1995). Towards the periphery of the rhizomorph, the cells become smaller, darker, and thicker walled. Extending outwards between the outer cells of the rhizomorph, there may be a growth of vegetative hyphae somewhat resembling the root-hair zone in a higher plant. Rhizomorphs may develop on monokaryotic mycelia derived from single basidiospores, or on dikaryotic

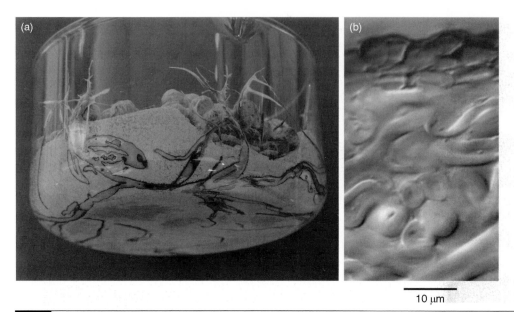

10 μm

Fig 1.15 Rhizomorphs of *Podosordaria tulasnei* (Ascomycota). (a) Subterranean rhizomorphs by which the fungus spreads through the soil. (b) T.S. showing the dark rind (1–2 cells thick) and a cortex consisting of thick-walled hyaline cells.

mycelia following fusion of compatible monokaryotic hyphae. Dikaryotic rhizomorphs of *Armillaria* do not possess clamp connections (Hintikka, 1973).

Rhizomorphs are also produced by other Basidiomycota and a few Ascomycota (Fig. 1.15; Webster & Weber, 2000). They are mainly formed in soil. An interesting exception is presented by tropical *Marasmius* spp., which form a network of aerial rhizomorphs capable of intercepting falling leaves before they reach the ground (Hedger *et al.*, 1993). Because these rhizomorphs have a rudimentary fruit body cap at their extending apex (Hedger *et al.*, 1993), they have been interpreted as indefinitely extending fruit body stipes (Moore, 1994). Mycelial strands and rhizomorphs represent extremes in a range of hyphal aggregations, and several intergrading forms can be recognized (Rayner *et al.*, 1985).

1.3.3 Sclerotia

Sclerotia are pseudoparenchymatous aggregations of hyphae embedded in an extracellular glucan matrix. A hard melanized rind may be present or absent. Sclerotia serve a survival function and contain intrahyphal storage reserves such as polyphosphate, glycogen,

protein, and lipid (Willetts & Bullock, 1992). The glucan matrix, too, may be utilized as a carbohydrate source during sclerotium germination (Backhouse & Willetts, 1985). Sclerotia may also have a reproductive role and are the only known means of reproduction in certain species. They are produced by a relatively small number of Asco- and Basidiomycota, especially plant-pathogenic species such as *Rhizoctonia* spp. (p. 595), *Sclerotinia* spp. (p. 429) and *Claviceps purpurea* (p. 349). The form of sclerotia is very variable (Butler, 1966). The subterranean sclerotium of the Australian *Polyporus mylittae* (see Figs. 18.13c,d) can reach the size of a football and is known as native bread or blackfellow's bread. At the other extreme, they may be of microscopic dimensions consisting of a few cells only. Several kinds of development in sclerotia have been distinguished (Townsend & Willetts, 1954; Willetts, 1972).

The loose type

This is exemplified by *Rhizoctonia* spp., which are sclerotial forms of fungi belonging to the Basidiomycota. Sclerotia of the loose type are readily seen as the thin brownish-black scurfy scales so common on the surface of

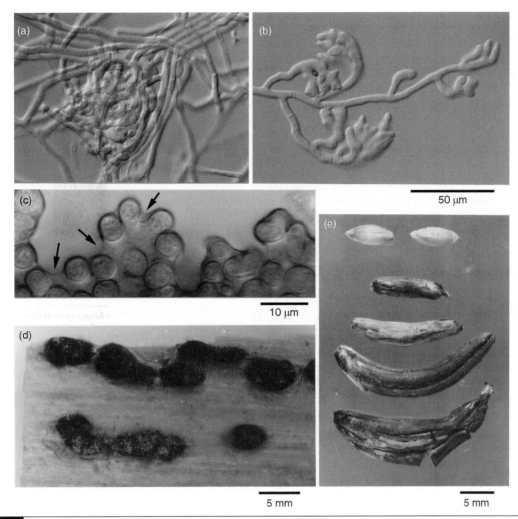

Fig 1.16 Development of sclerotia. (a) The loose type, as seen in *Rhizoctonia (Moniliopsis) solani*. (b) Hypha of *Botrytis cinerea* showing dichotomous branching on a glass coverslip to initiate the terminal type of sclerotium. (c) Later stage of sclerotium formation in *B. cinerea*. The hyphae have become melanized and are growing away from the glass surface. They are embedded in a glucan matrix (arrows). (d) Mature sclerotia of *B. cinerea* on a stem of *Conium*. Some sclerotia are germinating to produce tufts of conidiophores. (e) Sclerotia of *Claviceps purpurea* from an ear of rye (*Secale cereale*). Rye grains are shown for size comparison. (a) and (b) to same scale.

potato tubers. In pure culture, sclerotial initials arise by branching and septation of hyphae (Fig. 1.16a). These cells become filled with dense contents and numerous vacuoles, and darken to reddish-brown. The mature sclerotium does not show well-defined zones or 'tissues'. It is made up of a central part which is pseudoparenchymatous, although its hyphal nature can be seen. Towards the outside, the hyphae are more loosely arranged; a rind of thick-walled hyphae is absent (Willetts, 1969).

The terminal type

This form is characterized by a well-defined pattern of branching. It is produced, for example, by *Botrytis cinerea*, the cause of grey mould diseases on a wide range of plants, and by the saprotrophic *Pyronema domesticum* (see p. 415). Sclerotia of *B. cinerea* are found on overwintering stems of herbaceous plants, especially umbellifers such as *Angelica*, *Anthriscus*, *Conium* and *Heracleum*. They can also be induced to form in culture, especially on agar media with a high carbon/nitrogen ratio. When growing on host tissue, the sclerotia of *Botrytis* may include host cells, a feature shared also by sclerotia of *Sclerotinia* spp. to which *Botrytis* is related (see p. 429). Sclerotia arise by repeated dichotomous branching of hyphae, accompanied by cross-wall formation (Fig. 1.16b). The hyphae then aggregate, melanize and produce mucilage, giving the appearance of a solid tissue (Fig. 1.16c). A mature sclerotium may be about 10 mm long and 3–5 mm wide, and is usually flattened, measuring 1–3 mm in thickness. It is often orientated parallel to the long axis of the host plant (Fig. 1.16d). It is differentiated into a rind composed of several layers of rounded, dark cells, a narrow cortex of thin-walled pseudoparenchymatous cells with dense contents, and a medulla made up of loosely arranged filaments. Nutrient reserves are stored in the cortical and medullary regions (Willetts & Bullock, 1992).

The strand type

Sclerotinia gladioli, the causal agent of dry rot of corms of *Gladiolus*, *Crocus* and other plants, forms sclerotia of this type. Sclerotial initials commence with the formation of numerous side branches which arise from one or more main hyphae. Where several hyphae are involved, they lie parallel. They are thicker than normal vegetative hyphae, and become divided by septa into chains of short cells. These cells may give rise to short branches, some of which lie parallel to the parent hypha, whilst others grow out at right angles and branch again before coalescing. The hyphae at the margin continue to branch, and the whole structure darkens. The mature sclerotium is about 0.1–0.3 mm in diameter, and is differentiated into a rind of small, thick-walled cells and a medulla of large, thin-walled hyphae. More complex sclerotia are found in *Sclerotium rolfsii*, the sclerotial state of *Pellicularia rolfsii* (Basidiomycota). Here the mature sclerotium is differentiated into four zones: a fairly thick skin or cuticle, a rind made up of 2–4 layers of tangentially flattened cells, a cortex of thin-walled cells with densely staining contents, and a medulla of loose filamentous hyphae with dense contents. Chet *et al.* (1969) have shown that the skin or cuticle is made up of

the remnants of cell walls attached to the outside of the empty, melanized, thick-walled rind cells. All the cells of the strand-type sclerotium have thicker walls than those of vegetative hyphae. Cells of the outer cortex contain large storage bodies which consist of protein (Kohn & Grenville, 1989) and leave little room for cytoplasm or other organelles. The inner cortex is also densely packed with storage granules.

Other types

There is a great diversity of other types of sclerotia (Butler, 1966). The sclerotia of *Claviceps purpurea*, the 'ergots' of grasses and cereals (Fig. 1.16e; see also p. 349), develop from a pre-existing mass of mycelium which fills and replaces the cereal ovary, starting from the base and extending towards the apex. The outer layers form a violet, dark grey or black rind enclosing colourless, thick-walled cells. These contain abundant storage lipids which constitute 45% of the dry weight of a *C. purpurea* sclerotium (Kybal, 1964). *Cordyceps militaris*, an insect parasite, forms a dense mass of mycelium in the buried insect's body (p. 360). This mass of mycelium, from which fructifications develop, is enclosed by the exoskeleton of the host, not by a fungal rind. Many wood-rotting fungi enclose colonized woody tissue with a black zone-line of dark, thick-walled cells, and the whole structure may be regarded as a kind of sclerotium.

The giant sclerotium of *Polyporus mylittae* is marbled in structure, comprising white strata and translucent tissue. It has an outer, smooth, thin black rind. Three distinct types of hyphae make up the tissues: thin-walled, thick-walled and 'layered' hyphae. Thin- and thick-walled hyphae are abundant in the white strata but sparse in the translucent tissue, whereas the layered hyphae occur only in the translucent tissue. Detached sclerotia are capable of forming basidiocarps without wetting. It is believed that the translucent tissue functions as an extracellular nutrient and water store (Macfarlane *et al.*, 1978). The structure of the sclerotium appears to be related to its ability to fruit in dry conditions, such as occur in Western Australia.

Germination of sclerotia

Sclerotia can survive for long periods, sometimes for several years (Coley-Smith & Cooke, 1971; Willetts, 1971). Germination may take place in three ways – by the development of mycelium, asexual spores (conidia) or sexual fruit bodies (ascocarps or basidiocarps). Mycelial germination occurs in *Sclerotium cepivorum*, the cause of white-rot of onion, and is stimulated by volatile exudates from onion roots (see p. 434). Conidial development occurs in *Botrytis cinerea* and can be demonstrated by placing overwintered sclerotia in moist warm conditions (Fig. 1.16d; Weber & Webster, 2003). The development of ascocarps (i.e. carpogenic germination) is seen in *Sclerotinia*, where stalked cups or apothecia, bearing asci, arise from sclerotia under suitable conditions (Fig. 15.2), and in *Claviceps purpurea*, where the overwintered sclerotia give rise to a perithecial stroma (Fig. 12.26c). Depending on environmental conditions, the sclerotia of some species may respond by germinating in different ways.

1.3.4 The mantle of ectomycorrhiza

The root tips of many coniferous and deciduous trees with ectomycorrhizal associations, especially those growing in relatively infertile soils, are covered by a **mantle**. This is a continuous sheet of fungal hyphae, several layers thick (see Fig. 19.10). The mycelium extends outwards into the litter layer of the soil, and inwards as single hyphae growing intercellularly, i.e. between the outer cortical cells of the root, to form the so-called **Hartig net**. Hyphae growing outwards from the mantle effectively replace the root hairs as a system for the absorption of minerals from the soil, and there is good evidence that, in most normal forest soils of low to moderate fertility, the performance and nutrient status of mycorrhizal trees is superior to that of uninfected trees (Smith & Read, 1997). Most fungi causing ectomycorrhizal infections are Basidiomycota, especially members of the Homobasidiomycetes (pp. 526 and 581). Within the soil or in pure culture, mycelial strands may form, but the mycelium is not aggregated into the tissue-like structure of the mantle.

1.3.5 Fruit bodies of Ascomycota and imperfect fungi

In the higher fungi, hyphae may aggregate in a highly regulated fashion to form fruiting structures which are an important and often species-specific feature of identification. In the Ascomycota, the fruit bodies produce sexual spores (i.e. as the result of nuclear fusion and meiosis) which are termed **ascospores** and are contained in globose or cylindrical cells called **asci** (Lat. *ascus* = a sac, tube). In most cases, the asci can discharge their ascospores explosively. Asci, although occasionally naked, are usually enclosed in an aggregation of hyphae termed an **ascocarp** or **ascoma**. Ascocarps are very variable in form, and several types have been distinguished (see Fig. 8.16). Their features and development will be described more fully later. Forms in which the asci are totally enclosed, and in which the ascocarp has no special opening, are termed **cleistothecia**. In contrast, **gymnothecia** consist of a loose mesh of hyphae. Both are found in the Plectomycetes (Chapter 11). A modified cleistothecium is characteristic of the Erysiphales (Chapter 13). Cup fungi (Discomycetes, Chapters 14 and 15) possess saucer-shaped ascocarps termed **apothecia**, with a mass of non-fertile hyphae supporting a layer of asci lining the upper side of the fruit body. The non-fertile elements of the apothecium often show considerable differentiation of structure. The asci in apothecia are free to discharge their ascospores at the same time. In other Ascomycota, the asci are contained within ascocarps with a very narrow opening or **ostiole**, through which each ascus must discharge its spores separately. Ascocarps of this type are termed **perithecia** or **pseudothecia**. Perithecia are found in the Pyrenomycetes (Chapter 12) whilst pseudothecia occur in the Loculoascomycetes (Chapter 17). These two types of ascocarp develop in different ways. In many of the Pyrenomycetes, the perithecia are borne on or embedded in a mass of fungal tissue termed the **perithecial stroma**, and these are well shown by the Xylariales (p. 332), and by *Cordyceps* (p. 360) and *Claviceps* (p. 349). In some cases, in addition to the perithecial stroma, a fungus may develop a stromatic tissue on or within which asexual spores (conidia) develop. *Nectria cinnabarina* (p. 341), the coral spot fungus so common on freshly dead deciduous twigs, is such an example. It initially forms pink conidial stromata which later, under suitable conditions of humidity, become converted into perithecial stromata.

Among the imperfect (asexual) fungi, mycelial aggregations bearing conidia are seen in various genera. In some, there are tufts of parallel conidiophores termed **coremia** or **synnemata**, exemplified by *Penicillium claviforme* (see Fig. 11.19). In some imperfect fungi formerly called Coelomycetes, the conidia develop in flask-shaped cavities termed **pycnidia** (see Figs. 17.3–17.5). Various other kinds of mycelial fruiting aggregates are also known.

1.3.6   Fruit bodies of Basidiomycota

The fruit bodies of mushrooms, toadstools, bracket fungi, etc., are all examples of **basidiocarps** or **basidiomata** which bear the sexually produced spores (basidiospores) on basidia. Basidiocarps are almost invariably constructed from dikaryotic hyphae, but how vegetative hyphae aggregate to form a mushroom fruit body is still a mystery (Moore, 1994). Wessels (1997) has suggested that hydrophobins coating the surface of hyphae may confer adhesive properties, leading to their aggregation to form a fruit body initial as the first step in morphogenesis. Once an initial has been formed, its glucan matrix may provide an environment for the exchange of signalling molecules between hyphae. Moore (1994) speculated that morphogenesis might ultimately be determined by induction hyphae exerting a control over surrounding hyphae, leading to the development of morphogenetic units. This morphogenetic commitment must happen at a very early stage. For instance, in the ink-cap (*Coprinus cinereus*) an initial measuring only 1% of the final fruit body size is already differentiated into stipe and cap

(Moore *et al.*, 1979). Therefore, when a mushroom fruit body expands, this is due mainly to the enlargement of existing hyphae, whereas new apical growth is restricted mainly to branches filling up the space generated during expansion (Moore, 1994). Hyphae making up the mature basidiocarp may show considerable differentiation in structure and function. This is perhaps most highly developed in polypore-type basidiocarps, where a number of morphologically distinct hyphal types have been recognized (p. 517).

1.4 | Spores of fungi

The reproduction by means of small spores is a cornerstone in the ecology of fungi. Although a single spore may have a negligible chance of reaching a suitable substrate, spores may be produced in such quantities that even discrete substrates can be exploited by the species as a whole. Only a few fungi make do without spores, surviving solely by means of mycelium and sclerotia. Spores may be organs of sexual or asexual reproduction, and they are involved in dispersal and survival. Gregory (1966) distinguished between **xenospores** (Gr. *xenos* = a foreigner) for spores which are dispersed from their place of origin and **memnospores** (Gr. *mémnon* = steadfast, to persist), which stay where they were formed. Some spores are violently discharged from the organs which bear them, energy for dispersal being provided by the spore itself or the structure producing it (Ingold, 1971). However, many spores are dispersed passively by the action of gravity, air or water currents, rain splash, or by animals, especially insects. Dispersal may also occur by human traffic. Spores may be present in the outdoor air at such high concentrations (e.g. 100 *Cladosporium* spores l^{-1}) that they can cause allergic respiratory diseases when inhaled (Lacey, 1996). In freshwater, the asexually produced spores (conidia) of aquatic hyphomycetes, which colonize autumn-shed tree leaves, may reach concentrations of 10 000–20 000 spores l^{-1} (see p. 685). Long-range dispersal of air-borne spores over thousands of kilometres is known to occur in nature. For instance, the urediniospores of the coffee rust fungus, *Hemileia vastatrix*, are thought to have travelled from Africa to South America by wind at high altitudes, and the urediniospores of black stem rust of wheat (*Puccinia graminis*) undergo an annual migration from states bordering the Gulf of Mexico to the prairies of North America and Canada (Fig. 22.11). These spores are protected from the deleterious effects of UV irradiation in the upper atmosphere by pigments in the spore wall.

Some spores are not dispersed but survive *in situ*, e.g. the oospores of many soil-inhabiting Oomycota (Chapter 5), the zygospores of Zygomycota (Chapter 7) and the chlamydospores of Glomales (see p. 217) and other fungi. Fungal spores may remain dormant for many years, especially under dry and cold conditions (Sussman & Halvorson, 1966; Sussman, 1968). An extreme example of spore survival is shown by the recovery of viable spores of several fungi from glacial ice cores, including those of *Cladosporium cladosporioides* from ice samples 4500 years old (Ma *et al.*, 2000).

The morphology and structure of fungal spores show great variability, from unicellular to multicellular, branched or unbranched or sometimes spirally coiled, thin- or thick-walled with hyaline or pigmented walls, dry or sticky, smooth or ornamented by mucilaginous extensions, spines, folds or reticulations. A number of general descriptive terms have been applied to characterize spores in relation to the number of cells and septa which they contain. Single-celled spores are termed **amerospores** (Gr. *a* = not, *meros* = a part; i.e. not divided), two-celled spores are **didymospores** (Gr. *didymos* = double), spores with more than one transverse septum are **phragmospores** (Gr. *phragmos* = a hedge, barricade), and spores with transverse and longitudinal septa are **dictyospores** (Gr. *dictyon* = a net). These terms may be qualified by prefixes indicating spore pigmentation such as hyalo- for colourless (hyaline) spores and phaeo- for spores with dark-coloured (melanized) walls.

Special terms have also been used to refer to spore shape. **Scolecospores** (Gr. *skolex* = a worm) are worm-shaped, **helicospores** (Gr. *helix* = twisted or wound) are spores with a two- or three-dimensional spiral shape, whilst **staurospores** (Gr. *stauros* = a cross) have arms radiating from a central point or axis. Spore septation, colour and shape, along with other criteria such as the arrangement of structures which bear the spores, have been used in classification and identification, especially in conidial fungi which do not show sexual reproduction. These criteria rarely lead to natural systems of classification, but to 'form genera' or 'anamorph genera' made up of species unified by having similar spore forms.

Some of the more common spore types are described below. There are numerous other, less-common kinds of spore found in fungi, and they are described later, in relation to the particular fungal groups in which they occur.

1.4.1 Zoospores

These are spores which are self-propelled by means of flagella. Propulsion is often coupled with chemotactic movement, zoospores having the ability to sense chemicals diffusing from suitable substrata and to move towards them, or gametes detecting and following extremely low concentrations of hormones. In some cases oxygen or light are also stimuli for tactic movement. The fungal groups which possess flagella are mostly aquatic or, if terrestrial, rely on water for dispersal or infection. Their zoospores are of four kinds (see Fig. 1.17):

1. Posteriorly flagellate zoospores with flagella of the whiplash type are characteristic of the Chytridiomycota (Chapter 6). Each whiplash flagellum has 11 microtubules arranged in the 9 + 2 pattern typical of eukaryotes. The microtubules are enclosed in a smooth, membranous axoneme sheath continuous with the plasma membrane. In most members of the Chytridiomycota there is a single posterior flagellum (Fig. 1.17a), but in the rumen-inhabiting Neocallimastigales there may be up to 16 flagella (Fig. 1.17b). Such spores are driven forward by sinusoidal rhythmic beating of the flagellum. This type of zoospore

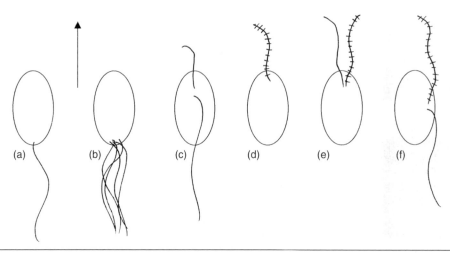

Fig 1.17 Zoospore types found in fungi, diagrammatic and not to scale. The arrow indicates the direction of movement of the zoospore. (a) Posteriorly uniflagellate (opisthokont) zoospore with a flagellum of the whiplash type found in many Chytridiomycota. (b) Posteriorly multiflagellate zoospore with numerous (up to 16) whiplash flagella which occur in certain anaerobic rumen-inhabiting Chytridiomycota (Neocallimastigales). (c) Zoospore with unequal (anisokont) whiplash flagella characteristic of the Myxomycota and the Plasmodiophoromycota. (d) Anteriorly uniflagellate zoospore with a flagellum of the tinsel type, the axoneme being clothed with rows of mastigonemes, typical of the Hyphochytriomycota. (e,f) Biflagellate zoospores with heterokont flagella, one of the whiplash and the other of the tinsel type, which are found in different groups of the Oomycota. For more details turn to the different fungal groups.

flagellation is termed **opisthokont** (Gr. *opisthen* = behind, at the back; *kontos* = a pole). Detailed descriptions of the fine structure of chytridiomycete zoospores are given on p. 129.

2. Biflagellate zoospores with two whiplash flagella of unequal length are called **anisokont** (Fig. 1.17c) and are found in some Myxomycota and the Plasmodiophoromycota, both now classified among the Protozoa (see Chapters 2 and 3).

3. Anteriorly uniflagellate zoospores with a flagellum of the tinsel type are characteristic of the Hyphochytriomycota (Chapter 4). The axoneme sheath of the tinsel or **straminipilous** flagellum (Lat. *stramen* = straw; *pilus* = hair) is adorned by two rows of fine hairs (Fig. 1.17d). These are called **tripartite tubular hairs** or **mastigonemes** (Gr. *mastigion* = a small whip; *nema* = a thread). Rhythmic sinusoidal beating of the tinsel type flagellum pulls the zoospore along, in contrast to the pushing action of whiplash flagellum. Details of the fine structure of this type of zoospore are given in Fig. 4.5.

4. Biflagellate zoospores with anteriorly or laterally attached flagella, one of which is of the whiplash type and the other of the tinsel type (Figs. 1.17e,f), are characteristic of the Oomycota (Chapter 5). Zoospores with the two different kinds of flagellum are **heterokont**. Where the two types of flagellum are attached anteriorly, as in the first-released zoospores of *Saprolegnia*, their propulsive actions tend to work against each other and the zoospore is a very poor swimmer (Fig. 1.17e). However, the secondary zoospore (termed the principal zoospore) in *Saprolegnia* and in many other Oomycota has laterally attached flagella, with the tinsel-type (pulling action) flagellum pointing forwards and the whiplash-type (pushing action) flagellum directed backwards and possibly acting as a rudder, jointly providing much more effective propulsion (Fig. 1.17f).

1.4.2 Sporangiospores

In the Zygomycota, and especially in the Mucorales (see p. 180), the asexual spores are contained in globose sporangia (Fig. 1.18) or cylindrical merosporangia. Because they are non-motile, the spores are sometimes termed **aplanospores** (Gr. *a* = not, *planos* = roaming).

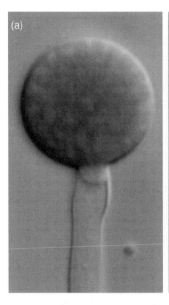
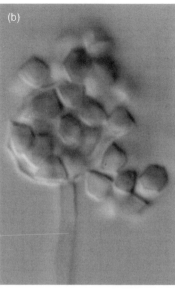

Fig 1.18 Sporangia in *Mortierella* (*Umbelopsis*) *vinacea*. (a) Maturing sporangium in which the cytoplasm is being cleaved into numerous sporangiospores. (b) Release of sporangiospores by breakdown of the sporangial wall. Unusually, in *M. vinacea* the sporangiospores are angular in shape.

5 μm

The spores may be uni- or multinucleate and are unicellular. They generally have thin, smooth walls and are almost always globose or ellipsoid in shape. They are formed by cleavage of the sporangial cytoplasm. They vary in colour from hyaline (colourless) to yellow, due to carotenoid pigments in the cytoplasm. When mature, they may be surrounded by mucilage, in which case they are usually dispersed by rain splash or insects, or they may be dry and dispersed by wind currents. In some genera, e.g. *Pilobolus*, entire sporangia become detached. The number of sporangiospores per sporangium may vary from several thousand to only one. The detachment and dispersal of intact sporangia containing a few sporangiospores or a single one is indicative of the way in which conidia may have evolved from one-spored sporangia.

1.4.3 Ascospores

Ascospores are the characteristic spores of the largest group of fungi, the Ascomycota or ascomycetes. They are meiospores and are formed in the developing ascus as a result of nuclear fusion immediately followed by meiosis. The four haploid daughter nuclei then divide mitotically to give eight haploid nuclei around which the ascospores are cut out. Details of ascospore development are described in Fig. 8.11. In most ascomycetes, the eight ascospores are contained within a cylindrical ascus, from which they are squirted out together with the ascus sap when the tip of the turgid ascus breaks down and the elastic ascus walls contract. The distance of discharge may be 1 cm or more. In some cases, for example, the Plectomycetes (Chapter 11) and in ascomycetes with subterranean fruit bodies, such as the false truffles (*Elaphomyces* spp.; Fig. 11.21) and truffles (*Tuber* spp. and their allies; p. 423), ascospore release is non-violent and their asci are not cylindrical but globose. Ascospores vary greatly in size, shape and colour. In size, the range is from about $4-5 \times 1\,\mu m$ in small-spored forms such as the minute cup fungus *Dasyscyphus*, to $130 \times 45\,\mu m$ in the lichen *Pertusaria pertusa*. The shape of ascospores varies from globose to oval, elliptical, lemon-shaped, sausage-shaped, cylindrical, or needle-shaped. Ascospores are often asymmetric in form with a wider, blunter, anterior part and a narrower, more tapering posterior. This shape increases their acceleration as they are squeezed out through the opening of the ascus. Ascospores may be uninucleate or multinucleate, unicellular or multicellular, divided up by transverse or by transverse and longitudinal septa. In some genera, e.g. *Hypocrea* (Fig. 12.15) or *Cordyceps* (Fig. 12.33), the multicellular ascospores may break up into part-spores within the ascus prior to discharge. The ascospore wall may be thin or thick, hyaline or coloured, smooth or rough, sometimes cast into reticulate folds or ornamented by ridges, and it may have a mucilaginous outer layer which is sometimes extended to form simple or branched appendages, especially in marine ascomycetes where they aid buoyancy and attachment. In many cases, ascospores are resting structures which survive adverse conditions. They may have extensive food reserves in the form of lipids and sugars such as trehalose. Because the formation of ascospores involves meiosis, they are important not only as a means of dispersal and survival but also in genetic recombination.

It is obvious that there is no such thing as a typical ascospore. *Neurospora tetrasperma* will serve as an example of an ascospore whose structure has been extensively studied (Lowry & Sussman, 1958, 1968). This fungus is somewhat unusual in that it has four-spored asci and the ascospores are binucleate. The spores are black, thick-walled and shaped rather like a rugby football, but with flattened ends. The name *Neurospora* refers to the ribbed spores, because the dark outer wall is made up of longitudinal raised ribs, separated by interrupted grooves. The structure of a spore in section is shown in Fig. 1.19. Within the cytoplasm of the spore are the two nuclei, fragments of endoplasmic

reticulum (not illustrated), swollen mitochondria and vacuoles, bounded by single unit membranes. The wall surrounding the protoplast is composed of several layers. The innermost layer is the **endospore**, outside of which is the **epispore**. The ribbed layer is termed the **perispore**. Between the ribs are lighter intercostal veins containing a material which is chemically distinct from the ribs. This material is continuous over the whole surface of the spore, giving it a relatively smooth surface. The spore germinates by the extrusion of germ tubes from a pre-existing **germ pore**, a thin area in the epispore at either end of the spore. In many ascomycetes a trigger is required for germination, e.g. heat shock in *Neurospora* or a chemical stimulus, for example in ascomycetes which grow and fruit on the dung of herbivorous mammals and whose spores are subjected to digestive treatment.

1.4.4 Basidiospores

Basidiospores are the sexual spores which characterize a large group of fungi, the Basidiomycota or basidiomycetes. In comparison with the morphological diversity of ascospores, basidiospores are more uniform. They also show a smaller size range, from about 3 to 20 μm, which is possibly related to their unique method of discharge. They are normally found in groups of four attached by tapering sterigmata to the cell which bears them, the **basidium**. At the time of their discharge all basidiospores

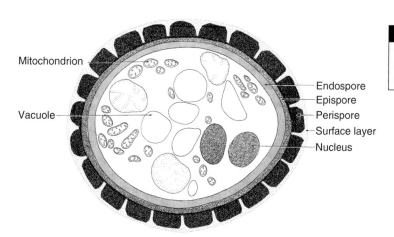

Mitochondrion

Vacuole

Endospore
Epispore
Perispore
Surface layer
Nucleus

Fig 1.19 *Neurospora tetrasperma*. T.S. ascospore. Simplified diagram based on an electron micrograph by Lowry in Sussman and Halvorson (1966).

are unicellular, but they may become septate after release in some members of the Heterobasidiomycetes (Chapter 21). In shape, basidiospores are asymmetric and vary from sub-globose, sausage-shaped, fusoid, to almond-shaped (i.e. flattened), and the wall may be smooth or ornamented with spines, ridges or folds. The colour of basidiospores is important for identification. They may be colourless, white, cream, yellowish, brown, pink, purple or black. The spore colour may be due to pigments in the spore cytoplasm or in the spore wall. The appearance of pigments in the wall occurs relatively late in spore development. This explains the change of colour of the gill

of a domestic mushroom (*Agaricus*) from pink, due to cytoplasmic spore pigments, to dark purplish-brown when mature, due to wall pigments.

The generalized structure of a basidiospore is illustrated in Fig. 1.20. Most basidiospores have a flatter adaxial face and a more curved abaxial face. The point of attachment of the spore to the sterigma is the **hilum**, which persists as a scar at the base of a discharged spore. Close to the hilum is a small projection, the **hilar appendix**. This is involved in the unique mechanism of basidiospore discharge, in which a drop of liquid perched on the hilar appendix coalesces with a second blob of liquid on the spore surface,

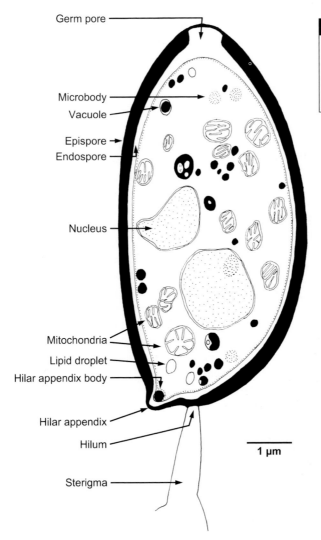

Germ pore

Microbody
Vacuole

Epispore
Endospore

Nucleus

Mitochondria
Lipid droplet
Hilar appendix body

Hilar appendix
Hilum

Sterigma

1 µm

Fig 1.20 Generalized view of a median vertical section through a basidiospore as seen by transmission electron microscopy. For clarity, structures such as endoplasmic reticulum and ribosomes are not illustrated. Diagram based on *Agrocybe acericola*, after Ruch and Nurtjahja (1996).

creating a momentum which leads to acceleration of the spore (Money, 1998; see p. 493). The spore is projected for a short distance (usually less than 2 mm) from the basidium. Violently projected spores are termed **ballistospores** (Lat. *ballista* = a military engine for throwing large stones), but whilst most basidiospores are ballistospores, some are not. For example, in the Gasteromycetes (Chapter 20), which include puffballs, stinkhorns and their allies, violent spore projection has been lost in the course of evolution from ancestors which possessed it. Likewise, the basidiospores of smut fungi (Ustilaginales, Chapter 23) are not violently discharged. The term **statismospore** (Lat. *statio* = standing still) is sometimes used for a spore which is not forcibly discharged.

The cytoplasm of basidiospores usually contains a single haploid nucleus resulting from meiotic division in the basidium; sometimes a post-meiotic division gives rise to two genetically identical nuclei. The structure of the wall is complex. In *Agrocybe acericola* there are two layers, a thicker, dark-pigmented, electron-dense outer layer or epispore, and a thinner, electron-transparent inner layer, the endospore (Ruch & Nurtjahja, 1996; see Fig. 1.20). The cultivated mushroom, *Agaricus bisporus*, has a three-layered wall making up some 35% of the dry weight of the spore (Rast & Hollenstein, 1977), whereas the wall of the *Coprinus cinereus* basidiospore comprises six distinct layers (McLaughlin, 1977). A histochemical feature of the walls of some basidiospores is that they are **amyloid**, i.e. they include starch-like material which stains bluish-purple with iodine-containing stains such as Melzer's reagent. This reaction is used as a taxonomic character. The amyloid reaction is due to the presence of unbranched, short-chain amylose molecules. It has been suggested that this 'fungal starch' may aid dormancy by creating a permeability barrier to oxygen in dry spores. When the amyloid material is dissolved as water becomes available, dormancy is lost and spore germination can proceed (Dodd & McCracken, 1972). In some basidiospores, e.g. those of *Coprinus cinereus* and *Agrocybe acericola*, the basidiospore has a distinct germ pore at the end opposite to the hilum

(see Fig. 1.20). In other basidiomycetes, e.g. *Oudemansiella mucida*, *Schizophyllum commune* and *Flammulina velutipes*, the basidiospores have no specialized pore.

The reserve contents of the spore may vary. In some species, lipid is the major storage product, and there is an apparent lack of insoluble polysaccharides such as glycogen (Ruch & Motta, 1987). In other spores, glycogen predominates. Where lipid is present, germination may be fuelled by its breakdown and utilization, but where it is absent spores are dependent on external nutrient supplies before germination and further development is possible. In addition to the usual organelle complement, microbodies are also prominent in basidiospores. These are single membrane-bound organelles often associated with mitochondria and lipid globules; they may function as glyoxisomes containing enzymes involved in the oxidation of lipids (Ruch & Nurtjahja, 1996).

1.4.5 Zygospores

Zygospores are sexually produced resting structures formed as a result of plasmogamy between gametangia which are usually equal in size (Fig. 1.21a). Nuclear fusion may occur early, or may be delayed until shortly before meiosis and zygospore germination. Zygospores are typical of Zygomycota (Chapter 7). They are often large, thick-walled, warty structures with abundant lipid reserves and are unsuitable for long-distance dispersal, usually remaining in the position in which they were formed and awaiting suitable conditions for further development. The gametangia which fuse to form the zygospore may be uninucleate or multinucleate, and correspondingly the zygospore may have one, two or many nuclei within it. Zygospore germination may be by a germ tube or by the formation of a germ sporangium.

1.4.6 Oospores

An oospore is a sexually produced spore which develops from unequal gametangial copulation or markedly unequal (oogamous) gametic fusion (Fig. 1.21b). It is the characteristic sexually

25 µm   10 µm

Fig 1.21 Sexual resting structures. (a) Zygospore of *Rhizopus sexualis*. The zygote has been produced by fusion of two gametangia and has laid down a thick wall with warty ornamentations. (b) Oospore of *Phytophthora erythroseptica*. The oogonium (o) has grown through the antheridium (a), and the oosphere has picked up a fertilization nucleus in the process. a kindly provided by H.-M. Ho; reprinted from Ho and Chen (1998) with permission of *Botanical Bulletin of Academia Sinica*.

produced spore of the Oomycota (Chapter 5), although oospores are also found in the Monoblepharidales (Chytridiomycota; Fig. 6.25). In the Oomycota, oospore development begins with the formation of one or more oospheres within the larger gametangium, the oogonium. After fertilization, i.e. the receipt of an antheridial nucleus by the oosphere, this lays down a thick wall and becomes the oospore. The number of oospores per oogonium may vary, and this is an important taxonomic criterion. Meiotic nuclear divisions precede oosphere and antheridial maturation in the Oomycota and nuclear fusion follows fertilization, so that the oospore is diploid. The oospore develops a thick outer wall and lays down food reserves, usually in the form of lipids. In the Peronosporales the outer wall of the oospore is surrounded by periplasm, the residual cytoplasm left in the oogonium after the oospheres have been cleaved out. Oospores are sedentary (memnospores) and are important in survival rather than dispersal. They often require a period of maturation before germination can occur and may remain dormant for long periods.

1.4.7 Chlamydospores

In most groups of fungi, terminal or intercalary segments of the mycelium may become packed with lipid reserves and develop thick walls within the original hyphal wall (Fig. 1.22). The new walls may be colourless or pigmented, and are often hydrophobic. Structures of this type have been termed chlamydospores (Gr. *chlamydos* = a thick cloak). They are formed asexually. Generally there is no mechanism for detachment and dispersal of chlamydospores, but they may become separated from each other by the collapse of the hyphae producing them. They are therefore typical memnospores, forming important organs of asexual survival, especially in soil fungi. Chlamydospores may develop within the sporangiophores of some species of the Mucorales, e.g. in *Mucor racemosus* (see Fig 7.14). The Glomales, which are fungal partners in symbiotic mycorrhizal associations with many vascular plants, reproduce primarily by large, thick-walled chlamydospores. These develop singly or in clusters (sporocarps) on coarse hyphae attached to their host plants. They are sedentary in soil but may be dispersed

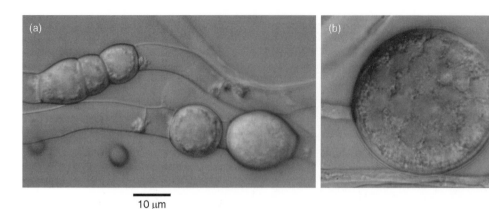

10 μm

Fig 1.22 Chlamydospores formed by soil-borne fungi. (a) Intercalary hyphal chlamydospores in *Mucor plumbeus* (Zygomycota). (b) Terminal chlamydospore in *Pythium undulatum* (Oomycota). Both images to same scale.

by wind or by burrowing rodents which eat the spores. Chlamydospores may also develop within the multicellular macroconidia of *Fusarium* spp. and may survive when other, thin-walled cells making up the spore are degraded by soil micro-organisms. Similar structures are found in old hyphae of the aquatic fungus *Saprolegnia* (see Fig. 5.6g), either singly or in chains. In this genus, the chlamydospores may break free from the mycelium and be dispersed in water currents. Chlamydospores which are dispersed in this way are termed **gemmae** (Lat. *gemma* = a jewel).

The term chlamydospore is also sometimes used to describe the thick-walled dikaryotic spore characteristic of smut fungi (Ustilaginales; Chapter 23) but the term teliospore is preferable in this context. Hughes (1985) has discussed the use of the term chlamydospore.

1.4.8 Conidia (conidiospores)

Conidiospores, commonly known as conidia, are asexual reproductive structures. The word is derived from the Greek *konidion*, a diminutive of *konis*, meaning dust (Sutton, 1986). Conidia are found in many different groups of fungi, but especially within Ascomycota and Basidiomycota. The term conidium has, unfortunately, been used in a number of different ways, so that it no longer has any precise meaning. It has been defined by Kirk *et al.* (2001) as 'a specialized non-motile (cf. zoospore) asexual spore, usually caducous (i.e. detached), not developed by cytoplasmic cleavage (cf.

sporangiospore) or free cell formation (cf. ascospore); in certain *Oomycota* produced through the incomplete development of zoosporangia which fall off and germinate to produce a germination tube'. In many fungi conidia represent a means of rapid spread and colonization from an initial focus of infection.

In general, conidia are dispersed passively, but in a few cases discharge is violent. For instance, in *Nigrospora* the conidia are discharged by a squirt mechanism (Webster, 1952), and in *Epicoccum* (Fig. 17.8) discharge is brought about by the rounding-off of a two-ply septum separating the conidium from its conidiogenous cell (Webster, 1966; Meredith, 1966). In the *Helminthosporium* conidial state of *Trichometasphaeria turcica*, drying and shrinkage of the conidiophore is associated with the sudden development of a gas phase, causing a jolt sufficient to project the conidium (Meredith, 1965; Leach, 1976).

There is great variation in conidial ontogeny. This topic will be dealt with more fully later when considering the conidial states of Ascomycota, and at this stage it is sufficient to distinguish between the major types of conidial development, which may be either **thallic** or **blastic**. Cells which produce conidia are conidiogenous cells. The term thallic is used to describe development where there is no enlargement of the conidium initial (Fig. 1.23a), i.e. the conidium arises by conversion of a pre-existing segment of the fungal thallus. An example of this kind is *Galactomyces candidus*, in which the conidia are

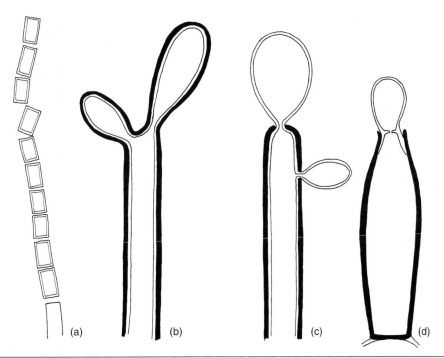

Fig 1.23 Diagrams to illustrate different kinds of conidial development. (a) Thallic development. There is no enlargement of the conidium initial. (b) Holoblastic development. All the wall layers of the conidiogenous cell balloon out to form a conidium initial recognizably larger than the conidiogenous cell. (c) Enteroblastic tretic development: only the inner wall layers of the conidiogenous cell are involved in conidium formation. The inner wall layers balloon out through a narrow channel in the outer wall. (d) Phialidic development: the conidiogenous cell is a phialide. The wall of the phialide is not continuous with the wall surrounding the conidium. The conidial wall arises de novo from newly synthesized material in the neck of the phialide. Diagrams based on Ellis (1971a).

formed by dissolution of septa along a hypha (Fig. 10.10). In most conidia, development is blastic, i.e. there is enlargement of the conidium initial before it is delimited by a septum. Two main kinds of blastic development have been distinguished:

1. **Holoblastic**, in which both the inner and outer wall layers of the conidiogenous cell contribute to conidium formation (Fig. 1.23b). An example of this kind of development is shown by the conidia of *Sclerotinia fructigena* (Fig. 15.3).

2. **Enteroblastic**, in which only the inner wall layers of the conidiogenous cell are involved in conidium formation. Where the inner wall layer balloons out through a narrow pore or channel in the outer wall layer, development is described as **tretic** (Fig. 1.23c). Examples of enteroblastic tretic development are found in *Helminthosporium velutinum* (Fig. 17.12) and *Pleospora herbarum* (Fig. 17.9d). Another important method of enteroblastic development is termed **phialidic**

development. Here the conidiogenous cell is a specialized cell termed the **phialide**. During the expansion of the first-formed conidium, the tip of the phialide is ruptured. Further conidia develop by the extension of cytoplasm enclosed by a new wall layer which is laid down in the neck of the phialide and is distinct from the phialide wall. The protoplast of the conidium is pinched off by the formation of an inwardly growing flange which closes to form a septum (Fig. 1.23d). New conidia develop beneath the earlier ones, so that a chain may develop with the oldest conidium at its apex and the youngest at its base. Details of phialidic development are discussed more fully in relation to *Aspergillus* and *Penicillium* (p. 299), which reproduce by means of chains of dry phialoconidia dispersed by wind. Sticky phialospores which accumulate in slimy droplets at the tips of the phialides are common in many genera; they are usually dispersed by insects, rain splash or other agencies.

As mentioned on p. 24, the term conidium is sometimes used for structures which are probably homologous to sporangia. A series can be erected in the Peronosporales in which there are forms with deciduous sporangia which release zoospores when in contact with water (e.g. *Phytophthora*), and other forms which germinate directly, i.e. by the formation of a germ tube (e.g. *Peronospora*). A similar series can be erected in the Mucorales where in some forms the number of sporangiospores per sporangium is reduced to several or even one (see Figs. 7.24, 7.26, 7.30). One-spored sporangia may be distinguished from conidia by being surrounded by two walls, i.e. that of the sporangium and that of the spore itself.

There are numerous other kinds of spore found in fungi, and they are described later in this book in relation to the particular groups in which they occur.

1.4.9 Anamorphs and teleomorphs

Fungi may exist in a range of forms or morphs, i.e. they may be **pleomorphic**. The morph which includes the sexually produced spore form, e.g. the ascocarp of an ascomycete or the basidiocarp of a basidiomycete, is termed the **teleomorph** (Gr. *teleios, teleos* = perfect, entire; *morphe* = shape, form) (Hennebert & Weresub, 1977). Many fungi also have a morph bearing asexually produced spores, e.g. conidiomata. These asexual morphs are termed **anamorphs** (Gr. *ana* = throughout, again, similar to). In the older literature, the term **perfect state** was used for the teleomorph and **imperfect state** for the anamorph. This is the origin of the name of the artificial group Fungi Imperfecti or Deuteromycetes, which included fungi believed to reproduce only by asexual means. The term **mitosporic fungi** is sometimes used alternatively for such fungi. The complete range of morphs belonging to any one fungus is termed the **holomorph** (Gr. *holos* = whole, entire) (see Sugiyama, 1987; Reynolds & Taylor, 1993; Seifert & Samuels, 2000). Some fungi have more than one anamorph as in the microconidia and macroconidia of some *Neurospora*, *Fusarium* and *Botrytis* species. These distinctive states are

synanamorphs and may play different roles in the biology of the fungus. The morph may have a purely sexual role as a fertilizing agent, e.g. in the case of spermatia of many ascomycetes and rust fungi. Such states have been termed **andromorphs** (Gr. *andros* = a man, male) (Parbery, 1996a).

The existence of different states in the life cycle of a fungus has nomenclatural consequences, because they had often been described separately and given different names before the genetic connection between them was established. Further, even after the proof of an anamorph—teleomorph relationship, usually achieved by pure-culture studies, the anamorphic name may still be in wide use, especially where it is the more common state encountered in nature or culture. For example, most fungal geneticists refer to *Aspergillus nidulans* (the name of the conidial state) instead of *Emericella nidulans* (the name for the ascosporic state; p. 308). Similarly, most plant pathologists use *Botrytis cinerea*, the name for the conidial state of the fungus causing the common grey mould disease of many plants, in preference to the rarely encountered *Sclerotinia* (*Botryotinia*) *fuckeliana*, the name given to the apothecial (ascus-bearing) state (see p. 434).

1.5 | Taxonomy of fungi

Taxonomy is the science of classification, i.e. the 'assigning of objects to defined categories' (Kirk *et al.*, 2001). Classification has three main functions: it provides a framework of recognizable features by which an organism under examination can be identified; it is an attempt to group together organisms that are related to each other; and it assists in the retrieval of information about the identified organism in the form of a list or catalogue.

All taxonomic concepts are man-made and therefore to a certain extent arbitrary. This is especially true of classical approaches relying on macroscopic or microscopic observations because it is a matter of opinion whether the difference in a particular character – say, a spore or the way in which it is formed – is significant to distinguish two fungi and, if so, at which taxonomic level. The great fungal taxonomist R. W. G. Dennis (1960) described taxonomy as 'the art of classifying organisms: not a science but an art, for its triumphs result not from experiment but from disciplined imagination guided by intuition'.

Recently, great efforts have been made at introducing a seemingly more objective set of criteria based directly on comparisons of selected DNA sequences encoding genes with a conserved biological function, instead of or in addition to phenotypic characters. The results of such comparisons are usually displayed as **phylogenetic trees** (see Fig. 1.26), which imply a common ancestry to all organisms situated above a given branch. Such a grouping is ideally 'monophyletic'. However, as we shall see later, quite different phylogenies may result if different genes are chosen for comparison. Further, a decision on the degree of sequence divergence required for a taxonomic distinction is based mainly on numerical parameters generated by elaborate computerized statistical treatments, occasionally at the expense of sound judgement. An excessive emphasis on such purely descriptive studies in the recent literature has led an eminent mycologist to characterize phylogenetic trees as 'the most noxious of all weeds'. Despite their limitations, these methods have led to a revolution in the taxonomy of fungi. At present, a new, more 'natural' classification is beginning to take shape, in which DNA sequence data are integrated with microscopic, ultrastructural and biochemical characters. However, many groups of fungi are still poorly defined, and many more trees will grow and fall before a comprehensive taxonomic framework can be expected to be in place. One of the core problems in fungal taxonomy is the seemingly seamless transition between the features of two taxa, and the question as to where to apply the cut-off point. To quote Dennis (1960) again, 'a taxonomic species cannot exist independently of the human race; for its constituent individuals can neither taxonomise themselves into a species, nor be taxonomised into a species by science in the abstract; they can only be grouped into species by individual taxonomisers'.

1.5.1 Traditional taxonomic methods

Early philosophers classified matter into three Kingdoms: Animal, Vegetable, and Mineral. Fungi were placed in the Vegetable Kingdom because of certain similarities to plants such as their lack of mobility, absorptive nutrition, and reproduction by spores. Indeed, it was at one time thought that fungi had evolved from algae by loss of photosynthetic pigmentation. This was indicated by the use of such taxonomic groups as Phycomycetes, literally meaning 'algal fungi'. This grouping, approximately synonymous with the loose term 'lower fungi', is no longer used because it includes taxa not now thought to be related to each other (chiefly Oomycota, Chytridiomycota, Zygomycota). Early systems of classification were based on morphological (macroscopic) similarity, but the invention of the light microscope revealed that structures such as fruit bodies which looked alike could be anatomically distinct and reproduce in fundamentally different ways, leading them to be classified apart.

Until the 1980s, the taxonomy of fungi was based mainly on light microscopic examination of typical morphological features, giving rise to classification schemes which are now known to be unnatural. Several examples of unnatural groups may be found by comparing the present edition with the previous edition of this textbook (Webster, 1980). Examples of traditional taxonomic features include the presence or absence of septa in hyphae, fine details of the type, formation and release mechanisms of spores (e.g. Kendrick, 1971), or aspects of the biology and ecology of fungi. Useful ultrastructural details, provided by transmission electron microscopy, concern the appearance of mitochondria, properties of the septal pore, details of the cell wall during spore formation or germination, or the arrangement of secretory vesicles in the apex of growing hyphae (Fig. 1.4). Biochemical methods have also made valuable contributions, especially in characterizing higher taxonomic levels. Examples include the chemical composition of the cell wall (Table 1.1), alternative pathways of lysine biosynthesis (see p. 67), the occurrence of pigments (Gill & Steglich, 1987) and the types and amounts of sugars or polyols (Pfyffer *et al.*, 1986; Rast & Pfyffer, 1989).

Microscopic features are still important today for recognizing fungi and making an initial identification which can then, if necessary, be backed up by molecular methods. Indeed, the comparison of DNA sequences obtained from fungi is meaningful only if these fungi have previously been characterized and named by conventional methods. It is therefore just as necessary today as it ever was to teach mycology students the art of examining and identifying fungi.

1.5.2 Molecular methods of fungal taxonomy

A detailed description of modern taxonomic methods is beyond the scope of this book, and the reader is referred to several in-depth reviews of the topic (e.g. Kohn, 1992; Clutterbuck, 1995). A particularly readable introduction to this subject has been written by Berbee and Taylor (1999). Only the most important molecular methods are outlined here. They are based either directly on the DNA sequences or on the properties of their protein products, especially enzymes.

Proteins extracted from the cultures of fungi can be separated by their differential migration in the electric field of an electrophoresis gel. The speed of migration is based on the charge and size of each molecule, resulting in a characteristic banding pattern. Numerous bands will be obtained if the electrophoresis gel is stained with a general protein dye such as Coomassie Blue. More selective information can be obtained by **isozyme analysis**, in which the gel is incubated in a solution containing a particular substrate which is converted into a coloured insoluble product by the appropriate enzyme, or in which an insoluble substrate such as starch is digested. In this way, the number and electrophoretic migration patterns of isoenzymes can be compared between different fungal isolates. Protein analysis is useful mainly for

distinguishing different strains of the same species or members of the same genus (Brasier, 1991a).

Gel electrophoresis can also be used for the separation of DNA fragments generated by various methods. One such method is called **RFLP** (restriction fragment length polymorphisms) and involves the digestion of a total DNA extract or a previously amplified target sequence with one or more restriction endonucleases, i.e. enzymes which cut DNA only at a particular target site defined by a specific oligonucleotide sequence. Fragments from this digest can be blotted from the gel onto a membrane; fragments belonging to a known gene can be visualized by hybridizing with a fluorescent or radioactively labelled DNA probe of the same gene. In this way, a banding pattern is obtained and can be compared with that of other fungal isolates prepared under identical experimental conditions.

A similar method, **RAPD** (random amplified polymorphic DNA), produces DNA bands not by digestion, but by the amplification of DNA sequences. For this purpose, a DNA extract is incubated with a DNA polymerase, deoxynucleoside triphosphates and one or more short oligonucleotides which act as primers for the polymerase by binding to complementary DNA sequences which should be scattered throughout the genome. Amplification is achieved by means of the **PCR** (polymerase chain reaction), in which the mixture is subjected to repeated cycles of different temperatures suitable for annealing of DNA and primer, polymerization, and dissociation of double-stranded DNA. The largest possible size of the amplification product depends on the polymerization time; bands visible on a gel will be produced only if two primer binding sites happen to be in close proximity to each other, so that the intervening stretch of DNA sequence can be amplified from both ends within the chosen polymerization time. The number and size of RAPD bands on electrophoresis gels can be compared between different fungi, provided that all samples have been produced under identical conditions.

Isozyme, RFLP and RAPD analyses all generate data which are useful mainly for comparing closely related isolates. Since the results strongly depend on the experimental conditions employed, there are no universal databases for these types of analysis. Further, they are unsuitable for comparisons of distantly related or unrelated organisms. A breakthrough in the taxonomy of fungi as well as other organisms was achieved when primers were developed which guided the PCR amplification of specific stretches of DNA universally present and fulfilling a homologous function in all life forms. Once amplified, the sequence of bases can be determined easily. Such methods were first applied to bacterial systematics with spectacular results (Woese, 1987). In eukaryotes, the most widely used target sequences are those encoding the 18S or 28S ribosomal RNA (rRNA) molecules, which fulfil a structural role in the small or large ribosomal subunits (respectively), or the non-coding DNA stretches (**ITS**, internal transcribed spacers), which physically separate these genes from each other and from the 5.8S rRNA sequence in the nuclear genome (Fig. 1.24; White *et al.*, 1990). The structural role which rRNA molecules play in the assembly of ribosomes requires them to take up a particular configuration which is stable because of intramolecular base-pairing. Since certain regions of each rRNA molecule hybridize with complementary regions within the same molecule or with other rRNA molecules, mutations in the DNA encoding these regions are rare because they would impair hybridization and thus the functioning of the rRNA molecule unless accompanied by a mutation at the complementary binding site. The non-pairing loop regions of the rRNA gene and the ITS sequences are not subjected to such a strong selective pressure and thus tend to show a higher rate of mutation. Nucleotide sequences therefore permit the comparison of closely related species or even strains of the same species (ITS sequences), as well as that of distantly related taxa or even members of different kingdoms (18S or 28S rRNA). Further, because extensive databases are now available, the sequence analysis of a single fungus can provide meaningful taxonomic information when compared with existing sequences. In addition to ribosomal DNA sequences, genes encoding cytochrome oxidase (*COX*), tubulins or other proteins with conserved functions are now used extensively for phylogenetic purposes.

Once comparative data have been obtained either by banding patterns or gene sequencing, they need to be evaluated. This is usually done by converting the data into a matrix, e.g. by scoring the absence or presence of a particular band. With comparisons of aligned DNA sequences, only informative positions are selected for the matrix, i.e. where variations in the nucleotides between different fungi under investigation are observed. When the matrix has been completed, it can be subjected to statistical treatments, and phylogenetic trees are drawn by a range of algorithms. In some, the degree of relatedness of taxa is indicated by the length of the branch separating them (see Figs. 1.25, 1.26). Such information is thought to be of evolutionary significance; the greater the number of differences between two organisms, the earlier the separation of their evolutionary lines should have occurred.

1.5.3 How old are fungi?

Several lines of evidence indicate that fungi are a very ancient group of organisms. Berbee and Taylor (2001) have attempted to add a timescale to phylogenetic trees by applying the concept of a 'molecular clock', i.e. the assumption that the rate of mutations leading to phylogenetic diversity is constant over time and in various groups of organisms. By calibrating their molecular clock against fossil evidence, Berbee and Taylor (2001) estimated that fungi may have separated from animals some 900 million years ago, i.e. long before the evolution of terrestrial organisms. This estimate is consistent with the discovery of fossilized anastomozing hypha-like structures in sediments about 1 billion years old (Butterfield, 2005). Fungi recognizable as Chytridiomycota, Zygomycota and Ascomycota have been discovered among fossils of early terrestrial plants from the Lower Devonian Rhynie chert, formed some 400 million years ago (Taylor *et al.*, 1992, 1999, 2005). It is apparent that these early terrestrial plants already entertained mycorrhizal symbiotic associations

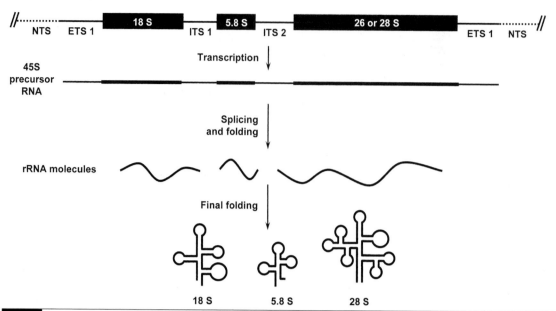

Fig 1.24 The spatial arrangement of a nuclear rRNA gene repeat unit. Each haploid fungal genome contains about 50–250 copies of this repeat, depending on the species (Vilgalys & Gonzalez, 1990). The three structural rRNA genes encoded by one repeat unit, i.e. 18S, 5.8S and 28S, are separated by internal and external transcribed spacers (ITS and ETS, respectively). Adjacent copies of the repeat unit are separated by a short non-transcribed spacer (NTS). The whole unit is transcribed into a 45S precursor RNA in one piece, followed by excision of the three structural RNA molecules from the spacers which are not used. The 5S rRNA gene is encoded at a separate locus. The 18S rRNA molecule is part of the small ribosomal subunit, whereas the other three contribute to the large subunit.

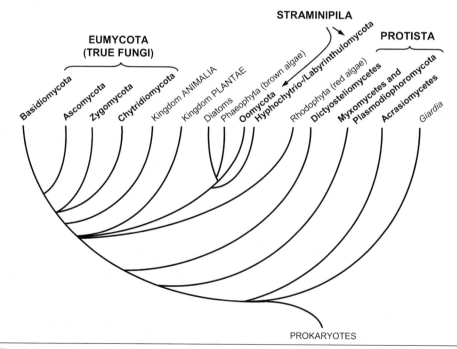

Fig 1.25 The phylogenetic relationships of Fungi and fungus-like organisms studied by mycologists (printed in bold), with other groups of Eukaryota. The analysis is based on comparisons of 18S rDNA sequences. Modified and redrawn from Bruns *et al.* (1991) and Berbee and Taylor (1999).

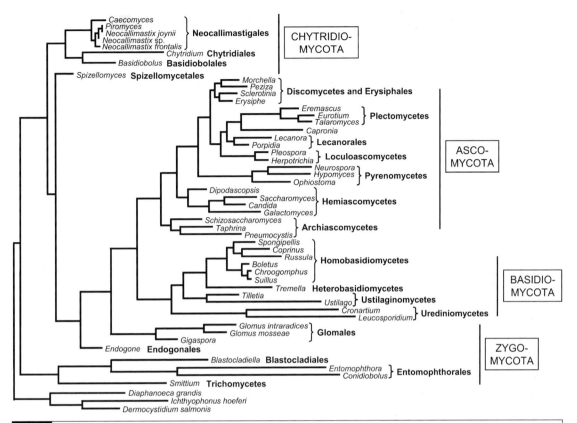

Fig 1.26 Phylogenetic relationships within the Eumycota, based on 18S rDNA comparisons. This tree illustrates the analytical power of molecular phylogenetic analyses; all four phyla of Eumycota are resolved. However, it also highlights problems in that *Basidiobolus* groups with the Chytridiomycota, although sharing essential biological features with the Zygomycota, and that conversely *Blastocladiella* groups with the Zygomycota instead of the Chytridiomycota. Modified and redrawn from Berbee and Taylor (2001), with kind permission of Springer Science and Business media.

with glomalean members of the Zygomycota (see p. 218).

1.5.4 The taxonomic system adopted in this book

The discipline of fungal taxonomy is evolving at an unprecedented speed at present due mainly to the contributions of molecular phylogeny. Numerous taxonomic systems exist, but this is not the place to discuss their relative merits (see Whittaker, 1969; Margulis *et al.*, 1990; Alexopoulos *et al.*, 1996; Cavalier-Smith, 2001; Kirk *et al.*, 2001). In this book we have tried to follow the classification proposed in *The Mycota* Volumes VIIA and VIIB (McLaughlin *et al.*, 2001), but even in these volumes the authors of different chapters have used their own favoured

systems of classification rather than adopting an imposed one. In cases of doubt, we have attempted to let clarity prevail over pedantry.

Fungi in the widest sense, as organisms traditionally studied by mycologists, currently fall into three kingdoms of Eukaryota, i.e. the Eumycota which contain only fungi, and the Protozoa and Chromista (= Straminipila), both of which contain mainly organisms not studied by mycologists and were formerly lumped together under the name Protoctista (Beakes, 1998; Kirk *et al.*, 2001). The Protozoa are notoriously difficult to resolve by phylogenetic means, and the only firm statement which can be made at present is that they are a diverse and ancient group somewhere between the higher Eukaryota ('crown eukaryotes') and the

Table 1.2. The classification scheme adopted in this book, showing mainly those groups treated in some detail.

KINGDOM PROTOZOA
Myxomycota (Chapter 2)
    Acrasiomycetes
    Dictyosteliomycetes
    Protosteliomycetes
    Myxomycetes
Plasmodiophoromycota (Chapter 3)
    Plasmodiophorales
    Haptoglossales (Oomycota?)

KINGDOM STRAMINIPILA
Hyphochytriomycota (Chapter 4)
Labyrinthulomycota (Chapter 4)
    Labyrinthulomycetes
    Thraustochytriomycetes
Oomycota (Chapter 5)
    Saprolegniales
    Pythiales
    Peronosporales

KINGDOM FUNGI (EUMYCOTA)
Chytridiomycota (Chapter 6)
    Chytridiomycetes
Zygomycota (Chapter 7)
    Zygomycetes
    Trichomycetes
Ascomycota (Chapter 8)
    Archiascomycetes (Chapter 9)
    Hemiascomycetes (Chapter 10)
    Plectomycetes (Chapter 11)
    Hymenoascomycetes
        Pyrenomycetes (Chapter 12)
        Erysiphales (Chapter 13)
        Pezizales (Chapter 14)
        Helotiales (Chapter 15)
        Lecanorales/lichens (Chapter 16)
    Loculoascomycetes (Chapter 17)
Basidiomycota (Chapter 18)
    Homobasidiomycetes (Chapter 19)
    Homobasidiomycetes: gasteromycetes
    (Chapter 20)
    Heterobasidiomycetes (Chapter 21)
    Urediniomycetes (Chapter 22)
    Ustilaginomycetes (Chapter 23)

prokaryotes (Kumar & Rzhetsky, 1996). An overview of eukaryotic organisms, in which those groups treated in this book are highlighted, is given in Fig. 1.25. Among the Protozoa, the Plasmodiophoromycota are given extensive treatment because of their role as pathogens of plants (Chapter 3), whereas the various forms of slime moulds are considered only briefly (Chapter 2). Similarly brief overviews will be given of most groups of Straminipila studied by mycologists (Chapter 4), except for the Oomycota which, despite their separate evolutionary origin, represent a major area of mycology (Chapter 5). All remaining chapters deal with members of the Eumycota (= Kingdom Fungi). The scheme is summarized in Table 1.2 and illustrated in Fig. 1.26. An overview of the nomenclature used for describing taxa within the Eumycota is given in Table 1.3.

In the past, fungi which solely or mainly reproduce asexually (Fungi Imperfecti, Deuteromycota, mitosporic fungi, anamorphic fungi) were considered separately from their sexually reproducing relatives the teleomorphs, and separate anamorph and teleomorph genera were erected. However, information from pure-culture studies and molecular phylogenetic approaches has linked many anamorphs with their teleomorphs. For instance, the conidial (imperfect) state of the common brown-rot fungus of apples and other fruits is called *Monilia fructigena*, whereas the sexual (perfect)

Table 1.3. Example of the hierarchy of taxonomic terms. The wheat stem rust fungus, *Puccinia graminis*, is used as an example.

Kingdom Fungi
  Subkingdom Eumycota
    Phylum Basidio**mycota**
      Class Uredinio**mycetes**
        Order Uredin**ales**
          Family Puccini**aceae**
            Genus *Puccinia*
              Species *Puccinia graminis*
                Race *Puccinia graminis*
                  f. sp. *tritici*

state is apothecial, being called *Sclerotinia* (*Monilinia*) *fructigena*. As far as is possible, we shall consider anamorphic states of fungi in the context of their known sexual state. Thus, an account of the brown-rot of fruits, although encountered predominantly as the conidial state, will be given in the chapter dealing with apothecial fungi (Helotiales, Chapter 15). Where practical, we have given the teleomorph name priority over the anamorph. As a long-term future goal, Seifert and Samuels (2000) and Seifert and Gams (2001) have outlined a unified taxonomy which might ultimately lead to the abolition of the names of anamorphic genera.

However, with certain ecological groups such as the Ingoldian aquatic fungi (Section 25.2) and nematophagous fungi (Section 25.1), which have diverse relationships, we have deliberately chosen to consider them in their ecological context rather than along with their varied taxonomic relatives.

Protozoa: Myxomycota (slime moulds)

2.1 Introduction

When the first slime moulds were described by Johann H.F. Link in 1833, they were given the term myxomycetes (Gr. *myxa* = slime). Link used the suffix *-mycetes* because of the superficial similarity of the fructifications of slime moulds with the fruit bodies of certain fungi, notably Gasteromycetes (see Chapter 20). Although it has been appreciated for some time that they lack any true relationship with the Eumycota (de Bary, 1887; Whittaker, 1969), slime moulds have none the less been studied mainly by mycologists rather than protozoologists, probably because they occur in the same habitats as fungi and are routinely encountered during fungus forays. Since slime moulds are only rarely covered by zoology courses even today, they are briefly described in this chapter, referring to more specialized literature as appropriate.

Slime moulds differ substantially from the Eumycota not only in phylogenetic terms, but also regarding their physiology and ecology. Their vegetative state is that of individual **amoebae** in the cellular slime moulds, or of a multinuclear (coenocytic) **plasmodium** in the plasmodial slime moulds. Motile stages bearing usually two anterior whiplash-type flagella may be present in the plasmodial slime moulds (Sections 2.4, 2.5) and in the Plasmodiophoromycota (Chapter 3). Amoebae or plasmodia feed by the ingestion (**phagocytosis**) of bacteria, yeast cells or other amoebae. This is followed by

intracellular digestion in vacuoles. The mode of nutrition in slime moulds is therefore fundamentally different from extracellular degradation and absorption as shown by Eumycota.

Numerous phylogenetic analyses of DNA sequences encoding rRNA molecules and various structural proteins or enzymes have been carried out, but the results obtained are difficult to interpret because the comparison of different genes have led to rather variable phylogenetic schemes. Of the four groups treated in this chapter, it seems that the Dictyosteliomycetes, Protosteliomycetes and Myxomycetes are related to each other whereas the Acrasiomycetes have a different evolutionary origin (Baldauf, 1999; Baldauf *et al.*, 2000). The general evolutionary background is, however, still rather diffuse in these lower eukaryotes.

2.2 Acrasiomycetes: acrasid cellular slime moulds

The Acrasiomycetes, or Acrasea as they are called in zoological classification schemes, are a small group currently comprising 12 species in six genera (Kirk *et al.*, 2001). Although appearing somewhat removed from the bulk of the slime moulds, they still clearly belong to the Protozoa (Roger *et al.*, 1996). The trophic stage consists of amoebae which are morphologically distinct from those of the dictyostelid cellular slime moulds (Section 2.3) in having a cylindrical, rather than flattened, body bearing a single

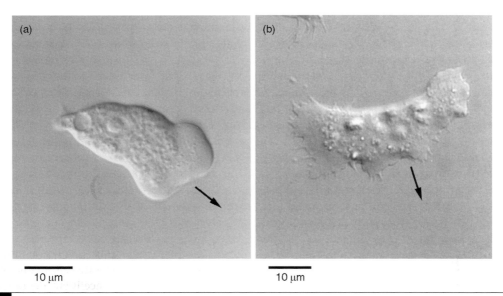

10 µm   10 µm

Fig 2.1 Amoebae of cellular slime moulds. The arrows indicate the direction of movement at the time when the photomicrographs were taken. (a) Limax-type amoeba of *Acrasis rosea*, an acrasid cellular slime mould. Note the absence of granular contents from the lobose pseudopodium at the tip of the amoeba. (b) Amoeba of *Protostelium mycophaga* with filose pseudopodia. Reproduced from Zuppinger and Roos (1997), with permission from Elsevier; original prints kindly supplied by C. Zuppinger.

large-lobed (**lobose**) anterior pseudopodium. The granular cellular contents trail behind the pseudopodium, which appears clear. The posterior end is knob-shaped and is called the uroid (Fig. 2.1a). Such amoebae are of the **limax** type because their movement resembles that of slugs of the genus *Limax*. Good accounts of the acrasids have been given by Olive (1975) and Blanton (1990).

Acrasid slime moulds are common on decaying plant matter, in soil, on dung and on rotting mushrooms, but they are rarely recorded because of their small size, which necessitates observations with a dissecting microscope. The most readily recognized species is *Acrasis rosea*, which has orange- or pink-coloured amoebae due to the presence of carotenoid pigments, including torulene (Fuller & Rakatansky, 1966). *Acrasis rosea* can be observed if dead twigs, leaves or fruits are incubated on weak nutrient agar for a few days. Spore-bearing structures called **sorocarps** (Gr. *sorus* = heap, *karpos* = fruit) will develop, and spores can be transferred to fresh agar with yeast cells as a food source (Blanton, 1990). The uninucleate amoebae feed on yeast cells, bacteria or fungal spores and can

encyst under unfavourable conditions, especially drought, to form **microcysts**. Each microcyst germinates again to release a single amoeba. Eventually amoebae aggregate to form a pseudoplasmodium, in which the individual amoebae retain their identity but are surrounded by a common sheath. The chemical signal for aggregation is unknown but it is not cyclic AMP (cAMP) as in the dictyostelid slime moulds (see below). The pseudoplasmodium develops into a branched sorocarp in which the amoebae align themselves in single rows and then round off, each forming a walled spore. Each spore germinates to release a single amoeba. The cells making up the stalk of the sorocarp also encyst and are capable of germination (Olive, 1975). Sexual reproduction in the acrasid slime moulds is unknown.

2.3 | Dictyosteliomycetes: dictyostelid slime moulds

The Dictyosteliomycetes (zool.: Dictyostelia) are a group of cellular slime moulds comprising

46 species in four genera (Kirk *et al.*, 2001). The best-known example is *Dictyostelium* which has been so named because the stalk of its multicellular sorocarp appears as a network, made up from cellulose walls secreted by the amoebae from which it is formed. *Dictyostelium* spp. are common in soil, on decaying plant material and on dung, and can be demonstrated by smearing non-nutrient agar with cells of a suitable bacterial food such as *Escherichia coli* or *Klebsiella aerogenes*, and adding a small crumb of moistened soil to the centre of the bacterial smear. Amoebae will creep out of the soil and consume the bacteria. At the end of the feeding phase, sorocarps develop and isolations can be made (Cavender, 1990). An axenic defined medium has been developed for *D. discoideum* and has greatly facilitated experimentation with this organism (Franke & Kessin, 1977). Good general accounts of the dictyostelids are those by K.B. Raper (1984), Cavender (1990) and Alexopoulos *et al.* (1996). The history of research on *Dictyostelium* has been recounted by Bonner (1999). Work on *D. discoideum* has contributed significantly to our understanding of the key features of eukaryotic cell biology, especially signalling events, phagocytosis, and the evolution of multicellularity in animals. Consequently, there is a vast literature on this organism. An excellent introduction to the impact of research on *D. discoideum* on general eukaryotic biology is the book by Kessin (2001), and challenging questions have been summarized by Ratner and Kessin (2000). Bonner (2001) has also provided a stimulating read.

The life cycle of *D. discoideum* is shown in Fig. 2.2. Amoebae of dictyostelids are morphologically different from those of acrasids in that they have **filose** (acutely pointed) rather than lobose pseudopodia (see Fig. 2.1b). Each spore from a sorocarp germinates to give rise to one uninucleate haploid amoeba which feeds by phagocytosis of bacteria. Amoebae reproduce asexually by division to form two haploid daughter amoebae. As with acrasid slime moulds, the amoebae of dictyostelids can form microcysts under unfavourable environmental conditions. Encystment may be triggered by the production of ammonia, which thus functions as a signal molecule (Cotter *et al.*, 1992). Sexual reproduction occurs by means of **macrocysts** and is initiated when two compatible amoebae meet and fuse. Both homothallic and heterothallic species and strains of *Dictyostelium* are known. In *D. discoideum*, fusion is inhibited by light and by the presence of cAMP, but is stimulated by ethylene (Amagai, 1992). The fusion cell is greatly enlarged relative to the two progenitor amoebae. This giant cell attracts unfused amoebae which aggregate and secrete a sheath (primary wall) around themselves and the zygote. Inside the primary wall, the giant cell undergoes karyogamy, and the resulting zygote feeds cannibalistically on the other amoebae by phagocytosis and eventually produces a secondary wall. Cellulose seems to be the main structural wall polymer. Meiosis is followed by mitotic divisions and cytoplasmic cleavage, and the macrocyst germinates to release numerous haploid uninucleate amoebae (Nickerson & Raper, 1973; Szabo *et al.*, 1982).

The most striking feature of *D. discoideum* is the **aggregation** of thousands of amoebae to form a pseudoplasmodium with radiating arms (Figs. 2.3a,b). This is a vegetative process not involving meiosis or mitosis. Aggregation is initiated when the bacterial food supply is exhausted, and follows the gradient of a hormone which causes directional (chemotactic) movement of starving amoebae (Konijn *et al.*, 1967; Swanson & Taylor, 1982). In the case of *D. discoideum*, the hormone is cAMP (Konijn *et al.*, 1967), but other molecules are implicated in this role in different dictyostelids. Upon exposure to a cAMP gradient, amoebae of *D. discoideum* change their shape from isodiametric to elongated, with the migrating tip pointing towards the highest cAMP concentration. Migration occurs in waves which correspond to the production of cAMP by starving amoebae, its detection and further synthesis by neighbouring amoebae, and its degradation by cAMP phosphodiesterase (Nagano, 2000; Weijer, 2004). In this way, waves of cAMP diffuse outwards, and waves of amoebae migrate inwards. During aggregation, amoebae migrate to the centre or one of the arms of the pseudoplasmodium. This is a highly co-ordinated effort in which hundreds of thousands of

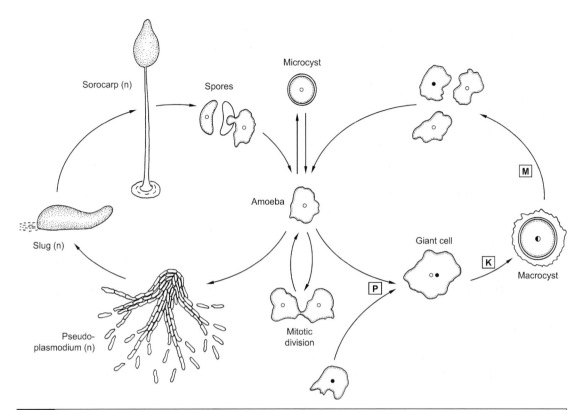

Fig 2.2 Life cycle of *Dictyostelium discoideum*. The central feature is the haploid amoeba which is free-living in the soil. It divides mitotically to produce two daughter amoebae or, under unfavourable conditions, may form a microcyst. If two amoebae of compatible mating type meet, a diploid macrocyst may be formed which can remain dormant for some time and eventually germinates by meiosis and then mitosis to release numerous haploid amoebae. Under certain circumstances, starvation may lead to aggregation of amoebae to form a slug and a sorocarp in which individual amoebae become converted into spores. These are purely asexual, and meiosis is not involved in their formation or germination. Open and closed circles represent haploid nuclei of opposite mating type; diploid nuclei are larger and half-filled. Key events in the life cycle are plasmogamy (P), karyogamy (K) and meiosis (M).

amoebae from an area of $1 \, cm^2$ of soil can be involved. Aggregating amoebae adhere to each other and secrete a common slime sheath (Figs. 2.3c,d). Eventually they pile up to form a compact bullet-shaped **slug** which flops over onto the substratum. In *D. discoideum* and some other species, the slug undergoes a period of **migration** towards the light (Figs. 2.3e–g). The individuality of amoebae is retained within the slug. As the slug moves along, it leaves behind a slime trail. Within the slug, the amoebae are divided into two functionally different populations, i.e. an anterior group of large, highly vacuolated cells (pre-stalk cells) and a posterior group of smaller ones, the pre-spore cells (Fig. 2.4). It is the pre-stalk group of cells which co-ordinates slug migration by secreting cAMP.

Various environmental stimuli can direct movement. For instance, the anterior end of the slug follows an oxygen gradient but is repelled by ammonia. Temperature as well as light can also act as triggers of directed movement. The end of the migration phase is marked by the rounding-off and erection of the pseudoplasmodium to form a flat-based, somewhat conical structure, which undergoes further development by differentiating into a multicellular stalk composed of the large anterior cells, and the sorus which rises up on the outside of the stalk (Figs. 2.3h–j, 2.4). This final stage of development is called **culmination**. About 80% of the amoebae become converted into spores, with the remainder being sacrificed for the formation of the fruit body structure.

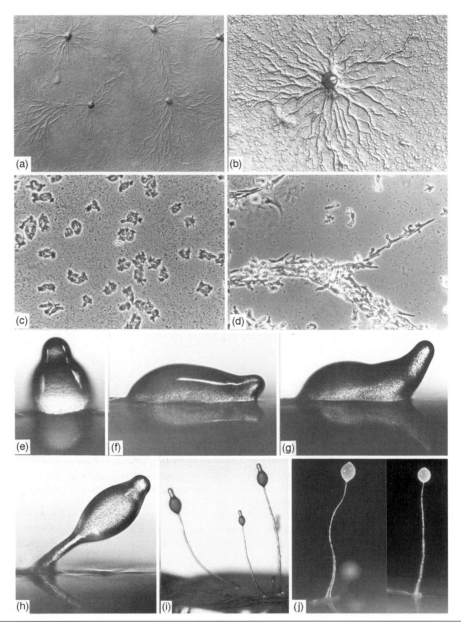

Fig 2.3 *Dictyostelium discoideum* development. (a) Aggregation of amoebae. (b) Aggregation, enlarged. (c) Amoebae feeding on bacteria; note their isodiametric shape. (d) Aggregating amoebae; note their elongated shape. (e) Late aggregation stage. (f,g) Migration stage. (h) Culmination; the spore mass is rising around the stalk. (i) Spore mass almost at the apex of the stalk. (j) Mature sorocarps.

The ability of free-living individual amoebae of *Dictyostelium* to aggregate into the multicellular slug has led to dictyostelid slime moulds being called social amoebae (Kessin, 2001). This phenomenon gives rise to interesting and fundamental questions. To give an example, since amoebae in the anterior end of the slug become stalk cells and are thus excluded from perpetuation as spores, cells skiving off to the rear of the slug and thereby avoiding self-sacrifice would have a selective advantage. 'Cheater strains' are indeed known from nature and the laboratory;

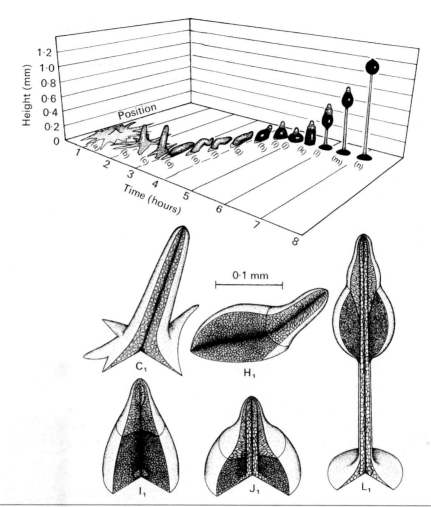

Fig 2.4 *Dictyostelium discoideum.* Development of sorocarp (after Bonner, 1944). (a)−(c) Aggregation. (d)−(h) Migration. (i)−(n) Culmination. C_1 End of aggregation. H_1 End of migration. I_1 Beginning of culmination and stalk formation. J_1 Flattened stage of culmination. I_1 A later stage of culmination.

some of them cheat only to a degree or only if altruistic non-cheater strains are present, whereas others are entirely unable to make a fruit body in the absence of wild-type amoebae prepared to form the pre-stalk cells (Dao *et al.*, 2000; Strassmann *et al.*, 2000). The cheater phenomenon has raised thought-provoking questions about the evolution and control of cheating in social systems (Hudson *et al.*, 2002).

Another interesting aspect involves the mode of nutrition of *Dictyostelium* by the phagocytosis of bacterial cells. Several bacteria pathogenic to humans and other animals, e.g. *Pseudomonas aeruginosa* and *Legionella pneumophila*, also kill *Dictyostelium* upon ingestion (Solomon *et al.*, 2000; Pukatzki *et al.*, 2002). The observation that interactions between *Dictyostelium* amoebae and phagocytosed bacterial pathogens are similar to those involving human phagocytes may stimulate further research on this fascinating slime mould (Steinert & Heuner, 2005).

2.4 | Protosteliomycetes: protostelid plasmodial slime moulds

This class of organisms (zool.: Protostelea) comprises 14 genera and 35 species (Kirk *et al.*, 2001).

Useful treatments of the group have been written by Olive (1967, 1975) and Spiegel (1990). Protostelids are ubiquitous on decaying plant parts in soil and humus, as well as on dung or in freshwater. They occur in all climatic zones from the tundra to tropical rainforests. Protostelids produce amoebae with filose pseudopodia (Fig. 2.1b), feeding phagocytotically on bacteria, yeast cells or spores of fungi. Some species also produce small plasmodia, thereby providing structural affinities to both the cellular and plasmodial slime moulds. Sporulation occurs by the conversion of a feeding amoeba or plasmodium into a round prespore cell which then rises at the tip of a delicate acellular stalk, ultimately forming one or several spores in a single sporangium. It is possible to isolate protostelids by transferring a spore from its stalk onto a weak nutrient agar plate with appropriate food organisms.

Protostelium is a typical member of the group (Fig. 2.5). The **sporocarp** consists of a long, slender stalk about 75 μm long, bearing a single spherical spore about 4–10 μm in diameter. The spore is deciduous and readily detached. Upon germination, a single uninucleate amoeba with thin pseudopodia emerges. The amoeboid stage feeds voraciously on yeast cells and may also feed cannibalistically on amoebae of the same species. Development of the sporocarp probably follows the generalized pattern described by Olive (1967) and summarized in Fig. 2.6. When feeding stops, the amoeba rounds off and heaps its protoplasm in the centre to form the 'hat-shaped' stage (Fig. 2.6b). A membranous, pliable, impermeable sheath develops over the surface of the cell. When the protoplast contracts into the central hump, the sheath collapses at the margins, forming the disc-like base to the stalk of the sporocarp. This may be the structural equivalent of the hypothallus of the Myxomycetes (see p. 48). Within the protoplast, a granular basal core, the **steliogen**, differentiates and begins to mould a hollow tube (Figs. 2.6d,e). As the tube extends at its tip, the protoplast migrates upwards, always seated on top of the growing tip. The entire structure remains covered by the sheath. Tube extension is an actin–myosin-driven process (Spiegel

et al., 1979). Ultimately, the steliogen is left behind at the tip of the stalk to form an **apophysis** (Fig. 2.5a), and the protoplast secretes a cell wall and becomes the spore.

Variations of this pattern occur within the protostelids. For instance, some species produce spores which are discharged forcibly (e.g. Spiegel, 1984). In *Ceratiomyxa fruticulosa*, a species which may or may not belong to the Protosteliomycetes (Spiegel, 1990; Kirk *et al.*, 2001; Clark *et al.*, 2004), numerous spores are formed externally on a sporocarp (Figs. 2.7a,b) and are the product of meiosis. They germinate to release a single quadrinucleate protoplast (Figs. 2.7c–e) which divides repeatedly to produce a clump of four and later eight haploid cells, the octette stage (Figs. 2.7f,g). Each of these cells releases a motile cell (a **swarmer**) which has one or two whiplash-type flagella (Fig. 2.7h).

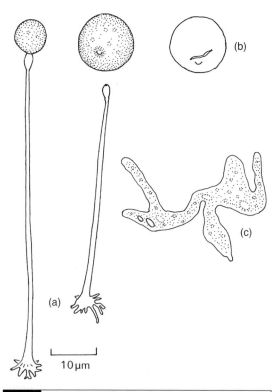

Fig 2.5 *Protostelium* sp. (a) Two sporocarps, one immature, the other with a detached spore. Note the apophysis beneath the spore. (b) Empty spore case after germination. (c) Amoeboid phase.

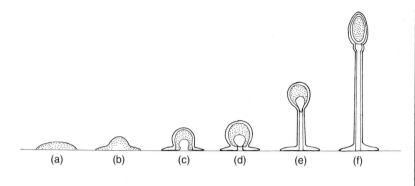

Fig 2.6 Sporogenesis in a protostelid (after Olive, 1967). (a) Early pre-spore stage. (b) Hat-shaped stage. (c) Appearance of the steliogen. (d) Beginning of stalk formation. (e) Later stage in stalk development, with steliogen extending into upper part of stalk tube. (f) Mature sporocarp showing terminal spore, with subtending apophysis, outer sheath, and inner stalk tube.

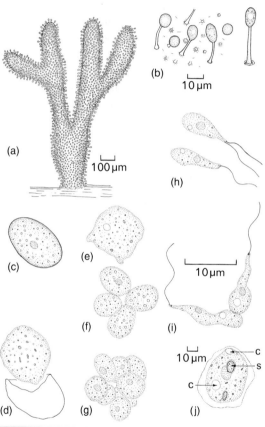

Fig 2.7 *Ceratiomyxa fruticulosa*. (a) Fruiting sporocarp bearing stalked spores. (b) Portion of the surface of the sporocarp showing spores and their attachment. (c) Spore. (d) Naked protoplast emerging from the spore at germination. (e) Naked protoplast before cleavage. (f) Cleavage of protoplast to form a tetrad of protoplasts. (g) Octette stage: a clump of eight protoplasts. (h) Uniflagellate and biflagellate swarmer released from the octette protoplasts. (i) Copulation of swarmers by their posterior ends. (j) Young plasmodium: c, contractile vacuole; s, ingested spore within food vacuole. (c–i) to same scale.

The swarmers eventually fuse to form a diploid zygote which initiates the plasmodial stage (Figs. 2.7i,j), from which the sporocarp develops (Spiegel, 1990). *Ceratiomyxa fruticulosa* thus shows features of both the Protosteliomycetes in producing its spores externally, and the Myxomycetes (see below) in having a flagellated stage in its life cycle. Its precise phylogenetic position remains to be established. This species is probably homothallic (Clark *et al.*, 2004). Its whitish semitransparent sporocarps are rather common on the surface of rotting wood (Plate 1a).

2.5 Myxomycetes: true (plasmodial) slime moulds

The Myxomycetes (zool.: Myxogastrea) are by far the largest group of slime moulds, comprising some 800 species in 62 genera which are currently divided into five orders (Kirk *et al.*, 2001). General accounts have been given by Frederick (1990), Stephenson and Stempen (1994) and Alexopoulos *et al.* (1996). A monograph of British species has been compiled by Ing (1999). These are the familiar slime moulds so common on moist, decaying wood and other organic substrata. They are also abundant in soil and may fulfil ecological functions which are as yet poorly understood (Madelin, 1984).

The vegetative phase is a free-living plasmodium, i.e. a multinucleate wall-less mass of protoplasm. This may or may not be covered by a slime sheath. Plasmodia vary in size and can be loosely grouped into three categories.

(1) **Protoplasmodia** are inconspicuous microscopic structures usually giving rise only to a single sporangium. They resemble the simple plasmodia of protostelids.

(2) **Aphanoplasmodia** (Gr. *aphanes* = invisible) are thin open networks of plasmodial strands. The aphanoplasmodium is transparent, with individual strands only 5–10 μm wide and the entire plasmodium about 100–200 μm in diameter. Most aphanoplasmodia are only seen with the aid of a dissection microscope.

(3) **Phaneroplasmodia** (Gr. *phaneros* = visible) are large sheets or networks with conspicuous veins (Fig. 2.8a) within which the protoplasm shows rhythmic and reversible streaming, each pulse lasting about 60–90 s. This striking phenomenon is readily observed with a dissection microscope and is probably due to interactions of Ca^{2+} ions with cytoskeletal elements lining the veins (see Section 2.5.3).

2.5.1 Life cycle of myxomycetes

The life cycle of *Physarum polycephalum*, a typical myxomycete, is summarized in Fig. 2.9. The plasmodium is diploid and feeds by phagocytosis of bacteria, yeasts or fungal mycelia or spores. It gives rise to a sporophore under appropriate conditions. The haploid spores are dispersed

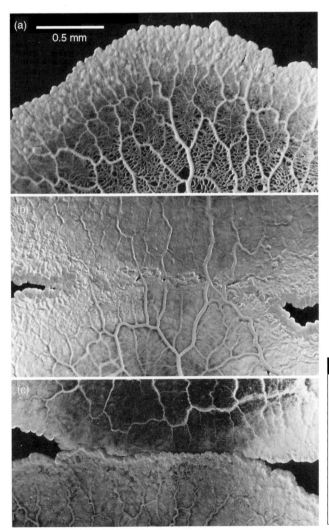

(a)
0.5 mm

(b)

(c)

Fig 2.8 Phaneroplasmodia of *Physarum polycephalum*. (a) Margin of extending plasmodium. The protoplasm is particularly dense at the advancing edge. Further behind, protoplasm is concentrated in large veins which show rhythmic pulsation. (b) Fusion between compatible plasmodia. Note the complete fusion of veins. (c) Lethal reaction following fusion between incompatible plasmodia. (a) from Carlile (1971), (b) and (c) from Carlile and Dee (1967), by permission of Academic Press (a) and Macmillan Journals (b,c). Original prints kindly supplied by M. J. Carlile.

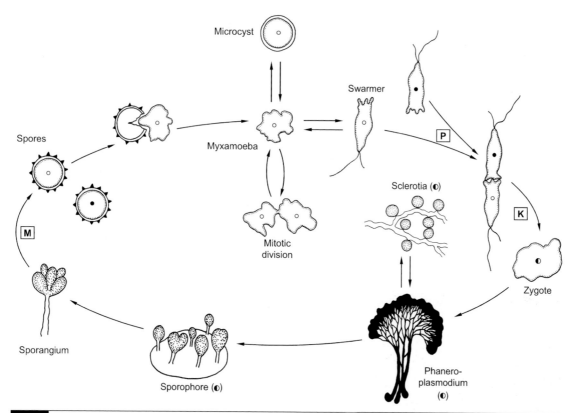

Fig 2.9 Life cycle of the myxomycete *Physarum polycephalum*. Spores released from the sporangium are haploid and can germinate by releasing either a single myxamoeba or a swarmer cell. These two cell types are interconvertible. The myxamoeba can divide mitotically. In *P. polycephalum*, plasmogamy (P) usually takes place between swarmers which must belong to different mating types. Karyogamy (K) follows, and the diploid zygote establishes a phaneroplasmodium. When nutrients become limiting, a sporophore is formed and differentiates sporangia in which meiosis (M) occurs. Unfavourable conditions can be overcome at the haploid stage when the myxamoeba forms a microcyst, or at the diploid stage when the plasmodium forms sclerotia. Open and closed circles represent haploid nuclei of opposite mating type; diploid nuclei are larger and half-filled.

by wind or insects and, depending on environmental conditions such as moisture, germinate by releasing either amoebae or zoospores (swarmers) with usually two anterior whiplash flagella, of which one is shorter than the other and is thus often invisible (Fig. 2.10). The amoebae are called **myxamoebae**, in order to distinguish them from the amoebae of cellular slime moulds which have a different function in the life cycle. Myxamoebae are capable of asexual reproduction by division. Swarmers cannot divide, but can readily and reversibly convert into myxamoebae. Under adverse conditions, myxamoebae secrete a wall to form microcysts. Both swarmers and myxamoebae form filose pseudopodia with which they engulf

their prey. Sexual reproduction is initiated when two haploid myxamoebae or swarmers of compatible mating type fuse to form a zygote from which the diploid plasmodium develops. The plasmodium can survive adverse conditions by turning into a resistant sclerotium in which numerous walled compartments (**spherules**), each containing several nuclei, are formed. Upon resumption of growth, the protoplasts emerge from their spherules and fuse to re-establish the plasmodium. When sexual reproduction ensues, the entire content of a plasmodium is converted into one or more sporangia in which meiosis takes place. Beneath the developing sporangia, the plasmodium deposits a specialized layer, the **hypothallus**, which is very variable in

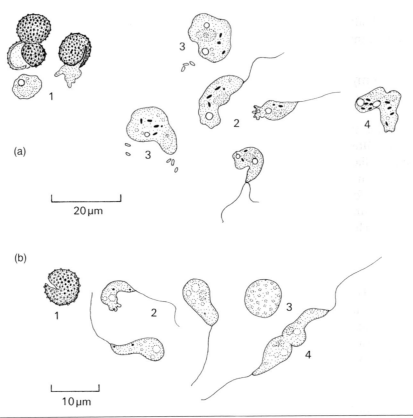

Fig 2.10 Spore germination and swarmers in *Physarum* and *Reticularia*. a. *Physarum polycephalum*: I, spores germinating to release myxamoebae; 2, uniflagellate and biflagellate swarmers, note the pseudopodia at the front end of one swarmer; 3, myxamoeba; 4, fusion between two myxamoebae. b. *Reticularia lycoperdon*: I, spore showing cracked wall; 2, swarmers, one with pseudopodia; 3, encystment stage; 4, fusion between two swarmers.

form (disc-like, membranous, horny or spongy). In *P. polycephalum*, sexual reproduction is triggered by environmental factors such as starvation and light, and by chemical factors, e.g. Ca^{2+} and malate (Renzel *et al.*, 2000).

Depending on species, the sporophores may take a range of shapes. Intermediates between these different types of sporophore are possible. The most common form is the **sporangium**, a vessel enclosed by a wall (**peridium**) within which the spores are contained (Plates 1e,f,h). Protoplasmodia produce only one sporangium each, but numerous sporangia may arise from phaneroplasmodia. Sporangia may be stalked or sessile. A second common sporophore is the **aethalium** (Gr. *aethes* = irregular, curious, unusual) in which the entire plasmodium becomes converted into a hemispherical or cushion-shaped structure (Plates 1b–d,g). This can comprise several sporangia, but these have usually lost their structural identity and are surrounded by one common peridium. In a **pseudoaethalium**, several sporangia are grouped together but are still recognized as structurally distinct. In the **plasmodiocarp**, the protoplasm accumulates in the main veins of the plasmodium, and spores are produced there.

Frederick (1990) has described methods for the isolation and cultivation of myxomycetes. Some species, such as *Physarum polycephalum*, can be grown in axenic culture and have become valuable systems for experimentation. Other species need to be fed with bacteria or sterile oat flakes. Plasmodia can be maintained for prolonged periods in a vegetative state, and sclerotia can be stored dry for months. Spores have been revived after more than 50 years' storage in a herbarium (Elliott, 1949).

2.5.2 Orders of myxomycetes

Myxomycetes are currently grouped into five orders, all of which are frequently found either in nature or upon incubating suitable plant material on moist filter paper.

The Echinosteliales (e.g. *Echinostelium*, *Clastoderma*) contain the smallest known myxomycetes. They form protoplasmodia, with each protoplasmodium giving rise to only one sporangium. The Echinosteliales resemble the protostelids from which they are probably derived (Frederick, 1990; Spiegel, 1991; see Fig. 2.5).

The Liceales (e.g. *Lycogala*, *Dictydium*, *Cribraria*, *Reticularia*) are common on the bark of dead trees. Some of the smaller species produce protoplasmodia, but most have phaneroplasmodia. Various types of sporophores are formed; the aethalia of *Lycogala epidendron* (Plate 1b) and *Reticularia* (= *Enteridum*) *lycoperdon* (Plates 1c,d) are particularly common.

The Trichiales (e.g. *Arcyria*, *Trichia*, *Hemitrichia*) are ubiquitous on fallen logs. The plasmodia are intermediate between aphanoplasmodia and phaneroplasmodia. Fructifications in *Trichia flori-forme* are well-defined sporangia which contain an internal meshwork of threads, collectively called the **capillitium**. The peridium breaks open at maturity, and the spores are released over time by the twisting of the capillitial threads which thus act as elaters (Fig. 2.11). *Arcyria denudata* produces reddish sporangia on rotting wood (Plate 1e). Another member, *Hemitrichia serpula*, produces plasmodiocarps.

The Physarales (e.g. *Physarum*, *Fuligo*) produce the largest plasmodia. *Physarum polycephalum* has been used extensively in fundamental research on cell biology, for example on the nature of protoplasmic streaming, or the synchrony of nuclear division in a large plasmodium comprising thousands of nuclei (see below). The plasmodia are typical phaneroplasmodia, each of which produces numerous sporangia at maturity (Plate 1f). *Fuligo septica* forms particularly large sporophores (aethalia) which are bright yellow and are frequently seen on decaying wood (Plate 1g).

The Stemonitales include such genera as *Comatricha* and *Stemonitis*. *Stemonitis* spp. produce clusters of stalked sporangia from aphanoplasmodia which are visible on rotting wood (Plate 1h).

2.5.3 *Physarum polycephalum* as an experimental tool

This species has been used to investigate several aspects of cell biology. The conspicuous cytoplasmic shuttle streaming in the veins of its large phaneroplasmodia is a fascinating phenomenon and has been examined extensively. The pulse is caused by actin–myosin interactions controlled by Ca^{2+} (Smith, 1994). It is brought about not by the direct binding of organelles to actin cables, but by the constriction and relaxation of an actin–myosin skeleton lining the veins. Several proteins interacting with actin and myosin are directly or indirectly regulated by Ca^{2+}, but the most important effect of Ca^{2+} is on one of the myosin light chains. This is a regulatory subunit which directly binds Ca^{2+}. In contrast to most animal actin–myosin systems which are stimulated by Ca^{2+}, that of *Physarum* is inhibited, i.e. contraction occurs at low Ca^{2+} concentrations, and relaxation at higher concentrations. Ca^{2+}-inhibited actin–myosin interaction also occur in plant cells where they are visible as cytoplasmic streaming. Nakamura and Kohama (1999) have written a thorough review of the actin–myosin system in *Physarum*.

Mitotic division of all nuclei throughout the plasmodium of *P. polycaphalum* occurs in a synchronized manner, and *Physarum* was one of the pioneer organisms in which the existence of the cell cycle was demonstrated. Synchrony of mitosis is regulated by a protein kinase which catalyses the phosphorylation of H1 histones, leading to the condensation of chromosomes at the onset of mitosis (Bradbury *et al.*, 1974; Inglis *et al.*, 1976). This protein kinase is now known to be homologous to the *cdc2* product in the fission yeast *Schizosaccharomyces pombe* (see Fig. 9.5; Langan *et al.*, 1989).

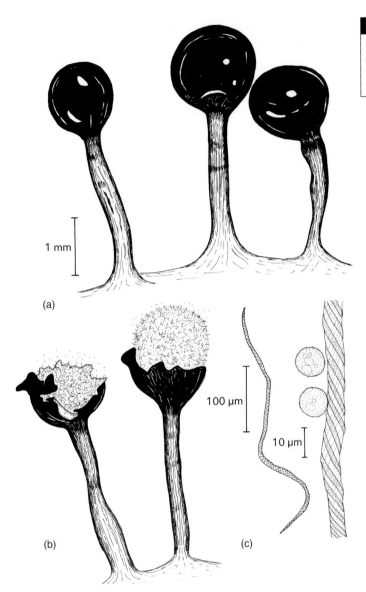

1 mm

(a)

100 µm

10 µm

(b) (c)

A further interesting feature of *P. polycephalum* is the behaviour of the plasmodium and the manner in which its actions are coordinated. Little work has been carried out beyond descriptions of striking phenomena. One is the ability of *P. polycephalum* plasmodia to find the shortest way to a food source through an artificially constructed maze (Nakagaki, 2001). Another is the pattern of veins which is established when different regions of a plasmodium are presented with food sources; the configuration of the plasmodium has been called a 'smart network' because it presents the shortest possible total length of veins to provide good interconnections while making allowances for blockage of individual veins (Nakagaki *et al.*, 2004).

When separate plasmodia of *P. polycephalum* or other species meet, two reactions are possible, i.e. a compatible reaction in which the plasmodia fuse and their veins coalesce (Fig. 2.8b) or an incompatible reaction in which the plasmodia fail to fuse and move away from each other, or fusion is attempted but stalls and is followed by death of the fusion regions of both plasmodia (Fig. 2.8c). This is called the **lethal reaction**.

Genetic studies have shown that fusion occurs between plasmodia of genetically closely related strains (Carlile & Dee, 1967). The type of incompatibility brought about by the interaction of genetically distinct plasmodia is an example of a widespread phenomenon called **vegetative incompatibility** which is found not only in slime moulds, but also in the Eumycota, vertebrates and other organisms. In humans, a similar phenomenon accounts for blood grouping or the failure of tissue transplantations. It is interesting to consider the paradox that fusion between genetically dissimilar myxamoebae is encouraged during sexual reproduction by the existence of different mating types, whereas it is discouraged during vegetative fusion of plasmodia.

3

Protozoa: Plasmodiophoromycota

3.1 | Introduction

The Plasmodiophoromycota are a group of obligate (i.e. biotrophic) parasites. The best-known examples attack higher plants, causing economically significant diseases such as club-root of brassicas (*Plasmodiophora brassicae*), powdery scab of potato (*Spongospora subterranea*; formerly *S. subterranea* f. sp. *subterranea*) and crook-root disease of watercress (*S. nasturtii*; formerly *S. subterranea* f. sp. *nasturtii*). In addition to damaging crops directly, some species (*S. subterranea, Polymyxa betae, P. graminis*) also act as vectors for important plant viruses (Adams, 1991; Campbell, 1996). Other species infect roots and shoots of non-cultivated plants, especially aquatic plants. Algae, diatoms and Oomycota are also attacked. If the nine species of *Haptoglossa*, which parasitize nematodes and rotifers, are included in the Plasmodiophoromycota, the phylum currently comprises 12 genera and 51 species (Dick, 2001a). Genera are separated from each other largely by the arrangement of resting spores in the host cell (Waterhouse, 1973). This feature has also been used for naming most genera; for instance, in *Polymyxa*, numerous resting spores are contained within each sorus, whereas in *Spongospora* the resting spores are grouped loosely in a sponge-like sorus (Fig. 3.6). Accounts of the Plasmodiophoromycota have been given by Sparrow (1960), Karling (1968), Dylewski (1990) and Braselton (1995, 2001).

3.1.1 Taxonomic considerations

Plasmodiophoromycota have traditionally been studied by mycologists and plant pathologists. Many general features of their biology and epidemiology are similar to those of certain members of the Chytridiomycota such as *Olpidium* (see p. 145). However, it is now clear from DNA sequence analysis and other criteria that *Plasmodiophora* is related neither to the Oomycota and other Straminipila (Chapters 4 and 5) nor to the true fungi (Eumycota). Instead, it is distantly related to the Myxomycota discussed in Chapter 2 but belongs to a different grouping within the Protozoa (Barr, 1992; Castlebury & Domier, 1998; Ward & Adams, 1998; Archibald & Keeling, 2004).

Some believe that *Haptoglossa* is related to the Oomycota rather than Protozoa, although no molecular data seem to be available as yet to support this claim. Since *Haptoglossa* strikingly resembles *Plasmodiophora* in its infection biology, we shall include it in this chapter. With the possible exception of *Haptoglossa*, the phylum Plasmodiophoromycota is monophyletic and contains a single class (Plasmodiophoromycetes). We consider two orders in this chapter, Plasmodiophorales and Haptoglossales.

3.2 | Plasmodiophorales

The zoospore of the Plasmodiophorales is biflagellate. The flagella are inserted laterally and are of unequal length, the anterior one being shorter. Both flagella are of the whiplash type (Fig. 1.17c). Zoospores of this type are said to be **anisokont**. Transmission electron microscopy (TEM) studies have shown that the tips of the flagella are tapered rather than blunt (Clay & Walsh, 1997). Like the zoospore, the main vegetative unit — the amoeba, which enlarges to become a plasmodium — is wall-less. It is present freely within host plant cells, its membrane being in direct contact with the host cytoplasm. The plasmodia possess amoeboid features because they can produce pseudopodia and engulf parts of the host cytoplasm by phagocytosis (Claxton *et al.*, 1996; Clay & Walsh, 1997). This has been interpreted as a primitive trait perhaps betraying a free-living amoeboid ancestor with a phagocytotic mode of nutrition (Buczacki, 1983). Some Plasmodiophorales can now be grown away from their host on artificial media for prolonged periods if bacteria are present. These are phagocytosed by amoeboid growth forms (Arnold *et al.*, 1996). In their hosts, amoeboid plasmodia can digest their way through plant cell walls, moving to adjacent uninfected cells and thus spreading the infection within an infected root (Mithen & Magrath, 1992; Claxton *et al.*, 1996).

The walled stages of Plasmodiophorales are confined to the zoospore cysts on the plant surface, and the zoosporangia and resting sporangia inside host plant cells. The wall of resting spores is particularly thick and has been shown to contain chitin (Moxham & Buczacki, 1983).

3.2.1 Life cycle of Plasmodiophorales

Certain details of the life cycle of the Plasmodiophorales are still doubtful (Fig. 3.1). However, the known stages show very little variation between different species, indicating that the life cycle is conserved throughout the order. A resting spore germinates by releasing a single haploid zoospore (**primary zoospore**) which encysts on a suitable surface by secreting a cell wall. After a while, an amoeba is injected from the cyst into a host cell such as a root hair where it enlarges to form a plasmodium, accompanied by mitotic nuclear divisions. Nuclear

divisions at this stage are **cruciform**; the nucleolus is prominently visible throughout the mitotic process, elongating in two directions to take up a cross-like shape when viewed in certain sections by transmission electron microscopy. This feature is unique to the Plasmodiophorales (Braselton, 2001). After a while, nuclei divide mitotically in a non-cruciform manner, and the contents of the plasmodium differentiate into zoospores. This type of plasmodium is termed the **primary plasmodium** or sporangial plasmodium because it produces zoospores. The zoospores are called **secondary zoospores** because they arise from a sporangium, not from a resting spore. Once released, secondary zoospores may re-infect the host to give rise to further primary plasmodia and zoosporangia. Eventually, however, a different type of plasmodium, the **secondary plasmodium** or sporogenic plasmodium, is formed which undergoes meiotic nuclear divisions and produces resting spores (Garber & Aist, 1979; Braselton, 1995). It is not known where plasmogamy and karyogamy occur in the life cycle of the Plasmodiophorales.

All developmental stages of *P. brassicae* can be produced readily in the laboratory. Clubbed roots should be collected from a field or garden and kept frozen at $-20°C$. Seedlings of brassicas, susceptible Chinese cabbage cultivars or *Arabidopsis thaliana* should be grown in a soil with a high peat content which must be kept well watered. Infections can be established by adding slices of infected root material or a resting spore suspension to the soil. Zoosporangia will be formed within a few days, and root galls should be visible within 3–7 weeks (Castlebury & Glawe, 1993). Potato or tomato plants can be infected with *Spongospora subterranea* using similar protocols. Cabbage callus cultures are occasionally used as a simplified experimental system for life cycle studies of *P. brassicae* (Tommerup & Ingram, 1971).

3.2.2 *Plasmodiophora brassicae*

Plasmodiophora brassicae is the causal organism of club root or finger-and-toe disease of brassicas (Fig. 3.2) and was first described by Woronin (1878). The disease is common in gardens where

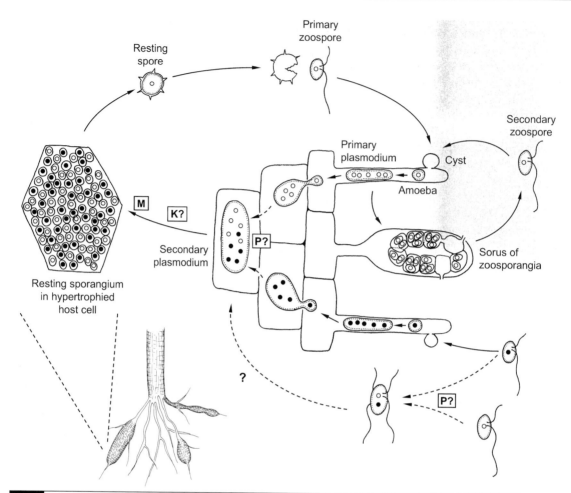

Fig 3.1 Probable life cycle of *Plasmodiophora brassicae*. A haploid resting spore forms a haploid primary zoospore giving rise to a multinucleate haploid primary plasmodium upon infection of a root hair. Secondary zoospores are also haploid, and the way in which they meet to form a secondary heterokaryotic plasmodium is not known for sure. Open and closed circles represent haploid nuclei of opposite mating type; the position of the diploid phase in the life cycle is unclear. Key events in the life cycle are plasmogamy (P), karyogamy (K) and meiosis (M). After Tommerup and Ingram (1971), Buczacki (1983) and Dylewski (1990).

brassicas are frequently grown, especially if the soil is acidic and poorly drained. A wide range of brassicaceous hosts is attacked, and root-hair infection of some non-brassicaceous hosts can also occur (Ludwig-Müller *et al.*, 1999). The disease is widely distributed throughout the world.

Club root symptoms

Infected crucifers usually have greatly swollen roots. Both tap roots and lateral roots may be affected. Occasionally, infection results in the formation of adventitious root buds which give rise to swollen stunted shoots. Above ground, however, infected plants may be difficult to distinguish from healthy ones. The first symptom is wilting of the leaves in warm weather, although such wilted leaves often recover at night. Later the rate of growth of infected plants is retarded so that they appear yellow and stunted. Plants infected at the seedling stage may be killed, but if infection is delayed the effect is much less severe and well-developed heads of cabbage, cauliflower, etc., can form on plants with quite extensive root **hypertrophy** (swelling of cells) and **hyperplasia** (enhanced

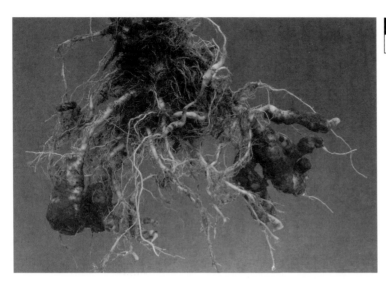

Fig 3.2 Club root of cabbage caused by *Plasmodiophora brassicae*.

division of cells). Microscopically, even infected root hairs are expanded at their tips to form club-shaped swellings which are sometimes lobed and branched (Fig. 3.3). Rausch *et al.* (1981) followed the growth of infected and uninfected seedlings of Chinese cabbage, a particularly susceptible host. Within the first 30 days, the growth rates of infected and control plants were almost identical, and clubs developed in proportion to shoot growth. Wilting of infected plants was observed beyond 30 days when the clubs developed at the expense of shoots. Plants growing in suboptimal conditions, e.g. in the shade, produced disproportionately smaller clubs. Generally, the root/shoot ratio is appreciably higher in infected plants, suggesting a diversion of photosynthetic product to the clubbed roots. The *P. brassicae* infection therefore acts as a new carbon sink.

The process of infection

Swollen roots contain a large number of small spherical resting spores, and when these roots decay the spores are released into the soil. Electron micrographs show that the resting spores have spiny walls (Yukawa & Tanaka, 1979). The resting spore germinates to produce a single zoospore with two flagella of unequal length, both of the whiplash type and with the usual 9 + 2 arrangement of microtubules (Aist & Williams, 1971). Germination of resting spores is stimulated by substances specific to Brassicaceae,

possibly allyl isothiocyanates, which diffuse from the cabbage roots into the soil (Macfarlane, 1970).

The primary zoospore (i.e. the first motile stage released from the resting spore) swims by means of its flagella, the long flagellum trailing and the short one pointing forward. The process of root hair infection has been followed in a classical study by Aist and Williams (1971). Since the first such study, on penetration by *Polymyxa betae*, was written in German (Keskin & Fuchs, 1969), the German terminology is still in use today. Primary zoospores of *P. brassicae* are released some 26–30 h after placing a suspension of resting spores close to seedling roots of cabbage. The zoospores may collide several times with a root hair before becoming attached, and appear to be attached at a point opposite to the origin of the flagella.

The flagella coil around the zoospore body, which becomes flattened against the host wall, and pseudopodium-like extensions of the zoospore develop, being continuously extended and withdrawn. The flagella are then withdrawn, and the zoospore encysts, attached to the root hair (Fig. 3.4). The zoospore cyst contains lipid bodies and a vacuole which enlarges during cyst maturation, which takes a few hours. The most conspicuous ultrastructural feature of mature cysts is a long **Rohr** (tube), with its outer end pointing towards the root hair wall. This end of the tube is occluded by a plug. Within the tube

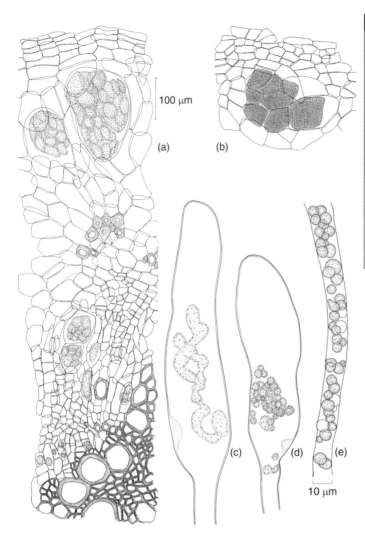

Fig 3.3 *Plasmodiophora brassicae.* (a) T.S. through young infected cabbage root showing secondary (sporogenic) plasmodia in the cortex. Note the hypertrophy of some of the host cells containing plasmodia, and the presence of young plasmodia in cells immediately outside the xylem. (b) T.S. cabbage root at a later stage of infection, showing the formation of resting spores. (c) Primary (zoosporangial) plasmodium in cabbage root hair 4 days after planting in a heavily contaminated soil. (d) Young primary zoosporangia in root hair. Note the club-shaped swelling of the infected root hair. (e) Mature and discharged primary zoosporangia. a and b to same scale; (c–e) to same scale.

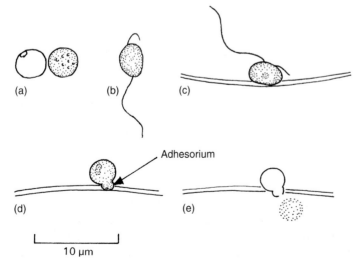

Fig 3.4 *Plasmodiophora brassicae.* (a) Resting spores, one full, one empty (showing a pore in the wall). (b) Zoospore. (c) Attachment of zoospore to root hair. (d) Zoospore cyst with adhesorium following withdrawal of flagellar axonemes. (e) Entry of amoeba into root hair. Based on Aist and Williams (1971).

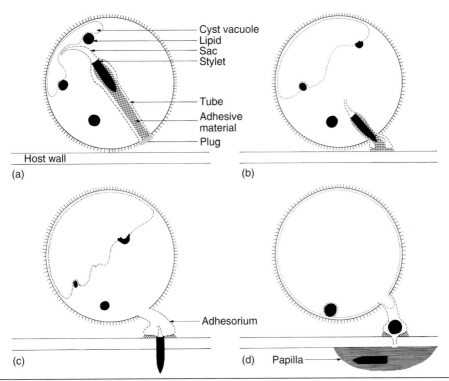

Fig 3.5 *Plasmodiophora brassicae.* Diagrammatic summary of penetration process (after Aist & Williams, 1971). The diagram shows a zoospore cyst attached to the wall of a root hair. (a) Cyst vacuole not yet enlarged. (b) About 3 h later, the cyst vacuole enlarges and a small adhesorium appears. (c) About 1 min later, the stylet punctures the host cell wall. (d) Penetration has occurred and the host protoplast has deposited a papilla at the penetration site.

is a bullet-shaped **Stachel** (stylet), the outer part of which is made up of parallel fibrils. Behind the blunt posterior end of the stylet, the tube narrows to form a **Schlauch** (sac).

Penetration of the root hair wall occurs about 3 h after encystment, as after this time the first empty vacuolated cysts are observed. The penetration process takes place rapidly, and an interpretation of it is shown in Fig. 3.5. Firm attachment of the tube to the root hair is brought about by the **adhesorium**, which may develop by partial evagination (i.e. turning inside out) of the tube (Fig. 3.5b). During evagination, an adhesive substance which has a fibrillar appearance in TEM micrographs is released onto the adhesorial surface from its storage site inside the tube. The enlargement of the vacuole is presumably the driving force which brings about complete evagination of the tube within 1 min, followed by thrusting the stylet through the host wall. The pathogen is injected into the

host cell as a small, spherical, wall-less amoeba which becomes caught up by cytoplasmic streaming. After penetration (Fig. 3.5d), a papilla of callose is deposited around the penetration point beneath the adhesorium, possibly as a wound-healing response. Similar penetration mechanisms have been described for other Plasmodiophorales, including *Spongospora subterranea* (Merz, 1997), *S. nasturtii* (Claxton *et al.*, 1996) and *Polymyxa betae* (Keskin & Fuchs, 1969). Details of the infection process by *P. betae* have been filmed (see Webster, 2006a). A yet more elaborate process of infection is found in *Haptoglossa*, which parasitizes nematodes and rotifers (see p. 65).

Development of zoosporangia

Within the infected root hair, the amoeba may divide into several uninucleate amoebae. Later the nuclei within each amoeba show cruciform divisions, giving rise to small multinucleate primary plasmodia. Each plasmodium divides up to form a group (**sorus**) of roughly spherical thin-walled zoosporangia lying packed together in the host cell (Fig. 3.3). Separate protoplasts might coalesce at this stage. Each zoosporangium finally contains 4–8 uninucleate zoospores. These are morphologically identical to primary zoospores. Some mature zoosporangia become attached to the host cell wall and an exit pore develops at this point through which the zoospores escape. The zoospores of other sporangia are released into those with an exit pore. Occasionally, zoospores escape into the lumen of the host cell. Liberated zoospores can re-infect plant roots, thereby completing an asexual cycle (Fig. 3.1).

Sexual reproduction

In *P. brassicae*, resting sporangia are not formed in root hairs after the first cycle of infection, but are located mainly in older infections in strongly hypertrophied regions of the root cortex. There is evidence that resting sporangia are involved in sexual reproduction (Fig. 3.1) because meiotic nuclear divisions with synaptonemal complexes have been observed in maturing resting sporangia (Garber & Aist, 1979). Further, each resting spore normally contains one haploid nucleus (Narisawa *et al.*, 1996). Thirdly, infection experiments have established that resting sporangia are formed only if two genetically dissimilar nuclei are present (Narisawa & Hashiba, 1998) which could be contributed either by two uninucleate zoospores or by a binucleate zoospore.

The positions of the preceding stages of sexual reproduction – plasmogamy and karyogamy – in the life cycle of *P. brassicae* are still a matter of doubt. One possibility is that secondary zoospores fuse to form a dikaryon, followed by karyogamy. Quadriflagellate binucleate swarmers have indeed been observed and can result from the fusion of zoospores (Tommerup & Ingram, 1971). However, it is not yet clear whether these quadriflagellate spores can infect plant cells from the outside. Quadriflagellate binucleate zoospores may also arise from incomplete cleavage of cytoplasm during zoospore formation.

Plasmodia of *P. brassicae* have been shown to break through plant cell walls, thereby spreading an infection from root hairs into deeper tissues of the root cortex (Mithen & Magrath, 1992). A conceivable alternative would be their migration through plasmodesmata. It is possible that two primary plasmodia or uninucleate amoebae arising from separate root hair infections fuse upon encountering each other deep inside the host plant. Such a fusion would produce a secondary plasmodium, and could be followed by karyogamy and meiosis, which would lead to the development of resting spores (Fig. 3.1).

Hypertrophy of infected host cells

As the plasmodia within a host cell enlarge, the host nucleus remains active and undergoes repeated divisions. Hypertrophy and an increased ploidy of the host nuclei result, at least in callus culture experiments, because the mechanism for host cell division is apparently blocked (Tommerup & Ingram, 1971).

Unsurprisingly, the grossly hypertrophied clubs contain enhanced levels of plant growth hormones. The concentration of auxins (especially indole-3-acetic acid, IAA) in clubbed roots was measured to be about 1.7 times as high as in uninfected roots (Ludwig-Müller *et al.*, 1993), and that of cytokinins was 2–3 times elevated (Dekhuijzen, 1980). Isolated secondary plasmodia of *P. brassicae* have been demonstrated to synthesize the cytokinin zeatin (Müller & Hilgenberg, 1986), and the amount of zeatin produced would be sufficient to establish a new carbon sink. The situation is more complicated with respect to auxins which are not synthesized by plasmodia. Instead, the pathogen interferes with the host's auxin metabolism, which is complex (Normanly, 1997). The tissues of healthy crucifers contain relatively large amounts of indole glucosinolates such as glucobrassicin (= indole-3-methylglucosinolate) which is converted by the enzyme myrosinase to 3-indoleacetonitrile (IAN), a direct IAA precursor. Conversion of IAN to IAA is catalysed by nitrilase. Increased concentrations of indole glucosinolates, IAN and IAA have been measured in clubbed roots (Ludwig-Müller, 1999), and the expression of nitrilase and myrosinase was also enhanced. Further, nitrilase protein was detectable by immunohistochemical methods only in cells containing sporulating plasmodia. The activities of the above enzymes might be regulated by the signalling molecule, jasmonic acid (Grsic *et al.*, 1999). However, these metabolic changes were confined to a narrow window of time, and other sources of IAA, such as its release from IAA–alanine conjugates by the activity of amidohydrolase, are likely to contribute (Ludwig-Müller *et al.*, 1996). The host–pathogen interactions leading to enhanced auxin levels in clubbed roots are therefore very intricate.

At first, only cortical cells of the young root are infected, but later small plasmodia can be found in the medullary ray cells and in the vascular cambium. Subsequently, tissues derived from the cambium are infected as they are formed. In large swollen roots, extensive wedge-shaped masses of hypertrophied medullary ray tissue may cause the xylem tissue to split. At this stage, the root tissue shows a distinctly mottled appearance. When the growth of the plasmodia is complete, they are transformed into masses of haploid resting spores. Only during the late stages of resting spore development do the host nuclei begin to degenerate. Eventually, the resting spores are released into the soil as the root tissues decay.

3.2.3 *Spongospora*

The life cycle of *S. subterranea*, the cause of powdery scab of potato, is similar to that of *P. brassicae* (Harrison *et al.*, 1997; Hutchison & Kawchuk, 1998). Diseased tubers show powdery pustules at their surface, containing masses of resting spores clumped into hollow balls. The resting spores release anisokont zoospores which can infect the root hairs of potato or tomato plants. In the root hairs, plasmodia form which develop into zoosporangia. Zoospores from such zoosporangia are capable of infection, resulting in a further crop of zoosporangia. Zoospores released from the zoosporangia have also been observed to fuse in pairs or occasionally in groups of three to form quadri- or hexaflagellate swarmers, but whether these represent true sexual fusion stages is uncertain. *Spongospora nasturtii* causes a disease of watercress in which the most obvious symptom is a coiling or bending of the roots. Zoosporangia and resting spore balls are found in infected root cells (Fig. 3.6), and plasmodia can migrate through the root tissue by breaking through host cell walls (Claxton *et al.*, 1996; Clay & Walsh, 1997). The encounter of two plasmodia might initiate sexual reproduction and thus complete the life cycle without any need for the parasite to leave the host (Heim, 1960).

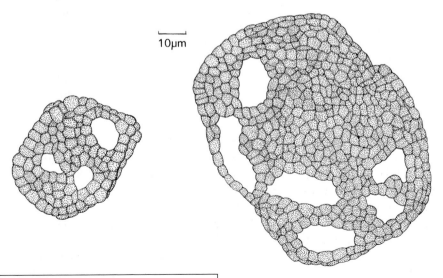

10μm

Fig 3.6 *Spongospora nasturtii.* Spore balls from watercress roots with crook root disease.

In addition to being the causal agent of powdery scab of potatoes, *S. subterranea* is also important as the vector of potato mop-top virus disease, which can reduce the yield of tubers by over 20% in some varieties (Campbell, 1996; Harrison *et al.*, 1997). The virus is transmitted by the zoospores and can also persist for several years in spore balls in the soil. It seems to be located inside the resting spores (Merz, 1997). Zoospores of *S. subterranea* can cause zoosporangial infections in the root hairs of a wide range of host plants outside the family Solanaceae, and can transmit viruses to them. Thus *S. subterranea* and numerous wild plants can provide a reservoir of infection for the potato mop-top virus even if potatoes have not been grown in a field for many years. Other members of the Plasmodiophorales also act as vectors for plant viruses, notably *Polymyxa betae* which transmits the beet necrotic yellow vein virus, and *P. graminis* which transmits several mosaic viruses on most major cereal crops.

3.3 | Control of diseases caused by Plasmodiophorales

3.3.1 Club root

The control of club root disease is difficult. Because resting spores retain their viability in the soil for up to 20 years, short-term crop rotation will not eradicate the disease. The fact that *Plasmodiophora brassicae* can infect brassicaceous weeds such as shepherd's purse (*Capsella bursa-pastoris*) or thalecress (*Arabidopsis thaliana*) suggests that the disease can be carried over on such hosts and that weed control is important. Moreover, it is known that root hair infection can also occur on many ubiquitous non-brassicaceous hosts such as *Papaver* and *Rumex*, or the grasses *Agrostis*, *Dactylis*, *Holcus* and *Lolium*. All infections of non-brassicaceous hosts are probably reduced to the zoosporangial cycle, and no root clubs are formed. Whether such infections play any part in maintaining the disease in the prolonged absence of a brassicaceous host is not known.

General measures aimed at mitigating the incidence of clubroot traditionally include improved drainage and the application of lime, which retards the primary infection of root hairs. Since the effect of liming does not persist, it is possible that it may simply delay the germination of resting spores and thus prolong their existence in the soil (Macfarlane, 1952). More recently, boron added at $10-20\,\mathrm{mg\,kg^{-1}}$ soil in conjunction with a high soil pH has been shown to suppress primary as well as secondary infections (M. A. Webster & Dixon, 1991). Early infection of seedlings can result in particularly severe symptoms, so it is important to raise seedlings in non-infected or steam-sterilized soil. The young plants can then be transplanted to infested soil. Since it is known that some resting spores survive animal digestion, manure from animals fed with diseased material should not be used for growing brassicas.

Infection can be retarded by the application of mercury-containing compounds or benomyl, but these are now banned in many countries. At present, no economically and ecologically acceptable fungicide appears to be available, although research efforts continue (Mitani *et al.*, 2003). Some attempts have been made to establish biological control methods for *P. brassicae* (Narisawa *et al.*, 1998; Tilston *et al.*, 2002), but it is doubtful whether such methods will gain full commercial viability in the near future.

In recent years, increasing emphasis has been placed on breeding club root resistant cultivars of crop plants. The weed *Arabidopsis thaliana*, which develops the full set of club root symptoms, has been used as a host for such studies because it is accessible by molecular biological methods. Natural resistance in *Arabidopsis* is based on a single gene and involves the **hypersensitive response**, in which infected plant cells die before the pathogen has had a chance to multiply. The resistance of susceptible cultivars can be enhanced by transformation with various resistance genes, e.g. a gene from mistletoe (*Viscum album*) encoding viscotoxin, a thionin-type cystein-rich polypeptide with antimicrobial activity (Holtorf *et al.*, 1998). Further, mutant lines with reduced levels of IAA precursors show reduced club development (Ludwig-Müller, 1999).

In contrast to *Arabidopsis*, natural resistance in cabbage is multigenic, with no obvious hypersensitive response (Ludwig-Müller, 1999). Breeding for resistance is difficult (Bradshaw *et al.*, 1997) and may not provide long-lasting success due to the development of new virulent races of *P. brassicae* on the resistant cultivars after a few years in the field. By 1975, 34 different physiological races of *P. brassicae* from Europe had already been differentiated based on infection experiments with *Brassica* cultivars varying in their degree of resistance (Buczacki *et al.*, 1975). Further, *P. brassicae* can still infect root hairs and reproduce by zoosporangia even in resistant cultivars.

3.3.2 Powdery scab and crook root

Powdery scab of potatoes is normally of relatively slight economic importance and amelioration of the disease can be brought about by good drainage. Potato mop-top virus infections can be more serious, however. Transgenic plants containing the viral coat protein gene have been shown to be completely resistant against infections by the virus (Reavy *et al.*, 1995), and it may be possible to produce transgenic crop plants in future.

Crook root of watercress can be controlled by application of zinc to the water supply. The zinc can be applied by dripping zinc sulphate into the irrigation water for watercress beds to give a final concentration of about 0.5 ppm, or by the

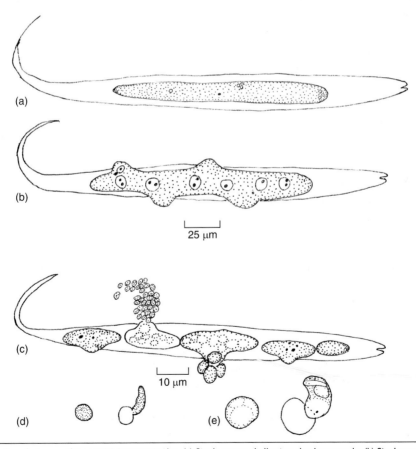

Fig 3.7 *Haptoglossa heteromorpha* parasitizing nematodes. (a) Single young thallus in a dead nematode. (b) Single maturing sporangium with developing dome-shaped exit papillae. (c) Nematode body containing several plasmodia and sporangia. One sporangium has released large aplanospores, and an adjacent one small ones. (d) Small aplanospores, one germinating to form a gun cell. (e) Large aplanospores, one germinating to form a gun cell. (a–c) to same scale; d,e to same scale. Redrawn from Glockling and Beakes (2000a).

addition of finely powdered glass containing zinc oxide (zinc frit) to the beds. The slow release of zinc from the frits maintains a sufficiently high concentration to inhibit infection (Tomlinson, 1958).

3.4 | *Haptoglossa* (Haptoglossales)

3.4.1 General biological features of *Haptoglossa*

If a slurry of soil or herbivore dung is spread on a weak medium such as tap water agar or cornmeal agar, the nematodes or rotifers contained within these samples may become parasitized and killed by fungi producing thalli within the cadavers. Although superficially resembling the plasmodia of *Plasmodiophora*, this term cannot be applied to *Haptoglossa* because its thalli are surrounded by a wall at all stages of development. One or several thalli may fill almost the entire body cavity of a nematode and become converted into sporangia upon maturity (Fig. 3.7). Sporangia of some species of *Haptoglossa* release zoospores which are anisokont, with both flagella of the whiplash type. Zoospore release occurs through one or several exit papillae (Barron, 1977). Zoospores of

Haptoglossa are weak swimmers and encyst within a few minutes in the vicinity of the host cadaver from which they were released. Other species of *Haptoglossa* do not release zoospores but produce non-motile spores (aplanospores) resembling cysts of the zoospore-forming species. Aplanospore release occurs by explosive rupture of the exit tube, followed by several further, progressively weaker bursts of discharge (Glockling & Beakes, 2000a). A few hours after their formation or release, cysts or aplanospores germinate to produce an elongated or glossoid (= tongue-shaped) cell, which is also often called a **gun cell** or an infection cell. This explosively injects a small amount of walled protoplasm (**sporidium**) containing a nucleus and a few organelles into a host passing by (see below). The sporidium enlarges to form a new thallus and, upon host death, a new sporangium. The mechanism of gun cell discharge is rather similar to that found in cysts of *Plasmodiophora* or *Polymyxa*. This, together with the occurrence of anisokont zoospores, has been taken as an indication that *Haptoglossa* should be included in the Plasmodiophoromycota (Beakes & Glockling, 1998; Dick, 2001a), whereas formerly the genus was thought to be related to the Oomycota.

The aplanosporic species of *Haptoglossa* produce spores of two distinctly different sizes,

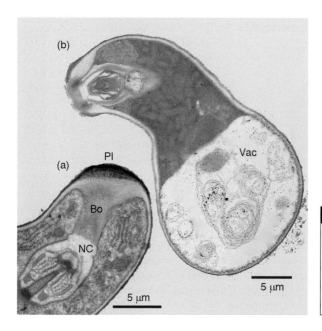

Fig 3.8 *Haptoglossa* sp. (a) Tip of a developing gun cell. The muzzle is still sealed by its plug (Pl). Bore (Bo) and needle chamber (NC) are visible. (b) Transmission electron micrograph of a mature gun cell. The basal part of the gun cell is entirely occupied by the enlarging posterior vacuole (Vac). Original prints kindly supplied by S. L. Glockling.

although any one sporangium produces propagules only of either size (Glockling & Beakes, 2000a; Fig. 3.7). In contrast to the Plasmodiophorales, sexual reproduction or resting stages have not yet been described for any species of *Haptoglossa*, and it is difficult at present to explain the occurrence of spores of different sizes. What appears clear is that each thallus is the result of a discrete infection event.

3.4.2 The gun cell of *Haptoglossa*

Germination of the spherical zoospore cyst or aplanospore of *Haptoglossa* occurs by means of a short germ tube which enlarges to form the elongated gun cell (Robb & Lee, 1986a). This remains attached to the cyst until maturity and is perched on top of it in many species. The mature gun cell (Figs. 3.8, 3.9a) shows strong ultrastructural similarities to the infection apparatus of *Plasmodiophora* (see Fig. 3.5) and is the object of considerable mycological curiosity. A tube leads into the pointed tip of the gun cell but its opening (**muzzle**) is separated from the exterior by a thin wall (**plug**) for most of its development (Fig. 3.8a). The formation of this internal tube from the tip of the gun cell backwards has been likened to inverted internal tip growth and is mediated by a scaffold of actin fibres against the turgor pressure of the gun cell (Beakes & Glockling, 1998). The inner (non-cytoplasmic) surface of the anterior part of the tube (**bore**) is lined with fibrillar material. A second wall separates the bore from a swollen section of the tube, the **needle chamber**. This contains a projectile (**needle**) resembling the bullet of *Plasmodiophora*, but terminating in a much finer tip, possibly reflecting the different properties of the host surface which it has to puncture. The needle is held in place by a complex set of cones and cylinders (Fig. 3.8a) which are thought to exercise a restraining function, fixing the needle against the high turgor pressure of the gun cell. The cones and cylinders may contain actin filaments. The shaft of the needle is much wider than its tip. The posterior (innermost) part of the tube (**tail**) coils around itself and the nucleus, almost touching the side of the needle chamber. The tail is walled,

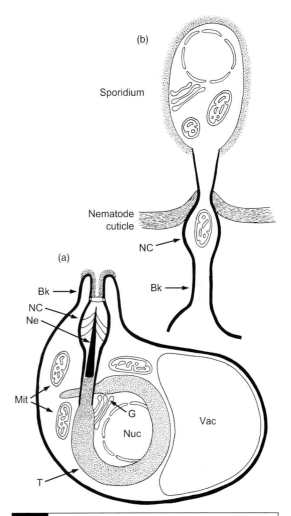

Fig 3.9 Schematic drawings of the nematode penetration mechanism in *Haptoglossa*. (a) Gun cell ready for discharge. The tube has already protruded to form a beak (Bk), the exterior of which is lined by a glue originating from the inside surface of the bore. This aids in the attachment of the gun cell to a passing nematode. The needle (Ne) is held in position by actin filaments inside the needle chamber (NC), which is separated from the outside by a wall. Behind the needle chamber is the coiled tail (T) which contains wall material in its lumen (dotted area). In fact, the tail is multi-layered, but this has not been illustrated here. The tail coils round the nucleus (Nuc) and a Golgi stack (G), and mitochondria (Mit) are also located in the vicinity. The posterior of the gun cell is filled by one large vacuole (Vac). (b) Tip of a fired gun cell showing the everted tail which has penetrated the nematode cuticle and has formed a sporidium inside the nematode body (above the cuticle). The wall material formerly located inside the tail has formed the sporidium wall. The detached needle is also visible inside the nematode body. For a more detailed description of the eversion process, see Glockling and Beakes (2000b).

and additional electron-dense cell wall precursor material is deposited within the lumen of the tail. Synthesis of the tube is mediated by one large Golgi stack which is always closely associated with the nucleus and faces the inward-growing tube tip, emitting vesicles towards it. As the tube extends and coils round the nucleus, the nucleus and Golgi stack turn like a dial by 360° (Beakes & Glockling, 1998, 2000). The turgor pressure of the gun cell is probably generated by a large posterior vacuole (Fig. 3.8b), similar to that found in cysts of *Plasmodiophora*. The osmotically active solutes required for turgor generation may originate from the degradation of lipid droplets within the enlarging vacuole.

Shortly before discharge, the increasing turgor pressure of the posterior vacuole is thought to push the tip of the gun cell forward; the wall sealing the muzzle is lost, and the bore shortens and extends a beak-like projection (Fig. 3.9a). The cell wall material from the interior of the bore now forms the external beak wall, and the needle is ready for injection. The nature of the discharge trigger probably varies between different species of *Haptoglossa* and may be chemical or mechanical. The beak wall is thought to act as an adhesive and immediately glues the gun cell to the cuticle of a passing nematode or rotifer. Firm attachment is necessary to provide resistance against the recoil of the needle attempting to penetrate the tough cuticle of the host, as it is for the penetrating bullet in adhesoria of *Plasmodiophora*.

Beakes and Glockling (1998) speculated that stretch-activated membrane channels (see p. 8) might be involved in triggering the launch of the needle. Following attachment, Ca^{2+} ions entering the needle chamber would cause the actin-rich cones and cylinders near the needle tip to contract and rupture. Once the constraints exercised by the cones and cylinders are broken, the high turgor pressure of the gun cell will immediately fire the needle, followed by explosive eversion of the entire tube which forms a syringe, conducting the nucleus, Golgi apparatus and mitochondria of the gun cell through the nematode cuticle (Fig. 3.9b). The infective propagule is called a sporidium because it is surrounded by a wall, the material for which is probably contributed by precursor material at the end of the tail section (Robb & Lee, 1986b; Glockling & Beakes, 2000b).

Straminipila: minor fungal phyla

4.1 | Introduction

The kingdom **Chromista** was erected by Cavalier-Smith (1981, 1986) to accommodate eukaryotic organisms which are distinguishable from the Protozoa by a combination of characters. Some of these are concerned with details of photosynthesis, such as the enclosure of chloroplasts in sheets of endoplasmic reticulum, and the absence of chlorophyll *b*, the latter feature being used for the naming of the kingdom. Other defining characters apply also to the non-photosynthetic members of the Chromista (Kirk *et al.*, 2001). These are as follows:

1. The structural cell wall polymer is cellulose, in contrast to walls of Eumycota which contain chitin.

2. The inner mitochondrial membrane is folded into tubular cristae (Fig. 4.1a) which are also found in plants. In contrast, mitochondrial cristae are generally lamellate in the kingdoms Eumycota (Fig. 4.1b) and Animalia.

3. Golgi stacks (dictyosomes) are present; these are also found in the Protozoa (see p. 64). In contrast, in the Eumycota the Golgi apparatus is usually reduced to single cisternae (see Figs. 1.3, 1.10).

4. Flagella are usually present during particular stages of the life cycle; they always include one **straminipilous** flagellum (Lat. *stramen* = straw, *pilus* = hair). Dick (2001a) considered this feature to be of such high phylogenetic significance that he has renamed the kingdom Chromista as **Straminipila**. The straminipilous flagellum is discussed in detail in the following section.

5. The amino acid lysine is synthesized via the α,ε-diaminopimelic acid (DAP) pathway. Diaminopimelic acid originates from aspartic semialdehyde and pyruvic acid and is present in terrestrial plants, green algae, Chromista and prokaryotes. The alternative route, the α-aminoadipic acid (AAA) pathway, draws on α-ketoglutaric acid and acetyl-CoA and is found almost exclusively in members of the Eumycota. Yet other organisms, including animals and Protozoa, are auxotrophic for lysine (Griffin, 1994). Lysine biosynthesis has been used as a chemotaxonomic marker for some time (Vogel, 1964; LéJohn, 1972).

The kingdom Chromista/Straminipila currently includes the diatoms, golden and brown algae, chrysophytes and cryptomonads, as well as three phyla of straminipilous organisms traditionally studied by mycologists, i.e. the Oomycota, Hyphochytriomycota and Labyrinthulomycota. The first two groups are also called **straminipilous fungi** because of the similarity of their mode of life to the fungal lifestyle (Dick, 2001a). The Oomycota are by far the more important of these, and are considered in detail in Chapter 5. The Hyphochytriomycota and Labyrinthulomycota are treated briefly in the present chapter. The Straminipila as circumscribed above are a diverse but natural

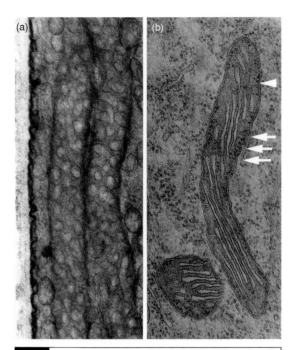

Fig 4.1 Mitochondrial ultrastructure observed by transmission electron microscopy. (a) Mitochondrion of *Phytophthora erythroseptica* (Oomycota). The inner mitochondrial membrane is folded into a complex tubular network. (b) Mitochondrion of *Sordaria fimicola* (Ascomycota) with the inner membrane appearing lamellate. Mitochondrial ribosomes (arrows) are also visible. Reprinted from Weber *et al.* (1998), with permission from Elsevier.

(monophyletic) grouping which has been confirmed by comparisons of the small-subunit (18S) ribosomal DNA sequences (e.g. Hausner *et al.*, 2000; Fig. 4.2).

4.2 | The straminipilous flagellum

The eukaryotic flagellum is a highly conserved structure. It is formed within the cytoplasm by a **kinetosome**, i.e. a microtubule-organizing centre resembling the centriole which co-ordinates the formation of the microtubular spindle during nuclear division. Like the centriole, the kinetosome contains an outer ring of nine triplets of microtubules surrounding two central microtubules (see Figs. 6.2 and 6.19). The flagellum extends outwards from the centriole as nine doublets of microtubules surrounding the two single central microtubules. This is the 9 + 2 arrangement. Where the eukaryotic flagellum protrudes beyond the cell surface, it is ensheathed by the plasma membrane. Within the flagellum, there are no obvious cytoplasmic features other than the microtubules which together are called the **axoneme**. Flagella which are entirely smooth or bear a coat of fine fibrillar surface material visible only by high-resolution electron microscopy (Fig. 4.3a; Andersen *et al.*, 1991) are commonly called **whiplash** flagella. Dick (2001a) has pointed out that whiplash flagella in a strict sense are pointed at their tip due to the fact that the two inner microtubules are longer than the nine outer doublets (Fig. 4.3a).

A second type of flagellum is decorated with hair-like structures 1–2 μm long (Fig. 4.3b). This is the **tinsel** or **straminipilous** flagellum (Dick, 1997). The hairs are called **tripartite tubular hairs** (TTHs) because they are divided into three parts. They were formerly called mastigonemes, thereby naming the fungi which produced them Mastigomycotina, but both terms are no longer used. Each TTH is attached to the flagellum by a conical base pointed towards the axoneme. The main part of the TTH is a long tubular shaft thought to consist of two fibres of different thickness coiled around each other (Domnas *et al.*, 1986). At the tip of the TTH, the two fibres separate from each other to form loose ends (Figs. 4.3b, 4.4). In the TTHs of some straminipilous organisms, only one loose end is visible (Fig. 4.7b). TTHs are assembled in anti-parallel arrays in Golgi-derived vesicles of the maturing zoospore, and are released by fusion of the vesicles with the plasma membrane (Fig. 4.5; Heath *et al.*, 1970; Cooney *et al.*, 1985). When a spore encysts, the flagellum may be withdrawn, shed or coiled around the spore. If it is withdrawn, the TTHs are sloughed off and left behind as a tuft on the surface of the cyst (Dick, 1990a).

TTHs are arranged in two rows along the axoneme. The cones of each row are adjacent to an outer microtubule doublet, and because there are nine such doublets, the two rows of TTHs are at an angle of about 160° rather than 180° to each other (Fig. 4.4a). In zoospores of

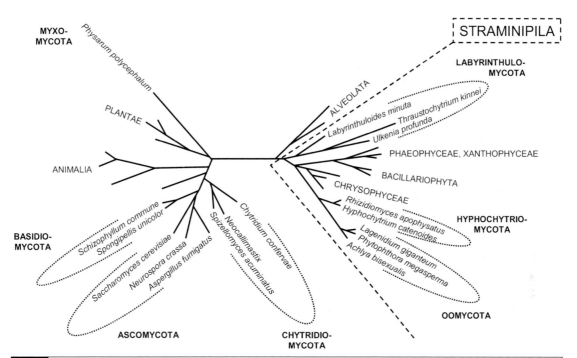

Fig 4.2 Unrooted phylogenetic tree of the Straminipila and members of other kingdoms, based on analyses of 18S rDNA sequences. Redrawn and modified from Hausner *et al.* (2000), by copyright permission of the National Research Council of Canada.

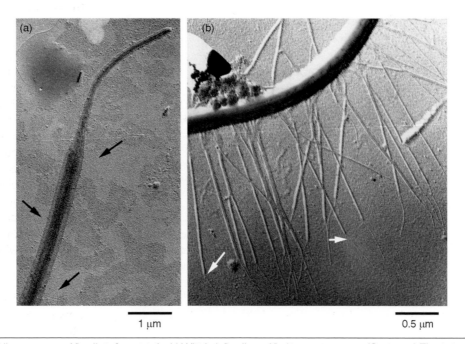

Fig 4.3 Ultrastructure of flagella in Straminipila. (a) Whiplash flagellum of *Pythium monospermum* (Oomycota). The tip is narrower than the main body of the flagellum because the two central microtubules are longer than the nine outer doublets. Arrows indicate the coating of the flagellum with very fine hairs. (b) Tinsel flagellum of *Achlya colorata* (Oomycota) with numerous TTHs. Each TTH ends in two fibres, one longer and thicker than the other (arrows). Original images kindly provided by M. W. Dick and I. C. Hallett.

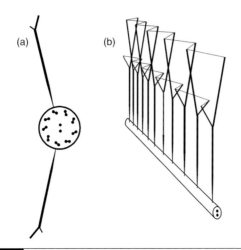

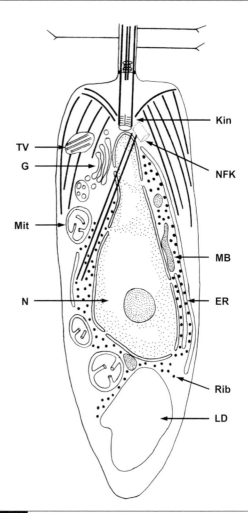

Fig 4.4 Organization of the straminipilous flagellum. (a) Postulated attachment of TTHs to the microtubule doublets I and 5 of the axoneme as seen in transverse section (after Dick, 2001a). (b) Longitudinal arrangement of TTHs along the axoneme of a straminipilous flagellum. Only one row of TTHs is drawn. The TTHs are thought to be arranged in an alternating fashion as regards the orientation of long and short fibres in adjacent TTHs. b redrawn from Dick (1990a). © 1990 Jones and Bartlett Publishers, Sudbury, MA. www.jbpub.com.

straminipilous fungi, the straminipilous flagellum always seems to point towards the direction of movement, and Dick (1990a, 2001a) has advanced a theory to explain how movement can be generated from a sinusoidal wave starting at the flagellar base, likening the straminipilous flagellum to 'a rowing eight with fixed oars and a flexible keel' (Fig. 4.4b; Dick, 2001a). An anterior straminipilous flagellum therefore pulls the spore through the water, whereas a backwardly directed whiplash flagellum pushes the spore.

The construction of the straminipilous flagellum is so elaborate that it is most unlikely to have arisen more than once during evolution (Dick, 2001a). The presence of a straminipilous flagellum, whether or not accompanied by another, smooth flagellum, therefore indicates membership in the Straminipila.

4.3 | Hyphochytriomycota

This group, formerly called Hyphochytridiomycetes probably due to the perpetuation of

Fig 4.5 Schematic drawing of a L.S. of a zoospore of the hyphochytrid *Hyphochytrium catenoides*. The elongated shape of the zoospore and of the nucleus (N) is maintained by a system of 'rootlets' consisting of parallel bundles of microtubules (thick lines). The straminipilous flagellum arises from a kinetosome (Kin). A second, non-functional kinetosome (NFK) is interpreted as the base of a whiplash flagellum lost in the course of evolution from a heterokont ancestor. Mitochondria (Mit), TTH-containing vesicles (TV), a Golgi stack (G), ER, ribosomes (Rib), a large basal lipid droplet (LD) and microbodies (MB) are also visible. Some organelles of unknown function, e.g. electron-opaque bodies and osmiophilic bodies, have been omitted from the original for improved clarity. Redrawn and modified from Cooney *et al.* (1985).

a typographical error (see Dick, 1983), is a very small phylum currently comprising 23 species in 6 genera (Kirk *et al.*, 2001). The Hyphochytriomycota (colloquially called hyphochytrids) are phylogenetically closely related to the Oomycota (van der Auwera *et al.*, 1995; Hausner *et al.*, 2000; see Fig. 4.2). Treatments of the group have been given by Karling (1977), Fuller (1990, 2001) and Dick (2001a). The diagnostic feature is the zoospore with its single anterior straminipilous flagellum (Fig. 4.5). This kind of zoospore is not found in any other known life form. The zoospore of hyphochytrids contains one prominent Golgi stack, one nucleus, and lipid droplets and microbodies (Barr & Allan, 1985; Cooney *et al.*, 1985). The latter are not arranged in a microbody–lipid complex like they are in chytrids (cf. Fig. 6.3). The TTHs are localized within Golgi-derived vesicles. The flagellum arises from a kinetosome, with microtubules rooting deeply within the spore and probably maintaining its shape. A second (dormant) kinetosome lies adjacent but at an angle, at the same position as that which gives rise to the backward-directed smooth flagellum in zoospores of Oomycota. This whiplash flagellum is missing in Hyphochytriomycota, and Barr and Allan (1985) have speculated that it could have been lost during evolution of the latter from the former. Like the Oomycota, hyphochytrids synthesize lysine by the α,ε-diaminopimelic acid (DAP) pathway (Vogel, 1964).

Hyphochytrids occur in the soil and in aquatic environments (both freshwater and marine) as saprotrophs or parasites of algae, oospores of Oomycota or azygospores of Glomales. *Hyphochytrium peniliae* was reported once as the cause of a devastating epidemic of marine crayfish (Artemchuk & Zelezinkaya, 1969), but no further cases have been observed since. Some species can be isolated into pure culture relatively easily (Fuller, 1990).

Zoospores encyst by withdrawing their flagellum and secreting a wall, leaving the TTHs dispersed on the surface of the cyst wall (Beakes, 1987). The cyst germinates by enlargement or by putting out rhizoids. Because of the similarity of their vegetative thalli with those of Chytridiomycota (see Chapter 6), hyphochytrids have been studied primarily by comparison with chytrids, and the same terminology has been used (see Fig. 6.1). Depending on the species, cysts germinate to develop in three different

ways, which have been used to subdivide the Hyphochytriomycota into families: (1) **Holocarpic** thalli are produced by simple enlargement of the cyst. The entire content of the sac-like thallus ultimately becomes converted into zoospores (Anisolpidiae, e.g. *Anisolpidium* which parasitizes marine algae; Canter, 1950). (2) In **eucarpic monocentric** thalli, the cyst produces a bunch of rhizoids at one end, which anchor the enlarging thallus to the substratum and/or absorb nutrients (Rhizidiomycetidae, e.g. *Rhizidiomyces*; Wynn & Epton, 1979). (3) In **eucarpic polycentric** thalli, a broad hypha-like germ tube emerges, branches and produces several zoosporangia (Hyphochytriaceae, e.g. *Hyphochytrium*; Ayers & Lumsden, 1977). The asexual life cycle is completed when a fresh crop of zoospores is released. Sexual reproduction has not yet been reliably described for the hyphochytrids.

4.4 | Labyrinthulomycota

Whereas the Hyphochytriomycota described in the previous section have a strong resemblance to true fungi (especially Chytridiomycota), the Labyrinthulomycota do not, and the only justification for mentioning them here is the fact that they have traditionally been studied by mycologists. They have been the subject of numerous taxonomic rearrangements, and are known under many different names such as Labyrinthomorpha, Labyrinthista and Labyrinthulea. Some 48 species are currently recognized (Kirk *et al.*, 2001). DNA sequence comparisons have placed them within the Straminipila (Fig. 4.2; Hausner *et al.*, 2000; Leander & Porter, 2001), and they are characterized by having **heterokont** flagellation, i.e. possessing a straminipilous and a whiplash flagellum with a pointed tip (Fig. 4.7). In addition, they have mitochondria with tubular cristae. Recent treatments of this group can be found in Moss (1986), Porter (1990) and Dick (2001a).

Labyrinthulomycota occur in freshwater and marine environments where they are attached to solid substrata by means of networks of slime within which individual vegetative cells are contained. For this reason, they are sometimes referred to as 'slime nets' (Porter, 1990). The vegetative cells possess a wall which, uniquely, is produced from Golgi-derived scales of a polymer of L-galactose (Dick, 2001a). These scales are located between the plasma membrane and the inner membrane of the slime net. The slime net is delimited by an inner and an outer membrane and is produced by specialized organelles termed **sagenogens** or bothrosomes; the net membranes are continuous with the plasma membrane at the sagenogen (Perkins, 1972). Labyrinthulomycota feed by absorption (osmotrophy) of nutrients. The nets contain degradative enzymes which can lyse plant material or microbial cells. Two orders are distinguished.

4.4.1 Labyrinthulales

Members of this order, especially of the genus *Labyrinthula*, can be readily isolated from marine angiosperms such as *Zostera* and *Spartina* or from seaweed by placing a small piece of one of these substrata directly on low-nutrient sea water agar augmented with penicillin and streptomycin (Porter, 1990). Within a few days, a fine network of strands can be seen extending over the agar surface (Fig. 4.6). *Labyrinthula* spp. can be kept in monoxenic culture with yeasts or bacteria as food source. These are presumably lysed by the enzymes contained in the slime net.

A closer examination shows that the network consists of branched slime tubes within which spindle-shaped cells move backwards and forwards (Fig. 4.7a; see Webster, 2006a). Movement of a speed up to $100\,\mu\text{m}\,\text{min}^{-1}$ has been reported and is due to a system of contractile actin-like proteins in the slime net (Nakatsuji & Bell, 1980). Cells occasionally aggregate to form sporangia containing numerous round cysts. Following meiosis, eight heterokont zoospores (Figs. 4.6a, 4.7b) are released by each cyst. These possess a pigmented eyespot not found in other types of heterokont zoospore (Porter, 1990). It is, however, unclear whether zoospores can establish new colonies (Porter, 1990). Asexual reproduction occurs by division of spindle cells within the slime net, and fragments of such a colony can establish new colonies (Porter, 1972). Further details of the life cycle appear to be unknown at present.

Labyrinthula spp. were implicated as pathogens in a wasting epidemic of eelgrass (*Zostera marina*) at the west coast of North America in the 1930s (Young, 1943; Muehlstein *et al.*, 1991), causing considerable disturbance to the littoral ecosystem and collateral damage to the local fisheries industry. However, although *Labyrinthula* spp. are still frequently associated with pieces of moribund *Zostera* shoots, no further epidemics seem to have occurred since. Instead, a new species, *L. terrestris*, has recently been identified as the cause of a rapid blight of

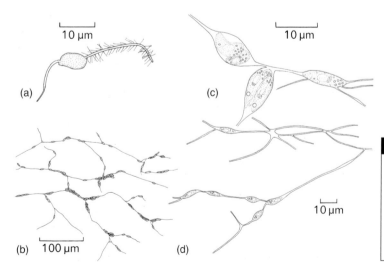

(a)

(c)

(b) 100 μm

(d)

10 μm

10 μm

10 μm

Fig 4.6 *Labyrinthula*. (a) Zoospore with long anterior straminipilous flagellum and a short posterior whiplash flagellum with a pointed tip (after Amon & Perkins, 1968). (b–d) Portions of colonies at different magnifications. In (c) spindle cells are seen in swellings in the slime tracks.

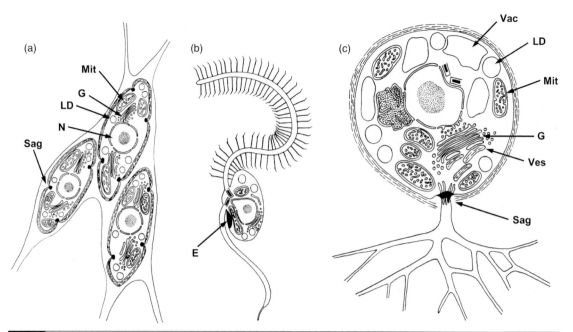

Fig 4.7 Ultrastructural features of Labyrinthulomycota. (a) Spindle-shaped cells of *Labyrinthula* within their slime net. Each cell has mitochondria with tubular cristae (Mit), Golgi stacks (G), a single nucleus (N), and cortical lipid droplets (LD). The slime net is produced by several sagenogens (Sag) in each cell. The plasma membrane is continuous with the inner membrane of the slime net. Wall scales are released at the sagenogen point and accumulate between the plasma membrane and the inner membrane of the slime net. (b) Biflagellate heterokont zoospore of *Labyrinthula* showing an eyespot (E) close to the base of the whiplash flagellum. Note that each TTH of the *Labyrinthula* zoospore produces only one terminal fibre. (c) Young thallus of *Thraustochytrium*. Mitochondria with tubular cristae, a Golgi stack, lipid droplets and larger vacuoles (Vac) are seen. The wall consists of scales pre-formed in Golgi-derived vesicles (Ves). The slime net is produced at the base of the thallus by a single sagenogen. All images schematic and not to scale; redrawn and modified from Porter (1990). © 1990 Jones and Bartlett Publishers, Sudbury, MA. www.jbpub.com.

turf-grass on golf courses, infection presumably being brought about by irrigation with contaminated water of unusually high salinity (Bigelow *et al.*, 2005).

4.4.2 Thraustochytriales
Thraustochytrids are probably ubiquitous in marine environments, occurring on organic debris as well as calcareous shells of invertebrates (Porter & Lingle, 1992). Like the labyrinthulids, they feed on organic matter, algae and bacteria (Raghukumar, 2002). Thraustochytrids can be baited by sprinkling pine pollen grains onto water samples or organic debris immersed in water. Within one to several days, the pollen grains become colonized by one or several thalli, the main bodies of which protrude beyond the grain surface (Figs. 4.8a,b). If colonized pollen grains are transferred to a suitable agar medium

containing sea salts, yeast extract and sugar (Yokochi *et al.*, 1998), thalli will grow on the agar surface and may be induced to release zoospores by mounting them in water. Thraustochytrids can be stored in pollen grain suspensions or on agar overlaid with sea water. They also possess the ability to survive in a dry state at room temperature for a year or longer (Porter, 1990).

The thallus of thraustochytrids superficially resembles that of an epibiotic monocentric chytrid in having a roughly spherical shape with 'rhizoids' at its base (Fig. 4.8c). These 'rhizoids' are, in fact, the slime net produced by one basal sagenogen (Fig. 4.7c). The thallus is surrounded by Golgi-derived scales forming a wall, but the slime net does not extend over the thallus. Sexual reproduction is unknown, but asexual biflagellate heterokont zoospores are released from the main body of the thallus,

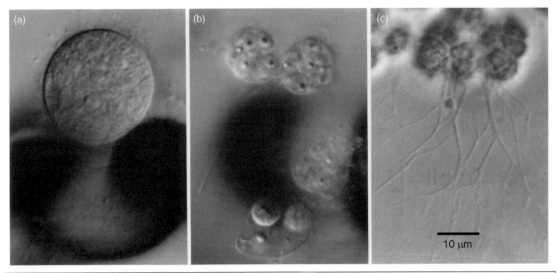

Fig 4.8 Thraustochytriales. (a) Thallus of *Thraustochytrium* sp. growing on a pollen grain sprinkled onto seawater. (b) Thalli of *Schizochytrium* sp. growing on a pollen grain. (c) Thalli of *Schizochytrium* sp. growing on agar medium. Note the slime net extending away from the thalli.

and these can settle onto a suitable substratum, giving rise to new thalli (Porter, 1990). Thus, these zoospores of Thraustochytriales are mitospores formed following mitosis, in contrast with those of Labyrinthulales which are meiospores, i.e. formed by meiosis. Although thraustochytrid zoospores lack a recognizable eye-spot, they are phototropic, reacting to light of blue wavelengths such as that produced by bioluminescent bacteria (Amon & French, 2004). Chemotropism has also been described for thraustochytrid zoospores (Fan *et al.*, 2002), and both sensual responses may enable zoospores to locate potential food sources.

Thraustochytrids, and especially the genera *Thraustochytrium* and *Schizochytrium*, have recently attracted attention as producers of polyunsaturated fatty acids (PUFAs). These are important as nutrient supplements, and thraustochytrid oils might eventually be able to compete with fish oils on the market (Yokochi *et al.*, 1998; Lewis *et al.*, 1999).

Straminipila: Oomycota

5.1 | Introduction

The phylum Oomycota, alternatively called Peronosporomycetes (Dick, 2001a), currently comprises some 800–1000 species (Kirk *et al.*, 2001). The Oomycota as a whole have been resolved as a monophyletic group within the kingdom Straminipila in recent phylogenetic studies (e.g. Riethmüller *et al.*, 1999; Hudspeth *et al.*, 2000; see Fig. 4.2), although considerable rearrangements are still being performed at the level of orders and families. A scholarly treatment of the Oomycota has been published by Dick (2001a) and will remain the reference work for many years to come. Because of the outstanding significance of Oomycota, especially in plant pathology, we give an extended treatment of this group.

5.1.1 The vegetative hypha

Although some members of the Oomycota grow as sac-like or branched thalli, most of them produce hyphae forming a mycelium. Oomycota are now known to be the result of convergent evolution with the true fungi (Eumycota), and their hyphae differ in certain details. However, the overall functional similarities are so great that they provide a persuasive argument for the fundamental importance of the hypha in the lifestyle of fungi (Barr, 1992; Carlile, 1995; Bartnicki-Garcia, 1996). Much physiological work has been carried out on hyphae of Oomycota (see Chapter 1), and the results have a direct bearing on our understanding of the biology of the Eumycota. Like them, the hyphae of Oomycota display apical growth and enzyme secretion, ramify throughout the substratum by branching to form a mycelium, and can show morphogenetic plasticity by differentiation into specialized structures such as appressoria or haustoria.

The hyphae of Oomycota are **coenocytic**, i.e. they generally do not form cross-walls (septa) except in old compartments or at the base of reproductive structures. The cytoplasm is generally coarsely granular and contains vacuoles, Golgi stacks, mitochondria and diploid nuclei. The apex is devoid of organelles other than numerous secretory vesicles. These are not, as in the Eumycota, arranged into a Spitzenkörper because the microvesicles which contain chitin synthase and make up the Spitzenkörper core are lacking. This is in line with the general absence, with a few exceptions, of chitin from the walls of Oomycota; instead, cellulose, a crystalline β-(1,4)-glucan, contributes the main fibrous component. As in the Eumycota, these structural fibres are cross-linked by branched β-(1,3)- and β-(1,6)-glucans, although the biochemical properties of the glucan synthases seem to differ fundamentally between those of Eumycota on the one hand and those of Oomycota and plants on the other (Antelo *et al.*, 1998). Other biochemical differences include the lysine synthetic pathway (DAP in plants and Oomycota; AAA in true Fungi; see p. 67) and details of sterol metabolism (Nes, 1990; Dick, 2001a).

The mitochondria of Oomycota are indistinguishable by light microscopy from those of the Eumycota, but when viewed with the transmission electron microscope they have tubular

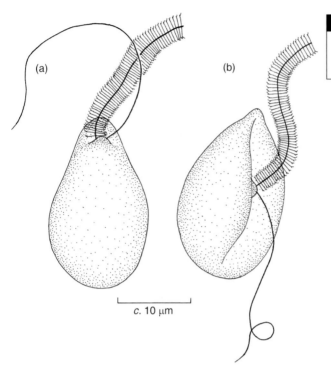

(a)

(b)

Fig 5.1 Asexual reproductive stages in *Saprolegnia*. (a) Auxiliary (primary) zoospore. (b) Principal (secondary) zoospore. Schematic drawings, based partly on Dick (2001a).

c. 10 μm

rather than lamellate cristae (see Fig. 4.1). The vacuolar system of Oomycota is also unusual in containing **dense-body vesicles** or '**fingerprint vacuoles**' (see Fig. 5.24b) which consist of deposits of a phosphorylated β-(1,3)-glucan polymer, mycolaminarin. Mycolaminarin may serve as a storage compound for carbohydrates as well as phosphate (Hemmes, 1983), and the polyphosphate storage deposits which are typically found within vacuoles of true Fungi are absent from vacuoles of Oomycota (Chilvers *et al.*, 1985). Apart from that, however, vacuoles of Oomycota share many features with those of true Fungi, including the membranous continuities which often link adjacent vacuoles and provide a means of transport by peristalsis (Rees *et al.*, 1994; see Fig. 1.9). Cytoplasmic glycogen granules, which are one of the major carbohydrate storage sites in Eumycota, are absent from hyphae of Oomycota (Bartnicki-Garcia & Wang, 1983).

5.1.2 The zoospore

The Oomycota are characterized by motile asexual spores (zoospores) which are produced in spherical or elongated zoosporangia. They are heterokont, possessing one straminipilous and one whiplash-type flagellum. Two types of zoospore may be produced and, if so, the **auxiliary zoospore** is the first formed. It is grapeseed-shaped, with both flagella inserted apically (Fig. 5.1a), and it encysts soon after its formation. Encystment is by withdrawal of the flagella, so that a tuft of tripartite tubular hairs (TTHs; see p. 68) is left behind on the surface of the developing cyst (Dick, 2001b). The cyst germinates to give rise to the **principal zoospore**, which is by far the more common type and also the more vigorous swimmer. This typical and readily recognized oomycete zoospore is uniform in appearance across the phylum (Lange & Olson, 1983; Dick, 2001a). In species lacking auxiliary zoospores, the principal zoospore is usually produced directly from a sporangium. It is kidney-shaped, with the flagella inserted laterally in a kinetosome boss which in turn is located within the lateral groove (Fig. 5.1b). Encystment is initiated by the shedding, rather than withdrawal, of the flagella; no tufts of TTHs are left on the cyst surface (Dick, 2001a). Fascinating insights into the cytology of zoospore encystment have been

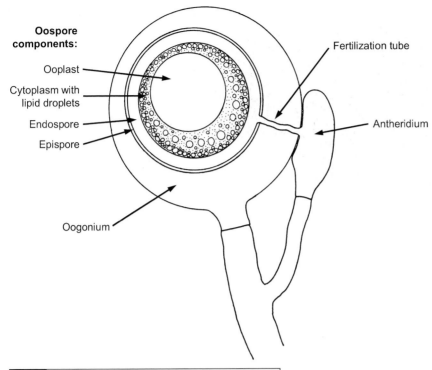

Oospore components:

Ooplast

Cytoplasm with lipid droplets

Endospore

Epispore

Oogonium

Fertilization tube

Antheridium

Fig 5.2 Schematic drawing and terminology of sexual reproductive organs in the Oomycota. Modified from Dick (1995).

obtained from several species (see Fig. 5.24). At the onset of encystment, adhesive and cell wall material is secreted by the synchronized fusion of pre-formed storage vesicles with the zoospore plasma membrane (Hardham *et al.*, 1991; Hardham, 1995), thereby providing a rare example of regulated secretion in fungi. Constitutive secretion by growing hyphal tips is more commonly associated with their mode of life.

Some members of the Oomycota have no motile spore stages but can be readily related to groups still producing them.

5.1.3 Sexual reproduction

The life cycle of the Oomycota is of the haplomitotic B type, i.e. mitosis occurs only between karyogamy and meiosis. All vegetative structures of Oomycota are therefore diploid (see Figs. 5.3 and 5.19). This is in contrast to the Eumycota in which vegetative nuclei are usually haploid, the first division after karyogamy being

meiotic. Sexual reproduction in Oomycota is **oogamous**, i.e. male and female gametangia are of different size and shape (Fig. 5.2). Meiosis occurs in the male **antheridia** and in the female **oogonia**, and is followed by plasmogamy (fusion between the protoplasts) and karyogamy (fusion of haploid nuclei). Numerous meioses can occur synchronously, so that true sexual reproduction can actually happen within the same protoplast (Dick, 1990a). Heterothallic species of Oomycota display **relative sexuality**, i.e. a strain can produce antheridia in combination with a second strain but oogonia when paired against a third (see pp. 86 and 95). Steroid hormones play an important role in sexual reproduction (see Fig. 5.11).

The mature oospore contains three major pools of storage compounds (Fig. 5.2; Dick, 1995). The oospore wall often appears stratified, and this is due in part to a polysaccharide reserve compartment, the **endospore**, which is located

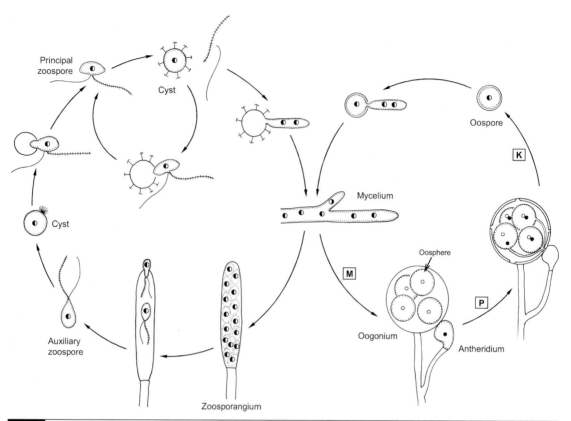

Fig 5.3 Life cycle of *Saprolegnia*. Vegetative hyphae are diploid and coenocytic. Asexual reproduction is by means of diplanetic (auxiliary and principal) zoospores. The principal zoospore state is polyplanetic. *Saprolegnia* is homothallic, and sexual reproduction is initiated by the formation of antheridia and oogonia. For simplicity, only a single nucleus is shown in each of the oospheres and in the antheridium. Each oogonium contains several oospheres. Karyogamy occurs soon after fertilization of an oosphere by an antheridial nucleus. The oospore may germinate by means of a germ sporangium (not shown) or a hyphal tip. Open and closed circles represent haploid nuclei of opposite mating type; diploid nuclei are larger and half-filled. Key events in the life cycle are meiosis (M), plasmogamy (P) and karyogamy (K).

between the plasma membrane and the outer spore wall (**epispore**). Upon germination, the endospore is thought to coat the emerging germ tube with wall material, and some material may also be taken up by endocytosis. A large storage vacuole inside the oospore protoplast is called the **ooplast**. It arises by fusion of dense-body vesicles and, like them, contains mycolaminarin and phosphate. Dick (1995, 2001a) speculated that the ooplast contributes membrane precursor material during the process of oospore germination. The third storage compartment consists of one or several lipid droplets which provide the endogenous energy supply required for germination. Ultrastructural changes during oospore

germination have been described by Beakes (1981).

5.1.4 Ecology and significance

Oomycota have a major impact on mankind as pathogens causing plant diseases of epidemic proportions. Two events have had particularly far-reaching political and social consequences, and have shaped and interlinked the young disciplines of mycology and plant pathology in the nineteenth century. These were the great Irish potato famine of 1845–1848 caused by *Phytophthora infestans* (Bourke, 1991), and the occurrence of downy mildew of grapes caused by *Plasmopara viticola* (Large, 1940). The former prepared the way for the then revolutionary theory that fungal infections can be the cause rather than the consequence of disease, whereas the latter stimulated research into chemical control of diseases which directly gave rise to the first fungicide, Bordeaux mixture (p. 119; Large, 1940).

Although all members of Oomycota depend on moist conditions for the dispersal of their zoospores, they are cosmopolitan and ubiquitous even in terrestrial situations. In species adapted to drier habitats, the sporangia often germinate directly to produce a germ tube, with zoospores released as an alternative germination method only in the presence of moisture, or lacking altogether. Oomycota occur in freshwater, the sea, in the soil and on above-ground plant organs. Most are obligate aerobes, although some tolerate anaerobic conditions (Emerson & Natvig, 1981; Voglmayr *et al.*, 1999), and one species (*Aqualinderella fermentans*) is obligately anaerobic and lacks mitochondria (Emerson & Weston, 1967). Oomycota live either saprotrophically on organic material, or they may be obligate (biotrophic) or facultative (necrotrophic) parasites of plants. Some can also cause diseases of animals, such as *Aphanomyces astaci* which has all but eliminated European crayfish from many rivers (p. 94), *Saprolegnia* spp. which cause serious infections of farmed fish, especially salmon (Plate 2a; Dick, 2003), or *Pythium insidiosum* causing equine phycomycosis (de Cock *et al.*, 1987). Yet other Oomycota, notably *Lagenidium giganteum*, parasitize insects and may prove valuable in the biological control of mosquito larvae (Dick, 1998).

5.1.5 Classification

As indicated above, the classification of Oomycota at the level below the phylum is still an ongoing process, and it is difficult at present to reconcile the different classification schemes that are being proposed. Kirk *et al.* (2001) listed eight orders in the phylum Oomycota, of which Dick (2001b) treated six within the class Peronosporomycetes, his equivalent to the Oomycota, considering the other two of

uncertain affinity (*incertae sedes*). These groups are summarized in Table 5.1.

5.2 | Saprolegniales

The order Saprolegniales is currently divided up into two families, the Saprolegniaceae (e.g. *Achlya*, *Brevilegnia*, *Dictyuchus*, *Saprolegnia*, *Thraustotheca*) and Leptolegniaceae (*Aphanomyces*, *Leptolegnia*, *Plectospira*), totalling 132 species in about 20 genera (Dick, 2001a; Kirk *et al.*, 2001). The Saprolegniales are the best-known group of aquatic fungi, often termed the water moulds. Members of this group are abundant in wet soils, lake margins and freshwater, mainly as saprotrophs on plant and animal debris. Whilst some Saprolegniales occur in brackish water, most are intolerant of it and thrive best in freshwater. A few species of *Saprolegnia* and *Achlya* are economically important as parasites of fish and their eggs (Willoughby, 1994). *Aphanomyces euteiches* causes a root rot of peas and some other plants, whilst *A. astaci* is a serious parasite of the European crayfish *Astacus* (Alderman *et al.*, 1990). Algae, fungi, rotifers and copepods may also be parasitized by members of the group, and occasional epidemics of disease among zooplankton have been reported.

Members of the Saprolegniales are characterized by coarse, stiff hyphae which branch to produce a typically fast-growing mycelium. The hyphae of Saprolegniales are coenocytic, containing a peripheral layer of cytoplasm surrounding a continuous central vacuole. Cytoplasmic streaming is readily observed in the peripheral cytoplasm. Numerous nuclei are present. Mitotic division is associated with the replication of paired centrioles and the development of an intranuclear mitotic spindle; the nuclear membrane remains intact throughout division (Dick, 1995). Filamentous mitochondria and lipid droplets can also be observed in vegetative hyphae. The mitochondria are orientated parallel to the long axis of the hypha and are sufficiently large to be seen in cytoplasmic streaming in living material. Important physiological work has been carried out on the

Table 5.1. Summary of the most important groups of Oomycota and their characteristic features. Only the last four groups are considered further in this book. Based on information provided by Dick (2001a,b) and Kirk *et al.* (2001).

Order	Number of species	Thallus and reproduction	Ecology
Myzocytiopsidales (*incertae sedes*)	74	Holocarpic,* later coralloid or breaking up into segments. Zoospores, oospores.	Parasites of invertebrates or algae.
Olpidiopsidales (*incertae sedes*)	21	Holocarpic,* becoming converted into a sporangium. Zoospores, oospores.	Biotrophic parasites of Oomycota, Chytridiomycota and algae.
Rhipidiales	12	Eucarpic* with rhizoids. Zoospores, oospores.	Freshwater saprotrophs, facultatively or obligately anaerobic.
Leptomitales	25	Constricted hyphae producing sporangia. Zoospores, oospores.	Freshwater saprotrophs or parasites of animals.
Saprolegniales (see Section 5.2)	132	Mycelium of wide stout hyphae. Zoospores, oospores.	Saprotrophs or necrotrophic pathogens of animals, plants and other organisms.
Pythiales (see Section 5.3)	>200	Mycelium of relatively narrow hyphae. Zoospores, oospores.	Saprotrophs or pathogens (often necrotrophic) of plants, fungi and animals.
Peronosporales (see Section 5.4)	252	Intercellular mycelium with haustoria. Differentiated sporangiophores. Zoospores or 'conidia', oospores.	Biotrophic plant pathogens, causing downy mildews and other diseases.
Sclerosporaceae (see Section 5.5)	22	Mycelium of very narrow hyphae. Differentiated sporangiophores. Zoospores or 'conidia', oospores.	Biotrophic pathogens of grasses, causing downy mildews.

*For thallus terminology, see Fig. 6.1.

mechanisms of hyphal polarity and growth regulation in *Achlya* and *Saprolegnia* (see Heath, 1995b; Hyde & Heath, 1997; Heath & Steinberg, 1999). Like other Oomycota but in contrast to the Eumycota (Pfyffer *et al.*, 1986; Rast & Pfyffer, 1989), these fungi are unable to synthesize compatible osmotically active solutes such as glycerol, mannitol and other polyols to maintain their intrahyphal turgor pressure against fluctuating external conditions. Under conditions of water stress, the turgor pressure in hyphae of *Achlya* and *Saprolegnia* approaches zero, yet hyphal growth can still occur at least under laboratory conditions because of the enhanced secretion of cell wall-softening enzymes and the role of the cytoskeleton in pushing forward the growing tip (see pp. 6–9; Money & Harold, 1992, 1993; Money, 1997; Money & Hill, 1997).

The Saprolegniales are the only order within the Oomycota to produce both auxiliary and principal zoospores, although both forms are not produced in all genera. The production of two distinct motile stages is termed **diplanetism**. It has also been called **dimorphism**, but this term has several different meanings and is best avoided in the current context. Depending on environmental conditions, the cysts of principal zoospores may germinate either by means of a germ tube developing into a hypha or by the emergence of a new principal zoospore. The repetition of the same type of motile spore is called **polyplanetism**.

Sexual reproduction in the Saprolegniales is oogamous, with a large, usually spherical oogonium containing one or several oospheres. Antheridial branches apply themselves to the wall of the oogonium and penetrate the wall by fertilization tubes through which a single nucleus is introduced into each oosphere. A feature of many Saprolegniales, especially when grown in culture, is the formation of thick-walled enlarged terminal or intercalary portions of hyphae which become packed with dense cytoplasm and are cut off from the rest of the mycelium by septa. These structures, which may occur singly or in chains (see Fig. 5.6g), are termed gemmae or **chlamydospores**, and their formation can be induced by manipulating the culture conditions. Morphologically less distinct but otherwise similar structures are frequently found in old cultures. Although it is known that chlamydospores cannot survive desiccation or prolonged freezing, they remain viable for long periods in less extreme conditions. They may function as female gametangia or as zoosporangia, but more frequently they germinate by means of a germ tube. Another feature of old cultures is the fragmentation of cylindrical pieces of mycelium cut off at each end by a septum.

Members of the Saprolegniales can be isolated readily from water, mud and soil by floating split boiled hemp seeds or dead house flies in dishes containing pond water, or by covering soil samples or waterlogged twigs with water (Stevens, 1974; Dick, 1990a). Within about 4 days the fungi can be recognized by their stiff, radiating, coarse hyphae bearing terminal sporangia, and cultures can be prepared by transferring hyphal tips or zoospores to cornmeal agar or other suitable media. The most commonly encountered genera are *Achlya*, *Dictyuchus*, *Saprolegnia*, *Thraustotheca* and *Aphanomyces*. With the exception of a few obligately parasitic species, most of the Saprolegniales will grow readily in pure culture even on chemically defined media, and extensive studies of their nutritional physiology have been undertaken (summarized by Cantino, 1955; Gleason, 1976; Jennings, 1995). Most species examined have no requirement for vitamins. Organic forms of sulphur such as cysteine, cystine, glutathione and methionine are preferred, and most species are unable to reduce sulphate. Organic nitrogen sources such as amino acids, peptone and casein are preferred to inorganic sources. Ammonium is widely utilized, but nitrate is not. Glucose is the most widely utilized carbon source, but many species also degrade maltose, starch and glycogen. In liquid culture, *Saprolegnia* can be maintained in the vegetative state indefinitely if supplied with organic nutrients in the form of broth. When the nutrients are replaced by water, the hyphal tips quickly develop into zoosporangia. The formation of sexual organs can similarly be affected by manipulating the external conditions in some species, and the concentration of salts in the medium may play a decisive role (Barksdale, 1962; Davey & Papavizas, 1962).

5.2.1 *Saprolegnia* (Saprolegniaceae)

Species of *Saprolegnia* are common in soil and in freshwater as saprotrophs on plant and animal remains. A few species such as *S. parasitica* and *S. polymorpha* cause disease in fish and their eggs (Plate 2a). Salmonid fish are particularly affected, and the disease can cause significant damage in fish farms around the world (Willoughby, 1994, 1998a). Control by fungicides is difficult but possible (Willoughby & Roberts, 1992). The disease is also seen in wild salmon and other fish (Söderhäll *et al.*, 1991; Bly *et al.*, 1992). Pathogenic strains or species may be closely related to non-pathogenic ones but can be distinguished by physiological characteristics, DNA sequencing (Yuasa & Hatai, 1996) and the length of the 'boat hook' appendages on the cysts of principal zoospores (Figs. 5.5b,c; Beakes, 1983; Burr & Beakes, 1994).

The life cycle of *Saprolegnia* is summarized in Fig. 5.3. A monographic treatment of the genus has been published by Seymour (1970).

Asexual reproduction in *Saprolegnia*

Sporangia of *Saprolegnia* develop when a hyphal tip, which is pointed in the vegetative condition, swells, rounds off and becomes club-shaped. It accumulates denser cytoplasm around the vacuole which remains clearly visible. A septum develops at the sporangial base and it is at first straight or convex with respect to the sporangium, i.e. it bulges into it (Figs. 5.4c,d). The sporangium contains numerous nuclei, and

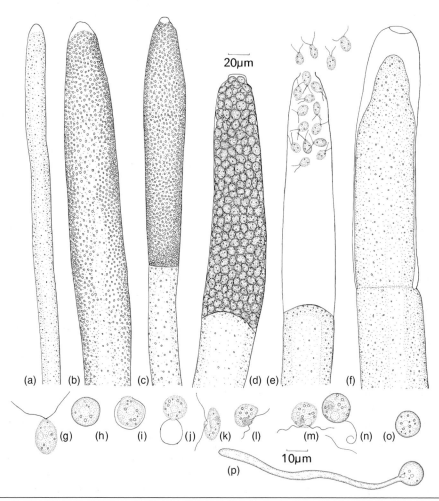

Fig 5.4 *Saprolegnia*. (a) Apex of vegetative hypha. (b–d) Stages in the development of zoosporangia. (e) Release of zoospores. (f) Proliferation of zoosporangium. A second zoosporangium is developing within the empty one. (g) Auxiliary zoospore (first motile stage). (h) Cyst formed at the end of the first motile stage (auxiliary cyst). (i,j) Germination of auxiliary cyst to release a second motile stage (principal zoospores). These have the typical reniform shape. (k–m) Principal zoospores. (n) Principal zoospore at the moment of encystment. Note the shed flagellum. (o) Principal cyst. (p) Principal cyst germinating by means of a germ tube. (a–f) to same scale; (g–p) to same scale. Note that the straminipilous flagellum cannot be distinguished from the whiplash flagellum at the magnification chosen.

cleavage furrows separate the cytoplasm into uninucleate pieces, each of which differentiates into an auxiliary zoospore. As the zoospores are cleaved, the central vacuole disappears. The tip of the cylindrical sporangium contains clearer cytoplasm and a flattened protuberance, the **papilla**, develops at the apex. As the sporangium ripens and the zoospores become fully differentiated, they show limited movement and change of shape (Figs. 5.4b–d). Shortly before discharge, there is evidence of a build-up of turgor pressure within the sporangium because the basal septum becomes concave, i.e. it is bent towards the lumen of the hypha beneath the sporangium. After cleavage, the positive turgor pressure is lost concomitantly with the loss of the sporangial plasma membrane which becomes part of the zoospore membranes, and the septum again bulges into the sporangium while the zoospores become fully differentiated. The sporangium undergoes a slight change of shape at this time and the sporangium wall breaks down at the papilla. The spores are released quickly, many zoospores escaping in a few seconds and moving as a column through the opening. Osmotic phenomena have been invoked to explain the rapidity of discharge, and the osmotically active substances must be large enough to be contained by the sporangial wall. Mycolaminarin, released from the central vacuole during zoospore differentiation, is the likely solute (Money & Webster, 1989). The whole process of sporangium differentiation takes about 90 min. The zoospores leave the sporangium backwards, with the blunt posterior end emerging first. The size of the zoospore is sometimes greater than the diameter of the sporangial opening so that the zoospores are squeezed through it. An occasional zoospore may be left behind, swimming about in the empty zoosporangium for a while before making its exit. Zoospores in partially empty sporangia orientate themselves in a linear fashion along the central axis of the sporangium.

A characteristic feature of *Saprolegnia* is that, following the discharge of a zoosporangium, growth is renewed from the septum at its base so that a new apex develops inside the old sporangial wall by **internal proliferation**. This in turn may develop into a zoosporangium, discharging its spores through the old pore (Fig. 5.4f). The process may be repeated so that several empty zoosporangial walls may be found inside, or partially inside, each other.

Upon release, the auxiliary zoospores slowly revolve and eventually swim somewhat sluggishly with the pointed end directed forwards. They are grapeseed- or pear-shaped ('Conference' pear; Dick, 2001a) and bear two apically attached flagella (see Figs. 5.1a, 5.4g). Each zoospore also contains a diploid nucleus, mitochondria, a contractile vacuole and numerous vesicles (Holloway & Heath, 1977a,b). The zoospores from a single sporangium show variation in their period of motility, the majority encysting within about a minute, but some remaining motile for over an hour. The zoospore then withdraws its flagella and encysts, i.e. the cytoplasm becomes surrounded by a distinct wall which is produced from pre-formed material stored in the cytoplasmic vesicles. Only the axonemes of the flagella are withdrawn, leaving the TTHs of the straminipilous flagellum at the surface of the cyst (see Fig. 5.5a). Following a period of rest (2–3 h in *S. dioica*), the cyst germinates to release a further zoospore, the principal zoospore (Figs. 5.4i,j). This differs in shape from the auxiliary zoospore in being bean-shaped, with the two flagella inserted laterally in a shallow groove running down one side of the zoospore (Fig. 5.1b). The principal zoospore may swim vigorously for several hours before encysting. Salvin (1941) compared the rates of movement of auxiliary and principal zoospores in *Saprolegnia* and found that the latter swam about three times more rapidly. The probable reason for this is that the lateral insertion of both flagella allows the straminipilous flagellum to point forward and the whiplash one to point backward, thereby improving the propulsion relative to the apical insertion in which both flagella point forward.

Movement of principal zoospores is chemotactic and zoospores can be stimulated to aggregate on parts of animal bodies such as the leg of a fly, or the surface of a fish (Fischer & Werner, 1958; Willoughby & Pickering, 1977). When principal zoospores encyst, they shed

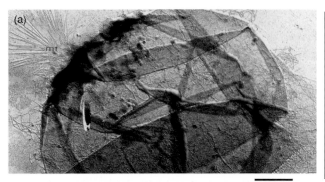

2 μm

Fig 5.5 Surface features of *Saprolegnia*. (a) Detail of an auxiliary zoospore cyst of *S. parasitica* showing the tuft of TTHs (mt) at the point where the straminipilous flagellum was withdrawn. (b) Surface of a principal zoospore cyst of *S. parasitica*; the long boat hook spines are arranged in fascicles. (c) Surface of a principal zoospore cyst of *S. hypogyna* with discrete boat hooks of intermediate length. All bars = 2 μm. All images kindly provided by M.W. Dick and I.C. Hallett; (b) reprinted from Hallett and Dick (1986), with permission from Elsevier.

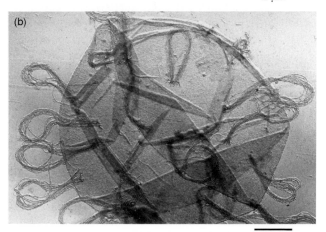

2 μm

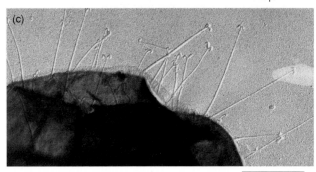

2 μm

rather than withdraw their flagella. The first step in encystment is the fusion of vesicles called K-bodies with the plasma membrane. These are so called because they are located near the kinetosome. The material they secrete is involved in attachment of the zoospore to a substratum, which occurs in the region of the groove near the flagellar bases, designated the ventral region (Lehnen & Powell, 1989). The cyst wall and pre-formed boat hook spines are secreted by fusion of encystment vesicles with the plasma membrane (Beakes, 1987; Burr & Beakes, 1994). The length and arrangement of spines on the surface of a mature principal cyst are characteristic features of individual species (Figs. 5.5b,c). They probably mediate attachment of the cyst to the host, and pathogenic isolates of *Saprolegnia* have much longer spines than saprotrophic ones (Burr & Beakes, 1994). Alternatively, the boat hooks may mediate attachment to the water meniscus.

Either way, attachment must be very effective because trout or char, placed in a water bath with principal zoospores of *S. parasitica* for 10 min and followed by 1 h in clean water, had an extremely high concentration of cysts attached to the skin (Willoughby & Pickering, 1977).

Principal zoospore cysts can germinate either by means of a germ tube (Fig. 5.4p) or by releasing a further principal zoospore which in turn may germinate directly or by releasing yet another motile stage. *Saprolegnia* is therefore polyplanetic. The auxiliary and principal zoospores, as well as the cysts they form, differ morphologically from each other, i.e. they are diplanetic.

Sexual reproduction in *Saprolegnia*

All members of the genus *Saprolegnia* characterized to date are homothallic, i.e. a culture derived from a single zoospore will give rise to a mycelium forming both oogonia and antheridia. In contrast, *Achlya* also contains heterothallic species in which sexual reproduction occurs only when two different strains are juxtaposed, one forming oogonia, the other antheridia (see Fig. 5.10).

Sexual reproduction follows a similar course in all members of the Saprolegniales. Oogonia containing one or several eggs are fertilized by antheridial branches. Fertilization is accomplished by the penetration of fertilization tubes into the oogonium. In some species, ripe oogonia are found without antheridia associated with them (Fig. 5.6f); this could be due either to the fusion of two haploid nuclei from adjacent meiotic events in a single oogonium (**apomixis**) or the formation of an oospore around a diploid nucleus that never underwent meiosis (**parthenogenesis**). Both processes are impossible to distinguish without detailed cytological evidence (Dick, 2001a). The typical arrangement of oogonia and antheridia in Saprolegnia is shown in Fig. 5.6. Antheridial branches arising from the stalk of the oogonium or the same hypha as the oogonium are said to be **monoclinous** whereas they are **diclinous** if they originate from different hyphae.

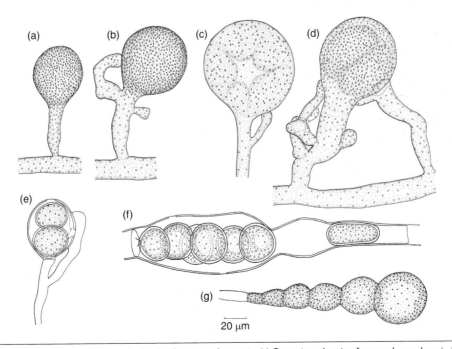

Fig 5.6 *Saprolegnia litoralis.* (a–d) Stages in the development of oogonia. (c) Oogonium showing furrowed cytoplasm indicative of centrifugal cleavage. (d) Outlines of two oospheres become visible. (e) Oogonium with two mature oospores. (f) Intercalary oogonium lacking antheridia. The oospores have developed by apomixis or parthogenesis. (g) Chain of chlamydospores.

The oogonial initial is multinucleate, and nuclear divisions continue as it enlarges. Eventually some of the nuclei degenerate, leaving only those nuclei which are included in the oospheres. From the central vacuole within the oogonium, cleavage furrows radiate outwards to divide the cytoplasm into uninucleate portions which round off to form oospheres. Oogonium differentiation is thus centrifugal, which is typical of the Saprolegniales. Cleavage of the oospheres from the cytoplasm is brought about by the coalescence of dense body vesicles which finally fuse with the plasma membrane of the oogonium so that the oospheres tumble into the centre of the oogonium (Dick, 2001a). The entire mass of cytoplasm within the oogonium is used up in the formation of oospheres and there is no residual cytoplasm (**periplasm**) as in the Peronosporales. The wall of the oogonium is often uniformly thick, but in some species it shows thin areas or pits through which fertilization tubes may enter (Fig. 5.6e). A septum at the base of the oogonium cuts it off from the subtending hypha.

The antheridia are also multinucleate. The antheridial branch grows towards the oogonium and attaches itself to the oogonial wall. The tip of the antheridial branch is cut off by a septum, and the resulting antheridium puts out a fertilization tube which penetrates the oogonial wall and may branch within the oogonium. After the tube has penetrated an oosphere wall, a male nucleus eventually fuses with the single oosphere nucleus. The fertilized oosphere (oospore) undergoes a series of changes described by Beakes and Gay (1978a,b). The wall of the oospore thickens and oil globules become obvious. Mature oospores contain a membrane-bound vacuole-like body, the ooplast, surrounded by cytoplasm containing various organelles, with lipid droplets particularly prominently visible. In *Saprolegnia*, the ooplast contains particles in Brownian motion. The position of the ooplast in the oospore is used for species identification, and four types of oospore have been distinguished (Fig. 5.7; Seymour, 1970; Howard, 1971). **Centric** oospores have a central ooplast surrounded by one or two peripheral layers of small lipid droplets (e.g. *S. hypogyna*, *S. ferax*). **Subcentric** oospores have several layers of small lipid droplets on one side of the ooplast and only one layer or none at all on the other (e.g. *S. unispora*, *S. terrestris*). In **subeccentric** oospores, the small lipid droplets have fused into several large ones all grouped to one side, with the ooplast contacting the plasma membrane on the opposite side (e.g. *S. eccentrica*). The **eccentric** type (found, for example, in *S. anisospora*) is similar to the subeccentric type except that there is only one very large lipid drop. These descriptive terms are also used for many other species of Oomycota.

5.2.2 *Achlya* (Saprolegniaceae)

Phylogenetic analyses have shown that the genera *Achlya* and *Saprolegnia* as well as minor genera of the Saprolegniales are closely related to each other, with possible overlaps which may necessitate the re-assignment of some species in future (Riethmüller *et al.*, 1999; Leclerc *et al.*, 2000; Dick, 2001a). Morphologically and ecologically, *Achlya* and *Saprolegnia* also share several key features. Both are common in soil and in waterlogged plant debris such as twigs, and certain species are pathogens of fish (Willoughby, 1994; Kitancharoen *et al.*, 1995). Unlike *Saprolegnia*, some species of *Achlya* are heterothallic, but their life cycle is otherwise similar to that of *Saprolegnia* given in Fig. 5.3. Heterothallic strains of *Achlya* have been the subject of classical

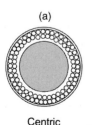
(a)

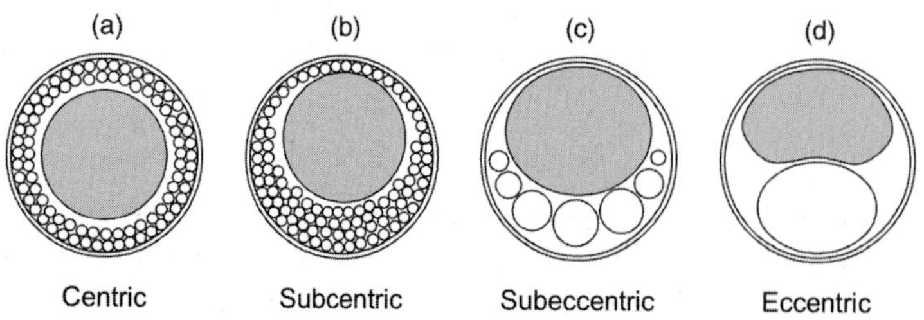

Centric

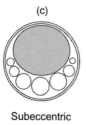
(b)

Subcentric

(c)

Subeccentric

(d)

Eccentric

Fig 5.7 Possible arrangements of the ooplast (shaded organelle) and lipid droplets (empty circles or ellipses) in oospores of *Saprolegnia*. (a) Centric. (b) Subcentric. (c) Subeccentric. (d) Eccentric.

studies on the nature of mating hormones (pheromones); additionally, more recent work has focused on zoospore release. Both aspects are described below.

Asexual reproduction in *Achlya*

The development of zoosporangia in *Achlya* is similar in all aspects to that in *Saprolegnia* but has been better researched. The central vacuole in the developing cylindrical sporangium is typical of the Saprolegniales and originates from the fusion of dense body vesicles containing mycolaminarin. The centrifugal cleavage of cytoplasm from the vacuole towards the plasma membrane, and the partitioning of individual spores, are controlled mainly by the actin cytoskeleton (Heath & Harold, 1992). In the Pythiales, vital roles of microtubules in the organization of differentiating cytoplasm have been described (see p. 102), and microtubules may have similar but as yet undescribed functions in the Saprolegniales. As the plasma membrane of the *Achlya* zoosporangium is breached, the zoosporangial volume decreases by about 10% due to the loss of turgor pressure. Since the membranes of the vacuole contribute to the zoospore plasma membrane, the vacuolar contents of water-soluble mycolaminarins (β-1,3-glucans) are released into the sporangium. These molecules are osmotically active but are too large to diffuse through the sporangial wall, thus causing the osmotic inward movement of water into the sporangium, which in turn pressurizes the sporangium and drives the rapid discharge of the auxiliary zoospores (Money & Webster, 1985, 1988; Money *et al.*, 1988).

On discharge, the zoospores do not swim away but cluster in a hollow ball at the mouth of the zoosporangium and encyst there (Fig. 5.8a). In fact, it is doubtful whether the term 'zoospore' is altogether appropriate as functional flagella are probably not formed. Partial fragmentation of the cyst ball frequently occurs and may have ecological significance in the dispersal of cysts prior to the release of principal zoospores. Unlike certain species of *Saprolegnia*, *Achlya* cysts are normally found at the bottom of culture dishes, and presumably also at the water/bottom sediment interface in natural environments. Cysts of

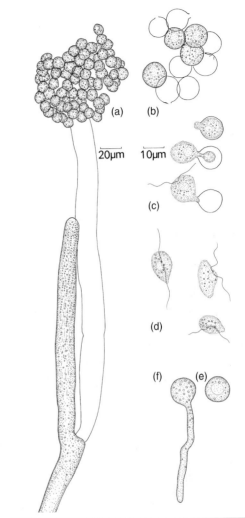

Fig 5.8 *Achlya colorata*. (a) Zoosporangium showing a clump of primary cysts at the mouth. Note the lateral proliferation of the hypha from beneath the old sporangium. (b) Full and empty auxiliary cysts. (c) Stages in the release of principal zoospores from an auxiliary cyst. (d) Principal zoospores. (e) Principal cyst. (f) Principal cyst germinating by means of a germ tube.

A. klebsiana may remain viable for at least two months when stored aseptically at 5°C (Reischer, 1951). However, most auxiliary cysts remain at the mouth of the sporangium for a few hours and then each cyst releases a principal zoospore through a small pore (Figs. 5.8b,c). After a period of swimming, principal zoospores encyst, and principal cysts germinate either by a germ tube or by releasing another principal zoospore. When the zoosporangium of *Achlya* has released its

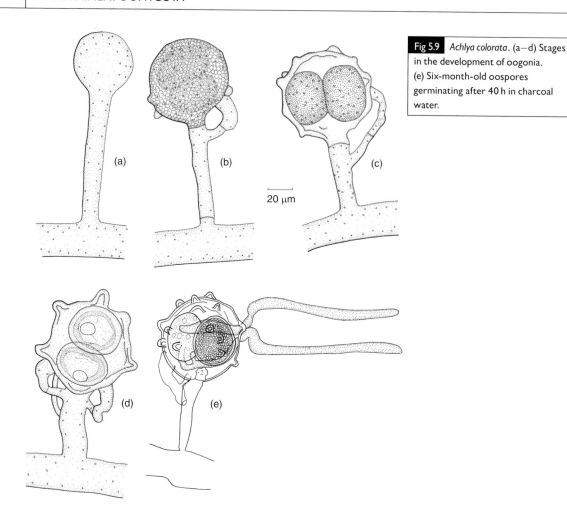

20 μm

Fig 5.9 *Achlya colorata*. (a–d) Stages in the development of oogonia. (e) Six-month-old oospores germinating after 40 h in charcoal water.

zoospores, growth is usually renewed laterally by the outgrowth of a new hyphal apex just beneath the first sporangium (Fig. 5.8a), rather than by internal proliferation.

Sexual reproduction in *Achlya*

Some species of *Achlya* are homothallic (Fig. 5.9) whereas others are heterothallic (Fig. 5.10). *Achlya colorata*, a homothallic species common in Britain, has oogonial walls which develop blunt, rounded projections so that the oogonium appears somewhat spiny (Fig. 5.9d). Otherwise, the process of sexual reproduction is similar to that of *Saprolegnia litoralis* (Fig. 5.6). Germination of oospores is often difficult to achieve with Oomycota, but can be stimulated in *A. colorata* by transferring mature oospores to freshly distilled water (preferably after shaking with charcoal and filtering). Germination occurs by means of a germ tube which grows out from the oospore through the oogonial wall. Here it may continue growth as a mycelium (Fig. 5.9e) or may give rise to a sporangium.

The study of heterothallic species of *Achlya* by John R. Raper quickly revealed that the formation of oogonia and antheridia by compatible strains must be under hormonal control (Raper, 1939, 1957). A particularly readable account of the classical series of experiments leading to the discovery of the steroid sex hormone, antheridiol (Fig. 5.11b), has been given by Carlile (1996b). Several reviews of the broader role of hormones in fungal reproduction have appeared recently (Gooday & Adams, 1992; Elliott, 1994). If isolates of *Achlya bisexualis*, *A. ambisexualis* or *A. heterosexualis* made from water or mud are grown singly

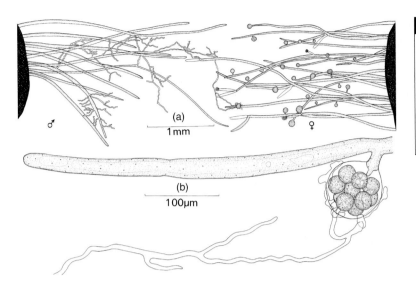

Fig 5.10 *Achlya ambisexualis.*
(a) Male and female mycelia grown on hemp seeds and placed together in water for 4 days. Note the formation of antheridial branches on the male and oogonial branches on the female. (b) Fertilization, showing the diclinous origin of the antheridial branch.

(a)

1 mm

(b)

100 μm

(a)

HO

(b)

HO

OH

O

O

(c)

HO

HO

OH

OH

O

Fig 5.11 Sterols from *Achlya* spp. Fucosterol (a) is one of the most abundant sterols in Oomycota and precursor to the sex hormones antheridiol (b) and oogoniol (c).

on hemp seed in water, reproduction is entirely asexual, but when certain of the isolates are grown together in the same dish, it becomes apparent within 2–3 days that one strain is forming oogonia, and the other antheridia. The development of oogonia and antheridia occurs even when the two strains are held apart in the water or separated by a cellophane membrane or by agar. This suggests that one or more diffusible substances are responsible for the phenomenon. As compatible colonies approach each other, the first observable reaction is the production of fine lateral branches behind the advancing tips of the male hyphae. These are antheridial branches.

By growing male (antheridial) strains in water in which a female (oogonial) strain had been grown previously, Raper (1939) showed that the vegetative female mycelium was capable of initiating the development of antheridial branches on the male. The reverse experiment showed no effect on female colonies in medium in which undifferentiated male colonies had been grown. The role of the vegetative female colony as initiator of the sequence of events leading to sexual reproduction was confirmed by ingenious experiments in micro-aquaria consisting of several consecutive chambers through which water flowed by means of small siphons. Male and female colonies were placed alternately in successive chambers, so that water from a male colony would flow over a female colony and so on. If a female colony was placed in the first chamber, the male colony in the second chamber reacted by developing antheridial hyphae. If, however, a male colony was placed in the first chamber, the male colony in the third chamber was the first to react. Raper (1939) postulated that the development of the antheridial branches was in response to a hormone, termed Hormone A, secreted by vegetative female colonies. By further experiments of this kind, he showed that the later steps in the sexual process were also regulated by means of diffusible substances. He postulated that the antheridial branches secreted a second substance, Hormone B, which resulted in the formation of oogonial initials on the female colony. The oogonial initials in their turn secreted a further substance called Hormone C, which stimulated the antheridial initials to grow towards the oogonial initials and also resulted in the antheridia being delimited. Having made contact with the oogonial initials, the antheridial branches secreted Hormone D which resulted in the formation of a septum cutting off the oogonium from its stalk, and in the formation of oospheres. The original scheme (Table 5.2) therefore implicated four hormones, but confusion arose subsequently because the effect of Hormone A can be modulated by amino acids and other metabolites released from the hemp seeds (Barksdale, 1970; Schreurs et al., 1989).

Since Hormone A is active at extremely low concentrations of 2×10^{-11} M (Barksdale, 1969), purification of this substance was extremely challenging, and 6000 l of culture fluid had to be extracted to obtain 20 mg crystalline Hormone A (Barksdale, 1967). It was eventually identified as the steroid antheridiol (Fig. 5.11b). Soon after, the structure of Hormone B was elucidated and found also to be a steroid, oogoniol (Fig. 5.11c), which is, in fact, present as three chemically closely related forms (McMorris et al., 1975). The effect postulated by Raper (1939) to be due to Hormone C is now thought to be mediated by antheridiol activity, whereas Hormone D may not exist (Carlile, 1996b). Both antheridiol and the oogoniols are derived from fucosterol (Fig. 5.11a), the principal sterol in Achlya (see Elliott, 1994).

The physiological roles of antheridiol and the oogoniols are several-fold and include induction or suppression of sexuality (Thomas & McMorris, 1987), directional growth of gametangial tips (McMorris, 1978), and stimulation of the production of cell wall-softening enzymes (especially cellulase) at points of branching and contact between gametangia (Mullins, 1973; Gow & Gooday, 1987). A cytoplasmic receptor protein for antheridiol has been detected (Riehl et al., 1984), and the hormone probably acts like its equivalents in mammalian cells, by the receptor–hormone complex moving to the nucleus and binding specifically to DNA, increasing transcription rates of certain genes (Elliott, 1994).

There is evidence that the co-ordination of sexual reproduction by hormonal control is not confined to heterothallic forms of Achlya, but also takes place in homothallic species. The fact that it is possible to initiate sexual reactions between homothallic and heterothallic species of Achlya shows that some of the hormones are common to more than one species, although

Hormone	Produced by	Affecting	Specific action
A	Vegetative hyphae	Vegetative hyphae	Induces formation of antheridial branches.
B	Antheridial branches	Vegetative hyphae	Initiates formation of oogonial initials.
C	Oogonial initials	Antheridial branches	(1) Attracts antheridial branches. (2) Induces thigmotropic response and delimitation of antheridia.
D	Antheridia	Oogonial initials	Induces delimitation of oogonium by formation of basal septum.

Table 5.2. Postulated effects of hormones on sexual reactions in *Achlya ambisexualis*.

After Raper (1939). So far, only hormones A and C (antheridiol) and B (oogoniol) have been shown to exist.

there is also evidence of some degree of specificity of the hormones of different species (Raper, 1950; Barksdale, 1965).

One further interesting phenomenon which has been discovered in relation to heterothallic *Achlya* spp. is relative sexuality. If isolates of *A. bisexualis* and *A. ambisexualis* from separate sources are paired in all possible combinations, it is found that certain strains show a capacity to react either as male or as female, depending on the particular partner to which they are apposed. Other strains remain invariably male or invariably female, and these are referred to as true or strong males or females. The strains can be arranged in a series with strong males and strong females at the extremes, and intermediate strains whose reaction may be either male or female depending on the strength of their mating partner. Similar interspecific responses between strains of *A. bisexualis* and *A. ambisexualis* are also possible. Further, some of the strains which appear heterothallic at room temperature are homothallic at lower temperatures. Barksdale (1960) has postulated that the heterothallic forms are derived from homothallic ones. She argued that the most notable difference between strong males and strong females lies in their differential antheridiol production and response. Very little of this substance is found in male cultures, and these are much more sensitive in their response to the hormone than female cultures. Another important difference is in the uptake of antheridiol. Certain strains appear capable of absorbing it much more readily than others, and it is the strains with a high ability to absorb antheridiol that produce antheridial branches during conjugation with other thalli (Barksdale, 1963). If one assumes that heterothallic forms have been derived from homothallic ones, this might have occurred by mutations leading to increased sensitivity to antheridiol and hence to maleness. Conversely, mutations leading to enhanced extracellular accumulation of antheridiol should lead to increasing femaleness.

Germination of the oospores of *A. ambisexualis* results in the formation of a multinucleate germ tube which develops into a germ sporangium if transferred to water, or into a coenocytic mycelium in the presence of nutrients. This mycelium can be induced to form zoosporangia when transferred to water. From zoosporangia of either source, single zoospore cultures can be obtained which can be mated with the parental male or female strains. All zoospores or germ tubes derived from a single oospore gave the same result with regard to their sexual interaction. This finding suggests that nuclear division on oospore germination is not meiotic, and is thus consistent with the idea that the life cycle is diploid (Mullins & Raper, 1965). Confirmation of these results, implying meiosis during gamete differentiation, has also been obtained with *A. ambisexualis* (Barksdale, 1966).

5.2.3 *Thraustotheca, Dictyuchus* and *Pythiopsis* (Saprolegniaceae)

In *Thraustotheca clavata* the sporangia are broadly club-shaped, and there is no free-swimming auxiliary zoospore stage. Encystment occurs within the sporangia and the auxiliary cysts are released by irregular rupture of the sporangial wall (Fig. 5.12a). After release, the angular cysts germinate to release bean-shaped principal zoospores with laterally attached flagella (Figs. 5.12c,d). After a period of swimming, further encystment occurs, followed by germination by a germ tube (Figs. 5.12e,f), or by emergence of a further principal zoospore. The zoospores are thus monomorphic and polyplanetic. Sexual reproduction is homothallic, but formation of gametangia is stimulated by *Achlya* sex hormones (Raper, 1950). Oospores germinate either by a germ tube or by a germ sporangium (Fig. 5.12g).

In *Dictyuchus*, there is again no free-swimming auxiliary zoospore stage. Commonly the entire zoosporangium is deciduous, and detached zoosporangia are capable of forming zoospores. Auxiliary zoospore initials are cleaved out but encystment occurs within the cylindrical sporangium. The cysts are tightly packed together and release their principal zoospores independently through separate pores in the sporangial wall (Fig. 5.13a). When zoospore release is complete, a network made up of the polygonal walls of the auxiliary cysts is left behind. After swimming, the laterally biflagellate zoospores

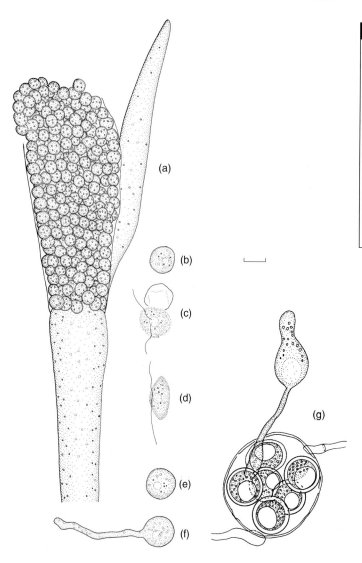

(a)

(b)

(c)

(d)

(e)

(f)

(g)

Fig 5.12 *Thraustotheca clavata*.
(a) Zoosporangium showing formation of auxiliary cysts within the sporangium. The auxiliary cysts are being released through breakdown of the sporangial wall.
(b) Auxiliary cyst. (c) Auxiliary cyst germinating to release a principal zoospore, the first motile stage in this species.
(d) Principal zoospore. (e) Principal cyst.
(f) Principal cyst germinating by means of a germ tube. (g) Sexual reproduction. Six-month-old oospore germinating after 17 h in charcoal water. The germ tube is terminated by a germ sporangium. Bar=20 μm (a) or 10 μm (b)−(g).

encyst (Figs. 5.13b,c). Electron micrographs have shown that the wall of the secondary cyst of *D. sterile* bears a series of long spines looking somewhat like the fruit of a horse chestnut (Fig. 5.14; Heath *et al.*, 1970). Following the formation of the first zoosporangium, a second may be produced immediately beneath it by the formation of a septum cutting off a subterminal segment of the original hypha, or growth may be renewed laterally to the first sporangium (Fig. 5.13a).

Because there is only one motile stage in *Thraustotheca* and *Dictyuchus* (i.e. a zoospore of the principal type), they are said to be mono-morphic. *Pythiopsis cymosa* (Figs. 5.13e−i) is also

monomorphic, but in this species the only motile stage is of the auxiliary type and princi-pal zoospores are not formed. After swim-ming, the zoospore encysts and then germinates directly by means of a germ tube (Figs. 5.13g−i).

5.2.4 Aplanetic forms

In certain cultures of Saprolegniaceae the zoo-sporangia produce cysts which do not release any motile stage. Instead, germ tubes are put out which penetrate the sporangial wall. Forms without motile spores are said to be **aplanetic**. The aplanetic condition is occasionally found in staling cultures of *Saprolegnia*, *Achlya* and

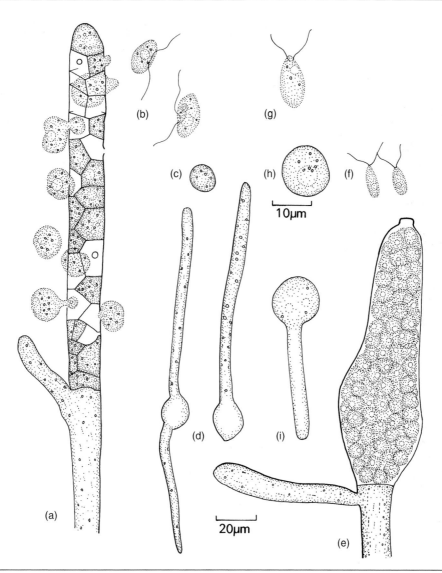

Fig 5.13 (a–d) *Dictyuchus sterile.* (a) Zoosporangium showing cysts within the sporangium and the release of principal zoospores through separate pores in the sporangium wall. Note the network of auxiliary cyst walls. (b) Principal zoospores. (c) Principal cyst. (d) Germination of principal cysts by means of germ tubes. (e–i) *Pythiopsis cymosa.* (e) Zoosporangium. (f,g) Auxiliary zoospores. (h) Auxiliary cyst. (i) Auxiliary cyst germinating by means of a germ tube. Principal zoospores have not been described. (a–c,e,f) to same scale; (g–i) to same scale.

Dictyuchus. Some species produce sporangia only rarely and the genus *Aplanes* has been erected for these forms. However, in very clean cultural conditions, all have been shown to behave as *Achlya*, and they are currently accommodated within that genus (Dick, 2001a). Two species of Saprolegniaceae are not known to form sporangia at all. They are common in soil, and have been placed in a separate genus, *Aplanopsis.* Another genus, *Geolegnia*, forms sporangia containing thick-walled aplanospores which never produce a flagellate stage. The final classification of these small genera of Saprolegniaceae will have to await the results of comparisons of suitable DNA sequences (see M. A. Spencer *et al.*, 2002).

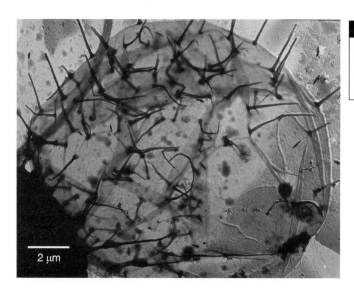

Fig 5.14 Surface of a principal cyst of *Dictyuchus sterile*. Note the spines covering the surface. Image kindly provided by M.W. Dick and I.C. Hallett; reprinted from Hallett and Dick (1986), with permission from Elsevier.

2 μm

5.2.5 *Aphanomyces* (Leptolegniaceae)

Aphanomyces is distinguished from *Achlya* by its thin, delicate hyphae and its narrow sporangia containing a single row of spores. Based on these morphological differences and DNA sequence analyses, the genus *Aphanomyces* has been removed from the Saprolegniaceae and classified in the family Leptolegniaceae, still within the Saprolegniales (Dick *et al.*, 1999; Hudspeth *et al.*, 2000; Dick, 2001a).

Asexual reproduction in *Aphanomyces* is variable. In *A. euteiches*, flagella do not develop on the first-formed spores. Protoplasts are cleaved out, move to the mouth of the sporangium, and encyst. Principal zoospores develop from the cysts and are the first true motile stage. *Aphanomyces euteiches* is thus monomorphic. In *A. patersonii*, the motility of the first-formed zoospore is controlled by variation in temperature. Below 20°C, encystment of the auxiliary zoospores at the mouth of the sporangium occurs in a manner typical of the genus, but above this temperature the auxiliary zoospores swim away and encyst some distance away from the zoosporangium.

The genus *Aphanomyces* has been monographed by Scott (1961). It has gained notoriety particularly because *A. astaci* is the cause of the plague of European crayfish. Having been introduced probably in the 1860s from America, where the local crayfish populations are fairly resistant to *A. astaci* infections, the fungus has now spread across Europe, severely damaging commercial production of the highly susceptible European crayfish, *Astacus fluviatilis* (Alderman & Polglase, 1986; Cerenius *et al.*, 1988; Alderman *et al.*, 1990). Although it would be possible to introduce resistant stock of American crayfish into European river systems affected by the disease, resistant crayfish still harbour the pathogen, thereby making it impossible to restore the native crayfish populations in the future (Dick, 2001a). The difference in resistance between North American and European crayfish lies in the melanization reaction which arrests hyphal growth from encysted zoospores (Nyhlén & Unestam, 1980; Cerenius *et al.*, 1988). In European crayfish, melanization occurs too slowly to prevent the spread of the fungus into the haemocoel which causes rapid death. *Aphanomyces astaci* can also cause epizootic ulcerative disease in fish, the symptoms often being very similar to those caused by *Saprolegnia* (Lilley & Roberts, 1997).

Aphanomyces euteiches is a significant pathogen of roots of peas and other terrestrial plants (Papavizas & Ayers, 1974; Persson *et al.*, 1997). Recently, methods have been developed to quantify the prevalence of the pathogen in infected plants by measuring the levels of specific fatty acids which are produced by *A. euteiches* but not by plants or pathogens belonging to the Eumycota (Larsen *et al.*, 2000). Other species of *Aphanomyces* are keratinophilic, occurring in the soil or in water on insect remains (Dick, 1970; Seymour & Johnson, 1973).

5.3 | Pythiales

The order Pythiales includes two families, the Pythiaceae and Pythiogetonaceae (Dick, 2001a; Kirk *et al.*, 2001). The Pythiogetonaceae are a small group of aquatic saprotrophs presently comprising one genus and six species. They occur in anoxic sediments at the bottom of freshwater lakes and are facultatively anaerobic as well as obligately fermentative, i.e. they break down sugars incompletely to give organic acids irrespective of the presence or absence of oxygen (Emerson & Natvig, 1981; Natvig & Gleason, 1983). Another member of the Pythiogetonaceae, *Pythiogeton zeae*, causes root and stalk rot in maize (Jee *et al.*, 2000). The Pythiogetonaceae are clearly related to the Pythiaceae by DNA sequence homology (Voglmayr *et al.*, 1999).

Only the Pythiaceae will be considered further in this book. This is a large family of over 200 species in approximately 10 genera, of which 2 are of outstanding significance: *Pythium* and *Phytophthora*. *Phytophthora* species are primarily pathogenic to plants from which they can be isolated and grown in pure culture. The genus *Pythium* is best known for its saprotrophic soil-inhabiting members, many of which are opportunistic pathogens especially in young plants. There are also obligately pathogenic *Pythium* spp. Generally, *Pythium* spp. parasitize a wider diversity of hosts than *Phytophthora*, including mammals, fungi and algae.

5.3.1 Life cycle of Pythiaceae
The life cycle of *Phytophthora infestans* is summarized in Fig. 5.19. Asexual reproduction in *Pythium* and *Phytophthora* is by means of sporangia which vary in shape from swollen hyphae or globose structures (*Pythium*) to lemon-shaped (*Phytophthora*). Sporangia are borne on more or less undifferentiated hyphae. In most cases, sporangia germinate to produce zoospores which are of the principal (kidney-shaped) type.

In many *Pythium* spp., the final stages of zoospore differentiation take place outside the sporangium in a walled **vesicle**, followed by breakdown of the soft wall and release of the zoospores. In *Phytophthora*, in contrast, zoospores differentiate within the sporangium and are released directly or via a very short-lived vesicle which is surrounded only by a membrane. About 20% of the total respiratory activity within a released zoospore is used up to fuel propulsion (Hölker *et al.*, 1993). The forward-directed straminipilous flagellum generates about 10 times more thrust than the posterior whiplash flagellum which acts mainly as a rudder (Erwin & Ribeiro, 1996). Zoospores can swim for several hours before they encyst. The process of encystment has been examined in great detail for *Phytophthora* (see p. 102). Cysts usually germinate by means of a germ tube, only rarely producing a further zoospore stage. In many species, sporangia can germinate either indirectly by releasing zoospores or directly by means of a germ tube, depending on environmental conditions and age of the sporangium.

Sexual reproduction is oogamous. Each oogonium contains a single oosphere (except for *Pythium multisporum* in which there are several). The antheridial and oogonial initials are commonly multinucleate at their inception and further nuclear divisions may occur during development. Meiosis eventually takes place in the gametangia so that karyogamy occurs between haploid antheridial and oogonial nuclei. In many forms, there is only one functional male and female nucleus, but in others multiple fusions occur. Oospores germinate either by producing a single germ sporangium, or by sending out vegetative hyphae.

Most members of the Pythiaceae are homothallic, although heterothallism and relative sexuality have been reported, e.g. for *Phytophthora infestans* (Fig. 5.19) and *Pythium sylvaticum*. Heterothallic species are thought to be derived from homothallic ones (Kroon *et al.*, 2004). The situation of mating in heterothallic strains is rather complex and still only incompletely understood. A system of two mating types (A1 and A2) seems to be superimposed on a hormonal control mechanism of mating akin to that described for *Achlya* (p. 86). When two strains of *Pythium* or *Phytophthora* were separated by a membrane preventing hyphal contact but permitting the exchange of diffusible metabolites, oospores were formed by either or both strains (Ko, 1980; Gall & Elliott, 1985). Because the mycelia were separated by a membrane, oospores formed by selfing, whereas in direct contact they may form by hybridization (Shattock *et al.*, 1986a,b). Oospore formation can also be induced by non-specific stimuli, such as volatile metabolites of the unrelated fungus *Trichoderma* stimulating reproduction in A2 but not A1 strains of *Phytophthora palmivora* (Brasier, 1975a). This 'Trichoderma effect' may well have ecological implications, since *Trichoderma* spp. are very common, especially in soil. Oospore formation may be a defence reaction against antibiotics commonly produced by *Trichoderma*, and the 'Trichoderma effect' may actually enhance the survival of *Phytophthora* spp. in soil, since it stimulates production of the long-lived oospore stage even in the absence of a compatible mating type (Brasier, 1975b). It is not known why *Trichoderma* spp. do not stimulate oosporogenesis in A1 strains.

Like *Achlya*, the Pythiaceae display relative sexuality, i.e. a strain can act as male in one pairing but as female in another. To complicate matters further, a given strain of *Phytophthora parasitica* can switch its mating type from predominantly male to predominantly female or vice versa, e.g. upon fungicide treatment (Ko *et al.*, 1986). Clearly, despite substantial research efforts over many years the genetic basis of sexual reproduction in the Pythiaceae still poses numerous unresolved questions!

By analogy with the hormones oogoniol and antheridiol of *Achlya*, a male strain needs to be induced to produce the oogonium-inducing hormone whereas female strains constitutively produce the antheridium-inducing hormone (Elliott, 1994). The ability of homothallic species to stimulate sexual reproduction in heterothallic species (Ko, 1980) indicates that these hormones may also fulfil a morphogenetic role in homothallic sexual reproduction. However, nothing seems to be known as yet about the chemical nature of these hormones.

Sterols are neither synthesized nor strictly required by vegetatively growing *Pythium* or *Phytophthora* spp. (Nes *et al.*, 1979). None the less, they are required for the formation of sexual reproductive organs (Elliott, 1994). It seems, therefore, that sterols – especially sitosterol and stigmasterol which are normally taken up from the host plant – are converted into as yet unidentified steroid hormones which initiate sexual morphogenetic events downstream of the action of the diffusible *Achlya*-like hormones (Elliott, 1994). An alternative hypothesis is that sterols interact with an as yet unknown membrane protein to transmit the hormonal signal and trigger the signalling cascade leading to sexual morphogenesis (Nes & Stafford, 1984). In *Lagenidium giganteum*, a member of the Pythiaceae parasitizing mosquito larvae (Cuda *et al.*, 1997), this cascade seems to be carried by Ca^{2+} and calmodulin (Kerwin & Washino, 1986).

5.3.2 *Pythium*

Species of *Pythium* grow in water and soil as saprotrophs, but under suitable conditions, e.g. where seedlings are grown crowded together in poorly drained soil, they can become parasitic, causing diseases such as pre-emergence killing, damping off and foot rot. Damping off of cress (*Lepidium sativum*) can be demonstrated by sowing seeds densely on heavy garden soil or garden compost which is kept liberally watered. Within 5–7 days some of the seedlings may show brown zones at the base of the hypocotyl, and the hypocotyl and cotyledons become water-soaked and flaccid. In this condition the seedling collapses. A collapsed seedling coming into contact with other seedlings will spread the disease (Plate 2b). The host cells separate from each other easily due to the breakdown of the middle lamella, probably brought about by pectic and possibly cellulolytic enzymes secreted by the fungus. The enzymes diffuse from their points of secretion at the hyphal tips, so that softening of the host tissue actually occurs ahead of the growing mycelium. Pure culture studies suggest that species of *Pythium* may also secrete heat-stable substances which are toxic to plants. Within the host the mycelium is coarse and

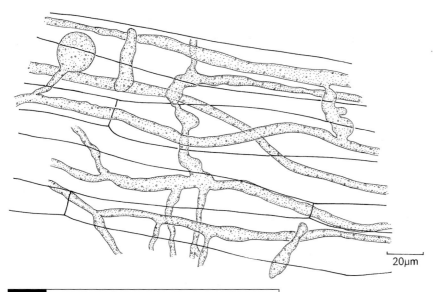

Fig 5.15 *Pythium* mycelium in the rotting tissue of a cress seedling hypocotyl. Note the spherical sporangium initial and the absence of haustoria.

coenocytic, with typically granular cytoplasmic contents (Fig. 5.15). At first there are no septa, but later cross walls may cut off empty portions of hyphae. Thick-walled chlamydospores may also be formed. There are no haustoria.

Several species are known to cause damping off, e.g. *P. debaryanum* and, perhaps more frequently, *P. ultimum*. *Pythium aphanidermatum* is associated with stem rot and damping off of cucumber, and the fungus may also cause rotting of mature cucumbers. *Pythium mamillatum* causes damping off of mustard and beet seedlings and is also associated with root rot in *Viola*. Many *Pythium* spp. have a very wide host range; e.g. *P. ultimum* parasitizes over 150 plant species belonging to many different families (Middleton, 1943; Hendrix & Campbell, 1973). Far from parasitizing only plant roots, several soil-borne species, e.g. *P. oligandrum*, *P. acanthicum* and *P. nunn*, are capable of attacking hyphae of filamentous fungi, including plant-pathogenic species and even other *Pythium* spp. (Foley & Deacon, 1986b; Deacon *et al.*, 1990). Attack may be mediated by the secretion of wall-degrading β-1,3-glucanase, chitinase and cellulase, or by inducing the host to undergo autolysis (Elad *et al.*, 1985; Laing & Deacon, 1991; Fang & Tsao,

1995). In contrast to plant-pathogenic *Pythium* spp., the mycoparasitic species require thiamine for growth and are unable to utilize inorganic nitrogen sources. These deficiencies may explain their mycoparasitic habit (Foley & Deacon, 1986a). Other species of *Pythium* parasitize freshwater and marine algae (Kerwin *et al.*, 1992).

The taxonomy of *Pythium* is somewhat confused at present due to the existence of numerous synonyms. Including a few varieties, Dick (2001a) listed 129 names in current use. Since the morphological characteristics traditionally used for diagnosis can be variable, the delimitation of species and their assignment to the genus *Pythium* will have to await the results of detailed molecular phylogenetic analyses which are in progress (Matsumoto *et al.*, 1999; Lévesque & de Cock, 2004). Keys and descriptions have been published by Waterhouse (1967, 1968), van der Plaats-Niterink (1981) and Dick (1990b).

Asexual reproduction

The mycelium within the host tissue or in culture usually produces sporangia, but their form varies. In some species, e.g. *P. gracile*,

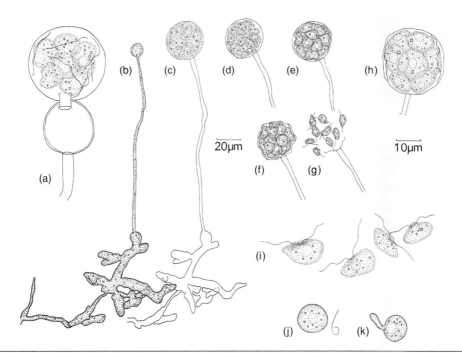

Fig 5.16 Sporangia and zoospores of *Pythium*. (a) *Pythium debaryanum*. Spherical sporangium with short tube and a vesicle containing zoospores. (b–k) *Pythium aphanidermatum*. (b) Lobed sporangium showing a long tube and the vesicle, which is beginning to expand. (c–g) Further stages in the enlargement of the vesicle, and differentiation of zoospores. Note the transfer of protoplasm from the sporangium to the vesicle in (c). The stages illustrated in (b–g) took place in 25 min. (h) Enlarged vesicle showing the zoospores. Flagella are also visible. (i) Zoospores. (j) Encystment of zoospore showing a shed flagellum. (k) Germination of a zoospore cyst. (b–g) to same scale; (a) and (h–k) to same scale.

the sporangia are filamentous and are scarcely distinguishable from vegetative hyphae. In *P. aphanidermatum*, the sporangia are formed from inflated lobed hyphae (Fig. 5.16b). In many species, however, e.g. *P. debaryanum*, the sporangia are globose (Fig. 5.16a). A terminal or intercalary portion of a hypha enlarges and assumes a spherical shape, then becomes cut off from the mycelium by a cross wall. The sporangia contain numerous nuclei. Cleavage of the cytoplasm to form zoospores begins in the sporangium, but is completed within a thin-walled vesicle which is extruded from the sporangium. This is a **homohylic** vesicle because its glucan wall is continuous with one layer of the sporangial wall (Dick, 2001a). Within the sporangium, cleavage vesicles begin to coalesce to separate the cytoplasm into uninucleate portions; membrane-bound packets of TTHs are already present within the cytoplasm of the sporangium. In *P. middletonii* (Fig. 5.17), the

fascinating process of differentiation from amorphous cytoplasm to motile zoospores takes about 30–45 min (Webster, 2006a) and is readily demonstrated in the laboratory (Weber *et al.*, 1999). The sporangium is extended into an apical papilla capped by a mass of fibrillar material which is lamellate in ultrastructure (Lunney & Bland, 1976). Shortly before sporangial discharge, there is an accumulation of cleavage vesicles behind the apical cap and at the periphery of the cytoplasm close to the sporangium wall. The cleavage vesicles around the sporangial cytoplasm discharge their contents to form a loose, fibrous interface between the cytoplasm and the sporangial wall.

Discharge of the sporangium occurs by the formation of a thin-walled vesicle at the tip of the papilla from the fibrillar material of the apical cap, and the partially differentiated zoospore mass is extruded into it. The movement of the cytoplasm from the sporangium into the

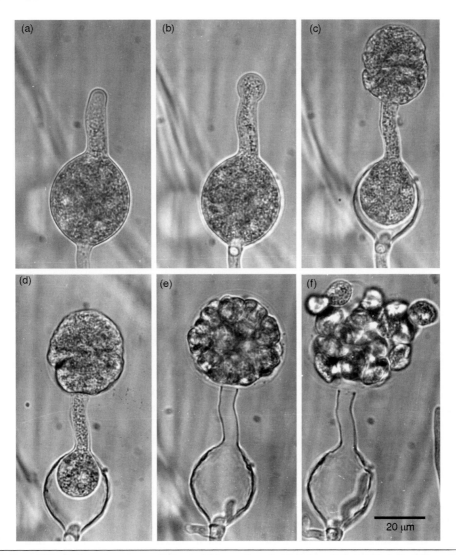

Fig 5.17 *Pythium middletonii*. Stages in zoospore discharge. (a) Sporangium shortly before discharge. Note the thickened tip of the papilla which consists of a cap of cell wall material. (b) Inflation of the vesicle begins. (c,d) Protoplasm is retreating from the sporangium. Note the shrinkage in sporangium diameter as compared with (a). (e) Zoospores have differentiated within the vesicle, with flagella visible between the vesicle wall and the zoospores. (f) Zoospores escape following the rupture of the vesicle wall. The whole process of discharge takes about 20 min.

vesicle is probably the result of several forces including the elastic contraction of the sporangium wall and possibly surface energy (Webster & Dennis, 1967). Lunney and Bland (1976) have also suggested that the fibrillar material extruded from the cleavage vesicles at the zoosporangium periphery may imbibe water, resulting in a build up of turgor pressure. The vesicle enlarges as cytoplasm from the sporangium is transferred to it, and during the next few minutes the cytoplasm cleaves into 8–20 uninucleate zoospores which jostle about inside the sporangium, causing the thin vesicle wall to bulge irregularly (Fig. 5.17). Finally, about 20 min after the inflation of the vesicle, its wall breaks down and the zoospores swim away. Internal sporangial proliferation, i.e. the formation of a new sporangium inside an old discharged one, occurs in certain species, e.g. *P. middletonii* and *P. undulatum*.

In some forms, e.g. *P. ultimum* var. *ultimum*, sporangia do not release zoospores but germinate directly by producing a germ tube. Sporangia of *P. ultimum* var. *ultimum* may survive in soil, whether moist or air-dry, for several months, and are stimulated to germinate within a few hours by sugar-containing exudates from seed coats. The germ tubes grow very rapidly so that a host in the vicinity may be penetrated within 24 h (Stanghellini & Hancock, 1971). The oospore of *P. ultimum* var. *ultimum* can germinate either by means of a germ tube or by forming a zoosporangium which releases zoospores (Figs. 5.18d,e).

The zoospore

Zoospores of *Pythium* spp. are always of the principal type. They can swim for several hours in a readily recognizable manner of helical forward movement. Donaldson and Deacon (1993) have provided evidence that the zoospore swimming pattern is regulated by Ca^{2+} and calmodulin; manipulations of Ca^{2+} concentrations cause aberrations such as circular, straight, spirally skidding or irregular movement. Zoospores of *Pythium* are attracted towards host surfaces, usually roots. The Ca^{2+}/calmodulin system may be the means by which the sensing of attractants is translated into directed movement. It is this *directed* movement (**taxis**), i.e. the ability to aim precisely at a suitable encystment site, rather than the ability to move per se, which represents the main benefit of zoospores to their producer (Deacon & Donaldson, 1993).

Chemotaxis to root exudates is often non-host-specific, being mediated by amino acids and other common metabolites (Jones *et al.*, 1991). Other tactic movements also occur, such as phototaxis, electrotaxis or negative geotaxis (Dick, 2001a). In general, zoospores of *Pythium* spp. accumulate around the root cap, root elongation zone or sites of injury.

Once the zoospore has alighted on a suitable surface, it encysts by shedding rather than withdrawing its flagella, and secreting a wall from pre-formed material. Much valuable ultrastructural work has been carried out on the encystment process of *Phytophthora* and is discussed on pp. 102–111. The cyst of *Pythium* spp. can germinate almost immediately, usually by emitting a germ tube which can directly penetrate the relatively soft root tissue. In *P. marinum*, which is parasitic on marine red algae, the germ tube forms a specialized infection structure termed an **appressorium** (Kerwin *et al.*, 1992); this is also commonly formed by leaf-infecting *Phytophthora* spp. The entire process from zoospore encystment to successful penetration is called **homing sequence** and may take place in as little as 30 min (Deacon & Donaldson, 1993). If a zoospore encysts on a non-host surface, the cyst may germinate by producing a further principal zoospore.

Sexual reproduction

Most species of *Pythium* are homothallic, i.e. oogonia and antheridia are readily formed in cultures derived from single zoospores. However,

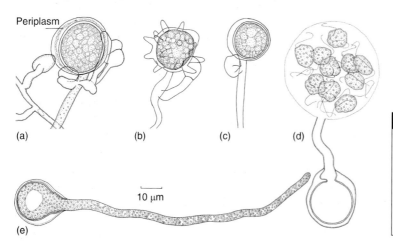

Periplasm

(a)   (b)   (c)   (d)

10 μm

(e)

Fig 5.18 Oogonia and oospores of *Pythium*. (a) *Pythium debaryanum*. Note that there are several antheridia. (b) *Pythium mamillatum*. Oogonium showing spiny outgrowths of oogonial wall. (c) *Pythium ultimum*. (d, e) Germination of oospores of *P. ultimum* (after Drechsler, 1960).

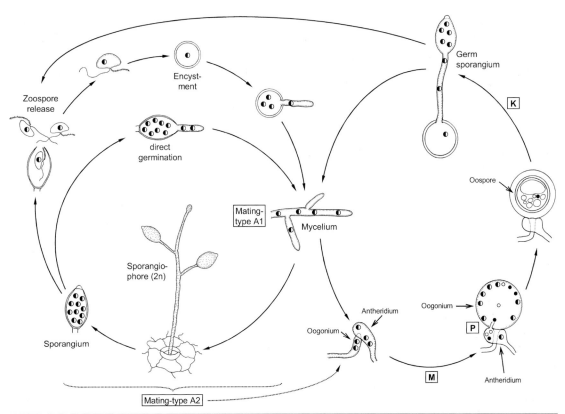

Fig 5.19 Life cycle of *Phytophthora infestans*. This fungus is heterothallic, and the asexual part of the life cycle (left of diagram) is shown only for one mating type (AI). Nuclei in vegetative states are diploid. When two compatible mycelia meet, multinucleate oogonia and antheridia are differentiated, and one meiotic event in each results in the transfer of one haploid nucleus from the gametangium to the oogonium. Karyogamy is delayed until shortly before oospore germination. Open and closed circles represent haploid nuclei of opposite mating type; diploid nuclei are larger and half-filled. Key events in the life cycle are meiosis (M), plasmogamy (P) and karyogamy (K).

some heterothallic species are known, e.g. *P. sylvaticum*, *P. heterothallicum* and *P. splendens*. In these cases, mating is a complicated affair under hormonal control, and with relative sexuality (see p. 95).

Oogonia arise as terminal or intercalary spherical swellings which become cut off from the adjacent mycelium by cross-wall formation. In some species, e.g. *P. mamillatum*, the oogonial wall is folded into long projections (Fig. 5.18b). The antheridia arise as club-shaped swollen hyphal tips, often as branches of the oogonial stalk (monoclinous) or sometimes from separate hyphae (diclinous). In some species, e.g. *P. ultimum*, there is typically only a single antheridium to each oogonium, whilst in others, e.g. *P. debaryanum*, there may be several (Fig. 5.18a).

The young oogonium is multinucleate and the cytoplasm within it differentiates into a multinucleate central mass, the ooplasm from which the oosphere develops, and a peripheral mass, the periplasm, also containing several nuclei. The periplasm does not contribute to the formation of the oosphere.

As soon as the gametangia have become delimited by the basal septum, mitotic divisions cease. Nuclei may be aborted at this stage, and in oogonia of *P. debaryanum* 1–8 nuclei undergo meiosis (Sansome, 1963). Meiotic divisions are synchronous in the antheridium and the oogonium, although no protoplasmic continuities exist at this stage (Dick, 1995). In the antheridium of *P. debaryanum* and *P. ultimum*, all nuclei but one degenerate prior to meiosis, so that four haploid nuclei are present in each antheridium just prior to plasmogamy (Sansome, 1963; Win-Tin & Dick, 1975). The antheridium then attaches itself to the oogonial wall and penetrates it by means of a fertilization tube. Following penetration, only three nuclei were counted in the antheridium, suggesting that one had entered the oogonium. Later still, empty antheridia were found, and it is presumed that the three remaining nuclei enter the oogonium and join the oogonial nuclei degenerating in the periplasm. Fusion between a single antheridial and oosphere nucleus has been described. The fertilized oosphere secretes a double wall, and the ooplast appears in the protoplasm. Material derived from the periplasm may also be deposited on the outside of the developing oospore. Such oospores may need a period of rest (after-ripening) of several weeks before they are capable of germinating. Germination may be by means of a germ tube, or by the formation of a vesicle in which zoospores are differentiated (Figs. 5.18d,e), or in some forms the germinating oospore produces a short germ tube terminating in a sporangium.

Ecological considerations

Pythium spp. can live saprotrophically and may survive in air-dry soil for several years. They are more common in cultivated than in natural soils (Foley & Deacon, 1985), and appear to be intolerant of highly acidic soils. As saprotrophs, species of *Pythium* are important primary colonizers, probably gaining initial advantage by virtue of their rapid growth rate. They do not, however, compete well with other fungi which have already colonized a substrate, and they appear to be rather intolerant of antibiotics.

The control of diseases caused by *Pythium* is obviously rendered difficult by its ability to survive saprotrophically and as oospores in soil. Its wide host range means that it is not possible to control diseases by means of crop rotation. The effects of disease can be reduced by improving drainage and avoiding overcrowding of seedlings. *Pythium* infections are particularly severe in greenhouses and nurseries, where some measure of control can be achieved by partial steam sterilization of soil. Recolonization

of the treated soil by *Pythium* is slow. The use of certain types of compost instead of peat in nurseries can provide good control (Craft & Nelson, 1996; Zhang *et al.*, 1996). The fungicide metalaxyl (see Fig. 5.27) also gives good control of seedling blight.

Pythium insidiosum

This species is associated with algae in stagnant freshwater in tropical and subtropical regions. When horses or cattle come into contact with *P. insidiosum*-contaminated water, zoospores are attracted to wounds and can infect them, causing severe open lesions of skin and subcutaneous tissues known as *pythiosis insidiosi* (Meireles *et al.*, 1993; Mendoza *et al.*, 1993). If contaminated water is consumed, gastrointestinal or systemic infections may also arise. In addition to grazing animals, infections in dogs and humans have been reported. *Pythium insidiosum* is keratinophilic and survives well at 37°C. Infections can be treated successfully by immunotherapy in which horses are injected with killed fungal material, the immune response leading to healing of infections (Mendoza *et al.*, 1992). *Pythium insidiosum* used to be known under different names, but its taxonomy has been clarified by de Cock *et al.* (1987).

5.3.3 Phytophthora

The name *Phytophthora* (Gr.: 'plant destroyer') is apt, most species being highly destructive plant pathogens. The best known is *P. infestans*, cause of late blight of potatoes (Plate 2e). This fungus is confined to solanaceous hosts (especially tomato and potato), but others have a much wider host range. For example, *P. cactorum* has been recorded from over 200 species belonging to 60 families of flowering plants, causing a variety of diseases such as damping off or rots of roots, fruits and shoots (Erwin & Ribeiro, 1996). *Phytophthora cinnamomi* has the widest host range of all species, being capable of infecting over 1000 plants and causing serious diseases especially on woody hosts, including conifers and *Eucalyptus* (Zentmyer, 1980). Several other *Phytophthora* spp. and related *Pythium* spp. can also cause diebacks and sudden-death symptoms of trees, with roots severely rotted by the time above-ground symptoms become apparent (Plate 2c,d). Other important pathogens are *P. erythroseptica* associated with pink rot of potato tubers (Plate 2f), *P. fragariae* causing red core of strawberries, and *P. palmivora* causing pod rot and canker of cocoa. The genus is cosmopolitan, although there are differences in the geographic distribution of individual species; for instance, *P. cactorum*, *P. nicotianae*, *P. cinnamomi* and *P. drechsleri* occur worldwide whereas *P. fragariae* and *P. erythroseptica* are found predominantly in Northern Europe and North America (Erwin & Ribeiro, 1996). Many *Phytophthora* spp. are spreading actively at present, e.g. *P. infestans* which has been spread worldwide by human activity (Fry & Goodwin, 1997) or *P. ramorum*, a serious pathogen of oak trees and other woody plants (Henricot & Prior, 2004). To make matters worse, different *Phytophthora* species may hybridize in nature, producing strains with new host spectra. An example is the recent outbreak of wilt of *Alnus glutinosa* in Europe caused by *P. alni*, a tetraploid hybrid of species resembling *P. cambivora* and *P. fragariae* (Brasier *et al.*, 2004).

In accordance with the great importance of the genus *Phytophthora* in mycology and plant pathology, a vast amount of literature has been published, and some of it has been summarized by Erwin & Ribeiro (1996) and Dick (2001a). Several books on the genus have appeared, including those edited by Erwin *et al.* (1983), Ingram and Williams (1991) and Lucas *et al.* (1991), and the masterly compendium by Erwin and Ribeiro (1996). Keys to the genus have been produced by Waterhouse (1963, 1970) and Stamps *et al.* (1990). Including *formae speciales*, Dick (2001a) listed 84 names in current use.

Phytophthora is closely related to *Pythium* and there are transitional species which may need to be re-assigned as more DNA sequences and other data become available (Panabières *et al.*, 1997). In general, the two genera can be distinguished morphologically in that the sporangia of *Phytophthora* spp. are typically pear- or lemon-shaped with an apical papilla (Fig. 5.20b), and ecologically by the predominantly saprotrophic existence of *Pythium* and the predominantly parasitic mode-of-life of *Phytophthora*. Probably all *Phytophthora* spp. are pathogenic on plants in some form, and they differ merely in the extent to which they have a free-living saprotrophic phase. All may survive in the soil at least in the form of oospores, or in infected host tissue. However, in contrast to the downy mildews (Peronosporales; Section 5.4), almost all pathogenic forms can be isolated from their hosts and can be grown in pure culture. Selective media, often incorporating antibiotics or fungicides such as pimaricin or benomyl, have been devised for the isolation of *Phytophthora* (Tsao, 1983; Erwin & Ribeiro, 1996).

Vegetative growth

Most species form an aseptate mycelium producing branches at right angles, often constricted at their point of origin. Septa may be present in older cultures. Within the host, the mycelium is intercellular, but **haustoria** may be formed. These are specialized hyphal branches which penetrate the wall of the host cell and invaginate its plasmalemma, thereby establishing a point of contact between pathogen and host membranes. Haustoria are typical of biotrophic pathogens such as the Peronosporales (see Fig. 5.29) but may also be formed during initial biotrophic phases of infections which subsequently turn necrotrophic. In *P. infestans* within potato tubers, the haustoria appear as finger-like protuberances (Fig. 5.20c). Electron micrographs of infected potato leaves show that the haustoria are not surrounded by host cell wall material, but by an encapsulation called the **extrahaustorial matrix** which is probably of fungal origin. This is delimited on the outside by the host plasma membrane, and on the inside by the wall and then the plasma membrane of the pathogen (Fig. 5.21; Coffey & Wilson, 1983; Coffey & Gees, 1991). Haustoria of *Phytophthora* do not normally contain nuclei, although one may be situated near the branching point within the intercellular hypha (Fig. 5.21a).

Asexual reproduction

The sporangia of *Phytophthora* spp. are usually pear-shaped or lemon-shaped (Fig. 5.22a) and arise on simple or branched sporangiophores which are more clearly differentiated than

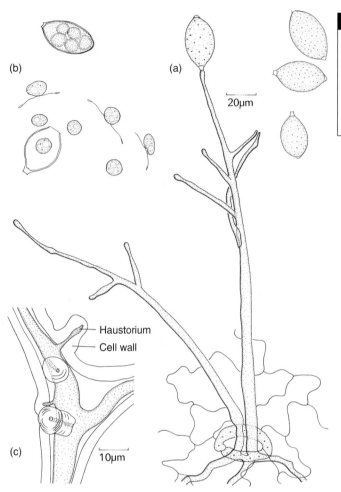

(b)

(a)

20µm

Fig 5.20 *Phytophthora infestans.*
(a) Sporangiophores penetrating a stoma of a potato leaf. (b) Zoospores and zoospore cysts, one formed inside a zoosporangium. (c) Intercellular mycelium from a potato tuber showing the finger-like haustoria penetrating the cell walls. Note the thickening of the cell walls around the haustorium.

Haustorium
Cell wall

(c)
10µm

those of *Pythium*. On the host plant, the sporangiophores may emerge through the stomata, as in *P. infestans* (Fig. 5.20a). The first sporangium is terminal, but the hypha bearing it may push it to one side and form further sporangia by sympodial growth. Mature sporangia of most species have a terminal papilla which appears as a plug because it consists of material different from the sporangial wall (Coffey & Gees, 1991).

In species of *Phytophthora* infecting aerial plant organs, the sporangia are detached, possibly aided by hygroscopic twisting of the sporangiophore on drying, and are dispersed by wind before germinating. In aquatic or soil-borne forms, zoospore release commonly occurs whilst the sporangia are still attached; internal proliferation of attached sporangia may occur.

Whether deciduous or not, sporangia may germinate either directly by means of a germ tube, or by releasing zoospores. The latter seems to be the original route because undifferentiated sporangia contain pre-formed flagella within their cytoplasm, and these are degraded under unfavourable conditions leading to direct germination (Hemmes, 1983; Erwin & Ribeiro, 1996). The mode of germination is dependent on environmental parameters. For example, in *P. infestans*, uninucleate zoospores are produced below 15°C whilst above 20°C multinucleate germ tubes arise. Further, with increasing age sporangia lose their capacity to produce zoospores and tend to germinate directly. In *P. cactorum*, sporangia have been preserved for several months under moderately dry conditions.

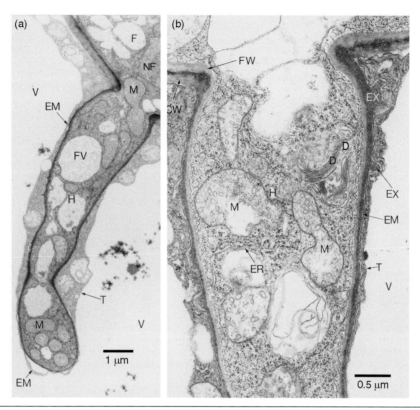

Fig 5.21 TEM images of haustoria of *P. infestans*. (a) Mature haustorium within a leaf cell of potato. (b) The basal region of a haustorium. The haustorium contains fungal vacuoles (FV) and mitochondria (M) but no nuclei. However, a nucleus (NF) is located within the intercellular hypha close to the branch point. The plant tonoplast (T), plant extrahaustorial membrane (EM), extrahaustorial matrix (EX) and fungal wall (FW) are visible. The seemingly empty space surrounding the haustorium is the plant vacuole (V). Both images reprinted from Coffey and Wilson (1983) by copyright permission of the National Research Council of Canada. Original prints kindly provided by M. D. Coffey.

When water becomes available again, such sporangia may germinate by the formation of a vegetative hypha, or a further sporangium.

Thick-walled asexual spherical chlamydospores have also been described for many *Phytophthora* spp. and can survive in soil for several years (Ribeiro, 1983; Erwin & Ribeiro, 1996). The morphological differences between sporangia, chlamydospores and oospores are illustrated in Fig. 5.22.

Once formed, mature sporangia may remain undifferentiated for several hours or even days, but zoospore differentiation can be induced by suspending mature sporangia in chilled water or soil extract. Detailed methods to trigger zoospore release have been established for many species (Erwin & Ribeiro, 1996). Once cold-shock has been received, differentiation can be completed in less than 60 min and probably involves cAMP-mediated signalling cascades (Yoshikawa & Masago, 1977). The processes of differentiation of sporangial protoplasm into zoospores differ in certain details between *Phytophthora* and the Saprolegniales (see Hardham & Hyde, 1997). For instance, in *Saprolegnia* the central vacuole is prominent and its membrane as well as the plasma membrane contribute to the plasma membranes of the developing zoospores (p. 81). In contrast, in *Phytophthora* the central vacuole disappears from the young sporangium before cleavage of the cytoplasm begins, and the plasma membrane remains intact even after zoospores have become fully differentiated. The zoospore plasma membranes therefore mostly originate from Golgi-derived cleavage cisternae (Hyde *et al.*, 1991). Detailed cytological studies

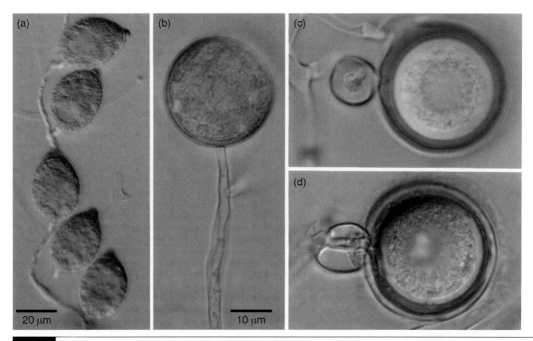

Fig 5.22 Reproductive structures in *Phytophthora cactorum*. (a) Sporangia. (b) Chlamydospore. (c) Oospore showing the paragynous mode of fertilization. (d) Oospore with amphigynous fertilization. (b–d) to same scale.

have revealed an important role of micro-tubules in organizing the distribution of nuclei during zoospore formation (Hyde & Hardham, 1992, 1993). Cleavage of the cytoplasm of a zoospore begins close to that end of the nucleus which subsequently points towards the ventral groove. At this stage, three types of vesicle which become important during zoospore encystment also move into their positions: large peripheral vesicles, dorsal vesicles, and small ventral vesicles. When the pre-formed flagella have been inserted, the zoospores acquire their mobility (Hardham, 1995). Zoospores are either discharged directly through the plug after this has dissolved, or they are transferred into a very transient membranous vesicle which forms outside the opened plug upon discharge and bursts one or a few seconds later (Gisi, 1983). Since the plasma membrane of the sporangium has not become part of the zoospore membranes, the membranous vesicle is probably continuous with the plasma membrane.

Encystment of zoospores

Zoospores of *Phytophthora* swim for several hours, travelling distances of a few centimetres in water

or wet soil, although they can be spread much further by passive movement within water currents (Newhook *et al.*, 1981). They are attracted chemotactically to plant roots by non-specific root exudates such as amino acids, host-specific substances, or the electrical field generated by plant roots (Carlile, 1983; Deacon & Donaldson, 1993; Tyler, 2002). No equivalent studies seem to have been carried out for zoospores of *Phytophthora* infecting leaves. The process of zoospore encystment described below for *Phytophthora* seems to apply also to *Pythium* (Hardham, 1995). It is an act of regulated secretion, i.e. the release of pre-formed contents by synchronous fusion of vesicles with the plasma membrane. Regulated secretion is common in animal cells, e.g. in epithelial or neuronal systems, but in fungi it is probably confined to encysting zoospores.

Zoospores of *Phytophthora* are kidney-shaped; both flagella arise from the kinetosome boss protruding from within the longitudinal groove at the ventral surface. The anterior end of the spore is indicated externally by the straminipilous flagellum and internally by the water expulsion vacuole; the nucleus is located in

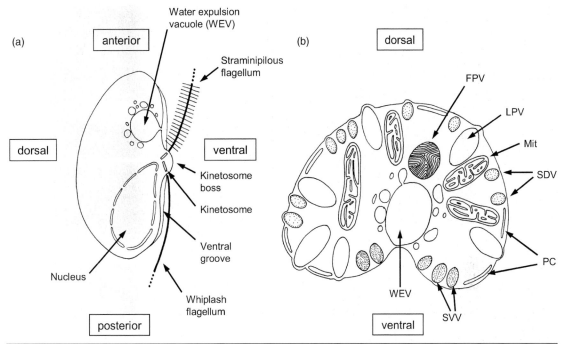

Fig 5.23 Schematic drawings of a zoospore of *Phytophthora* (not to scale). (a) Longitudinal section. (b) Transverse section of the anterior region showing several types of vesicle, namely the water-expulsion vacuole (WEV), fingerprint vacuole (FPV), large peripheral vesicles (LPV), small ventral vesicles (SVV), small dorsal vesicles (SDV) and peripheral cisternae (PC). Mitochondria (Mit) with unusually lamellate cristae are also indicated. a modified from Dick (2001b); b based on the ultrastructural work of Hardham *et al.* (1991).

the posterior half of the spore (Fig. 5.23a). The nucleus is associated with the microtubular roots of the flagella which force it into a somewhat conical shape, the pointed end pointing towards the kinetosome boss. Zoospores contain several vesicular compartments. Their positions are drawn schematically in Fig. 5.23, and electron micrographs are provided in Fig. 5.24. Fingerprint vacuoles, equivalent to the dense-body vesicles of *Saprolegnia* and *Achlya*, are defined by the lamellate structure of their contents, presumably deposits of β-1,3-glucan (mycolaminarin) and phosphate. Fingerprint vacuoles are located mainly in the interior of the zoospore and play no part in the encystment process but are thought to provide carbon and energy reserves during subsequent germination of the cyst (Gubler & Hardham, 1990). In zoospores of *Phytophthora cinnamomi*, there are several kinds of peripheral vesicle which have been distinguished morphologically (Fig. 5.23) and by labelling with specific antibodies. When

zoospores approach a root, the groove of the ventral surface faces the root surface, initial contact presumably being made by the flagella. Attachment of the zoospore is achieved by means of a glue discharged by the synchronous fusion of the small ventral vesicles with the ventral plasma membrane (Hardham & Gubler, 1990). At the same time, the small dorsal vesicles also secrete their contents, leading to the deposition of the first cyst wall (Figs. 5.24c,d; Gubler & Hardham, 1988). The process of exocytosis is complete within 2 min of receiving the encystment trigger. In contrast, the large peripheral vesicles do not fuse with the plasma membrane but withdraw to the centre of the cyst. Their contents are proteinaceous and probably serve as reserves for the germination process. Peripheral cisternae, ultrastructurally distinct from the ER, line the inside of the zoospore plasma membrane and disappear during encystment (Hardham *et al.*, 1991; Hardham, 1995).

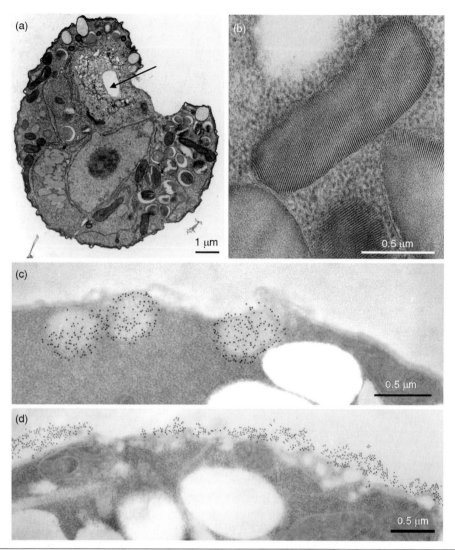

Fig 5.24 Ultrastructure of *Phytophthora cinnamomi* zoospores as seen with the TEM. (a) Oblique section through a zoospore. Several kinds of vesicle are visible, as are mitochondria, the water-expulsion vacuole (arrow) and the conical nucleus with its prominent nucleolus. (b) Fingerprint vacuoles. (c,d) Immunogold labelling of wall material located within dorsal vesicles before (c) and in the cyst wall 1 min after (d) encystment of the zoospore. (a,b) reproduced from Hardham and Hyde (1997), with permission from Elsevier; (c,d) previously unpublished work. All images kindly provided by F. Gubler and A. R. Hardham.

Zoospore encystment can be triggered by several stimuli, e.g. contact with host cell surface polysaccharides, change in medium composition, or presence of root exudates. Commitment to encystment occurs within 20–30 s of receiving the stimulus (Paktitis *et al.*, 1986). Complex signalling cascades involving Ca^{2+} and phospholipase D are involved (Zhang *et al.*, 1992), and commitment to several future developmental processes is made before the onset of encystment, including the point of germ tube emergence (Hardham & Gubler, 1990).

Zoospore cysts germinate quite rapidly after their formation, usually by means of a germ tube which infects the plant roots directly. In the case of hard surfaces such as leaves, the germ tube may form an appressorium which mediates infection (see pp. 378–381).

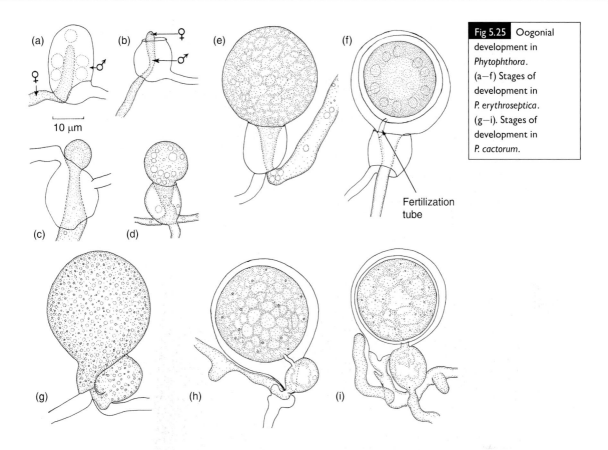

Fertilization tube

Fig 5.25 Oogonial development in *Phytophthora*. (a–f) Stages of development in *P. erythroseptica*. (g–i). Stages of development in *P. cactorum*.

10 μm

Sexual reproduction

Oospore formation is dependent on sterols and mating hormones (p. 95) and may be homo- or heterothallic. Phylogenetic studies have indicated that the former is ancestral, heterothallism having arisen repeatedly within the genus *Phytophthora* (Kroon *et al.*, 2004). Two distinct types of antheridial arrangement are found. In *P. fragariae*, *P. megasperma* and a number of other species, antheridia are attached laterally to the oogonium and are described as **paragynous** meaning 'beside the female' (Figs. 5.22c, 5.25g–i). In other *Phytophthora* species such as *P. infestans*, *P. cinnamomi* and *P. erythroseptica*, the oogonium, during its development, penetrates and grows through the antheridium (Hemmes, 1983). The oogonial hypha emerges above the antheridium and inflates to form a spherical oogonium, with the antheridium persisting as a collar around its base (Figs. 5.25a–f). This arrangement of the antheridium is termed **amphigynous** ('around the female'). In some species (e.g. *P. cactorum*,

P. clandestina, *P. medicaginis*), both types of arrangement may be found (Figs. 5.22c,d); one or the other may predominate, depending on strain and culture conditions (Erwin & Ribeiro, 1996).

Both the oogonia and antheridia contain several diploid nuclei, but as the oosphere matures only a single nucleus remains at the centre while the remaining nuclei are included in the periplasm, i.e. the space between the oosphere and the oogonial walls (see Fig. 5.2). Meiosis occurs in the antheridium and oogonium (Shaw, 1983). Fertilization tubes have been observed and a single haploid nucleus is introduced from the antheridium into the oosphere (Fig. 5.26). Fusion between the oosphere nucleus and the antheridial nucleus is delayed. Even mature, dormant oospores may still be binucleate, karyogamy usually occurring after breakage of dormancy as a first step towards germination (Jiang *et al.*, 1989).

Following fertilization, the physiology and ultrastructure of the oospore change to

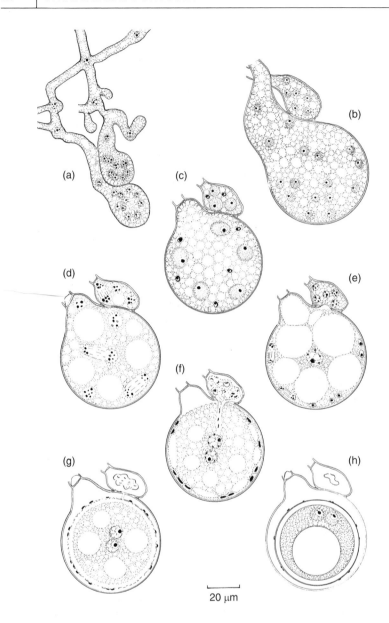

Fig 5.26 *Phytophthora cactorum.* Development of oogonium, antheridium and oospore. (a) Initials of oogonium and antheridium. (b) Oogonium and antheridium grown to full size: the oogonium has about 24 nuclei and the antheridium about 9. (c) Development of a septum at the base of each, and degeneration of some nuclei in each until the oogonium has 8 or 9 nuclei and the antheridium 4 or 5. (d) A simultaneous division of the surviving nuclei in oogonium and antheridium. The protoplast has large vacuoles. (e) Separation of oosphere from periplasm. Nuclei divide in the periplasm prior to degeneration. The oogonium presses into the antheridium. (f) Entry of one antheridial nucleus by a fertilization tube. The protoplasm and remaining nuclei of the antheridium degenerate. (g) Development of oospore wall. (h) The oospore enters its dormant period with exospore formed from dead periplasm, endospore deposited inside it, and paired nuclei in association but not yet fused. (a—h) are composite drawings of eight stages in sequence (after Blackwell, 1943).

20 μm

a resting state. Oospore differentiation proceeds from the outside inwards (centripetal development). The oospore has a thin outer wall (epispore) which is derived from the periplasm and appears to consist of pectic substances. The inner oospore wall (endospore) is rich in β-1,3-glucans which form a major storage reserve and are mobilized by glucanases just prior to germination (Erwin & Ribeiro, 1996). Within the developing oospore, the numerous small lipid droplets coalesce into a few large ones. Lipids are

undoubtedly the major endogenous storage reserve in the spores of Oomycota (Dick, 1995) and many other fungi. Later, the dense body vesicles which are rich in mycolaminarin and phosphate fuse together, giving one large structure, the ooplast. Like the endospore, the ooplast is consumed during germination whereas some lipid droplets are saved and are translocated into the germ tube (Hemmes, 1983). Considering their thick walls and abundant storage reserves, it is not surprising that oospores are the longest-lived propagule of *Phytophthora*, being capable of surviving in soil for many years.

5.3.4 *Phytophthora infestans*, cause of potato late blight

Late blight of potato caused by *P. infestans* is a notorious disease. In the period between 1845 and 1848 it resulted in famine across much of Europe, and especially in Ireland where most people had come to depend on the potato as their major source of food. In Ireland alone, the population size dropped from over 8 million in 1841 to 6.5 million in 1851 (Salaman, 1949). The history of the Great Famine has been ably documented by Large (1940), Woodham-Smith (1962) and Schumann (1991). The social and political repercussions of this tragedy have been immense and still reverberate today.

An enormous amount of literature about *P. infestans* has been published over the past 150 years, including several books (Ingram & Williams, 1991; Lucas *et al.*, 1991; Dowley *et al.*, 1995). It has been estimated that about 10% of the entire phytopathological literature is concerned just with this one species. None the less, there are many uncomfortable gaps in our knowledge, and the fungus continues to provide unpleasant surprises to this day.

Origin and spread

The probable centre of evolution of most *Solanum* spp. and hence also their pathogens, notably *P. infestans*, lies in Mexico (Niederhauser, 1991), although the potato (*S. tuberosum*) was first cultivated in South America. There are several theories accounting for the spread of *P. infestans* round the world (Ristaino, 2002). In the early 1840s *P. infestans* rapidly spread to North America, and it is generally assumed that it was introduced to Europe (Belgium) in June 1845 with a shipment of contaminated potatoes (Bourke, 1991). *Phytophthora infestans* is heterothallic, and there is good evidence that in the first wave of migration in 1845 only the A1 mating type reached Europe (Goodwin *et al.*, 1994a). Over the next century or more, the fungus probably survived entirely on an asexual life cycle, overwintering in tubers infected during the previous

season and discarded together with shoots and other debris in the field. Despite the absence of sexual reproduction, *P. infestans* showed a considerable genetic adaptability, as documented by its ability to break the resistance bred into new potato cultivars (p. 114), and also the rapid emergence of strains resistant against newly introduced fungicides (p. 112).

A second wave of *P. infestans* migration brought the A2 mating type from central Mexico to North America and Europe where it was first isolated in 1981 (Hohl & Iselin, 1984). It is now established worldwide (Spielman *et al.*, 1991; Fry *et al.*, 1993; Gillis, 1993; Goodwin *et al.*, 1994b). The enhanced genetic recombination brought about by sexual reproduction is catalysing a change in the genetic make up of *P. infestans*, which may be leading to an explosive evolution of new *P. infestans* strains (Fry *et al.*, 1993; Goodwin *et al.*, 1995). This situation is seen as the biggest threat posed by *P. infestans* since the 1840s (Fry & Goodwin, 1997).

Epidemiology

There is clear genetic evidence of sexual reproduction taking place in the field, and it is also possible that oospores contribute to the survival of *P. infestans* in soil during the winter (Andrivon, 1995). Additionally, the fungus has a good capacity to survive the winter without oospores. A very low proportion of infected tubers left on the field gives rise to infected 'volunteer' plants in the following spring. In experimental plots, the proportion of infected plants developing from naturally or artificially infected tubers was found to be less than 1% (Hirst & Stedman, 1960). Nevertheless, such infected shoots form foci within the crop from which the disease spreads. The sporangia of *P. infestans* are deciduous, and they are blown from diseased shoots to healthy leaves where they germinate either by the formation of germ tubes or zoospores. Zoospore production is favoured by lower temperatures (9−15°C). After swimming for a time, the zoospores encyst and then form germ tubes which usually penetrate the epidermal walls of the potato leaf, or occasionally enter the stomata. An appressorium is formed at the tip of the germ tube, attaching the zoospore cyst firmly to the leaf. Penetration of the cell wall is probably achieved by a combination of mechanical and enzymatic action and can occur within 2 h. Within the leaf tissue, an intercellular mycelium develops and haustoria are formed where hyphae contact host cell walls (Fig. 5.21). The resulting lesion acquires a dark green water-soaked appearance associated with tissue disintegration (Plate 2e). Such lesions are visible within 3–5 days of infection under suitable conditions of temperature and humidity. Around the margin of the advancing lesion on the lower surface of the leaf, a zone of sporulation is found in which sporangiophores emerge through the stomata (Fig. 5.20a). Sporulation is most prolific during periods of high humidity and commonly occurs at night following the deposition of dew. In potato crops, as the leaf canopy closes over between the rows to cover the soil, a humid microclimate is established which may result in extensive sporulation. As the foliage dries during the morning, the sporangiophore undergoes hygroscopic twisting which results in the flicking-off of sporangia. Thus the concentration of sporangia in the air usually shows a characteristic diurnal fluctuation, with a peak around 10 a.m. Although sporangia can survive drying if they are rehydrated slowly (Minogue & Fry, 1981), in practice the long-range spread of inoculum is probably by sporangia in contact with water drops (Warren & Colhoun, 1975).

The destructive action of *P. infestans* is directly associated with the killing of photosynthetically active foliage. When about 75% of the leaf tissue has been destroyed, further increase in the weight of the crop ceases (Cox & Large, 1960). Thus, the earlier the onset of the epidemic, the more serious the consequences. To a certain extent, the crop reduction may be offset by the fact that epidemics are more common in rainy cool seasons which are conducive to higher crop yields.

Phytophthora infestans can also cause severe post-harvest crop losses because tubers can be infected by sporangia falling onto them, either during growth or lifting. Such infected tubers may rot in storage, and the diseased tissue is susceptible to secondary bacterial and fungal infections.

Chemical control

By spraying with suitable fungicides, epidemic spread of the disease can be delayed. This results in a prolongation of photosynthetic activity of the potato foliage and hence an increase in yield. Fungicides developed against the Eumycota are often ineffective against Oomycota such as *Phytophthora* because the latter differ in fundamental biochemical principles, including many of the molecular targets of fungicides active against Eumycota (Bruin & Edgington, 1983; Griffith *et al.*, 1992). In 1991, about 20% of the total amount of money spent on chemicals for controlling plant diseases worldwide was used for the control of Oomycota (Schwinn & Staub, 1995).

The first of all fungicides was Bordeaux mixture, an inorganic formulation containing copper sulphate and calcium oxide which was found to be effective against downy mildew of vines caused by *Plasmopara viticola*, another member of the Oomycota (see p. 119; Large, 1940; Erwin & Ribeiro, 1996). Oomycota in general are extremely sensitive to copper ions, and Bordeaux mixture is still widely used (Agrios, 2005).

The **dithiocarbamates** such as zineb or maneb (Fig. 5.27a) were among the first organic fungicides to be developed. They act against a wide range of fungi, including Oomycota, because of their non-selective mode of action. The molecule is sufficiently apolar to diffuse across the fungal plasma membrane; once inside, it is metabolized, and the released isothiocyanate radical (Fig. 5.27b) reacts with the sulphydryl groups of amino acids (Agrios, 2005).

The most important agrochemicals against Oomycota are the **phenylamides** such as metalaxyl (Fig. 5.27c) which are **systemic fungicides**, i.e. they can enter the plant and are translocated throughout it. Metalaxyl appears to inhibit the transcription of ribosomal RNA in Oomycota but not Eumycota (Davidse *et al.*, 1983). This is an inhibition of a specific biochemical target, and the immense genetic variability of *P. infestans* enabled it to develop resistance against metalaxyl in the early 1980s shortly after this was released for agricultural use (Davidse *et al.*, 1991). Resistance is now widespread and has serious implications for future control of

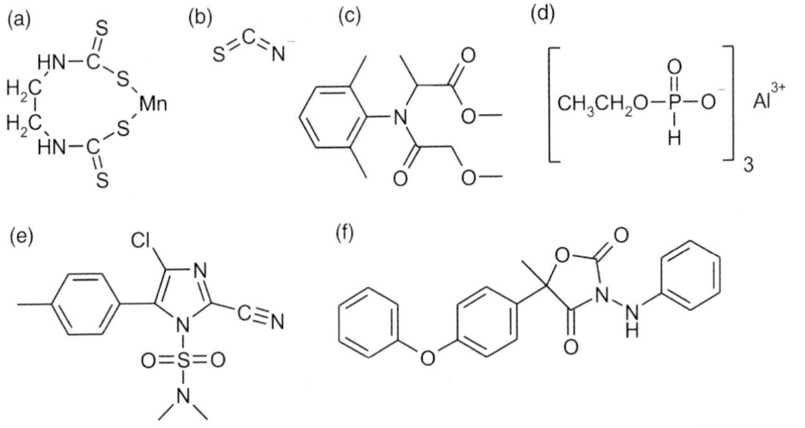

Fig 5.27 Fungicides against *P. infestans*. (a) The dithiocarbamate maneb which is active against Oomycota and Eumycota. (b) The isothiocyanate radical released by metabolism of dithiocarbamates by fungal hyphae. (c) The phenylamide metalaxyl which is active only against Oomycota. (d) Aluminium ethyl phosphonate (fosetyl-Al). (e) Cyazofamid, a new fungicide specific against Oomycota. (f) Famoxadone, a new fungicide active against Oomycota and Eumycota.

Phytophthora spp. (Erwin & Ribeiro, 1996). Phenylamides are now protected by being used in a cocktail, e.g. with the less-specific dithiocarbamates, and tailor-made application regimes are recommended for each year and each region (Staub, 1991).

The **phosphonates** are a different type of fungicide against *Phytophthora* spp. Fosetyl–Al (aluminium ethyl phosphonate; Fig. 5.27d) is readily taken up by plants in which it is broken down to release phosphorous acid (= phosphonate), which seems to be the active principle (Griffith *et al.*, 1992). Fosetyl–Al as well as phosphorous acid can move downwards through the phloem and upwards in the xylem, showing similar transport characteristics as sucrose (Ouimette & Coffey, 1990; Erwin & Ribeiro, 1996). The mode of action of phosphonates is not known but is likely to be complex, with a stimulatory effect also on the host plant immune system (Molina *et al.*, 1998). Although active only against potato tuber blight but not foliar blight caused by *P. infestans* (L. R. Cooke & Little, 2002), phosphonates are effective against a wide range of root-infecting *Phytophthora* spp. and even show good curative properties (Erwin & Ribeiro, 1996).

A useful introduction to current fungicides and their modes of action has been provided by Uesugi (1998). Because of the enormous economic significance of *P. infestans* and other Oomycota, new fungicide candidates are continually being developed and introduced into the market. Two recent examples are cyazofamid (Fig. 5.27e) and famoxadone (Fig. 5.27f). Both inhibit mitochondrial respiration. However, whilst the former is specific against Oomycota (Sternberg *et al.*, 2001), famoxadone inhibits both Oomycota and Eumycota (Mitani *et al.*, 2002). Its molecular target is different from that of cyazofamid but probably the same as that of the strobilurins (see Figs. 13.15e,f), as indicated by the development of cross-resistance in fungal pathogens against famoxadone and strobilurins.

Disease forecast

To avoid unnecessary spraying and to ensure that timely spray applications are made, it has proven possible to provide forecasts of the incidence of potato blight epidemics for certain countries. Beaumont (1947) analysed the incidence of blight epidemics in south Devon (England) and established that a 'temperature–humidity rule' controls the relationship between blight epidemics and weather. After a certain date (which varies with the locality) and assuming that inoculum on volunteer plants is always present, Beaumont (1947) predicted that blight would follow within 15–22 days of a period of at least 48 h during which the minimum temperature was not less than 10°C and the relative humidity was over 75%. The warm humid weather during this **Beaumont period** provides conditions suitable for sporulation and the initiation of new infections. Modified in the light of experience and adapted to regional climates, computerized forecasting systems are now used worldwide, limiting fungicide applications to situations in which they are necessary (Doster & Fry, 1991; Erwin & Ribeiro, 1996). After receipt of a blight warning, fungicide sprays are applied prophylactically by the farmer, irrespective of whether *P. infestans* is actually present in his field or not.

Haulm destruction

The danger of infection of tubers by sporangia falling onto them from foliage at lifting time can be minimized by ensuring that all the foliage is destroyed before lifting. This is achieved by spraying the foliage with herbicides 2–3 weeks before harvest time. The ridging of potato tubers also helps to protect the tubers from infection. Although sporangia may survive in the soil for several weeks, they do not penetrate deeply into it.

Crop sanitation

In principle, one infected volunteer plant per hectare is sufficent to initiate an epidemic. This is because late blight is a typical multicyclic disease, with numerous cycles of reproduction occurring in a single growing season under favourable conditions, leading to the rapid build up of inoculum. Crop sanitation, which is effective against single-cycle diseases, therefore has only limited value in the control of *P. infestans* (van der Plank, 1963).

Breeding for major gene resistance

A worldwide screening of *Solanum* spp. showed that a number of them have natural resistance to *P. infestans*. One species which has proven to be an important source of resistance is *S. demissum* which grows in Mexico, the presumed centre of origin of *P. infestans*. Although this species is

valueless in itself for commercial cultivation, it is possible to cross it with *S. tuberosum*, and some of the progeny are resistant to the disease. *Solanum demissum* contains at least four major genes for resistance (R_1, R_2, R_3 and R_4), together with a number of minor genes which determine the degree of susceptibility in susceptible varieties (Black, 1952). The four genes may be absent from a particular host strain, or they may be present singly (e.g. R_1), in pairs, in threes, or all together, so that 16 host genotypes are possible representing different combinations of R genes. The identification of the R gene complex was dependent on the discovery that the fungus itself exists in a number of strains or **physiological races**. For each host R gene, the pathogen was assumed to carry a gene which enables it to overcome the effect of the R gene. This is the basis of the **gene-for-gene** hypothesis, and gene-for-gene interactions are common in many host–pathogen interactions (Flor, 1971). Assuming a gene-for-gene situation for the interaction of *P. infestans* with *S. tuberosum*, 16 races of *P. infestans* should theoretically be demonstrable. If the corresponding genes of the fungus are termed 1, 2, 3 and 4, then the different races can be labelled (0), (1), (2), etc., (1.1), (1.2), etc., (1.2.3), (1.2.4), etc., and (1.2.3.4). By 1953, 13 of the 16 races had been identified, the prevalent race being Race 4. By 1969, 11 R genes had been recognized in Britain (Malcolmson, 1969). Resistance based on a small number of defined genes of major effect has been termed **major gene resistance** or **race-specific resistance**. Because of the uncanny ability of *P. infestans* to break major gene resistance even before the arrival of the A2 mating type in Europe and North America, attempts at breeding fully resistant potato cultivars have now been abandoned (Wastie, 1991).

The origin of physiological races is difficult to determine. The occurrence and spread of resistance genes before the arrival of the A2 mating type may have been due to mutation followed by selection imposed by the monoculture of a resistant host. Another possibility is that the mycelium of *P. infestans* is heterokaryotic, carrying nuclei of more than one race. Yet another scenario is vegetative hybridization followed by parasexual recombination (see p. 230); by mixing sporangia of two different races, new races with a different pattern of virulence towards potato varieties have been obtained after several cycles of inoculation (Malcolmson, 1970). The parasexual cycle has been experimentally demonstrated for *P. parasitica* using fungicide resistance as a genetic marker (Gu & Ko, 1998).

Within 1–2 days of infection, tissues of resistant hosts undergo necrosis so rapidly that sporulation and further growth of the fungus cannot occur. Such a reaction is sometimes termed **hypersensitivity**, and the function of the *R* genes is to accelerate this host reaction. When potato tubers are inoculated with an avirulent race of *P. infestans*, they respond by secreting antifungal substances called **phytoalexins**. Two of the phytoalexins formed by resistant tubers are rishitin and phytuberin. Rishitin, originally isolated from the potato variety Rishiri, is a bicyclic sesquiterpene. Tomiyama *et al.* (1968) showed that R_1 tuber tissue inoculated with an avirulent race of *P. infestans* produced over 270 times the amount of rishitin than when inoculated with a virulent race. The *R* genes of the potato probably determine the ability of host tissue to recognize and respond to avirulent races of *P. infestans* (Day, 1974). The detailed molecular interactions which determine race specificity are, however, complex and still only incompletely understood at present (Friend, 1991).

Breeding for field resistance
In addition to the major genes for resistance in potato, numerous other genes also exist which, although individually of small effect, may contribute to resistance if present together. Resistance of this kind is known as **general resistance** or **field resistance**, and some potato breeding programmes aim at producing varieties possessing it (Niederhauser, 1991). This is preferable to single-gene resistance because *P. infestans* is less likely to overcome the combined resistance of numerous minor genes simultaneously. Field resistance retards the infection process, e.g. by production of a particularly thick cuticle or by a leaf architecture unfavourable to infection, lowers the number of sporangia produced, and extends the time needed by the pathogen to initiate new infections (Wastie, 1991). Field resistance is equally effective against all physiological races of *P. infestans*, and it reduces the severity of an epidemic and consequently the need to apply fungicides (Erwin & Ribeiro, 1996).

Tomato late blight
P. infestans also causes significant worldwide crop losses of tomato (*Lycopersicon esculentum*) which, like potato, belongs to the Solanaceae. The general principles of control of tomato late blight are similar to those described above for potato, including fungicides used and blight forecasting (Erwin & Ribeiro, 1996). Many strains of *P. infestans* are capable of infecting both tomato and potato. However, since the resistance gene systems are different in these two hosts, correlations between virulence of a given strain on potato and tomato cannot be drawn (Legard *et al.*, 1995).

5.4 | Peronosporales

The Peronosporales are obligately biotrophic pathogens of a few groups of higher plants and are responsible for diseases mainly of aerial plant organs known collectively as **downy mildews**. The order currently comprises two families, the Peronosporaceae (*Peronospora*, *Plasmopara*, *Bremia*) and Albuginaceae (*Albugo*). There are about 250 species (Kirk *et al.*, 2001). DNA sequencing data (Cooke *et al.*, 2000; Riethmüller *et al.*, 2002) are confusing at present because species of *Phytophthora* (Pythiales) and *Peronospora* (Peronosporales) seem to intergrade in phylogenetic analyses. *Peronospora* seems more closely related to *Phytophthora* than to other members of the Peronosporales such as *Albugo*, which in turn may have affinity with *Pythium*. Considerable rearrangements between the Peronosporales and Pythiales will therefore have to be carried out at some point in the future. However, we prefer to retain the conventional system for the time being because the downy mildews (Peronosporales) represent a convincing biological entity (Dick, 2001a). The key features distinguishing them from the Pythiales are as follows.

First, they are obligate biotrophs and cannot be grown apart from their living host. The mycelium in the host tissues is coenocytic and intercellular, with haustoria of various types penetrating the cell walls. No member of the Peronosporales has as yet been grown in axenic culture, although some can be propagated in dual culture with callus tissues of their plant hosts. None the less, some species (e.g. *Plasmopara viticola*) can cause cell damage to their hosts which leads to the leakage of cytoplasm (Lafon & Bulit, 1981). This is similar to the rots caused, for example, by *Phytophthora erythroseptica* (Plate 2f) and suggests an incomplete adaptation to the biotrophic habit, tying in with the likely origin of Peronosporales from within the Pythiales (Dick, 2001a).

Second, whereas *Pythium* and *Phytophthora* spp. are typically able to attack a very wide range of host plants, Dick (2001a) has pointed out that Peronosporales parasitize a narrow range of angiosperm families, usually dicotyledons, and especially herbaceous plants which are either highly evolved or accumulate large amounts of secondary metabolites such as essential oils or alkaloids. Any one species of downy mildew is specific to only one or a few related host genera. Dick (2001a, 2002) has speculated that a co-evolution of the downy mildews with herbaceous angiosperms occurred mainly in the Tertiary period, and as several independent events, whereby *Phytophthora* and downy mildews share common ancestors. The Peronosporaceae are relatively recent; *Peronospora*, along with its host plants, may have arisen in the mid to late Tertiary in the vicinity of Armenia and Iran. *Plasmopara* is probably of South American origin and dates back to the early Tertiary, whereas *Bremia lactucae* is a central European species. In contrast, the Albuginaceae (*Albugo*) are more ancient, with a late Cretaceous origin possibly in South America (Dick, 2002).

A third major feature of the Peronosporales is the tendency of their sporangia to germinate directly, rather than by releasing zoospores. Many species have lost the ability to produce zoospores altogether, their sporangia being functional 'conidia' which are disseminated by wind. The sporangiophores are well-differentiated, showing determinate growth and branching patterns which provide characteristic features for identification. The production of directly germinating sporangia on well-defined sporangiophores represents an adaptation to the terrestrial lifestyle and supports the postulated origin of the Peronosporales in the drier Tertiary period (Dick, 2002). The life cycle of Peronosporales is similar to that of *Phytophthora* (see Fig. 5.19). Sporangia infect directly or produce infective zoospores, leading to a new crop of sporangiophores and sporangia, and this asexual cycle spreads the disease during the vegetation period. Sexual reproduction is by means of oospores which are formed within the host tissue and survive adverse conditions after host death.

Peronosporales cause economically significant diseases, and one of them – *Plasmopara viticola* – has had a major impact on agriculture and plant pathology because it led to the discovery of Bordeaux mixture (see p. 119). Overviews of the Peronosporales have been given by Spencer (1981), Smith *et al.* (1988) and Dick (2002).

5.4.1 *Peronospora* (Peronosporaceae)

Peronospora destructor causes a serious disease of onions and shallots whilst *P. farinosa* causes downy mildew of sugar beet, beetroot and spinach, but can also be found on weeds such as *Atriplex* and *Chenopodium*. *Peronospora tabacina* causes blue mould of tobacco. This name refers to the bluish purple colour of the sporangia, which is actually a feature of many species of *Peronospora*. Crop losses associated with *P. tabacina* can be up to 95%. This species was introduced into Europe in 1958 and has spread rapidly since (Smith *et al.*, 1988).

Peronospora parasitica attacks members of the Brassicaceae. Although many specific names have been applied to forms of this fungus on different host genera, it is now customary to regard them all as belonging to a single species (Dickinson & Greenhalgh, 1977; Kluczewski & Lucas, 1983). Turnips, swede, cauliflower, Brussels sprouts and wallflowers (*Cheiranthus*)

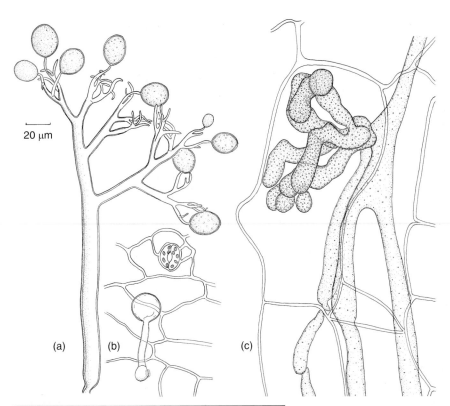

20 μm

(a)   (b)   (c)

Fig 5.28 *Peronospora parasitica* on *Capsella bursa-pastoris.* (a) Sporangiophore. (b) Sporangium germinating by means of a germ tube. (c) L.S. of host stem showing intercellular mycelium and coarse lobed haustoria.

are commonly attacked, and the fungus is found particularly frequently on shepherd's purse (*Capsella bursa-pastoris*). Diseased plants stand out by their swollen and distorted stems bearing a white 'fur' of sporangiophores (Plate 2g). On leaves the fungus is associated with yellowish patches on the upper surface and the formation of white sporangiophores beneath. Sections of diseased tissue show a coenocytic intercellular mycelium and branched lobed haustoria in certain host cells (Fig. 5.28c; Fraymouth, 1956).

Following penetration of the host cell by *P. parasitica*, reactions are set up between the host protoplasm and the invading fungus. The haustorium becomes ensheathed by a layer of callose which is visible as a thickened collar around the haustorial base in susceptible host plants, whereas the entire haustorium may be coated by thick callose deposits in interactions showing a resistance response (Donofrio &

Delaney, 2001). The general appearance of haustoria of *Peronospora* is very similar to that of *Phytophthora* shown in Fig. 5.21; the main body of the haustorium is surrounded by host cytoplasm, the host plasma membrane, an extrahaustorial matrix, the fungus cell wall, and the fungal plasma membrane (Fig. 5.29). Although the haustoria undoubtedly play a major role in the nutrient uptake of the fungus from the host plant, it should be noted that intercellular hyphae are also capable of assimilating nutrients *in planta* (Clark & Spencer-Phillips, 1993; Spencer-Phillips, 1997).

The sporangiophores emerge singly or in groups from stomata. There is a stout main axis which branches dichotomously to bear egg-shaped sporangia at the tips of incurved branches (Fig. 5.28a). Detachment of sporangia is possibly caused by hygroscopic twisting of the sporangiophores related to changes in humidity.

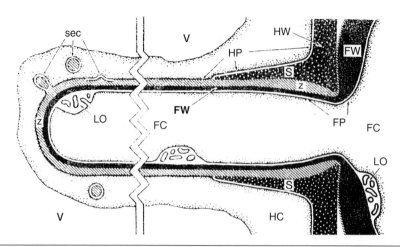

Fig 5.29 *Peronospora manshurica*. Diagram of host–pathogen interface in the haustorial region. Fungal cytoplasm (FC) is bounded by the fungal plasma membrane (FP), lomasomes (LO) and the fungal cell wall (FW) in both the intercellular hyphae (right) and the haustorium (centre). The relative positions of the host cell vacuole (V), host cytoplasm (HC) and host plasmalemma (HP) are indicated. The host cell wall (HW) terminates in a sheath (S). The zone of apposition (Z) separates the haustorium from the host plasmalemma. Invaginations of the host plasmalemma and vesicular host cytoplasm are considered evidence for host secretory activity (sec). After Peyton and Bowen (1963).

In *P. tabacina*, however, it has been suggested that changes in turgor pressure of the sporangiophores occur which parallel changes in the water content of the tobacco leaf. Sporangia may be discharged actively by application of energy at their point of attachment to the sporangiophore. In the Sclerosporaceae (see Section 5.5), violent sporangial discharge also occurs. Upon alighting on a suitable host, sporangia of *P. parasitica* germinate by the formation of a germ tube rather than zoospores. The germ tube penetrates the wall of the epidermis by means of an appressorium (Fig. 5.28b).

Oospores of *P. parasitica*, like those of most other Peronosporales, are embedded in senescent leaf tissues and are found throughout the season. There is evidence that some strains of the fungus are heterothallic whilst others are homothallic (McMeekin, 1960). Both the antheridium and oogonium are at first multinucleate. Nuclear division precedes fertilization, and meiosis occurs in the oogonium and antheridium (Sansome & Sansome, 1974). Fusion between two nuclei is delayed at least until the oospore wall is partly formed.

The wall of the oospore of *P. parasitica* is very tough, and it is difficult to induce germination. In *P. destructor* and some other species, germination

occurs by means of a germ tube but in *P. tabacina* zoospores have been described. It is probable that oospores overwinter in soil and give rise to infection in subsequent seasons. Although oospores of *P. destructor* have been germinated after 25 years, it has not proven possible to infect onions from such material. Possibly in this case the disease is carried over by means of systemic infection of volunteer onion bulbs (Smith *et al.*, 1988).

Peronospora parasitica and *Arabidopsis thaliana*

The chance discovery of a *P. parasitica* infection in an *Arabidopsis thaliana* weed population in a Zurich garden showing haustoria, sporangia and oospores (Koch & Slusarenko, 1990) opened up the possibility of using this genetically well-characterized 'model plant' to investigate plant–pathogen interactions involving downy mildews. The interaction between *Arabidopsis* and *Peronospora* is governed by a gene-for-gene relationship, i.e. it is a form of major gene resistance based on specific recognition of a pathogen avirulence gene (*avr*) product by the product of a matching host resistance (R) gene (e.g. Botella *et al.*, 1998). Molecular aspects of the *Arabidopsis* immune response to infections by

P. parasitica and other pathogens have been investigated in some detail. Infection of one leaf triggers a localized reaction, the hypersensitive response, leading to death of the plant cells in the vicinity of infection. Additionally, a systemic response is initiated, i.e. plant organs distal to the infected leaf become resistant against further attack. This phenomenon is called **systemic acquired resistance** and is active against attacks by the same as well as many other pathogens. It is triggered at the site of initial infection by various **elicitor** molecules of pathogen origin, e.g. fatty acids such as arachidonic acid, or by other substances. The signal is transmitted by signalling molecules such as salicylic acid (Lawton *et al.*, 1995; Ton *et al.*, 2002) which itself has no antimicrobial activity. Salicylic acid-independent signalling events are probably also involved (McDowell *et al.*, 2000). Salicylic acid is produced at sites of infection, diffuses through the plant and interacts with a signalling chain, leading to the expression of a set of pathogenesis-related (*PR*) genes. A whole subset of *PR* genes involved in resistance to *P. parasitica* (*RPP* genes) is now known (McDowell *et al.*, 2000). The function of many *PR* genes is still obscure; those whose functions are known encode chitinases, β-1,3-glucanases, proteinases, peroxidases or enzymes involved in toxin biosynthesis (Kombrink & Somssich, 1997). By creating mutants of *Arabidopsis* or of crop plants which overexpress their own regulatory genes or *PR* genes, or express introduced genes encoding elicitor molecules of pathogen origin, constitutive resistance against pathogen attack may be generated. This is considered to hold great potential for agriculture (Cao *et al.*, 1998; Maleck *et al.*, 2002).

Control of *Peronospora*

Downy mildew infections caused by *Peronospora* spp. are controlled mainly by fungicide applications. Metalaxyl is very effective against all downy mildews, but resistance has arisen in several species, and thus this fungicide is now applied in a cocktail with dithiocarbamates (Smith *et al.*, 1988). Fosetyl−Al is also now widely used as a foliar spray, root dip or soil amendment (Agrios, 2005).

The breeding of cultivars with resistance against *Peronospora* spp. has been successful in certain crops, e.g. in lucerne (*Medicago sativa*) against *P. trifoliorum* (Stuteville, 1981). In tobacco plants attacked by *P. tabacina*, this strategy is a useful component of integrated control but is not sufficient on its own to afford complete control (Schiltz, 1981). In the tobacco−*P. tabacina* system, a disease warning system is also in operation in Europe; subscribing tobacco growers are informed of the occurrence of the pathogen, so that preventative measures can be taken (Smith *et al.*, 1988). This is profitable because tobacco is a high-value crop.

Because downy mildews infect aerial plant parts and produce air-borne propagules in large numbers, crop sanitation measures are generally not very effective. However, in the case of *P. destructor* which overwinters systemically in volunteer onion bulbs, removal of volunteers is essential. In *P. viciae* on peas and beans, deep ploughing of the crop residue is important as the pathogen survives on infected haulms (Smith *et al.*, 1988).

5.4.2 *Plasmopara* (Peronosporaceae)

Although downy mildews caused by species of *Plasmopara* are rarely serious in temperate climates, *P. viticola* is potentially a very destructive pathogen of the grapevine. The disease, which was endemic in North America and not particularly destructive on the local vines, was introduced into France during the nineteenth century with disastrous results on the French vines which had never been exposed to the disease and were highly susceptible. Large (1940) has vividly recounted the moment when Alexis Millardet, walking past a heavily infected vineyard in 1882, noticed that vines close to the road appeared healthy and had been sprayed with a mixture of lime and copper sulphate to discourage passers-by from pilfering fruit. This led to the discovery of Bordeaux mixture, one of the world's first fungicides and still effective against *P. viticola* and other foliar pathogens belonging to the Oomycota.

Plasmopara nivea is occasionally reported in Britain on umbelliferous crops such as carrot and parsnip, and it is also found on *Aegopodium podagraria*. *Plasmopara pygmaea* is found on yellowish patches on the leaves of *Anemone nemorosa* (Fig. 5.30b), whilst *P. pusilla* is similarly associated with *Geranium pratense* (Fig. 5.30a). The haustoria of *Plasmopara* are knob-like, the sporangiophores are branched monopodially and the sporangia are hyaline (Fig. 5.30). Two types of sporangial germination have been reported. In *P. pygmaea* there are no zoospores but the entire sporangium detaches and later produces a germ-tube. In other species the sporangia germinate by means of zoospores which encyst and penetrate the host stomata. Oospore germination in *P. viticola* is also by means of zoospores.

Because the grapevine is such a high-value crop, the fungicide market is lucrative. Bordeaux mixtures are still used today, and similar fungicide applications to those described for *Peronospora* are made. Resistance to metalaxyl

has been observed in *P. viticola*. Disease forecasting systems are being developed (Lafon & Bulit, 1981; Smith *et al.*, 1988). Breeding for resistant cultivars is being carried out, but because of the long generation times of the crop, this will be a prolonged effort.

5.4.3   *Bremia* (Peronosporaceae)

Bremia lactucae causes downy mildew of lettuce (*Lactuca sativa*) and strains of it can be found on 36 genera of the Asteraceae including *Sonchus* and *Senecio* (Crute & Dixon, 1981). Cross-inoculation experiments using sporangia from these hosts have failed to result in infection of lettuce and it seems that the fungus exists as a number of host-specific strains (*formae speciales*). Although wild species of *Lactuca* can carry strains capable of infecting lettuce, these hosts are not sufficiently common to provide a serious source of infection. The disease can be troublesome both in lettuce grown in the open and under frames,

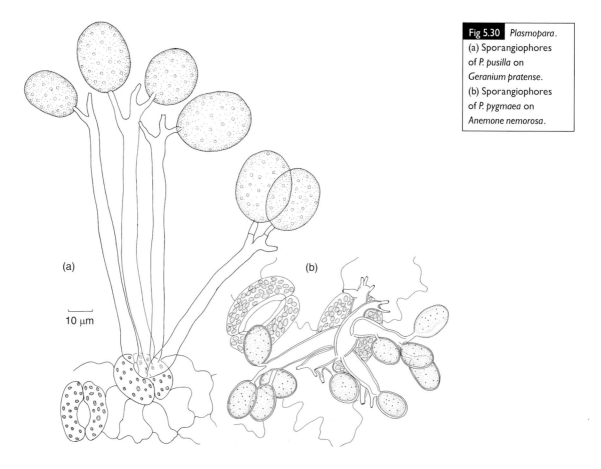

(a)

(b)

10 μm

Fig 5.30 *Plasmopara*. (a) Sporangiophores of *P. pusilla* on *Geranium pratense*. (b) Sporangiophores of *P. pygmaea* on *Anemone nemorosa*.

and in market gardens there may be sufficient overlap in the growing of lettuce for the disease to be carried over from one sowing to the next. The damage to the crop caused by *Bremia* may not in itself be severe, but infected plants are prone to secondary infection by the more serious grey mould, *Botrytis cinerea*. Systemic infections can occur. The intercellular mycelium is coarse, and the haustoria are sac-shaped, often several of them being present in each host cell (Fig. 5.31d). The sporangiophores emerge singly or in small groups through the stomata and branch dichotomously. The tip of each branch expands to form a cup-shaped disc bearing short cylindrical

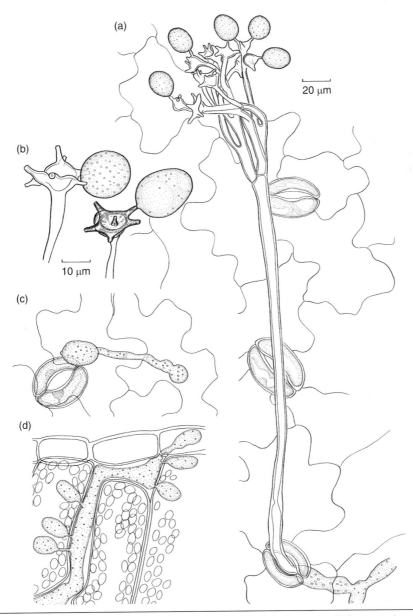

Fig 5.31 *Bremia lactucae* from *Senecio vulgaris*. (a) Sporangiophore protruding through a stoma. (b) Sporangiophore apex. (c) Sporangium germinating by means of a germ tube which has produced an appressorium at its apex. (d) Cells of epidermis and palisade mesophyll, showing intercellular mycelium and haustoria. (a,c,d) to same scale.

sterigmata at the margin and occasionally in the centre, and from these the hyaline sporangia arise (Figs. 5.31a,b). Germination of the sporangia is usually by means of a germ tube which forms an appressorium to penetrate epidermal cells (Fig. 5.31c), or it enters through a stoma. Zoospore formation has been reported but not confirmed. Sexual reproduction is usually heterothallic, although homothallic strains also exist. The oospores are formed in leaf tissue and remain viable for 12 months (Michelmore & Ingram, 1980; Morgan, 1983).

Chemical control of *B. lactucae* on lettuce is certainly possible although not necessarily desirable; hence, intensive efforts for major gene resistance breeding have been made. Integrated control based on resistant cultivars and fungicide applications using metalaxyl and dithiocarbamates is successful (Crute, 1984). However, resistance against metalaxyl arose in Britain as early as 1983. Fosetyl–Al is not as effective as metalaxyl (Smith *et al.*, 1988).

5.4.4 *Albugo* (Albuginaceae)

This family has only a single genus, *Albugo*, with about 40–50 species of biotrophic parasites of flowering plants which cause diseases known as white blisters or white rusts. The commonest British species is *A. candida* causing white blisters of crucifers such as cabbage, turnip, swede, horseradish, etc. (Plate 2h). It is particularly frequent on shepherd's purse (*Capsella bursapastoris*). There is some degree of physiological specialization in the races of this fungus on different host genera. *Albugo candida* can infect *Arabidopsis thaliana*, and the host defence response is governed by resistance genes involved in the recognition of the pathogen (Holub *et al.*, 1995). The principle is similar to, although not as well researched as, the *Arabidopsis–Peronospora* interaction described earlier (p. 116). It is also now possible to establish callus cultures of mustard plants (*Brassica juncea*) containing balanced infections of *A. candida* (Nath *et al.*, 2001). This experimental system should facilitate studies of the physiology of host–pathogen interactions. A less common species is *A. tragopogonis*, causing white blisters of salsify (*Tragopogon porrifolius*), goatbeard (*T. pratensis*) and *Senecio squalidus*.

In *A. candida* on shepherd's purse, diseased plants may be detected by the distorted stems and the shining white raised blisters on the stem, leaves and pods before the host epidermis is ruptured (Plate 2h). Later, when the epidermis has burst open, a white powdery pustule is visible. The distortion is possibly associated with altered auxin levels. The host plant may be infected simultaneously with *Peronospora parasitica*, but the two fungi are easily distinguishable microscopically both in the structure of the sporangiophores and by their different haustoria. In *Albugo*, the mycelium in the host tissues is intercellular with only small spherical haustoria (Fig. 5.32) which contrast sharply with the coarsely lobed haustoria of *P. parasitica*. The fine structure of *A. candida* haustoria has been described by Coffey (1975) and Soylu *et al.* (2003). They are spherical or somewhat flattened and about 4 μm in diameter, connected to the intercellular mycelium by a narrow stalk about 0.5 μm wide. Inside the plasma membrane of the haustorium, lomasomes, i.e. tubules and vesicles apparently formed by invagination of the plasma membrane, are more numerous than in the intercellular hyphae. The cytoplasm of the haustorial head is densely packed with mitochondria, ribosomes, endoplasmic reticulum and occasional lipid droplets, but nuclei have not been observed. Since nuclei of *Albugo* are about 2.5 μm in diameter, they may be unable to traverse the constriction which links the haustorium to the intercellular hypha. Nuclei may (e.g. *Peronospora pisi*) or may not be present in the haustoria of other Oomycota. The base of the haustorium of *A. candida* is surrounded by a collar-like sheath which is an extension of the host cell wall, but this wall does not normally extend to the main body of the haustorium. Between the haustorium and the host plasma membrane is an encapsulation. Host cytoplasm reacts to infection by an increase in the number of ribosomes and Golgi complexes. In the vicinity of the haustorium the host cytoplasm contains numerous vesicular and tubular elements not found in uninfected cells. These structures have been interpreted

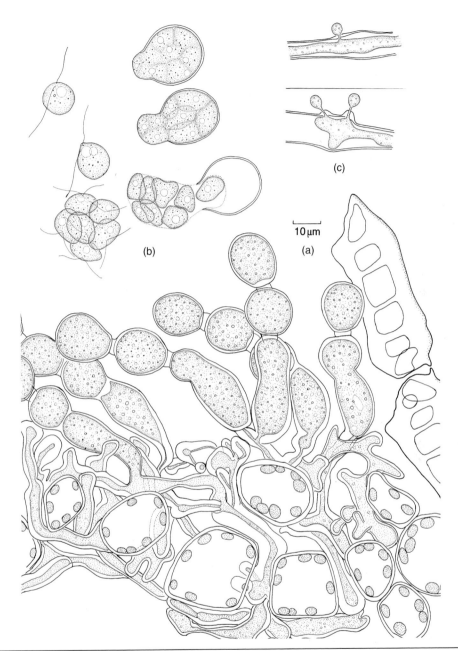

Fig 5.32 *Albugo candida* on *Capsella bursa-pastoris*. (a) Mycelium, sporangiophores and chains of sporangia formed beneath the ruptured epidermis (right). (b) Germination of sporangia showing the release of eight biflagellate zoospores. The stages illustrated took place within 2 min. (c) Haustoria.

as evidence of secretory processes induced in the host cell by the presence of the pathogen.

The intercellular mycelium aggregates beneath the host epidermis to form a palisade of cylindrical or skittle-shaped sporangiophores which give rise to chains of spherical sporangia in basipetal succession – i.e. new sporangia are formed at the base of the chain. The pressure of the developing chains of sporangia raises the host epidermis and finally ruptures it.

The sporangia are then visible externally as a white powdery mass dispersed by the wind. Sporangia reaching a suitable host leaf will germinate within a few hours in films of water to form biflagellate zoospores of the principal type, about eight per sporangium (Fig. 5.32b). After swimming for a time, a zoospore encysts and then forms a germ tube which penetrates the host epidermis. The asexual disease cycle may be completed within 10 days. Infections may be localized or systemic. Gametangia are formed in the intercellular spaces of infected stems and leaves. Both the antheridium and the oogonium are multinucleate at their inception, and during development two further nuclear divisions occur so that the oogonium may contain over 200 nuclei. However, there is only one functional male and one functional female nucleus. In the oogonium all the nuclei except one migrate to the periphery and are included in the periplasm. Following nuclear fusion a thin membrane first develops around the oospore. Division of the zygote nucleus takes place and is repeated, so that at maturity the oospore may contain as many as 32 diploid nuclei. Sansome and Sansome (1974) reported that meiosis occurs within the gametangia. They also suggested

that *A. candida* is heterothallic. The high incidence of oospores of *Albugo* in *Capsella* stems simultaneously infected with *Peronospora parasitica* may result from some stimulus towards self-fertilization in *Albugo* produced by *Peronospora*, a situation analogous to the *Trichoderma*-induced sexual reproduction in heterothallic species of *Phytophthora* (see p. 95).

The mature oospore is surrounded by a brown exospore, thrown into warty folds (Fig. 5.33a). Germination of the oospores takes place only after a resting period of several months. Under suitable conditions the outer wall of the oospore bursts and the endospore is extruded as a thin, spherical vesicle, which may be sessile or formed at the end of a wide cylindrical tube. Within the thin vesicle 40–60 zoospores are differentiated and are released on its breakdown (Figs. 5.33b,c).

The cytology of oospore development in some other species of *Albugo* differs from that of *A. candida*. In *A. bliti*, a pathogen of *Portulaca* in North America and Europe, the oogonia and antheridia are also multinucleate and two nuclear divisions take place during their development. Numerous male nuclei fuse with numerous female nuclei and the fusion nuclei

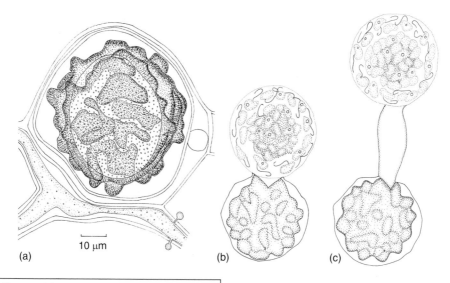

10 µm

(a)   (b)   (c)

Fig 5.33 *Albugo candida* oospores. (a) Oogonium and oospore from *Capsella* leaf. (b,c) Two methods of oospore germination (after Vanterpool, 1959).

pass the winter without further change. In *A. tragopogonis*, a multinucleate oospore develops and again there are two nuclear divisions involved in the development of the oogonium and antheridium, but finally there is a single nuclear fusion between one male and one female nucleus. This fusion nucleus undergoes repeated divisions so that the overwintering oospore is multinucleate.

Albugo candida alone or in combination with co-infecting *Peronospora parasitica* can occasionally cause significant crop losses in cabbage cultivation. Fungicide treatment is possible, with copper-based or dithiocarbamate-type fungicides commonly used (Smith *et al.*, 1988).

5.5 | Sclerosporaceae

This family comprises the downy mildews of grasses and cereals. Although it is well defined as a biological group, its phylogenetic position is unclear, recent ribosomal DNA-based studies placing its members among the Peronosporales (Riethmüller *et al.*, 2002). For reasons of their distinctly different biological features, we consider them briefly here. The principal genera are *Sclerospora*, with sporangia capable of germinating by releasing zoospores, and *Peronosclerospora*, whose sporangia show direct germination by germ tubes and are thus, functionally speaking, 'conidia'. Sporangia or conidia are produced on repeatedly branching aerial structures which resemble those of *Peronospora* spp. In *Peronosclerospora*, the conidiophores project through stomata of the host and branch at their apices to produce up to 20 finger-like tapering extensions which expand to form conidia (Figs. 5.34a–c). The conidia are oval and hyaline. Unlike those of other Oomycota, conidia of Sclerosporaceae are projected actively by a sudden rounding-off of the conidiophore tip and conidial base, and this is visible as a

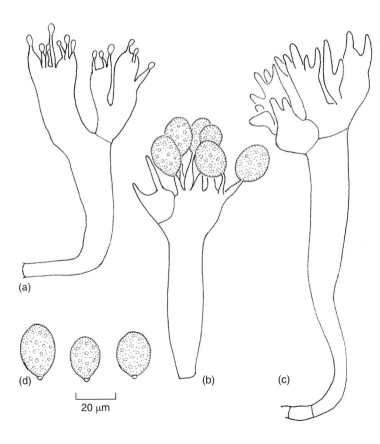

(a)

(d) (b) (c)

20 μm

Fig 5.34 *Peronosclerospora sorghi.* (a) Immature conidiophore showing conidium initials. (b) Mature conidiophore from which two conidia have become detached. (c) Old conidiophore; all conidia have become detached. (d) Discharged conidia. Note the small basal projection. Drawn from material kindly provided by K. Mathur.

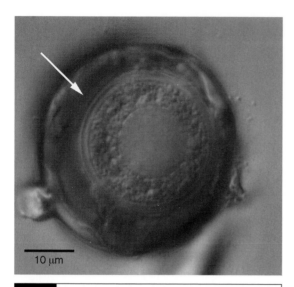

Fig 5.35 Oospore of *Peronosclerospora sorghi*. Note the thickened oogonial wall (arrow), within which the spherical oospore with its wall and ooplast is clearly visible.

small projection at the base of discharged conidia (Fig. 5.34d). Oospores of Sclerosporaceae are distinctive in being surrounded by a thickened oogonial wall (Fig. 5.35), and this feature may enhance the longevity of the oospore. The most important species are *Sclerospora graminicola* infecting pearl millet (*Pennisetum americanum*), and *Peronosclerospora sorghi* pathogenic on sorghum and maize. Because of their similar biological features and great economic importance, these two species are often considered together. Thorough reviews have been written by R. J. Williams (1984) and Jeger *et al.* (1998).

Downy mildews of grasses cause serious crop losses especially in dry subtropical and tropical zones in Africa, their putative centre of evolution, as well as Asia and, to a lesser extent, North and South America. The thick-walled oospores can survive on plant debris and in the soil for up to 10 years, and infections are usually initiated from oospores which germinate directly by means of a germ tube. The plant root may be the initial route of entry, although both *S. graminicola* and *P. sorghi* may also become seed-borne. Later infections are through the shoot surface, either by direct penetration of the epidermis by means of appressoria, or through stomata. Infections of host plants are obligately biotrophic and can become systemic if they reach the apical meristem. Sporangia or conidia are formed only on freshly infected living host tissues under moist conditions, and infections are therefore polycyclic only when sufficient moisture is available. In dry regions, infections may be carried exclusively by oospores, confining the pathogen to one disease cycle per growing season. Oospore production is buffered against environmental extremes by taking place within the tissue of aerial host organs. Like sporangia or conidia, oospores can be blown about by wind.

Control of downy mildew of grasses is difficult. Metalaxyl gives good control both as a seed dressing and as a foliar spray but may not always be available. Numerous cultivars of sorghum and pearl millet show resistance against downy mildews, but this is usually based on one or a few major genes and can therefore be overcome by the pathogens if single cultivars are grown in large coherent areas. On small-scale farms, it may be possible to remove individual infected plants prior to the onset of sporulation (Gilijamse *et al.*, 1997).

Chytridiomycota

6.1 | Introduction

The phylum Chytridiomycota comprises over 900 species in five orders (D. J. S. Barr, 2001; Kirk et al., 2001). Fungi included here are colloquially called 'chytrids'. Most chytrids grow aerobically in soil, mud or water and reproduce by zoospores with a single posterior flagellum of the whiplash type, although the zoospores of some members of the Neocallimastigales are multiflagellate. Some species inhabit estuaries and others the sea. Sparrow (1960) has given an extensive account of aquatic forms, Karling (1977) a compendium of illustrations, and Powell (1993) has provided examples of the importance of the group. Many members are saprotrophs, utilizing cellulose, chitin, keratin, etc., from decaying plant and animal debris in soil and mud, whilst species of *Caulochytrium* grow as mycoparasites on the mycelium and conidia of terrestrial fungi (Voos, 1969). Saprotrophs can be obtained in crude culture by floating baits such as cellophane, hair, shrimp exoskeleton, boiled grass leaves and pollen on the surface of water overlying samples of soil, mud or pieces of aquatic plant material (Sparrow, 1960; Stevens, 1974; Willoughby, 2001). From such crude material, pure cultures may be prepared by streaking or pipetting zoospores onto agar containing suitable nutrients and antibiotics to limit contamination from bacteria. The growth and appearance of chytrids in pure culture is variable and often differs significantly from their natural habit. This has led to problems in classification systems based on thallus morphology (Barr, 1990, 2001). The availability of cultures has, however, facilitated studies on chytrid nutrition and physiology (Gleason, 1976).

Some chytrids are biotrophic parasites of filamentous algae and diatoms and may severely deplete the population of freshwater phytoplankton (see p. 139). Two-membered axenic cultures of diatom host and parasite have been prepared, making possible detailed ultrastructural studies of comparative morphology, zoospores, infection processes and reproduction. Other chytrids such as species of *Synchytrium* and *Olpidium* are biotrophic parasites of vascular plants. *Synchytrium endobioticum* is the agent of the potentially serious black wart disease of potato. *Olpidium brassicae*, common in the roots of many plants, is relatively harmless, but its zoospores are vectors of viruses such as that causing big vein disease of lettuce. *Coelomomyces* spp. are pathogens of freshwater invertebrates including copepods and the larvae of mosquitoes. The possibility of using them in the biological control of mosquitoes has been explored. The most unusual group are the Neocallimastigales, which grow in the guts of herbivorous mammals, are obligately anaerobic and subsist on ingested herbage.

The cell walls of some chytrids have been examined microchemically by X-ray diffraction and other techniques. Chitin has been detected in many species (Bartnicki-Garcia, 1968, 1987), and in *Gonapodya* cellulose is also present (Fuller & Clay, 1993). The composition of the wall is of interest because chitin, a polymer of

N-acetylglucosamine, is also found in the walls of other Eumycota (i.e. Zygomycota, Ascomycota and Basidiomycota), whilst the cell walls of members of the Oomycota contain cellulose. Cellulose and chitin occur together in the walls of species of *Hyphochytrium* and *Rhizidiomyces*, members of the Hyphochytriomycota (Fuller, 2001; see Section 4.3).

The form of the thallus in the Chytridiomycota is varied. In biotrophic species such as *Olpidium* and *Synchytrium*, where the whole thallus is contained within the host cell, there is no differentiation into a vegetative and a reproductive part. At maturity the entire structure, except for the wall which surrounds it, is converted into reproductive units, i.e. zoospores, gametes or resting sporangia. Such thalli are termed **holocarpic** (Fig. 6.1). More usually, the thallus is differentiated into organs of reproduction (sporangia and resting sporangia) arising from a vegetative part which often consists of **rhizoids**. These serve in the exploitation of the substratum and the assimilation of nutrients. Thalli of this type are **eucarpic**. Eucarpic thalli may have one or several sporangia and are then termed **monocentric** or **polycentric**, respectively (Fig. 6.1). In some species there are both monocentric and polycentric thalli, so that these terms have descriptive rather than taxonomic significance. A further distinction has been made, especially in monocentric forms, between those in which only the rhizoids are inside the host cell whilst the sporangium is external (**epibiotic**), in contrast with the **endobiotic** condition in which the entire thallus is inside the host cell (Fig. 6.1). In monocentric thalli, the rhizoids usually radiate from a single position on the sporangium wall, but in polycentric forms a more extensive, branched rhizoidal system, the **rhizomycelium**, develops.

The zoosporangium is generally a spherical or pear-shaped sac bearing one or more discharge tubes or exit papillae. The method of zoospore release has been used in classification.

Holocarpic

| Fig 6.1 | Types of thallus structure in the Chytridiales, diagrammatic and not to scale. |

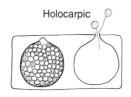

Eucarpic

Monocentric   Polycentric

Endobiotic

Epibiotic

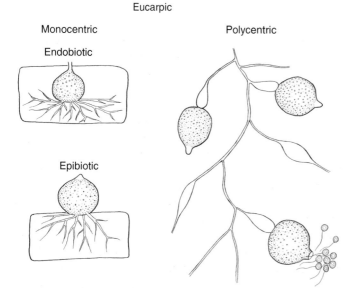

In the **inoperculate** chytrids such as *Olpidium*, *Diplophlyctis* and *Cladochytrium*, the sporangium forms a discharge tube which penetrates to the exterior of the host cell and its tip becomes gelatinous and dissolves away. In **operculate** chytrids such as *Chytridium* and *Nowakowskiella*, the tip of the discharge tube breaks open at a special line of weakness and is seen as a special cap or **operculum** after discharge (see Fig. 6.4b).

6.1.1 The zoospore

The number of zoospores formed inside zoosporangia of chytrids varies with the size of the spore and sporangium. Although the zoospore size is roughly constant for a given species, the size of the sporangium may be very variable. In *Rhizophlyctis rosea*, tiny sporangia containing only one or two zoospores have been reported from culture media deficient in carbohydrate, whereas on cellulose-rich media large sporangia containing many hundred spores are formed. The release of zoospores is brought about by internal pressure which causes the exit papillae to burst open. In studies of the fine structure of mature sporangia of *R. rosea* and *Nowakowskiella profusa* (Chambers & Willoughby, 1964; Chambers *et al.*, 1967), it has been shown that the single flagellum is coiled round the zoospore like a watch spring. The zoospores are separated by a matrix of spongy material which may absorb water and swell rapidly at the final stages of sporangial maturation. When the internal pressure has been relieved by the ejection of some zoospores, those remaining inside the sporangium escape by swimming or wriggling through the exit tube. In some species the spores are discharged in a mass which later separates into single zoospores, but in others the zoospores make their escape individually.

The form of the zoospore is similar in all chytrids (with the exception of the multiflagellate members of the Neocallimastigales). There is a spherical or ellipsoidal body which in some forms is capable of plastic changes in shape, and a long trailing flagellum. When swimming, the zoospores show characteristic jerky or 'hopping' movements; additionally, abrupt changes in direction are sometimes made. The internal structure of the zoospore as revealed by light and electron microscopy is variable, but characteristic of particular genera (Lange & Olson, 1979). In view of the plasticity in morphology of the thallus under different growth conditions, zoospore ultrastructure is regarded as a more satisfactory basis of classification (D. J. S. Barr, 1990, 2001). Two features are of taxonomic importance, the flagellar apparatus and an assemblage of organelles termed the **microbody–lipid globule complex** (MLC) (D. J. S. Barr, 2001).

The flagellar apparatus

The whiplash flagellum resembles that of other eukaryotes, with a smooth membrane enclosing a cylindrical shaft, the axoneme, made up internally of nine doublet pairs of microtubules surrounding two central microtubules. As shown in Fig. 6.2, the base of the axoneme comprises three regions, the flagellum proper, the transitional zone and the kinetosome. The function of the kinetosome is to generate the flagellum. An interesting feature found in several species is a second kinetosome or the remainder of one, the **dormant kinetosome**. Its presence has led to the suggestion that the ancestors of the Chytridiomycota may have had biflagellate zoospores, the second flagellum having been lost in the course of evolution (Olson & Fuller, 1968).

In section, the kinetosome resembles a cartwheel (Fig. 6.2f), because to each of the nine outer microtubule doublets seen in the flagellum proper, a third microtubule is attached. This is called the C-tubule; in the doublets, that tubule with extended dynein arms is the A-tubule, and its partner is labelled B. These flagellar microtubules radiate as kinetosome props into the zoospore, perhaps providing structural support and anchorage of the flagellum (D. J. S. Barr, 2001). Microtubules may also be attached laterally to the kinetosome, contributing to the flagellar root system (Figs. 6.2c, 6.19). In the innermost (proximal) part of the transitional zone, the nine microtubule triplets of the kinetosome are converted into the doublets of the flagellum proper; concentric fibres, possibly arranged helically, surround the nine doublet pairs. Also within the transitional zone,

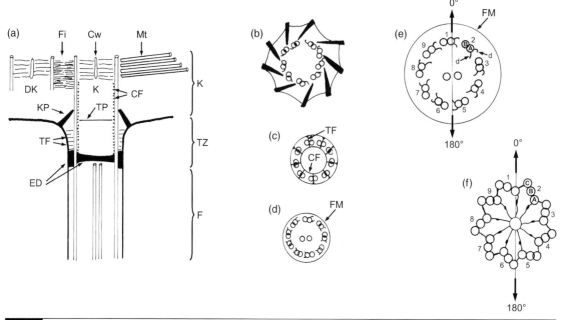

Fig 6.2 Flagellar apparatus typical of zoospores of Chytridiomycota. (a) Median longitudinal section of the junction of the flagellum with the body of the zoospore. The labels indicate the flagellum proper (F), transitional zone (TZ), kinetosome (K), electron-dense region (ED), concentric fibres (CF), transitional fibres (TF), kinetosome props (KP), terminal plate (TP), kinetosome (K) showing a cartwheel-like organization (Cw), dormant kinetosome (DK), fibrillar material (Fi) found in some taxa, and microtubular roots (Mt) extending from the side or end of the kinetosome into the body of the zoospore. (b) Transverse section near the terminal plate showing nine kinetosome props extending from doublet microtubules to the cell membrane. (c) Transverse section in the lower part of the transition zone showing concentric and transitional fibres. (d) Transverse section of the flagellum proper showing two central microtubules and nine peripheral doublet microtubules enclosed in the flagellar membrane (FM). (e) Schematic drawing of the flagellum proper in transverse section. The arrowed line 0°–180° shows an imaginary plane which coincides with the plane of undulation of the flagellum, passing through doublet pair I and between the central microtubules and doublet pairs 5 and 6. The convention used in labelling the outer doublet pairs of microtubules is shown: the microtubule with dynein arms (d) is the A microtubule and its partner is the B microtubule. (f) Kinetosome in transverse section showing the triplet arrangement of the peripheral microtubules by the addition of a third microtubule (C). Redrawn from Barr and Désaulniers (1988) by copyright permission of the National Research Council of Canada, Barr (1992). ©The Mycological Society of America, and D. J. S. Barr (2001) with kind permission of Springer Science and Business Media.

the two central microtubules arise near a terminal plate. The structure of the flagellum and kinetosome in transverse section is shown in Figs. 6.2e and f (Barr & Désaulniers, 1988).

The microbody–lipid complex

The MLC (Fig. 6.3) is made up of a microbody which is often closely appressed to a large lipid globule and to simple membrane cisternae or a tubular membrane system, the **rumposome**. This is defined as a cisterna in which there is an area with hexagonally arranged, honeycomb-like pores called fenestrae (Fuller, 1976; Powell &

Roychoudhury, 1992). The rumposome may be involved in signal transduction from the plasma membrane to the flagellum because it is known that this organelle sequesters calcium. Regulation of external calcium concentrations has an effect on the symmetry of flagellar beat and hence on the direction of zoospore movement (Powell, 1983).

There are several distinct types of MLC (Powell & Roychoudhury, 1992) and Fig. 6.3 illustrates diagrammatically just one of them, that described for *Rhizophlyctis harderi*. In this species, the MLC includes several (3–5) lipid globules.

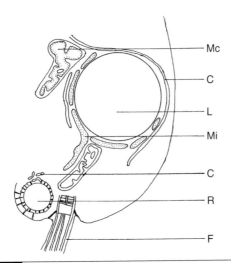

Fig 6.3 Schematic diagram of the microbody—lipid complex of the zoospore of *Rhizophlyctis harderi* as seen in a longitudinal section through the base of the zoospore and flagellum. The following organelles are drawn: mitochondrion (Mc), simple cisterna (C), lipid globule (L), microbody (Mi), flagellum (F) and rumposome (R). Redrawn from Powell and Roychoudhury (1992), by copyright permission of the National Research Council of Canada.

Those at the anterior of the cell are embedded in an aggregation of ribosomes. The surfaces of lipid globules close to the plasma membrane are partially covered by one to several simple cisternae, sometimes with irregularly scattered pores. Towards the centre of the cell the lipid bodies are clasped by cup-shaped microbodies. At the posterior of the zoospore near the kinetosome, 1–3 smaller lipid globules are partially covered by a rumposome, linked to the plasma membrane by short bridges and to the kinetosome by a microtubule root.

Other features

Patches of glycogen are located in the peripheral cytoplasm of the zoospore and it is likely that these and the lipid globules represent sources of energy used in respiration and propulsion. Mitochondria tend to be concentrated in the posterior of the zoospore close to the kinetosome; in *Allomyces* and *Blastocladiella* (Blastocladiales), the base of the flagellum passes through the perforation of a single large mitochondrion (see Fig. 6.19).

Most zoospores are uninucleate. The nucleus is surrounded in many cases (but not all) by a nuclear cap of uneven thickness. The nuclear cap is especially prominent in zoospores of members of Blastocladiales such as *Allomyces* and *Blastocladiella* (Fig. 6.19). It is rich in RNA and protein and also contains ribosomes.

6.1.2 Zoospore encystment and germination

The period of zoospore movement varies. Some flagellate zoospores seem to be incapable of active swimming and amoeboid crawling may take place instead, or swimming may last for only a few minutes. In other spores, motility may be prolonged for several hours. Prior to germination, the zoospore comes to rest and encysts. The flagellum may contract, it may be completely withdrawn or it may be cast off, but the precise details are often difficult to follow. The subsequent behaviour also differs in different species. In holocarpic parasites the zoospore encysts on the host surface and the cytoplasmic contents of the zoospore are injected into the host cell. In many monocentric chytrids rhizoids develop from one point on the zoospore cyst and the cyst itself enlarges to form the zoosporangium, but there are variants of this type of development in which the cyst enlarges into a **prosporangium** from which the zoosporangium later develops. In the polycentric types, the zoospore on germination may form a limited rhizomycelium on which a swollen cell arises, giving off further branches of rhizomycelium. Germination may be from a single point on the wall of the zoospore cyst (**monopolar** germination) or from two points, enabling growth to take place in two directions (**bipolar** germination). The mode of germination is an important character in distinguishing, for example, the Chytridiales (monopolar) from the Blastocladiales (bipolar).

6.1.3 Life cycles of the Chytridiomycota

Most chytrids have haploid zoospores and thalli but some Blastocladiales show an alternation of haploid (gametothallic) and diploid (sporothallic) generations. Apart from differences in the reproductive organs, the morphology of the two types of thallus is very similar, a phenomenon known as **isomorphic alternation of generations**.

Sexual reproduction, i.e. a life cycle which includes nuclear fusion and meiosis, may occur in several different ways (e.g. Figs. 6.6 and 6.22). In some chytrids it is by **gametogamy**, the fusion of gametes which are posteriorly uniflagellate. **Isogamous conjugation** occurs if there is no morphological distinction between the two fusing partners, but in some Blastocladiales (e.g. *Allomyces*) **anisogamy** takes place by fusion between a smaller, more actively motile male gamete with a larger, sluggish female gamete. **Oogamy**, fusion between an actively motile male gamete and a much larger, non-flagellate, immobile globose egg, is characteristic of Monoblepharidales. **Somatogamy**, the fusion of undifferentiated hyphae or rhizoids, has been well documented in cultures of the fresh-water fungus *Chytriomyces hyalinus* by Moore and Miller (1973) and Miller and Dylewski (1981, 1987). As shown in Fig. 6.4, zoospores of *C. hyalinus* are released from the zoosporangium by the opening of a lid-like operculum. They germinate to form uninucleate rhizoidal thalli (contributory thalli) and the tips of the rhizoids from adjacent thalli, which are apparently not genetically distinct from each other, may fuse (Fig. 6.4c). At the point of fusion an incipient resting body develops (Fig. 6.4d) and swells while cytoplasm and a nucleus migrate into it from each contributory thallus. Nuclear fusion occurs in the resting body to form a diploid zygote nucleus. The resting body continues to enlarge and develops a thick wall. This type of sexual reproduction by somatogamous conjugation probably occurs in several genera of inoperculate and operculate chytrids (Moore & Miller, 1973).

Fusion of gametangia (**gametangio-gametangiogamy**) has been reported by Doggett and Porter (1996) for *Zygorhizidium planktonicum*, a parasite of the diatom *Synedra*. This species reproduces asexually by epibiotic zoosporangia. Germinating zoospores develop either new zoosporangial thalli or gametangial thalli of two sizes with globose uninucleate gametangia.

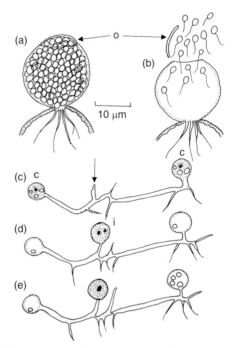

Fig 6.4 *Chytriomyces hyalinus* somatogamy. (a,b) Epibiotic fruiting thallus seated on a pollen grain into which rhizoids have penetrated. In (a) the zoosporangium, containing numerous zoospores, is seen shortly before discharge with a bulging operculum (o). In (b) the operculum has lifted off and the zoospores are escaping. (c–e) Stages in somatogamy. (c) Rhizoids from two uninucleate contributory thalli (c) have undergone anastomosis (arrow). (d) Cytoplasm and a nucleus from each contributory thallus have migrated towards the point of anastomosis, where the thallus swells to form a globose incipient resting body (i) which is binucleate and packed with cytoplasm, leaving the contributory thalli empty. (e) The two nuclei in the incipient resting body have fused. After C. E. Miller and Dylewski (1981).

Conjugation occurs when a conjugation tube grows from the smaller donor to the larger recipient gametangium (Fig. 6.5a). Following nuclear fusion, the larger gametangium develops a thick wall and functions as a diploid resting spore. After a period of maturation the resting spore acts as a prosporangium, giving rise to a thin-walled meiosporangium. Meiosis, as evidenced by the presence of synaptonemal complexes, occurs here, followed by mitosis and cytoplasmic cleavage to form zoospores (Fig. 6.5b). A variant of this form of sexual differentiation (**gametangio-gametogamy**) has

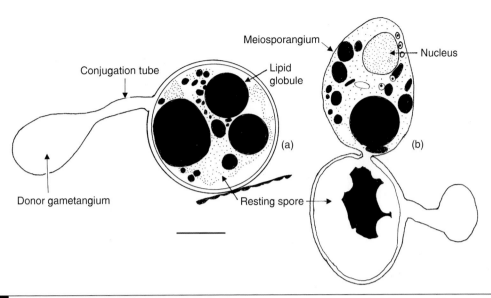

Fig 6.5 Sexual reproduction in *Zygorhizidium planktonicum*. (a) Empty donor gametangium to the left connected by a conjugation tube to a mature resting spore. (b) Near-median section of a fully formed meiosporangium which has developed from a germinating resting spore. The donor gametangium is on the right. Scale bar = 4 μm. After Doggett and Porter (1996).

been reported in species of *Rhizophydium* (Karling, 1977); this involves copulation between the gametangium of a rhizoid-forming thallus and a motile gamete that encysts directly on the gametangium.

Generally the product of sexual reproduction is a resting spore or resting sporangium with thick walls, but it is known that thick-walled sporangia may also develop asexually and in many chytrids sexual reproduction has not been described and possibly does not occur. Resting sporangia of some chytrids may remain viable for many years.

6.1.4 Classification and evolution

Fossil chytrids have been reported from the 400 million-year-old Rhynie chert, a site known for the discovery of fossil remains of the earliest known vascular land plants. Clusters of holocarpic, endobiotic thalli resembling the present day *Olpidium* have been found inside cells of a coenobial alga preserved within the hollow axes of a vascular plant, and epibiotic sporangia with endobiotic rhizoids have been seen attached to meiospores of a vascular plant, much like those of extant chytrids like

Rhizophydium which grow on pollen grains (Taylor *et al.*, 1992). Chytrid-like fossils have also been found in strata of the 340 million-year-old Pennsylvanian (Carboniferous) era (Millay & Taylor, 1978) and from the more recent Eocene strata (Bradley, 1967).

Formerly thought to have an affinity for the Oomycota, Hyphochytriomycota or protists, the Chytridiomycota are now accepted as members of the true fungi, the Eumycota. They are probably ancestral to other groups of true fungi, especially the Zygomycota (Cavalier-Smith, 1987, 2001; D. J. S. Barr, 2001). The inclusion of the chytrids in the Eumycota is supported by several DNA-based phylogenetic analyses (e.g. Bowman *et al.*, 1992; James *et al.*, 2000), but the delimitation of orders within the Chytridiomycota is still problematic. Particularly puzzling is the grouping of the Blastocladiales with the Zygomycota on the basis of 18S ribosomal DNA sequences (see Fig. 1.26).

D. J. S. Barr (2001) and Kirk *et al.* (2001) have classified the Chytridiomycota into five orders (Table 6.1) but the details of their distinguishing features need not concern us here. We shall study examples from each order.

Table 6.1. Orders of Chytridiomycota following D. J. S. Barr (2001) and Kirk *et al.* (2001).

Order	Number of described taxa	Examples
Chytridiales (see Section 6.2)	80 genera 600 spp.	*Cladochytrium, Nowakowskiella, Rhizophydium, Synchytrium*
Spizellomycetales (see Section 6.3)	13 genera 86 spp.	*Olpidium, Rhizophlyctis*
Neocallimastigales (see Section 6.4)	5 genera 16 spp.	*Anaeromyces, Caecomyces, Neocallimastix, Orpinomyces, Piromyces*
Blastocladiales (see Section 6.5)	14 genera 179 spp.	*Allomyces, Blastocladiella, Coelomomyces, Physoderma*
Monoblepharidales (see Section 6.6)	4 genera 19 spp.	*Gonapodya, Monoblepharella, Monoblepharis*

6.2 | Chytridiales

This is by far the largest order, comprising more than 50% of the total number of chytrids described to date. It is difficult to characterize members of the Chytridiales because they lack any specific features by which species have been assigned to the other four orders. The classification of the Chytridiales has traditionally been based on thallus morphology (Sparrow, 1973) but, as pointed out by D. J. S. Barr (2001), this is unsatisfactory because of the great variability in thallus organization shown by the same fungus growing on its natural substratum and in culture. Future systems of classification will be based on zoospore ultrastructure and the comparison of several different types of DNA sequences, but too few examples have yet been studied to provide a definitive framework. Because of this we shall study genera which illustrate the range of morphology, life cycles and ecology of the Chytridiales without attempting to place them into families.

6.2.1 *Synchytrium*

In this genus the thallus is endobiotic and holocarpic, and at reproduction it may become converted directly into a group (sorus) of sporangia, or to a **prosorus** which later gives rise to a sorus of sporangia. Alternatively the thallus may turn into a resting spore which can function either directly as a sporangium and give rise to zoospores, or as a prosorus. The zoospores are of the characteristic chytrid type (Lange & Olson, 1978). Sexual reproduction is by copulation of isogametes, resulting in the formation of thalli which develop into thick-walled resting spores. *Synchytrium* includes about 120 species which are biotrophic parasites of flowering plants. Some species parasitize only a narrow range of hosts, e.g. *S. endobioticum* on Solanaceae, but others, e.g. *S. macrosporum*, have a wide host range (Karling, 1964). Most species are not very destructive to the host plant but stimulate the formation of galls on leaves, stems and fruits.

Synchytrium endobioticum

This is the cause of wart disease affecting cultivated potatoes and some wild species of *Solanum*. It is a biotrophic pathogen which has not yet been successfully cultured outside living host cells. Wart disease is now distributed throughout the main potato-growing regions of the world, especially in mountainous areas and those with a cool, moist climate. Lange (1987) has given practical details of techniques for studying the fungus but in most European countries handling of living material by unlicensed workers is illegal. Diseased tubers bear dark brown cauliflower-like excrescences. Galls may also be formed on the aerial shoots, and they are then green with convoluted leaf-like masses of tissue (the leafy gall stage; Plates 3a,b). Heavily infected tubers may have a considerable proportion of their tissues converted to warts. The yield of saleable potatoes from a heavily infected crop may be less than the actual weight of the seed potatoes planted. The disease is thus potentially a serious one, but fortunately varieties of potatoes are available which are immune from the disease, so that control is practicable. The possible life cycle of *S. endobioticum* is summarized in Fig. 6.6.

The dark warts on the tubers are galls in which the host cells have been stimulated to divide by the presence of the fungus. Many of the host cells contain resting spores which are more or less spherical cells with thick dark brown walls folded into plate-like extensions (see Fig. 6.7a). The resting spores are released by the decay of the warts and may remain alive in the soil for over 40 years (Laidlaw, 1985). The outer wall (exospore) bursts open by an irregular aperture and the endospore balloons out to form a vesicle within which a single sporangium differentiates (Kole, 1965; Sharma & Cammack, 1976; Hampson *et al.*, 1994). Thus the resting spore functions as a **prosporangium** on germination. Germination of the resting spore may occur spontaneously but can be stimulated by passage through snails. It is presumed that abrasion and digestion of the spore wall

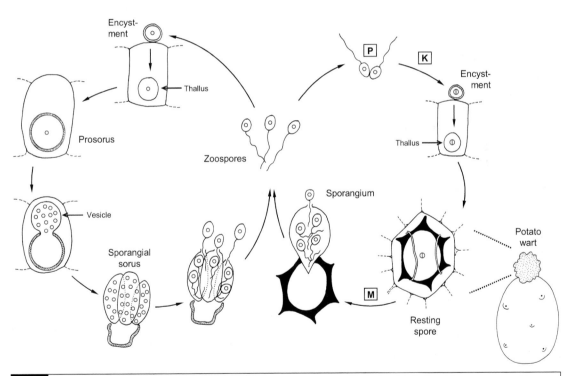

Fig 6.6 Schematic outline of the probable life cycle of *Synchytrium endobioticum*. Haploid and diploid nuclei are represented by small empty and larger split circles, respectively. Key events in the life cycle are plasmogamy (P), karyogamy (K) and meiosis (M). Resting spores within a warted potato contain a single nucleus which undergoes meiosis upon germination. Haploid zoospores are released from a single sporangium. If two zoospores pair up, a zygote is formed and penetration of a potato cell gives rise to a diploid thallus and, ultimately, a resting spore. Diploid infections cause host hyperplasia visible as the potato wart symptoms. If a zoospore infects in the haploid state, a haploid prosorus (summer spore) is formed, and hypertrophy of the infected and adjacent host cells ensues. A sorus of several sporangia is ultimately produced, with each sporangium releasing a fresh crop of haploid zoospores. *Synchytrium endobioticum* appears to be homothallic.

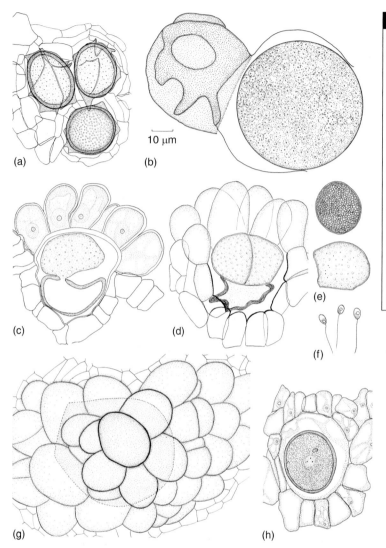

Fig 6.7 *Synchytrium endobioticum.*
(a) Resting spores in section of wart.
(b) Germinating resting spore showing the formation of a vesicle containing a single globose sporangium (after Kole, 1965). (c) Section of infected host cell containing a prosorus. The prosorus is extruding a vesicle. Note the hypertrophy of the infected cell and adjacent uninfected cells. (d) Cleavage of vesicle contents to form zoosporangia. (e) Two extruded zoosporangia. (f) Zoospores. (g) Rosette of hypertrophied potato cells as seen from the surface. The outline of the infected host cell is shown dotted. (h) Young resting sporangium resulting from infection by a zygote. Note that the infected cell lies beneath the epidermis due to division of the host cells.

10 µm

(a)   (b)   (c)   (d)   (e)   (f)   (g)   (h)

in the snail gut causes breakdown of the thick wall which contains chitin and branched-chain wax esters, so overcoming dormancy related to the impermeability of the wall (Hampson *et al.*, 1994).

The zoospores are capable of swimming for about two hours in the soil water. If they alight on the surface of a potato 'eye' or some other part of the potato shoot such as a stolon or a young tuber before its epidermis is suberized, they come to rest and withdraw their flagellum. During penetration, the contents of the zoospore cyst are transferred to the host cell whilst the cyst wall remains attached to the outside. When

a dormant 'eye' is infected, dormancy may be broken and the tuber may begin to sprout. If the potato variety is susceptible to the disease, the small fungal thallus inside the host cell will enlarge. The infected host cell as well as surrounding cells also enlarge so that a rosette of hypertrophied cells surrounds a central infected cell (Fig. 6.7c). The walls of these cells adjacent to the infected cell are often thickened and assume a dark brown colour. The infected cell remains alive for some time but eventually it dies. The pathogen thallus passes to the bottom of the host cell, enlarges and becomes spherical. A double-layered chitinous wall which is golden brown in colour is secreted around the thallus, now termed a prosorus or summer spore. Further development of the prosorus involves the protrusion of the inner wall through a pore in the outer wall, and its expansion as a vesicle which enlarges upwards and fills the upper half of the host cell (Fig. 6.7c). The cytoplasmic contents of the prosorus including the single nucleus are transferred to the vesicle. The process is quite rapid and can be completed in about 4 h. During its passage into the vesicle the nucleus may divide, and mitoses continue so that the vesicle contains about 32 nuclei. At this stage the cytoplasmic contents of the vesicle become cleaved into about 4−9 sporangia (Fig. 6.7d), forming a sorus. After the deposition of sporangial walls, further nuclear divisions occur in each sporangium, and finally each nucleus with its surrounding mass of cytoplasm becomes differentiated to form a zoospore. As the sporangia ripen, they absorb water and swell, causing the host cell containing them to burst open. Meanwhile, division of the host cells underlying the rosette has been taking place, and enlargement of these cells pushes the sporangia out onto the surface of the host tissue (Fig. 6.7e). The sporangia swell if water is available and burst open by means of a small slit through which the zoospores escape. There may be as many as 500−600 zoospores in a single large sporangium. The zoospores resemble those derived from resting sporangia and are capable of initiating further asexual cycles of reproduction throughout spring and early summer. Sometimes several zoospores succeed in penetrating a single cell so that it contains several fungal protoplasts.

Alternatively, zoospores may function as gametes, fusing in pairs (or occasionally in groups of three or four) to form zygotes which retain their flagella and swim actively for a time. Since zoospores acting as gametes do not differ in size and shape, copulation can be described as isogamous. However, the gametes may differ physiologically. Curtis (1921) has suggested that fusion may not occur between zoospores derived from a single sporangium, but only between zoospores from separate sporangia. Köhler (1956) has claimed that the zoospores are at first sexually neutral. Later they mature and become capable of copulation. Maturation may occur either outside the sporangia or within, so that in over-ripe sporangia the zoospores are capable of copulation on release. At first the zoospores are 'male', and swim actively. Later the swarmers become quiescent ('female') and probably secrete a substance which attracts 'male' gametes. After swimming by means of its two flagella, the zygote encysts on the surface of the host epidermis and penetration may then follow by a process essentially similar to zoospore penetration. Multiple infections by several zygotes penetrating a single host cell can also occur. Nuclear fusion occurs in the young zygote before penetration.

The results of zygote infections differ from infection by zoospores. The host cell reacts to zoospore infection by undergoing hypertrophy, i.e. increase in cell volume, and adjacent cells also enlarge to form the characteristic rosette which surrounds the resulting prosorus. In contrast, when a zygote infects, the host cell undergoes hyperplasia, i.e. repeated cell division. The pathogen lies towards the bottom of the host cell, adjacent to the host nucleus, and cell division occurs in such a way that the fungal protoplast is located in the innermost daughter cell. As a result of repeated divisions of the host cells, the typical gall-like potato warts are formed and fungal protoplasts may be buried several cell layers deep beneath the epidermis (see Fig. 6.7h). During these divisions of the host tissue the zygote enlarges and becomes surrounded by a two-layered wall, a thick outer layer which eventually becomes dark brown in colour and is thrown into folds or ridges which appear as spines in section, and a thin hyaline inner wall surrounding the granular cytoplasm (Lange & Olson, 1981). The host cell eventually dies and some of its contents are deposited on the outer wall of the resting sporangium, forming the characteristic brown ridges. During its development the resting spore remains uninucleate. Resting spores are released into the soil and are capable of germination within about 2 months. Before germination, the nucleus divides repeatedly to form the nuclei of the zoospores whose further development has already been described. It has been claimed that the zygote and the young resting spore are diploid, and it has been assumed that meiosis occurs during germination of the resting sporangia prior to the formation of zoospores, so that these zoospores, the prosori and the soral zoospores are also believed to be haploid. These assumptions seem plausible in the light of knowledge of the life history and cytology of other species (e.g. Lingappa, 1958b), and an essentially similar life cycle has been described for *S. lagenariae* and *S. trichosanthidis*, parasitic on Cucurbitaceae, which differ from *S. endobioticum* in that their resting spores function as prosori instead of prosporangia (Raghavendra Rao & Pavgi, 1993).

Control of wart disease

Control is based largely on the breeding of resistant varieties of potato. It was discovered that certain varieties such as Snowdrop were immune from the disease and could be planted on land heavily infected with *Synchytrium* without developing warts. Following this discovery, plant breeders have developed a number of immune varieties such as Maris Piper. However, some potato varieties that are susceptible to the disease are still widely grown, including the popular King Edward. In most countries where wart disease occurs, legislation has been introduced requiring that only approved immune varieties be planted on land where wart disease has been known to occur, and prohibiting the movement and sale of diseased material. Within the British Isles, the growing of immune varieties on infested land has prevented the spread of the disease, and it is now confined to a small number of foci in the West Midlands, northwest England and mid and south Scotland. It has also persisted in Newfoundland. The majority of the outbreaks are found in allotments, gardens and smallholdings.

The reaction of immune varieties to infection varies (Noble & Glynne, 1970). In some cases when 'immune' varieties are exposed to a heavy inoculum load of *S. endobioticum* in the laboratory, they may become slightly infected, but infection is often confined to the superficial tissues which are soon sloughed off. In the field such slight infections would probably pass unnoticed. Occasionally infections of certain potato varieties may result in the formation of resting spores, but without the formation of noticeable galls. Penetration of the parasite seems to occur in all potato varieties, but when a cell of an immune variety is penetrated it may die within a few hours, and since the fungus is a biotrophic parasite, further development is checked. In other cases the parasite may persist in the host cell for up to 2–3 days, apparently showing normal development, but after this time the fungal thallus undergoes disorganization and disappears from the host cell.

Unfortunately, it has been discovered that new physiological races (or pathotypes) of the pathogen have arisen, capable of attacking potato varieties previously thought to be immune. About 20 pathotypes are now known, and the implications are obvious. Unless their spread can be prevented, much of the work of potato plant breeders over the past century will have to be started all over again.

Other methods of control are less satisfactory. Attempts to kill the resting spores of the fungus in the soil have been made, but this is a costly and difficult process, requiring large-scale fungicide applications to the soil. Copper sulphate or ammonium thiocyanate have been applied in the past at amounts of up to 1 ton acre^{-1}, and local treatment with mercuric chloride or with formaldehyde and steam has been used to eradicate foci of infection (Hampson, 1988). Control measures based on the use of resistant varieties seem more satisfactory. An interesting method of control developed in Newfoundland is the use of crabshell meal placed above seed tubers at the time of planting. This technique has resulted in significant and sometimes complete control (Hampson & Coombes, 1991) which may be due to selective enhancement of chitinolytic soil micro-organisms degrading the chitinous walls of the resting spores of *S. endobioticum*.

Other species of *Synchytrium*

Not all species of *Synchytrium* show the same kind of life cycle as *S. endobioticum*. *Synchytrium fulgens*, a parasite of *Oenothera*, resembles *S. endobioticum* in that both summer spores and resting spores are formed (Lingappa, 1958a,b), but in this species the zoospores from resting sporangia can also function as gametes and give rise directly to zygote infections from which further resting spores arise (Lingappa, 1958b). It has been suggested that the same phenomenon may occasionally occur in *S. endobioticum*. In *S. taraxaci* parasitic on *Taraxacum* (Fig. 6.8; Plate 3c), as well as a number of other *Synchytrium* spp., the mature thallus does not function as a prosorus but cleaves directly to form a sorus of sporangia, and the resting spore also gives rise to zoospores directly. In some species, e.g. *S. aecidioides*, resting sporangia are unknown, whilst in others, e.g. *S. mercurialis*, a common parasite on leaves and stems of *Mercurialis perennis* (Fig. 6.9), only resting sporangia are known and summer sporangial sori do not occur. *Mercurialis* plants collected from March to June often show yellowish blisters on leaves and stems. The blisters are galls made up of one or two layers of hypertrophied cells mostly lacking chlorophyll, surrounding the *Synchytrium* thallus during its maturation to form a resting sporangium. In this species the resting sporangium functions as a prosorus during the following spring. The undivided contents are extruded into a spherical sac which becomes cleaved into a sorus containing as many as 120 sporangia from which zoospores arise. The variations in the life histories of the various species of *Synchytrium* form a useful basis for classifying the genus (Karling, 1964).

6.2.2 *Rhizophydium*

Rhizophydium is a large, cosmopolitan genus of about 100 species (Sparrow, 1960) which grow in soil, freshwater and the sea. The thallus is eucarpic, with a globose epibiotic zoosporangium which develops from the zoospore cyst, and endobiotic rhizoids which penetrate the host. Whilst some species are saprotrophic, others are biotrophic pathogens of algae and can cause severe epidemics of freshwater phytoplankton. Saprotrophic forms such as *R. pollinispini* and *R. sphaerocarpon* colonize pollen grains and are easily isolated by sprinkling pollen onto the surface of water overlying soil (Fig. 6.10). Within 3 days, sporangia with exit papillae are

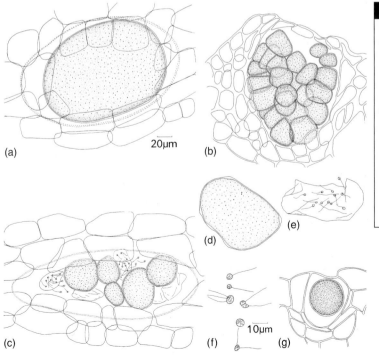

(a)

20μm

(b)

(c)

(d)

(e)

(f)

10μm

(g)

Fig 6.8 *Synchytrium taraxaci.*
(a) Undivided thallus in epidermal cell of scape of *Taraxacum*. Outline of host cell shown dotted. (b) Section of *Taraxacum* scape showing thallus divided into a sorus of sporangia. (c) A sorus of sporangia seen from above. Two sporangia are releasing zoospores. (d) A ripe sporangium. (e) Sporangium releasing zoospores. (f) Zoospores and zygotes. The triflagellate zoospore probably arose by incomplete separation of zoospore initials. (g) Section of host leaf showing a resting sporangium. (a—e) and (g) to same scale.

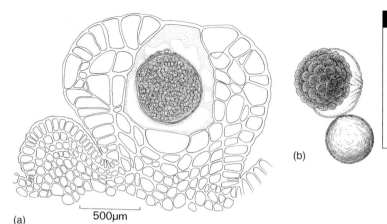

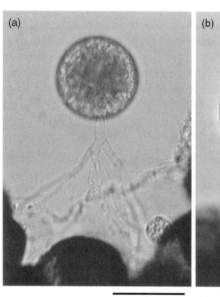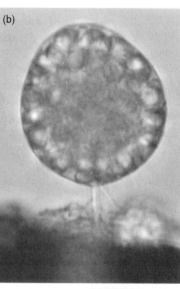

500µm

(a)

(b)

Fig 6.9 *Synchytrium mercurialis.* (a) Section of stem of *Mercurialis perennis* showing hypertrophied cells surrounding a resting sporangium. (b) Germination of a resting sporangium to release a sorus of zoosporangia. Thus in *S. mercurialis* the resting sporangium functions as a prosorus (after Fischer, 1892).

(a)

(b)

20 µm

10 µm

Fig 6.10 Pine pollen grains colonized by *Rhizophydium* sp. (a) The rhizoid system attaching the epibiotic sporangium to the colonized pollen grain. (b) Mature sporangium; the cytoplasm has become cleaved into numerous zoospores.

found in crude cultures on pine pollen. The zoospores are at first released into a hyaline vesicle which soon dissolves, allowing them to swim away. Gauriloff and Fuller (1987) have outlined techniques for growing *R. sphaerocarpon* in pure culture. This species can also grow parasitically on filaments of the green alga *Spirogyra*.

Douglas Lake (Michigan, USA) is surrounded by conifers shedding pollen which floats on the lake and becomes colonized by *Rhizophydium* spp. Using the MPN (most probable number) technique, Ulken and Sparrow (1968) have estimated that the number of chytrid propagules in the surface waters (epilimnion) can rise to over 9001^{-1} by late June. Some infected pollen grains sink through the hypolimnion to the mud at the floor of the lake. It is thought likely that these develop resting sporangia which survive the winter and provide inoculum to start off colonization of new pollen deposits in the following season.

Rhizophydium planktonicum

This species is the best-studied chytrid phytoplankton parasite. It is a biotrophic pathogen of the diatom *Asterionella formosa*, an inhabitant of eutrophic lakes. This alga forms cartwheel-like colonies, the diatom frustules making up the spokes, cemented together by mucilage pads at the hub of the wheel. *Rhizophydium planktonicum* may form one to many thalli on each host cell (Fig. 6.11a). Dual cultures of the host and parasite have been established (Canter & Jaworski, 1978) and from such cultures a detailed picture of infection, development and zoospore structure has been built up (Beakes *et al.*, 1993). Zoospores are attracted to the alga and encyst on it, forming monocentric rhizoidal thalli. The rhizoids penetrate between the girdle lamellae of the host (Fig. 6.11b). The rhizoids may extend throughout the whole length of the host cell and infection is often accompanied by loss of photosynthetic pigment, failure of cells to divide, and ultimately early death of the host cell. The zoospore is uninucleate and the nucleus is retained within the zoospore cyst, the rhizoids being devoid of nuclei. The zoospore cyst enlarges to form the sporangium. Synchronous nuclear divisions result in the formation of several nuclei lying within the cytoplasm, followed by the development of cleavage furrows which divide up the sporangial contents into zoospores. A septum develops at the base of the sporangium and, prior to cleavage, the upper part of the sporangium wall develops a thickened apical papilla which balloons out to form a vesicle into which the immobile zoospores are released. The complete cycle from infection to zoospore release depends on temperature and can be as short as 2–3 days. About 1–30 zoospores may be formed in a sporangium depending on the state of the host cells, in turn affected by external physical and chemical conditions. Breakdown of the vesicle allows the zoospores to swim away. No resting stage has been described for *R. planktonicum*.

A striking feature of the zoospore ultrastructure is the presence of several paracrystalline bodies near the nucleus in the peripheral part of the cytoplasm (Beakes *et al.*, 1993). They consist of parallel arrays of regularly arranged crystals interconnected to each other with fibrous material. They appear late in sporangial development but disappear following encystment of zoospores. Similar structures have been reported from the zoospores of a few other Chytridiomycota, but their composition and function are unknown.

There have been several studies on the ecology of *Asterionella* subjected to parasitism by *R. planktonicum* (see Canter & Lund, 1948, 1953;

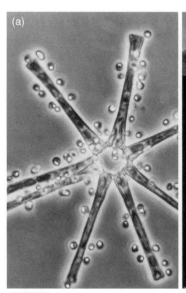

5 μm

1 μm

Fig 6.11 *Rhizophydium planktonicum* growing parasitically on the frustules of the colonial diatom *Asterionella formosa*. (a) Heavily infected colony from a dual-clone culture showing encysted zoospores. (b) Scanning electron micrograph of *Asterionella* cells showing heavy infection and zoospore cysts which have germinated and penetrated the host cells via the girdle lamellae. From Beakes *et al.* (1993), with permission from Elsevier; original images kindly provided by G.W. Beakes.

Canter & Jaworski, 1981; van Donk & Bruning, 1992). *Asterionella* is also parasitized by two other chytrids, *Zygorhizidium planktonicum* and *Z. affluens*, and some of the early studies in fresh-water lakes may well have included a mixture of species.

Studies on the epidemiology of infection of *Asterionella* by *R. planktonicum* in lakes have shown that there are peak periods of *Asterionella* population density both in spring and in autumn, related to the availability of dissolved nutrients, water temperature, thermal stratification and its breakdown, daylength and light intensity. *Asterionella* cells infected with *Rhizophydium* can occur throughout the year, but epidemics in which a high proportion of cells are infected only occur at concentrations of around 10 host cells ml^{-1} (Holfeld, 1998). Interpretation of the conditions conducive to the occurrence of epidemics has been aided by experiments using dual cultures of pathogen and host in which effects such as light intensity, temperature and phosphorus concentration have been varied (van Donk & Bruning, 1992). The effects of light are complex. Although *Rhizophydium* zoospores are not photo-tropic, they are quiescent and incapable of infection in the dark or at low light intensity. Experiments by Canter and Jaworski (1981) have indicated that a light intensity below 200 lx is inadequate for zoospore settlement on host cells. In light-limited cultures of *Asterionella*, the sporangia of the pathogen and hence the number of zoospores produced are smaller than when light is not limiting (Bruning, 1991a). Similarly, zoospore production is also reduced when the concentration of phosphorus limits growth of the host (Bruning, 1991b). Temperature affects the rate of sporangium development and the size of sporangia, with maximum dimensions at 2°C at fairly high light intensities (Bruning, 1991a). It also affects the duration of swimming of zoospores and therefore their infective lifetime which can vary from about 10 days at 3°C to only 2 days at 20°C. Epidemic development may result from a combination of factors and there is a remarkable interaction between the effects of light intensity and temperature (Bruning, 1991c). At higher temperatures, optimal conditions for epidemic development occur at high light intensities, but at temperatures below 5–6°C epidemic development is encouraged by lower light intensities. This may explain why, in nature, epidemics can occur both in summer (high light intensity, high temperature) and winter (low light intensity, low temperature).

Rhizophydium planktonicum is a specialized parasite infecting only *Asterionella*. It is more compatible with certain clones of host cells than others, and cells from incompatible clones show hypersensitivity, undergoing rapid death following infection (Canter & Jaworski, 1979).

6.2.3 *Cladochytrium*

There are about a dozen species of *Cladochytrium* (Sparrow, 1960) which are widespread sapro-trophs, mostly of aquatic plant debris. The thallus is eucarpic and polycentric and the vegetative system may bear intercalary swellings and septate **turbinate cells** (sometimes termed **spindle organs**). The sporangia are inoperculate. *Cladochytrium replicatum* is a common representative in decaying pieces of aquatic vegetation and can be distinguished from other chytrids by the bright orange lipid droplets found in the sporangia. It is frequently isolated if moribund aquatic vegetation is placed in a dish of water and baited with boiled grass leaves or cellulosic materials such as dialysis tubing. Lucarotti (1987) has given details of its isolation and growth in culture. The bright orange sporangia which are visible under a dissecting microscope appear on baits within about 5 days, arising from an extensively branched hyaline rhizomycelium bearing two-celled intercalary swellings. Sporangium development is encouraged by exposure to light. On release from the sporangium, the zoospores each contain a single orange lipid droplet and bear a single posterior flagellum. Lucarotti (1981) has described the fine structure of the zoospore. After swimming for a short time, the zoospore attaches itself to the surface of the substratum and puts out usually a single germ tube which can penetrate the tissues of the host plant. The germ tube expands to form an elliptical or cylindrical turbinate cell which is often later divided into two by a transverse septum (Fig. 6.12d). The zoospore is uninucleate and during germination the single nucleus is transferred to the swollen turbinate cell which becomes a vegetative centre from which rhizoids are put out which in turn produce further turbinate cells (see Figs. 6.12b,d). Nuclear division is apparently confined to the turbinate cells, and although nuclei are transported through the rhizoidal system they are not resident there. The thallus so established branches profusely, and at certain points spherical zoosporangia form, either terminally or in intercalary positions.

Sometimes one of the cells of a pair of turbinate cells swells and becomes transformed into a sporangium. In culture, both cells may be modified in this way. The spherical to pear-shaped zoosporangium undergoes progressive nuclear division, and the contents of the sporangium acquire a bright orange colour due to accumulation of lipid droplets containing the carotenoid lycopene. These lipid reserves are later found in the zoospores. Cleavage of the cytoplasm to form uninucleate zoospore initials follows. The zoospores escape through a narrow

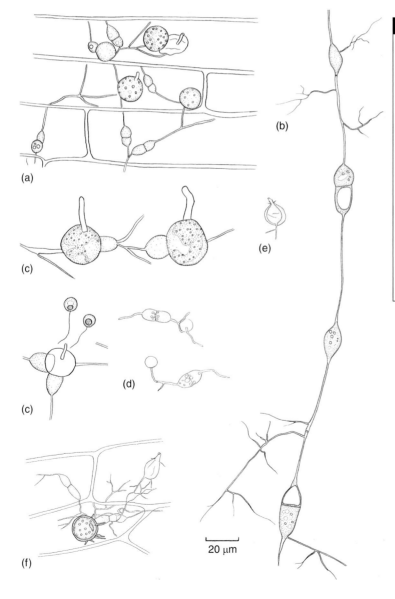

(a)

(b)

(c)

(c)

(d)

(e)

(f)

20 μm

Fig 6.12 *Cladochytrium replicatum.*
(a) Rhizomycelium within the epidermis of an aquatic plant bearing the two-celled hyaline turbinate cells and globose orange zoosporangia. (b) Rhizomycelium and turbinate cells from a culture. (c) Zoosporangia from a two-week-old culture. One zoosporangium has released zoospores, each of which contains a bright orange-coloured globule. (d) Germinating zoospores on boiled wheat leaves. The empty zoospore cysts are spherical. The germ tubes have expanded to form turbinate cells. (e) A zoosporangium which has proliferated internally to form a second sporangium. (f) Rhizomycelium within a boiled wheat leaf bearing a thick-walled, spiny resting sporangium.

exit tube which penetrates to the exterior of the substratum and becomes mucilaginous at the tip. There is no operculum. Sometimes zoosporangia may proliferate internally, a new zoosporangium being formed inside the wall of an empty one. Resting sporangia with thicker walls and a more hyaline cytoplasm are also formed either terminally or in an intercalary position on the rhizomycelium. In some cases the wall of the resting sporangium is reported to be smooth and in others spiny, and it has been suggested (Sparrow, 1960) that the two kinds of resting sporangia may belong to different species. However, studies by Willoughby (1962) of a number of single-spore isolates have shown that the presence or absence of spines is a variable character. The contents of the resting sporangia divide to form zoospores which also have a conspicuous orange droplet, and escape by means of an exit tube as in the thin-walled zoosporangia. Whether the resting sporangia are formed as a result of a sexual process is not known. Pure cultures of *C. replicatum* have been studied by Willoughby (1962), Goldstein (1960) and Lucarotti (1981). The fungus is heterotrophic for thiamine. Biotin, while not absolutely required, stimulates growth. Nitrate and sulphate are utilized, as are a number of different carbohydrates; a limited amount of growth takes place on cellulose.

6.2.4 Nowakowskiella

Species of *Nowakowskiella* are widespread saprotrophs in soil and on decaying aquatic plant debris, and can be obtained by baiting aquatic plant remains in water with boiled grass leaves, cellophane, dialysis tubing and the like. *Nowakowskiella elegans* is often encountered in such material, and pure cultures can be obtained and grown on cellulosic materials overlying agar, or directly in liquid culture media (Emerson, 1958; Johnson, 1977; Lucarotti, 1981; Lucarotti & Wilson, 1987). In culture, considerable variation in growth habit and morphology can result from changing the concentration of nutrients and the availability of water (Johnson, 1977). In boiled grass leaves the fungus forms an extensive rhizomycelium with turbinate cells

(Fig. 6.13c). Zoosporangia are formed terminally or in an intercalary position (Fig. 6.13c) and are globose or pear-shaped with a subsporangial swelling (apophysis), and granular or refractile hyaline contents. At maturity some sporangia develop a prominent beak, but in others this is not present. When an operculum becomes detached, zoospores escape and initially remain clumped together at the mouth of the sporangium (Figs. 6.13b,c). The fine structure of the zoospore is very similar to that of *Rhizophydium* but paracrystalline bodies have not been observed (Lucarotti, 1981). It also has close resemblance to the zoospore ultrastructure of the inoperculate, polycentric *Cladochytrium replicatum*.

Yellowish resting sporangia (Fig. 6.13e) have been described (Emerson, 1958; Johnson, 1977; Lucarotti & Wilson, 1987). They develop as spherical to fusiform swellings in the rhizomycelium which become delimited by septa, develop thick walls and a large central vacuole surrounded by dense cytoplasm with small spherical lipid droplets. The resting sporangium is at first binucleate. After nuclear fusion the diploid nucleus divides meiotically. Further nuclear divisions are mitotic and the contents of the resting sporangium cleave into zoospores which may be released through a papilla in the sporangium wall. Alternatively, the resting sporangium may give rise to a thin-walled zoosporangium from which the zoospores are released, i.e. the resting sporangium may function as a prosporangium as in some other chytrids (Johnson, 1977).

In *N. profusa*, which is probably synonymous with *N. elegans* (Johnson, 1977), three kinds of sporangial dehiscence have been described: exo-operculate, in which the operculum breaks away to the outside of the sporangium; endo-operculate, in which the operculum remains within the sporangium; and inoperculate, where the exit papilla opens without any clearly defined operculum (Chambers *et al.*, 1967; Johnson, 1973). Such variations within a single chytrid strain add emphasis to criticisms of the value of dehiscence as a primary criterion in classification.

Goldstein (1961) has reported that *N. elegans* requires thiamine and can utilize nitrate,

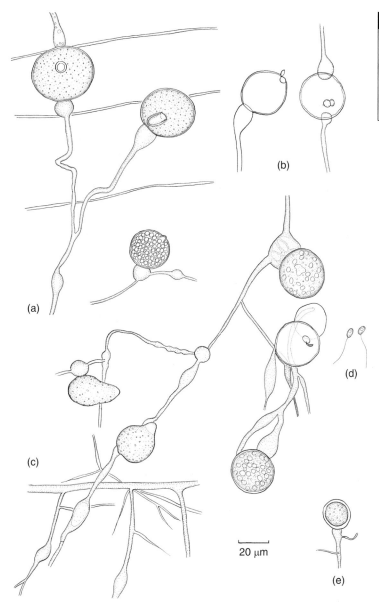

Fig 6.13 *Nowakowskiella elegans*. (a) Polycentric mycelium bearing zoosporangia. (b) Empty zoosporangia showing opercula. (c) Mycelium showing turbinate cells and zoosporangia. (d) Zoospores from culture. (e) Resting sporangium from culture.

sulphate and a number of carbohydrates including cellulose, but cannot utilize starch.

6.3 | Spizellomycetales

Members of this order differ from the Chytridiales in possessing zoospores which contain more than one lipid droplet and are capable of limited amoeboid movement. Thalli are generally monocentric. The order takes its name from the genus *Spizellomyces* which in turn was named in honour of the chytrid pioneer F.K. Sparrow after *Spizella*, a genus of North American sparrows (Barr, 1980). Some 86 species of Spizellomycetales are currently recognized.

6.3.1 *Olpidium*

About 30 species of *Olpidium* are known, but the genus is in need of revision and possibly some of the species should be classified else-where. Typical species are holocarpic. Some are parasitic on fungi and aquatic plants or algae, or saprotrophic on pollen (Sparrow, 1960). Others parasitize rotifers (Glockling, 1998), nematodes and their eggs (Tribe, 1977; Barron & Szijarto, 1986), moss protonemata or leaves and roots of higher plants (Macfarlane, 1968; Johnson, 1969). *Olpidium bornovanus* (= *O. radicale*) develops on various monocotyledonous and dicotyledonous plant roots following inoculation (Lange & Insunza, 1977). *Olpidium brassicae* is common on the roots of cabbages, especially when growing in wet soils, and is also found on a wide range of unrelated hosts, but some host specialization has been reported. Both *O. bornovanus* and *O. brassicae* are vectors of a number of plant viruses (Barr, 1988; Adams, 1991; Hiruki, 1994; Campbell, 1996) and this topic is discussed more fully below. Weber and Webster (2000a) have given practical details of how to grow *O. brassicae* for observation on *Brassica* seedlings. A film featuring *O. brassicae* is also available (Webster, 2006a).

Epidermal cells and root hairs of infected cabbage roots contain one or more spherical or cylindrical thalli, sometimes filling the whole cell (Fig. 6.14a). The cytoplasm of the thallus is granular and the entire contents divide into numerous posteriorly uniflagellate zoospores that escape through one or more discharge tubes which penetrate the outer wall of the host cell (Temmink & Campbell, 1968). Release of the zoospores takes place within a few minutes of washing the roots free from soil. The tip of the discharge tube breaks down and zoospores rush out and swim actively in the water. The zoospores are very small, tadpole-like, with

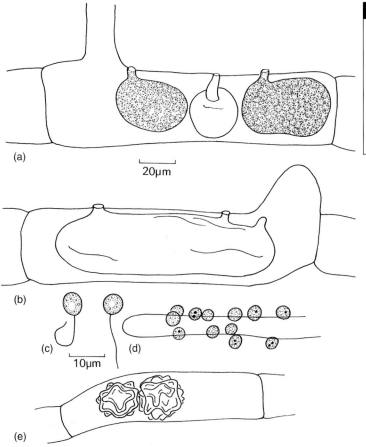

Fig 6.14 *Olpidium brassicae* in cabbage roots. (a) Two ripe sporangia and one empty sporangium in an epidermal cell. Each sporangium has a single exit tube. (b) Empty sporangium showing three exit tubes. (c) Zoospores. (d) Zoospore cysts on a root hair. Note that some cysts are uninucleate and some are binucleate. (e) Resting sporangia. (a,b,d,e) to same scale.

20µm

10µm

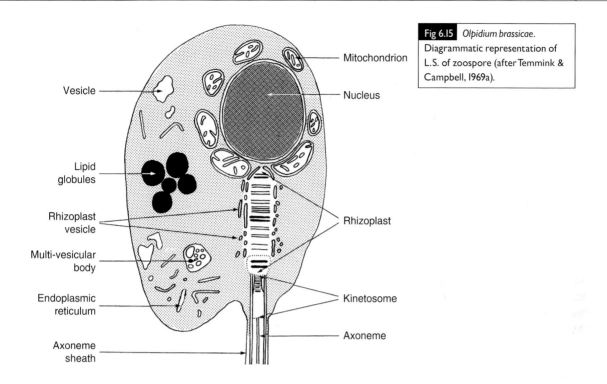

Fig 6.15 *Olpidium brassicae.* Diagrammatic representation of L.S. of zoospore (after Temmink & Campbell, 1969a).

a spherical head and a long trailing flagellum. The fine structure of the zoospore is summarized in Fig 6.15. A distinctive feature is the banded rhizoplast which connects the kinetosome to the nucleus (Temmink & Campbell, 1969a; Lange & Olson, 1976a,b; Barr & Hartmann, 1977). This structure has also been reported from the zoospore of the eucarpic chytrid *Rhizophlyctis rosea* (see p. 148; Barr & Hartmann, 1977).

The zoospores swim actively in water for about 20 min. If roots of cabbage seedlings are placed in a suspension of zoospores, these settle on the root hairs and epidermal cells, withdraw their flagella and encyst. The cysts are attached by a slime-like adhesive (Temmink & Campbell, 1969b). The cyst wall and the root cell wall at the point of attachment are dissolved and the root cell is penetrated. The cyst contents are transferred to the inside of the host cell, probably by the enlargement of a vacuole which develops inside the cyst, whilst the empty cyst remains attached to the outside. The process of penetration can take place in less than one hour (Aist & Israel, 1977). Within 2 days of infection, small spherical thalli can be seen in the root hairs and epidermal cells of the root, carried around the cell by cytoplasmic streaming. The thalli enlarge and become multinucleate. Within 4–5 days discharge tubes develop and the thalli are ready to release zoospores.

In some infected roots, stellate bodies with thick folded walls, lacking discharge tubes, are also found (Fig. 6.14e). These are resting sporangia. There is no evidence that they are formed as a result of sexual fusion either in *O. brassicae* or in *O. bornovanus* (Barr, 1988). Although biflagellate zoospores may occur in *O. brassicae*, these possibly result from incomplete cleavage (Temmink & Campbell, 1968) and zoospores with as many as 6 flagella have been observed (Garrett & Tomlinson, 1967). The resting sporangia are capable of germination 7–10 days after they mature, and germinate by the formation of one or two exit papillae through which the zoospores escape.

Virus transmission by *Olpidium*

Several plant viruses are transmitted by zoospores of *Olpidium*. By analogy with plant virus transmission by aphids, Adams (1991) arbitrarily distinguished viruses with non-persistent and persistent transmission by fungi, although Campbell (1996) objected to the use of these terms, distinguishing instead between viruses which can be acquired in vitro (i.e. outside the plant) and those that can only be acquired in vivo (within the host cell).

Tobacco necrosis virus (TNV) and cucumber necrosis virus (CNV) are non-persistent viruses which can be acquired in vitro by zoospores of *O. brassicae* or *O. bornovanus* (respectively). Virus particles (virions) are adsorbed onto the plasmalemma of the zoospore and onto the flagellar axonemal sheath which is continuous with it (Temmink *et al.*, 1970). Binding seems to occur between the virus coat and specific molecules at the zoospore surface, possibly oligosaccharide side chains of proteins (Kakani *et al.*, 2003; Rochon *et al.*, 2004). When the flagellum is withdrawn into the body of the zoospore at encystment, virus particles are introduced into the fungal cytoplasm and are then transmitted into the plant upon infection. Air-dried roots containing TNV virus and *O. brassicae* resting sporangia, or living virus-infected roots with resting sporangia treated with 5N HCl, were incapable of transmitting virus even though the resting sporangia survived these treatments, indicating that TNV is not carried inside the resting sporangia (Campbell & Fry, 1966).

Lettuce big vein virus, LBVV, in contrast, is an example of the persistent type (Grogan *et al.*, 1958). In this case it has been shown that the virus can persist in air-dried resting sporangia for 18–20 years (Campbell, 1985). Here the virions are acquired in vivo and they are present inside the zoospores which emerge from sporangia and resting sporangia (Campbell, 1996).

Classification of *Olpidium*

Although previously classified within the family Olpidiaceae in the order Chytridiales, D. J. S. Barr (2001) has placed *Olpidium* in the order Spizellomycetales along with *Rhizophlyctis* on the basis of similarities in zoospore structure. Ribosomal DNA sequence comparisons are inconclusive in that they do not show any close

similarity between *Olpidium* and either *Chytridium* or *Spizellomyces* (Ward & Adams, 1998).

6.3.2 *Rhizophlyctis*

There are about 10 known species of *Rhizophlyctis* with monocentric eucarpic thalli, growing as saprotrophs on a variety of substrata in soil, freshwater and the sea. *Rhizophlyctis rosea* grows on cellulose-rich substrata in soil, and it probably plays an active but currently underestimated role in cellulose decay (Powell, 1993). It can survive for prolonged periods in dry soil, even when this is heated to 90°C for two days (Gleason *et al.*, 2004) and, in fact, the recovery of *R. rosea* is greatly enhanced if soil samples are air-dried prior to isolation experiments (Willoughby, 2001). Willoughby (1998b) has estimated that over 1000 thallus-forming units could be recovered per gram of air-dry soil or leaf humus fragments from Provence, France. These numbers may arise from one or a few sporangia, since a single sporangium about 100 μm in diameter may discharge up to 30 000 zoospores. Mitchell and Deacon (1986) have shown that zoospores of *R. rosea* accumulate preferentially on cellulosic materials.

The fungus is readily isolated and grown in culture, and details of techniques have been provided by Stanier (1942), Barr (1987), Willoughby (1998b) and Weber and Webster (2000a). The placing of a small crumb of soil onto moist tissue paper or cellophane overlying agar containing mineral salts, or the floating of squares of cellophane on water containing a soil sample, are followed within a few days by the development of thalli with bright pink sporangia. The sporangia are attached to coarse rhizoids which arise at several points on the sporangial wall and extend throughout the cellulosic substratum, tapering to fine points. Extensive corrosion of the substrate underneath the thallus and rhizoids points at the secretion of powerful cellulases (Fig. 6.17).

Although the fungus is usually monocentric, there are also records of some polycentric isolates. When ripe, the sporangia have pink granular contents which differentiate into numerous uninucleate posteriorly uniflagellate zoospores (Fig. 6.16a). One to several discharge tubes are formed, and the tip of each tube contains a clear mucilaginous plug which, prior to discharge, is exuded in a mass from the tip of the tube (Fig. 6.16c). While the plug of mucilage dissolves, the zoospores within the sporangium show active movement and then escape by swimming through the tube. In some specimens of *R. rosea* it has been found that a membrane may form over the cytoplasm at the base of the discharge tubes. If the sporangia do not discharge their spores immediately, the membrane may thicken. When spore discharge occurs, these thickened membranes can be seen floating free within the sporangia, and the term endo-operculum has been applied to them. The genus *Karlingia* was erected for forms possessing such endo-opercula, including *R. rosea*, which is therefore sometimes referred to as *Karlingia rosea*, but the validity of this separation is questionable because the presence or absence

of endo-opercula is a variable character (Blackwell & Powell, 1999).

Zoospores of *R. rosea* are capable of swimming for several hours. The head of the zoospore is often globose, but can become pear-shaped or show amoeboid changes in shape. It contains a prominent lipid body, several bright refringent globules, and bears a single trailing flagellum. Ultrastructural details resemble those of *Olpidium brassicae* in the presence of a striated rhizoplast connecting kinetosome and nucleus (Barr & Hartmann, 1977). On coming to rest on a suitable substratum, the flagellum is withdrawn and the body of the zoospore enlarges to form the rudiment of the sporangium, whilst rhizoids appear at various points on its surface. Within the sporangium, the flagella are tightly wrapped around the zoospores (Chambers & Willoughby, 1964).

Resting sporangia are also found. They are brown, globose or angular and have a thickened

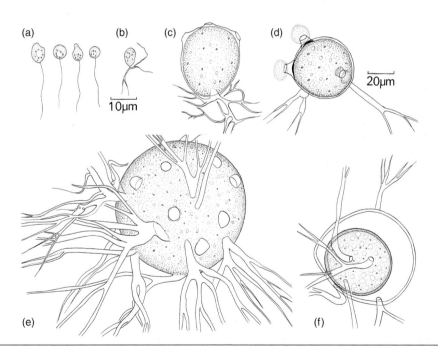

Fig 6.16 *Rhizophlyctis rosea*. (a) Zoospores. (b) Young thallus formed on germination of zoospore. The zoospore cyst has enlarged and will form the sporangium. (c) Older sporangium showing three discharge tubes. (d) Sporangium showing mucilage plugs at the tips of the discharge tubes and thickenings of the cell membrane at the bases of the tubes. Such thickenings are termed endo-opercula. (e) Globose sporangium and seven visible papillae. (f) Resting sporangium formed inside an empty zoosporangium. (a,b) to same scale; (c–f) to same scale.

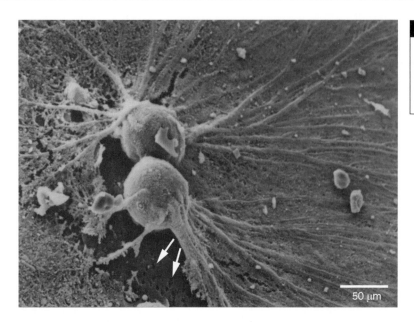

Fig 6.17 Scanning electron micrograph of two thalli of *Rhizophlyctis rosea* on a cellophane membrane. Pit corrosion is visible where a thallus has been lifted from the substratum (arrows).

50 μm

wall (Fig. 6.16f). Whether they are formed sexually in *R. rosea* is not known. Couch (1939) has, however, put forward evidence that the fungus is heterothallic because single isolates grown in culture failed to produce resting sporangia whereas these structures did form when certain cultures were paired. Stanier (1942) has reported the occurrence of biflagellate zoospores, but whether these represented zygotes seemed doubtful. In the homothallic chitinophilic fungus *Rhizophlyctis oceanis*, Karling (1969) has described frequent fusions between zoospores. These fusions are possibly sexual, but unfortunately Karling was unable to cultivate the resulting thalli to the stage of resting spore development.

On germination, the resting sporangium of *R. rosea* functions as a prosporangium, although it is uncertain whether resting sporangia are important for survival in nature. Willoughby (2001) has shown that *R. rosea* could be recovered from cellophane baits in as little as 5–6 h after placing air-dried soil samples in water, and it was concluded that these zoospores were derived from sporangia instead of resting spores which need a longer time to produce zoospores.

The nutritional requirements of *R. rosea* are simple. It shows vigorous growth on cellulose as the sole carbon source but it can utilize a range of carbohydrates such as glucose, cellobiose and starch. The pink colour of the sporangia is due to the presence of carotenoid pigments such as γ-carotene, lycopene and a xanthophyll.

6.4 | Neocallimastigales (rumen fungi)

A very interesting and unusual group of zoosporic fungi inhabits the rumens (foreguts) of ruminants (herbivorous mammals which regurgitate and masticate previously ingested food) like cows, sheep and deer. They have also been found in some non-ruminants such as horses and elephants and probably occur in the guts of many large herbivores. These fungi are obligate anaerobes which can flourish in the rumen because oxygen is depleted there by the intense respiratory activity of a dense population of protozoa and bacteria, some of which are facultative anaerobes capable of scavenging free oxygen. Their zoospores were at first thought to be protozoa and were not recognized as belonging to fungi because obligately anaerobic fungi were not believed to exist. Further, microbiologists working on microbes from the ruminant gut studied only strained rumen fluid and therefore failed to see the thalli of fungi attached to herbage fragments. The view that the motile cells swimming in rumen fluid belonged to flagellates was challenged by Orpin (1974), who observed that there was an enormous increase in the concentration of 'flagellates' in the rumen of sheep within a short time of feeding. The ratio of minimum (pre-feeding) to maximum concentration of motile cells could vary between 1:15 and 1:296 (average 1:47), and if these were organisms reproducing by binary fission it would be necessary for them to undergo six successive cell divisions in 15 min. The explanation for the rapid increase in motile cells is that sedentary fungal thalli, anchored by rhizoids to partially digested food fragments floating in the rumen, are stimulated to release zoospores by soluble substances such as haems released from the newly ingested food material. The zoospores attach themselves in large numbers to the herbage fragments, and germinate to form rhizoidal or rhizomycelial thalli with sporangia capable of releasing further zoospores within about 30 h.

Some 5 genera and 15 species have now been distinguished (Theodorou *et al.*, 1992, 1996; Trinci *et al.*, 1994). They include *Caecomyces* which has mono- and polycentric thalli, *Anaeromyces* and *Orpinomyces* with polycentric thalli, and *Piromyces* and *Neocallimastix* which are monocentric. The zoospores of *Anaeromyces*, *Caecomyces* and *Piromyces* are uniflagellate whilst those of *Neocallimastix* and *Orpinomyces* are multiflagellate (see Fig. 6.18). They were classified within the order Spizellomycetales, family Callimasticaceae by Heath *et al.* (1983) and Barr *et al.* (1989) but are now placed in a separate order Neocallimastigales (Li *et al.*, 1993; D. J. S. Barr, 2001). Special techniques and media are needed for isolating and handling anaerobic fungi, but the life cycle details of several have now been followed in pure culture. One of the best known is *N. hurleyensis*, isolated from sheep (Fig. 6.18). Minutes after the arrival of fresh food, globose ripe zoosporangia on previously colonized grass fragments release zoospores through an apical pore and these attach themselves to herbage fragments and germinate to produce rhizoids which penetrate and digest the ingested plant material. The walls of the thallus contain chitin. A single zoosporangium develops and is cut off from the rhizoidal system by a septum. The rhizoidal part is devoid of nuclei, but within the zoosporangium repeated nuclear divisions occur before the cytoplasm cleaves to form 64–128 zoospores. The life cycle of *N. hurleyensis* from zoospore germination to the release of a fresh crop of zoospores lasts about 29–31 h at 39°C (Lowe *et al.*, 1987a). The zoospores bear 8–16 whiplash flagella inserted posteriorly in two rows.

The ultrastructure of zoospores has been described for several species of *Neocallimastix*, including *N. patriciarum* (Orpin & Munn, 1986), *N. frontalis* (Munn *et al.*, 1981; Heath *et al.*, 1983) and *N. hurleyensis* (Webb & Theodorou, 1988, 1991). There are differences in detail. For example, the zoospore of *N. frontalis* has a waist-like constriction, with the majority of the cytoplasmic organelles concentrated in the posterior portion near the insertion of the flagella. Characteristic organelles known from zoospores of aerobic chytridiomycetes such as mitochondria, Golgi bodies, lipid droplets or gamma particles (seen in zoospores of *Blastocladiella emersonii*; Fig. 6.19) are absent. In the posterior portion of the zoospore of *N. hurleyensis* near the point of insertion of the flagella, an irregularly shaped complex structure interpreted as a **hydrogenosome** has been reported in place of a mitochondrion. In zoospores of *N. patriciarum* there are many presumed hydrogenosomes concentrated around the region of flagellar insertion. Hydrogenosomes are organelles capable of the anaerobic metabolism of hexoses to acetic and formic acids. Protons (H^+) act as electron acceptors, so that gaseous H_2 is released by the activity of the enzyme hydrogenase (Müller, 1993; Boxma *et al.*, 2004). The hydrogen, in turn, is used by anaerobic methanogenic bacteria to reduce CO_2 to CH_4 (methane) which escapes in profusion through the front and hind exits of the ruminant digestive tracts. Hydrogenosomes are found in several anaerobic lower eukaryotes and are believed to be derived

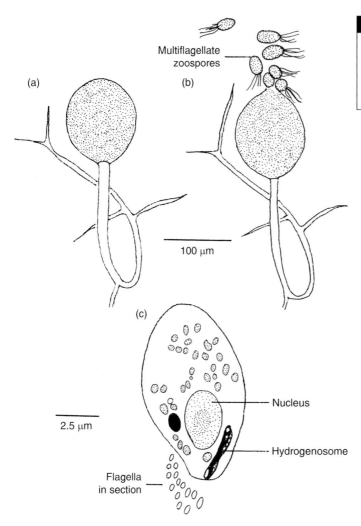

Multiflagellate
zoospores

(a)

(b)

100 μm

Fig 6.18 *Neocallimastix hurleyensis*. (a) Rhizoidal thallus with zoosporangium. (b) Release of zoospores. (c) Tracing of T.E.M. of zoospore with 14 flagella in longitudinal section. Diagrams based on Webb and Theodorou (1991).

(c)

Nucleus

2.5 μm

Hydrogenosome

Flagella
in section

from mitochondria (Embley *et al.*, 2002). Whereas mitochondria of most fungi contain a limited amount of DNA, hydrogenosomes of rumen chytrids seem to have lost their genome altogether (Bullerwell & Lang, 2005).

Granular inclusion bodies which contain aggregates of ribosome-like particles and also free ribosome-like arrays are found anterior to the nucleus. Rosettes of glycogen represent the energy reserve of the zoospore. The shafts of the flagella contain the familiar eukaryotic 9 + 2 arrangement of microtubules, but in *N. frontalis* the microtubules do not extend into the tips of the flagella which are narrower than the proximal part.

Ecologically, these anaerobic fungi play an important role in the early colonization of ingested herbage and have a wide range of enzymes which enable them to utilize monosaccharides, disaccharides and polysaccharides such as xylan, cellulose, starch and glycogen (Theodorou *et al.*, 1992). They may play an active role in fibre breakdown. It is likely that colonization of straw particles by these fungi aids further attack by bacteria. The survival of anaerobic fungi outside the unusual and protective environment of the herbivore gut occurs in dried faeces in the form of cysts or as melanized thick-walled thalli whilst transmission to young animals takes place in saliva during licking and grooming (Lowe *et al.*, 1987b; Wubah *et al.*, 1991; Theodorou *et al.*, 1992).

6.5 | Blastocladiales

6.5.1 Introduction

Species belonging to the Blastocladiales are mostly saprotrophs in soil, water, mud or aquatic plant and animal debris, and some are pathogens of plants, invertebrate animals or fungi. Most are obligate aerobes, but *Blastocladia* spp. are facultatively anaerobic, requiring a fermentable substrate and growing on submerged fleshy fruits, twigs or other plant materials rich in soluble carbohydrates (Emerson & Robertson, 1974). The life cycles of Blastocladiales show great variations and in some forms there is an alternation of distinct haploid **gametothallic** and diploid **sporothallic** generations. These terms are used in preference to the botanical terms gametophytic and sporophytic. Species of *Physoderma*, previously grouped with the Chytridiales (Lange & Olson, 1980), are biotrophic parasites of higher plants (Karling, 1950). They include *P. maydis*, the cause of brown spot of maize, and *P. alfalfae* (Lange *et al.*, 1987). One genus, *Coelomomyces*, consists of obligate parasites of insects, usually mosquito larvae (Couch & Bland, 1985). This genus is unusual in that the vegetative thallus is a wall-less plasmodium-like structure lacking rhizoids. The life cycle is completed in unrelated alternate animal hosts, sporothalli occurring in mosquito larvae (Insecta) and gametothalli in a copepod (Crustacea) (Whisler *et al.*, 1975; Federici, 1977). Attempts are being made to use *Coelomomyces* in the biological control of mosquitoes. *Catenaria anguillulae*, a facultative parasite of nematodes and their eggs, liver fluke eggs and some other invertebrates, can be grown in culture (Couch, 1945; Barron, 1977; Barstow, 1987), whilst *Catenaria allomycis* is a biotrophic parasite of *Allomyces* (Couch, 1945; Sykes & Porter, 1980).

With the exception of *Coelomomyces*, the thallus of members of the Blastocladiales is eucarpic. The morphologically simpler forms such as *Blastocladiella* (Fig. 6.22) are monocentric, with a spherical or sac-like zoosporangium or resting sporangium arising directly or on a short one-celled stalk from a tuft of radiating rhizoids. These simpler types show considerable similarity to monocentric Chytridiales of other orders such as *Rhizophlyctis rosea* (Figs. 6.16 and 6.17), and in the vegetative state they may be difficult to distinguish. The more complex organisms such as *Allomyces* are poly-centric, and the thallus is differentiated into a trunk-like portion which has rhizoids below whilst branching above, often dichotomously, and bearing sporangia of various kinds at the tips of the branches. Chitin has been demon-strated in the walls of *Allomyces* and *Blastocladiella* (Porter & Jaworski, 1966; Youatt, 1977; Maia, 1994).

The zoospore of Blastocladiales

The zoospore of Blastocladiales has a single posterior flagellum of the whiplash type. Details of the fine structure of this kind of zoospore have been reviewed by Fuller (1976) and Lange and Olson (1979). The best known are *Blastocladiella emersonii* (Cantino *et al.*, 1963; Reichle & Fuller, 1967) and *Allomyces macrogynus* (Fuller & Olson, 1971). The structure of the zoospore of *B. emersonii* is summarized diagram-matically in Fig. 6.19. The zoospore is tadpole-like with a pear-shaped head about $7 \times 9 \mu m$ and a single, trailing flagellum about $20 \mu m$ long. Under the light microscope, the most conspic-uous internal structure is the dense crescent-shaped nuclear cap which surrounds the more transparent nucleus. The nuclear cap is rich in RNA and protein, and is filled with ribosomes. The zoospore of *B. emersonii* is unusual in that it contains only a single large mitochondrion, situated near the flagellar kinetosome.

The organization of the flagellum is essen-tially as described on p. 129. The nine triplet microtubules extend in a funnel-shaped manner from the proximal end of the kinetosome towards the nucleus and nuclear cap, maintain-ing its conical shape. Extending into the mito-chondrion and linking up the kinetosome

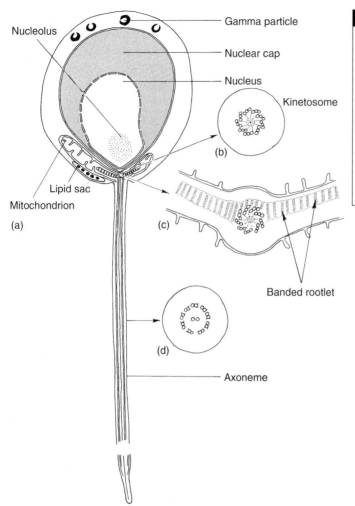

Nucleolus

Gamma particle

Nuclear cap

Nucleus

Kinetosome

(b)

Lipid sac

Mitochondrion

(a)

(c)

Banded rootlet

(d)

Axoneme

Fig 6.19 *Blastocladiella emersonii* zoospore, fine structure, diagrammatic and not to scale. (a) L.S. of zoospore along the axis of the flagellum. (b) T.S. of kinetosome showing nine triplets of microtubules. (c) T.S. of kinetosome at a slightly lower level showing the origin of two of the banded rootlets which extend into the mitochondrion. The cristae of the mitochondrion are close to the membrane which surrounds the banded rootlets. (d) T.S. of axoneme showing the nine paired peripheral microtubules and the two central microtubules.

with it are three striated bodies variously referred to as flagellar rootlets, striated rootlets or banded rootlets. They are contained within separate channels, and each is surrounded by a unit membrane. Since the energy for propulsion is generated within the mitochondrion, it is possible that the banded rootlets are, in some way, responsible for transmitting energy to the base of the axoneme. It is also possible that the banded rootlets serve to anchor the flagellum within the body of the zoospore.

There are two other obvious kinds of organelle within the body of the *Blastocladiella* zoospore. The **lipid sac** attached to the mitochondrion contains a group of lipid droplets which is surrounded by a unit membrane. It is not known whether lipid forms the energy reserve

used in swimming, cytoplasmic glycogen deposits being a more plausible alternative (Cantino *et al.*, 1968). In the anterior of the zoospore between the nuclear cap and the plasma membrane, there is a group of granules about 0.5 μm in diameter, called **gamma particles**. They consist of an inner core, shaped like an elongated cup and bearing two unequal openings at opposite sides of the cup. This cup-shaped structure is enveloped in a unit membrane (Myers & Cantino, 1974). Gamma particles are only present in developing and motile zoospores but disappear as the zoospore encysts. Formerly thought to represent the chytrid equivalent of the chitosome found in higher fungi (see p. 6), this notion has now been discarded (Hohn *et al.*, 1984).

The zoospore of *Allomyces macrogynus* broadly resembles that of *B. emersonii* (Fuller & Olson, 1971). Gamma particles are present in the zoospore and, during encystment, these form vesicles which fuse with the plasma membrane. Fusion coincides with the appearance of wall material around the cyst (Barstow & Pommerville, 1980). The zoospore of *Allomyces* differs from that of *Blastocladiella* in some other ways. Although there is a large basal mitochondrion, many smaller mitochondria are also present, generally located along the membrane of the nuclear cap in the anterior part of the cell. A complex structure situated laterally at the base of the body of the zoospore, between the nucleus and the zoospore membrane, has been termed the **side body complex** by Fuller and Olson (1971). It consists of two closely appressed membranes separated by an electron-opaque material. These membranes subtend numerous electron-opaque, membrane-bound bodies, lipid bodies and a portion of the basal mitochondrion. In addition, there are membrane-bound non-lipid bodies termed **Stüben bodies** by Fuller and Olson (1971), whose function and composition are uncertain.

The zoospore is propelled forward by rhythmic lashing of the flagellum, and it can swim for a period even under anaerobic conditions. It is also capable of amoeboid changes of shape. On coming to rest, the flagellum is retracted into the body of the zoospore. There are different interpretations of the manner in which flagellar retraction is achieved. Cantino *et al.* (1968) have suggested that the flagellum is retracted by a revolving action of the nucleus, whereas in the 'lash-around' mechanism the flagellum coils around the body of the spore, the flagellar membrane fuses with the plasmalemma of the spore and the axoneme enters the spore cytoplasm (Olson, 1984). In *Allomyces*, the zoospore cyst produces, at one point, a narrow germ-tube which branches to form the rhizoidal system. At the opposite pole, the zoospore cyst forms a wider germ tube which gives rise to hyphae which branch and later bear sporangia. This **bipolar germination** pattern is a point of difference between the Blastocladiales and the Chytridiales, in which germination is typically unipolar. The rhizoids are strongly chemotropic and specialize in nutrient uptake and transport. An inwardly directed electrical current has been detected around the rhizoids, and an outwardly directed current around the hyphae and hyphal tips. The inward current at the rhizoids may be the consequence of localized proton-driven solute transport (de Silva *et al.*, 1992).

Life cycles of Blastocladiales

A number of distinct life history patterns are found. In *Allomyces arbuscula*, for example, isomorphic alternation of haploid gametothallic and diploid sporothallic generations has been demonstrated. In *A. neo-moniliformis* (= *A. cystogenes*) there is no free-living sexual generation, but this stage is represented by a cyst (see below). In *A. anomalus*, only the asexual stage has been found in normal cultures, but experimental treatments may result in the development of sexual thalli. Similar variations in life cycles have been found in other genera such as *Blastocladiella*. A characteristic feature of the asexual thalli of the Blastocladiales is the presence of resting sporangia with chitinous, pitted walls impregnated with a dark brown, melanin-type pigment. The pits are inwardly directed conical pores in the wall. The inner ends of the pores abut against a smooth, colourless inner layer of wall material surrounding the cytoplasm (Skucas, 1967, 1968). The resting sporangia of *Allomyces* can remain viable for up to 30 years in dried soil. The ease with which certain members of the group can be grown in culture has facilitated extensive studies of their nutrition and physiology, and the results of some of these investigations are discussed below.

Four families have been recognized – Coelomomycetaceae, Catenariaceae, Physodermataceae and Blastocladiaceae – but of these we shall study only *Allomyces* and *Blastocladiella*, both representatives of the Blastocladiaceae.

6.5.2 *Allomyces*

Species of *Allomyces* are found in mud or soil of the tropics or subtropics, including desert soil, and if dried samples of soil are placed in water and 'baited' with boiled hemp seeds, the baits may become colonized by zoospores. From such material, it is possible to obtain pure cultures by streaking or pipetting zoospores onto suitable agar media and to follow the complete life history of these fungi in the laboratory. Olson (1984) has given a full account of the taxonomy, life cycles, morphogenesis and genetics of different species of *Allomyces*, with practical details of how to grow and handle them. Good growth occurs on a medium containing yeast extract, peptone and soluble starch (YPsS), but chemically defined media have also been used. There is a requirement for thiamine and organic nitrogen in the form of amino acids.

Emerson (1941) isolated species of *Allomyces* from soil samples from all over the world. He distinguished three types of life history, represented by three subgenera.

Sub-genus *Eu-Allomyces*

The *Eu-Allomyces* type of life history is exemplified by *A. arbuscula* and *A. macrogynus* (Fig. 6.21; for a film, see Webster & Hard, 1998a). Resting sporangia are formed on asexual diploid thalli. They contain about 12 nuclei which undergo meiosis during the early stages of germination (Olson, 1974). The cytoplasm cleaves around the 48 haploid nuclei to form the zoospores. Since meiosis occurs in the resting sporangia, these have been termed **meiosporangia**, and the haploid zoospores **meiospores**. The meiospores are released when the outer wall of the brown pitted resting sporangium cracks open by a slit and the inner wall balloons outwards and eventually opens by one or more pores. The meiospores swim by movement of the trailing flagellum and, on coming to rest, encyst and germinate as described above to form a rhizoidal system and a trunk-like region which bears dichotomous branches.

The tips of the branches have been claimed to resemble the Spitzenkörper of higher fungi in being actin-rich, although secretory vesicles and/or microvesicles (chitosomes) have not been clearly shown (Srinivasan *et al.*, 1996). Repeated nuclear division occurs to form a coenocytic structure, and finger-like ingrowths from the

walls of the trunk-region and branches form incomplete septa, sometimes termed pseudo-septa, with a pore in the centre through which cytoplasmic connections can be seen (Fig. 6.20d; Meyer & Fuller, 1985). The haploid thalli which develop from the meiospores are gametothallic, i.e. sexual. They are monoecious, and the tips of their branches swell to form paired sacs – the male and female gametangia. The male gametangia can be identified by the presence of a bright orange pigment, γ-carotene, whilst the female gametangia are colourless. In *A. arbuscula* the male gametangium is subterminal or **hypogynous**, i.e. beneath the terminal female gametangium, but in *A. macrogynus* the positions are reversed and the male gametangium is terminal or **epigynous** (Figs. 6.20e,i). The gametangia bear a number of colourless papillae on their walls, blocked by pulley-shaped plugs which eventually dissolve.

The contents of the gametangia differentiate into uninucleate gametes which differ in size and pigmentation. The female gametangium forms larger, colourless motile gametes (**swarmers**) whilst the male gametangium releases smaller, more active, orange-coloured swarmers. After escaping through the papillae in the walls of the gametangia, the gametes swim for a time and then pair off. A female gamete which fails to pair can function as a zoospore by germinating to form a new sexual thallus. A hormone, sirenin, is secreted by female gametangia during gametogenesis and by the released female gametes, and this stimulates a chemotactic response in male gametes at the extremely low concentration of 8×10^{-11} M (Machlis, 1972; Carlile, 1996a). The chemical structure of sirenin has been determined (Fig. 6.22), and both D- and L-forms have been synthesized. Only L-sirenin is active. It is a bicyclic sesquiterpene, probably derived from the parent hydrocarbon sesquicarene (Nutting *et al.*, 1968; Plattner & Rapoport, 1971). A second hormone, parisin, which attracts female gametes, is secreted by male gametes. Its structure has not been determined, although it may well be related to sirenin (Pommerville & Olson, 1987).

The biflagellate zygote resulting from the fusion of two gametes may swim for a while

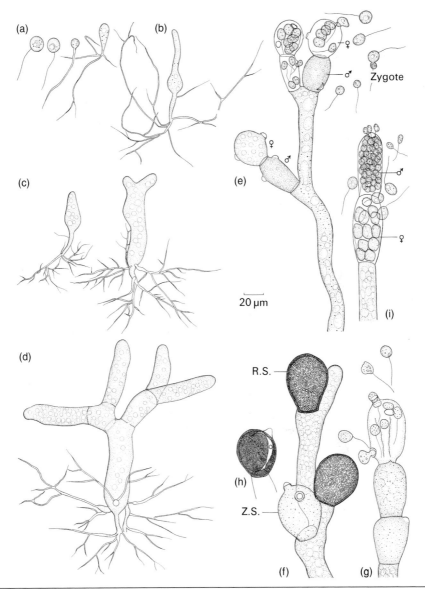

Fig 6.20 (a–h) *Allomyces arbuscula*. (a) Zoospores (haploid meiospores). (b) Young gametothalli, 24 h old. (c) Young sporothalli, 18 h old. (d) Sporothallus, 30 h old. Perforations are visible in some of the septa. (e) Gametangia at the tips of the branches of the gametothallus. Note the disparity in the size of the gametes (anisogamy). The smaller male gametes are orange in colour whilst the larger female gametes are colourless. Compare the hypogynous arrangement of the male gametangia with the epigynous arrangement in *A. macrogynus* shown at (i). (f) Meiosporangia (resting sporangia, R.S.) and mitosporangia (zoosporangia, Z.S.) on a sporothallus. (g) Release of mitospores from zoosporangia (= mitosporangia) on sporothallus. (h) Rupture of meiosporangium (= resting sporangium). (i) *Allomyces macrogynus*. Branch tip from gametothallus showing the arrangement of gametangia with terminal, epigynous male gametangia and anisogamous gametes.

before it encysts and casts off the flagella. Nuclear fusion then follows (Pommerville & Fuller, 1976). The zygote develops immediately into a diploid asexual thallus which differs from gametothalli in bearing two types of zoosporangia instead of gametangia. The first formed are thin-walled papillate zoosporangia formed singly or in rows at the tips of the

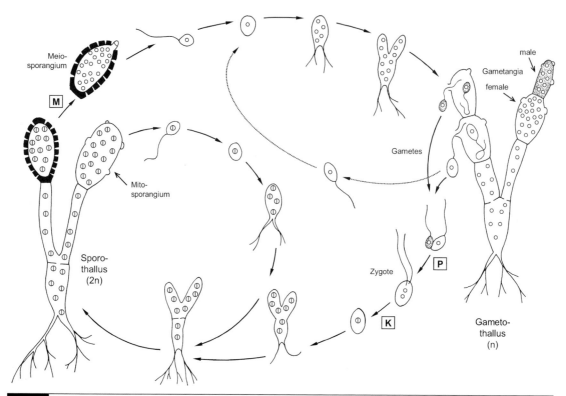

Fig 6.21 Life cycle diagram of *Eu-Allomyces* as exemplified by *A. macrogynus*. A diploid sporothallus may produce diploid mitospores from a colourless, thin-walled papillate mitosporangium, and haploid meiospores from a thick-walled pitted meiosporangium in which meiosis occurs. Meiospores germinate to form a haploid gamethothallus which produces two different gametangia and releases haploid gametes of two kinds, small carotenoid-rich (shaded) 'male' gametes and larger colourless 'female' ones. Upon copulation, a diploid zygote gives rise to a sporothallus. Alternatively, if failing to pair up, the female gametes may function as zoospores, in which case they give rise to a new gametothallus. Small open circles represent haploid nuclei whereas diploid nuclei are drawn larger and split. It should be noted that many field strains of *A. macrogynus* have a higher ploidy level, e.g. alternating between diploid (small circles) and tetraploid (large split circles) conditions. Key events in the life cycle are plasmogamy (P), karyogamy (K) and meiosis (M).

branches (Fig. 6.20g). Within these thin-walled sporangia the nuclei undergo mitosis. Initially the nuclei are arranged in the cortical region of the cytoplasm, but later they migrate and become uniformly spaced apart. Movement of the nuclei is controlled by forces generated by actin microfilaments whilst their spacing and positioning is controlled by microtubules (Lowry et al., 1998). Cleavage of the cytoplasm around the nuclei to form diploid colourless zoospores is initiated by the formation of membranes seen first at the plasmalemma, then extending into the cortex to form a complex membranous network (Fisher et al., 2000). The process of cytokinesis, i.e. the extension and fusion of

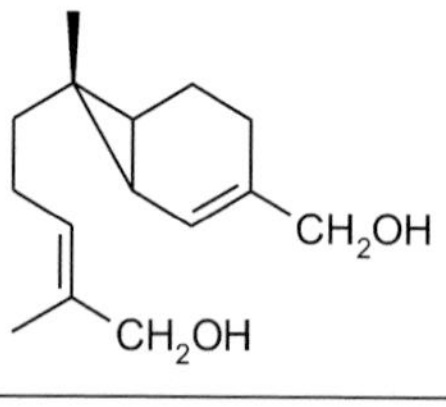

Fig 6.22 Chemical structure of the hormone L-sirenin which attracts male gametes of *Allomyces macrogynus*. The structure of parisin, attractive to female gametes, does not seem to have been elucidated as yet.

membranes, seems to be mediated principally by the actin component of the cytoskeleton (Lowry et al., 2004). According to Barron and Hill (1974), the development of the cleavage membranes is induced by the availability of free water. Zoospores are released from the sporangia after dissolution of the plugs blocking the exit papillae. Since nuclear division in the thin-walled sporangia is mitotic, these are termed **mitosporangia**, and the diploid swarmers they release are **mitospores**. The mitospores, after a swimming phase, encyst and are capable of immediate germination, developing into a further diploid asexual thallus.

The second type of zoosporangium is the dark brown, thick-walled, pitted resting sporangium (meiosporangium), formed at the tips of the branches. Meiotic divisions within these sporangia result in the formation of the haploid meiospores, which develop into sexual thalli. The life cycle of a member of the subgenus *Eu-Allomyces* is thus an isomorphic alternation of gametothallic and sporothallic generations (Fig. 6.21). Comparisons of the nutrition and physiology of the two generations show no essential distinction between them up to the point of production of gametangia or sporangia.

Emerson and Wilson (1954) have made cyto-logical and genetic studies of a number of collections of *Allomyces*. Interspecific hybrids between *A. arbuscula* and *A. macrogynus* have been produced in the laboratory, and it has been shown that the fungus earlier described as *A. javanicus* is a naturally occurring hybrid between these two species. Cytological examination of the two parent species and of artificial and natural hybrids showed a great variation in chromosome number. In *A. arbuscula* the basic haploid chromosome number is 8, but strains with 16, 24 and 32 chromosomes have been found. In *A. macrogynus* the lowest haploid number encountered is 14, but strains with 28 and 56 chromosomes are also known. The demonstration that these two species each represent a polyploid series was the first to be made in fungi. The wild-type strain of *A. macrogynus* appears to be an autotetraploid which, after meiosis, produces diploid gameto-thalli (Olson & Reichle, 1978).

The behaviour of the hybrid strains is of considerable interest. As seen above, the parent species differ in the arrangement of the primary pairs of gametangia, *A. arbuscula* being hypogynous whilst *A. macrogynus* is epigynous. Following fusion of gametes derived from different parents, zygotes formed, germinated and gave rise to sporothalli. The meiospores from the hybrid sporothalli had a low viability (0.1–3.2%), as compared with a viability of about 63% for *A. arbuscula* meiospores, but some germinated to form gametothalli. The arrangement of the gametangia on these F$_1$ gametothalli showed a complete range from 100% epigyny to 100% hypogyny. Also, in certain gametothalli the ratio of male to female gametangia (normally about 1:1) was very high, with less than one female per 1000 male gametangia. It was concluded from these experiments that, since intermediate gametangial arrangements are found in hybrid haploids, this arrangement is not under the control of a single pair of non-duplicated allelic genes, but that a fairly large number of genes must be involved. Hybridization in some way upsets the mechanism which controls the arrangement of gametangia in the parental species. By treating meiospores of *A. macrogynus* with DNA extracted from gametothallic cultures of *A. arbuscula*, Ojha and Turian (1971) have demonstrated an inversion of the normal gamet-angial arrangement, i.e. a proportion of the DNA-treated meiospores developed colonies with hypogynous antheridia instead of the normal epigynous arrangement. Similar inversions were also obtained in converse experiments. In an isolate of the naturally occurring hybrid *A. javanicus*, Ji and Dayal (1971) have shown that although copulation between anisog-amous gametes results in the formation of sporothalli bearing thin-walled and thick-walled sporangia, the swarmers from the thick-walled sporangia rarely develop into gametothalli, but into sporothalli. This is not surprising for a hybrid, and is possibly due to a failure of meiosis in the thick-walled sporangia.

Sub-genus *Cystogenes*

A life cycle different from *Eu-Allomyces* is found in *Allomyces moniliformis* and *A. neo-moniliformis*. There is no independent gametothallic genera-tion, but this stage is probably represented by a cyst (C. M. Wilson, 1952). The asexual thalli resemble those of subgenus *Eu-Allomyces*, bearing both thin-walled mitosporangia and brown, thick-walled, pitted meiosporangia. The mitospores encyst and germinate to form a further crop of asexual thalli. In the meiosporangium, meiosis takes place, but before cytoplasmic cleavage occurs, the haploid nuclei pair, the paired nuclei being united by a common nuclear cap. When cleavage does occur it therefore results in the formation of some 30 binucleate cells. When the meiosporangial wall cracks open, the binucleate cells are released as amoeboid bodies which may or may not bear flagella, and it is these cells which form the cysts. A mitotic division in each cyst results in four haploid nuclei, and cytoplasmic cleavage gives rise to four colourless uniflagellate isogametes. These copulate to form biflagellate zygotes, each of which can develop into an asexual sporothallus. In the *Cystogenes* life cycle there is thus a free-living diploid asexual sporothallic generation, whereas the haploid generation is reduced to the cysts and gametes.

Sub-genus Brachy-Allomyces

In certain isolates of *Allomyces* which have been placed in a 'form species' *A. anomalus*, there are neither sexual thalli nor cysts. Asexual thalli bear mitosporangia and brown resting sporangia. The spores from the resting sporangia develop directly to give asexual thalli again. The cytological explanation proposed by C.M. Wilson (1952) for this unusual behaviour is that, due to complete or partial failure of chromosome pairing in the resting sporangia, meiosis does not occur and nuclear divisions are mitotic. Consequently the zoospores produced from resting sporangia are diploid, like their parent thalli and, on germination, give rise to diploid asexual thalli again. Similar failures in chromosome pairing were also encountered in the hybrids between *A. arbuscula* and *A. macrogynus* leading to very low meiospore viability from certain crosses. In view of this it seemed possible that some of the forms of *A. anomalus* might have arisen through natural hybridization. In a later study, Wilson and Flanagan (1968) showed that there is a second way in which the life cycle of this fungus is maintained without a sexual phase. In certain isolates, meiosis does occur in

the resting sporangia, followed by **apomixis**, i.e. the fusion of two meiosis-derived nuclei in the same thallus. Propagules from the resting sporangia are therefore diploid and the cysts develop into sporothalli. By germinating resting sporangia in dilute K_2HPO_4, a small percentage of zoospores were produced which developed into gametothalli, some of which were identified as *A. macrogynus* and some as *A. arbuscula*. No hybrids were found. Thus *A. anomalus* is not a single species, but represents sporothalli of these two species in which the normal alternation of generations has been upset by cytological deviations.

6.5.3 *Blastocladiella*

About a dozen species of *Blastocladiella* have been isolated from soil or water, and one is parasitic on the cyanobacterium *Anabaena* (Canter & Willoughby, 1964). The form of the thallus is comparatively simple, resembling that of some monocentric chytrids. There is an extensive branched rhizoidal system which is attached either to a sac-like sporangium or to a cylindrical trunk-like region bearing a single sporangium at the tip. In *B. emersonii* it has been shown that the rhizoids are chemotropic and function not only in attachment, but in absorption and selective translocation of nutrients (Kropf & Harold, 1982).

Different species of *Blastocladiella* have life cycles resembling those of the three subgenera of *Allomyces*, and Karling (1973) has proposed that *Blastocladiella* should similarly be divided into three subgenera, i.e. *Eucladiella* corresponding to *Eu-Allomyces*, *Cystocladiella* corresponding to *Cystogenes*, and *Blastocladiella* corresponding to *Brachy-Allomyces*. In some species there is an isomorphic alternation of generations, probably matching in essential features the *Eu-Allomyces* pattern, but cytological details are needed to confirm this. For example, in *Blastocladiella variabilis* two kinds of asexual thallus are found. One bears thin-walled zoosporangia which release posteriorly uniflagellate swarmers. These swarmers may develop to form thalli resembling their parents or may give rise to the second type of asexual thallus bearing a thick-walled dark-brown sculptured resting sporangium within the terminal sac. The resting sporangium releases posteriorly uniflagellate swarmers which, after swimming, germinate to form sexual thalli of two kinds. About half of the sexual thalli are colourless ('female'), and about half are orange-coloured ('male'). However, in contrast to the anisogamy of *Eu-Allomyces*, in *Blastocladiella* there is no distinction in size between the gametes. The orange and colourless gametes pair to produce zygotes, which germinate directly to produce asexual thalli. In other species (e.g. *B. cystogena*) the life cycle is of the *Cystogenes* type, i.e. there are no gametothalli.

In yet other species there is no clear evidence of sexual fusion. In *B. emersonii* (Fig. 6.23), the resting sporangial thallus contains a single globose, dark reddish brown resting sporangium with a dimpled wall. Meiosis occurs during development of the resting sporangium (Olson & Reichle, 1978). After a resting period, the wall cracks open and one to four papillae protrude from which swarmers are released. The swarmers germinate to form two types of thallus bearing thin-walled zoosporangia. About 98% of the swarmers give rise to thalli bearing colourless sporangia (Fig. 6.23a), and about 2% to thalli with sporangia coloured orange due to the presence of γ-carotene. The colourless thalli develop rapidly and are ready to discharge zoospores within 24 h. These have about twice the DNA content as the swarmers released from resting sporangia (Horgen *et al.*, 1985). Thus young colourless thalli are at first haploid, but release diploid zoospores. The manner in which the diploid state of the colourless thalli or of the resting sporangia is brought about is not known. The life cycle of *B. emersonii* thus corresponds to that of the sub-genus *Brachyallomyces*.

Blastocladiella emersonii has a number of other unusual features. If zoospore suspensions are pipetted onto yeast−peptone−glucose (YPG) agar, the majority of thalli which develop will be of the thin-walled colourless type. On the same medium containing 10 mM bicarbonate, resting sporangial thalli develop. The addition of 40−80 mM KCl, NaCl or NH₄Cl, or exposure of cultures to ultra-violet light, will similarly induce the formation of resting sporangia (Horgen & Griffin, 1969). Thus, by means of simple manipulation of the environment it is possible to switch the metabolic activities of the fungus into one of two morphogenetic

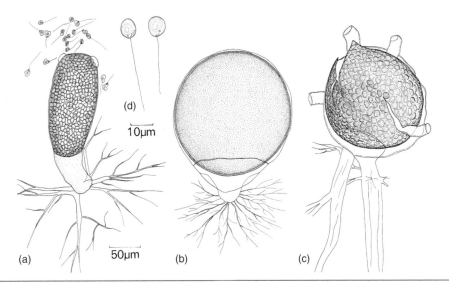

Fig 6.23 *Blastocladiella emersonii*. (a) Thin-walled thallus releasing zoospores. (b) Three-day-old thallus with immature resting sporangium. (c) Thallus with germinating resting sporangium showing the cracked wall and four exit tubes. (d) Zoospores from thin-walled thallus.

pathways. There are important differences in the activities of certain enzymes (Cantino et al., 1968; Lovett, 1975). In the absence of bicarbonate, there is evidence for the operation of a tricarboxylic acid cycle, whereas in the presence of bicarbonate, part of this cycle is reversed, leading to alternative pathways of primary carbon metabolism. In addition, a polyphenol oxidase, absent in the thin-walled thallus, replaces the normal cytochrome oxidase. There is also increased synthesis of melanin and of chitin in the presence of bicarbonate. The effect of bicarbonate can be brought about by increased levels of CO_2.

Another unusual feature is that B. emersonii fixes CO_2 more rapidly in the light than in the dark. In the presence of CO_2, light-grown thalli show a number of differences when compared with dark-grown controls. Illuminated thalli take about three hours longer to mature, and are larger than dark-grown thalli. They also have an increased rate of nuclear division and a higher nucleic acid content. The most effective wavelengths for this increased CO_2 fixation (or lumisynthesis) lie between 400 and 500 nm, i.e. at the blue end of the spectrum. This suggests that the photoreceptor should be a yellowish substance. Attempts to identify the photoreceptor have as yet been unsuccessful, but it is known not to be a carotenoid.

6.6 | Monoblepharidales

This group includes about 20 species and is represented by 5 genera, namely Monoblepharis, Monoblepharella, Gonapodya, Oedogoniomyces and Harpochytrium. Fungi belonging to this order can be isolated from soil samples or from twigs or fruits submerged in freshwater, sometimes under anoxic conditions (Karling, 1977; Fuller & Clay, 1993). Whisler (1987) has given details of isolation techniques. Most species are saprotrophs and several are available in culture. In all genera the thallus is eucarpic either with rhizoids or a holdfast, and with branched or unbranched filaments. The walls contain microfibrils of chitin (Bartnicki-Garcia, 1968), but the walls of G. prolifera also contain cellulose (Fuller & Clay, 1993). A characteristic feature is the frothy or alveolate appearance of the cytoplasm caused by the presence of numerous vacuoles often arranged in a regular pattern. Asexual reproduction is by posteriorly uniflagellate zoospores which are borne in terminal, cylindrical or flask-shaped sporangia. Sexual reproduction, where known, is unique for fungi in being oogamous with a large egg and a smaller, posteriorly flagellate spermatozoid. The egg may be retained within the oogonium or may move to its mouth by amoeboid movement in some species of Monoblepharis, or propelled by the lashing of the flagellum of the spermatozoid in Monoblepharella and Gonapodya.

6.6.1 The zoospore

The fine structure of zoospores is similar in representatives of all five genera (Fig. 6.24; see Mollicone & Longcore, 1994, 1999). In all cases the body of the zoospore is oval, the narrow part facing forward and with a long whiplash flagellum trailing from the wider posterior. Amoeboid changes of shape may occur and swimming zoospores may develop pseudopodia anteriorly. The body of the zoospore is differentiated into three regions: an anterior region which is often devoid of organelles apart from lipid globules, a few vacuoles and tubular cisternae; a central region which contains the nucleus, surrounded by ribosomal aggregations (sometimes termed the nuclear cap), microbodies and spherical mitochondria with flattened cristae; and a posterior 'foamy' region at the base of which are the functional kinetosome, a non-functional kinetosome and a rumposomal complex. The functional kinetosome is surrounded by a striated disc, apparently anchored to annular cisternae. From an electron-dense region of the striated disc, about 31–34 microtubules extend outwards into the body of the zoospore. Water expulsion vacuoles have been identified in the anterior part of the zoospore of G. prolifera. Another distinctive feature in this fungus is the presence of a pair of paraxonemal structures, solid cylindrical fibres which are

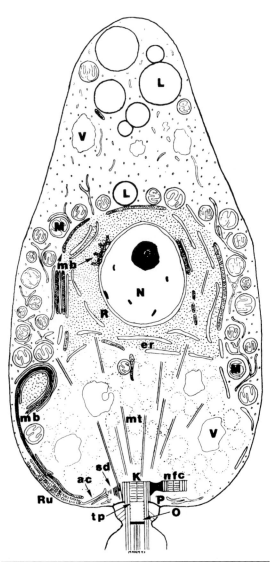

smaller in diameter than the axonemal microtubules, running parallel to them within the axoneme and connected at intervals to doublets 3 and 8 (Mollicone & Longcore, 1999).

6.6.2 *Monoblepharis*

Species of *Monoblepharis* occur in quiet silt-free pools containing neutral or slightly alkaline water (i.e. pH 6.4–7.5) on waterlogged twigs on which the bark is still present. Twigs of birch, ash, elm and especially oak are suitable substrata, and although samples taken at varying times throughout the year may yield growths of the fungus, there are two main periods of vegetative growth, one in spring and another in autumn, with resting periods during the summer and winter months. Low temperatures appear to favour asexual development and good growth can be obtained on twigs incubated in dishes of distilled water at temperatures around 3 °C. The mycelium is delicate and vacuolate. The hyphae are multinucleate. During the formation of a sporangium, a multinucleate tip is cut off by a septum. The cytoplasm cleaves around the nuclei to form zoospore initials which are at first angular and then later pear-shaped. The ripe sporangium is cylindrical or club-shaped and may not be much wider than the hypha bearing it. A pore is formed at the tip of the sporangium through which the zoospores escape by amoeboid crawling. The free zoospores swim away. On coming to rest, a zoospore encysts and germinates by emitting a germ tube. The single nucleus of the zoospore cyst divides and further nuclear divisions occur as the germ tube elongates.

Sexual reproduction can be induced by incubating twigs at room temperature. Light also affects reproduction in *M. macrandra*. Cultures of this fungus incubated in the dark produced only gametothalli whilst those grown in light formed only sporothalli (Marek, 1984). In *M. polymorpha* and related species, the antheridia are epigynous, becoming cut off by a basal septum. Beneath the antheridium the hypha becomes swollen somewhat asymmetrically so that the antheridium is displaced into a lateral position. The swollen subterminal part becomes spherical and is then cut off by a basal septum to form the oogonium. In *M. sphaerica* and some other species, the arrangement of the sex organs is the reverse of that in *M. polymorpha*, i.e. hypogynous. In *M. macrandra* the antheridia and oogonia may grow as solitary organs at the

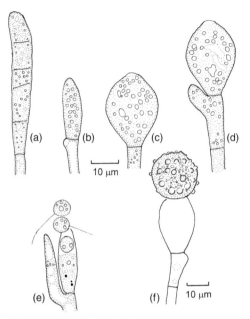

10 μm

10 μm

Fig 6.25 *Monoblepharis macrandra* reproduction. (a) Terminal zoosporangium containing cleaved zoospores. (b) Solitary terminal antheridium. (c) Solitary terminal oogonium with apical receptive area. (d) Oogonium with hypogynous antheridium. (e) Spermatozoid release from solitary terminal antheridium. (f) Exogenous oospore on empty oogonium with bullations on the wall of the oogonium and lipid inclusions in the cytoplasm. (a–e) to same scale. Traced from Whisler and Marek (1987), with permission by Southeastern Publishing Corporation.

tips of the hyphae (Figs. 6.25b,c) or in pairs, with the antheridium in a hypogynous position (Fig. 6.25d). The antheridium often releases sperm before the adjacent oogonium is ripe. Each antheridium forms about four to eight posteriorly uniflagellate swarmers which resemble, but are somewhat smaller than, the zoospores. The oogonium contains a single spherical uninucleate oosphere, and when this is mature an apical receptive papilla on the oogonial wall breaks down. A spermatozoid approaching the receptive papilla of the oogonium becomes caught up in mucus and fusion with the oosphere then follows, the flagellum of the spermatozoid being absorbed within a few minutes. Following plasmogamy, the oospore secretes a golden-brown wall around itself and nuclear fusion later occurs. In some species, e.g. *M. sphaerica*, the oospore remains within the oogonium (**endogenous**) but in others, e.g. *M. macrandra* and *M. polymorpha*, the oospore begins to move towards the mouth of the oogonium within a few minutes of fertilization, and remains **exogenous**, i.e. attached to it (Fig. 6.25f). In the exogenous species, nuclear fusion is delayed but finally fusion occurs and the oospore becomes uninucleate. In some species the oospore wall remains smooth, but in others such as *M. macrandra* the wall may be ornamented by hemispherical warts or bullations (Fig. 6.25f). The oospore germinates after a resting period which coincides with frozen winter conditions or summer drought by producing a single hypha which branches to form a mycelium. The cytological details of the life cycle are not fully known but it seems likely that reduction division occurs during the germination of the overwintered oospores.

Zygomycota

7.1 | Introduction

The phylum Zygomycota comprises the first group of fungi considered in this book which lacks any motile stage. Asexual reproduction is by spores which are called **aplanospores** because they are non-motile, and **sporangiospores** because they are typically contained within sporangia. They are dispersed passively by wind, insects and rain splash, although violent liberation of entire sporangia (e.g. *Pilobolus*) or individual spores (e.g. *Basidiobolus, Entomophthora*) can also occur. Sexual reproduction is by gametangial copulation which is typically isogamous and results in the formation of a **zygospore**. The mycelial organization is coenocytic, and the cell wall contains chitin and its deacetylated derivative, chitosan (Bartnicki-Garcia, 1968, 1987; see Fig. 1.5). As in the Chytridio-, Asco- and Basidiomycota, the mitochondria possess lamellate cristae, and the Golgi system is reduced to single cisternae. Lysine is synthesized by the α-aminoadipic acid (AAA) route, as it appears to be in all Eumycota.

General accounts of the Zygomycota have been given by Benjamin (1979), Benny (2001) and Benny *et al.* (2001). Molecular evidence indicates that the group may have diverged from the Chytridiomycota early in the history of terrestrial life. The Zygomycota, in turn, probably gave rise to the Asco- and Basidiomycota, i.e. the 'higher fungi' (Jensen *et al.*, 1998; Schüssler *et al.*, 2001). Two classes are included in the Zygomycota, namely Zygomycetes comprising

870 species in 10 orders, and Trichomycetes with 218 species in 3 orders (Kirk *et al.*, 2001). The most prominent orders of the Zygomycetes are the Mucorales, Entomophthorales and Glomales. Mucorales are ubiquitous in soil and dung mostly as saprotrophs, although a few are parasitic on plants and animals. Entomophthorales include a number of insect parasites, but some saprotrophic forms also exist. Glomales are mutualistic symbionts associated with almost all kinds of terrestrial plants as arbuscular and vesicular–arbuscular mycorrhiza. Trichomycetes are mostly commensal in the guts of arthropods, e.g. millipedes and the larvae of aquatic insects.

The Zygomycetes are almost certainly polyphyletic, but the precise evolutionary relationships within this class are still controversial (O'Donnell *et al.*, 2001; Schüssler *et al.*, 2001; Tanabe *et al.*, 2004, 2005), and comparisons of numerous representative organisms with several different DNA sequences, e.g. genes encoding ribosomal RNA, cytochrome oxidase or cytoskeletal proteins, will be required before a satisfactory natural arrangement can be found. A recent phylogenetic scheme is presented in Fig. 7.1.

7.2 | Zygomycetes: Mucorales

In most members of the Mucorales, numerous spores are contained in globose sporangia borne at the tips of aerial sporangiophores (Fig. 7.2). Within the sporangium the spores may surround

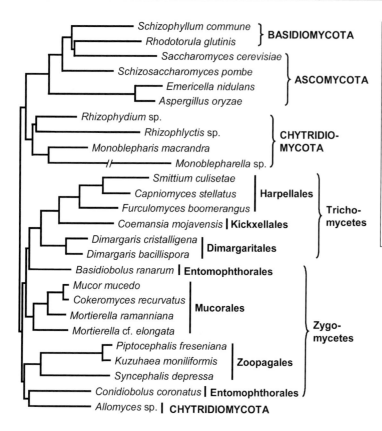

Fig 7.1 Recent phylogenetic scheme of the Zygomycota based on partial sequences of the gene encoding a subunit of RNA polymerase II. Orders discussed in detail in this book include Mucorales (Sections 7.2–7.3), Zoopagales (Section 7.4), Entomophthorales (Section 7.5) in the Zygomycetes, and Harpellales (Section 7.7) in the Trichomycetes. The Glomales (Section 7.6), only distantly related to other Zygomycota, were not included in this analysis. Redrawn and modified from Tanabe *et al.* (2004), with permission from Elsevier.

a central core or **columella**, although in some species (e.g. *Mortierella* spp.) the columella is greatly reduced. Some species possess few-spored sporangia, termed **sporangiola**, which are often dispersed as a unit, and in some groups the spores are arranged as a single row inside a cylindrical sac termed a **merosporangium**. Yet other Mucorales reproduce by means of unicellular propagules which are sometimes termed **conidia**, but Benjamin (1979), in his consideration of asexual propagules formed in the Zygomycota, has recommended the use of the term 'sporangiolum' instead of 'conidium' in this context. It is believed that 'conidia' may have evolved several times within different groups of Mucorales from forms with monosporous sporangiola. A distinction between sporangiospores and conidia is that germinating sporangiospores lay down a new wall, continuous with the germ tube, within their original spore wall, whilst within germinating conidia there is no new wall layer formed.

The Mucorales are mostly saprotrophic and are abundant in soil, on dung and on other organic matter in contact with the soil. They may play an important role in the early colonization of substrata. Sometimes, however, they can behave as weak pathogens of soft plant tissues, e.g. *Rhizopus stolonifer* can cause a rot of sweet potatoes or fruits such as apples, tomatoes and strawberries (Plate 3d). Such infections may cause spoilage of food (Samson *et al.*, 2002). Some species are parasitic on other fungi, a common example being *Spinellus fusiger* which forms a tuft of sporangiophores on the caps of moribund fruit bodies of *Mycena* spp. (Plate 3e). Others cause diseases of animals including man, especially patients suffering from diabetes, leukaemia and cancer. Lesions may be localized in the brain, lungs or other organs, or may be disseminated, e.g. at various points in the vascular system (Kwon-Chung & Bennett, 1992). Species of *Rhizopus* and *Mucor* are reported from human lesions, and these genera together with species of *Absidia* may also infect domestic animals.

A number of species have been used in the production of oriental foods such as sufu, tempeh and ragi (Nout & Aidoo, 2002) and some are used as starters in the saccharification of starchy materials before fermentation to alcohol (Hesseltine, 1991). In modern biotechnology, many mucoralean fungi are employed in biotransformation processes (for references, see Kieslich, 1997). Further, a number of species are oleaginous, i.e. they are able to synthesize and accumulate lipids to over 20% (dry weight) of their biomass. Because these lipids (principally triacylglycerides) may be enriched in polyunsaturated fatty acids (PUFAs), oleaginous members of the Mucorales are of current biotechnological interest (Certik & Shimizu, 1999).

Extensive studies of nutrition and physiology have been made. A wide variety of sugars can be used, and whilst starch can be decomposed by some species, cellulose is generally not utilized. Under anaerobic conditions, ethanol and numerous organic acids are produced. Many Mucorales need an external supply of vitamins for growth in synthetic culture. Thiamine is a common requirement, and the amount of growth of *Phycomyces* has been used as an assay for the concentration of thiamine.

Zycha *et al.* (1969) have given a general account of the taxonomy of the Mucorales, including keys to genera and species. Benny *et al.* (2001) recognized 13 families and 57 genera. Classification and identification are based largely on the morphology of the anamorph. However, DNA sequence comparisons indicate that several families and even some larger genera are polyphyletic (O'Donnell *et al.*, 2001), meaning that the traditional family-level classification scheme is artificial. We retain it here because it presents an accessible framework of morphological features within which the Mucorales can be understood, and because convincing alternative schemes have not yet been put forward.

7.2.1 Growth and asexual reproduction

The mycelium is coarse, coenocytic and richly branched, the branches tapering to fine

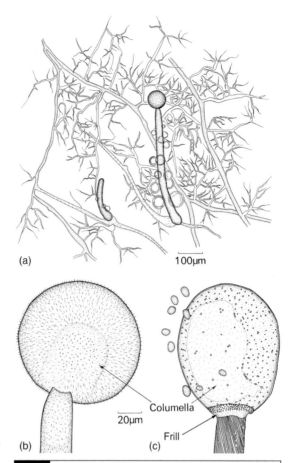

(a) 100μm

(b) (c) 20μm Columella Frill

Fig 7.2 *Mucor mucedo*. (a) Mycelium and young sporangiophores with globules of liquid attached. (b) Immature sporangium with the columella visible through the sporangial wall. (c) Dehisced sporangium showing the columella, the frill representing the remains of the sporangial wall, and sporangiospores.

points (Fig. 7.2). Later, septa may appear. Thick-walled mycelial segments (chlamydospores) may be cut off by such septa (Benjamin, 1979) and in certain species, e.g. *Mucor racemosus*, the presence of chlamydospores in sporangiophores may be a useful diagnostic feature (Fig. 7.14b). In anaerobic liquid culture, especially in the presence of CO_2, several species of *Mucor* (e.g. *M. rouxii*) grow in a yeast-like instead of a filamentous form (Fig. 7.3) but revert to filamentous growth in the renewed presence of O_2. The cell walls of Mucorales are chemically complex (Ruiz-Herrera, 1992; Gooday, 1995). Chitin microfibrils are present but are often

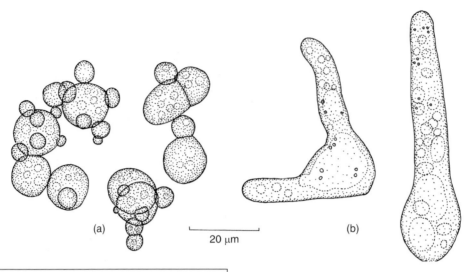

Fig 7.3 *Mucor rouxii.* (a) Yeast-like growth in liquid medium under anaerobic conditions 24 h after inoculation with spores. (b) Filamentous growth from spores in liquid medium under aerobic conditions 4 h after inoculation.

deacetylated to chitosan. Other compounds such as poly-D-glucuronides, polyphosphates, proteins, lipids, purines, pyrimidines, magnesium and calcium have also been detected. Comparison of the structure and composition of yeast-like and filamentous cells of *Mucor rouxii* shows that the yeast-like cells have much thicker walls. They also have a mannose content about five times as great as that of filamentous cell walls. The synthesis of chitin microfibrils takes place within chitosomes which have been described from sporangiophores of *Phycomyces* (Herrera-Estrella *et al.*, 1982; Gamow *et al.*, 1987) and from a number of other fungi with chitinous walls (see p. 6).

Asexual reproduction is by aplanospores (sporangiospores) contained in globose or pear-shaped sporangia, which are borne singly at the tip of a sporangiophore or on branched sporangiophores. In *Absidia* (Fig. 7.17), the sporangia are arranged in whorls on aerial branches, and in many species of *Rhizopus* the sporangiophores arise in groups from a clump of rhizoids (Fig. 7.16). Sporangiophores are often phototropic, and several studies on the phototropism of the strikingly large sporangiophores of *Phycomyces blakesleeanus* have been

undertaken (see p. 169) and have been summarized in elegant and stimulating reviews of the biology of this fungus (Bergman *et al.*, 1969; Cerdá-Olmedo & Lipson, 1987; Cerdá-Olmedo, 2001). 'The sporangiophore of the fungus *Phycomyces* is a gigantic, single-celled, erect, cylindrical aerial hypha. It is sensitive to at least four distinct stimuli: light, gravity, stretch, and some unknown stimulus by which it avoids solid objects. These stimuli control a common output, the growth rate, producing either temporal changes in the growth rate or tropic responses' (Bergman *et al.*, 1969). The avoidance by the sporangiophore of solid objects is termed the avoidance response or fugitropism. Despite its obvious fascination, the mechanisms behind the avoidance response are still not understood. Because, under certain conditions, *P. blakesleeanus* may also develop much smaller sporangiophores (microsporangiophores or **microphores**), the larger sporangiophores are sometimes termed **macrophores**. Despite their remarkable height (Fig. 7.4), for much of their length the macrophores are a constant 100 μm in diameter. The wall, about 0.6 μm thick, encloses a peripheral layer of cytoplasm of about 30 μm surrounding a central vacuole about 40 μm in diameter (Fig. 7.5). The mature sporangium is spherical and some 500 μm across.

The sporangiophore of *Phycomyces* develops as a conical outgrowth from the vegetative mycelium. Elongation of the sporangiophore is confined to a yellow-pigmented growing zone (about 1 cm long) beneath the apex. In the absence of light, sporangiophores grow vertically as a negative response to gravity. Schimek *et al.* (1999) have suggested that gravity may be detected by a combination of at least two mechanisms. Proteinaceous crystals located inside vacuoles have a higher density than the vacuolar sap and therefore sediment in response to gravity, whereas a cluster of buoyant lipid droplets less dense than the cytoplasm floats to the apex of the sporangiophore. Both mechanisms would be different from that found in the fruit bodies of basidiomycetes such as *Flammulina*, in which nuclei denser than the surrounding cytoplasm seem to be the organelles involved in graviperception (see p. 546).

7.2.2 Phototropism in *Phycomyces*

If a sporangiophore is subjected to unilateral illumination it bends towards the light, especially blue light. Phototropism in *Phycomyces* is extremely sensitive, the lower threshold being $1\,\mathrm{nW\,m^{-2}}$, which is equivalent to the light emitted by a single star at night (Cerdá-Olmedo, 2001). Bending is the consequence of a deceleration of about 6% in the growth rate of the side proximal to the direction of light, and an increase by the same rate on the distal side (Fig. 7.4). Because the refractive index of the sporangiophore contents exceeds that of air, the sporangiophore functions as a cylindrical lens, focusing unilateral light on the distal wall of the sporangiophore, resulting in more intense illumination of that side. Evidence in support of the lens effect is the demonstration that sporangiophores immersed in mineral oil with a higher refractive index than that of the sporangiophore contents function as a diverging lens and bend *away* from the light. The illumination of the edge of a sporangiophore by a narrow beam of light from a laser is followed by bending of the sporangiophore in a direction perpendicular to the light beam (Meistrich *et al.*, 1970). Photoreceptors are located in the plasma membrane (Fukshansky, 1993), and the transmission of the signal leads to localized wall softening and the synthesis of new cell wall

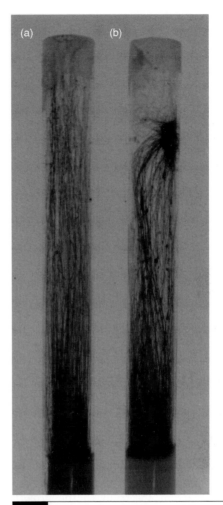

Fig 7.4 Sporangiophore development of *Phycomyces blakesleeanus* in standard test tubes (about 1.5 cm diameter). The tubes were wrapped except for the tip of tube (a) or a square on the right-hand side near the top of tube (b) In tube (a), the sporangiophores have grown straight towards the light, whereas in tube (b) they have bent towards the lateral light source.

material (Herrera-Estrella & Ruiz-Herrera, 1983; Ortega, 1990).

A central problem in studies of photoresponses is the nature of the photoreceptor(s). Two photoreceptors – one for low and the other for high light intensities – are involved in determining the phototropism in *Phycomyces*, and there are also two receptors each for light-induced microphore formation, macrophore formation, and carotenoid biosynthesis (Cerdá-Olmedo, 2001). A clue to the possible

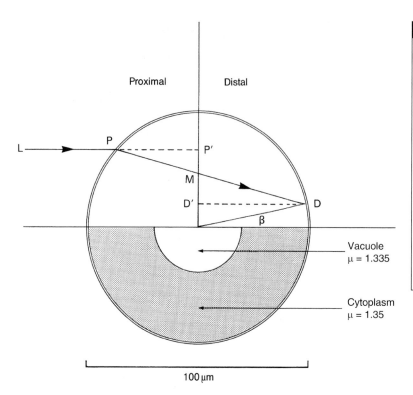

Proximal   Distal

L

P

P'

M

D'

D

β

Vacuole
μ = 1.335

Cytoplasm
μ = 1.35

100 μm

Fig 7.5 Sporangiophore of *Phycomyces* as a cylindrical lens. Upper portion: light ray L impinges from the left and is refracted at the first surface. The ratio of 'path length' of the light ray in the proximal part of the sporangiophore to the path length in the distal part is PM/DM. The maximum value of the angle β is about 20°. Lower portion: sporangiophore in section to show the peripheral layer of cytoplasm surrounding the central vacuole. The values are estimates of the refractive index of cytoplasm and vacuolar sap. Diagram modified from Bergman *et al.* (1969).

nature of the photoreceptors can be obtained by studying the action spectrum of the response over a range of light wavelengths. The phototropic curvature of *Phycomyces* sporangiophores has a similar action spectrum to the growth response of the vegetative mycelium, which is also stimulated by light. There are several clearly defined peaks at 485, 455, 385 and 280 nm, i.e. mostly in the blue part of the spectrum. Although β-carotene is present in large amounts in the growing zone of the sporangiophore, the photoreceptor system is more likely to comprise a flavin-type molecule and a pterin-type protein (Flores *et al.*, 1999; Galland & Tölle, 2003). Mutants with less than 0.1% of the wild-type β-carotene content remain fully photosensitive. However, β-carotene is involved in the other light-induced responses. The signalling chains involved in transduction of the light signal are only partially unravelled at present (Cerdá-Olmedo, 2001).

Moss and Baker (2002) have described a technique for demonstrating the phototropic response of *P. blakesleeanus* in the laboratory.

7.2.3 Sporangiophore development in *Phycomyces*

As the sporangiophore of *Phycomyces* develops it rotates. Castle (1942) followed the growth and rotation of the sporangiophore by attaching *Lycopodium* spores as markers and tracking the displacement of the markers. His findings are illustrated in Fig. 7.6. After a period of apical growth of the tubular sporangiophore (stage I), the sporangium appears as a terminal swelling and growth ceases. During this period (stage II), growth is limited to sporangial enlargement. In the next period (stage III) no further enlargement of the sporangium occurs, and elongation is also at a standstill. During stages IVA and IVB, elongation of the sporangiophore is resumed and growth is mainly localized in a zone somewhat below the sporangium. During stage I the tip of the sporangiophore rotates clockwise (as seen from above looking down) through a maximum angle of about 90°. There is no rotary movement during stages II and III. When sporangiophore elongation recommences in stage IVA, the direction of rotation is now *anti-clockwise* (as seen from

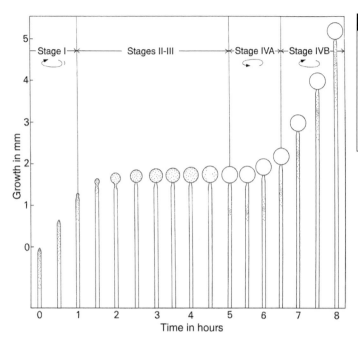

Fig 7.6 Diagram of developmental stages of the sporangiophore of *Phycomyces*. Regions in which growth is taking place are stippled. The rotary component of growth is indicated. During stage I the axis of growth is directed sinistrally, in stages II and III growth is unoriented. In Stage IVA dextral spiralling occurs and in Stage IVB sinistral spiralling again takes place.

above). During this stage, which lasts about an hour, markers attached to the growth zone may make up to two complete revolutions around the axis. During stage IVB the direction of rotation reverses once more. The reasons for the spiral growth are far from clear (see Ortega *et al.*, 2003). It is known that the chitin microfibrils which make up the wall of the sporangiophore show a right-handed or Z-spiral orientation. One possible explanation is that the laying down of the fibrils in this way is responsible for the rotation. A second is that the extension due to turgor pressure of a cylinder whose walls are composed of spirally arranged fibrils would naturally result in a passive rotation. The phenomenon of spiral growth is not peculiar to *Phycomyces*, occurring also during elongation of the sporangiophores of other members of the Mucorales such as *Thamnidium* and *Pilobolus*, and in various cylindrical plant cells.

The mechanical properties of the sporangiophore of *Phycomyces* change during development. During stage II, when no elongation of the sporangiophore is taking place, the sporangiophore shows elastic deformation when small loads are applied to it, i.e. the fractional change in length is directly proportional to the applied load, and on removal of the load, the

sporangiophore returns to its original length. During stage IV, although the sporangiophore changes in length in response to applied loads, upon unloading the sporangiophore does not return to its original length.

There is evidence that the spores secrete one or more unknown substances which control elongation of the sporangiophore. If mature sporangia are removed, growth of the sporangiophore ceases. Replacement of the detached sporangium with a substitute sporangium, with a suspension of spores, or with a drop of supernatant liquid from a centrifuged spore suspension, results in resumption of growth. Another effect of the removal of a ripe sporangium is that branching is induced in the sporangiophore, and this phenomenon has been likened to the breaking of apical dominance upon removal of the terminal bud in shoots of angiosperms.

7.2.4 Sporangium development

The tip of the sporangiophore expands to form the sporangium initial containing numerous nuclei which continue to divide. A dome-shaped septum is laid down and cuts off a distal portion which will contain the spores, from a cylindrical or subglobose spore-free core, the columella.

The columella is curved from its inception. Cleavage planes separate the nuclei within the sporangium, and finally the spores are cleaved out. They may be uninucleate or multinucleate according to the species, e.g. *M. hiemalis* and *Absidia glauca* have predominantly uninucleate spores (Storck & Morrill, 1977) whilst *M. mucedo*, *P. blakesleeanus*, *Rhizopus stolonifer* and *Syzygites megalocarpus* have multinucleate spores (Hammill & Secor, 1983). The number of spores formed is very variable. On nutrient-poor media minute sporangia containing very few spores may be formed, but in *P. blakesleeanus* the number of spores may be as high as 50 000–100 000 in a single sporangium of normal size.

The ultrastructure of developing sporangia of *Gilbertella* has been studied by Bracker (1968a) and is essentially similar in multisporous columellate sporangia of other Mucorales, e.g. *M. mucedo* (Hammill, 1981) and *Zygorhynchus heterogamus* (Edelman & Klomparens, 1994). One difference is that in *M. mucedo* mitotic division continues during sporangial development, which has not been reported from the other species

studied. The cleavage of the sporangial cytoplasm to form spores is accomplished by the fusion of membranous cleavage vesicles lined by electron-opaque granules. The vesicles are at first globose but coalesce and become flattened to form cleavage furrows. A three-dimensional network of cleavage furrows envelops the individual spore protoplasts, radiating outwards until they fuse with the sporangial plasma membrane. After cleavage, the flattened membrane which bounded the cleavage vesicle persists as the plasma membrane of the sporangiospore, whereas the electron-opaque granules make up part of the spore wall (Fig. 7.7). The columella is delimited from the rest of the sporangium by a process similar to that which cuts out the spores. Edelman and Klomparens (1994) noted that in *Z. heterogamus* the wall of the sporangium contains chitin, but the walls of the sporangiospores and columella do not. They have suggested that the columella may be a source of chitinase which causes enzymatic degradation of the mature sporangial wall whilst not affecting the walls of the spores or the columella itself.

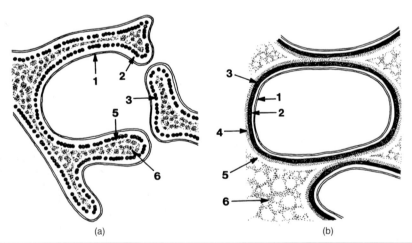

(a) (b)

Fig 7.7 *Zygorhynchus heterogamus*. Diagrammatic interpretation of ultrastructural transition from cleavage furrows isolating spore protoplasts (a) to post-cleavage spores (b). (a) shows the cleavage furrow membrane (1), an electron-translucent layer (2), a layer of fusing electron-opaque granules (3), an electron-transparent zone (5) and the electron-translucent matrix of the cleavage furrow (6). (b) shows the spore plasma membrane which was initially the cleavage furrow membrane (1), an electron-translucent layer (2), the electron-opaque layer which originated from the fusing granules of the cleavage furrows (3), an additional electron-translucent layer which had no corresponding layer within the cleavage furrows (4), a uniform zone of electron transparency, similarly placed in the cleavage furrows (5), and a non-uniform granular matrix separating the spores (6). Redrawn with permission from Edelmann and Klomparens (1994), *Mycologia*. ©The Mycological Society of America.

The sporangial wall is sometimes colourless or yellow, but it often darkens and develops a spiny surface due to the formation of crystals of calcium oxalate dihydrate (weddellite) beneath the surface layer (Fig. 7.14c; Jones *et al.*, 1976; Urbanus *et al.*, 1978; Whitney & Arnott, 1986). In some species similar crystals may develop on the sporangiophore. A possible function ascribed to these structures is that they form a barrier against grazing arthropods.

Despite the apparent similarity in sporangial structure across members of the Mucorales, spore liberation may be brought about by two different mechanisms (Ingold & Zoberi, 1963; Zoberi, 1985). In many of the commonest species of *Mucor* (e.g. *M. hiemalis*), the sporangium wall dissolves and the sporangium becomes converted at maturity into a 'sporangial drop' adhering to the columella. Sporangial walls which dissolve in this way are said to be **diffluent**. In large sporangia, for example of *M. plasmaticus*, *M. mucedo* and *Phycomyces*, the spores are embedded in mucilage. The sporangial wall does not break open spontaneously, but the slimy contents exude when the wall is touched. Such sticky spore masses are distributed by insects or rain splash, or by wind after drying. In the second spore liberation mechanism, the sporangial wall breaks into pieces, and here air currents or mechanical agitation readily liberate spores. An example of this is *Mucor plumbeus* (Figs. 7.14c,d).

In *M. plumbeus* the columella terminates in one or more finger-like or spiny projections (Fig. 7.14d), and in some *Absidia* spp. the columella may also bear a single nipple-like projection (Fig. 7.17b). In *Rhizopus stolonifer* the columella is large, and as the sporangium dries the columella collapses so that it appears like a basin balanced at the end of the sporangiophore (Fig. 7.16d). Associated with these changes in columella shape, the sporangium wall breaks up into many fragments and the dry spores can escape in air currents.

7.2.5 Sexual reproduction

The Mucorales reproduce sexually by a process of gametangial conjugation resulting in the formation of zygospores. By strict definition, what we describe as a zygospore is actually a **zygosporangium**, the dark warty ornamentation representing its outer wall (Benny *et al.*, 2001). According to this definition the zygosporangium contains a single globose zygospore, sometimes referred to as the zygospore proper. For convenience, we continue to use the term 'zygospore' in a wide sense.

Some species are homothallic, zygospores being formed in cultures derived from a single sporangiospore (e.g. *Rhizopus sexualis*, *Syzygites megalocarpus*, *Zygorhynchus moelleri* and *Absidia spinosa*). However, the majority of species are heterothallic and only form zygospores when compatible strains are mated together. It is believed that homothallic species are derived from heterothallic ancestors (O'Donnell *et al.*, 2001). There is, in reality, no absolute distinction between the homothallic and heterothallic conditions because some species normally homothallic or heterothallic are ambivalent, i.e. they can change their mating behaviour under certain conditions (Schipper & Stalpers, 1980). Zygospore formation is affected by environmental conditions, being generally favoured by darkness (Hesseltine & Rogers, 1987; Schipper, 1987). The effects of temperature are variable. In *Mucor piriformis* lower temperatures (0–15°C, optimum 10°C) favour zygospore formation, whilst for *Choanephora cucurbitarum* the optimum is 20°C (Michailides *et al.*, 1997).

In heterothallic species, if the appropriate strains are inoculated at opposite sides of a Petri dish, the mycelia grow out and a line of zygospores develops where they meet (Fig. 7.8). The two compatible strains rarely differ from each other in any obvious morphological or physiological features, although there may be slight differences in growth rate and carotenoid content. Because it was not possible to designate one strain as male and the other as female, Blakeslee (1906) labelled them (+) and (−). The two compatible strains are said to differ in mating type. The morphological events preceding zygospore formation are sufficiently similar to allow a general description of the process.

When two compatible strains approach each other, three reactions can be distinguished.

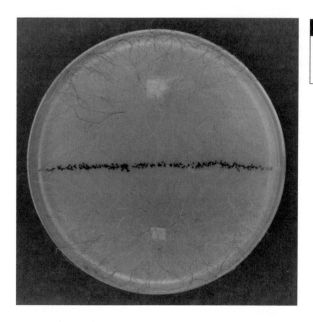

Fig 7.8 *Phycomyces blakesleeanus.* Malt agar plate 5 days after inoculation with a (+) and a (−) strain. A line of black zygospores has been produced where the two mycelia have met.

(1) A 'telemorphotic reaction' which involves the formation of aerial (or occasionally submerged) swollen hyphal tips. These are called **zygophores** or, when they have made contact with each other, **progametangia**. They are often coloured yellow due to a high β-carotene content. (2) A 'zygotropic reaction', in which directed growth of zygophores of (+) and (−) mating partners towards each other is observed. (3) A 'thigmotropic reaction', i.e. a touch response involving the events which occur after contact of the respective zygophores, such as gametangial fusion and septation of the progametangia to form gametangia and suspensors.

Hormonal control of sexual reproduction

The mating process is under the control of mating hormones (sex hormones, gamones, pheromones), and the hormones involved are effective in all members of the Mucorales studied (see Gooday, 1994; Gooday & Carlile, 1997). Early evidence of the involvement of pheromones was the demonstration that in *Mucor mucedo* the mating process can be initiated between mycelia of different mating types separated by a collodion membrane. The effect of the mating hormone is to switch the vegetative mycelium from asexual to sexual development. Other effects are the accumulation of carotenoids in cultures containing both mating types and, in *Phycomyces*, of a marked reduction in the growth rate of the vegetative mycelia as they approach each other (Drinkard *et al.*, 1982). The mating hormones have been identified as trisporic acid, actually a family of structurally related molecules, and its precursors. Trisporic acid was so named after *Blakeslea trispora* (see Fig. 7.27) from which this substance was first isolated (Austin *et al.*, 1969; Sutter, 1987). Liquid media inoculated with a mixture of (+) and (−) spores of *B. trispora* developed more intense yellow pigmentation than unmated cultures due to a massive stimulation of β-carotene synthesis.

Trisporic acid itself is derived from β-carotene (see Fig. 7.9) and is synthesized by collaborative metabolism of the two different mating type strains. Each strain has an incomplete enzyme pathway for the synthesis of trisporic acid so that intermediates accumulate which can only be metabolized further by mycelium of the opposite mating type. As shown in Fig. 7.9, the enzymatic steps in the conversion of β-carotene (**I**) to 4-dihydrotrisporol (**III**) via retinal (**II**) are common to both mating types. The (+) strain can convert 4-dihydrotrisporol to methyl-4-dihydrosporate (**IV**), whereas the (−) strain converts 4-dihydrotrisporol to trisporol (**V**). Thus **IV** and **V** function as two complementary prohormones, each of which is inactive in its own mycelium but is converted to the active hormone trisporic acid after diffusion into the mycelium of the complementary strain (Gooday, 1994). Trisporic acid stimulates further synthesis of β-carotene and of the two prohormones, leading to amplification of its own synthesis by a 'cascade' mechanism. The 15–20-fold enhanced synthesis of β-carotene upon mating of two compatible strains of *B. trispora* holds potential for commercial production of this substance (Lampila *et al.*, 1985; Sandmann & Misawa, 2002).

The role of the trisporic acid in inducing β-carotene synthesis and zygophore formation is widespread in the Mucorales, having been characterized in *Blakeslea, Phycomyces, Mucor* and even *Mortierella* (Schimek *et al.*, 2003). It is also known that trisporic acid is involved in the sexual response of some homothallic Mucorales such as *Zygorhynchus moelleri, Mucor genevensis* and *Syzygites megalocarpus* (Lampila *et al.*, 1985). The common nature of the hormones of homothallic

and heterothallic species could also be inferred from earlier observations of attempted matings between such forms, either at the interspecific or intergeneric level.

Zygophores show directional growth towards each other in response to volatile hormones. Gooday (1994) has suggested that these may be the mating type-specific prohormones, methyl-4-dihydrosporate of the (+) strain and trisporol of the (−) strain.

Thigmotropic reactions

When compatible zygophores make contact, they become firmly attached to each other and develop into progametangia. In *Mucor mucedo* there is evidence that the cell wall chemistry of the zygophores is distinct from that of the vegetative mycelium and that the (+) and (−) zygophores are bound together by **lectins**, i.e. glycoproteins exhibiting specific binding for polysaccharides (Jones & Gooday, 1977). In

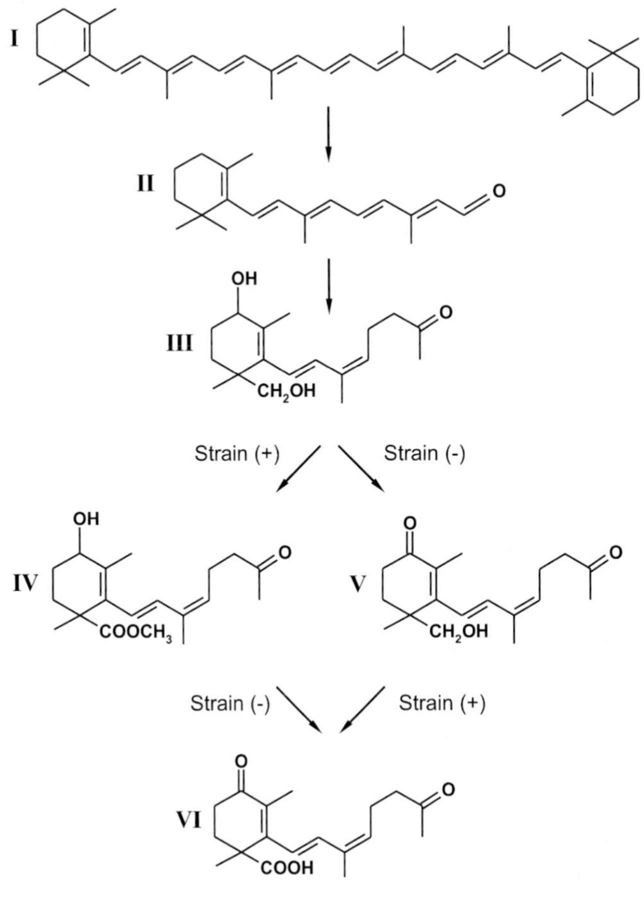

Fig 7.9 Collaborative biosynthesis of trisporic acid by cross-feeding of intermediates between (+) and (−) mating types of *Blakeslea trispora*. β-Carotene (**I**) is metabolized by both (+) and (−) mating-types via retinal (**II**) to 4-dihydrotrisporol (**III**). This is metabolized by (+) strains to 4-dehydrosporic acid and its methyl ester (**IV**) and by (−) strains to trisporol (**V**). These are converted to trisporic acid (**VI**) only after diffusing to the (−) and (+) strains, respectively. Redrawn from Gooday (1994), with kind permission of Springer Science and Business Media.

Phycomyces, after arrest of the growth of vegetative hyphae, certain submerged hyphal tips develop short branches called 'knobbly knots' (Fig. 7.18) which break through the surface of the agar and become progametangia (O'Donnell *et al.*, 1976; Sutter, 1987). The progametangia become tightly appressed and their close contact is enhanced by the formation of extracellular fimbriae whose presence appears to be essential for further development in *Phycomyces* (Yamakazi & Ootaki, 1996) as well as in other groups of fungi (see p. 652). Fimbriae may be the lectin-bearing structures. In other Mucorales the zygophores are aerial and club-shaped. The tip of each progametangium becomes cut off by a septum to separate a distal multinucleate gametangium from the subterminal suspensor (see Fig. 7.10).

The walls separating the two gametangia break down so that the numerous nuclei from each cell become surrounded by a common cytoplasm. The fusion cell, or zygote, swells and develops a dark warty outer layer to become the zygospore.

Cytology of zygospore formation

There have been numerous accounts of the cytology of zygospore formation. Meiosis usually occurs before zygospore germination, so that the zygospore can be regarded as a meiosporangium. Four main types of nuclear behaviour can be distinguished.

(1) In *Mucor hiemalis*, *Absidia spinosa* and some other species, all the nuclei fuse in pairs within a few days, then quickly undergo meiosis so that the mature zygospore contains only haploid nuclei.

(2) In *Rhizopus stolonifer* and *Absidia glauca*, some of the nuclei entering the zygospore do not pair, but degenerate. The remainder fuse in pairs, but meiosis is delayed until germination of the zygospore.

(3) In *Phycomyces blakesleeanus*, the haploid nuclei continue to divide mitotically in the young zygospore and then become associated in groups, with occasional single nuclei also present. Before germination some of the nuclei pair up, and in the germ sporangium diploid nuclei and also haploid nuclei are found; some of

these may be products of meiosis, others may represent the scattered solitary nuclei which failed to pair up.

(4) In *Syzygites megalocarpus* mitotic nuclear divisions continue in the young zygospore, but nuclear fusion and meiosis apparently do not occur. This fungus can therefore be described as **amictic** (Burnett, 1965).

The fine structure of zygospore development has been studied in the homothallic *Rhizopus sexualis* (Hawker & Beckett, 1971; Ho & Chen, 1998). Following contact of the tips of the two zygophores, their walls adhere to each other and become flattened (Figs. 7.10a–c) to form the **fusion septum**. On either side of the fusion septum, each cell becomes distended to form a progametangium. In each progametangium, an oblique septum, concave to the developing zygospore, develops by gradual inward extension mediated by the coalescence of vesicles. When the septum is complete, it separates the terminal gametangium from the progametangial base now called the **suspensor**. However, cytoplasmic continuity between the suspensor and the gametangium persists through a series of pores which probably enable nutrients to flow into the developing zygospore from the surrounding mycelium. Numerous nuclei congregate on either side of the fusion septum. It has been estimated that there may be over 150 nuclei in a pair of progametangia, but the number may rise to over 300 in a pair of completely delimited gametangia, reflecting further nuclear divisions.

The breakdown of the fusion septum is associated with an accumulation of vesicles in the vicinity of the dissolving wall. These may contain wall-degrading enzymes. The fusion septum is completely dissolved, and once the cytoplasmic contents of the two gametangia are continuous, the nuclei become arranged in the periphery of the cytoplasm. In *R. sexualis*, it is probable that most of the gametangial nuclei fuse in pairs immediately, and that the fusion nuclei then quickly divide.

Even before dissolution of the fusion wall is complete, the primary outer wall of the zygote thickens, and beneath this original wall, the warts (which will eventually ornament the wall

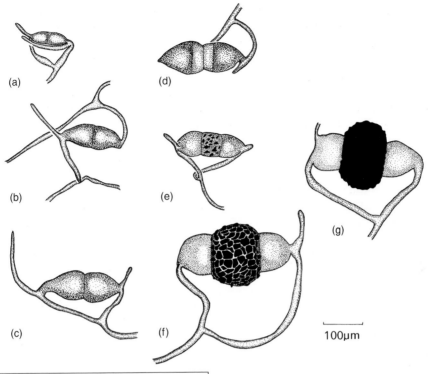

Fig 7.10 *Rhizopus sexualis.* (a–g) Successive stages in the formation of zygospores. The fungus is homothallic.

of the mature zygospore) are initiated as widely separated patches shaped like inverted saucers (Fig. 7.11c). The cytoplasm fills the domes of the 'saucers' and also balloons out between them, enveloped by the plasmalemma. As the zygospore continues to enlarge, the saucers change in shape and size to resemble inverted flower pots which increase in size by the addition of new material at their rims until they are contiguous. From this moment onwards, electron microscopy fixatives can no longer penetrate, explaining why cytological studies of later stages of zygospore development have proven difficult. Eventually, the tips of the warts become pushed through the original primary wall. At least three wall layers are deposited beneath the original primary wall (Fig. 7.11g). The darkening of the wall is probably due to the deposition of melanin. The sculpturing of the zygospore wall of other members of the Mucorales, as seen by scanning electron microscopy, shows different patterns, ranging from circular or conical warts to branched stellate warts. Essentially similar zygospore development has been reported in the heterothallic *Gilbertella persicaria* (O'Donnell *et al.*, 1977a).

In *M. mucedo* and *P. blakesleeanus*, the wall of the zygospore is rich in sporopollenin (Gooday *et al.*, 1973; Furch & Gooday, 1978). This substance, which is also present in the walls of pollen grains, is extremely resistant to degradation and enables zygospores to remain dormant but undamaged in the soil for long periods. Sporopollenin is formed by oxidative polymerization of β-carotene, and this may explain the high content of this pigment in developing zygophores. However, β-carotene and sporopollenin appear to be absent from the zygospore walls of *R. sexualis* (Hocking, 1963).

Zygospore investment

In *Phycomyces* and *Absidia*, the suspensors may bear appendages which arch over the zygospore. In *Phycomyces* the suspensor appendages are black

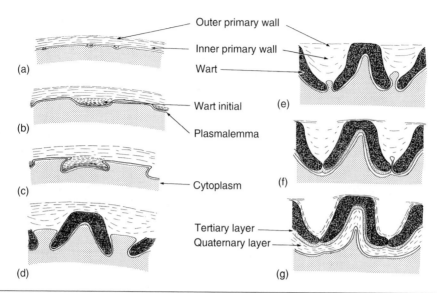

Outer primary wall
Inner primary wall
Wart
Wart initial
Plasmalemma
Cytoplasm
Tertiary layer
Quaternary layer

(a) (b) (c) (d) (e) (f) (g)

Fig 7.11 Development of the zygospore wall in *Rhizopus sexualis* (diagrammatic, after Hawker & Beckett, 1971). (a) Primary wall before inflation of zygospore, showing thin electron-dense outer layer, thicker less electron-dense inner one, and scattered lomasome-like bodies. (b) Blocks of secondary material (wart initials) developing locally on inner surface of primary wall. (c) Wart initials growing by deposition of secondary material at the rims to give saucer-shaped pigmented masses. (d) Warts becoming flower pot shaped by further growth at rims, inner layer of primary wall becoming gelatinous and swollen. Note pockets of cytoplasm between warts. (e) Rims of warts nearly touching, inner layer of primary wall showing stress lines, pockets of cytoplasm between warts much reduced. (f) Edges of warts touching, warts lined with tertiary smoothing layer, outer layer of primary wall torn. (g) Thick stratified impermeable layer of quaternary material laid down inside smoothing layer, inner gelatinous layer of primary wall has collapsed as a horny skin enveloping the warts.

and forked, whilst in *Absidia* they are hyaline and coiled or curved inwards (see Figs. 7.17, 7.18). The function of such appendages is unknown; possibly they assist in attaching zygospores to passing animals. The forked appendage tips of *Phycomyces* bear a drop of liquid, and they have been interpreted as hydathodes (i.e. water-secreting structures). In the homothallic species *A. spinosa* the appendages arise on only one suspensor.

Mating behaviour
Analysis of the results of crosses involving several genes suggest that there is a single mating type locus with two alternative alleles, (+) and (−), which segregate at meiosis. However, no DNA sequences of this locus have as yet been published, and there are also a number of anomalous results for which a full cytological explanation is still awaited.

Hybridization experiments have been conducted between different species and genera of Mucorales, and in some cases imperfect zygospores are formed. Attempted copulation has also been observed between homothallic and heterothallic strains. An unusual type of mating behaviour has been discovered in *Mucor pusillus* which is predominantly heterothallic but in which homothallic strains are known. It has been possible to induce a (+) strain to mutate to a (−) strain, and also to a homothallic strain by γ-irradiation (Nielsen, 1978).

7.2.6 Zygospore germination
After a resting period the zygospore may germinate by developing a **germ sporangium** which resembles an ordinary sporangium and contains sporangiospores of the normal type. In some cases vegetative mycelium develops from the germinating zygospore. The conditions for zygospore germination are, in many cases, imperfectly known, but a protocol for germination has been established for *Phycomyces blakesleeanus* (Eslava & Alvarez, 1987). Mature zygospores

collected from 6-week-old cultures and placed on moist filter paper at 22°C under alternating light/dark illumination will germinate after about 8 days, reaching maximum germination after a further 8–10 days. In *Mucor piriformis*, the germination rate is highest in fresh zygospores (Guo & Michailides, 1998), a vigorous germ tube emerging through one of the suspensors or through a crack in the zygospore wall. The germ tube may continue development as mycelium or grow into the air and form a germ sporangium at the tips of single or branched sporangiophores.

Mating-types represented in germ sporangia
The distribution of mating types amongst the germ spores which are present in germ sporangia falls into three categories.

1. Pure germinations in which all the spores are homothallic, e.g. in *Mucor genevensis*, *Zygorhynchus dangeardi* and *Syzygites megalocarpus*.

2. Pure germinations in which all sporangiospores are of one mating type, i.e. all (+) or all (−). *Mucor mucedo*, *M. hiemalis* and *P. blakesleeanus* generally behave in this way. In *P. blakesleeanus*, the analysis of progeny from crosses involving up to four unlinked factors which included mating type were best explained on the basis of the survival of a single diploid nucleus from the thousands which are present in the young zygospore (Cerdá-Olmedo, 1975; Eslava *et al.*, 1975a,b). This single diploid nucleus undergoes meiosis and one or more of the resultant nuclei divide mitotically to provide nuclei for the germ sporangium (Fig. 7.12). Occasionally two or three diploid nuclei may survive and undergo meiosis. In some germ sporangia heterokaryotic spores are present. If these are heterokaryotic for mating type, the mycelium which develops from them may be abnormal and 'neuter', i.e. it is unable to mate with (+) as well as (−) strains.

3. Mixed germinations. In *Phycomyces nitens*, the same germ sporangium sometimes contains (+), (−) and homothallic (i.e. self-fertile) spores. The finding that diploid nuclei enter the germ sporangium may be the explanation for the presence of homothallic spores which should properly be described as **secondarily** homothallic. Mixed germinations have also been reported by Gauger (1961) for *Rhizopus stolonifer* in which both (+) and (−) spores were present in some germ sporangia, whereas others contained spores of either mating type. For this type of mixed germination to occur, it would be necessary only to postulate the survival of more than one meiotic product so that both mating types are represented in the sporangium. 'Neuter' spores were found in some germ sporangia. Thus, in *Choanephora cucurbitarum* mixed germinations have been reported in which the majority (usually all) of the germ sporangia contained only either (+) or (−) spores, but a low proportion gave heterokaryotic spores of mating-type (+/−). A characteristic feature of *C. cucurbitarum* cultures derived from heterokaryotic (+/−) germ spores or fusion of (+) with (−) protoplasts is that they produce azygospores (Yu & Ko, 1996, 1999; see below).

7.2.7 Azygospores

In some Mucorales, if gametangial copulation fails to take place normally, one or both gametangia may give rise parthenogenetically to a structure morphologically similar to the zygospore, termed an **azygospore** (azygosporangium). Azygospores therefore usually appear as warty spherical structures borne on a single suspensor-like cell, or occasionally on a sporangiophore. They are formed regularly in cultures of *Mucor bainieri* and *M. azygospora* (Fig. 7.13), both of which are obligately azygosporic and do not form true zygospores (Benjamin & Mehrotra, 1963), and they have also been reported in *Rhizopus azygosporus* (Yuan & Yong, 1984). The development of azygospores of *M. azygospora* resembles that of normal zygospores in other Mucorales (O'Donnell *et al.*, 1977b; Ginman & Young, 1989). Azygospore formation may occur in intergeneric and interspecific crosses, for example in crosses between a (+) strain of *Gilbertella persicaria* and a (−) strain of *Rhizopus stolonifer* (O'Donnell *et al.*, 1977c) and between different species of *Rhizopus* (Schipper, 1987). Azygospore development has also been seen in intraspecific crosses, e.g. in certain isolates of *M. hiemalis* (Gauger, 1966, 1975). These azygosporic isolates of *M. hiemalis* were derived from

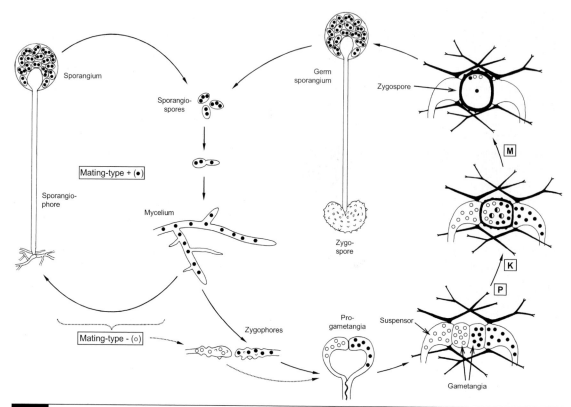

Fig 7.12 Life cycle of *Phycomyces blakesleeanus* (diagrammatic and not to scale). From a coenocytic haploid mycelium of either mating type (+) or (−), sporangiophores develop. Sporangia are columellate and contain numerous sporangiospores which, in *P. blakesleeanus*, are multinucleate. When hyphae of both mating types meet, sexual reproduction is initiated by the formation of knobbly zygophores which develop into progametangia. Each progametangium divides into a gametangium and a suspensor, the latter ornamented by black forked appendages. Plasmogamy (P) occurs by lysis of the wall separating the two multinucleate gametangia. This is followed by mass karyogamy (K), but only one of the numerous diploid fusion nuclei seems to undergo meiosis (M), and only one of the resulting tetrad nuclei survives in the zygospore during dormancy, so that the sporangiospores in the germ sporangium are usually of either one or the other mating type. Open and closed circles represent haploid nuclei of opposite mating type; diploid nuclei are larger and half-filled.

spores of germ sporangia developed from normal zygospores. If the azygosporic strains are subcultured, either from single sporangiospores or by mass transfer, they show a tendency to 'break down' to strains of (+) or (−) mating type of normal appearance. It seems that azygosporic strains of *M. hiemalis* are typically diploid and heterozygous for mating type, i.e. the diploid nucleus carries both (+) and (−) mating type alleles. The breakdown to the normal (+) or (−) mating type condition may be brought about by somatic (i.e. non-meiotic) reduction leading to aneuploid intermediates, and finally to haploids. The germination of azygospores is unknown.

7.3 | Examples of Mucorales

As mentioned before, the traditional family classification within the Mucorales is artificial (see Benny *et al.*, 2001; Tanabe *et al.*, 2004), and we use it here solely for convenience of presentation.

7.3.1 Mucoraceae
Mucor
About 50 species of *Mucor* are currently known (Kirk *et al.*, 2001). The genus is cosmopolitan, with

20 μm

Fig 7.13 Azygospore of *Mucor azygospora*. Original image kindly provided by T. W. K. Young.

many species being widespread in soil or on substrates in contact with soil. Most species are mesophilic (growing at 10–40°C with an optimum 20–35°C), but some, e.g. *M. miehei* or *M. pusillus* (sometimes classified as species of *Rhizomucor*; see Mouchacca, 1997, 2000) are thermophilic, with a minimum growth temperature of about 20°C and a maximum extending up to 60°C (Cooney & Emerson, 1964; Maheshwari *et al.*, 2000). *Mucor indicus* and *M. circinelloides* are used as starters in food processing to break down starchy polysaccharides in rice, cassava and sorghum, releasing simple sugars for the preparation of fermented foods or alcohol production (Hesseltine, 1991).

Most species of *Mucor* grow rapidly on agar at room temperature, filling a Petri dish in 2–3 days with their coarse aerial mycelium. When incubated in liquid culture under semi-anaerobic conditions, several species grow in a yeast-like state. The ability to switch between the yeast-like and filamentous state is termed **dimorphism**, a phenomenon which has been studied in greatest detail in *M. rouxii* (see Fig. 7.14), but also occurs in *M. circinelloides*, *M. fragilis*, *M. hiemalis*, *M. lusitanicus* and in other Mucorales (Orlowski, 1991, 1995). Sporangia are globose and borne on branched and unbranched sporangiophores growing into the air. The columella is large and typically elongated (Figs. 7.2 and 7.3). Zygospores are rarely formed in agar culture because most species are heterothallic. Amongst the most common species from soil are *M. hiemalis*, *M. racemosus* and *M. spinosus* (Domsch *et al.*, 1980). Several species of Mucor, e.g. *M. mucedo*, fruit on dung (Ellis & Ellis, 1998; Richardson & Watling, 1997), and they are the earliest fungi to appear in the succession of fungal fruit bodies on this substrate (Dix & Webster, 1995). The sporangiospores of coprophilous *Mucor* spp. survive digestion by herbivorous mammals.

A few species of *Mucor* are human pathogens. The term **mucormycosis**, however, usually refers to conditions caused by Mucorales generally rather than the genus *Mucor* (Rinaldi, 1989; Eucker *et al.*, 2001) because it is not possible to identify species by the microscopic appearance of their coenocytic mycelium within diseased tissue. Diagnosis is dependent on the isolation and identification of the suspected pathogen in culture, sometimes *post mortem*. By these means, several ubiquitous species of *Mucor* have been associated with disease symptoms, including *M. circinelloides*, *M. hiemalis* and *M. racemosus*. Infections are opportunistic, derived from sporangiospores present in the soil or air, and are usually associated with patients suffering from other diseases such as diabetes, leukaemia, AIDS and post-operative conditions. There are no records of person-to-person transmission. Mucormycoses are serious, even fatal in immunocompromised patients, although some can be successfully treated by surgery and antibiotics such as amphotericin B (Kwon-Chung & Bennett, 1992).

Schipper (1978) has given a key to 49 species of *Mucor*, and Watanabe (1994) has described the six homothallic and two azygosporic species.

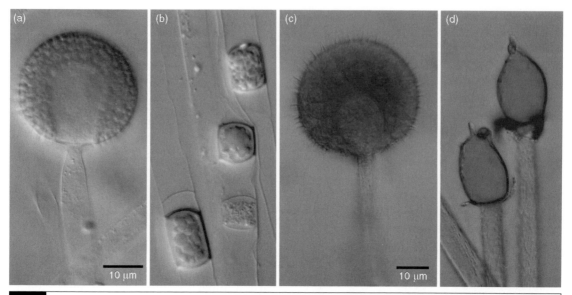

Fig 7.14 *Mucor racemosus* (a,b) and *M. plumbeus* (c,d). (a) Tip of a sporangiophore which has formed a sporangium. The columella (arrow) is visible. (b) Lower region of sporangiophores showing intercalary thick-walled chlamydospores which are typical of the species. (c) Sporangium with a spiny surface of calcium oxalate crystals. (d) Exposed columellae with finger-like projections. (a,b) to same scale; (c,d) to same scale.

Zygorhynchus

There are about six species, mostly reported from soil, often from considerable depth (Hesseltine *et al.*, 1959). All species are homothallic and, unusually, form heterogametangic zygospores (Fig. 7.15). The sporangiophores are commonly branched and the columella is often broader than high. The most frequently encountered species is *Z. moelleri*, which has been isolated worldwide from a range of soils and from the rhizosphere of numerous plants (Domsch *et al.*, 1980). Most species are mesophilic, but *Z. psychrophilus* forms zygospores readily at 5°C. Sporangium development in *Z. heterogamus* has been studied by Edelmann and Klomparens (1994) (see p. 171). Zygospore development and structure in several species of *Zygorhynchus* have been described by O'Donnell *et al.* (1978a). The warts on the outside of the zygosporangia often appear as interlocking, starfish-like pointed thickenings. The inner wall of the zygosporangium is ornamented by a network of ridges and grooves radiating from centres corresponding to the points of the warts. The outer wall of the zygospore proper, lying within the zygosporangium, is similarly ornamented by a pattern of radiating grooves and ridges which are a template of the lining of the zygosporangial wall.

Rhizopus

There are about 10 species which grow in soil (Domsch *et al.*, 1980) and on fruits, other foods and all kinds of decaying materials. *Rhizopus* spp. also occur frequently as laboratory contaminants. *Rhizopus stolonifer* (syn. *R. nigricans*) grows rapidly. It is often found on ripe fruits, especially if these are incubated in a moist atmosphere (see Plate 3d). Characteristic features of *Rhizopus* are the presence of rhizoids at the base of the sporangiophores (which may grow in clusters), and the stoloniferous habit (Fig. 7.16). An aerial hypha grows out, and where it touches on the substratum it bears rhizoids and sporangiophores. Growth in this manner is repeated. The sporangium wall is brittle and the sporangiospores are dry and wind-dispersed. Some species of *Rhizopus*, e.g. *R. oryzae*, *R. microsporus* and its allies, are used as starters in ragi fermentations of rice (Hesseltine, 1991). Several species (*R. arrhizus*, *R. microsporus*, *R. rhizopodiformis*) are

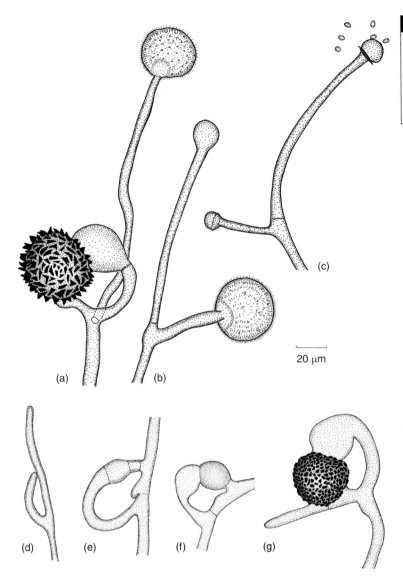

Fig 7.15 *Zygorhynchus moelleri.*
(a) Zygospore and sporangium.
(b) Young sporangiophores.
(c) Dehisced sporangia. (d—g) Stages in zygospore formation. Note that the fungus is homothallic and that the suspensors are unequal.

20 μm

(a) (b) (c)

(d) (e) (f) (g)

human pathogens associated with mucormycosis (Rinaldi, 1989). Most species of *Rhizopus* are heterothallic, but *R. sexualis* is homothallic and forms zygospores freely within 2 days in the laboratory (see Fig. 7.10).

Rhizopus microsporus causes rice seedling blight in which root growth is strongly impaired by a toxin, rhizoxin, excreted by the soil-borne pathogen. The toxin binds to β-tubulin, thereby interfering with mitosis. Intriguingly, it is synthesized not by *R. microsporus* but by bacteria (*Burkholderia* spp.) living endosymbiotically within the cytoplasm of *Rhizopus* hyphae (Partida-Martinez & Hertweck, 2005). Bacterial endosymbionts have been reported only rarely from fungi, e.g. in the zygomycete *Geosiphon pyriforme* (p. 221), or within hyphae of the ascomycete *Morchella elata* (p. 427) and the basidiomycete *Laccaria bicolor* (p. 552).

Absidia

There are some 20 species growing in soil (Domsch *et al.*, 1980). Characteristic features are pear-shaped sporangia arising in partial whorls along stolon-like branches which produce rhizoids at intervals but not opposite the

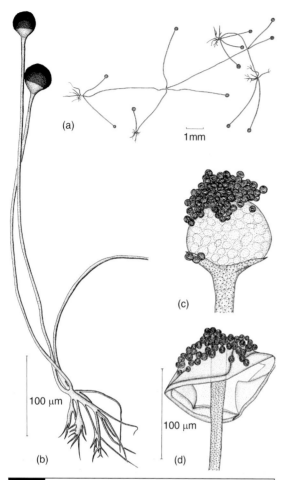

Fig 7.16 *Rhizopus stolonifer.* (a) Habit sketch, showing stolon-like branches which develop rhizoids and tufts of sporangiophores. (b) Two sporangiophores showing basal rhizoids. (c) Dehisced sporangium showing the columella with attached spores. (d) Invaginated columella.

sporangiophores. The zygospores are surrounded by curved unbranched suspensor appendages which may arise from either or both suspensors (Fig. 7.17). Most species are heterothallic but *A. spinosa* is homothallic. *Absidia glauca* and *A. spinosa* are amongst the most commonly isolated species. *Absidia corymbifera* is a human pathogen.

Phycomyces

The two best-known species are *P. blakesleeanus* and *P. nitens.* The sporangiospores of *P. nitens* are larger than those of *P. blakesleeanus*, but it is likely that many workers confused the two and much of the early literature on *P. nitens* probably refers to *P. blakesleeanus*. Neither species is particularly common, but likely substrata are fatty products and empty oil casks. Bread, dung and decaying hops are other recorded substrata. The zygospores are unusual in that they are overarched by black, forked suspensor appendages (see Fig. 7.18; O'Donnell *et al.*, 1976, 1978b). Great interest has been focused on the development and sensory perception of the spectacularly large sporangiophore, especially to light (see pp. 169–171), and on the genetics of *Phycomyces* (Eslava & Alvarez, 1987). The genus has been classified in the family Phycomycetaceae by Benny *et al.* (2001).

Syzygites

Syzygites megalocarpus (= *Sporodinia grandis*; see Hesseltine, 1957) is found on the decaying basidiocarps of various toadstools, especially *Boletus, Lactarius* and *Russula*. It grows readily in culture and is homothallic. Probably because of this fact it was the first member of the Mucorales for which sexual reproduction was described in detail (Davis, 1967). The sporangiophores are dichotomous and bear thin-walled sporangia (Fig. 7.19a). In culture, light is essential for the development of sporangiophores but has no effect on zygospore formation. Lowering of the osmotic potential markedly stimulates zygospore development and this is possibly significant ecologically in that the drying out of basidiocarp tissue on which the fungus grows might induce the development of the zygosporic (resting) state (Kaplan & Goos, 1982).

7.3.2 Pilobolaceae

There are two common genera, *Pilobolus* and *Pilaira*, which grow on the dung of herbivores. Both genera have evolved mechanisms to ensure that their sporangia escape from the vicinity of the dung patch on which they were produced. In *Pilobolus* the sporangiophore is swollen and the sporangium is shot away violently by a jet of liquid, whilst in *Pilaira* the sporangiophore is elongated and the sporangium becomes converted into a sporangial drop, breaking off

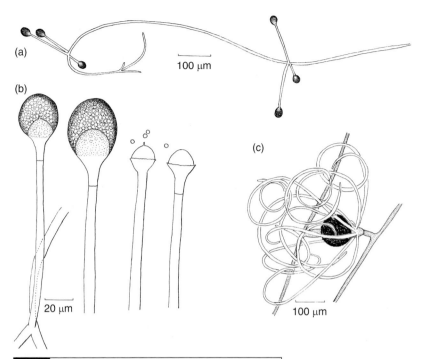

Fig 7.17 *Absidia glauca.* (a) Habit showing whorls of pear-shaped sporangia. (b) Intact and dehisced sporangia. Note the single pointed projection on certain columellae. (c) Zygospore showing the arching suspensor appendages.

upon contact with an object. The sporangia are black and melanized, presumably as a protection against UV irradiation. Grove (1934) has written a monograph of the family which has stood the test of time.

Pilobolus

The generic name means literally the 'hat thrower', referring to the sporangial discharge mechanism. If fresh herbivore (e.g. rabbit, sheep, deer, horse) dung is incubated in light, the characteristic bulbous sporangiophores of *Pilobolus* appear after a preliminary phase of fruiting of *Mucor* which may last for 4–7 days (Fig. 7.20, Plate 3f). Nine *Pilobolus* spp. previously recognized have been reduced to five, but including a number of varieties (Hu *et al.*, 1989). Common species are *P. crystallinus*, *P. kleinii* (= *P. crystallinus* var. *kleinii*), and *P. umbonatus* (= *P. roridus* var. *umbonatus*). Here we adopt the nomenclature proposed by Grove (1934). As far as

is known all members of the Pilobolaceae are heterothallic.

A full account of the development and discharge of the sporangium has been given by Buller (1934) and Ingold (1971). Discharged sporangia of *Pilobolus* become attached to vegetation surrounding the dung on which they were produced. When the vegetation is eaten by a herbivore, the spores are released into the gut. In the voided faeces, the spores germinate to form a mycelium. After about 4 days, the mycelium near the surface of the dung pellet forms **trophocysts**, swollen hyphal segments coloured yellow by carotenoids (Fig. 7.20). Sporangiophores develop from the trophocysts in a regular daily sequence, and the stage of development can be correlated with the time of day. During the late afternoon the sporangiophore grows away from the trophocyst towards the light and during the night its tip enlarges to become the sporangium. The swelling of the subsporangial vesicle takes place mainly between midnight and the early morning. Young sporangiophores are highly phototropic even before their sporangia are differentiated, and the clear tip of the developing sporangiophore is

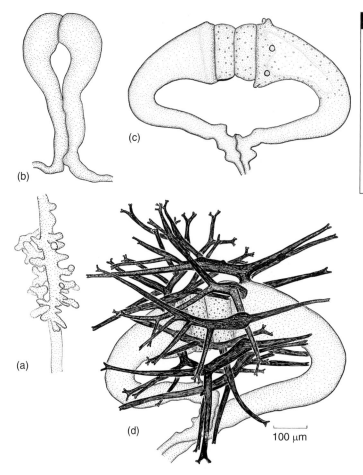

(a)

(b)

(c)

(d)

100 μm

Fig 7.18 *Phycomyces blakesleeanus.* Stages in zygospore formation. The fungus is heterothallic. (a) Zygophores consisting of knobbly hyphal branch tips which become closely appressed. (b) Paired club-shaped progametangia which develop from the appressed zygophores. (c) Septation of the progametangia to form terminal gametangia and subterminal suspensors. Appendages are developing on the suspensor to the right. (d) Young zygospore overarched by dichotomous suspensor appendages.

the sensitive region. Despite the bright yellow carotenoid deposits in the trophocysts and young sporangiophores, studies of the phototropic response to light of various wavelengths suggest that the photoreceptor in the sporangiophore is more likely to be a flavonoid than a carotenoid (Page & Curry, 1966). Fully developed sporangiophores are also highly phototropic. Light projected along the axis of the sporangiophore is brought to a focus at a point beneath the swollen vesicle termed the **ocellus**. In this region, there is an accumulation of carotenoid-rich cytoplasm which glows orange when illuminated (Plate 3f). When light falls asymmetrically onto the sporangiophore, it is focused onto the back of the subsporangial vesicle near its base, and some stimulus is probably transmitted to the cylindrical part of the sporangiophore, resulting in more rapid growth of the wall facing away from the light. Curvature of the whole sporangiophore thus occurs until it is again orientated parallel to the incident light (see Fig. 7.21).

The structure of the sporangium differs in a number of ways from that of the Mucoraceae. The sporangium is hemispherical, and its wall is dark black, shiny, tough and unwettable. At the base of the sporangium is a conical columella, which is separated from the spores by a pad of mucilage. During late morning the sporangium cracks open by a suture running around the base, just above the columella. The spores are prevented from escaping by the mucilaginous pad which protrudes through the crack in the sporangium wall as a ring of mucilage (Figs. 7.20e,f). The subsporangial vesicle is turgid, and the osmotic pressure of the liquid has been estimated to be around 5.5 bars (Buller, 1934). Drops of liquid decorate the outside of

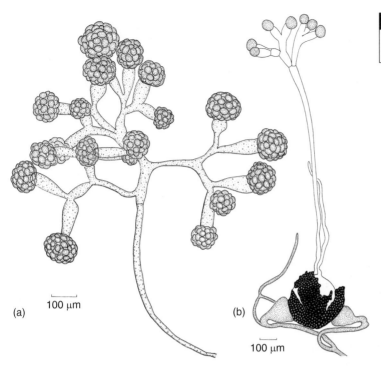

(a)

100 μm

(b)

100 μm

Fig 7.19 *Syzygites megalocarpus.*
(a) Sporangiophore. (b) Germinating zygospore. The fungus is homothallic.

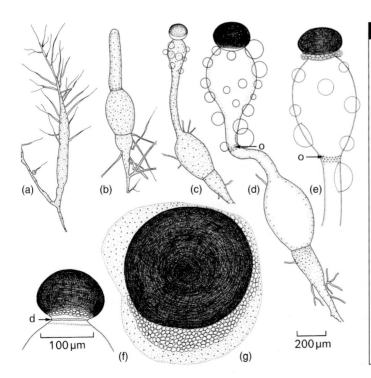

(a)   (b)   (c)   (d) o   (e) o

d

100 μm

(f)   (g)

200 μm

Fig 7.20 Asexual reproduction in *Pilobolus kleinii*. (a) Developing trophocyst which is becoming distended by carotenoid-rich cytoplasm. (b) Trophocyst with immature sporangiophore. The clear tip of the sporangiophore is light-sensitive.
(c) Trophocyst bearing a developing sporangium. The upper part of the sporangium is beginning to darken. Globules of liquid accumulate on the sporangiophore surface (9.00 p.m.). (d) Trophocyst with sporangium which has not yet dehisced (9.00 a.m.). The arrow (o) points to a carotenoid-rich band of cytoplasm called the ocellus.
(e) Sporangiophore bearing a sporangium which has dehisced near its base. Spores have extruded and are held in place by a ring of mucilage (11.30 a.m.). (f) Sporangium showing dehiscence line at its base (d). (g) Discharged sporangium surrounded by dried-out vesicular sap. The spores are enclosed in mucilage. (a−e) to same scale; (f,g) to same scale.

the sporangiophore of *Pilobolus*, as they do in many zygomycetes. Eventually, usually about midday, the sporangial vesicle explodes at a line of weakness just beneath the columella. Due to the elasticity of the vesicle wall the liquid contents are squirted out, projecting the entire sporangium forward in the direction of the light. Photographs of the jet show that it is at first cylindrical but eventually breaks up into fine droplets (Fig. 7.22c; Page, 1964). In *P. kleinii*, the velocity of projection varies between wide limits of $4.7–27.5\,\mathrm{m\,s^{-1}}$ with a mean of $10.8\,\mathrm{m\,s^{-1}}$ (Page & Kennedy, 1964). The sporangia can be projected vertically upwards for as much as 2 m and horizontally for up to 2.5 m. On striking a grass blade or other herbage, the sporangium becomes attached in such a way that the mucilaginous ring adheres to it, with the black sporangium wall facing outwards. Buller (1934) has suggested that the projectile contains a drop of liquid attached to the sporangium (Fig. 7.22a). When the projectile strikes an object the liquid flows around the sporangium, but because the sporangium wall is hydrophobic and the base of the sporangium is surrounded by the wettable mucilaginous ring, the sporangium turns round in the liquid so that its wall faces outwards (Fig. 7.22b). The non-wettable nature of the sporangial wall may be related to the presence on its surface of hollow, blunt-tipped spines and crystals as seen by electron microscopy (Bland & Charles, 1972). As the mucilage dries, the sporangium becomes cemented onto the surface which it struck. The spores of *Pilobolus* are released only after the sporangium has been ingested by an animal. They survive gut passage and are voided with the faeces. A film featuring the life cycle of *Pilobolus* has been made (Webster & Hard, 1999). An unexpected consequence of the attachment of *Pilobolus* sporangia to herbage is that the sporangia may act as vectors for parasitic nematodes such as *Dictyocaulus* spp., which multiply on dung and, when ingested, cause lungworm disease in sheep, cattle and some wild mammals.

The physiology of *Pilobolus* shows a number of interesting features possibly related to its coprophilous habit. Spores germinate best above

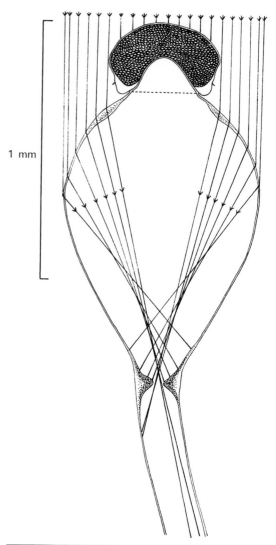

1 mm

Fig 7.21 *Pilobolus kleinii*. Diagrammatic L.S. of sporangiophore showing the path of light rays falling parallel to the axis of the sporangiophore which are brought to a focus beneath the subsporangial vesicle. The sporangiophore illustrated is orientated symmetrically with respect to the incident light. Note the mucilaginous ring extruded through the sporangial wall at its base (after Buller, 1934).

pH 6.5, and can be induced to germinate by treatment with alkaline pancreatin. Germination can also be triggered by hexoses such as glucose and mannose. Mycelial growth occurs over a wide range of temperatures, with optimum temperatures at 25–35°C. Growth on synthetic media with asparagine and acetic acid

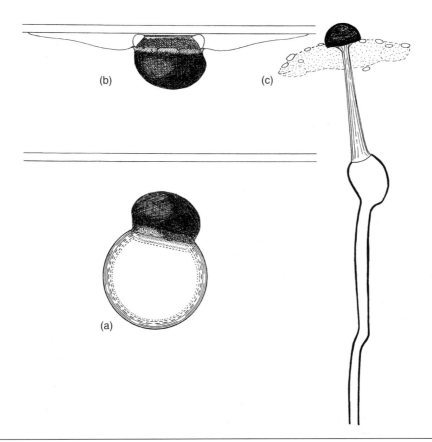

Fig 7.22 Projectiles of *Pilobolus* (diagrammatic). (a) Sporangium with adherent drop of sporangiophore sap about to strike an obstacle. (b) Sporangium after striking the obstacle. The sporangiophore sap has flowed round the sporangium which has turned outwards so that the mucilage ring adheres to the surface of the obstacle (after Buller, 1934). (c) Sporangiophore releasing a sporangium. Note the jet of liquid and the bending of the narrow base of the sporangiophore under the recoil of the discharge (after Page, 1964).

as nutrients is stimulated by the addition of thiamine, haemin and coprogen, an organo-iron compound produced by various fungi and bacteria (Hesseltine *et al.*, 1953; Page, 1960; Levetin & Caroselli, 1976). Sporangium formation is stimulated by ammonia, and in dual cultures *Mucor plumbeus* may release sufficient gaseous ammonia to induce asexual reproduction in *Pilobolus* spp. (Page, 1959, 1960).

Pilaira

Pilaira (Fig. 7.23) also appears early in the succession of coprophilous fungi, i.e. the order in which their fruit bodies appear on herbivore dung incubated under moist conditions. It has not been found in the tropics (Kirk, 1993). The structure of the melanized sporangium closely resembles that of *Pilobolus* in that the spores are separated from the columella by a mucilaginous ring which extrudes from the base of the sporangium. There are, however, no trophocysts or subsporangial vesicles, and sporangial release is non-violent. The cylindrical sporangiophores are phototropic, and when mature, especially under moist conditions, they elongate rapidly to a length of several centimetres (H. J. Fletcher, 1969, 1973). Their development essentially resembles that of *Phycomyces*. In a moist atmosphere, the mucilaginous ring may absorb water and swell considerably so that a large sporangial drop is formed (Ingold & Zoberi, 1963). When the mucilaginous ring at the base of the sporangium makes contact with adjacent herbage it becomes firmly attached to it. The sporangium slips off its columella and dries down onto the herbage. *Pilaira anomala* forms zygospores resembling those of *Pilobolus*. On germination, a germ sporangium is produced (Fig. 7.23g).

Nutritional studies on *P. anomala* indicate a preference for NH_4^+ and urea over NH_3 or asparagine as nitrogen sources, and a biotin and thiamine requirement, but no requirement of haem compounds for either growth or fruiting, in contrast to the related genus *Pilobolus*. In culture, of the simple carbon sources tested only glucose and fructose supported good growth, and there was no evidence of enzymatic ability to degrade starch, cellulose, pectin or proteins (Wood & Cooke, 1987). These nutritional characteristics support the idea that *P. anomala* is a

typical ruderal fungus, its activities constrained by the availability of simple soluble nutrients which would be depleted rapidly in decomposing dung.

7.3.3 Thamnidiaceae

In this family two kinds of asexual reproductive structure are found, namely columellate sporangia of the *Mucor* type and smaller, few-spored, usually non-columellate sporangia termed sporangiola, which are often borne in whorls or at the tips of branches. The branches bearing the sporangiola may be borne laterally on the columellate sporangiophores or may arise separately. In some cases the branch system bearing the sporangiola is terminated by a spine. Benny *et al.* (2001) have recognized 10 genera but we shall consider only *Thamnidium*.

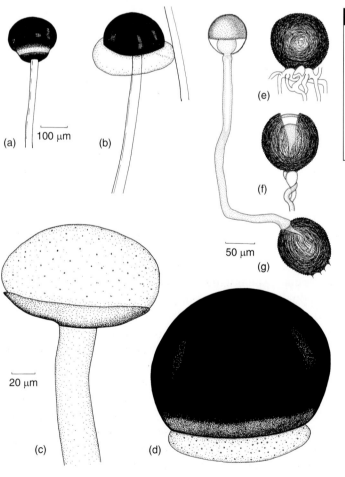

Fig 7.23 *Pilaira anomala*. (a) Sporangiophore from rabbit dung showing rupture of the sporangial wall at the base of the sporangium. (b) Sporangium with extruded mucilage ring adhering to an adjacent hypha. (c) Columella after sporangium has been detached. (d) Detached sporangium showing basal mucilage ring. (e) Zygospore. (f,g) Stages in zygospore germination (e–g) after Brefeld, 1881).

Thamnidium

The only species is *T. elegans* (Fig. 7.24), which grows in soil in cold and temperate regions, and on the dung of many different animals (Benny, 1992). It is psychrophilic, continuing to grow at 1–2°C, with an optimum 18°C and a maximum at 27–31°C (Domsch *et al.*, 1980). It has been reported from meat in cold storage. In culture, large terminal columellate sporangia are produced on tall sporangiophores which may also have repeatedly dichotomous lateral branches bearing fewer-spored columellate or non-columellate sporangiola. The sporangiola may also be borne on separate branch systems. Low temperature and light induce the formation of sporangia as opposed to sporangiola. During the development of the sporangiophores, spiral growth occurs as in *Phycomyces* (see Fig. 7.6).

Electron microscopy studies of the development of sporangia and sporangiola show that they develop in essentially the same way (J. Fletcher, 1973a,b). At maturity the columellate sporangia become converted into sticky sporangial drops. In contrast, the sporangiola are easily detached in wind tunnel experiments. A change from damp to dry air leads to increased liberation of sporangiola (Ingold & Zoberi, 1963). *Thamnidium elegans* is heterothallic and forms zygospores resembling those of *Mucor* or *Rhizopus*, but they are produced best at low temperatures such as 6–7°C and not at 20°C (Hesseltine & Anderson, 1956).

7.3.4 Chaetocladiaceae

The family Chaetocladiaceae contains two genera, the facultatively mycoparasitic *Chaetocladium* and the saprotrophic *Dichotomocladium*. Their fertile hyphae are branched and bear monosporous sporangiola on fertile vesicles. The main branches terminate in sterile spines (Benny & Benjamin, 1993). Whilst species of *Chaetocladium* are believed to be psychrophilic and are rarely collected within the tropics, all known species of *Dichotomocladium* have been recorded only in tropical areas (Kirk, 1993).

Chaetocladium

In *Chaetocladium* (Fig. 7.25) there are no *Mucor*-like sporangia. Sporangiola, each containing a single spore, are borne on lateral branches which end in spines. Such monosporous sporangiola are sometimes termed conidia. There are two species, *C. jonesii* and *C. brefeldii*, both parasitic on other Mucorales (Benny & Benjamin, 1976), especially on *Mucor* or *Pilaira* growing on dung. At the point of attachment to the host there are numerous yellow galls. These are unique bladder-like outgrowths which contain nuclei of both the host and the parasite in a common cytoplasm. *Chaetocladium* does not form haustoria and has been described as a **fusion biotroph** (Jeffries & Young, 1994). Both *Chaetocladium* spp. can, however, be cultured on standard agar media in the absence of a host. They are heterothallic. *Chaetocladium brefeldii* is heterogametangic, forming zygospores resembling *Zygorhynchus*. Burgeff (1920, 1924) has claimed that a given strain of *Chaetocladium* can only parasitize one of the two mating type strains of heterothallic *Mucor* spp., suggesting that the parasitic habit of fungi such as *Chaetocladium* may have originated from attempted copulation with other members of the Mucorales. Jeffries and Young (1994) believed that contact is truly mycoparasitic and not pseudosexual.

7.3.5 Choanephoraceae

This is probably the only current family in the Mucorales to be monophyletic (O'Donnell *et al.*, 2001). Members of the Choanephoraceae are essentially tropical in their distribution. There are three genera of which the best-known are *Blakeslea* and *Choanephora* (Kirk, 1984). Asexual reproduction is by sporangia and sporangiola. The sporangia which have brown persistent walls are usually columellate and often hang downwards. They contain dark brown sporangiospores with a striate wall and bristle-like appendages at each end. The sporangiola contain one or a few spores, also with brown striate walls and with (*Blakeslea*) or without (*Choanephora*) polar appendages. The dark sporangium walls and the dark walls of the sporangiospores (due to melanin and carotenoid pigments), both unusual features in the Mucorales, may have evolved as a protection against the mutagenic and oxidizing UV light and may help to explain the tropical

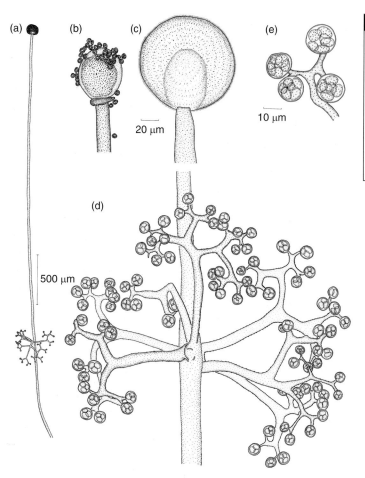

Fig 7.24 *Thamnidium elegans.*
(a) Sporangiophore showing terminal sporangium and lateral branches bearing sporangiola. (b) Dehisced sporangium showing columella and spores. (c) Immature terminal sporangium showing the columella. (d) Base of sporangiophore with dichotomous branches bearing sporangiola. (e) Sporangiola. Note the absence of a columella in these sporangiola. (b–d) to same scale.

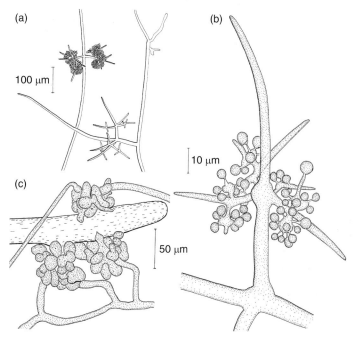

Fig 7.25 *Chaetocladium brefeldii.* (a) Habit sketch to show branches ending in spines and bearing lateral sporangiola. (b) Branch showing spine and sporangiola. (c) Hypha of *Pilaira anomala* bearing bladder-like outgrowths following parasitism by *Chaetocladium.*

distribution of these fungi (Kirk, 1993). The zygospores proper, when extruded from their zygosporangia, also have striate walls.

Choanephora

Choanephora cucurbitarum is a weak pathogen causing soft rot and wet rot diseases of a wide range of tropical and subtropical plants such as okra, chilli pepper, cowpea and *Amaranthus*. It also grows on decaying flowers of various kinds. Infection of male inflorescences of *Artocarpus integer* (Moraceae) by *Choanephora* attracts gall midges which feed on the mycelium and build up large populations on the decaying flesh of the inflorescence. The gall midges are probably involved in pollination of the female inflorescences of *Artocarpus* (Sakai *et al.*, 2000).

Asexual reproduction is by two types of structure, drooping multisporous sporangia and monosporous sporangiola ('conidia') borne on separate sporangiophores (Fig. 7.26). The development of sporangia is stimulated by growth on carbon-limited media and temperatures around 30°C, whilst the optimum temperature for sporangiolum formation is around 25°C. Light is essential for sporulation. The sporangia are columellate or non-columellate and dehisce into two halves along a line of weakness. The sporangiospores have brown walls with longitudinal grooves appearing as striations, and bear a group of hyaline tapering appendages at each pole. These appendages may play a role in the dispersal of the spores in water films since they only become extended if the sporangium dehisces in water (Higham & Cole, 1982). Sporangiola develop on globose vesicles at the tips of separate sporangiophores, each of which may bear about 100 sporangiola. The sporangiolum is multinucleate and the spore within it develops a separate thick, brown, ridged

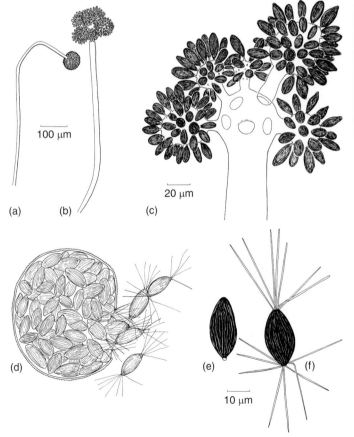

Fig 7.26 *Choanephora cucurbitarum.*
(a) Sporangiophore with drooping sporangium.
(b) Sporangiophore ('conidiophore') with numerous monosporous sporangiola ('conidia').
(c) Apex of conidiophore showing swollen vesicles bearing conidia. (d) Dehisced sporangium showing striate spores with terminal appendages. (e) Conidium.
(f) Sporangiospore. (c) and (d) to same scale.

100 μm

20 μm

(a)   (b)   (c)

(d)   (e)   (f)

10 μm

(i.e. striate) wall inside the thin sporangiolum wall which clings to it and conforms to its shape, making it difficult to discern that the spore wall is distinct (Higham & Cole, 1982). The spore inside the sporangiolum has no appendages. *Choanephora cucurbitarum* is heterothallic. Its zygosporangia develop from intertwined zygo-phores and are held in place between tongs-like suspensors (Kirk, 1977; Chang *et al.*, 1984). The zygosporangial wall is thin and may flake off or fracture to reveal the striate wall of the enclosed zygospore.

Blakeslea

Blakeslea trispora, which has been isolated from cowpeas, tobacco and cucumber leaves, forms two kinds of asexual reproductive structure in culture: nodding columellate or non-columellate sporangia with brown, faintly striate spores which usually bear bristle-like appendages, and non-columellate sporangiola borne in large numbers on globose vesicles (Fig. 7.27). The sporangiola contain 2−5 (typically 3) distinctly striate, dark brown spores which also have bristle-like appendages. The production of the

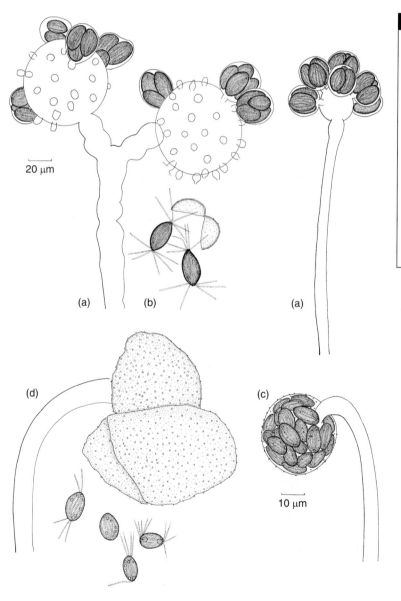

Fig 7.27 *Blakeslea trispora.* (a) Sporangiophores with globose terminal vesicles bearing sporangiola containing three or four spores. (b) A dehisced sporangiolum showing two spores released from it. Note the striate wall, the polar spore appendages and the splitting of the sporangiolum wall into two halves. (c) Sporangiophore with a drooping sporangium. No columella was observed. (d) Dehisced sporangium also lacking a columella. Note the split sporangial wall and the sporangiospores with striate walls and polar appendages.

20 μm

(a) (b) (a)

(d) (c)

10 μm

nodding sporangia is enhanced in culture by growth at a temperature of 30°C and that of sporangiola by a temperature of 26°C, with mixed sporulation at 28°C (Tereshina & Feofilova, 1995). The number of spores within sporangiola is affected by nutrition, and when grown on media with limiting nutrient content the sporangiola may contain only a single spore, thus resembling *Choanephora*. In *B. unispora* the sporangiola generally contain only one spore, rarely two. The sporangiola of *B. trispora* are readily detached by wind and break open in water like the two halves of a bivalve shell to release the spores which are carried by insects from one plant to another (Fig. 7.27b). *Blakeslea trispora* is heterothallic and has brown striate zygospores resembling those of *Choanephora* (Mistry, 1977).

The Choanephoraceae have been the subject of physiological investigations. An interesting phenomenon observed in intra- and inter-specific crosses is that the production of β-carotene is markedly enhanced when (+) and (−) strains are mated on liquid media, as compared with production from either strain grown singly. Commercial production of β-carotene and lycopene from fermentations of mixed cultures of (+) and (−) strains of *B. trispora* is possible (Mehta *et al.*, 2003). The discovery that β-carotene production can be stimulated by an acid fraction of culture filtrates from mixed cultures of *B. trispora* led to the discovery of trisporic acid as the sex hormone of Mucorales (see p. 173).

7.3.6   Syncephalastraceae

A characteristic feature of this family is that asexual reproduction occurs by means of cylindrical sporangia containing typically a single row of sporangiospores. Such sporangia are termed merosporangia and are formed in groups on inflated vesicles (Benjamin, 1966). Merosporangia appear to have evolved independently in the Piptocephalidaceae (Zoopagales) (see p. 201). There is only a single genus, *Syncephalastrum*, in the Syncephalastraceae. DNA sequence analysis indicates close relationships with certain genera traditionally classified in Mucoraceae and Thamnidiaceae (O'Donnell *et al.*, 2001).

Syncephalastrum

Syncephalastrum racemosum (Fig. 7.28) can be isolated from soil and dung in tropical and subtropical areas (Domsch *et al.*, 1980). It grows rapidly in culture over a wide range of temperatures (7–40°C) and is mainly saprotrophic, but has been implicated in mucormycosis in human and animal hosts. It has also been isolated from foodstuff, cereal grains, other seeds and spices. In culture it forms aerial branches terminating in club-shaped or spherical vesicles. The vesicles are multinucleate and bud out all over their surface to form cylindrical outgrowths, the merosporangial primordia. Into these outgrowths one or perhaps several nuclei pass, and nuclear division continues. The cytoplasm in the merosporangium cleaves into a single row of 5–10

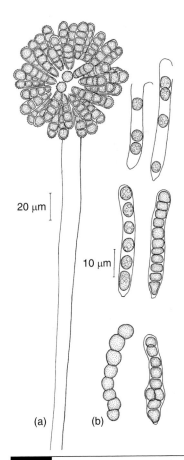

20 μm

10 μm

(a)   (b)

Fig 7.28   *Syncephalastrum racemosum*. (a) Sporangiophore bearing a vesicle and numerous merosporangia. (b) Merosporangia and merospores.

sporangiospores, each with 1–3 nuclei. The cleavage process is similar to that found in other Mucorales (Fletcher, 1972). The sporangial wall shrinks at maturity so that the spores appear in chains reminiscent of *Aspergillus*. Occasionally the merospores may lie in more than a single row. The spore heads remain dry and entire rows of spores (spore rods) are detached by wind (Ingold & Zoberi, 1963). *Syncephalastrum racemosum* is heterothallic and forms zygospores resembling those of other Mucorales.

7.3.7 Cunninghamellaceae

In this family, asexual reproduction is entirely by means of monosporous sporangiola. Sporangia are not formed. There is a single genus.

Cunninghamella

There are about 12 species of *Cunninghamella*, found in soil in the warmer regions of the world, e.g. the Mediterranean and subtropics (Domsch *et al.* 1980; Zheng & Chen, 2001). *Cunninghamella elegans* and *C. echinulata* are saprotrophs but *C. bertholettiae* is a serious, sometimes fatal, human pathogen. *Cunninghamella echinulata* may also be a destructive mycoparasite of *Rhizopus arrhizus*. *Cunninghamella elegans* and *C. echinulata* have been used in a wide range of biotransformations of pharmaceutical products (Kieslich, 1997). DNA sequence studies have grouped *C. echinulata* with some of the genera traditionally classified in Mucoraceae (O'Donnell *et al.*, 2001), but comparisons of fatty acid and cell wall composition of *Cunninghamella japonica* and *Blakeslea trispora* have suggested that *Cunninghamella* is related to members of the Choanephoraceae, a conclusion reached also on morphological criteria by some other workers.

The sporangiola of *Cunninghamella* are hyaline and clustered on globose vesicles (ampoules) on branched or unbranched sporangiophores (Fig. 7.29). They are sometimes referred to as conidia, but details of their development indicate that they are best interpreted as one-spored sporangiola. Khan and Talbot (1975) have studied sporangiolum development in *C. echinulata*. The ampoules are club-shaped, globose or pear-shaped and bear spherical sporangiola, each arising from a tubular denticle. Localized areas of weakness in the ampoule wall, yielding to turgor pressure, blow out to form the denticles. The wall is two-layered in the ampoule and denticle, but single in the developing sporangiolum where it develops hollow spines all over the surface. Within the sporangiolum wall a two-layered wall develops around the multinucleate protoplast. Hawker *et al.* (1970) have studied structure and germination of sporangiola in a species of *Cunninghamella*. Here, too, the wall

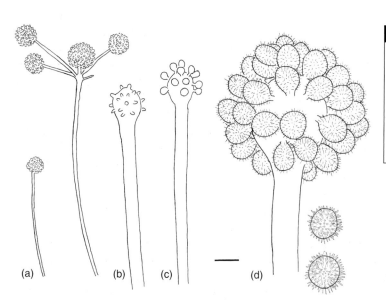

Fig 7.29 *Cunninghamella echinulata*. (a) A simple and a branched sporangiophore. (b,c) Apices of sporangiophores showing expanded vesicles developing sporangiola. (d) Apex of mature sporangiophore with cluster of attached and two detached spiny-walled sporangiola. Scale bar (a) = 25 μm, (b–d) = 10 μm.

(a)   (b)   (c)   (d)

of the ungerminated sporangiolum consisted of at least two layers, the outermost layer relatively thin, enclosing a distinct thicker inner layer. On germination only the inner layer extends as a germ tube.

The zygospores of *Cunninghamella* resemble those of *Mucor*.

7.3.8 Mortierellaceae

The distinctive feature of this family is that the sporangiophore produces only a rudimentary columella or lacks it altogether. In the most frequently encountered genus, *Mortierella*, zygospores are often heterogametangic and may be naked or enclosed in a weft of mycelium. The family, which includes about seven genera, has been monographed by Zycha *et al.* (1969). DNA sequence analyses indicate relationships to certain genera usually placed in Mucoraceae (O'Donnell *et al.*, 2001; Tanabe *et al.*, 2004; see Fig. 7.1), although many authors regard *Mortierella* and its allies as a separate order, Mortierellales.

Mortierella

About 90 species of *Mortierella* are known mainly from soil, the rhizosphere, and plant or animal remains in contact with soil (Gams, 1977; Domsch *et al.*, 1980). These fungi can be isolated readily on nutrient-poor media which prevent the growth of more vigorous moulds. Many species are psychrophilic and may comprise the bulk of fungal isolates from soil if the isolation media are incubated near 0°C (Carreiro & Koske, 1992). *Mortierella wolfii* is associated with mycotic abortion in cattle and can be isolated from the placenta and foetal stomach contents and from liver. In nature it grows in warm soils, over-heated silage and rotten hay and can grow well at 40–42°C (Austwick, 1976; Domsch *et al.*, 1980). Certain species of *Mortierella*, e.g. *M. alpina*, have been used in fermentations as catalysts of biotransformations in the production of pharmaceuticals (Kieslich, 1997). Another focus of biotechnological interest is their accumulation of lipid, notably polyunsaturated fatty acids (PUFAs) which are of nutritional value (Dyal & Narine, 2005). These are also produced

by thraustochytrids (see p. 73). The genus *Mortierella* is polyphyletic, and many of the best-known species, including *M. isabellina*, *M. ramanniana* and *M. vinacea*, are now placed in other genera such as *Micromucor* or *Umbelopsis* (Meyer & Gams, 2003).

The mycelium of most species of *Mortierella* is fine and, in agar culture, often shows a characteristic series of fan-like zones. Cultures frequently have a garlic-like odour. The sporangia are borne on branched or unbranched tapering sporangiophores (Fig. 7.30a). The sporangium wall is delicate and may collapse around the spores. There is no protruding columella (Fig. 7.30b). Frequently the entire sporangium is detached. In a number of species, and also dependent upon environmental conditions, there may be only one or a few spores per sporangium (Figs. 7.30c,d). Asexual reproduction may also include the formation of sessile, intercalary chlamydospores which are not dispersed but remain in the soil when their subtending mycelium breaks down (Fig. 7.30e). **Stylospores** are also produced; unfortunately this term has been used for two different, non-homologous structures. In its original application by van Tieghem, it referred to aerial chlamydospores, i.e. relatively thick-walled, stalked spores as seen, for example, in *M. polycephala* (Domsch *et al.*, 1980). In other species, classified in the section *Stylospora*, e.g. *M. humilis* and *M. zonata* (Gams, 1977), single-spored sporangiola (Fig. 7.31a) have been termed stylospores. On detachment of the sporangiolum, the remnants of the sporangiolum wall can often be seen at the tip of the sporangiophore (Domsch *et al.*, 1980). In some species, e.g. *M. stylospora* and *M. zonata*, only sporangiola are present and true sporangia are lacking. *Mortierella chlamydospora* also lacks true sporangia, reproducing asexually by intercalary smooth or stalked echinulate chlamydospores and sexually by zygospores (Ansell & Young, 1982).

The zygospores of *Mortierella* spp. may be naked or surrounded by a partial or complete investment of sterile hyphae (see Figs. 7.31b,c). Of the 90 species, 26 are known to form zygospores, half of which are homothallic (Watanabe *et al.*, 2001). It is likely that the majority of the remaining species will prove to be heterothallic.

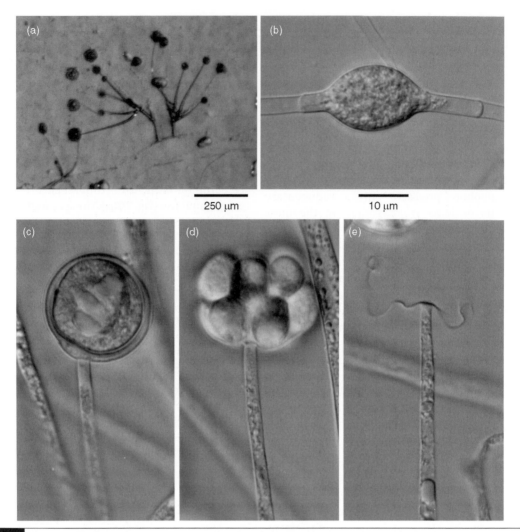

250 μm   10 μm

Fig 7.30 *Mortierella hyalina.* (a) Branched sporangiophore as viewed with the dissection microscope. (b) Intercalary chlamydospore. (c) One-spored sporangium (sporangiolum) showing separate walls, the outer belonging to the sporangium and the inner to the sporangiospore. (d) Multi-spored sporangium with the sporangial wall disintegrating. (e) Apex of a sporangiophore in which the sporangium has dehisced, leaving fragments of the sporangial wall as a frill. Note the absence of a bulging columella. (b–e) to same scale. Reprinted from Weber and Tribe (2003), with permission from Elsevier.

Zygospore production takes place in culture, often embedded in the agar, on media with a relatively poor nutrient content which discourages profuse development of aerial mycelium. A common feature is that zygospores are heterogametangic, one suspensor being considerably larger than the other. The smaller progametangium or gametangium does not enlarge and may disappear soon after plasmogamy. The early development of such heterogametangic zygospores is illustrated in the heterothallic

M. umbellata (Fig. 7.32; Degawa & Tokumasu, 1998). In this species, hyphal coiling occurs at the point of contact of compatible mycelia, followed by the development of club-shaped progametangia which grow parallel and closely appressed to each other. One, the macroprogametangium, soon becomes much larger than the other, the microprogametangium. In each progametangium a septum delimits a terminal gametangium from a suspensor. The macrogametangium and macrosuspensor both enlarge considerably,

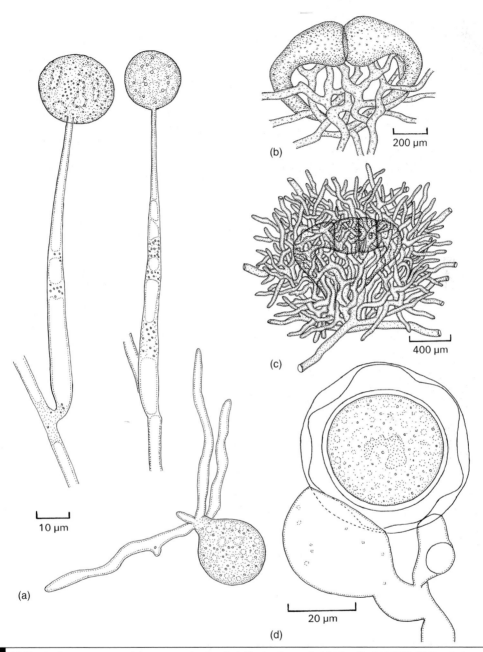

Fig 7.31 (a) *Mortierella zonata* sporangiola, one germinating. (b,c) *Mortierella rostafinskii* (after Brefeld, 1876). (b) Developing zygospore. (c) Older zygospore surrounded by a weft of hyphae. (d) *Mortierella epigama* zygospore with unequal suspensors arising from a common branch showing that this fungus is homothallic and heterogametangic. The zygospore proper, lying within the zygosporangium, has a thick undulating wall.

appearing as two contiguous spheres (Figs. 7.32e,f). Eventually the macrogametangium becomes a zygosporangium, its wall ornamented with small warts, and containing a smooth, thick-walled zygospore.

The heterothallic *M. indohi* is also heterogametangic, one progametangium being blunter and more rounded than the other. A cross wall develops only in this larger progametangium to cut off a gametangium which enlarges and

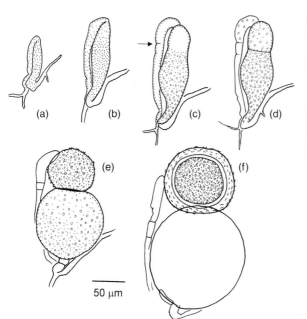

(a)   (b)   (c)   (d)

(e)   (f)

50 μm

Fig 7.32 *Mortierella umbellata*, stages in zygospore development (traced from Degawa & Tokumasu, 1998). (a) Developing progametangia with micro-progametangium on the left. (b) Swelling of progametangia. (c) Septum formation in micro-progametangium, arrowed. (d) Septa have formed in both progametangia to delimit the terminal gametangia from sub-terminal suspensors. (e) The macrogametangium and macrosuspensor have enlarged. (f) Mature zygosporangium containing a thick-walled zygospore.

becomes converted into a subspherical zygosporangium with a dimpled wall. The narrower progametangium does not enlarge appreciably. It is not divided by a septum and remains as a lateral attachment to the zygosporangium (Ansell & Young, 1983, 1988). Delimitation of the zygosporangium by means of a single wall in only one of the fusing progametangia occurs in several other species of *Mortierella* (Kuhlman, 1972).

Mortierella capitata shows an unusual mode of zygospore development (Degawa & Tokumasu, 1997). It is heterothallic and heterogametangic, with two mating types designated A and B. When compatible vegetative hyphae meet in culture, their tips swell to form progametangia. The hyphal tips from strain B are always larger than those from A and are designated as macroprogametangia. The narrower hyphae from strain A (microprogametangia) coil around the macroprogametangia, branch dichotomously and become septate, resulting in the formation of microsuspensors and microgametangia. A septum divides the terminal macrogametangium from its macrosuspensor. The macrogametangium becomes the zygosporangium and eventually contains a thick-walled hyaline zygospore. The macrosuspensor elongates and persists so that the mature zygosporangium appears at the

end of a long stalk surrounded at its base by the coiled microsuspensors. Apart from the other unusual features of this developmental process, *M. capitata* is distinctive in that the morphology of its gametangia is linked to mating type, i.e. the formation of macrogametangia occurs only in the B strain and microgametangia in the A strain. This condition, termed **morphological heterothallism**, is comparatively rare in Zygomycota and in fungi generally.

Complete investment of the zygospore by branching hyphae is a feature of *M. rostafinskii* (see Figs. 7.31b,c) and *M. ericetorum* (Kuhlman, 1972). The zygospores proper in *Mortierella* are hyaline with thick smooth walls, sometimes showing coarse, undulating folds (see Fig. 7.31d). Little is known about the germination of zygospores.

7.4 | Zoopagales

The order Zoopagales contains soil- and dung-inhabiting parasites of fungi and small terrestrial animals such as protozoa and nematodes. Reproduction is by conidia, merosporangia and zygospores. Benny *et al.* (2001) recognized

5 families and 20 genera but we shall study only the Piptocephalidaceae, earlier classified in the Mucorales.

7.4.1 Piptocephalidaceae

This family includes *Piptocephalis* and *Syncephalis*, both mycoparasites. DNA sequence analysis suggest that these two genera are not closely related (Tanabe *et al.*, 2000). *Piptocephalis* is a biotrophic haustorial parasite which needs the presence of a susceptible host for good growth and reproduction (Manocha, 1975), although on certain agar media *Piptocephalis* spores will germinate and give rise to a limited mycelium producing dwarf sporangiophores. The spores so formed are unable to germinate if transferred to fresh agar, but they do germinate and infect a suitable host fungus if one is present. *Syncephalis* develops intrahyphal hyphae within the host mycelium and can be grown more readily in culture if supplied with appropriate nutrients (Jeffries & Young, 1994).

Piptocephalis

Most of the 20 or so known species of *Piptocephalis* (Gr. *pipto* = to fall, *kephale* = head) parasitize the mycelium of Mucorales, with *P. xenophila* exceptional in its ability to infect members of the Ascomycota. Species of *Piptocephalis* are most abundant in the surface layers of soils where there is a rapid recycling of organic matter, such as in woodland and in grazed grassland (Richardson & Leadbeater, 1972). They also parasitize Mucorales on dung. A characteristic habitat for *P. freseniana* is herbivore dung towards the end of the fruiting phase of *Mucor* and *Pilaira*. From an infected host mycelium *Piptocephalis* develops an erect dichotomous sporangiophore (Fig. 7.33a). Swollen nodulose (knobbly) **head cells** form at the tips of the branches (see Fig. 7.33c), and from these cylindrical merosporangia radiate outwards. The merosporangia are thin-walled and usually contain from one to several multinucleate merospores, arranged in a single row. *Piptocephalis unispora* is unusual in that its merosporangia contain only a single sporangiospore. Its merosporangial wall encloses the sporangiospore which has a two-layered wall and may contain 1–3 nuclei (Jeffries & Young, 1975). At maturity *Piptocephalis* merosporangia behave in two different ways (Ingold & Zoberi, 1963). In some species the thin sporangial wall collapses around the spores which remain attached together as spore rods, appearing as short chains (see Fig. 7.33c). Alternatively, as in *P. freseniana*, the merosporangial wall becomes diffluent and all the spores in a head collapse to form a spore drop. In some species the whole head cell with its attached merosporangia becomes detached at maturity. All types of propagule can be dispersed by wind.

On germination sporangiospores swell and emit one to several germ tubes (McDaniell & Hindal, 1982). There is a chemotropic attraction

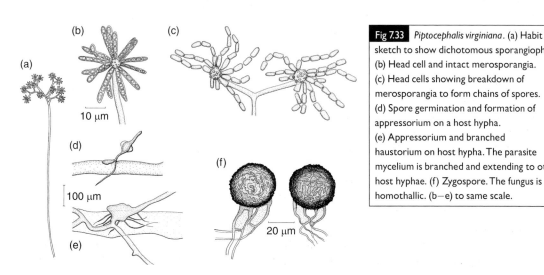

Fig 7.33 *Piptocephalis virginiana.* (a) Habit sketch to show dichotomous sporangiophore. (b) Head cell and intact merosporangia. (c) Head cells showing breakdown of merosporangia to form chains of spores. (d) Spore germination and formation of appressorium on a host hypha. (e) Appressorium and branched haustorium on host hypha. The parasite mycelium is branched and extending to other host hyphae. (f) Zygospore. The fungus is homothallic. (b–e) to same scale.

of germ tubes towards host hyphae (Fig. 7.33d), with preferential growth towards the hyphal tips. On agar the chemotropic stimulus can be detected over distances as great as 5 mm (Evans & Cooke, 1982). Fimbriae extending outwards for up to 25 μm from the cell walls of potential host fungi may play a role in directing the growth of *Piptocephalis* germ tubes towards the host hyphae (Rghei *et al.*, 1992). At the point of contact an appressorium develops, but in some combinations the parasite hyphae may coil around those of their host and several appressoria form. In successful host–parasite combinations, the host wall is penetrated beneath the appressorium by mechanical and possibly also enzymatic means. An infection peg penetrates the host wall. Enclosed by the plasmalemma of the host cell, the tip of the penetration peg expands to form a haustorium which may branch inside the host hypha. The haustoria of *Piptocephalis* have close similarity to those of biotrophic haustorial parasites of plants (Manocha & Lee, 1971; Jeffries & Young, 1976). Nutrients taken up by the haustorium are translocated to the germinating spore and its germ tubes may then grow out to form a mycelium which extends over the host hypha, producing further haustoria. The distinctive biochemical features of the Mucorales which are correlated with their ability to support the growth of these mycoparasites are that their walls contain chitosan and that their cytoplasm is rich in the polyunsaturated fatty acid γ-linolenic acid which is essential for growth of the mycoparasite (Manocha, 1975, 1981; Manocha & Deven, 1975).

Recognition between the mycoparasite and its hosts operates on at least two levels, the cell wall and the protoplast surface (Manocha *et al.*, 1990). There are qualitative and quantitative differences in the carbohydrates present at the hyphal surface of host and non-host species. Attachment is favoured by the presence of two distinctive glycoproteins in the wall of susceptible host hyphae. These two glycoproteins act as subunits of an agglutinin which may serve as receptor to a complementary protein in the mycoparasite (Manocha *et al.*, 1997).

Piptocephalis virginiana readily infects young but not old cultures of *Choanephora cucurbitarum*. This is correlated with the fact that the wall of young hyphae of *C. cucurbitarum* is single-layered whilst that of older hyphae is double-layered. Although appressoria and penetration pegs develop on older hyphae, penetration of the inner layer of the cell wall is rarely successful. The inner wall layer develops a papilla opposite the point of attempted penetration (Manocha, 1981). Similar findings were made when *P. virginiana* failed to penetrate the resistant species *P. articulosus* (Manocha & Golesorkhi, 1981). Where successful penetration of a susceptible host occurs, the mycoparasite *P. virginiana* can suppress wall synthesis by the host in the vicinity of infection points, so overcoming one of its defence reactions (Manocha & McCullough, 1985; Manocha & Zhonghua, 1997).

The effects of *Piptocephalis* spp. on the growth of their hosts are very variable (Curtis *et al.*, 1978). In some combinations the rate of growth of dual cultures was not significantly different from that of uninfected hosts, in others it was reduced, whilst in yet others it was enhanced. These effects are temperature-dependent. Growth and sporulation of the coprophilous fungus *Pilaira anomala* were reduced in culture when infected by *P. fimbriata* or *P. freseniana* (Wood & Cooke, 1986). A curious effect was found in culture when *P. fimbriata* challenged its normally susceptible host *Mycotypha microspora*. In the presence of *P. fimbriata* the host grew in a yeast-like state which was not infected. In contrast, the mycelial state of this fungus is readily infected (Evans *et al.*, 1978).

Most species of *Piptocephalis* are homothallic (Leadbeater & Mercer, 1957). In culture zygospores are usually formed within the agar. The mature zygospore is a spherical dark brown sculptured globose cell held between two tong-shaped suspensors.

7.5 | Entomophthorales

Many Entomophthorales are parasites of insects and other animals, whilst some parasitize desmids, nematodes or fern prothalli, or grow saprotrophically in plant litter, dung or soil.

An illustrated account of entomopathogenic species has been provided by Samson *et al.* (1988) and a key to genera by Humber (1997). The major entomopathogenic genera are *Batkoa, Conidiobolus, Entomophaga, Entomophthora, Erynia, Furia, Massospora, Neozygites, Pandora* and *Zoophthora*. Some of these insect pathogens hold promise for the control of insect pests, not least because many of them can be grown in culture, albeit on complex media containing ingredients such as sugars, egg yolk, yeast extract and milk (Wolf, 1981; Papierok & Hajek, 1997). The cells of Entomophthorales are uninucleate or coenocytic with chitinous walls, or they may exist in the bodies of insects as wall-less protoplasts. The absence of a wall presumably reduces the elicitation of immune responses in their hosts (Dunphy & Nolan, 1982). Asexual reproduction in most genera is by means of forcibly discharged conidia, and on germination such conidia may develop a variety of secondary conidia. Sexual reproduction is by isogamous or anisogamous conjugation between uni- or multinucleate gametangia, to give a thick-walled zygospore. Azygospores may also be formed without conjugation, but it is likely that nuclear fusion and reduction division occur during their development (McCabe *et al.*, 1984).

Fossil evidence indicates that, as insect pathogens, members of the group were extant at least 25 million years ago. A well-preserved specimen of a winged termite probably infected with a species of *Entomophthora* has been found embedded in amber dated around the Oligocene–Miocene border in the Dominican Republic (Poinar & Thomas, 1982).

According to Benny *et al.* (2001), the order Entomophthorales consists of six families including the Basidiobolaceae. If this family is excluded, the remaining Entomophthorales appear to be monophyletic by DNA-based analysis (Jensen *et al.*, 1998). In many phylogenetic schemes, *Basidiobolus ranarum* seems to be more closely related to Chytridiales and Neomastigales than to Entomophthorales (see Figs. 1.26, 7.1), and Cavalier-Smith (1998) has placed it in a separate order, the Basidiobolales. In the current context, we sacrifice these taxonomic details in favour of a better understanding of the Zygomycota as a whole, and therefore retain *Basidiobolus* in the Entomophthorales.

Important criteria in the classification of Entomophthorales are the branched or unbranched nature of the conidiophores, whether the conidia are uninucleate or multinucleate, whether the wall of the conidium is single (**unitunicate**) or separates into two layers (**bitunicate**), and the presence or absence of secondary conidia and their morphology (Humber, 1989).

7.5.1 Basidiobolaceae: *Basidiobolus*

Basidiobolus is the only genus in the Basidiobolaceae. The best-known species is *B. ranarum*, which has a worldwide distribution. It fruits on the dung of frogs, toads, lizards, some insectivorous fish and mammals such as bats. It has also been found on the dung of kangaroos and wallabies (Speare & Thomas, 1985). If a frog is captured and placed in a jar with a little water, it will defaecate in due course and its dung can be filtered off. If the damp filter paper is placed in the lid of an inverted Petri dish containing a suitable agar medium (e.g. 1% peptone agar, potato-dextrose agar, or cornmeal agar), conidia of *B. ranarum* will be shot upwards from the dung onto the agar surface, and within a few days coarsely septate colonies will become visible on the agar (Weber & Webster, 1998a). The presence of *Basidiobolus* and other ballistosporic fungi such as *Conidiobolus* in surface soil and litter can also be disclosed by the 'canopy' technique. A suspension of soil is filtered and the filter paper, bearing a thin layer of soil, is placed in the lid of a Petri dish facing downwards over a suitable agar medium. The dish is illuminated from below and this encourages the discharge of conidia onto the agar (Smith & Callaghan, 1987; Callaghan, 2004).

In agar cultures of *Basidiobolus*, the cytoplasm in the mycelium moves towards the hyphal apex so that only a few terminal segments contain cytoplasm and a single large, prominent nucleus whilst the older segments are empty, being isolated by **retraction septa** (Fig. 7.34d). The cytoplasm-filled mycelial segments are termed **hyphal bodies**. Branching occurs immediately

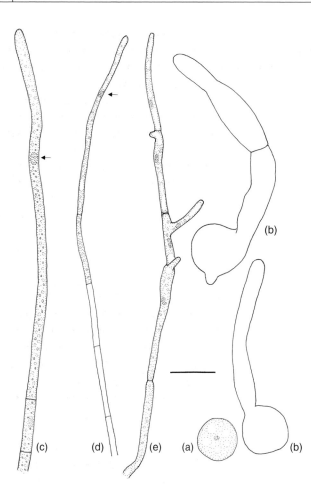

Fig 7.34 *Basidiobolus ranarum.* (a) Gut-stage cell from fresh frog dung. (b) Germination of gut-stage cells producing a coarse septate mycelium. (c) Hyphal apex with the terminal cells full of cytoplasm. The prominent nucleus in the apical cell is arrowed. (d) Apex of an older hypha in which only the two terminal cells contain cytoplasm; those behind are empty. (e) Branches arising beneath the septa of three subterminal cells. Scale bar: (a,b) = 25 μm, (c) = 50 μm, (d,e) = 100 μm.

behind the septum delimiting the apical segment, after mitotic division of the nucleus (Fig. 7.34e). The conidiophores, which develop in a few days, are phototropic and resemble the sporangiophores of *Pilobolus* but bear a colourless pear-shaped to globose ballistosporic conidium (Fig. 7.35a). O'Donnell (1979) interpreted the conidium as a monosporous sporangiolum but since it can, under certain conditions, cleave to form endospores it may also be regarded as a modified sporangium. The conidium is uninucleate. A conical columella projects into it. Beneath is a swollen sub-conidial vesicle containing liquid under turgor pressure. This is probably generated by a single large vacuole which fills most of the sub-conidial vesicle at maturity. A line of weakness can be detected as a slight constriction around the base of the vesicle, and when this ruptures the conidium and vesicle fly forward for

a distance of 1–2 cm. The elastic upper portion of the vesicle contracts and the vacuolar sap within it squirts out backwards, so that it behaves as a minute rocket (Ingold, 1971). During their flight the conidium and the rocket motor (i.e. the vesicle) may be separated or the two parts remain attached to each other until landing (Figs. 7.35c,d).

Conidium germination in *B. ranarum*

Primary ballistosporic conidia can germinate in a number of different ways depending on external conditions (Zahari & Shipton, 1988; Waters & Callaghan, 1999).

(1) By direct germination, producing one to several germ tubes from which the vegetative mycelium develops (Fig. 7.35e). Germination of this type requires a nutrient concentration above that of 0.1% malt extract agar.

(2) Germination by repetition to form a secondary conidiophore with a ballistosporic conidium. This is essentially similar to the primary conidium (Fig. 7.35a) and is produced under conditions of high water availability and low nutrient concentration. Secondary conidia may germinate by further repetition or in other ways.

(3) Discharged ballistosporic conidia formed in culture on certain media or located within the gut of the frog may cleave to form many endospores (sporangiospores, sometimes termed **meristospores**), and these are released by dissolution of the original conidial wall (Fig. 7.36a; Dykstra, 1994).

(4) Germination under somewhat drier conditions with a water activity at or below 0.995 stimulates the development of **capilliconidia** or **capillispores** (Fig. 7.36c). The body of the capilliconidium may cleave by transverse and longitudinal septa to form endogenous segments (endospores, meristospores) which are released by breakdown of the wall of the capilliconidium (Fig. 7.36d; Drechsler, 1956).

Capilliconidia are so called because they are formed on long (over 0.3 mm), slender conidiophores. The conidia themselves are spindle-shaped and apically beaked with a terminal globose adhesive droplet or **haptor**. The material making up the haptor is extruded through a narrow channel within the beak of the conidium. The droplet has unusual properties because it is not affected by water but rapidly spreads out to form a film when in contact with a solid surface (Dykstra & Bradley-Kerr, 1994). The capilliconidia are easily detached from their conidiophores and may be dispersed by mites (Blackwell &

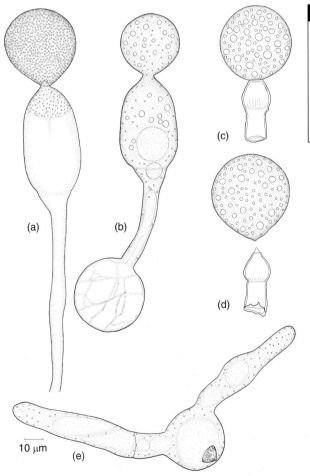

Fig 7.35 *Basidiobolus ranarum*. (a) Conidiophore from culture. Note the conical columella and the swollen vesicle with a line of weakness around its base. (b) Primary conidium germinating to produce a secondary conidiophore and ballistosporic conidium. (c) Discharged conidium with remnant of the vesicle attached. (d) Discharged conidium separated from the remnant of the vesicle. (e) Conidium germinating directly to form a septate mycelium.

(a)

(b)

(c)

(d)

(e)

10 µm

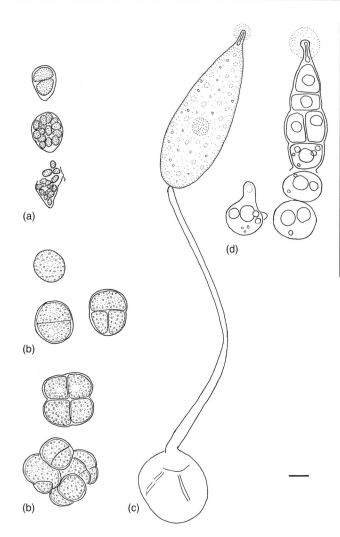

Fig 7.36 *Basidiobolus ranarum.* (a) Stages in the development of endospores by primary ballistoconidia. Above: cytoplasmic contents cleaving to form two protoplasts. Centre: sporangium containing over a dozen sporangiospores. Below: sporangium showing breakdown of sporangium wall (traced from Dykstra, 1994). (b) Successive cleavage of protoplasts from a primary ballistosporic conidium placed on a rich agar medium to form the 'Palmella' stage. (c) Germination of a primary ballistoconidium to form a uninucleate secondary capilliconidium with a terminal beak which has extruded a sticky haptor. (d) Capilliconidium which has divided to produce several endospores, some of which have been released following breakdown of its wall. One of the endospores is germinating. Scale bars: (a,b) = 20 μm, (c,d) = 12.5 μm.

Malloch, 1989). Mites are ingested by beetles, the main vectors of *B. ranarum*, but other insects, spiders, millipedes, woodlice, worms and snails may also acquire conidia. These are ingested by vertebrate insectivores. Within the vertebrate gut, endospores are released from ballistosporic conidia and capilliconidia.

Endospores germinate in the vertebrate gut to form spherical, large, uninucleate, hyaline cells up to 20 μm. These were called the 'Darm-Form' (gut-stage) of *B. ranarum* by Levisohn (1927) who showed that a single ingested primary conidium can give rise to 50–60 division products (meristospores) forming gut-stage cells (Fig. 7.34a). The concentration of *Basidiobolus*

propagules builds up in the guts of the vertebrate vectors, and the fungus can be isolated from faeces of lizards up to 18 days after the animals are deprived of infected prey (Coremans-Pelseneer, 1973; Okafor *et al.*, 1984). The gut-stage cells are voided with the faeces and can survive for several months under dry conditions. Under moist warm conditions they germinate to form a mycelium from which ballistosporic conidiophores develop, whereas on certain media they enlarge and their contents undergo successive binary fission to form globose thick-walled cells. This state is sometimes termed the 'Palmella' state (Fig. 7.36b) because of its superficial resemblance of a genus of green algae.

Basidiobolus microsporus, which grows in deserts in California, has a method of asexual reproduction not found in *B. ranarum*. Primary conidia can germinate directly or by repetition as in *B. ranarum*, but capilliconidia have not been found. However, under relatively dry conditions primary conidia may produce large numbers of exogenous obclavate spores (**micro-spores**) each attached by a separate pedicel to the wall of the primary conidium. They have been interpreted as modified sporangiospores (Benjamin, 1962).

In culture it has been found that light, especially blue light of wavelength 440–480 nm, stimulates conidial development and discharge in *B. ranarum*. The effect of light is to stimulate aerial growth from hyphal bodies within the medium, and the aerial hyphae which develop in the light become modified as conidiophores (Callaghan, 1969a,b).

Sexual reproduction in *B. ranarum*

Zygospores are formed following conjugation. The fungus is homothallic and development can be seen on certain agar media (e.g. Czapek-Dox agar) within 4–5 days in cultures derived from a single conidium. Zygospore development appears to occur most readily in the dark, and under these conditions the hyphal bodies become bicellular prior to developing into zygospores (Callaghan, 1969b). On either side of a septum, beak-like projections develop, and the single nucleus within each hyphal segment migrates into the tip and divides there. One daughter nucleus is cut off by a septum in the terminal cell of the beak and later disintegrates, whereas the second nucleus migrates back into the parent cell. Following this, one of the parent cells enlarges to several times the volume of the adjacent cell and a pore is formed connecting the two cells through the original septum separating them. A nucleus from the smaller cell passes through the pore and lies close to the nucleus of the larger cell. Nuclear fusion may occur directly or after a further division. The enlarged parent cell forms the zygospore which has a thick wall when mature (Fig. 7.38). Meiosis occurs within the mature zygospore to give four haploid nuclei, of which three usually degenerate. The mature

zygospores of some isolates of *B. ranarum* have thick undulating walls of variable thickness, but in others the wall may be smooth. On germination the zygospore forms a germ tube or a conidiophore terminated by a ballistosporic conidium. Capilliconidia may also develop from germinating zygospores (Dykstra & Bradley-Kerr, 1994). The complicated and unusual life cycle of *B. ranarum* is illustrated in Fig. 7.37.

Basidiobolus ranarum is an atypical zygomycete in that its mycelium becomes divided into uni-nucleate segments. The nucleus is also unusually large, up to 25 µm, and this fact has led to several investigations of its cytology (e.g. Robinow, 1963; Tanaka, 1970; Sun & Bowen, 1972). The number of chromosomes has been estimated to be as high as 900, and the nucleus may be polyploid.

Pathogenicity of *B. ranarum*

Basidiobolus is probably not harmful to most insects and mites, although it has been isolated as a mass infection of mosquitoes, from termites and from larvae of *Galleria* (Krejzová, 1978). It was earlier thought to be harmless to reptiles and amphibians and there is no evidence of intestinal lesions in them. However, an epizootic cutaneous infection caused by *B. ranarum* has been reported from the dwarf African clawed frog, *Hymenochirus curtipes* (Groff *et al.*, 1991). There are many reports of the isolation of *B. ranarum* from man and domestic animals such as horses (see Gugnani, 1999; Ribes *et al.*, 2000). Although several specific names have been applied to isolates pathogenic to humans and other mammals, the consensus is that they should be regarded as synonyms of *B. ranarum* (McGinnis, 1980). This view is supported by ribosomal DNA analysis (Nelson *et al.*, 1990). Isolates from humans, unsurprisingly, are capable of growing at 37°C (Cochrane *et al.*, 1989). Human disease caused by *B. ranarum* is more common in tropical and subtropical regions than in temperate zones. Infection is associated with subcutaneous swellings of affected areas of the lower limbs but rare intestinal infections are also known. It is assumed that the inoculum is usually soil-borne, and the use of fallen leaves in place of toilet paper has sometimes been implicated as the cause of

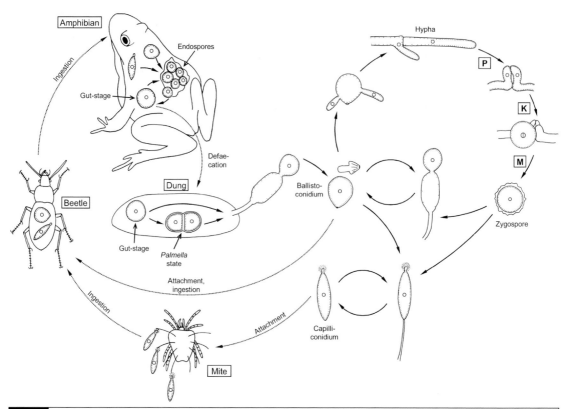

Fig 7.37 The eventful life cycle of *Basidiobolus ranarum*, not to scale. A beetle with attached or ingested ballistoconidia and capilliconidia is eaten by a frog. In the gut of the frog, both conidial types can undergo cleavage to form endospores, which germinate by enlargement to form the gut stage. After defaecation, gut-stage cells germinate to produce ballistoconidia or cleave to give the *Palmella* stage. Discharged ballistoconidia may germinate by repetition, by forming capilliconidia, or by emitting a hypha. Zygospore formation is initiated by conjugation between two adjacent hyphal cells. Small open circles represent haploid nuclei; diploid nuclei are larger and split. *Basidiobolus ranarum* is homothallic. Key events in the life cycle are plasmogamy (P), karyogamy (K) and meiosis (M).

infections. In horses, infection of the nasal mucosa is again most probably from soil.

7.5.2 Ancylistaceae: *Conidiobolus*

There are about 30 species of *Conidiobolus*, the 'conidium thrower'. King (1977) has given keys and descriptions. Most of them grow saprotrophically in soil and litter and can be readily isolated by the canopy technique described for *Basidiobolus* (see p. 203) because they forcefully project their conidia. Several species have been isolated from basidiocarps of the Jew's Ear fungus, *Auricularia auricula-judae*. Some are pathogenic to insects such as aphids and termites, and certain species cause disease in mammals including man. The characteristic feature of

Conidiobolus is the formation of globose multinucleate primary conidia surrounded by a two-layered wall whose layers do not separate (Latgé *et al.*, 1989). The primary conidia are projected by an **eversion mechanism** in which the inner wall of the double-walled conidiophore apex ('columella') suddenly rounds off because of different mechanical properties of the two wall layers. The force responsible is turgor pressure. After discharge, the tip of the columella can be seen projecting outwards as a conical papilla, and the base of the discharged conidium ends in a similar projection (Fig. 7.39b). Spores may be shot away for up to 4 cm. Germination of primary conidia may be by repetition producing the same spore type, by germ tubes (direct

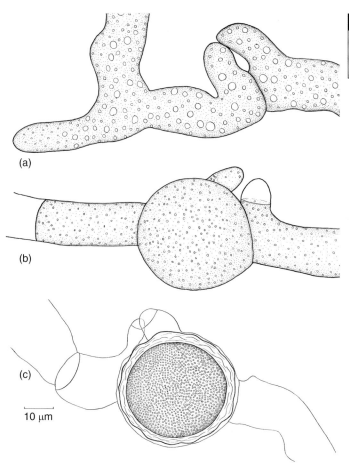

Fig 7.38 *Basidiobolus ranarum.* Successive stages in the formation of zygospores. (a) Progametangia. (b) Young zygospore. (c) Mature zygospore.

(a)

(b)

(c)

10 μm

germination), by the formation, in some species, of numerous microconidia, or by the development of capilliconidia resembling those of *Basidiobolus*. Zygospores or azygospores have been reported and all species which reproduce sexually in this way are homothallic. It is believed that the cytological condition of the nuclei is haploid, as it is in other Zygomycota, and that karyogamy and meiosis are involved in the formation of zygospores and azygospores (McCabe *et al.*, 1984).

The best-known species of *Conidiobolus* is the cosmopolitan *C. coronatus*, a fungus which has been referred to under various names, e.g. *Entomophthora coronata, Delacroixia coronata* and *Conidiobolus villosus*. It grows readily and rapidly in agar culture, forming a septate mycelium and numerous phototropic conidiophores (Fig. 7.39) which shoot off conidia onto the lid of the Petri dish. Conidial discharge takes place both in the

light and in the dark, but is enhanced by light (Callaghan, 1969a).

The behaviour of a conidium on germination depends on pH, humidity, availability of light, and nutrients. If the conidium falls on a medium containing nutrients, it germinates by means of a germ tube, but on nutrient-poor media, such as water agar, it may develop into a secondary conidiophore, forming a slightly smaller conidium (Fig. 7.39c). The secondary conidiophore develops from the illuminated side of a primary conidium, and the conidiophore which develops is phototropically orientated, but not very precisely (Page & Humber, 1973). Under conditions of reduced humidity the primary conidia may develop a cluster of globose microconidia (Fig. 7.39f). The entire cytoplasm of the primary conidium is evacuated by the expansion of a large vacuole into numerous buds formed by localized softening of the primary conidium wall. At

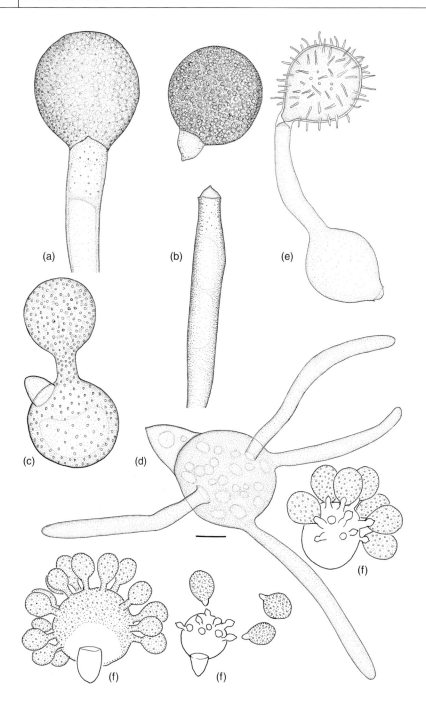

Fig 7.39 *Conidiobolus coronatus.* (a) Conidiophore with attached primary conidium. Note the columella which protrudes into the body of the conidium. (b) Apex of conidiophore and conidium after discharge by eversion of the columella from inside the spore. (c) Primary conidium germinating by repetition to produce a secondary conidium of similar type. (d) Conidium germinating directly, forming several germ tubes. (e) Conidium germinating to produce a villose secondary conidium. (f) Conidia germinating to produce numerous secondary microconidia which are discharged by columellar eversion.

maturity each microconidium is supported on a two-ply columella and projected by the eversion mechanism. In older cultures, primary conidia may form short conidiophores terminating in pear-shaped spiny-walled (villose) conidia, which are also projected by columellar eversion. The precise conditions under which spiny conidia are formed are not known. They have been inter-preted as resting spores, and they germinate to form a coarse septate mycelium. The possession of both microconidia and villose conidia is a combination unique to *C. coronatus*. Another type of resting spore may develop from ungermi-nated primary conidia which swell and form 2–3 layered thickened walls. Such conidia have been termed **loriconidia** (Gindin & Ben-Ze'ev, 1994). No zygospores have been reported for *C. coronatus*, structures resembling zygospores being interpreted as aerial chlamydospores.

Conidiobolus coronatus is a parasite of aphids, termites and whiteflies attacking tobacco, cotton and sweet potato, as well as of waxmoths and some other insects (Gindin & Ben-Ze'ev, 1994; Bogus & Szczepanik, 2000). It should be regarded as a relatively primitive opportunistic pathogen. Infection of termites can occur by penetration of germ tubes through the exoskeleton, or via the oesophagus after ingestion of germinated conidia (Yendol & Paschke, 1965). Following infection of insects, death can occur within 2 days, probably by the production of toxins (Evans, 1989). A highly insecticidal 30 kDa protein has been found in mycelium and culture filtrates of *C. coronatus* (Bogus & Scheller, 2002). This and possibly other toxins induce damage to blood cells or early death in several insects when injected into the haemocoel. In artificially infected waxmoths (*Galleria mellonella*), infection is followed by melanization of the host cuticle and damage to the Malpighian tubules with no evidence of tissue penetration (Bogus & Szczepanik, 2000). *Conidiobolus coronatus* and *C. obscurus* are being investigated as potential agents of biological control of insect pests. *Conidiobolus coronatus* is also pathogenic to mammals such as horses, llama, chimpanzee and man. Human infections are most common in the moist tropics and subtropics, especially in male outdoor workers from the rain forests of West Africa. Although the mode of transmission has not been established it is probably by inhalation of spores which germinate in the nasal mucosa. Other species of *Conidiobolus* known to infect vertebrates are *C. incongruus* and *C. lamprauges*. Isolates pathogenic to vertebrates grow readily at 37°C (Gugnani, 1992; Ribes *et al.*, 2000).

7.5.3 Entomophthoraceae
Benny *et al.* (2001) included 12 entomopathogenic genera in the family Entomophthoraceae (Gr. 'insect destroyer'), of which we shall study the three most important, i.e. *Erynia*, *Entomophthora* and *Furia*.

Erynia
There are about 12 species of *Erynia* parasitic on terrestrial insects such as aphids and Lepidoptera, but some attack the aquatic larval stages of Diptera such as *Simulium* spp. (river blackflies), stone flies and caddis flies. Characteristic features of the genus are branched conidiophores bearing uninucleate, bitunicate primary conidia which are discharged by septal eversion. Germination of primary conidia is by the production of various types of secondary conidia. Tertiary conidia may also develop. **Resting bodies** (azygospores and zygospores) occur in some, but not all species. Attempts are being made to use *Erynia neoaphidis* to control aphid populations in field crops (Pell *et al.*, 2001).

Erynia neoaphidis
This species, synonymous with *Entomophthora aphidis*, *Pandora neoaphidis* and *Zoophthora neoaphidis*, is the most widespread aphid pathogen of temperate regions and has been found on over 70 species of aphids on annual and perennial crops. It also attacks aphids on non-cultivated plants, a common example being the nettle aphid, *Microlophium* (Fig. 7.40). Infected aphids are cream to brown in colour. They are attached on the ventral side to their plant host by fungal rhizoids and their bodies are distended. Within the body of an infected aphid there are numerous closely packed, wide, septate hyphal bodies (Fig. 7.41a). Widely spaced, thick-walled, long, awl-shaped **pseudocystidia** (Fig. 7.41b) pierce the cuticle and, surrounding them, numerous tightly packed, branched conidiophores emerge, usually made up of uninucleate segments (Figs. 7.41b,d; Brobyn & Wilding, 1977). The tip of the conidiophore is cut off by a two-ply septum to form a uninucleate primary conidium with a two-layered wall (Figs. 7.41d,e).

Under humid conditions (relative humidity >95%), primary conidia are discharged by septal eversion for a distance of about 1 cm, and detached conidia show a bulging papilla at their base. Violent discharge projects the conidia through the boundary layer of still air

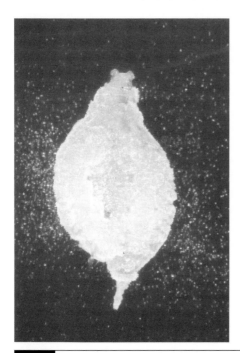

Fig 7.40 Carcass of a nettle aphid (*Microlophium*) incubated on tap water agar for 12 h. Note the halo of discharged conidia of *Erynia neoaphidis*.

globose appressoria develop at the tips of the germ tubes. After penetrating the cuticle and epidermis, the hyphal tips branch and fragment to form multinucleate protoplasts which become rapidly dispersed throughout the haemocoel within about 12–24 h. The protoplasts may increase in number by budding (Butt *et al.*, 1981). It is believed that the switch to the protoplast form is in response to contact with the nutrient-rich haemolymph. Later, as the nutrients in the haemolymph are exhausted, the protoplasts develop cell walls and are transformed into hyphal bodies. Protoplasts and hyphal bodies colonize fat bodies, nerve ganglia and muscle tissue. Infected aphids die some 72 h after inoculation, and shortly before death rhizoids develop from enlarged hyphal bodies and emerge from the ventral side of the abdomen, making contact with the leaf on which the aphid has been feeding, then branching by bifurcation to form digitate holdfasts. About 15–30 rhizoids may develop from a single aphid before it dies. Soon after death, pseudocystidia and conidiophores emerge. Under natural conditions, infected aphids die in the late afternoon and sporulation begins at night. The moist conditions and dew formation after sunset play a role in enhancing spore production and discharge. Hemmati *et al.* (2001) found that concentrations of air-borne conidia among wheat crops were usually highest at night and in the early morning and relatively low during the day, peak concentrations being correlated with high relative humidity.

It is rare for an infected aphid to produce both conidia and resting bodies. Germinating resting bodies form branched or unbranched germ tubes bearing retraction septa. They are terminated by an apical conidium, followed by one or two lateral conidia. These conidia closely resemble those which develop on infected aphids, and they are discharged by septal eversion (Tyrell & MacLeod, 1975). The germination of resting bodies is markedly stimulated by long-day conditions of more than 14 h of light per day (Wallace *et al.*, 1976). In the pea aphid, *E. neoaphidis* does not form resting bodies, surviving instead as hyphal bodies in aphid cadavers. Artificially infected cadavers can be stored and

surrounding the host so that they come under the influence of wind and gravity, falling at a velocity of about $1 \, \mathrm{cm \, s^{-1}}$ (Hemmati *et al.*, 2002). About 200 000 conidia are produced per cadaver of adult pea aphid over a period of 2–3 days. Immediately after discharge, primary conidia may germinate to produce secondary conidia which are wider and more ovate than the primary conidia (see Figs. 7.41e,f). Direct germination by the production of a germ tube from one or both ends of primary and secondary conidia also occurs (Fig. 7.41c). As in many Entomophthorales, cytoplasm is concentrated into a few terminal cells, leaving empty intercalary segments cut off by retraction septa (Fig. 7.41c).

Brobyn and Wilding (1977) and Butt *et al.* (1990) have described the process of infection of the pea aphid *Acyrthosiphon pisum*. Conidia adhere to any point on the aphid cuticle, often in clumps, and may germinate by forming secondary conidia or germ tubes. The tip of a germ tube can penetrate any part of the cuticle. Clavate or

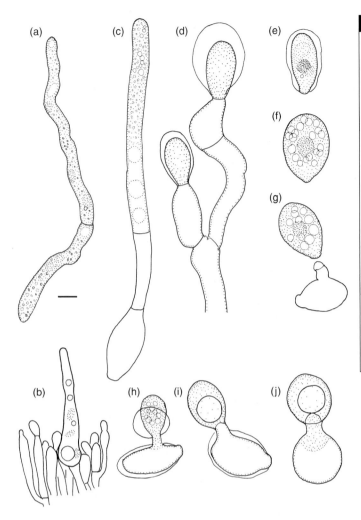

Fig 7.41 *Erynia neoaphidis.* (a) Hyphal body from within an infected aphid. (b) Pointed pseudocystidium projecting above a layer of conidiophores at the surface of a dead aphid. (c) Direct germination of a secondary conidium. Cytoplasm is confined by a retraction septum to the tip of the germ tube. (d) Branched conidiophore with terminal primary conidia. The wall surrounding the conidium is bitunicate with a thin outer envelope. (e) Discharged uninucleate primary conidium. (f) Discharged secondary conidium. Compare its more ovoid shape with the shape of the primary conidium. (g) A discharged secondary conidium has germinated by repetition to form a further conidium of the same type. Note the bulging septum on the empty conidium and at the base of the newly developed conidium. (h) Primary conidium germinating to form a secondary conidium. Both are bitunicate. (i) Secondary conidium germinating to form a tertiary conidium with a single large lipid body. (j) Tertiary conidium germinating by repetition. (a,b) = 20 µm; (c−j) = 12.5 µm.

used as inoculum to introduce the parasite into field populations of aphids pathogenic to crops, such as *Aphis fabae*, the common blackfly of broad beans. It is also possible to grow *E. neoaphidis* in agar culture and to introduce inoculum into aphid-infested crops in this form (Shah *et al.*, 2000).

Erynia conica

Whilst *E. neoaphidis* shows some versatility in its asexual reproduction, a more extreme example is *E. conica*, which forms four distinct types of conidium, some of them primary and some secondary (Descals *et al.*, 1981; Hywel-Jones & Webster, 1986a). *Erynia conica* is a parasite of the blackfly *Simulium* (Diptera) and some other insect hosts associated with aquatic habitats. *Simulium*

spp. have aquatic larval stages generally found in rapidly flowing streams. The larvae are attached to stones, twigs and aquatic plants, feeding by the ingestion of particulate matter collected by modified branched mouthparts (head rakes). After pupation, winged adults emerge, mate, and the females take a blood meal from a mammal. Gravid females lay egg masses amongst algae and mosses on water-splashed boulders kept continuously wet by trickling water. At such sites the white swollen bodies of dead females infected by *E. conica*, attached by rhizoids, may sometimes be found in large numbers. Conidiophores project from the carcass, bearing conidia of two types. From conidiophores which develop in air, i.e. at the surface of the insect's body projecting out of water, boat-shaped bitunicate conidia develop. These are termed *primary cornute* (type 1) conidia and are illustrated in Fig. 7.42a. They are discharged from the conidiophores by septal eversion. If a type 1 conidium floats on the water surface, it will germinate to produce a balloon-shaped *secondary globose* (type 2) conidium on its upper side, projecting into the air. The type 2 conidium contains a single large lipid body and a two-ply septum capable of discharge by eversion (see Fig. 7.42b).

When primary cornute conidia become submerged in water, they germinate to produce a secondary conidium of a different type. Its body has four branches (i.e. it is **tetraradiate**) and the position of attachment of this conidium to the short conidiophore is the central point from where the four arms radiate (see Fig. 7.42c). This type of spore is termed a *secondary stellate* (type 3) conidium. If a dead infected insect is continuously bathed in water or is submerged,

the conidiophores emerging from it will develop primary conidia which are also tetraradiate, but these are attached at the tip of the main arm from which three upper arms radiate (see Fig. 7.42d). This type of conidium is a *primary coronate* (type 4) conidium. So types 1 and 2 are aerial conidia, formed and discharged into air, whilst types 3 and 4 are aquatic conidia, formed and released under water. Tetraradiate conidia are a typical adaptation of fungi to dispersal in aquatic environments and are produced also by aquatic hyphomycetes (see p. 685).

Most of the four types of conidium can germinate by repetition, by germinating to form conidia of one of the other types, or by the formation of germ tubes. For example, a type 1 conidium can germinate by repetition to form a secondary conidium morphologically identical to itself, i.e. another type 1 conidium. This is described in shorthand as 1−1

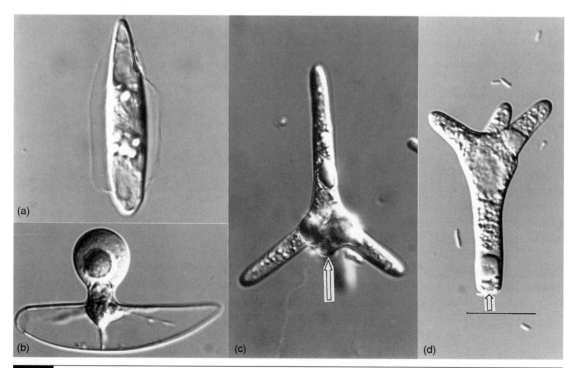

Fig 7.42 *Erynia conica.* The four types of conidia. (a) Primary cornute conidium (type 1). Note the bitunicate wall. (b) Primary cornute conidium germinating to produce a secondary globose conidium (type 2). (c) Secondary stellate conidium (type 3) which has developed from a submerged primary cornute conidium. The point of attachment of the conidium is between the three backwardly directed arms (arrow). (d) Primary coronate conidium (type 4) with the point of attachment at the end of the main, vertical, arm (arrow). The single large nucleus is visible below the point of branching. Bar = 20 μm, all images to same scale. From Webster (1992), with kind permission of Springer Science and Business media.

germination (Webster *et al.*, 1978). Type 1 conidia may show 1–1, 1–2 and 1–3 germination. Of the 16 (i.e. 4 × 4) possible interconversions, 12 have been observed so far (Webster, 1987). The only type of conidium shown to be infective is the secondary globose, i.e. type 2, conidium (Hywel-Jones & Webster, 1986b). It is sometimes termed an invasive conidium. This kind of conidium only develops from cornute conidia, although these may be primary or secondary. When a secondary globose conidium is in contact with an insect cuticle, a short germ tube develops with an appressorium at its tip. Penetration of the cuticle seems to be mainly by enzymatic means and is followed by the formation of multinucleate, branched hyphal bodies in the haemocoel. It is presumed that the other kinds of conidia function as dispersal rather than infection units, and they can be found in appreciable numbers trapped in foam near infected flies. It appears that only adult flies are infected through the cuticle. Although all types of conidia are known to be present in larval guts, there is no evidence that larvae are infected from ingested conidia. Survival over winter, when adult insects are not available, is by globose, thick-walled zygospores which are formed within the dead body of an insect, surrounded by a network of brown hyphae.

The precise physical conditions associated with the different types of germination in *E. conica* are not known and most attention has been devoted to the germination of the primary cornute (type 1) and secondary globose (type 2) conidia (Nadeau *et al.*, 1995, 1996). Germination of the latter, resulting in appressorium formation and cuticular penetration on wings of the susceptible host *S. rostratum*, occurs over the temperature range of 15–25°C with an optimum at 20°C. Germination occurs within 2 h and penetration within 9 h. The development of appressoria is related to the presence of a coating of lipid on the host cuticle. In experiments in which lipids were removed from susceptible blackfly wings, there was no discernible appressorium formation or cuticular penetration. On the non-susceptible host *S. decorum*, germination is delayed and appressorium formation and cuticular penetration do not occur. Instead, a

high level (26%) germination of the 2–1 type takes place.

The plasticity of asexual reproduction shown by *E. conica* is not unique. Similar versatility is shown by some other members of the Entomophthoraceae which grow on insects with aquatic larval stages such stoneflies (Plecoptera) and crane flies (Tipulidae) (Descals & Webster, 1984).

Entomophthora muscae

There are about a dozen species of *Entomophthora*, occurring as widespread insect pathogens (Samson *et al.*, 1988; Humber, 1997). They are characterized by unbranched conidiophores and multinucleate primary conidia which are projected by a squirt mechanism. Secondary conidia may form on germination of the primary conidia, but these are discharged by a septal eversion mechanism similar to that described above for *Erynia* and *Conidiobolus*. Sexual reproduction is by the formation of zygospores and azygospores.

The best-known taxon is *E. muscae* which is, in fact, a complex of about five species with similar morphology and spore dimensions (MacLeod *et al.*, 1976). This fungus is a parasite of houseflies and other Diptera. Disease is apparent in summer to autumn and is more frequent in wet weather. In the field, epizootics occur in places where there are dense populations of potential hosts, for example dung flies (*Scatophaga* spp.) on farms, or hoverflies (*Melanostoma* spp.) attracted to the honeydew secreted by *Claviceps* (see Fig. 12.26b) on the moor grass *Molinia*. Diseased flies can occasionally be found attached to the glass of a window pane surrounded by a white halo about 2 cm in diameter made up of discharged conidia (Plate 3g).

The dead fly shows a distended abdomen with white bands of conidiophores projecting between the segments of the exoskeleton. The unbranched multinucleate conidiophores arise from the coenocytic mycelium which plugs the body of the dead fly. The conidia are also multinucleate (Fig. 7.43b). They are projected by a forwardly directed jet of cytoplasm from the elastic conidiophores. On impact, the bitunicate nature of the wall of the primary conidium becomes apparent (Fig. 7.43e). Recently discharged conidia have a dried out drop around them which represents the cytoplasm squirted from the conidiophore (Figs. 7.43d–f). This cytoplasmic coating may act as a protective agent against desiccation and may possibly help in attaching the primary conidium to the cuticle of an insect. If the conidium impinges on the body of a fly, it develops an adhesive pad which attaches it firmly to the cuticle (Fig. 7.43h). Penetration of the cuticle is probably brought about by a combination of mechanical and enzymatic means (Brobyn & Wilding, 1983). A few hours after infection, tri-radiate fissures can be seen in the cuticle beneath attached conidia. When the cuticle in such a region is examined from the inside, a thin-walled bladder-like expansion can be seen. From this cell mycelial branches develop. The hyphae grow towards the fatty tissues, and as these are consumed the hyphae break up to form wall-less protoplasts which are carried by the circulatory system to all parts of the body. Eventually the protoplasts secrete walls and become converted into hyphal bodies (Fig. 7.43c). Infected flies show behavioural changes, often crawling to the top of a grass stem and clasping it or adhering to walls or window panes by the proboscis (Maitland, 1994). The sexual behaviour of the host may also be affected (Moller, 1993). Males attempting to mate with diseased females may themselves

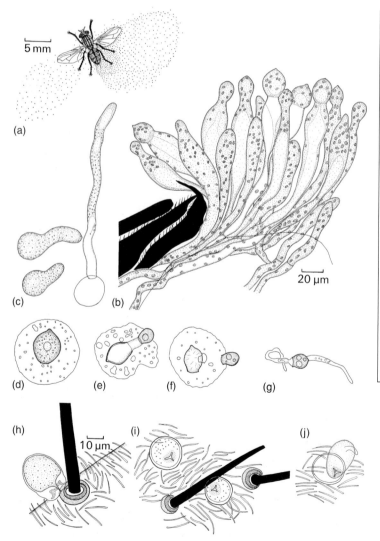

Fig 7.43 *Entomophthora muscae.*
(a) House fly adhering to a window pane, surrounded by a halo of discharged conidia. (b) L.S. house fly showing palisade of unbranched conidiophores projecting between segments of the exoskeleton. The conidiophores and conidia are multinucleate. (c) Hyphal bodies from recently dead fly extending to form conidiophores. (d) Primary conidium immediately after discharge surrounded by cytoplasm from the conidiophore. (e,f) Germination of primary conidia to form secondary conidia which are discharged by bouncing off (septal eversion). (g) Germination of secondary conidium by germ tubes. (h) Attachment of primary conidium to integument of a fly. (i) Two primary conidia attached to integument and penetrating it by a tri-radiate fissure. (j) View of penetration from within the integument. Note the bladder-like expansion within the tri-radiate fissure. (b–g) to same scale, (h–j) to same scale.

become infected with *E. muscae*, making it a sexually transmitted pathogen. Although it was previously generally accepted that there are no rhizoids in *E. muscae*, Balazy (1984) has shown that rhizoids do develop from hyphal bodies within the head, growing through the proboscis and forming a network of branched hyphae with short irregular holdfasts. A few days after infection the fly dies and the hyphal bodies within the abdomen then grow out into coenocytic hyphae which penetrate between the abdominal segments and develop into conidiophores. Discharge of primary conidia begins within about 5 h, reaching a maximum about 10–12 h after death. Over 8000 conidia may develop from a single cadaver (Mullens & Rodriguez, 1985). The primary conidia remain viable for only 3–5 days. If they fail to penetrate a fly, they may produce secondary conidia within 3 h. The secondary conidia are released from the tips of short conidiophores by septal eversion. They may germinate by a germ tube or by producing the same type of conidium by repetition.

Within the body of the dead fly, multinucleate spherical resting bodies (azygospores) are formed. In the wheat bulb fly *Leptohylemia coarctata* it has been observed that a much higher proportion of infected female flies contain resting bodies as compared with infected males. This is probably associated with the longer lifespan of females than males (Wilding & Lauckner, 1974). Resting bodies may develop terminally or in an intercalary position from short hyphae, or by budding from hyphal bodies. They germinate by developing a germ conidiophore. Germination is stimulated by the action of chitin-decomposing bacteria on the resting spore wall. It is from such resting bodies that infection probably begins each year (Goldstein, 1923).

The onion fly *Delia antiqua* has maggots which pupate in the soil and overwinter there. Adults become infected as they emerge through the soil the following season, presumably from germ conidia which develop from resting spores (Carruthers *et al.*, 1985). In some members of the Entomophthorales, e.g. *Entomophthora sepulchralis*, zygospores develop following conjugation between hyphal bodies (see Fig. 7.44). *Entomophthora muscae*, like many other entomopathogenic Entomophthorales, can be grown in complex media such as those used in tissue culture (Wolf, 1981; Papierok & Hajek, 1997). Yeast extract and ingredients of animal origin such as egg yolk, fat and serum or blood are also used. Growth is markedly stimulated by glucosamine, a breakdown product of chitin. Successful cultures have also been established on a medium containing wheat grain extract, peptone, yeast extract and glycerol (Srinivasan *et al.*, 1964).

Furia

Some of the species formerly placed in the genus *Entomophthora* have features distinct from *E. muscae* and have been re-classified into different genera. An example is *Furia americana* (Plate 3h, Fig. 7.45), a fungus found on blowflies in the autumn, especially around corpses of dead animals or stinkhorns. In wet weather severe epidemics may occur, greatly affecting the blowfly population. Distinctive features are the conidiophores which branch close to the conidiogenous cells; uninucleate, bitunicate clavate conidia with a rounded apex and basal papilla; and discharge by septal eversion. Dead flies are often attached to adjacent plants by filamentous rhizoid-like hyphae. The conidiophores form yellowish pustules between the abdominal segments and the branched tips bear conidia. The two layers of the conidium wall are frequently separated from each other by liquid (Figs. 7.45a–c). These conidia are projected for several centimetres from the host and, on germination, may form germ tubes, or may produce secondary conidia which are projected by the rounding off of a two-ply septum. Within the dry body of the dead fly numerous smooth hyaline thick-walled resting spores (azygospores) are formed by budding from the lateral walls of parent hyphae (Fig. 7.45f).

7.6 | Glomales

The roots of most terrestrial plants grow in a mutualistic symbiosis with fungi, i.e. an association in which both partners benefit.

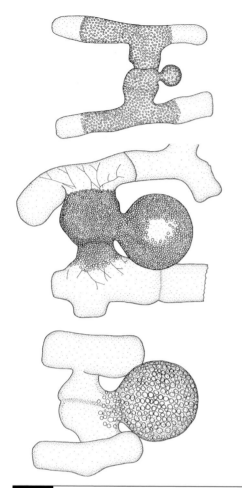

Such symbiotic associations are termed **mycorrhiza** (Gr. 'fungus root'). There are several different kinds of mycorrhiza, including vesicular and arbuscular mycorrhiza, ectomycorrhiza (sheathing mycorrhiza, pp. 21 and 526), ericoid mycorrhiza (p. 442), and orchid mycorrhiza (p. 596) (Smith & Read, 1997; Peterson *et al.*, 2004). It is important to realize that the nature of the relationships between the fungi and their host plants in these distinct types of association is not the same. In this section we shall look at the Glomales, a group of zygomycetous fungi causing the development of vesicular arbuscular mycorrhiza (VAM) and arbuscular mycorrhiza (AM). These fungi are particularly well-known as mycorrhizal associates of herbaceous plants, but they may also associate with trees, especially in the tropics.

7.6.1 General features of VAM and AM

A coarse, intercellular, aseptate coenocytic mycelium within the root tissues may develop large, balloon-shaped intercalary or terminal thick-walled **vesicles** (intraradical vesicles) which are multinucleate and contain large amounts of lipid (Figs. 7.46c,d). In some plants, e.g. the roots of *Paris*, the mycelium emits branches which penetrate the cortical root cells, forming extensive intracellular coils. More commonly, hyphae penetrating host cells fork repeatedly to form richly branched **arbuscules** (Fig. 7.46c) which invaginate the plasmalemma. Plant and fungal plasma membranes are separated by an apoplastic compartment, the periarbuscular space. The arbuscule is therefore a type of haustorium, and there is an interchange of nutrients and water across the periarbuscular space. Arbuscules have a relatively short active life, lasting only a few days. After this time the fine tips of the arbuscules are digested by the host cell so that only irregular clumps of fungal material remain (Fig. 7.46c).

A coarse, angular and often thick-walled mycelium extends outwards from infected roots, sometimes for several cm, and penetrates into the surrounding soil. It may bear large (>100 μm dia.) globose multinucleate thick-walled spores which are sometimes termed chlamydospores. These spores contain thousands of nuclei as well as energy reserves including lipid droplets, glycogen, protein and trehalose. These spores may be borne singly or in clusters and are often naked, but in some species, e.g. *Glomus mosseae*, they are enveloped in a weft of hyphae to form a **sporocarp** (Figs. 7.46a,b). Chlamydospores are asexual reproductive structures and are known to survive in dry soil for many years. For most members of the group only asexual reproduction is known, but in *Gigaspora decipiens* zygospores and azygospores have been reported in addition to chlamydospores. This species is heterothallic (Tommerup &

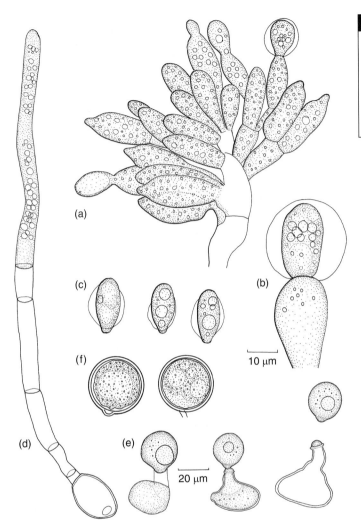

Fig 7.45 *Furia americana* from blowfly.
(a) Branched conidiophore. (b) Single conidiophore and conidium. Note that the wall of the conidium is bitunicate. (c) Conidia after discharge. (d) Conidium germinating by means of a germ tube. (e) Primary conidia germinating to produce secondary conidia. (f) Spherical resting bodies from dead fly.

Sivasithamparam, 1990). Sporocarps may form part of the diet of some mammals and chlamydospores can be dispersed by soil animals, including invertebrates and some rodents as well as larger hoofed mammals. Chlamydospores can survive in their faeces and are also dispersed in wind-borne soil dust (Allen, 1991; Allen *et al.*, 1997; Linderman, 1997).

The spores of Glomales can be extracted from soil by wet sieving and decanting from soil slurries using a series of sieves in the 2000–60 μm size range (Gerdemann & Nicolson, 1963). After surface sterilization, single chlamydospores placed near the roots of susceptible host plants such as *Trifolium* and *Sorghum* germinate, produce hyphae which make contact with the root surface and form appressoria before infecting the root (Hepper, 1984; Menge, 1984). In this way, dual cultures have been established and can be maintained by the addition of freshly extracted spores or infected root pieces to pots containing a suitable host plant. Viable spores derived from dual cultures maintained on potted plants are available from the International Collection of Vesicular–Arbuscular Mycorrhizal Fungi (Morton *et al.*, 1993). Spores have also been produced under aseptic conditions in association with hairy root cultures (Mugnier & Mosse, 1987; Bécard & Fortin, 1988). Limited extension of germ tubes takes place after germination in vitro, but sustained growth in the absence of living root tissues does not occur, so the fungi causing this

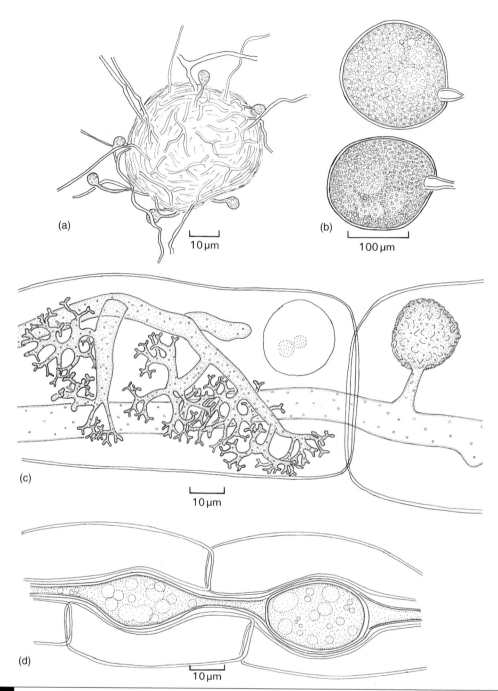

Fig 7.46 Vesicular–arbuscular mycorrhiza. (a) *Glomus mosseae* sporocarp in which chlamydospores are embedded. There are also naked chlamydospores attached to external hyphae. (b) Chlamydospores dissected from a sporocarp, borne on single subtended hyphae. (c) Onion root cells infected with *Glomus mosseae*. The cell to the left contains a nucleus with two nucleoli and a branched haustorium or arbuscule. In the cell to the right the arbuscule has degenerated. (d) Vesicles from roots of *Arum maculatum*.

kind of mycorrhiza are obligate mutualistic symbionts. Glomalean fungi are generally non-specific in their host range. The roots of most groups of vascular land plants are associated with this type of mycorrhiza, as are the gametophytic stages of Bryophyta and Pteridophyta.

7.6.2 Taxonomy and evolution of Glomales

Fungi in this group were originally classified in the Endogonales but are currently placed in a separate order, the Glomales, with the Endogonales now reduced to a single genus, *Endogone*, with subterranean fleshy sporocarps which contain zygospores. Each zygospore is formed after conjugation of two gametangia (Pegler *et al.*, 1993). One species, *E. flammicorona*, forms ectomycorrhizae with some Pinaceae (Fassi *et al.*, 1969). The fruit bodies of *Endogone* spp. are colloquially called 'pea truffles' (Plate 3i; Pegler *et al.*, 1993).

The order Glomales was proposed by Morton and Benny (1990) to include all soil-borne fungi which form arbuscules in obligate mutualistic associations with terrestrial plants. Sexual reproduction is rare. There are about 150 species and 6 genera in 2 suborders, the Glomineae with 2 families (Glomaceae and Acaulosporaceae), and the Gigasporineae with a single family (Gigasporaceae). Members of the Glomineae (such as *Glomus*, *Acaulospora*) form intraradical vesicles (VAM type), whilst members of the Gigasporineae have no intraradical vesicles and are of the AM type. The separation of genera within the Glomales is based partly on different patterns of chlamydospore development, and partly on the structure of the spore wall which may be complex and multilayered (Hall, 1984; Morton & Bentivenga, 1994). Schüssler *et al.* (2001) have suggested that the Glomales are not closely related to the Zygomycota and should be considered as a separate phylum, the Glomeromycota.

VAM and AM associations are very ancient, and structures resembling extant arbuscules have been discovered in the fossilized rhizome tissues of early vascular plants, including Devonian psilophytes such as *Rhynia* (Pirozynski & Dalpé, 1989; Taylor *et al.*, 1995). Even older are the fossilized chlamydospores found among bryophytes of the Ordovician period (some 460 million years old; Redecker *et al.*, 2000a). It is believed that the origin and evolution of land plants was dependent on symbiotic associations of the VAM and AM type (Pirozynski & Malloch, 1975; Malloch, 1987; Simon *et al.*, 1993).

An interesting non-mycorrhizal relative of the Glomales is the fungus *Geosiphon pyriforme*, which is unusual in harbouring a mutualistic endosymbiont, the cyanobacterium *Nostoc*. When the hyphal tip of *Geosiphon* encounters a suitable symbiont, this is taken up and the hypha swells to form a so-called bladder cell. The *Geosiphon*–*Nostoc* symbiosis resembles cyanolichens (see p. 451) in being autotrophic both for carbon and nitrogen (Schüssler & Kluge, 2001). *Geosiphon* reproduces by forming chlamydospores similar to those of *Glomus*. Molecular studies show that *Geosiphon* is closely related to the Glomales and may be ancestral to the group (Gehrig *et al.*, 1996; Redecker *et al.*, 2000b).

7.6.3 Physiological and ecological studies

The immense current interest in AM and VAM mycorrhiza has its origins in the demonstration of the improved growth of mycorrhiza-infected host plants compared to uninfected controls. Literature on the physiology of this relationship has been reviewed by Hause and Fester (2005). The arbuscule is the main interface for nutrient exchange between the plant and its fungal partner, although the latter may also be able to take up nutrients through intercellular hyphae. The periarbuscular space is a highly acidic compartment (Guttenberger, 2000) due to the outward-directed pumping of protons by H^+ ATPases located in the plasma membranes of both partners. This sets up proton gradients which may be used for active uptake of sucrose hydrolysis products (fructose and glucose) by the fungus, and phosphate and other mineral nutrients by the plant. Proton-dependent transport proteins have been localized in both plant and fungal perihaustorial membranes (see Hause & Fester, 2005).

The ecology of VAM and AM fungi in crop plants and natural communities is of particular interest (Allen, 1991; Smith & Read, 1997; Leake et al., 2004). There are numerous reports of significant improvements in growth rate, dry weight and mineral content following infection especially of plants growing on nutrient-deficient soils. Emphasis has been placed on phosphate nutrition. The supply of phosphate (as HPO_4^{2-} or $H_2PO_4^-$, depending on soil pH) is often a limiting factor to plants growing in natural soils. It is usually present in low concentrations and diffuses through soil very slowly. Its influx may increase 3–4-fold in infected plant roots but there are also significant increases in other minerals such as Zn, Cu, and ammonium. The water relations and resistance of infected plants to infections by pathogens may also be improved. Increased uptake of minerals is largely due to the exploration of larger volumes of soil by the extramatrical hyphae which can extend beyond the depletion zone surrounding plant roots. The depletion zone is a region in which minerals are taken up by plant roots at a rate greater than can be replenished by diffusion through the soil. For many plants the depletion zone is only 1–2 mm wide whilst the extramatrical hyphae may extend for several centimetres and can penetrate into soil cavities too fine to be explored by roots. Moreover, phosphate can be translocated through fungal hyphae towards the host root at much faster rates than is possible by diffusion through soil. The improved growth of host plants associated with increased supply of minerals obtained through the hyphae of the mycorrhizal symbiont is achieved at a cost to the plant, i.e. the drain of photosynthate taken up by the fungus whose biomass, achieved largely at the expense of the host, may amount to 3–20% of the root weight. In experiments in which $^{14}CO_2$ was supplied to the shoots of young cucumber plants infected by G. fasciculatum, as much as 20% of the radioactive carbon fixed by the plant was used by the fungus (Jakobsen & Rosendahl, 1990).

In nutrient-deficient soils such as sand dunes, recently disturbed soil, spoil heaps, areas covered by volcanic ash, etc., successful colonization by plants appears to be correlated with root infection by Glomales (Allen, 1991). In closed vegetation such as mature grassland which contains a diversity of plants, spore extraction reveals a wide diversity of AM and VAM species. The roots of different plant species making up the community are in close contact and may also be connected by a hyphal network (Newman, 1988). There is experimental evidence using isotopically labelled ^{15}N, ^{32}P, and ^{14}C that there may be an interchange of mineral nutrients and carbon between unrelated plant species mediated by VAM mycelia, but Newman (1988) has cautioned against the conclusion that any increases in labelled materials necessarily imply net gains to receiver plants at the expense of donors. There is also experimental evidence using soil microcosms seeded with a mixture of grassland grasses and dicotyledons and inoculated with Glomus constrictum that mycorrhizal infection may increase species diversity by selectively enhancing the performance of less dominant dicotyledons. This results largely from a reduction in relative abundance of canopy dominants such as Festuca ovina (Grime et al., 1987).

7.7 | Trichomycetes

The Trichomycetes are a group of fungi which grow commensally in the guts of terrestrial, freshwater and marine arthropods such as insects, millipedes and crustaceans. In most cases there is little evidence that the host is harmed by their presence, although it has been shown that some species may extend parasitically into the ovarian tissue to form chlamydospores (cysts) in place of eggs. These are deposited amongst egg masses laid by uninfected females. McCreadie et al. (2005) have documented an element of plasticity in the association of a given trichomycete species, Smittium culisetae, with its blackfly host which may vary from commensalistic in well-fed larvae to mutualistic under starvation conditions to parasitic if the ovaries of adult females are infected.

More than 50 genera and over 200 species have been described but doubtless many more await discovery (Lichtwardt, 1986, 1996; Misra, 1998; Misra & Lichtwardt, 2000). Members of the group have a worldwide distribution and are especially common in the guts of larvae of aquatic insects. A few species belonging to one order (Harpellales) have been grown in culture and appear to have no unusual nutritional requirements. The term trichomycete (Gr. 'hairy fungus') refers to the fuzzy appearance of heavily infested gut linings. Branched or unbranched thalli are attached by a holdfast to the hindgut cuticle or to the peritrophic membrane, a transparent membranous sleeve which surrounds digested food material in the mid-gut of certain insects. Asexual reproduction is by various types of spore, including trichospores, chlamydospores, arthrospores or sporangiospores. Sexual reproduction by the formation of zygospores is known in the Harpellales. The occurrence of zygospores, the presence of chitin in the walls of *Smittium culisetae* (Sangar & Dugan, 1973) and molecular studies (O'Donnell *et al.*, 1998; Gottlieb & Lichtwardt, 2001) all provide evidence linking Trichomycetes with the Zygomycota. It is possible that the class Trichomycetes is polyphyletic, and it is therefore preferable to refer to the gut fungi as a biological group, trichomycetes with a lower-case 't' (Lichtwardt, 1986).

Three orders have been distinguished, namely the Harpellales, Asellariales and Eccrinales, of which we shall consider only the first. The Amoebidiales, previously included, are now classified with the protozoa.

7.7.1 Harpellales

Harpella melusinae is one of the most common and abundant trichomycetes with a worldwide distribution in temperate regions. It is found in larval blackflies (*Simulium* spp.) which live attached to stones, twigs and aquatic vegetation submerged in rapidly flowing streams. The dissection of larval guts reveals the peritrophic membrane, to the inner wall of which unbranched cylindrical thalli are attached. Developing thalli receive nutrients from the material passing through the gut. The peritrophic membrane is continuously secreted by endothelial cells lining the upper part of the mid-gut, i.e. new membrane material is added at the upper end. Young thalli

are present here and progressively older thalli further down. Attachment is by a simple holdfast (Fig. 7.47a). The holdfast consists of a chamber at the base of the thallus from which numerous finger-like projections protrude, cemented to but not penetrating the peritrophic membrane (Reichle & Lichtwardt, 1972). The cylindrical part of the thallus is divided by septa into 2–12 or more uninucleate segments called generative cells. The septal ultrastructure consists of a flared pore associated with a plug, somewhat resembling the bordered pit of a conifer xylem tracheid. This feature is characteristic of trichomycetes (Moss, 1975).

Asexual reproduction

The entire contents of the thallus are converted to reproductive cells. Reproduction begins at the terminal generative cell and progresses basipetally (towards the holdfast) by production of **trichospores** (Figs. 7.47a–c). Trichospores are really monosporous sporangia. They have been defined as 'exogenous, deciduous sporangia containing a single uninucleate sporangiospore and normally having one to several basally attached filamentous appendages' (Lichtwardt, 1986). The trichospores, which are usually coiled but sometimes straight, develop at the upper end of a generative cell (Fig. 7.47a). The nucleus of the generative cell divides mitotically, one daughter nucleus remaining in the generative cell, the other entering the developing trichospore. In *H. melusinae* there are four basal appendages which, before trichospore release, are spirally coiled inside the upper part of the generative cell (Figs. 7.47b,c; Reichle & Lichtwardt, 1972). At the distal end of the trichospore within the cytoplasm is an elongated apical spore body (Fig. 7.47a). This contains holdfast material which is released after extrusion of the sporangiospore on its germination within the insect gut, cementing the holdfast to the gut wall (Moss & Lichtwardt, 1976; Horn, 1989a). Trichospores are separated from the generative cell by a septum and are released by breakdown of the wall beneath it. After release, the appendages uncoil and extend up to 10 times their original length. The released trichospores are passed out from the larval gut with faecal material and the

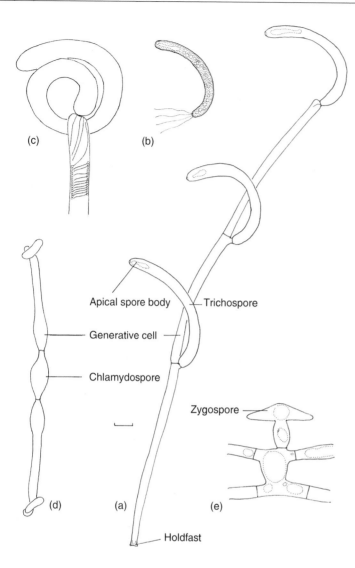

(c)

(b)

Apical spore body — Trichospore

— Generative cell

— Chlamydospore

Zygospore —

(d)   (a)   (e)

— Holdfast

Fig 7.47 *Harpella melusinae.* (a) Thallus attached by a holdfast to the peritrophic membrane from the mid-gut of a larva of a blackfly, *Simulium* sp. The thallus is divided by septa into three generative cells, each of which is producing a curved trichospore at its upper end. The apical spore body at the upper end of the trichospore contains material extruded to cement the holdfast to the peritrophic membrane. (b) Detached trichospore bearing four filamentous appendages at its base. (c) Developing trichospore showing the coiled filamentous appendages wrapped around inside the wall of the generative cell. (d) Chlamydospore which has germinated to form two generative cells, each of which is developing a trichospore. (e) Zygospore development. Two adjacent thalli have conjugated and from one of the conjugating cells a zygosporophore has been produced, terminating in a biconical zygospore. Bars: (a,b) = 10 µm, (c,e) = 5 µm, (d) = 15 µm. (d) after Moss and Descals (1986); (e) after Lichtwardt (1967).

appendages cause the trichospores to be entangled with faeces and other particulate material. Further development of trichospores, i.e. the extrusion of a sporangiospore from its sporangium and the formation of a holdfast, only occurs after ingestion and is stimulated by conditions in the larval gut.

Smittium culisetae (Harpellales) inhabits the midgut and hindgut of larval mosquitoes and can be grown in culture. Horn (1989a,b, 1990) has investigated in vitro the conditions which trigger sporangiospore extrusion in this and related species of *Smittium*. The trigger for extrusion in *S. culisetae* is a two-stage process. Phase 1, which simulates mid-gut conditions, involves exposure to 20 mM KCl at pH 10 followed by phase 2, in which the pH is reduced to 7, simulating hind-gut conditions. Following this sequence of treatments sporangiospores (= trichospores sensu stricto) are rapidly extruded, a process in which they increase in size by the uptake of water and by vacuolation, generating turgor pressure which aids extrusion. Spore germination quickly follows with the secretion of holdfast material from the apical spore body through canals in the distal wall of the sporangiospore (Horn, 1989a).

A second asexual, free-living stage in the life cycle of *H. melusinae* is the chlamydospore (sometimes termed ovarian cyst or cystospore), masses of which are deposited by adult female

Simulium in the place of eggs (Moss & Descals, 1986; Lichtwardt, 1996). A similar stage has been reported from *Simulium* infected with *Genistellospora homothallica* (Labeyrie *et al.*, 1996). Ovarian tissue is invaded by the fungus, growing in a parasitic mode. Cysts of *H. melusinae* dissected from ovaries are surrounded by a membranous sheath. They are ellipsoidal and, on germination, form two germ tubes, one at each pole, ending as a spherical knob, the generative cell initial. A single generative cell develops from each initial and forms a terminal trichospore (Fig. 7.47d). The chlamydospores are deposited among egg masses and infection of young larvae results from ingestion of trichospores produced by them. The ovarian chlamydospores therefore represent a 'missing link' in the life cycle of Harpellales. Adult blackflies do not contain trichomycete thalli because at the final ecdysis (moult) before pupation, the cuticular lining of the larval gut is shed.

Sexual reproduction

This occurs by the production of zygospores and has so far been reported only in Harpellales. In *H. melusinae* zygospores are rarely detected, possibly because they are associated with the last stage of development of the *Simulium* larval host before pupation and are shed at ecdysis. Zygospore formation is preceded by conjugation between swollen cells on adjacent thalli (see Fig. 7.47e; Lichtwardt, 1967). From one of the conjugating cells a zygosporophore grows out, and from this a biconical zygospore develops.

The biconical shape, which is characteristic of the Harpellales, is possibly adapted to passage through the insect gut. In some members of the group zygospores bear polar filamentous appendages, but these are absent in *Harpella* (Moss & Lichtwardt, 1977). The cytological details of zygospore formation have not been fully worked out but in *H. melusinae* the zygospore, zygosporophore and the two conjugant cells each contain a single nucleus. Moss and Lichtwardt (1977) have speculated that the four nuclei might have been derived from meiotic division of a diploid zygote nucleus within the fused conjugants. On this hypothesis the conjugant cells would be interpreted as gametangia, a situation markedly different from that found in other zygomycetes.

Relationships

On the basis of similarities in serological reactions, septal pore structure and in sporangial morphology, it has been suggested that the Harpellales are related to the Kickxellales, an order of mostly dung- and soil-inhabiting saprotrophic zygomycetes (Moss & Young, 1978). However, the evidence for a phylogenetic relationship between these two groups is conflicting. K.L. O'Donnell *et al.* (1998) and Gottlieb and Lichtwardt (2001) have attempted to correlate morphological criteria with molecular data (18S rDNA) but found only poor support, whereas Tanabe *et al.* (2004), comparing a range of DNA sequences, found a strong link between Harpellales and Kickxellales (see Fig. 7.1).

8

Ascomycota (ascomycetes)

8.1 | Introduction

The phylum Ascomycota (colloquially called ascomycetes) is by far the largest group of fungi, estimated to include more than 32 000 described species in 3400 genera (Kirk *et al.*, 2001). It is assumed that the majority of ascomycetes has yet to be discovered, and the total number of species may well be higher by a factor of 10–20 or even more (see Hawksworth, 2001). The name is derived from the Greek words *askos* (a leather bottle, bag or bladder) and *mykes* (a fungus), so ascomycetes are sac fungi. The characteristic feature of the group is that the sexually produced spores, the ascospores (see p. 25), are contained within a sac, the ascus. In most ascomycetes the ascus contains eight ascospores and is turgid, ejecting its spores by a squirt mechanism.

There is a very wide range of lifestyles. Some ascomycetes are saprotrophs, others necrotrophic or biotrophic parasites of plants and animals, including humans. Examples of biotrophic parasites are the Erysiphales, the cause of many powdery mildew diseases of plants (Chapter 13), the Taphrinales (p. 251) causing a range of plant diseases associated with growth abnormalities, and the Laboulbeniales, relatively harmless ectoparasites of beetles and some other insects (Blackwell, 1994; Weir & Blackwell, 2001). Many ascomycetes grow as endophytes in symptomless associations with plants. Some are mutualistic symbionts, for example the lichens (Chapter 16) which make up

about 40% of the described species of ascomycetes. Lichens are dual organisms consisting of a fungus (usually an ascomycete) and a photosynthetic alga or cyanobacterium living in close association. This type of association has evolved independently in several unrelated groups of ascomycetes and indeed it has been claimed that several major fungal lineages are derived from lichen-symbiotic ancestors (Lutzoni *et al.*, 2001), although this hypothesis is under dispute (Liu & Hall, 2004; see Fig. 8.17). Symbiotic mycorrhizal relationships also exist between true truffles (e.g. *Tuber* spp.) or false truffles (e.g. *Elaphomyces* spp.) and trees such as oak and beech (see pp. 423 and 313). The range of habitats is wide, as would be expected of such a large and diverse group of fungi. Ascomycetes grow in soil, are common on the above-ground parts of plants, and are also found in freshwater and in the sea.

Most ascomycetes are recognized by their fruit bodies or ascocarps, i.e. the structures which surround the asci. These will be described more fully later (see Fig. 8.16).

8.2 | Vegetative structures

Ascomycetes may grow either as yeasts, i.e. unicells multiplying by budding or fission, or as mycelia consisting of septate hyphae (Fig. 8.1a). Some fungi may switch from the yeast to the filamentous state or vice versa, i.e. they are **dimorphic**. A good example of a dimorphic fungus is *Candida* (see Fig. 8.1b).

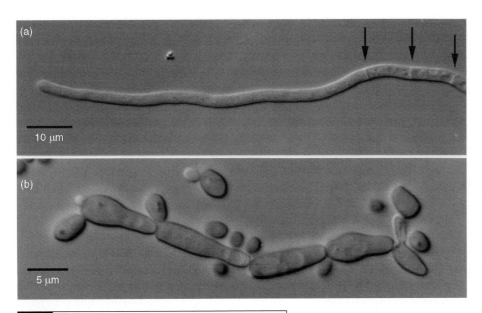

(a)

10 μm

(b)

5 μm

Fig 8.1 (a) Hypha of *Hormonema dematioides*. The positions of the first septa are indicated by arrows. (b) *Candida parapsilosis*. Pseudohypha budding off cells which continue to bud in a yeast-like manner.

Candida albicans is the cause of diseases such as thrush in mammals, including man.

The mycelial septa of ascomycetes are usually incomplete, developing as transverse centripetal flange-like ingrowths from the cylindrical wall of a hypha, which fail to meet at the centre so that in most ascomycete septa there is one central pore permitting cytoplasmic continuity and streaming between adjacent segments of mycelium (Buller, 1933; Gull, 1978). This means that organelles such as mitochondria and nuclei are free to travel from cell to cell; the large nuclei are constricted as they pass through the pore (Fig. 8.2). Individual cells may be uni- or multinucleate and the cytoplasmic continuity between the cells means that the mycelium of an ascomycete is effectively coenocytic. Proteinaceous organelles termed **Woronin bodies** (Buller, 1933) may be closely grouped near the central pore (Fig. 8.3). Woronin bodies are globose structures or 'hexagonal' (polyhedral) crystals made up essentially of one protein (Tenney *et al.*, 2000), and surrounded by a unit membrane. They measure 150–500 nm in width and are sufficiently large to block the septal pore. They rapidly do so near regions where

a hypha is physically damaged. Usually one Woronin body blocks one pore. The blockage in the septal pore is consolidated by deposition of further material (for references see Markham & Collinge, 1987; Markham, 1994; Momany *et al.*, 2002). Woronin bodies are formed near the hyphal apex and are transported to more distal regions of the hypha as septa develop. Woronin bodies have been recorded from ascomycetes and their related conidial fungi, but there are no reliable reports from other fungal phyla.

The mycelium of many ascomycetes is **homokaryotic** (Gr. *homoios* = like, resembling; *karyon* = a nut, meaning nucleus), i.e. all nuclei in a given mycelium are genetically identical. **Heterokaryotic** mycelia also occur and generally arise through **anastomosis**, i.e. the cytoplasmic fusion of vegetative hyphae. Following anastomosis between homokaryons of differing genotypes, nuclei, other organelles and plasmids may be transferred between one mycelium and another so that a given mycelium or even a single cell may contain nuclei of different kinds. However, the ability to form heterokaryons is under genetic control and a degree of genetic similarity between homokaryons is necessary for it to occur. Failure to establish a heterokaryon is a phenomenon known as heterokaryon incompatibility or **vegetative incompatibility** (Caten & Jinks, 1966; see pp. 320 and 594).

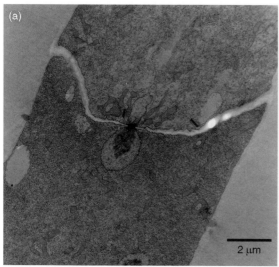

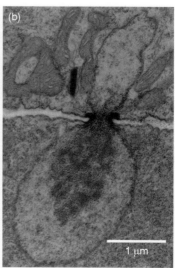

Fig 8.2 Nucleus of *Botrytis cinerea* passing through a septal pore. (a) View of the entire hyphal diameter. (b) Close up. Note the constricted appearance of the nucleus.

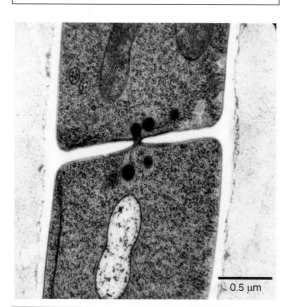

Fig 8.3 Transmission electron micrograph of a transverse septum in the hypha of *Emericella nidulans* showing five Woronin bodies near the central septal pore. Scale bar = 0.25 μm. Reprinted from Momany *et al.* (2002), *Mycologia*, with permission. ©The Mycological Society of America.

Some ascospores, e.g. those of *Neurospora tetra-sperma*, are heterokaryotic, and multinucleate conidia can also be heterokaryotic. Following

mycelial anastomosis between homokaryons, nuclear division succeeded by migration may result in the rapid spread of an introduced nucleus into a mycelium, thus transforming a homokaryon into a heterokaryon. An ascomycete mycelium may thus consist of a mosaic of cells, some of which are homokaryotic and others heterokaryotic. Because the different types of nuclei do not always divide at the same rate, the ratio of nuclear types in a heterokaryotic mycelium may change with time and respond to changes in external conditions such as nutrient availability. This gives the mycelium a degree of genetic flexibility which sometimes manifests itself as the formation of sectors in a mycelium in agar culture (Fig. 8.4). Another important source of genetic variability which may arise within a heterokaryon is the parasexual cycle, a process in which genetic recombination is brought about in the absence of meiosis. This is discussed more fully below.

8.3 | Life cycles of ascomycetes

8.3.1 Sexual life cycles
Sexual life cycles in the strict sense, i.e. involving nuclear fusion and meiosis, occur only in those ascomycetes which possess asci, because it is within the young ascus that these events occur.

Fig 8.4 Sectoring of a mycelial colony of *Pseudeurotium* sp. The slow-growing wild-type mycelium appears dark due to the formation of melanized cleistothecia. Two non-fruiting sectors have formed which show faster vegetative growth but no sporulation on the rich agar medium.

Ascospores of most ascomycetes contain one or more haploid nuclei, and therefore most (but by no means all) ascomycetes have a haploid vegetative mycelium. The mycelium is often capable of asexual reproduction, e.g. by fragmentation, budding or by the formation of conidia, chlamydospores, sclerotia, etc. The structure and formation of conidia is described below. Some yeasts, e.g. *Saccharomyces cerevisiae*, show an alternation of diploid and haploid yeast-like states and here the diploid state is the commonly encountered form (p. 265), in contrast to *Schizosaccharomyces* in which the vegetative cells are haploid (p. 253).

The mating behaviour of ascomycetes may be homothallic or heterothallic. In homothallic ascomycetes the mycelium derived from a single ascospore is capable of reproducing sexually, i.e. by developing asci. Examples are *Emericella nidulans*, *Pyronema domesticum* and *Sordaria fimicola*. However, the homothallic condition does not preclude outcrossing as is shown by the formation of hybrid asci containing black (wild type) and white (mutant) ascospores in crosses between different strains of *Sordaria*

fimicola (see Fig. 12.2). In heterothallic ascomycetes the ascus usually contains four ascospores of one mating type and four of the other. The two mating types differ at a single allele and the mating types may be designated A and a, a and α, or $(+)$ and $(-)$. Sexual reproduction occurs following plasmogamy between cells of the two mating types. Plasmogamy is of three main types:

1. Gametangio-gametangiogamy. Fusion occurs between differentiated gametangia. An example is *Pyronema domesticum* where fusion is between the **trichogyne**, a filamentous extension of the large, swollen 'female' gametangium (the **ascogonium**) and a less swollen 'male' gametangium, the **antheridium**, which donates nuclei to the trichogyne and thereby to the ascogonium (see p. 416).

2. Gameto-gametangiogamy. Fusion takes place between a small unicellular male gamete (**spermatium**) and a differentiated female gametangium (ascogonium). The spermatium is rarely capable of independent germination and growth and may only germinate to produce a short conjugation tube which fuses with the wall of the ascogonium. An example is *Neurospora crassa* in which the spermatium fuses with a trichogyne (see Fig. 12.7).

3. Somatogamy. Fusion takes place between undifferentiated hyphae, i.e. there are no recognizable sexual organs. This type of sexual behaviour is shown by *Coprobia granulata*, whose orange ascocarps are common on cattle dung.

8.3.2 Asexual life cycles

Most fungi which were formerly classified in the artificial group Deuteromycotina or Fungi Imperfecti are conidial forms (anamorphs) of Ascomycota, although a few have affinities with Basidiomycota. Evidence for a relationship to Ascomycota comes from morphological similarity and from DNA sequence comparisons. Morphological similarities include the structure of the mycelium, the layering of the hyphal wall as seen by electron microscopy, the fine-structure of nuclear division, and also close resemblances of conidial structure and development. Some genera contain species which reproduce by asexual means only, whilst closely similar forms have sexual as well as asexual reproduction. Examples include *Aspergillus* and *Penicillium*, which are anamorphs of several genera of Ascomycota (Trichocomaceae; see pp. 308–313) and *Fusarium* which is the anamorph of *Gibberella* and *Nectria* (members of the Hypocreales; see p. 343). It is presumed that fungi which reproduce only by conidia have lost the capacity to form ascocarps in the course of evolution.

8.3.3 Parasexual reproduction

This is a process in which genetic recombination can occur through nuclear fusion and crossing-over of chromosomes during mitosis. Meiosis does not occur, and instead haploidization takes place by the successive loss of chromosomes during mitotic divisions. It is believed that the necessary cytological steps take place in a regular sequence which Pontecorvo (1956) has termed the **parasexual cycle**. The essential steps include (i) nuclear fusion between genetically distinct haploid nuclei in a heterokaryon to form diploid nuclei; (ii) multiplication of the diploid nuclei along with the original haploid nuclei; (iii) the development of a diploid homokaryon; (iv) genetic recombination by crossing-over during mitosis in some of the diploid nuclei; and (v) haploidization of some of the diploid nuclei by progressive loss of chromosomes (aneuploidy) during mitosis.

This process was discovered in *Emericella* (*Aspergillus*) *nidulans*, which can reproduce sexually by forming asci and asexually by forming conidia (see Fig. 11.17). By changing the nutrient content of the medium on which the fungus is grown, the development of asci and therefore of normal sexual reproduction can be prevented. Genetic mapping based on gene recombination following conventional sexual reproduction has been compared with mapping based on parasexual recombination and has yielded identical results.

Parasexual recombination is known to occur not only in Ascomycota but also in Oomycota and Basidiomycota. It makes possible genetic recombination in organisms not known to reproduce by sexual means and helps us to understand why purely asexual fungi such as many species of *Aspergillus* and *Penicillium* have achieved success and have continued to flourish in the course of evolution. However, because parasexual reproduction is comparatively rare in nature, it is probably only a partial substitute for sexual reproduction, so that purely asexual species are more prone to accumulating deleterious mutations (Geiser *et al.*, 1996).

8.4 | Conidia of ascomycetes

The asexual spores or conidia of ascomycetes are remarkably diverse in form, structure and modes of dispersal, but their development or **conidiogenesis** occurs in a limited number of ways (see below). The cell from which a conidium develops is the **conidiogenous cell** and usually one or more such cells are borne on a stalk, the **conidiophore**. Conidiophores which are narrow and not differentiated from the vegetative mycelium are said to be **micronematous** (Gr. *nema* = a thread) whilst those that are clearly differentiated are **macronematous**. Conidiophores frequently arise singly as in *Eurotium repens* (Fig. 11.16), *Emericella nidulans* (Fig. 11.17) and in many species of *Penicillium* (Fig. 11.18).

However, in certain fungi the conidiophores may aggregate to form a **conidioma**. Descriptive terms have been given to different types of conidioma. Conidiophores aggregated into parallel bundles (fascicles) are termed **coremia** (Gr. *korema* = a brush) or **synnemata** (Gr. prefix *syn* = together). Examples are *Penicillium claviforme* (Fig. 11.19) and *Cephalotrichum* (*Doratomyces*) *stemonitis* (Fig. 12.39). Seifert (1985) has distinguished several types of synnema, some of which are simple, some compound, some made up of parallel conidiophores, and others where the hyphae making up the synnema are intricately interwoven (see Kirk *et al.*, 2001). In many ascomycetes the conidiophores develop on or in a **stroma** (Gr. *stroma* = bed, cushion), an aggregation of pseudoparenchymatous cells. A good example of a conidial stroma is seen in the wood-rotting candle-snuff fungus, *Xylaria hypoxylon* (see Fig. 12.11a). Here, powdery white conidia develop at the tips of the branches of the conidial stroma and, later, asci develop in flask-shaped perithecia at the base of the old stroma. The term **sporodochium** (Gr. *spora* = a seed; *doche* = a receptacle) is used for the cushion-like conidiomata bearing a layer of short conidiophores. An example is the conidial (*Tubercularia*) state of *Nectria cinnabarina* (see Fig. 12.20c).

Another type of conidioma is the **acervulus** (Lat. *acervulus* = a little heap), a saucer-shaped fructification which may develop inside the tissues of a host plant or may be superficial. Subepidermal acervuli develop from a pseudo-parenchymatous stroma, and as the acervulus matures the overlying epidermis of the host becomes ruptured to expose conidia formed from conidiogenous cells lining the base of the saucer. The conidia are held together in slime and are chiefly dispersed by rain splash. A good example of an acervular fungus is *Colletotrichum* (see Fig. 12.51). The teleomorphs of *Colletotrichum*, where known, are species of *Glomerella*, many of which are serious plant pathogens. In many ascomycetes and their allies, the conidia are borne inside flask-shaped conidiomata termed **pycnidia** (Gr. diminutive of *pyknos* = dense, packed, concentrated). Traditionally, fungi with pycnidial and acervular states have been grouped together in the artificial taxon **coelomycetes** (Sutton, 1980), in contrast to **hyphomycetes** in which the conidiogenous cells are exposed on single conidiophores or in synnemata, coremia or sporodochia (see above). Pycnidia may be superficial or embedded in host tissue. The opening of the pycnidium is generally by means of a circular ostiole. Conidia formed from conidiogenous cells lining the inner wall of the pycnidium are held together in slimy masses which ooze out through the ostiole, sometimes as spore tendrils. They are generally dispersed by splash or in water films. In some cases the pycnidia, instead of producing conidia with an asexual function, produce **spermatia** which are involved in fertilization. Examples of fungi with pycnidial anamorphs are *Leptosphaeria acuta* with a *Phoma* anamorph (Fig. 17.3), and *Phaeosphaeria nodorum* (anamorph *Stagonospora nodorum*; see Fig. 17.4).

8.5 | Conidium production in ascomycetes

There are several steps in the production and release of conidia, namely (1) conidiogenesis, i.e. conidial initiation; (2) maturation; (3) delimitation; (4) secession, i.e. separation from the conidiogenous cell; (5) proliferation of the conidiogenous cell or conidiophore to form further conidia. Many of the current ideas on conidiogenesis stem from a seminal paper by Hughes (1953) based on light microscopy studies of conidial development in a range of hyphomycetes. Hughes classified the development of conidia in a limited number of ways. His ideas were extended by other workers, and advances were also made possible by the use of electron microscopy and time-lapse cinephotomicrography (Cole & Samson, 1979). An excellent review of these aspects of conidiogenesis has been written by Cole (1986). The descriptions which follow are based on the account by de Hoog *et al.* (2000a). Conidiogenesis occurs in two ways which appear to be distinct at first glance: **blastic** and **thallic** (see Fig. 8.5). In reality, when surveying conidium formation and release in a range of fungi, there is a continuum of development of which these two concepts represent extremes (Minter *et al.*, 1982, 1983a,b; Minter, 1984).

8.5.1 Blastic conidiogenesis
The conidium develops by the blowing-out of the wall of a cell, usually from the tip of a hypha, sometimes laterally as in *Aureobasidium* (conidial

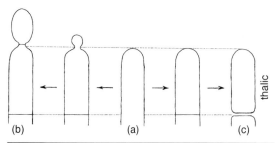

Fig 8.5 Two basic modes of development during conidiogenesis of a hyphal apex (a): blastic (b) and thallic (c). From de Hoog *et al.* (2000a), with kind permission of Centraalbureau voor Schimmelcultures.

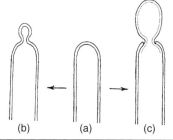

Fig 8.6 Two alternative types of conidiogenesis starting from an undifferentiated hyphal apex (a). In holoblastic conidiogenesis (b), the entire wall becomes inflated to form the conidium initial. In enteroblastic conidiogenesis (c), the conidium initial develops through a hole in the rigid outer wall. From de Hoog *et al.* (2000a), with kind permission of Centraalbureau voor Schimmelcultures.

state of *Discosphaerina*; see Fig. 17.25). In certain yeasts including *Saccharomyces*, the blastic development of new daughter cells is known as budding (see Fig. 10.3). Two kinds of blastic development have been distinguished:

1. **Holoblastic**. All the wall layers of the conidiogenous cell contribute to the wall of the newly formed conidium (see Fig. 8.6b). *Aureobasidium pullulans* (Fig. 17.25) and *Tricladium splendens* (conidial *Hymenoscyphus*; see Fig. 25.12) are examples. In some genera with dark (i.e. melanized), relatively thick-walled conidiophores such as *Stemphylium* and *Alternaria* (anamorphs of *Pleospora*), the conidia develop holoblastically, but a narrow channel persists in the wall of the conidiogenous cell through which cytoplasm had passed as the spore expanded. This type of development has been described as **porogenous** (Luttrell, 1963) or **tretic** (Ellis, 1971a) and the conidia are sometimes termed **porospores** or **poroconidia** (see Figs. 17.10–17.13; Carroll & Carroll, 1971; Ellis, 1971b).

2. **Enteroblastic**. The wall of the conidiogenous cell is rigid and breaks open. The initial of the conidium is pushed through the opening and is surrounded by a newly formed wall (Fig. 8.6c). Two types of enteroblastic development have been distinguished, **phialidic** and **annellidic** (see Fig. 8.7). In phialidic development a basipetal succession of conidia (phialospores, phialoconidia) develops from a specialized conidiogenous cell, the **phialide** (Gr. diminutive of *phialis* = flask), usually shaped like a bottle with a narrow neck. Phialides are formed singly

or in clusters at the tip of a conidiophore or, more rarely, laterally. There may be one or several nuclei in a phialide. As shown in Fig. 8.8 for *Thielaviopsis basicola*, the initial of the first-formed phialoconidium is surrounded by the apical wall of the phialide and is, in reality, holoblastic. The phialide wall breaks transversely near its tip and the first conidium, surrounded by a newly formed wall and capped by the wall from the broken tip of the phialide, is pushed out (Hawes & Beckett, 1977; Ingold, 1981). The new wall material which encases the phialoconidium is secreted in the form of a cylinder from the surface of the cytoplasm deep within the phialide, a process known as **ring wall building** (Minter *et al.*, 1983a). Before the conidium is extruded, a septum develops within the phialide below its neck, at the base of the conidium. The upper part of the wall of the now open-ended phialide persists as a small collar, the **collarette**. The nucleus or nuclei within the phialide continue to divide mitotically. A second conidium develops below the first, and is surrounded by newly secreted wall material. This conidium is also cut off by a septum and pushed out. Part of the newly secreted wall material may persist around the inside of the neck of the phialide as **periclinal thickening**. The process is repeated so that many phialoconidia may develop from a single phialide. In phialides which have developed several conidia, the periclinal

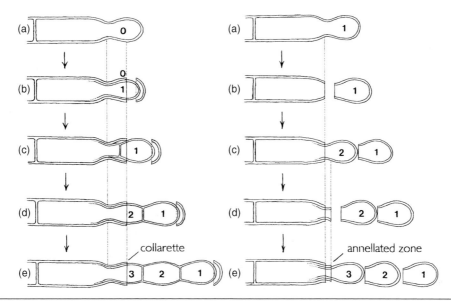

Fig 8.7 Enteroblastic phialidic (left) and enteroblastic (right) annellidic conidiogenesis. From de Hoog *et al.* (2000a), with kind permission of Centraalbureau voor Schimmelcultures.

Phialidic conidiogenesis	Annellidic conidiogenesis
a. Apex of conidiophore expands to form a phialide with a blown-out holoblastic conidium initial (o).	a. Apex of conidiophore differentiates to form a conidiogenous cell (annellide) and the initial of a holoblastic first conidium (1).
b. The first-formed conidium (1), surrounded by a new wall secreted inside the phialide, is pushed out and breaks the outer wall of the phialide whose tip persists as a cap.	b. The first conidium is cut off by a septum.
c. The first conidium is cut off by a septum.	c. A second conidium (2) develops beneath the first.
d. A second conidium develops below the first, also surrounded by new wall secreted inside the phialide.	d. The septum cutting off the second conidium is formed beyond the point at which the original annellide wall was ruptured and persists as an annellation.
e. A third conidium develops in basipetal succession adding to the length of the conidial chain. The lower part of the broken original wall of the phialide has persisted as a collarette, the extent of which is shown by vertical dashed lines. Successive layers of wall material may accrete in the neck of the phialide to form a periclinal thickening.	e. The development of further conidia results in the addition of more annellations so that an annellated zone, marked by vertical dashed lines, increases in length.

thickening may be seen even with the light microscope, but in others it is less obvious.

During maturation of the phialoconidium, the spore may increase in size, its wall may become thickened and ornamented by spines and may become pigmented by melanin and other materials. In some genera of ascomycetes and their conidial derivatives, the phialospores are dry and appear in chains. Dry-spored conidial chains are often persistent and are typical of *Aspergillus* and *Penicillium* (see Figs. 11.16–11.18).

The hydrophobic nature of the spore wall is due to incorporation of hydrophobin rodlets. The adherence of the conidia in chains depends on the strength of the septum between adjacent spores. Secession of the conidia into separate spores occurs by breakage of the septum. Where the spore wall is wet, the succession of conidia may briefly persist in the form of a short chain (false chain) or may collapse into slimy balls at the tips of the phialides (no-chain phialides). An example of the latter is *Trichoderma* (conidial

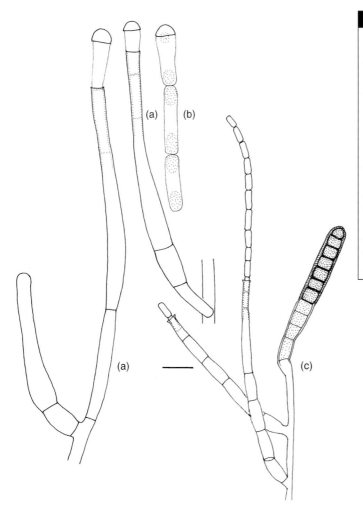

(a) (b)

(a)

(c)

Fig 8.8 Conidiogenesis in *Thielaviopsis basicola*. (a) Phialides and phialoconidia. To the left is a branched conidiophore with a short phialide which has not yet formed conidia and a longer phialide in which the tip has broken transversely and the first phialoconidium is being extruded. This spore is capped by the remnants of the phialide tip. Developing phialoconidia can be seen in the necks of the phialides. (b) Three end spores from a spore chain. The capped terminal spore is more bulbous than the cylindrical spores which succeed it. (c) A branched conidiophore bearing two phialides with chains of hyaline thin-walled phialoconidia and a dark, thick-walled transversely septate chlamydospore. These two distinct conidial states are synanamorphs. Scale bar: (a) = 10 μm, (b) = 20 μm.

Hypocrea; see Fig. 12.16). Phialoconidia are, in general, unicellular but multicellular conidia are found in certain genera such as in the transversely septate conidium of *Sporoschisma* (conidial *Melanochaeta*; for references see Sivichai *et al.*, 2000).

Annellidic conidiogenesis (Fig. 8.7) in many ways resembles phialidic, and indeed the term annellidic phialide is sometimes used for this type of conidiogenous cell. These are also termed **annellides** (Lat. *annulus* = little ring) or **annellophores**, and the spores which develop from them are **annelloconidia**. As in phialidic development, the first-formed annelloconidium is holoblastic. The difference between the two modes of development is that new wall material which is secreted within the annellide protrudes beyond

its neck and the septum which cuts off the newly formed conidium also forms beyond the neck. As each new conidium develops in basipetal fashion, a small ring of wall material (annellation) is left at the neck of the annellide, which thus grows in length as successive conidia develop. This accumulation of short collars of wall material is the **annellated zone** (see Fig. 8.7). With normal light microscopy annellation may be difficult to see, but detection is improved by interference contrast or phase contrast optics. Examples of fungi reproducing by annelloconidia are *Scopulariopsis brevicaulis* (see Fig. 12.38; Cole & Kendrick, 1969a) and *Cephalotrichum* (*Doratomyces*) *stemonitis* (Fig. 12.39). Both genera contain species which are conidial forms of *Microascus*.

Secession of conidia, irrespective of their mode of development, is in most cases by dissolution of the septum or septa which separate them from the conidiogenous cell or from adjacent spores. This process is termed **schizolytic secession** (Gr. *schizo* = to split, divide; *lyticos* = able to loosen). In some other cases secession is brought about by the collapse of a special **separating cell** beneath the terminal conidium. This is termed **rhexolytic secession** (Gr. *rhexis* = a rupture, breaking).

8.5.2 Thallic conidiogenesis

Thallic conidiogenesis (Gr. *thallos* = a branch) occurs by conversion of a pre-existing hyphal element in which terminal or intercalary cells of a hypha become cut off by septa (see Fig. 8.9). Two kinds of thallic development have been distinguished: **holothallic** (Gr. *holos* = whole, entire) and **thallic-arthric** (Gr. *arthron* = a joint). In holothallic development a hyphal element, e.g. a terminal segment of a hypha, is converted as a whole into a single conidium (see Fig. 8.9).

Secession of such conidia may be schizolytic or rhexolytic. *Microsporum* spp. (anamorphic *Arthroderma*), which are skin pathogens (dermatophytes) of mammals, provide examples of this holothallic development (see Fig. 11.6). During thallic-arthric conidiogenesis, septa develop in a hypha and divide it up into segments which separate into individual cells by dissolution of the septa (see Fig. 8.9). *Geotrichum candidum* (anamorphic *Galactomyces*), a common soil fungus and frequent contaminant of milk and milk products, develops conidia in this way (see Fig. 10.10; Cole, 1975).

The proliferation of the conidiogenous cell or the conidiophore may occur in various ways, for example by the formation of a new growing point in the region of the conidiophore beneath the point at which the first conidium was formed. The new apex extends beyond the point of origin of the first conidium and develops a new conidiogenous cell. These methods of conidiophore regeneration are discussed more fully in relation to some of the different genera.

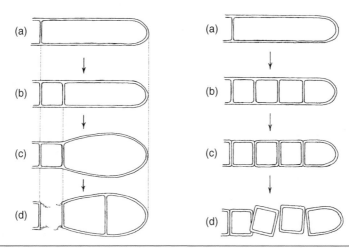

Fig 8.9 Holothallic and thallic—arthric conidiogenesis. From de Hoog *et al.* (2000a), with kind permission of Centraalbureau voor Schimmelcultures.

Holothallic conidiogenesis with rhexolytic secession	Thallic—arthric conidiogenesis with schizolytic secession
a. The terminal portion of a hypha is cut off by a septum.	a. A terminal segment of a hypha.
b. A second septum laid down near the first cuts off a subterminal segment, the separating cell.	b. Septa develop, dividing the segment into several cells.
c. The terminal cell enlarges to form the conidium.	c. The septa divide, each separating into two layers.
d. Collapse of the separating cell causes conidium secession.	d. The daughter cells separate.

8.6 | Development of asci

The morphogenesis of asci and ascospores has been reviewed by Read and Beckett (1996). In yeasts and related fungi, the ascus arises directly from a single cell, but in most other ascomycetes it develops from a specialized hypha, the **asco-genous hypha**, which in turn develops from an ascogonium (Fig. 8.10a). The ascogenous hypha of many ascomycetes is multinucleate, and its tip is recurved to form a **crozier** (shepherd's crook). Within the ascogenous hypha, nuclear division occurs simultaneously. Two septa at the tip of the crozier cut off a terminal uninucleate cell and a penultimate binucleate cell (Fig. 8.10c) destined to become an ascus. The ante-penulti-mate cell beneath the penultimate cell is termed

the stalk cell. The terminal cell of the crozier curves round and fuses with the stalk cell, and this region of the ascogenous hypha may grow on to form a new crozier in which the same sequence of events is repeated. Repeated prolif-eration of the tip of the crozier can result in a tight cluster of asci in many ascomycetes or a succession of well-separated asci as in *Daldinia concentrica* (see Fig. 12.10c). Specialized septal plugs, more elaborate than normal Woronin bodies, block the pores in the septa at the base of the ascus (Kimbrough, 1994). The septal pore plugs probably aid in retaining the high turgor pressure which develops in asci shortly before ascospore discharge.

In the ascus initial the two nuclei fuse and the diploid fusion nucleus undergoes meiosis to form four haploid daughter nuclei (Figs. 8.10d,e).

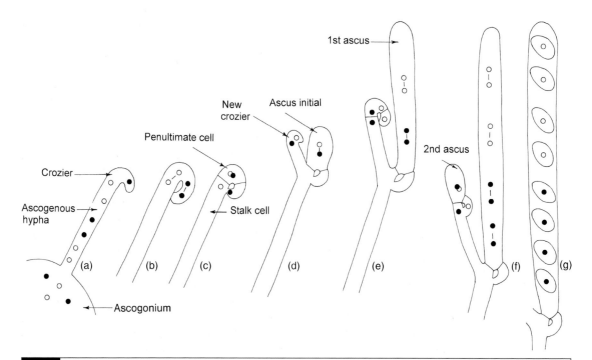

Fig 8.10 Diagrammatic representation of cytological features during ascus development. (a) Ascogenous hypha with a crozier at its tip developing from an ascogonium. (b) Conjugate nuclear division of the two nuclei in the crozier. (c) Two septa have cut off a binucleate penultimate cell. The two nuclei fuse to form a diploid nucleus. The uninucleate terminal segment of the ascogenous hypha has recurved and fused with the ascogenous hypha to form the stalk cell. (d) The penultimate cell enlarges to become an ascus initial within which the fusion nucleus begins to divide meiotically. A new crozier is developing from the stalk cell. (e) Second division of meiosis has occurred in the developing ascus. The behaviour of the new crozier repeats that of the first. (f) Mitotic division of the four haploid nuclei resulting from meiosis in the first ascus. (g) Ascospores formed.

These nuclei then undergo a mitotic division so that eight haploid nuclei result (Fig. 8.10f). The eight nuclei may divide further mitotically so that each ascospore is binucleate, or, if still more mitoses follow, the ascospore becomes multinucleate. For example, a single mitosis occurs in the immature ascospores of *Neurospora crassa*, and the spores remain binucleate for 2−3 days after they have been delimited. Later, a series of four or more synchronous mitoses occur after the spores have become pigmented so that they contain 32 or more nuclei when they are mature (Raju, 1992a). Where the ascospores are multicellular, there are repeated nuclear divisions accompanied by the formation of septa which divide up the spore. In some ascomycetes more than eight ascospores are formed, usually in numbers which are a multiple of eight, e.g. in the coprophilous genera *Podospora* and *Thelebolus*. In others the eight multicellular ascospores break up into part-spores, e.g. in *Hypocrea* (Fig. 12.15c) and *Cordyceps* spp. (Fig. 12.33b). In *Taphrina* ascospores may bud mitotically within the ascus so that the mature ascus contains numerous yeast cells (Fig. 9.2c). Asci with fewer than eight spores are also known, e.g. in *Neurospora tetrasperma* where the four ascospores are binucleate, in *Phyllactinia guttata* where there are two ascospores (Fig. 13.14e), or in *Monosporascus cannonballus* which has a single ascospore.

8.6.1 Cleavage of ascospores

In many ascomycetes, studies of the fine structure of asci during cleavage of the ascospores have shown that a system of double membranes continuous with the endoplasmic reticulum extends from the envelope of the diploid fusion nucleus (Fig. 8.11). The double membrane develops to form a cylindrical envelope lining the young ascus. This peripheral membrane cylinder or lining layer is termed the **ascus vesicle** or **ascospore-delimiting membrane**. The ascospores are cut out from the cytoplasm within the ascus by infolding and fusion of the inner edges of the double membrane around a portion of cytoplasm and a nucleus (Fig. 8.11d). In some ascomycetes, e.g. *Taphrina*, a peripheral

membrane cylinder has not been observed and the nuclei within the ascus become enveloped by ascospore-delimiting membranes formed by direct invagination of discrete parts of the ascus plasma membrane.

Between the two layers of the ascospore-delimiting membrane enclosing the ascospores, the primary spore wall is secreted. The inner membrane forms the plasma membrane of the ascospore and the outer membrane becomes the spore-investing membrane. Secondary wall material is secreted within the primary wall. There may be several such layers. In *Sordaria humana* a total of four spore wall layers have been distinguished, a primary wall layer and three secondary layers (Read & Beckett, 1996). The secondary wall layers are often quite thick, and in dark-walled ascospores the pigment is usually laid down within the secondary wall layers. The spore wall may be smooth or extended to form a variety of ornamentations such as spines, ridges or reticulations. The ascus epiplasm, i.e. the residual cytoplasm remaining outside the spores after these have become cleaved out, may continue to play a part in the formation of the ascospore wall. For example, in *Ascobolus immersus*, the outer leaf of the spore-delimiting membrane may extend irregularly outwards into the surrounding epiplasm to form a perisporic sac within which secondary wall material is deposited, derived from the epiplasm, and passing as globular bodies through the membrane of the perisporic sac. This secondary wall material ornaments the ascospore wall but is not involved in the formation of the purple pigment characteristic of *Ascobolus* ascospores (Wu & Kimbrough, 1992).

In many ascomycetes the outermost layer of the ascospore wall, the **perispore**, is mucilaginous, as seen for example in *Ascobolus immersus* (Fig. 14.5), *Sordaria fimicola* (Fig. 12.1c) and *Pleospora herbarum* (Fig. 17.10a). The properties of this outer wall layer may aid in the lubrication of the spore and also enable it to be compressed as it emerges from the ascus. Further, it may aid in the attachment of ascospores to substrata. It may also cause ascospores to stick together to form multisporous projectiles, an adaptation which results in an increased distance of

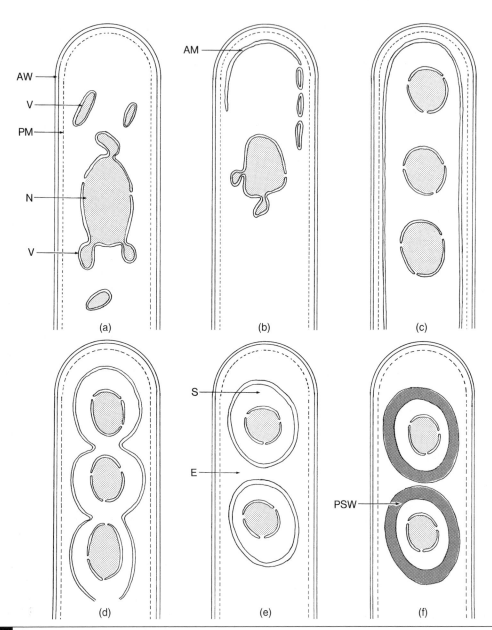

Fig 8.11 Ascus development in *Ascobolus* (after Oso, 1969). (a) Young ascus showing the formation of membrane-bounded vesicles (V) from the nucleus (N). The ascus wall (AW) is lined by the plasmalemma (PM). (b) Appearance of the ascospore membrane (AM) at the tip of the ascus and the arrangement of vesicles along the periphery of the ascus. (c) Ascospore membrane now in the form of a peripheral tube open at the lower end. The diploid nucleus has divided. (d) Invagination of the ascospore membrane between the haploid nuclei. (e) Young ascospores (S) delimited by the ascospore membrane from the epiplasm (E). (f) Separation of the two layers of the ascospore membrane due to the formation of the primary spore wall (PSW) between them.

ascospore discharge as compared with single-spored projectiles (Ingold & Hadland, 1959). This adaptation is especially common in coprophilous fungi, i.e. those which grow and fruit on herbivore dung, such as *Ascobolus* (Fig. 14.5) and *Sordaria*. A special adaptation occurs in another coprophilous fungus, *Podospora*, in which the basal part of the spore proper develops as a primary appendage, whilst other parts of the perispore extend as mucilaginous secondary appendages (see Figs. 12.3, 12.4; Beckett *et al.*, 1968). The appendages of adjacent spores intertwine so that the spores are discharged strung together in the manner of a slingshot (Ingold, 1971). In some aquatic ascomycetes the ascospores have extensions of the spore wall which aid in attachment. *Pleospora scirpicola*, which forms ascocarps on the submerged parts of culms of *Schoenoplectus lacustris*, an inhabitant of the shoreline of freshwater lakes, canals and slow-moving rivers, has long, mucilaginous, tapering extensions from each end of the ascospore (Fig. 17.1d). Appendaged ascospores are especially common in marine ascomycetes. The appendages develop in a variety of ways and unfurl in sea water, slowing down their rate of sedimentation and increasing the likelihood of their attachment to underwater substrata such as wood (Hyde & Jones, 1989; Hyde *et al.*, 1989; Jones, 1994).

Germ pores or germ slits, through which germ tubes emerge on spore germination, are found in many ascomycetes, especially those with thick dark-pigmented walls. Germ pores, representing thin areas in the spore wall, occur at each end of the spore in *Neurospora* and germination may occur at either or at both ends. In *Sordaria humana* there is a single germ pore at the lower end of the ascospore plugged by a pore plug (Read & Beckett, 1996). The ascospores of Xylariaceae, e.g. *Xylaria*, *Hypoxylon* and *Daldinia*, have black walls with a hyaline germ slit running along the length of the spore (Figs. 12.10 and 12.14).

Because the division which follows the four-nucleate stage is mitotic and because the division plane is usually parallel to the length of the ascus, adjacent pairs of spores starting from the tip of an ascus are normally sister spores and are thus genetically identical. Rare exceptions to this situation are occasionally found where the division planes are oblique, or for other reasons (see Raju, 1992a).

8.6.2 The ascus wall

The wall of the ascus consists of several distinguishable layers. The outer layer is laid down first and inside it is a succession of later-formed layers so that the mature wall may consist of four or more layers (Bellemère, 1994; Read & Beckett, 1996). The wall material includes chitin, polysaccharides and proteins, but there is no evidence of lipid. The ascus wall is elastic. All or parts of it may stretch considerably during ascospore liberation, and contraction of the elastic wall provides the force for ascospore discharge. During discharge all the layers of the ascus wall may remain attached to each other, thus appearing as a single layer. Such asci are termed **unitunicate** (Lat. *tunica* = a garment). Despite the term unitunicate which refers to the behaviour (i.e. function) of the ascus wall during ascus dehiscence, the wall of unitunicate asci is often composed of two superposed tunicae, a thin, single-layered or double-layered exoascus and a thicker endoascus. The endoascus may be fibrillar, or at first granular and then with parallel or reticulate fibrils (Parguey-Leduc & Janex-Favre, 1984). During ascospore discharge the two layers of the ascus wall remain attached, i.e. they do not glide over each other.

A variant of the unitunicate type of ascus dehiscence is found in the lichenized ascomycetes *Lecanora* and *Physcia* (the *Lecanora* or **rostrate** type of dehiscence). In *Physcia stellaris* the ascus has a prominent amyloid dome. The ripening ascospores push against this dome and on ascospore discharge it is extruded to form a rostrum (Lat. *rostrum* = beak) which extends upwards to the surface, whilst its base remains attached to the upper part of the wall of the ascus (see Figs. 8.12e,f; Honegger, 1978).

In other ascomycetes, the ascus wall appears distinctly two-layered (**bitunicate**) when viewed with the light microscope (Luttrell, 1951; Reynolds, 1971, 1989). The layers of many (but not all) bitunicate asci separate at ascospore discharge into two functionally distinct layers (see Fig. 8.12), and such asci are termed **fissitunicate** (Dughi, 1956). Fissitunicate asci are particularly common in the Loculoascomycetes. Development of the wall of a bitunicate ascus takes place in two stages prior to ascospore formation. The first stage involves the growth of the ascus initial and the expansion of the

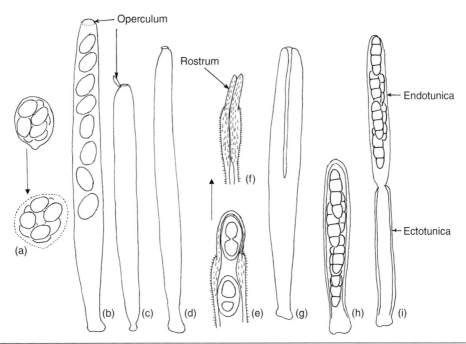

Fig 8.12 Types of ascus dehiscence. (a) Prototunicate ascus; the wall dissolves to release the ascospores passively. (b,c) Operculate asci before and after discharge; the ascus opens by means of a lid or operculum. (d) Discharged inoperculate ascus which has opened through a pore. (e,f) Rostrate ascus as seen in *Physcia*. In (e) a thickened part of the upper wall of the ascus is being extruded and is visible in a discharged ascus (f) as an extension of the inner part of the ascus wall, the rostrum. (g) Discharged bilabiate ascus showing the longitudinal slit by which the ascus opens. (h,i) Bitunicate ascus before and after the first stage of spore release. Rupture of the ectotunica has allowed the endotunica to expand. (e,f) after Honegger (1978).

ascus mother cell. During this stage the outer layers of the wall making up the **ectotunica** (= ectoascus) are deposited. In the second stage, secondary wall layers making up the **endotunica** (= endoascus) are laid down within the primary wall. The development of bitunicate asci has been studied by Reynolds (1971) and by Parguey-Leduc and Janex-Favre (1982). At the beginning of development, asci are surrounded by a single homogeneous layer which is sometimes granular, bearing externally a loose network (a fuzzy coat) of interascal material. The ascus wall becomes divided into a densely granular external layer (the ectoascus) and a clearer, but equally granular, inner layer (the endoascus). A clear space then separates these two layers. The granular material of the endoascus rearranges itself into lines of fibrils at first following a wavy pattern as seen in transverse sections of developing asci. Later the fibrils become strongly

folded into pointed zigzag shapes, a development which progresses from the inside towards the outside of the endoascus. The folds of the zigzags are closely pressed against the pointed teeth which mark out the plasmalemma. The density of the fibrils increases considerably throughout the thickness of the endoascus. Finally the two layers of the ascal wall, separated from each other by a clear space, appear as a double-layered ectoascus and a single-layered endoascus within which the fibrils are strongly pleated into accordion-like folds. The pointed crests of the pleats lie parallel to each other and perpendicular to the plasmalemma of the ascus. The folding of the layers of the endoascus and plasmalemma permit the rapid expansion of the ascus prior to spore discharge, i.e. by providing material which can unfold rapidly. Towards the tip of the non-discharged ascus the crests of the pleated folds of the endoascus may converge and appear as a kind of apical basket which has been termed by Chadefaud (1942) the **nasse apicale** (Fr. *nasse* = keep net, eel trap).

In some asci the ascus wall does not function in ascospore discharge, but dissolves or disintegrates at maturity, the spores being released passively. Such asci are termed **prototunicate** (see Fig. 8.12a). This type of ascus is characteristic of certain groups of ascomycetes such as the Eurotiales and Onygenales but they are also found in unrelated groups (Currah, 1994). Examples are *Eurotium* (Fig. 11.16) and *Gymnoascus* (Fig. 11.9).

8.6.3 The apical apparatus of asci

The apical dome of the ascus may be modified in various ways. In certain types of discomycete with an open saucer-like fruit body or apothecium, the ascus is capped by a wall which has an annulus of thinner wall material forming a lid or **operculum** (Lat. *operculum* = a cover, lid) (van Brummelen, 1981). When the ascus explodes to discharge its ascospores, the operculum may be lifted off completely or may hinge to one side (see Figs. 14.5, 14.6). Such asci are **operculate** (Figs. 8.12b,c). However, the majority of ascomycetes have no ascus lid; they are **inoperculate** and when ascospore discharge occurs, the tip of the ascus opens by a pore (Fig. 8.12d). The presence or absence of an operculum is a character used in the classification of discomycetes. Operculate asci are characteristic of Pezizales including genera such as *Aleuria*, *Ascobolus* and *Pyronema* (see Chapter 14). Inoperculate discomycetes include Helotiales such as *Sclerotinia* (Figs. 15.1, 15.2) and many other orders. In a few cases, e.g the lichenized ascomycete *Pertusaria* and the coprophilous fungus *Ascozonus*, the ascus may burst by one or two longitudinal slits at the apex (see Fig. 8.12g). Such asci are described as **bilabiate** (i.e. two-lipped).

Other kinds of specialized structures found in ascus tips are generally referred to as the **apical apparatus**. Their functions relate to the mechanism of discharge (see below). In many perithecial fungi the tip of the ascus contains an **apical ring** or **annulus**. This is a specially thickened inward extension of the apical wall of the ascus, arranged in the form of a cylindrical flange (Fig. 8.13). In some fungi (e.g. *Xylaria*) the annulus is amyloid, i.e. it stains blue with Melzer's iodine, an aqueous solution of I_2 in KI (Beckett & Crawford, 1973). In other ascomycetes, reddish-brown (dextrinoid) staining may be observed, whereas in yet others (e.g. *Sordaria*) the annulus does not stain with iodine (Read & Beckett, 1996). When ascospores are discharged the annulus is everted, i.e. turned inside out like a sleeve (see Fig. 8.13b). The cylindrical opening of the annulus is considerably less than the diameter of the ascospores which pass through it as shown in Fig. 8.13a for *Xylaria longipes* and Fig. 12.1 for *Sordaria fimicola*, so that the annulus must be sufficiently elastic to expand and contract as an ascospore passes through it. The function of the annulus is to act as a sphincter, minimizing the decrease in hydrostatic pressure inside the ascus as spore discharge proceeds. It may also separate the spores from each other as they pass through, and by gripping the tapering rear portion of an ascospore impart some force which helps to expel it (Ingold, 1954a). Filiform (i.e. needle-shaped) ascospores are discharged singly and not in groups. This is well shown in the ergot fungus *Claviceps purpurea* and its allies such as *Cordyceps* (Figs. 12.27, 12.33). Their ascus apices are capped by a swollen plug of wall material pierced by a narrow pore. Ascospores are squeezed out through the pore, sometimes with an interval of several seconds between successive discharge events (Ingold, 1971). Other elaborations of the upper part of the ascus have been reported (Beckett, 1981; Bellemère, 1994).

The actual form of the mature ascus is very variable. In prototunicate forms, i.e. with non-explosive ascospore release, the ascus is often a globose sac, but in the majority of ascomycetes the ascus is cylindrical, and the spores are expelled from the ascus explosively.

8.6.4 Hamathecium

In many cases the asci are surrounded by packing tissue in the form of **paraphyses**

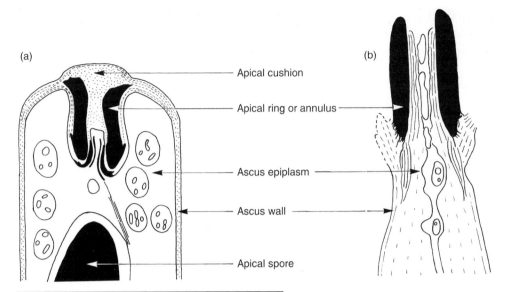

(a)

Apical cushion

Apical ring or annulus

Ascus epiplasm

Ascus wall

Apical spore

(b)

Fig 8.13 *Xylaria longipes*. Fine structure of the ascus apex (after Beckett & Crawford, 1973). (a) L.S. undischarged ascus showing the apical ring. (b) L.S. discharged ascus showing the eversion of the apical ring.

(Gr. *para-* = near, beside, parallel; *physis* = growth), or **pseudoparaphyses**. The general term for such sterile inter-ascal tissue is the **hamathecium** (Gr. *hama* = all together, at the same time) (Eriksson, 1981). Paraphyses are filaments which are attached to the ascocarp near the bases of the asci and are free at their upper ends as in *Pyronema* (Figs. 14.2a,c) and *Ascobolus* (Fig. 14.6). Pseudoparaphyses are hyphae which usually arise above the level of the asci and grow downwards between them. They may become attached at their lower ends as in *Pleospora* (Fig. 17.9). Because the paraphyses and pseudoparaphyses pack tightly around the asci, the latter cannot expand laterally but are forced to elongate. A hamathecium is lacking in certain groups of ascomycetes, e.g. the Eurotiales and Clavicipitales, and also in *Mycosphaerella* (Fig. 17.19). The sum of all contents of the ascoma (i.e. the hamathecium plus asci) but excluding the ascoma wall is called the **centrum**.

8.6.5 The mechanism of ascospore discharge

Explosive release of ascospores follows increased turgor pressure, caused by water uptake by the ascus. In the young ascus, after the spores have been cut out, the epiplasm remains lining the ascus wall, and this surrounds a large central vacuole containing ascus sap, within which the ascospores are suspended. The epiplasm is rich in the polysaccharide glycogen which can be visualized cytochemically by its reddish-brown staining with the I_2/KI stain. As the ascus matures, the red stain diminishes in intensity due to the conversion of the polysaccharide to osmolytes of lower molecular weight. This brings about an increased osmotic concentration of the ascus sap, followed by increased water uptake. The resulting increase in turgor pressure causes the ascus to stretch and, eventually, to burst open, squirting out the ascospores. The osmotic pressure of the sap in mature asci extending from apothecia of *Ascobolus immersus* has been determined to be up to 3 bar (0.3 MPa), with glycerol being the main organic osmolyte (Fischer *et al.*, 2004). In *Gibberella zeae*, the turgor pressure required for ascus discharge (1.54 MPa) seems to be caused mainly by a K^+ and Cl^- influx across the plasma membrane, with the most abundant organic osmolyte (mannitol) making only a small contribution (Trail *et al.*, 2005). Higher turgor pressures have been recorded when asci are mounted in water (Ingold, 1939, 1966).

In cup fungi (discomycetes), as the asci mature they elongate and project above the

Fig 8.14 Ascospore puffing in *Aleuria aurantia*. A thick white cloud of ascospores has been released by a cluster of apothecia. Reprinted from Fuhrer (2005), with permission by Bloomings Books Pty Ltd. Original image kindly provided by B. Fuhrer.

general surface of the hymenium. Their tips may be phototropic, as in the coprophilous fungus *Ascobolus*, and this ensures that the ascospores are directed upwards, towards the light. In some discomycetes, especially those with operculate asci (e.g. *Ascobolus*, *Peziza*), large numbers of ripe asci may discharge their spores simultaneously, a phenomenon known as **puffing** (Fig. 8.14). This may also occur, but less obviously, in forms with inoperculate asci, e.g. *Sclerotinia* and *Rhytisma*. Puffing results in a cloud of ascospores being discharged for greater distances than with spores discharged from a single ascus (Buller, 1934; Ingold, 1971).

In the flask fungi (pyrenomycetes), such as *Sordaria* or *Podospora*, as an ascus ripens it elongates and takes up a position inside the ostiole, often gripped in position by a lining layer of hairs, **periphyses** (Gr. prefix *peri* = near, around, roundabout). In this case the asci discharge their spores in turn. The necks of the perithecia in *Sordaria* and *Podospora* are phototropic so that the ascospores are shot towards the light. A variant of this method of discharge is found in fungi whose perithecia have necks very much longer than the length of the asci, such as *Ceratostomella* and *Gnomonia*. Here, before discharge, the asci break at their bases and detached asci move up a canal inside the perithecial neck and are held in place within the ostiole by periphyses as they discharge their spores (Ingold, 1971).

The behaviour of the bitunicate type of ascus during discharge has been described as the Jack-in-the-box mechanism (Ingold, 1971). The outer wall is relatively rigid and inextensible. As the ascus expands, the outer wall ruptures laterally or apically (see Figs. 8.12h,i) and the inner wall then stretches before the ascus explodes. Ascus discharge is thus a two-stage process. This type of mechanism is found in Loculascomycetes such as *Sporormiella* (Fig. 17.18), *Leptosphaeria* (Fig. 17.3) and *Pleospora* (Figs. 17.1, 17.9).

In *Cochliobolus* the ascus is bitunicate but the endotunica is incomplete at its base, i.e. vestigially or partially bitunicate. In *C. cymbopogonis* the eight ascospores are spirally coiled around each other in the ascus and each has a recurved tip (Figs. 8.15a,b). The ectoascus bursts open near its tip and the sheaf of ascospores is expelled, the incomplete endoascus forming a thimble-like cap over the tips of the ascospores as they pass through the long pseudothecial neck. The spores are not explosively discharged but are extruded, *en masse*, from the neck of the pseudothecium in a long tendril from which they are dispersed by rain splash. In water the spores separate from each other and push away the endotunica which earlier capped their tips together (see Fig. 8.15d; El-Shafie & Webster, 1980; Alcorn, 1981).

It is likely that in most cases the spores are spatially separated from each other as they are

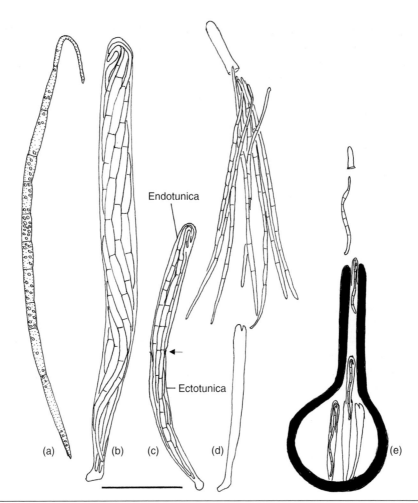

Fig 8.15 Ascospore liberation in *Cochliobolus cymbopogonis*. (a) Ascospore with a recurved tip. (b) Ascus containing a sheaf of eight spirally coiled ascospores. The ascus is bitunicate but the endoascus is not shown. (c) Ascus during the first stage of discharge. The ectoascus has broken (arrow), the endoascus has broken at its base and has extended, remaining as a thimble-like cap over the sheaf of ascospores. (d) An ascus after the release of the ascospores which have straightened out and pushed the broken endoascus aside. (e) Diagrammatic representation of a section through a pseudothecium showing stages in ascospore release. Scale bar, (a,b) = 50 µm; (c,d) = 200 µm. After El-Shafie and Webster (1980).

constricted on passing through the ascus pore. This has been neatly demonstrated by spinning a transparent disc over the surface of a culture of *Sordaria* discharging spores (Ingold & Hadland, 1959). The ascus contents are laid out on the disc in the order in which they are released. Various patterns of spore clumping and separation are visible, and although in many asci the eight spores are well separated from each other, in others there is a tendency for spores to stick together. Calculations made from measurements of the length of the ascospore deposit and the speed of rotation of the disc, as well as by other methods, have revealed ascospores to be the fastest-accelerating biological objects (Trail *et al.*, 2005; Vogel, 2005). The actual time taken for ascus discharge was estimated by the rotating disc method to be 0.000024 s (Ingold & Hadland, 1959). When ascospores stick together, they are discharged further than single-spored projectiles. In many coprophilous ascomycetes (e.g. *Ascobolus*, *Saccobolus*, *Podospora*) the spores may be attached together by mucilaginous secretions and may be projected for distances of 30 cm in *Ascobolus immersus* and 50 cm in *Podospora fimicola*. The distances to which individual ascospores are discharged vary, but are often in the range of 1–2 cm.

In some ascomycetes the ascospores are not discharged violently, and in such cases the asci are often globose instead of cylindrical. The Hemiascomycetes (Chapter 10), Plectomycetes (Chapter 11) and several other groups have asci of this type. In *Ophiostoma* (Fig. 12.36) and *Sphaeronaemella fimicola* (Fig. 12.42) the ascus walls dissolve to release a mass of sticky spores which ooze out as a drop held in place by a ring of hairs surrounding the ostiole at the tip of a cylindrical neck which surmounts the perithecium. They are dispersed by insects. Breakdown of asci within the fruit body is also found in *Chaetomium* (Fig. 12.9). Ripe ascospores are extruded from the neck of the perithecium in a tendril. Possibly they are dispersed by jerking movements generated as the rough-walled perithecial hairs twist around each other. Tendrils of ascospores are sometimes found in ascomycetes which normally discharge their spores violently, e.g. *Daldinia concentrica* (Plate 5a), *Hypocrea pulvinata* (Plate 5c) and *Nectria*. In many marine ascomycetes the ascus walls are evanescent and dissolve to release the ascospores passively. In ascomycetes with subterranean fruit bodies, e.g. in the truffle *Tuber* and its relatives (Figs. 14.7, 14.8), the ascospores are not discharged violently, but are dispersed when the fruit bodies are eaten by rodents and other animals attracted by their characteristic odour.

8.7 | Types of fruit body

The main types of ascomycete fruit body have been listed earlier (p. 21) and are drawn in Fig. 8.16. In yeasts and related fungi the asci are not enclosed by hyphae, but in most ascomycetes they are surrounded by hyphae to form an **ascocarp** (i.e. an ascus fruit body) or **ascoma**. An old term for ascus is theca (Gr. *theca* = a case), and although this word is not now in general use, it is still found as a suffix in terms for different types of ascocarp. *Byssochlamys* forms clusters of naked asci (Fig. 11.15). In *Gymnoascus* there is a loose open network of peridial hyphae

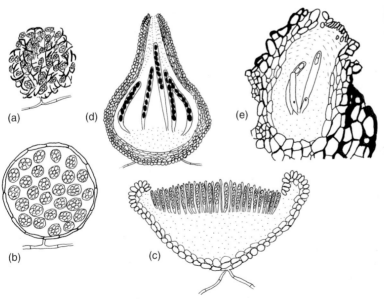

Fig 8.16 Different types of ascocarp, diagrammatic and not to scale. (a) Gymnothecium made up of branched hyphae which do not completely enclose the asci. (b) Cleistothecium completely enclosing the asci which are formed throughout the ascocarp. There is no opening. (c) Apothecium, an open cup lined by a layer of asci and associated structures forming the hymenium. (d) Perithecium with a layer of asci at the base. It opens by a pore or ostiole. Its wall or peridium is made up of flattened cells. (e) Pseudothecium. The asci are formed within locules in a pseudoparenchymatous ascostroma. There is no peridium. (d) after Ingold (1971), (e) after Luttrell (1981).

forming a **gymnothecium** (Gr. *gymnos* = naked) and the asci can be seen through the network (Fig. 11.8). Gymnothecia are also seen in *Myxotrichum* (Fig. 11.10) and *Ctenomyces* (Fig. 11.5) where certain peridial hyphae extend as hooked hairs. In most species of *Aspergillus* and *Penicillium* which possess ascocarps, the asci are enclosed in a globose fructification with no special opening to the outside. Such ascocarps are termed **cleistocarps** or **cleistothecia** (Gr. *kleistos* = enclosed). A modified cleistothecium capable of cracking open along a line of weakness is found in the Erysiphales (powdery mildews), and this is called a **chasmothecium** (Gr. *chasma* = an open mouth). In the cup fungi (Pezizales and Helotiales) as well as in many lichenized ascomycetes the asci are borne in open saucer-shaped ascocarps, and at maturity the tips of the asci are freely exposed (see Plates 6 and 7). Such fruit bodies are termed **apothecia** (Gr. *apo-* = away from, separate). The Pyrenomycetes (e.g. Sphaeriales and Hypocreales) have **perithecia** (Gr. *peri-* = around) which are flask-shaped fruit bodies opening by a pore or ostiole (see Fig. 12.1, *Sordaria fimicola*). The perithecial wall is formed from sterile cells derived from hyphae which surrounded the ascogonium during development. Perithecia are often single, as in *Sordaria* and *Neurospora*, but in some genera they are embedded in or seated on a mass of tissue forming a **perithecial stroma** (for examples, see Plate 5). The development of **pseudothecia** differs from that of perithecia in that the asci are contained in one or several cavities (locules) formed within a pre-existing **ascostroma** (Gr. *stroma* = mattress, bed) (Luttrell, 1981). Examples are *Leptosphaeria* (Fig. 17.3) and *Sporormiella* (Fig. 17.18). Although the structure and development of perithecia and pseudothecia are essentially different, the term perithecium is often loosely applied to both.

8.8 | Fossil ascomycetes

Ascomycetes are an ancient group of fungi, and fossilized structures possibly representing ascocarps made up of septate, anastomosing hyphae have been described from the Proterozoic period about 1 billion years ago (Butterfield, 2005). Lichen-like associations between fungi and cyanobacteria or algae may have existed some 600 million years ago (Yuan *et al.*, 2005). What are believed to be the remains of perithecia have been reported from beneath the epidermis of stems and rhizomes of one of the earliest known land plants, *Asteroxylon*, in the Rhynie chert of the Devonian period about 400 million years ago (Taylor *et al.*, 1999, 2005). Fossil cleistothecia containing asci and ascospores resembling those of present-day Trichocomaceae have been found in coal balls of the Carboniferous age (Stubblefield & Taylor, 1983; Stubblefield *et al.*, 1983). Stalked ascocarps with well-preserved ascospores have been found in amber, the fossilized resin of a conifer. They have been assigned to an extant genus *Chaenothecopsis* (Mycocaliciaceae). Their close resemblance of present-day species which are also associated with resin indicate little evolutionary change during the past 20 million years (Rikkinen & Poinar, 2000).

On morphological grounds, Barr (1983) suggested that the ancestors of Ascomycota should be sought among the Chytridiomycota. Confirmation of this view has since been obtained by comparison of DNA sequence data. Ascomycota are also closely related to Basidiomycota, each being a derived monophyletic group (Bruns *et al.*, 1992; Berbee & Taylor, 2001). Berbee and Taylor (2001) have estimated that these two groups evolved from a common ancestor about 600 million years ago, well before the development of vascular terrestrial plants.

8.9 | Scientific and economic significance of ascomycetes

The study of ascomycetes is of considerable scientific importance. *Neurospora crassa* has been the subject of intensive genetical research related to its relatively simple nutrient requirements, rapid growth, its capacity to produce mutants and the ease with which it can be grown and cross-mated in culture. The dissection of ascospores from its asci by micromanipulation has enabled tetrad analysis to be performed. Research on this fungus led to the important one-gene–one-enzyme concept. Budding yeast (*Saccharomyces cerevisiae*) and a fission yeast (*Schizosaccharomyces pombe*) were amongst the first eukaryotes for which the entire genome was sequenced. Studies on *S. cerevisiae* were basic to the understanding of the biochemistry of anaerobic respiration whilst studies of *S. pombe* have provided key facts by which to interpret the fundamental process of cell division, which in turn has a bearing on the understanding of the apparently uncontrolled growth of cancerous cells. The economic significance of fermentation processes involving ascomycetes and their conidial relatives is immense. Examples include alcoholic fermentations by yeasts as the basis of the wine and brewing industries, antibacterial antibiotics such as penicillin from *Penicillium chrysogenum* and cephalosporin from *Acremonium* spp., and organic acids such as citric acid from *Aspergillus niger*. The immunosuppressant drug cyclosporin, which reduces the tissue rejection response and thus facilitates organ transplants, is a metabolite of *Tolypocladium inflatum*. Some ascomycetes are important in food production as in bread-making by yeast, cheese ripening by *Penicillium roqueforti* and *P. camemberti* and the fermentation of soybeans and wheat by *Aspergillus*, yeasts and bacteria to produce soy sauce. The mycoprotein Quorn is produced from mycelial biomass of *Fusarium venenatum*. Examples of the direct use of ascocarps as food or food flavourings are morels (*Morchella* spp.) and truffles (*Tuber* spp.).

However, food spoilage may result from ascomycete contamination. A well-known example is contamination of cereal grains and grass by sclerotia of the ergot fungus *Claviceps purpurea*, which can cause severe, sometimes fatal, neurological, muscular and circulatory diseases such as gangrene or abortion in cattle and man. Studies on the alkaloid toxins contained in ergot sclerotia led to the discovery of drugs useful in obstetrics and the treatment of migraine, and in the identification of the

hallucinogen lysergic acid. Another potentially serious mycotoxin is aflatoxin produced in groundnuts, cereals and other foodstuffs infected by *Aspergillus flavus*. Aflatoxins are highly carcinogenic in poultry and mammals, including man. Other mycotoxins include zearalenone from *Gibberella zeae*, which causes infertility in cattle and pigs, and trichothecenes from *Trichothecium roseum* and *Fusarium* spp., which cause aleukia in farm animals and man. A family of plant growth hormones, the gibberellins, now produced commercially, were discovered in an investigation of Bakanae (foolish seedling) disease of rice.

It is not surprising that such a large group as the Ascomycota should contain numerous pathogens of plants and animals. Lifestyles are similarly varied, including biotrophic, hemibiotrophic and necrotrophic associations. Many ascomycote pathogens are of considerable economic importance.

8.10 | Classification

It is impractical to attempt a detailed classification of ascomycetes which could include around 55 orders and 291 families (Kirk *et al.*, 2001). We shall adopt the simplified classification outlined by M. E. Barr (2001) and Kurtzman and Sugiyama (2001). Based on a wealth of microscopic data, and especially the results of several phylogenetic analyses, five major groups (classes) of Ascomycota have been proposed, namely Archiascomycetes (Chapter 9), Hemiascomycetes (Chapter 10), Plectomycetes (Chapter 11), Hymenoascomycetes (Chapters 12–16) and Loculoascomycetes (Chapter 17). The latter three are sometimes called 'higher ascomycetes' or Euascomycetes. The class Hymenoascomycetes contains ascomycetes producing asci in a hymenium, i.e. in a fertile layer around which the ascocarp develops. This is in contrast to the Loculoascomycetes, where the asci develop in a pre-formed stroma. Since the Hymenoascomycetes are a very large and diverse group, we have subdivided them in this book.

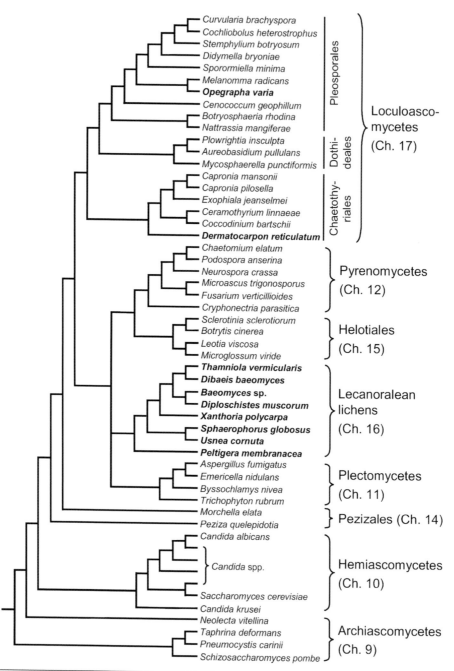

Fig 8.17 Outline of a possible ascomycete phylogeny, presented as a consensus tree based on sequences of the RNA polymerase II gene. Lichenized fungi are printed in bold. Redrawn from Liu and Hall (2004), with permission. © 2004 National Academy of Sciences, U.S.A.

An outline of current phylogenetic relationships among the Ascomycota has been given by Liu and Hall (2004) and is shown in Fig. 8.17. More detailed delimitations of the individual groups will be discussed in subsequent chapters. In our treatment of the ascomycetes, we have attempted to consider purely asexual forms in the taxonomic context of their ascomycete state wherever practical and where known with certainty.

Archiascomycetes

9.1 | Introduction

Several independent phylogenetic analyses of DNA sequence data (e.g. Berbee & Taylor, 1993; Sjamsuridzal et al., 1997; Liu & Hall, 2004) have grouped together a range of seemingly very diverse genera of ascomycetes. This group is considered to be the oldest of three broad evolutionary lineages of Ascomycota and has thus been named Archiascomycetes (Nishida & Sugiyama, 1994). The core of the Archiascomycetes consists of the genera Taphrina and Protomyces, which are facultative biotrophic plant pathogens, and the saprotrophic fission yeast Schizosaccharomyces. Also now included are the yeast-like Pneumocystis, which causes pneumonia in immunocompromised patients (see p. 259); the filamentous fungus Neolecta, which parasitizes the roots of higher plants (Redhead, 1977; Landvik et al., 2003); and the anamorphic yeast Saitoella. Yet other genera are included as possible members because even though their appropriate DNA sequences have not yet been obtained, they are known to be related to confirmed members. In total, the class Archiascomycetes currently contains some 150 species in 10 genera.

Because of their diverse morphological appearances and modes of life, it is difficult to describe common characters typical of the Archiascomycetes. With the exception of Neolecta, which produces apothecia, ascocarps are lacking and asci are produced individually by yeast cells or by conversion of hyphal tips. There are no differentiated ascogenous hyphae. Asexual reproduction is usually by simple division of vegetative yeast cells by budding or fission. Even in Neolecta, the apothecia are highly unusual in that they lack ascogenous hyphae and paraphyses, and in that the ascospores are capable of producing yeast-like conidia by budding while still within the ascus or after discharge (Redhead, 1977). This phenomenon is also found in Taphrina (see Fig. 9.2c).

The presence of Neolecta in the most basal group of ascomycetes indicates that the capacity to produce fruit bodies is probably an ancient trait. The inclusion of both yeasts and mycelial forms among the Archiascomycetes makes it impossible to decide the chicken-and-egg question as to which of these states is ancestral and which is derived. It is significant that all recent molecular studies have placed Taphrina within the Archiascomycetes because this genus has long been suspected to be close to the origin of both the higher ascomycetes and the basidiomycetes (Savile, 1968; Alexopoulos et al., 1996). Further, the position of Pneumocystis has been unclear until recently, oscillating between Basidiomycota (Wakefield et al., 1993) and Archiascomycetes (Sjamsuridzal et al., 1997; Kurtzman & Sugiyama, 2001), and thereby further supporting the suspected ancestral status of the organisms included among the Archiascomycetes.

Two genera — Taphrina and Schizosaccharomyces — are of special significance to mycology and will be discussed more fully below, together with a brief account of Pneumocystis.

9.2 | Taphrinales

The Taphrinales are ecologically biotrophic parasites mainly of flowering plants, causing a wide variety of disorders which often lead to strikingly abnormal development of the infected host tissue to form witches' brooms, galls or leaf curls. About six genera are known of which *Protomyces* (10 species) and *Taphrina* (95 species) are the most important. Both *Protomyces* and *Taphrina* can be isolated from their hosts as ascospores, and these germinate in pure culture by budding to form saprotrophic haploid yeast cells. In the host plant, however, a mycelium of intercellular septate hyphae is produced. In *Protomyces*, hyphae are diploid, whereas they are dikaryotic in *Taphrina*. Dikaryotic hyphae are most unusual among ascomycetes but are typical of basidiomycetes. Biotrophic infection of the host plant culminates in individual hyphal tips undergoing meiosis (preceded by karyogamy in *Taphrina*), producing usually eight haploid ascospores which are discharged violently. We will discuss only *Taphrina* here; for *Protomyces* and related genera, species descriptions are given by Reddy and Kramer (1975).

9.2.1 *Taphrina*

Species of *Taphrina* are mostly parasitic on Fagaceae and Rosaceae (Mix, 1949), causing diseases of three main kinds. (1) Leaf curl or blister diseases, e.g. *Taphrina deformans*, the cause of peach leaf curl (Fig. 9.1; Plate 4a); *T. tosquinetii*, the cause of leaf blister of alder; and *T. populina*, the cause of yellow leaf blister of poplar. (2) Diseases of above-ground plant organs in which the infected twig undergoes repeated branching to form dense tufts of twigs called witches' brooms. Examples are *T. betulina*, causing witches' brooms of birch (Plate 4b), *T. insititiae* causes witches' brooms of plum and damson, and *T. wiesneri* causes witches' brooms and leaf curl of cherry. However, not all witches' brooms are caused by *Taphrina*, and similar twig proliferation is also associated with infection by mites. (3) Diseases of fruits, e.g. *T. pruni*, which causes the condition known as pocket plums in which the fruit is wrinkled and shrivelled and has a cavity in the centre in place of the stone. *Taphrina amentorum* causes conspicuous tongue-like outgrowths on female catkins of alder, *Alnus glutinosa* (Plate 4c).

Taphrina deformans

Peach leaf curl is common on leaves and twigs on peach and almond, especially after a cool and moist spring. Towards the end of May, infected peach leaves show raised reddish puckered blisters which eventually acquire a waxy bloom (Fig. 9.1). Sections of leaves in this condition show an extensive septate mycelium growing between cells of the mesophyll and between the cuticle and epidermis, where the hyphae end in

Fig 9.1 *Taphrina deformans.* Peach leaf showing leaf curl.

swollen tips which have been termed chlamydo-spores by Martin (1940) but are, in fact, ascus initials or ascogenous cells (see Fig. 9.2). The interface between the parasitic fungus and the host takes the form of contact between their walls. No specialized haustoria have been found in *T. deformans* (Syrop, 1975), although they have been reported in some other species. Cytological studies (Martin, 1940; Kramer, 1961; Syrop & Beckett, 1976) have revealed that the segments of the mycelium and the young ascogenous cells are mostly binucleate. If the cells are multi-nucleate, then the nuclei are at least arranged in pairs (Syrop & Beckett, 1976). In the ascogenous cell, the two nuclei fuse and the diploid nucleus divides mitotically. The upper of the two daughter nuclei then undergoes meiosis followed by a mitosis so that eight nuclei result, which form the nuclei of the eight ascospores. The lower daughter nucleus remains in the lower part of the ascogenous cell and is often separated from the upper nucleus by a cross wall. During these nuclear divisions, the wall of the ascogenous

cell has stretched to form an ascus. Delimitation of the ascospores occurs at the eight-nucleate stage. The individual nuclei become enclosed by double-delimiting membranes which do not arise from the nuclear envelope as in most ascomycetes, but by invagination of the plasma-lemma of the developing ascus (Syrop & Beckett, 1972). Within the ascus, the ascospores may bud so that ripe asci may contain numerous yeast cells (see Fig. 9.2c). These yeast cells can be regarded as the anamorphic state of *Taphrina*, and they have been named *Lalaria* (Moore, 1990; Inácio et al., 2004). The asci form a palisade-like layer above the epidermis, and it is their presence which gives the infected leaf its waxy bloom.

The ascospores and yeast cells are projected from the ascus which often opens by a characteristic slit (Fig. 9.2c). Yarwood (1941) has shown that there is a diurnal cycle of ascus development and discharge in *T. deformans*. Nuclear fusion takes place during the afternoon or evening; nuclear divisions are complete by

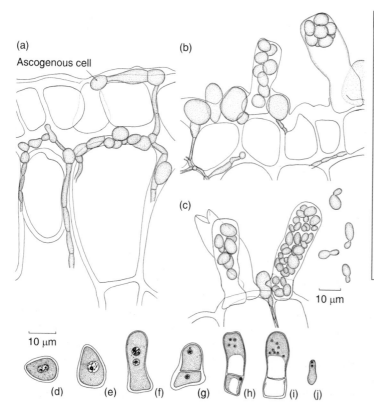

(a) Ascogenous cell
(b)
(c)

10 μm

Fig 9.2 *Taphrina deformans.* (a) T.S. peach leaf showing intercellular mycelium and subcuticular ascogenous cells. (b) T.S. peach leaf showing ascogenous cells and asci, containing eight ascospores. (c) T.S. leaf showing a dehisced ascus, an eight-spored ascus and an ascus in which the ascospores are budding. Ascospores budding outside the ascus are also shown. (d–j) Cytology of ascus formation (after Martin, 1940). (d,e). Fusion of nuclei in ascogenous cell. (f) Elongating ascogenous cell containing two nuclei formed by mitosis from the fusion nucleus. The upper nucleus has begun to divide meiotically. (g) Uninucleate ascus with uninucleate basal cell. (h,i) Four- and eight-nucleate asci. (j) Binucleate germ tube in germinating ascospore.

10 μm

(d) (e) (f) (g) (h) (i) (j)

about 5 a.m. and the spores appear mature by 8 a.m. However, maximum spore discharge does not occur until about 8 p.m. Outside the ascus, ascospores or yeast cells may continue budding and the fungus can be grown saprotrophically as a yeast in agar or liquid culture. The yeast cells are often pigmented due to the presence of β-carotene and several other carotenoid pigments (van Eijik & Roeymans, 1982). Young leaves can be infected from such yeast cells and it has been shown that a culture derived from a single ascospore can cause infection resulting in the formation of a fresh crop of asci, so that *T. deformans* is homothallic. In this respect it differs from some other species, e.g. *T. epiphylla*, where the fusion of yeast cells, presumably of different mating types, is necessary before infection can occur (Kramer, 1987). In *T. deformans* the binucleate condition is established at the first nuclear division of a yeast cell placed on a peach leaf, and the two daughter nuclei remain associated in the germ tube which penetrates the cuticle (Fig. 9.2j). In other *Taphrina* species, the germ tube penetrates through stomata but is unable to breach the intact cuticle (Taylor & Birdwell, 2000).

9.2.2 Growth hormones

In infections of peach leaves with *T. deformans*, the distortions of the host tissue are associated with division and hypertrophy of the cells of the palisade mesophyll. In liquid cultures, especially on media containing tryptophane, considerable quantities of the auxin-type phytohormone indole acetic acid (IAA) have been demonstrated. A number of different cytokinins are also produced by several species of *Taphrina* in culture (Kern & Naef-Roth, 1975; Tudzynski, 1997). Together, these hormones promote processes of cell division, enlargement and differentiation in plants, and leaves infected with *T. deformans* show higher levels of auxins and cytokinins than uninfected leaves (Sziráki *et al.*, 1975). It is therefore tempting to assume that the fungus produces these substances also *in planta*. However, this has not been formally proven yet, and the *Taphrina*–peach system seems to have been less thoroughly examined than the

interaction between *Plasmodiophora brassicae* and cabbage plants (see p. 63).

9.2.3 Control of *Taphrina deformans*

Taphrina deformans is by far the most serious pathogen among the Taphrinales, and it occurs wherever peach or almond trees grow. It is not yet entirely clear how *T. deformans* overwinters; Butler and Jones (1949) and Smith *et al.* (1988) considered it unlikely that the mycelial form is involved because leaves harbouring mycelium are shed in the autumn. It is more probable that yeast cells arising from discharged ascospores survive saprotrophically on the surface of twigs or in bud scales. Between November and March, the yeast cells develop thick walls and in spring, as the peach buds open, they produce germ tubes which penetrate the young leaves. The first symptoms of infection can be seen as soon as the buds break, but no further infection occurs from about early July onwards. This may be because *T. deformans* has a relatively low temperature maximum of 26–30°C (Butler & Jones, 1949).

Good chemical control of *T. deformans* can be achieved by spraying with Bordeaux mixture in autumn after leaf fall, in order to reduce the population of yeast cells on the twigs. Another spray in early spring, at the time of bud swelling, will give improved control of infection because the time span in which *Taphrina* can infect is limited. Dithiocarbamates or other simple fungicides are commonly used in spring (Smith *et al.*, 1988). Other diseases caused by *Taphrina* can be controlled in a similar way if necessary.

9.3 | Schizosaccharomycetales

The classification of the Schizosaccharomycetales has been the subject of controversial discussions, but the emerging consensus is that there is only one genus with three species, *S. japonicus*, *S. octosporus* and *S. pombe* (Kurtzman & Robnett, 1998; Vaughan-Martini & Martini, 1998a; Barnett *et al.*, 2000). All three species grow as saprotrophic yeasts which reproduce asexually by fission, i.e. by division of a vegetative cell into two daughter cells of equal size (Fig. 9.3a). *Schizosaccharomyces* is therefore called the **fission yeast**. Occasionally, especially in *S. japonicus*, true septate hyphae can be formed, and these may fragment into arthrospores. Sexual reproduction is by conjugation of two haploid vegetative yeast cells, followed by karyogamy and meiosis which gives rise to four or, more usually, eight ascospores (Figs. 9.3a,b).

Schizosaccharomyces can be isolated from substrates rich in soluble carbon sources, e.g. tree exudates, fruits, honey and fruit juices. The best-known species are *S. octosporus* and *S. pombe*. The latter is the fermenting agent of African millet beer (pombe) and arak in Java. It can tolerate ethanol levels up to 7% (v/v). Both species grow well in liquid culture or on solid media such as malt extract agar, developing ripe asci within 3 days at 25°C. All stages of the life cycle can be readily seen if a preparation of cells from an agar culture of *S. octosporus* is made in water (Fig. 9.3). Individual cells are globose to cylindrical,

uninucleate and haploid. Cell division is preceded by intranuclear mitosis, towards the end of which the nucleus constricts and becomes dumb-bell shaped (Tanaka & Kanbe, 1986). The division of the cell into two daughter cells is brought about by the centripetal development of a septum which cuts the cytoplasm into two. The two sister cells may remain attached to each other for a while, or may separate by breakdown of a layer of material in the middle of the septum (Sipiczki & Bozsik, 2000).

Ascus formation in *S. pombe* is preceded by copulation. *Schizosaccharomyces octosporus* is homothallic, and quite often adjacent sister cells may fuse together. In the case of *S. pombe*, both homothallic and heterothallic strains are known, the latter with a bipolar mating system (h+ and h− mating types). When cells of opposite mating type of *S. pombe* are grown together in liquid culture, especially under conditions of nitrogen starvation, a strong sexual agglutination occurs. This clumping together of the cells becomes visible as a flocculation of the culture. Changes in cell surface properties are

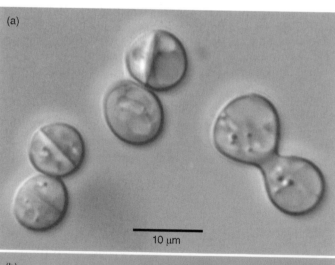

(a)

10 µm

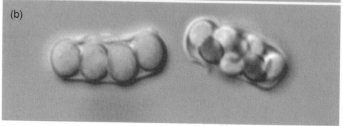

(b)

Fig 9.3 *Schizosaccharomyces octosporus.* (a) Vegetative cells, three of which showing transverse division. Two cells to the right of the picture are conjugating. (b) Four- and eight-spored asci.

important in agglutination, and fimbriae have been observed at the surface of cells stimulated by the appropriate pheromone (Johnson *et al.*, 1989). Using stable heterothallic haploid strains, the purification of the pheromones of *S. pombe* has been achieved; both are linear oligopeptide hormones (Davey, 1992; Imai & Yamamoto, 1994). The binding of a hormone released by cells of one mating type to receptors in the membranes of cells of opposite mating type initiates a signalling chain which in turn triggers the cellular response leading to agglutination and conjugation (Davey, 1998). The principle of mating factors will be discussed in more detail for *Saccharomyces cerevisiae* (p. 266).

During agglutination, two cells come into contact by a portion of their cell wall. In homothallic strains, the fusing cells are often sister cells formed by preceding mitotic division. A pore is formed in the centre of the attachment area and this widens and elongates to form a conjugation canal (Fig. 9.3a). During this process the nuclei, one from each cell, migrate towards each other and fuse. Vacuoles may appear in the young ascus following nuclear fusion. The fused nucleus elongates and may reach half the length of the ascus, and then divides by constriction, the nuclear membrane remaining intact during division. The two daughter nuclei migrate to opposite ends of the ascus and divide further. These two divisions constitute meiosis. A single mitotic division usually follows so that eight haploid nuclei result, and eight ascospores are finally differentiated (Tanaka & Hirata, 1982).

Four-spored asci are also common. The ascospores are released passively following disintegration of the ascus wall.

The life cycle of *Schizosaccharomyces* (Fig. 9.4) is thus interpreted as being based on haploid vegetative cells which fuse to form asci, the only diploid cells. Meiosis in the ascus restores the haploid condition. Some variation in this pattern may occur. For example, in *S. japonicus* and *S. pombe*, limited division of the zygote in the diploid state before ascospore formation may take place, and it is possible to select diploid strains (Tange & Niwa, 1995).

Physical and chemical analyses of the cell walls of *Schizosaccharomyces* show that they are principally composed of a β-(1,3)-glucan with β-(1,6)-branches making up 50–54% of the total cell wall carbohydrates, and of an α-(1,3)-glucan with α-(1,4)-branches contributing 28–32%. There are also trace amounts of a branched β-(1,6)-glucan (Manners & Meyer, 1977; Kopecka *et al.*, 1995). The β-(1,3)-glucan synthase is involved in all aspects of wall synthesis, including polarized growth, septum formation, and the formation and germination of ascospores (Cortés *et al.*, 2002). A galactomannan linked to wall matrix glycoprotein makes up about 9–14% of the cell wall polysaccharides (Manners & Meyer, 1977). As in other Archiascomycetes, chitin is present only in traces (Sietsma & Wessels, 1990), but it seems to play an important role in ascospore formation (Arellano *et al.*, 2000). The walls of ascospores contain amylose and give a blue reaction with iodine.

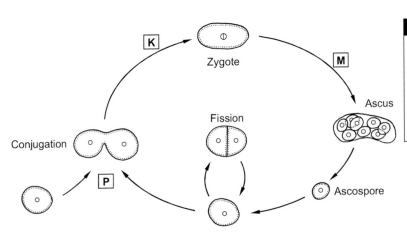

Fig 9.4 The life cycle of the homothallic yeast *Schizosaccharomyces octosporus*. Small open circles represent haploid nuclei; diploid nuclei are larger and split. Key events in the life cycle are plasmogamy (P), karyogamy (K) and meiosis (M).

Conjugation   Fission   Zygote   Ascus   Ascospore

In the following sections we give brief summaries of areas in which research on *S. pombe* is of outstanding significance for the discipline of biology as a whole. We anticipate that the relevance of this yeast for fundamental research will further increase in the future. Whilst we do not believe in the concept of a 'model organism' or even a 'model fungus', it is becoming clear that *S. pombe*, based on its more ancestral position in the phylogenetic system, is in many ways more relevant to the study of eukaryotic biology than its great rival, the more derived *Saccharomyces cerevisiae* (see p. 270). The entire genomes of both yeasts have been sequenced, and research is under way with *S. pombe* to find out the minimum number of genes (approximately 17.5% of all genes) required for the basic functioning of this organism (Decottignies *et al.*, 2003).

9.3.1 *Schizosaccharomyces pombe* and the cell cycle

The term 'cell cycle' denotes a carefully controlled sequence of regulatory and biosynthetic processes which guide a cell arising from mitosis towards its division into two daughter cells. Research on *S. pombe* has given us a fundamental understanding of the cell cycle. The literature on this topic is vast, and it is beyond the scope of this book to give more than the briefest of summaries. Our account borrows heavily from the textbook by Lewin (2000), which also provided the basis of the diagrammatic summary (Fig. 9.5).

A young cell arising from mitotic division starts its life in the G1 phase (G = gap) and may synthesize RNA, protein and other cellular constituents. It may grow in size but it does not duplicate its DNA at this point. The first crucial control point of the cell cycle is the START point, located in G1. At this point, the cell becomes committed to mitosis, and other options – notably sexual reproduction – are no longer available, i.e. beyond the START point the cell becomes insensitive to mating pheromones. When DNA duplication is actually initiated, the cell moves from the G1 to the S (synthesis of DNA) phase. After DNA replication has been

completed, the G2 phase follows, during which the *S. pombe* cell further enlarges in size and produces all organelles and macromolecules which are required to support two daughter cells. A second control point is the boundary between G2 and the M (mitotic) phase; when this has been passed, the cell stops elongating. Condensation and separation of the chromosomes occur, followed by septation and physical separation of the two daughter cells. The identification of genes whose products are involved in the regulation of the cell cycle was possible by analysing temperature-sensitive mutants, i.e. mutants which grow normally at reduced (permissive) temperature (e.g. 25°C) but are blocked at some stage of the cell cycle at a higher (restrictive) temperature (e.g. 37°C).

The most fundamental gene involved in the cell cycle is *cdc2* because its product – a protein kinase – is involved at both the START and G2/M control points, and it is now known to fulfil the same universal role in all eukaryotes, including humans (Lee & Nurse, 1987; Nurse, 1990). In order to act in such a way, the *cdc2* protein (written as Cdc2) combines with different proteins at specific stages of the cell cycle. These proteins are termed cyclins because their levels in the yeast cell show one peak in each cell cycle, followed by their degradation or inactivity. There are G1 cyclins and G2 cyclins which have different properties in combination with Cdc2. However, the activity of Cdc2 is modulated not only by the binding of cyclins, but also by kinases or phosphatases which, respectively, phosphorylate or dephosphorylate the Cdc2 protein. These respond to environmental stimuli and often antagonize each other in their effects on Cdc2. This allows a fine-tuning of the cell cycle in response to environmental factors such as the presence of pheromones which would prevent progression through START, or nutrient availability. Whilst the regulation of Cdc2 is relatively well understood, few of its substrates have been identified as yet, and this is an area of ongoing research.

An understanding of the cell cycle of *S. pombe* is of significance far beyond mycology because the principles are conserved across all eukaryotes. In mammals, one regulatory factor

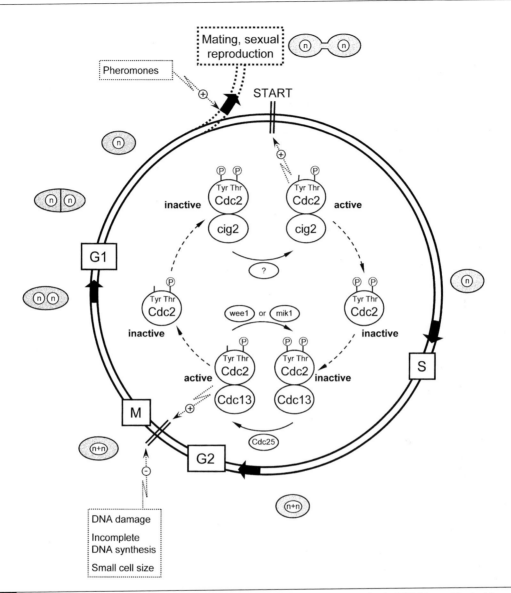

Fig 9.5 The central role of Cdc2 in the cell cycle of *S. pombe*. The START checkpoint is passed only if Cdc2 is combined with a GI cyclin (cig2) *and* is phosphorylated at a threonine residue at position 161 (Thr) but dephosphorylated at a tyrosine residue at position 15 (Tyr). The second major checkpoint is between the G2 and M phases; here Cdc2 needs to be coupled with a G2 cyclin (Cdcl3) *and* must be phosphorylated at Thr but dephosphorylated at Tyr. The cell cycle is therefore controlled by the type of cyclin available for coupling with Cdc2, and by the action of kinases such as weel or mikl and phosphatases (e.g. Cdc25). Adapted from Lewin (2000).

which stops the cell cycle at the G2/M control point and commits the cell to apoptosis (self-destruction) is DNA damage. This recognition mechanism is a most important protection against uncontrolled cell growth (cancer). These and other implications of the work on the cell cycle of *S. pombe* have resulted in the award of the Nobel Prize for Medicine and Physiology, among others, to Sir Paul Nurse in 2001. His Nobel lecture (Nurse, 2002) is a stimulating and readable account of the unravelling of the cell cycle in *S. pombe*.

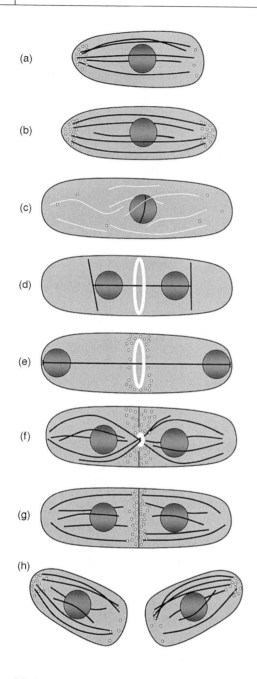

(a)

(b)

(c)

(d)

(e)

(f)

(g)

(h)

Fig 9.6 The cytoskeleton and the cell cycle in *Schizosaccharomyces pombe*. A new cell initially grows only at the old end (a) before bipolar growth is assumed (b). In growing cells, microtubules (dark lines) are orientated longitudinally, forming a basket around the nucleus (large sphere) and projecting into the poles. Actin patches (white dots) are located in the growing tips, as they are in filamentous fungi. At mitosis (c), microtubules form the intranuclear spindle while actin aggregates to form a cortical ring in the vicinity of the dividing nucleus (d,e). The ring constricts as the wall of the septum is laid down; upon completion of nuclear migration into the poles, the nuclear spindle breaks down and the basket of cytoplasmic microtubules reforms (f). After cell fission has been completed (g), actin relocates into the old end (h) at which growth is resumed. Redrawn from Brunner and Nurse (2000), *Philosophical Transactions of the Royal Society*, by copyright permission of The Royal Society.

9.3.2 Morphogenesis in *S. pombe*

Many aspects of the cell biology and ultrastructure of *S. pombe* have been studied extensively, and the results pertaining to the cytoskeleton are of particular relevance to filamentous fungi. Freshly divided cells of *S. pombe* grow only at one end (the 'old end' opposite the septum), i.e. growth is initially monopolar (Fig. 9.6a) before bipolar growth starts in early G2 phase

(Fig. 9.6b). In growing cells, microtubules are located in the cytoplasm and are orientated longitudinally, enclosing the nucleus like a basket. Mutant and inhibitor studies have revealed a crucial role for microtubules in co-ordinating polarized growth, i.e. in focusing cell wall extension to either or both of the two opposite poles (Brunner & Nurse, 2000; Hayles & Nurse, 2001). Microtubules seem to be involved in transporting the Tea1 protein to the poles. This protein acts as a termination signal for microtubule elongation, and by attracting microtubules it effectively controls its own transport (Mata & Nurse, 1998; Sawin & Nurse, 1998). Interactions between Tea1p and other proteins mark the cell end, i.e. the site of Tea1p accumulation, thereby fixing the growth direction in *S. pombe* (Niccoli *et al.*, 2003; Sawin & Snaith, 2004). Actin is also located in the growing poles of *S. pombe*. We can speculate that actin filaments and microtubules fulfil a similar role in *S. pombe* as in the apices of filamentous fungi, with microtubules mediating long-distance transport and actin fine-tuning the fusion of secretory vesicles with the plasma membrane.

When mitosis starts, there is a complete remodelling of the cytoskeleton, with the disappearance of cytoplasmic microtubules and the formation of the intranuclear spindle (Hagan, 1998). No further cell wall extension takes place during mitosis, presumably because no cytoplasmic microtubules are available to drive it. By contrast, actin relocates from the poles to the centre of the cell, forming a ring around the equator in close proximity to the nucleus (Figs. 9.6c,d). Actin aggregation is co-ordinated by a protein (Mid1p) emitted from the nucleus to form a broad cortical band, which in turn is controlled by the activity of the nuclear protein kinases Plo1p and Pdk1p (Bähler et al., 1998; Brunner & Nurse, 2000; Bimbó et al., 2005). Since the positioning of the nucleus in the cell is determined by microtubules, these are ultimately responsible for morphogenesis in S. pombe. Microtubules pull apart the two daughter nuclei (Figs. 9.6d,e), and when these have reached the two opposite poles (Fig. 9.6e), the spindle breaks down and the basket of cytoplasmic microtubules is re-established (Fig. 9.6f). Meanwhile, the actin ring co-ordinates the inward growth of the septum wall. When the septum has been completed, actin relocates to the two old ends, one in each daughter cell, which resume growth (Figs. 9.6g,h). It is remarkable that the septum is laid down precisely in the middle of a cell which was itself synthesized by asymmetric elongation of its two ends.

If a cell of S. pombe comes into close proximity to a cell of opposite mating type and is in the G1 phase prior to the START point, it will conjugate. The formation of a projection tip during conjugation requires the presence of actin for localized cell wall synthesis and lysis, and microtubules to localize actin towards this new if transient growing point (Petersen et al., 1998). Cell-to-cell fusion also requires the accumulation of actin (Kurahashi et al., 2002).

9.4 | Pneumocystis

This is an unusual but appropriately named group of organisms living as cyst-like cells in the lungs of mammalian hosts, where they cause pneumonia in immunocompromised individuals. The identity of these organisms as fungi was established beyond doubt only relatively recently, following DNA sequence comparisons. Recently, these techniques have also permitted the distinction between different taxonomic entities within Pneumocystis which correlate with the taxonomy of their hosts. Thus, the original name P. carinii is now applied by most workers only to a species infecting rats, with the human pathogen called P. jirovecii (Stringer et al., 2002).

Little is known about the life cycle of Pneumocystis because this organism cannot be grown satisfactorily outside its host. Cushion (2004) summarized evidence indicating that there is probably a haploid trophic phase in which cells divide by binary fission in an amoeba-like way, i.e. by constriction. Trophic cells of Pneumocystis are firmly attached to mammalian pneumocyte I cells in the alveolar regions of lung tissue. The cell wall of the trophic stage is unusually thin and flexible. If two compatible trophic cells fuse, a diploid zygote is formed and undergoes meiosis, producing a thick-walled cyst containing eight ascospores which are presumed to develop into a fresh crop of trophic cells upon germination. Pneumocystis can cause lethal pneumonia in immunocompromised hosts and the pneumonia it causes is regarded as an AIDS-defining illness. Infections have also been linked with the sudden infant death syndrome, reflecting contact of children with Pneumocystis very soon after birth. Exposure to inoculum seems to have little effect on immunocompetent individuals, although there is evidence that they may carry latent Pneumocystis infections of limited duration. Considerable uncertainty exists about the epidemiology of Pneumocystis. Although there have been occasional reports of the detection of P. jirovecii DNA in nature, there is no convincing evidence of any external reservoir of inoculum which could represent a source of human infections. The fungus therefore seems to be spread mainly or exclusively between humans. One way by which Pneumocystis may avoid the mammalian immune response is its ability to alter the properties of surface glycoproteins acting as antigens.

There are many oddities about *Pneumocystis* (Stringer, 1996, 2002). One is that this fungus lacks ergosterol, utilizing cholesterol instead as its major membrane sterol. This explains the insensitivity of *Pneumocystis* to amphotericin B, the most important drug against fungal infections of humans (see Fig. 10.9). In contrast, although the cell wall of trophic cells of *Pneumocystis* is unusually thin, this fungus is susceptible to inhibitors of β-(1,3)-glucan synthesis such as echinocandins (Schmatz *et al.*, 1990). Thirdly, there are only two rRNA gene repeat units, in contrast to other fungi which contain some 50−250 copies of it (see Fig. 1.24).

Hemiascomycetes

10.1 | Introduction

The class Hemiascomycetes contains the classical ascomycete yeasts, exclusive of those which belong to the Archiascomycetes (see the preceding chapter) and the 'black yeasts' such as *Aureobasidium* (see p. 486). Detailed descriptions of the individual yeast genera and species are given in Kurtzman and Fell (1998) and Barnett *et al.* (2000). A useful taxonomic overview is that by Kurtzman and Sugiyama (2001). There is only one order, the Saccharomycetales, which has been divided into 11 families and 276 species (Kirk *et al.*, 2001; Kurtzman & Sugiyama, 2001). However, detailed phylogenetic analyses of the Hemiascomycetes (Kurtzman & Robnett, 1998, 2003) indicate that this family arrangement is likely to be modified in the future, and for this reason we shall focus on selected genera.

The key feature that distinguishes the Hemi- and Archiascomycetes from the higher ascomycetes (Euascomycetes) is that ascogenous hyphae and an ascocarp, i.e. an investment of sterile hyphae surrounding the asci, are lacking in the first two groups. Instead, the asci are formed freely and singly, either directly following karyogamy or more rarely after a prolonged diploid phase. Another distinguishing feature is the composition of the cell wall, which contains very little chitin in the Hemi- and Archiascomycetes. Chitin is often confined to a small ring around the site where the daughter cell is produced (the bud scar). An ultrastructural feature of distinction concerns the septal pore of any hypha that may be produced. One or several pores may be present, and these are usually very small or plugged. They lack Woronin bodies, in contrast to Euascomycete septa which usually have only one large pore with associated Woronin bodies (Alexopoulos *et al.*, 1996; M.E. Barr, 2001; see Fig. 8.3). Hence, the Euascomycete septal pore permits passage of organelles including nuclei (see Fig. 8.2), whereas the micropore of the Hemiascomycete septum does not. Cytoplasmic communication between adjacent hyphal cells therefore does not seem to be possible.

It is impossible to give a watertight set of criteria by which Hemiascomycetes can be distinguished from Archiascomycetes. The predominant growth form of Hemiascomycetes in culture as well as in nature is the yeast state, although a limited mycelium or pseudomycelium may also be present. Archiascomycetes may grow as a mycelium in nature but as yeasts in the laboratory (*Taphrina*, *Protomyces*). In Archiascomycetes, asci may (*Taphrina*, *Protomyces*) or may not (*Pneumocystis*, *Schizosaccharomyces*) forcibly discharge their spores, whereas asci of Hemiascomycetes generally have evanescent walls, i.e. they release their ascospores passively.

In the absence of asci and ascospores, the microscopic identification of yeasts is difficult or impossible, and other methods have to be employed, e.g. physiological tests based on the ability of test species to grow on any of a standard set of carbon or nitrogen sources

(Yarrow, 1998; Barnett *et al.*, 2000). The analysis of DNA sequences (e.g. 18S rDNA) is now performed routinely in many laboratories, and a comparison with the extensive databases of appropriate sequences should afford identification at least to genus level. In this way, hemiascomycete yeasts can be distinguished from Archiascomycetes and also from basidiomycete yeasts. Such a distinction should be unequivocal since it utilizes the very same characters by which the classes Hemi- and Archiascomycetes were established.

10.1.1 Occurrence and isolation of Hemiascomycetes

Hemiascomycete yeasts are prominent as epiphytic saprotrophic colonizers of plant organs, especially where sugars are present, e.g. in the nectar of flowers, on fruits, and on wounded or exposed surfaces of plants. Between 10^5 and 10^7 yeast cells g^{-1} plant material (fresh weight) may be present (Phaff & Starmer, 1987). Yeasts also occur in the soil, although only a few exclusively soil-borne species have been described. Most yeasts are probably introduced into the soil with the plant material with which they were originally associated (Phaff & Starmer, 1987). Yeasts also occur in freshwater and marine situations. Some species are associated with insects and other animals, including the guts of vertebrates which have a thriving yeast mycota. Yeasts may grow on skin surfaces and one species – *Candida albicans* – can, under certain circumstances, turn into a mild or severe pathogen of humans, especially of immunocompromised patients (see p. 276). Hemiascomycetes are of little importance as plant pathogens with the exception of *Eremothecium* spp. which cause lesions on citrus fruits, cotton and other crop plants, and are spread by sucking insects (see p. 284).

Many species of Hemiascomycetes can grow under conditions of reduced water availability corresponding to about 50% glucose or a near-saturated NaCl solution. Consequently, they can colonize most types of preserved foods, whereby the type of preservative determines the species composition (Pitt & Hocking, 1985; Fleet, 1990).

Fortunately, food spoilage by yeasts does not normally result in the production of toxins, in contrast to bacteria or certain filamentous fungi. However, the economic losses of food spoilage due to yeasts are still considerable.

Hemiascomycete yeasts are easily isolated onto most standard agar media augmented with a suitable antibiotic to suppress bacteria, e.g. a mixture of penicillin G and streptomycin sulphate ($100–200\,mg\,l^{-1}$ each), added to the cooling agar after autoclaving. Plant or soil samples can be plated either directly, or the yeasts can be suspended by shaking the sample in sterile distilled water containing a detergent such as 0.01% (v/v) Triton X-100 or Tween 80. The undiluted sample or a dilution series in water can be plated out, and the density of colony-forming units (CFU)g^{-1} soil or leaves can be calculated. Yeasts are just large enough to be resolved as individual cells when a Petri dish is inverted and viewed with a ×10 objective, whereas bacterial cells are not resolved at that magnification.

10.1.2 The importance of Hemiascomycetes

A very small number of species is of immense importance to biotechnology, and an adequate discussion is beyond the scope of this book. Below is a mention of the most important aspects; some further applications and the yeast species involved have been summarized by J.F.T. Spencer *et al.* (2002).

1. Alcoholic fermentation mainly by *Saccharomyces cerevisiae*. This is the oldest and yet still the most important area of biotechnology, with about $10^{11}\,l$ of beer and $3 \times 10^{10}\,l$ of wine produced worldwide each year (Oliver, 1991; Kurtzman & Sugiyama, 2001). The discovery of alcoholic fermentations has been made several times independently in the history of mankind. Details of fermentation processes are given in on pp. 274–276. Industrial alcohol (ethanol) is often obtained from fermentations of corn starch hydrolysate by *S. cerevisiae*, but there is an ongoing interest in using other yeasts (*Pachysolen tamophilus*, *Pichia stipitis*) for the

production of ethanol from pentose sugars in wastes of industrial processes (Jeffries & Kurtzman, 1994).

2. Bread-making. About 1.5×10^6 tons of fresh cells of *S. cerevisiae* are produced worldwide per annum for use in the production of bread dough (see p. 274).

3. Single-cell protein (SCP). This term describes the conversion of low-cost substrates into protein-rich biomass of unicellular organisms. Yeasts have a high nutritional value to animals and man because they are rich in vitamins and protein, and because they do not generally produce mycotoxins. Since they also have very simple growth requirements, yeasts can be used to convert low-cost substrates such as wastes from industrial processes into high-value products for human or animal consumption. While the use of mineral oil as a substrate was a somewhat predictable failure, other substrates such as whey wastes from cheese production, molasses from sugar cane or pentose-containing wastes from paper production are promising (Tuse, 1984; Scrimshaw & Murray, 1995; Paul *et al.*, 2002). Currently, about 800 000 tons of fodder yeasts are produced per annum (Kurtzman & Sugiyama, 2001), but an extended application of single-cell protein technology is hampered by the low current cost of alternative protein sources such as soy meal or fish meal (Harrison, 1993; Scrimshaw & Murray, 1995).

4. Vitamin production. Riboflavin (vitamin B2) is produced industrially by *Eremothecium* spp. (see p. 284).

5. Production of recombinant proteins, e.g. enzymes, or clinically relevant molecules such as antigens, insulin and epidermal growth factor. Expression systems for heterologous proteins, i.e. proteins of interest whose gene has been linked to the promoter sequence of the producing organism, include *S. cerevisiae* and *Pichia pastoris*. The latter holds advantages because proteins of interest are secreted more efficiently. Further, the glycosylation (sugar) chains which are added to the polypeptide during and after its translation in the rough endoplasmic reticulum are more similar between *Pichia* and mammals than either is to *S. cerevisiae*.

Therefore, *Pichia* proteins cause fewer immunological problems in clinical use (see p. 281).

6. Biological control. Because yeasts do not produce mycotoxins and because of their ability to colonize the skin of fruits, they are being developed as biological control agents against postharvest losses in fruit crops. *Pichia guilliermondii* sprayed onto fruits selectively reduces development of moulds caused by *Penicillium* spp. (Chalutz & Wilson, 1990; McLaughlin *et al.*, 1990).

10.2 | *Saccharomyces (Saccharomycetaceae)*

About 10–16 species of *Saccharomyces* are currently recognized (Vaughan-Martini & Martini, 1998b; Barnett *et al.*, 2000; Kirk *et al.*, 2001). We will focus on *S. cerevisiae*, which in many ways is the most important fungus yet discovered. About 25 strains of *S. cerevisiae* exist, and these have different physiological properties which are relevant to their biotechnological applications. Many were formerly regarded as different species (Vaughan-Martini & Martini, 1998b; Rainieri *et al.*, 2003). *Saccharomyces cerevisiae* is the brewer's and baker's yeast (see below), although some of the best brewing yeasts in current use belong to *S. pastorianus* (= *S. carlsbergensis*). In nature, *S. cerevisiae* is found on ripe fruits, like many other yeasts. Grape and fruit wines are still often made by relying on spontaneous fermentations by yeasts which happen to be growing on the skin of the fruits used.

The relatively small size of yeast cells (about $6–8 \times 5–6\,\mu m$) has limited their investigation by light microscopy, but great progress has been made recently by the use of fluorescent dyes. Further, by fusing the green fluorescent protein (GFP) gene to the promoters of diverse yeast proteins, it has become possible to locate the site of a defined gene product within the yeast cell (Kohlwein, 2000). The availability of freeze-substitution fixation for transmission electron microscopy has led to the production of highly resolved 'natural' images of organelles and cellular processes (Baba & Osumi, 1987), although many issues of yeast cytology have remained controversial. A vegetative yeast cell (Fig. 10.1) is densely packed with organelles, including one nucleus, a large central vacuole, and mitochondria. Mitochondria are very dynamic in shape, showing a pronounced tendency to fuse into one or a few large reticulate organelles when the energy demand is high, or to fragment into numerous small promitochondria during anaerobic fermentation or metabolic inactivity (Jensen *et al.*, 2000; Okamoto & Shaw, 2005).

An immense amount of literature exists on *S. cerevisiae*, covering aspects of genetics and molecular biology (Pringle *et al.*, 1997), physiology (see Jennings, 1995), cytology (Baba & Osumi, 1987) and biotechnology (Spencer & Spencer, 1990; Walker, 1998). We can only broach a few selected topics here to give an impression of

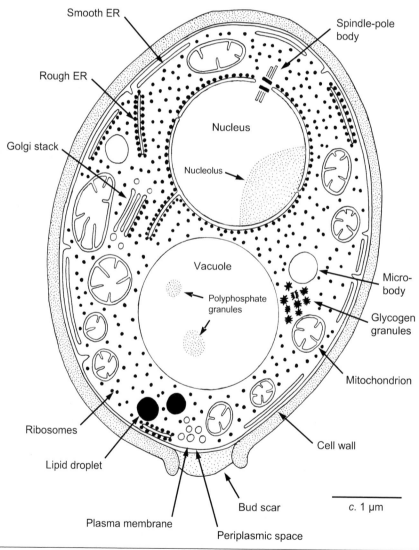

Fig 10.1 *Saccharomyces cerevisiae.* Sketch of a budding yeast cell as seen by transmission electron microscopy using material fixed by freeze-substitution. The presence of a morphologically recognizable Golgi stack is unusual among Eumycota. Modified and redrawn from Baba and Osumi (1987).

the enormous importance of *S. cerevisiae* for fundamental cell biology. This fungus was the first eukaryote (in 1996) to have its complete genome sequenced, and this — together with the ease of genetic manipulation — has further enhanced the status of *S. cerevisiae* as a work-horse, if not 'model organism', for eukaryote research.

10.2.1 The life cycle of *S. cerevisiae*

Vegetative cells of *S. cerevisiae* are generally diploid in nature, although tetraploid or aneu-ploid strains also occur. Strains may be homo- or heterothallic. The chromosome number is 16 (Cherry *et al.*, 1997). The life cycle of *S. cerevisiae* is presented in Figs. 10.2 and 10.3. The haploid ascospores often fuse within the ascus where they were formed (Fig. 10.3d), or shortly after release. However, if individual ascospores become isolated, they can germinate and repro-duce as haploid cells by budding. Where two haploid cells of opposite mating type are in close contact, they secrete peptide hormones and pro-duce plasma membrane-bound receptors which recognize the hormone of opposite mating type. The binding of a hormone molecule to the matching receptor sets a signalling chain in motion (reviewed by Bardwell, 2004) which arrests the mitotic cell cycle at G1, stimulates transcription of mating-specific genes and causes polarized growth of the two cells towards each other (Leberer *et al.*, 1997). Mating initially involves an increased ability of the surfaces of two cells to adhere to each other. This so-called **sexual agglutination** is mediated by glycopro-teins. It seems that these are components of fimbriae, i.e. long filaments radiating outwards from the cell wall (see Fig. 23.15). Agglutination is followed by co-ordinated digestion of the walls separating the two cells. Plasmogamy and karyogamy follow swiftly (Gammie *et al.*, 1998). The resulting diploid cell (Fig. 10.3e) can carry on reproducing asexually by budding. In contrast to *Schizosaccharomyces pombe*, there are therefore two mitotic cycles in the life cycle of *S. cerevisiae*. Under optimum conditions, the culture doubling time by mitosis is about 100 min.

Diploid strains of *S. cerevisiae* can be induced to form ascospores by suitable treatment, and this yeast is therefore termed an **ascosporoge-nous** yeast, in contrast to **asporogenous** yeasts in which ascospores have not been observed. Meiosis can be induced by growing the yeast on a nutrient-rich presporulation medium contain-ing an assimilable sugar, a suitable nitrogen source for good growth (nitrate is not utilized),

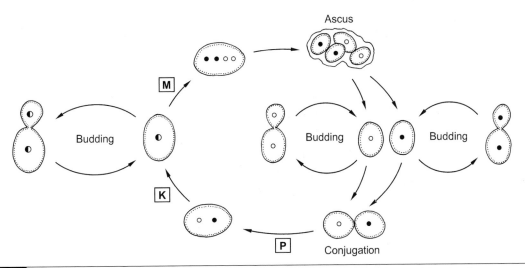

Fig 10.2 The life cycle of *S. cerevisiae*. Both haploid and diploid cells can reproduce by budding. Open and closed circles represent haploid nuclei of opposite mating type; diploid nuclei are larger and half-filled. Key events in the life cycle are plasmogamy (P), karyogamy (K) and meiosis (M).

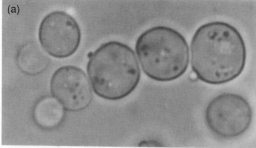

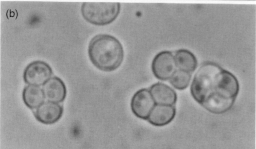

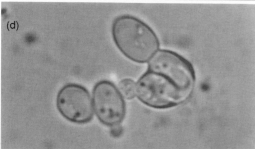

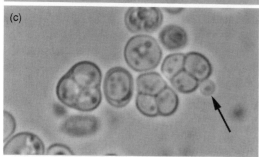

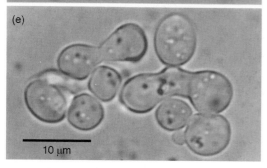

10 μm

Fig 10.3 *Saccharomyces cerevisiae.* (a) Vegetative yeast cells reproducing by budding. (b) Yeast asci containing mostly four spores, sometimes with only three spores in focus. (c) Ascus showing a budding ascospore (arrow). (d) Ascus in which two spores have fused together and are budding. (e) Two ascospores fusing (top left), and two fused ascospores forming a diploid bud (right).

and vitamins of the B group. Such a medium results in well-grown cells which will sporulate on transfer to a sporulation medium. Sporulation occurs best on media in which budding is inhibited. Low concentrations of an assimilable carbon source are necessary to provide energy for the sporulation process. Acetate agar (1 g glucose, 8.2 g Na acetate·3H$_2$O and 2.5 g yeast extract l^{-1}) is a good sporulation medium (Yarrow, 1998).

Diploid yeast cells convert directly into asci within 12−24 h of incubation in a suitable sporulation medium. The frequency of ascus formation in most isolates is quite low, usually less than 10% (Vaughan-Martini & Martini, 1998b). The cytoplasm differentiates into four thick-walled spherical spores, although the number of spores may be fewer (see Fig. 10.3d). The nuclear divisions which precede ascus formation are meiotic. As is also the case in mitosis, the nuclear envelope remains intact during meiosis, taking up a lobed shape as the nuclear spindles draw the chromosomes into two and then four corners of the envelope (Fig. 10.4). The mechanism of ascospore formation has been extensively reviewed by Neiman (2005). It is very similar to that of higher ascomycetes, the main difference being that there is no common vesicle enclosing all nuclei prior to ascospore delimitation. Instead, a cup-shaped double-membrane, the prospore membrane, associates with each of the four spindle-pole bodies, and this gradually encapsulates its nuclear lobe until the four nuclei separate (Figs. 10.4g,h). As in other ascomycetes, the ascospore wall is then laid down in the lumen between the two membranes surrounding the developing ascospore.

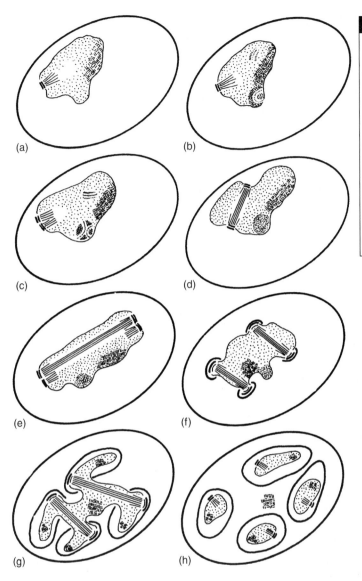

(a)

(b)

(c)

(d)

(e)

(f)

(g)

(h)

Fig 10.4 *Saccharomyces cerevisiae.*
Diagrammatic summary of the processes of meiosis and ascospore delimitation (from Beckett *et al.*, 1974). (a–d) The spindle pole body replicates and the two new spindle pole bodies move to opposite poles of the nucleus. The nuclear membrane remains intact. (e,f) Further replication of the spindle pole bodies and rearrangement. The nuclear envelope is still intact. New membranes, the ascospore-delimiting membranes, form outside the spindle pole bodies. (g,h) Envelopment of the lobes of the dividing nucleus by the ascospore-delimiting membranes results in the formation of four haploid uninucleate ascospores.

10.2.2 Mating in *S. cerevisiae*

Many strains of *S. cerevisiae* are heterothallic, and the ascospores are of two mating types. Mating type specificity is controlled by a single genetic locus which exists in two allelic states, α and **a**, and segregation at the meiosis preceding ascospore formation results in two α and two **a** ascospores. Fusion normally occurs only between cells of opposite mating type, and this has been termed legitimate copulation. Such fusions result in diploid cells which can, under appropriate conditions, form asci with viable ascospores.

The mating type (*MAT*) alleles are rather small and structurally similar to each other, but differ in their central region which comprises about 650 base pairs (bp) in *MATa* and 750 bp in *MATα* (Fig. 10.5). Because of this difference, mating type alleles are often called **idiomorphs**. In haploid α-cells, the *MATα* locus expresses two genes, both of which encode regulatory proteins. The α1 gene product interacts with a constitutively expressed protein not encoded by the *MAT* locus, Mcm1p, to activate several α-specific genes outside the *MAT* locus, notably those encoding the α-pheromone which is secreted by α-cells, and

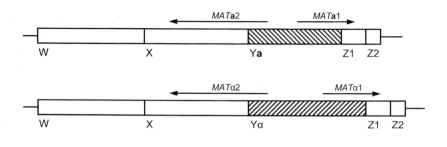

c. 500 nt

Fig 10.5 The structure of the mating type idiomorphs **a** (top) and α (bottom) of *Saccharomyces cerevisiae*. The two alleles differ only in their central (Y) regions which contain parts of two genes **a**l and **a**2 or αl and α2. The entire lengths of these genes and their directions of transcription are indicated by arrows. The function of **a**2 is unknown. In diploid cells, the lack of expression of αl and the formation of a dimeric α2/**a**l protein suppresses the expression of mating type-specific proteins including hormones and their receptors. nt = nucleotides. Redrawn from Haber (1998) *Annual Reviews of Genetics* **32**, with permission. © 1998 Annual Reviews, www.annualreviews.org.

the plasma membrane receptor Ste2p, which can bind **a**-pheromone from the environment. In diploid cells, expression of α1 and thus of α-specific proteins is repressed. The α2 gene encodes a repressor protein which interacts with several other regulatory proteins, including Mcm1p, to repress the expression of **a**-specific genes, including those encoding the **a**-pheromone and the α-factor receptor Ste3p. In the absence of the α1 and α2 gene products, haploid cells have an **a**-phenotype with respect to mating behaviour because **a**-genes are constitutively expressed. The function of **a**2 is unknown, and the **a**1 gene product is active only in diploid cells, combining with the α2 protein to repress haploid-specific genes including those encoding the two pheromones and their receptors. Another gene repressed by the α2/**a**1 dimer is *RME1*, which encodes a repressor of meiosis. Meiosis can therefore only take place if a diploid α/**a** nucleus exists in which Rme1p is repressed by the α2/**a**1 dimer, and if nutrient conditions are limiting. The signal for nutrient limitation is sensed and transmitted by a cyclic AMP-dependent signalling chain (Klein *et al.*, 1994).

Therefore, the most important difference between α/**a** diploids and homozygous diploids or haploids is that only the α/**a** diploids can initiate meiosis under nutrient-limiting conditions, leading to the production of ascospores which are more resistant to adverse conditions than vegetative cells. It is likely that this enhanced survival of ascospores is the reason why haploid or homozygous populations of *S. cerevisiae* and some other ascomycetes (e.g. *Schizosaccharomyces pombe*) possess the intriguing ability to switch their mating type, thereby acquiring the ability to undergo meiosis. An excellent account of the experiments and ideas leading to the unravelling of the mating factor switch in *S. cerevisiae* has been given by Haber (1998), and we borrow heavily from it in the following summary.

Strathern and Herskowitz (1979) observed that the ability to switch their mating type is acquired only by cells which have previously divided at least once. The pattern established by a single germinating **a**-type haploid ascospore is shown in Fig. 10.6: the first daughter cell has the mating type **a**, but then the mother cell switches its mating type prior to its second division, so that two α-cells result. Meanwhile, the first daughter (**a**-type) cell undergoes its first division so that a cluster of two **a**- and two α-type cells is produced. Conjugation can occur, and the two zygotes can carry on dividing as diploid yeast cells with the additional option to undergo meiosis and produce asci if required. The mating type switch is brought about by the

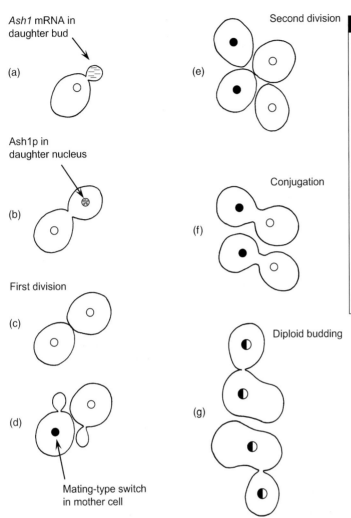

Ash1 mRNA in daughter bud

(a)

Ash1p in daughter nucleus

(b)

First division

(c)

(d)

Mating-type switch in mother cell

Second division

(e)

Conjugation

(f)

Diploid budding

(g)

Fig 10.6 Mating type switch in *Saccharomyces cerevisiae*. A germinating **a**-type ascospore denoted by its white nucleus produces a bud which has the same mating type. *Ash1* mRNA is selectively translocated into the bud (a), and its protein Ash1p is expressed in the daughter nucleus (b), preventing it from switching its mating type. However, the mother cell is now competent to switch its mating type to α because it has divided once before (c); when it undergoes its second division (d,e), it produces an α-type daughter cell. The first-formed daughter cell cannot switch its mating type because it has no previous history of cell division. Consequently it produces a daughter of **a**-type. Mating (f) occurs between one α- and one **a**-type cell apiece, and conjugation results in two diploid cells which can reproduce by further budding (g) or by meiosis and ascus formation. Based on Haber (1998).

HO endonuclease, the gene of which is only expressed in mother cells which have divided at least once (Nasmyth, 1983). Expression of the HO endonuclease gene is controlled by a series of repressor proteins, and one of them – Ash1p – is confined to the daughter cell upon division. This is due to the selective transport of its mRNA molecule into the bud prior to cytokinesis (Long *et al.*, 1997).

Hicks *et al.* (1977) proposed the cassette model to account for the mating type switch, and this has been confirmed by subsequent experimentation. In addition to the mating type locus which is active in a given haploid cell, each haploid genome possesses two further complete copies, one to the left of the active locus which usually contains the α allele (i.e. *HML*α) and the other to the right, encoding a (i.e. *HMR*a). These genes are silenced, i.e. they are not transcribed because their DNA is coated by histones and other proteins encoded and regulated by numerous other genes. Silencing is determined by the location of these gene copies in the proximity of silencing sequences (Haber, 1998). The mating type switch is perfomed when the HO endonuclease cleaves the currently active locus at the Y/Z boundary (see Fig. 10.5), followed by the digestion of one of the two DNA strands of the Z region by an exonuclease. A new sequence is then copied into that gap from either of the two silent genes, using the one-stranded Z region as a template. The integrity of the silent gene which acts as template is unaffected during that process (for details, see Haber, 1998). Hence, yeast cells can repeatedly switch their mating type. There is an element of selectivity in the mating type switch because, for example, α-cells choose the silent **a**-locus 85–90% of the time (Weiler & Broach, 1992). In the case of α-cells, preference for the switch to the **a** mating type is brought about by the α2 protein (for details, see Haber, 1998).

10.2.3 The cell wall of *S. cerevisiae*

The wall of *S. cerevisiae* represents a considerable biochemical investment, making up 15–30% of the dry weight of vegetative cells. Up to three wall layers can be distinguished by electron microscopy. They differ in their chemical composition, and the relatively simple architecture of the wall of *S. cerevisiae* is considered a model for other fungi (Molina *et al.*, 2000; de Nobel *et al.*, 2001). The middle layer is electron-translucent and consists of the main structural scaffold of branched β-(1,3)-glucan molecules which bind β-(1,6)-glucans and chitin. The latter, however, is present only in low quantities (1–2% of the total wall material) and it is unevenly distributed, being concentrated in a ring around the region where the bud emerges. The outer wall layer of *S. cerevisiae* is electron-dense because it consists mainly of proteins. These determine the cell surface properties, including the porosity (pore size) of the cell wall (Zlotnik *et al.*, 1984) and adhesiveness to other cells (flocculation; see p. 274). The outer wall proteins may be highly glycosylated in *S. cerevisiae* by the addition of large mannose chains. In pathogenic yeasts such as *Candida albicans*, this outer layer is also important because of its involvement in attachment of the fungus to its host, and because it conveys antigenic properties. There are two main groups of outer cell wall proteins (CWPs) in *S. cerevisiae*. The members of one are modified by a glycosylphosphatidylinositol (GPI) chain which is indirectly linked to the β-(1,3)-glucans of the central wall layer via the β-(1,6)-glucans. These are called GPI–CWPs. The second type of outer cell wall protein is called Pir–CWP (Pir = protein with internal repeats) and is

linked directly to the β-(1,3)-glucan component (Kapteyn *et al.*, 1999; de Nobel *et al.*, 2001). Both GPI–CWPs and Pir–CWPs are structural proteins. The innermost layer (**periplasmic space**) is also electron-dense and consists of proteins, but these are mostly enzymes which are too large to pass through the central layer. They are therefore restrained by the glucan layer (de Nobel & Barnett, 1991).

The polarity of wall synthesis in *S. cerevisiae* is controlled by the localization of the plasma membrane-bound enzymes (glucan synthetases, chitin synthetases) which produce the elements of the middle layer, and by the secretion of the structural outer wall proteins as well as enzymes which cross-link the various elements of the cell wall. Cell wall synthesis is thus regulated spatially by the polarity of the yeast cell, and temporally by the cell cycle; the transcription of many genes involved in cell wall synthesis is cell cycle-dependent (Molina *et al.*, 2000; Rodríguez-Peña *et al.*, 2000).

10.2.4 Morphogenesis and the cell cycle of *S. cerevisiae*

One oddity about *S. cerevisiae* is that it does not require microtubules for the maintenance of its cellular polarity, as shown by mutant and inhibitor studies. In other eukaryotes, including filamentous fungi and also the fission yeast *Schizosaccharomyces pombe*, microtubules are employed for long-distance transport processes. It is possible that they are dispensable in *S. cerevisiae* simply because of the small distance between the mother cell and the growing bud. Of course, microtubules are required in *S. cerevisiae* as in all other eukaryotes for nuclear division. Actin, in contrast, is crucial for cell polarity and cell viability in *S. cerevisiae* (Pruyne & Bretscher, 2000b; Pruyne *et al.*, 2004).

The cell cycle of *S. cerevisiae* is similar to that of *Schizosaccharomyces pombe* (see p. 256) and other eukaryotes in its regulatory mechanisms (Lewin, 2000; Alberts *et al.*, 2002), except that the G2 phase is lacking, and that cell division (**cytokinesis**) is initiated early in the cycle, the bud being already present during the S phase. A summary of the budding process is given in Fig. 10.7.

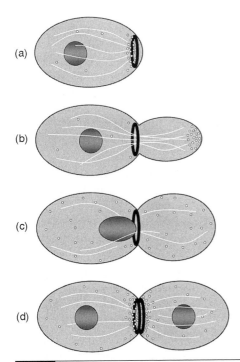

Fig 10.7 The mitotic cell cycle of *Saccharomyces cerevisiae*. Actin patches and cables are drawn in white; the septin ring is black. Secretory vesicles are not drawn, but their distribution follows essentially that of actin patches. (a) Initiation of the bud site occurs during the S phase by the assembly of actin patches around a septin ring. (b) Bud extension during late S phase. The cap co-ordinates apical bud growth by establishing actin cables which mediate the transport of secretory vesicles into the bud, and their fusion in the region of the cap. (c) A later stage of bud growth; the cap components become distributed more evenly over the bud membrane surface, and growth is non-polarized (isodiametric). Meanwhile, the dividing nucleus is drawn, via cytoplasmic microtubules (not shown), to an actin/myosin ring superimposed onto the septin ring. The nucleus divides so that each cell receives one daughter nucleus. (d) Re-establishment of polarized growth by the formation of two septin rings, one on either side of the bud site. Growth leads to closure of the pore between the mother and daughter cell. Redrawn from Sheu and Snyder (2001), with kind permission of Springer Science and Business Media.

Useful and comprehensive reviews are those by Pruyne and Bretscher (2000a,b) and Sheu and Snyder (2001). The bud site is determined by the assembly of a protein cap at the inner surface of the plasma membrane. This cap contains an essential regulatory protein, Cdc42p, which is controlled directly by the cell cycle and in turn determines the sequestration of numerous other scaffold and regulatory proteins by the cap. Budding differs between haploid and diploid cells, the former initiating new buds adjacent to the previous one, and the latter budding in a bipolar fashion (Pruyne *et al.*, 2004).

One important group of proteins are the **septins**, which form a ring around the bud site (Fig. 10.7a). Actin filaments are initiated from the centre of the cap. As the bud extends in a polarized fashion, the septin ring remains at the site of bud emergence whereas the cap – which governs bud extension – migrates with the bud, staying at the apex and controlling bud extension (Fig. 10.7b). Later, the cap components and actin filament attachment points become distributed diffusely over the bud surface. This leads to the fusion of secretory vesicles over the entire bud surface, and to a change in the growth pattern from polarized to isodiametric (Fig. 10.7c).

The septin ring binds numerous proteins, including important regulatory ones (Versele & Thorner, 2005). Actin and myosin are attracted early during bud emergence, and a contractile actin ring is superimposed on the septin ring. Later, during nuclear division, cytoplasmic microtubules are also captured; these, in turn, are in contact with the microtubules of the intranuclear mitotic spindle, and thus the dividing nucleus is drawn towards the ring, with one daughter nucleus apiece ending up in each of the two cells (Fig. 10.7c; Kusch *et al.*, 2002). Cytokinesis is brought about as the actin ring contracts (Lippincott & Li, 1998). At this point, septins appear as a double ring, sandwiching the constricting actin ring. The septin double ring assembles two caps, and these serve as a nucleation centre for actin filaments which direct secretory vesicles to the bud site, closing the wall between mother and daughter cell (Fig. 10.7d). The regulatory mechanisms are immensely complex but are beginning to be unravelled (Pruyne & Bretscher, 2000a,b; Pruyne *et al.*, 2004; Versele & Thorner, 2005) and are likely to be of fundamental significance because the division of cells by a constricting actin ring is found also in other fungi and in animal systems, although apparently not in plants.

Following separation of the daughter cell, a circular, crater-like **bud scar** is left as a permanent mark on the surface of the mother cell (see Fig. 10.12). The maximum number of scars that could be accommodated on the surface of a yeast cell is about 100, suggesting that individual yeast cells are not capable of unlimited budding. Individual yeast cells age just like other organisms, although the timing of death is determined by complicated genetic factors and the sum of metabolic energy expended throughout the life of the yeast cell, rather than the number of the bud scars per se (Jazwinski, 2002).

Under certain environmental conditions (notably nutrient deficiency) diploid and, to a lesser extent, haploid cells of *S. cerevisiae* can change their growth pattern from budding, which produces heaps of cells only on the agar surface, to the formation of pseudohyphae which can grow into the agar. Pseudohyphae may be of significance in the ecology of *S. cerevisiae* because they allow the organism to spread over and penetrate into substrates, and to assimilate nutrients more readily (Gimeno *et al.*, 1992). Formation of pseudohyphae requires an enhanced adhesion of the cells to each other and an enhanced polarity of daughter cell growth. Not surprisingly, the signalling events leading to pseudohyphal growth are rather complex (Palecek *et al.*, 2002; Ceccato-Antonini & Sudbery, 2004).

10.2.5 Membrane cycling in *S. cerevisiae*

An enormous amount of work has been done to elucidate the secretory route in *S. cerevisiae*, and a sizeable collection of temperature-sensitive mutants with defects at different points of the secretory route has been assembled (Schekman, 1992). Further, individual stepwise modifications to proteins travelling the secretory route can be identified, especially with respect to their glycosylation pattern and proteolytic cleavage of parts of the original polypeptide chain (Graham & Emr, 1991). The export of proteins starts with their synthesis in the rough endoplasmic reticulum and continues with their processing in a Golgi system. Along this route,

the proteins are modified by the addition of glycosylation chains, and by the proteolytic cleavage of signal sequences. Transport has long been thought to occur by means of vesicle-like carriers, and the biochemical events leading to the budding of a vesicle from its source and its fusion with the destination membrane (e.g. ER → Golgi) have been extensively characterized (Rothman & Orci, 1992). However, it is still unclear whether discrete vesicular carriers are an obligate transport system in vegetative yeast cells. An alternative is the dynamic maturation model in which sheets of ER become transformed into Golgi compartments which gradually dilate and fragment into secretory vesicles (Rambourg *et al.*, 2001). Whatever their initial history, secretory vesicles emerge from the Golgi system (Baba & Osumi, 1987) and migrate to the growing bud along actin cables (Finger & Novick, 1998).

The cell membrane shows a high capacity for endocytosis, i.e. the removal of excess membrane material and the uptake of specific molecules by membrane-bound receptors from the liquid medium of the environment. The occurrence of endocytosis has been controversial in filamentous fungi, but it has been obvious for some time that this must take place in *S. cerevisiae* as it is the route through which mating hormones are internalized and transported to the vacuole for degradation. While actin is certainly involved in endocytosis, it is still unclear whether the actin patches long known to exist inside the plasma membrane of *S. cerevisiae* cells are the scaffold around which the inward-budding of the plasma membrane is moulded (Shaw *et al.*, 2001). Endocytosis occurs when pits are formed at the plasma membrane and bud inwards to form small vesicles (endocytotic vesicles) which fuse to form a tubular early endosome. From there, material is transported via a late endosome to the vacuole in which it is degraded (Munn, 2000; Shaw *et al.*, 2001). The protein ubiquitin plays a vital role as a tag for endocytosis at the plasma membrane and for transport of endosomes to the vacuole (Horák, 2003). The purposes of endocytosis could include the removal of excess membrane material, the removal of nutrient uptake systems no longer required, and the removal of mating type receptors, e.g. after the fusion of two haploid cells.

10.2.6 The yeast vacuole

The vacuole is the central destination of membrane trafficking in *S. cerevisiae*. It receives input directly from the secretory route in the form of most vacuolar proteins which are separated at the Golgi stage from those bound for secretion. Material also reaches the vacuole from the endocytotic route (see above), and from the cytoplasm, especially during starvation. Cytoplasmic material may be engulfed directly by the tonoplast (microautophagy) or redundant material may first be surrounded by a double membrane to form an autophagosome whose outer membrane then fuses with the tonoplast (macroautophagy). Details of these processes have been described by Klionsky (1997) and Thumm (2000). Degradation of protein is a major function of the vacuole in starvation situations, and about 40% of the total protein content of a yeast cell can be degraded within 24 h (Teichert *et al.*, 1989). Not surprisingly, the vacuole contains a large set of powerful hydrolytic enzymes, especially proteases (Klionsky *et al.*, 1990).

When nitrogen is abundant, it is stored in the vacuole as arginine at concentrations of up to 400 mM, and this can be re-released into the cytoplasm if nitrogen becomes limiting (Kitamoto *et al.*, 1988). Likewise, phosphate can be stored and released (Castro *et al.*, 1999), as can many other ionic nutrients (Jennings, 1995). Toxic ions and metabolites may be stored in the vacuole (e.g. Ramsay & Gadd, 1997). Vacuoles thus fulfil a crucial function in maintaining the homeostasis of the yeast cytoplasm against changing external conditions. In order to fulfil such functions, the vacuolar morphology can change dramatically, e.g. by fragmentation of one large central vacuole into numerous small ones (Çakar *et al.*, 2000).

10.2.7 Killer yeasts and killer toxins

Killer yeasts are strains which produce toxins capable of killing other strains belonging to the same or to closely related species. Toxin producers are resistant against their own toxin, but may be susceptible to toxins produced by other strains. Three important virus-encoded killer toxins (K1, K2, K28) are known to exist in *S. cerevisiae*; all three are polypeptides and are encoded by double-stranded RNA encapsulated in virus-like particles (VLPs). Another group of double-stranded viruses (the L-A viruses) belonging to the genus *Totivirus* is necessary for the replication of the killer toxin VLPs. The subject of killer yeasts has been reviewed by Magliani *et al.* (1997) and Marquina *et al.* (2002).

The best-researched killer toxin is K1. It is encoded by a single open reading frame and is synthesized as a single polypeptide which is initially localized in the ER membrane. As the membrane-bound polypeptide travels the secretory route, it is modified by glycosylation and proteolytic cleavage, much like other secretory proteins. In the Golgi system, the polypeptide is cleaved into two parts which are held together by disulphide bonds, and a third part, the glycosylated region, which is not part of the active toxin. The active toxin is secreted and diffuses into the growth medium. The two parts of the active molecule fulfil two different functions; the β-chain binds the molecule to its receptor site which is the β-(1,6)-glucan component of the cell wall. Following binding to the wall, the toxin is thought to be transferred to the plasma membrane where the α-chain forms a trans-membrane pore. Death of the target cell occurs because the trans-plasma membrane proton and ionic gradients are disrupted. The toxin can bind to the wall of the producing cell but not to its plasma membrane; presumably a membrane receptor is altered, masked or destroyed. Self-immunity is conveyed by a precursor molecule of the mature toxin (Boone *et al.*, 1986).

The K1 toxin has been an important instrument in elucidating the processing of proteins along the secretory route, and the mechanism of cell wall synthesis in *S. cerevisiae*. Additionally, there are biotechnological implications. The possession of a killer toxin conveys a selective advantage upon a yeast strain, and killer yeasts are particularly common (25% of all isolates) in habitats which contain abundant nutrients, such as the surface of ripe fruits (Starmer *et al.*, 1987). Not surprisingly, contaminations by killer yeasts can be a problem in long-term fermentation processes, e.g. wine production (van Vuuren & Jacobs, 1992), and attempts have been made to incorporate the killer virus into yeast strains used for biotechnological purposes (Javadekar *et al.*, 1995). Killer toxins cannot themselves be used against clinically relevant yeasts because the molecules are so large that they would elicit an immune response in the patient. However, it is possible to create antibodies which mimic the membrane-disrupting action of killer toxins (Polonelli *et al.*, 1991). Whether these will become useful in medicine, e.g. against *Candida* infections, remains to be seen, but the existence of natural human antibodies with a killer toxin effect on *Candida* points to potential applications of this strategy (Magliani *et al.*, 1997, 2005).

10.2.8 Bread-making

The principle behind the leavening of bread dough by baker's yeast is the same as in brewing, i.e. the anaerobic metabolism of glucose and other reducing sugars via pyruvic acid into ethanol and CO_2. The difference is that the released CO_2 is the important product in bread-making because it is responsible for the texture of the bread. Ethanol may, however, contribute to the flavour of fresh bread. Originally, a portion of the risen dough medium was retained as a starter for the next baking session, or surplus yeast from brewing processes was used (Jenson, 1998). Specific yeasts for baking were first produced in Vienna in 1846, and baker's yeast is now produced commercially under aerobic conditions because the yield of biomass can be maximized (Caron, 1995). In the bread dough the yeast cells are subjected to anoxic or anaerobic conditions and must be able to release CO_2 quickly. The carbon sources available to yeast cells in bread dough are hexoses, especially glucose, and the disaccharides maltose and sucrose, all of which are present at fairly low concentrations. Starch is not utilized by *S. cerevisiae* but can be hydrolysed by amylases present in the flour, and the glucose thereby released may be available to the yeast (Oliver, 1991; Jakobsen *et al.*, 2002). A very thorough account of the microbiology and processes of baking is that by Spicher and Brümmer (1995).

10.2.9 Beer brewing

Several good accounts of the process of beer brewing have been given (e.g. Oliver, 1991; Russell & Stewart, 1995; Hartmeier & Reiss, 2002), and there are numerous popular books exploring the diversity of beers worldwide. In his masterful history of beer, Hornsey (2003) has summarized evidence of the first known records and recipes of beer which date back 6000 years or more and originate from Mesopotamia and ancient Egypt, where beer was more widely consumed than wine. The art of brewing may be almost as ancient as the cultivation of cereals, and indeed some historians believe that brewing was a major incentive for the development of agriculture around 6000 BC (Hornsey, 2003). Brewing has remained the most important area of biotechnology to this day. Astonishing quantities of beer are being consumed, with several sources agreeing on the Czech Republic as the top beer-drinking nation at around 160 l per person per annum, followed by the Republic of Ireland (155 l) and Germany (128 l). These values appear frugal when put into the historical context, e.g. of one gallon (3.8 l) as the daily personal allowance of ale for monks in medieval England (Hornsey, 2003).

Two fundamentally different types of fermentation exist, and these differ in the strains of yeast used. In bottom-fermenting beers, especially lager beers, the yeast settles as a sludge at the bottom of the brewing vessel at the end of the fermentation, whereas it floats at the top in top-fermenting beers, especially the English ales, porters, stouts, and the German Altbier. In Germany, a purity law was passed in 1516 which banned the use of ingredients other than water, yeast, malted barley and hops. Although now formally abolished by the European Union, most brewers still abide by it. In contrast, ale and lager brewers outside Germany often add other ingredients to their beer, e.g. cereals other than barley, other fermentable sugar sources such as syrups or enzymatic starch digests, fruits or interesting spices.

Since *S. cerevisiae* cannot utilize starch, this has to be hydrolysed into fermentable sugars first. The starch reserves in the endosperm of barley are hydrolysed naturally by endogenous amylases when the grains germinate. At a suitable time, the germination process is terminated by drying and heating (kilning), and the degree to which the malt is roasted determines the colour of the beer. For instance, heavily roasted malts are used for the dark milds, porter ales and stouts. During mashing, the ground malt is heated in water to about 65°C; surviving endogenous enzymes or added enzymes continue with starch hydrolysis, and with the degradation of proteins to amino acids. Hops are added to the liquid (wort) which is then boiled, traditionally in a copper vessel, and after filtration and cooling the yeast is added. Top fermentation of ales takes a few days at 15–22°C whereas lager beers are fermented for up to 2 weeks at 8–15°C. In addition to ethanol, yeast metabolites which impart flavour to the beer are esters of higher alcohols (e.g. isobutanol), diketones, diacetyl, isobutyraldehyde and methylglyoxal. Sulphur-containing metabolites may also be important. There are several ways in which the flavour profile can be modified to give a desirable taste. For example, a yeast strain with the appropriate ester profile can be used, or new strains expressing the required enzymes can be engineered. A re-fermentation, often with *Brettanomyces* spp., may be performed in order to alter the flavour profile (Vanderhaegen *et al.*, 2003).

Towards the end of the fermentation, the yeast cells should flocculate, i.e. form aggregates. Flocculation is dependent on the expression of a number of surface proteins (flocculins) which recognize and bind to the mannose residues on mannoproteins located in the outermost wall layer of other yeast cells (Verstrepen *et al.*, 2003). These flocculins are probably located in fimbriae, short hairs (0.5 μm long) on the cell surface which have been observed by ultrastructural studies (Day *et al.*, 1975). Numerous environmental factors such as carbon and nitrogen deficiency, ethanol levels and cell age contribute to efficient flocculation, the control of which is one of the most difficult tasks in brewing. Whether flocculated yeast accumulates at the top or bottom of the vessel seems to depend on the flocculins as well as other wall surface properties (Dengis & Rouxhet, 1997). Attempts are being made to engineer improved brewer's yeast strains with respect to their flocculation behaviour (Verstrepen *et al.*, 2003) and also their ability to utilize other carbon sources, including starch (Hammond, 1995).

Most beer in Europe is brewed in batch fermentations, but in other countries continuous cultures following the chemostat principle are performed. Either way, the ale or lager must be stored for a while before it is sterilized and filled into barrels or bottles. In the case of cask-conditioned Real Ales, the beer is filled directly into the casks and stored until it is ready to be sold to public houses; cask-conditioned ale is therefore in direct contact with the yeast until it is served to the customer. It requires skilled brewers and publicans to keep cask-conditioned Real Ale, but it does, in the opinion of many, result in a superior pint.

10.2.10 Wine production

Wine is the fermentation product of starting materials which already contain high levels of monosaccharides, i.e. typically fruit juices and especially the must of grapes. Wine is at least as ancient as beer and seems to have originated in Transcaucasia and the Near East in the early Neolithic, around or before 6000 BC (McGovern, 2003). One of the pre-requisites for wine-making was the invention of pottery, since the fermentation process requires anaerobic conditions. Wine has been given a special place in many civilizations by its association with religious ceremonies, e.g. in ancient Egypt, Greece, Rome and in Christianity. In France, Portugal, Luxemburg and Italy, more than 50 l wine are consumed per person per annum.

Red wine takes its colour (and high tannin content) from the skin of the grapes which are macerated, and the juice ('must') is left in contact with the solid parts for some time. In contrast, in white wine and rosé wine, the must is extracted rapidly from the white or red grapes, respectively. Once the must has been obtained and filtered, subsequent treatment is similar for red, rosé and white wines. The must is either fermented directly, relying on the natural yeast flora of the grapes ('spontaneous fermentation'), or a pure yeast starter culture is added at such high concentrations that this strain suppresses the wild yeasts. Spontaneous fermentations still account for 80% of the worldwide wine production, and up to 100 000 wild yeast cells – mostly not belonging to *S. cerevisiae* – may be found on the surface of one berry or in 1 ml must (Dittrich, 1995). The diversity of yeasts changes rapidly during the initial stages of the wine fermentation, with *S. cerevisiae* displacing the obligately aerobic species as oxygen becomes depleted. The total fructose and glucose content of musts may be as high as $150 \, \mathrm{g \, l^{-1}}$. In principle, fermentation is completed when no further release of CO_2 occurs, either due to exhaustion of sugars or due to ethanol poisoning of the yeast, but in practice fermentations are often terminated artificially by addition of sulphite, especially if a sweet wine is desired.

Fermentation may carry on for up to 1 year with white wine; red wine develops faster but is often stored in barrels for prolonged periods to permit maturation. If red wine is stored in oak wood, it is called 'barrique' wine and it acquires a characteristic additional flavour. Good concise accounts of the processes and microbiology of wine making are those of Dittrich (1995) and Hartmeier and Reiss (2002).

10.2.11 Production of saké

Saké production (for a review, see Oliver, 1991) involves the conversion of rice starch into monosaccharides which are then fermented into ethanol. Saké is thus technically a beer rather than a wine. It has been produced for several thousand years in China, but its current production principles based on the synergistic action of two fungi date back to the fifth century AD. Saké production relies on the degradation of starch in cleaned and boiled rice by a filamentous fungus, *Aspergillus oryzae*, which produces several different amylases as well as proteases

and other enzymes (see p. 302). Koji, a solid culture of *A. oryzae* on steamed rice, is used as a starter for starch hydrolysis.

Fermentation (moromi) is carried out in a large volume of water to which successive quantities of boiled rice, koji and the *S. cerevisiae* starter culture (moto) are added. Stepwise addition and a highly ethanol-tolerant yeast strain ensure that saké is the most strongly alcoholic beverage produced by fermentation without distillation, containing up to 20% (v/v) ethanol. Fermentation takes about 25 days and is followed by storage, maturation and filtration. In order to avoid contamination by lactic acid bacteria, saké is pasteurized. It is interesting to note that this practice was introduced in the sixteenth century, 300 years before Pasteur.

10.3 | *Candida* (anamorphic Saccharomycetales)

Candida is a very large genus of anamorphic Saccharomycetales, currently comprising some 165 accepted species (Meyer *et al.*, 1998; Kirk *et al.*, 2001), with new ones being described at a high frequency. The genus is polyphyletic (Kurtzman & Robnett, 1998). By far the best-known species is *C. albicans*, which is associated with human disease, and on which we will focus here. A very similar species, and possibly one which has been misdiagnosed as *C. albicans* in the past, is *C. dubliniensis* (Martinez *et al.*, 2002). Other species (*C. glabrata*, *C. inconspicua*, *C. krusei*) may also cause opportunistic infections of man. In contrast, *Candida utilis* (now called *Pichia jadinii*; see p. 281) has been used for food and fodder production for over 80 years, and other *Candida* spp. are also suitable for this purpose (Boze *et al.*, 1995; Scrimshaw & Murray, 1995).

Candida spp. are cosmopolitan and can be found in many ecological situations (Meyer *et al.*, 1998), e.g. the surface of fruits and other plant organs, rotting wood, the soil, sea water, or associations with mammals and insects (especially bees). *Candida* spp. can contaminate grape musts during the early stages of wine making but are usually displaced by *S. cerevisiae* later. *Candida albicans* is slightly atypical of the genus in that it does not appear to be distributed widely in the environment and can be considered a commensal of humans and other warm-blooded animals.

10.3.1 Dimorphism in *Candida albicans*

Candida albicans can grow as yeast cells, true septate hyphae, or pseudohyphae which are an intermediate form between these two extremes (see Fig. 1.1d). Thick-walled chlamydospores may be formed by hyphae or pseudohyphae. This dimorphism – or polymorphism – has long been thought to represent an important pathogenicity determinant, pathogenicity commonly being associated with hyphal growth whereas yeasts are indicative of saprotrophic commensal growth. The switch between yeast and hyphal states is reversible and is determined by an interplay of several factors, e.g. temperature (hyphae at 37°C, yeasts below), pH (hyphae at neutral pH, yeasts at acid pH), nutrient abundance (yeast growth) or deficiency (hyphal growth), and presence (hyphal growth) or absence (yeast growth) of blood serum. Thus, conditions which mimic the bloodstream encourage hyphal growth, whereas conditions as found on the skin or in mucosal linings tend to promote yeast growth. *Candida albicans* is a commensal colonist of most humans, occasionally causing skin lesions, but under exceptional circumstances it turns into a serious pathogen causing deep-seated or systemic mycoses, especially when the host's immune system is weakened, e.g. in AIDS sufferers or patients who have undergone an organ transplantation. From such infections, *C. albicans* is usually recovered in the hyphal form.

The yeast and hyphal forms differ in many features which have a bearing on their ability to cause disease (Odds, 1994). For instance, hyphae are coated with mannoproteins which adhere strongly to mammalian proteins found in the membranes of cell surfaces. Such adhesive proteins (**adhesins**) take the shape of fimbriae projecting beyond the cell wall (Yu *et al.*, 1994; Vitkov *et al.*, 2002; see Fig 23.15). Enhanced adhesion may play a role in pathogenesis, especially when coupled with the invasive mode of growth displayed by hyphae (Gow *et al.*, 1999). Further, hyphae secrete aspartyl proteases and lipases capable of degrading host tissue (Hube & Naglik, 2001). Mannoproteins as well as proteases are potential targets for new anti-*Candida* drugs.

Yeast-hyphal dimorphism in *C. albicans* has been investigated in some detail. The signalling chains leading to the formation of a hypha are extremely complex, involving cyclic AMP as well as mitogen-activated protein kinase (MAP kinase) pathways. Both are also involved in the switch from yeast cells to pseudohyphal growth in *S. cerevisiae* (Brown & Gow, 2002). An extensive cross-talk between different signalling pathways is not surprising, since the switch from yeast to hypha responds to many different environmental signals which need to be integrated. The control mechanisms determining the switch from yeast to (pseudo)hyphal growth may also be similar between *S. cerevisiae* (see Section 10.2.4) and *C. albicans*.

10.3.2 Mating and switching in *Candida albicans*

Whereas *C. albicans* is permanently diploid, other *Candida* species such as *C. glabrata* are haploid. An exclusively diploid vegetative phase is very unusual among true fungi, although it is found in *Protomyces* (Archiascomycetes; see p. 251) or *Xanthophyllomyces* (Heterobasidiomycetes; see Fig. 24.3) and, of course, in the Oomycota (see Chapter 5). Until recently, *C. albicans* was thought to reproduce strictly asexually. However, when the genome sequence of *C. albicans* became available and was examined closely, a complete set of genes relevant to mating, homologous with those known for *S. cerevisiae*, was detected, and it was found that the fungus is heterozygous for the two mating type idiomorphs α and **a**, similar to the diploid cells of *S. cerevisiae* but unable to sporulate. The signalling processes involved in mating are likely to be similar between *S. cerevisiae* and *C. albicans* (Bennett & Johnson, 2005), and conjugation in *C. albicans* has now been observed between diploid strains each

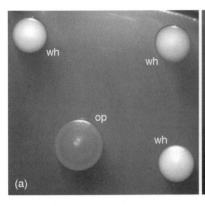
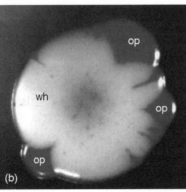

Fig 10.8 Examples of white−opaque switching in *Candida albicans*. (a) Cells from a white colony plated at low density. A switch has occurred from white (wh) to opaque (op). (b) A white colony which has been aged on a plate with limited gas exchange. This has caused increased rates of switching from white (wh) to opaque (op) at the colony edge. Original images kindly provided by D. R. Soll and K. Daniels.

containing only either mating type, although karyogamy was doubtful and meiosis was not observed (Magee & Magee, 2000; Lockhart *et al.*, 2003). If tetraploid strains result from karyogamy in nature, these may undergo meiosis or random loss of chromosomes by a parasexual cycle, i.e. *C. albicans* may have a cryptic sexual phase in its life cycle which has eluded mycologists for over a century (Gow, 2002). Wong *et al.* (2003) have suggested that *Candida glabrata* has a similarly cryptic sexual cycle.

In contrast to *S. cerevisiae*, there are no silent additional copies of mating type idiomorphs in the genome of *C. albicans*. Before a diploid strain of *C. albicans* heterozygous for mating type idiomorph (i.e. α/a) can mate, it will therefore have to convert to a/a or α/α by a mechanism different from the mating type switch based on a cassette system as found in *S. cerevisiae* (see p. 266).

A remarkable phenomenon that has been known for some time is the spontaneous and reversible switching of yeast colony phenotypes in *C. albicans*. A given strain can switch its colony morphology between smooth, wrinkled or star-like, and white or opaque (Fig. 10.8). The latter switch is particularly well-characterized (Slutsky *et al.*, 1987; Soll, 2002) and occurs at an unusually high frequency (about one colony in 1000–10 000). The different morphological appearances of white and opaque colonies are due to differences in size, shape and surface properties of the yeast cells. Genetically, the switch is accompanied by the co-ordinated up- or down-regulation of numerous genes, some of them potentially involved in pathogenesis

(Soll, 2003). Examples are secretory proteases or an ABC transporter involved in drug resistance (see p. 278). Not surprisingly, these two different colony types have widely differing pathogenic properties, the white-phase cells appearing to be better adapted to colonization of internal organs and opaque-phase cells superior in colonizing external skin regions.

An interesting link between mating and switching is that the latter is suppressed in strains heterozygous for the mating type idiomorphs a and α. This may be mediated by the regulatory heterodimer presumably formed by protein products of the α and a idiomorphs. Another noteworthy observation involves sexual reproduction: opaque cells mate about 10^6 times more efficiently than white cells (Miller & Johnson, 2002). Mating competence in *C. albicans* is therefore regulated at two levels, namely the requirement for a given cell to be homozygous for mating type (either α or a) and to be in the opaque state (Soll *et al.*, 2003). It is furthermore possible that some of the phenotypes characteristic of opaque cells are required for mating, in addition to or instead of being pathogenicity factors. In this context, it is of interest that there is a mass switch from opaque to white at 37°C, and that mating between opaque cells of opposite mating type is strongly stimulated on skin surfaces which have a lower temperature (Lachke *et al.*, 2003). Clearly, the ability of *C. albicans* to adapt to different situations by changing between several pre-programmed cell types, e.g. the mating-competent opaque and the invasive white cells, contributes to the success of this organism as a pathogen (Staib *et al.*, 2001; Bennett & Johnson, 2005).

10.3.3 Treatment of candidiasis and resistance mechanisms

Candida infections often occur following treatment with antibacterial antibiotics which also kill the benign bacteria which compete against *Candida*. Such superficial infections are especially common in mucosal linings of the mouth cavity, vagina or on the skin. They are collectively called 'thrush'. Infections of the oesophagus occur in patients with weakened immune system and are considered an AIDS-defining illness. The mucous membranes and skin are usually effective as primary barriers against infection, and *Candida* cells within the human body are vigorously attacked by the immune system (Murphy, 1996; Magliani *et al.*, 2005). If all these barriers are broken or weakened, deep invasive candidiasis can occur. Yeast cells (conidia) can be disseminated in the blood stream, and individual organs can become colonized by hyphae. Contaminated catheters are also an important entry point for *Candida*. An extended account of candidiasis in all its forms has been given by Kwon-Chung and Bennett (1992).

Generally, treatment of *Candida* infections is difficult because of the relative genetic similarity between *Candida* and humans, which greatly reduces the range of available targets as compared to the treatment of bacterial infections. Consequently, certain anti-*Candida* drugs have severe side effects. None the less, drugs belonging to several different classes are in current clinical use, as reviewed by Georgopapadakou (1998), Cowen *et al.* (2002) and Sanglard and Bille (2002). Lucid accounts of the fascinating array of mechanisms by which *Candida* achieves resistance against the various drugs are those by Ghannoum and Rice (1999), Sanglard (2002) and Akins (2005).

Many drugs target ergosterol, a fungal membrane sterol which is not found in animals. Amphotericin B (Fig. 10.9a) or nystatin are polyene antibiotics which associate with ergosterol in the membranes of *Candida*, forming pores in the plasma membrane and thereby rendering it leaky. Amphotericin B has severe side effects but has to be used especially against deep-seated infections (Lemke *et al.*, 2005). Resistance is usually based on the replacement of ergosterol by a precursor molecule, or a general reduction of the sterol content in the plasma membrane.

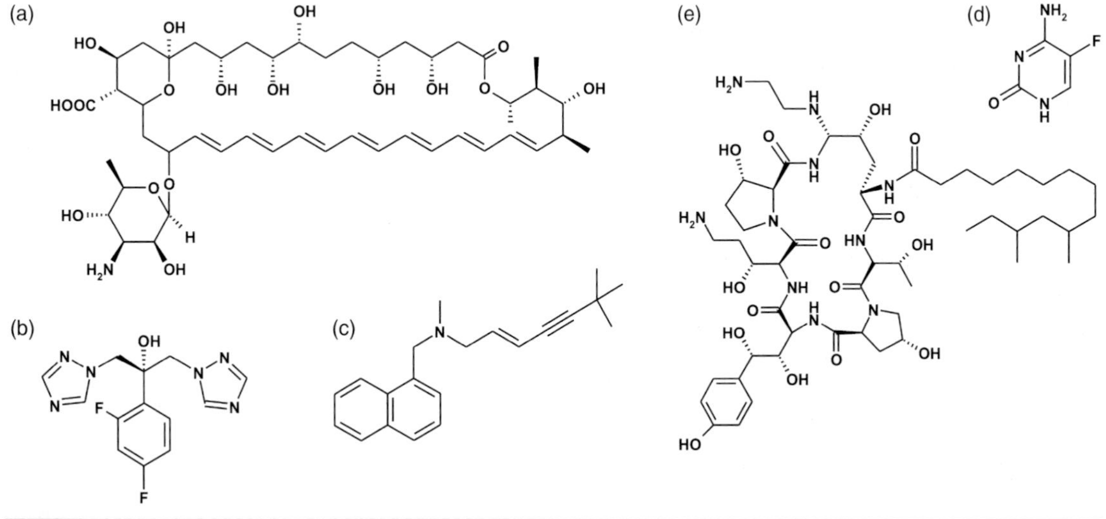

Fig 10.9 The most important anti-*Candida* drugs in current use. (a) The polyene compound amphotericin B. (b) The triazole compound fluconazole. (c) The allylamine terbinafine. (d) The fluoropyrimidine compound 5-fluorocytosine. (e) The echinocandin caspofungin.

The azole-type fungicides inhibit the enzyme lanosterol demethylase which is involved in ergosterol biosynthesis (see Fig. 13.16). A much-used example is fluconazole (Fig. 10.9b) which is free from severe side effects. There are several resistance mechanisms in *C. albicans*. The most common, found in 85% of all resistant isolates (Perea *et al.*, 2001), is based on active exclusion of the drug by means of ABC (ATP binding cassette) transporters or similar mechanisms. These are plasma membrane proteins with numerous (usually 12) transmembrane domains and two cytoplasmic ATP-binding domains. Alternating binding and hydrolysis of ATP changes the conformation of these proteins, enabling them to open and close membrane pores. ABC transporters are often capable of transporting a group of different metabolites, thereby producing cross-resistance. The natural role of ABC transporters probably lies in the exclusion of endogenous antibiotics, toxins or other substances, e.g. mating factors in *S. cerevisiae*. The second most important type of resistance against azole-type drugs (65% of all isolates) is a mutation of the cellular target, i.e. the azole binding site on the enzyme lanosterol demethylase. The fact that the occurrences of these two types of resistance add up to more than 100% indicates that many clinical *C. albicans* isolates (about 75%) possess both resistance mechanisms. A third resistance mechanism against azoles is the overexpression of the gene *ERG11* encoding lanosterol demethylase, which occurs in about 35% of resistant isolates (Perea *et al.*, 2001).

A third group of compounds, the allylamines (e.g. terbinafine; Fig. 10.9c) act against a different enzyme involved in ergosterol biosynthesis, squalene epoxidase (see Fig. 13.16). Terbinafine is not used extensively on its own, but it is useful in combination with other drugs in order to treat infections by resistant *Candida* strains.

Another target, nucleic acid biosynthesis, is attacked by 5-fluorocytosine (Fig. 10.9d). Following uptake, it is deaminated to 5-fluorouracil and converted to 5-fluoro-UTP or a corresponding deoxynucleotide, which inhibit RNA and DNA biosynthesis, respectively. Resistance is associated with a reduced capacity of the fungus to metabolize 5-fluorocytosine. Mammalian cells do not efficiently metabolize this drug but intestinal bacteria can, which precludes the oral use of this antibiotic.

A recently described group are the echinocandins which inhibit β-(1,3)-glucan synthesis. There is preliminary evidence that echinocandins such as caspofungin (Fig. 10.9e) do not act directly on β-(1,3)-synthase but in an indirect manner by interfering with upstream regulatory proteins (Edlind & Katiyar, 2004). No drugs against mannoproteins or aspartic proteases are as yet commercially available, although treatment against HIV uses protease inhibitors which also affect *C. albicans* (Dupont *et al.*, 2000). Research efforts into new anti-*Candida* drugs are intensive, given that the incidence of *Candida* infections is strongly on the increase, few substances are currently available, and resistance of *Candida* against them is becoming a problem.

10.3.4 Ecology and drug resistance of *Candida albicans*

Numerous investigations of the distribution of *Candida* spp. on their hosts have been carried out. Generally, *C. albicans* is by far the most frequent species, followed by *C. parapsilosis*. Other species such as *C. glabrata*, *C. krusei* and *C. tropicalis* are very much less frequent. Within the species *C. albicans*, many different strains exist, and their colonization pattern has been followed on the same human host over time (Xu *et al.*, 1999; Kam & Xu, 2002). Each human being can be colonized by a diversity of strains. Displacement of one strain by another is possible, as is the transfer of strains between humans. Appropriate analyses of allelic distributions have shown that the mode of genetic inheritance is predominantly clonal, i.e. sexual reproduction and the exchange of genetic material between different *Candida* strains do not seem to play an important role (Lott *et al.*, 1999). There are no significant differences in the *Candida* populations between healthy individuals and AIDS patients unless, of course, the population dynamics are shaken up by anti-*Candida* drug treatments. In the course of a prolonged treatment of patients against oral candidiasis, Martinez *et al.* (2002) reported the displacement of the initially predominant, fluconazole-sensitive *C. albicans* flora by fluconazole-resistant *C. dubliniensis* strains especially in those patients where *C. albicans* had failed to develop resistance. Resistance may arise spontaneously after prolonged treatment, and the spread of resistant clones in hospitals may not be as important with *Candida* as, for example, with multiple drug-resistant bacteria (Taylor *et al.*, 2003).

10.4 | *Pichia* (Saccharomycetaceae)

The genus *Pichia* contains 94 species (Kurtzman, 1998; Kirk *et al.*, 2001) and is characterized by budding cells, with only a few species also producing arthroconidia, pseudohyphae and hyphae. Sexual reproduction is by ascospores (1−4 per ascus) which are often hat-shaped (galeate). Molecular characterization of the genus is still in progress (Suzuki & Nakase, 1999) and will undoubtedly lead to rearrangements in future.

Pichia is cosmopolitan and ubiquitous. A surprising number of species has been isolated from the frass of wood-attacking beetles (Kurtzman, 1998); others grow on the exudates (slime fluxes) of trees or on decaying cacti, or they occur as contaminants of industrial fermentations. Two species are of particular biotechnological interest: *Pichia jadinii* (anamorph *Candida utilis*), formerly called *Torula* yeast, has been developed since World War I as a food yeast for single-cell protein. It can utilize the pentoses of pulping-waste liquors from the paper industry and is also grown on other biological wastes (Boze *et al.*, 1995).

Pichia pastoris is interesting for a different reason; it can utilize methanol by expressing and secreting large quantities of alcohol oxidase. Since the protein glycosylation chains of *P. pastoris* are similar to those of humans, and because the products of heterologous genes are secreted efficiently, *P. pastoris* has advantages over *S. cerevisiae* in the industrial production

processes of proteins of pharmacological interest (Daly & Hearn, 2005).

10.5 | *Galactomyces* (Dipodascaceae)

The genus *Galactomyces* (formerly called *Endomyces*) is characterized by true hyphae which

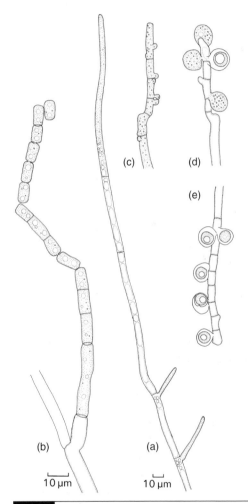

Fig 10.10 *Galactomyces candidus.* (a) Vegetative hyphal apex. The two lateral branches near the base are developing conidiophores. (b) Conidiophore showing the development and separation of arthroconidia. (c) Gametangia developing as lateral bulges of the hyphae on either side of a septum. (d) Fusion of gametangia to form asci. In one ascus, a single ascospore is differentiated. (e) Mature asci, each containing a single ascospore.

form septa and quickly fragment into arthro-spores (Fig. 10.10), giving the colonies a creamy appearance. The septa have micropores, like those of true hyphae of *C. albicans*. Six species are now known; by far the most important is *G. candidus* (anamorph *Geotrichum candidum*), formerly known as *Galactomyces geotrichum* (de Hoog & Smith, 2004). It is a ubiquitous mould which is common in soil, dairy products, sewage and other substrata. It is also thought to be a common constituent of the skin and gut flora of humans and animals, although reports of it being a human pathogen have generally remained unsubstantiated (Kwon-Chung & Bennett, 1992). In addition, *G. candidus* is a well-known cause of post-harvest rot in ripe fruits and vegetables especially when these are kept in plastic bags. Infected plant organs become soft and eventually, upon puncturing, exude a creamy mass of decaying tissue which has a sour smell; hence the name 'sour rot' (Agrios, 2005).

The fungus grows readily in culture, forming broad hyphae with finer lateral branches. The vegetative cells contain 1−4 nuclei. Branching is of two kinds, pseudodichotomous near the apex, and lateral immediately behind a septum. It is from such lateral branches that conidia develop (Figs. 10.10a,b). Conidiophores are diffi-cult to differentiate from vegetative hyphae. Prior to conidium formation, apical growth of a hypha ceases, then septa are laid down in the tip region. The septa are two-ply, and separa-tion of the two layers making up the septum leads to the disarticulation of the terminal part of a hypha into cylindrical segments termed arthrospores or arthroconidia (Cole & Kendrick, 1969b). Conidia of other *Galactomyces* spp. are virtually indistinguishable from those of *G. candidus*.

Galactomyces candidus may be homo- or hetero-thallic, but the sexual state is not frequently seen. After mating, fertile hyphae are produced and gametangia arise in pairs on either side of a septum, in the broad main hyphae or short side branches (Figs. 10.10c−e). Fusion of the gametangia gives rise to a globose fusion cell which becomes transformed directly into an ascus. The ascus contains only a single ascospore which has two wall layers, a smooth inner layer and a furrowed outer layer. Each ascospore contains 1−2 nuclei. Whether and when meiosis occurs is not yet known.

The *Geotrichum* arthroconidial state is found also in the only other genus of the Dipodasca-ceae, *Dipodascus*, which produces multispored asci with 4−128 spores. A superficially similar state, *Saprochaete*, is formed by a genus of phylo-genetically unrelated fungi now called *Magnusio-myces*. Species descriptions and a key of *Galactomyces*, *Dipodascus* and *Magnusiomyces* have been provided by de Hoog and Smith (2004).

10.6 | *Saccharomycopsis* (Saccharomycopsidaceae)

Saccharomycopsis (formerly *Endomycopsis*) is a myce-lial yeast which reproduces by buds (blastospores or yeast cells) and also forms asci parthenoge-netically or following isogamous fusion. About 10 species are known (Kurtzman & Smith, 1998; Barnett *et al.*, 2000). *Saccharomycopsis fibuligera* grows in flour, bread, macaroni and other starchy substrates, and produces a complex of numerous active extracellular amylases, an unusual property in yeasts (Hostinová, 2002). This has been used to develop *S. fibuligera* as a food yeast for cattle feed which can be grown on potato starch processing wastes (Jarl, 1969). This species is also used extensively for starch hydrolysis by starter cultures in Far Eastern fermented food (Beuchat, 1995).

In culture, *S. fibuligera* may form budding yeast cells and branched septate hyphae which produce blastospores laterally and terminally (Fig. 10.11). Arthrospore formation has also been demonstrated. Ascus formation in this homothallic species can be induced by growing the yeast for a few days on malt extract agar and transferring it to distilled water. The asci are mostly four-spored, and the spores are hat-shaped (Fig. 10.11d), having a flange-like extension of the wall.

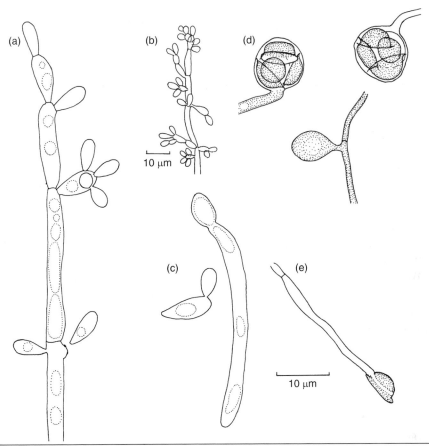

Fig 10.11 *Saccharomycopsis fibuligera.* (a,b) Mycelium from three-day-old culture showing blastospore formation. (c) Blastospores germinating by germ tube, or budding to form a further blastospore. (d) A young ascus and two mature asci containing four hat-shaped ascospores. (e) Germinating ascospore. (a,c–e) to same scale.

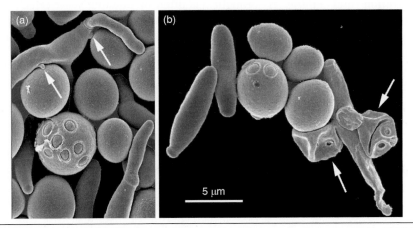

Fig 10.12 Scanning electron micrographs of *Saccharomycopsis javanensis* preying upon *Saccharomyces cerevisiae.* (a) Points of penetration (arrowheads) at an early stage. Also note the bud-scars on an older *S. cerevisiae* cell in the centre of the picture. (b) Collapsed cells of penetrated prey (arrows) at a later stage. Original images kindly provided by M.-A. Lachance. Reprinted from Lachance *et al.* (2000) by copyright permission of the National Research Council of Canada.

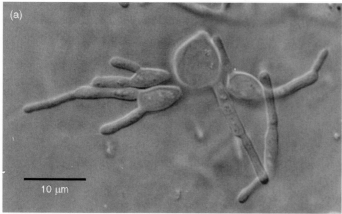

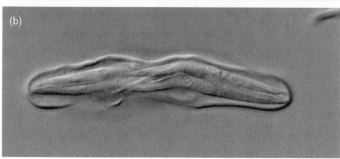

Fig 10.13 *Eremothecium coryli*. (a) Vegetative growth as a mass of hyphae and yeast cells. (b) Ascus containing eight ascospores. Both images to same scale.

Most members of the genus *Saccharomycopsis* have been observed to behave as predacious yeasts on leaf surfaces, i.e. they attack, penetrate and digest the cells of other yeasts (Fig. 10.12). This phenomenon is different from that of killer yeasts because no toxins appear to be involved. Instead, penetration and killing is brought about mainly by cell wall-degrading enzymes, notably β-(1,3)-glucanase (Lachance *et al.*, 2000). It is interesting to note that all those yeasts capable of preying on others are incapable of utilizing sulphate as a source of sulphur, although not all yeasts deficient in sulphate transport are predacious. Predation can be cannibalistic, but many other yeasts belonging to the Asco- and Basidiomycota are also attacked. Clearly, the phylloplane is a highly competitive environment.

10.7 | *Eremothecium* (Eremotheciaceae)

This genus contains five species which were formerly classified in different genera but have now been shown to be closely related by DNA sequence analyses (Kurtzman, 1995; de Hoog *et al.*, 1998). Pseudohyphae and true hyphae are present in culture, and vegetative reproduction is often by budding yeast cells. Asci are formed terminally or in intercalary positions, and they contain 8–32 needle-shaped ascospores (Fig. 10.13).

Eremothecium spp. are the only important plant pathogens among the Hemiascomycetes and can infect numerous plant species, causing damage especially on cotton (*Gossypium* spp.). They are transmitted by hemipteran insects which may harbour inoculum in their stylet pouches. The route of entry into the plant is often via the stigma of the flower (Batra, 1973). *Eremothecium* (formerly *Nematospora*) *coryli* (Fig. 10.13) causes a disease called stigmatomycosis on a wide range of plants, including hazel (*Corylus*).

In biotechnology, *E. ashbyi* and *E.* (*Ashbya*) *gossypii* are used in fermentations for the

Plectomycetes

11.1 | Introduction

The class Plectomycetes originally contained all ascomycetes which produce their asci within a cleistothecium, i.e. a 'closed case'. DNA sequence comparisons have revealed that this character was a fairly good one because, with the major exception of the powdery mildews (Erysiphales; see Chapter 13) and few scattered examples in the Pyrenomycetes (Chapter 12) and Helotiales (Chapter 15), most cleistothecium-forming fungi and the anamorphs associated with them have been found to be monophyletic (Berbee & Taylor, 1992a; Geiser & LoBuglio, 2001). As they stand now, the Plectomycetes can be defined by the following set of characters (Alexopoulos *et al.*, 1996; Geiser & LoBuglio, 2001).

(1) A cleistothecium or gymnothecium is usually present; a cleistothecium proper has a thick and continuous (pseudoparenchymatous) wall, whereas in the gymnothecium the wall consists of an open cage-like construction of hyphae, the reticuloperidium (Greif & Currah, 2003). Naked asci are produced only in rare cases.

(2) Ascogenous hyphae are usually not conspicuous.

(3) Asci are scattered throughout the cleistothecium, not produced by a fertile layer (hymenium).

(4) Asci are mostly globose and thin-walled, and the ascospores are released passively after disintegration of the ascus wall, not by active discharge.

(5) Ascospores are small, unicellular and usually spherical or ovoid.

(6) Conidia are commonly produced from phialides (in Eurotiales) or as arthroconidia, which are typically formed as chains of conidia alternating with sterile cells. An arthroconidium becomes released when the neighbouring cells disintegrate. This **rhexolytic secession** is typical of the microconidia of Onygenales and Ascosphaerales. The alternative is schizolytic secession in which adjacent cells separate when the septum joining them splits into two (see Figs. 8.9, 10.10), but this is not found in the Plectomycetes. However, terminal thick-walled chlamydospores and multicellular blastic macroconidia may be produced by some Plectomycetes.

Plectomycetes are predominantly saprotrophic fungi associated with the soil. Many have a capacity to degrade complex biopolymers, e.g. starch and cellulose, while others degrade proteins such as keratin which makes up hair, horn and feathers. If proteolytic fungi can grow at 37°C, they are potentially pathogenic to mammals, and some of them are indeed among the most dangerous fungal pathogens of man. Many other Plectomycetes produce important secondary metabolites, e.g. antibiotics and mycotoxins.

Several taxonomic arrangements have been proposed for the Plectomycetes (e.g. Kirk *et al.*, 2001; Eriksson *et al.*, 2003), but we have chosen that of Geiser and LoBuglio (2001) because of its clarity. This divides the Plectomycetes into three orders, the Ascosphaerales, Onygenales, and Eurotiales (Table 11.1). We will consider

Table 11.1. Classification of Plectomycetes following Geiser and LoBuglio (2001) and Kirk *et al.* (2001).

Order	Family	No. of taxa	Examples of teleomorphs	Examples of anamorphs
Ascosphaerales (see Section 11.2)	Ascosphaeraceae Eremascaceae	3 gen., 13 spp. 1 gen., 2 spp.	*Ascosphaera* *Eremascus*	(mostly unknown) (unknown)
Onygenales (see Section 11.3)	Onygenaceae (see p. 290)	22 gen., 57 spp.	*Ajellomyces, Auxarthron, Amauroascus, Onygena*	*Malbranchea, Chrysosporium, Coccidioides, Histoplasma, Paracoccidioides, Blastomyces*
	Arthrodermataceae (see p. 293)	2 gen., 48 spp.	*Ctenomyces, Arthroderma*	*Chrysosporium, Microsporum, Epidermophyton, Trichophyton*
	Gymnoascaceae (see p. 295)	10 gen., 23 spp.	*Gymnoascus*	(mostly unknown)
	Myxotrichaceae (see p. 295)	4 gen., 12 spp.	*Myxotrichum*	*Geomyces, Malbranchea, Oidiodendron*
Eurotiales (see Section 11.4)	Trichocomaceae (see p. 297)	20 gen., >500 spp.	*Byssochlamys, Emericella, Eupenicillium, Eurotium, Talaromyces*	*Aspergillus, Paecilomyces, Penicillium*
	Monascaceae Elaphomycetaceae (see p. 313)	2 gen., 7 spp. 1 gen., 20 spp.	*Monascus* *Elaphomyces*	*Basipetospora* (unknown)

representatives from all three orders, with a particular emphasis on the Eurotiales which contain the important anamorphic genera *Aspergillus* and *Penicillium*.

11.2 | Ascosphaerales

This small order currently comprises the 4 teleomorphic genera *Arrhenosphaera* (1 species), *Ascosphaera* (11 species), *Bettsia* (1 species) and *Eremascus* (2 species). The first three genera are associated with beehives whereas *Eremascus* is a food-spoilage fungus. All genera can grow on, and sometimes require, substrates rich in sugar

or salt. Some species are truly xerophilic, i.e. they can grow at water activities (a_W) lower than 0.85, which is equivalent to a solution containing 60% glucose. Ascosphaerales are atypical of the Plectomycetes because they do not produce true cleistothecia, but DNA-based phylogenetic studies have shown that they belong here (Berbee *et al.*, 1995).

11.2.1 *Eremascus*

In mycology as in many other areas of biology, it is virtually impossible to establish a rule without having to qualify it almost immediately by giving exceptions and modifications to it. *Eremascus* is a member of the Euascomycete clade with free asci which are not organized into ascocarps, and it is thus an exception to the generalization which places such fungi in the Archiascomycetes (Chapter 9) or Hemiascomycetes (see *Eremothecium, Galactomyces, Saccharomycopsis*; Chapter 10). Support for the inclusion of *Eremascus* in the Euascomycetes comes not only from molecular data (Berbee & Taylor, 1992a; Anderson *et al.*, 1998) but also from the presence of typical Euascomycete septa with one central pore and associated Woronin bodies (Kreger-van Rij *et al.*, 1974). Further, the arthroconidia are delimited by a double-septum (Harrold, 1950) and are released by rhexolytic secession, which is typical of certain Plectomycetes (see Fig. 11.3f).

Two species are known, *E. albus* and *E. fertilis* (Fig. 11.1). Both are associated with sugary substrates such as mouldy jam, but several collections of *E. albus* have been made from powdered mustard. Harrold (1950) has shown that both fungi grow best on media with a high sugar content (e.g. 40% sucrose), but do not grow well in a water-saturated atmosphere. The mature mycelium consists of uninucleate segments. Both species are homothallic. On either side of a septum, short gametangial branches arise which are swollen at their tips and, in the case of *E. albus*, coil around each other. The gametangial tips of *E. albus* are usually uninucleate and, following breakdown of the wall separating the tips of adjacent gametangia, nuclear fusion occurs. This is followed by meiosis and mitosis so that eight nuclei result, each one being surrounded by cytoplasm to form a uninucleate ascospore (Fig. 11.1m). The ascospores are dispersed passively following breakdown of the ascus wall. On germination a multinucleate germ tube emerges, but the uninucleate condition is soon established by the formation of septa.

11.2.2 *Ascosphaera*

Ascosphaera spp. are associated with bees and related insects, growing saprotrophically in their nests on the gathered pollen and nectar. They can be maintained in pure culture but commonly

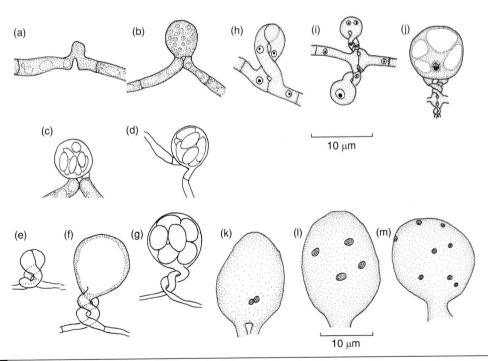

Fig 11.1 *Eremascus.* (a–d) *Eremascus fertilis*, stages in the development of asci. (e–g) *Eremascus albus*, stages in the development of asci. Note the coiling of the gametangia and the globose ascospores of *E. albus*. (h–m) *Eremascus albus*, nuclear behaviour during ascus formation (after Harrold, 1950). (h) Uninucleate gametangia. (i) Plasmogamy and karyogamy. (k–m) Nuclear divisions preceding ascospore formation.

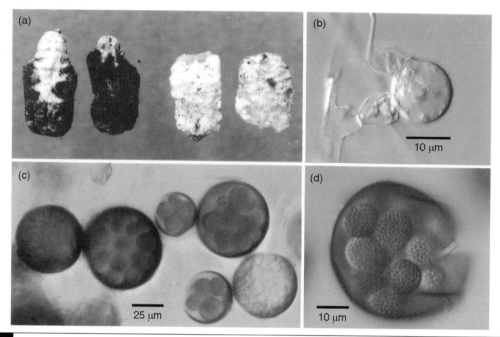

Fig 11.2 *Ascosphaera apis.* (a) Dead infected bee larvae. (b) Young sporocyst. (c) Several intact mature sporocysts. (d) Ruptured sporocyst with released ascospore balls.

require up to 40% glucose in the medium. On lower-strength agar media they may still grow but often fail to produce ascospores. *Ascosphaera apis* is a pathogenic species causing 'chalk brood' disease of honey bees (*Apis mellifera*). Dead infected larvae appear white and hard like chalk (Fig. 11.2a), and black spore balls may break through the integuments (Skou, 1972, 1975, 1988). The disease occurs as an epidemic in some years and can seriously weaken bee colonies, especially if accompanied by pests such as *Varroa* mite infestations.

Some species are homothallic but *A. apis* is heterothallic. Ascospores are produced in a unique structure termed a **sporocyst** (Skou, 1982). The 'female' colony produces an ascogonium terminating in a trichogyne, and plasmogamy occurs between the trichogyne and an undifferentiated hypha of the opposite mating type. Following plasmogamy, the trichogyne grows backwards into the ascogonium, the wall of which swells greatly to form the sporocyst (Fig. 11.2b; Spiltoir, 1955). When it is mature, the sporocyst acquires a brown pigmentation (Figs. 11.2c,d). Within the sporocyst, a system of binucleate cells with croziers forms eight-spored

asci in clusters. The ascus walls are evanescent, and the ascospores from the asci of any one cluster stick together as spore balls which are released when the sporocyst wall breaks (Figs. 11.2c,d). The sporocyst is not homologous with a cleistothecium because it arises from a single cell which enlarges prior to formation of the asci, whereas a cleistothecium is multicellular and grows around the developing asci.

In order to produce ascospores on infected bee larvae, *A. apis* requires a slight reduction of temperature (normally around 33−36°C in intact hives) to about 30°C. Infections by *A. apis* are usually most severe in cool weather, especially in spring. Interestingly, bee colonies have been found to respond to *A. apis* infections by elevating their temperature, and this so-called 'behavioural fever' may retard the outbreak of the disease (Starks *et al.*, 2000). A further way for bees to control the disease is hygiene, i.e. they uncap brood cells and remove dead larvae before *A. apis* can sporulate on them. Bees can be bred for hygiene, and the basis of this is thought to be an enhanced sensitivity to the odour of infected larvae rather than hygienic behaviour per se which is instinctive (Masterman *et al.*, 2001). Larvae become infected by *A. apis* by ingestion. Many types of commercially available honey contain viable spores of *A. apis* (Anderson *et al.*, 1997).

11.3 | Onygenales

This order of the Plectomycetes is of utmost significance to medical mycologists because it contains most of the true human pathogens, i.e. fungi able to cause disease in otherwise healthy and immunocompetent individuals. Some taxonomic confusion has arisen because many of the serious pathogens have been known for a long time only in their anamorphic form and continue to be called by their anamorphic names. The current *Dictionary of the Fungi* (Kirk *et al.*, 2001) recognizes three families – Arthrodermataceae, Gymnoascaceae and Onygenaceae – but it

excludes the family Myxotrichaceae which is of uncertain placement (*incertae sedis*), possibly belonging to the Helotiales (Tsuneda & Currah, 2004). Since members of this last family have many features in common with the other three, we will consider them briefly here. Including the Myxotrichaceae, there are some 120 species in the Onygenales.

Defining features of the Onygenales are that their ascoma consists of loosely interwoven and often thick-walled hyphae which sometimes bear complex and species-characteristic appendages (Figs. 11.3a, 11.5, 11.8, 11.9). Such a cage-like ascoma is termed **gymnothecium**, and the meshwork of hyphae making up the basket (peridium) is called **reticuloperidium**. Greif and Currah (2003) have shown that the reticuloperidium can be pierced by the stiff hairs of arthropods such as flies, and gymnothecial appendages may also be caught by the limbs of flies during grooming. Movements by the animals shake the ascospores out of the

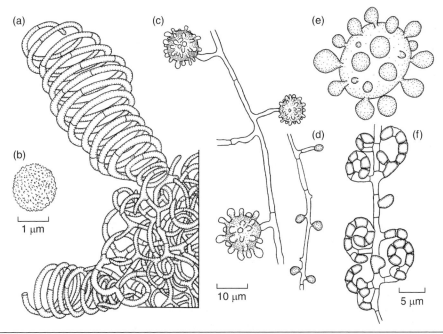

Fig 11.3 Onygenaceae. (a) Quarter-segment of a gymnothecium of *Ajellomyces capsulatus* showing coiled appendages. (b) Ascospore of *A. capsulatus*. (c) Tuberculate macroconidia of *Histoplasma capsulatum*. (d) Microconidia of *H. capsulatum*. (e) The 'pilot wheel' stage of *Paracoccidioides brasiliensis*. One giant yeast cell is producing several buds. (f) *Malbranchea*-type arthroconidia. The conidia are released by rhexolytic secession, i.e. conidia are spaced apart by sterile cells which eventually disintegrate. (a,c–e) to same scale. Redrawn and modified from de Hoog *et al.* (2000a), with kind permission of Centraalbureau voor Schimmelcultures.

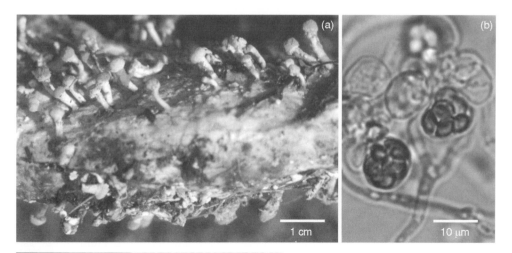

1 cm   10 μm

Fig 11.4 *Onygena*. (a) Stalked gymnothecial stromata of *O. equina* on a cast sheep's horn. (b) Ascogenous hyphae and asci of *O. corvina*.

gymnothecium and distribute them. Thus, the gymnothecium may be an adaptation to dispersal by arthropods.

The asci are formed loosely throughout the gymnothecium. Asci are eight-spored and evanescent, releasing their ascospores passively. Ascus development is similar to that in other higher ascomycetes, with the cytoplasm being delimited by two membranes between which the ascospore wall is laid down. The inner membrane eventually becomes the ascospore plasma membrane (Ito *et al.*, 1998). The anamorphic states are usually more readily seen than the teleomorph and typically consist of rhexolytic arthrospores, although thick-walled chlamydospores are also sometimes present.

Members of the Onygenales are cosmopolitan, although many individual species have a very limited distribution. Thankfully, this is true especially of many of the human pathogens. Most species, including the pathogenic ones, are soil-borne and associated with keratin-containing substrates such as hair, hooves, feathers and the dung of carnivores (Hubalek, 2000). An excellent review of the order has been compiled by Currah (1985); Geiser and LoBuglio (2001) and Sugiyama *et al.* (2002) have discussed phylogenetic aspects.

11.3.1 Onygenaceae

This family contains 22 genera and 57 species and includes the most important human pathogens. The anamorphic states are arthrosporic with rhexolytic secession (e.g. *Malbranchea*; Fig. 11.3f), or solitary terminal spores are produced which may be unicellular (*Chrysosporium*-like) or multicellular. The ascospores carry ornamentations (spines, pits or reticulations). The gymnothecia often have a few conspicuously large coiled hyphae (see Fig. 11.3a). Kwon-Chung and Bennett (1992), de Hoog *et al.* (2000a) and Sigler (2003) have given accounts of the most important pathogens; these are associated with the teleomorph genus *Ajellomyces* (Guého *et al.*, 1997), although gymnothecia are seldom formed and the species are better known by their anamorphic names. *Histoplasma capsulatum*, *Blastomyces dermatitidis* and *Paracoccidioides brasiliensis* are particularly closely related to each other, and this grouping has been given family status by Untereiner *et al.* (2004), with *Coccidioides immitis* being less closely related and retained in the Onygenaceae. We shall consider these four pathogenic species together because of their medical importance. It is not permitted to work with them in standard laboratories because they are among the handful of fungi currently listed in hazard category 3 (Kirk *et al.*, 2001). Teleomorph genera other than *Ajellomyces* are *Auxarthron*, *Amauroascus* and *Onygena*; the latter, being the type of the family, is also briefly considered (p. 293).

Onygenaceae as human pathogens

The salient features of the four important human pathogens are listed below. The points at which *C. immitis* differs from the other three are indicated.

1. All four species are probably mainly saprotrophic in the soil, having become serious pathogens mainly because of their ability to grow at 37°C, evade the human immune system, bind to human tissue, and produce proteases. Pathogenicity is probably coincidental and represents a dead end in the life cycle of these fungi because the transmission of inoculum from infected humans to the environment or to other humans is negligible (Berbee, 2001). Infection of humans occurs by inhalation of microconidia produced in the soil. These are sufficiently small to penetrate into the alveoli of the lung. There, yeast-like stages are formed which are the agents of disease. This is in contrast to *Candida albicans* where hyphae rather than yeast cells represent the invasive stage.

2. All species are dimorphic, with a temperature-dependent switch from hyphae (27°C) to yeast (37°C). In *C. immitis*, instead of producing yeast cells at 37°C, the conidium swells to produce an endospore-forming cyst or **spherule**. In all four species, however, the switch is relatively simple because the temperature shift is sufficient to trigger it. This differs from *C. albicans* in which the switch from yeast-like to hyphal growth is influenced by a complexity of environmental factors (see p. 277).

3. Pulmonary infections may take the form of influenza-like symptoms in the majority of immunocompetent patients but sometimes develop into more severe tuberculosis-like illnesses. Following initial infection, yeast cells (or endospores) can be disseminated, causing systemic mycoses in other organs. In the case of mild infections, patients may make a complete recovery and may then possess lifelong immunity. This observation raises the possibility that vaccines may be developed against these pathogens (Cox & Magee, 2004; Nosanchuk, 2005) and also against *Candida albicans* and other fungi (Magliani *et al.*, 2005).

4. The yeast cells of *P. brasiliensis*, *H. capsulatum* and *B. dermatitidis* are internalized by macrophages of their human hosts, but they have a remarkable ability to survive and even reproduce inside the lytic vacuoles by raising the intravacuolar pH and withstanding the attack of the lytic enzymes and the 'oxidative burst' created by the macrophages. Yeast cells inside macrophages represent latent inoculum which can cause relapses many years after the initial infection, especially when the host's immune system becomes weakened by other causes. Thus, these three species have been likened, in terms of their pathology, to the bacterium *Mycobacterium tuberculosis* (Borges-Walmsley *et al.*, 2002; Woods, 2002).

5. Even prolonged chemotherapy may not altogether eliminate the pathogens. The drugs in common current use are similar to those applied against *C. albicans* (p. 278) and include amphotericin B and azole-type compounds (Harrison & Levitz, 1996). Since long treatment periods are required to control these diseases effectively, the side effects of the drugs in current use are problematic. The anti-*Candida* drug caspofungin (see Fig. 10.9e) also shows promise against onygenalean pathogens (Letscher-Bru & Herbrecht, 2003).

6. Diseases caused by all four pathogens are much more prevalent in men than in women, often by a ratio of 10:1 or higher. This is due to the inhibitory effects of oestrogen and other female steroid hormones on the conidium–yeast transition (Hogan *et al.*, 1996; Aristizabal *et al.*, 1998).

Ajellomyces capsulatus (anamorph *Histoplasma capsulatum*)

Gymnothecia of this species (Fig. 11.3a) are easily recognized, with a few conspicuous coiled appendages and very small ascospores (<1.5 μm diameter) which have a minutely roughened surface visible only with the scanning electron microscope (Fig. 11.3b). The fungus is heterothallic, and the teleomorph is very rare (Kwon-Chung, 1973). The fungus produces two types of conidia in culture. Macroconidia are spherical and tuberculate, i.e. they have finger-like projections. They are borne on short lateral hyphae (Fig. 11.3c). Microconidia are sessile, globose and smooth-walled (Fig. 11.3d).

Histoplasma capsulatum grows as a mycelium at room temperature, but at 37°C it develops small budding yeast cells (3–5 μm) which can spread throughout the patient under favourable conditions. The normal route of infection is by inhalation, especially of the small microconidia. One common way to become infected is by cleaning buildings of bird or bat excreta, or by exploring caves in which these animals dwell; histoplasmosis is therefore also called the 'spelunker's disease' (Woods, 2002). Three strains of *H. capsulatum* are distinguished, and they are present in soil, especially when contaminated with bird or bat guano. Endemic areas are North America (var. *capsulatum*) or equatorial Africa (var. *duboisii*). *Histoplasma capsulatum* var. *farciminosum* infects horses sporadically in Africa, Asia and Eastern Europe (Weeks *et al.*, 1985).

Ajellomyces dermatitidis (anamorph *Blastomyces dermatitidis*)

The gymnothecia of *A. dermatitidis* are similar to those of *A. capsulatus*, but asexual reproduction is by means of stalked or sessile conidia which are smooth or spiny, but not tuberculate. Yeast cells are formed at 37°C, and these are much larger (10–12 μm in diameter) and have a thicker wall than those of *H. capsulatum*. The disease (blastomycosis) starts as an infection of the lung which can spread systemically to other sites, especially skin and bones. As in *H. capsulatum*, considerable efforts are currently being made to characterize the cell surface properties in *B. dermatitidis* (Hogan *et al.*, 1996; Brandhorst *et al.*, 2002). Crucial roles are probably played by the presence or absence of α-(1,3)-glucan in the cell wall, and by surface adhesins which are proteins that mediate the recognition of yeast cells by the host's immune system.

The fungus is soil-borne, especially in moist soil such as the banks of rivers. It is of North American origin, being particularly prevalent in the Mississippi and Ohio river valleys. Isolates from Africa represent a genetically different subpopulation (Guého *et al.*, 1997).

Paracoccidioides brasiliensis

No teleomorph has been described as yet, but DNA sequence data predict that it will be an *Ajellomyces* if it is found (Guého *et al.*, 1997). Conidia (arthroconidia) are seldom formed in culture, but the fungus is readily recognizable in the yeast state at 37°C because it produces a large, thick-walled central cell (30 μm in diameter) from which several smaller daughter cells bud off, often several at the same time. This type of budding is therefore called the 'pilot-wheel stage' (Fig. 11.3e). The first organ affected by paracoccidioidomycosis is the lung, but the infection may spread, causing grossly deforming lesions on mouth, nose and in gastrointestinal regions. The fungus is thought to occur in forest soils in areas with heavy rainfall in Central and South America. As with the other serious pathogens described in this section, the precise ecological niche of *P. brasiliensis* is still obscure, but it is possibly spread by the nine-banded armadillo, *Dasypus novemcinctus* (Restrepo *et al.*, 2001).

Coccidioides immitis

In culture, *C. immitis* reproduces by *Malbranchea*-type arthroconidia (Fig. 11.3f). In infected tissues, arthrospores swell and give rise to large, thick-walled spherical cysts or spherules (50–100 μm in diameter) which produce endospores. Endospores are very small (3 × 4 μm) and are readily disseminated in the bloodstream. Each endospore can develop into a new spherule. The disease (coccidioidomycosis) can be benign with influenza-like symptoms, but infections of the lung may spread to other organs such as the skin, brain and bones. Coccidioidomycosis can be fatal even to immunocompetent humans (Dixon, 2001).

DNA sequence comparisons indicate a relationship with the teleomorphic genus *Uncinocarpus* (Sigler *et al.*, 1998). Further, detailed studies of the distribution of specific DNA sequences in the genomes of isolates from various patients have revealed that sexual recombination must occur in nature, even though no teleomorph is known (Burt *et al.*, 1996). *Coccidioides immitis* is a soil-borne fungus present as arthrospores, especially in arid regions of the south-western United States and in localized places in South America. It can infect wild mammals, but outbreaks occur especially among farmers and building workers, and after dust storms, earthquakes and other events that disturb the soil. One hot spot of infection is the San Joaquín Valley in Southern California, where coccidioidomycosis is known as 'valley fever'. The Californian population of *C. immitis* is reproductively isolated from populations elsewhere and has recently been given the status of a separate species, *C. posadasii* (Fisher *et al.*, 2002). A readable account of the history of *C. immitis* has been given by Odds (2003), and Cox and Magee (2004) have covered general aspects of its interactions with the mammalian host.

Onygena

This is perhaps the most unusual yet least researched member of the Onygenaceae. Its fructification (Fig. 11.4a) is interpreted as a stalked ascostroma, i.e. an aggregate of several gymnothecia at the tip of a sterile stalk which may be 1 cm or more in length. The stalks are phototropic during growth. At maturity, the peridium of the stroma ruptures, thereby exposing the ascospores. There are two species, *O. corvina* which is associated with animal hair and bird feathers, and *O. equina* growing on the hooves and horns of herbivorous mammals (Currah, 1985). Both species are strongly keratinolytic, and although little work has been published on their biological features, we can assume that keratinolysis proceeds as in other Onygenales. According to Kunert (2000), the key feature is the ability to use keratin as the sole source of both carbon and nitrogen. The vast surplus of nitrogen is released into the environment as ammonia, thereby generating an alkaline pH of 9.0 or higher. The cystein-rich keratin also contains sulphur in excess of the growth requirements of keratinolytic fungi. This is often released as sulphate, thereby buffering the pH increase caused by the release of ammonia.

Like many protein-degrading fungi, *Onygena* produces a cadaverous smell in culture, and Currah (1985) has suggested that this might attract carrion flies if produced in nature. This would make *Onygena* an insect-dispersed fungus.

11.3.2 Arthrodermataceae

There are only two genera in this small family. *Ctenomyces* (Fig. 11.5), with only one species

(*C. serratus*), has a *Chrysosporium* anamorph in which one-celled microconidia are formed as terminal or intercalary cells of hyphae. Species of *Arthroderma* also sometimes produce *Chrysosporium*-like microconidia (Fig. 11.6b) but are better known by their macroconidial synanamorphs *Epidermophyton*, *Microsporum* and *Trichophyton*. These are multicellular with transverse septa, and are spindle-shaped or cylindrical. They are typical and readily recognized (Fig. 11.6a). The perfect state has not been found for many of these anamorphs, but they are suspected to be phylogenetically close to *Arthroderma* (Gräser *et al.*, 1999; Hirai *et al.*, 2003). Howard *et al.* (2003) have given a useful summary of *Arthroderma* and its associated anamorphs, listing 47 species. The teleomorph of the Arthrodermataceae is a typical gymnothecium with a basket of branching and anastomosing hyphae enclosing spherical asci which release their spores passively. Characteristic appendages are often present, e.g. in *Ctenomyces* (Fig. 11.5), a keratinolytic species associated with feathers. The combed appendages may serve to attach the gymnothecium to bird feathers for dispersal (Currah, 1985).

Members of the Arthrodermataceae are generally keratinolytic, i.e. they degrade skin and hair. Howard *et al.* (2003) distinguished between species primarily associated with man (anthropophilic), animals (zoophilic) or the soil (geophilic). Because of their ability to grow on the skin, hair and nails of animals, the Arthrodermataceae are collectively called **dermatophytes**. Diseases caused by dermatophytes are colloquially known as 'ringworm', whereas they are called *tinea* within the medical profession, with descriptive terms such as *capitis*, *barbae*, *corporis* and *pedis* added to describe mycoses of the scalp, beard, general body, or feet, respectively (Howard *et al.*, 2003). These infections are usually confined to the outer (dead) skin regions and are relatively easily controlled either by the superficial (topical) application of creams containing a wide variety of drugs, or by oral treatment especially with triazoles or terbinafine (see p. 278; Weitzman & Summerbell, 1995; Gupta *et al.*, 1998). Griseofulvin, produced by *Penicillium griseofulvum* (see p. 302), was one of the first oral and topical drugs and is still in use today, especially in

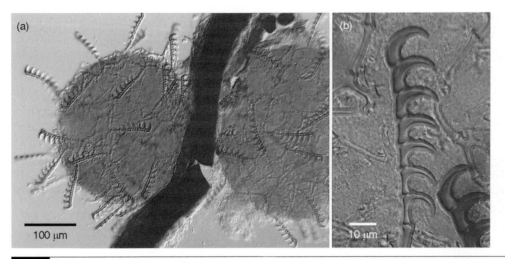

Fig 11.5 *Ctenomyces serratus*. (a) Gymnothecia attached to a horse hair with which the fungus was baited. (b) Gymnothecial appendage. Note the thick walls. Micrographs taken from material kindly provided by R. Sharma.

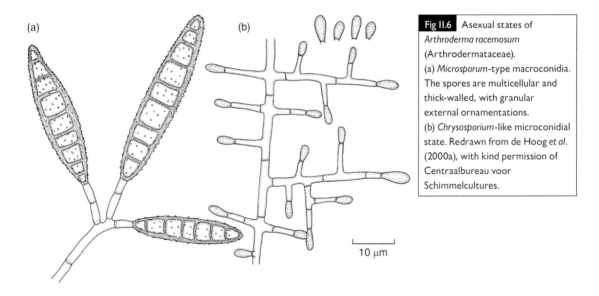

Fig 11.6 Asexual states of *Arthroderma racemosum* (Arthrodermataceae).
(a) *Microsporum*-type macroconidia. The spores are multicellular and thick-walled, with granular external ornamentations.
(b) *Chrysosporium*-like microconidial state. Redrawn from de Hoog *et al.* (2000a), with kind permission of Centraalbureau voor Schimmelcultures.

the treatment of dermatomycoses in children, although resistance has developed among some strains of dermatophytes. Transmission of *tinea* infections to humans is often from pets, especially dogs, but also herbivores such as cows and horses. One species particularly common on cows is *T. verrucosum* (Fig. 11.7). The main transmission route for athlete's foot and infections of toenails, both caused mainly by *T. mentagrophytes* and *T. rubrum*, is the use of communal facilities such as swimming pools and dressing rooms. Extensive listings of the

various skin mycoses and the species causing them have been compiled by Weitzman and Summerbell (1995) and Howard *et al.* (2003).

Members of the Arthrodermataceae are of historical interest because they were among the first fungi recognized as causing disease. In 1842, Robert Remak succeeded in experimentally infecting himself with a fungus now known as *T. schoenleinii* (Ainsworth, 1976). Another milestone was laid by Raymond Sabouraud who pioneered the identification of dermatophytes on the basis of their appearance on

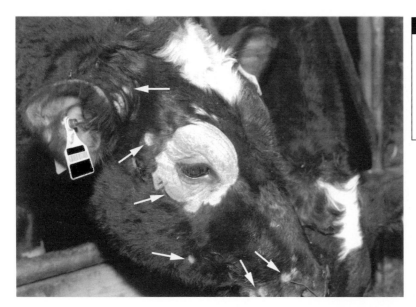

Fig 11.7 Friesian cattle showing typical symptoms of *Trichophyton verrucosum* infections. Lesions are indicated by arrows. The circular lesion around the eye is particularly large because the animal has spread the infection by rubbing.

culture media, in addition to the clinical symptoms (see Weitzman & Summerbell, 1995). All Arthrodermataceae appear to be heterothallic. Different strains of *A. simii* can induce enhanced growth but not complete mating and gymnothecium formation in strains of opposite mating type of a range of other species (Stockdale, 1968). Using this feature as a test system, it has been found for most anthropophilic species that all known isolates are of only one mating type, e.g. (−) for *T. rubrum*. Since the *Arthroderma* state of dermatophytes is found only on substrates in contact with soil (Summerbell, 2000), the lack of sexual reproduction in most anthropophilic species may be the result of adaptation to a highly patchy and specialized habitat (Howard *et al.*, 2003).

Geophilic members of the Arthrodermataceae are best isolated from soil or other substrata enriched in hair, skin or feathers (e.g. rodent burrows or birds' nests) using horse hair or human hair as bait. Gymnothecia can be picked up with the aid of a dissection microscope, and transferred to a suitable medium such as Sabouraud agar (Sharma *et al.*, 2002).

11.3.3 Gymnoascaceae

The Gymnoascaceae are a small family (10 genera, 23 species; Kirk *et al.*, 2001). A recent phylogenetic study is that by Sugiyama *et al.* (2002). Members of the Gymnoascaceae are isolated mainly from soil and are saprotrophic, degrading keratin and also cellulose. The genus *Gymnoascus* has been described by von Arx (1986). *Gymnoascus reessii* is a species commonly encountered on herbivore dung, producing strikingly coloured gymnothecia which are at first yellow, then red and finally brown as the ascospores mature. The reticulo peridium consists of branched, recurved, thick-walled hyphae loosely enclosing a mass of asci (Fig. 11.8). Ascocarp development can be followed readily in culture and begins from paired gametangia which arise from the same or different hyphae. The antheridium is club-shaped and the ascogonium coils around it (Fig. 11.9a). The ascogonium then becomes septate and its cells give rise to ascogenous hyphae (see Fig. 11.9b), whose tips develop into croziers. Asci develop from the penultimate cells of the croziers (Kuehn, 1956). The branched reticulo peridial hyphae arise from vegetative hyphae in the region of the gametangium (Figs. 11.9b,c). The asci do not discharge violently; the ascus wall disappears and the spores escape through the loose envelope. There is no conidial stage.

11.3.4 Myxotrichaceae

This family contains 4 genera (12 species). Anamorphs are *Oidiodendron* or *Malbranchea*. The genus *Oidiodendron* may have important

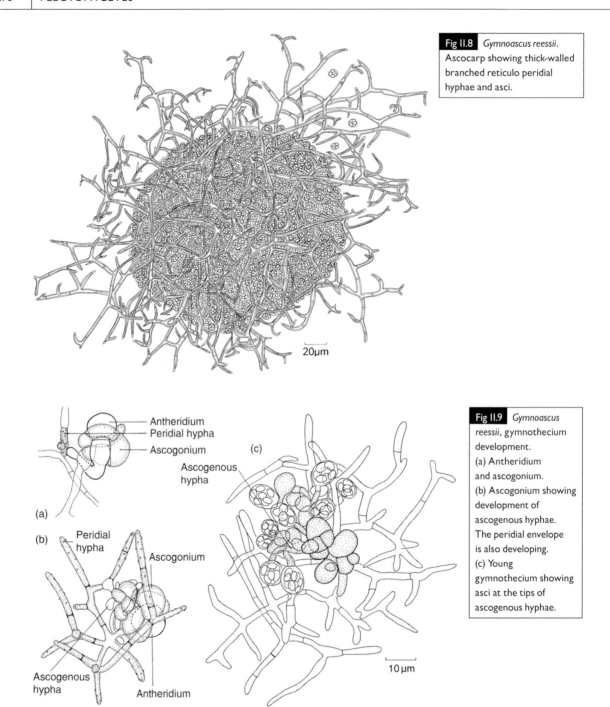

Fig 11.8 *Gymnoascus reessii.* Ascocarp showing thick-walled branched reticulo peridial hyphae and asci.

20μm

Antheridium
Peridial hypha
Ascogonium
Ascogenous hypha
(a)
(b)
Peridial hypha
Ascogonium
(c)
Ascogenous hypha
Antheridium

Fig 11.9 *Gymnoascus reessii*, gymnothecium development. (a) Antheridium and ascogonium. (b) Ascogonium showing development of ascogenous hyphae. The peridial envelope is also developing. (c) Young gymnothecium showing asci at the tips of ascogenous hyphae.

10 μm

functions in soil ecosystems, and keys and species descriptions have been provided by Calduch *et al.* (2004) and Rice and Currah (2005). *Myxotrichum chartarum* is cellulolytic and has been known as a paper spoilage organism since the early nineteenth century. Gymnothecia form a thick chocolate-brown layer on paper or cardboard (Fig. 11.10a) and are also readily formed in culture. They are typical of the Onygenales, consisting of a reticuloperidium of anastomosing hyphae with beautiful curved appendages (Fig. 11.10b). Other *Myxotrichum* spp.

have gymnothecia with simpler appendages, or these lack appendages altogether (Currah, 1985).

The taxonomic position of Myxotrichaceae is uncertain, recent phylogenetic studies suggesting an affinity with inoperculate discomycetes (Helotiales; Tsuneda & Currah, 2004). Support for this proposal comes from ecological surveys in which *Oidiodendron* spp. as well as members of the Helotiales have been isolated as mycorrhizal symbionts of ericaceous plants (Berch *et al.*, 2002; Peterson *et al.*, 2004; see also p. 442). Since both the teleomorphs and anamorphs of Myxotrichaceae are clearly referable to the Onygenales, a connection with the Helotiales would be surprising, rendering the concept of a gymnothecium with evanescent asci one of the most striking examples of convergent evolution among the fungi (Greif & Currah, 2003).

11.4 | Eurotiales

To the practical mycology student, the order Eurotiales is among the most important groups of fungi because it contains many ubiquitous and readily recognized species, notably in the anamorphic genera *Aspergillus* and *Penicillium*. Virtually any environmental sample – soil, water, rhizosphere, air, and indoor or food contaminations – will yield viable spores. One important feature of many species of *Aspergillus* and *Penicillium* is that

(a)

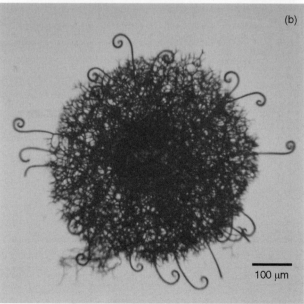

(b)

Fig 11.10 *Myxotrichum chartarum.* (a) Thick layer of gymnothecia on cardboard from a damp cellar. (b) Gymnothecium with dark curved appendages. Reprinted from Tribe and Weber (2002), with permission from Elsevier.

they are xerophilic, i.e. capable of growing at a water potential (a_W) at or below 0.85. Thus, they are major food spoilage organisms, growing on stored cereals, spices, nuts, bread, dried and cured ham, pickles, jams and preserves (Lacey, 1994; Filtenborg *et al.*, 2002). Colonization of food and feedstuff can result in its contamination by serious mycotoxins (see pp. 304–306).

Colonies typically spread slowly but quickly assume a greenish-blue pigmentation due to abundant conidium formation. The pigmentation of mature conidia is at least partly due to melanin (see Fig. 12.46; Plate 4d). The conidial state is more commonly observed than the teleomorph, and indeed many species have lost their capacity of sexual reproduction altogether (Geiser *et al.*, 1996). Loss of sexual reproduction seems to have occurred independently on several occasions within the Eurotiales. When present, the ascocarps range from gymnothecial structures with loose mesh-like reticuloperidia (e.g. *Talaromyces*) to the hard sclerotium-like fructifications of certain species of *Eupenicillium*. The conidial states are generally phialidic. Microscopically,

the conidiophores of *Aspergillus* and *Penicillium* can be categorized according to the arrangement of phialides on the conidiophore (Fig. 11.11). The phylogenetic value of these structures may be limited, but they are very useful for species identification. In *Aspergillus* (Fig. 11.11a), the conidiophore tip is swollen into a hemispherical or club-shaped structure, the **vesicle**. Phialides may be formed directly at the vesicle surface in which case the conidiophore is said to be **uniseriate** (Figs. 11.11a, 11.16a; Raper & Fennell, 1965). Alternatively, a palisade of sterile cells (**metulae**) is formed by the vesicle, and the tips of the metulae give rise to the phialides (**biseriate** conidiophores; Fig. 11.17a). No vesicle is produced in *Penicillium* (Figs. 11.11b and 11.18), and instead the conidiophore tip either gives rise to phialides directly in the **monoverticillate** arrangement (Pitt, 1979), or it produces one series of metulae (**biverticillate**) or further branching layers. In **terverticillate** penicilli, the conidiophore tip produces one or several **rami**, each of which develops several metulae which in turn produce several phialides each. In **quaterverticillate**

10 μm

Fig 11.11 Examples of conidiophores of three important anamorphic genera of the Trichocomaceae. (a) *Aspergillus penicillioides*. (b) *Penicillium notatum* (= *P. chrysogenum*), the original penicillin-producing strain isolated by Sir Alexander Fleming in 1928. (c) *Paecilomyces marquandii*. All images to same scale.

penicilli, a further branch, the **ramulus**, is inserted between the ramus and the metula. *Paecilomyces*, another important anamorphic genus in the Eurotiales, has conidiophores similar to those of *Penicillium*, except that the phialides are more loosely arranged and have a different shape, being more elongated with a narrow drawn-out tip (Fig. 11.11c). Conidia of *Paecilomyces* are usually pale rather than pigmented green or blue.

The order Eurotiales is thought by some to consist of only a single family, Trichocomaceae, and these two terms are often used interchangeably. However, Geiser and LoBuglio (2001) also included the Monascaceae (not discussed further here) and Elaphomycetaceae (see p. 313). Many species of the Trichocomaceae are much better known by their anamorphic names, and these will continue to be used by most mycologists, especially those working in applied fields. Correlations of anamorphic and teleomorphic taxa are given in Table 11.2 (Pitt *et al.*, 2000). These data show that a teleomorph has been found only for about 40% of all *Aspergillus* species described to date, and 31% of *Penicillium* spp. The taxonomy of the Eurotiales is still in a considerable state of confusion because it is not possible unequivocally to correlate the various anamorphs with the appearance of cleistothecia and other features such as DNA sequence data and biochemical features (Ogawa *et al.*, 1997; Ogawa & Sugiyama, 2000).

11.4.1 Aspects of morphogenesis in *Aspergillus*

To recapitulate on our discussion of conidiogenesis on p. 235, a phialide, according to Kendrick (1971), is:

a conidiogenous cell in which at least the first conidium initial is produced within an apical extension of the cell, but is liberated sooner or later by the rupture or dissolution of the upper wall of the parent cell. Thereafter, from a fixed conidiogenous locus, a basipetal succession of enteroblastic conidia is produced, each clad in a newly laid-down wall to which the wall of the conidiogenous cell does not contribute... The length of the phialide does not change during the production of a succession of conidia...

The development of phialoconidia in *Aspergillus niger* is illustrated in Fig. 11.12. Young phialides are somewhat club-shaped in outline. In *A. niger*, *A. nidulans* and many other species of *Aspergillus* and *Penicillium*, the phialoconidia are uninucleate, but in some species they are multinucleate. The tip of the phialide expands to form a spherical knob which is the initial of the first-formed spore. Meanwhile, the single nucleus in the phialide divides, and a daughter nucleus passes into the spore which begins to be separated by the formation of a septum at the phialide tip. The expansion of the first conidium leads to the rupture of the phialide wall near its tip, and the remnants of the broken phialide wall persist as a cap around the first-formed conidium. Before the rupture of the phialide wall, a layer of cell wall material is laid down (Fig. 11.12d). This layer becomes the outer wall

Table 11.2. | A summary of some anamorphic states found in the Trichocomaceae, and their associated teleomorphs.

Anamorphic name	Teleomorphic name
Aspergillus (218)	*Chaetosartorya* (3)
	Emericella (33)
	Eurotium (24)
	Fennellia (3)
	Hemicarpenteles (2)
	Neosartorya (19)
	Petromyces (3)
	Sclerocleista (2)
Geosmithia (10)	**Talaromyces** (3)
Paecilomyces (48)	*Byssochlamys* (4)
	Talaromyces (3)
	Thermoascus (4)
Penicillium (249)	**Eupenicillium** (44)
	Talaromyces (33)

For each taxon, the numbers of species are indicated in brackets. Numerically important genera are highlighted in bold. Note that only a few of the 48 *Paecilomyces* species are referable to the Trichocomaceae, many others belong to Pyrenomycetes (see p. 360). Summarized from Pitt *et al.* (2000) and Samson (2000).

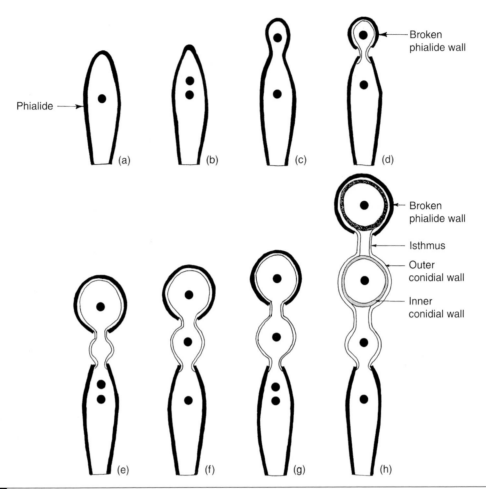

Phialide

Broken phialide wall

Broken phialide wall

Isthmus

Outer conidial wall

Inner conidial wall

(a) (b) (c) (d) (e) (f) (g) (h)

Fig 11.12 Phialoconidium ontogeny in *Aspergillus niger* (modified from Subramanian, 1971). (a) Young phialide. (b) Mitosis in phialide. (c) Conidium initial with daughter nucleus. (d) Breakage of phialide wall and formation of a new wall layer surrounding the conidial cytoplasm. (e,f) Mitosis in the phialide and extrusion of the second phialospore. Note that the phialide has not increased in length. (g,h) Further mitosis in the phialide and formation of the third spore. Note the formation of an inner conidial wall within the outer, and the isthmus connecting mature adjacent spores within the chain.

layer of the conidium within which an inner wall develops later (Fig. 11.12h). Nuclear division continues within the phialide, and cytoplasm and a wall are laid down around a daughter nucleus to form a second conidium which is extruded from the broken tip of the phialide (Figs. 11.12e,f). The second and all subsequent conidia differ from the first in that they are not enveloped by remnants of the broken phialide wall. The cytoplasm of the second conidium is initially continuous with that of the first by means of a cylindrical **isthmus**. The formation of an inner wall layer by the conidia severs this cytoplasmic connection. The surviving empty isthmus is sometimes termed the **connective**. Each phialide can produce 100 spores or more, so that the total crop from one conidiophore may be more than 10 000 conidia (Adams *et al.*, 1998). The fine structural details of conidium production in *A. nidulans* have been described by Mims *et al.* (1988).

In the phialide, a transition from the polarized growth pattern of the mycelial hypha, conidiophore and phialide to yeast-like (isodiametric) growth of the nascent conidium takes place. Results obtained from experiments with the biseriate species *Aspergillus* (*Emericella*) *nidulans* can be compared with those summarized earlier for the two 'model yeasts', *Schizosaccharomyces pombe* (pp. 256–259) and *Saccharomyces cerevisiae* (p. 270). While the cell cycle (nuclear division) is always followed by cell division (cytokinesis) in these yeasts, this is not necessarily the case in *A. nidulans*. For instance, when a uninucleate conidium of *A. nidulans* germinates, several mitotic divisions take place prior to and during the emergence of the germ tube before the first septum is laid down near the base of the germ tube. Subsequent septa are formed at regular intervals along the hypha, and each hyphal segment contains about 3–4 nuclei (Fiddy & Trinci, 1976). Only the nuclei in the apical cell continue to divide whereas those in intercalary compartments are arrested at the G1 stage but become reactivated later if a lateral branch is formed (Kaminskyj & Hamer, 1998). As the mycelium develops and spreads horizontally on an agar surface, conidiophores begin to grow out vertically after about 16 h in optimal conditions, with the first conidia being produced about 8 h later. Thus the time span from the germination of a conidium to the production of the new crop of conidia is only 24 h. It is not surprising that *Aspergillus* and *Penicillium*-type moulds are omnipresent in our environment!

The development of the conidiophore tip has been described in detail by Mims *et al.* (1988) and is shown in Fig. 11.13. After vertical growth for a limited distance (about 100 μm in *A. nidulans*), the tip of the conidiophore swells to form the vesicle (Fig. 11.13a). In the conidiophore, cytokinesis is suppressed whereas nuclear division continues, resulting in a multinucleate cell. In contrast, nuclear division and cytokinesis are tightly controlled beyond the vesicle stage because a single nucleus migrates into each of the 60 or so metulae which are formed by the vesicle of *A. nidulans* (Fig. 11.13b). Each metula then produces about two uninucleate phialides (Fig. 11.13c). At the phialide tip, conidia are formed (Fig. 11.13d), and here the transition from polarized to yeast-like growth takes place. The nucleus divides as the conidium enlarges, and one daughter nucleus migrates along microtubules into the spore whereas the other remains

in the phialide, ready to divide again. The signalling events which co-ordinate the interactions between the nuclear division cycle and cytokinesis are beginning to be unravelled (Ye *et al.*, 1999).

As in the case of *S. cerevisiae* and *S. pombe*, septin-type proteins play a crucial role in organizing morphogenesis in *A. nidulans* and most probably other members of the Eurotiales, although detailed studies are lacking. In *A. nidulans*, septin rings with superimposed constricting actin rings are found at the sites of septum formation, the points of origin of metulae at the vesicle surface, the point of origin of phialides, and at the phialide tip where conidia are budded off (Momany & Hamer, 1997; Westfall & Momany, 2002). All rings except for the last-mentioned are transient, disappearing as soon as the septum is complete.

11.4.2 Morphogenesis in *Penicillium*

Little work has been carried out on conidiogenesis in *Penicillium*, although it is reasonable to assume that the general principles will be similar to those described above for *Aspergillus* (Borneman *et al.*, 2000). One interesting case is *P. marneffei*, which is the only species displaying a switch from hyphae to yeast cells, growing as a mycelium at 25°C and as yeast cells at 37°C. Upon transfer of a mycelial colony to 37°C, nuclear division becomes tightly coupled with cytokinesis so that uninucleate hyphal segments are formed which fragment into arthrospores (Garrison & Boyd, 1973; Andrianopoulos, 2002). These form yeast cells which reproduce by fission at 37°C. Interestingly, the same regulatory mechanism is responsible for phialidic conidiogenesis at 25°C and for the switch from multinucleate hyphae to uninucleate yeast cells (Borneman *et al.*, 2000), supporting the notion that conidiogenesis by phialides should be regarded as yeast-like growth.

In *Penicillium cyclopium*, conidiophore formation is thought to be induced by a hormone called conidiogenone, which is permanently produced by growing colonies. When the concentration of conidiogenone exceeds a particular threshold (10^{-7}–10^{-8} M), conidiogenesis, i.e. growth and differentiation of the conidiophore, is triggered (Roncal *et al.*, 2002). For conidiogenesis to occur, hyphae must usually be exposed to air. This may

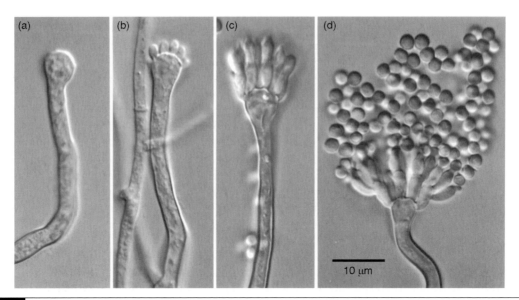

(a) (b) (c) (d)

10 μm

Fig 11.13 Conidiophore development in *Aspergillus nidulans*. (a) Tip of a conidiophore which has swollen to produce a vesicle. The conidiophore is multinucleate (nuclei not visible). (b) Development of metulae. Each metula contains one nucleus. (c) Production of phialides, two from each metula. Each phialide contains one nucleus. (d) Production of uninucleate conidia. All images to same scale.

be because the hormone is concentrated in the cell wall, rather than diffusing into the medium. In liquid cultures, conidiogenesis in *P. cyclopium* can be triggered by the addition of Ca^{2+} ions which are thought to act merely by enhancing the sensitivity of the fungus to its own hormone (Roncal *et al.*, 2002). It is not known whether a similar hormonal system exists in *Aspergillus*.

11.4.3 The roles of *Aspergillus* and *Penicillium* in biotechnology

Species of *Aspergillus* and *Penicillium* are among the most important organisms used in biotechnology, second only to *S. cerevisiae*. Their applications are diverse, including the production of enzymes or primary and secondary metabolites, and direct colonization and modification of foodstuff. We will briefly consider examples of each of these applications.

Food production

Species of *Aspergillus* have been used in the Far East for food production for many centuries, and we have already mentioned the role of *A. oryzae* in the degradation of rice starch as a first step in saké production (see p. 276). Similar two-stage fermentation processes were developed for soy

sauces, although in this case the raw material consists of a mixture of soy beans and wheat, and proteolytic as well as amylolytic enzymes are relevant. The degradation of this substrate is called the koji process, and it is one of the best examples of a solid-substrate fermentation. As such, it requires great skill to find the optimum moisture level of the soybean cake because excessive moisture will limit aeration, whereas low moisture limits growth. The koji fermentation is completed within 72 h, and it utilizes mainly *A. oryzae* and *A. sojae*. Interestingly, on the basis of morphological as well as molecular data (Geiser *et al.*, 2000), these appear to be domesticated forms of the potent mycotoxin producers *A. flavus* and *A. parasiticus*, respectively (see p. 304). The partially degraded substrate enriched in fungal extracellular enzymes is then suspended in brine, and the main fermentation (moromi) is carried out with a consortium of bacteria and yeasts. The *Aspergillus* enzymes continue to be active during the moromi fermentation, thus releasing a steady supply of degradation products. A readable account of soy sauce production is that by Aidoo *et al.* (1994). Other Far Eastern food types produced with the aid of *Aspergillus* spp. have been summarized by Nout and Aidoo (2002).

Cheese production involves complex consortia of bacteria and numerous different yeast species, but two *Penicillium* spp. are important in specialized cheeses. These are *P. roqueforti* for blue-veined cheeses, and *P. camemberti* for white mould cheeses. The subject has been summarized by Jakobsen *et al.* (2002). The moulds are inoculated together with enzymes or bacterial starters after the cooling of the pasteurized milk, and gradually colonize the maturing cheese. They contribute significantly to the texture as well as the flavours, with the characteristic blue cheese or camembert flavours being due mainly to the activity of extracellular lipases which break down short-chain fatty acids. A substantial contribution is also made by protein and peptide degradation products resulting from the activities of fungal proteinases and peptidases (Jakobsen *et al.*, 2002).

Production of enzymes

The prominent role of *Aspergillus* and *Penicillium* species in food production is, of course, due to their ability to produce large quantities of extracellular enzymes. This feature has also been harnessed for industrial purposes, with the majority of all commercial fungal enzymes produced by *Aspergillus* spp. (Oxenbøll, 1994). Proteases, amylases, lipases and pectinases are important in many industrial processes, including the manufacture of dairy, bakery, distillery and brewery products, juices and leather, and in the starch industry.

Citric acid fermentation

Citric acid is found in many fruits, and it is used for flavouring and pH control of food and beverages. In combination with carbonates and bicarbonates, it is also used to create the effervescent effect when medications such as vitamin preparations or aspirin are dissolved in water. Initially extracted from citrus fruits, citric acid has been produced commercially by *Aspergillus niger* since about 1923 and this fungus remains the world's most important producer. The total current annual world production of citric acid is about 9×10^6 tons. Curiously, although production of citric acid by *A. niger* is one of the most efficient biotechnological fermentations with conversion of up to 95% (by weight) of the sugar substrate, the biochemistry of it is still only poorly understood. The uptake of sugar (as hexose) is followed by glycolysis in the cytosol, the tricarboxylic acid cycle in the mitochondrion, and export of citric acid into the cytosol and thence into the extracellular medium where it accumulates, creating a pH below 3. Citric acid production proceeds optimally when an excess of sugar and aeration is provided, whereas phosphate and trace elements, especially manganese, must be limiting. Excellent reviews of citric acid production have been published by Brooke (1994) and Karaffa and Kubicek (2003).

Production of antibiotics

The accidental discovery of penicillin by a contamination of *Penicillium notatum* growing on a bacterial agar culture (Fleming, 1929, 1944) followed by the re-discovery of penicillin by Florey and Chain and its development into an antibiotic against Gram-negative bacteria has been told many times, and the original 1945 Nobel Lectures by Fleming, Florey and Chain can be found at http://www.nobel.se. In nature, penicillin (Fig. 11.14a) is produced by *P. notatum*, the closely related or identical *P. chrysogenum*, by *A. nidulans* and a few other conidial fungi, whereas the chemically related cephalosporins are produced by *Acremonium chrysogenum* (formerly *Cephalosporium chrysogenum*), which probably belongs to the Pyrenomycetes (p. 348). Together, penicillin- and cephalosporin-type antibiotics take a staggering 50% share (approximately 11 billion US$) of the total worldwide sales of antibiotics (Schmidt, 2002). *Aspergillus nidulans* has been useful for studies of the genetics and biosynthesis of penicillin production because it is easily manipulated by molecular biological methods. However, for commercial production *P. chrysogenum* has been used traditionally. The first penicillins were produced commercially by purification of the final product from static liquid cultures. Over several decades, high-producing mutants were generated, resulting in a 50 000-fold enhanced penicillin yield relative to that of the original strain, and current yields are as high as 50 g penicillin l^{-1} of liquid culture (Schmidt, 2002). A wide range of penicillin (and cephalosporin) derivatives has been produced,

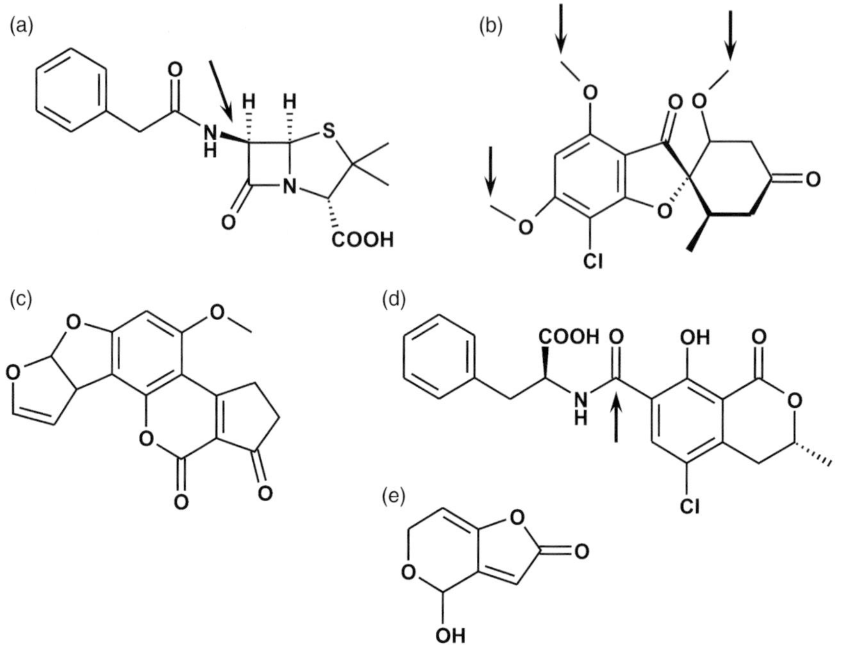

Fig 11.14 Important metabolites produced by Trichocomaceae. (a) Penicillin G, an antibiotic against Gram-positive bacteria which is synthesized from three amino acids. The cleavage point of penicillin acylases is indicated by an arrow. (b) Griseofulvin, an antifungal antibiotic which is synthesized as a heptaketide, with three methyl groups (arrows) added subsequently by methylation. (c) The polyketide aflatoxin B$_1$. (d) Ochratoxin A. This is a pentaketide to which the amino acid phenylalanine is linked via a previously added one-carbon group (arrow). (e) The derived tetraketide patulin.

partly to counter bacterial resistance and partly to broaden the range of applications or reduce allergic responses by patients. Current production seems to be mainly semi-synthetic; penicillin G is produced by *P. chrysogenum*, followed by the removal of the side-chain to give 6-aminopenicillanic acid, which is then derivatized chemically. There are tendencies to use microbial enzymes (penicillin acylases) for the removal of the side chain (see Fig. 11.14a) and for subsequent synthetic steps (Arroyo *et al.*, 2003; Bruggink *et al.*, 2003). Good general reviews of the history of penicillin biotechnology have been written by Rolinson (1998) and Demain and Elander (1999). The biosynthesis of penicillins and cephalosporins in fungi has been described in detail by Martin *et al.* (1997).

Another important secondary metabolite of the Trichocomaceae is griseofulvin (Fig. 11.14b), which is used as a systemic antifungal drug, especially against dermatophytes (see p. 293). Griseofulvin was first detected in *P. griseofulvum* (Oxford *et al.*, 1939) and was then re-discovered in *P. janczewskii* (see Brian, 1960). It is now known to be produced by a wide range of *Penicillium* spp. as well as *Aspergillus versicolor* and by the Hemiascomycete *Eremothecium coryli* (Bérdy, 1986). Commercial production is still achieved by means of fungal fermentations.

11.4.4 Mycotoxins

Members of the genera *Aspergillus* and *Penicillium* are notorious for their production of secondary metabolites which are highly toxic against many different organisms and are therefore collectively called **mycotoxins**. Since these fungi are often found as food contaminants, their mycotoxins present a major health hazard and, consequently, have been thoroughly investigated. We summarize the major groups of substances here, i.e. aflatoxins, ochratoxin A and patulin (Figs. 11.14c−e). These are derived at least in part from the **polyketide pathway** in which acetyl-coenzyme A or malonyl-CoA units are fused head-to-tail in a stepwise fashion. The principle has similarities to the synthesis of fatty acids from acetyl-CoA. Synthesis proceeds in cycles, with one addition in each cycle which is followed by modification of the side chain (initially a keto group). Many other mycotoxins arising from diverse biochemical pathways are produced by *Aspergillus* and *Penicillium*, and an excellent introduction to the biochemical diversity of mycotoxins has been given by Moss (1994). Good reviews of ecological aspects and health implications of these mycotoxins have been written by Scudamore (1994), Bhatnagar *et al.* (2002) and Pitt (2002). The ability to produce mycotoxins such as aflatoxin, ochratoxin or patulin is found in diverse groups of *Aspergillus* and *Penicillium*. Since all or most of the genes involved in the biosynthesis of a given mycotoxin tend to be clustered in the genome, it is possible that sporadic horizontal gene transfer has occurred between different species, thus explaining the lack of correlation between mycotoxin production and phylogenetic placement (Varga *et al.*, 2003).

Aflatoxins

These polyketide-type metabolites are produced by strains of *Aspergillus flavus* but not, apparently, by the closely related *A. oryzae* (Bayman & Cotty, 1993). The most common is aflatoxin B_1 (Fig. 11.14c), which is so named because it fluoresces blue on a thin-layer chromatography plate under UV light (aflatoxins G fluoresce blue–green). The fluorescence of aflatoxins is so strong that heavily contaminated food samples, e.g. the kernels of Brazil nuts, will fluoresce under UV light. Aflatoxin B_1 is one of the most potent carcinogens known, being capable of inducing liver cancer at concentrations below $1 \, \mu g \, kg^{-1}$ body weight (Cotty *et al.*, 1994). Consequently, stringent regulations concerning maximum permissible aflatoxin levels are in place in many countries. However, these toxins may still present a health hazard to consumers, and also to agricultural workers because spores of *A. flavus* contain such high toxin levels that their inhalation may pose a risk of liver cancer (Olsen *et al.*, 1988). Although the crop may well become contaminated on the field, *A. flavus* infections become visible only during the storage of agricultural produce. Since the fungus is xerophilic, it can colonize even dry products (Lacey, 1994). Nuts, peanuts and spices are particularly susceptible, but almost any food can be contaminated. An infamous outbreak of aflatoxicosis, turkey-X disease, occurred in the UK in 1960 when about 100 000 turkeys were killed by contaminated groundnut meal. Further, cows eating contaminated feed will produce milk containing the slightly modified aflatoxins M (Scudamore, 1994). Because of their ubiquity and extreme toxicity, aflatoxins must be considered the most important food-borne mycotoxins worldwide. The biosynthetic pathways of aflatoxins are well characterized (Klich & Cleveland, 2000). The mycotoxin sterigmatocystin, produced by various *Aspergillus* spp., is a precursor of aflatoxin and is also carcinogenic, although it is comparatively rare in food and feed (Moss, 1994). The production of secondary metabolites usually occurs only when vegetative growth has ceased and when conidium formation ensues; the regulatory mechanisms coupling conidiation with secondary metabolism are beginning to be unravelled for *A. nidulans* (Adams & Yu, 1998).

Ochratoxin A

This is a pentaketide/amino acid hybrid molecule (Fig. 11.14d) which is produced by numerous species of *Aspergillus* and *Penicillium*, especially *P. verrucosum*, which is common on cereals in temperate climates, and *A. ochraceus* and *A. carbonarius*, which grow on the flesh of coffee berries during drying. Coffee can, therefore, be contaminated with ochratoxin A, but mercifully much of it is destroyed during roasting (Viani, 2002). Ochratoxin A consumed with contaminated cereals or meat has a long residence time (half-life 35 days) in the human body. It is highly nephrotoxic and has been implicated in a degenerative human kidney disorder called 'Balkan endemic nephropathy'; it is also strongly suspected to cause cancer of the gall bladder (Stoev, 1998; O'Brien & Dietrich, 2005). Further, ochratoxin A causes a renal degenerative disorder of farm animals, especially pigs. A chemically closely related mycotoxin is citrinin.

Patulin

Although patulin is a small (tetraketide-derived) molecule (Fig. 11.14e), its biosynthesis is complex, involving the formation and subsequent cleavage of an aromatic ring (Moss, 1994). Patulin is produced by several species of *Aspergillus* and *Penicillium* as well as *Byssochlamys nivea* (see p. 307), but the most important producer by far is *P. expansum*, a cause of brown rot of apples. Patulin is often detected in apple juices, sometimes at concentrations greatly exceeding safety limits set at or below $50\,\mu g\,l^{-1}$. It is, however, destroyed during alcoholic fermentation to wine or cider (Moss & Long, 2002), or by adding sulphite. It is also formed by *A. clavatus* in spent barley from beer brewing which is often fed to cattle. Patulin may be carcinogenic; it also reacts with the sulphydryl groups of proteins, thereby inactivating enzymes (Mahfoud *et al.*, 2002). A review of safety issues and methods for analysis and control of patulin levels in food has been written by Moake *et al.* (2005).

11.4.5 Pathogenic species

In principle, all species of *Aspergillus* and *Penicillium* and indeed many other types of fungi can cause health hazards because of the potential of their spores to act as allergens to those suffering from hay fever or asthma. Further, many species of *Aspergillus* and *Penicillium* produce mycotoxins (see above). In the present section we will consider only those species which cause mycoses, i.e. infections which require chemotherapy. Good general reviews have been written by Kwon-Chung and Bennett (1992) and Summerbell (2003).

Aspergillus

Two species cause most of the mycotic infections associated with *Aspergillus*. These are *A. fumigatus* (69% of all reports) and *A. flavus* (17% of reports) (Summerbell, 2003). Both produce similar diseases. Like many other fungal pathogens of humans, these species primarily cause infections of the respiratory tract and the lung, although wound infection can also occur occasionally. In immunocompetent patients, non-spreading 'fungus balls' (aspergillomas) may be formed

in the lung in cavities caused, for example, by previous tuberculosis. In immunocompromised patients, invasive aspergillosis may arise, i.e. the infection spreads throughout the lung and even to other organs. Aspergillosis is a major cause of death among cancer patients and is strongly on the increase among AIDS sufferers. One reason why *A. fumigatus* is a more frequent cause of infection than *A. flavus* may be that its conidia are smaller ($3\,\mu m$ diameter or less) and can penetrate more deeply into the lung. They are also more buoyant in the air, and Chazalet *et al.* (1998) have routinely measured concentrations above $1\,\mathrm{conidium\,m^{-3}}$ air even in protected hospital environments. This means that every human normally inhales several hundred conidia of *A. fumigatus* every day. One disease caused almost solely by *A. fumigatus* is allergic bronchopulmonary aspergillosis, in which infections occur in patients already suffering from chronic irritation of the lung, e.g. due to asthma or cystic fibrosis. The disease can lead to fatal destruction of the lung tissue. Treatment by chemotherapy is possible, with amphotericin B and the triazole itraconazole being the major current drugs. In-depth reviews on all aspects of diseases caused by *A. fumigatus* have been written by Latgé (1999, 2001).

Aspergillus fumigatus is a particularly thermotolerant species with an upper growth limit at 52°C (Dix & Webster, 1995), although it can survive 80°C for up to 60 min (Jesenská *et al.*, 1993). It is one of the most abundant moulds found in compost heaps and other situations in which the decay of vegetation generates heat. Workers at compost sites are therefore subjected to a massive spore inoculum, although the incidence of aspergillosis does not seem to be generally higher among them. This indicates the opportunistic nature of aspergillosis in man. In fact, the spores of thermophilic actinomycetes seem to cause most of the problems associated with 'compost worker's lung' (van den Bogart *et al.*, 1993).

Penicillium

In general, species of *Penicillium* are not as thermotolerant as *Aspergillus*, with only relatively few species capable of growing at 37°C. Consequently, clinical reports of *Penicillium* infections are uncommon, with the major exception of *P. marneffei*. As already noted, this species grows as a fission yeast at 37°C, and it causes systemic and disseminated infections in South East Asia which have increased dramatically with the spread of AIDS there. The disease symptoms are similar to those of *Histoplasma capsulatum* (see p. 290), including the predominance of the disease in male patients (Harrison & Levitz, 1996). Disseminated infections are most commonly found in lung, liver and skin and can be treated with amphotericin B and itraconazole (Harrison & Levitz, 1996). *Penicillium marneffei* can be isolated with high frequency from the internal organs of bamboo rats as well as their burrows in South East Asia. However, since contact between these rodents and humans is probably infrequent, there may be unknown sources of inoculum in the environment to which both rats and humans are exposed (Vanittanakom *et al.*, 2006).

11.4.6 *Byssochlamys*

Byssochlamys is a small genus of soil fungi currently comprising four species (Pitt *et al.*, 2000) which are noteworthy because of their thermotolerance. The most tolerant structures are the ascospores which may survive heating to 90°C for 25 min, especially in the presence of high sucrose concentrations (Beuchat & Toledo, 1977; Bayne & Michener, 1979). *Byssochlamys* spp. are important contaminants of canned fruits or bottled fruit juices (Tournas, 1994) because of their heat tolerance, ability to produce pectolytic enzymes, and tolerance of conditions of low oxygen tension. Contamination can be dangerous because *Byssochlamys* spp. can produce several mycotoxins, including patulin (Rice *et al.*, 1977). Another habitat associated with human activity is silage in which *Byssochlamys* is a common contaminant (Inglis *et al.*, 1999). In nature, *Byssochlamys* is ubiquitous in the soil. In orchards, it may be splashed onto the fruit prior to or during harvesting.

In culture, *Byssochlamys* spp. reproduce asexually by the formation of chains of hyaline conidia derived from tapering open-ended phialides (Fig. 11.15a) which have been assigned to the *Paecilomyces* type. Terminal, thick-walled

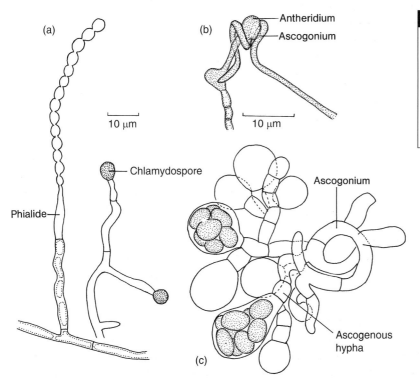

(a)

10 μm

(b)

Antheridium
Ascogonium

10 μm

Phialide

Chlamydospore

Ascogonium

Ascogenous hypha

(c)

Fig 11.15 *Byssochlamys nivea.* (a) Phialospores and chlamydospores. (b) Coiled ascogonium surrounding an antheridium. (c) Ascogonium bearing ascogenous hyphae which in turn produce asci. Note the absence of sterile investing hyphae.

unicellular chlamydospores are also found. The asci of *Byssochlamys* develop best in cultures incubated around 30°C. In *B. nivea*, a club-shaped antheridium becomes encoiled by an ascogonium (Fig. 11.15b). Later, the coiled ascogonium develops short branches (ascogenous hyphae) which bear globose, eight-spored asci either terminally or laterally, so that eventually clusters of asci can be found. There is no sign of any sterile hyphae enclosing them (Fig. 11.15c). Like most if not all members of the Eurotiales, *Byssochlamys* spp. appear to be homothallic.

11.4.7 *Aspergillus* and its teleomorphic states

Keys to *Aspergillus* may be found in Raper and Fennell (1965), Domsch *et al.* (1980) and Klich (2002). There are about eight teleomorphic genera which have an *Aspergillus* conidial state (Table 11.2), although many *Aspergillus* species (about 60%) have no known sexual state. Among the purely asexual aspergilli are some of the most important species such as *A. parasiticus, A. flavus, A. oryzae, A. niger* and *A. fumigatus*. Geiser *et al.* (1996) proposed that the ability to reproduce sexually has been lost on many separate occasions, because many strictly mitotic *Aspergillus* species have teleomorphic species as their closest relatives. The loss of the teleomorph seems to have occurred very recently on an evolutionary timescale, which is also indicated by the fact that many *Aspergillus* species still produce sterile structures (e.g. sclerotia) or cells (e.g. Hülle cells; see below) which are similar to those found in cleistothecia. Such defects hint at the deletion of one or more of the many genes whose products are necessary for cleistothecium formation and meiosis. It is possible that purely asexual species are more likely to become extinct because they accumulate mutations without the possibility of meiotic recombination. Whilst the parasexual cycle can certainly be used to bring about efficient genetic recombination in the laboratory (Bradshaw *et al.*, 1983; see Plate 4d), it is doubtful whether it occurs sufficiently frequently in nature to present a viable alternative to meiosis (Geiser *et al.*, 1996).

It is a contentious question as to how to name purely mitotic species. To bring current efforts at unifying anamorphic and teleomorphic taxonomy to an extreme, these *Aspergillus* species would have to be given the name of a teleomorph which does not exist. An additional problem is that the anamorphic features by which the genus *Aspergillus* is divided into sections do not correspond to the groupings obtainable with phylogenetic analyses (Peterson, 2000b). The nomenclature and taxonomy of *Aspergillus* are thus in a horrible state of flux, and for the time being we have a clean conscience in continuing to use the name *Aspergillus*, especially for those species with no teleomorph. One desirable consequence of this approach is that it considerably reduces the degree of confusion in mycology courses. The same approach, of course, applies to *Penicillium*. For both taxa, we will now introduce examples of the most common teleomorph forms.

Eurotium

Members of this genus are widely distributed in nature, especially in soil. They are responsible for the spoilage of foodstuffs, especially those with high osmotic concentrations. A typical example is *E. repens* (Fig. 11.16), which is common on mouldy jam. In culture, conidia are formed on agar media low in sugar content (e.g. 2% malt extract). The hyphal segment from which the conidiophore arises persists as a swollen **foot cell** (Fig. 11.16a). The tip of the conidiophore swells to form a club-shaped vesicle bearing directly on its surface a cluster of bottle-shaped phialides which give rise to chains of green conidia in basipetal succession. On agar media with a high sugar content (e.g. 2% malt extract with 20% sucrose), yellow spherical ascocarps also develop and conidia are sparse, so that the entire Petri dish may look bright yellow instead of olive green. Aerial hyphae develop coiled ascogonia (Fig. 11.16b), and although there are reports of associated antheridia, these are not always seen. The ascogonium becomes invested by sterile hyphae which grow up from the stalk of the ascogonium. The ascogonium becomes septate, and from its segments ascogenous hyphae develop which penetrate and dissolve the

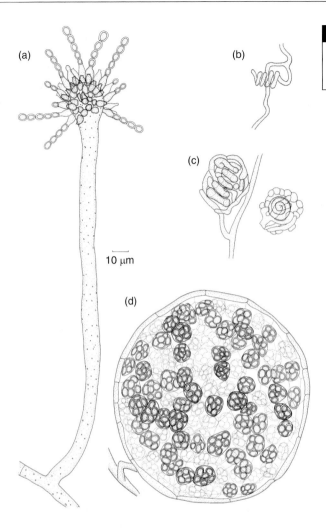

surrounding pseudoparenchyma derived from the investing hyphae. Globose asci develop from croziers at the tips of the ascogenous hyphae and, when ripe, the ascocarp consists of clusters of asci surrounded by a single-layered, yellow-coloured peridium. The peridium breaks open irregularly. The asci do not discharge violently, but the ascospores escape as the ascus wall breaks down. The ascospores are broadly lenti-cular, and are without obvious surface ornamen-tation in *E. repens*, although they may bear an equatorial furrow in other species.

Eurotium repens, like nearly all *Eurotium* spp., is homothallic, and cultures can be transferred by means of conidia, ascospores or hyphal tips. If successive conidial transfers are made, the ability to form ascospores declines with each transfer. Ascospore production can be restored to the initial level by making one subculture from an ascospore (Mather & Jinks, 1958). This suggests that the formation of cleistothecia and conidio-phores is partially controlled by cytoplasmic determinants, and it also indicates a way in which purely anamorphic forms may have evolved from *Eurotium*. Little recent work seems to have been carried out on sexual reproduction in *Eurotium*.

Emericella

Emericella differs from *Eurotium* in a number of features. Whilst in *Eurotium* the ascocarp is even-tually surrounded by a single-layered peridial envelope, that of *Emericella* is enclosed by chains of very thick-walled cells termed **Hülle cells**

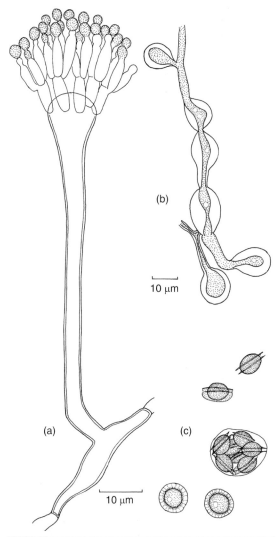

(b)

10 μm

(a)

(c)

10 μm

Fig 11.17 *Emericella nidulans.* (a) Conidiophore. Note that the phialides are not borne directly on the vesicle. (b) Hülle cells, thick-walled cells surrounding the ascocarp. (c) Ascus and ascospores. Note that the ascospores bear a double flange. (a) and (c) to same scale.

of the nest-like arrangement of cleistothecia surrounded by Hülle cells. This species has been used widely in genetic studies on sexual and parasexual recombination (Roper, 1966; Bos & Swart, 1995) and, more recently, for experiments on morphogenetic aspects (see p. 299) as well as the biosynthetic pathways of penicillin and mycotoxins (pp. 302 and 304).

11.4.8 *Penicillium* and its teleomorphic states

Keys to *Penicillium* have been provided by Pitt (1979, 2000) and Ramírez (1982). The anamorphic genus *Penicillium* presents the same kind of taxonomic and nomenclatural problems as *Aspergillus*. The classical conidial apparatus is a branched conidiophore, bearing successive whorls of branches which terminate in clusters of phialides (Figs. 11.11, 11.18). Members of the subgenus *Aspergilloides* produce phialides directly on the conidiophore and are thus superficially similar to *Aspergillus*. An example is *P. spinulosum* (Fig. 11.18a). More commonly, the phialides are borne on a further whorl of branches, the metulae, and such species are grouped by their phialide shape and depending on whether the penicilli are symmetrical (subgenus *Biverticillium*) or more irregular (subgenus *Furcatum*). Examples are given in Figs. 11.18b,c. A third possibility is that the metulae may in turn arise from a further verticil of branches, the rami (subgenus *Penicillium*; e.g. *P. expansum*, Fig. 11.18d). In some species, e.g. *P. claviforme*, the individual conidiophores may be aggregated together into club-shaped fructifications or coremia (Fig. 11.19).

As in the case of *Aspergillus*, the morphology of the conidiophore unfortunately does not correlate with DNA sequencing data (Peterson, 2000a), and it is possible that *Penicillium* will eventually be broken up into several new genera. However, whilst of only limited taxonomic value, the conidiophore architecture will continue to be used for identification purposes. Some species of *Penicillium* have teleomorphs which can be assigned to *Talaromyces* or *Eupenicillium*. The different kinds of ascocarp represented by these generic names can be correlated weakly with conidiophore branching (see below). As for

(Fig. 11.17b). Whereas the ascospores of *Eurotium* are colourless and without conspicuous ornamentations, those of *Emericella* are red and bear a prominent double equatorial flange, so that the spores resemble pulley wheels (Fig. 11.17c). The conidiophores also differ in that the phialides are not borne directly on the vesicle but on a series of cylindrical cells termed metulae.

The best-known species is the soil fungus *Emericella* (*Aspergillus*) *nidulans*, so called because

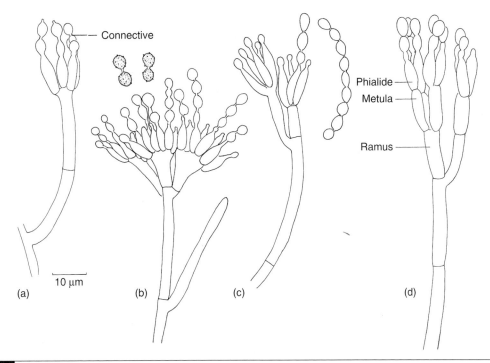

Fig 11.18 Conidiophores of the four anamorphic subgenera of *Penicillium*. (a) *Penicillium spinulosum* (subgenus *Aspergilloides*). (b) *Penicillium verruculosum* (subgenus *Biverticillium*). (c) *Penicillium citrinum* (subgenus *Furcatum*), possibly 'one of the most common eukaryotic life forms on earth' (Pitt, 1979). (d) *Penicillium expansum* (subgenus *Penicillium*).

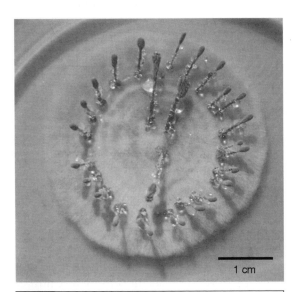

Fig 11.19 *Penicillium claviforme* producing coremia on an agar plate.

Aspergillus, there is evidence that the ability to reproduce sexually has been lost on several independent occasions in *Penicillium* (LoBuglio *et al.*, 1993).

Penicillium is one of the most ubiquitous groups of fungi, occurring on all kinds of decaying materials. The conidia are universally present in air, so that *Penicillium* colonies are frequent contaminants of cultures. It was such a chance contaminant which led to the discovery of penicillin (see p. 302). *Penicillium italicum* and *P. digitatum* cause rotting of citrus fruits (Plate 4e) whilst *P. expansum* causes a brown rot of apples (see p. 304).

Eupenicillium

Members of the genus *Eupenicillium* produce conidiophores which may be mono-, bi- or terverticillate. The terverticillate species (subgenus *Penicillium*) have been particularly well studied because many of them are relevant to man as producers of antibiotics (*P. griseofulvum*), in food production (e.g. *P. camemberti* and *P. roqueforti*) or food spoilage (e.g. *P. aurantiogriseum*, *P. digitatum*, *P. italicum*, *P. expansum*), or as mycotoxin producers (*P. expansum*). A superbly illustrated key to terverticillate *Penicillium* spp. has been produced by Frisvad and Samson (2004), accompanied by phylogenetic analyses in which the grouping of these species into sections has been correlated with morphological features and the production of mycotoxins and other metabolites (Samson *et al.*, 2004). An overview of secondary metabolites within *Penicillium* subgenus *Penicillum* has been compiled by Frisvad *et al.* (2004).

In phylogenetic analyses, species of *Penicillium* subgenus *Penicillium* aggregate in a well-resolved

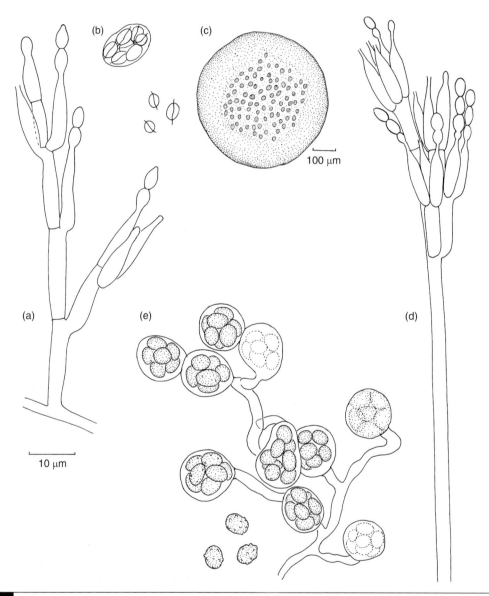

Fig 11.20   *Talaromyces*. (a) *Talaromyces stipitatus*, conidiophore. (b) Ascus and ascospores. Note the equatorial frill. (c) *Talaromyces vermiculatus*, ascocarp. (d) Conidiophore. Note the long tapering phialides characteristic of the subgenus *Biverticillium*. (e) Ascogenous hyphae and asci. Note that some asci arise in chains.

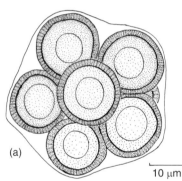

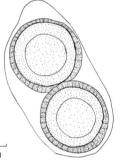

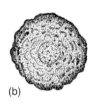

(a)

(b)

10 μm

cluster around *E. crustosum* (Peterson, 2000a). *Eupenicillium* spp. produce cleistothecia with very tough peridia. The ascospores have pulley wheel-like flanges. *Penicillium* spp. associated with *Eupenicillium* often produce thick-walled sclerotia, and it is likely that these sclerotial forms are cleistothecial forms which have lost their ability to complete meiosis.

Talaromyces

The ascocarp of *Talaromyces* is rather different from that of *Eupenicillium* in having a peridium with soft cottony hyphae. It is thus a gymnothecium rather than a cleistothecium. *Penicillium* anamorphs of *Talaromyces* all belong to the subgenus *Biverticillium* (LoBuglio *et al.*, 1993), with long tapering phialides which are closely appressed to each other, rather than

divergent (Fig. 11.20). In this they resemble *Paecilomyces* (Fig. 11.11c), some species of which are also associated with *Talaromyces* (see Table 11.2).

II.4.9 Elaphomycetaceae

Like so many other apparently diagnostic structures of fungi, the hypogeous (= subterranean) habit of the truffle has evolved independently several times. Whereas the best-known edible truffles belong to the Pezizales (see p. 423), the Elaphomycetaceae are firmly included among the Eurotiales (Geiser & LoBuglio, 2001). The genus *Elaphomyces* contains the most common hypogeous fungi of temperate climates, and *E. granulatus* (Figs. 11.21, 11.22) and *E. muricatus* can be collected throughout the year beneath the litter layer under various trees, but especially beech with which they form ectomycorrhizal associations. *Elaphomyces muricatus* is often parasitized by *Cordyceps ophioglossoides* forming yellow mycelium around the subterranean fruit bodies, and a club-shaped perithecial stroma above ground (Plate 4f). There are 5 British species (Pegler *et al.*, 1993) and about 20 species worldwide. *Elaphomyces* spp., along with other truffles and epigeous fungi, form an important part of the winter diet of squirrels (Currah *et al.*, 2000). The common name, hart's truffle, indicates that the fruit bodies of *Elaphomyces* spp. are dug up and consumed by deer.

The fruit bodies of *Elaphomyces* vary in size (about 1–4 cm in diameter) and are regarded as cleistothecia. When cut open, a two-layered rind (peridium) can be distinguished from a central mass containing the globose asci, traversed by lighter sterile 'veins'. The asci in *E. granulatus* usually contain six spores and in *E. muricatus* two to four. The spores are dark brown and thick-walled when mature, and the conditions necessary for their germination are not known. There are no anamorphic states.

Further evidence of the remarkable plasticity of fruit body morphology in the fungi comes in the shape of an unusual member of the Elaphomycetaceae from tropical South America. In *Pseudotulostoma*, a subterranean initial produces a sizeable stalk (up to 7 cm high) with a head which looks much like the gleba of a puffball but is, in fact, a cleistothecium. At maturity, the peridial layers disintegrate, leaving the ascospores to be distributed by the wind (Miller *et al.*, 2001).

Hymenoascomycetes: Pyrenomycetes

12.1 | Introduction

The Pyrenomycetes are defined here according to Samuels and Blackwell (2001) as fungi which produce non-fissitunicate or occasionally proto-tunicate asci usually in flask-shaped ascomata (perithecia), less frequently in cleistothecia. The sub-class Pyrenomycetes is one of several groups belonging to the huge and heterogeneous class Hymenoascomycetes. The characteristic feature of this class is that the asci develop in an ascohymenial way, i.e. the ascoma is formed *after* plasmogamy and the pairing of nuclei have occurred, and the asci therefore arise from a hymenium. This is in contrast to asci being formed singly (Archiascomycetes, Hemiascomycetes), scattered throughout the fruit body (Plectomycetes), or formed in a locule within a pre-formed fruit body (Loculoascomycetes). Although the term 'Pyrenomycetes' is not generally understood in a taxonomic sense at the present, Samuels and Blackwell (2001) pointed out the monophyly of a core group of orders, including all those which we shall describe in this chapter (summarized in Table 12.1).

The development of the perithecium follows several different schemes defined by Luttrell (1951), which are described in more detail for the different orders. Following fertilization and plasmogamy, the ascogonium gives rise to ascogenous hyphae while the perithecial wall is formed by hyphae arising from the ascogonial stalk or elsewhere. Sterile hyphae growing up from the basal fertile region (**paraphyses**) and

periphyses which line the inner surface of the ostiole, may be present. The development of the opening of the perithecium is typically **schizogenous**, i.e. it is formed by the pushing apart of tissue by the periphyses at the apex of the perithecium. This is in contrast to **lysigenous** development, e.g. in the pseudothecial neck of Loculoascomycetes (see p. 459). Perithecia may be formed singly or in a perithecial stroma. Ascospore discharge is by active turgor-driven liberation or occurs passively, with the ascospores oozing out of the perithecial ostiole as a tendril (**cirrhus**).

Although the core orders belonging to the Pyrenomycetes appear to be monophyletic, the fungi considered here follow numerous different lifestyles. Many species grow saprotrophically in terrestrial habitats, while others are associated with plants, covering a wide range from mutualistic or commensalistic endophytes over biotrophic pathogens through to hemibiotrophic and necrotrophic pathogens. Animals, especially insects, are also parasitized. Numerous biologically active metabolites are produced, including alkaloids, antibiotics and phytotoxins.

12.2 | Sordariales

The order Sordariales is a substantial group of ascomycetes containing some 7 families, 115 genera and over 500 species. We shall study representatives of only 2 families, the Sordariaceae (6 genera, 37 spp.) and Chaetomiaceae (15 genera, 150 spp.).

Table 12.1. Core groups of Pyrenomycetes as treated in this chapter. Species numbers (in brackets) are based mostly on Kirk *et al.* (2001).

Taxon	Examples of teleomorphs	Examples of anamorph
Sordariales (see p. 315)	*Sordaria* (14)	
	Podospora and *Schizothecium* (80)	
	Neurospora (12)	*Chrysonilia*
	Chaetomium (80)	
Xylariales (see p. 332)	*Daldinia* (13)	*Nodulisporium*
	Xylaria (100)	*Nodulisporium*
	Biscogniauxia (25)	*Nodulisporium*
	Hypoxylon (120)	*Nodulisporium, Geniculosporium*
	Kretzschmaria (5)	*Nodulisporium*
	Hypocopra (30)	
	Podosordaria (17)	*Lindquistia*
	Poronia (2)	
	Rosellinia (100)	*Dematophora*
Hypocreales (see p. 337)	*Hypocrea* (100)	*Trichoderma*
	Hypomyces (30)	*Cladobotryum, Verticillium*
	Nectria (28)	*Acremonium, Cylindrocarpon, Fusarium*
	Gibberella (10)	*Fusarium*
	Sphaerostilbella (4)	*Gliocladium*
Clavicipitales (see p. 348)	*Claviceps* (36)	*Sphacelia*
	Cordyceps (100)	*Beauveria, Metarhizium, Tolypocladium*
	Epichloe (8)	*Neotyphodium*
	Balansia (20)	*Ephelis*
Ophiostomatales (see p. 364)	*Ophiostoma* (100)	*Graphium, Leptographium, Sporothrix, Pesotum*
Microascales (see p. 368)	*Microascus* (13)	*Cephalotrichum, Scopulariopsis, Wardomyces*
	Ceratocystis (14)	*Thielaviopsis*
	Sphaeronaemella (5)	*Gabarnaudia*
Diaporthales (see p. 373)	*Diaporthe* (75)	*Phomopsis*
	Cryphonectria (6)	*Endothiella*
	Apiognomonia, Gnomoniella (18)	*Discula*
Magnaporthaceae (see p. 377)	*Magnaporthe* (4)	*Pyricularia*
	Gaeumannomyces (5)	*Phialophora*
Glomerellaceae (see p. 386)	*Glomerella* (5)	*Colletotrichum*

The best-known genera of Sordariaceae are *Sordaria*, *Podospora* and *Neurospora*. Many *Sordaria* and *Podospora* spp. are coprophilous, fruiting on the dung of herbivores, but species growing on wood or in soil are also known. *Neurospora* occurs in nature on burnt soil and vegetation, especially in warmer countries. All three genera have been extensively used in genetic studies.

The dark-coloured perithecia usually have an ostiole lined by periphyses, but some genera are astomous, i.e. they have fruit bodies lacking ostioles, thus forming cleistothecia. Stromata are not produced. The asci are unitunicate and thin-walled, and the apical apparatus of the ascus is in the form of a thickened annulus or apical plate which does not stain blue with iodine. Free-ended paraphyses are often present but may dissolve at ascus maturity. The ascospores are black and sometimes surrounded by a mucil-aginous epispore, or they have mucilaginous appendages. The spores are mostly unicellular and germinate through a germ pore.

In the Chaetomiaceae, the ascomata (peri-thecia or sometimes cleistothecia) are generally clothed with thick-walled ornamented hairs. The club-shaped asci are thin-walled and without apical apparatus. If a hamathecium is formed, it does not persist. The ascus wall dissolves and, in perithecial forms, the ascospores are extruded as a cirrhus. The ascospores are grey to brown in colour, mostly unicellular and with a single germ pore. Molecular data indicate a close relationship between the Chaetomiaceae and Sordariaceae (Huhndorf et al., 2004).

12.2.1 Sordaria (Sordariaceae)

Most species of Sordaria are cellulolytic. Perithecia are common on the dung of herbi-vores and occasionally on other substrata such as seeds and plant remains, while a few species are reported from soil. Guarro and von Arx (1987) have given a key to 14 species, 5 of which are coprophilous. Lundqvist (1972) has described and illustrated Nordic species. Sordaria fimicola is especially common on horse dung and has been widely used in experiments on nutrition, the physiology of fruiting, spore liberation and genetics. It is homothallic, and perithecial devel-opment occurs within 10 days on a wide range of media. A longitudinal section of a perithecium (Fig. 12.1) shows a basal tuft of asci at different stages of development. The asci elongate in turn so that only one ascus can occupy the ostiole at a time. Because each spore is about 13 μm wide and the diameter of the apical apparatus of the

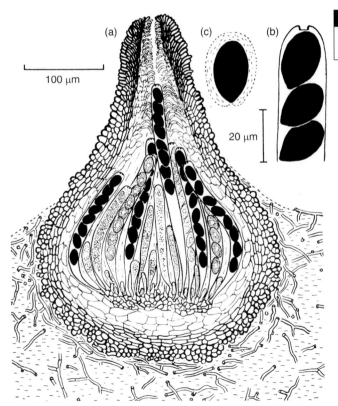

100 μm

20 μm

Fig 12.1 *Sordaria fimicola.* (a) L.S. perithecium. (b) Ascus apex. (c) Ascospore showing mucilaginous epispore. After Ingold (1971).

ascus is only about 4 μm, the latter acts as a sphincter, gripping the spores as they leave the ascus. Projectiles which vary in size from one to eight spores may be formed. The larger the number of spores in the projectile, the greater the distance of discharge due to the fact that the surface-to-volume ratio of single spores is greater than that of multiple-spored projectiles, so that the effects of wind resistance are disproportionately high (Ingold & Hadland, 1959). Thus, single-spored projectiles have a mean discharge distance of about 1.5 cm whilst for eight-spored projectiles the distance is about 6 cm. The necks of the perithecia are phototropic and, as in many other coprophilous fungi, this adaptation ensures that the spores are projected away from the dung substratum. The ascospores of *S. fimicola* have a distinct mucilage envelope which enables them to adhere to herbage, but in *S. humana* the sheath is much reduced or absent. The ascospores of *S. fimicola* can survive for long periods. On drying, gas vacuoles (de Bary bubbles) may appear in the spore cytoplasm, but despite this the spores remain viable, and the bubbles disappear upon rehydration (Ingold, 1956; Milburn, 1970). Ascospore germination is enhanced by digestive treatment in the herbivore gut and this effect can be simulated in the laboratory by treatment with pancreatin or sodium acetate.

Perithecium development

There have been numerous studies on perithecial development and structure in *Sordaria*, e.g. in *S. fimicola* (Mai, 1977), *S. humana* (Uecker, 1976; Read & Beckett, 1985) and *S. macrospora* (Hock *et al.*, 1978). Intercalary multinucleate cells of vegetative hyphae give rise to multinucleate, spirally coiled, septate ascogonia which are not associated with antheridia. In these species, there is no trichogyne and there are no microconidia (spermatia). The ascogonium is enveloped by branched investing hyphae which originate from the ascogonial stalk or from adjacent vegetative hyphae to form a spherical **protoperithecium**, i.e. an immature ascoma which does not, at this stage, show differentiation into a neck or ostiole. The outer region of the protoperithecium is made up of about five

layers of rounded cells which have thick, pigmented (melanized) walls. Lying inside them are several layers of flattened, thinner, non-pigmented cells (see Fig. 12.1). Differentiation of the innermost cells of the protoperithecium gives rise to the centrum consisting of elongate, free-ended, thin-walled, septate paraphyses and ascogenous cells. As the ascoma matures, it changes in outline from spherical to pear-shaped with an elongate neck, thus developing into a perithecium. Neck development is associated with meristematic activity of the cells at the base of the neck and by the appearance of short, tapering periphyses which line it. By pushing against each other the periphyses create the ostiole, the opening to the outside through which the asci will discharge their spores. Thus, the cells are pushed apart rather than lysed, and this process of ostiole development is called schizogenous. Phototropism of the perithecial neck is associated with differential enlargement of the periphyses.

The ascogenous hyphae arise on a placenta-like mound at the base of the centrum and elongate upwards. They show the usual type of crozier found in many ascomycetes with a binucleate penultimate cell (the mother cell of an ascus), a terminal cell and a stalk cell (see Fig. 8.10). Proliferation of the ascogenous hypha occurs by fusion of the recurved terminal cell with the stalk cell. Nuclear fusion in the ascus mother cell is followed by meiosis, yielding four haploid nuclei. Two further mitotic divisions produce 16 nuclei. Cleavage of the cytoplasm accompanied by wall formation results in eight binucleate ascospores. The fine structure of ascospore development has been studied by Mainwaring (1972).

Perithecium development is affected by environmental factors such as temperature and nutrient supply. For *S. macrospora*, Hock *et al.* (1978) have shown that in pure culture on a defined nutrient medium, perithecium development requires a simultaneous supply of biotin and arginine. In the presence of certain other fungi such as *Armillaria* spp. and *Mortierella* spp., *S. fimicola* is stimulated to increased production of perithecia and ascospores and it is likely that this effect is related to the production of vitamins by the stimulating fungi (Watanabe, 1997).

Mating systems of *Sordaria*

Although *S. fimicola* is homothallic, it has the ability to hybridize. Wild-type strains have black ascospores, but mutants are known with pale or colourless spores. If a wild-type strain and a white-spored mutant strain are inoculated on opposite sides of a Petri dish, hybrid perithecia develop from heterokaryotic parts of the mycelium. The asci from hybrid perithecia usually have four black and four white ascospores. Six different arrangements of the ascospores are found in such hybrid asci (Fig. 12.2). Asci with four black or four white ascospores at the tip of the ascus are those in which the gene for spore colour segregated at the first meiotic nuclear division separating the paired chromosomes. In those with two black or two white ascospores at the tip of the ascus, segregation of the gene for spore colour occurred at the second meiotic division separating the two sister chromatids of each chromosome. First-division segregation results from the absence of a cross-over between the gene for spore colour and the centromere of the chromosome, whilst second-division segregation results from a single cross-over between gene and centromere. Since the likelihood of crossing-over depends on the distance between gene and centromere, the frequency of the two kinds of segregation pattern can be used for determining the distance of the gene for spore colour relative to the centromere.

A low proportion of hybrid asci shows 5:3, 6:2 or (very rarely) 7:1 colour segregation patterns. These findings are explained in terms of **gene conversion**, a non-reciprocal process where one allele of a gene converts another allele at the same locus to its own type (Lamb, 1996). Similar patterns have been reported in other ascomycetes, e.g. *Ascobolus immersus* (p. 423). A model to account for the molecular basis of gene conversion was developed with the smut fungus *Ustilago maydis* (see p. 652). In this so-called 'Holliday model' DNA strands are exchanged at 'Holliday junctions' between two paired DNA double helices, which can result in the generation of hybrid or heteroduplex DNA. Following separation of the two double helices from each other, non-matching DNA will be excised and the undamaged strand will be used as template to synthesize the second, damaged strand (Holliday, 1964; Lewin, 2000). This can then give rise to the phenomenon of gene conversion which, if it happens in the ascus, is manifested as deviations from the 4:4 gene segregation patterns.

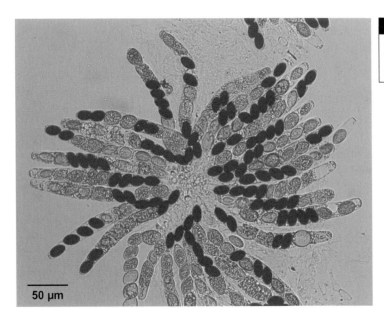

Fig 12.2 *Sordaria fimicola*. Squash preparation from a hybrid perithecium. Most ripe asci contain four black and four white ascospores.

50 μm

Sordaria brevicollis and *S. heterothallis* are heterothallic. Both species form minute spermatia which are involved in the fertilization of mycelia of the opposite mating type. *Sordaria brevicollis* is a heterothallic relative of *S. fimicola* according to Guarro and von Arx (1987). In *S. brevicollis* it has been discovered that perithecium development can occur in unmated cultures of mating type *A*. In about 30% of such perithecia one or two asci with viable ascospores may form, with all spores being mating type *A* (Robertson *et al.*, 1998).

The molecular basis of homothallism has been elucidated in *S. macrospora* by Pöggeler *et al.* (1997). The molecular configuration of the genes conferring the ability to mate is similar to that in two heterothallic members of the Sordariaceae, *Podospora anserina* and *Neurospora crassa*. In *Neurospora crassa* (see Fig. 12.7) the haploid mating types are designated *A* and *a*. The 'alleles' which confer mating competence in *P. anserina* and *N. crassa* consist of dissimilar DNA sequences termed **idiomorphs** which are present at the homologous loci in the mating partners. This term has been introduced to denote sequences like those of mating types *A* and *a*, which occupy the same locus in different strains but are related neither in sequence nor (probably) by common descent (Metzenberg & Glass, 1990). In *S. macrospora*, the two idiomorphs are contiguous (i.e. adjoin each other), and they have been used in experiments to transform (+) and (−) strains of *P. anserina* in order to induce them to form perithecia.

12.2.2 *Podospora* and *Schizothecium* (Sordariaceae)

The perithecia of *Podospora* and *Schizothecium* develop on herbivore dung. In *Podospora*, the upper part of the perithecium wall is often ornamented by various kinds of 'vestiture' (Lundqvist, 1972) such as short or long single hairs or long, pointed setae which may be aggregated into a tuft to one side of the perithecial neck as seen in *P. anserina* and *P. curvicolla*. In *Schizothecium*, the hairs are composed of swollen cells and agglutinate together to form short scale-like tufts (Bell &

Mahoney, 1995). The separation between these two genera has been confirmed by phylogenetic analyses (Cai *et al.*, 2005). Taking *Schizothecium* and *Podospora* together, about 80 species are known (Mirza & Cain, 1969; Lundqvist, 1972). Different species show a degree of substrate specificity. For instance, perithecia of *P. curvicolla*, *P. pleiospora* and *S. vesticola* are especially common on the dung of lagomorphs (rabbits and hares) whilst *P. curvula* fruits commonly on horse dung (Lundqvist, 1972; M. Richardson, 1972, 2001). The reasons for these preferences are not known. Some species have semi-transparent perithecia within which the outline of the club-shaped asci can be seen and the sequential development and discharge of individual asci can be followed (Fig. 12.3a). The number of spores in the ascus varies from 4 to 512. Spore number has been used as a taxonomic criterion in the past, although the species concept has been widened to include forms with a range of spore numbers. For example, 8-, 16-, 32- and 64-spored forms of *Podospora decipiens* are recognized. The name *Podospora* (Gr. *podos* = foot, *spora* = seed) refers to the mucilaginous appendage attached to one or both ends of the black ascospore (Fig. 12.3b). This character is also found in *Schizothecium*. In some of the commonest species, *P. curvula* and *S. tetrasporum*, the upper spore appendages are attached to the cap of the ascus, and when the ascus explodes, the spores, roped together by their appendages, are propelled as a single slingshot projectile (Fig. 12.3c). As in *Sordaria*, it has been shown that multi-spored projectiles are discharged further than single spores (Walkey & Harvey, 1966a). The ascus wall breaks across, just beneath the cap, and in contrast to *Sordaria* there is usually no distinctive apical apparatus.

Developmental aspects

Many morphological and experimental studies have been made on *P. anserina*, which Lundqvist (1972) treated as a synonym of *P. pauciseta*. This fungus was originally described from goose dung but also fruits on the dung of sheep, horse, cattle, mice, grouse and zoo animals. Perithecium development has been described for *P. anserina* (Beckett & Wilson, 1968; Mai, 1976) and *Schizothecium* spp. (Bell & Mahoney,

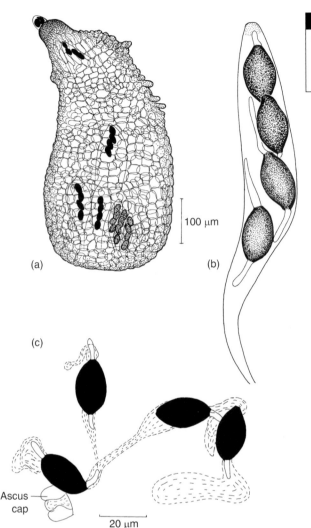

Fig 12.3 *Schizothecium tetrasporum.* (a) Perithecium showing asci through the semi-transparent wall. (b) Ascus. (c) Projectile consisting of four spores attached to the ascus cap and to each other by means of mucilaginous appendages.

100 μm

(a)

(b)

(c)

Ascus cap

20 μm

1996). It is similar to that described above for *Sordaria*.

The development of the ascospores of *P. anserina* has been studied by Beckett *et al.* (1968). The ascospores (four in *P. anserina*, but see below) are delimited by a double membrane system as in other ascomycetes (Fig. 12.4). The primary spore wall develops between the two membranes and gradually pushes them apart. The inner membrane continues to function as the plasma membrane of the spore, whilst the outer functions as the spore-investing membrane. As the primary spore wall widens, secondary wall material is laid down towards the inside of the primary wall. These primary and secondary walls enclose the whole of the spore, including the spore head and the tail. A tertiary wall representing the pigmented layer of the spore head is laid down to the inside of the secondary layer (Figs. 12.4e–g). The elongated tail of the spore is cut off from the spore head by a septum. The tertiary wall layer does not extend into the spore tail, which therefore remains colourless. Its contents degenerate. This part of the spore persists as the primary appendage, sometimes termed the pedicel. Secondary appendages develop at the apex of the spore head and at the primary appendage. They arise by localized evaginations of the spore-investing membrane. A thinner area in the tertiary wall at

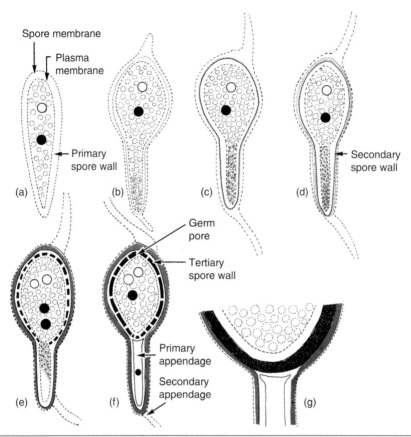

Fig 12.4 *Podospora anserina*, ascospore development (based on Beckett *et al.*, 1968). (a) Binucleate ascospore initial enclosed by two membranes, between which the primary spore wall develops. (b–d) The secondary spore wall develops within the primary wall, and the secondary appendages develop towards each end of the spore by outpushing of the spore membrane. (e–g) Development of tertiary, pigmented wall layer. A further mitotic nuclear division occurs. The uninucleate tail of the spore is cut off from the body of the spore, and its cytoplasm degenerates, but the tail persists as the primary appendage. Note that the tertiary wall layer does not extend into the primary appendage. At the opposite pole, a thinner area in the tertiary wall marks the position of the germ pore.

the end of the spore opposite the primary appendage marks the position of the germ pore (Fig. 12.4f). The secondary appendages of many species of *Podospora* are very elaborate branched structures.

Mating systems of *Podospora*

Anamorphs of *Podospora* species, where known, consist of dark-celled phialides which produce small, sticky, unicellular, uninucleate, hyaline phialoconidia assigned to the anamorph genera *Phialophora* and *Cladorrhinum*. These do not germinate and are presumed to function as spermatia (Bell & Mahoney, 1997; Lundqvist *et al.*, 1999).

The sexual compatibility within the genus varies. Most species for which information is available are homothallic, including species with eight-spored asci such as *P. decipiens*, and species with more than eight spores in the ascus, e.g. *P. pleiospora* (Lundqvist *et al.*, 1999). *Podospora anserina* and *S. tetrasporum* are **pseudohomothallic** (Esser, 1974; Raju & Perkins, 1994). *Podospora anserina* normally has four-spored asci, each ascospore eventually becoming quadrinucleate following two post-meiotic mitotic nuclear divisions (see Fig. 12.6). During the final stages of development three nuclei remain in the main body of the spore and one nucleus passes into the primary spore appendage, where it degenerates. Cultures derived from single asco-spores form perithecia readily. Occasionally, however, smaller uninucleate ascospores may occur in some asci and when such spores are germinated, the resulting mycelium does not fruit. Instead, perithecia only develop when certain strains derived from uninucleate asco-spore cultures are paired together. On each such strain, spermatia and ascogonia bearing tricho-gynes are formed, but these are self-incompati-ble; perithecia only develop if trichogynes of one strain are spermatized by spermatia of a genetically distinct strain. Thus, although the behaviour of the large ascospores suggests that *P. anserina* is homothallic, it is clear that the underlying mechanism controlling perithecial development is a heterothallic one of the usual bipolar type (i.e. with (+) and (−) strains). Most of the large ascospores (about 97%) contain nuclei of the two distinct mating types (Esser, 1974).

The fact that such a high proportion of the binucleate ascospores in four-spored asci contain nuclei of two distinct mating types implies some regulated process of nuclear movement and arrangement. The normal sequence of nuclear divisions occurs during ascus development, including the two nuclear divisions of meiosis and a post-meiotic mitosis (**PMM**). The plane of the two meiotic nuclear divisions lies parallel to the long axis of the developing ascus, but the spindles formed during **PMM** lie at a right angle to it (Fig. 12.5). Delimitation of the ascospores (closure) caused by invagination of the ascospore-delimiting membrane is associated with a 'cage' of microfilaments surrounding each spore initial. At the same time a 'rope' of microfila-ments running along the whole length of the ascus develops, and the cage of each ascospore initial becomes attached to it (Figs. 12.5a,b). Pairing between nuclei of differing mating types occurs during the cleavage of the ascos-pores and this appears to be mediated by astral microtubules radiating from their closely asso-ciated spindle pole bodies (SPBs) and pulling the two nuclei together (Thompson-Coffe & Zickler, 1994). This pairing between genetically different nuclei during ascospore formation is similar to the recognition mechanism leading to

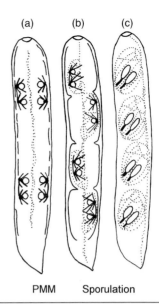

PMM   Sporulation

Fig 12.5 Nuclear alignment in the pseudohomothallic four-spored fungi *Podospora anserina* and *Neurospora tetrasperma*. Fine continuous lines represent the spore cell membrane (omitted in c for clarity); dotted lines represent microfibrils; thicker continuous lines represent microtubules. Spindle pole bodies (SPBs) are shown as black bars and nuclei as open circles or ovals. (a) Following post-meiotic mitosis (PMM) the eight nuclei are seen in pairs linked to each other by microtubules which radiate from the SPBs. A rope of actin—myosin is forming along the centre line of the ascus. (b) The nuclei become rearranged and move to a staggered formation along the microfibrillar rope as microfibrillar cages form and microfibrils extend upwards towards the SPBs. The spore membranes begin to invaginate. (c) Nuclei are re-aligned in a row, presumably by the actin—myosin rope and cage assembly. Redrawn from Thompson-Coffe and Zickler (1994), with permission from Elsevier.

karyogamy in the crozier prior to meiosis. Where the SPBs are not closely associated, uninucleate spores develop.

It is interesting to compare nuclear behaviour within the asci of the two pseudohomothallic four-spored ascomycetes *Podospora anserina* and *Neurospora tetrasperma* (see Fig. 12.6). In *N. tetra-sperma*, the mating type idiomorphs *A* and *a* lie close to the centromere so that *A* and *a* almost invariably go with the centromeres to opposite poles of the first meiotic division spindle (M_I), i.e. there is first-division segregation of the alleles for mating type. The second division spindles (M_{II}) then overlap. The third nuclear

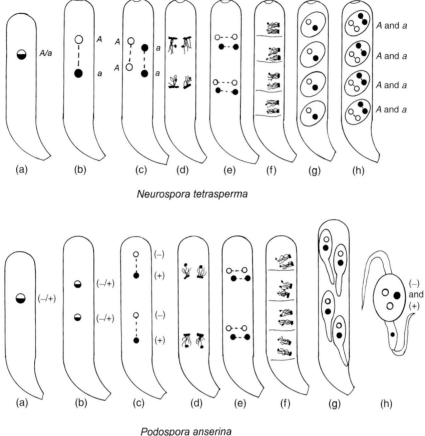

Neurospora tetrasperma

Podospora anserina

Fig 12.6 Schematic diagram of ascus development in the two pseudohomothallic ascomycetes *Neurospora tetrasperma* and *Podospora anserina* (based on Raju & Perkins, 1994). *Neurospora tetrasperma*: (a,b) First meiotic nuclear division M_I. The mating type alleles *A* and *a* are closely linked to the centromere and segregate at the first division. (c) During the second meiotic nuclear division M_{II} the spindles of the dividing nuclei lie in tandem parallel to the long axis of the ascus. (d) Interphase of M_{II}; the nuclei are aligned in pairs. (e) Telophase of a post-meiotic mitosis PMM_I; the spindles are paired and lie transversely to the long axis of the ascus. (f) Interphase of PMM_I; the four pairs of nuclei are realigned more or less regularly to one side of the ascus prior to ascospore delimitation. (g) Young ascospores are binucleate and heterokaryotic, containing both kinds of mating type allele. (h) Following a second post-meiotic mitosis PMM_{II}, the ascospores contain four nuclei, two of each mating type. *Podospora anserina*: the alleles for mating type are not tightly linked to the centromere and do not segregate at M_I (b) but at M_{II} (c). (d–f) as for *N. tetrasperma*. (g) Following ascospore delimitation, four binucleate, heterokaryotic ascospores are formed. (h) A second post-meiotic division has occurred. Three of the four resulting nuclei, two of one mating type and one of the other, remain in the body of the ascospore whilst the fourth nucleus migrates to the basal appendage and degenerates.

division (PMM) brings *A* and *a* nuclei close together in four pairs, and walls develop around the nuclei and cytoplasm so that the binucleate ascospores each contain one *A* and one *a* nucleus. In *P. anserina*, the same result is achieved by different means. The mating type idiomorphs (+) and (−) are sufficiently distant from the centromere for a single cross-over to occur between the centromere and the locus of the idiomorphs. Second-division spindles do not overlap and, by the mechanism outlined above, nuclei of opposite mating types are brought close together at the PMM stage and become enclosed in walls as binucleate ascospores. A further mitosis brings the number of nuclei in each ascospore to four, and three nuclei remain in the body of the ascospore whilst one migrates to the tail (Raju & Perkins, 1994).

Mating type factors in *Podospora anserina*

As explained above, wild-type *P. anserina* is pseudohomothallic, each binucleate ascospore normally containing both distinct mating type idiomorphs *mat+* and *mat−*, whilst uninucleate ascospores contain either *mat+* or *mat−*. The genes which control mating type specificity have been labelled *FPR1* and *FPR2* (fertilization plus and minus regulators). The molecular structure of both genes has been determined (Debuchy & Coppin, 1992; Coppin *et al.*, 1997). The *mat+* locus contains 3800 ± 200 base pairs (bp) with the *FPR1* gene within it, whilst the *mat−* locus is larger and contains 4700 ± 200 bp, enclosing *FPR2* and three regulatory genes *SMR2*, *SMR2* and *FMR1*. There is a close similarity between the structure of the mating type genes in *P. anserina* and *Neurospora crassa*.

Studies on incompatibility in *Podospora anserina*

Incompatibility is usually defined as the genetic control of mating competence, but this concept extends beyond the sexual phase to the vegetative phase. Two different systems provide genetic control, namely homogenic and heterogenic incompatibility (Esser & Blaich, 1994). **Homogenic** incompatibility is caused by the sexual incompatibility of nuclei carrying identical idiomorphs, and it thus favours outbreeding. In contrast, in **heterogenic** incompatibility (also known as heterokaryon, somatic or vegetative incompatibility), the coexistence of nuclei in a common cytoplasm is inhibited by the genetic difference in one or more genes. Thus heterogenic incompatibility restricts outbreeding. It may also play a role in speciation. Another consequence is the reduced risk of transmission, following hyphal anastomosis, of the spread of infectious cytoplasmic elements such as mycoviruses or transposons (e.g. in *Cryphonectria*, p. 375).

Heterogenic incompatibility was discovered when attempts were made to cross strains of *P. anserina* of different geographic origin. When two mycelia grow towards each other and intermingle, hyphal anastomosis occurs. Nuclear exchange is not inhibited but is followed by an antagonistic reaction, sometimes accompanied by death of the fusing cells and by profuse branching of adjacent cells. This **barrage phenomenon**, observed as a white or colourless zone between two mycelia, occurs irrespective of mating type. Perithecium formation may occur in inter-racial crosses of differing mating types but the number of perithecia is much reduced. In some pairings, one or both of the reciprocal crosses between the different mating partners are unsuccessful.

Nine unlinked loci are now known to be involved in the control of heterokaryon incompatibility, and these are termed *het* loci. A *het* locus can be defined as a locus in which heteroallelism cannot be tolerated in a heterokaryon. The nine *het* loci comprise five allelic systems (in which different alleles of the same gene provoke vegetative incompatibility) and three non-allelic systems (involving the interactions of two specific alleles from different loci). One locus (*het*-V) is simultaneously involved in an allelic and a non-allelic interaction (Saupe, 2000). The molecular structures of the genes at some of these loci have been characterized. The different genes encode very different products in the form of HET polypeptides. Complexes between the different HET polypeptides may function in the recognition process between self and non-self and may act as the trigger to mediate biochemical events causing vegetative incompatibility. Alternatively, HET heterocomplexes may function to poison the cell and thus may directly mediate growth inhibition and death (Glass *et al.*, 2000).

The *het-s/het-S* allelic system of *P. anserina* is of particular interest. Both alleles encode polypeptides differing only in a few amino acids, and incompatibility results if a heterodimer is formed. However, whilst the HET-S protein is immediately active, HET-s is initially translated as an inactive form, HET-s*, which is present in a soluble form in the cytoplasm. Biologically active HET-s molecules may arise by a rare spontaneous rearrangement to another tertiary conformation, and HET-s molecules have the ability to convert HET-s* to their own state by catalysing this conformational change. This interaction may lead to the formation of aggregates. Once initiated, the conversion of HET-s* to

HET-s spreads like an infection throughout a mycelium at a rate of several $mm\,h^{-1}$. Further, transmission can occur from a HET-s containing hypha to a HET-s* mycelium by anastomosis. The ability of a protein to convert others to its own state in an infectious transcription-independent manner, accompanied by the formation of cytoplasmic aggregates, has the hallmarks of a prion disease (Coustou et al., 1997; Coustou-Linares et al., 2001).

Senescence in *Podospora anserina*

Podospora anserina has been the subject of research into senescence and has been treated as a model of the ageing phenomenon in more complex organisms (Griffiths, 1992; Bertrand, 2000; Silar et al., 2001). In *P. anserina* senescence is defined as a diminution in the ability of cells to proliferate and/or differentiate. This may or may not culminate in cell death. In pure cultures of *P. anserina*, senescence is marked by a progressive reduction in growth rate and loss of ability to form perithecia. Eventually it proves impossible to transfer viable sub-cultures so that a given isolate has a limited lifespan. Different isolates of *P. anserina* have characteristic lengths of growth before growth ceases, and the mean lengths of growth can be used as a convenient indicator of lifespan. For example, two races A and S grown on cornmeal agar in glass tubes ($20 \times 150\,mm$) at 26°C in the dark had mean lengths of 15 and 170 cm, respectively (Smith & Rubenstein, 1973). A hypothesis to explain the phenomenon of senescence in *P. anserina* is that, after a period of growth characteristic of a given race of the fungus, a senescence factor appears in a culture and is presumed to be produced, or to reproduce itself, more rapidly than other cellular components. The factor is transmissible through hyphal anastomosis, i.e. fusion between a senescent hypha and a non-senescent hypha results in the non-senescent hypha acquiring the factor controlling senescence. Senescence is inherited maternally; it can be transmitted to 90% of the progeny of a senescent protoperithecial strain but to none of the progeny of a spermatial parent. No nuclear mixing is involved.

A large number of genes (between 600 and 3000) can modulate lifespan; 50% increase it and 50% diminish it (Rossignol & Silar, 1996). Senescence is a complex process affected by environmental factors and is also controlled by the interactions between nuclear and mitochondrial DNA (Osiewacz & Kimpel, 1999; Osiewacz, 2002). The onset of senescence is marked by the appearance of dysfunctional mitochondria and of circular plasmid-like senility DNAs derived from the mitochondria.

During the respiratory activities of mitochondria, reactive oxygen species (ROS) are generated as by-products and these molecules are able to damage all cellular components, leading to cellular dysfunctions such as the cytochrome oxidase pathway (Osiewacz, 2002). To compensate for these dysfunctions, ROS scavengers can be produced which reduce the level of ROS. Alternative oxidative pathways may also be induced which may help in reducing the adverse effects of ROS (Dufour et al., 2000; Lorin et al., 2001). It is interesting that different mutants with defective mitochondrial DNA associated with growth arrest can be restored to wild-type mitochondrial DNA by crossing, indicating an important role of sexual reproduction in this pseudohomothallic fungus (Silliker et al., 1997).

12.2.3 *Neurospora* (Sordariaceae)

There are about 12 species of *Neurospora*, mostly growing on soil. A key has been provided by Frederick et al. (1969). Many species grow in humid tropical and subtropical countries but others have been reported from temperate areas (Perkins & Turner, 1988; Turner et al., 2001). In nature, the most conspicuous species colonize burnt ground and charred vegetation following fire caused by volcanic eruptions or deliberate burning to clear vegetation (slash and burn) or crop residues such as sugar cane. Within a few days the burnt areas are covered by an orange or pink powdery mass of macroconidia. The commonest species here is *N. intermedia* (Perkins & Turner, 1988). The association with burnt ground is related to the fact that dormant ascospores in the soil are stimulated to germinate by heat. *Neurospora* species also grow in warm humid environments such as wood-drying kilns and bakeries where they cause serious trouble because of their rapid growth and sporulation. For this reason *N. sitophila* is sometimes called the red bread mould. Strains of *N. intermedia* are used in the preparation of the fermented food 'omchom' (ontjom) in which conidia are used to inoculate soybean or peanut solids from which oil and protein have been extracted by pressing.

Neurospora as a genetic tool

Neurospora has been widely used in genetic and biochemical studies (Perkins, 1992; Davis, 1995). The development of auxotrophic mutants deficient in successive steps of arginine biosynthesis led to the proposition of the 'one-gene–one-enzyme' hypothesis by Beadle and Tatum (1941) who were awarded the Nobel Prize in 1958. There is a very extensive literature relating to the genus. The loci of over 1000 genes have been mapped (Perkins *et al.*, 2000), and the complete genome of *N. crassa* has now been sequenced. It is haploid and has seven chromosomes. The best-known species are *N. crassa* and *N. sitophila*, both of which are eight-spored and heterothallic. *Neurospora tetrasperma* is four-spored and pseudohomothallic. Some other species, e.g. *N. africana, N. dodgei* and *N. terricola*, are homothallic (Coppin *et al.*, 1997). The homothallic species do not form conidia.

The reasons why *Neurospora* has proven so useful as a tool in biochemical and genetic research are: (1) that it is haploid; (2) that wild-type strains have simple nutritional requirements, namely a carbon source, simple mineral salts and one vitamin, biotin; (3) that mutations can be induced readily by the use of chemical mutagens or UV irradiation of conidia; (4) that growth and sexual reproduction is rapid; and (5) that tetrad analysis by dissecting asci is straightforward. By the use of marked strains it is now not even necessary to dissect asci because tetrad analysis can be performed on octets of projected ascospores. Another advantage of *Neurospora* is that cultures can be preserved for long periods in suspended animation as spores stored over silica gel, or following lyophilization or freezing.

The life cycle of *Neurospora*

The life cycle of *N. crassa* is illustrated diagrammatically in Fig. 12.7. The name *Neurospora* is derived from the characteristically ribbed ascospores. The ascospore walls bear dark, raised, thicker ribs separated by thinner, paler, branched or unbranched **inter-costal veins** (Figs. 1.19 and 12.8). The spores are multinucleate and have reserves of lipids and the carbohydrate trehalose. The dark ascospores of *N. crassa* are viable for many years and do not germinate readily unless treated chemically (e.g. by furfural) or by heat shock (e.g. 60°C for 20–40 min). In contrast, the conidia are killed by such heat treatment. Following treatment, the ascospores germinate through a germ pore at either or both ends, forming at first a globose, inflated vesicle and then a coarse, incompletely septate, rapidly growing mycelium, each segment of which is multinucleate. Cytoplasm and organelles including nuclei pass freely through the septa.

Within 24 h, the mycelium can begin asexual reproduction. The cues inducing conidial development include light, desiccation, and nutrient deprivation. Upright branches develop which, instead of continuing to grow by hyphal tip elongation, undergo repeated apical budding. The resulting cells are separated from each other by incomplete septa with a wide central pore. These cells have been termed **proconidia** (Springer & Yanofsky, 1989). The proconidia continue to bud apically, forming multinucleate macroconidia (blastoconidia) separated from each other by septa with narrower pores. The septa thicken and develop a centripetal furrow which widens, leaving a central strand of material, the connective, by which the spores remain attached to each other. Further conidia develop by budding of the terminal conidium on a chain and, when the terminal conidium gives rise to two buds, the chain branches (Figs. 12.8d,e; Hashmi *et al.*, 1972). Conidia of this type belong to the form genus *Chrysonilia* (formerly *Monilia*). The individual segments of the spore chain break apart and are readily dispersed by wind. The surface of macroconidia in wild-type strains is made up of hydrophobin rodlets which impregnate it, but the conidia of the *easily wettable* mutant *eas* are devoid of rodlets

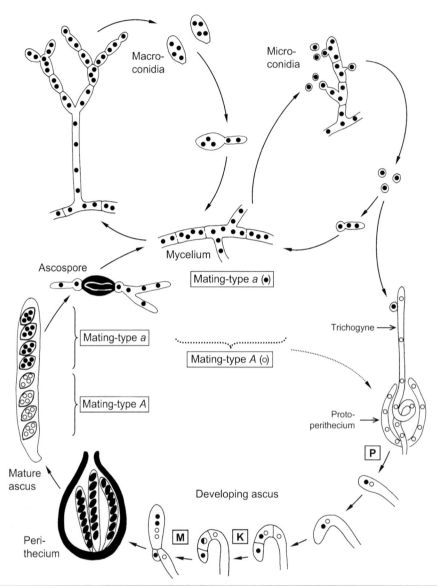

Fig 12.7 The life cycle of *Neurospora crassa*, diagrammatic and not to scale (based on Fincham & Day, 1971). The fungus is heterothallic and a mature ascus (left of diagram) contains four haploid multinucleate ascospores of mating type *a* and four of mating type *A*. Ascospores germinate to form a branched, incompletely septate mycelium with multinucleate segments. Multinucleate macroconidia and uninucleate microconidia develop, and both types of conidium can germinate to form a new mycelium. Protoperithecia, consisting of a coiled ascogonium and a trichogyne, and surrounded by sterile hyphae, also develop on the haploid mycelium (right of diagram). When microconidia or macroconidia of opposite mating type are transferred to a trichogyne, plasmogamy (P) occurs and the protoperithecium develops into a perithecium. Ascogenous hyphae containing paired *A* and *a* nuclei grow out from the ascogonium. At the tip of the ascogenous hypha a crozier develops (bottom of diagram) and, within the penultimate cell, karyogamy (K) occurs between nuclei of the two mating types. Within this diploid cell, which is the ascus initial, meiosis (M) takes place, followed by mitoses (not shown). Ascospores are cleaved around the nuclei (not shown) and the mature asci discharge their ascospores through the neck of the perithecium.

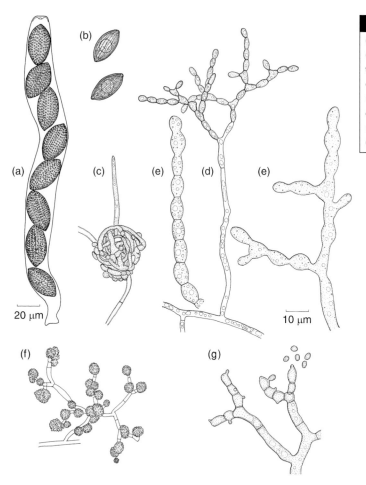

Fig 12.8 *Neurospora crassa.* (a) Ascus. (b) Ascospores showing ribbed surface. (c) Protoperithecium showing projecting trichogyne. (d) Macroconidia from one-day-old culture. (e) Enlarged view of developing macroconidia. (f) Microconidia forming sticky clusters. (g) Enlarged view showing origin of microconidia. (a–d,f) to same scale; (e,g) to same scale.

and cohere in sticky masses (Beever *et al.*, 1978). The pink colour of the conidia is due to the presence of a carotenoid pigment, neurosporoxanthin, which is stimulated to develop by light. The spores are formed in vast numbers. In laboratories they can cause serious contamination of other cultures, partly because of the rapid growth of the mycelium (up to 5 mm h^{-1}), and partly because macroconidia can develop in profusion beyond the rims of closed Petri dishes. In humid environments, the mycelium may grow through the cotton wool plugs of testtube cultures and sporulate on the plugs. Shaw (1993) has reported that macroconidia formed on steamed pine logs in Queensland, Australia, are collected by honeybees in their pollen baskets.

Cultures derived from a single ascospore also develop two other types of reproductive structure. In contrast to the large, dry, wind-dispersed macroconidia which are formed within 1–2 days, clumps of smaller, oval, uninucleate, sticky microconidia develop after about 12–15 days (Figs. 12.8f,g; Maheshwari, 1999). The conidiogenous cells from which the microconidia develop have been interpreted as reduced phialides. The microconidia are capable of slow and variable germination but function primarily as spermatia. Coiled ascogonia, terminated by long tapering trichogynes and surrounded at the base by hyphae, also develop (Fig. 12.8c). Such structures are termed protoperithecia or bulbils and each often forms several trichogynes. In *N. crassa* and *N. sitophila*, no further development occurs in single ascospore cultures, i.e. each strain is selfincompatible. Incompatibility is controlled by a pair of mating type idiomorphs, *A* and *a*, and if two compatible strains are grown together in a Petri dish for a few days, microconidia of one strain can be transferred to trichogynes of the opposite strain by flooding with sterile water. In nature it is possible that mites or insects are involved in transfer. There is evidence that a microconidium produces a pheromone which induces directional growth (positive chemotropism) of a trichogyne of opposite mating type towards it before plasmogamy occurs (Bistis, 1983). The transfer of macroconidia or hyphae of the opposite strain to a trichogyne can also effect fertilization. Fusion between the trichogyne and the fertilizing cell is followed by the migration of one or more nuclei from the fertilizing cell down the trichogyne into the ascogonium. The development of ripe perithecia occurs within 7–10 days and follows the typical general ascomycete pattern. This has been described by Nelson and Backus (1968) in two homothallic species. Much is now known of the genetic control of sexual development in *Neurospora* and several genes have been identified which control steps in the process. The cytological details of ascus development have also been worked out. After the eight-nucleate stage in the developing ascus of *N. crassa*, several mitoses ensue so that a fully developed ascospore may contain as many as 32 nuclei (Raju, 1992a).

Mating type genes in *Neurospora*

In *N. crassa*, heterokaryons are not normally formed between mycelia of opposite mating types, and this implies that plasmogamy usually occurs only between a trichogyne of one strain and a fertilizing agent (e.g. a microconidium, macroconidium or hypha) of the opposite strain. This condition is termed **restricted heterokaryosis**. Even within one mating type, the ability to form heterokaryons is under genetic control, i.e. there is heterokaryon incompatibility as in *Podospora anserina* and heterokaryons generally develop only if the *het* genes which control compatibility are homoallelic (see p. 325). Several *het* genes have been identified (Mylyk, 1976; Perkins, 1992), and Micali and Smith (2003) have provided evidence of a yet more complex regulation of heterokaryon incompatibility in the shape of suppressor genes which modify the effect of *het* and mating type genes. When the

mycelia of unlike genotype anastomose, cytoplasmic incompatibility results in vacuolation and disorganization of cell contents in the region of the anastomosis. Similar cytoplasmic reactions are visible when anastomosis occurs between the hyphae of wild-type strains differing in mating types. In contrast to *N. crassa*, heterokaryons are readily formed between different mating type strains of *N. tetrasperma*, which thus exhibits **unrestricted heterokaryosis**.

The molecular structure of the *A* and *a* idiomorphs has been elucidated in *N. crassa*. They are strikingly dissimilar (Glass *et al.*, 1988). The *A* idiomorph is composed of a region of 5301 bp bearing little similarity to the *a* idiomorph comprising 3235 bp (Glass *et al.*, 1990; Staben & Yanofsky, 1990). The *A* idiomorph of *N. crassa* is also involved in heterokaryon incompatibility. There are similarities in the structure and functions of the mating type idiomorphs between *N. crassa*, *Podospora anserina* and the yeasts *Saccharomyces cerevisiae* and *Schizosaccharomyces pombe*, but there are also differences (p. 266; Glass & Lorimer, 1991). The idiomorphs of *N. crassa* are larger than those of the yeasts. Mating type idiomorphs are present in several homothallic species of *Neurospora* which hybridize with the *A* DNA probe of *N. crassa* (Coppin *et al.*, 1997). This and other lines of evidence suggest that the heterothallic condition was primitive and that the homothallic condition has probably arisen several times in the course of evolution and may have a selective advantage (Metzenberg & Glass, 1990).

The pseudohomothallic condition, represented by *N. tetrasperma*, is of interest. This species is functionally homothallic in that the mycelium from a single heterokaryotic ascospore containing nuclei of two distinct mating types can develop perithecia directly. However, homokaryotic mycelia can develop from uninucleate ascospores and also from about 20% of macroconidia. Such mycelia are capable of outcrossing. Raju (1992b) has summarized,

Thus *N. tetrasperma* appears to have the best of both worlds. On one hand the single-mating-type homokaryotic cultures offer *N. tetrasperma* the advantages of outbreeding. On the other hand the heterokaryotic ($A + a$) cultures have the potential advantage of hybrid vigour. They can also enter the sexual cycle and produce ascospores without waiting for a sexual partner. This is very useful in situations where rapid completion of the life cycle is advantageous.

Neurospora and the biological clock

Many organisms show a daily (circadian) rhythm in their activities, entrained by a combination of light and temperature (Dunlap, 1999). When transferred to a uniform environment, they may continue to display the same rhythm, suggesting that it is controlled by some internal clock. Important research helping to interpret the molecular basis of circadian rhythmicity in fungi as well as other eukaryotes is being performed on *N. crassa* (Davis, 1995; Liu, 2003; Dunlap & Loros, 2004). A *band* mutant (*bd*) was discovered which, when grown in culture in long tubes, formed alternating bands of macroconidia interspersed by non-sporulating bands. In continuous darkness, bands continued to form, with a periodicity of 21.5 h, little affected by temperature. From the *bd* mutant, further mutants were developed by mutagenesis in which the *frequency* of the circadian rhythm was affected. The *frq* locus was identified as a key control element of the frequency of sporulation, with partial loss-of-function mutations capable of shortening the frequency to as little as 16 h or extending it to 29 h. The effect of temperature on the circadian rhythm seems to be controlled by the temperature-dependent alternative splicing of the *frq* mRNA, giving either of two major FRQ proteins (Colot *et al.*, 2005). Several other loci are also involved in the integration of the rhythm with temperature and light. The *white collar wc-1* and *wc-2* gene products are especially important, forming a heterodimer (WCC = white collar complex) which stimulates *frq* transcription. The FRQ dimer, in turn, inhibits existing WCC and lowers the continued expression of WC-1, thus introducing a circadian pattern into the cycle such that the levels of FRQ proteins hit their lowest point late at night, and WC-1 (and WCC activity) late in the day (Dunlap & Loros, 2004). It should be noted that regulation is partly at the level of gene expression, partly by translation of existing mRNA molecules, and partly by protein phosphorylation/dephosphorylation. Entrainment of the rhythm, i.e. the switching on of the clock, is sensitive to blue light and the photo-receptor has been identified as the WC-1 component of WCC coupled with the chromophore FAD (flavin-adenine dinucleotide).

12.2.4 *Chaetomium* (Chaetomiaceae)

There are over 80 species of *Chaetomium* (von Arx *et al.*, 1986), many of which are cosmopolitan, growing in soil and fruiting on cellulose-rich substrata such as seeds, textiles in contact with soil, straw, sacking and dung. Wood infected by *Chaetomium* spp. may undergo a superficial decay known as soft rot. Wood inside buildings damaged by flooding or water used in fire control is particularly susceptible. Most species are saprotrophic and cellulolytic, but some have been isolated from human lesions. Some produce mycotoxins (Ugadawa, 1984) and others have been used in biological control because of their competitive ability to colonize cereal stubble, thus displacing plant-pathogenic fungi (Dhingra *et al.*, 2003). Many potentially valuable chemical compounds such as enzymes (cellulases, xylanases), pharmaceutical products and antifungal substances have been extracted from *Chaetomium* in culture. *Chaetomium thermophile* is thermophilic and has potential for use in composting palm oil fibre for recycling biomass (Suyanto *et al.*, 2003).

The perithecia of *Chaetomium* are superficial, barrel-shaped, thin-walled and, in most species, clothed with projecting, dark, stiff hairs. In *C. elatum*, one of the commonest species, the hairs are dichotomously branched. In others, e.g. *C. cochliodes*, the body of the perithecium bears straight or slightly wavy, unbranched hairs, whilst the apex bears a group of spirally coiled hairs. The hairs are roughened or ornamented, and the type of ornamentation is an aid to identification (Guarro & Figueras, 1989). It is likely that the perithecial hairs have special functions. Jerking caused by the movement of the hairs over each other on drying may help in ascospore dispersal. Another possible function is to deter grazing of the perithecia by insects and

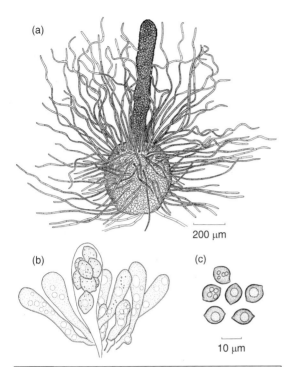

(a)

200 μm

(b)   (c)

10 μm

Fig 12.9 *Chaetomium globosum.* (a) Perithecium showing cirrhus of ascospores. (b) Asci. (c) Ascospores.

other arthropods (Wicklow, 1979). When the perithecia are ripe, a column-like mass (cirrhus) of dark ascospores is extruded through the ostiole, supported by the perithecial hairs which surround it (Fig. 12.9a). The spore column results from the breakdown of the asci within the body of the perithecium, i.e. the asci do not discharge their spores violently. The young asci are cylindrical to club-shaped, but this stage is very evanescent and is only found in young perithecia (Fig. 12.9b), often before the spores have become pigmented and sometimes before the perithecium has developed an ostiole. Spores of most species are pale brown to grey and lemon-shaped, with a single terminal germ pore.

The development of perithecia in *Chaetomium* shows some variation between species (Whiteside, 1957, 1961). The ascogonia are coiled and lack antheridia. Investing hyphae arise from the ascogonial stalk or from vegetative cells surrounding the ascogonium. Perithecial hairs develop early from the external

cell layer. The centrum is at first filled with hyaline pseudoparenchyma. At the apex of the perithecium, certain of these cells become meristematic and give rise to the elongate periphyses which line the inside of the ostiole. Ascogenous hyphae develop from the ascogonium at the base of the centrum and, at about this time, the surrounding pseudoparenchyma cells of the centrum deliquesce to form a cavity. In some species croziers develop, but in others they are absent. Paraphyses of two types have been described. In *C. brasiliense*, lateral paraphyses arise from the pseudoparenchyma cells outside the hymenium, whilst hymenial paraphyses have been illustrated in *C. globosum* (Whiteside, 1961). The paraphyses are evanescent and disappear as the asci mature.

Most species of *Chaetomium* are homothallic, but a few, e.g. *C. cochliodes*, are heterothallic. Both homo- and heterothallic strains of *C. elatum* have been reported. Conidial states are rare in *Chaetomium*, but simple phialides and phialospores occur in *C. elatum* and *C. globosum*, whilst *C. piluliferum* forms both phialospores and globose thalloconidia of the *Botryotrichum* type (Daniels, 1961).

12.3 | Xylariales

The order Xylariales is probably polyphyletic and comprises about 800 species in 8 families (Kirk *et al.*, 2001). Members of this group produce dark perithecial stromata with asci which have an iodine-positive apical apparatus. Ascospores are typically melanized. Xylariales are saprotrophs or plant pathogens and are associated especially with the bark and wood of trees.

We shall only consider one family, the Xylariaceae, which is by far the most important. The dark-coloured ascospores are generally smooth-walled and inaequilateral (one face being more strongly curved than the other), with a hyaline germ slit. Anamorphs are sometimes also stromatic, producing hyaline or lightly pigmented conidia holoblastically from a sympodially proliferating conidiogenous region, the **sympodula** (Fig. 12.13). They have been placed in several anamorph genera, including *Geniculosporium* and *Nodulisporium* (Rogers, 1979, 2000; Whalley, 1996). Some develop as a thin covering of the perithecial stroma. The conidia are dry and are dispersed by air currents, rain splash or insects.

There are about 40 genera in the Xylariales. Most species (e.g. *Xylaria, Hypoxylon, Daldinia*) are hemi-saprotrophic or saprotrophic, fruiting on wood. These forms are ligninolytic and cause white-rot of their substrate (Rayner & Boddy, 1988; Eaton & Hale, 1993). The limits of the mycelial colonies in decaying wood are demarcated by black zone lines (pseudosclerotial plates) made up of brown bladder-like fungal cells which fill the wood tissue (Campbell, 1933). Other genera fruit on herbivore dung, e.g. *Hypocopra, Poronia, Podosordaria* and *Wawelia*. Some are serious plant pathogens; for example, *Hypoxylon mammatum* (*H. pruinatum*) causes canker of aspen (*Populus tremuloides* and other *Populus* species) in North America (Manion & Griffin, 1986), *Biscogniauxia mediterranea* causes canker of cork oak (*Quercus suber*), and *B. nummularia* causes strip canker of beech (*Fagus sylvatica*). *Kretzschmaria* (*Ustulina*) *deusta* causes a fatal butt rot of beech, elm (*Ulmus*) and horse chestnut (*Aesculus hippocastanum*), whereas *Rosellinia necatrix* is a plurivorous root pathogen known to attack over 130 plant species. Members of the Xylariaceae have been isolated as endophytes (symptomless symbionts or commensals) from a range of plants on which they do not fruit so that their host range and distribution may extend far beyond that which would have been inferred from the occurrence of their ascocarps (Petrini & Petrini, 1985; Petrini *et al.*, 1995). Although some species are plurivorous, others have a restricted host range, e.g. *H. fragiforme* which generally fruits on dead beech twigs (*Fagus*) and *Xylaria carpophila* which fruits only on fallen beech cupules.

The development of perithecia in the family conforms to what Luttrell (1951) has termed the '*Xylaria*' type:

> The ascogonia are produced free upon the mycelium or more commonly within a stroma. Branches from the stalk cells of the ascogonium or from neighbouring vegetative hyphae surround the ascogonium and form the perithecial wall. Hyphal branches with free tips (paraphyses) grow upward and inward from the inner surface of the wall over the base and sides of the perithecium. Pressure exerted by the growth of opposed paraphyses expands the perithecium and creates a central cavity. The perithecium becomes pyriform as a result of growth of hyphae in the apical region of the wall to form a neck. The layer of inward-growing hyphae is continuous up the sides and into the perithecial neck. Growth of these hyphae within the neck produces a schizogenous ostiole lined with free hyphal tips (periphyses). The ascogonium produces ascogenous hyphae which typically grow out along the inner wall over the base and sides of the perithecium. Asci derived from the ascogenous hyphae grow among the paraphyses to form a continuous hymenium of asci and more or less persistent paraphyses lining the perithecial cavity. In some forms the paraphyses are evanescent, and the ascogenous hyphae form a plexus in the base of the perithecium. The asci then arise in a single aparaphysate cluster.

The cytology of ascus development in *Xylaria* (Beckett & Crawford, 1973; Rogers, 1975a) and in the related genus *Hypoxylon* (Rogers, 1965, 1975b) follows the usual pattern in most cases. The ascospores may be uninucleate, binucleate, or occasionally multinucleate (Rogers, 1979). In *X. polymorpha* and *H. serpens*, immature ascospores are divided by a septum near the base which cuts off a small appendage. The appendage disappears in the mature ascospore, leaving a truncate base. The ascospore wall of *Daldinia concentrica*, as seen with the transmission electron microscope, consists of five recognizable layers numbered progressively from the outside inwards as wall layers W1–W5 (Beckett, 1976a). The thin, non-pigmented outer layer W1 acts as a sheath which completely encloses the spore. Before germination this layer is sloughed off (see below). The wall of the mature ascospore is dark, probably due to the pigment melanin located in an inner wall layer (W4). There is a hyaline germ slit in W4 running the length of the ascospore (Fig. 12.10a). Germ slits are microfibrillar in construction but their structure varies within the family (Beckett, 1979a,b). The germ slits create a weak point in the ascospore wall, causing the spore to gape wide open as it swells and permitting the germ tubes to emerge (Beckett, 1976b; Chapela *et al.*, 1990; Read & Beckett, 1996). The ascus tip contains a cylindrical apical apparatus which is amyloid (i.e. it stains bright blue with iodine). The apical apparatus is pierced by a narrow pore and is everted as the ascus explodes (Greenhalgh & Evans, 1967; Beckett & Crawford, 1973).

12.3.1 *Daldinia*

There are about 13 species of *Daldinia* worldwide (Ju *et al.*, 1997), 5 of which grow in Northern Europe (Johannesson *et al.*, 2000). They form concentrically zoned perithecial stromata on wood. The best known is *D. concentrica*, with stromata mostly on ash (*Fraxinus excelsior*) but occasionally on other hosts. *Daldinia loculata* (*D. vernicosa*) and *D. fissa* form stromata on charred branches of birch (*Betula*), gorse (*Ulex*) and some other hosts following fire.

Daldinia concentrica colonizes the living branches of ash as a symptomless endophyte but can continue to grow saprotrophically, fruiting on dead branches and trunks. Recently infected wood (calico wood) has a dark speckled appearance caused by the presence of a dark-coloured mycelium in the vessels of the spring wood. Boddy *et al.* (1985) isolated *D. concentrica* from attached branches of ash and regarded it as a primary colonizer. Only a small number of individual mycelia were isolated from any one branch, but often these occupied extensive volumes of wood. Individual mycelia are detected by their reactions in culture to other individuals of the same species. Identical mycelia intermingle freely, showing no obvious reaction, but mycelia of different genotype show vegetative incompatibility when confronted. Such observations have led to the suggestion that the colonization of attached branches is by a process of latent invasion (Boddy & Rayner, 1983) whereby the fungus might be distributed in the sap stream of the living tree host as mycelial fragments or spores. It is likely that *D. concentrica* is heterothallic (Sharland & Rayner, 1986) like *D. loculata* which grows on fire-scorched birch

branches (Johannesson *et al.*, 2001). In this fungus, the *Nodulisporium* conidial stroma develops in spring beneath the birch bark before the perithecial stroma emerges. Pyrophilous (fire-loving) insects feed on the conidia and also disperse them. Conidia of different mating types and different genotypes may be spread in this way, and since some perithecial stromata are known to contain several genetically distinct perithecia, multiple mating events are possibly involved.

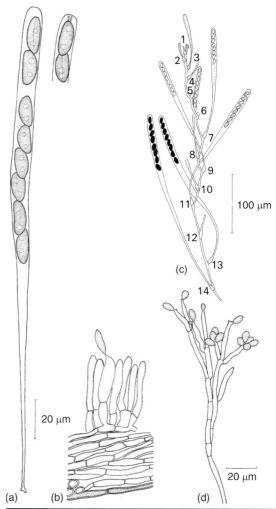

Fig 12.10 (a,b) *Xylaria hypoxylon*. (a) Ascus. The ascus tip to the right has been stained with iodine to reveal the apical apparatus. (b) Conidiophores. (c,d) *Daldinia concentrica*. (c) Ascogenous hypha (after Ingold, 1954b). The numbers represent successive asci working backwards from the apex. (d) *Nodulisporium*-type conidiophore.

Daldinia concentrica forms large (5–10 cm diameter) hemispherical purplish-brown annual stromata called 'cramp balls' or 'King Alfred's cakes'. They contain ripe asci between May and October. In cross section (Plate 5a), the stromata show a concentric zonation of alternating light and dark bands. The surface of young stromata may be covered with a pale fawn powdery mass of conidia. The conidia are dry and ovoid in shape, developing successively at the tips of branched conidiophores by the outgrowth of the wall and, when detached, leave a small scar (Fig. 12.10d). Conidia of this type have been named *Nodulisporium tulasnei*. Perithecia develop in the outer layers of the stroma. The perithecial wall is lined by ascogenous hyphae which are unusual in that there is often a considerable distance separating successive asci (Ingold, 1954b) (Fig. 12.10c). The stroma of *Daldinia* apparently functions as a water reservoir and detached stromata will continue to discharge ascospores for about 3 weeks even if placed in a desiccator (Ingold, 1946). Spore discharge is nocturnal and the rhythm of spore discharge is maintained for several days if detached stromata are kept in continuous dark. In continuous light, periodic spore discharge ceases after about three days but is restored on return to alternating light and dark (Ingold & Cox, 1955). The output of spores from a single stroma of average size is about 10 million a night. The ability of perithecial stromata to store water enables the fungus to continue sporulating on dry branches. Mycelial growth is also possible at lower water potentials (−10 MPa) than by competing fungi (Boddy *et al.*, 1985).

12.3.2 *Xylaria*

There are over 100 species of *Xylaria*, most of which are lignicolous, but some are endophytic and others grow on fallen fruits (Whalley, 1985, 1987). One of the best known is *Xylaria hypoxylon*, the candle-snuff fungus. Stromata are common on stumps and fallen branches of deciduous trees. As in most Xylariaceae growing on wood, the boundaries of individual mycelia within infected tissues are visible as conspicuous black demarcation lines. The stromata are branched

and cylindrical or flattened. At the upper end, the stroma is covered by a white powdery mass of conidia (Figs. 12.10b, 12.11a). Perithecia develop later at the base of the stroma and are visible externally as swellings at the surface (Fig. 12.11b). The apical apparatus of the ascus is visible even in immature asci after staining in iodine as a bright blue cylindrical collar pierced by a narrow pore (Fig. 12.10a). *Xylaria polymorpha* ('dead men's fingers') fruits in late summer and autumn at the base of old tree stumps. The stromata are swollen, finger-like and clustered (Plate 5b). The surface is at first covered by an inconspicuous conidial layer, but eventually perithecia develop beneath the surface of the whole stroma, and are not restricted to the basal region as in *X. hypoxylon*. Both species are active wood-rotting fungi causing decay of the white-rot type. Other common species with a more restricted host range are *X. longipes* on fallen branches of sycamore (*Acer pseudoplatanus*) and *X. carpophila* on fallen cupules of beech (*Fagus sylvatica*).

12.3.3 *Hypoxylon*

This is a large genus of over 120 species (Whalley & Greenhalgh, 1973; Ju & Rogers, 1996) forming stromata which are often hemispherical or sometimes flattened on the surface of wood and bark. Some show a preference for a particular host. Common species are *H. fragiforme* almost confined to branches and trunks of *Fagus* (Fig. 12.12a), *H. multiforme* on *Betula* (Fig. 12.12b), and *H. rubiginosum* which forms flat stromata on decorticated wood of *Fraxinus*. The young stromata of all these species bear a conidial felt of the *Nodulisporium* or *Geniculosporium* type (Fig. 12.13; Chesters & Greenhalgh, 1964). Most species show nocturnal spore discharge (Walkey & Harvey, 1966b).

Freshly cut lengths of healthy beech branches incubated under water-saturated conditions show no evidence of the presence of *H. fragiforme*, but if similar sections are incubated under conditions in which the branches are allowed to dry, characteristic patches of stained or discoloured wood become apparent within 21 days (Chapela & Boddy, 1988a,b). Isolations from such areas produce several genetically

Fig 12.11 *Xylaria hypoxylon.* (a) Conidial stroma. Conidia are borne on the white tips of the branches. (b) Perithecial stromata which have developed at the base of the old conidial stroma. The knobbly swellings are the perithecia.

distinct mycelia of *H. fragiforme*, indicating that the infections are derived from a number of separate ascospores. The fungus is present in apparently healthy sapwood as a number of inconspicuous pockets of 'inoculum units' from which mycelial growth is held in check by the high water content of the wood. On drying, these infections can spread rapidly. This phenomenon is another example of latent invasion (Boddy & Rayner, 1983). Chapela (1989) used the term 'xylotropic endophyte' for fungi such as *H. fragiforme* which occur within living trees and have the capacity to extend into secondary xylem upon drying of the wood.

Ascospore eclosion
The results of experimental studies on ascospore germination in *H. fragiforme* provide a partial explanation of its host specificity. Although known to fruit on several other woody hosts, perithecial stromata are most regularly associated with drying branches and trunks of European beech (*Fagus sylvatica*) and American beech (*F. grandiflora*). Ascospores suspended in extracts from living *F. sylvatica* twigs react by a dramatic germination process termed **eclosion**, an entomological term for the emergence of a pupa from its case or a larva from an egg (Chapela *et al.*, 1990, 1993). In *H. fragiforme* this appears to be a host-specific recognition system.

Exposure to aqueous extracts of beech twigs for a period of about 10 min activates the eclosion mechanism, which is a two-stage process. In the first stage the spore swells slightly and its germ slit widens until the pigmented spore wall opens abruptly along the germ slit, forcing the outer transparent sheath (corresponding to the W1 layer of *D. concentrica*) to crack open transversely. This stage is a millisecond event. The second stage, lasting about 10 s, involves the widening of the germ slit up to about 7 μm, and expansion of the coloured wall of the ascospore. This forces the transparent outer sheath away from the rest of the spore, which then escapes and is free to germinate by the formation of a germ tube. Eclosion is readily observed with the light microscope, as summarized in Fig. 12.14 (see Webster & Weber, 2004).

The triggers which stimulate eclosion have been identified as two monolignol glucosides, Z-syringin and Z-isoconiferin, which are mobile transport intermediates of lignin biosynthesis. Both induce eclosion at micromolar concentrations. However, the presence of these germination activators in beech extracts cannot fully explain the host specificity of *H. fragiforme* because eclosion is induced by extracts from a range of trees which do not normally support fruiting of this fungus, such as *Abies*, *Corylus* and *Populus* (Chapela *et al.*, 1991).

Fig 12.12 *Hypoxylon.* (a) Perithecial stromata of *H. fragiforme* on beech (*Fagus sylvatica*). One stroma has been broken open to show the perithecia embedded in the outer layer.
(b) Perithecial stromata of *H. multiforme* on birch (*Betula pendula*).

Fig 12.13 The conidial (*Geniculosporium*) state of *Hypoxylon serpens*.

12.4 | Hypocreales

The Hypocreales are a large group of fungi, although estimates vary as to the number of taxa contained in it. Kirk *et al.* (2001) included 117 genera and 654 species, whilst Rossman (1996) stated that there may be 2000–5000 holomorphic taxa. Some are known only as anamorphs. Hypocreales are characterized by pale or brightly coloured perithecia (or rarely cleistothecia) which may be single or borne on or embedded in a fleshy stroma. The asci are unitunicate, with or without a well-defined apical apparatus.

Perithecial development conforms to the 'Nectria' type of Luttrell (1951). The ascogonia, which are formed within a stroma, become surrounded by concentric layers of vegetative hyphae which form a true perithecial wall. The cells of the inner wall layer in the apical region of the young perithecium produce a palisade of inward-growing hyphal branches. These hyphal branches grow downward to form a vertically arranged mass of hyphae with free ends termed **apical paraphyses** (Luttrell, 1965). Pressure exerted by the elongation of the apical paraphyses, accompanied by expansion of the wall, creates a central cavity within the perithecium. The free tips of the apical paraphyses ultimately push into the lower portion of the wall so that they become attached at both the top and bottom of the perithecial cavity. Ascogenous hyphae arising from the ascogonium spread out across the floor and sides of the cavity and produce asci by means of croziers. The asci grow upward among the apical paraphyses and form

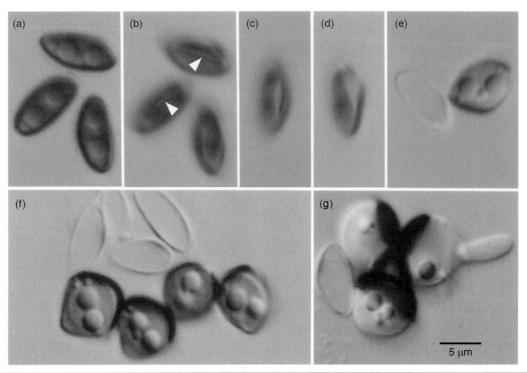

Fig 12.14 The course of events during eclosion of *Hypoxylon fragiforme* ascospores. (a,b) Dormant spores as seen in median (a) and surface view (b), the latter showing the germ slits (arrowheads). (c–e) Rupture of wall layer WI along a longitudinal seam, gaping of germ fissure of ascospore, and escape of the spore from the perispore shell formed by wall layer WI. The sequence (c–e) can be observed some 10 min after exposure of ascospores to the beech extract. It occurs abruptly, taking about 3 s. (f) Further distension of valves composed of wall layers W2–W4, 30 min after addition of beech twig extract. The perispores are still visible as ghosts. (g) Formation of a germ tube some 8 h after addition of beech twig extract. All images to same scale. Reprinted from Webster and Weber (2004), with permission from Elsevier.

a concave layer lining the inner surface of the wall in the basal region of the perithecium. At the upper end of the perithecium a schizogenous ostiole develops in the wall. It is lined with periphyses, i.e. hyphae with free apices which are attached at their bases to the inner wall of the neck. At maturity, the perithecia may protrude from the stroma and appear to be seated on its surface. Perithecial development of this type has been described for *Nectria* (Strickmann & Chadefaud, 1961; Hanlin, 1971), and for *Hypocrea* (Canham, 1969).

There is an exceptionally wide range of anamorphs (Rossman, 2000; Seifert & Gams, 2001). In most cases conidiogenous cells are phialides, and these may be terminal or lateral, single or grouped in synnemata, sporodochia or, more rarely, pycnidia. However, other developmental types of conidia may occur and in many species there are synanamorphs. The conidia are generally light-coloured and produced in slimy masses. The names of some anamorph genera related to Hypocreales are listed in Table 12.2. There is a tendency in modern classification, with support of molecular evidence, to link anamorph states with teleomorph genera even if they have, as yet, no proven connection from pure culture studies (Rossman, 2000).

Many Hypocreales are saprotrophs active in the decay of plant substrata above ground, in soil or fresh water. Others are serious plant pathogens (e.g. *Nectria* and *Fusarium* spp.) or mycoparasites, especially of agaric, bolete or polypore basidiocarps (e.g. *Hypomyces*, *Apiocrea* and *Hypocrea*; Põldmaa, 2000) but also of cultivated mushroom mycelium. Some are lichenicolous. The ability to parasitize other fungi has been employed for the biological control of fungal pathogens of plants, using species of *Trichoderma*, *Gliocladium* and *Clonostachys* (see below).

Several antifungal compounds, e.g. gliotoxin or viridin, are produced as secondary metabolites by anamorphs of *Hypocrea* spp. (*Trichoderma* and *Gliocladium*; see below). Other pharmacologically important secondary metabolites synthesized by anamorphic Hypocreales include the antibacterial antibiotic cephalosporin from *Acremonium* spp. The gibberellins are growth hormones ubiquitous in all higher plants, but they were first detected as a secondary metabolite produced by *Gibberella fujikuroi* (anamorph *Fusarium moniliforme*). This species causes the 'foolish seedling disease' in rice, in which seedlings show excessive stem elongation and eventually keel over due to the production of gibberellic acid by the infecting fungus (Tudzynski, 1997). Mycotoxins poisonous to farm animals and humans are produced by some species of *Fusarium* (see below) and other species have been reported as human pathogens. The edible fungus food 'Quorn' is

manufactured by large-scale fermentation of *F. venenatum* (previously identified as *F. graminearum*) (Trinci, 1991; Moore & Chiu, 2001). Cellulase production by *Trichoderma reesei* is exploited commercially.

General accounts of the Hypocreales have been given by Rogerson (1970) with keys to genera, and by Rossman (1996). The taxonomic treatment, i.e. the division into orders and families, varies between different authors, some placing *Claviceps* and its allies as a family (Clavicipitaceae) of the Hypocreales, whilst others, e.g. M. Barr (2001), placed them in a separate order (Clavicipitales), a disposition followed in this book. However, all are agreed that the two groups are closely related, a view based on morphological and molecular evidence (Spatafora & Blackwell, 1993). Samuels and Blackwell (2001) included four families in the Hypocreales, i.e. Hypocreaceae, Nectriaceae, Bionectriaceae and Niessliaceae, but we shall consider representatives only of the first two.

12.4.1 *Hypocrea* (Hypocreaceae)

Species of *Hypocrea*, of which about 100 are known, usually fruit on decaying wood or occasionally on herbaceous plant material, forming brightly coloured fleshy perithecial stromata, with perithecia embedded in the outer layers. The thin-walled asci contain 8 two-celled ascospores, and in many species the spores separate into 2 part-spores before ascus discharge, so that 16 part-ascospores are released (Fig. 12.15c). The ascospores may be colourless (hyaline) or green in colour. *Hypocrea pulvinata* forms bright yellow stromata on the underside of dead overwintered basidiocarps of *Piptoporus betulinus*, the birch polypore. It is possible that this fungus grows parasitically on the basidiocarp. The ascospores are often visible as white tendrils issuing from the ostioles of the perithecia (Plate 5c). In culture, conidia are formed in sticky masses at the tip of single phialides (Fig. 12.15b; Rifai & Webster, 1966). Conidia of this type are described as *Acremonium*-like. Some species of *Hypocrea* have conidia of the *Trichoderma* type, in which whorls of phialides give rise to separate, sticky green or white

| Table 12.2. | Some teleomorph−anamorph connections in the Hypocreales. |

Teleomorph genus	Anamorph genus
Nectria sensu lato	*Tubercularia, Fusarium, Cylindrocarpon, Verticillium, Heliscus, Flagellospora*
Bionectria	*Clonostachys*
Calonectria	*Cylindrocladium*
Gibberella	*Fusarium*
Hypocrea	*Trichoderma, Gliocladium, Acremonium*-like
Sphaerostilbella	*Gliocladium*
Hypomyces	*Mycogone, Cladobotryum*
Apiocrea	*Sepedonium*
Melanopsamma	*Stachybotrys*

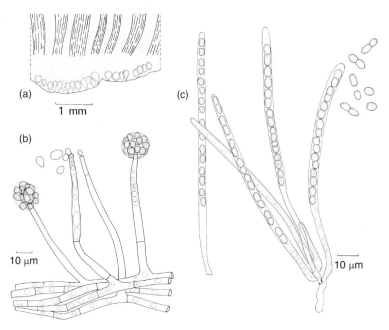

Fig 12.15 *Hypocrea pulvinata.* (a) L.S. lower part of fruit body of *Piptoporus betulinus* showing perithecial stromata of *Hypocrea* in section (see also Plate 5c). (b) *Acremonium*-like conidia produced from upright phialides. (c) Asci and ascospores. Note how the two-celled ascospores may break up into unicellular part-spores.

spore masses. Detailed descriptions of *Hypocrea–Trichoderma* connections have been given by Chaverri and Samuels (2003). *Hypocrea rufa* forms conidia referable to *T. viride* (Fig. 12.16d), a species with globose, warty conidia. Conidia of the *Gliocladium* type, in which conidia derived from individual phialides coalesce in a single slimy mass, are found in cultures of *H. gelatinosa* (Figs. 12.16a,b; Webster, 1964). The distinction between *Trichoderma* and *Gliocladium* anamorphic states can be difficult and some species, e.g. *G. virens*, have been transferred to *Trichoderma* as *T. virens* (Fig. 12.17a), the anamorph of *H. virens* (Chaverri *et al.*, 2001).

The biology and taxonomy of *Trichoderma* have been reviewed by Samuels (1996, 2006), ecological aspects by Klein and Eveleigh (1998), and keys to identification have been given by Gams and Bissett (1998). It is a large genus and species identification is difficult using morphological criteria. DNA-based and biochemical techniques are now widely used to support identification. Such studies indicate that genus is a monophyletic group within the Hypocreaceae which evolved about 110 million years ago (Kullnig-Gradinger *et al.*, 2002). *Trichoderma* spp. are cosmopolitan in soil and on decaying woody substrata. On soil isolation plates, *Trichoderma* spp. are often the most

rapidly growing and dominant fungi, smothering the plates and forming clusters of sticky green phialoconidia within a few days. Certain species, especially *T. harzianum* and *T. virens*, have been used in the biological control of soil-borne fungal plant pathogens such as *Rhizoctonia solani* and *Pythium ultimum* (Papavizas, 1985; Chet, 1987; Lumsden, 1992; Tang *et al.*, 2001). Antagonism to the fungal pathogen may be associated with coiling of *Trichoderma* around the host hypha, followed by penetration and parasitism (Figs. 12.17b–d). Volatile and non-volatile antifungal antibiotics may also be produced. Several species of *Trichoderma* are troublesome contaminants and parasites of mycelia and basidiocarps of cultivated mushrooms in Europe and North America (Castle *et al.*, 1998; Samuels *et al.*, 2002).

Trichoderma reesei is used commercially for cellulase production, secreting 20 g or more of cellulase l^{-1} culture fluid (Durand *et al.*, 1988; Kubicek *et al.*, 1990). Until recently, this interesting species had been isolated only once, during the Second World War on the Solomon Islands from canvas in contact with soil. Molecular studies have revealed that *T. reesei* is the anamorph of *Hypocrea jecorina*, an uncommon tropical species (Kuhls *et al.*, 1996; Lieckfeldt *et al.*, 2000).

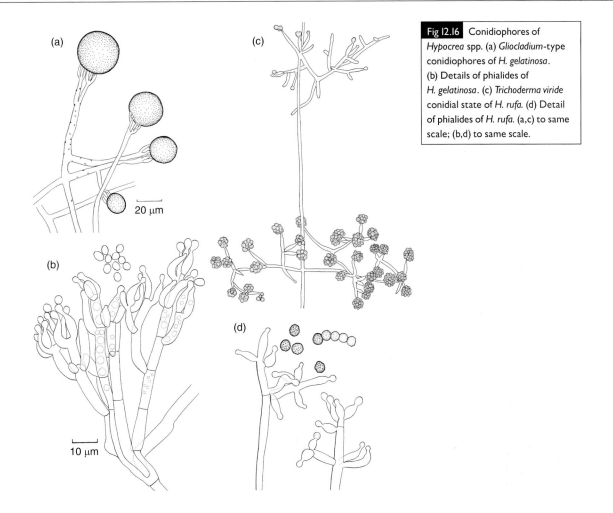

Fig 12.16 Conidiophores of *Hypocrea* spp. (a) *Gliocladium*-type conidiophores of *H. gelatinosa*. (b) Details of phialides of *H. gelatinosa*. (c) *Trichoderma viride* conidial state of *H. rufa*. (d) Detail of phialides of *H. rufa*. (a,c) to same scale; (b,d) to same scale.

The mating behaviour of *Hypocrea* spp. is poorly understood because perithecial development in culture occurs only rarely, except for the tropical species *H. jecorina*. This species shows bipolar heterothallism, half of its ascospores being of one and half of the opposite mating type (Lieckfeldt *et al.*, 2000). Another tropical species, *H. poronioidea*, also forms perithecia in culture but the genetic basis of sexual reproduction is not clear. Cultures from eight part-ascospores in each of several asci were self-fertile (i.e. homothallic), whilst the other eight part-ascospores produced only conidia and were self-sterile. Another unusual feature of this fungus is that, in addition to having a *Trichoderma* anamorph, it has an *Acremonium*-like synanamorph. The *Acremonium*-like conidia can germinate, i.e. they have an asexual function, but it is

possible that they also have a spermatial role (Samuels & Lodge, 1996).

12.4.2 *Nectria* (Nectriaceae)

The concept of the genus *Nectria* has changed in recent years. Previously considered to include about 200 species, the genus has now been narrowed to about 30, centred around *N. cinnabarina* (Rossman, 1983; Rossman *et al.*, 1999). Perithecia of *Nectria* are common on twigs and branches of woody hosts. Many are saprotrophic or weakly parasitic but some cause economically important diseases, e.g. *N. coccinea* causes bark disease of beech (*Fagus sylvatica*), whereas *N. galligena* infects apple and pear trees. Although there are some reports that *N. galligena* may have a prolonged latent (endophytic) phase, the fungus acts as a wound pathogen in most

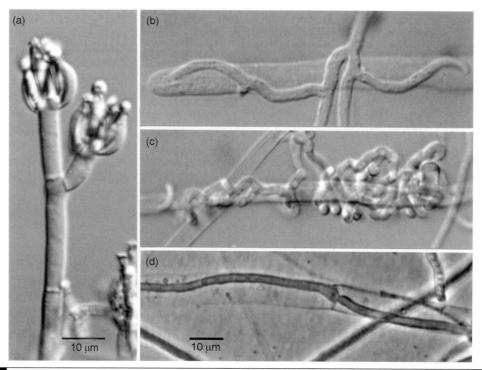

Fig 12.17 Mycoparasitism by *Trichoderma* spp. (a) Tip of conidiophore of *T. virens* with phialides. (b) Contact of a hypha of *T. harzianum* with one of *Rhizoctonia cerealis*. (c) Coiling of *T. harzianum* around a hypha of *R. cerealis*. (d) Intrahyphal growth of *T. virens* in a hypha of *Fusarium* sp.

situations, entering through pruning wounds, cracks, leaf scars and sites of branch breakage. The colonization of the bark and cambium tissues results in a necrotic lesion, often with the dead heartwood visible in the centre, and surrounded by raised callus-like bark tissue where the tree launches a defence response. This type of symptom is called a **canker** (Fig. 12.18). Conidia and, later, perithecia are produced in the bark of cankers. **Girdling** results if a canker surrounds an entire branch, and this leads to the death of all shoot tissue distal to the lesion.

The conidial states of *Nectria* include some terrestrial species of *Fusarium*, *Cylindrocarpon* and *Tubercularia*, whilst in freshwater streams they include the aquatic hyphomycetes *Cylindrocarpon*, *Flagellospora* and *Heliscus* (Section 25.2; Webster, 1992). Some species have synanamorphic states with conidia of distinctive size and shape, e.g. macro- and micro-conidia (e.g. *N. haematococca*; Fig. 12.19).

Nectria cinnabarina, the cause of coral spot, is common on freshly cut twigs of hardwood trees and shrubs, but may occasionally be a wound parasite on these hosts. The name coral spot refers to the pale pink conidial pustules, about 1−2 mm in diameter, which burst through the bark (Plate 5d). Before the connection with *Nectria* was understood, these conidial pustules had been named *Tubercularia vulgaris*. They consist of a column of pseudoparenchyma bearing a dense tuft of conidiophores. These are long slender hyphae producing intercalary phialides at intervals along their length (Figs. 12.20b,d). Conidial pustules of this type are termed **sporodochia**. The conidia are sticky and form a slimy mass at the surface of the sporodochium. They are dispersed very effectively by rain splash. Around the base of the old conidial pustule perithecia arise (Plate 5d), and eventually the pustule may bear perithecia over its entire surface. Perithecial pustules develop in damp conditions in late summer and autumn and are

Fig 12.18 Canker caused by *Nectria galligena* on the twig of an old apple tree.

readily distinguished from conidial pustules by their bright red colour and their granular appearance. The pustules are regarded as perithecial stromata, bearing as many as 30 perithecia. Ripe perithecia contain numerous club-shaped asci, each with eight two-celled hyaline ascospores (Fig. 12.20c). These are somewhat unusual in being multinucleate (El-Ani, 1971). There is no obvious apical apparatus to the ascus (Strickmann & Chadefaud, 1961).

Perithecial development begins by the formation of ascogonial primordia beneath the surface of the conidial pustule. The details of development are as described on p. 337. An important feature is the development of apical paraphyses which grow downwards from the upper part of the perithecial cavity. As the asci develop from ascogenous hyphae lining the base of the perithecial cavity, they grow upwards through the mass of apical paraphyses, which are difficult to find in mature perithecia (Strickmann & Chadefaud, 1961).

Other species of *Nectria* differ from *N. cinnabarina* in a number of ways. In many, the perithecia are not grouped together on a stroma, but occur singly (e.g. in *N. galligena*). The asci of some species, e.g. *N. mammoidea*, have a well-defined apical apparatus (Fig. 12.21a).

12.4.3 *Fusarium* (Nectriaceae)

Fusarium-type conidia are known in several species of *Nectria* and also in the related genus *Gibberella* (Samuels *et al.*, 2001). For many *Fusarium* species a teleomorph has not yet been found. The number of species recognized by morphological characters in pure culture varies from one authority to another. Booth (1971) included 44 species and 7 varieties; Gerlach and Nirenberg (1982) included 73 species and 26 varieties, while Nelson *et al.* (1983) distinguished 30 species. However, using molecular techniques a large number of species which cannot be distinguished by morphological characters are now being defined (O'Donnell, 1996), and a publically accessible database of diagnostic DNA sequences has been established as an aid to identification (Geiser *et al.*, 2004). Species of *Fusarium* are cosmopolitan in soils from the permafrost of the Arctic to sand in the Sahara desert (Booth, 1971). They are particularly common in cultivated soil and are associated with a wide range of plant diseases as indicated in Table 12.3 (Nelson *et al.*, 1981). Some species are also known to cause diseases of humans such as onychomycosis (nail infections), keratomycosis of the cornea, ulcers, necroses, skin infections and fatal infections of internal organs, especially in immunocompromised patients (Joffe, 1986; de Hoog *et al.*, 2000a).

A characteristic feature in the asexual reproduction of *Fusarium* is the development from phialides of fusoid, transversely septate macroconidia with a basal contracted foot cell (Fig. 12.22b). Microconidia, also formed from phialides, develop in some species. On plant hosts macroconidia often accumulate in orange−pink sporodochia (Plate 5e, Fig. 12.22a). Macroconidia may be dispersed by rain splash

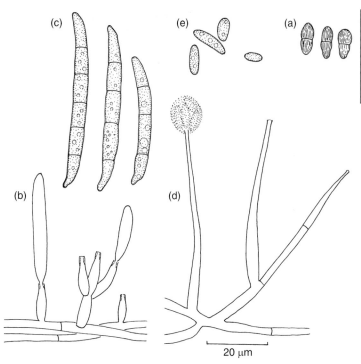

(c) (e) (a)

Fig 12.19 *Nectria haematococca.*
(a) Ascospores. (b) Phialides bearing macroconidia. (c) Macroconidia of the *Fusarium* type. (d) Phialides producing microconidia which accumulate in a drop of mucilage. (e) Microconidia.

(b) (d)

20 µm

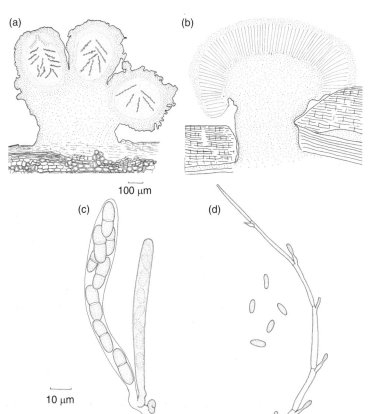

(a) (b)

Fig 12.20 *Nectria cinnabarina.* (a) V.S. perithecial stroma. (b) V.S. conidial stroma (sporodochium). (c) Asci. (d) Conidiophore, phialides and conidia.

100 µm

(c) (d)

10 µm

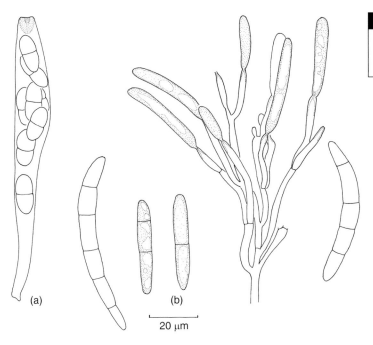

Fig 12.21 *Nectria mammoidea*. (a) Ascus; note the apical apparatus. (b) *Cylindrocarpon*-type conidiophore with phialides and conidia.

20 μm

from such pustules (see Jenkinson & Parry, 1994). In culture, greasy accumulations of macro-conidia are termed **pionnotes**. Survival in soil is probably in the form of dormant chlamydospores which may be formed as a result of energy deprivation or in response to bacterial secretions. They are thick-walled and develop by modification of segments of the vegetative mycelium or by enlargement and modification of a segment of a macroconidium (Schippers & van Eck, 1981) (Figs. 12.22c,d). Chlamydospores are induced to germinate in the presence of root secretions (for references see Griffiths, 1974).

One of the most commonly isolated species from soil is *F. oxysporum*, which often grows in association with roots. Many isolates are non-pathogenic but this fungus may also be a serious plant pathogen, reported from a very wide range of plant hosts. Although no sexual state is known, molecular studies show that it is mono-phyletic with the *Gibberella*–*Fusarium* complex and can be regarded as an asexual *Gibberella* (Samuels *et al.*, 2001). In this species over 80 *formae speciales* characteristic of different host genera have been recognized, in addition to a large number of vegetative compatibility types and races (Gordon, 1993; Gordon & Martyn, 1997). The large amount of genetic variation in

F. oxysporum in the absence of conventional sexual reproduction is possibly explained by the demonstration of parasexual recombination even between members of different vegetative compatibility groups (Molnár *et al.*, 1990).

Fusarium wilts

As can be seen from Table 12.3, *formae speciales* of *F. oxysporum* cause wilt diseases of numerous crop plants. The range of affected hosts is so wide that it is easier to mention those plants unaffected by *Fusarium* wilts, i.e. grasses and many tree species. Although *F. oxysporum* has no known sexual state, strains of this species complex show huge genetic variation which may come about by the presence of transposable elements (Daboussi & Capy, 2003) and the parasexual cycle (Teunissen *et al.*, 2002). The variability of *F. oxysporum* is such that a given disease, e.g. Panama wilt of banana, is caused by several unrelated strains (K. O'Donnell *et al.*, 1998), and that strains with different host specificities may be closely related (Hua-Van *et al.*, 2001).

Wilting is associated with the presence of fungal hyphae in the xylem vessels. *Fusarium oxysporum* can persist in the soil as chlamydo-spore inoculum for many years and infects the roots of hosts by direct penetration and

Table 12.3. Some species of *Fusarium* causing plant diseases of economic importance. From Holliday (1998).

Anamorph	Teleomorph	Disease caused and name of host
F. culmorum	not known	cortical rot, foot rot and pre-emergence blight of temperate cereals; head blight of barley, wheat and rye
F. oxysporum	not known	a wide range of wilts, yellows and foot rots
F. oxysporum f. sp. *apii*	not known	celery yellows
F. oxysporum f. sp. *cepae*	not known	basal rot and storage rot of onion
F. oxysporum f. sp. *cubense*	not known	Panama wilt of banana
F. oxysporum f. sp. *dianthi*	not known	wilt of carnation and pinks
F. oxysporum f. sp. *elaeidis*	not known	oil palm wilt
F. oxysporum f. sp. *lycopersici*	not known	tomato wilt
F. oxysporum f. sp. *pisi*	not known	pea wilt
F. oxysporum f. sp. *vasinfectum*	not known	cotton wilt
F. solani	*Nectria haematococca*	root and collar rot of many plants; canker of woody crops
F. solani f. sp. *cucurbitae*	not known	foot rot of cucumber and melon
F. solani f. sp. *coeruleum*	not known	storage rot and dry rot of potato
F. avenaceum	*Gibberella avenacea*	damping off and root damage to cereals, conifers and legumes
F. sulphureum	*G. cyanogena*	storage rot and dry rot of potato tubers
F. moniliforme	*G. fujikuroi*	diseases of many plants, e.g. banana black heart, cotton boll rot, maize and sorghum stalk rots, rice seedling bakanae disease
F. sambucinum	*G. pulicaris*	hop canker, potato storage rot, root rot of many crops
F. graminearum	*G. zeae*	numerous diseases of temperate and tropical cereals, e.g. pre- and post-emergence blights, root and foot rot, culm decay, head or kernel blight, ear scab and stalk rot.

intercellular growth through the root tip into the xylem vessels (Bishop & Cooper, 1983). *In planta*, *F. oxysporum* exists mainly as microconidia which can spread in the xylem vessels. Sieve plates are overcome by germination and penetration, followed by the production of further microconidia. Wilting is caused by blockage of the xylem by fungal biomass and also by the accumulation of gums of plant origin, some of them released by fungal pectinase activity (Pegg, 1985; Beckman, 1987). In cross section, infected stem bases show a browning of the vascular bundles. The host plant eventually dies because of water shortage (wilting), and after host death extensive colonization of host tissue and sporulation of *F. oxysporum* ensues. Extensive work is being carried out especially on molecular biological aspects of wilt diseases caused by

Fig 12.22 *Fusarium culmorum.* (a) Part of a sporodochium with clusters of phialides producing macroconidia. (b) Mature macroconidium from a two-week-old culture. The foot cell is indicated by an arrowhead. (c) Older macroconidia (eight-week-old) in which some cells have collapsed and the others have turned into chlamydospores. (d) Chlamydospores formed by a hypha submerged in agar. (b–d) to same scale.

F. oxysporum, and this has been summarized by Di Pietro *et al.* (2003).

The control of diseases caused by *Fusarium* is mainly through the use of resistant host plant cultivars (Beckman, 1987). The disastrous economic consequences of the outbreak and spread of Panama wilt of banana caused by *F. oxysporum* f. sp. *cubense* which attacked the popular banana variety 'Gros Michel' were overcome by switching production to clones of the resistant variety 'Cavendish' which could be planted into soils heavily contaminated with *F. oxysporum* f. sp. *cubense* (Moore *et al.*, 2001). Another well-characterized example is tomato wilt caused by *F. oxysporum* f. sp. *lycopersici*. Resistance seems to be based mainly on major-gene resistance, and numerous resistance genes have been crossed into the cultivated tomato plant from wild relatives. These occur in several discrete gene clusters (Sela-Buurlage *et al.*, 2001). A range of tomato cultivars with resistance against one or more of the three races of *F. oxysporum* f. sp. *lycopersici* is available for planting.

Another approach actively pursued at present is the biological control of *F. oxysporum*. An interesting strategy is the use of apathogenic strains of *F. oxysporum* which colonize the root cortex but not the vascular system, and therefore have an endophytic lifestyle. They inhibit infections by pathogenic strains. Various explanations have been offered, including competition for nutrients, the induction of systemic acquired resistance in plants pre-infected with the apathogenic strains, and direct competition between apathogenic and pathogenic strains in the root cortex (Larkin & Fravel, 1999; Bao & Lasarovits, 2001). Other soil micro-organisms such as *Pseudomonas* spp. are also being developed as biocontrol agents.

Mycotoxins

Some species of *Fusarium* (e.g. *F. graminearum* and *F. moniliforme*) are seed-borne, with mycelium being present inside the grain and often producing macroconidia on the surface. Infected seed not only ensures carry-over of the disease to following seasons but can be a source of mycotoxins if fed to livestock or consumed by humans. Mycotoxins such as zearalenone, fumonisin, trichothecenes and vomitoxin (Fig. 12.23) are produced by *Fusarium* spp. present on animal and human feedstock (Creppy, 2002; Moss, 2002). Zearalenone, an oestrogenic hormone from *F. graminearum*, causes vulvovaginitis and infertility in cattle and pigs, and trichothecenes such as T-2 toxin from several other *Fusarium* species cause toxic aleukia (reduction in white blood cell count) in farm animals and humans (Joffe, 1986; Moss, 2002).

12.4.4 *Cylindrocarpon* (Nectriaceae)

Fungi with *Cylindrocarpon*-type conidia are anamorphic states of species of *Nectria sensu lato*, now placed in a distinct genus *Neonectria* (Rossman *et al.*, 1999; Mantiri *et al.*, 2001). The macroconidia of *Cylindrocarpon* are hyaline, curved, transversely septate phialoconidia which resemble those of *Fusarium* but do not have the constricted basal foot cell characteristic of the latter (compare Figs. 12.21b and 12.20c). Some species also have microconidia and some have chlamydospores. About 120 taxa (species and varieties) have been described and about 50 have been linked to *Nectria sensu lato*. Most species of *Cylindrocarpon* grow in soil and are saprotrophic or weakly parasitic. *Cylindrocarpon destructans* (formerly known as *C. radicicola*), whose teleomorph is *Nectria radicicola* (Samuels & Brayford, 1990), causes seedling blights and basal rots of bulbs, as well as root rots of various plants. *Cylindrocarpon heteronema* is the anamorph of *Nectria galligena*, the cause of apple and pear canker.

12.5 | Clavicipitales

Fungi in this group have been and still are assigned to various groups (Spatafora & Blackwell, 1993; Rossman, 1996; Stensrud *et al.*, 2005), notably a family (Clavicipitaceae) of the Hypocreales or a separate order (Clavicipitales). Whatever its taxonomic rank, this group contains fungi with several distinguishing characteristics. Perithecia develop on a fleshy stroma. The perithecial centrum contains a central basal mound from which the asci arise. Any paraphyses which develop are obliterated by crushing as the asci enlarge (White, 1997; Rossman *et al.*, 1999). The asci have a well-defined

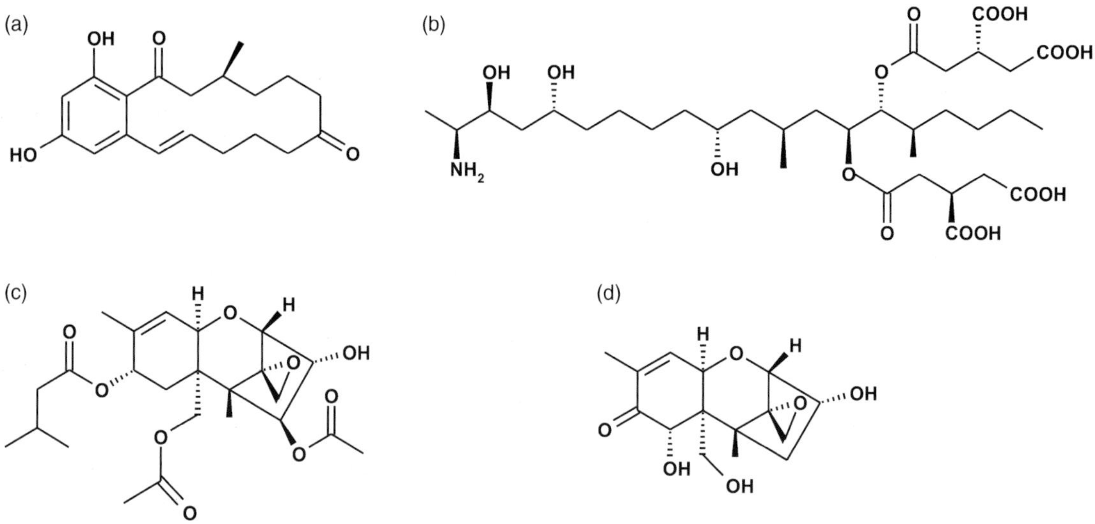

Fig 12.23 Common mycotoxins produced by *Fusarium* spp. (a) The nonaketide zearalenone, a suspected carcinogen and oestrogen analogue produced by *F. graminearum* in maize and cattle feed. (b) Fumonisin B$_1$, a suspected cause of oesophageal cancer, produced by *F. moniliforme* infecting maize. (c) T-2 toxin, a highly cytotoxic trichothecene (sesquiterpene derivative) produced by *F. graminearum* in various cereals. (d) Vomitoxin (desoxynivalenol), a trichothecene produced by *F. culmorum* and *F. graminearum* in maize and wheat.

thick apical cap perforated by a narrow pore through which the ascospores are discharged, singly and successively. The ascospores are long, narrow and often multi-septate, breaking up into part-spores. Most members (e.g. *Claviceps*, *Balansia*, *Epichloe*) are pathogens or endophytes of grasses, whereas *Cordyceps* parasitizes insects or fruit bodies of the hypogeous ascomycete *Elaphomyces*. *Claviceps* sclerotia are the source of toxic alkaloids, and grasses infected with endophytic *Balansia* may also be toxic to herbivorous insects and mammals. The grass host may thus be at least partially protected against insect herbivory, and the effects of alkaloids on grazing mammals can also be severe. The conidia of certain species of *Cordyceps* show promise as agents of biological control of insect pests (Evans, 2003). Some members of the Clavicipitales attack nematodes (Gams & Zare, 2003). For example, *Atricordyceps* (now called *Podocrella*) is the teleomorph of *Harposporium anguillulae* (Samuels, 1983) (see Fig. 25.8), and phylogenetic analyses of nuclear ribosomal DNA have also placed the nematophagous fungus *Drechmeria coniospora* (see Fig. 25.8) in the Clavicipitaceae (Gernandt & Stone, 1999). The immunosuppressant drug cyclosporin A (Fig. 12.24a) is produced by *Tolypocladium inflatum*, anamorph of *Cordyceps subsessilis* (Hodge *et al.*, 1996). A group of β-lactam antibiotics, the cephalosporins (Fig. 12.24b; see p. 302), are derived from the anamorphic *Acremonium chrysogenum* and *A. salmosynnematum*. Many other valuable secondary metabolites have been isolated from members of the Clavicipitales (Isaka *et al.*, 2003).

12.5.1 *Claviceps*

There are over 40 species of *Claviceps*, all of which are parasitic on grasses, rushes and, occasionally, sedges (Alderman, 2003). The best known species is *C. purpurea*, the cause of ergot of grasses and cereals. Other economically important species are *C. sorghi* and *C. africana* which cause ergot of sorghum (Frederickson *et al.*, 1991), *C. paspali* on *Paspalum* and *C. fusiformis* on pearl millet (*Pennisetum typhoides*). *Claviceps purpurea* occurs in temperate regions and has an exceptionally wide

host range for a biotrophic pathogen, infecting over 400 grass species. The course of infection has been described by Luttrell (1980), Tenberge (1999) and Oeser *et al.* (2002).

Life cycle

The life cycle of *C. purpurea* is summarized in Fig. 12.25. The primary inoculum is an ascospore shot away from a perithecium which has developed from an overwintered sclerotium. The time of ascospore release coincides with anthesis in a susceptible host. Ascospores germinate on a grass stigma to form an intercellular mycelium which grows down to the base of the ovary towards the vascular bundle of the floret stalk (rachilla), thus gaining access to the photosynthetic products of the host. Subsequent growth is upwards and within a few days a conidial stroma develops beneath the ovary. A palisade of phialides lining labyrinthine chambers is formed from which a succession of unicellular, uninucleate conidia develops in a sugary syrup. This becomes visible on the grass florets as beads of liquid termed honeydew (Fig. 12.26b). The conidial stage was given the separate name *Sphacelia segetum* before its connection with ergot was understood. Honeydew contains glucose, fructose, sucrose and other sugars (Mower & Hancock, 1975a), and is attractive to insects, which feed on it and in so doing disperse conidia to healthy grass flowers, thus causing secondary infection. Infection of a grass flower by *Claviceps* results in increased translocation of water and sucrose towards the diseased flower, and infected flowers are more effective at acquiring photosynthetic products from the host than uninfected flowers (Parbery, 1996b). Within the infected host tissue, conversion of host-derived sucrose to mono-, di- and oligo-saccharides by the fungus creates a continuing sink for sucrose translocation, and evaporation at the surface of the diseased grain results in increased osmotic concentration of the sugars, possibly accelerating the rate of translocation (Mower & Hancock, 1975b). The high osmotic concentration prevents conidial germination until the honeydew has been diluted.

As infection proceeds, the entire ovary is pushed upwards by the developing fungal tissue

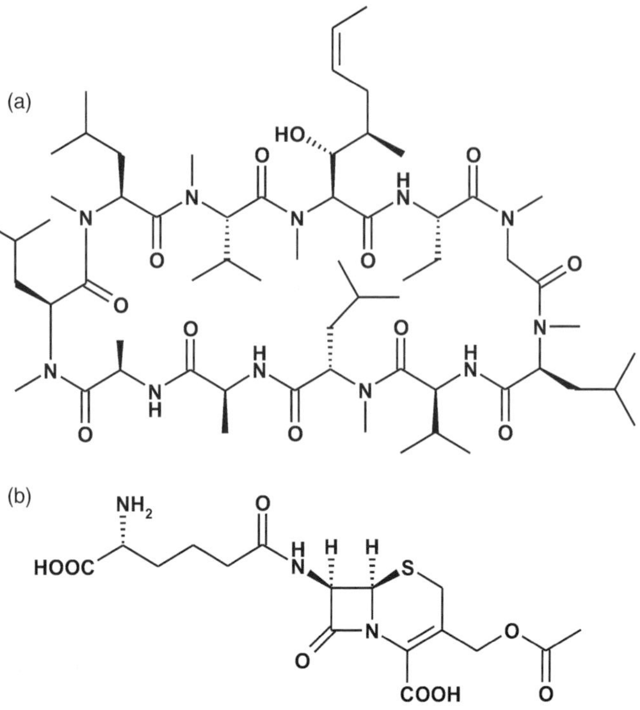

Fig 12.24 Important metabolites from Clavicipitales. (a) Cyclosporin A, produced by *Tolypocladium inflatum*. This is used extensively as an immunosuppressive drug, e.g. after organ transplantations. (b) Cephalosporin C, a β-lactam antibiotic with activity against Gram-positive and Gram-negative bacteria.

and sits like a cap over it. The ovary, which would normally develop into a caryopsis filled with grain, becomes replaced by fungal tissue. For this reason the disease caused by *C. purpurea* has been termed a replacement disease (Luttrell, 1980). The fungal structure which develops is considerably longer than the ovary which it replaces and it differentiates into a sclerotium up to 3 cm in length, the foot of which continues to obtain nutrients via the vascular connection in the host rachilla. The sclerotium is made up of three distinct layers: a thin purplish-brown rind, a discontinuous layer of mealy white tissue, and a central layer of translucent gelatinous tissue (Luttrell, 1980).

Ergot is the French name for a cock's spur, referring to the curved, banana-shaped sclerotia which project from the inflorescence of infected grasses and cereals in late summer (Fig. 12.26a). The sclerotia fall to the ground and overwinter near the surface of the soil. They need a period of low temperature before they can develop further. A chilling period of 0°C for at least 25 days is optimal for further development. The main reserve substance of the sclerotium is lipid, which may account for 50% of the dry weight.

It is likely that the chilling period is necessary before enzymes capable of mobilizing the lipid reserves develop (Cooke & Mitchell, 1970). Sclerotia do not remain viable in soil in the field for more than a few months, often being invaded by fungi, bacteria, mites and insects (Cunfer & Seckinger, 1977). The following summer, sclerotia develop one or more perithecial stromata (clavae) about 1–2 cm high, shaped like miniature drumsticks (Fig. 12.26c). The perithecial stromata are positively phototropic (Hadley, 1968).

The enlarged spherical head or capitulum contains a number of perithecia which are embedded in the stroma, each surrounded by a distinct perithecial wall (Fig. 12.27a). The cytological details of perithecial development have been studied in *C. purpurea* by Killian (1919) and in *C. microcephala* (regarded by some as a form of *C. purpurea*) by Kulkarni (1963). In the outer layers of the head of the perithecial stroma, club-shaped multinucleate antheridia and ascogonia undergo plasmogamy. Ascogenous hyphae made up of predominantly binucleate segments develop from the base of the ascogonium, and the tips of the ascogenous hyphae form croziers

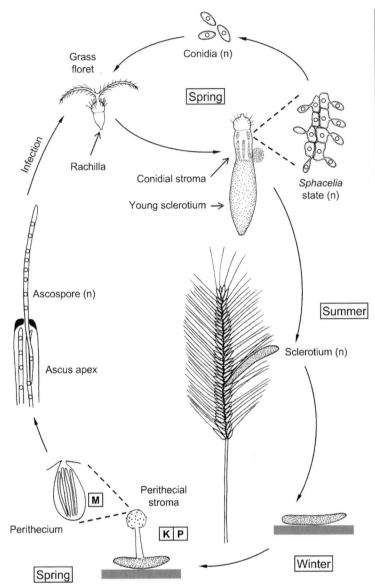

Fig 12.25 The homothallic life cycle of *Claviceps purpurea* growing on rye (*Secale cereale*). A sclerotium fallen to the ground will produce a perithecial stroma after overwintering. In the perithecia, plasmogamy (P), karyogamy (K) and meiosis (M) give rise to filamentous ascospores, each with numerous haploid nuclei (small open circles). Ascospores germinate on the stigmata of the rye ovary, and mycelium penetrates to the ovary stalk (rachilla). Infections develop both as conidial regions and, lower down, as a sclerotium initial. The developing sclerotium pushes the ovary upwards, and this remains as a cap above the conidial stroma. The stroma contains cavities lined by phialides which produce phialoconidia of the *Sphacelia segetum* type. These accumulate in beads of a sugary liquid (honeydew). Conidia are carried by insects to fresh rye stigmata and initiate secondary infections. Later in the season, the sclerotia enlarge and become visible as ergots. Some images redrawn from Luttrell (1980).

with binucleate penultimate segments. The penultimate cell elongates to form the ascus, and fusion between the two nuclei occurs. There are numerous asci in each perithecium, each containing a bundle of eight filiform ascospores. The ascus bears a conspicuous perforated cap at its tip (Fig. 12.27c).

Successful infection of rye (*Secale cereale*) from cultures derived from a single ascospore show that *C. purpurea* is homothallic. Despite this, genetic recombination is possible through heterokaryosis and parasexual reproduction (Tudzynski, 1999). Curiously, sclerotia are frequently formed from heterokaryotic mycelia, indicating multiple infection of the grass flower.

Although common on rye and some other cereals in Europe and North America, *C. purpurea* is not usually troublesome on cereals in Britain. In the occasional years in which its incidence is high, there is a correlation with high relative humidity and low maximum temperature in June, which probably prolongs the period during

Fig 12.26 *Claviceps purpurea.* (a) Head of rye (*Secale cereale*) bearing several sclerotia (ergots) of *Claviceps purpurea*. (b) Rye inflorescence at anthesis bearing two drops (arrowed) of the honeydew or *Sphacelia* conidial state of *C. purpurea*. A fly has landed near these drops. (c) Germinated sclerotium showing several stalked perithecial stromata.

which the host grass flowers are open and therefore susceptible to infection.

Ergotism

The effect of infection on the host can result in yield reduction by as much as 80% of seeds. Severe though this reduction in crop yield is, the consequences of consumption of ergot-contaminated grain can be disastrous to herbivorous animals and humans alike. The purple sclerotia contain a number of toxic alkaloids (Buchta & Cvak, 1999) and if they are eaten they can cause severe illness and sometimes death. Even at relatively low concentration there may be severe effects on feed refusal, lack of weight gain in farm animals and on reduced fertility, resulting in part from agalactia, the inability to produce sufficient milk to nourish the young (Shelby, 1999). One effect of the toxins is to constrict the blood vessels, and the impaired circulation may result in gangrene or loss of limbs. Gruesome descriptions of the symptoms on humans have been related by several contemporary authors, e.g. Sidney (1846) who wrote:

> The medical effects of ergot, in small doses, have already been noticed as being extremely powerful, but if taken to any extent its results on the animal frame are truly awful. This has been proved by numerous experiments, of which Professor Henslow gives a most striking account in his most valuable notice of this disease; to which he adds a proper caution against their repetition now the question is settled. Animals which refused ergot mixed with their food have been compelled to swallow it, and it reduced them to a wretched condition. It was tried upon pigs, and also upon poultry, and the consequences were sickness, gangrene, and inflammatory action so intense, that the flesh actually sloughed away. In some cases, the limbs rotted off, and no description of animal suffering has ever exceeded the direful ills thus inflicted. These experiments were made with a view to determine whether the ergot of rye, constantly ground up with the flour in some parts of France, might not be the cause of the gangrenous disease so prevalent amongst the poor in certain districts. The symptoms of these epidemic diseases are

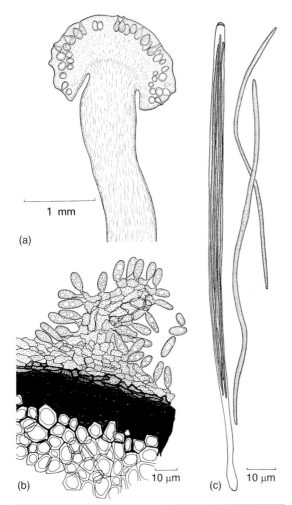

and dumb, and, besides, lost a limb which actually rotted off, precisely in the same way as the limbs of the animals which were compelled to swallow the experimental ergot.

Another effect is on the nervous system, resulting in convulsions, hallucinations and burning sensations. In the Middle Ages the symptoms of ergotism were called 'St Anthony's fire' and there are numerous records of out-breaks of the disease (see Ramsbottom, 1953; Fuller, 1969; Matossian, 1989). Matossian (1989) has outlined some of the social consequences of ergotism, e.g. in depressing population growth after the plague outbreaks in Europe, and in provoking witch trials in North America when women were accused of and executed for bewitching people who were probably suffering from ergot-induced food poisoning and hallucinations.

With improved grain-cleaning techniques and a switch in carbohydrate consumption from rye to wheat, maize and potatoes, the disease is now rare in humans. Cattle and sheep which have eaten sclerotia from pasture grasses, or pigs and horses fed on ergot-contaminated grain, are still affected and if pregnant animals are involved there is a risk of abortion. However, these problems are now relatively rare.

Alkaloids of *Claviceps purpurea*

Ergot alkaloids are used in human medicine. 'No other class of compounds exhibits such a wide spectrum of structural diversity, biological activity and therapeutic uses as ergot derivatives' (Křen & Cvak, 1999). This is because the tetra-cyclic ergoline ring structure (Fig. 12.28) mimics several neurotransmitter molecules such as noradrenaline, dopamine and serotonin (Mantegani *et al.*, 1999). Two alkaloids, ergometrine and ergotamine, are of special importance and are produced by field strains of *C. purpurea* (Pažoutová *et al.*, 2000). Ergometrine causes constriction of smooth muscle tissues. The name ergometrine is derived from endometrium, the lining of the uterus, because the drug is used to stimulate uterine contraction. Ergotamine can similarly accelerate uterine contraction and is used as a vasoconstrictor

Fig 12.27 *Claviceps purpurea.* (a) L.S. perithecial stroma. (b) T.S. young sclerotium showing the formation of phialoconidia on the surface. (c) Ascus and ascospores. Note the cap of the ascus.

dreadful, and there seems to be very little doubt that the suspicions as to their originating from ergotted flour of rye are correct. Tessier, who has paid great attention to the subject, mentions a case which came under his own observation. A family were in a state of great destitution, and the father begged of a neighbouring farmer a quantity of ergotted rye to supply the urgent calls of his distressed family for food. The farmer gave it him, but added that he was afraid it was not wholesome. Still the calls of hunger prevailed, and in the face of this caution it was eaten. The result was the death of the father, mother, and five of the children out of seven. Two survived, but one of them became subsequently deaf

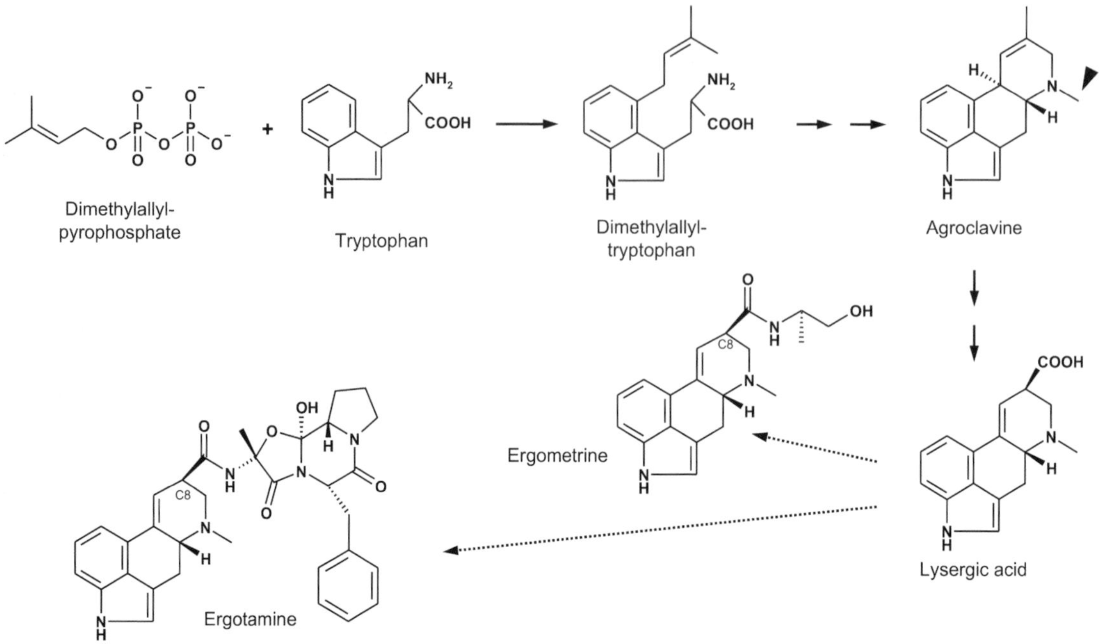

Fig 12.28 Biosynthesis of alkaloids in ergots of *Claviceps purpurea*. Starting from tryptophan and the mevalonic acid-derived metabolite dimethylallylpyrophosphate, dimethylallyltryptophan is synthesized. Ring closure, decarboxylation and addition of a methyl group from methionine (arrowhead) gives the simplest alkaloids, agroclavine and D-lysergic acid. The latter is then derivatized at its C-8 position to give more complex alkaloids such as ergometrine and ergotamine. LSD (D-lysergic acid diethylamide) is a semisynthetic derivative of D-lysergic acid.

and a haemostatic drug. It was also the first medication available against migraine (Eadie, 2004). A third ergot-derived drug is lysergic acid diethylamide (LSD), a synthetic derivative of lysergic acid. As is well known, this drug has hallucinogenic properties and is associated with euphoria (Minghetti & Crespi-Perellino, 1999). Numerous chemical derivatives have been synthesized, and ergot alkaloids are lead structures for drugs against a range of illnesses, including Parkinson's disease and Alzheimer's disease.

The biosynthesis of ergot alkaloids in *C. purpurea* from tryptophan, mevalonic acid and methyl groups donated by methionine leads to the ergoline ring structure (Fig. 12.28). The simplest alkaloids are the clavine alkaloids such as agroclavine, but most of the pharmaceutically important ones are based on D-lysergic acid because it is the carboxylic acid group at C-8 in D-lysergic acid which is further substituted. Simple substituted D-lysergic acid derivatives

are ergometrine and the semi-synthetic LSD (Fig. 12.28). In the most complex ergot alkaloids, the ergopeptines, three amino acids are added to D-lysergic acid by a non-ribosomal peptide synthase, followed by ring closure. Ergotamine is a commonly found example of this type of alkaloid in *C. purpurea* (Fig. 12.28). Reviews of alkaloid biosynthesis have been written by Flieger *et al.* (1997) and Tudzynski *et al.* (2001). A concise general treatment is that by Lohmeyer and Tudzynski (1997).

The ergot of commerce is produced by cultivating the fungus on rye, and profitable crops of ergot sclerotia are obtained, often on specially bred strains of male-sterile rye in Eastern Europe, Spain and Portugal (Németh, 1999). Rye flowers are infected with a suspension of conidia from strains of *C. purpurea* selected for high yield of certain alkaloids. Alkaloids are extracted from harvested sclerotia. Alkaloids are also extracted from special strains of *C. purpurea* grown saprotrophically in deep fermentation. About 60% of alkaloid production is currently derived from fermentations, the rest from harvested sclerotia. The advantages of saprotrophic fermentation are that the process can be closely controlled and the alkaloids produced are less variable than those from harvested ergots. A disadvantage is that the ability to produce alkaloids in economically significant amounts is variable and may be lost on prolonged cultivation. Nevertheless, the market share of alkaloids produced by fermentation is currently increasing (Tudzynski et al., 2001).

Control of *Claviceps*

The control of ergot in cereals is difficult. Although several techniques are available, none is completely effective. Use of ergot-free seed would reduce infection, but inoculum may survive from a previous crop. It can also be provided by wild grasses bordering the field because *C. purpurea* strains have wide host ranges. Deep ploughing, which buries the sclerotia, and crop rotation involving a non-cereal are also helpful. Systemic fungicides would need to be applied in sufficient amounts to produce an effective concentration at the surface of the ovary, and they have been used to control ergot in seed crops of Kentucky bluegrass, *Poa pratensis* (Schulz et al., 1993). Fungicide sprays are used at present to control *C. africana* on sorghum in Australia (Ryley et al., 2003) but not against *C. purpurea* on cereals.

Other *Claviceps* spp.

Some other species of *Claviceps* differ in significant ways from *C. purpurea*. For example, *C. fusiformis* has two synanamorphs, a macro- and a micro-conidial state. *Claviceps africana* and *C. paspali* may produce secondary phialoconidia and these may develop in sufficient quantity on the surface of the conidial stroma to be capable of dispersal by wind (Luttrell, 1977; Frederickson & Mantle, 1989; Alderman, 2003). *Claviceps paspali* which infects dallisgrass (*Paspalum dilatatum*) is the source of alkaloids such as paspalic acid and its derivatives. Ingestion of its sclerotia causes paspalum staggers in sheep. The salt marsh grass *Spartina anglica* often shows heavy infection by a specialized variety, *C. purpurea* var. *spartinae*

(Plate 5e). This fungus appears to be adapted to an aquatic environment. Its unusually slender sclerotia float on the surface of sea water whilst those of other forms of *C. purpurea* sink. The high levels of infection may be related to the fact that *S. anglica*, an allopolyploid grass of recent origin, is genetically uniform. Despite the heavy infection, seed production by the host plant is not severely affected (Raybould et al., 1998; Duncan et al., 2002). In contrast, *C. phalaridis*, which is endemic in Australia on the introduced pasture grass *Phalaris tuberosa*, is systemic and when its mycelium penetrates the inflorescence, sclerotia are formed in all the florets, rendering the host plant sterile (Walker, 2004). Its systemic habit is shared by other clavicipitaceous endophytes such as *Epichloe* (see below).

12.5.2 *Epichloe*

There are about 10 biological species (i.e. mating populations) of *Epichloe* (Gr. *epi* = on, upon; *chloë* = young shoots of grass) mainly infecting cool-season grasses with the C$_3$ photosynthetic pathway (Leuchtmann, 2003). They grow in nature as biotrophic, systemic, parasitic or symbiotic endophytes in grass shoots, forming at first conidial, then perithecial stromata around the uppermost leaf sheaths of tillers containing floral primordia. Anamorphic relatives, now classified as species of *Neotyphodium* (previously *Acremonium* Section *Albo-lanosum*; Glenn et al., 1996), are symptomless endophytes. *Neotyphodium*-infected grasses contain ergot alkaloids and other mycotoxins which are injurious to herbivorous insects and mammals and cause economic damage (Schardl, 1996; Kuldau & Bacon, 2001; Clay & Schardl, 2002).

Epichloe typhina (*sensu lato*) causes 'choke' of pasture grasses and is common on grasses such as *Dactylis*, *Holcus* and *Agrostis*. However, forms of *Epichloe* on certain hosts are distinct from *E. typhina* in dimensions of stromata, ascospores, ascospore septation, and in molecular characteristics. They have now been accorded different species names. The specific name *typhina* should be applied to the forms on eight genera of grasses including *Dactylis*, whilst the name

E. baconii has been given to the fungus on *Agrostis stolonifera*, and *E. clarkii* to that on *Holcus lanatus* (White, 1993). The form on *Festuca rubra* and *F. valesiaca* is *E. festucae* (Leuchtmann *et al.*, 1994). The uppermost leaf sheath of flowering tillers becomes surrounded by a white mass of mycelium 2 cm or more in length, and at the surface small unicellular phialoconidia are produced (Fig. 12.29b). These conidia function as spermatia (see below). Later, the conidial stroma becomes thicker and turns orange in colour as perithecia are formed (Plate 5f). The perithecia produce numerous asci, each with a well-defined apical cap, and containing eight long narrow ascospores which may break up within the ascus to form part-spores (Fig. 12.29d). The mycelium is for the most part intercellular, unbranched and mainly located in the pith, although intracellular penetration of the vascular bundles is found in the region of the inflorescence primordium. Perithecial stromata are formed only on tillers containing inflorescence primordia, and by manipulating incubation conditions it has been shown that the formation of stromata is correlated directly with the presence of an inflorescence primordium rather than with external conditions (Kirby, 1961).

Epichloe typhina is heterothallic with a unifactorial (bipolar) mating system (White & Bultman, 1987). Throughout its range *Epichloe* is attacked by a parasitic fly, *Botanophila phrenione* (*Phorbia phrenione*), which feeds on conidia, conidiophores and hyphae. The relationship is a symbiotic one. Before laying eggs, female flies feed on conidia and hyphae from conidial stromata. Possibly they are attracted by the white colour and distinctive smell of the stroma (Leuchtmann, 2003). The conidia remain viable after passing through the gut of the flies. After laying an egg in a fresh conidial stroma an ovipositing female shows an unusual but characteristic pattern of behaviour, walking in a linear or spiral path around the conidial stroma whilst dragging its abdomen and depositing a trail of faecal material with viable conidia on the receptive hyphae of its surface, so spermatizing them. This track is later marked by the development of perithecia (Bultman *et al.*, 1998).

Perithecial ontogeny has been studied by White (1997). The perithecial primordium develops as a cavity lined by inwardly directed branched hyphae. In *E. typhina*, at the base of the cavity a mound of ascogenous tissue appears. This is made up of ascogenous hyphae with croziers and with lateral paraphyses, but the paraphyses do not persist as the asci mature. The perithecial ostioles protrude above the surface of the stroma and are lined by curved periphyses (Fig. 12.29c). The apical apparatus of the ascus consists of a thickened ring pierced by a narrow canal continuous with the cytoplasm of the ascus (Figs. 12.29d, 12.30a) and through the canal the ascospores are discharged singly, one after another.

Ascospores may segment into part-spores within the ascus or after discharge. It has been claimed that they never germinate directly (i.e. by germ tube), but only by the production of conidia from narrow tapering phialides (Figs. 12.30b,c; Bacon & Hinton, 1988). However, our own observations on ascospores of *E. typhina* from *D. glomerata* show that direct germination may occur (Fig. 12.30d). Primary conidia may germinate directly or by forming secondary conidia, a process described as microcyclic conidiation (Bacon & Hinton, 1991). Tertiary conidia may also develop from secondary conidia.

Attempts to infect grasses from ascospores have generally been unsuccessful, so it is likely that infection is by conidia. After allowing ascospores to be discharged close to emerging inflorescences of uninfected *Lolium perenne* plants, about 12% of the seeds gave rise to infected progeny, but whether infection was directly from ascospores or from primary or secondary conidia was not determined (Chung & Schardl, 1997a). Experimentally, it has not been possible to infect developing seeds of *Dactylis*, and the only effective method of infection is by application of ascospores or conidia to the cut ends of green stubble (Western & Cavett, 1959). If this is the natural route of infection in other grasses, it may explain the greater incidence of the disease in *Agrostis* in heavily grazed pastures (Bradshaw, 1959).

The time of release of ascospores coincides with the emergence of larvae of the parasitic fly

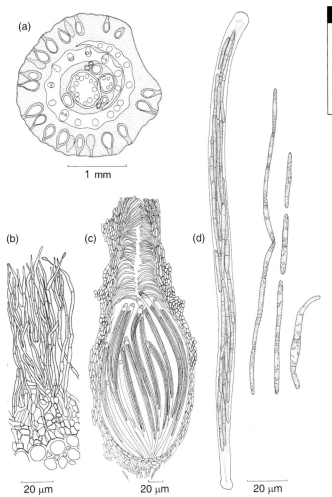

Fig 12.29 *Epichloe baconii*. (a) T.S. stem and leaf sheath of *Agrostis* surrounded by a perithecial stroma. Note the axillary shoots between the leaf sheath and the stem. (b) Part of the conidial stroma. (c) A single perithecium. Note the periphyses lining the ostiole. (d) Ascus and ascospores. Note the apical apparatus of the ascus.

which may feed on perithecia and so reduce ascospore production (Welch & Bultman, 1993).

12.5.3 *Epichloe*-related grass endophytes

White (1988) has classified the relationships between *Epichloe* or related endophytes and their hosts into three types. In type I associations, perithecial stromata are formed on the inflorescences of most if not all infected individuals so that flowering of the host is suppressed. *Dactylis glomerata* and *Agrostis tenuis* harbour this type of association which should be regarded as parasitic. In type II associations, stromata are formed on only a few (1–10%) of infected individuals in a population, although 50–75% of the population may contain the infection. This type of association has been found only in the sub-family Festucoideae. *Agrostis hiemalis*, *Bromus anomalus* and *Elymus canadensis* have associations of this type. The endophyte is probably spread by clonal (i.e. vegetative) growth of the infected host, by ascospores (contagious or horizontal transmission) and by seed transmission (vertical transmission). In type III associations, stromata are not formed on infected plants and apparently are never produced. Such associations have been found only in festucoid grasses including tall fescue (*Festuca arundinacea*) and perennial ryegrass (*Lolium perenne*). In many of these grasses over 90% of individuals are infected. In this type of association the endophytes rely on vertical transmission which involves mycelial growth from the parent plant to the embryo within the seed (White *et al.*, 1991). Since the host plant is

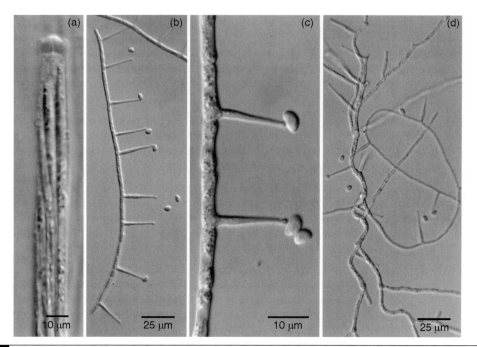

Fig 12.30 *Epichloe typhina*. (a) Tip of ascus showing the thickened apical ring pierced by a cytoplasmic canal. (b) Discharged ascospore which has germinated to produce several phialides. (c) Phialides with primary conidia. (d) Ascospore showing direct germination by hyphal growth. Some phialides are also visible.

not detrimentally affected and indeed may benefit by experiencing reduced herbivory, the relationship can be regarded as mutualistic.

The anamorphic states of *Epichloe* spp., now classified as species of *Neotyphodium*, grow as endophytes, i.e. symptomless symbionts, of many grasses. Most species of *Neotyphodium* produce phialoconidia in laboratory culture and *N. typhinum*, the anamorph of *E. typhina*, has been reported to form conidiophores on the phylloplane (leaf surfaces) of *Poa rigidifolia* and *Agrostis hiemalis*, although it is unclear whether or not they are involved in horizontal transmission in nature (White *et al.*, 1996). What is certain is that these endophytes are seed-borne and are vertically transmitted. The leaf sheaths of infected plants show a characteristic fine, intercellular, infrequently branched, contorted mycelium with lipid contents, running parallel to the vascular bundles, and following the longitudinal cell walls of the inner epidermis of the lower leaf sheaths (Fig. 12.31). Endophyte mycelium is less extensive in leaf blades and it is believed that the ligule, lacking intercellular spaces,

may be a barrier to spread (Hinton & Bacon, 1985; Christensen *et al.*, 2002). The mycelium of *Neotyphodium* appears indistinguishable from that of *Epichloe*. At flowering, the *Neotyphodium* mycelium grows upwards through the stem, extending into the inflorescence and infecting the embryos of the developing seeds so that a high proportion of them are infected (White *et al.*, 1991). In natural infections of *Lolium perenne* by *N. lolii*, the mycelium is scanty in vascular tissues but is occasionally found in smaller vascular bundles (Christensen *et al.*, 2001).

Alkaloids and endophytism

Intense interest in *Neotyphodium* has developed since the discovery that consumption of endophyte-infected grass is associated with disorders of grazing livestock caused by mycotoxins, including ergot alkaloids. Two associations (symbiota) have been particularly well investigated. Ingestion of tall fescue (*Festuca arundinacea*) infected with *N. coenophialum* causes fescue toxicosis in cattle and horses in the Southeastern USA (Bacon *et al.*, 1977; Blodgett, 2001). The

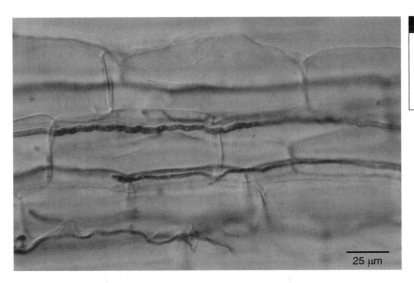

Fig 12.31 Endophytic hyphae of *Neotyphodium lolii* growing intercellularly between epidermal cells of *Lolium perenne*. Photographed from material kindly provided by P. J. Fisher.

25 μm

economic impact of fescue toxicosis to livestock producers in the USA has been estimated at $50–200 million annually (Siegel *et al.*, 1984). Perennial ryegrass (*Lolium perenne*) containing *N. lolii* is associated with ryegrass staggers in sheep in New Zealand (Fletcher & Harvey, 1981).

Various alkaloids are present in *Neotyphodium*-infected grasses. These belong to several groups, including the ergopeptide-type ergot alkaloids which we have already encountered in *Claviceps purpurea* (Figs. 12.28, 12.32a), the lolitrems (Fig. 12.32b), the lolines (Fig. 12.32c) and peramine (Fig. 12.32d). The primary causal agent of ryegrass staggers is the neurotoxin lolitrem B (Siegel & Bush, 1997; Kuldau & Bacon, 2001). In addition to their toxicity to mammals, endophyte-infected grasses have deleterious effects on insects feeding on them. The loline alkaloids are primarily insecticidal. The endophytes are therefore regarded as mutualistic symbionts with their grass hosts, protecting them against mammalian and insect herbivory whilst themselves gaining nutrients and a means of dissemination (Clay, 1988; Schardl & Clay, 1997). Endophyte-infected grasses are also more resistant than uninfected hosts against attack by certain fungal pathogens (Christensen, 1996) and nematodes.

There are other effects of infection on the grass hosts. Infection causes enhanced production of biomass, increased tillering, drought resistance and competitiveness. These are valuable attributes of turf grasses and attempts have been made to enhance the agronomic value of turf grasses by deliberately infecting them with endophytic fungi (Bacon *et al.*, 1997), including genetically modified strains. Attempts have also been made to free seeds and seedlings of pasture grasses from endophytes by fungicidal or heat treatment. Prolonged storage may also achieve this goal because the endophytes do not retain viability for as long as the seeds. However, endophyte-free plants are often less competitive than their infected counterparts.

The origin of grass endophytes

The most convincing evidence that *Neotyphodium* species are the anamorphic state of *Epichloe* is molecular (Glenn *et al.*, 1996; Kuldau *et al.*, 1997). Some of the endophytic *Neotyphodium* species are believed to be directly related to a species of *Epichloe*, e.g. *N. lolii* which is apparently derived from *E. festucae*, whilst others are of hybrid origin (Tsai *et al.*, 1994; Schardl & Moon, 2003). Interspecific heterokaryon formation has been observed in culture (Chung & Schardl, 1997b). Hyphal anastomosis and heterokaryosis may take place within grass shoots infected with more than one species of *Epichloe*, and the parasexual origin of some *Neotyphodium* species is a possibility (Tredway *et al.*, 1999).

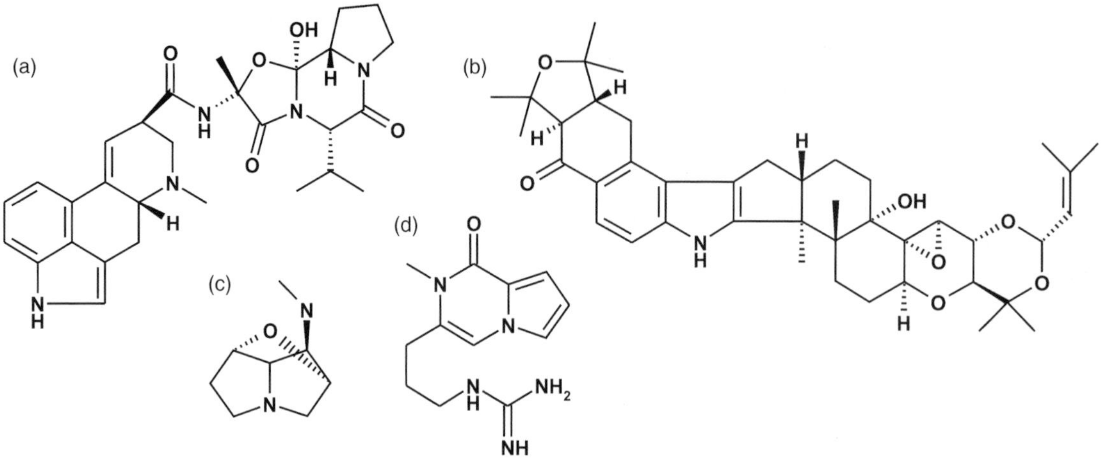

Fig 12.32 Alkaloids produced by *Neotyphodium* endophytes in grasses. (a) Ergovaline, an ergopeptide similar to ergotamine (see Fig. 12.28). (b) Lolitrem B, an alkaloid derived from geranylgeranylpyrophosphate and tryptophan. This substance is neurotoxic to grazing animals, causing ryegrass staggers. (c) Loline, an insecticidal pyrrolizidine alkaloid derived from ornithine and S-adenosylmethionine. (d) Peramine, an insecticidal pyrrolopyrazine-type alkaloid.

Other evidence comes from the distribution of alkaloids in species of *Epichloe* and *Neotyphodium*. *Epichloe festucae* is the only sexual species known to synthesize representatives of three classes of alkaloids, namely ergovaline, lolitrem B and lolines (Fig. 12.32). *Neotyphodium lolii* produces the same three alkaloids and it seems reasonable to assume that *E. festucae* was the ancestor that contributed the genes for synthesis of these three alkaloids (Clay & Schardl, 2002).

12.5.4 *Cordyceps* and its anamorphs

There are about 400–500 species of *Cordyceps* (Kobayasi, 1982; Liu *et al.*, 2002) with an epicentre of species diversity in northeastern Asia and Japan. There are 29 species in the United States and Canada, and 18 species in Europe (Humber, 2000). Most species are necrotrophic parasites of insect adults, larvae or pupae, but several species grow on the ascocarps of *Elaphomyces* (Mains, 1957). There is a wide range of insect hosts including moths, ants, beetles, and cicadas. Spiders are also attacked. Within the dead body of an infected insect a mass of mycelium or a sclerotium develops, and from this a perithecial stroma grows out. The perithecial stroma is usually fleshy and brightly coloured. It may

bear perithecia over its entire surface or they may be restricted to an upper or lateral portion so that there is a stalk region lacking perithecia. The perithecia are embedded in stromatal tissue and tightly packed together. They are elongate, with protruding ostioles. They contain numerous narrowly cylindrical asci, each with a conspicuous swollen cap pierced by a narrow canal. There are 4–8 long cylindrical ascospores which are typically divided by transverse septa into cylindrical or fusoid part-spores which may number 16, 32, 64 or 128 (Hywel-Jones, 2002). The ascospores escape singly through the narrow pore at the tip of the ascus and usually, but not invariably, break up into constituent part-spores outside the ascus.

There is an exceptionally wide range of conidial forms including the anamorph genera *Paecilomyces*, *Hirsutella*, *Hymenostilbe*, *Beauveria*, *Metarhizium* and *Tolypocladium* (Hodge, 2003; Stensrud *et al.*, 2005). Some of these are shown in Fig. 12.35. Evidence for the connection between anamorphs and their *Cordyceps* teleomorphs has been obtained in various ways, e.g. by the development of anamorphs in cultures derived from ascospores or from hyphal bodies within a parasitized insect, by the development of perithecial stromata on insect larvae artificially infected with conidia, or by molecular comparison of gene sequences (Liu *et al.*, 2002). Teleomorphic states may be rare and some of the fungi listed above are far better known as anamorphs than as teleomorphs. This is especially true of *Beauveria bassiana* and *Metarhizium anisopliae*, both of which are widely distributed in soil, probably growing as saprotrophs. They have been used in the biological control of insect pests.

Cordyceps

Cordyceps militaris forms club-shaped orange- or red-coloured stromata (Plate 5g) which project above the ground in autumn from buried lepidopteran larvae and pupae (Winterstein, 2001). Species from several genera of Lepidoptera and some Hymenoptera are susceptible. If ascospore segments or conidia come into contact with the integument of a pupa, germination occurs and is followed by penetration of the cuticle, aided by the secretion of chitinolytic enzymes. Soon after penetration, cylindrical hyphal bodies appear in the haemocoel. The hyphal bodies increase in number by budding, and become distributed within the insect's body. Death of the insect is probably associated with the secretion of the toxin cordycepin (3′-deoxyadenosine) which also accumulates in the perithecial stroma (Yu *et al.*, 2001; Kim *et al.*, 2002). After death some 5 days after infection, mycelial growth takes place and the body of the dead insect becomes transformed into a sclerotium. Under suitable conditions one or more perithecial stromata develop above ground, some 45–60 days post infection. Perithecial stromata can also form in pure culture on rice grain supplemented with haemoglobin or casein (Basith & Madelin, 1968). In pure cultures derived from single ascospores, a phialidic conidial state called *Paecilomyces militaris* is formed (Fig 12.33c).

It is not known whether *C. militaris* is homo- or heterothallic. Perithecial development has been studied in *C. militaris* by Varitchak (1931). Coiled septate ascogonia arise in the peripheral layers of the perithecial stroma. The segments of the ascogonium become multinucleate and give rise to ascogenous hyphae from which asci develop in a single cluster at the base of the

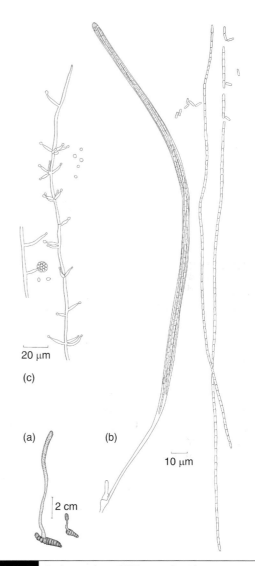

Fig 12.33 *Cordyceps militaris.* (a) Two perithecial stromata attached to pupae. (b) Ascus and multiseptate ascospores. Note the ascus cap. (c) Conidiophores and conidia.

perithecium. The perithecial wall is derived from hyphae which develop from the stalk of the ascogonium or from surrounding hyphae. Paraphysis-like hyphae grow inwards from the perithecial wall, but at maturity these hyphae dissolve and disappear.

Cordyceps sinensis (Fig. 12.34) is highly prized in traditional Chinese medicine (Pegler *et al.*, 1994). It grows on the larvae of hepialid moths at 3600–5000 m in mountainous areas in southern

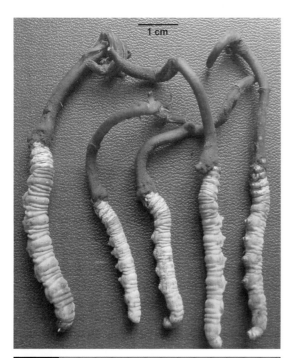

Fig 12.34 Stromata of *Cordyceps sinensis* growing from insect larvae. Image kindly supplied by Y.-J. Yao.

and western China, Tibet and Nepal (Jiang & Yao, 2002). The cylindrical perithecial stromata may be 10 cm or more in length and bundles of dried stromata are sold commercially. The anamorph is *Hirsutella sinensis* (Liu *et al.*, 2001). The conidiogenous cells are tightly packed together on the surface of a stroma.

The ecological role of *Cordyceps* spp. on insects is difficult to evaluate. Evans and Samson (1982, 1984), judging by the large numbers of stromata collected on worker ants in tropical forests, were of the opinion that they exerted a controlling effect on population size.

Cordyceps ophioglossoides grows in woods on the subterranean ascocarps of the hart's truffle *Elaphomyces*, forming bright yellow mycelial stands over its surface. Brown club-shaped perithecial stromata grow above ground in autumn (Plate 4f). *Cordyceps capitata* grows in similar situations. Phylogenetic analyses using nuclear and mitochondrial ribosomal DNA derived from 22 species of *Cordyceps* and their known anamorphs have shown that four species of *Cordyceps* parasitic on *Elaphomyces* have very

close similarity to two species which parasitize the nymphs of cicadas, and have probably evolved from them, an example of 'interkingdom host-jumping'. These related species have been grouped together in a 'truffle–cicada clade' and it has been suggested that the truffle–cicada clade separated from other *Cordyceps* clades about 43 $\pm$ 13 million years ago (Nikoh & Fukatsu, 2000). Cicada nymphs develop for several years underground, feeding on xylem sap from the host trees with which they are associated. In this respect their physiology is similar to that of *Elaphomyces* spp., which are mycorrhizal partners of trees, obtaining nutrients from them. Nikoh and Fukatsu (2000) have speculated that the overlapping niches of the species of *Cordyceps* which parasitize cicadas and those which parasitize hart's truffles may have promoted this interkingdom host jumping.

Species of *Cordyceps* reproduce asexually, some having more than one anamorphic state, i.e. with synanamorphs (Hodge, 2003). Considerable research attention has been focused on the three anamorphs *Beauveria bassiana*, *Metarhizium anisopliae* and *Tolypocladium inflatum* (Figs. 12.35a–c). The first two have potential for the biological control of insect pests, whereas *Tolypocladium inflatum* is the source of the immunosuppressant drug cyclosporin A (Fig. 12.24a).

Beauveria bassiana

This is the conidial form of *Cordyceps bassiana* (Huang *et al.*, 2002) and causes the serious white muscardine disease of silkworm (*Bombyx mori*) larvae, which is a threat to silk production. This species has been known for well over a century and is of historical interest because Agostino Bassi, who studied the fungus around 1835, proposed the germ theory of disease on the basis of his results. This preceded by several years the publications of Robert Koch, who is usually given credit for the formal proof of pathogenesis by micro-organisms. *Beauveria bassiana* is widely distributed in the soil, usually associated with diseased insects (Domsch *et al.*, 1980). Species of Coleoptera, Diptera, Lepidoptera and other groups, including Arachnida, are covered by white dusty raised tufts of hyphae bearing conidia. The association has had a long history. A worker ant covered with a fungus similar to the present-day *B. bassiana* has been discovered embedded in 25 million year-old amber (Poinar & Thomas, 1984). *Beauveria bassiana* grows readily in culture, forming dry conidia and, occasionally, synnemata. The conidiophores form densely clustered whorls of conidiogenous cells which are swollen at the base and extend into a zigzag shaped rachis forming small, globose, smooth, hyaline conidia (Fig. 12.35a).

Several mycotoxins have been obtained from cultures of *B. bassiana* including beauvericin, a cyclic depsipeptide, oosporein and bassianolide, all of which are toxic to insect larvae (see Boucias & Pendland, 1998). Numerous attempts have been made to use commercial preparations of conidia in the biological control of insect pests, such as the Colorado beetle on potatoes or the codling moth on apples. Conidia are either applied alone or, in integrated control, in conjunction with a chemical insecticide. However, dependable success in biocontrol using *B. bassiana* is still awaited. Whilst it is possible to produce large quantities of inoculum, this cannot be stored for very long, and there are also problems in maintaining a reproducibly high level of biocontrol of insect

pathogens in outdoor situations (Boucias & Pendland, 1998).

Metarhizium anisopliae

This fungus is the cause of green muscardine disease of insects. There are three varieties, var. *anisopliae*, var. *acridum* and var. *major*. The teleomorph of *M. anisopliae* var. *major* is *C. brittlebankisoides* (Liu *et al.*, 2001, 2002). *Metarhizium anisopliae* grows in soil (Domsch *et al.*, 1980), but it is also one of the most important insect pathogens. *Metarhizium anisopliae* var. *anisopliae* has a wide host range, attacking members of the Coleoptera, Orthoptera, Hemiptera and Hymenoptera as well as Arachnida. *Metarhizium anisopliae* var. *major* is more host-specific, mostly infecting soil-inhabiting scarabeid beetles (Boucias & Pendland, 1998). In culture it grows slowly, forming columns of green, shortly cylindrical, uninucleate phialoconidia which are rich in lipid droplets (Fig. 12.35b). The wall of the conidium is three-layered and the outermost layer is highly hydrophobic due to impregnation by a hydrophobin. There have been extensive and detailed studies of the physiology and enzymology of germination and penetration by the germ tubes, mainly carried out by Charnley and St Leger (1991). The brief account which follows

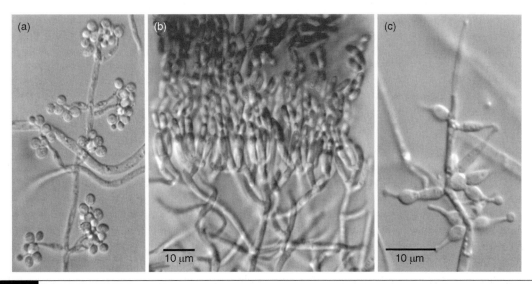

Fig 12.35 Anamorphic states associated with *Cordyceps*. (a) *Beauveria bassiana*. (b) *Metarhizium anisopliae*. (c) *Tolypocladium inflatum*. The phialides have a swollen base and a long neck. (a,b) to same scale.

draws heavily on the summary in Boucias and Pendland (1998). Water is required for germination, which is followed by attachment to the host cuticle by means of an appressorium. Appressorium development is stimulated by contact with a hard surface and can take place on glass or polystyrene, so long as complex nitrogenous substances (e.g. yeast extract or peptone) are present. In nature it is presumed that such compounds are derived from the cuticle. The wall of the appressorium is surrounded by a coat of mucilage which tightly attaches the appressorium to the host integument. An infection peg from the appressorium penetrates the surface layer of the integument, the epicuticle, but when it reaches the procuticle (the layer beneath it), the infection peg expands to form a penetration plate which grows out parallel to the surface of the integument. From the penetration plate, penetration hyphal bodies develop and, from these, vertical penetration hyphae grow through the innermost layer, the procuticle, to the hypodermis and the body cavity. From the vertical penetration hyphae, hyphal bodies in turn develop which become dispersed in the haemolymph. The hyphal bodies come to rest in the fat bodies of the insect and then give rise to mycelium, by which time, 48–72 h post infection, the host is dead. Death is probably brought about by the action of secondary metabolites which function as toxins. These include a series of depsipeptides (destruxins A–E), a hydrophobin, cytochalasins and alkaloids.

Metarhizium anisopliae var. *acridum* (formerly known as *M. flavoviride*) has been used in the biological control of grasshoppers and locusts by suspending conidia in oil for low volume application. As for *B. bassiana*, however, the commercial viability of *M. anisopliae* as a mainstream insecticide remains to be established (Hajek *et al.*, 2001).

Tolypocladium inflatum

This is the anamorph of *Cordyceps subsessilis* which fruits on the larvae of scarabeid beetles (Hodge *et al.*, 1996). It is the commercial source of the immuno-suppressant cyclosporin A, which has become a crucial drug in the treatment of the rejection reaction after organ transplantation (Dreyfuss *et al.*, 1976; Borel, 1986). Other secondary metabolites are the efrapeptins, compounds with anti-fungal and insecticidal properties (Krasnoff & Gupta, 1992). *Tolypocladium inflatum* grows in soil. In culture it first forms *Acremonium*-like conidia on single slender phialides, but later copious hyaline conidia are produced in slime from clusters of phialides with a globose base and a long, narrow tapering neck (Fig. 12.35c).

12.6 | Ophiostomatales

This group contains about 6 genera (110 species) of perithecial ascomycetes which are mainly saprotrophic or parasitic on woody hosts. The perithecia are non-stromatic, generally long-necked and solitary. We shall consider only *Ophiostoma*.

12.6.1 Ophiostoma

There are about 100 species of *Ophiostoma* (Grylls & Seifert, 1993; Seifert *et al.*, 1993), largely confined to wood and bark and associated with bark-boring beetles which disperse their ascospores and conidia. Some cause fatal diseases of trees, notably Dutch elm disease (see below). Others, e.g. *O. piceae* (Fig. 12.36), cause blue-stain (= sap-stain) of conifer wood (Seifert 1993; Gibbs, 1993). The anamorphic fungus *Sporothrix schenkii*, which is closely related to *Ophiostoma* (Berbee & Taylor, 1992b), is a human pathogen (Summerbell *et al.*, 1993).

The perithecia of *Ophiostoma* are black in colour with a bulbous base and a long cylindrical neck, the ostiole of which is surmounted by a ring of stiff tapering hairs which hold the hyaline unicellular ascospores in a mucilaginous blob. The asci have thin evanescent walls and the ascospores are released into the body of the perithecium and move up the narrow tube inside the neck. The perithecia closely resemble those of *Ceratocystis* and the two genera are sometimes synonymized (see de Hoog & Scheffer, 1984). However, molecular and morphological comparisons have indicated that, despite their

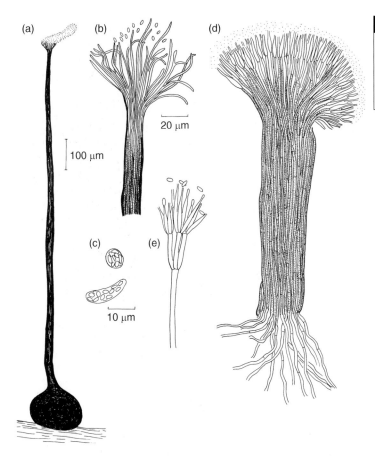

(a)

(b)

20 μm

100 μm

(c)

(e)

10 μm

(d)

Fig 12.36 *Ophiostoma piceae.*
(a) Perithecium showing spore drop at tip of neck. (b) Details of ostiole with ring of setae. (c) Asci. (d) Synnematous conidiophore. (e) Details of apex of conidiophore.

similarity, the two genera are not closely related, *Ophiostoma* being in a sister clade to the Diaporthales (p. 373) whilst *Ceratocystis* is allied to the Microascales (p. 368; Hausner *et al.*, 1993; Spatafora & Blackwell, 1994). The similarity in perithecial morphology suggests parallel evolution in adaptation to the insect dispersal of ascospores (Malloch & Blackwell, 1993). There are differences in anamorphs between the two groups. In *Ceratocystis sensu stricto* the anamorphs are phialidic, of the *Thielaviopsis* type (see Figs. 8.8, 12.42e), whereas the anamorphs of *Ophiostoma* are annellidic (de Hoog, 1974). Other points of difference are listed on p. 369.

Most species of *Ophiostoma* are heterothallic and the ascospores are of two mating types, A and B. Although ascogonia have been described, there are no antheridia. When isolates of different mating types from the same ascocarp of *O. ulmi* or *O. piceae* are inoculated near each

other in agar cultures, the two mycelia intermingle and perithecia develop. However, when cultures of different origin are opposed to each other, a line of barrage analogous to that seen in *Podospora anserina* may develop between the approaching mycelia, a phenomenon associated with vegetative incompatibility (Brasier, 1993).

Perithecium development in *O. ulmi* has been described by Rosinski (1961) and the ultrastructure of the ascogenous hyphae and ascosporogenesis by Jeng and Hubbes (1980). The multinucleate coiled ascogonium, differentiated by its greater width from the hyphae which subtend it, becomes surrounded by a mantle of branched hyphae probably derived from the ascogonial stalk. Ascogenous hyphae arise as buds from the ascogonium which is positioned near the base of the ascocarp. A central cavity arises as the result of enlargement of the outermost cells of the ascocarp, and the ascogenous hyphae form a lining layer around this cavity and develop centripetally towards the centre of the cavity. Croziers develop at the tips of the ascogenous hyphae and produce a succession of asci, indicating the ascohymenial nature of the fungus. The asci have extremely thin walls. There are no paraphyses and no periphyses.

Ascospores released into the cavity of the perithecium are extruded through the narrow neck and accumulate in a mucilaginous blob held in place by the ring of ostiolar hairs. Bark-boring beetles which feed on the ascospores (and also conidia) help in their dispersal. The beetles are attracted by 'fruity' odours emitted from the mycelium. These are volatile metabolites, mainly short-chain alcohols and esters, as well as monoterpenes and sesquiterpenes (Hanssen, 1993).

Some species of *Ophiostoma* have two or more synanamorphs. These vary in structure from yeast-like to mononematous or synnematous. They have been assigned to several anamorph genera including *Leptographium*, *Sporothrix*, *Pesotum* and *Graphium*. For example, *O. ulmi* has a yeast-like anamorph, a mononematous anamorph referred to *Pesotum* and a synnematous anamorph, *Graphium ulmi*. Older hyphae may also produce endoconidia (Fig. 12.37). Wingfield *et al.* (1991) have placed *Pesotum* in synonymy with *Graphium*.

12.6.2 Dutch elm disease

Dutch elm disease is a vascular wilt disease of *Ulmus* spp. caused by *O. ulmi*, *O. novo-ulmi* and *O. himal-ulmi*. It is best regarded as a disease complex because it is invariably associated with the activities of bark-boring scolytid beetles such as *Scolytus scolytus*, *S. multistriatus* and *Hylurgopinus rufipes*. These are the vectors for the disease. Elm populations worldwide have been ravaged, causing the death of millions of trees in Europe and North America and changing the appearance of the landscape, particularly where hedgerow elms have been killed. Diseased tree leaves wilt in dry weather and rapidly turn brown and brittle. Defoliation and death of the twigs ensues, and eventually the whole tree dies (Plate 5h). The fungus persists as a saprotroph

in the bark of dead trees. Infected twigs show a characteristic brown flecking in the sapwood. This is associated with the development of brown-coloured bladder-like inflated cells of the xylem parenchyma called tyloses, which invade the xylem vessels and block them. Gums released partly by the action of cell wall-degrading enzymes of the pathogen impede water flow. Wilting is also associated with the production of the wilt toxin cerato-ulmin, which is a hydrophobin (Richards, 1993). However, since mutants of *O. novo-ulmi* with low cerato-ulmin production still retain pathogenicity, an alternative role for it has been sought. Temple *et al.* (1997) have shown that cerato-ulmin can enhance the adhesiveness of yeast-like propagules of the pathogen and also protect them from desiccation.

Dutch elm disease was first described in Holland in the late 1920s and spread to the rest of north-western Europe, North America and to parts of Asia. The disease declined in severity in Europe in the 1940s but persisted in North America, possibly because the American elms were more susceptible. This first pandemic was relatively mild in effect and did not destroy the elm tree populations. When first discovered, the disease was associated with the synnematal conidial state, *Graphium ulmi*. Later the teleomorphic state was discovered and named *Ceratostomella ulmi*, then *Ceratocystis ulmi*, now *O. ulmi*. In the mid-1960s simultaneous outbreaks of a more severe and aggressive form of the disease occurred, centred around ports in southern England and originating from infected elm logs from North America. These outbreaks spread rapidly from the original infection foci into much of Britain and mainland Europe.

Isolations from trees affected by the aggressive form of the disease yielded a strain of *Ophiostoma* distinguished by its fluffy appearance and more rapid growth in culture, in contrast with the waxy appearance and slower growth of the non-aggressive strain. In many places where *O. ulmi* was present, it has now been replaced by the more aggressive form (Brasier *et al.*, 1998). The aggressive strain is regarded as a distinct species, *O. novo-ulmi* (Brasier, 1991b). Closer investigation of isolates of *O. novo-ulmi*,

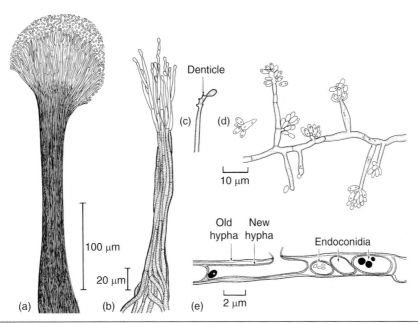

Fig 12.37 Asexual reproduction in *Ophiostoma ulmi*. (a,b) Synnematous conidiophores (*Graphium ulmi*). Parallel bundles of dark hyphae branch at their tips to produce holoblastic conidia which accumulate in a sticky drop. (c,d) Mononematous synanamorph terminating in conidiogenous cells with a succession of holoblastic conidia which, on detachment, leave a protruding scar or denticle. In d some of the conidia show yeast-like budding. (e) An older hypha within which a new hypha and endoconidia have developed (from Harris & Taber, 1973).

based on their vegetative incompatibility characteristics, showed that they can be separated into two distinct vc (vegetative compatibility) supergroups. One biotype, centred on the Romania–Moldova–Ukraine region of Europe, was designated the Eurasian or EAN vc supergroup, whilst the other, centred on the Southern Great Lakes region of North America, has been designated the North American or NAN vc supergroup. They have been formally recognized as distinct subspecies, *O. novo-ulmi* subsp. *novo-ulmi* for the EAN strains and subsp. *americana* for the NAN strains. There are morphological differences between the perithecia of the two subspecies, those of subspecies *novo-ulmi* having longer necks than those of subspecies *americana* (Brasier & Kirk, 2001). Naturally occurring hybrids between the two subspecies have been detected. Rare interspecific hybrids between *O. ulmi* and *O. novo-ulmi* also occur where the former species is being replaced by the latter (Brasier *et al.*, 1998). Hybridization between introduced pathogens permits their rapid evolution

(Brasier, 2001). *Ophiostoma himal-ulmi* is endemic to the Himalayas and has been distinguished as a third species (Brasier & Mehrotra, 1995).

Fertilized females of bark-boring beetles, possibly carrying ascospores or conidia, excavate tunnels beneath the bark of living trees, often those already weakened by the disease, to lay a cluster of eggs. After hatching, the developing larvae also make tunnels beneath the bark radiating outwards from the egg chamber. They feed on the infected wood, and the galleries which they excavate are often lined by synnemata or perithecia, so that conidia and ascospores become attached to their mouthparts and bodies. Young, sexually immature adults emerge in the spring and summer. They fly to the crotches (branch points) of young twigs where they feed on bark before maturation and mating. During this twig-feeding stage spores of the fungal pathogen may be introduced into the host sapwood, and from these multiple inoculation points the fungus may spread into host tissues by mycelial growth or by movement of conidia in xylem vessels (Webber & Brasier, 1984; Webber & Gibbs, 1989). By the latter method it has been estimated that movement can be as much as $10\,\mathrm{cm\,day}^{-1}$. In addition to being transmitted by insect vectors, the pathogen can be passed from tree to tree by natural root contact.

The control of Dutch elm disease has been attempted by the combined use of fungicides and insecticides but is now based largely on the breeding of resistant cultivars. Major genes for resistance have been identified in a group of Asian species. By crossing some of these with European elms, cultivars have been bred and released for commercial sale (Smalley & Guries, 1993). Resistance may be correlated with the production of phytoalexins (mansonones) by the host in response to infection (Smalley et al., 1993). In U. minor a correlation has also been found between vessel diameter and susceptibility, trees having vessels of large diameter being more susceptible to the disease than those with smaller diameter vessels (Solla & Gil, 2002). An interesting novel potential method of control is by the use of hypovirulent strains of the pathogen infected by cytoplasmically transmissible virus-like agents called d-factors (d for disease), now known to consist of double-stranded mitochondrial RNA elements. Twelve d-factors have been characterized. Strains carrying d-factors are hypovirulent and this is correlated with a reduced capacity in vitro to produce cerato-ulmin (Rogers et al., 1986; Sutherland & Brasier, 1995).

12.7 | Microascales

The Microascales are a small order currently containing 67 species (Kirk et al., 2001). The taxonomy of the Microascales has seen a turbulent history. Species accommodated here are characterized by perithecia (rarely cleistothecia) which usually have a long neck. The fact that asci are scattered throughout the perithecial cavity and are not produced by croziers is one reason why members of the Microascales have been considered in the past to belong to the Plectomycetes. Another moot point has been the great similarity of perithecia and general ecological features between Ceratocystis (Microascales) and Ophiostoma (Ophiostomatales), which will be discussed in more detail below. In its current shape, the order Microascales is monophyletic by DNA analyses (Hausner et al., 1993). Because the two major lineages, the Ceratocystidaceae and Microascaceae, show considerable differences in their biology and ecology, we will briefly discuss members of both groups.

12.7.1 Microascus (Microascaceae)

The family Microascaceae currently contains 43 species in 8 genera. The conidial forms are hyphomycetous, with conidia produced on annellides (see Fig. 8.7). The most important genus is Microascus with 14 species, which have been described in detail by Barron et al. (1961). Simple conidiogenous forms of Microascus are referable to Scopulariopsis (Fig. 12.38), whereas more complex, synnematous forms belong to Cephalotrichum (formerly called Doratomyces; Fig. 12.39) or Trichurus (Fig. 12.40). In agar culture, both Cephalotrichum and Trichurus produce simple Scopulariopsis-like forms in addition to the striking synnemata. The connection between these annellidic forms and Microascus has now been confirmed by phylogenetic analyses (Issakainen et al., 2003). Just to confuse matters, Cephalotrichum stemonitis produces, in addition to annelloconidia, a distinct Echinobotryum conidial state (Fig. 12.39e). Graphium is another form-genus producing synnemata which form annelloconidia at their apex (Fig. 12.41), but here the conidia are held in a drop of mucilage at the tip of the synnema. Certain Graphium-like forms, e.g. G. penicillioides (Fig. 12.41a), have affinity with Microascus, although others seem to belong to the Ophiostomatales (Okada et al., 2000).

Microascus spp. and members of related genera (e.g. Pseudoallescheria, Petriella) are capable of metabolizing a wide range of carbon sources, including keratin, crude oil, cellulose and even phenol, and they are tolerant of wide-ranging environmental conditions. Consequently, they are found from the Arctic to hot desert soil, in salt marshes, bat guano, herbivore dung, food,

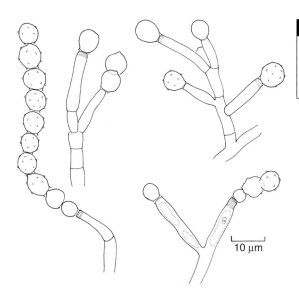

Fig 12.38 *Scopulariopsis brevicaulis*. Conidiophores terminating in conidiogenous cells (annellides) from which chains of conidia develop. The stippled areas at the tips of the annellides indicate the region of growth associated with the development of successive conidia.

10 µm

silage, and on keratin-rich substrates. Because of their ability to degrade keratin and grow at 37°C, some *Scopulariopsis* spp. can be pathogenic to humans, occasionally causing deep-seated mycoses (de Hoog et al., 2000a). Not surprisingly for such versatile and adaptable fungi, they quickly develop resistance against commonly used antifungal drugs (Cuenca-Estrella et al., 2003). This feature, rather than their pathogenicity per se, seems to be the main reason why they can be troublesome in clinical situations.

Several well-known *Scopulariopsis* states have only recently been assigned to heterothallic *Microascus* species (Abbott & Sigler, 2001). Detailed developmental studies of perithecium formation in homothallic species have been carried out by Corlett (1966). The vegetative hyphae of *Microascus* consist of uninucleate segments. Sexual reproduction is initiated when a coiled ascogonium forms and becomes ensheathed by hyphae arising from the ascogonial base as well as the surrounding vegetative mycelium. At this primordium stage, the outermost layer already becomes melanized, and it surrounds the developing pseudoparenchyma. The primordium enlarges by further growth of the pseudoparenchyma. The ascogonium is positioned initially in a cavity in the centre of the primordium, but soon sterile hyphae grow inwards from the surrounding pseudoparenchyma, raising the ascogonium towards the

apex of the perithecium. From the ascogonium, ascogenous hyphae grow between the sterile, inwardly radiating hyphae and form asci without croziers from their tips which have become binucleate. Simultaneously, at the tip of the perithecium pseudoparenchymatous hyphae begin to grow inwards and then upwards, rupturing the apical wall and extending a perithecial neck with an ostiole. During maturation of the perithecium, the sterile hyphae in the perithecial centre lyse, followed by disintegration of the ascus walls, so that the mature perithecium contains an outer melanized layer surrounding a pseudoparenchyma which encloses masses of free ascospores. These ooze out through the ostiole as a tendril.

12.7.2 *Ceratocystis* and *Sphaeronaemella* (Ceratocystidaceae)

The most important genus of this small family is *Ceratocystis* (14 species) associated with living trees. Some species cause vascular wilts, perhaps the most important pathogen being *C. fagacearum*, the cause of oak wilt, which is especially destructive in the United States. *Ceratocystis fimbriata* attacks an exceptionally wide range of plants, including the European plane (*Platanus acerifolia*), *Eucalyptus* spp., mango, coffee, sweet potato and rubber (Kile, 1993). However, infections by *Ceratocystis* spp. need not be destructive to the host, and some species produce benign

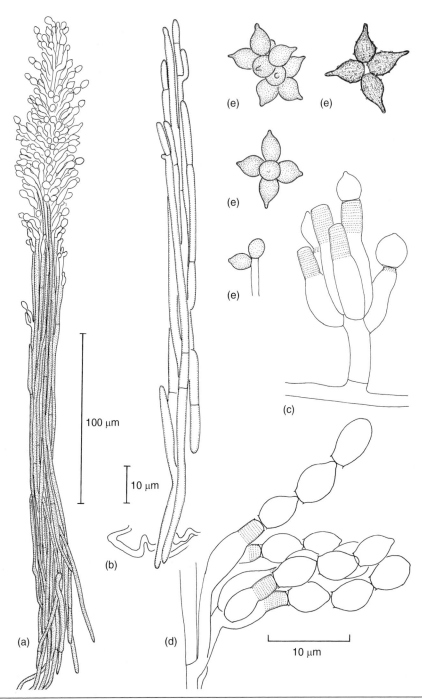

Fig 12.39 *Cephalotrichum* (*Doratomyces*) *stemonitis*. (a) Synnema or macronematous conidiophore consisting of a parallel bundle of dark hyphae branching at their tips to form conidiogenous cells (annellides) and chains of conidia. (b) Developing synnema. (c) Micronematous conidiophore bearing six annellides. The stippled region is the cylinder of annellide growth. (d) Annellides from a synnema showing chains of annelloconidia. The stippled zone represents the growth cylinder of the annellide. (e) *Echinobotryum* conidial state. (b) and (e) to same scale, (c) and (d) to same scale.

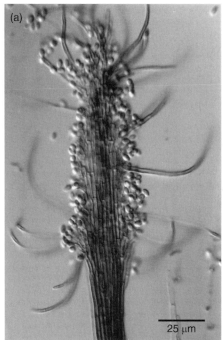

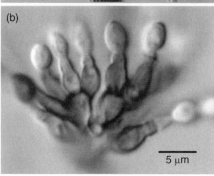

Fig 12.40 *Trichurus* sp. on agar. (a) Macronematous fructification. Note the melanized curved setae typical of the form-genus. Conidia are produced from annellides arising from the tips of the parallel hyphae making up the synnema. (b) Micronematous fructification with annellidic conidiogenesis.

infections which merely stain the wood. This so-called sap-stain can none the less be of economic importance as it reduces the market value of the infected wood.

The asci of the Ceratocystidaceae (Fig. 12.42c) deliquesce early in the development of the perithecium. The perithecium has a very long neck (Fig. 12.42a) through which the ascospores are exuded single file, accumulating in a sticky drop at the tip of the perithecial neck

(Fig. 12.42b). This is thought to be an adaptation to dispersal by insects which may pick up the slime containing ascospores upon browsing, or by feeding directly on it. Many *Ceratocystis* spp. produce fruity smells (pineapple, banana, pear) which attract insects. The substances responsible are complex mixtures of alcohols, esters and other volatile substances (Soares *et al.*, 2000), and industrial processes are being developed to produce them under fermentation conditions (Bluemke & Schrader, 2001). The conidial forms of *Ceratocystis* are phialidic, and the conidia are typically barrel-shaped or cylindrical and are produced deep inside the phialide (Figs. 8.8, 12.42e). This anamorph used to be called *Chalara* but has now been renamed *Thielaviopsis* (Paulin-Mahady *et al.*, 2002). Additionally, dark melanized chlamydospores may be present.

In addition to *Ceratocystis* spp., which are mainly associated with trees, other members of this group are soil-borne saprotrophs or weak pathogens attacking the roots of herbaceous plants. Examples are *Thielaviopsis basicola* and *T. thielavioides* (Fig. 12.42e), which can cause significant post-harvest rots in stored carrots (Punja *et al.*, 1992). The genus *Sphaeronaemella* probably also belongs to the Ceratocystidaceae. Its perithecium is typical of the family in being long-necked and topped by a frill of hyphae which hold the ascospore drop in place (Figs. 12.42a,b). *Sphaeronaemella fimicola* grows on dung, and its distinctive perithecia are commonly found in association with the fructifications of other coprophilous fungi. Although *S. fimicola* can be a weak hyphal parasite, it also seems to require diffusible metabolites and, in turn, supplies other metabolites to fellow coprophilous fungi (Weber & Webster, 1998b). There is also an interaction of *S. fimicola* with animals because its ascospore drop appears to stick to mites brushing past the perithecial neck, and the mites in turn hijack rides on flies (Malloch & Blackwell, 1993). In contrast, *S. helvellae* is a mycoparasite infecting the fruit bodies of *Helvella* spp. The conidial state of *Sphaeronaemella* is called *Gabarnaudia* and differs from *Thielaviopsis* mainly in that the conidia are formed at the tip of the phialide (Fig. 12.42d), not inside.

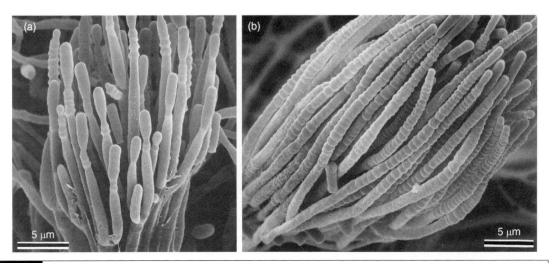

Fig 12.41 Scanning EM of synnemata of *Graphium*. The annellations are prominently visible. (a) *Graphium penicillioides*. (b) *Graphium* sp. Both images reprinted from Okada *et al.* (2000) with kind permission of Centraalbureau voor Schimmelcultures (Utrecht). Original prints kindly provided by G. Okada.

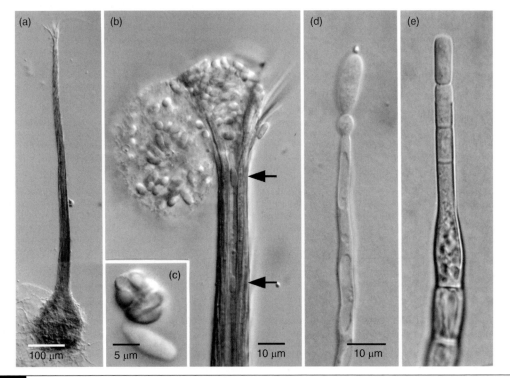

Fig 12.42 Ceratocystidaceae. (a) Perithecium of *Sphaeronaemella fimicola*. Note the extremely long neck and the crown of ostiolar hairs. (b) Enlargement of a mature perithecium apex, with the ostiolar hairs supporting a mucilaginous blob containing ascospores. The mucilage has a frothy appearance. Arrows indicate two ascospores ascending in the neck canal. (c) Ascus and conidium of *S. fimicola*. (d) The *Gabarnaudia* state of *S. fimicola*. The conidia are delimited at the apex of the phialide. (e) Phialide of *Thielaviopsis basicola*. Note that the cylindrical conidia are delimited deep inside the phialide neck. (d,e) to same scale.

Aspects of pathogenicity

Although *Ceratocystis* spp. are spread by insects, they are not normally associated with the tunnels of bark-boring beetles. Instead, they infect through pruning wounds or the bark, and may be introduced by sap-feeding insects. Transmission from infected to adjacent healthy trees may also occur via root contact. *Ceratocystis fagacearum* causes oak wilt symptoms by invading the xylem and blocking the conducting vessels with mycelium and a gel- or gum-like substance. Additionally, toxins are produced by *C. fagacearum* and other wilt-causing species, and these can reproduce many wilt symptoms in the absence of the fungus (Pazzagli *et al.*, 1999). Upon death of the host, *C. fagacearum* forms compact mycelial mats underneath the bark. The hyphae of the mycelial mat produce conidia which are dispersed by beetles attracted by the fruity smell. The fungus is heterothallic and the conidia double up as spermatia. When spermatia of compatible mating types are carried to a mycelial mat, perithecia are formed.

There are several *Ceratocystis* spp. which are associated with coniferous trees, partly as pathogens but also as sap-staining fungi (Harrington & Wingfield, 1998). These belong to a phylogenetically clearly defined group around *C. coerulescens* (Witthuhn *et al.*, 1998). Because conifers are unusual hosts for *Ceratocystis* and all known species associated with them are closely related, it is assumed that *Ceratocystis* has only recently switched hosts from broad-leaved trees.

Ceratocystis versus *Ophiostoma*

Much confusion has arisen in the past between the genera *Ceratocystis* and *Ophiostoma* which are now known not to be closely related phylogenetically (Spatafora & Blackwell, 1994). Alexopoulos *et al.* (1996) have summarized the main differences, which are as follows. (1) The anamorphic state of *Ophiostoma* is annellidic (*Graphium*; see Fig. 12.41) whereas that of *Ceratocystis* and its allies is phialidic (*Thielaviopsis*, *Gabarnaudia*; Fig. 12.42). (2) *Ophiostoma* spp. infect through insect tunnels whereas *Ceratocystis* spp. infect through wounds. (3) The cell wall of *Ophiostoma* spp. contains rhamnose-based polymers and cellulose, but

neither kind is found in *Ceratocystis*. (4) *Ophiostoma* spp. are insensitive to cycloheximide whereas *Ceratocystis* spp. are sensitive.

12.8 | Diaporthales

The order Diaporthales currently includes some 447 species in 94 genera (Kirk *et al.*, 2001) and is well separated phylogenetically from the other orders of perithecial ascomycetes (Zhang & Blackwell, 2001). Several clades can be resolved within the Diaporthales, although the naming of families is problematic at present (Castlebury *et al.*, 2002). Members of the Diaporthales grow mainly in the bark of trees as saprotrophs or parasites. *Discula destructiva* infects all above-ground organs of dogwood (*Cornus* spp.) causing anthracnose (leaf blight, twig dieback, stem cankers) which is devastating the native *Cornus* populations of North America (Redlin, 1991). Two other important plant-pathogenic genera are *Diaporthe* and *Cryphonectria*, and these will be considered in more detail below.

Members of the order Diaporthales are characterized by perithecia which are produced in clusters or stromata, often embedded in the host tissue with their long necks protruding beyond the surface. The asci are unitunicate and contain eight ascospores which have one or more septa. Diaporthales may be homothallic or heterothallic, in the latter case with a bipolar mating type system. The conidia are produced by phialide-like structures lining the inside surface of pycnidia, i.e. anamorphic Diaporthales were formerly classified as coelomycetes. Pycnidia are usually dark-walled (melanized) and have one or more openings through which the conidia exude in slimy drops. They are more commonly encountered than the sexual state (see Fig. 12.43).

12.8.1 *Diaporthe* and its anamorph *Phomopsis*

Developmental aspects

Perhaps the only detailed developmental study in *Diaporthe* was carried out by Jensen (1983) on *D. phaseolorum* var. *sojae*, which is homothallic

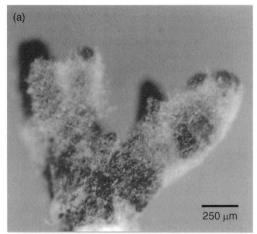

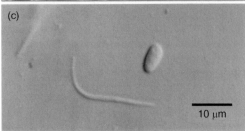

Fig 12.43 *Phomopsis phaseoli.* (a) Surface view of a conidium-producing stroma formed in agar culture. (b) Phialides producing β-conidia. (c) Ovoid α-conidium and elongated β-conidium. Both types of conidium are produced from similar phialides. (b) and (c) to same scale.

Diaporthe state which is most often seen on dead plant material. Conidial locules form in the upper region of the stroma (ectostroma) when two opposing palisades of hyphae press against each other, accompanied by lysis of hyphae bordering the developing slit. In consequence, the slit becomes convoluted and lined by hymenium. Ectostromatic cells proliferate to form a neck, resulting in the typical shape of the pycnidium consisting of a more or less globose structure within which the lobed or folded hymenium is located, and one or several necks through which conidia are exuded (Fig. 12.43a). The conidiogenous cells are interpreted as phialides. They are awl-shaped, 20 μm long and tapering from 2–3 μm at their base to 1 μm at their apex (Fig. 12.43b). Two types of conidia are produced, often within the same conidioma (Fig. 12.43c). The ovoid α-conidia contain two lipid droplets and germinate readily in culture, whereas β-conidia are highly elongated and do not usually germinate in *P. phaseoli* or other *Phomopsis* spp. They are therefore interpreted as spermatia (Jensen, 1983).

According to Uecker (1988), *Phomopsis* typically has a dark stroma producing α- and β-conidia, the teleomorph being *Diaporthe*. Unfortunately, few typical *Phomopsis* species exist since many produce only either α- or β-conidia, and/or lack the *Diaporthe* state. Further, there are few distinguishing features in conidial shape between different *Phomopsis* species, and species identification is based mainly on the host species with which a given strain is associated. This poses problems because the delimitations of host ranges are not precisely characterized. Thus, the taxonomy of *Phomopsis* is in a state of confusion, with many synonyms probably in existence. Uecker (1988) has compiled over 800 *Phomopsis* names in current use, but Kirk *et al.* (2001) have estimated that only about 100 species of *Phomopsis* exist. The genus is thus in urgent need of an up-to-date monographic treatment, Grove (1935) having been the last mycologist to rise to this formidable task.

In *D. phaseolorum*, sexual reproduction is initiated by the formation of ascogonial coils in the lower region of the stroma, termed entostroma by Jensen (1983). These coils become

like all species of *Diaporthe* examined to date. In culture, the mycelium consists of narrow hyphae containing 3–4 nuclei per segment, and wider ones with up to 15 nuclei per segment. Hyphae aggregate and swell to form a pseudo-parenchymatous stroma which produces both pycnidia and perithecia. In nature, the pycnidial *Phomopsis* state is produced earlier than the enveloped by hyphae which proliferate to form the wall (peridium) of the perithecium, and by others which form the pseudoparenchymatous centrum and paraphyses. A neck lined by periphyses is formed relatively early in perithecium development. The ascogenous hyphae arising from the ascogonial coils form a bowl-shaped hymenium in the base of the perithecium. Shortly before maturity of the perithecium, the ascogenous hyphae produce croziers. Karyogamy, meiosis and mitosis all occur in the usual way, giving rise to eight nuclei which divide once more, so that eight bicellular ascospores are produced in each ascus. A typical feature of *Diaporthe* is that the asci often become detached from the hymenium before or after the ascospores are ripe, so that the cavity of the perithecium is filled with loose asci or free ascospores. Ascospores are usually discharged non-violently as sticky tendrils exuding from the ostiole of the perithecium. The ascus has a prominent apical apparatus which stains with iodine.

Ecology

Species of *Phomopsis* and *Diaporthe* cause serious plant diseases of commercial significance, such as pod and stem blight of soybeans and other pulse fruits (*D. phaseolorum*, anamorph *P. phaseoli*) or stem canker and leaf necrosis of sunflower (*D. helianthi*, anamorph *P. helianthi*). Many *Phomopsis* diseases are caused by species complexes, i.e. different *Phomopsis* spp. can be isolated from the same diseased plant in the field. All species associated with a given disease may not be equally pathogenic. A good example is dead arm of vines, which is caused primarily by *P. viticola*. Several other *Phomopsis* species which can also be isolated from vines are only weakly pathogenic or entirely non-pathogenic (Mostert *et al.*, 2001). Fungi colonizing the living plant host as permanently asymptomatic infections are termed endophytes. The example of *P. viticola* illustrates the diffuse boundary between endophytism and parasitism; some *Phomopsis* strains are entirely endophytic whereas others initially cause latent infections but later become pathogenic (Mostert *et al.*, 2000). Another example is provided by *Diaporthe toxica* (anamorph formerly called *P. leptostromiformis*) which infects living lupins as coralloid hyphae with a limited spread beneath the cuticle. Such latent infections can persist for many months until death of the colonized host organs occurs by natural causes, e.g. senescence at the end of the growing season. Within 1−2 days of host death, large-scale colonization of the infected tissue occurs (Shankar *et al.*, 1998). Colonization by saprotrophic and pathogenic *Phomopsis* spp. can be accompanied by the secretion of large amounts of cell wall-degrading enzymes which indiscriminately macerate the plant tissue (Heller & Gierth, 2001). This accounts for the strongly necrotic nature of many *Phomopsis* diseases. Toxins may also be produced by phytopathogenic species, and these may facilitate rapid colonization by killing host cells and preventing an immune response. The *Phomopsis* state of *D. toxica* produces phomopsins, which are cyclic peptides comprising six unusual amino acids. These can accumulate in lupin stems and seeds to such high levels that they cause a serious toxicosis called lupinosis in sheep grazing on lupin stubble, or fed with lupin seeds (Culvenor *et al.*, 1977). They seem to act mainly on the microtubular cytoskeleton, and their primary effect is on the liver of affected sheep (Edgar *et al.*, 1986). Another example is the toxin phomozin which is produced by *P. helianthi* and has been shown to be capable of killing host tissue (Mazars *et al.*, 1991).

Recently, *Diaporthe ambigua*, the cause of cankers on roots and stems of fruit trees, has been found to be infected by a mycovirus which reduces the ability of its fungal host to cause disease on plants (Preisig *et al.*, 2000). Hypovirulence-causing fungal viruses and their implications for biological control are discussed in the following section on *Cryphonectria parasitica*.

12.8.2 *Cryphonectria parasitica*

Readable accounts of various aspects of *C. parasitica* have been written by Nuss (1992), Heiniger and Rigling (1994) and Dawe and Nuss (2001). The origin of *C. parasitica* is uncertain. It was first reported in 1904 as a sudden and rapidly spreading infection of *Castanea dentata* (American edible chestnut) in North America, and in 1938 on *C. sativa* (European edible chestnut) in Italy, from where it spread rapidly to other countries. It is also known in Asia, the putative centre of origin of *C. sativa* which was brought from the Near East to Western Europe by the Romans. It is possible that the pathogen was accidentally introduced into the USA and Europe with seedlings of Asian *C. crenata* (Anagnostakis, 1987). Other tree species such as oak (*Quercus* spp.) are occasionally attacked, although only with minor commercial damage. The fungus is a wound pathogen, colonizing the bark and cambium tissues as a spreading mycelium. The host defence reaction produces cankers (see Fig. 12.18), and branches die if cankers girdle their circumference. Cankers are often coloured reddish-brown due to the abundant formation of conidia which ooze out in sticky tendrils from the pycnidia producing them. Conidia are spread by animals and rain-splash. Perithecia are also formed, embedded in the bark tissue. The fungus is heterothallic, with a bipolar mating system. Conidia function as spermatia but can also germinate directly to cause fresh infections. The roots are not normally affected, and sucker shoots may grow from intact rootstocks or from points proximal to girdling cankers, but these new shoots also become infected in due course. Following the establishment of *C. parasitica*, the American chestnut tree has become reduced to a shrub-like habit, not unlike the way in which the elm tree in Europe has been affected by Dutch elm disease (see p. 366). The disease caused by *C. parasitica* is known as chestnut blight and has had a dramatic effect especially in the eastern United States, where *C. dentata* once accounted for 50% of the value of hardwood timber (Agrios, 2005).

Hypovirulence in *Cryphonectria parasitica*

Similarly severe outbreaks to those in the United States were noted in Europe, but self-healing cankers were observed about 15 years after the first sighting of the disease. Self-healing was shown by Grente and Sauret (1969) to be associated with the presence of hypovirulent *C. parasitica* isolates, i.e. strains with a much reduced virulence. Moreover, when the hyphae of a fully virulent colony were allowed to fuse with those of a hypovirulent strain by anastomosis, the former became hypovirulent, too. Eventually it was discovered that hypovirulence is associated with the presence of an unusual fungal virus consisting of double-stranded RNA surrounded by membranes but without a protein coat (Anagnostakis & Day, 1979). This new type of virus was named *Hypovirus* (see Dawe & Nuss, 2001). There are now three species of *Hypovirus* known from *Cryphonectria* (Smart *et al.*, 2000), the best-examined being CHV-1 (*Cryphonectria Hypovirus* 1). *Cryphonectria* is also a repository for mycoviruses belonging to several other taxonomic groups (Hillman & Suzuki, 2004).

Strains of *C. parasitica* infected by *Hypovirus* have several phenotypic characteristics that distinguish them from uninfected *C. parasitica*. In addition to their hypovirulence on chestnut trees, these include a pale rather than orange colony colour on agar, reduced production of an extracellular laccase and of a cell surface hydrophobin protein due to reduced transcription of their genes, poor asexual sporulation, and lack of sexual reproduction because of female sterility. Several genes were thus found to be affected by the presence of the virus. This pleiotropic effect is thought to be due to the action of the virus on various signalling cascades, whereby the stimulation of the cAMP pathway due to an inhibition of the inhibitory α subunit (CPG-1) of a large trimeric G protein has been particularly strongly implicated (Chen *et al.*, 1996).

The CHV-1 virus genome consists of two open reading frames which produce altogether three proteins (Shapira *et al.*, 1991). It is not yet clear how these act to cause the observed symptoms, except that the p29 protein is responsible for reducing the pigmentation and asexual sporulation as well as transcription of the laccase gene. However, p29 does not seem to cause hypovirulence (Nuss, 1996). In contrast, when the CPG-1 protein levels were reduced by the presence of the wild-type virus or by genetic manipulation of uninfected *C. parasitica*, or when cAMP levels were raised by chemical treatment, much of the phenotype of hypovirus infection including hypovirulence was induced. This hints at an interplay between several signalling pathways, as outlined for *Magnaporthe grisea* (Fig. 12.48).

Experiments with CHV-1 have also provided an insight into the sexual reproductive system of *C. parasitica* because the virus suppresses the genes encoding the mating pheromones which could thus be identified (Zhang *et al.*, 1998). Both male and female pheromones were suppressed by CHV-1 infections. The reason why infection leads selectively to female sterility is unclear but may be because a certain proportion of conidia, which double up as spermatia, remains uninfected by the virus during conidiogenesis.

Biological control of *Cryphonectria parasitica* by *Hypovirus*

The transmission of *Hypovirus* from one strain of *C. parasitica* to another is mediated by anastomosis, which requires vegetative compatibility. In *C. parasitica* as in *Podospora anserina* (p. 320), there are several genetic loci controlling vegetative compatibility, and anastomosis as well as virus transmission occur readily between two strains possessing identical alleles at all six vegetative incompatibility (*vic*) loci. Heteroallelism at one or more loci restricts anastomosis and reduces the percentage of virus transmission, whereby mismatches at certain loci have a more restrictive effect than those at others (Cortesi *et al.*, 2001). Of course, virus transmission in the field will be higher in populations containing a low diversity of *vic* alleles. Such is the case in Europe, where hypovirulent strains were observed to spread rapidly after their discovery. There are different strains of *Hypovirus* which may be mildly hypovirulent, i.e. only moderately restricting canker development and asexual sporulation in *Cryphonectria*, or may be aggressive. The latter permit *Cryphonectria* to form only very small cankers, but also greatly reduce asexual sporulation. Since the virus can be disseminated at least to a certain extent in conidia of *C. parasitica*, mild virus strains may spread faster in nature than aggressive ones.

In addition to relying on the natural spread of hypovirulent strains of *C. parasitica* in Europe, it has proven possible to implement a biological control strategy by inoculating active cankers with a paste containing a mixture of hypovirulent strains differing in their *vic* alleles. Success has been obtained especially in chestnut orchards or in regions where hypovirulent strains were rare in the field (Heininger & Rigling, 1994). Active cankers can be converted into healing cankers if one of the inoculated hypovirulent strains can anastomose with the fully pathogenic strain. In addition, the natural spread of hypovirulence was observed around sites of release (Heininger & Rigling, 1994).

In contrast, in the eastern USA where a great diversity of *vic* alleles exists in the wild, biological control measures have not generally been successful except in isolated forests. Another reason for the difficulties may be that an aggressive hypovirus strain was chosen for initial release experiments, which strongly reduced the ability of *C. parasitica* to produce conidia (Nuss, 1992). Current strategies are using the fact that cDNA of *Hypovirus* can be stably integrated into the genome of *C. parasitica*, and produces double-stranded viral RNA in the fungus. Whereas the RNA of the virus is not transmissible via sexual reproduction into ascospores, the integrated genomic DNA, of course, is transmitted so long as the viral RNA, which causes female sterility, is absent from the fungal cytoplasm. This strategy promises to be more successful than previous release experiments because the *vic* alleles are re-mixed during sexual reproduction, thereby facilitating the introduction of the virus into a population with diverse *vic* alleles (Dawe & Nuss, 2001).

12.9 | Magnaporthaceae

This small family (9 genera, 26 species) is currently homeless, having been excluded from the Diaporthales (see p. 373) with which it was formerly thought to be associated (Berbee, 2001; Castlebury *et al.*, 2002). We include it here because two members, *Magnaporthe grisea* and *Gaeumannomyces graminis*, are important plant pathogens. Seminal work on developmental and molecular aspects of plant pathogenesis has been done with *M. grisea* which will be the main focus of this section.

Magnaporthe grisea causes rice blast disease in which individual infections give rise to spindle-shaped necrotic lesions on rice leaves. The fungus is particularly common in South East Asia, its probable centre of origin where it has been known for centuries (Rao, 1994). Rice blast has spread to virtually all rice-growing areas, although it is more severe in cooler climates. Considerable research efforts are being directed at controlling rice blast, although the output in the shape of fungicides against *M. grisea* or the seeds of resistant rice cultivars may well be beyond the financial means of the small-scale agricultural systems found in many countries in Asia and Africa where rice blast is a problem.

In the field, the fungus is encountered mainly in the anamorph state which used to be called *Pyricularia oryzae* if growing on rice. In addition to rice, *M. grisea* can attack wheat, barley and various wild grasses on which the asexual state is called *P. grisea*. Accordingly, the rice pathogen should perhaps be renamed *M. oryzae*, but we feel bound by convention to retain *M. grisea* since that name is universally used. There are several strains with different host spectra, and e.g. the wheat blast strain in Brazil is genetically distinct from rice blast strains present in the same regions (Urashima *et al.*, 1993). This is consistent with the suggestion by Couch *et al.* (2005) that the rice-infecting lineage of *M. grisea* arose from a single host-switching event from *Setaria* millet early in the history of rice cultivation which began around 5000 BC. Sexual reproduction is rare in the field especially with the rice strains, so that discrete clones of *M. grisea* populations are typically formed in many rice-growing areas (Zeigler, 1998). In tropical climates with up to three cropping seasons each year, the fungus can continuously infect fresh green foliage, whereas in Southern Europe it overwinters on rice stubble. *Magnaporthe grisea* is

haploid, with each nucleus containing about seven chromosomes (Valent, 1997).

12.9.1 Conidium germination and appressorium formation in *Magnaporthe grisea*

A summary of the disease cycle is given in Fig. 12.44. Good reviews have been written by Howard and Valent (1996), Valent (1997) and Tucker and Talbot (2001). The conidia of *M. grisea* are three-celled, with each cell containing a single nucleus. A mature conidium swells upon hydration, and this causes the breakage of the wall at the tip of the spore, releasing a drop of mucilage stored in the periplasmic space (Fig. 12.45a). This may already occur while the spore is still attached to the conidiophore. The exact chemical nature of this mucilage is unknown, although it probably contains glycoproteins. It attaches the spore firmly to the wax of the host cuticle or other hydrophobic surfaces (Hamer *et al.*, 1988; Howard, 1994). Mucilage release is a purely physical phenomenon which does not require any de novo metabolic activities.

Under suitable conditions, a single conidial cell — usually the basal or apical cell, more rarely the central one — emits a germ tube. Numerous conflicting reports have been published on the requirements for germination, but in our experience conidia germinate readily in aqueous suspension or in contact with any inert surface, provided that they have been washed by centrifugation and resuspension in water. Washing removes an auto-inhibitor which prevents germination of spores in dense suspensions (Kono *et al.*, 1991). Germination on the plant cuticle may appear to be stimulated simply because the inhibitor is lipophilic and dissolves into the cuticular waxes, thereby becoming diluted from the spore (Hedge & Kolattukudy, 1997).

In contrast to spore germination, commitment to appressorium formation requires the presence of specific environmental signals. These are perceived after the germ tube has formed a hook-like appressorium initial (Bourett & Howard, 1990; deZwaan *et al.*, 1999). Appressoria are formed from hooks upon contact

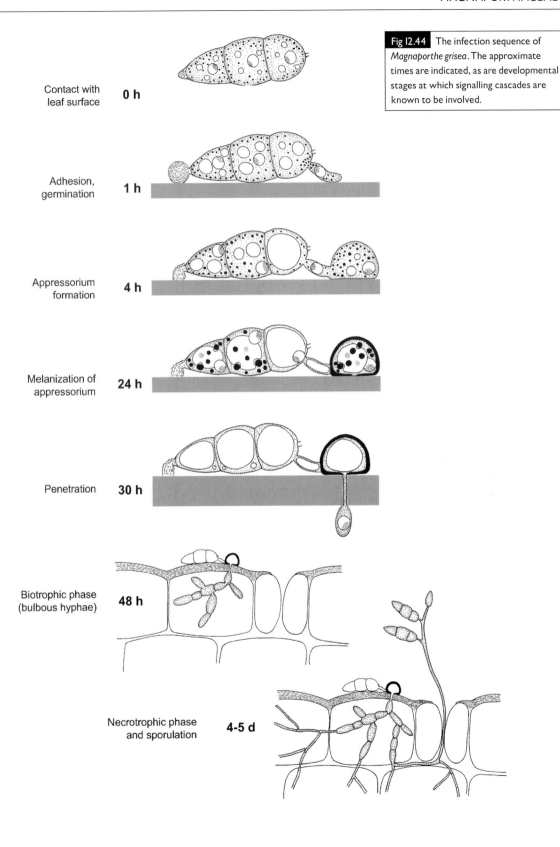

Contact with
leaf surface   **0 h**

Adhesion,
germination   **1 h**

Appressorium
formation   **4 h**

Melanization of
appressorium   **24 h**

Penetration   **30 h**

Biotrophic phase
(bulbous hyphae)   **48 h**

Necrotrophic phase
and sporulation   **4-5 d**

Fig 12.44 The infection sequence of *Magnaporthe grisea*. The approximate times are indicated, as are developmental stages at which signalling cascades are known to be involved.

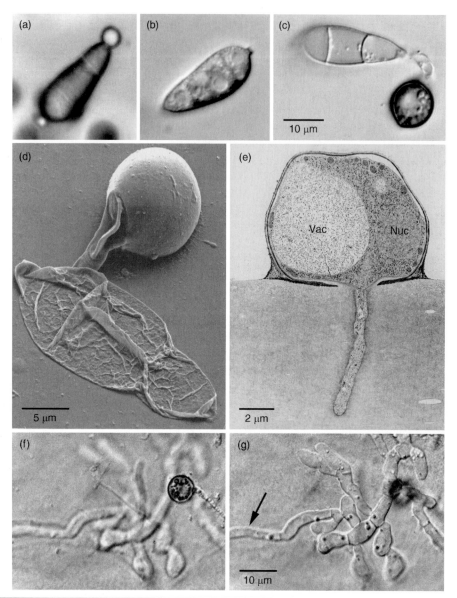

Fig 12.45 Microscopy of *Magnaporthe grisea*. (a) Hydrated conidium of *M. grisea* showing the drop of mucilage exuded at the apex. (b) Ungerminated conidium with dense cytoplasmic contents and developing vacuoles. (c) Conidium on a hydrophobic surface about 18 h after germination. Most cytoplasmic contents have been translocated into the appressorium which is laying down a dark melanized wall. (d) SEM of a mature appressorium of *M. grisea*. Whereas the conidial cells have collapsed due to the loss of turgor pressure, the appressorium is still turgid. (e) TEM of an appressorium penetrating a cellophane membrane. The appressorial wall is heavily melanized with the exception of the basal region through which the penetration peg has emerged. A large vacuole (Vac) and nucleus (Nuc) are visible inside the appressorium. (f,g) Formation of bulbous hyphae from an appressorium, some 36 h after penetration of an onion epidermis. A thinner secondary hypha has already been produced (arrow). Micrographs in (d) and (e) kindly provided by R. J. Howard; (d) reprinted from Valent (1997) with kind permission of Springer Science and Business Media; (e) reprinted from Bourett and Howard (1990), by copyright permission of the National Research Council of Canada. Original micrographs in f and g kindly supplied by A. J. Foster. (a–c) to same scale; (f,g) to same scale.

with a hydrophobic surface, or by means of chemical cues such as cutin monomers (especially 1,16-hexadecanediol). Major carbon and energy storage products in spores of *M. grisea* are cytoplasmic glycogen deposits and lipid droplets. During and after germ tube emergence, the glycogen becomes hydrolysed and enters the germ tube as soluble sugars. In contrast, lipid droplets become mobilized and migrate intact into the germ tube (Thines *et al.*, 2000), most probably along elements of the cytoskeleton. Lipid droplets accumulate in the developing appressorium (Figs. 12.45b,c), and their exit is prevented by the formation of a septum which cuts off the appressorial cytoplasm from that of the germ tube. The single nucleus present in the germinating conidial cell divides once, and one nucleus enters the appressorium whereas the other remains in the germ tube. After the septum has been completed, the germ tube and conidial cell collapse (Fig. 12.45d).

12.9.2 Appressorium maturation and host penetration by *Magnaporthe grisea*

The maturation of the appressorium is a rapid process, and within 12 h of conidia being placed on a suitable surface, mature appressoria will have formed (Fig. 12.44). Penetration takes place about 24 h after spore germination. Several processes can be observed with the light microscope, e.g. when drops of a conidial suspension are placed on plastic coverslips or onion epidermis. The lipid droplets which have accumulated in the appressorium aggregate into larger drops. In the centre of the appressorium, a vacuole forms and enlarges, and the lipid droplets enter the vacuole by microautophagocytosis and are degraded there (Weber *et al.*, 2001). This presumably provides the energy needed for the penetration events which follow. While the lipid droplets are being degraded, a thick brown wall forms on the outside of the initially hyaline wall. This contains melanin. The melanin of *M. grisea*, like that of most ascomycetes, is a polymer of dihydroxynaphthalene (DHN), although other pathways exist in other groups of fungi, plants and animals. DHN is a

pentaketide, i.e. it is formed by the head-to-tail condensation of five acetate units (Fig. 12.46) which may well arise directly from lipid oxidation. Melanin biosynthesis in fungi has been reviewed by Bell and Wheeler (1986) and Butler and Day (1998). In addition to giving strength to the appressorial wall, melanin also reduces the pore size to less than 1 nm so that molecules larger than water cannot traverse the melanized wall.

Soon after the melanin layer has been deposited, the appressorium begins to synthesize large quantities of solutes, especially glycerol (de Jong *et al.*, 1997). Since these solutes cannot escape, water moves inwards by osmosis, resulting in the generation of a turgor pressure of up to 8 MPa (= 80 bar). This is one of the highest pressures recorded in any living cell (Howard *et al.*, 1991). At its base, the appressorium is cemented tightly to the surface of the cuticle by a very effective adhesive which consists mainly of glycoproteins and lipids (Ohtake *et al.*, 1999). In the basal plate of the appressorium, there is a small non-melanized pore, so that eventually the turgor pressure is relieved by driving a penetration peg through the surface (Fig. 12.45e). The penetration peg contains a conspicuous internal skeleton of actin (Bourett & Howard, 1992), which is almost certainly involved in maintaining its shape, akin to the internal skeleton at the tip of a growing hypha (see Fig. 1.8).

12.9.3 Colonization of the host and sporulation of *Magnaporthe grisea*

Once the cuticle has been penetrated, the infection peg enlarges and branches to form hyphae which initially look swollen and are therefore called bulbous hyphae. Interestingly, at this stage the ultrastructure of an infection by *M. grisea* is similar to that of haustoria produced by biotrophic pathogens such as rusts or powdery mildews in that the bulbous hyphae do not pierce but invaginate the host plasmalemma (Heath *et al.*, 1992). This initial biotrophic phase is confined to the first epidermal cell encountered; hyphae reaching adjacent cells quickly

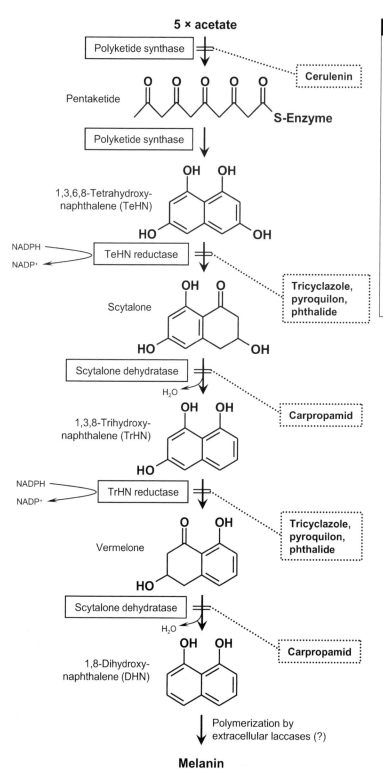

5 × acetate

Polyketide synthase — — — — Cerulenin

Pentaketide — — S-Enzyme

Polyketide synthase

1,3,6,8-Tetrahydroxy-naphthalene (TeHN)

NADPH — TeHN reductase — — — Tricyclazole, pyroquilon, phthalide
NADP+

Scytalone

Scytalone dehydratase — — — Carpropamid
H_2O

1,3,8-Trihydroxy-naphthalene (TrHN)

NADPH — TrHN reductase — — — Tricyclazole, pyroquilon, phthalide
NADP+

Vermelone

Scytalone dehydratase — — — Carpropamid
H_2O

1,8-Dihydroxy-naphthalene (DHN)

Polymerization by extracellular laccases (?)

Melanin

Fig 12.46 The biosynthetic pathway of fungal melanin via DHN (dihydroxynaphthalene). A polyketide synthase fuses five acetate units head-to-tail followed by cyclization, resulting in the pentaketide 1,3,6,8-tetrahydroxynaphthalene. This is modified by two reductases and two dehydratases to 1,8-dihydroxynaphthalene. This melanin precursor, which is presumably synthesized in the cytoplasm, is released into the wall and then polymerized to melanin by wall-bound oxidases, presumably laccases. Melanin biosynthesis is an excellent target for fungicides, and the points of inhibition of several of them are indicated. Based partly on Uesugi (1998) and Thompson et al. (2000).

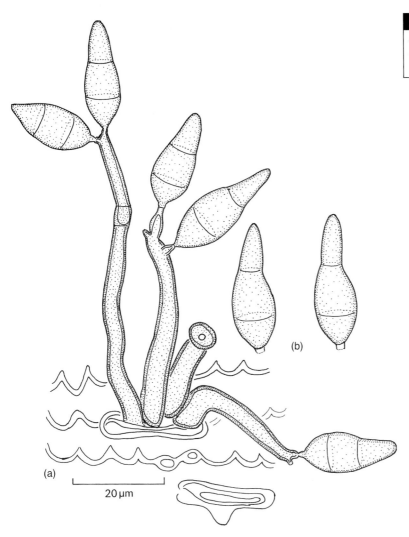

Fig 12.47 (a) Conidiophores of *Magnaporthe grisea* emerging from a stoma on a rice leaf. Detached conidia are also shown (b).

(b)

(a)

20 µm

colonize the plant tissue. These secondary colonizing hyphae are thinner and straighter than the bulbous hyphae (Figs. 12.45f,g). Within 5 days of infection, diamond-shaped lesions of dead tissue develop, and these emit conidiophores which often grow out through dead stomata (Fig. 12.47).

The interaction between *M. grisea* and rice is governed by a gene-for-gene relationship (Valent, 1997; see p. 619). In compatible interactions, the host response is delayed because the fungus is not immediately recognized by the plant. However, even virulent strains of *M. grisea* will eventually be exposed to the toxic products (phytoalexins) of the host's delayed immune response, and *M. grisea* seems to have evolved mechanisms to deal with phytoalexins by exclusion. It is becoming clear that ABC transporters play a crucial role, and we have already come across them as a resistance mechanism developed by *Candida albicans* against clinical drugs (see p. 278). The report by Urban *et al.* (1999) on *M. grisea* was one of the first to implicate a phytoalexin-excluding ABC transporter as an important factor in the colonization of a plant host by a fungal pathogen. ABC transporters have now been shown to play a crucial role in the exclusion of toxic substances as well as fungicides in several plant-pathogenic fungi (Hayashi *et al.*, 2002; Stergiopoulos *et al.*, 2002).

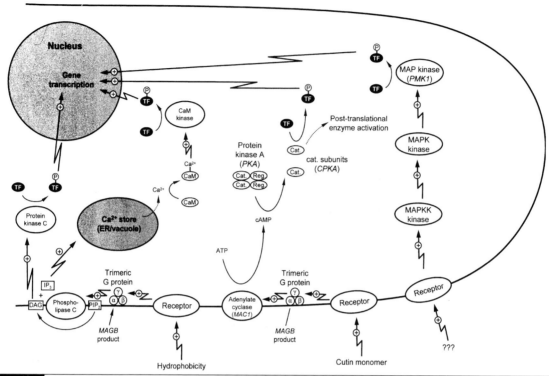

Fig 12.48 The three principal signalling cascades involved in appressorium differentiation in *Magnaporthe grisea* and in many other differentiation processes in other eukaryotes. Cross-talk occurs between the individual pathways, but this is not shown here. Abbreviations are as follows: CaM (calmodulin), cAMP (cyclic adenosine monophosphate), Cat. (catalytic), DAG (diacylglycerol), GTP (guanosine triphosphate), IP$_3$ (inositol trisphosphate), MAPK (mitogen-activated protein kinase), PIP$_2$ (phosphatidylinositol bisphosphate), Reg. (regulatory), TF (transcription factor).

12.9.4 Signalling and pathogenesis in *Magnaporthe grisea*

Both signals for appressorium initiation – surface hydrophobicity and cutin monomers – are probably perceived at the plasma membrane of the *Magnaporthe* germ tube. The actual receptors are unknown at present, but the Pth11p plasma membrane protein is likely to be one of them (deZwaan *et al.*, 1999). The transmission of signals from the plasma membrane to the nucleus occurs along several different routes (Fig. 12.48; Dean, 1997; Tucker & Talbot, 2001) which are briefly outlined below:

1. One membrane receptor receiving the chemical stimulus 1,16-hexadecanediol acts via a trimeric GTP-binding protein to activate an adenylate cyclase which converts ATP into the second messenger, cyclic AMP (cAMP). This, in

turn, activates a protein kinase A by releasing its monomeric catalytic subunits from the inactive tetramer, and the catalytic subunits then phosphorylate regulatory proteins (transcription factors) which enter the nucleus and activate the genes required for specific developmental steps.

2. The involvement of a second common eukaryotic signalling cascade in appressorium formation was suggested by Thines *et al.* (1997) who noted that diacylglycerols could trigger appressorium formation on normally non-inductive hydrophilic surfaces such as glass. The initial signal (hydrophobicity) is thus likely to be transduced via phospholipase C which hydrolyses the membrane lipid phosphatidylinositol into inositol-triphosphate (IP$_3$) and diacylglycerol (DAG). IP$_3$ acts by releasing Ca^{2+} from

intracellular stores such as the endoplasmic reticulum; this forms a complex with the calcium-binding protein calmodulin, and the calcium–calmodulin complex activates a calmodulin-dependent protein kinase. The DAG component directly activates protein kinase C (Fig. 12.48). Both the IP$_3$ and DAG branches of this signalling pathway thereby lead to the activation of protein kinases which phosphorylate transcription factors.

3. Mitogen-activated protein kinase (MAPK) pathways have been proposed to be involved at various points in appressorium induction and maturation in *M. grisea* (Xu & Hamer, 1996; Thines *et al.*, 2000). A mitogen is an extracellular substance which stimulates nuclear division or cell differentiation; here it is a stimulus for appressorium formation. The MAP kinase encoded by *PMK1* responds to a surface signal and interacts with the cAMP pathway in a manner not yet entirely understood, to initiate appressorium formation (see Fig. 12.48). A second MAP kinase, *MPS1*, is involved in penetration of the epidermis from mature appressoria. A third MAP kinase, *OSM1*, is involved in turgor regulation during osmotic stress but plays no role in appressorium functioning (Dixon *et al.*, 1999). MAP kinases are very highly conserved between different fungi, to the extent that they are functional when their genes are exchanged,

e.g. between *M. grisea*, *Candida albicans* and *Saccharomyces cerevisiae* (Xu, 2000).

It should be noted that any one of the above principal signalling cascades may act repeatedly in the course of appressorium development, as indicated in Fig. 12.44, and in other events such as production of conidiophores and conidia. Signal cascades acting repeatedly in the life cycle of *M. grisea* use many shared components and only a few specific ones at any one time point. There is also considerable cross-talk between the three types of signalling cascade mentioned here (see Kronstad *et al.*, 1998), and equivalent signalling pathways are involved in infection processes of many other fungal pathogens and in other fundamental processes such as yeast–hyphal dimorphism, mating and osmoregulation in most fungi examined to date (Xu, 2000). They are also fundamentally conserved across other eukaryotic life forms, and some are even found in prokaryotes. For this reason, signalling cascades are unlikely to provide suitably specific targets for fungicides, and interest in this aspect of signalling seems to have waned somewhat in recent years.

12.9.5 *Gaeumannomyces graminis*

Gaeumannomyces graminis var. *tritici* causes take-all disease of wheat, barley, rye and numerous wild grasses. The pathogen is soil-borne and

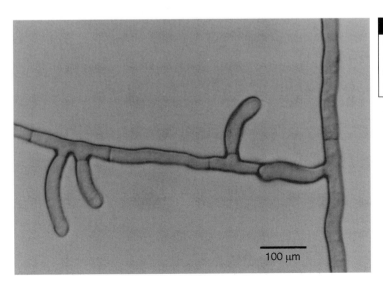

Fig 12.49 Melanized hyphopodia of *Gaeumannomyces graminis* arising from an equally strongly melanized runner hypha formed on the hydrophobic surface of a plastic coverslip.

100 μm

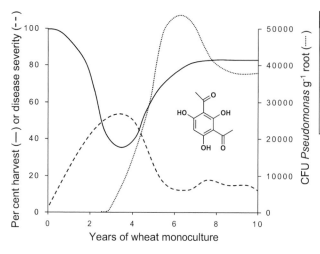

Fig 12.50 Diagrammatic representation of the relationship between wheat yield (solid line), the severity of take all infection (dashed line) and the occurrence of 2,4-diacetylphloroglucinol-producing *Pseudomonas* spp. (dotted line) on a field with successive wheat cultivation. The insert shows the molecular structure of 2,4-diacetylphloroglucinol. Based partly on Parry (1990) and Weller *et al.* (2002).

infects the roots, causing blackening and decay. As a result, the cereal plants appear stunted and produce small, non-fertile heads which appear bleached and are therefore called 'white heads'. The fungus overwinters on cereal stubble and infects the new crop mainly in spring. The infection mechanism is not entirely clear, and the fungus may enter host roots directly from dark melanized hyphae (runner hyphae) or melanized **hyphopodia** which may be aggregated to form infection cushions or mycelial mats (Butler & Jones, 1949). A hyphopodium is defined as an appressorium produced from a vegetative hypha rather than a germinating spore (Fig. 12.49). The turgor pressure in hyphopodia is around 1.5 MPa (= 15 bar), i.e. considerably less than in appressoria of *M. grisea* (Money *et al.*, 1998).

Like *M. grisea*, *G. graminis* is a complex of strains with different but overlapping host ranges, and varieties *tritici*, *avenae*, *graminis* and *maydis* are distinguished. The conidial state of *Gaeumannomyces* is a *Phialophora*, but since the assignments of anamorphs is difficult, it may be best to speak of the *Gaeumannomyces–Phialophora* complex (Bryan *et al.*, 1995). *Gaeumannomyces* is closely related to *Magnaporthe* (Bryan *et al.*, 1995).

Take-all is considered to be the most important cereal disease in temperate climates. If wheat is grown on a field for about four successive years, the disease will build up to very high levels, causing crop losses in excess of 50%

(Polley & Clarkson, 1980), but if cultivation is continued the disease will decline to an acceptable base level (Fig. 12.50). If a crop rotation is carried out, the antagonistic properties of the soil are lost. The reasons for this remarkable phenomenon are now beginning to be understood, and it is clear that the establishment of an antagonistic soil microflora plays an important role. In particular, fluorescent *Pseudomonas* spp., i.e. pseudomonads which produce water-soluble substances which fluoresce green or yellow, have been implicated as agents of take-all decline (Weller *et al.*, 2002). The metabolite 2,4-diacetylphloroglucinol (Fig. 12.50) produced by them seems to be responsible for suppression of *G. graminis* in the rhizosphere (Raajmakers & Weller, 1998). Wheat roots injured by *Gaeumannomyces* are colonized extensively by *Pseudomonas* spp., thereby explaining why a severe take-all attack must occur in order to render the soil suppressive.

12.10 | Glomerellaceae

Like the Magnaporthaceae, the family Glomerellaceae is a group of plant-pathogenic fungi which cannot be assigned to any order at present, although it is certain that it belongs to the core group of Pyrenomycetes like all other groups described in this chapter (Wanderlei-Silva *et al.*, 2003). The family comprises only one

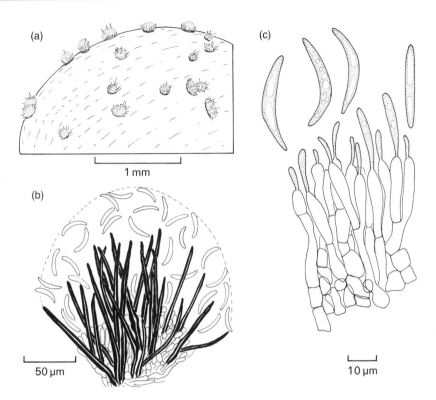

Fig 12.51 *Colletotrichum graminicola*. (a) Portion of *Sorghum* grain bearing acervuli. (b) An acervulus. (c) Phialides and phialoconidia.

teleomorph genus, *Glomerella*, which consists of five species, and numerous species belonging to the form-genus *Colletotrichum*. Early workers defined *Colletotrichum* spp. mainly by their association with host plants, so that approximately 750 'species' were described before von Arx (1957) tidied them up by synonymizing them into 11 taxa. For *C. gloeosporioides* alone, von Arx (1957) recognized 600 synonyms. A more recent treatment is that by Sutton (1992) in which 39 species are described. Many of these are probably species complexes, and for example six *formae speciales* have been described for *C. gloeosporioides*.

Glomerella forms dark-walled perithecia which occur singly. They are long-necked and release ascospores passively. The *Colletotrichum* state is an **acervulus**, i.e. a saucer-shaped conidioma which develops within the host tissue, rupturing the cuticle at maturity (Fig. 12.51a). The conidiogenous cells form a closely packed palisade of phialides (Fig. 12.51c). The curved elongated phialoconidia are produced in a slimy droplet which is held in place by stiff dark setae surrounding the acervulus (Fig. 12.51b). This mucilage contains the autoinhibitor mycosporine-alanine which prevents the germination of conidia until it has been diluted out (Leite & Nicholson, 1992). Other germination autoinhibitors are localized within the conidial cytoplasm of diverse *Colletotrichum* spp. (García-Pajón & Collado, 2003). The presence of such substances explains why the spores of many fungi germinate better after they have been washed.

Mating systems are complex and variable in *Glomerella*. For example, in *G. cingulata* (anamorph *C. gloeosporioides*), both homothallic and heterothallic strains occur, and there may be more than two mating type idiomorphs (Cisar & TeBeest, 1999). The occurrence of multiple alleles at one mating type locus is common in the Basidiomycota but has apparently not been found in any ascomycete other than *Glomerella*. Many strains belonging to a given species complex appear to be reproductively isolated; for instance, the sexual state of *C. lindemuthianum* (teleomorph *G. cingulata* f. sp. *phaseoli*) has not been found in nature but can be readily induced in agar culture by the pairing of compatible isolates (Roca *et al.*, 2003).

Species of *Colletotrichum* cause serious diseases on a wide range of plants. These are often referred to as anthracnose because of the appearance of sunken necrotic lesions, as exemplified by *C. lindemuthianum* which causes anthracnose of beans, peas and other legumes (Plate 5i). These necrotic lesions contain the acervuli. Other important pathogenic species are *C. gloeosporioides* (anthracnose of a range of tropical fruits and many other plants), *C. coffeanum* (coffee berry disease), *C. gossypii* (boll rot and anthracnose of cotton), *C. musae* (post-harvest fruit anthracnose on banana), *C. graminicola* (anthracnose of maize and sorghum), and *C. coccodes* (anthracnose of tomato, black dot disease of potato). Many of these produce phytotoxic substances which are involved in causing disease symptoms (García-Pajón & Collado, 2003). Whilst diseases in the field are important, post-harvest rots probably cause even greater economic damage, especially in the tropics. This comes about because *Colletotrichum* spp. can cause latent infections which give rise to disease symptoms only during fruit ripening in storage. Good descriptions of anthracnose diseases may be found in two volumes dedicated to *Colletotrichum*, which is one of the most important genera of fungal plant pathogens worldwide (Bailey & Jeger, 1992; Prusky *et al.*, 2000). Several *Colletotrichum* spp. have been developed with limited success as biocontrol agents against weeds (Watson *et al.*, 2000).

12.10.1 Infection strategies in *Colletotrichum*

Most species infect their host from germinating conidia which emit germ tubes terminating in a melanized appressorium. The details of appressorium formation and function are very similar to those in *Magnaporthe grisea* (see pp. 378–381), as far as they are known (Bailey *et al.*, 1992). One of the fascinations of the genus *Colletotrichum* lies in the range of post-appressorial infection strategies which various

species have evolved. These have been well described by O'Connell *et al.* (2000) and Latunde-Dada (2001), and are briefly summarized below.

Necrotrophic pathogens

Colletotrichum capsici is a necrotrophic pathogen infecting red peppers. It shows intercellular growth which commences immediately after penetration and is accompanied by the secretion of cell wall-degrading enzymes (Pring *et al.*, 1995).

Hemibiotrophic pathogens

The pattern of infection by numerous species (exemplified by *C. lindemuthianum*) is very similar to that in *Magnaporthe grisea* (p. 381). The penetration peg arising from an appressorium forms a vesicle which emits swollen primary hyphae inside the first-colonized epidermal cell and in adjacent cells (see Figs. 12.45f,g). The primary hyphae do not breach the plant plasma membrane but invaginate it, and there is evidence of a distinct matrix between the fungal hypha and the host membrane, in analogy to the haustorium formed by biotrophic fungal pathogens (O'Connell *et al.*, 1985; Mendgen & Hahn, 2002). After 1–3 days, the biotrophic phase breaks down and thinner secondary hyphae are formed which spread the infection, forming a necrotrophic lesion. Secondary hyphae differ fundamentally in surface properties from primary hyphae and are not surrounded by an extracellular matrix (Perfect *et al.*, 2001). A modification of this pattern is observed, e.g. in *C. destructivum* on cowpea (*Vigna unguiculata*), in which the biotrophic stage is confined to one epidermal cell containing a multi-lobed vesicle, from which secondary hyphae initiate the necrotrophic phase (Latunde-Dada *et al.*, 1996).

Post-harvest pathogens

This group of *Colletotrichum* spp. is responsible for most of the diseases of ripe fruits, especially in tropical areas. An example is *C. musae* on banana fruits. Spores alighting on unripe fruits prior to harvest may germinate and form appressoria and even penetration pegs, but there the infection process stalls. Infection is continued on the ripened fruits after a period of storage. This infection strategy enables the pathogen to avoid the high levels of phytoalexins in unripe fruits (Latunde-Dada, 2001). Resumption of development is triggered by the strong increase in ethylene levels associated with fruit ripening (Flaishman & Kolattukudy, 1994).

Endophytic colonization

Several anthracnose pathogens including *C. musae*, *C. coccodes* and strains of *C. gloeosporioides* develop a prolonged latent phase inside their host plant (Rodriguez & Redman, 1997). Colonization during the endophytic phase may begin by infection through stomata rather than direct penetration of the epidermis, and intercellular colonization of host tissue. During senescence of the host plant, the host's immune system degenerates and active colonization may ensue, resulting in the development of sporulating anthracnose-type lesions. Freeman and Rodriguez (1993) and Redman *et al.* (1999a) have shown that the deletion of a single gene can convert an aggressive pathogen (*C. magna*) to a permanently symptomless endophyte. This finding emphasizes the possibility that 'endophytic mutualism is only a gene away from pathogenicity' (Latunde-Dada, 2001), and also indicates one mechanism by which endophytes may evolve in nature. A further fundamental point of interest is that colonization by such symptomless endophytes renders the host plant resistant against infection by pathogenic strains of the same species, and also other fungal pathogens (Redman *et al.*, 1999b).

Hymenoascomycetes: Erysiphales

13.1 | Introduction

The Erysiphales are a clearly defined, monophyletic order of about 500 species, all of which are obligately biotrophic pathogens of plants. Braun (1987, 1995) has provided thorough monographic treatments of this group, including descriptions of most known species. There is only one family, the Erysiphaceae. The species grouped here cause symptoms readily recognized as 'powdery mildews' because the conidia produced in abundance on the shoots of infected host plants give them a whitish powdery appearance. The term 'powdery mildews' is often also applied to the organisms causing them. Only angiosperms are attacked, almost always belonging to the dicotyledons. One notable exception is the powdery mildew of cereals and grasses caused by *Blumeria graminis* (formerly *Erysiphe graminis*). Other economically relevant species are *Podosphaera leucotricha* causing apple mildew, *Podosphaera* (formerly *Sphaerotheca*) *mors-uvae* causing American gooseberry mildew, and *Uncinula necator* (now *Erysiphe necator*) causing the powdery mildew of grapes. Many other species produce less destructive infections and are ubiquitous in nature, e.g. *Microsphaera alphitoides* (now *Erysiphe alphitoides*) on the leaves of oak (*Quercus* spp.) and *Phyllactinia guttata* on hazel and many other broad-leaved trees. Being obligate biotrophs, powdery mildews cannot be kept in axenic culture, although Arabi and Jawhar (2002) have recently grown *B. graminis* on agar augmented with shredded barley leaves. The possibility of

infecting *Arabidopsis thaliana* with several different powdery mildew species holds promise for future investigations (Vogel & Somerville, 2002).

In the Erysiphales, the mycelium generally consists of uninucleate haploid segments. It is almost always confined to the leaf surface, and infections are limited to the epidermal cells which are penetrated from the outside following the formation of appressoria. Inside the host cell, haustoria are formed which provide a large area of contact with the host. Only relatively few powdery mildews (e.g. *Phyllactinia* spp.) are able to penetrate more deeply into the host tissue. Either way, soon after infection conidia are produced at the infected leaf surface from a foot cell, either in basipetal chains or singly as in *Erysiphe* if a conidium is released before the next one is formed. Conidial states of the Erysiphales are referable to the anamorphic genus *Oidium*, barring a few specialized genera such as the *Ovulariopsis* state of *Phyllactinia*. A division of *Oidium* into subgenera has been proposed (Cook et al., 1997; Braun et al., 2002; see Fig. 13.1). The conidia of Erysiphales are generally described as meristem arthroconidia because the conidiogenous cell is not homologous to a rudimentary phialide (Hughes, 1953). Numerous crops of conidia can be produced in a growing season, and they are the main carriers of infection.

The conidia of Erysiphales are unusual because unlike most fungal spores they are fully hydrated. Germination does not require the uptake of exogenous water and can proceed even in atmospheres with low relative humidity (Somers & Horsfall, 1966). In fact, germination

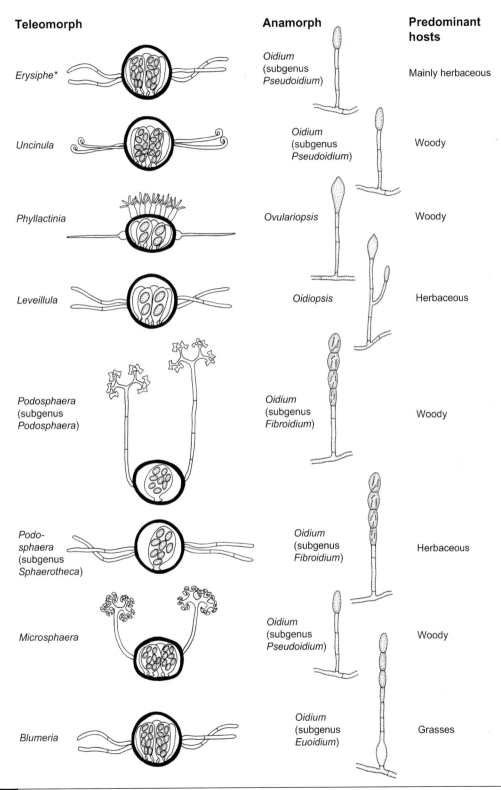

Fig 13.1 Summary of the key features of the most important genera of Erysiphales. Based on the results of Saenz and Taylor (1999a) and Braun *et al.* (2002). * Note that *Erysiphe* is a polyphyletic genus at present.

in some species is inhibited by free water. Ultra-structurally, conidia are highly vacuolated (see Fig. 13.11). The major carbon and energy reserve seems to be glycogen (McKeen *et al.*, 1967; Roberts *et al.*, 1996), although lipid droplets have also been reported, and lipids may contribute about 10% of the dry weight of conidia (Lösel, 1988). The conidia of Erysiphales are uninucleate.

Ascocarps are usually formed late in the vegetation period. These have traditionally been termed cleistothecia, although they differ fundamentally from those of the Plectomycetes because the asci of Erysiphales are club-shaped, not globose, and are formed at one level at the bottom of the ascocarp rather than being scattered throughout. Further, at maturity the ascocarp breaks open by a pre-determined line of weakness, exposing the asci which forcibly discharge their spores by a squirt mechanism. In contrast, the ascospores of Plectomycetes are released passively when the ascus wall disintegrates. Braun *et al.* (2002) have proposed the term **chasmothecium** (Gr. *chasma* = an opening, open mouth) for the ascocarp of the Erysiphales. Chasmothecia are brown globose bodies which have no ostiole. Depending on species, they may contain one or several asci, and their line of weakness may run around the equator of the chasmothecium, or through its apex. Chasmothecia are often ornamented by highly characteristic appendages (Fig. 13.1) which are usually sufficient to permit unambiguous species identification, together with the number of asci in the ascocarp (one or several), the number of ascospores per ascus, and the identity of the host plant. However, the phylogenetic value of chasmothecial appendages appears limited (see Fig. 13.1). In Northern European climates, the asci are usually fully formed in late autumn, but chasmothecia do not open until the following spring when the host plants begin to grow. They are therefore thought of primarily as overwintering structures, even if their viability may be low. In countries with dry hot summers, chasmothecia may serve as oversummering structures, being formed in late spring and releasing ascospores in the autumn.

The developmental events taking place during chasmothecium formation are immensely complex and have been described by Luttrell (1951) and Gordon (1966). They are probably similar in most species. Initially, two superficial uninucleate hyphae meet and one encircles the other. The central cell receives a nucleus and enlarges somewhat. The central cell has been termed the pseudoascogonial cell because it does not seem to play any direct role in ascus formation. The pseudoantheridial cells which encircle the pseudoascogonium divide to form the peripheral cells of the ascocarp. Some outer peripheral cells ('mother cells') develop short septate receptive hyphae which make contact with vegetative hyphae on the host surface. Following plasmogamy, one nucleus is taken up by the receptive hypha and divides in each segment of the receptive hypha until one nucleus derived from the vegetative hypha reaches the mother cell. The mother cell then divides repeatedly.

At this stage, the immature ascocarp consists of a pseudoparenchymatous centrum composed largely of binucleate cells derived from the mother cells intermixed with some uninucleate cells, and surrounded by a peridium, some 4–6 cell layers thick. The peridium becomes darkly pigmented. Uninucleate and binucleate cells above the middle part of the centrum lyse. Karyogamy occurs only within certain of the binucleate cells which are more or less isolated from the surrounding cells by lysis. These cells then enlarge to form asci. The asci appear to grow at the expense of the uninucleate and binucleate cells of the centrum, so that eventually the asci (or a single ascus, depending on the genus) occupy almost the entire centrum. Meiosis of the fusion nucleus in developing asci is usually delayed until the centrum cells have all been absorbed, although it tends to be completed before the winter dormancy.

13.2 | Phylogenetic aspects

The Erysiphales are clearly delimited and defined as a group, but the question where to position this order within the Ascomycota has aroused considerable controversy over the past 150 years or so and is still undecided. Braun *et al.* (2002) have given an overview of the taxonomic history of the Erysiphales. Affinities with Pyrenomycetes or Plectomycetes have been proposed in the past but are not now thought to be true. Instead, the phylogenetic position of the Erysiphales is fairly isolated, possibly associated weakly with the Helotiales (Saenz & Taylor, 1999b).

For many years, the taxonomy of genera and species was based on the system proposed by Léveillé (1851) who emphasized the features of the chasmothecium, especially the number of asci (one or several) and the type of appendage which may be simple, uncinate (= recurved or hooked), dichotomously branched, or bulbous (see Fig. 13.1). Recent phylogenetic studies based on a variety of DNA sequences have revealed that in the Erysiphales, unlike most other groups of organisms, the features of asexual reproduction are more clearly diagnostic than those associated with sexual reproduction. There is a good correlation between the anamorphic state and the groupings obtained by DNA sequence analysis (Saenz & Taylor, 1999a), and the gross anamorphic features also correlate with the surface ornamentations of the conidia as seen by scanning electron microscopy (Fig. 13.3; Cook et al., 1997). Thus, the genera of Erysiphales are currently defined mainly by their anamorphs. In contrast, the striking chasmothecial appendages do not correspond well to the individual groups because notably the dichotomous and simple mycelioid appendages are found in more than one taxon (Saenz & Taylor, 1999a). One casualty of these findings is the genus *Sphaerotheca* with simple appendages and one ascus per chasmothecium, which has been incorporated into *Podosphaera*, formerly comprising only species with chasmothecia containing one ascus but bearing appendages with dichotomously branched tips (Braun et al., 2002).

We have encountered a great plasticity of appendages on ascocarps before, especially in the Plectomycetes where they may be involved in insect dispersal (see p. 289). In contrast, in the Erysiphales the chasmothecial appendages seem to be related to the type of host infected. According to the analyses of Mori et al. (2000), the most basal species among the Erysiphales is *Parauncinula septata* which has uncinate (hooked) appendages. This species occurs on oak (Fagaceae),

and Mori et al. (2000) have speculated that the Erysiphales originated on members of the Fagaceae because that host family hosts by far the greatest diversity of powdery mildews (Amano, 1986; Braun, 1987, 1995). Together with another ancestral species possessing chasmothecia with uncinate appendages, *Caespitotheca forestalis*, *P. septata* may have diverged from other powdery mildews some 80–90 million years ago (Takamatsu et al., 2005).

Many of the powdery mildews associated with the leaves of deciduous trees have hooked or branched appendages, whereas those on evergreen or herbaceous plants often have simple (mycelioid) appendages. It is known that complex appendages facilitate the attachment of chasmothecia to twigs or the bark of host trees for overwintering (see p. 407; Gadoury & Pearson, 1988; Cortesi et al., 1995), and it must be advantageous for the primary inoculum in spring to be close to the budding leaves, rather than falling to the ground. Mori et al. (2000) proposed that no such selection pressure may hold for species parasitizing herbaceous plants, which might have permitted a reduction in the complexity of appendages to simple mycelioid ones. Such changes from woody to herbaceous hosts accompanied by the reduction in the complexity of appendages may have occurred several times independently within the Erysiphales (Takamatsu et al., 2000).

13.3 | *Blumeria graminis*

Although *Blumeria graminis* is somewhat atypical of a powdery mildew in being a pathogen of grasses and cereals, we shall describe it in detail because more is known about it than any other member of the Erysiphales. *Blumeria* is separated from *Erysiphe* on several grounds. The haustorium of *B. graminis* is digitate, i.e. it has striking finger-like projections (Fig. 13.2c), whereas the haustoria are knob-like in most members of the Erysiphales. Upon closer inspection, they also have lobes, but these are tightly folded round the main haustorial body, giving it a globose appearance (Bushnell & Gay, 1978). A second distinguishing feature of *B. graminis*

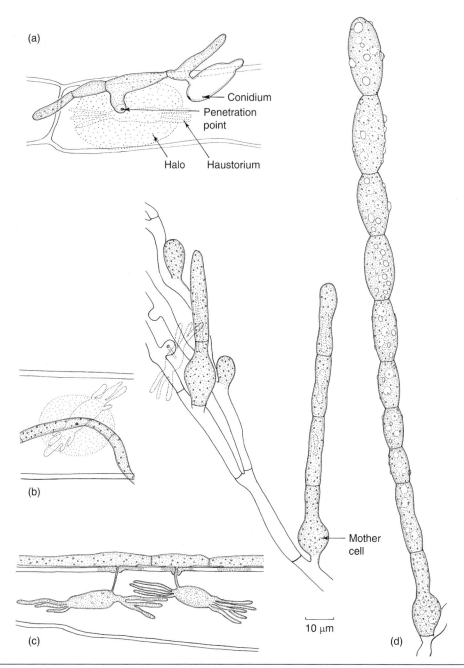

Fig 13.2 *Blumeria graminis.* (a) Two-day-old germinating conidium on wheat leaf, showing penetration point surrounded by a 'halo' (stippled). A haustorium has developed beneath the penetration point. (b) Penetration from an established mycelium. (c) Section of an epidermal cell showing two penetration points and two haustoria. Note the thickening of the epidermal cell beneath the penetration point. (d) Mycelium and conidia, showing the swollen flask-shaped foot cell ('mother cell').

is that the conidial foot cell is swollen (Fig. 13.2d). Thirdly, scanning electron microscopy studies by Cook *et al.* (1997) have shown that the surface of the conidium of *B. graminis* is distinct from that of other Erysiphales because of the spiny wall and the raised septum surrounded by a depressed ring (Figs. 13.3a,b). These differences are matched by DNA sequence data (Saenz & Taylor, 1999a).

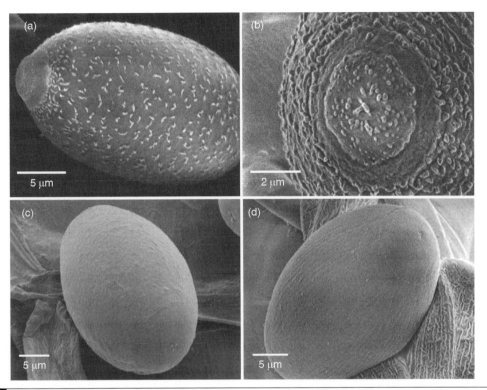

Fig 13.3 An example of the use of scanning electron microscopy in fungal taxonomy, separating *Blumeria graminis* (a,b) from other members of the Erysiphales (c,d). (a) Whole conidium of *B. graminis* showing the spiny surface of the wall. (b) The 'annular' septum of the conidium. The actual septum is raised and warty, but it is surrounded by a depressed annulus. (c) Conidium of *Golovinomyces* (*Erysiphe*) *cynoglossi* showing a slightly roughened wall. (d) Conidium of *Neoerysiphe* (*Erysiphe*) *galeopsidis* with its striate conidial wall. Micrographs kindly provided by R. T. A. Cook. Reprinted from Cook *et al.* (1997), British Crown copyright 1997.

The genus *Blumeria* contains only one species, *B. graminis*, which is separable into several **formae speciales** distinguishable by their different grass (*Agropyron*, *Bromus*, *Poa*) or cereal hosts (barley, oat, rye, wheat). Thus, *B. graminis* f. sp. *tritici* infects wheat (*Triticum*) but not barley, whilst *B. graminis* f. sp. *hordei* infects barley (*Hordeum*) but not wheat. *Blumeria graminis* is heterothallic and hybrids between certain *formae speciales* may arise, especially if the hosts themselves can hybridize. For example, it has been shown that hybridization between *B. graminis* ff. spp. *agropyri*, *tritici* and *secalis* can occur, resulting in viable ascospores. The three host genera *Agropyron* (couch grass), *Triticum* (wheat) and *Secale* (rye) can also hybridize (Hiura, 1978). Recently, it has become obvious that the definition of *formae speciales* is less clear-cut than previously thought. Extensive overlaps in host spectra occur especially on wild hosts in the Middle East where *Hordeum spontaneum*, the presumed ancestor species of cultivated barley, is endemic and where *B. graminis* probably originated (Clarke & Akhkha, 2002; Wyand & Brown, 2003). Further, within a given *forma specialis*, numerous **races** can be distinguished by specific features such as resistance to fungicides or ability to infect a given host cultivar.

13.3.1 The infection process

A conidium of *B. graminis* alighting on the surface of a grass or cereal leaf will initially come to rest by the tips of its spines. Within seconds of contact with hydrophobic (but not with hydrophilic) surfaces, an extracellular matrix is released by those spines touching the surface. The matrix contains proteins, including enzymes such as cutinases and non-specific esterases, and it may serve in the initial attachment of the conidium to the surface (Carver *et al.*, 1999; Nielsen *et al.*, 2000).

About 15 min later, a further batch of matrix material is released from the conidium, but unlike the initial secretion this second wave requires de novo protein biosynthesis. Within 2 h, a primary germ tube emerges. In all powdery mildews except *B. graminis*, this develops an appressorium under suitable conditions, but in *B. graminis* the primary germ tube grows only to a limited distance (up to 10 μm). An appressorium is formed by a separate secondary germ tube which is much longer than the primary germ tube (up to 40 μm) and becomes septate. If the primary germ tube fails to make contact with a suitable surface, further short germ tubes may be emitted. Upon contact with the host surface, the primary germ tube secretes an adhesive pad which provides more secure anchorage to the surface. By suspending conidia on a spider's thread over different kinds of surface, Carver and Ingerson (1987) observed that a long appressorial germ tube is emitted only if the primary germ tube has made contact with an inductive surface such as a host leaf. The primary germ tube therefore functions as a probe. Signals perceived by the primary germ tube may be the hydrophobicity of the surface, or minute quantities of cutin or cellulose degradation products released by appropriate hydrolytic enzymes which are secreted by the primary germ tube (Green *et al.*, 2002). The signals are probably transduced by cascades containing cAMP and cAMP-dependent protein kinase A (cPKA), similar to those described in more detail for *Magnaporthe grisea* (Fig. 12.48). The primary germ tube may penetrate the surface of the epidermis to a limited extent but does not achieve successful infection of the host cell. It may, however, take up water and dissolved substances from the plant surface (Kunoh & Ishizaki, 1981; Carver & Bushnell, 1983).

Under optimal conditions, the appressorium-forming germ tube emerges about 3 h after the primary germ tube. It elongates and its tip swells to form an appressorium which is lobed (Fig. 13.4b) and non-melanized, in contrast to the hemispherical melanized appressorium of *M. grisea* described on p. 381. Appressoria of other species of the Erysiphales may have different shapes, and these are of taxonomic significance (see Braun *et al.*, 2002). The sequence

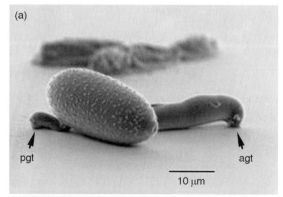

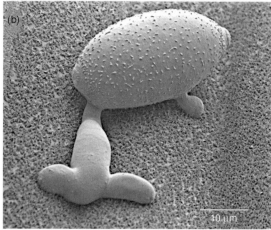

Fig 13.4 Penetration events in *Blumeria graminis*. (a) Conidium incubated on a cellulose membrane for 12 h. After contact of the primary germ tube (pgt) with the surface, the appressorial germ tube (agt) was emitted and has formed an incipient appressorium. (b) A lobed mature appressorium on the surface of a wheat leaf. (a) reprinted from Carver *et al.* (1999), with permission from Elsevier; original print kindly provided by T. L. W. Carver. (b) reprinted from Howard (1997), with kind permission of Springer Science and Business Media; original print kindly provided by R. J. Howard.

of infection-related morphogenetic events up to this stage is shown in Fig. 13.4.

Penetration is achieved by means of a thin penetration peg which originates from the underside of the appressorium. In all probability, both the activity of secreted wall-degrading enzymes (cutinases and cellulases) and appressorial turgor pressure contribute to successful penetration events, with the former predominating (Edwards & Allen, 1970). Penetration can be achieved within about 12 h of the conidium landing on the leaf surface. The tip of the penetration peg then enlarges to form the haustorium initial which differentiates in the course of 2–3 days, invaginating but not breaching the host plasmalemma. As a nutrient supply is established, surface hyphae grow from the appressorium or the appressorial germ tube and further epidermis cells are penetrated. It is also possible for the same host cell to be penetrated repeatedly so that it contains several haustoria. Haustorium formation represents the end-point of *in planta* growth; further penetration occurs from epidermal hyphae.

13.3.2 Self-defence of the host plant against infection

Good summaries of this vast topic may be found in Carver *et al.* (1995), Giese *et al.* (1997) and Zeyen *et al.* (2002). Although the ungerminated conidium already emits signals perceived by the epidermis cell, it is the contact of the primary germ tube with the host cell surface which elicits initial defence reactions. These are visible as a dramatic re-organization of the cytoplasm of the attacked epidermal cell, with dense cytoplasm aggregating beneath the point of contact of the primary germ tube with the leaf surface. This is followed by a modification of the epidermal cell wall by secreted substances. The altered cell wall region is visible as a **halo** (Fig. 13.2a). Phenolic substances and hydrolytic enzymes become incorporated into the cell wall within the halo region. This is a non-specific defence reaction because it is elicited by both virulent and avirulent strains of the pathogen. A further such halo is produced by the epidermal cell when the appressorial penetration peg attempts to penetrate. A **papilla** is then formed around the penetration peg between the host cell wall and the host plasma membrane, and this may succeed in plugging the peg and preventing infection. Papillae are easily seen with an epifluorescence microscope because they show strong autofluorescence, a general indicator of an attempt by the host plant to resist infection. Autofluorescence is due mainly to the accumulation of phenolic substances which have antimicrobial activity (von Röpenack *et al.*, 1998). Additionally, hydrolytic enzymes, callose and silica may be deposited in the papilla. Papilla

formation is an important mechanism of general resistance, although it is unknown why some strains of *B. graminis* can penetrate the papillae and others fail. Even if a susceptible host is infected, only about 70% of the penetration events succeed beyond the papilla stage. In certain barley cultivars, particularly thick papillae are formed because of mutations in which restrictions of the resistance response are lifted; only 0.5% of infection pegs get through the epidermis of these *mlo* mutants, and barley cultivars homozygous for the *mlo* allele show broad-spectrum resistance to all strains of *B. graminis* f. sp. *hordei* (Jørgensen, 1994; Collins *et al.*, 2002). It should be noted here that papillae are not particularly prominent in infections of dicotyledons by other powdery mildews.

In cereals attacked by *B. graminis*, most strain-specific resistance mechanisms are initiated later, when the tip of the penetration peg enlarges and begins to differentiate into the first haustorium. At this point the infected epidermal cell of a resistant cultivar displays an oxidative burst, i.e. it releases H_2O_2 and various enzymes into its own cytoplasm and dies (Zhou *et al.*, 1998). This phenomenon is known as the **hypersensitive response**. Since *B. graminis* is an obligate biotroph, penetration ending in a haustorium inside a dead cell is a wasted effort. The haustorium itself may also be directly affected by the oxidative burst, with first signs of degeneration appearing in the mitochondria (Hippe-Sanwald *et al.*, 1992). If sufficient nutrient reserves are present in the appressorial germ tube, further appressoria may be formed, each resulting in failure to establish a functional haustorium in resistant cultivars. Successful infection results if the host cell tolerates the establishment of the initial haustorium. Curiously, therefore, it is sensitivity rather than tolerance which leads to resistance. Numerous genes are involved in the various lines of defence which barley plants possess against *B. graminis* f. sp. *hordei* (Collinge *et al.*, 2002).

13.3.3 Genetics of plant resistance against *B. graminis*

The fact that only certain strains of *B. graminis* f. sp. *hordei* can elicit the hypersensitive response in a given barley cultivar indicates that specific recognition mechanisms between host and pathogen must be involved. The molecular basis of recognition is still obscure, but the genetics are well understood. They are based on the **gene-for-gene** concept first formulated by Flor (1955) for an interaction between the rust fungus *Melampsora lini* and flax, but soon after also demonstrated for *B. graminis* and cereal hosts (see Moseman, 1966). Given that in strain-specific interactions recognition leads to resistance via the hypersensitive response, every avirulence gene of the pathogen (e.g. encoding a surface protein recognized by the host) is matched by a specific resistance gene of the host (e.g. a receptor). It follows that avirulence alleles should be dominant to virulence alleles in diploid or dikaryotic pathogens (not, of course, applicable to the haploid *B. graminis*), whereas resistance should be dominant to susceptibility in the host. Successful infection occurs only if the avirulence gene is modified so that the host can no longer recognize the pathogen. Numerous resistance genes have been identified especially in barley and are being used for breeding programmes, although it is relatively easy for the pathogen to overcome such single-gene resistance (Brown *et al.*, 1993; Collins *et al.*, 2002). It should be noted that there are deviations from the classical gene-for-gene concept in the interaction between *B. graminis* f. sp. *hordei* and barley. Further, interactions between a given race of *B. graminis* and its host can take various courses with intermediates between complete resistance and full development of symptoms, due to the influence of minor genes depending on the host's genetic make-up, and also due to environmental parameters. A good summary of this complicated topic has been written by Brown (2002). In general terms, research on the molecular biology of *B. graminis* would benefit greatly from the availability of a reliable DNA transformation method for this important pathogen.

13.3.4 The haustorium of *B. graminis*

In compatible interactions, a functional haustorium is formed by the enlarging tip of the penetration peg. The papilla remains as a collar around the peg at the point where it penetrated the epidermis wall. The host plasmalemma is invaginated around the haustorium, but it is not in direct contact with the plasma membrane of the haustorium. Instead, the two membranes are separated by the haustorial wall, and surrounding it by the **extrahaustorial matrix** which is of host origin. It is a compartment with a gelatinous texture, sealed by the host and pathogen plasma membranes, and at the epidermal wall by a collar (Manners, 1989). The host plasmalemma is strongly modified and seems to lack H^+ ATPases, so that the host cell may not be able to control leakage of solutes into the extrahaustorial matrix (Gay *et al.*, 1987). The escape of solutes from the extrahaustorial matrix to the cell surface is prevented by the collar seal. Haustoria of the Erysiphales contain a full complement of organelles including a single nucleus. The haustorium is separated from the surface hypha by a septum which is perforated, thus permitting nutrient transfer. Ultrastructural details are summarized in Fig. 13.5 and have been described by Bracker (1968b) and Hippe-Sanwald *et al.* (1992).

Whereas in other biotrophic plant pathogens nutrients can, to a certain extent, be absorbed by intercellular hyphae in addition to haustoria, in the Erysiphales the haustorium seems to be the sole means of nutrient uptake. Not surprisingly, the haustorial membrane differs from that of surface hyphae in terms of protein composition and physiology (Manners, 1989; Mendgen & Deising, 1993; Green *et al.*, 2002). Nutrient uptake has traditionally been studied with the haustoria of *E. pisi* which can be isolated intact from infected pea plants (Gil & Gay, 1977). Sutton *et al.* (1999) have shown that glucose is the sugar which is taken up by haustoria of *B. graminis* f. sp. *tritici*, and that the plant hydrolyses the transport sugar sucrose before this reaches the epidermal cells and the haustoria contained within them. Glucose probably diffuses passively into the extrahaustorial matrix and is then taken up across the haustorial membrane by a proton uniport mechanism (Sutton *et al.*, 1999). This is in line with the situation in most other fungi examined to date, which generally take up glucose but not sucrose (Jennings, 1995).

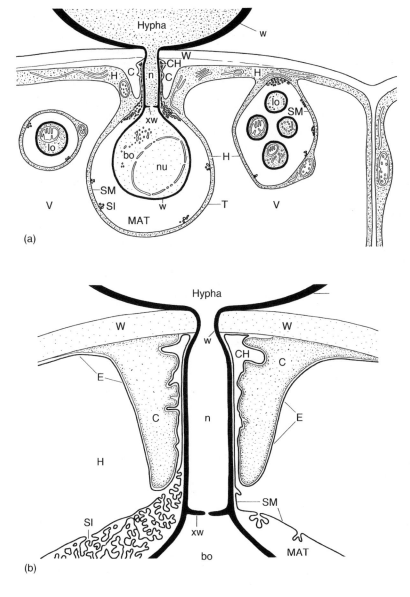

Fig 13.5 Interpretation of the fine structure of the haustorium of B. graminis f. sp. hordei (redrawn from Bracker, 1968b). (a) Section of host leaf at point of penetration. The body of the haustorium (bo) containing a single nucleus (nu) lies immediately beneath the point of penetration. The body of the haustorium is enclosed in a sheath with extensive matrix (MAT). The sheath membrane (SM) is close to the host tonoplast (T). The sheath membrane bears invaginations (SI). A single lobe (lo) of another haustorium enclosed in an extension of the sheath is shown to the left of the diagram, and four lobes enclosed in a common sheath to the right. (b) Enlargement of the neck of a haustorium. Note the thickened collar (C) deposited on the host cell wall (W). The sheath membrane (SM) is continuous with the host plasmalemma (E). Other abbreviations: XW (cross-wall), CH (channel), H (host cytoplasm).

The difference is that other fungi are capable of hydrolysing sucrose externally by the activity of their own secreted invertases.

13.3.5 Life cycle and epidemiology of B. graminis

About 7–10 days after infection by B. graminis, the haploid surface mycelium can produce conidia under field conditions. Conidia develop from a flask-shaped foot cell within which mitotic nuclear division occurs. The foot cell elongates away from the host leaf and a cross-wall cuts off the hyphal tip. Further cross-walls develop so that a chain of cells is formed, increasing in length at its base by further divisions of the foot cell. Such conidia are said to be **catenate** (Fig. 13.2d). Each conidium is uninucleate. In successful infections, a dense stand of conidiophores is produced so that the lesion appears as a white powdery pustule (Fig. 13.6a). Aust (1981) has estimated that a single pustule can release about 1.5×10^4 conidia day^{-1}. A high spore inoculum can be built up very quickly, and numerous infection cycles can occur during one growing season. Dispersal of powdery mildew conidia is mainly by wind, whereby short gusts of wind are ideal for spore dispersal (Hammett & Manners, 1971, 1974). This ties in with the duration of conidium production (about 3 h for each conidium) and the fact that usually only the terminal, most mature spore becomes detached from the conidial chain. Thus, the observed diurnal rhythm of spore abundance in the air, with a peak usually in early afternoon, can probably be explained by factors affecting release, rather than formation, of spores (Hammett & Manners, 1971, 1974). Both the formation and release of spores are strongly inhibited by rain (Hirst, 1953).

The conidia of *B. graminis* can travel considerable distances with the prevailing wind. For instance, Hermansen *et al.* (1978) have demonstrated the migration of spores from Northeastern England and Scotland to Denmark, a journey which would have taken approximately 48 h. In an extensive survey undertaken across the whole of Europe, Limpert *et al.* (1999) found that *B. graminis* is a nomadic species, with the prevailing westerly winds driving waves of populations eastwards at a rate of about 100 km year^{-1}. Since such populations encounter hosts with different spectra of resistance genes, they face a selective pressure to adapt, and this may explain why the complexity of virulence alleles in *B. graminis* populations was found to increase from west to east by about one virulence factor every 1000 km. Ultra-long distance (intercontinental) dispersal of viable conidia, as found e.g. for urediniospores of rust fungi (see p. 632) or teliospores of smut fungi (p. 639), has not been reported for powdery mildews.

Whereas conidium formation and release by *B. graminis* are favoured by dry windy conditions, infection proceeds best at 98–100% relative humidity but is inhibited by free water. The optimum infection temperature is about 15–20°C. Thus, the conditions in Northwestern Europe are ideal for *B. graminis* in terms of rapidly changing weather conditions (Smith *et al.*, 1988). Crop losses of up to 40% have been reported, with barley the most seriously affected cereal. The timing of infection is important, early infections reducing the number of ears and later infections reducing the size of the grain. Crop damage is due to a decrease in photosynthesis in infected leaves, meaning that they are unable to

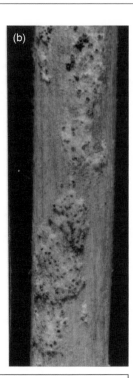

Fig 13.6 *Blumeria graminis.* (a) Conidial pustules on wheat. (b) Dark spherical chasmothecia nestling in a felt of surface hyphae on a wheat leaf sheath.

export carbohydrates to the developing grains. Heavily infected leaves may even act as a sink for photosynthetic product (sucrose) from other leaves.

The chasmothecia of *B. graminis* are dark brown and globose, and nestle in a dense mass of mycelium formed on the basal leaves and leaf-sheaths of cereals (Fig. 13.6b). In contrast to the chasmothecia of most Erysiphales, those of *B. graminis* do not bear any conspicuous thick-walled appendages (Fig. 13.7a). Each chasmothecium has a wall made up of several layers of cells, surrounding a number of asci. At maturity, it cracks open by swelling of the contents, and the asci discharge their spores. In Europe, chasmothecia are formed in late summer and it is unclear what their exact role in the disease cycle is. Ascospores of *B. graminis* formed in the current season are capable of infection, and dried chasmothecia can survive under herbarium conditions for up to 13 years (Moseman & Powers, 1957). However, it is generally assumed that *B. graminis* survives mainly in the vegetative state on overwintering host plants, especially winter varieties of cereals (Jenkyn & Bainbridge, 1978). To survive a single growing season, *B. graminis* may have to switch four times between winter barley and summer barley main crops and their volunteers germinating after harvest (Limpert *et al.*, 1999). The availability of living shoots for infection throughout the year is called the 'green bridge'. In contrast, in hot and arid regions such as the Mediterranean, chasmothecia play a role in the oversummering of the pathogen on wild grasses such as *Hordeum spontaneum*, discharging ascospores when the host seeds germinate in autumn (Clarke & Akhkhra, 2002). Since *B. graminis* seems to have its centre of origin in the Middle East, it is possible that the chasmothecia are primarily oversummering rather than overwintering structures.

13.4 | *Erysiphe*

The genus *Erysiphe* is currently subject to major taxonomic rearrangements. *Erysiphe* has traditionally been defined by chasmothecia with undifferentiated mycelioid appendages (Fig. 13.8a), and was later distinguished from *Blumeria* by its knob-like haustoria and a cylindrical (non-swollen) conidial foot cell (Fig. 13.8b). This anamorph gives rise to one conidium at a time, i.e. it is non-catenate. It is called *Pseudoidium*. *Erysiphe* spp. now fall into several branches of phylogenetic trees, and they are interspersed by species which have similar anamorphs but chasmothecia with uncinate or lobed appendages (Saenz & Taylor, 1999a; Mori *et al.*, 2000). Braun *et al.* (2002) therefore proposed the absorption of genera such as *Microsphaera* and *Uncinula* into *Erysiphe*. The most important species of *Erysiphe* in an agricultural context (Spencer, 1978; Smith *et al.*, 1988) are mentioned below. They seem to share the fundamental biological principles of infecting by wind-dispersed conidia which do not require or even tolerate free water, and causing disease symptoms as a superficial mycelium producing haustoria only in epidermal cells.

13.4.1 *Erysiphe sensu stricto*

Erysiphe cruciferarum (formerly part of *E. polygoni sensu lato* or *E. communis*) is a pathogen of

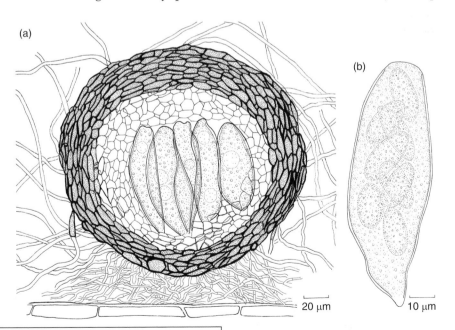

(a)

(b)

20 µm

10 µm

Fig 13.7 *Blumeria graminis.* (a) Mat of superficial hyphae with a sectioned chasmothecium containing several asci. (b) Ascus.

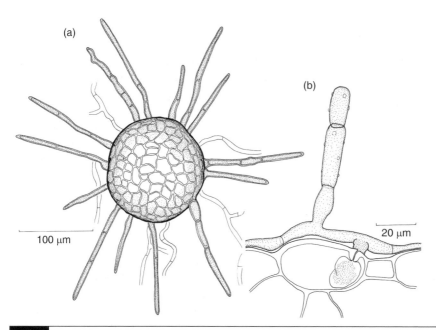

(a)

(b)

100 μm

20 μm

Fig 13.8 *Erysiphe polygoni.* (a) Chasmothecium showing dark, free-ended equatorial appendages, and the hyaline superficial mycelium anchoring the chasmothecium to the host leaf. (b) T.S. host leaf showing simple haustorium, superficial mycelium and conidial chain arising from a foot cell which does not appear bulbous.

crucifers, causing severe infections on cabbage and Brussels sprouts. It is morphologically similar to *E. betae* which is a specialized pathogen of sugar beet and related hosts (*Beta* spp.). *Erysiphe polygoni* is now considered a separate species of no commercial interest, forming powdery mildew on *Rumex* and *Polygonum*. *Erysiphe pisi* is a cosmopolitan pathogen of leguminous plants including peas and lucerne. There are several *formae speciales*. This species is of particular interest because much physiological work has been done on its haustoria which can be isolated from infected leaves (Gil & Gay, 1977). Another powdery mildew fungus on Leguminosae is *Erysiphe trifolii* infecting mainly clover and with *formae speciales* on *Trifolium*, *Lathyrus*, *Melilotus* and *Lotus*. *Erysiphe heraclei* (= *E. umbelliferarum*) infects umbellifers such as carrot and celery and is of limited importance in winter crops in the Mediterranean. It is exceedingly common on hogweed (*Heracleum sphondylium*).

All the above species belong to the redescribed genus *Erysiphe*, possess a *Pseudoidium* anamorph and cluster together in phylogenetic analyses (Saenz & Taylor, 1999a; Mori *et al.*, 2000).

13.4.2 *Microsphaera* and *Uncinula*

Both these genera group together with the main *Erysiphe* cluster around *E. pisi*, *E. cruciferarum* and other species described above (Saenz & Taylor, 1999a) and will in due course be called *Erysiphe* (Braun *et al.*, 2002) possibly with the addition of further names reflecting distinct phylogenetic sub-groups rather than characteristics of chasmothecial appendages. Anamorphic states belong to *Pseudoidium*.

Microsphaera

Chasmothecia of *Microsphaera* (now *Erysiphe* sect. *Uncinula*) contain several asci, but they carry appendages with tips showing a highly diagnostic dichotomous branching (see Fig. 13.10a). *Microsphaera* (now *Erysiphe*) *alphitoides* is the cause of oak mildew which is extremely common, especially on *Quercus robur*. It appears from June onwards, infecting mainly leaves which are produced on sucker shoots and seedlings, and causing distortions to growing shoots until growth finally stalls for the rest of the growing season (Fig. 13.9). The chasmothecial stage is not always present but may be more

Fig 13.9 Summer shoot of oak (*Quercus*) infected by oak powdery mildew, *Microsphaera* (now *Erysiphe*) *alphitoides*. Contrast the stunted appearance of the infected foliage with the healthy basal leaves formed in spring.

common during hot summers or in hot climatic zones.

Uncinula

Members of the genus *Uncinula* (now *Erysiphe* sect. *Uncinula*) also produce several asci in each chasmothecium but are distinguished by their chasmothecial appendages which are uncinate, i.e. they have recurved tips. Strikingly similar appendages are found in the unrelated genus *Sawadaea* (see Fig. 13.10b). Undoubtedly the best-known species is *U. necator*, the cause of powdery mildew of vines. Chasmothecia are quite rare, and when the fungus first appeared in Europe in a glasshouse in Kent in 1845 it was known only in its anamorphic state and given the name *Oidium tuckeri*, after Mr Tucker, the gardener who discovered it. An accidental introduction from North America, *U. necator* quickly spread throughout the major vine-growing regions of Europe, causing such severe damage especially in

France that the entire wine industry was threatened. By 1854, the French wine production had fallen from 54 million to 10 million hectolitres. Fortunately, Mr Tucker had also discovered that a mixture of sulphur dust and lime provided good control of the disease, and this simple fungicide is still used today (Smith *et al.*, 1988). An excellent account of the arrival of vine powdery mildew (and many other fungal diseases) has been given by Large (1940).

Because of the extremely high value of the vine crop, extensive epidemiological data have been compiled and are used for disease forecasting models (Jarvis *et al.*, 2002). Conidia are, of course, the major inoculum, and they are dispersed by wind movements and also by the air currents generated by high-pressure spraying equipment (Willocquet & Clerjeau, 1998). Infections are mainly found on the upper (adaxial) surface of leaves, in contrast to the grape downy mildew (*Plasmopara viticola*; see p. 120) which is more common on the underside (abaxial surface). The grapes themselves are also readily infected by *U. necator*; if they are young, they will be aborted altogether whereas older grapes suffer skin damage, and moulds such as *Botrytis cinerea* can easily infect through these cracks. *Uncinula necator* overwinters as mycelium in dormant buds (Rugner *et al.*, 2002), although chasmothecia formed in autumn can survive the winter on twigs and give rise to sizeable quantities of ascospores capable of causing new infections in spring (Jailloux *et al.*, 1998, 1999).

13.4.3 Species previously attributed to *Erysiphe*

Species of *Golovinomyces* (formerly *Erysiphe cichoracearum sensu lato*) are not closely related to the *Erysiphe* as circumscribed above. They have an anamorph (*Oidium* subgenus *Reticuloidium*; Cook *et al.*, 1997) that has catenate conidia with slightly roughened conidial surfaces as seen under the scanning electron microscope (see Fig. 13.13c). Braun (1987) has placed the forms affecting Asteraceae (including lettuce) in *G. cichoracearum* and most of the plurivorous forms in the morphologically very similar *G. orontii*. Important crops

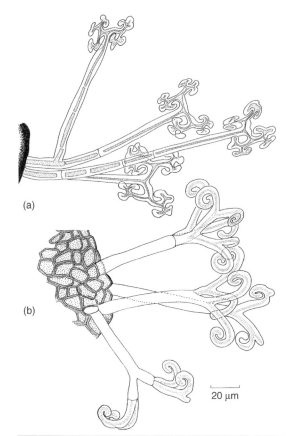

These two genera, although easily distinguished from each other with the light microscope by their chasmothecial appendages, have now been grouped together, and *Podosphaera* takes precedence over *Sphaerotheca* (Braun *et al.*, 2002). Both produce chasmothecia containing only a single ascus. The important features of the conidia of all species grouped here are that they are catenate and contain conspicuous fibrosin bodies. These are also seen in some other genera, e.g. *Sawadaea* (see p. 405). Fibrosin bodies are filamentous organelles up to 8 μm long which appear highly light-refractile or sparkling when viewed with the light microscope (Fig. 13.11). Their biochemical composition and function appear to be unknown.

13.5.1 *Sphaerotheca* (now *Podosphaera*)

Common species are *S. fuliginea* on dandelion (*Taraxacum officinale*) and other Asteraceae, *S. macularis* causing powdery mildew of hops (*Humulus lupulus*), *S. mors-uvae* (American gooseberry mildew) and *S. pannosa*, the common rose mildew. The chasmothecial structure of *Sphaerotheca* closely resembles that of *Erysiphe* with its simple appendages, the only notable microscopic difference being that the chasmothecium contains only one ascus in *Sphaerotheca* (Fig. 13.12). The fine structure of developing and mature chasmothecia of *S. mors-uvae* has been studied by Martin *et al.* (1976). They have shown that the darker melanized cells forming the peridium are, like those of vegetative hyphae, uninucleate. Most of the inner cells of the chasmothecium are binucleate, suggesting that they may have arisen from a binucleate ascogonial fusion cell. Another interesting discovery was that fibrosin bodies, previously reported from conidia, are also present in the ascospores.

The mycelium and conidia of *S. pannosa* are common on leaves and shoots of cultivated and wild roses. Chasmothecia are formed on twigs, embedded in a dense mycelial felt. Overwintering is not only by means of ascospores, but particularly as mycelium within dormant

---

20 μm

Fig 13.10 Chasmothecial appendages of Erysiphales. (a) Flattened dichotomously branched appendages of *Podosphaera clandestina*. (b) Branched appendages of *Sawadaea* (formerly *Uncinula*), *bicornis* with recurved tips.

stated to be attacked by the plurivorous species are those belonging to Solanaceae (e.g. tobacco) and Cucurbitaceae (especially melon, cucumber, squash and pumpkin). However, it is unclear whether the same species attacks both families. On cucurbits it has been described as *G. cucurbitacearum* but here it seems to be of lesser significance than *Sphaerotheca fuliginea* (= *Podosphaera xanthii*) and might have been confused with it in the past (Jahn *et al.*, 2002). Molecular data indicate that both *Golovinomyces* spp. are polyphyletic, and that the clades do not obviously relate to their host ranges. Another catenate species, *Neoerysiphe* (formerly *Erysiphe*) *galeopsidis* affecting mainly Lamiaceae differs from *Golovinomyces* in having conidia with a minutely striated conidial surface (Fig. 13.3d).

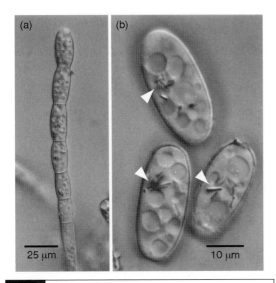

Fig 13.11 *Sphaerotheca* (now *Podosphaera*) *pannosa*. (a) Conidial foot cell producing a chain of conidia. (b) Close-up of mature conidia. Note the large vacuoles and the elongated light-refractile fibrosin bodies (arrowheads).

buds, and the chasmothecia may not play any significant role in perennation. In the case of *S. mors-uvae* from blackcurrants, only a small proportion (<0.1%) of overwintered chasmothecia seems to be functional. Loss of viability is associated with degeneration of asci and ascospores. Chasmothecia overwintering on the soil surface are invaded by chitinolytic microorganisms which may play a role in degrading the walls of the chasmothecium prior to cracking open as the ascus expands (Jackson & Gay, 1976).

13.5.2 Podosphaera

The chasmothecia of *Podosphaera* contain a single ascus and bear characteristic flattened, dichotomously branched appendages (Fig. 13.10a), much like those of *Microsphaera*. Apart from the presence of only one ascus, the chasmothecium of *Podosphaera* is thus very similar to that of *Microsphaera*, but the anamorph is quite distinct in having catenate conidia with fibrosin bodies. *Podosphaera leucotricha* is the cause of apple mildew, and the mycelium and conidia are visible in spring on the expanding foliage and young shoots, probably developing from hyphae overwintering in the buds. This is the most common

mode of survival for most powdery mildews on woody plants (Jarvis *et al.*, 2002). Chasmothecia are formed on the young branches. Woodward (1927) has reported that the asci themselves are projected from the chasmothecia. The ascocarp gapes open as the asci expand by absorbing water, but the wall is elastic and eventually the asci are thrown out by the snapping together of the 'jaws' of the chasmothecium, for a distance of several centimetres. If the ascus alights in water it continues to expand and explodes, shooting out its ascospores. A similar double discharge has been reported for *B. graminis* (Ingold, 1939).

Another common species of *Podosphaera* is *P. clandestina* (= *P. oxyacanthae*) on hawthorn (*Crataegus monogyna*). The incidence of disease is associated with hedge-clipping during the summer when conidia are abundant in the air. Clipping removes the apical buds of the host, thus destroying apical dominance and leading to the enlargement of axillary buds. The fungus can penetrate these lateral buds and overwinter there (Khairi & Preece, 1978).

13.6 | Sawadaea

One very common species is *S. bicornis* (formerly *Uncinula bicornis*) which causes powdery mildew of sycamore (*Acer pseudoplatanus*) and produces chasmothecia on the underside of the leaves. The uncinate appendages of the chasmothecia (Fig. 13.10b) are very similar to *Uncinula*, but the anamorph is quite distinct having micro- as well as macroconidia that are octagonal in outline, catenate and contain fibrosin bodies (Braun, 1987).

13.7 | Phyllactinia and Leveillula

These genera have distinct anamorphs, quite unlike the *Oidium* anamorph of all the powdery mildews described above. Unusually for powdery mildews, *Phyllactinia* and *Leveillula* spp. produce internal (endophytic) as well as the usual external (ectophytic) superficial mycelium. The latter

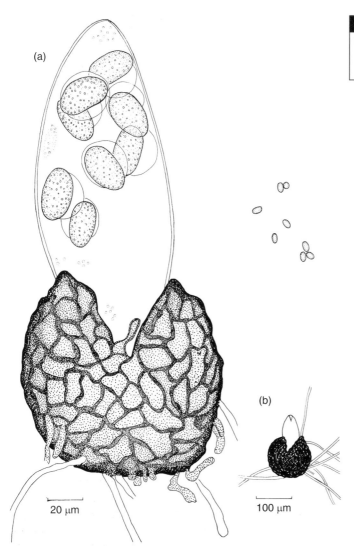

(a)

(b)

Fig 13.12 *Sphaerotheca* (now *Podosphaera*) *pannosa*. (a) Chasmothecium crushed to show a single ascus. (b) Chasmothecium with discharged ascospores.

20 μm

100 μm

gives rise to very large and distinctly shaped conidia.

13.7.1 Phyllactinia

Phyllactinia guttata (= *P. corylea*) grows on the leaves of hazel (*Corylus avellana*) and other woody plants, forming chasmothecia on the lower leaf surface in late summer and autumn. In this species the superficial mycelium produces short lateral appressoria which emit penetration hyphae through the stomata into the mesophyll (Fig. 13.13b). Directly beneath the stoma, the penetration peg swells to produce a substomatal

vesicle from which septate hyphae grow into the leaf mesophyll, penetrating mesophyll cells and forming haustoria within them. The conidia are club-shaped and are formed singly (Fig. 13.13a). However, in a damp chamber pseudo-chains of up to four conidia may hang together. Conidia of this type are classified in the anamorphic genus *Ovulariopsis*.

Phyllactinia guttata presents a most intriguing spore dispersal mechanism which is easily demonstrated in mycology classes because the same hazel trees are reliably infected each year and, once a source has been located, leaves can be

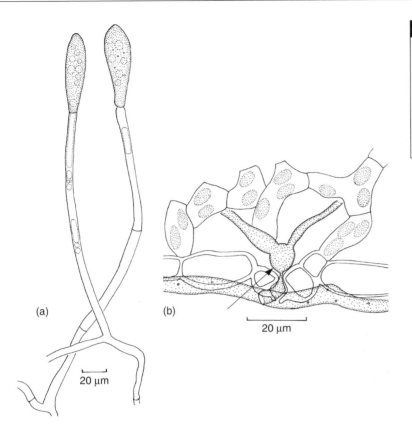

Fig 13.13 *Phyllactinia guttata.* (a) Conidiophores showing the single terminal conidium. (b) T.S. leaf of *Corylus avellana* showing penetration of stoma in lower epidermis, formation of a substomatal vesicle (arrow) and extension of the mycelium into the mesophyll.

(a)

(b)

20 μm

20 μm

collected and kept dry until required for classes (Weber & Webster, 2001a). Chasmothecia are formed only on the lower leaf surface. They bear two types of appendage, an equatorial group of radiating unbranched appendages with bulbous bases, and a crown of highly branched appendages which secrete mucilage (Fig. 13.14). The base of the bulbous appendage is thin-walled in the region facing the leaf surface, but thick-walled facing away from the leaf. On drying, the appendage bends towards the leaf surface as the thin part buckles inwards, and the pressure of the appendage tip levers the chasmothecium free from the superficial mycelium (for a video sequence, see Webster, 2006b). The bulbous appendages now function as 'flights' or vanes, and the chasmothecium plummets downwards like a shuttlecock, orientated during its fall with the sticky mucilage on the lower side. The blob of mucilage between the apical crown of branched appendages helps to stick the chasmothecium

onto twigs and leaves. The asci usually contain only two spores (Fig. 13.14e). Overwintered ascocarps open by means of an equatorial line of dehiscence, and the base of the ascocarp hinges back to place the asci in a suitable position for ascus discharge to occur (Cullum & Webster, 1977; Weber & Webster, 2001a).

For historical interest, we may note that in 1861 the Tulasne brothers, working on *P. guttata* and other powdery mildews, were among the first to demonstrate and illustrate in timeless beauty the connection between a conidial state and a morphologically very different-looking sexual state (see Tulasne & Tulasne, 1931).

13.7.2 *Leveillula*

This genus is related to *Phyllactinia*. Its members are common in warmer countries and occasionally in glasshouses in cooler areas. *Leveillula taurica* attacks a wide range of mainly herbaceous plants including tomato and pepper. It is distinguished from *Phyllactinia* by its rare chasmothecia bearing only mycelioid appendages and by its spear-shaped conidia (*Oidiopsis* anamorph).

13.8 | Control of powdery mildew diseases

Several different strategies are employed to control powdery mildew diseases in various crops, chemical control and breeding for resistance being by far the most important in commercial terms. We describe them here in detail because the fundamental principles apply to diseases caused by many other groups of fungi within the Eumycota, as mentioned elsewhere in this book.

13.8.1 Breeding for resistance

Wheat and barley were among the first crop plants for which resistance breeding programmes were initiated almost a century ago, following Biffen's (1905) seminal work on the genetics of resistance to wheat stripe rust, *Puccinia striiformis*. Initial attempts at resistance breeding against *B. graminis* as well as other biotrophic plant pathogens used major gene resistance based on introducing single resistance genes into the cultivar of choice. Numerous such major resistance genes originating from cereal cultivars as well as wild grasses related to wheat or barley are now available for breeding purposes (Hsam & Zeller, 2002). Breeding for major gene resistance has also been pursued in many other crops susceptible to powdery mildew diseases, such as melons and cucumbers (*Podosphaera xanthii* and *Golovinomyces cichoracearum*; Jahn et al., 2002), or clover (*Erysiphe trifolii*), hops (*Sphaerotheca macularis*) and gooseberries (*Sphaerotheca mors-uvae*) (see Smith et al., 1988). Breeding for resistance in slow-growing perennial crops such as apple or vines is obviously rather more difficult.

Major gene resistance is usually based on a recognition mechanism involving the hypersensitive response (see pp. 397–398). The danger inherent in major gene resistance breeding is that the pathogen can overcome resistance by mutation of its corresponding avirulence allele to virulence. This is a frequent occurrence, e.g. in *B. graminis*. On the other hand, the frequency of virulence alleles in the field may decrease again in pathogen populations no longer exposed to the cultivar and its resistance gene. Further, by co-ordinating the release of resistant cultivars, the effect of major gene resistance in crop protection can be maximized. For instance, the 'green bridge' of *B. graminis* (see p. 399) can be broken if the summer and winter cereal cultivars sown in any one year carry different resistance genes. Further, it is possible to combine several resistance genes in one host variety, a process known as 'pyramiding'. This requires the pathogen to develop multiple virulence alleles before it can infect such a crop variety (Hsam & Zeller, 2002). The danger, of course, lies in the creation of multiply resistant 'super-races' of *B. graminis* or other powdery mildews.

An interesting type of resistance in barley is mediated by the *mlo* allele which, as mentioned on p. 397, mediates the formation of very thick papillae, restricting infection by all races of *B. graminis* f. sp. *hordei*. This resistance is unusual for a broad-spectrum resistance in giving almost total control, and it is exceptional for a single-gene resistance in that it has remained stable in barley cultivars since it was introduced about 20 years ago (Jørgensen, 1994; Collins et al., 2002). In 1990, 30% of the spring barley acreage was sown with *mlo* resistant barley cultivars (Jørgensen, 1992). More commonly, broad-spectrum resistance is based on the combined effects of several minor genes and it is only partial, i.e. it retards infection and sporulation of the pathogen but does not altogether prevent disease. It is sometimes termed **horizontal resistance** because it controls many different races of the pathogen to the same limited level, in contrast to **vertical resistance** mediated by major resistance genes in which individual races are controlled totally, and others not at all. Although the molecular basis of most resistance genes is still unknown, it is noteworthy that some genes involved in horizontal resistance become more effective in

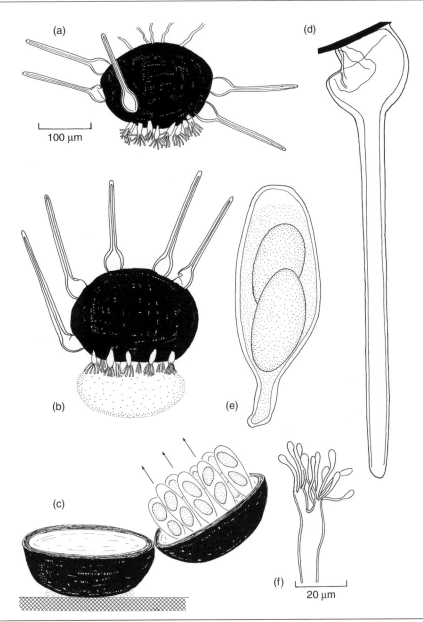

Fig 13.14 *Phyllactinia guttata*. (a) Chasmothecium on lower side of hazel (*Corylus*) leaf. The radiating bulbous appendages are horizontal. The branched secretory appendages which crown the chasmothecium are on the morphologically upper side, i.e. the side to which the ascus apices point. (b) Position of chasmothecium during its fall from the host leaf. The bulbous appendages are now folded to form flight vanes, ensuring that the sticky mass of mucilage faces downwards. (c) Diagrammatic representation of an open chasmothecium. The chasmothecium is shown attached by mucilage to a surface. The chasmothecium has opened by a circumscissile line of weakness, and has hinged back so that the apices of the asci now point outwards. Arrows indicate the direction of ascospore discharge. (d) A bulbous appendage, showing the differential thickening of the wall of the bulb. Collapse of the thinner walls result in movement of the appendage. (e) Two-spored ascus. (f) Branched secretory appendage. (a,b) to same scale; (d–f) to same scale.

adult plants. This phenomenon is called **adult plant resistance** (Hsam & Zeller, 2002). Thus, it can be envisaged that the products of horizontal resistance genes may be involved, for example, in producing a thicker cuticle, cell wall or papilla.

Resistance breeding will certainly remain one of the key methods for controlling powdery mildews, and it is likely that major and minor genes will be combined to produce cultivars with more durable and effective resistance to *B. graminis* and other powdery mildews. Further, genetic engineering may enable breeders to produce cultivars with overexpresssed pathogenesis-related (*PR*) genes (Salmeron *et al.*, 2002). The principles of *PR* genes are explained briefly on p. 116.

13.8.2 Chemical control

A good overview of fungicides against powdery mildews has been given by Holloman and Wheeler (2002). By far the oldest remedy against powdery mildew is powdered elemental sulphur, which was mentioned by Homer in about 1000 BC (Agrios, 2005) and was rediscovered in the nineteenth century and combined with lime for enhanced efficacy (Large, 1940). Sulphur–lime mixtures saved the French wine industry from ruin in the mid nineteenth century when

U. necator appeared in Europe, just as Bordeaux mixture (based on copper sulphate and lime) played a crucial role in protecting vines against *Plasmopara viticola* later that century (see p. 120). In the early twentieth century, as described above, efforts to control *B. graminis* on cereals shifted to breeding for resistance. Dithiocarbamate (see Fig. 5.27) was released in 1934 and, like sulphur, had a purely protectant activity.

The first systemic fungicide against powdery mildews was the **benzimidazole** benomyl (Fig. 13.15a) which is converted to the active molecule carbendazim inside the plant. Carbendazim binds to tubulin proteins, interfering with their assembly into microtubules and especially with the nuclear spindle during mitosis (Davidse & Ishii, 1995). Resistance against benomyl is common among plant-pathogenic fungi and is usually due to a mutation in the β-tubulin gene at the site where carbendazim binds, thereby reducing its affinity (Davidse & Ishii, 1995).

The next two important groups of systemic fungicides to be released against powdery mildews were the **morpholines** (Pommer, 1995) and **2-aminopyrimidines** introduced in the 1960s (Hollomon & Schmidt, 1995). Both are used more or less exclusively against powdery mildews.

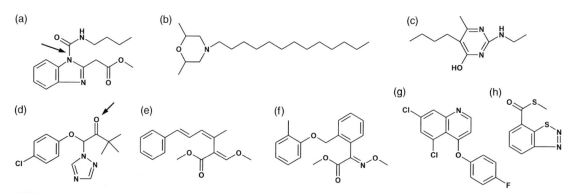

Fig 13.15 Fungicides against powdery mildews. (a) The benzimidazole benomyl. The active substance, carbendazim, is produced in planta by removal of the side chain at the position indicated by the arrow. (b) The morpholine tridemorph. (c) The 2-aminopyrimidine ethirimol. (d) The triazole triadimefon. The molecule becomes reduced to its more active alcohol in planta and by fungi at the position indicated by the arrow. (e) Strobilurin A, an antifungal substance from *Strobilurus tenacellus*. (f) Kresoxim methyl, the first strobilurin-based fungicide. (g) Quinoxyfen. (h) Benzothiadiazole (= benzo(1,2,3)thiadiazole-7-carbothioic acid *S*-methyl ester), an inducer of systemic acquired resistance.

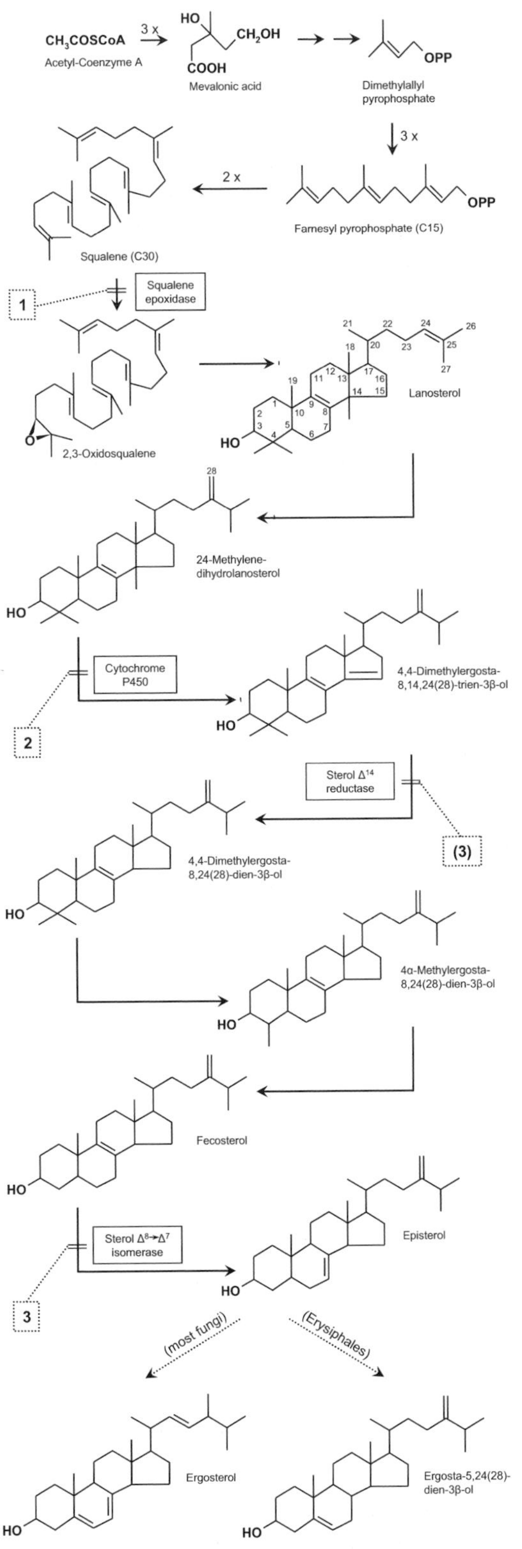

Fig 13.16 One of several possible sterol biosynthesis pathways in fungi. Most fungi have ergosterol as their major membrane sterol, but in the Erysiphales it is ergosta-5,24(28)-dien-3β-ol (Loeffler et al., 1992). Putative targets of various sterol biosynthesis inhibitors are indicated as follows. (1) Allylamines, e.g. terbinafine (against *Candida albicans*; see Fig. 10.9). (2) Triazoles (against fungi pathogenic to humans as well as plants). A specific demethylation step is inhibited at C14. (3) Morpholines, e.g. fenpropimorph (against plant-pathogenic fungi). These compounds seem to inhibit two different steps in sterol biosynthesis. Re-drawn and modified from Kerkenaar (1995).

Morpholines (e.g. tridemorph; Fig. 13.15b) were the first sterol biosynthesis inhibitors, acting by inhibiting two sterol-modifying enzymes in fungi (Fig. 13.16) whilst not affecting sterol biosynthesis in plants (Kerkenaar, 1995; Uesugi, 1998). A more recent fungicide, fenpropimorph, is derived from tridemorph and has a similar mode of action but is used against other fungi as well. In contrast, the 2-aminopyrimidines (Fig. 13.15c) act by inhibiting the incorporation of adenine into nucleic acids, thereby halting DNA synthesis. The effects are visible early, usually at the stage of formation of the first haustorium (Holloman & Schmidt, 1995).

Triazoles, which we have already encountered as drugs in the treatment of *Candida* infections (Fig. 10.9), are also extensively used in agriculture because of their selective action in sterol biosynthesis. There is a huge diversity of structures (Kuck et al., 1995; Uesugi, 1998), but they all seem to act as inhibitors of a cytochrome P-450 involved in demethylating sterols at the C-14 position (Fig. 13.16). These compounds are therefore collectively called **demethylation inhibitors** (DMIs). The target organism is unable to produce a functional membrane. Triazoles are active against a very wide range of fungal pathogens and are systemic fungicides with curative properties. One of the first important examples against powdery mildews and other biotrophic plant pathogens was triadimefon, which is reduced in plants and fungi to its more active alcohol (Fig. 13.15d). However, resistance has arisen on numerous occasions, and various resistance mechanisms have been implicated, such as exclusion or reduced uptake of the fungicides, altered membrane lipid composition with reduced sterol content, or mutation of the fungicide-binding site on the cytochrome P-450 enzyme.

One of the most important recently introduced classes of fungicides are the **strobilurins**. They are based on strobilurin A (Fig. 13.15e), a natural product from the basidiomycete *Strobilurus tenacellus* initially described by Anke *et al.* (1977) as a strongly and selectively antifungal substance. This was derivatized to give kresoxim methyl, the first commercial strobilurin-type fungicide (Fig. 13.15f). The mode of action is based on an inhibition of complex III of the mitochondrial respiratory chain (Anke, 1997). Resistance of *Strobilurus tenacellus* to its own product is based on a mutation of three amino acids in the target polypeptide, and resistance based on similar mechanisms has arisen in many fungal plant pathogens. In consequence, strobilurins are not currently recommended for control of powdery mildews on cereals and cucumber/melon crops (Hollomon & Wheeler, 2002).

A recently released compound with exclusive activity against powdery mildews is **quinoxyfen** (Fig. 13.15g). The mode of action is interesting because this compound selectively inhibits morphogenetic events related to infection, such as germination or appressorium formation, but not vegetative growth. It is therefore non-toxic to other fungi. Quinoxyfen is not systemic but is distributed in the vapour phase and binds to the surface waxes of the epidermis. It is thus ideally placed for activity against the germinating powdery mildew conidium (Hollomon & Wheeler, 2002). Non-fungicidal compounds with such a highly specific mode of action have fewer side effects against non-target organisms such as saprotrophic or mycorrhizal fungi, and are likely to increase in importance in future.

Activators of systemic acquired resistance (SAR; see p. 116), especially the salicylic acid derivative **benzothiadiazole** (Fig. 13.15h), may prove their worth in the prevention of powdery mildew infections as well as many other plant diseases (Salmeron *et al.*, 2002), although they cannot cure existing infections. Since SAR activators trigger the expression of a multitude of defence-related proteins and other mechanisms in the crop plant, powdery mildews are unlikely to develop resistance against them, like they have done against most of the currently used fungicides, many of which target a single site in the physiology of the pathogen. Therefore, fungicides must be used as cocktails containing chemicals with different modes of action, and following strict recommendations (Hollomon & Wheeler, 2002).

13.8.3 Biological control

Their exposed habitat on the leaf surface renders powdery mildews potentially susceptible to parasitism by phylloplane fungi. Perhaps the best-known of them is *Ampelomyces quisqualis*, which produces conidia within pycnidial fruit bodies associated with the hyphae, conidia and, less frequently, chasmothecia of powdery mildews. It is commonly found in nature (Falk *et al.*, 1995; Kiss, 1997). Attempts are being made to develop it as a biological control agent (see Bélanger & Labbé, 2002), but success is limited by the ability of powdery mildews to grow at lower humidities than *A. quisqualis*. This may be alleviated by the application of the fungus in paraffin oil (Bélanger & Labbé, 2002). One positive aspect is that *A. quisqualis* is tolerant of several fungicides used against powdery mildews, so that integrated control is possible in principle (Sundheim, 1982).

In general, however, the biological control of powdery mildews, like that of most other airborne pathogens, would appear to be limited to greenhouses because there the environment can be controlled to a certain extent. With powdery mildews, humidity appears to be the crucial parameter not only in interactions with *A. quisqualis*, but also other potential control fungi such as *Verticillium lecanii* or basidiomycete yeasts, e.g. *Tilletiopsis* and *Pseudozyma* (summarized in Bélanger & Labbé, 2002). All these, in contrast to *A. quisqualis*, control powdery mildews through the production of biologically active substances rather than by direct parasitism. Fatty acid derivatives seem to be the main active compounds, and these may act by

Plate I Slime moulds (Myxomycota). (a) White coralloid sporocarps of *Ceratiomyxa fruticulosa* (Protosteliomycetes) on rotting wood. (b) Developing aethalia of *Lycogala epidendron* on rotting wood. (c) Aethalium of *Reticularia lycoperdon* with a silvery grey peridium. (d) Aethalium of *R. lycoperdon* with its peridium ruptured to reveal a dark brown powdery mass of spores. (e) *Arcyria denudata* on rotting wood. The sporangia have opened up, releasing their dull red spores. The capillitium network is exposed. (f) Phaneroplasmodium of *Physarum polycephalum* producing numerous stalked sporangia on an agar surface. (g) Slightly immature aethalium of *Fuligo septica*. (h) Clustered stalked sporangia of *Stemonitis axifera*. (e) and (h) kindly provided by G. L. Barron.

Plate 2 Oomycota. (a) Salmon infected by *Saprolegnia* sp. (b) Damping-off of cress caused by *Pythium* sp. (c) Sudden death of a 20-year-old *Pseudotsuga menziesii* plantation caused by *Pythium undulatum* (= *Phytophthora undulata*) due to root rot following heavy summer rains. (d) *Phytophthora* root rot symptoms of *Abies procera*. (e) Late blight symptoms on potato leaf caused by *Phytophthora infestans*. (f) Pink rot of potato tuber caused by *Phytophthora erythroseptica*. Killed tissue shows a pink discoloration after cutting and exposure to air for 30 min. (g) *Peronospora parasitica* on wallflower (*Cheiranthus cheiri*) showing the distorted host shoot and downy appearance of sporulating regions. (h) Cauliflower leaf infected with *Albugo candida*, showing the typical white blister rust symptoms caused by the eruption of sporangia through the epidermis.

Plate 3 Chytridiomycota (a–c) and Zygomycota (d–i). (a,b) *Synchytrium endobioticum*. (a) Excavated potato plant showing a gall at the base of a shoot. (b) Leafy gall stage at the soil surface. (c) *Synchytrium taraxaci*, sporangial sori on dandelion (*Taraxacum officinale*). (d) Strawberry infected by *Rhizopus stolonifer*. (e) Sporangiophores of *Spinellus fusiger* on an old basidiocarp of *Mycena pura*. (f) Sporangiophore of *Pilobolus crystallinus* producing a hemispherical black sporangium. A yellow band below the base of the swollen subsporangial vesicle, the ocellus, is enriched in carotenoids. The stalk and vesicle bear droplets of liquid. (g) *Entomophthora muscae*. Dead fly attached to a window pane, surrounded by a halo of discharged conidia. (h) *Furia americana*. Dead blowflies attached to a leaf. Conidiophores have penetrated between the abdominal segments. (i) Sporocarp of the pea truffle *Endogone lactiflua*, cut open to reveal the zygospores.

Plate 4 Archiascomycetes (a–c) and Plectomycetes (d–f). (a) Peach leaf infected with *Taphrina deformans*. (b) *Taphrina betulina* causing witches' broom disease on birch. (c) *Taphrina amentorum* causing abnormal enlargement of individual catkin segments on *Alnus glutinosa*. (d) *Aspergillus parasiticus*. Petri dish with parasexual recombinants deficient in melanin biosynthesis leading to colourless spores, and/or in aflatoxin biosynthesis which leads to the accumulation of bright yellow or orange pigments. (e) Orange rotted by *Penicillium digitatum*. (f) Excavated cleistothecium of *Elaphomyces* infected by *Cordyceps ophioglossoides*. Germinating ascospores have accumulated as whitish pustules around the perithecial ostioles. (d) kindly provided by J. F. Peberdy, (f) by J. Benn.

Plate 5 Pyrenomycetes. (a) *Daldinia concentrica* on an old log of ash. One stroma has been cut open. (b) Stromata of *Xylaria longipes* on a buried sycamore trunk. (c) *Hypocrea pulvinata*, perithecial crust on the underside of an old fruit body of *Piptoporus betulinus*. (d) *Nectria cinnabarina* with conidial pustules and dark red perithecial stromata. (e) Orange-coloured sporodochia of *Fusarium heterosporum* parasitizing sclerotia of *Claviceps purpurea* var. *spartinae* on *Spartina anglica*. (f) Perithecial stroma of *Epichloe typhina* 'choking' a shoot of *Dactylis glomerata*. (g) *Cordyceps militaris*, perithecial stromata emerging from a buried insect pupa. (h) Dutch elm disease. (i) Anthracnose on broad bean caused by *Colletotrichum lindemuthianum*. (c) kindly provided by J. Benn.

Plate 6 Apothecia of operculate discomycetes (Pezizales). (a) *Pyronema domesticum* fruiting on autoclaved pottery. (b) *Aleuria aurantia*. (c) *Peziza vesiculosa* on freshly manured garden soil. (d) *Ascobolus furfuraceus* on agar. The purple dots represent ripe asci. (e) *Sarcoscypha australis* on an old log of ash (*Fraxinus*). (f) *Helvella crispa*. (g) *Morchella esculenta*. (g) kindly provided by P. Davoli.

Plate 7 Inoperculate Discomycetes (Helotiales). (a,b) *Sclerotinia curreyana*. (a) Sclerotia formed inside culms of *Juncus effusus*. (b) Stalked apothecia of *S. curreyana* arising from a sclerotium in spring. (c) *Rutstroemia echinophila* on the shell of edible chestnuts. (d) *Mollisia cinerea*, a common saprotroph on twigs. (e) *Dasyscyphus virgineus* producing small apothecia with a hairy margin on rotting wood. (f) *Leotia lubrica*, a saprotroph soil fungus with stalked apothecia. (g) *Chlorociboria aeruginascens* on a decaying sycamore twig. Sections of the colonized wood show its green discolouration. (h) *Bisporella citrina* on a beech twig. (i) *Bulgaria inquinans* forming black gelatinous apothecia on freshly felled oak trunks. (j) Apothecial stromata of *Cyttaria darwinii* on living twigs of the southern beech *Nothofagus pumilio*. (d) and (e) kindly provided by H. Anke, (f) by P. Davoli. (a,b) reprinted from Weber and Webster (2003), with permission from Elsevier.

Plate 8 Thalli of lichens. (a) The crustose lichen *Lecanora muralis*. Numerous apothecia are seen in the centre of the thallus. (b) The crustose *Rhizocarpon geographicum*. Its mosaic-like appearance is due to greenish-yellow vegetative thalli with small dark apothecial areas, and larger black prothalli. (c) The crustose−foliose lichen *Xanthoria parietina*. The yellow colour is due to the pigment parietin. Numerous apothecia are seen in the centre of the thallus. (d) *Peltigera canina* on mossy boulders. The foliose fleshy thallus is coloured dark blue−green by the cyanobacterium *Nostoc*. Note the white rhizinae and the reddish apothecial areas. (e) *Cladonia floerkiana*. Fruticose upright podetia arise from a squamulose horizontal thallus. (f) The reindeer lichen *Cladina rangiferina* growing among ground vegetation. (g) *Usnea florida*, a fruticose lichen hanging down from tree branches. (b) kindly provided by B. Büdel.

Plate 9 Fruit bodies of Homobasidiomycetes: euagarics and boletoid clades. (a) *Amanita muscaria*, probably the best-known of all mushrooms. (b) *Amanita caesarea*, a prized delicacy in Mediterranean countries. (c) The scarlet waxcap, *Hygrocybe coccinea*. (d) The winter fungus, *Flammulina velutipes*. (e) *Pholiota squarrosa*. (f) *Boletus erythropus*. (g) *Suillus granulatus*. (h) *Boletus badius*. Healthy fruit body (left) and others attacked by the mycoparasitic mould, *Sepedonium chrysospermum*. (e) kindly provided by P. Davoli.

Plate 10 Fruit bodies of Homobasidiomycetes: other clades. (a) *Trametes versicolor*, the artist's fungus or turkey tail on a birch branch. Note the bleached (white-rot) appearance of the wood at the broken end. (b) *Laetiporus sulphureus*, the chicken of the woods. (c) *Lactarius deliciosus*. (d) *Chondrostereum purpureum*, cause of silver-leaf disease on plum trees. (e) *Thelephora terrestris*, an ectomycorrhizal species. (f) *Phellinus igniarius* fruiting on an old willow tree. (g) *Cantharellus cibarius*, the chanterelle. (h) *Ramaria botrytis*.

Plate II Gasteromycetes (a–f) and Heterobasidiomycetes (g–i). (a) *Calvatia excipuliformis*. (b) *Pisolithus tinctorius*, one gasterocarp cut open to reveal the peridioles. (c) *Rhizopogon* sp.; excavated gasterocarp which has been cut open. (d) *Phallus impudicus*, the common stinkhorn. (e) *Clathrus ruber*. (f) *Aseroe rubra*. (g) *Calocera viscosa*. (h) *Auricularia auricula-judae*. (i) *Tremella mesenterica*. (f) reprinted from Fuhrer (2005), with permission by Bloomings Books Pty Ltd. Image kindly provided by B. Fuhrer.

Plate 12 Urediniomycetes (a–g) and Ustilaginomycetes (h–j). (a) Aeciospore of *Puccinia distincta*. Lipid droplets have been displayed by the two nuclei in the upper spore. (b) Aecial infection of *Puccinia caricina* on stinging nettle. (c) Uredinia (orange pustules) and telia (dark purple lesions) of *Phragmidium violaceum* on the underside of a leaf of bramble. (d–f) *Gymnosporangium fuscum*. (d) Telial horns on a swollen canker on *Juniperus* in spring. (e) Spermogonial infection on a pear leaf in midsummer. (f) Roestelioid aecia on the underside of a pear leaf in autumn. The aecial caps are connected to the aecial base by trellis-like threads. (g) Uredinia of *Melampsora* sp. on *Populus tremula*. (h) Maize smut caused by *Ustilago maydis*. Swollen kernels have become converted into teliospore-bearing tumours. (i) *Exobasidium* sp. on an ornamental *Azalea*. Infected leaves are strongly hypertrophied. (j) Systemic *Exobasidium vaccinii* infection of blueberry (*Vaccinium myrtillus*). The infected shoot (left) shows a reddish discolouration. (d) and (i) kindly provided by H. Weber.

inserting themselves into the powdery mildew plasma membrane, disrupting its structural integrity (Avis & Bélanger, 2002; Urquhart & Punja, 2002). Other explanations for the strongly inhibitory effect of specific fatty acids (especially *cis*-monounsaturated ones) against powdery mildews are, however, also possible (Wang *et al.*, 2002). Unsaturated fatty acids could also be the basis for the anti-powdery mildew activity of cow's milk described by Bettiol (1999).

Hymenoascomycetes: Pezizales (operculate discomycetes)

14.1 | Introduction

The order Pezizales contains the operculate discomycetes which are the most readily recognized cup fungi. The order is large, containing some 15 families, about 160 genera and over 1100 species (Kirk *et al.*, 2001). Most are terrestrial and saprotrophic on soil, burnt ground, decaying wood, compost or dung, but some form sheathing mycorrhiza (ectomycorrhiza) with trees (Maia *et al.*, 1996). A somewhat exceptional case is *Rhizina undulata*, which causes root rot of conifers in plantation situations, usually starting from areas affected by recent fires (Callan, 1993). Whilst most species of Pezizales produce **epigeous** fruit bodies above ground level and have active ascus discharge mechanisms with wind-dispersed ascospores, the truffles (e.g. *Tuber* and *Terfezia*) form subterranean (**hypogeous**) ascomata. The dispersal of truffles relies on the ripe ascomata being eaten by rodents and other mammals attracted by their strong odour. The ascospores survive digestion and defaecation. There are also aquatic Pezizales, growing on wood in streams or other wet places. An overview of the Pezizales may be found in Pfister and Kimbrough (2001). Keys to genera are given by Korf (1972) and Dissing *et al.* (2000).

The ascocarp is generally an apothecium (p. 245) which can range in diameter from less than one millimetre to several centimetres. It is often cup-shaped or disc-like, fleshy, sometimes stalked, and frequently brightly coloured. The asci are, in most cases, cylindrical with a well-defined lid called **operculum** (see pp. 239–241) which is the characteristic feature of the Pezizales. Members of this order are therefore often referred to as 'operculate discomycetes'. The asci are interspersed by filamentous paraphyses, the tips of which often contain carotenoids giving the apothecia their striking yellow, orange or red colours. Several unusual carotenoids are known in nature only from apothecia of Pezizales (Gill & Steglich, 1987). The ascus wall appears distinctly two-layered under the light microscope, but the two layers do not separate during ascospore discharge as they do in functionally bitunicate (i.e. fissitunicate) ascomycetes (see p. 239). The ascus of the Pezizales is thus bitunicate but non-fissitunicate. In many Pezizales the ascus wall is **amyloid**, i.e. it stains blue or purple with Melzer's iodine (an aqueous solution of iodine and KI). The blue-staining properties are associated with an outer mucilaginous layer which may extend for the whole length of the ascus or may be confined to an apical region (Samuelson, 1978a,b). The presence or absence of the amyloid staining property may help in distinguishing certain genera (see Hansen *et al.*, 2001). The septal pore formed at the base of the ascus has characteristic features which may be useful in classification (Kimbrough, 1994). The ascospores are colourless to reddish-brown, globose to ellipsoidal, unicellular (i.e. non-septate) and may be uninucleate,

Table 14.1. | Families of the Pezizales which are commonly encountered in nature. Data from Kirk *et al.* (2001). Families printed in bold are considered in more detail in this chapter.

Family	Number of species	Examples
Ascobolaceae (p. 419)	118	*Ascobolus, Saccobolus*
Discinaceae	25	*Gyromitra*
Helvellaceae (p. 423)	68	*Helvella*
Morchellaceae (p. 427)	38	*Morchella, Verpa*
Pezizaceae (p. 419)	160	*Peziza*
Pyronemataceae (p. 415)	462	*Aleuria, Otidea, Pyronema, Scutellinia*
Rhizinaceae	2	*Rhizina*
Sarcoscyphaceae	36	*Sarcoscypha*
Sarcosomataceae	31	*Galiella, Urnula*
Terfeziaceae	15	*Terfezia*
Tuberaceae (p. 423)	87	*Tuber*

quadrinucleate or multinucleate. They contain one or several large lipid globules and may have smooth or ornamented walls. Individual asci may discharge their spores asynchronously, or large numbers of asci may shoot off their spores simultaneously. In this case, the spores may be released in a visible cloud in a process known as 'puffing' (see Fig. 8.14). Buller (1934) investigated this phenomenon in detail and reported that ascospore puffing produces an audible hissing sound. The asci of truffles, in contrast, are sac-like or globose, with no functional operculum, and the spores are released passively.

The fruit bodies of some members of the group are highly prized culinary delicacies, notably truffles (*Tuber* spp.) and morels (*Morchella* spp.). *Gyromitra esculenta*, the false morel, was formerly also widely consumed but, following a series of often fatal mushroom poisonings, was eventually found to contain the heat-labile toxin gyromitrin. This is readily converted into hydrazine derivatives such as the rocket fuel methylhydrazine, which are highly toxic and carcinogenic (Bresinsky & Besl, 1990). Illustrations of the ascomata of commonly occurring Pezizales have been provided by Breitenbach and Kränzlin (1984) and Dennis (1981). Selected examples are presented on Plate 6.

Molecular phylogenetic analyses indicate that the Pezizales are probably a primitive group ancestral to other Euascomycete orders (Fig. 8.17; Gargas & Taylor, 1995; Landvik *et al.*, 1997). This implies that apothecia are an ancient type of ascoma, with cleistothecia and perithecia representing later developments. Whilst the Pezizales as a whole are monophyletic, the arrangement into families within this order is still tentative (Harrington *et al.*, 1999). The more important of the currently recognized families are listed in Table 14.1. Since the morphological and ecological features of pezizalean fungi often transcend the family boundaries, we will describe a few characteristic genera in more detail by their biological features, merely indicating their family assignment where appropriate.

14.2 | *Pyronema* (Pyronemataceae)

The apothecia of *Pyronema* develop on burnt soil and on heat-sterilized composts in glasshouses. There are two species, *P. omphalodes* (= *P. confluens*) and *P. domesticum* (Moore & Korf, 1963). In *P. omphalodes* the apothecia are confluent and lack marginal hairs, whilst in *P. domesticum* the apothecia are more discrete, and surrounded by tapering hairs (Fig. 14.1a). *Pyronema domesticum* forms sclerotia in culture (Moore, 1962), whilst *P. omphalodes* does not. In earlier studies the distinction between the two species was sometimes not appreciated and

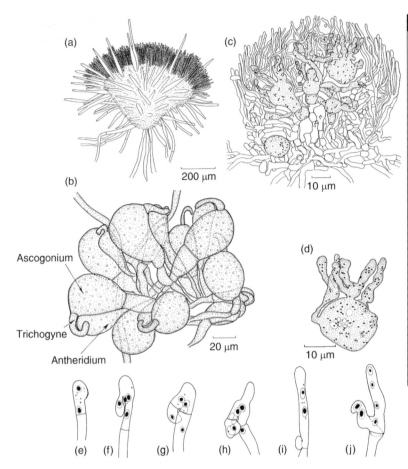

Fig 14.1 *Pyronema domesticum.*
(a) Apothecium showing hymenium and excipular hairs. (b) Group of ascogonia and antheridia. (c) V.S. through developing apothecium showing several ascogonia producing ascogenous hyphae, and the development of paraphyses and excipulum from the ascogonial stalks. (d) Enlarged view of the ascogonium and developing ascogenous hyphae. (e–j) Stages in the development of asci. (e) Binucleate tip of ascogenous hypha beginning to form a crozier. (f) Quadrinucleate stage. (g) Septation of crozier to form a binucleate penultimate cell. (h) Development of ascus from binucleate cell. (i) Completion of first meiotic division. (j) Completion of second meiotic division. Note the proliferation of a new ascogenous hypha from the stalk cell.

some reports purporting to be on *P. confluens* may well have been based on *P. domesticum*. These earlier studies include classical accounts of the cytology of the development of ascogenous hyphae, croziers and asci (see Moore, 1963).

14.2.1 Development of asci in *Pyronema*

Both species of *Pyronema* are homothallic and grow rapidly in agar culture or on sterilized soil and within 4–5 days form pink apothecia about 1–2 mm in diameter (Plate 6a; Webster & Weber, 2001). Apothecia of *P. domesticum* arise from clusters of ascogonia and antheridia formed by repeated dichotomy of a single hypha. The ascogonia are fatter than the antheridia and each ascogonium is surmounted by a tubular recurved trichogyne which grows to make contact with the tip of an antheridium (Fig. 14.1b). Both antheridia and ascogonia are multinucleate and, following fusion of the trichogyne with the

antheridium by breakdown of the walls separating them (plasmogamy), antheridial nuclei stream into the ascogonium and each antheridial nucleus becomes paired with an ascogonial nucleus. Nuclear fusion (karyogamy) does not occur at this stage, but the paired nuclei remain associated with each other. Branched investing hyphae develop from the ascogonial stalks and envelop the cluster of fertilized ascogonia, ultimately making up the tissues of the apothecium, i.e. the medullary and ectal excipulum. Several ascogenous hyphae extend from each ascogonium and grow between the surrounding investing hyphae (Figs. 14.1c,d).

Further development follows the common ascomycete pattern outlined in Fig. 8.10 (I. M. Wilson, 1952; Hung & Wells, 1971). The ascogenous hyphae are branched and septate at their tips, which recurve to form croziers. The tip of the crozier is binucleate and the two nuclei simultaneously divide by mitosis (conjugate mitosis). Two septa cut off a uninucleate terminal cell, a binucleate penultimate cell and a uninucleate antepenultimate cell (the stalk cell). The binucleate penultimate cell is the ascus mother cell and the two nuclei within it fuse, i.e. karyogamy now occurs. The diploid fusion nucleus undergoes meiosis and the four resulting haploid nuclei then divide mitotically so that eight haploid nuclei result around which the eight ascospores are subsequently cleaved (Reeves, 1967). No further nuclear divisions occur so that each ascospore contains one haploid nucleus. No special inclusions are seen in the septa of the crozier, but electron-dense plugs are formed at the base of the ascus (Hung & Wells, 1971; Kimbrough, 1994). When the uninucleate terminal cell grows backwards and makes contact with the stalk cell, their walls break down and a new binucleate cell is formed which grows on to form a further crozier and another ascus, a process which is repeated so that a single ascogenous hypha may produce several asci (Fig. 14.2b). The ascus mother cell elongates and acquires a cylindrical shape. It is surrounded by filamentous paraphyses. These develop from the stalks of the ascogonia (I. M. Wilson, 1952) and also appear to arise from ascogenous hyphae (Fig. 14.2c). As the ascus matures it extends above the layer of paraphyses and explodes, throwing out its ascospores. The operculum may persist as a hinged lid (Fig. 14.2c) or may be blown off. During the development of asci, before the ascospores are cleaved out, the operculum becomes apparent as a thickened rim of wall material at the upper end of the ascus.

Mature ascospores have three wall layers, a thicker, electron-transparent inner layer (the **endospore**), a thinner electron-opaque **epispore**, and an outer fibrous **perispore** of variable thickness. The perispore lies immediately within the ascospore-investing membrane, and, as ascospores mature, this membrane continues to produce vesicles, leading to degradation of the perispore. At discharge, the ascospores do not stick together but remain separate from each other (Hung, 1977). Merkus (1976) has described the development and structure of the ascospores of *P. omphalodes* in similar terms.

Much is known about the conditions under which *P. domesticum* forms apothecia and sclerotia (Moore-Landecker, 1975, 1992). Light is required for apothecium development, with white, blue and far-red light being particularly effective. Sclerotium formation is inhibited by intense blue light.

14.2.2 Ecology of *Pyronema*

In nature, the apothecia of both species of *Pyronema* are among the first to appear on burnt ground following volcanic eruptions, wildfires, controlled burns and bonfires. *Pyronema* forms part of a characteristic group of '**phoenicoid fungi**', i.e. fungi arising from ashes. Many other operculate discomycetes are also phoenicoid (Carpenter & Trappe, 1985; Dix & Webster, 1995). The ascospores of *P. domesticum* germinate readily at 20°C, although a short exposure to 50°C enhances germination. Apothecium formation is inhibited by the presence of other soil-inhabiting organisms and it is possible that the preference for burnt ground and steam-sterilized soil is associated with its rapid growth and inability to compete with other soil biota (El-Abyad & Webster, 1968a,b).

Unnatural (i.e. man-made) situations in which *P. domesticum* fruits are steam-sterilized soils and composts used in horticulture, and plaster prepared by slaking lime, a process which generates heat. Supposedly sterile surgical gauzes manufactured from Chinese cotton have been found to be contaminated with *P. domesticum* due to insufficient radiation treatment during manufacture. The source is likely to be raw cotton materials possibly already contaminated in the field (Yan, 1998). Laboratory experiments have shown that the γ-irradiation resistance of *P. domesticum* ascospores is higher even than that of *Bacillus* endospores (Richter & Barnard, 2002).

14.3 | *Aleuria* (Pyronemataceae)

Aleuria seems to be closely related to *Pyronema* (Landvik *et al.*, 1997). There are about 10 species of *Aleuria*, growing especially on forest soil.

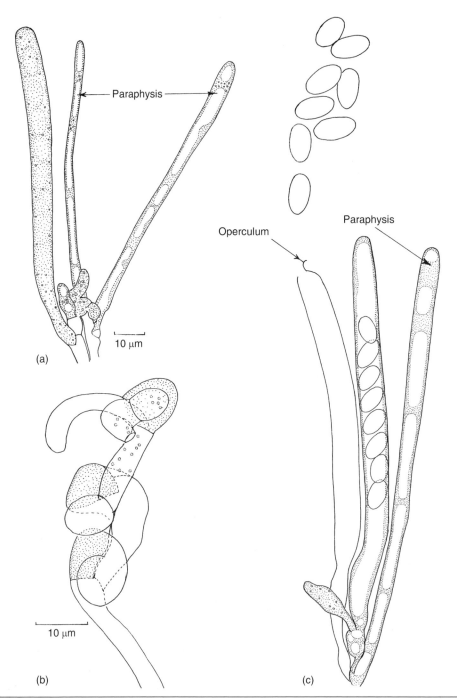

Fig 14.2 *Pyronema domesticum.* (a) An immature ascus (left). The ascogenous hypha from which it developed continues to proliferate. (b) More magnified view of the tip of an ascogenous hypha showing repeated proliferation. The three stippled cells represent penultimate cells of croziers probably destined to develop into asci. (c) Mature asci, one discharged and showing an operculum. A paraphysis is also shown apparently arising from the ascogenous hypha.

Aleuria is similar in appearance to *Peziza* (see below) but is distinguished from it in having non-amyloid asci. The best-known species is *A. aurantia*, the so-called orange-peel fungus which forms strikingly orange-coloured cup-shaped apothecia from about 1–10 cm in diameter in woodland and grassland soil during autumn (Plate 6b). Isotopic analyses of ascocarps using ^{15}N and ^{13}C isotopes indicate that the fungus may be mycorrhizal (Hobbie *et al.*, 2001). The orange colour of the hymenium is due to carotenoid-enriched granules in the club-shaped tips of the paraphyses (Fig. 14.3). The main carotenoids are β-carotene, γ-carotene and aleuriaxanthin (Gill & Steglich, 1987).

The ascospores contain two lipid globules and the ascospore wall is ornamented by a honeycomb-like series of raised ridges which represent secondary wall material. It is derived from granules in the perisporic sac which condense into larger spherical dense bodies. These accumulate and become attached to the epispore (Merkus, 1976; Wu & Kimbrough, 1993).

14.4 | *Peziza* (Pezizaceae)

Peziza is a large genus containing around 100 species. Analyses based on a combination of morphological and molecular evidence indicate that this genus is polyphyletic, i.e. it contains a number of unrelated taxa which have been grouped together artificially. *Peziza* in its traditional sense can be differentiated into at least eight clades (Hansen *et al.*, 2005). Hohmeyer (1986) has provided a key to European species. For descriptions and illustrations see Dennis (1981), Breitenbach and Kränzlin (1984) and Dissing *et al.* (2000). The apothecia are cup-shaped, often large (2–5 cm or more), usually pale brown and fleshy (Plate 6c). They are commonly encountered in a very wide range of habitats including soil, manure heaps, dung, rotting wood or straw, burnt ground and sand dunes. About six species have hypogeous ascocarps (Trappe, 1979). The ascus wall is, in general, amyloid. In *P. succosa* the blue-staining by I_2/KI is confined to the ascus apex (Samuelson, 1978a).

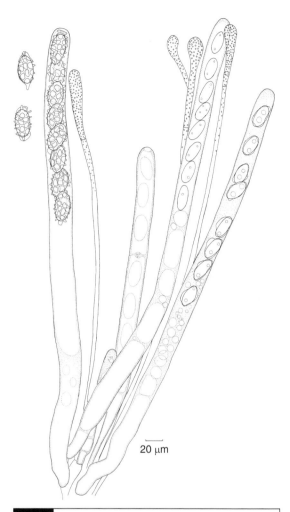

20 μm

Fig 14.3 *Aleuria aurantia*. Asci, ascospores and paraphyses. The tips of the paraphyses are filled with granules containing orange-coloured carotenoids.

The conidial states of *Peziza* spp., where known, have been classified in the anamorph genus *Oedocephalum* (Fig. 14.4).

A genus with more strikingly coloured apothecia is *Sarcoscypha* (Plate 6e). Although similar in appearance to *Peziza*, these two genera are not closely related (Landvik *et al.*, 1997).

14.5 | *Ascobolus* (Ascobolaceae)

There are about 80 species of *Ascobolus* (van Brummelen, 1967). Most of them are

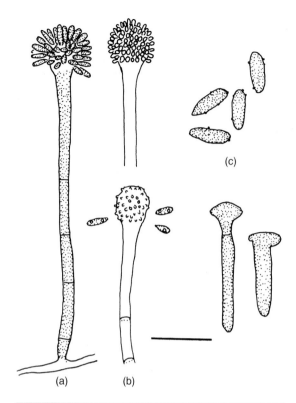

(c)

(a)　　　(b)

Fig 14.4 *Oedocephalum* conidial state of *Peziza subviolacea* (= *P. praetervisa*). (a) Conidiophores terminating in a club-shaped vesicle bearing numerous dry blastoconidia. (b) Details of developing conidia (top) and a vesicle from which the conidia have been detached (bottom). (c) Conidia, two of which are germinating. Scale bar = 40 μm (a,b) and 20 μm (c). From Webster *et al.* (1964), with permission from Elsevier.

coprophilous, growing on the dung of herbivorous animals, but *A. carbonarius* grows on old bonfire sites. Common coprophilous species are *A. furfuraceus* (= *A. stercorarius*) which is very commonly found on old cattle dung, often along with *A. immersus* (Figs. 14.5 and 14.6, respectively). Whilst these species are heterothallic, some others, e.g. *A. crenulatus* (= *A. viridulus*), are homothallic. Characteristic features of all species are the purple colour of the ascospores and the protruding, operculate asci. *Ascobolus furfuraceus* forms yellowish saucer-shaped apothecia up to 5 mm in diameter, and when mature the surface of the apothecium is studded with purple dots which mark the ripe asci (Plate 6d). As the asci mature they elongate above the general level of the hymenium. The ascus tips are phototropic and this ensures that when they explode the spores are thrown away from the dung. The ascospores have a mucilaginous perispore which aids attachment. *Ascobolus immersus* has yellow globose apothecia about 1–2 mm in diameter, with very large ascospores (about 70 × 30 μm). The perispores cause all the eight ascospores to adhere to form a single projectile about 250 μm long, capable of being discharged for up to 30 cm horizontally. In general, multi-spored projectiles have a lower surface-to-volume ratio and are projected further than single spores (see p. 317 for *Sordaria*). There is a general trend among coprophilous fungi towards multi-spored projectiles. In the genus *Saccobolus*, which also belongs to the Ascobolaceae, all eight spores are firmly cemented together by their perispores.

The spores of *Ascobolus* become attached to herbage and, when eaten by a herbivore, germinate in the faeces. It is likely that digestion stimulates spore germination. Most spores fail to germinate on nutrient media but can be triggered to do so by treatment with 0.4% NaOH or bile salts, and incubation at 37°C. The purple pigment in the spore wall develops late and is deposited within the perispore from the ascus epiplasm. Immature spores are colourless. The spore wall bears longitudinal colourless striations in some species, e.g. *A. immersus* (Fig. 14.5) and *A. furfuraceus* (Fig. 14.6). Both species can be grown and induced to form apothecia in culture (see Webster & Weber, 2001).

14.5.1 Mating behaviour

There is variation in the mating behaviour of different species of *Ascobolus*. A single ascospore culture of *A. scatigenus* (= *A. magnificus*) does not produce apothecia. Sex organs (coiled ascogonia and antheridia) are formed only when mycelia of different mating types are grown together. Each strain is hermaphroditic, i.e. is capable of developing both ascogonia and antheridia. However, *A. scatigenus* is self-incompatible, i.e. the antheridia of one strain do not fertilize the ascogonia borne on the same mycelium. The ascospores of this fungus are of two types, *A*

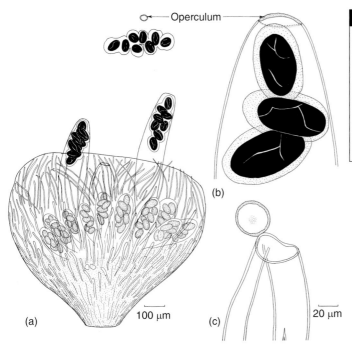

— Operculum —

(b)

(a)   100 μm

(c)   20 μm

Fig 14.5 *Ascobolus immersus.* (a) Apothecium showing two projecting asci. Immature asci can be seen below the general level of the surface. A single projectile consisting of eight adhering ascospores is shown above the apothecium. Note the operculum which has also been projected. (b) Tip of ripe ascus showing the operculum. (c) Tip of discharged ascus. In this case the operculum has remained attached to the ascus tip.

and *a*, and fertilization can only occur between an *A* ascogonium and an *a* antheridium, or vice versa. There is thus a gene for mating type represented in two idiomorphs *A* and *a*, and incompatibility is controlled by this gene irrespective of the presence of both types of sex organ on each strain. There is no morphological difference between the two different mating type strains.

A similar situation occurs in *A. furfuraceus*, but here there are no antheridia. Instead, each strain at first produces chains of arthrospores or oidia (see Fig. 14.6c). The oidia can germinate to form a fresh mycelium, i.e. they can function asexually as conidia, but they also play a part in sexual reproduction. Mites and flies may transport oidia of one strain to the mycelium of the opposite strain, and following this, apothecia develop. The process of fertilization has been studied by Bistis (1956, 1957) and Bistis and Raper (1963). If an *A* oidium is transferred to an *a* mycelium, the oidium fails to germinate and within 10 h an ascogonial primordium appears on the *a* mycelium (Fig. 14.6d). The ascogonium consists of a broad coiled base and a narrow apical trichogyne which shows chemotropic growth towards the oidium and eventually

fuses with it. There is evidence that this sequence of events is under hormonal control, and it has been suggested that a fresh *A* oidium is not immediately capable of inducing development of ascogonial primordia, but must itself at first be sexually activated by a messenger secreted by the *a* mycelium. Following activation, the oidium can induce ascogonial development. By substitution experiments, it has been shown that an *A* ascogonium can be induced to fuse with an *A* oidium, i.e. an oidium of the same mating type, but apothecia fail to develop from such fusions. In compatible crosses, fertile apothecia develop within about 10 days of fertilization, each ascus producing four *A* and four *a* spores.

In *A. immersus* there are no morphologically distinguishable antheridia and there are no oidia. When *A* and *a* mating type mycelia are grown together in culture, multinucleate ascogonia develop which are fertilized by fusion with slender multinucleate hyphae of the opposite strain. Experimentally, fertilization can also be achieved using homogenized mycelial fragments of one strain to spermatize a strain of opposite mating type (Lewis & Decaris, 1974).

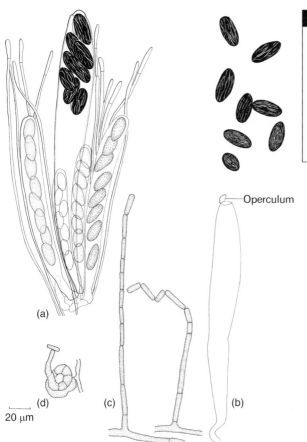

Operculum

Fig 14.6 *Ascobolus furfuraceus.* (a) Group of asci and paraphyses. One ascus is mature and contains purple-pigmented ascospores. (b) The same ascus as shown in a after discharge. The ascus has decreased in size during discharge. Note the operculum. (c) Arthrospores (oidia) developed in five-day-old culture. (d) Coiled ascogonium formed in a single ascospore culture within 48 h of adding oidia of the opposite mating type. The trichogyne of the ascogonium has grown towards the oidium and has fused with it.

(a)

(d)

(c)

(b)

20 μm

14.5.2 Apothecium development

The development of apothecia of *A. furfuraceus* has been studied by various authors, including Wells (1970, 1972) and O'Donnell *et al.* (1974). The fertilized ascogonium becomes surrounded by sheath hyphae which develop from the ascogonial stalk, and the paraphyses and excipular tissues develop from the sheath hyphae. The ascogonium gives rise to numerous ascogenous hyphae. Van Brummelen (1967) has distinguished two kinds of ascocarp development in *Ascobolus*. In **gymnohymenial** forms, the hymenium is exposed from the beginning until the maturation of the asci. In **cleistohymenial** forms the hymenium is enclosed during its early development. *Ascobolus furfuraceus* and *A. immersus* are examples of cleistohymenial development.

14.5.3 Ascosporogenesis

The details of ascosporogenesis have been studied in *A. immersus* by Wu and Kimbrough (1992) and in *A. stictoideus* by Wu and Kimbrough (2001). Although *A. immersus* has smooth ascospores and *A. stictoideus* has ornamented spores, development is very similar. As in most ascomycetes, invagination of the ascus vesicle encloses the young ascospores by a two-layered spore-delimiting membrane. The inner layer forms the plasmalemma of the ascospore, whilst the outer membrane forms a perisporic sac. The primary spore wall is laid down between these two membranes. It consists of an electron-transparent endospore and a laminated epispore. The perisporic sac is variable in thickness and projects into the epiplasm of the ascus. At points of contact with the perisporic sac of neighbouring ascospores, the perisporic sac may be lined by vesicles. Secondary wall formation results from deposition of material derived from the epiplasm onto the epispore. Dense, vesicle-bound bodies originally present in the epiplasm are responsible for the purple coloration of the secondary spore wall. The pigmented secondary wall layer is not uniformly thick and the hyaline striations in the spore wall represent crevices in the material deposited.

14.5.4 Studies on gene recombination

Ascobolus immersus has proved a useful tool for interpreting the mechanism of gene recombination through crossing-over during meiosis using ascospore colour as a marker. Recombination is detected by the simple technique of scoring the ratio of the spores of different colour in octads of spores shot away from hybrid apothecia. Although the wild-type strains have purple ascospores, several series of mutants with pale spores have been found. When crosses are made using certain non-allelic spore colour mutants, recombinants develop resulting from two types of event: (1) crossing-over, giving reciprocal recombinants and (2) gene conversion, yielding non-reciprocal recombinants corresponding to only one of the four products of meiosis. Gene conversion is a process where one allele of a gene converts another allele at the same locus to its own type. Crossing-over results in a 4:4 ratio of coloured to colourless spores whilst a conversion is detected by the presence of coloured and colourless spores in different ratios e.g. 7:1, 6:2 or 5:3. Conversions have been interpreted as implying a double-strand replication of one part of a chromatid whilst the corresponding part of the other is not replicated (Lamb, 1996).

14.6 | *Helvella* (Helvellaceae)

Helvella, a genus containing about 40 species, is mainly distributed in northern temperate areas (Dissing, 1986; Abbott & Currah, 1997). *Helvella* spp. are known as saddle fungi because the ascocarp is differentiated into a hollow, non-fertile stipe and a curved, saddle-shaped fertile part folded over it. They are also called false morels. Common species are *H. crispa* (Plate 6f) and *H. lacunosa*. Some species form ectomycorrhiza, and e.g. *H. crispa* is a mycorrhizal partner with beech, *Fagus sylvatica*. A characteristic feature of the Helvellaceae is that their ascospores are quadrinucleate. This feature, also found in certain truffle-like fungi with hypogeous ascocarps, has been used as evidence placing genera such as *Hydnotrya*, species of which also form sheathing mycorrhizae, into the Pezizales instead of the Tuberales where they were previously classified (Trappe, 1979). More recent molecular phylogenetic studies have confirmed a close relationship between the Helvellaceae and the Tuberaceae, the main truffle-containing family (Percudani *et al.*, 1999).

14.7 | *Tuber* (Tuberaceae)

About 100 species of *Tuber* are known. They are called the 'true truffles', ascomycetes which form subterranean fruit bodies in which the hymenium is not open to the exterior. It has been suggested that hypogeous fruiting is an adaptation for surviving drought and that the genera of fungi which have adopted this strategy (including ascomycetes and basidiomycetes) have evolved from ancestors with epigeous fruit bodies. According to this view the subterranean ascocarps of *Tuber* and other ascomycetous truffle genera are modified apothecia. The term **stereothecium**, defined as a more or less solid fleshy ascoma with asci which are either solitary, i.e. scattered relatively evenly throughout the medullary excipulum, or grouped in dispersed pockets, has been proposed for this type of ascoma (O'Donnell *et al.*, 1997; Hansen *et al.*, 2001). Supporting evidence that stereothecia are modified apothecia is that, during their ontogeny, ascocarps are at first open to the exterior and close over later as the hymenium develops (Barry *et al.*, 1993; Callot, 1999; Janex-Favre & Parguey-Leduc, 2002). This morphological evidence has been corroborated by molecular phylogenetic data (Spatafora, 1995; Percudani *et al.*, 1999; Hansen *et al.*, 2001). Although *Tuber* and other ascomycetes were formerly classified in a separate order (Tuberales), they are now placed in the Pezizales (Trappe, 1979) as a separate family, the Tuberaceae. This family is closely related to the Helvellaceae (see above).

14.7.1 The truffle ascocarp

The ascocarp is generally globose, varying in size from about 1 to 8 cm in diameter and, exceptionally, may weigh up to 1000 g. It is differentiated into an outer, usually dark **peridium** which in some species, e.g. *T. melanosporum* or *T. aestivum*, may bear pyramidal scales, and an inner, fertile **gleba**. The appearance of the gleba is marbled because it is traversed by light- and dark-coloured veins (Fig. 14.7a). The light-coloured veins are sterile, consisting of a loose network of hyphae and air, whilst the darker veins are fertile, made up of more closely packed hyphae, paraphyses and asci (Parguey-Leduc *et al.*, 1991; Barry *et al.*, 1995; Callot, 1999; Janex-Favre & Parguey-Leduc, 2002).

The asci are unitunicate, subglobose and contain 2−6 ascospores. They do not discharge their spores violently, and lack a specialized apical apparatus or operculum. The ascospores are at first hyaline, but later develop yellow to dark brown (melanized) thick walls which may be spiny or thrown into reticulate, honeycomb-like folds (Figs. 14.7b, 14.8). Many of the fruit bodies have a strong smell and flavour, and are excavated and eaten by animals such as badgers, wild boar, mice, moles, shrews, squirrels and rabbits (Trappe & Maser, 1977). Hypogeous ascocarps (and basidiocarps) form an important component of their diet. Spore dispersal is brought about in this way and ascospore germination is probably enhanced by passage through the gut of the mammal. Several different volatile chemical substances have been detected from truffle fruit bodies, but the most common and abundant is dimethyl sulphide, attractive to 'truffle flies' (*Suillia* spp.) which lay their eggs on the ascocarps (Pacioni *et al.*, 1990, 1991). Similar substances are emitted by stinkhorns (*Phallus* spp.), which likewise attract flies (see p. 590). Claus *et al.* (1981) have shown that the steroid hormone 5α-androst-16-en-3α-ol is also produced by *Tuber* spp. Since this is the main sex hormone produced by boars, its presence in truffles may account for the enthusiasm and efficiency with which sows locate and excavate truffles. It is possible that some of the odoriferous substances are produced by the activity of microbes associated with ascocarp.

14.7.2 The life cycle of true truffles

Surprise discoveries may happen even in seemingly well-studied life cycles such as those of *Tuber* spp., in which a sympodulosporic conidial state somewhat resembling *Geniculosporium* (see Fig. 12.13) has been described recently (Urban *et al.*, 2004). It is as yet unclear how frequent this state is in nature or among other *Tuber* spp. and which role, if any, it might play in their ecology.

The traditional life cycle of *Tuber* is based solely on sexual reproduction (see Giovannetti *et al.*, 1994). The haploid ascospores germinate to form hyphae with monokaryotic segments, and the mycelium grows towards the roots of potential mycorrhizal partners, usually trees, but is unable to form mycorrhiza. Anastomosis between monokaryotic mycelia derived from different ascospores results in the formation of a dikaryotic mycelium which forms sheathing mycorrhiza with suitable hosts (Fasolo-Bonfante & Brunel, 1972). Ascocarp development is initiated by the aggregation and differentiation of hyphae which at this stage remain attached to long roots and obtain nutrients from the host tree. According to Janex-Favre and Parguey-Leduc (2002), in *T. melanosporum* the primordium of the ascocarp consists of an ascogonium with its trichogyne, surrounded at the base by an envelope of sterile investing hyphae. Within the glebal tissues to the inside of the primordium, fertile cells develop. At the tips of these fertile cells, the two nuclei of the dikaryon fuse to give a diploid nucleus. This is followed by meiosis and one or more mitoses so that the ascospores may be uni- or multinucleate (Delmas, 1978). Ultrastructural studies

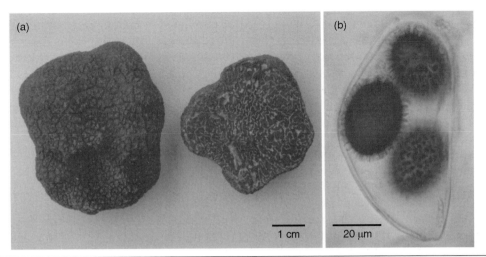

Fig 14.7 *Tuber melanosporum*, the black or Périgord truffle. (a) Two fruit bodies, weighing 60 g (left) and 40 g (right). One has been cut open to reveal the black fertile (ascospore-containing) regions interspersed by white (sterile) veins. (b) Ascus containing three mature ascospores.

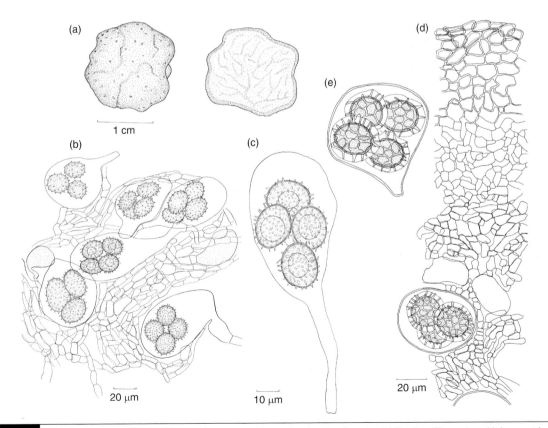

Fig 14.8 (a–c) *Tuber rufum*. (a) Fruit body in surface view and in section showing the veins. (b) Portion of hymenium. (c) Ascus with four spiny-walled ascospores. (d,e) *Tuber puberulum*. (d) V.S. ascocarp showing the structure of the peridium and developing asci. (e) Mature four-spored ascus showing ascospores with reticulate walls.

have shown that ascosporogenesis resembles that found in most other Euascomycetes, but instead of being delimited by the invagination of an ascus vesicle the ascospores become delimited individually by vesicles formed by material derived from lomasomes or from invaginations of the ascus plasma membrane (Berta & Fusconi, 1983).

At first, the developing ascocarp is attached by hyphal connections to a host tree, i.e. it grows symbiotically. Eventually it breaks free and continues to develop independently during a saprotrophic phase of growth. In *T. melanosporum*, tufts of hyphae extend into the surrounding soil from the pyramidal scales on the outside of the peridium. Application of radioactive tracers (e.g. $^{32}PO_4$, $^{3}H_2O$ and ^{14}C-labelled mannose) to these hyphae is rapidly followed by the appearance of radioactive material in the inner portions of the gleba and especially in the fertile veins (i.e. the dark regions of the ascocarp interior; see Fig. 14.7a) at rates and in patterns which could not be accounted for by simple diffusion (Barry *et al.*, 1994, 1995). This finding and reports that fruit bodies of truffles may be found a considerable distance away from living tree roots suggest that mature ascocarps may be autonomous and can obtain water and nutrients directly from the soil, from decaying roots and faecal deposits from the soil fauna (Callot, 1999).

The species of *Tuber* which have been investigated have a wide mycorrhizal host range, including Angiosperms and some Gymnosperms. The host ranges of four species of *Tuber* are shown in Table 14.2.

14.7.3 Truffle collecting

The truffles of greatest commercial value are *Tuber melanosporum* (the black truffle of Périgord) and *T. magnatum* (the white truffle of Piedmont), which can command prices of up to €2000 kg^{-1} in Continental Europe. Several other species are also traded. The only common British truffle which can be used for culinary purposes is *T. aestivum*. Périgord and Piedmont truffles are most abundant in Southern Europe (France, Spain and Italy) within latitudes 40° and 50° North in well-drained calcareous soils (Callot, 1999). Both are collected with the aid of pigs and dogs trained to detect them by smell. The Périgord truffle can also be detected by the presence of a 'burnt' ring-like zone (brûlé)

Table 14.2. Host species relationships of four species of *Tuber*. The main hosts are indicated by black symbols. Marks in brackets indicate the formation of mycorrhiza with some, but not all members of the host genus tested. Data summarized from Giovannetti *et al.* (1994).

	Tuber magnatum	T. melanosporum	T. aestivum group	T. albidum group
Alnus cordata	○	○	○	○
Carpinus betulus	○		●	○
Castanea sativa		○	○	○
Cistus (2 spp.)		●	○	○
Corylus avellana	●	○	●	○
Fagus sylvatica			●	●
Ostrya carpinifolia	●	●	●	○
Populus (2 spp.)	●		○	○
Quercus (6 spp.)	(●)	●	●	●
Salix (2 spp.)	●		○	○
Tilia (3 spp.)	●	○	○	○
Abies alba			○	○
Cedrus (2 spp.)			○	○
Pinus (7 spp.)	(○)	(○)	(○)	●

surrounding a tree with roots with mycorrhizal connections, in which associated herbaceous plants are wilting or dead. This is partly due to deleterious volatile metabolites from the *Tuber* mycelium (Pacioni, 1991) and possibly also to parasitic attack by the mycelium on roots of herbs. The Périgord truffle is associated in the wild with the roots of oaks (*Quercus* spp.) in France. Truffles are cultivated there in plantations (truffières) of appropriate species of oak or on hazel (*Corylus avellana*). Clonal material of suitable hazel cultivars may be used to ensure greater yield and uniformity of cropping (Mamoun & Olivier, 1996). The seedling roots of potential hosts are inoculated by dipping them in a suspension of ascospores, or seedlings can be naturally infected by growing them close to mature mycorrhizal trees where infection occurs by mycelial contact. Despite this, truffle yields have fallen continuously, with about 1000 tons harvested annually in France around the year 1900 but less than one-tenth of that yield collected 100 years later (Hall *et al.*, 2003). A method to cultivate truffles on a large scale under axenic conditions would be a marvellous achievement, but this is not yet in sight.

The literature on truffles, stimulated by their gourmet and high commercial value, is enormous. Guides to identification have been provided by Gross (1987), Pegler *et al.* (1993), Riousset *et al.* (2001), and a computer-based interactive key by Zambonelli *et al.* (2000). More general accounts of truffle biology and cultivation have been written by Delmas (1978), Hall *et al.* (1994) and Callot (1999).

The 'desert truffles', *Terfezia* spp., are not closely related to *Tuber* but may instead have an affinity with the Pezizaceae (Percudani *et al.*, 1999). *Terfezia* occurs as a mycorrhizal associate of shrubs in arid regions of Southern Europe and the Middle East, where it is consumed as food and traded on markets. The association of *Terfezia* with the roots of shrubs such as *Helianthemum almeriense* can greatly improve the ability of the plant to withstand drought stress and may play an important role in mediterranean ecosystems (Morte *et al.*, 2000).

14.8 | *Morchella* (Morchellaceae)

The fruit bodies of *Morchella* spp., the true morels, are among the most popular and highly prized edible fungi (Plate 6g). They appear for a few weeks in spring in cold-temperate regions soon after snow-melt as the soil becomes warmer and drier, but they are not confined to such areas. *Morchella* spp. have two ecological strategies; as saprotrophic ruderals, fruiting for a relatively short period (a few years) on disturbed or burnt ground, or in mycorrhizal association with tree roots, fruiting over a longer period. The validity of the alternative lifestyles has been confirmed by comparative analyses of the isotopes ^{15}N and ^{13}C, showing some populations to be saprotrophs whilst others are mycorrhizal (Hobbie *et al.*, 2001). Opinions on taxonomy vary, with some authors recognizing about 50 species and others as few as 3–5 species showing wide phenotypic variation. Three broad groups of species have been distinguished, i.e. the half-free morel (*M. semilibera*), the black morels (*M. elata*, *M. conica* and *M. angusticeps*), and the common or yellow morels (*M. esculenta*, *M. crassipes* and *M. deliciosa*). Molecular analysis indicates that the black morels and yellow morels are separate taxonomic groups (Bunyard *et al.*, 1995; Gessner, 1995). The ascoma of a *Morchella* consists of a hollow stipe and a fertile cap thrown into shallow cup-like depressions or **alveoli**. The alveoli are lined by asci and paraphyses but the ridges or ribs which separate the alveoli are sterile, containing only paraphyses (Janex-Favre *et al.*, 1998).

The cylindrical, unitunicate, operculate asci contain eight unicellular ascospores. Karyogamy occurs prior to ascus formation, but croziers are apparently absent. Following meiosis in the ascus, there are four successive mitoses so that the ascospores are multinucleate (Volk & Leonard, 1990). The tips of the asci are phototropic and are directed towards the opening of the alveolus. The ascospores are often discharged simultaneously by puffing, generating air currents which carry clouds of spores well away from the fruit body (Buller, 1934). They germinate soon after discharge to form a septate mycelium with multinucleate segments. Frequent anastomosis may result in the formation of heterokaryons. Later in the season sclerotia develop, and it is in this form that the fungus survives the winter. Ascospores themselves do not remain viable in the soil for very long (Schmidt, 1983). The mycelium may also develop 'muffs' around the roots of various hosts, mostly young trees, and within such muffs the mycelium may penetrate as far as the phloem, an association which is probably non-mycorrhizal (Buscot & Roux, 1987; Buscot, 1989). However, in association with spruce (*Picea abies*) rootlets already infected with basidiomycetous mycorrhizal fungi, the mycelium of *Morchella* is weakly mycorrhizal, forming a Hartig net of intercellular mycelium around one layer of cortical cells of the host (Buscot & Kottke, 1990). Several morphologically distinctive types of such secondary ectomycorrhiza have been observed, often in association with endobacteria which invade the hyphae of *M. elata* making up the Hartig net and also the host plant cells (Buscot, 1994). Pure culture synthesis of sheathing mycorrhizae between *Morchella* spp. and four species of Pinaceae has been reported (Dahlstrom *et al.*, 2000).

The mating system of *Morchella* is not fully understood, and it is as yet uncertain whether it is homo- or heterothallic. A conidial state, *Costantinella cristata*, has been reported to develop following ascospore germination of *M. esculenta*. The conidiogenous cells (phialides?) are formed in verticils arising from lateral branches of the erect conidiophores. They give rise to minute spherical conidia which do not germinate readily (Costantin, 1936). Possibly they function as spermatia. When certain mycelial isolates derived from single ascospores are confronted in pure culture, a barrage phenomenon occurs (Hervey *et al.*, 1978), and this may be due to heterokaryon incompatibility (Volk & Leonard, 1989). The assumption that sexual reproduction occurs, in the sense that different parental genomes are involved in ascocarp formation, is supported by electrophoretic data confirming that allelic variation exists in natural populations. It is also possible that the fertilization events leading to the production of different asci within one fruit body may involve several individuals (Gessner *et al.*, 1987).

The possibility of cultivating morel ascocarps commercially is being explored and patents have been taken out to protect the techniques involved. They are based on the observations by Ower (1982) that sclerotial development followed by ascocarp differentiation can be encouraged by growth of the mycelium on sterilized wheat grain.

Hymenoascomycetes: Helotiales (inoperculate discomycetes)

15.1 | Introduction

In contrast to the Pezizales (see preceding chapter) which produce apothecia with asci discharging their spores through a detachable lid at their apex, the asci of inoperculate discomycetes liberate their spores either through a valve or a slit. In the inoperculate as well as operculate discomycetes, the asci may contain two or more layers, i.e. they are often described as bitunicate. However, these layers do not separate during ascus discharge, i.e. they are non-fissitunicate. Fine structural details of the asci of Helotiales have been described by Verkley (1993, 1994, 1996). Two large ecological groups of inoperculate discomycetes can be distinguished: the lichenized and non-lichenized species. This feature correlates approximately with the taxonomy at the level of orders, and here we shall discuss the Helotiales (sometimes alternatively called Leotiales) which contain mostly non-lichenized fungi. The Lecanorales and other orders with mainly or exclusively lichen-forming fungi are described in Chapter 16.

Those relatively few phylogenetic studies that have so far been performed on the Helotiales lack the necessary power of resolution to delimit natural groups. Thus, it is not clear at present whether this order is monophyletic or not, and several different classification schemes are in use (Gernandt *et al.*, 2001; Kirk *et al.*, 2001; Pfister & Kimbrough, 2001). The families currently associated with the Helotiales are listed in Table 15.1. Thus circumscribed, the order Helotiales contains some 2300 species. The Orbiliaceae, formerly included here (Pfister, 1997), are now considered to be more closely related to the Pezizales. Since the conidial forms of some of them are of interest as nematode-trapping and aquatic fungi, they will be considered in Chapter 25.

Fungi belonging to the Helotiales have adapted to several different ecological situations. Many species are necrotrophic, hemibiotrophic or biotrophic plant pathogens, and some can cause considerable damage in economically important crops. Other species are saprotrophic, colonizing dead leaves and shoots of herbaceous and woody plants. Endophytic species are also known, and it would not be too surprising if many of the inoperculate discomycetes known as saprotrophs were found to be already present in the living plant as endophytes. Some species fruit on plant debris submerged in freshwater streams and have adapted to this habitat by producing conidia of unusual shapes (Chapter 25). Other species form ericoid mycorrhizal associations with the roots of Ericaceae (p. 442). The Thelebolaceae are a group of coprophilous species.

15.2 | Sclerotiniaceae

The Sclerotiniaceae and Rutstroemiaceae are closely related but can be separated by

Table 15.1. | The families currently placed in or near the Helotiales. Data from Gernandt *et al.* (2001), Kirk *et al.* (2001) and Pfister and Kimbrough (2001). Three families printed in bold are considered in more detail in this chapter.

Family	No. of species	Examples
Ascocorticiaceae	3	
Bulgariaceae	1	*Bulgaria inquinans*
Cudoniaceae	10	
Cyttariaceae	11	*Cyttaria*
Dermateaceae (p. 439)	385	*Mollisia, Pyrenopeziza, Rhynchosporium, Tapesia*
Geoglossaceae	48	*Geoglossum, Trichoglossum*
Helotiaceae	623	*Ascocoryne, Hymenoscyphus, Neobulgaria*
Hemiphacidiaceae	12	
Hyaloscyphaceae	541	*Hyaloscypha, Lachnum* (= *Dasyscyphus*)
Leotiaceae	13	*Leotia*
Loramycetaceae	2	
Phacidiaceae	3	*Phacidium*
Rhytismataceae (p. 440)	219	*Rhytisma*
Rutstroemiaceae	100	*Rutstroemia*
Sclerotiniaceae (p. 429)	124	*Sclerotinia*
Thelebolaceae	15	*Thelebolus*
Vibrisseaceae	14	

DNA-based phylogenetic analyses (Holst-Jensen *et al.*, 1997). Members of both families produce stalked apothecia which grow from **stromata** located within the colonized host plant tissue. The apothecia usually develop in spring from overwintered stromata. The stroma is a food storage organ and is usually differentiated into two parts, a rind (cortex) of dark, thick-walled cells and a medulla of hyaline cells. Two generalized types of stroma have been distinguished. To quote from Whetzel (1945),

> The **sclerotial stroma** (commonly called the **sclerotium**) has a more or less characteristic form and a strictly hyphal structure under the natural conditions of its development. While elements of the substrate may be embedded in its medulla, they occur there only incidentally and do not constitute part of the reserve food supply. The **substratal stroma** is of a diffuse or indefinite form, its medulla being composed of a loose hyphal weft or network permeating and preserving as a food supply a portion of the suscept or other substrate (e.g. culture media).

It is now clear that the sclerotial stroma is typical of the Sclerotiniaceae where it is often conspicuous (Plates 7a,b). Fungi belonging to the Rutstroemiaceae (Plate 7c) produce the less obvious substratal stroma. The Rutstroemiaceae grow mainly as saprotrophs, but the Sclerotiniaceae include some important plant-pathogenic species. We shall only consider the latter family further because much more is known about it.

Various types of macroconidia with or without accompanying microconidia are formed within the genus *Sclerotinia* in its widest sense, and species with different types of conidia are regarded by many mycologists as belonging to distinct genera. Since these are very closely related, *Sclerotinia* is a good example to illustrate the flexibility of asexual reproduction in fungi (Weber & Webster, 2003). For instance, *Sclerotinia fuckeliana* has *Botrytis cinerea* with polyblastic conidia as its asexual state (Figs. 15.5a,b) and is thus currently called *Botryotinia fuckeliana*. It also produces microconidia from clustered phialides (Fig. 15.5c). *Sclerotinia* (*Monilinia*) *fructigena* produces *Monilia*-type blastoconidia in chains (Fig. 15.3b) and lacks a microconidial state, whereas *S.* (*Myriosclerotinia*) *curreyana* produces only *Myrioconium*-like microconidia (Figs. 15.1d–f) but has no macroconidial

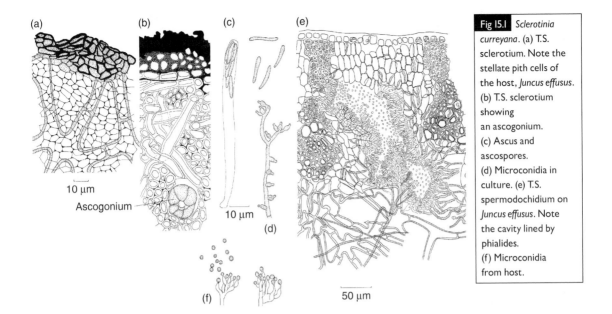

Fig 15.1 *Sclerotinia curreyana.* (a) T.S. sclerotium. Note the stellate pith cells of the host, *Juncus effusus*. (b) T.S. sclerotium showing an ascogonium. (c) Ascus and ascospores. (d) Microconidia in culture. (e) T.S. spermodochidium on *Juncus effusus*. Note the cavity lined by phialides. (f) Microconidia from host.

state. All of these species can produce apothecia. This is also true of *Sclerotinia sclerotiorum*, which is considered to represent *Sclerotinia sensu stricto* (Kohn, 1979). *Sclerotium cepivorum* produces neither functional conidia nor apothecia, and the sclerotia function purely as vegetative propagules, germinating by hyphal growth. Microconidia are sometimes produced by germinating sclerotia, but these do not appear to have any function. The relationship of *S. cepivorum* with the Sclerotiniaceae has been deduced from DNA-based studies (Carbone & Kohn, 1993). Whereas macroconidia generally germinate readily and play important roles in the spread of diseases, microconidia may or may not germinate and are considered to function mainly as **spermatia**, i.e. agents of fertilization in sexual reproduction.

15.2.1 *Sclerotinia curreyana* and *S. tuberosa*

The apothecia of *Sclerotinia* (*Myriosclerotinia*) *curreyana*, a pathogen of the rush *Juncus effusus*, are common in May. They arise from black sclerotia in the pith at the base of the *Juncus* stem (Plates 7a,b). Infected stems look paler than healthy stems, and by feeling down to the base of an infected stem the sclerotium can be felt as a swelling between finger and thumb. The sclerotium has an outer layer of dark cells and a pink interior which includes some of the stellate pith cells of the host (Fig. 15.1a; Plate 7a). One or several apothecia may grow from a single sclerotium. The ascospores are released in late spring and infect the new season's stems. In culture, germinated ascospores form a mycelium which produces microconidia from small phialides (Fig. 15.1d). Similar clusters of microconidia can be found on infected *Juncus* later in the season (Fig. 15.1f) where they line cavities beneath the epidermis in the upper part of infected culms. Whetzel (1946) has used the term **spermodochidium** for these microconidial fructifications (Fig. 15.1e). It is probable that microconidia play a role in fertilization.

The apothecia of *S.* (*Dumontinia*) *tuberosa* (Fig. 15.2) are about 2 cm in diameter and arise from sclerotia within rhizomes of *Anemone nemorosa* (Pepin, 1980). They may also occur on garden *Anemone* where they are associated with black rot disease. Microconidia are formed in culture. Electron microscopy studies of the ascus wall show that it has a two-layered wall, but the two layers do not separate from each other, i.e. the ascus is non-fissitunicate. The ascus apex contains a thickened dome of wall material with a central canal. As the ascus explodes, the apical apparatus is everted (Verkley, 1993).

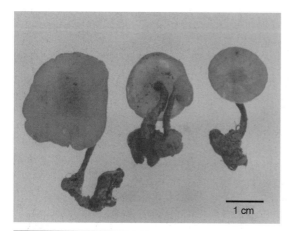

Fig 15.2 Apothecia of *Sclerotinia tuberosa* rising from subterranean sclerotia formed on rhizomes of *Anemone nemorosa*.

15.2.2 *Monilinia fructigena* and *M. laxa*

Monilinia fructigena is the cause of a brown fruit rot of apples, pears, plums and other stone fruits (Byrde & Willetts, 1977). Although the apothecial state is only rarely formed, the disease is common and is transmitted by means of conidia. Apples and pears showing brown rot bear buff-coloured pustules of conidia often in concentric zones (Fig. 15.3a). Sporulation is stimulated by light, and adjacent zones correspond to daily periods of illumination. The conidia are blasto-conidia formed in chains which extend in length at their apices by budding of the terminal conidium. Occasionally more than one bud is formed, and this results in branched chains (Fig. 15.3b). Conidiogenesis of this type is characteristic of the anamorphic genus *Monilia*. Infection of the fruit is commonly through wounds caused mechanically or by insects such as codling moth, wasps or earwigs (Croxall *et al.*, 1951; Xu & Robinson, 2000). Fruits left lying on the ground are the source of infection in the following season. During the winter, infected fruits become mummified, and the shrivelled fruit thoroughly colonized by mycelium is interpreted as the sclerotium. In the following year the sclerotium may produce further conidial pustules. Infections can develop as a post-harvest disease in stored apples, and in some varieties a twig infection (spur canker) may also occur.

A similar group of diseases of apple and plum is caused by *Monilinia laxa* which also has a *Monilia* conidial state. In addition to fruit rot, this species causes blossom and shoot blight, in which infected fresh shoots wilt and become coated by conidial pustules. *Monilinia fructicola* causes brown rot especially of peaches and nectarines in North and South America, South Africa, Australia and the Far East, but has not been reported from Europe. It produces the apothecial state more readily than the other species (Holtz *et al.*, 1998), and ascospores released from overwintered mummified fruits can be a source of inoculum in the field (Tate & Wood, 2000). These and a fourth species of the brown fruit rot complex, *M. polystroma*, can be distinguished by means of morphological features and DNA sequences (van Leeuwen *et al.*, 2002).

15.2.3 *Sclerotinia sclerotiorum*

This species causes a range of diseases (*Sclerotinia* rot, white mould, stalk break) in over 400 cultivated and wild plant species belonging to some 75 different families (Boland & Hall, 1994). The most important crop plants affected are sunflower, soybean and oilseed rape, with crop losses approaching 100% under conditions favourable to the disease. Sclerotia are formed on decaying crop debris and remain viable in a dormant state in the soil for many years, especially if deep-ploughed. Sclerotia located within the top 3 cm of soil germinate to produce hyphae or apothecia in spring. Plants can be infected either from mycelium or from ascospores; there is no macroconidial state. Phialidic microconidia are formed and probably serve as spermatia. Infection by *S. sclerotiorum* is often initiated by a saprotrophic phase on dead leaves or petals during which mycelial biomass is generated, prior to the attack on the living plant tissue. Above-ground shoots and, to a lesser extent, roots can be infected, and infections of living tissues are strongly necrotrophic. This necrotrophic phase is followed by further saprotrophic growth and the formation of sclerotia. The infection cycle in *S. sclerotiorum* is therefore tri-phasic. Reviews of the general biology and pathology of *S. sclerotiorum* have been written by

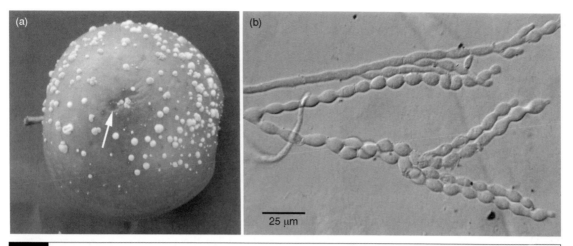

Fig 15.3 *Monilinia fructigena*. (a) Apple showing brown rot caused by this fungus, and bearing conidial pustules. The wound serving as entry point is indicated by an arrow. (b) Blastoconidia of the *Monilia* type.

Purdy (1979), Hegedus and Rimmer (2005) and Bolton *et al.* (2006).

Sclerotinia sclerotiorum can infect pollen grains of crop plants but can also become genuinely seed-borne, surviving systemically in the embryo for several years and replacing the rotten tissues with mycelium and sclerotia (Tu, 1988). Such seed-borne infections are readily eliminated by fungicidal seed-dressings, but infections of growing crops are less easily controlled. Hence, biological control of *S. sclerotiorum* has been attempted. A promising biocontrol agent is the pycnidial fungus *Coniothyrium minitans* which infects and parasitizes hyphae and sclerotia of *S. sclerotiorum* in the soil and in plant tissue (Tribe, 1957; de Vrije *et al.*, 2001). The ability of *C. minitans* to destroy dormant *S. sclerotiorum* sclerotia in the soil is particularly interesting, as it offers a chance to decontaminate infected soil on which susceptible crop plants could not otherwise be grown for several years. Gerlagh *et al.* (2003) have demonstrated that one or two conidia of *C. minitans* are sufficient to initiate infection of a sclerotium. Conidia of *C. minitans* can be spread rapidly by the activity of soil invertebrates, including mites (Williams *et al.*, 1998). These properties have led to the registration of *C. minitans* as a commercial biocontrol agent against *S. sclerotiorum*. Since *C. minitans* colonizing the soil can tolerate many fungicides used against *S. sclerotiorum*, the integrated

control of *Sclerotinia* rot is also possible in some crops (Budge & Whipps, 2001).

Oxalic acid and pH regulation

Like other fungi such as *Botrytis cinerea* (see p. 435) and brown-rot basidiomycetes (p. 527), *S. sclerotiorum* releases large amounts of oxalic acid into the infected plant tissue, and this is an important pathogenicity factor (Godoy *et al.*, 1990). Oxalic acid may chelate Ca^{2+} ions released from cell wall degradation, and it also suppresses the host's hypersensitive response (Cessna *et al.*, 2000). Most importantly, however, it acidifies the infected plant tissue. There is good evidence that *S. sclerotiorum* can sense the pH of its environment, and that it can adjust the production rate of oxalic acid accordingly. In this way, optimum conditions are created for the activity of its pectin-degrading enzymes, especially endopolygalacturonases, which macerate colonized host tissues (Rollins & Dickman, 2001). A transcription factor encoded by the *pac1* gene seems to be involved in regulating the expression of genes controlled by external pH (Rollins, 2003).

Transgenic sunflower or soybean plants containing an oxalate oxidase gene from cereals show good resistance to infection by *S. sclerotiorum*. This appears to be a promising control strategy for the future, but is currently still slow in gaining public acceptance (Lu, 2003). Conventional resistance breeding is also possible, although resistance is not usually due to major genes and is only partial. The underlying principle in runner bean (*Phaseolus vulgaris*) seems to be an enhanced tolerance of oxalic acid or a restriction of its diffusion through the infected tissue (Tu, 1985).

Sclerotia

Sclerotia of *S. sclerotiorum* form readily in culture and have been the subject of investigations into the physiology of their development (Willetts & Wong, 1980; Willetts & Bullock, 1992). They are of the terminal type. General aspects of sclerotial development have been summarized on pp. 18–21. The regulation of sclerotium formation is interesting because it, too, is stimulated by acid pH, and the Pac1 transcription factor is involved (Rollins, 2003). Another signal known to be a trigger of sclerotium development is oxidative stress, e.g. lipid peroxidation or irradiation with light. Excessive oxidation is prevented by the synthesis of antioxidants such as β-carotene or ascorbic acid (vitamin C), and if high concentrations of these are added to cultures of *S. sclerotiorum*, sclerotium formation is inhibited (Georgiou & Petropoulou, 2001; Georgiou *et al.*, 2001).

15.2.4   *Sclerotium cepivorum*

This fungus causes white rot, the most serious disease, of *Allium* spp., especially onions and garlic. Sclerotia germinate by emitting hyphae which grow towards the roots of the host plant and cause necrotrophic infections with maceration of the root and bulb base tissue. Large quantities of various pectinolytic enzymes are secreted (Metcalf & Wilson, 1999). New sclerotia are formed on and in the decaying bulb tissues. Sclerotia of *S. cepivorum* can survive in the soil for many years or even decades (Coley-Smith, 1959), and as little as one sclerotium per kg of soil can cause serious disease losses (Crowe *et al.*, 1980). Apart from soil fumigation, no effective treatment of infections or contaminated soil is available (but see below), and fields may need to be abandoned for *Allium* cultivation once *S. cepivorum* has become established.

The root exudates of *Allium* spp. have long been known to trigger germination of sclerotia of *S. cepivorum*. Substances such as alkyl-cysteine sulphoxides are themselves inactive but are probably metabolized by soil microbes to release volatile compounds which act as the stimulants (Coley-Smith & King, 1969; King & Coley-Smith, 1969). One such substance, which is also produced directly by *Allium* spp., is diallyl disulphide (Storsberg *et al.*, 2003). If this is sprayed onto an infested field, it will trigger the germination of sclerotia which is followed by their death if no host plants are present. This idea seems to hold potential for the control of *S. cepivorum* (Coley-Smith, 1986; Coley-Smith & Parfitt, 1986). Garlic powder worked into the soil seems to have similar effects, most probably because of the release of volatile substances from the non-volatile water-soluble alkyl cysteine sulphonates, catalysed by soil bacteria (Fig. 15.4). Biological control using *Trichoderma* spp., which secrete chitinases capable of lysing hyphae of *S. cepivorum*, may also be possible (Metcalf & Wilson, 2001).

15.2.5   The life cycle of *Botryotinia* (*Sclerotinia*) *fuckeliana*, anamorph *Botrytis cinerea*

Because the apothecia of *B. fuckeliana* are not commonly seen, the fungus is better known by its macroconidial state, *Botrytis cinerea*. This is ubiquitous on all kinds of moribund plant material and is also associated with a wide range of diseases often referred to as grey mould. The name *Botrytis cinerea* is now known to be a collective name used to describe a number of closely similar, but genetically distinct, strains (Giraud *et al.*, 1999; Beever & Weeds, 2004). For this reason some authors prefer to write of a *Botrytis* of the *cinerea* type. In-depth treatments of the biology of *Botrytis cinerea* have been compiled by Coley-Smith *et al.* (1980) and Elad *et al.* (2004). In addition to *B. cinerea*, there are some 20 other *Botrytis* spp. causing diseases on a wide range of host plants (Staats *et al.*, 2005).

Macroconidia of *B. cinerea* are formed on infected host tissue from dark-coloured branched conidiophores. The tips of the branches are

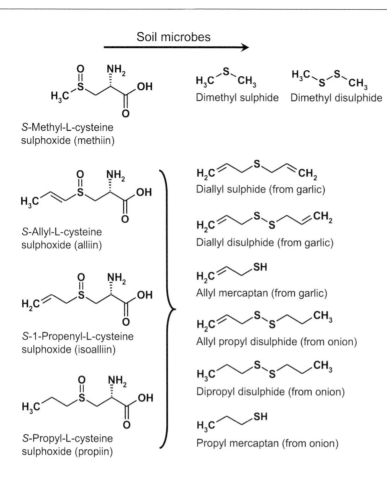

Soil microbes →

S-Methyl-L-cysteine
sulphoxide (methiin)

Dimethyl sulphide   Dimethyl disulphide

S-Allyl-L-cysteine
sulphoxide (alliin)

Diallyl sulphide (from garlic)

Diallyl disulphide (from garlic)

Allyl mercaptan (from garlic)

S-1-Propenyl-L-cysteine
sulphoxide (isoalliin)

Allyl propyl disulphide (from onion)

Dipropyl disulphide (from onion)

S-Propyl-L-cysteine
sulphoxide (propiin)

Propyl mercaptan (from onion)

Fig 15.4 Non-volatile water-soluble *Allium* metabolites (left) and their volatile breakdown products (right). S-Methyl-L-cysteine sulphoxide (methiin) is common in many plants, and its sulphide breakdown products do not trigger sclerotium germination in *Sclerotium cepivorum*. Alliin (S-2-propenyl-L-cysteine sulphoxide = S-allyl-L-cysteine sulphoxide), isoalliin and propiin are typical of members of the genus *Allium*. Their volatile breakdown products, especially mercaptans, sulphides and disulphides, are potent triggers of sclerotium germination. Diallyl disulphide is the major flavour component of garlic.

thin-walled and bud out to form numerous elliptical multinucleate conidia which are blastospores (Hughes, 1953). These are easily detached by the wind, or are thrown off as the conidiophores twist hygroscopically (Figs. 15.5a,b). Conidia can also be dispersed by the fruitfly *Drosophila melanogaster* (Louis *et al.*, 1996) and other insect vectors. Uninucleate microconidia are formed by clusters of phialides which arise directly from the mycelium (Weber & Webster, 2003) or from germinating macroconidia (Fig. 15.5c). The microconidia have been claimed to be capable of germination (Brierley, 1918) but do not do so in our experience. They are probably mainly involved in sexual reproduction, i.e. they function as spermatia. Sclerotia are formed at the surface of infected tissues and the fungus overwinters in this form. In spring the sclerotia may develop to give rise to tufts of macroconidia or, much less commonly, to apothecia. One or several stalked apothecia may

arise from one sclerotium, with the stalk 1 cm or more in length and the apothecial disc a few millimeters in diameter. *Botryotinia fuckeliana* is heterothallic with a bipolar mating system. In a single-ascospore culture, macroconidia, microconidia and sclerotia can be formed on the same agar plate (Weber & Webster, 2003), but apothecia never develop. However, apothecia will form if microconidia of one mating type are applied to sclerotia of the opposite mating type (Faretra *et al.*, 1988). Like the great majority of ascomycetes (Bistis, 1998), *B. fuckeliana* therefore shows **physiological heterothallism**. The life cycle of this fungus is summarized in Fig. 15.6.

15.2.6 Other life cycles in *Sclerotinia*

A deviation from the typical ascomycete life cycle of *B. cinerea* (Fig. 15.6) is found in *Sclerotinia* (*Stromatinia*) *narcissi* (Drayton & Groves, 1952). Of the eight spores formed in the asci of this fungus, four germinate to produce mycelia

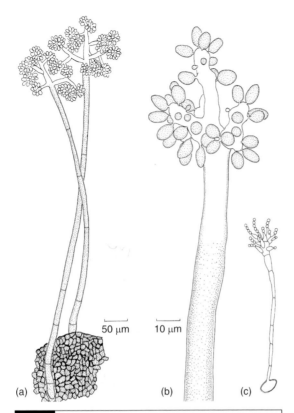

Fig 15.5 *Botrytis cinerea*. (a) Conidiophores developing from a sclerotium. (b) Apex of conidiophore showing origin of conidia as blastospores. (c) Conidium germinating to produce phialides and microconidia (after Brierley, 1918).

bearing microconidia but no sclerotia, whilst the other four produce mycelia bearing sclerotial stromata. Apothecia develop on the strains forming sclerotia if microconidia are transferred to them. Thus the mating behaviour of *S. narcissi* differs from that of *B. fuckeliana* and we can say that *S. narcissi* is sexually dimorphic. This type of behaviour is not common in ascomycetes and it is possible that such an incompatibility system has been derived from the more usual system exemplified by *B. fuckeliana* by aberrations which prevent the normal sequence of development of sexual organs in basically hermaphrodite forms (Raper, 1959).

A somewhat related phenomenon is found in *Sclerotinia trifoliorum*, in which each ascus contains ascospores of two different sizes. The four large ascospores germinate to give rise to homothallic (self-fertile) mycelia, whereas the four smaller ascospores produce self-sterile mycelia (Uhm & Fujii, 1983a,b). As discussed earlier in detail for *Saccharomyces cerevisiae* (see Fig. 10.5), the two mating type alleles of ascomycetes differ strongly in the genes they encode and are thus termed idiomorphs. Although no detailed studies seem to have been carried out on the Sclerotiniaceae, in other filamentous ascomycetes such as *Pyrenopeziza brassicae* (see p. 439), heterothallic strains carry either one or the other of the two idiomorphs whereas in homothallic species both are fused together and expressed simultaneously. In *Sclerotinia trifoliorum*, the formation of small ascospores may be preceded by a unidirectional switch from homothallic to heterothallic, i.e. the deletion of one of the two mating types during meiosis (Harrington & McNew, 1997).

Yet another kind of mating behaviour is seen in *Sclerotinia sclerotiorum*, which is homothallic. A single ascospore culture produces microconidia and sclerotia which bear ascogonial coils beneath the rind. The transfer of microconidia to the sclerotia on the same mycelium results in the formation of apothecia (Drayton & Groves, 1952). A similar process of self-fertilization also occurs in *Sclerotinia* (*Botryotinia*) *porri*.

15.2.7 Pathogenicity of *Botrytis cinerea*

Botrytis cinerea is pathogenic on over 200 species of plants. Serious diseases of crops are grey mould of lettuce, tomato, strawberry and raspberry, die-back of gooseberry and damping-off of conifer seedlings. A special case is bunch rot of grapes. Under normal circumstances, infected grapes shrivel and ultimately fall to the ground, forming mummies in which the fungus can survive the winter. Mummies give rise to infective macroconidia in the following spring. This type of bunch rot causes serious crop losses in both white and red grapes. Under certain circumstances and with certain grape varieties, however, *Botrytis* causes the 'noble rot' in which infections take a milder course and allow the grape to dry out gently, concentrating its sugar and flavours in the process. The resulting wine is much sweeter and richer than normal table wine and is consumed

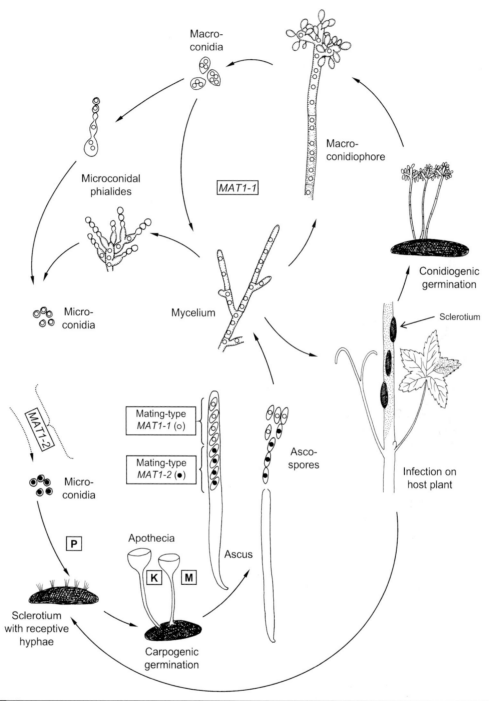

Fig 15.6 Life cycle of *Botryotinia fuckeliana* (anamorph *Botrytis cinerea*). In nature, this fungus overwinters by means of sclerotia which may germinate in either of two ways. Conidiogenic germination gives rise to the macroconidial state which can be formed also from mycelium. The blastic macroconidia are multinucleate, as are mycelial segments. Phialidic microconidia are formed from vegetative mycelium or from macroconidia. They are uninucleate, serving mainly as fertilizing agents for sclerotia of opposite mating type. Fertilization leads to carpogenic germination of a sclerotium, i.e. to the formation of apothecia, resulting in eight uninucleate ascospores of either of two mating types. The diploid state is confined to the tip of the ascogenous hypha (not shown; see Fig. 8.10). Open and closed circles represent haploid nuclei of opposite mating type. Key events in the life cycle are plasmogamy (P), karyogamy (K) and meiosis (M).

as a dessert wine. Probably the most famous example is produced in the Sauternes region of France, using the thin-skinned Sémillon grape which is particularly susceptible to *B. cinerea*.

Botrytis cinerea can kill its host's tissue rapidly and then carries on growing on the dead remains. It is thus a classical necrotrophic pathogen. Several research groups have examined factors which may be involved in the pathogenicity of *B. cinerea*, but it is as yet impossible to say which ones are the most important. Quite possibly *B. cinerea* employs different strategies for the colonization and killing of different hosts. This subject has been reviewed by Prins *et al.* (2000) and Kars and van Kan (2004) and is summarized below.

Attachment

Macroconidia of *B. cinerea* have a hydrophobic surface, but this is apparently not due to the presence of hydrophobin-type proteins (Doss *et al.*, 1997). Initial attachment of the macroconidium to the host surface is by weak hydrophobic interactions. When the germ tube emerges, it secretes a polysaccharide-based matrix which acts as a much stronger glue (Doss *et al.*, 1995). This polysaccharide may be the same as cinerean, a β-(1,3)-glucan with frequent β-(1,6) cross-linkages which is produced by *B. cinerea* from excess glucose in liquid culture and in infected grapes (Dubourdieu *et al.*, 1978b; Monschau *et al.*, 1997). When free glucose becomes scarce, cinerean is hydrolysed again by extracellular glucanases (Stahmann *et al.*, 1993). On the plant surface, the glucan matrix may thus serve in attachment, as an external carbohydrate reservoir, and as a matrix for hydrolytic enzymes (Doss, 1999).

Lytic enzymes

After a short period of growth, the germ tube terminates in a slightly swollen infection structure which may be considered a rudimentary appressorium. This is non-melanized, and thus penetration of the cuticle is probably mediated mainly by lytic enzymes rather than turgor pressure (see pp. 381 and 395). Cutin-degrading enzymes are secreted by *B. cinerea* during the initial infection stages (Comménil *et al.*, 1998), and proteases may also play a role in pathogenesis (Movahedi & Heale,

1990). Later, a battery of cell wall-degrading enzymes (especially pectinolytic enzymes) is produced during the colonization of the host tissue beyond the initial necrotic lesion. Pectin seems to be a major carbohydrate source for *B. cinerea* (Prins *et al.*, 2000). The degradation of pectin from the middle lamella may also be a contributing factor to host cell death (Tribe, 1955) and causes rapid and widespread maceration of host tissue (Kapat *et al.*, 1998; Kars & van Kan, 2004), which is typical of the necrotrophic appearance of *B. cinerea* infections. Oxalic acid is secreted by *B. cinerea* as it is by many other fungi, and its presence is also correlated with tissue necrosis. However, rather than acting directly as a toxin, it is more likely to enhance the activity of the pectinolytic enzymes which have an acidic pH optimum, and to chelate Ca^{2+} ions (Prins *et al.*, 2000). Substantial quantities of Ca^{2+} ions can be released during pectin degradation from the carboxylic acid groups of the monomers, galacturonic acid, which often form calcium salts.

Hypersensitive response

Biotrophic pathogens such as downy or powdery mildews or rust fungi fail to infect incompatible host plants because these recognize their presence. One important mechanism of defence is the hypersensitive response (see pp. 115 and 397) in which epidermal cells in the vicinity of the infection site undergo programmed cell death (Mayer *et al.*, 2001). The hypersensitive response is accompanied by an 'oxidative burst' followed by the synthesis of phytoalexins. With biotrophic pathogens which require living host cells for their nutrition, the hypersensitive response is often sufficient to kill the infection unit. If the necrotrophic *B. cinerea* attempts to infect a host plant, the hypersensitive response also takes place, but it fails to control the infection because *B. cinerea* can exploit the dead cells for nutrition and initial growth (Govrin & Levine, 2000). The reactive oxygen intermediates (especially superoxide and H_2O_2) released during the oxidative burst may be detoxified by the enzymes superoxide dismutase and catalase, respectively, which are secreted by *B. cinerea* and are probably localized in the glucan matrix surrounding the infection hypha (Gil-ad *et al.*, 2001).

Further, *B. cinerea* is known to produce laccase and other enzymes which can degrade or detoxify phytoalexins (Prins *et al.*, 2000). ABC transporters capable of excluding phytoalexins from the hyphal cytoplasm have also been reported from *B. cinerea* (Schoonbeek *et al.*, 2001; see also p. 278). Hence, Govrin and Levine (2000) have suggested that the hypersensitive response launched by the host actually facilitates, rather than represses, infection by *B. cinerea*.

Fungicide resistance

Although biological control strategies against *B. cinerea* are being attempted, especially in the greenhouse and in post-harvest storage of certain fruit crops, control in agricultural situations relies chiefly on the application of fungicides. This is especially the case for the control of grey mould on grapevines. *Botrytis cinerea* has developed resistance against almost all fungicides in current use, and this may be due to several factors, e.g. the occurrence of sexual reproduction in the field, the existence of at least two genetically distinct 'species', and the presence and spread of transposable genetic elements in one of them (Giraud *et al.*, 1999). All of these factors enhance the genetic diversity of populations of the pathogen, and thus the chances of development of fungicide resistance. Mechanisms of resistance of *B. cinerea* to fungicides have been discussed by Leroux *et al.* (2002) and seem to involve strategies also described from other fungi, i.e. reduced fungicide penetration into or enhanced export from the hyphae by means of ABC transporters, enzyme-mediated detoxification and degradation of fungicides, and mutations leading to a reduced binding of the fungicide to its modified target protein.

15.3 | Dermateaceae

This family (385 species) is almost certainly polyphyletic and it will take some time and numerous further name changes before the genera are circumscribed to the phylogeneticists' satisfaction. The species included here produce their apothecia directly on the substratum. Stromata are absent. The apothecia are small (less than 1 mm in diameter) and rather inconspicuous, being coloured in grey, brown or black tones. The development of apothecia has been described by Gilles *et al.* (2001) for *Pyrenopeziza brassicae* (Fig. 15.7). Apothecia are formed from hyphae aggregating into small globular structures resembling sclerotia or cleistothecia. Later a pore develops at the apex (Fig. 15.7a), and this increases in diameter by lateral expansion of the basal disc (Fig. 15.7b) Meanwhile the asci mature in the hymenium. Ultimately, a flat apothecium is formed which possesses a clearly defined margin typical of the Dermateaceae (Fig. 15.7c). This developmental pattern has been termed **hemiangiocarpic** by Corner (1929). The anamorphs of Dermateaceae are variable. One very common form (*Cadophora*) is *Phialophora*-like, i.e. the phialides bear an apical collarette (Harrington & McNew, 2003). Other forms do not have phialides, and instead long and transversely septate conidia are produced more or less directly from vegetative hyphae.

One large genus (*Mollisia*) is chiefly saprotrophic and forms apothecia on dead leaves and fallen twigs, as exemplified by the ubiquitous *Mollisia cinerea* which fruits on dead wood (Plate 7d). Other members of the family are hemibiotrophic plant pathogens causing limited lesions on agricultural crops. *Pyrenopeziza brassicae* (anamorph *Cylindrosporium concentricum*) causes light leaf spot on winter oilseed rape (Fig. 15.7) whereas *Tapesia yallundae* (anamorph *Pseudocercospora herpotrichoides*) is the cause of eyespot at the base of cereal stems, especially winter wheat, and its sister species, *T. acuformis*, causes a similar disease especially on rye. The conidial *Rhynchosporium secalis* is the agent of leaf blotch on a range of cereals. All of these pathogenic species are phylogenetically closely related (Goodwin, 2002).

15.3.1 *Tapesia yallundae* and *T. acuformis*

Apothecia have been found only recently for both *Tapesia* species (see Lucas *et al.*, 2000) and *Pyrenopeziza brassicae* (see Gilles *et al.*, 2001). They have not yet been found for *Rhynchosporium secalis*, although the high genetic diversity of field isolates of this species indicates that sexual reproduction should occur in nature

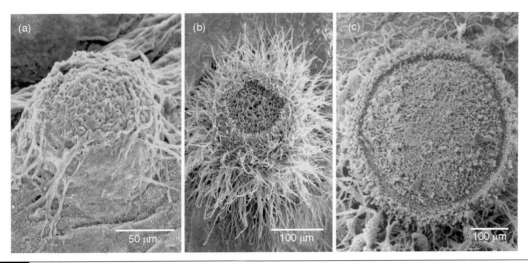

Fig 15.7 Development of *Pyrenopeziza brassicae* ascocarps. (a) Immature apothecium, 10 days old on oilseed rape. The apical pore has just formed. (b) 14-day-old apothecium on oilseed rape. The opening is widening due to the expansion of the basal disc. (c) Mature apothecium 46 days after inoculation onto agar. Reprinted from Gilles *et al.* (2001), with permission from Elsevier. Images kindly provided by N. Evans.

(Salamati *et al.*, 2000). The mating type idiomorphs have been characterized for all species except *Tapesia acuformis* (Foster & Fitt, 2004), and they are of the usual heterothallic/bipolar type.

The biology of the two sister species *Tapesia yallundae* on wheat and *T. acuformis* on rye (formerly called *T. yallundae* W and R pathotypes, respectively) is very similar and has been reviewed by Fitt *et al.* (1988) and Lucas *et al.* (2000). Eyespot is a major disease in winter cereals growing in cool climates. Infection is probably mainly by the needle-shaped conidia which are formed on overwintered stubble and spread by rain splash. However, ascospores released from apothecia (Fig. 15.8a) in early spring are also infectious. If a spore lands on the coleoptile of a host plant, it germinates and produces an aggregate of hyphae termed an **infection plaque** (Fig. 15.8b). Numerous melanized appressoria are formed at the interface of this structure with the host epidermis, so that infection of susceptible hosts occurs at several points (Fig. 15.8c). Penetration is probably mediated by a combination of turgor pressure and hydrolytic enzymes. The typical eyespot (Fig. 15.8d) develops as a greyish-brown lesion around clusters of infection plaques which may be visible as the 'pupil' of the eyespot. Detailed studies of infection mechanisms have been published by Daniels *et al.* (1991, 1995). The presence of eyespots at the haulm bases renders the cereal shoots prone to collapsing. Further infections can affect the vascular system, resulting in poorly developed 'whiteheads' containing inferior grain.

Resistance breeding seems to be a promising strategy for the control of eyespot in cereals (Lucas *et al.*, 2000). Chemical control is also practised, but *Tapesia* spp. have developed resistance against several types of fungicide (Leroux & Gredt, 1997).

15.4 | Rhytismataceae

The taxonomy of this family is still in a state of flux (Gernandt *et al.*, 2001). It is sometimes given ordinal status (Rhytismatales or Phacidiales). The apothecia are immersed in host tissue or embedded in a flat stroma. Individual apothecia become evident when the upper surface breaks open to reveal the hymenium. There are 219 species in this group at present (Kirk *et al.*, 2001). Most of them are associated with broad-leaved trees or conifers (Cannon & Minter, 1986; Johnston, 1997). Particularly difficult genera

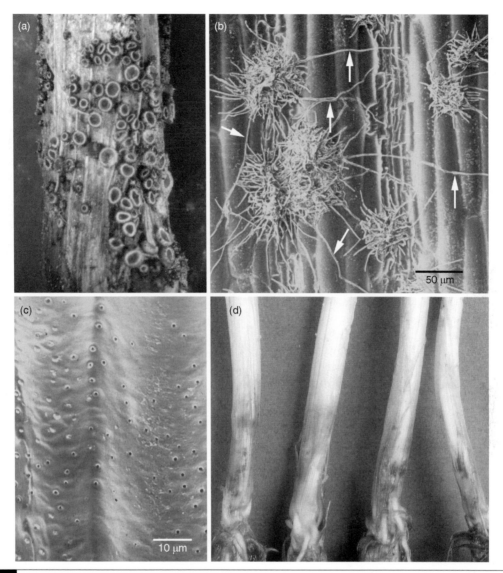

Fig 15.8 Infection biology of *Tapesia yallundae* (a,c,d) and *T. acuformis* (b). (a) Production of apothecia on overwintered wheat stubble. (b) SEM of infection plaques on a rye leaf. Runner hyphae (arrows) extend from established plaques onto the surrounding leaf surface. (c) SEM view of a wheat leaf after removal of an infection plaque. Penetration has occurred at numerous points. (d) Eyespot lesions at the stem bases of wheat plants. (a) kindly provided by P. S. Dyer. (b) and (c) reprinted from Daniels *et al.* (1991) with permission from Elsevier; original images of (b−d) kindly provided by J. A. Lucas.

in taxonomic terms are *Lophodermium* and *Lophodermiella* which cause needlecast diseases of *Pinus* spp. As with many other fungi, there is a gradient of interactions within the Rhytismataceae, ranging from the purely endophytic way of life (Deckert *et al.*, 2001) through saprotrophy to severely pathogenic species. Ortiz-García *et al.* (2003) have suggested that at least some pathogenic species have evolved from endophytic ancestors.

15.4.1 Rhytisma acerinum

Rhytisma acerinum is common on the leaves of sycamore, *Acer pseudoplatanus*, forming black shiny lesions (tar spots) about 1−2 cm wide (Fig. 15.9). The lesions arise from infections by ascospores released from apothecia on overwintered leaves. Lesions become visible to the naked eye in June or July, some 2 months after infection, as yellowish spots which eventually turn black. Sections of the leaf at this stage show an extensive mycelium filling the cells of the mesophyll, and especially the cells of the upper epidermis. Between the epidermal cells, a conidial state called *Melasmia acerina* develops. This consists of flask-shaped cavities (spermogonia) which give rise to uninucleate curved club-shaped conidia (spermatia) measuring about 6 × 1 μm (Figs. 15.10a,b). The spermatia are exuded from the upper surface of the centre of the lesion through ostioles in the spermogonial wall. The spermatia do not germinate, even on sycamore leaves, and it is believed that they play a sexual role (Jones, 1925), although this has not yet been proven. Apothecia begin development in the portion previously occupied by spermogonia, and the hymenium is roofed over by several layers of dark cells formed within the upper epidermis. The asci complete their development on the fallen leaves and are ripe about March to April when sycamore leaves of the new season unfold. The hymenium is exposed by means of cracks in the surface layer of the fungal stroma (Fig. 15.9b) and the asci discharge their spores, sometimes by puffing. Since the ascospores are very large, their discharge can be viewed with a dissecting microscope. Although the ascospores are only projected to a height of about 1 mm above the surface of the stroma, they are carried by air currents to leaves several metres above the ground. The ascospores are needle-shaped and have a mucilaginous epispore which is especially well developed at the upper end (Fig. 15.10d). This probably helps in attaching them to leaves. Infection occurs by penetration of the germ tubes through stomata on the lower epidermis.

Rhytisma acerinum is absent from densely populated areas, probably because the germination of ascospores is inhibited by sulphur dioxide. Greenhalgh and Bevan (1978) have suggested that the incidence and frequency of colonization of sycamore leaves by the tar spot fungus can be used as an accurate visual index of air pollution, although other interpretations are possible, such as the removal of fallen leaves from municipal parks or the drier microclimate in city centres (Leith & Fowler, 1988).

15.5 | Other representatives of the Helotiales

Many members of the Helotiales are encountered during fungus forays because they produce

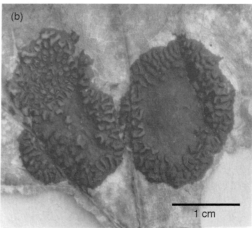

1 cm

Fig 15.9 *Rhytisma acerinum.* (a) Leaf of sycamore (*Acer pseudoplatanus*) with developing tar spot lesions. (b) Tar spot from an overwintered leaf showing cracking of the surface to reveal the hymenia of the apothecia. The flat central zones indicate areas where spermogonia had been formed during the previous summer.

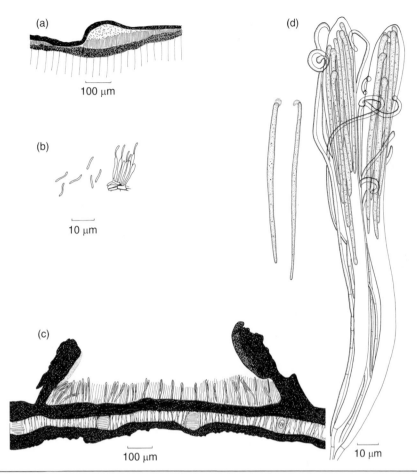

(a)

100 μm

(b)

10 μm

(c)

(d)

100 μm

10 μm

Fig 15.10 *Rhytisma acerinum*. (a) T.S. living leaf of *Acer pseudoplatanus* in June showing spermogonium. (b) Details of cells forming spermatia. (c) T.S. overwintered leaf of *Acer* showing the opening of lips of the stromatal surface to reveal the apothecial hymenium. (d) Asci, paraphyses and ascospores. Note the mucilaginous appendage at the upper end of the ascospore.

unusually shaped or brightly coloured fruit bodies. Good images and keys are given in Dennis (1981), Breitenbach and Kränzlin (1984) and Hansen and Knudsen (2000). Very little is known about Helotiales with small or inconspicuous apothecia, such as the Hyaloscyphaceae (Plate 7e).

15.5.1 Geoglossaceae

Trichoglossum is a representative of the Geoglossaceae (earth-tongues) which form club-shaped stalked apothecia. Members of this family grow saprotrophically on the ground, but sometimes also on dead leaves or amongst *Sphagnum* (e.g. *Mitrula*). An account of the family has been given by Nannfeldt (1942). *Trichoglossum hirsutum*

has black, somewhat flattened fruit bodies up to 8 cm high, and grows in pastures and lawns. The ascospores are long, dark and septate, and the asci are interspersed by black, thick-walled, pointed hymenial setae whose function is not known (Fig. 15.11b). The presence of hymenial setae separates *Trichoglossum* from *Geoglossum* which grows in similar habitats. The elongated ascospores of *Geoglossum* and *Trichoglossum* are discharged singly through a minute pore at the tip of the ascus. When the ascus is ripe, the pore bursts and one ascospore is squeezed into it, blocking it. The pressure of the ascus sap behind the spore causes the spore to protrude, at first slowly. When about half the spore is projecting, the spore gathers velocity and is

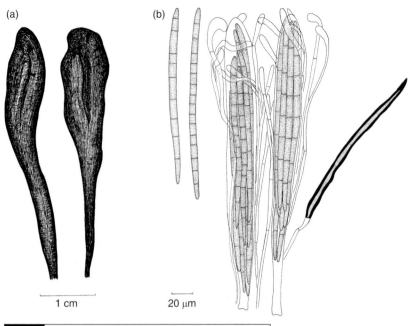

Fig 15.11 *Trichoglossum hirsutum.* (a) Apothecia. (b) Asci, ascospores, paraphyses and a hymenial seta.

rapidly discharged. Another ascospore immediately takes the place of the first spore and the process of discharge is continued until all eight ascospores have been released in single-file (Ingold, 1953).

15.5.2 Leotiaceae

Following the separation of the Helotiaceae (see below), the Leotiaceae now represent only a small group (13 species) of saprotrophic fungi. They can be distinguished from the Geoglossaceae by their brightly coloured ascocarps and hyaline ascospores (Lizoň *et al.*, 1998). A well-known example is *Leotia lubrica* (Plate 7f), a species colonizing woodland humus and known colloquially as 'jelly babies'.

15.5.3 Helotiaceae

Even after the separation of the Leotiaceae, this is still a very large (>600 spp.) and probably polyphyletic group. Well-known and widely distributed saprotrophic genera are *Ascocoryne* and *Neobulgaria* which produce gelatinous pinkish apothecia on relatively fresh dead wood, or *Chlorociboria* with its bright green apothecia

(Plate 7g). *Chlorociboria* spp. stain the colonized wood, and this is sometimes used in furniture making for ornamental inlays. *Bisporella citrina* (Plate 7h) is another commonly encountered species on relatively freshly fallen twigs.

Some members of the Helotiaceae, notably *Hymenoscyphus ericae*, as well as some other ascomycetes belonging to the Plectomycetes (e.g. *Pseudogymnoascus*, *Myxotrichum*, *Oidiodendron*; see p. 295), can form mycorrhizal associations with ericaceous plants such as *Erica* and *Vaccinium*. This association is called **ericoid mycorrhiza** and has fundamentally different properties from the vesicular−arbuscular (p. 202), ectomycorrhizal (pp. 525 and 581) and orchid mycorrhizal types (p. 596). Ericaceous plants form numerous small lateral roots called hair roots which consist of a narrow vascular bundle surrounded by a thin cortex and a thick epidermal monolayer. When *H. ericae* infects individual epidermal cells of its host's hair roots, it invaginates the plasmalemma and forms hyphal coils which superficially resemble those seen in orchid mycorrhiza (see Fig. 21.2). *Hymenoscyphus ericae* is credited with making nitrogen and phosphorus available to its host plants which typically grow in situations characterized by poor soils with acid pH. Good accounts of ericoid mycorrhiza have been given by Read (1996), Smith and Read (1997), Berch *et al.* (2002) and Peterson *et al.* (2004).

15.5.4 Bulgariaceae

Bulgaria inquinans forms gelatinous black apothecia on the bark of recently felled trees (Plate 7i), especially oak (*Quercus*), chestnut (*Castanea*) and beech (*Fagus*). Most unusually, the ripe ascus always seems to contain melanized as well as hyaline spores (Verkley, 1992). This species is cosmopolitan, and it is possible that it pre-colonizes the bark of the living tree as an endophyte.

15.5.5 Cyttariaceae

This family contains some of the most unusual and striking members of the Helotiales. *Cyttaria* spp. live biotrophically on the southern beech (*Nothofagus*). Orange-coloured apothecial stromata which can attain the size of golf balls arise singly or in clusters from galls on living tree branches (Plate 7j). Each of the dimples at the surface of the stroma represents a single apothecium. The ascospores are dark grey to black and continue to be discharged in great numbers even after several days of storage of detached stromata in dry conditions. *Cyttaria* spp. occur wherever *Nothofagus* grows, especially in South America, Australia and New Zealand. The fruit bodies of some species are edible (Minter *et al.*, 1987). A review of this enigmatic family of fungi has been given by Gamundí (1991).

Lichenized fungi (chiefly Hymenoascomycetes: Lecanorales)

16.1 | Introduction

The dual nature of lichens was first hinted at by de Bary (1866) and clearly recognized by Schwendener (1867). A lichen is now defined as a 'self-supporting association of a fungus (**mycobiont**) and a green alga or cyanobacterium (**photobiont**)' (Kirk *et al.*, 2001), 'resulting in a stable thallus of specific structure' (Ahmadjian, 1993). The fungal partner usually contributes most of the biomass to this symbiosis, including the external surface. It is thus termed the **exhabitant**, whereas the unicellular or filamentous photobiont cells are collectively called the **inhabitant** because they are located inside the lichen thallus (see Ahmadjian, 1993). Most lichens have a characteristic appearance which permits their identification if suitable keys are available (e.g. Purvis *et al.*, 1992; Wirth, 1995a,b; Brodo *et al.*, 2001). Since the structure of lichens is almost entirely due to the fungal partner, lichen taxonomy is synonymous with the taxonomy of the mycobiont.

It is possible to grow the algal and fungal partners of many lichens separately in pure culture (Ahmadjian, 1993; Crittenden *et al.*, 1995). Whereas most photobionts multiply readily in pure culture, the fungal partner, if it grows at all, typically shows slow growth as a sterile leathery mycelium but does not produce the characteristic lichen thallus. This is in marked contrast to the natural thallus where the

mycobiont displays its full sexual and asexual cycle, whereas the photobiont cells often appear swollen and are arrested in their cell cycle, i.e. their cell division is controlled by the mycobiont. The nature of the morphogenetic signals exchanged between the symbiotic partners is as yet unknown (Honegger, 2001).

Some 13 500 species of lichenized fungi have been described to date. Since lichens are often conspicuous and have been relatively well researched over the past 200 years, this number is not far below the estimated worldwide total of some 18 000 species (Sipman & Aptroot, 2001). In contrast, only about 100 species (40 genera) of photobionts are known, although this number may rise because photobionts are rarely formally and fully identified by lichenologists. The most common photobiont genera are the green algae *Trebouxia* (found in about 50% of all lichens) and *Trentepohlia*, and the cyanobacterium *Nostoc*. About 85% of lichenized fungi have a green algal photobiont, and 10% are associated with a cyanobacterium. The remaining lichens contain both a green alga and a cyanobacterium (Honegger, 2001). Most lichenized fungi (>98%) belong to the Euascomycetes, with only a few imperfect fungi and some 20 species of Basidiomycota also entering this type of symbiosis.

Because of their ability to tolerate repeated cycles of drying and rehydration and to survive extreme temperatures, high solar irradiation and other adverse conditions, lichens can colonize a range of terrestrial habitats not accessible to higher plants. Lichens are classical pioneer organisms, e.g. on bare rocks or infertile soils. Lichens can cause the weathering of rocks by secreting oxalic acid which reacts chemically with the rock surface; the rate of degradation may be 0.5–3.0 mm century^{-1} (Hale, 1983). *Dirina massiliensis* f. *sorediata* has been shown to cause much more rapid weathering of limestone surfaces, including those of historical monuments, at a rate of up to 2 mm in 12 years (Seaward & Edwards, 1997). Extensive lichen communities also exist on the bark and foliage of trees (corticolous lichens). Additionally, freshwater and marine species have been described. Lichens occur in all climatic zones from the Arctic and Antarctica, where they provide the dominant vegetation (Seppelt, 1995), to the tropics.

Some lichen thalli live for over 1000 years and can be used for determining the age of rock surfaces because of their slow growth rate. This discipline is known as **lichenometry** (Hale, 1983; Innes, 1988). It has been applied, for example, to date the standing stones on Easter Island or the time point of exposure of rock surfaces caused by avalanches or earthquakes. Crustose lichens are commonly used for lichenometry because they have the slowest growth rate. An example is the 'map lichen', *Rhizocarpon geographicum* (Plate 8b; O'Neal & Schoenenberger, 2003).

A wide range of lichens has been examined by different research groups. Therefore, in the present chapter we will give an introduction to the general features of lichen biology, followed by brief profiles of common examples taken from the Lecanorales, which is by far the largest order of lichenized fungi. Good general textbooks on lichens are those by Hale (1983), Ahmadjian (1993) and Nash (1996a). Richardson (1975) has written a stimulating account of the importance of lichens to mankind and in natural ecosystems.

16.2 | General aspects of lichen biology

16.2.1 Morphology of the lichen thallus

Lichen thalli come in three basic shapes – crustose (crust-like), fruticose (shaped like a miniature shrub) or foliose (leaf-like). It should be noted that these are purely descriptive terms which have no taxonomic meaning. Intermediate forms also exist. Good summaries are those by Büdel and Scheidegger (1996) and Honegger (2001).

By far the most common type is the **crustose** thallus which forms a thin spreading crust firmly attached to the substratum by its entire lower surface (Plate 8a,b). In the morphologically simplest crustose lichens, fungal hyphae are loosely associated with photobiont cells but do not form a protective upper cortex. Such lichen thalli appear powdery and are referred to as **leprose**. They often have a highly hydrophobic surface. Other crustose lichens produce a thicker thallus often held together by mucilage, as in the gelatinous lichens. In more highly differentiated crustose thalli, the photobiont cells are positioned in a defined layer located underneath an upper cortex formed exclusively by the mycobiont. The photobiont cells are thus protected from adverse environmental factors. Air spaces in the photobiont layer and the medulla underneath permit gas exchange (see Fig. 16.1). The differentiation of the lichen thallus into horizontal layers is called **stratification**. In **squamulose** lichens (Plate 8e), the crustose thallus forms small scales (squamules) which become partially raised from the substratum, giving the surface a scurfy appearance.

The stratification is developed further in the second thallus type, the **foliose** lichens (Plate 8c,d) by the development of a lower cortex. Attachment to the substratum is often by bundles of hyphae termed **rhizinae**. As a result, the thallus appears leaf-like or lobed and can be detached from the substratum without being damaged.

The third thallus type is called **fruticose**. Here the thallus has a shrub-like or branched appearance and is raised from the substratum (Plate 8f) or hangs down from it (Plate 8g). In some cases a fruticose thallus may develop from a basal crustose or foliose thallus (Plate 8e). Stratification in fruticose thalli is often tubular/concentric rather than horizontal, and it resembles the more complex crustose types in possessing an outer cortex overlying a photobiont layer

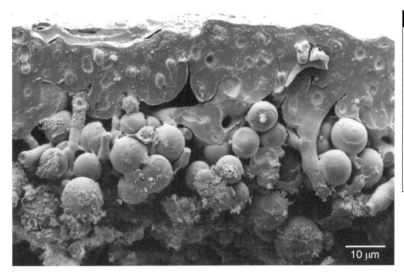

Fig 16.1 SEM view of a section of the stratified fruticose thallus of *Anaptychia ciliaris*. The globose cells of the photobiont (*Trebouxia*) are located in a loose layer underneath the thick mucilaginous cortex produced exclusively by the mycobiont hyphae. From Büdel & Scheidegger (1996), by permission of Cambridge University Press. Original image kindly provided by C. Scheidegger.

10 μm

and a medulla, but lacking an inner cortex (Fig. 16.1).

16.2.2 Reproduction of lichens

Sexual reproductive features of lichenized ascomycetes are largely equivalent to those found in non-lichenized groups. Both homothallic and heterothallic mating systems are known. Apothecia with inoperculate asci are especially common, but perithecia and pseudothecia also occur. In contrast, no cleistothecia or apothecia with operculate asci are known among lichenized fungi. The taxonomic affinities of the various orders of lichenized ascomycetes are summarized in Table 16.1. Conidial states, if produced, are often pycnidial. In most cases, the photobiont is excluded from the fertile regions of the mycobiont so that ascospores or conidia are released without photobiont cells. A new lichen thallus can be established from fungal spores only by re-lichenization with suitable photobiont cells. It is unclear how frequent this is in nature.

Most lichen thalli therefore produce vegetative propagules containing both symbionts. These can be very variable, and an extensive terminology has evolved to describe them (see Büdel & Scheidegger, 1996). The most common vegetative propagules are **soredia**, i.e. small clumps of hyphae enclosing a few algal cells (Fig. 16.2). They are produced over the entire surface of the thallus (Plate 8e; Fig. 16.8) or in differentiated structures called **soralia**. Soredia are usually hydrophobic and are dispersed by wind, perhaps following their initial detachment by the impact of a rain drop. **Isidia** are larger, upright cylindrical structures which contain both symbionts. They serve to increase the surface area of the lichen thallus but can also become detached and then function as vegetative propagules. In some lichens such as *Cladonia*, squamules broken off a vegetative thallus are capable of establishing a new thallus. Animals can also play a role in dispersing lichens. Meier *et al.* (2002) have shown for *Xanthoria parietina* (Plate 8c) that mites feeding on lichen thalli can spread both the mycobiont and photobiont via their faecal pellets. Since this lichen does not produce soredia or isidia, dispersal by invertebrates could be significant.

16.2.3 Establishment of a lichen thallus

A germinated mycobiont spore in nature may be able to survive for a while as a mycelium of limited spread, or it can undergo a loose association with free-living algae not suitable for an intimate and permanent lichen symbiosis (Ahmadjian & Jacobs, 1981; Ott, 1987). The initial stage of the lichen symbiosis is therefore non-specific. It results in mycobiont hyphae making contact with and growing around individual photobiont cells (Figs. 16.3a,b).

Table 16.1. Summary of the most important orders of lichenized ascomycetes and their characteristic features. Orders have been grouped approximately according to the phylogenetic summary by Grube and Winka (2002).

Order	Number of species	Features of sexual reproduction	Lichen thallus	Photobiont	Taxonomic reference
Lichinales	237 (all lichenized)	ascohymenial/apothecial with prototunicate asci	crustose, foliose or fruticose (often gelatinous)	cyanobacteria	Schultz et al. (2001)
Agyriales	98 (mostly lichenized)	ascohymenial/apothecial with bitunicate (non-fissitunicate) asci	crustose or squamulose	green algae (with cyanobacteria in cephalodia)	Lumbsch et al. (2001)
Gyalectales	108 (all lichenized)	ascohymenial/apothecial with unitunicate asci	mainly crustose	mainly green algae (especially *Trentepohlia*)	Lumbsch et al. (2004)
Ostropales (incl. Graphidales)	1854 (mainly non-lichenized)	ascohymenial/apothecial with unitunicate asci	crustose	green algae	Lumbsch et al. (2004)
Pertusariales	47 (all lichenized)	ascohymenial/apothecial with unitunicate asci	mainly crustose	green algae	Stenroos and DePriest (1998)
Lecanorales	7108 (mostly lichenized)	ascohymenial/apothecial with rostrate (non-fissitunicate) asci	all shapes	green algae and/or cyanobacteria	(see Section 16.3)
Pyrenulales	286 (two-thirds lichenized)	ascohymenial/perithecial with bitunicate (fissitunicate) asci	crustose	green algae	
Verrucariales	720 (mostly lichenized)	ascohymenial/perithecial with bitunicate (non-fissitunicate) asci	crustose or foliose	green algae	Wedin et al. (2005)
Arthoniales	1200 (mostly lichenized)	ascohymenial/apothecial with bitunicate (fissitunicate) asci	mainly crustose	green algae	Myllys et al. (1998)

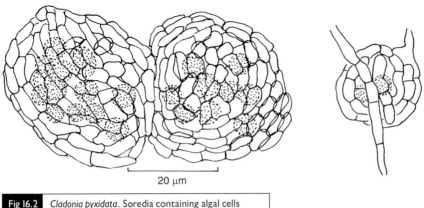

Fig 16.2 *Cladonia pyxidata.* Soredia containing algal cells surrounded by fungal hyphae.

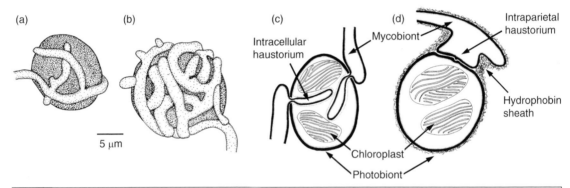

Fig 16.3 Interactions between photobiont cells (*Trebouxia*) and mycobiont hyphae. (a,b) Early stages in the establishment of a lichen thallus in *Cladonia cristatella*. (a) Single algal cell at point of contact with the mycobiont. (b) Ensheathment of algal cell by the mycobiont. Both images redrawn and modified from SEM images of Ahmadjian and Jacobs (1981). (c) Formation of tubular intracellular haustoria by the mycobiont of a simple crustose lichen (e.g. *Lecanora conizaeoides*). The same photobiont cell may be penetrated repeatedly. (d) Formation of an intraparietal haustorium by the mycobiont of a stratified lichen. Exchange of nutrients occurs apoplastically; the walls of both mycobiont and photobiont are surrounded by a common hydrophobin sheath produced by the former. (c,d) schematic drawings based on the results of Honegger (1986).

If partners are compatible, a **pre-thallus** is formed, i.e. a crustose non-stratified cluster of photobiont cells ensheathed by hyphae of the mycobiont (Trembley *et al.*, 2002). A pre-thallus is also formed by germinating isidia or soredia. Under suitable conditions, the pre-thallus enlarges and stratifies into a mature thallus; the first sign of this step is the secretion of mucilage by hyphae at the periphery of the pre-thallus (Honegger, 1993).

The factors determining photobiont–mycobiont specificity and thus permitting the development of a pre-thallus and mature thallus are not yet known. In addition to genetic determinants, environmental factors must also play a crucial role as indicated by the difficulties encountered when trying to grow pre-thalli into fully differentiated thalli in the laboratory (Ahmadjian, 1993). Recognition is probably a continuous and multi-step process mediated by surface molecules such as lectins, and facilitated by the embedding of both bionts in a gelatinous matrix of fungal origin (Ahmadjian, 1993).

A pre-thallus can be formed by fusion of several genetically distinct isidia, soredia or photo- and mycobionts, and likewise several pre-thalli can fuse in the process of thallus maturation. Thus, a mature thallus may contain a jigsaw of genetically heterogeneous myco- and photobionts (Fahselt, 1996). In other cases, e.g. the map lichen *Rhizocarpon geographicum*, the borders between incompatible thalli are demarcated by black barrage lines. In some lichens with a green alga as photobiont, the situation is further complicated by the inclusion of a second (cyanobacterial) photobiont. This is then usually confined to discrete regions termed **cephalodia** which often differ in their morphology from the thallus containing the green photobiont (Fig. 16.4).

16.2.4 Lichenicolous fungi

The capture of a compatible photobiont partner by a germinating fungus spore can be problematic, especially if the photobiont belongs to the genus *Trebouxia* which does not seem to be widespread as a free-living organism. One solution to the problem is the recovery of photobionts from the propagules of other lichen species. Certain photobiont strains are favoured by many taxonomically unrelated mycobionts (Rikkinen *et al.*, 2002). If these lichens grow in similar ecological situations, communities are

formed which have a high mycobiont diversity but share the same or closely related photobiont strains. Thus, the chances for a germinating spore to salvage compatible photobionts from soredia or isidia of other lichen species may be quite high (Rikkinen, 2003).

Several mycobionts have taken the ultimate step of poaching their photobiont from an existing lichen thallus in order to establish their own independent thallus (Ott *et al.*, 1995). Such organisms are called **lichenicolous lichens**, and the phenomenon has been aptly named 'cleptobiosis' (Honegger, 1993). Numerous other fungi feed on the photosynthetic products of a lichenized photobiont without ever establishing an independent thallus, while yet others destructively parasitize the host lichen (Rambold & Triebel, 1992; Richardson, 1999; Lawrey & Diederich, 2003). Such fungi are collectively called **lichenicolous fungi**, and they are often taxonomically related to lichenized fungi.

16.2.5 The nutritional basis of the lichen symbiosis

Hill and Smith (1972) devised a simple and elegant method termed the 'inhibition technique' which has permitted the identification of carbohydrates secreted by the photobiont. Lichen thalli are exposed to radiolabelled CO_2, and after a while an excess of a single unlabelled

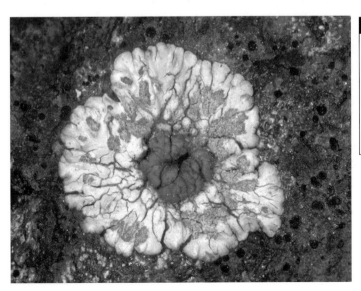

Fig 16.4 *Placopsis gelida*, a foliose lichen. The brightly coloured main thallus containing the primary (algal) photobiont has produced a dark central gall-like cephalodium in which the secondary (cyanobacterial) photobiont is localized. From Büdel and Scheidegger (1996), by permission of Cambridge University Press. Original image kindly provided by C. Scheidegger.

carbohydrate is added. This will saturate the uptake system of the mycobiont; only if it is identical to the radiolabelled mobile carbohydrate exported by the photobiont will the latter accumulate in the incubation medium, where its radioactivity can be measured. Such studies have shown that cyanobacterial photobionts export glucose to the mycobiont, whereas green algae export polyols such as erythritol (*Trentepohlia*), sorbitol (*Hyalococcus*, *Stichococcus*) or ribitol (*Trebouxia*, *Coccomyxa*, *Myrmecia*) (Ahmadjian, 1993). Tapper (1981) estimated that at least 70% of the total photosynthetically fixed carbon is transferred from the photobiont to the mycobiont in *Cladonia convoluta*. Once taken up by the fungus, the transport carbohydrate is rapidly converted to mannitol (Lines *et al.*, 1989; Ahmadjian, 1993).

Cultured algal photobionts do not secrete polyols into the medium, and if they are separated from a fresh lichen thallus, polyol secretion ceases within a few hours. Nothing appears to be known about the mechanism by which carbohydrate export from the photobiont is regulated (Ahmadjian, 1993). Contact of the mycobiont with its photobiont partner takes various shapes. The gelatinous extracellular sheath of cyanobacteria is penetrated by hyphal protrusions, whereas direct wall-to-wall contact occurs between mycobionts and green algal photobionts. Appressorium-like structures presumably facilitate attachment, and haustoria may also be formed within the photobiont cells especially in simple, non-stratified crustose lichens (Fig. 16.3c). In more highly differentiated lichens, haustoria are often reduced to a pad-like infection peg appressed to but not breaking the algal wall (Fig. 16.3d). Such structures have been called intraparietal haustoria (Honegger, 1986).

Fungal hyphae and attached photobiont cells are often coated by hydrophobin-type proteins and other hydrophobic molecules secreted by the mycobiont (Honegger, 1997; Scherrer *et al.*, 2000). Thus, the main transport route from the photobiont to the mycobiont, in the absence of large haustorial interfaces, must be through the apoplast by cell wall contact (Ahmadjian, 1993). This is different from the elaborate membrane-to-membrane contact as found in the haustorial complexes of arbuscular mycorrhizal fungi (see Fig. 7.46c) and biotrophic parasites (see Fig. 13.5). The reason for the reduction in membrane contact in the lichen symbiosis may lie in the fact that membranes are among the most easily damaged structures during the drying and rehydration cycles to which lichens are exposed. Large haustoria might thus reduce cell viability. Presumably sufficient carbohydrate leaks out of the photobiont cells during the frequent drying−rehydration cycles without the need for intracellular haustoria (Honegger, 1997).

Fungi with a cyanobacterium as their primary or secondary photobiont benefit by receiving nitrogen in addition to carbohydrates. Nitrogen fixed by cyanobacterial photobionts is released to the mycobiont as ammonium (NH_4^+) and is incorporated into the amino acid pool as glutamate (Nash, 1996b).

Whilst the advantages of the lichen symbiosis to the mycobiont are obviously nutritional, there is no clear evidence of the transfer of any minerals or other nutrients from the mycobiont to the photobiont. The benefits to the photobiont may include the buffering against adverse environmental conditions such as high solar irradiation. The upper cortex of many lichens is brightly coloured due to the presence of pigments which screen out UV light (see Plate 8b,c). In fact, cortical pigments may filter out as much as 50% of the incoming light, and this is particularly important with *Trebouxia* spp. as photobiont because these algae favour low light intensities (Masuch, 1993). Lichen thalli growing at higher altitudes or on surfaces facing the sun often contain higher pigment concentrations than less-exposed thalli. An example of a light-screen pigment is the polyketide usnic acid (Fig. 16.5) which is also toxic against bacteria, fungi and other organisms (Elix, 1996; Cocchietto *et al.*, 2002). This substance is produced by several taxonomically unrelated lichens, but has not yet been isolated from any non-lichenized fungus. Another example is the pulvinic acid derivative vulpinic acid (Fig. 16.5) produced by the wolf's lichen, *Letharia vulpina*. This species is so toxic that its thalli have been used in the past to poison foxes and wolves, by laying out animal carcasses spiked with ground lichen thalli

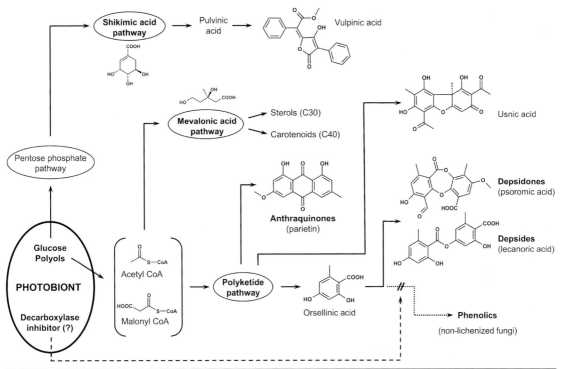

Fig 16.5 The main routes of secondary metabolism in lichenized fungi. The shikimic acid pathway (top) gives rise to vulpinic acid (= pulvinic acid methyl ester) and other pulvinic acid-derived metabolites (see also Fig. 19.22). The mevalonic acid pathway gives rise to triterpenes such as sterols (see Fig. 13.16) and tetraterpenes such as carotenoids (see Fig. 24.8). The most typical lichen metabolites are polyketides, especially those synthesized by polymerization of phenolic acids (orsellinic acid) or orcinols derived from them, giving rise to a wide range of depsides (e.g. lecanoric acid) or depsidones (e.g. psoromic acid). This biosynthetic route, although taking place in the mycobiont, is thought to be encouraged by the production of an orsellinic acid decarboxylase inhibitor produced by the photobiont. Usnic acid is also the product of oxidative coupling of two phenolic-type rings, although the biosynthetic route does not proceed via orsellinic acid and orcinol. Yet other lichen polyketides (e.g. parietin) arise by cyclization of a single long polyketide chain; this metabolic pattern is also common in non-lichenized fungi (see Fig. 12.46). Modified from Hale (1983), Masuch (1993) and Elix (1996).

(Richardson, 1988; Brodo *et al.*, 2001). Not surprisingly, lichens containing these and other toxic substances appear to be avoided by lichen-grazing animals (Masuch, 1993). Many secondary metabolites of lichens have acidic properties and are therefore sometimes collectively called **lichen acids**. Most of them are produced by the mycobiont only in the intact lichen thallus but not in isolation, indicating that the photobiont may exert a subtle influence on the secondary metabolism of the mycobiont. As an example, Culberson and Ahmadjian (1980) have proposed that a putative decarboxylase inhibitor secreted by lichen algae inhibits the conversion of orsellinic acid to phenolic substances which is

common in free-living fungi, leading instead to esterification of orsellinic acid or orcinol, and the accumulation of depside-type lichen acids (see Fig. 16.5). The topic of secondary metabolism in lichens is vast and has been reviewed by Lawrey (1986), Fahselt (1994) and Elix (1996). The main biosynthetic routes towards secondary metabolism in lichens are the shikimic acid, terpenoid and polyketide pathways, of which the polyketide route is particularly important (Fig. 16.5).

The benefit conveyed by the mycobiont may thus be the provision of a 'photobiont cultivation chamber' (Honegger, 2001) which permits the growth of photobionts in situations which might be too hostile for free-living forms. The lichen symbiosis can thus be considered an alternative adaptation to terrestrial life as compared to higher plants.

16.2.6 Lichens and pollution

Lichens are particularly sensitive to aerial pollutants, and especially to sulphur dioxide, SO_2 (Seaward, 1993; Gries, 1996; Nash & Gries, 2002). The photobiont appears to be generally more sensitive than the mycobiont. The disappearance of lichens from the centres of urban and industrial areas was first recognized by Nylander (1866), who had already correlated this phenomenon with aerial pollution. Because different lichens show a differential sensitivity to SO_2, the presence or absence of key species can be used as an index of the level of air pollution (Hawksworth & Rose, 1970). The most SO_2-tolerant lichen, *Lecanora conizaeoides*, may have evolved in SO_2-polluted areas and went on to become Northern Europe's most abundant lichen by the 1950s (Richardson, 1975). This lichen may actually *require* elevated SO_2 levels for good growth (Nash & Gries, 2002), as shown by its disappearance from some areas after the implementation of legislation to curb SO_2 emissions. At the same time, formerly polluted areas are being re-colonized by many SO_2-sensitive species (Rose & Hawksworth, 1981; Seaward, 1993). An example of this trend has been given by Masuch (1993) for the city of Munich. Between 1891 and 1956, the 'lichen desert' (i.e. lichen-free zone) in the city centre increased from 8 to 56 km^2, and then it decreased again, disappearing altogether by 1983. The size of the 'lichen desert' has been correlated with the degree of SO_2 pollution in the air. Careful studies of lichen population dynamics have revealed that lichen species re-colonizing a lichen desert may be different from those initially present. This phenomenon has been explained by the eutrophication of urban habitats, i.e. their enrichment especially with nitrogen (Seaward, 1997; Seaward & Coppins, 2004). One of these newcomers in urban lichen deserts is *Dirina massiliensis* f. *sorediata*, which is the cause of a rapid decay of limestone monuments (Seaward & Edwards, 1997).

Lichens obtain most of their mineral nutrients from the air and rainwater in which these are present only in very low concentrations. Not surprisingly, therefore, lichens can accumulate dissolved substances from very dilute solutions. For instance, lichens concentrate radioactive nuclides which enter the food chain lichen→reindeer→man, leading to their accumulation in human tissues (Richardson, 1991). Lichens are also being used to monitor the radioactive contamination resulting, for example, from the explosion of the Chernobyl nuclear reactor in 1986 (Seaward, 2004).

16.2.7 Taxonomy of lichens

The discovery of a fossilized cyanolichen in the Rhynie chert sediments (Taylor *et al.*, 1997) indicates that lichens were present some 400 million years ago when the terrestrial habitat was first colonized. Indeed, there is evidence of even older lichen-like associations (Yuan *et al.*, 2005). A huge diversity of lichens exists today, and there is good phylogenetic evidence that the lichenized habit has been developed and lost independently on several occasions in the course of evolution (Gargas *et al.*, 1995; Lutzoni *et al.*, 2001). This is also evident from the scattered placement of lichenized fungi in a wider ascomycete context. Lutzoni *et al.* (2001) even suggested that some of today's groups consisting entirely of non-lichenized species, notably the Plectomycete lineage (Chapter 11), originated from lichenized ancestors. Part of the proposed argument is a chemotaxonomic one, i.e. the presence of numerous secondary metabolites (especially polyketides) in the lichens and Plectomycetes, but their absence or less-frequent occurrence in certain other groups of fungi. As with many DNA-based analyses, the phylogenetic arrangement of taxa may vary with the kinds of sequences used, and other schemes showing a less scattered distribution of lichenized fungi within the Ascomycota have been put forward (Fig. 8.17; Liu & Hall, 2004).

Whereas much work remains to be done on the taxonomy of ascomycetes in general and lichenized ascomycetes in particular, some orders are beginning to take shape. These are listed in Table 16.1, summarizing data from Tehler (1996), Kirk *et al.* (2001), Ott and Lumbsch (2001) and Grube and Winka (2002). The orders Dothideales (Section 17.3), Hypocreales (Section 12.4) and Helotiales (Chapter 15) are not considered here because they contain only a small proportion of lichenized species. Some small families of lichenized fungi (e.g. Baeomycetaceae, Icmadophilaceae, Umbilicariaceae) are *incertae sedes* at the moment, and their accurate placement will be determined in further studies. Such studies will have to be based on the combined analysis of several different genes in order to obtain a greater degree of confidence in the resulting phylogenetic trees (Myllys *et al.*, 2002). Most results so far have been obtained with the small subunit (18S) ribosomal RNA sequence.

The data summarized in Table 16.1 are too diffuse to be fully understood at present, but we note in passing that the occurrence of fissitunicate asci is not always correlated with ascolocular development (see p. 459 for an explanation), i.e. that fissitunicate asci can be found in perithecia and apothecia, not just pseudothecia (see Chapter 17). Further, many orders, and especially the Lecanorales, produce bitunicate but non-fissitunicate asci, as do some other Ascomycota, e.g. Helotiales and Pezizales (see pp. 414. and 429).

16.3 | Lecanorales

Members of the Lecanorales produce inoperculate asci in apothecia. The asci are bitunicate but non-fissitunicate. The ascus apex is thickened, and ascospores are discharged when the outermost wall layer breaks and the innermost layer protrudes through the pore thus generated, to produce an apical beak called **rostrum** (Figs. 8.12e,f). Rostrate ascus dehiscence is typical of the Lecanorales.

Over 75% of all lichenized fungi belong to this order, making it one of the largest in the Ascomycota. Most of the best-known and most readily collected lichens belong to the Lecanorales. Only a few non-lichenized members of the Lecanorales are known; these are usually lichenicolous. This order has been divided into 42 families (Kirk *et al.*, 2001), of which many still have an uncertain phylogenetic position and are currently being circumscribed by DNA sequence comparisons. Stenroos and DePriest (1998) have identified five suborders which together make up a monophyletic order. We shall consider just a few representatives to indicate the astonishing morphological and ecological variability of lecanoralean lichens.

16.3.1 *Lecanora*

About 300 species of *Lecanora* have been described, mainly from temperate climates. Thalli are usually crustose and are very common on rock surfaces, including ancient monuments, dry stone walls and roof tiles, as well as on the bark of trees. The photobiont usually belongs to the genus *Trentepohlia*. We have already come across *L. conizaeoides* as a particularly SO_2-tolerant species (p. 454). Because of their exposed habitats, *Lecanora* spp. often deposit light-screen pigments in their upper cortex which give them a bright yellow coloration. An example is the xanthone lichexanthone; usnic acid and pulvinic acid derivatives have also been detected. The quantity and diversity of pigments may be greater in thalli exposed to higher levels of irradiation (Obermayer & Poelt, 1992). A common example of the genus is *L. muralis*, showing a typical crustose thallus with apothecia (Plate 8a).

The thalli of *Lecanora* (*Sphaerothallia*) *esculenta*, a species found from northern Africa to western Central Asia, may roll up and become detached upon maturity, being blown about by the wind. They are said to be edible. On occasions, windborne lichen thalli have been so abundant that the common name 'manna lichen' has been coined for *L. esculenta* (Richardson, 1988).

16.3.2 *Xanthoria*

The most abundant species is *X. parietina*, which forms bright yellow foliose thalli (Plate 8c) on the surface of rocks, roofs, trees and farm buildings, especially near the sea. It is particularly common in places enriched by manure, e.g. dust from cattle yards, or from birds. The thallus is lobed and is attached to the substratum by short rhizinae. The photobiont is the green alga

Trebouxia, which forms single globose cells in a defined layer beneath the upper cortex of the stratified thallus (Fig. 16.6c). The apothecia are saucer-shaped and about 2–3 mm in diameter. They are located on the upper surface of the thallus, and the algal zone extends into the apothecial margin (Fig. 16.6a). The ascospores are at first one-celled, but ingrowth from the wall of the ascospore eventually divides the contents of the spore into two. The yellow colour of the thallus is due to the presence of the anthraquinone parietin (Fig. 16.5) in the upper cortex. Unusual carotenoids are also produced by *Xanthoria* spp. (Czeczuga, 1983). *Xanthoria parietina* does not produce soredia, but it is very abundant none the less. One reason for this may be that germinating ascospores display a tendency towards cleptobiosis, i.e. the theft of photobiont cells from soredia or mature thalli of other lichens (Ott, 1987). Another means of dispersal may be by mite browsing and feeding as mentioned earlier.

16.3.3 *Peltigera*

About 45–60 species are known, and the genus has been thoroughly examined by Miadlikowska and Lutzoni (2004). They and many other workers now consider it to be part of a separate order, Peltigerales, which is closely related to Lecanorales. Species of *Peltigera* form large lobed leaf-like thalli attached to the ground or to rocks by groups of white rhizinae. The thallus is rather fleshy and is highly stratified (Fig. 16.7a). The commonest species are *P. polydactyla* and *P. canina* (Plate 8d), both of which have now been split up into several species. They grow among grass on heaths, on sand dunes and on rocks amongst moss. The usual photobiont is the cyanobacterium *Nostoc*. In some species, however, the primary photobiont can be either a *Nostoc*, giving rise to the usual greyish-black thallus, or a green alga (*Coccomyxa*), in which case the thallus is vividly green, with *Nostoc* sometimes present as a secondary photobiont in cephalodia (Brodo & Richardson, 1978). The two different

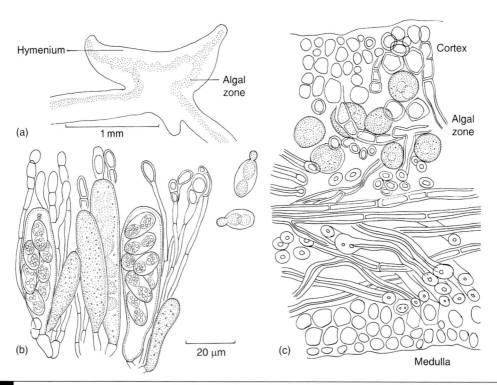

Fig 16.6 *Xanthoria parietina.* (a) V.S. thallus and apothecium showing the extension of the algal zone into the apothecium. (b) Asci, paraphyses and two germinating ascospores. (c) V.S. thallus.

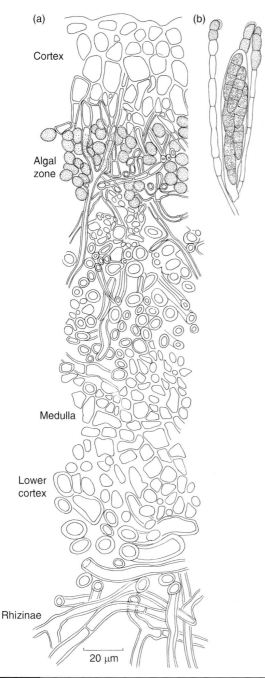

Fig 16.8 *Cladonia pyxidata*. Primary squamulose thallus bearing funnel-shaped podetia. Note the granular soredia outside and inside the podetia.

Fig 16.7 *Peltigera polydactyla*. (a) V.S. thallus. (b) Ascus, ascospores and paraphyses.

forms of the same lichen are then termed a **morphotype pair** or 'lichen chimera'. The apothecia of *Peltigera* are reddish-brown, folded extensions of the thallus (Plate 8d) which do not contain algal cells. The red colour of the apothecia is due to pigments in the tips of the paraphyses (Fig. 16.7b).

16.3.4 *Cladonia*

There are about 400 species of *Cladonia*, some of them extremely common, growing in heaths, moors and elsewhere on rocks and walls. There are two kinds of thallus. The primary thallus is squamulose, and the secondary thallus is upright and cylindrical (fruticose), often consisting of a hollow stalk which bears the apothecium at its tip. Such an apothecium-bearing vertical thallus arising from a horizontal primary thallus is called a **podetium** and is typical of the genus *Cladonia*. In *C. pyxidata*, the podetium opens out into a cup (Fig. 16.8), and the apothecia ultimately develop at the rim of the podetium. In *C. floerkiana* (Plate 8e), the podetia are shrub-like and bear their red apothecia as terminal heads. The colloquial name for this species is 'British soldiers'. The podetia frequently bear the granular soredia which contain algal cells surrounded by fungal hyphae (Fig. 16.2). In wind-tunnel experiments using *C. pyxidata*, Brodie and Gregory (1953) showed that soredia were blown away from the funnel-shaped podetia at wind speeds as low as $5.4-7.2 \, \mathrm{km \, h^{-1}}$ although they were not removed from horizontal glass slides at the same wind speeds. They suggested that funnel-shaped structures generate eddy currents when placed in a windstream and that eddy currents effectively remove soredia.

Usnic acids (Fig. 16.5) are particularly common in *Cladonia* spp., and their concentration in the lichen thallus has been shown to increase linearly with the intensity of UV light, supporting a possible role as a light screen (Rundel, 1969). Further, sun-exposed thalli of *Cladonia* spp. appear yellowish whereas those in more shaded habitats are greyish-green, although the pigment responsible is not usnic acid.

The reindeer lichen genus *Cladina* is closely related to *Cladonia* (Stenroos & DePriest, 1998).

The most common species are *C. rangiferina* (Plate 8f) and *C. stellaris*, which are a major winter food for grazing animals such as reindeer or caribou in northern boreal forests (Richardson, 1988). These lichens provide an important component of the ground cover grazed by animals, and are also used by Laplanders to make hay for their animals. Reindeer lichens are popular in Germany as decorations on wreaths and are also well-known among model railway enthusiasts and architects who use the highly branched fruticose thalli as miniature trees (Kauppi, 1979; Richardson, 1988).

Loculoascomycetes

17.1 Introduction

The characteristic feature of this group is that the ascus is bitunicate and fissitunicate; it has two separable walls (see p. 240). The outer wall (**ectotunica** or ectoascus) does not stretch readily, but ruptures laterally or at its apex to allow the stretching of the thinner inner layer, the **endotunica** or endoascus (Figs. 17.1a−c). Asci are generally non-amyloid. The fruit body with asci is regarded as an **ascostroma**, and each cavity in which asci develop is termed a **locule**. In contrast to the Hymenoascomycetes, in which ascocarps develop *following* plasmogamy and the pairing up of two genetically dissimilar nuclei (**ascohymenial** development), in the Loculoascomycetes the ascoma is already present *before* the compatible nuclei are brought together (Barr & Huhndorf, 2001). The development of asci in pre-formed locules is called **ascolocular**. The ascostroma has therefore been defined as an aggregation of vegetative hyphae not resulting from a sexual stimulus (Wehmeyer, 1926). However, Holm (1959) has questioned the accuracy of this definition, since examples are known where the ascocarps do develop following the pairing of nuclei (Shoemaker, 1955). Within the developing ascocarp, one or more locules are formed by the downgrowth of pseudoparaphyses (see below) and the development of asci. One or more ostioles then develop by the breakdown (lysis) of a pre-formed mass of tissue (**lysigenous development**). Where a single locule develops, a structure resembling a perithecium results and,

although this term is commonly used for such loculoascomycete fruit bodies, they should strictly be called **pseudothecia** (see p. 245). Although a mature ascostroma with several locules can be superficially similar to the perithecial stroma of Pyrenomycetes such as *Hypoxylon* (Fig. 12.11) or *Claviceps* (Fig. 12.27), the difference is that in the pyrenomycete stroma each fertile region (perithecium) is surrounded by a wall whereas the locules of the ascostroma are not (Alexopoulos *et al.*, 1996). However, the ascocarp of Loculoascomycetes does not always take the form of a pseudothecium. In some groups it may be an apothecium, a **hysterothecium** (an elongate ascoma with a slit-like opening) or a cleistothecium (Barr & Huhndorf, 2001). These deviations from the classical pseudothecium are found for example in lichenized Loculoascomycetes (see Table 16.1).

The name Loculoascomycetes was coined by Luttrell (1955) and corresponds to the Ascoloculares of Nannfeldt (1932). The group has also been named the Dothideomycetidae (see Kirk *et al.*, 2001). It is very large, with about 900 genera and over 7000 species. Most members are terrestrial, growing as saprotrophs, endophytes or parasites on the shoots and leaves of herbaceous or woody plants and may cause diseases of economic significance, but some grow in freshwater or the sea and others in dung or soil. There is a very wide range of anamorphs, some hyphomycetous, others pycnidial (Sivanesan, 1984; Seifert & Gams, 2001).

It is most unlikely that the Loculoascomycetes, as currently classified, are monophyletic

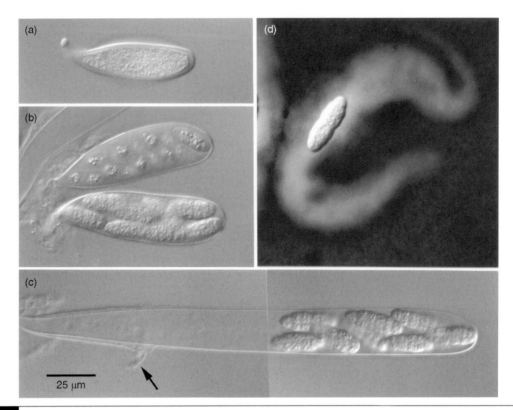

Fig 17.1 Sexual reproduction in *Pleospora scirpicola*. (a) Young ascus with undifferentiated cytoplasm. Note the thick ascus wall. (b) Maturing asci with cytoplasm cleaved into ascospores. (c) Mature ascus with muriform ascospores. The ascus is shown a few seconds before discharge; the ectotunica has ruptured (arrow), and the more flexible endotunica has been forced to expand by the increasing turgor pressure. (d) Discharged ascospore mounted in Indian ink to show the mucilaginous extensions. All images to same scale.

(Berbee, 1996; Dong *et al.*, 1998; Silva-Hanlin & Hanlin, 1999). Molecular analyses based on a relatively small number of representatives of ascomycete groups suggest that the most closely related groups are Pyrenomycetes (perithecioid fungi) belonging to Diaporthales, Hypocreales, Microascales and Sordariales (see Chapter 12; Spatafora, 1995). Several orders are currently included in the Loculoascomycetes (Barr & Huhndorf, 2001; Kirk *et al.*, 2001) but we shall consider representatives only of the two most important, the Pleosporales and Dothideales. Because the arrangement of orders and especially families is still in a state of flux, we shall not discuss it.

17.2 | Pleosporales

This is a large group of ascomycetes believed to be monophyletic on the basis of molecular evidence (Berbee, 1996). Several economically important genera of plant pathogens are included, such as *Cochliobolus*, *Phaeosphaeria* and *Pyrenophora*, parasitic on grasses and cereals. *Pleospora*, *Lewia* and *Leptosphaeria* are common endophytes, saprotrophs or parasites of other herbaceous plants. They and their anamorphs may also have significance as human allergens, as human pathogens and in the production of mycotoxins. *Sporormiella* fruits on herbivore dung.

The development of pseudothecia in these forms conforms to the *Pleospora* type of Luttrell (1951). Ascogonia arise within a stroma and, in the region of the ascogonia, a group of vertically arranged septate hyphae appears, each hypha arising as an outgrowth from a stromatal cell. These hyphae are capable of elongating by intercalary growth and are termed **pseudoparaphyses** (Luttrell, 1965).

Pseudoparaphyses arise near the upper end of the cavity and grow downwards. Their tips soon intertwine and push between the other cells of the stroma so that free ends are seldom found. They may thus be distinguished from the true paraphyses of other fungi (e.g. *Sordaria*) which are formed from hyphae attached at the base of the cavity, extend upwards and are free at their upper ends. They may also be distinguished from apical paraphyses which are attached above, arising from a clearly defined meristem near the apex of a perithecium, and form a well-defined palisade of hyphae free at their lower ends (see the *Nectria* type of development, p. 337). In the *Pleospora* type of development, asci arise amongst the pseudoparaphyses at the base of the cavity and grow up between them. The ostiole develops lysigenously, i.e. by breakdown of pre-existing cells. Development of this general type has been described in *P. herbarum* (Wehmeyer, 1955; Corlett, 1973), *Leptosphaeria* (Dodge, 1937), *Sporormiella* (Arnold, 1928) and other fungi (see Luttrell, 1951, 1973). A more recent discussion of pseudoparaphyses development in the Pleosporales and its taxonomic implications has been written by Liew *et al.* (2000).

17.2.1 *Leptosphaeria*

Leptosphaeria species fruit on moribund leaves and stems of herbaceous plants. There are probably some 100 species, many growing on a wide range of hosts, but others are confined to one host plant. Although most are saprotrophic or only weakly pathogenic, some are troublesome pathogens, e.g. *L. coniothyrium*, the cause of cane blight of raspberry, and *L. maculans* which causes blackleg of oilseed rape and other brassicas. Characteristic features are the fusoid, yellow or pale brown ascospores with two or more transverse septa. Anamorphic states are pycnidial (see Table 17.1).

Leptosphaeria acuta fruits in abundance in spring at the base of overwintered, decorticated stems of stinging nettles (*Urtica dioica*). The black shining pseudothecia are somewhat conical and flattened at the base (Fig. 17.2a). Bitunicate asci elongate within a pre-existing group of branching pseudoparaphyses, and close examination of the direction of growth and branching indicates that the pseudoparaphyses may be ascending and descending. The ostiole of the perithecium is formed lysigenously by breakdown of a pre-existing mass of thin-walled cells (Fig. 17.3a). The bitunicate structure of mature asci is difficult to discern because, as the ascus expands, the inner wall protrudes through a thin area in the outer wall at the ascus tip (Fig. 17.3e) and then the inner wall extends. Thus the ascus tip in expanded asci is single-walled. The ascospores have about 11 transverse septa and are discharged successively at intervals of about 5 s.

Associated with the thick-walled conical pseudothecia on the nettle stem are thinner-walled, slightly smaller, globose pycnidia with cylindrical necks (Figs. 17.2b and 17.3b). The cavity of the pycnidia is lined by small spherical cells which give rise to numerous rod-shaped conidia (Fig. 17.3c). These are dispersed by rain splash or in water films and are capable of germination, which suggests that they do not function as spermatia. Such pycnidia have been

Table 17.1.	Anamorphic states of some species of *Leptosphaeria* and *Phaeosphaeria*.	
Teleomorph	Anamorph	Disease
L. bicolor	*Stagonospora* sp.	Sugarcane leaf scorch
L. coniothyrium	*Coniothyrium fuckelii*	Raspberry cane blight
L. maculans	*Phoma lingam*	Blackleg of oilseed rape and other brassicas
P. avenaria	*Stagonospora avenae*	Oat leaf blotch
P. microscopica	*Phaeoseptoria festucae*	Leaf spot of fescue and other grasses
P. nodorum	*Stagonospora nodorum*	Glume and leaf blotch of wheat and barley

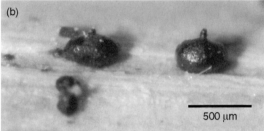

Fig 17.2 *Leptosphaeria acuta* on overwintered nettle stem. (a) Pseudothecia. (b) *Phoma* anamorph. Both images to same scale.

named *Phoma acuta*, and culture studies have confirmed that this stage is the conidial state of *L. acuta* (Müller & Tomasevic, 1957). Pseudothecia of *L. acuta* ripen in the spring and discharge ascospores as the new season's nettle shoots are elongating. There are no obvious disease symptoms on infected plants during the summer and the fungus persists as a symptomless endophyte, with the pseudothecia and pycnidia developing during the winter on dead stems of the past season's growth.

17.2.2 *Leptosphaeria* and *Phaeosphaeria*

An unusually high degree of taxonomic confusion has arisen in attempts to delimit and define several important plant-pathogenic species which clearly belong to the Pleosporales but show immense variations in their anamorph and teleomorph features. One example is a group of coelomycetous species comprising *Septoria* with multiseptate conidia more than 10 times longer than wide, *Stagonospora* with multiseptate conidia less than 10 times longer than wide (Fig. 17.4), and *Phoma* with aseptate globose or slightly elongated conidia (Figs. 17.6 and 17.7). The most common sexual states associated with *Septoria* and *Stagonospora* are *Leptosphaeria*

and *Phaeosphaeria* (Leuchtmann, 1984). As with the anamorphs, the morphological features of these teleomorphs show transitions, but a broad distinction can be made by *Leptosphaeria* being associated mainly with dicotyledons whereas *Phaeosphaeria* is pathogenic on monocotyledons (Cunfer & Ueng, 1999). These two genera have also been separated by critical DNA sequence analyses (Câmara *et al.*, 2002).

Among the cereal pathogens belonging here, *Phaeosphaeria* (*Stagonospora*) *nodorum* has two *formae speciales*, one on wheat and the other on barley, whereas *Phaeosphaeria avenaria* (*Stagonospora avenae*) is a more diverse species complex (Ueng *et al.*, 1998) but is not as serious a pathogen in most agricultural situations. These fungi cause leaf blotch diseases which are very common in most cereal-growing regions and often become the major cereal leaf disease in wet and cool conditions. Epidemics are slowed during periods of dry weather. Serious infections can cause heavy crop losses if they start early and affect the uppermost leaves, which contribute most photosynthetic product to the developing grains. *Stagonospora nodorum* also infects the heads of wheat, causing glume blotch. Lesions develop as small brown spots which enlarge into irregular brownish necrotic areas, giving the leaf a speckled appearance. Within the necrotic areas, pycnidia develop beneath the epidermis which they eventually pierce, releasing a tendril of cylindrical, three-septate pycnidiospores (Fig. 17.4d). These can spread the disease to neighbouring plants by rainsplash. In *S. nodorum*, a second pycnidial state containing minute unicellular conidia has also been discovered (Harrower, 1976). These infect host leaves via germ tubes which penetrate the stomata, whereas penetration from *Stagonospora*-type conidia is directly through the cuticle (Karjalainen & Lounatmaa, 1986). There is no evidence that either type of conidium plays a sexual (spermatial) role. Low temperatures (5–10°C) and irradiation with ultraviolet light favour conidium production. Both *S. avenae* and *S. nodorum* survive the winter either on living volunteer plants or as pycnidia and pseudothecia on stubble. Epidemics can also be started from infected seeds, and from ascospores which are released from pseudothecia

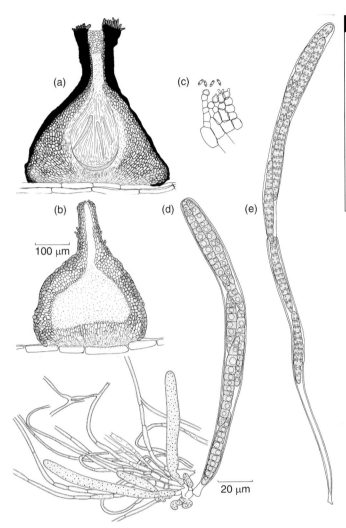

100 µm

20 µm

Fig 17.3 *Leptosphaeria acuta.* (a) L.S. immature pseudothecium. Note the young asci (stippled) elongating between a pre-formed mass of pseudoparaphyses, and the thin-walled cells which block the ostiole at this stage but later dissolve. The centrum subsequently enlarges, dissolving the pseudoparenchyma surrounding it. (b) L.S. pycnidium. (c) High-power drawing of cells lining the pycnidium, showing the origin of the conidia. (d) Cluster of developing asci from a young pseudothecium. Note the branching of the pseudoparaphyses. (e) Stretched bitunicate ascus showing rupture of the outer wall at its apex.

in the spring and autumn. Good reviews of *S. avenae* and *S. nodorum* have been written by Eyal (1999) and Cunfer (2000).

An important disease caused by *Leptosphaeria maculans* (anamorph *Phoma lingam*) is blackleg disease of winter oilseed rape (*Brassica napus*) and other brassicas. As the name suggests, the disease symptoms are seen mainly at the stem bases and main roots, although the foliage can also be affected. Infection may occur through leaves, followed by an endophytic phase lasting several months, before the blackleg symptoms manifest themselves. Lesions eventually turn pale and necrotic, and pycnidia are produced within them. Where the cortex of infected stems cracks open, cankers develop. Rain-splashed

pycnidiospores spread the disease during the growing season. The fungus overwinters on stubble and infects the new crop by ascospores released from pseudothecia, although it can also be seed-borne. The biology of this fungus has been reviewed by Rouxel and Balesdent (2005).

Leptosphaeria maculans comprises several morphologically similar species which can be assigned to two groups, the highly destructive A group which forms cankers and a less damaging B group which is confined to infections of leaves and the pith (Williams & Fitt, 1999). The B pathotype has recently been named *L. biglobosa* (Shoemaker & Brun, 2001). Some of the disease symptoms are probably caused by toxins produced by the A pathotype in planta

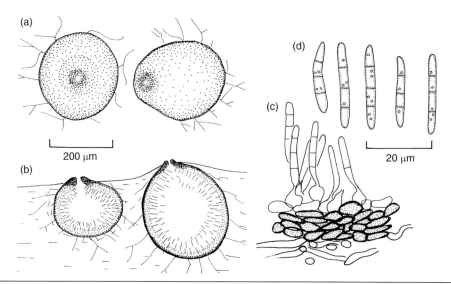

Fig 17.4 *Stagonospora nodorum.* (a) Pycnidia seen from above. (b) Pycnidia in section in agar culture. (c) Portion of wall of pycnidium showing origin of conidia. (d) Conidia. (a,b) to same scale; (c,d) to same scale.

(Pedras & Biesenthal, 1998; Howlett *et al.*, 2001). In addition, *L. maculans* is able to detoxify phytoalexins produced by the host plant (Pedras *et al.*, 2000). Since *P. lingam* is not closely related to most other *Phoma* spp., it has been re-named *Plenodomus* (Reddy *et al.*, 1998).

Diverse anamorphic states are known in other species of *Leptosphaeria* and *Phaeosphaeria*, and some examples of important pathogens are summarized in Table 17.1.

17.2.3 *Ascochyta* and *Phoma*

Several related *Ascochyta*-type anamorphs infect legumes, causing diseases such as blight of chickpea (*A. rabiei*), foot rot and blight of peas (*A. pinodes*), and leaf- and pod-spot of broad beans (*A. fabae*) and peas (*A. pisi*). The pseudothecial state, where present, is now called *Didymella* (formerly *Mycosphaerella*). These species usually overwinter on crop residues, infecting new crops in spring as ascospores, but they are also seedborne. During the growing season, the infection is spread by rainsplash of pycnidiospores. An excellent review of the biology of *Ascochyta*, highlighting the severity of diseases which can be caused, has been written by Pande *et al.* (2005) for *A. rabiei*. Pycnidia contain hyaline, two-celled conidia which, according to Brewer and Boerema

(1965), arise by septation from the conidiogenous cell. A few conidia with no, two or three septa may be produced occasionally (Fig. 17.5). Extensive work has been carried out on characterizing the mating type idiomorphs of *Ascochyta* spp. These data and analyses of other sequences have shown that *Ascochyta* spp. belong to the Pleosporales and are not closely related to *Mycosphaerella*, which is in the Dothideales (Barve *et al.*, 2003).

Phoma medicaginis (Fig. 17.6) has also been shown to be related to *Ascochyta* spp. (Fatehi *et al.*, 2003). Unfortunately, little information is available to circumscribe the very large genus *Phoma* which is almost certainly polyphyletic, being associated with several different teleomorphs. However, a valuable and much-needed monograph of this difficult group has been written by Boerema *et al.* (2004). The conidial fructification of *Phoma* is a pycnidium which usually has only one ostiole, rarely two (Figs. 17.6a and 17.7b). Like most pycnidia, its outermost wall layer is pigmented and consists of cells of distinct shape, which is a useful feature in species identification. In the case of *Phoma* spp., it is called a *textura angularis* (Fig. 17.6a). The cavity of the pycnidium is lined by a hymenium of conidiogenous cells from which one-celled hyaline pycnidiospores develop. The absence of filiform

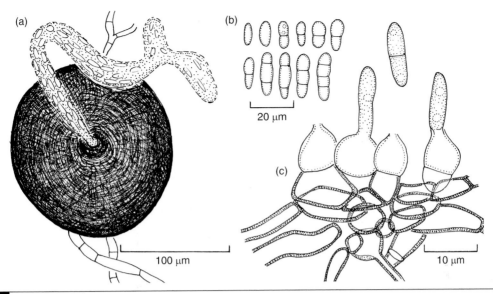

Fig 17.5 *Ascochyta pisi.* (a) Pycnidium seen from above, showing a cirrhus of conidia oozing from the ostiole. (b) Conidia (pycnidiospores). (c) Portion of pycnidium wall in section, showing origin of pycnidiospores.

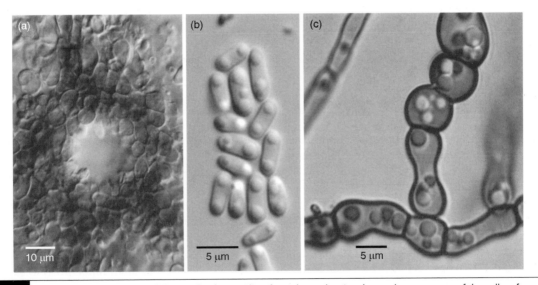

Fig 17.6 *Phoma medicaginis.* (a) View of the ostiole of a pycnidium from above, showing the angular appearance of the wall surface comprising a *textura angularis.* (b) Conidia (pycnidiospores), most of them containing two lipid droplets. (c) Pigmented chlamydospores formed by old hyphae embedded in agar. The light-refractive globose bodies inside the chlamydospores are lipid droplets.

β-conidia distinguishes *Phoma* from *Phomopsis*, which belongs to an altogether different group (see p. 373). The pycnidiospores often ooze out from the ostiole as a tendril (cirrhus).

The conidiogenous cells of *Phoma* are very small, and details of conidium development are difficult to discern with the light microscope (Fig. 17.7b). Brewer and Boerema (1965) have therefore studied spore development with the electron microscope. They described the process of spore formation as a monopolar, repetitive budding of the small, undifferentiated cells of the pycnidial wall. As repeated spore formation occurs, the apex of the conidiogenous cell

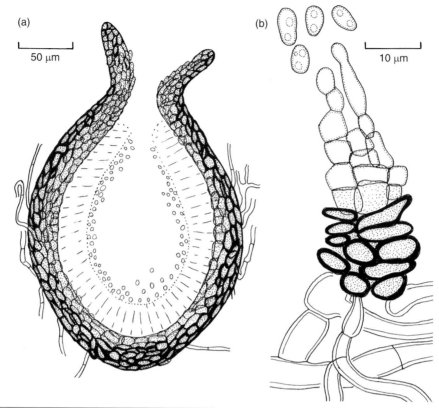

Fig 17.7 *Phoma betae.* (a) L.S. pycnidium. (b) Portion of pycnidium wall, showing conidiogenous cells and conidia.

develops a thickened rim which resembles a phialide or an annellophore (annellide). Sutton (1980) has interpreted these spores as phialoconidia.

Phoma epicoccina is an unusual species because it possesses a hyphomycetous synanamorph, *Epicoccum nigrum* (Fig. 17.8), which is very common on decaying plant material and in the soil. The *Epicoccum* state takes the form of cushion-shaped sporodochia, which are black or purplish-red in colour and covered with rough-walled, warted, segmented, brownish-red conidia (Fig. 17.8a). The conidia may be violently projected from the sporodochium, probably by the rounding off of the two turgid cells on either side of the septum which separates the conidium from its conidiophore (Fig. 17.8c; Webster, 1966). It is possible that conidial discharge is stimulated by drying because the peak concentration of *Epicoccum* spores in air occurs shortly before noon (Meredith, 1966). The synanamorphic

nature of *P. epicoccina* and *E. nigrum* has been conclusively demonstrated only relatively recently (Arenal *et al.*, 2000).

17.2.4 *Pleospora*

Estimates of the number of species of *Pleospora* vary, and Kirk *et al.* (2001) suggested that there are about 50. Most species form fruit bodies on moribund herbaceous stems apparently as saprotrophs, but some are weak pathogens. Of these, *P. bjoerlingii* (= *P. betae*) is the cause of blackleg of sugar beet. *Pleospora scirpicola* forms its pseudothecia on the underwater parts of the culms of the bulrush, *Schoenoplectus lacustris*, and was the first fungus in which the 'jack-in-the-box' mechanism of discharge of bitunicate asci was illustrated (see Fig. 17.1).

Pleospora herbarum attacks a wide range of cultivated hosts, causing such diseases as net blotch of broad bean and leaf spot of clover, lucerne and other hosts. It may be seed-borne.

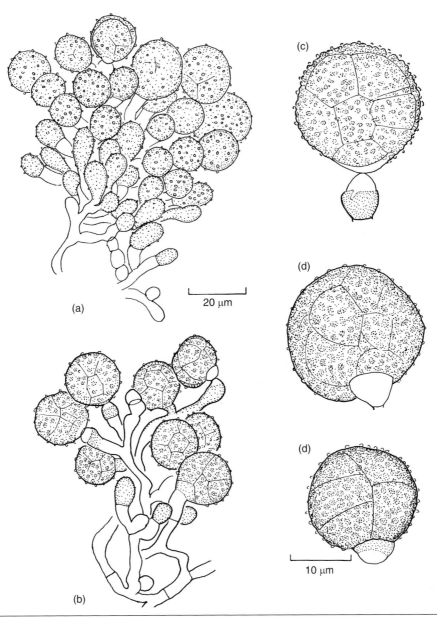

Fig 17.8 *Epicoccum nigrum.* (a) Young sporodochium. (b) Conidiophores and conidia. (c) Conidium almost separated from the conidiophore. Note the bulging septum of the conidiophore. (d) Two detached conidia.

The pseudothecia are common on overwintered stems of herbaceous maritime plants. The large, black, somewhat flattened pseudothecia contain broad, sac-like bitunicate asci, with eight yellowish-brown, slipper-shaped ascospores with transverse and longitudinal septa (Fig. 17.9a). The anamorphic state, *Stemphylium*, produces pigmented muriform conidia and is often associated with the pseudothecia. The connection between the teleomorphic and anamorphic states is readily demonstrated by shooting ascospores onto an agar surface where they germinate and the resulting mycelium develops conidia within a few days. Conversely, cultures started from a single conidium develop pseudothecia within a few weeks

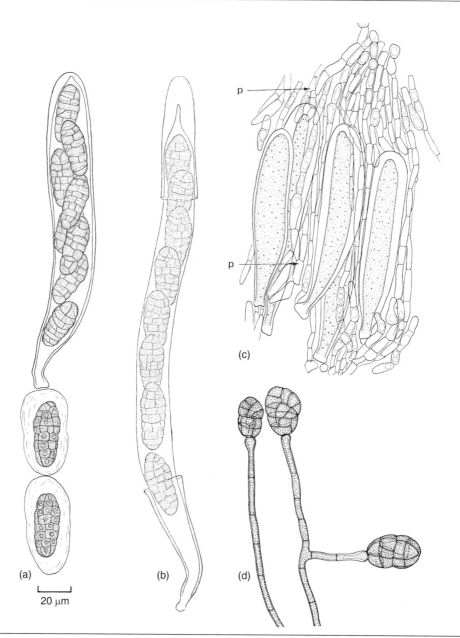

Fig 17.9 *Pleospora herbarum.* (a) Ascus and ascospores showing mucilaginous epispore. (b) Stretched bitunicate ascus showing rupture of outer wall. (c) Developing asci and pseudoparaphyses. The arrows (p) indicate points of branching of ascending and descending pseudoparaphyses. (d) Conidia of *Stemphylium* type.

(Weber & Webster, 2000b). The *P. herbarum* complex includes a number of similar species forming conidia which are critically different from each other in morphology and dimensions (Simmons, 1969). A distinctive form, *S. vesicarium*, is associated with leaf blight of onions and garlic.

Stemphylium conidia develop singly from the tips of conidiophores swollen at their apices, as blown-out ends. A narrow neck of cytoplasm connects the developing spore to its conidiophore through a pore, and Hughes (1953) has termed conidia of this type **porospores** (Fig. 17.9d), but the term poroconidium is also used (Ellis, 1971b). Electron microscopy studies (Carroll & Carroll, 1971) have shown that conidial development is blastic, involving the whole of the wall at the apex of the conidiogenous cell. The cytoplasmic connection between the conidiogenous cell and the conidium is narrow, and is surrounded by two layers of thickened wall material. Following the detachment of the first-formed conidium, the conidiophore may grow out through the detachment scar to form a second conidium, a process described as **percurrent** conidiogenesis. The conidia of *P. herbarum* are formed more readily in cultures illuminated by near-UV light (Leach, 1968), whereas daylight and low temperature stimulate pseudothecial development (Leach, 1971). The fungus is homothallic. According to Meredith (1965) the conidia are violently jolted from the tip of the conidiophores.

17.2.5 *Lewia*

The genus *Lewia* was named in honour of L. E. Wehmeyer by Simmons (1986) for *Pleospora*-like fungi with *Alternaria* anamorphs. Six ascocarpic species have been recognized, fruiting on grasses (including cereals) and on dicotyledonous hosts (including *Brassica* and *Pastinaca*) (Kwasna & Kosiak, 2003). The separation of *Lewia* from *Pleospora* is supported by molecular evidence (Pryor & Gilbertson, 2000).

Lewia infectoria (= *Pleospora infectoria*) forms black, shining, subepidermal pseudothecia on overwintered grass and cereal culms. It has golden-brown muriform ascospores with up to five transverse septa. The central cells of the ascospores also contain one or rarely two longitudinal septa (Fig. 17.10a). In culture, this fungus forms branching chains of obclavate, brown-coloured (melanized), muriform, beaked spores (**dictyospores** or **dictyoconidia**) and new spores are formed at the tip of the chain (Fig. 17.10c). A darkly pigmented thickened annulus is visible at the base of the conidium and at the apex of the conidiophore surrounding the point of spore separation and, if the spore has occupied an intercalary position on the spore chain, there is also an annulus at the opposite end. Chain branching occurs where a conidium

produces more than one spore. Conidia of this type have been classified in the anamorph genus *Alternaria*, and are poroconidia. The conidial state of *L. infectoria* is *A. infectoria*.

17.2.6 *Alternaria*

About 50 species of *Alternaria* are known which have not been connected to a teleomorph (Neergaard, 1945; Joly, 1964; Simmons, 1986; Kirk *et al.*, 2001). The taxonomy of *Alternaria* is difficult. Simmons (1992) has given a key to 10 species-groups and Ellis (1971a, 1976) has described and figured some common species. Despite the absence of formal evidence for sexual reproduction in many species of *Alternaria*, Berbee *et al.* (2003) have shown that three species of *Alternaria* not known to have sexual states, *A. brassicae*, *A. brassicicola* and *A. tenuissima*, have mating type gene sequences. In any one isolate of these species, only one mating type idiomorph was found, but in other isolates of the same species the opposite idiomorph was detected. This suggests that these currently asexual species were derived from sexually reproducing ancestors.

The fine structure of conidial development from a pre-existing conidium in *A. brassicae* has been studied by Campbell (1968). The mature spore has a two-layered wall, the outer of which is melanized. A pore develops in the outer wall, probably by enzymatic activity, and the inner wall layer expands through the pore to become the primary wall of the new conidium. Later, this in turn becomes two-layered. Transverse and longitudinal septa develop within the spore, but these are incomplete; a pore in each septum allows cytoplasmic continuity between adjacent cells and flow of cytoplasm through the spore to provide material for the formation of new spores at the tip of the chain.

The shape of the conidium in *Alternaria* affects its aerodynamic properties. Several species have conidia with long beaks, e.g. *A. solani* and *A. brassicae* (Fig. 17.11). It has been suggested that the long beaks increase the chance of wind-dispersal as compared to species with smaller, non-beaked conidia (Chou & Wu, 2002). Long beaks also increase the drag on the

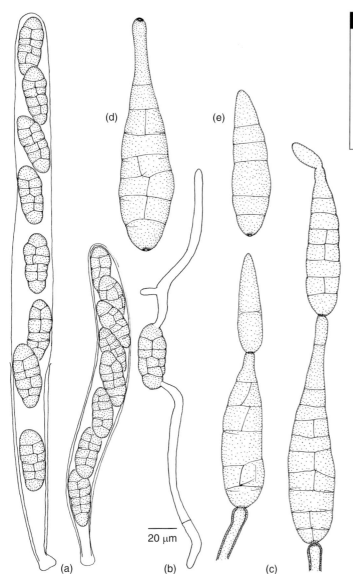

(d)

(e)

20 μm

(a) (b) (c)

Fig 17.10 *Lewia infectoria.* (a) Asci, one intact, the other stretched by expansion of the endotunica. (b) Germinating ascospore. (c) *Alternaria infectoria* conidial state developed in culture from a germinating ascospore. (d) A conidium from an intercalary position in a spore chain showing a scar at each end of the conidium. (e) A conidium from the end of a spore chain with only one basal scar.

spore, reducing its settling velocity (McCartney *et al.*, 1993).

Some species of *Alternaria* are of considerable economic significance (Chelkowski & Visconti, 1992; Rotem, 1994). *Alternaria alternata* (= *A. tenuis*) is the name given to a widespread and cosmopolitan opportunistic saprotroph, reported on all kinds of senescent plant material. Unfortunately, because of difficulties in identification, it is probable that the name encompasses several distinct taxa (Roberts *et al.*, 2000). *Alternaria* spp. are associated with diseases of crops, often showing a degree of host specificity

as indicated in Table 17.2. Most of these diseases are seed-borne; seeds become infected from the flowering stage onwards (Rotem, 1994). The host-specific pathogens may have evolved from non-specific saprotrophic forms of 'A. *alternata*' by the selection and multiplication of mutants capable of secreting host-specific toxins (Scheffer, 1992; Rotem, 1994; Thomma, 2003). Many of these host-specific toxins have been characterized chemically (Otani & Kohmoto, 1992). Numerous mycotoxins are also produced by *Alternaria* spp. (Montemurro & Visconti, 1992), and some have severe and fatal consequences if they accumulate

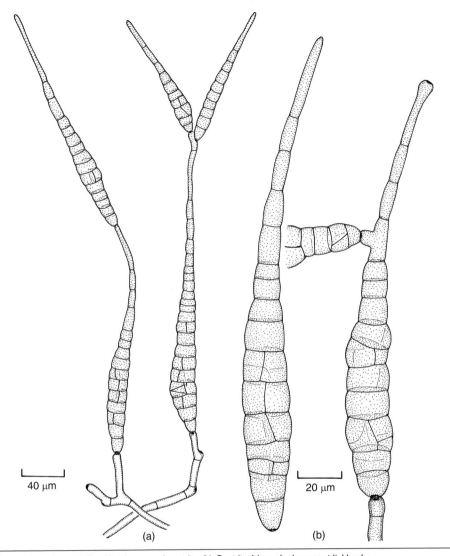

Fig 17.11 *Alternaria brassicae.* (a) Conidiophores and conidia. (b) Conidia. Note the long conidial beaks.

in human food and animal feed. Spores of *Alternaria* spp. are abundant in the air in late summer and autumn and may be a cause of inhalant allergy causing asthma. Some species, e.g. '*A. alternata*', *A. infectoria* and *A. longipes* are rare opportunistic human pathogens, associated with diseases of bone, cutaneous tissue, ears, eyes, nose and the urinary tract (Schell, 2003). Other species are insect pathogens.

17.2.7 *Cochliobolus*

In *Cochliobolus*, the pseudothecium is long-necked and contains elongate, transversely septate

ascospores spirally coiled around each other within a vestigially bitunicate ascus (see Figs. 8.15b–d).

Anamorphs associated with *Cochliobolus*

The genus *Cochliobolus* contains some of the best-studied and most highly damaging plant pathogens. As in other Loculoascomycetes, the taxonomic history of this genus has been tortuous (Sivanesan, 1987; Alcorn, 1988); the anamorphs were formerly classified in *Helminthosporium*, then transferred to *Drechslera*, and are now called *Bipolaris* (Fig. 17.13) and

Table 17.2.	Selected plant-pathogenic species of *Alternaria* (from Holliday, 1998).	
Species of *Alternaria*	Disease	Comments
'A. alternata'	Wide range of hosts and diseases	
'A. alternata' f. sp. *lycopersici*	Tomato stem canker	
A. brassicae	Brassica grey leaf spot	Seed-borne
A. brassicicola	Brassica black leaf spot	Seed-borne
A. carthami	Safflower leaf spot	Seed-borne
A. dauci	Carrot leaf blight	Seed-borne
A. dianthi	Carnation leaf blight	Seed-borne
A. linicola	Linseed seedling blight	Seed-borne
A. macrospora	Cotton leaf spot	Seed-borne
A. mali	Apple core rot	
A. porri	Onion purple blotch	Seed-borne
A. radicina	Carrot black rot; also infects celery and parsnip	
A. solani	Potato early blight and tuber rot	
	Tomato early blight and fruit rot	

Curvularia (Fig. 17.14). The revised genus *Helminthosporium* is now rather small, with about 20 species and *H. velutinum* as the type-species (Fig. 17.12). It has affinities to *Leptosphaeria* (Olivier *et al.*, 2000). *Drechslera*, in contrast, is the anamorphic state of *Pyrenophora* (Fig. 17.16; see p. 477). All these anamorphic forms produce pigmented (melanized) spores with only transverse septa, but they differ in their pattern of conidiogenesis and in the ultrastructural appearance of their cell walls.

Sporogenesis by *Helminthosporium* is tretic, i.e. porospores are formed as described for *Stemphylium* by the digestion of a small hole into the conidiophore cell wall, followed by an extension of the inner wall to form the conidium which then lays down its outer wall (Fig. 17.12c). In contrast to *Pleospora*, each site can produce only one conidium, and conidia can develop apically or laterally on the conidiophore. If the conidium is produced apically, the conidiophore cannot grow further. The sequence of conidium development in *Bipolaris* and *Curvularia* is not of this type. In *Bipolaris*, the first conidium always develops apically, and subsequent conidia are formed either by growing through the scar left by the first conidium, or by the conidiophore growing past the first conidium to form a new apex producing the second conidium

(*B. sorokiniana*; Fig. 17.13b). *Bipolaris* is so named because the conidia germinate by emitting two germ tubes, one at either end (Fig. 17.13d), and these grow as extensions of the long axis of the spore. This is in contrast to *Helminthosporium* conidia in which each cell is principally capable of germination, and the germ tubes grow perpendicular to the long axis of the spore.

In *Curvularia*, the conidia are curved because of an unevenly swollen central cell. The end cells are usually less strongly pigmented than the central cell (Figs. 17.14b,c). The first conidium also develops at the apex of the conidiophore as a poroconidium extending through a tiny pore, and then the conidiophore develops a new subterminal growing point from which a second conidium initial arises. The process is repeated so that a succession of new apices, each terminated by a conidium, is formed (Fig. 17.14a). The term **sympodula** has been applied to such a conidiophore producing conidia sympodially (Kendrick & Cole, 1968). In some *Curvularia* spp., such as *C. cymbopogonis* (Fig. 17.14c), the base of the conidium bears a protuberant hilum.

A feature of the conidia of *Drechslera*, *Helminthosporium* and *Bipolaris*, but not *Curvularia*, is that they are **distoseptate** (Luttrell, 1963). This means that the wall separating adjacent conidial segments is visibly

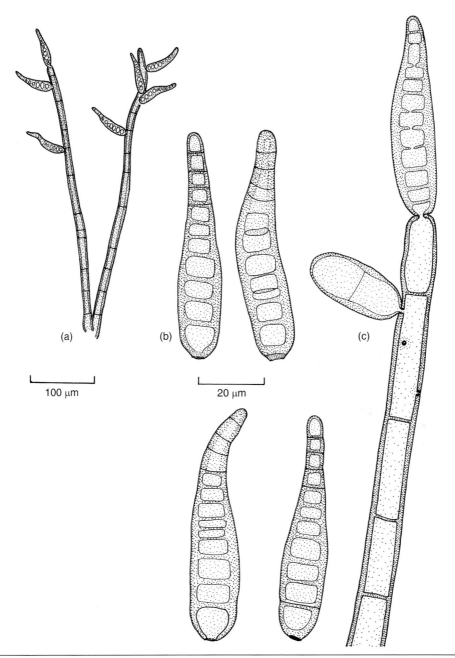

Fig 17.12 *Helminthosporium velutinum.* (a) Conidiophores and conidia. (b) Detached conidia. (c) Details of conidial development. Note the narrow channels in the wall through which cytoplasm passes to the developing conidia. This type of development is tretic. (b) and (c) to same scale.

different from the outer wall surrounding the entire conidium (White *et al.*, 1973). This can be readily seen even with the light microscope because the individual conidial cells of *Drechslera*, *Helminthosporium* and *Bipolaris* appear like peas in a pod (Figs. 17.12, 17.13, 17.16). In *Curvularia*, the conidial septa are of the normal (**euseptate**) type (Fig. 17.14).

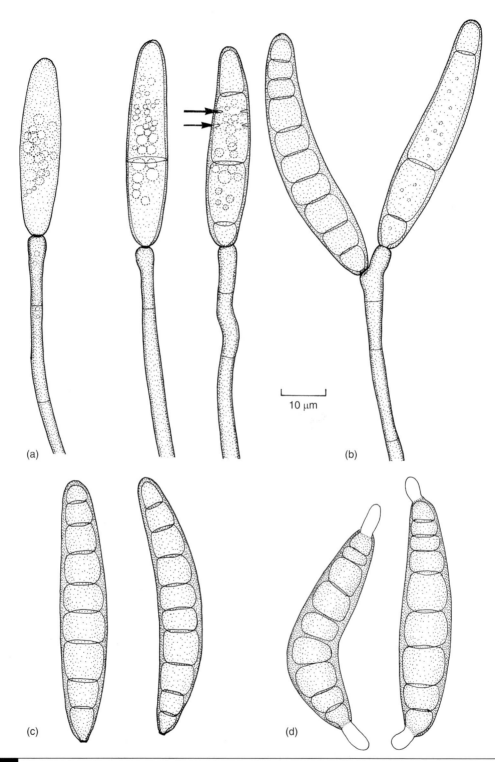

Fig 17.13 *Bipolaris sorokiniana*, the conidial state of *Cochliobolus sativus*. (a) Developing conidia. The arrows point to developing septa. (b) Conidiophore showing the development of a second conidium lateral to the first. (c) Mature conidia. (d) Two germinating conidia showing emergence of germ tubes from each end of the conidium.

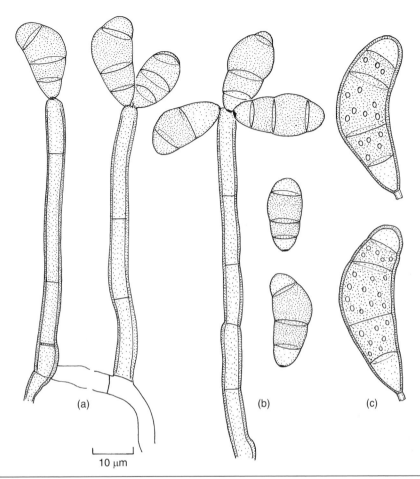

Fig 17.14 *Curvularia* spp. (a) *Curvularia lunata*, the conidial state of *Cochliobolus lunatus*. Conidiophores showing sequence of conidial development. (b) Mature detached conidia of *C. lunata*. Note the paler end cells. (c) Mature detached conidia of *C. cymbopogonis* showing the protuberant hilum.

The pathology of *Cochliobolus*

Cochliobolus has been examined by DNA sequencing methods (Berbee *et al.*, 1999), which revealed two separate groups. All important pathogens belong to one group and have *Bipolaris* anamorphs. In general, the teleomorphs are rare or absent in nature, and the diseases are carried mainly by the thick-walled conidia, which can survive in the soil but can also infect seeds. *Cochliobolus sativus* (*B. sorokiniana*) is the cause of a variety of root and leaf necroses of cereals (especially wheat and barley) in warm humid climates of South East Asia, Australia, North and South America. A good review of its biology has been written by Kumar *et al.* (2002).

It is a typical hemibiotrophic pathogen which forms quiescent infections in the first-infected epidermal cell, followed by a necrotrophic phase in which the surrounding tissue is aggressively invaded. This pathogen shares with *Botrytis cinerea* (see p. 435) the ability to evade the oxidative burst launched as a defence response by the newly infected host (Kumar *et al.*, 2001). Several sesquiterpene-type toxins are produced and contribute to a weakening of the host; prehelminthosporol (Fig. 17.15a) seems to be the most potent. These toxins act in a non-specific manner on several different cellular processes. Interestingly, barley cultivars possessing the *mlo*-type resistance against powdery

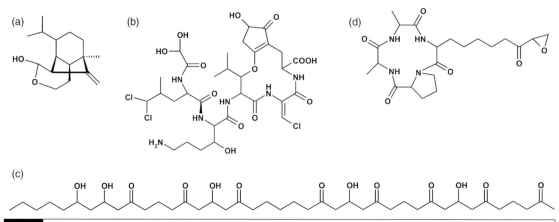

Fig 17.15 Toxins produced by *Cochliobolus* spp. a. The sesquiterpene prehelminthosporol produced by *C. sativus*. b. The chlorinated pentapeptide victorin C produced by *C. victoriae*. c. The dominant form of T-toxin, a polyketide produced by *C. heterostrophus*. d. The major form of HC-toxin, a tetrapeptide produced by *C. carbonum*.

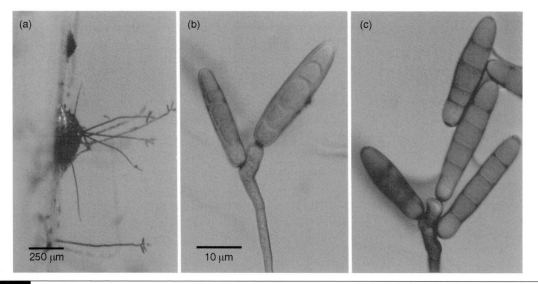

Fig 17.16 *Pyrenophora tritici-repentis*. (a) Immature pseudothecial stroma on overwintered barley stubble, with conidia produced on the setae and on separate conidiophores. (b,c) Conidia of *Drechslera tritici-repentis* mounted in lactic acid (b) showing the typical 'peas-in-a-pod' appearance, and mounted in water (c). (b) and (c) to same scale.

mildew (see p. 408) are highly susceptible to *C. sativus*. A hypersensitive response which effectively suppresses biotrophic pathogens may thus actually enhance pathogenicity by others with a necrotrophic potential.

A similar feature was observed with *C. victoriae*, which became pathogenic on the Victoria cultivar of oats bred for resistance against the crown rust, *Puccinia coronata* (Meehan & Murphy, 1946). The main toxin produced by *C. victoriae*, victorin C (Fig. 17.15b), is one of several related cyclic pentapeptides which are absolutely essential and sufficient for pathogenesis, i.e. the purified toxin can reproduce the disease symptoms which consist of widespread leaf necrosis, and strains of *C. victoriae* not synthesizing this toxin do not cause the disease. It now seems that victorin triggers a form of hypersensitive response, i.e. widespread host cell death (Navarre & Wolpert,

1999). Victorin binds to mitochondrial proteins, and it is possible that it initiates cell death via mitochondrial dysfunction (Curtis & Wolpert, 2002). Intriguingly, in the mammalian equivalent of the hypersensitive response (i.e. apoptosis), mitochondria are also among the first organelles to break down.

Mitochondria are also the target of the specific toxin produced by *C. heterostrophus* (*B. maydis*), which caused the well-described southern leaf blight of corn in the United States in 1970 (Ullstrup, 1972; Schuman, 1991). Prior to that epidemic, maize breeders had relied heavily on male-sterile cultivars, i.e. cultivars which do not produce viable pollen and are therefore dependent on pollen from another cultivar for seed production. This facilitated the production of hybrids, i.e. the F1 progeny of genetically dissimilar parents. These often produce particularly high yields, a phenomenon known as hybrid vigour. Male sterility was achieved by a mitochondrial mutation called *cms-T* (cytoplasmic male sterility). The plant line used as the pollen donor did not contain the *cms-T* mutation, so that the hybrid seeds grew into plants capable of producing pollen in the field of the farmers. In 1970, about 80% of the maize crop contained *cms-T* cytoplasm. At about the same time a mutation in a minor leaf spot pathogen, *C. heterostrophus* race O, spread in the field. This new race produced several related polyketides collectively called T-toxin (Fig. 17.15c), which specifically affected the mitochondria of hybrid maize. The epidemic of 1970 resulted in crop losses totalling over US $1 billion for that year. The target site of T-toxin in *cms-T* maize mitochondria is now known to reside in a small protein present as a tetramer in the outer mitochondrial membrane. T-toxin binds directly to this protein, causing conformational changes which open up pores and render mitochondria leaky (Levings *et al.*, 1995; Wolpert *et al.*, 2002). Race O causes only minor leaf spots on *cms-T* as well as other cultivars of maize.

Cochliobolus carbonum is the cause of northern leaf spot and ear rot of maize. There are three races, of which only race 1 is a serious pathogen on all those cultivars of maize containing two recessive alleles of the *Hm1* resistance gene. This strongly enhanced pathogenicity is caused by a group of cyclic tetrapeptides collectively called HC-toxin (Fig. 17.15d), where HC stands for *Helminthosporium carbonum*, the former name of the anamorph. Resistant cultivars produce an enzyme which detoxifies HC-toxin. The susceptibility of certain maize cultivars was caused by a simultaneous inactivation of both duplicate genes encoding the enzyme, HC-toxin reductase (Multani *et al.*, 1998). The mode of action of HC-toxin is not yet clear; it seems to inhibit, rather than induce, defence responses (Wolpert *et al.*, 2002).

As we have seen, certain strains of *C. carbonum*, *C. heterostrophus* and *C. victoriae* have caused catastrophic epidemics on particular cereal hosts employing biochemically unrelated toxins. This specificity of action − i.e. a specific pathogen race being pathogenic only against certain host cultivars − paved the way towards an understanding of the gene-for-gene concept (see pp. 112 and 397). A further question of interest is the origin of these highly aggressive strains. A partial answer was found somewhat fortuitously by an examination of the mating type genes in all three pathogens. Both idiomorphs *MAT-1* and *MAT-2* have been found in various field isolates of *C. heterostrophus* and *C. carbonum*, but all known isolates of *C. victoriae* belong to *MAT-2*. Further, *C. victoriae* and *MAT-1* strains of *C. carbonum* are interfertile. These observations have led to the suggestion that *C. victoriae* arose from a *MAT-2* strain of *C. carbonum* which received the gene cluster for pathogenicity on oats (i.e. the genes encoding the enzymes necessary for toxin synthesis) by horizontal gene transfer (Christiansen *et al.*, 1998). The integration of this gene cluster must have been close to the *MAT-2* locus, so that it did not spread to *MAT-1* strains of *C. carbonum* by crossing-over.

17.2.8 *Pyrenophora* (anamorph *Drechslera*)

In its taxonomically restricted use, the *Drechslera* state (Fig. 17.16b) is the conidial form of *Pyrenophora* and is clearly defined as a monophyletic group (Zhang & Berbee, 2001). *Drechslera* is commonly found in the field, whereas *Pyrenophora*-type pseudothecia are uncommon. Species belonging to this group are pathogens of cereals and grasses, and some of them cause significant diseases in agricultural situations. The disease symptoms are similar to those caused by other members of the Pleosporales, and phytotoxic substances are produced by several members of the genus.

Pyrenophora tritici-repentis (anamorph *Drechslera tritici-repentis*) occurs on a range of grasses, including, as the name suggests, *Agropyron* (formerly *Triticum*) *repens* and wheat (*Triticum aestivum*). The disease caused is known as yellow leaf spot or tan spot of wheat (De Wolf *et al.*, 1998). *Pyrenophora tritici-repentis* is spread as seed-borne infections but also overwinters on infected stubble, which is the most important source of inoculum, giving rise to pseudothecia and conidia in spring. Pseudothecia can be identified by their large size and by the presence of dark setae around the pseudothecial neck (Fig. 17.16a). The ascospores are transversely septate, with a longitudinal septum also present in one of the central cells. The conidia of *Drechslera tritici-repentis* are very large and have a variable number of distosepta (Fig. 17.16b). They are sometimes produced on the pseudothecial setae, or they arise directly from stubble or from necrotic leaf lesions. Phytotoxins are involved in causing disease symptoms. Most unusually, they consist of at least two extracellular proteins synthesized by ribosomes (Wolpert *et al.*, 2002). They are a critical factor in determining the host specificity of infections.

Pyrenophora teres is common wherever barley is grown, and is the major barley pathogen, especially in humid regions. This species exists in two forms which are distinguished by their symptoms, *P. teres* f. *teres* causing net blotch on barley leaves and *P. teres* f. *maculata* causing brown leaf spots. These two forms can hybridize in the laboratory and also in the field (Campbell & Crous, 2003). In addition to ascospores and *Drechslera*-type macroconidia similar to those of *P. tritici-repentis*, a pycnidial state producing unicellular conidia is also apparently associated with *P. teres*, although its role in the disease cycle is uncertain (Smith *et al.*, 1988). The epidemiology of the disease is similar to *P. tritici-repentis*, as is the involvement of phytotoxins in causing leaf necrosis. Leaf chlorosis and necrosis can be reproduced by phytotoxins purified from cultures of *P. teres*. However, biochemically the phytotoxins involved are rather different, the most potent of them being the aspartic acid derivative aspergillomarasmine A (Weiergang *et al.*, 2002).

17.2.9 *Venturia*

The genus *Venturia* contains some 50 species which cause scabs, i.e. limited lesions with a scurfy appearance, on the leaves and fruits of various trees. The genus is an unusual member of the Pleosporales in producing asci with one-septate ascospores, but Silva-Hanlin and Hanlin (1999) have confirmed its position within this order. The most important species is *V. inaequalis* which parasitizes apple (*Malus* spp.) and hosts related to it. This fungus is cosmopolitan and extremely common on apple fruits and leaves if fungicide treatments are not carried out (Fig. 17.17a). In many regions, scab is the most serious apple disease. The fungus overwinters on fallen leaves which, in spring, give rise to pseudothecia (Fig. 17.17b) releasing ascospores during periods of wetness. The ascospores require surface wetness in order to infect apple leaves. Infection is mediated by appressoria, but the invasion is limited to the space between the cuticle and the epidermis; the latter is not pierced, and haustoria are not formed. In this way, the fungus persists for several weeks. Conidia are eventually produced from such subcuticular stromata, and they spread the disease during the growing season. Invasion of host tissue takes place only on dead leaves in the autumn when *V. inaequalis* switches to a saprotrophic growth phase and produces pseudothecial initials. The biology of *Venturia*, which has been summarized by MacHardy *et al.* (2001), is thus very unusual for members of the Pleosporales.

From the above summary of the infection biology of *Venturia* it is apparent that the key to apple scab management lies in controlling the ascospore inoculum in spring. One commonly

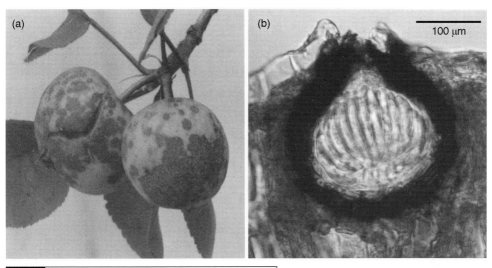

Fig 17.17 *Venturia inaequalis.* (a) Apples showing scab symptoms. (b) Section through a pseudothecium in an overwintered apple leaf.

practised approach is disease forecasting, based on the knowledge that ascospore discharge occurs within 1–2 h of wetting ripe pseudothecia, and that infection requires leaf surface wetness for some 25 h at 6°C or 9 h at 16–24°C (Smith *et al.*, 1988). Under certain circumstances, e.g. after a prolonged dry period, some time will elapse before the pseudothecia have produced a fresh crop of ascospores after the onset of rain, and this can be integrated into forecasting systems (Stensvand *et al.*, 2005). Protective fungicide sprays have to be applied as soon as possible after the onset of conditions conducive to infection, and especially if a high density of air-borne ascospores has already been detected by spore traps or other means (Kollar, 1998). Curative fungicides can be applied one or a few days after infection. Numerous fungicides are in use against apple scab. Protective agents include copper-based formulations which are registered in some countries even for organic farming, or the thiol reactant captan. Important curative fungicides include strobilurins (respiration inhibitors), and myclobutanil and imidazoles (demethylation inhibitors of ergosterol biosynthesis).

A different control strategy is to reduce the available ascospore inoculum in the spring by encouraging the decomposition of leaves during winter. This can be achieved by applying urea to the leaf litter, or by using a flail mower to shred the leaves (Sutton *et al.*, 2000). It may also prove possible to spray leaves before leaf-fall with spores of fungi antagonistic to *Venturia* (Carisse *et al.*, 2000). Yet another approach is the breeding of resistant cultivars.

17.2.10 *Sporormiella*

There are about 70 species of *Sporormiella* (Ahmed & Cain, 1972). Molecular studies indicate that the genus has affinities with Pleosporales (Liu *et al.*, 1999). Most species form pseudothecia on the dung of herbivores, but some are isolated from the soil or as endophytes. Characteristic features are dark, transversely septate ascospores which may disarticulate into separate part-spores, each of which is capable of germination, and whose walls are often marked by a hyaline longitudinal or oblique germ slit. *Sporormiella intermedia* is one of the common species and has thin transparent pseudothecial walls through which asci can be seen (Fig. 17.18a). The ascospores of *S. intermedia* are four-celled and surrounded by a mucilaginous envelope. Spore discharge is nocturnal (Walkey & Harvey, 1966b).

Because of the unmistakable shape of its ascospores and its association with dung, *Sporormiella* has been used in an archaeological

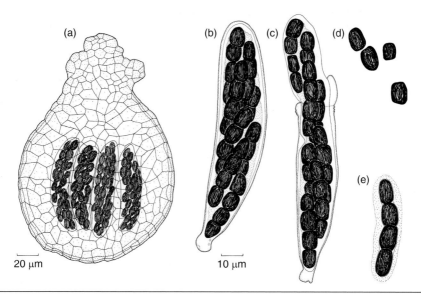

20 μm

10 μm

Fig 17.18 *Sporormiella intermedia*. (a) Pseudothecium with asci visible through the transparent wall. (b) Ripe unextended ascus showing the double wall. (c) Elongating ascus showing rupture of the outer wall (ectotunica) and extension of the inner (endotunica). (d) Ascospore separated into its four component cells. (e) Intact ascospore.

context as an indicator of changes in vegetation and land management. Thus, Burney *et al.* (2003) have shown for the island of Madagascar that *Sporormiella* was very common before the settlement of humans which occurred after AD 200 and then declined in abundance along with the extinction of several groups of large herbivores. After AD 1100, spore densities of *Sporormiella* in sediments showed an increase, coinciding with the introduction of grazing domesticated livestock.

17.3 | Dothideales

This is a large group of ascomycetes containing an enormous variety of conidial forms, both hyphomycetous and pycnidial. Where present, the teleomorph consists of dark-celled pseudothecial ascomata, usually developing as locules within an ascostroma. The asci are bitunicate (fissitunicate). In contrast to the Pleosporales, inter-ascal tissue (hamathecium with pseudoparaphyses) is generally lacking. A relationship between the Dothideales and the Pleosporales has been suggested on the basis of

molecular data (Silva-Hanlin & Hanlin, 1999; Lumbsch & Lindemuth, 2001). Many members are saprotrophic on dead plant material, but some grow as endophytes and some are plant pathogens. The single ascocarpic example which we shall study is *Mycosphaerella*.

17.3.1 *Mycosphaerella*

Mycosphaerella is one of the largest genera of ascomycetes, containing over 2000 described species (Corlett, 1991). However, many of the names are based mainly on the association of a *Mycosphaerella* with a particular host plant. Given the lack of critical inoculation experiments to clarify their host range, mating experiments or DNA sequence comparisons with similar forms on other plants, it is likely that many names are synonyms. Some species are plurivorous, growing on a broad range of monocotyledonous and dicotyledonous hosts. Many species of *Mycosphaerella* cause diseases of economic significance, and some of them are listed in Table 17.3. Most of these diseases involve the necrosis of host plant tissue, and the toxins produced by the pathogens are commonly associated with the disease symptoms (e.g. *Cercospora beticola*; see p. 481).

Table 17.3. Some anamorph genera with *Mycosphaerella* teleomorphs.

Mycosphaerella	Anamorph	Diseases caused
M. graminicola	*Septoria tritici*	Leaf blotch of wheat
M. brassicicola	*Asteromella brassicae*	Ring spot of brassicas
M. tassiana	*Cladosporium herbarum*	
M. berkeleyi	*Passalora personata*	Groundnut defoliation
(unknown)	*Cercospora beticola*	Leaf spot of sugar beet
M. fijiensis	*Paracercospora fijiensis*	Leaf spot of banana
M. musicola	*Pseudocercospora musae*	Sigatoka disease of banana

Crous *et al.* (2000, 2001) and Goodwin and Zismann (2001) listed and discussed the bewildering diversity of anamorph genera (about 23) connected with *Mycosphaerella* teleomorphs, and performed phylogenetic analyses on representatives of most of them. The list includes pycnidial forms such as *Septoria* and *Asteromella*, but also numerous hyphomycetous form-genera such as *Cercospora*, *Pseudocercospora* and *Cladosporium* (Table 17.3). Despite this range of anamorphs, molecular evidence has somewhat surprisingly indicated that the genus *Mycosphaerella* is monophyletic (Crous *et al.*, 2000, 2001; Goodwin & Zismann, 2001). Pseudothecia of *Mycosphaerella* are globose and small, rarely more than 100 μm in diameter. Because they show relatively little variation, they are difficult to identify to species level. Pseudothecia develop subepidermally, usually on leaves. The asci develop in a basal fascicle. The ascospores are hyaline with a single transverse septum (Fig. 17.19). In these features *Mycosphaerella* is similar to *Venturia*, although these two genera are not closely related.

17.3.2 *Mycosphaerella graminicola* (anamorph *Septoria tritici*)

The leaf blotch disease of wheat caused by *M. graminicola* (Fig. 17.20) is very similar to that caused by the wheat strain of *Phaeosphaeria nodorum* (see p. 17.2.2), and the two diseases often co-occur on wheat crops and are controlled in the same way, especially by the application of fungicides. The most important fungicides are strobilurin-type compounds and ergosterol biosynthesis inhibitors. The ascospores of *M. graminicola*, produced from pseudothecia

initially on overwintering stubble and later from infected leaves, are the main source of inoculum, and conidia are thought to be of lesser importance as propagules of the disease (Eyal, 1999). In consequence, the genetic diversity of *M. graminicola* in the field is often very high, with one square metre of infected wheat shown to contain about 70 genetically different strains (Zhan *et al.*, 2001). Infection by germinating ascospores and conidia of *M. graminicola* is almost always through stomata (Duncan & Howard, 2000), in contrast to *Phaeosphaeria nodorum* where it occurs directly through the cuticle. Following penetration, intercellular colonization of the surrounding leaf tissue by hyphae of *M. graminicola* occurs, but the onset of symptom development is delayed. Recent reviews of *M. graminicola* have been written by Eyal (1999) and Palmer and Skinner (2002).

17.3.3 *Cercospora*

This very large form-genus (>1000 species) contains numerous important plant pathogens associated with a wide range of host plants (Farr *et al.*, 1989). Examples are *C. beticola* causing leaf spot of sugar beet, *C. zea-maydis* (grey leaf spot of corn), and *C. coffeicola* (brown eyespot of coffee). It is difficult to estimate the real number of species; Johnson and Valleau (1949) isolated *Cercospora* from 28 host plants in 16 families, and all seemed to belong to the same species. Further, the dimensions of conidia and conidiophores can vary in response to changes in humidity. Where known, the teleomorphs of *Cercospora* spp. belong to *Mycosphaerella* (Goodwin *et al.*, 2001). The conidia of *Cercospora* are

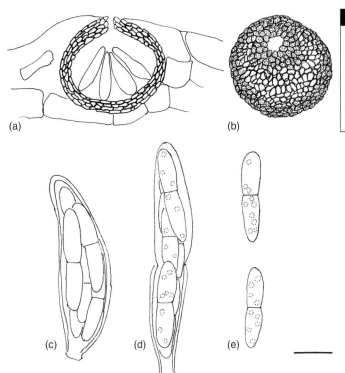

Fig 17.19 *Mycosphaerella brassicicola*. (a) Section of a brassica leaf with a subepidermal pseudothecium containing asci. Note that there is no hamathecium. (b) Pseudothecium as seen from above, showing the ostiole. (c) Intact bitunicate ascus. (d) Stretched ascus resulting from extension of the endotunica. (e) Ascospores. Scale bar: 40 μm (a,b) and 10 μm (c–e).

hyphomycetous, being produced from pigmented aerial hyphae. They are long and tapering, as shown by *C. beticola* (Fig. 17.21), which is seed-borne.

It has long been known that *Cercospora* infections are much less severe in shaded plants as compared to those growing in direct sunlight. The reason for this lies in the production of a potent photosensitizing toxin, cercosporin (Fig. 17.22), by many *Cercospora* spp. A fascinating and lucid account of cercosporin has been written by Daub and Ehrenshaft (2000). Cercosporin, upon absorption of light energy, is activated to an energized state in which it reacts with molecular oxygen, converting it into various radicals but especially into singlet oxygen ($^1O_2$). This energized form of oxygen is highly reactive, rapidly destroying organic molecules and especially membrane lipids. Cercosporin is a non-selective toxin, affecting bacteria, plants, fungi and animals unless these produce protective antioxidants such as carotenoids. It is not yet entirely clear how *Cercospora* spp. protect themselves against their own toxin,

but part of the answer may lie in keeping it in a non-reactive (reduced) state. Another mechanism is the synthesis of vitamin B_6 (pyridoxine) and its derivatives which can act as antioxidants. Cercosporin has been found only in *Cercospora* species (Goodwin *et al.*, 2001), but biosynthetically related photosensitizers are produced by a range of plant-pathogenic fungi, especially in the Loculoascomycetes (e.g. *Cladosporium*, *Alternaria*, *Stemphylium* spp.).

17.3.4 *Cladosporium*

The anamorph genus *Cladosporium* is large, containing about 60 species. A key to species in culture collections has been published by Ho *et al.* (1999). Several species have *Mycosphaerella* teleomorphs, e.g. *C. herbarum* is the anamorph of *M. tassiana* (von Arx, 1950; Barr, 1958), *C. echinulatum* is the anamorph of *M. dianthicola* and *C. humile* the anamorph of *M. macrospora*. The association of *Cladosporium* with *Mycosphaerella* has been supported by molecular studies (Crous *et al.*, 2001). However, most species are without known teleomorphs. *Mycosphaerella tassiana* forms

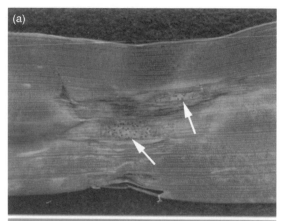

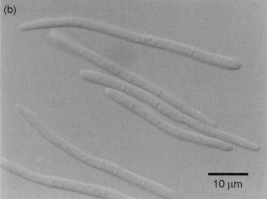

Fig 17.20 Wheat leaf blotch symptoms caused by *Mycosphaerella graminicola*. (a) Infected leaf showing a necrotic lesion, in the centre of which pycnidia of *Septoria tritici* have formed (arrows). (b) Septate conidia of *S. tritici*, more than 10 times longer than wide.

its pseudothecia on overwintered stalks and leaves of numerous monocotyledons and dicotyledons in subarctic and subalpine regions, and a period of cold is required for ascocarp initiation in culture (Barr, 1958; Corlett, 1991). In contrast, *C. herbarum* is ubiquitous and common in temperate regions on senescent and dead plant material, and in soil. Conidia of this and other *Cladosporium* spp. are the most abundant component of the fungal air spora (Gregory, 1973), and they are probably the most frequent contaminant of foodstuffs, textiles and paintwork. They also frequently contaminate cultures of other fungi in the laboratory. The conidia of *C. herbarum* and other common moulds such as *Alternaria alternata* and *Aspergillus fumigatus* are associated with severe asthma (Zureik *et al.*,

2002). Over 30 antigens causing mould allergy have been described from *C. herbarum*, and most of them are secretory or cytoplasmic glycoproteins, often representing common enzymes such as enolase or aldehyde dehydrogenase (Breitenbach & Simon-Nobbe, 2002).

Colonies of *C. herbarum* are dull olive green to black in colour, and appear as a network of hyphae or a plate-like mass (stroma) of tightly packed, dark, thick-walled cells (McKemy & Morgan-Jones, 1991). The conidiophores are branched or unbranched and conidiogenesis is holoblastic (Fig. 17.23a). The tip of the conidiophore bulges out to form the first conidium and it is presumed that all the wall layers of the apex are involved (Hashmi *et al.*, 1973). The first conidium buds to form a further conidium and this process continues so that a chain of conidia develops in acropetal succession, the youngest conidium at the end of the chain. Most conidia have a scar (hilum) at each end, but occasionally a conidium may form two daughter conidia at its tip so that, as further conidial development proceeds, a branched chain develops (see Fig. 17.23a). Such branch-point conidia have been termed **ramoconidia** (Lat. *ramus* = branch) and are marked by having a single scar at the base and two scars at the apex. The conidia of *C. herbarum* have dark (melanized) walls which are slightly roughened. They may remain unicellular or develop 1–3 transverse septa. Another common species of *Cladosporium* is *C. cladosporioides*, which has smooth conidium walls.

Cladosporium fulvum (also called *Fulvia fulva* or *Mycovellosiella fulva*) is probably not closely related to other *Cladosporium* spp. No sexual state has been reported, but other *Mycovellosiella* spp., like *Cladosporium* spp., belong to *Mycosphaerella* (Crous *et al.*, 2001). In contrast to most other plant-pathogenic members of the Loculoascomycetes, *C. fulvum* is biotrophic. It is a pathogen of tomato plants, especially in greenhouses. Conidia infect their host through stomata, and hyphae spread in the leaf apoplast in close contact with mesophyll cells, but without producing haustoria. Sucrose, the major plant transport sugar, is hydrolysed and taken up, being converted to mannitol by the fungus (Joosten *et al.*, 1990). Conspicuous yellow

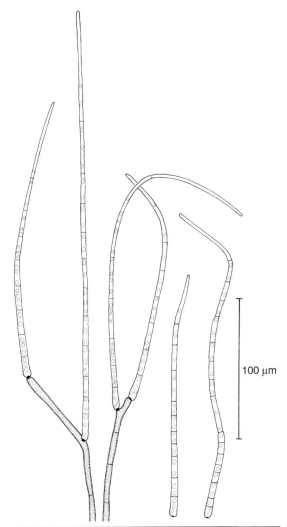

100 µm

Fig 17.21 *Cercospora beticola*. Conidiophores and conidia from sugar beet seed.

(chlorotic) leaf areas are produced as a result of such systemic infections (Fig. 17.24), and eventually conidiophores are emitted through stomata especially on the lower leaf surface, forming a lawn of spores resembling powdery mildews but being light brown in colour.

The interaction between *C. fulvum* and its tomato host is governed by a classical gene-for-gene relationship based on dominant host resistance genes (*Cf* genes, for *C. fulvum*) and dominant avirulence (*Avr*) genes in *C. fulvum*, i.e. virulence is a recessive trait (Joosten *et al.*, 1997). The hypersensitive response of incompatible interactions results if the product of the *Avr* gene is recognized by the host plant. Several *Avr* genes have been characterized, and their products are usually small proteins secreted by *C. fulvum* into the apoplast. Many of the corresponding *Cf* genes of tomato are also known; they encode proteins anchored in the plasma membrane of tomato cells, with large extracellular domains. The examination of the products of matching avirulence and resistance genes should provide an opportunity to examine their interactions, and thus the molecular basis of recognition events involved in specific resistance (Rivas & Thomas, 2005). This work is still ongoing.

17.3.5 *Aureobasidium* and black yeasts

Aureobasidium pullulans is a ubiquitous saprotroph whose main habitats are the phylloplane and other surfaces of living and senescent plants, but it can also occur as a symptomless endophyte. It grows in soil, but is often not recorded because it is temperature-sensitive and is not seen on soil plates prepared with warm agar. It has been isolated from fresh water, estuarine and marine sediments, sea water, sewage and other liquid waste (Domsch *et al.*, 1980). Its teleomorph is *Discosphaerina fulvida*, a relative of *Mycosphaerella* (Yurlova *et al.*, 1999). The fungus can be readily isolated from leaf washings. It is pleomorphic, and in culture it forms a rapidly growing mycelium with wide, septate hyphae from which intercalary and occasionally terminal cells give rise to single or clustered hyaline blastoconidia by a process of budding. Budding is associated with local lysis of the wall of the conidiogenous cell, the inner wall of which then balloons out and forms the wall of the conidium (Ramos *et al.*, 1975). Repeated conidium development from the same or closely adjacent conidiogenous loci results in the formation of slimy clusters of hyaline conidia (Fig. 17.25a). Intercalary cells may enlarge and develop thicker, dark, melanized walls to become chlamydospores (Fig. 17.25b). Conidia bud in a yeast-like manner when the fungus is grown in liquid culture with high inoculum densities. The yeast cells can continue to bud

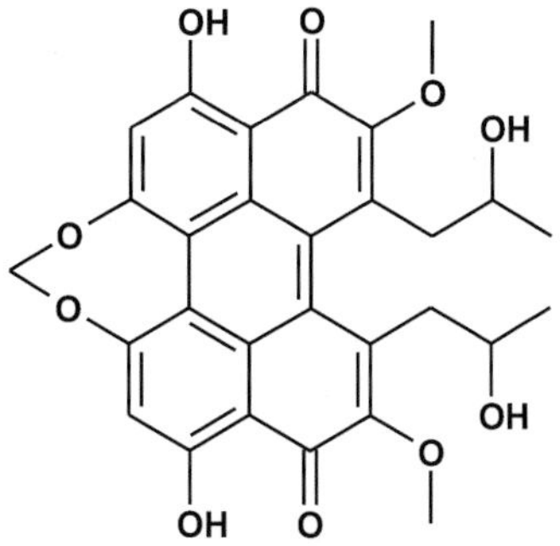

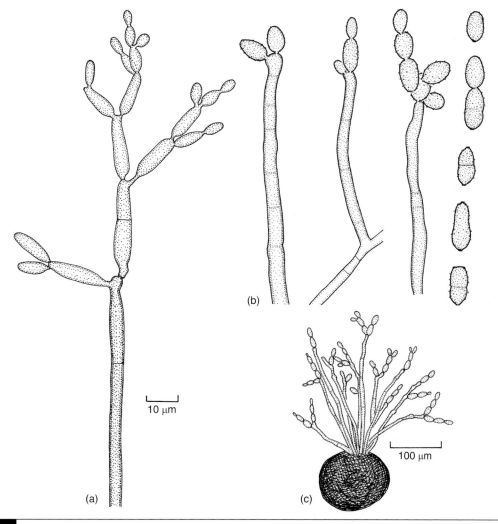

Fig 17.22 Cercosporin. This compound is synthesized as two polyketide halves which are then fused, accounting for the symmetric appearance of the molecule.

or may germinate by germ tube. Conidia are dispersed by rain splash and by air currents. The fungus may be involved in the biodeterioration of paint.

Aureobasidium pullulans has a number of potential applications. It is being investigated as a possible biocontrol agent against fungi like *Botrytis* and *Monilia* which cause post-harvest storage rots of fruit such as grapes, cherries and strawberries. It is a source of gluconic acid and of the dextran pullulan, an extracellular polysaccharide which has uses as an adhesive in laminates and in fabrics. Pullulan is also used as

10 μm

(a)

(b)

(c)

100 μm

Fig 17.23 *Cladosporium.* (a) *Cladosporium herbarum*; conidiophore with branching chain of blastoconidia. The terminal spores of the chain continue to develop blastoconidia. (b,c) *Cladosporium macrocarpum.* (b) Conidiophores and conidia. (c) Conidiophores developing from a sclerotium.

Fig 17.24 *Cladosporium fulvum.* Symptoms on greenhouse-grown tomato leaves as seen from above. The pale leaf areas are chlorotic spots, at the underside of which a pale greyish-brown felt of conidiophores is produced.

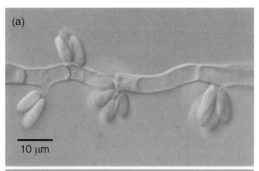

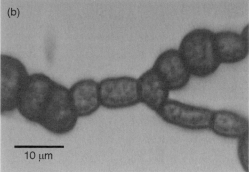

Fig 17.25 *Aureobasidium pullulans.* (a) Blastoconidia developing from an undifferentiated hypha. Several conidia can form from the same point, i.e. conidiogenesis is polyblastic. (b) Thick-walled melanized chlamydospores formed from hyphal segments in an older agar culture.

a low-calorie ingredient of foodstuffs and in pharmaceutical applications (Deshpande *et al.*, 1992; Leathers, 2003).

Aureobasidium is a member of a group loosely called **black yeasts**. This is not a taxonomic term but simply describes melanized fungi which produce yeast states, especially in culture. Several species of black yeasts, notably species of *Exophiala*, *Cladophialophora* and *Ramichloridium*, are known as opportunistic human pathogens causing infections of the brain and other organs which can be fatal (Horré & de Hoog, 1999; de Hoog *et al.*, 2000b). The microscopic features of these species have been described by de Hoog (1977) and, together with supplementary information, in an excellent compendium by de Hoog *et al.* (2000a). Although formerly considered to belong to the Loculoascomycetes, they are now recognized to be related to *Capronia* which is grouped in the Chaetothyriales (Haase *et al.*, 1999; Untereiner, 2000), an order quite remote from the Loculoascomycetes but with possible affinities with Plectomycetes or Lecanorales (Winka *et al.*, 1998). Black yeasts belonging to the Chaetothyriales are similar to those of the Dothideales not only in microscopic features, but also in their ecology, commonly occurring on living and decaying vegetation.

Basidiomycota

18.1 | Introduction

The Basidiomycota (colloquially basidiomycetes) are a large group of fungi with over 30 000 species. They include many familiar mushrooms and toadstools, bracket fungi, puffballs, earth balls, earth stars, stinkhorns, false truffles, jelly fungi and some less familiar forms. Also classified here are the rust and smut fungi, which are pathogens of higher plants and may cause serious crop diseases. Most basidiomycetes are terrestrial with wind-dispersed spores, but some grow in freshwater or marine habitats. Many are saprotrophic and are involved in litter and wood decay, but there are also pathogens of trees such as the honey fungus, *Armillaria*, which attacks numerous tree species, and *Heterobasidion annosum*, which can seriously damage conifer plantations. Common woodland mushrooms such as species of *Amanita*, *Boletus* and their allies grow in a mutually symbiotic relationship with the roots of trees, forming ectotrophic (sheathing) mycorrhiza. Species of *Rhizoctonia*, representing mycelial forms of basidiomycetes, behave as pathogens towards a wide range of plants but are mycorrhizal associates of orchids. As saprotrophs, basidiomycetes play a vital role in recycling nutrients but they also cause severe damage as agents of timber decay, e.g. dry rot of house timbers by *Serpula lacrymans*. The fruit bodies (basidiocarps) of many mushrooms are edible, and some are grown commercially for food, notably *Agaricus bisporus* (= *A. brunnescens*, the white button mushroom), *Pleurotus* spp.

(oyster mushrooms) and *Lentinula edodes* (shiitake). It is also well known that the basidiocarps of certain mushrooms are poisonous to eat, e.g. *Amanita phalloides* (the death cap). Some species have basidiocarps which are hallucinogenic, e.g. *Amanita muscaria* (the fly agaric) and *Psilocybe* spp. ('magic mushrooms').

The mycelium of basidiomycetes may be very long-lived. Estimates based on the rate of growth and the diameter of circles of the fairy ring fungus *Marasmius oreades* growing in permanent pasture show that they may be centuries old. It has been estimated that the age of an individual mycelium of *Armillaria* in a Canadian forest is at least 1500 years, with an extent of 15 hectares and a probable biomass in excess of 10 tonnes, making it one of the largest organisms on earth (Smith *et al.*, 1992).

Not all basidiomycetes grow in the mycelial form; some are yeast-like and others are dimorphic, i.e. capable of switching between mycelial and yeast-like growth. A dimorphic species which is a dangerous human pathogen to immunocompromised patients is *Filobasidiella* (= *Cryptococcus*) *neoformans* causing cryptococcosis, a fatal disease of the brain (see pp. 661–664).

18.2 | Basidium morphology

The characteristic structure of sexually reproducing basidiomycetes is the basidium. It is a spore-bearing cell which produces basidiospores externally through curved, tapering sterigmata

(Figs. 18.1d,e). Usually there are four spores but in some cases there are one, two or more than four basidiospores per basidium. *Itersonilia perplexans* has one-spored basidium-like structures (see Fig. 18.6a), the cultivated mushroom (*Agaricus bisporus*) has two-spored basidia, whilst basidia of the stinkhorn (*Phallus impudicus*) have as many as nine spores (see Fig. 20.9b). The form of the basidium varies, and this has taxonomic significance, different groups of basidiomycetes having distinctive types of basidium. In the mushrooms and their allies, the basidium is a cylindrical cell, undivided by septa (Fig. 18.1). Such basidia are termed **holobasidia**. In *Dacrymyces* and *Calocera* (Dacrymycetales) the basidium is undivided by septa, but the body of the basidium is forked into two, with each arm of the fork developing a single basidiospore. The basidia of the Jew's ear fungus, *Auricularia auricula-judae*, and related species are divided by transverse septa, whilst *Tremella* and its relatives have basidia with longitudinal septa. Basidia divided by septa are

termed **phragmobasidia** or **heterobasidia** (Gr. *phragmos* = a hedge or barricade; *heteros* = other, different). In rust fungi (Uredinales) and smut fungi (Ustilaginales) the basidia develop from thick-walled, originally dikaryotic resting cells termed **teliospores** or **chlamydospores**. A thin-walled tubular structure, the **promycelium**, develops from this resting cell and becomes divided by transverse septa, each of the resulting cells producing one basidiospore (in rusts) or several basidiospores (sporidia) in smuts. Some of these different kinds of basidia are illustrated in Fig. 18.2.

18.3 | Development of basidia

The development of a holobasidium is readily observed with the light microscope in the gill-bearing fungus *Oudemansiella radicata* which fruits on dead tree stumps (Figs. 18.1 and 19.18c)

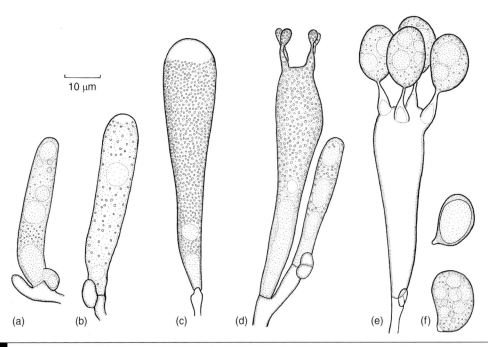

10 μm

(a) (b) (c) (d) (e) (f)

Fig 18.1 *Oudemansiella radicata.* Stages in the development of basidia. (a) Young basidium with numerous vacuoles. Note the clamp connection at its base and the formation of a further basidial initial. (b) A later stage showing the appearance of a clear apical cap. (c) Localization of vacuoles towards the base of the basidium. (d) Development of sterigmata and spore initials. A basal vacuole is enlarging. (e) Fully developed basidium. The spores are full of cytoplasm, whilst the body of the basidium contains only a thin lining of cytoplasm surrounding an enlarged vacuole. (f) Discharged basidiospores.

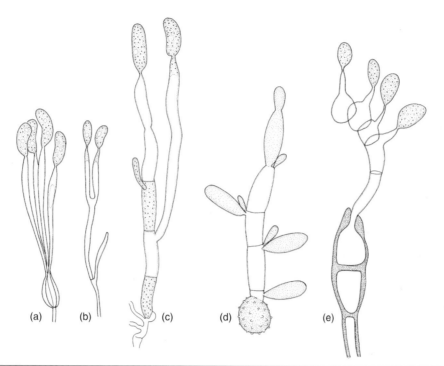

Some different kinds of basidia. (a) Longitudinally divided basidium of *Exidia glandulosa* (Tremellales). (b) Tuning-fork type of basidium of *Calocera viscosa* (Dacrymycetales). (c) Transversely divided basidium of *Auricularia auricula-judae* (Auriculariales). (d) Germinating chlamydospore of *Ustilago avenae* (Ustilaginales). A transversely septate promycelium has developed and each segment is forming sporidia. (e) Germinating teliospore of *Puccinia graminis* (Uredinales). A transversely septate promycelium has developed and each segment is producing a single basidiospore.

and has particularly large basidia. A detailed description of the process has been given by Corner (1948) for *O. canarii*. Ultrastructural studies have also been made on a range of fungi with holobasidia, e.g. the split-gill (*Schizophyllum commune*; Wells, 1965), the agarics *Coprinus cinereus* (McLaughlin, 1973, 1977, 1982) and *Panellus stypticus* (Lingle *et al.*, 1992), the clavarioid fungus *Clavicorona pyxidata* (Berbee & Wells, 1989) and the bolete *Boletus rubinellus* (McLaughlin, 1973; Yoon & McLaughlin, 1979, 1984).

18.3.1 Cytological aspects

In *Oudemansiella radicata* the basidium arises as the terminal cell of a hypha making up the gill tissue on the underside of the cap of the fruit body (basidiocarp). Basidia are packed tightly together in the hymenium at the surface of the gill. A basidium is at first filled with dense cytoplasm, but soon several small vacuoles appear near its

base. Later these coalesce into a single large vacuole at the base of the basidium and, by enlargement of this vacuole, cytoplasm is pushed towards the apex of the basidium. A clear cap is differentiated at the tip and it is in this region that the sterigmata develop. Corner (1948) postulated that there must be four elastic areas in the upper part of the basidium wall from which the sterigmata extend. The wall of the upper part of the basidium consists of two layers, the outer of which is mucilaginous, the inner firmer. In the areas where the sterigmata are about to bulge out, a new layer of wall material is deposited between these two original layers. The outer mucilaginous layer of the basidial wall bursts and the apex of the sterigma grows out. It is surrounded by two wall layers, the inner of which is continuous with the inner wall layer of the basidium (Clémençon, 2004). Ultrastructural studies of *Boletus rubinellus* show that, in the region where the sterigmata appear, the basidial wall is indeed thinner and differs in structure from the other parts of the wall (Yoon & McLaughlin, 1984). Beneath these areas there is some evidence of cytoplasmic differentiation, such as the presence of microtubules and vesicles. The growth of sterigmata can be compared to hyphal tip growth, except that a Spitzenkörper is absent (McLaughlin, 1973). At the tips of the sterigmata of *Coprinus cinereus*, vesicles apparently fusing with the plasma membrane have been observed by transmission electron microscopy studies, and vesicles of similar size were also found in the basidium (McLaughlin, 1973). Corner (1948) suggested that the force for the development of basidia comes from the expansion of the basal vacuole which acts as a piston, ramming the cytoplasm into the spores. Ripe basidia thus contain a large vacuole but very little cytoplasm (Fig. 18.1e). The vacuole, which is filled with liquid, keeps the basidium turgid until spore discharge has occurred.

The tip of the sterigma expands to form a small spherical knob, the **apophysis** (Gr. *apo-* = away from, separate; *physis* = growth). Further development of basidiospores is asymmetric, expansion being more rapid towards the outside of the long axis of the basidium. The narrow point of attachment of the spore at the tip of the sterigma is the eventual point of spore separation and is termed the **hilum**. The non-expanded part of the apophysis persists as the **hilar appendix** (Fig. 18.3a).

18.3.2 Nuclear events

Typically, a basidium is at first binucleate; it is formed on a dikaryotic mycelium, i.e. a mycelium with segments containing two haploid nuclei which are usually genetically different (see Section 18.9). In this cell nuclear fusion (karyogamy) occurs (see Fig. 18.4) and is followed immediately by meiosis, giving rise to four haploid nuclei. As in most fungi, division is intranuclear; the nuclear membrane remains intact. Meiosis occurs in the upper part of the basidium. In narrow basidia the plane of the second meiotic nuclear division lies parallel to the long axis of the basidium. This type of nuclear division is termed **chiastic** (Gr. *chiastos* = crossed, arranged

diagonally) and basidia with nuclear division of this sort are termed **chiastobasidia**. In contrast, in broader basidia, the plane of both meiotic nuclear divisions is transverse to the long axis. Nuclear divisions of this type are **stichic** (Gr. *stichos* = a line or row of things) and the corresponding term for basidia with such division is **stichobasidium**. The plane of nuclear division has relevance to taxonomy; chiastobasidia are found in mushrooms and toadstools whilst stichobasidia are more characteristic of certain genera of bracket fungi in the polyporoid clade. Terms used to define different parts of a basidium are **probasidium**, the part within which karyogamy occurs, and **metabasidium**, the part within which meiosis occurs (see Kirk *et al.*, 2001). The four haploid nuclei formed during meiosis move into the basidiospores which are therefore usually four in number. As they pass through the sterigmata, the nuclei are often elongated and tapered apically. They may be led through the sterigmata by microtubules attached to the nuclear spindle pole bodies (Thielke, 1982; Lingle *et al.*, 1992).

In many basidiomycetes, meiosis is followed by a post-meiotic mitosis (Duncan & Galbraith, 1972; Clémençon, 2004) which may happen in different places: (1) In the upper part of the basidium. Four of the eight nuclei enter the spores, and those remaining in the basidium abort. The ripe spores are thus uninucleate, e.g. in *Cantharellus cibarius*. (2) At the base of or inside the sterigmata. One nucleus enters each spore and the other four remain in the basidium and degenerate, e.g. in *Collybia butyracea*. (3) In the young spore. Four of the daughter nuclei migrate back into the basidium where they degenerate, e.g. in *Paxillus involutus*. (4) In the young spore, but all eight nuclei remain in the four spores and none abort; the basidium is left devoid of nuclei.

18.4 | Basidiospore development

On the adaxial side of the apophysis, an electron-dense cytoplasmic region appears at the moment

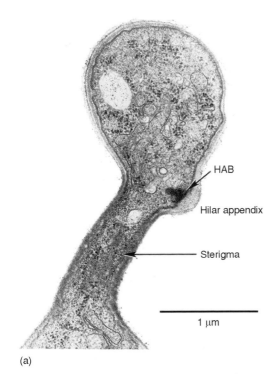

HAB

Hilar appendix

Sterigma

1 µm

(a)

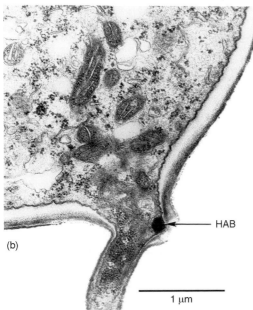

HAB

(b)

1 µm

Fig 18.3 *Coprinus cinereus.* Transmission electron micrographs of sections illustrating development of basidiospores. (a) Stage 4 basidiospore expanding at the tip of its sterigma. The hilar appendix body (HAB) is appressed to a wall thickening. (b) An early stage 4 basidiospore with the conical hilar appendix body in firm contact with the spore plasmalemma. For further explanation see p. 490. Photomicrographs kindly provided by D. J. McLaughlin.

of its initiation and, in most cases, persists throughout the development of the basidiospore, disappearing only shortly before spore discharge. It is hemispherical or conical, lying immediately within the plasma membrane of the apophysis and closely appressed to the wall of the hilar appendix (see Figs. 18.3a,b). This structure is the **hilar appendix body** and has been reported from several basidiomycetes, including *Boletus rubinellus*, *Coprinus cinereus* (McLaughlin, 1973, 1977), *Lactarius lignyotellus* (Miller, 1988) and *Panellus stypticus* (Lingle *et al.*, 1992). It is probably present in all ballistosporic species. Its function is not understood. Possibly it is involved in the softening of the closely adjoining wall layers of the hilar appendix or in the extrusion of material related to the expansion of Buller's drop (see Fig. 18.8). Another possibility is that it blocks the movement of wall vesicles into the adaxial side of the spore, so possibly contributing to the asymmetric shape of the spore. During further development, the wall of the hilar appendix thickens considerably, and uneven expansion of the wall of the developing spore occurs. Expansion of the abaxial face is more rapid, leaving the spore asymmetrically perched on the sterigma, attached at the hilum and with the remains of the apophysis forming the hilar appendix (see Fig. 1.20). In *Coprinus cinereus*, McLaughlin (1977) has distinguished four successive stages of basidiospore expansion (see Fig. 18.5a).

Stage 1, inception. This is characterized by the spherical enlargement of the sterigma apex to form a basidiospore primordium 0.6–0.8 µm in diameter. The hilar appendix body is already differentiated. The thin basidiospore wall is three-layered at first. Microtubules are occasionally present in the sterigma, being orientated parallel to its long axis.

Stage 2, asymmetric growth. The basidiospore initial grows asymmetrically on its abaxial side, and the hilar appendix develops. The hilar appendix body becomes conical and appressed to the plasma membrane of the spore initial. The hilar appendix is initiated adjacent to the hilar appendix body. The basidiospore wall thickens, being thickest at the apex of the spore, and is six-layered.

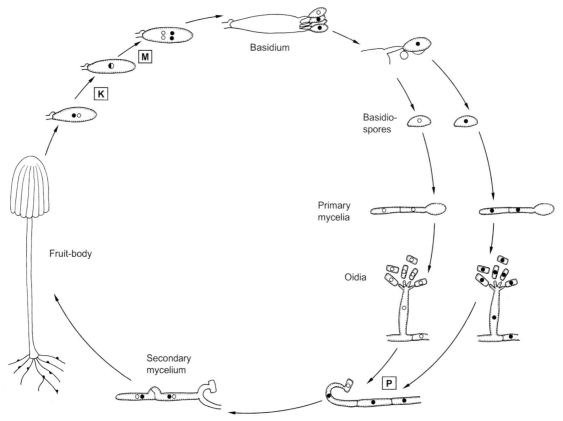

Fig 18.4 Life cycle of the basidiomycete *Coprinus* (diagrammatic and not to scale). The basidiocarp develops from a dikaryotic secondary mycelium and produces numerous basidia on the surface of its gills beneath the cap. Progressive stages of basidium maturation involving karyogamy (K) and meiosis (M) are indicated. Eventually, each basidium forms four basidiospores, each containing a single haploid nucleus. In many basidiomycetes there is a post-meiotic mitosis, giving rise to two identical haploid nuclei in each basidiospore (not shown here). Basidiospores are discharged by the surface-tension catapult mechanism involving Buller's drop. Discharged basidiospores germinate to form haploid (monokaryotic) mycelia with simple transverse septa. In *Coprinus*, these often produce upright conidiophores which form numerous sticky haploid oidia. The apex of a monokaryotic hypha in the vicinity of an oidium of compatible mating type will respond chemotropically by growing towards the compatible oidium (homing reaction). Fusion (plasmogamy, P) between the hypha and the oidium initiates the formation of a dikaryotic mycelium bearing clamp connections. Nuclear fusion does not occur at this stage. The dikaryotic mycelium can develop basidiocarps under appropriate environmental conditions. Open and closed circles represent haploid nuclei of opposite mating type; the diploid nucleus is drawn larger and half-filled.

Stage 3, equal enlargement. This is characterized by spherical enlargement of the basidiospore. Growth is at an angle of about 45° to the sterigma apex (see Fig. 18.5a). The hilar appendix body projects further into the spore wall. It is conical or hemispherical, with the apex of the cone or base of the hemisphere projecting towards the hilar appendix. The outermost layer of the basidiospore wall is sticky.

Stage 4, elongation. Basidiospores grow in length and a pore cap is formed at the upper end of the spore. The spore wall becomes darkly pigmented, starting at the upper end.

Similar asymmetric changes in spore expansion have been reported by Yoon and McLaughlin (1984) for basidiospores of *Boletus rubinellus* (Fig. 18.5b). Elongation of the spore occurs mainly at later stages of development.

In certain basidiomycetes, e.g. *Amanita vaginata*, the basidiospore is spherical. This involves even wall expansion during development and implies that the wall structure is

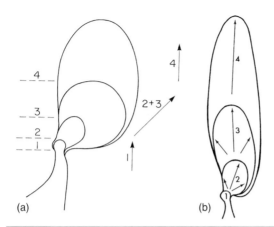

homogeneous. Corner (1948) believed that the non-spherical shape of basidiospores was the result of differential setting of the wall material. This may well be true, but more recent studies have shown that the basidiospore wall also varies in thickness. Many basidiospores have smooth outer walls, but others have characteristic ornamentations, e.g. spines, folds or ridges. The ornamentations generally develop by extension of the outer wall layer of the basidiospore (Pegler & Young, 1971; Clémençon, 2004). In such spores, e.g. those of *Lactarius* and *Russula*, only the main body of the spore is ornamented, whereas the region of the adaxial face immediately above the hilar appendix is smooth. This region is termed the **suprahilar plage, disc** or **depression** (Pegler & Young, 1971, 1979). It plays an important part in the spore discharge mechanism (see below).

Abscission of the basidiospore from its sterigma is preceded by the formation of a plug of material, the **hilar plug** which blocks the spore hilum, and a **sterigmal plug** of wall material immediately below the hilum. Between the two plugs a septum appears which represents the

point of abscission (Yoon & McLaughlin, 1986; Money, 1998).

18.5 | The mechanism of basidiospore discharge

With the exception of the gasteromycetes (see Chapter 20), in most terrestrial basidiomycetes basidiospores are ballistospores, i.e. they are actively projected from basidia. Various suggestions have been made as to the mechanism of ballistospore discharge (see Webster & Chien, 1990), but the one which we now accept is the **surface tension catapult**, originally suggested by Buller (1922) and Ingold (1939). Discussions of the various theories of basidiospore discharge have been written by Webster *et al.* (1988) and Money (1998). Shortly before discharge, dissolution of the abscission layer occurs, indicated by a slight wobble in the position of the spore. Then a spherical drop of liquid, Buller's drop, forms at the hilar appendix and a shallower liquid deposit, the adaxial drop (adaxial blob), appears on the face of the spore above the hilar appendix (Fig. 18.7). Cinephotography has been used to illustrate these events (Webster & Hard, 1998b; Webster, 2006b). Both drops increase in size until they eventually coalesce, and spore discharge then immediately occurs (Pringle *et al.*, 2005).

Experimental investigations on the phenomenon of ballistospore discharge have focused on *Itersonilia perplexans*, an unusual heterobasidiomycete with large ballistospores. This fungus is a weak plant pathogen and is commonly associated with lesions caused by other pathogens such as rust and smut fungi. It also grows in basidiocarps of certain jelly fungi and can be readily isolated by allowing it to shoot off its ballistospores from the basidiocarps of *Dacrymyces stillatus* or *Auricularia auricula-judae* (Ingold, 1983a, 1984a). In culture it forms a clamped dikaryotic mycelium, the tips of whose branches swell to form clamped sporogenous cells, each with a single ballistospore (Fig. 18.6a). The sporogenous cells do not fully match the definition of basidia because nuclear fusion and meiosis do not occur in them; instead the dikaryotic cell forms a dikaryotic ballistospore directly. A discharged primary ballistospore may germinate by a germ tube or by repetition to form a secondary ballistospore (Fig. 18.6b). Yeast-like growth can also occur, especially on rich media.

Working with *Itersonilia*, several observations were made which provided clues to the mechanism of discharge. (1) Ballistospores can be detached from their sterigmata with a micromanipulator needle and spores so detached still develop Buller's drop. This shows that the liquid in the drops does not originate from liquid transported through sterigmata. (2) During normal discharge, although the volume of Buller's drop may attain 60% of that of the spore, there is no decrease in the dimensions of the ballistospore, indicating that Buller's drop does not come from within the spore. The same observation has been made on other basidiomycetes and has led to the suggestion that the liquid in Buller's drop and also in the adaxial drop is formed by condensation of water vapour around a hygroscopic substance extruded from the hilar appendix and through the spore wall (Webster *et al.*, 1984a,b, 1989). Washings from the spores of basidiomycetes belonging to several different taxonomic groups were analysed by gas–liquid chromatography, and all gave a positive result for the presence of mannitol. Glucose was also sometimes detected. The presence of mannitol and hexose in liquid drawn off by a micropipette from Buller's drops in *Itersonilia* was confirmed by microscope fluorimetry, and measurements of the solute concentrations in Buller's drop corresponded closely to the calculated concentrations which would be necessary to drive the uptake of water from a saturated atmosphere at the rates observed (Webster *et al.*, 1995).

The surface tension catapult mechanism postulates that, as Buller's drop develops, the centre of mass of the spore plus drop moves towards the hilar appendix (Fig. 18.8b). The coalescence of Buller's drop with the adaxial drop causes a rapid redistribution of mass away from the hilar appendix, resulting in a momentum which carries the spore plus drop away from the sterigma (Fig. 18.8c; Webster *et al.*, 1988). Pringle *et al.* (2005) have suggested that an even

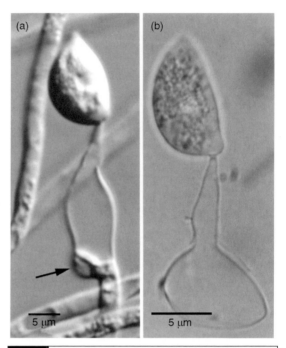

Fig 18.6 *Itersonilia perplexans.* (a) Basidium (sporogenous cell) bearing a single ballistospore. Note the clamp connection at the base of the basidium (arrow). (b) A primary ballistospore has germinated by repetition to form a secondary ballistospore.

greater momentum may be generated by the fusion drop moving towards the basidiospore apex and coming to an abrupt halt upon reaching it. For this mechanism to be effective, a rigid sterigma is required, and it is likely that the turgor pressure of the vacuolated basidium contributes to the required rigidity (Money, 1998).

The requirement of high humidity for effective operation of the mechanism is a likely explanation for observations that basidiospore concentrations in the air peak at night (Kramer, 1982). High humidity develops in the space between agaric gills or inside the hymenial tubes of toadstools and bracket fungi. The presence of free water would, of course, prevent operation of the surface tension catapult, and this may be the reason why agaric basidiocarps are often umbrella-shaped. The impossibility of the surface tension catapult mechanism operating under water also explains why, although some basidiomycetes grow vegetatively in fresh

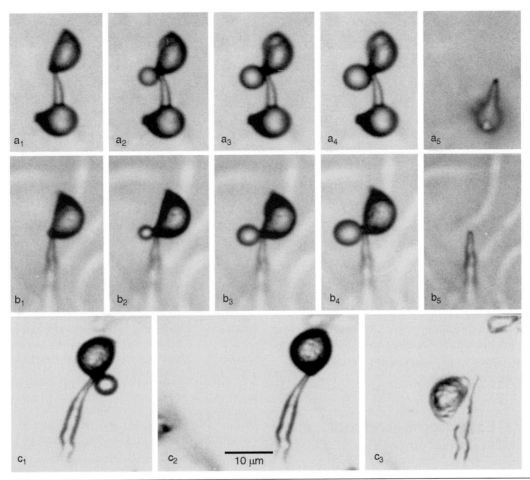

Fig 18.7 *Itersonilia perplexans.* Appearance of Buller's drop and adaxial drop. a_1–a_5 show events immediately preceding and following the discharge of a secondary ballistospore. An adaxial drop is clearly seen. b_1–b_5 show drops developing on a primary ballistospore. Note that there is no decrease in the size of the ballistospore as the drops develop. c_1–c_3 illustrate failure of ballistospore discharge. Here coalescence of Buller's drop and the adaxial drop is not accompanied by separation of the spore from its sterigma and the spore soon topples from its perch. Reprinted from Webster *et al.* (1984a), with permission from Elsevier.

water or the sea, they do not produce ballistospores there.

The spore of *Itersonilia* is subjected to considerable acceleration and moves away from the sterigma at a velocity of over $1\,\mathrm{m\,s^{-1}}$. However, the large surface/mass ratio of a relatively small object like a ballistospore results in high wind resistance and therefore rapid deceleration and loss of momentum, so that the spore soon falls under the preponderant influence of gravitational force. The trajectories of ballistospores of most fungi follow a short horizontal path for a distance of about 0.1–0.3 mm and then turn through a right angle so that spores in still

air drift downwards at a steady sedimentation velocity. This characteristic trajectory, termed a **sporabola**, ensures that basidiospores projected into the space between the gills or into the lumen of a pore turn vertically downwards before hitting the opposing hymenial surface.

18.6 | Numbers of basidiospores

The number of basidiospores produced by a single basidiocarp can be extremely high. Buller (1909) estimated that the detached cap of the

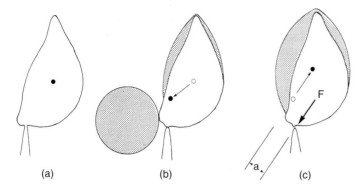

Fig 18.8 Representation of events associated with ballistospore discharge. (a) Ballistospore attached to its sterigma before drop formation. The closed circle within the spore indicates the centre of mass of the spore. (b) Buller's drop appears at the hilar appendix. The adaxial drop emerges on the spore wall above it and extends downwards as it increases in size. The centre of mass of spore plus drop moves to a position nearer the hilar appendix. (c) Contact between the two drops is followed by immediate coalescence and the combined mass of liquid moves rapidly up the adaxial face of the spore away from the hilar appendix. The centre of mass moves very rapidly in the direction of the thin arrow and the spore−drop system gains kinetic energy and momentum in the same direction, simultaneously exerting an opposite force F on the sterigma at the hilum (thick arrow). Some angular momentum is also exerted on the spore, related to the distance a between the hilum and the hilar appendix. Reprinted from Webster et al. (1988), with permission from Elsevier.

mushroom *Agaricus campestris* produced 1.8×10^9 spores over 2 days at an average rate of 40 million h^{-1}. Estimates for some other basidiomycetes are given in Table 18.1. The mycelia of all these fungi are perennial and an individual mycelium may produce numerous basidiocarps over a period of many years. It is therefore clear that an individual basidiospore has an infinitesimal chance of successfully establishing a fruiting mycelium.

18.7 | Basidiospore germination and hyphal growth

18.7.1 Germination

Basidiospores may remain dormant and retain viability for several months or even for a few years if conditions are unsuitable for germination. Dormancy is frequently exogenous, i.e. the spores require some external chemical or physical stimulus before germination can occur. Germination may be direct by production of a germ tube, by repetition (i.e. the formation of a secondary ballistospore), or by the formation of conidia. Repetitious germination is common in certain jelly fungi (Tremellales) (see Figs. 18.6b, 21.7b and 21.13d). Germination by the formation of conidia is illustrated for *Dacrymyces stillatus* (Fig. 21.4a), *Auricularia auricula-judae* (Fig. 21.6c) and by yeast-like budding in *Tremella frondosa* (Fig. 21.13c). During direct germination, germ tubes usually emerge through a special germ pore at the hilum or, as in *Coprinus*, through a pore at the opposite end of the spore.

18.7.2 Monokaryotic and dikaryotic hyphae

Because the nuclear divisions involved in basidiospore formation are meiotic, basidiospores are haploid. Since the post-meiotic nuclear divisions are mitotic, if there are several nuclei in a single basidiospore these are usually genetically identical. They are said to be **homokaryotic** (Gr. *homos* = equal, alike; *karyon* = a nut, here meaning nucleus). At germination, repeated mitotic nuclear divisions occur and the early germ tubes may consequently be multinucleate and coenocytic. Transverse septa are laid down behind the growing hyphal tip and eventually divide the hypha into segments which contain only a single nucleus. The uninucleate segments and the hyphae which contain them are said to be **monokaryotic**. The terms homokaryon and monokaryon or **primary mycelium** have also been applied to such haploid mycelia. As part of the sexual cycle, monokaryotic hyphae of genetically distinct mating type undergo plasmogamy (somatogamy), i.e. they fuse together and initiate the formation of a mycelium made up of

Table 18.1. Numbers of basidiospores produced from single basidiocarps of selected basidiomycetes. From Buller (1922).

Species	Total number of spores	Spore fall period	Spores discharged per day
Calvatia gigantea	7×10^{12}	—	—
Ganoderma applanatum	5.5×10^{12}	6 months	3×10^{10}
Polyporus squamosus	5×10^{10}	14 days	3.5×10^{9}
Agaricus campestris	1.6×10^{10}	6 days	2.6×10^{9}
Coprinus comatus	5.2×10^{9}	2 days	2.6×10^{9}

segments, each of which contains two genetically distinct nuclei. Such mycelia are said to be **dikaryotic** and **heterokaryotic**. The term **secondary mycelium** is also used.

18.7.3 Dolipore septa

The transverse septa which divide both monokaryotic and dikaryotic hyphae into segments are incomplete; they contain a central pore which permits cytoplasmic continuity between adjacent segments. The septal pore is surrounded by a barrel-shaped flange of thickened wall material. Such septa, which are characteristic of basidiomycete mycelia, are known as **dolipore septa** (Lat. *dolium* = large jar, cask). They are discernible by light microscopy, especially with ammoniacal Congo red staining, but interpretation of their structure is only possible in sections or freeze-fractured cells viewed by electron microscopy (Figs. 18.9 and 18.10). Septal development begins by centripetal ingrowth of a membrane on which wall material (glucan and chitin) is deposited at both faces from associated vesicles (for references see Moore, 1985). The thickening surrounding the pore results from more rapid deposition of wall material, but there is evidence that very thick pore rims may be an artefact associated with chemical fixation. The pore itself may be blocked by an occlusion shaped like two champagne corks attached end to end (Fig. 18.9), but blockage of the pore is not a permanent feature. In some cases there is a transverse central plate in the pore canal (Fig. 18.10b).

Overarching the septal pore on each side of the septum is a specialized portion of endoplasmic reticulum known as the **septal pore cap** or **parenthesome** (parenthesis = round bracket, Gr. *soma* = body). In some cases a second parenthesome (outer cap) has been reported. In many basidiomycetes with holobasidia (i.e. Homobasidiomycetes), the parenthesomes are perforated (fenestrated), but in many Heterobasidiomycetes, e.g. *Auricularia*, there is only a single perforation or none (Lü & McLaughlin, 1991; Wells, 1994; Wells & Bandoni, 2001). Other variations in ultrastructure are known and can be characteristic of different groups of basidiomycetes, so that the dolipore/parenthesome complex is considered to be of taxonomic significance (Khan & Kimbrough, 1982; Moore, 1985, 1996; McLaughlin *et al.*, 1995; Müller *et al.*, 1998a).

An important role of the dolipore/parenthesome complex is to secure the integrity of hyphal cells and to maintain intercellular communication and transport of some organelles. A variety of cytoplasmic structures has been reported from within the pores of *Rhizoctonia solani*. These include small tubular and filamentous structures, small vesicles, tubular endoplasmic reticulum and other plugging material. The movement of mitochondria through the septal pore cap has also been documented (Müller *et al.*, 2000). Whilst the movement of most organelles through the septal pore is permitted, the passage of nuclei is not, and this is possibly a consequence of their larger size. The migration of nuclei following plasmogamy between two sexually compatible monokaryotic mycelia is associated with enzymatic dissolution of the dolipore (see below). Another important function of dolipores is the repair of hyphal damage, the septal pore being rapidly plugged by electron-dense material in the compartment of a hypha

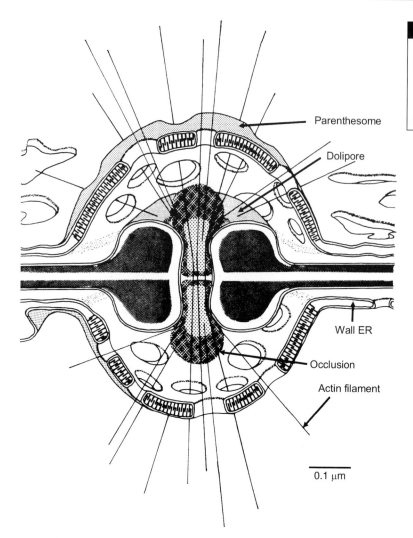

Parenthesome

Dolipore

Wall ER

Occlusion

Actin filament

0.1 μm

Fig 18.9 Diagrammatic interpretation of a basidiomycete dolipore/parenthesome septum. Reprinted from Moore and Marchant (1972), by copyright permission of the National Research Council of Canada.

adjacent to the damaged segment (Aylmore *et al.*, 1984; Markham, 1994).

18.7.4 Plasmogamy, dikaryotization, clamp connections

When two compatible monokaryotic mycelia make contact, the hyphal walls separating them break down and cytoplasmic continuity (i.e. plasmogamy) of the two monokaryons is established. In *Coprinus cinereus* and some other fungi, plasmogamy may also occur following fusion of an oidium of one mating type with a hyphal tip of a compatible monokaryon. Nuclear migration follows and is associated with the breakdown of the dolipore/parenthesome complex to permit transfer of a compatible

nucleus from one compartment to another (Giesy & Day, 1965; Marchant & Wessels, 1974). The mycelium which develops after plasmogamy is dikaryotic and the process of conversion of a monokaryon to a dikaryon is termed **dikaryotization**.

Nuclear fusion (i.e. karyogamy) is delayed until the basidia have formed, and is thus preceded by a prolonged dikaryotic state. As a result of nuclear migration, the tip of a dikaryon contains two nuclei which are of different mating types in heterothallic basidiomycetes. The speed of nuclear migration can be much higher than the hyphal growth rate. Nuclear migration rates have been measured in a number of fungi, e.g. *Coprinus cinereus*

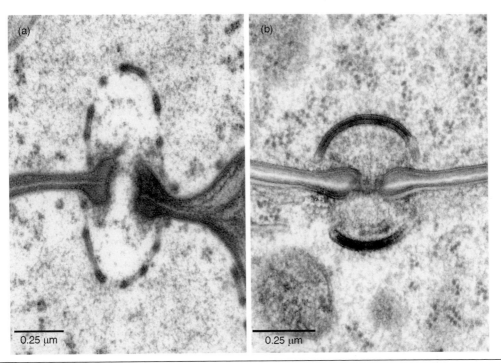

Fig 18.10 Longitudinal sections of the dolipore/parenthesome septum in two basidiomycetes as viewed by transmission electron microscopy. (a) *Auriscalpium vulgare*, a homobasidiomycete. The parenthesome is perforated. (b) *Auricularia auricula-judae*, a heterobasidiomycete. The parenthesome is imperforate. Photographs kindly provided by D. J. McLaughlin.

$(0.5-1.0\,\mathrm{mm\,h^{-1}})$, *Coprinus congregatus* $(4\,\mathrm{cm\,h^{-1}})$ and *Schizophyllum commune* $(1.5-5.4\,\mathrm{mm\,h^{-1}})$ (Snider, 1965, 1968; Raper, 1966; Ross, 1976). Microtubules are associated with migrating nuclei in *S. commune* (Raudaskoski, 1972) and in *Trametes versicolor* (Girbardt, 1968). In the tip cell and also in the subapical segments (Fig. 18.12) of a dikaryon, the two nuclei maintain a constant distance apart from each other, indicating that they are paired together by microtubules. They also move forward together at a fixed distance from the apex (Kamada *et al.*, 1993; Torralba *et al.*, 2004).

The two nuclei in a dikaryotic hyphal tip divide simultaneously, a process termed **conjugate nuclear division**. In most but not all basidiomycetes, division is accompanied by nuclear rearrangement involving the formation of **clamp connections**, visible as a lateral bulge in the hyphal wall adjacent to a transverse septum. The events connected with the development of a clamp connection are set out diagrammatically in Fig. 18.11. A clamp connection develops near the position of the pair of nuclei in the terminal segment of a dikaryotic hypha (Fig. 18.11b). A backwardly directed hyphal branch (hook) develops and one daughter nucleus migrates into it and divides there mitotically at the same time as mitotic nuclear division is taking place in the subterminal nucleus (Fig. 18.11c). A transverse septum develops in the main hypha between the two daughters of the subterminal nucleus and an oblique septum also forms at the base of the hook (Fig. 18.11d). The hook, containing a single daughter nucleus, grows round the transverse septum of the main hypha and its tip fuses with the wall of the subterminal cell. Plasmogamy occurs and the nucleus from the hook migrates into the subterminal cell (Fig. 18.11e). The two pairs of nuclei then move away from the transverse septum (Fig. 18.11f). The clamp connection is a device which ensures that each segment of a dikaryotic hypha contains two genetically distinct nuclei. In the absence of clamps or of some other mechanism for rearrangement of

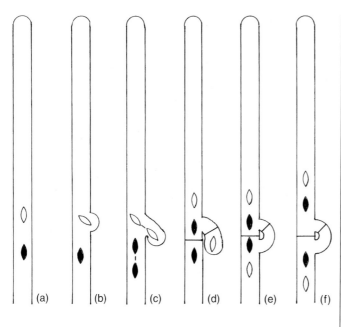

Fig 18.11 Diagrammatic representation of the sequence of events associated with the formation of a clamp connection in the dikaryotic hypha of a basidiomycete. (a) Terminal segment of a hypha with two genetically dissimilar nuclei. (b) A lateral bulge in the hypha appears near the paired nuclei. The leading nucleus moves into the bulge. (c) The two nuclei undergo conjugate (i.e. simultaneous) nuclear division. Mitosis of the leading nucleus occurs within the bulge. (d) The lateral bulge has developed into a backwardly directed branch or hook with one daughter of the leading nucleus at its tip. One of the daughter subterminal nuclei moves forward. Transverse septa have developed simultaneously in the main hypha and at the base of the hook. (e) The tip of the hook has fused with the wall of the main hypha. Its nucleus has moved into the main hypha and taken up position behind the daughter of the subterminal nucleus. (f) The pairs of nuclei move away from the transverse septum in the main hypha, the terminal pair moving nearer the hyphal tip and the subterminal pair distally. Note that both segments contain dissimilar nuclei but that their arrangement has been reversed.

nuclei, there would be a tendency for dikaryotic mycelia to break down into homokaryotic segments. Whilst it is reasonable to infer that mycelia with clamps at the septa are dikaryotic, the converse is not true because there are numerous basidiomycetes in which the dikaryotic mycelium does not bear clamps, and this is especially true of hyphae making up basidiocarps. For a fuller discussion of dikaryon formation see Casselton and Economou (1985).

18.7.5 Aggregates of vegetative hyphae

Basidiomycete hyphae can aggregate to form complex vegetative structures such as hyphal strands, mycelial cords, mycelial sheets and rhizomorphs (Butler, 1966; Watkinson, 1979; Rayner *et al.*, 1985). These often grow more rapidly than individual hyphae, connect food bases such as litter fragments in soil, fallen logs and tree stumps, and are capable of rapid two-way conduction of water and nutrients.

Mycelial cords

Mycelial cords have been defined by Boddy (1993) as aggregations of predominantly parallel, longitudinally aligned hyphae (see Fig. 1.13). They are especially common in wood-decaying basidiomycetes but are also involved in the underground spread of basidiomycetes forming ectomycorrhiza. Wood-decaying mycelial cord formers are a very successful ecological group, using their cords in a variety of strategies to secure resources, e.g. by displacing other fungi from food bases or exploring new resources in the soil. The cords show relatively little differentiation in structure and pigmentation. Examples of fungi with mycelial cords are the stinkhorn (*Phallus impudicus*) and the agaric *Megacollybia platyphylla*. Their white cords can extend for many metres in the soil and can be traced back from the fruit bodies to tree stumps or other food bases if highly motivated students and digging tools are available (Fig. 18.13a).

Rhizomorphs

The term rhizomorph refers to the root-like form of these mycelial aggregates which are more highly differentiated than mycelial cords and are often pigmented brown or black due to the presence of melanin in the walls of small,

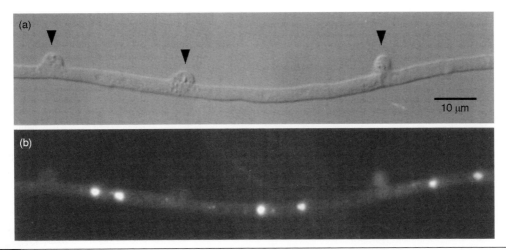

Fig 18.12 Subapical hypha of a dikaryon of *Omphalotus olearius*. (a) Interference contrast image showing the segments delimited by clamp connections (arrowheads). (b) DAPI fluorescence staining of the same hypha showing that each segment contains two paired nuclei (bright fluorescent objects).

thick-walled cells making up the rind (Townsend, 1954; Cairney, 1991). The best-known example of a rhizomorph is that of the honey fungus *Armillaria mellea*, a serious tree pathogen whose flat, black, bootlace-like strands often persist for long periods beneath the bark of trees killed by the fungus (see pp. 16–18 and Fig. 18.13b). Many ectomycorrhizal fungi especially in the boletoid clade form hyphal aggregates which show intermediate features between mycelial cords and rhizomorphs.

Sclerotia

Sclerotia, an adaptation to prolonged survival and propagation, develop in some basidiomycetes (see pp. 18–20). They vary in size from 50 μm to several centimetres and in weight from 10 μg to several kilogrammes. They also vary in organization from loose aggregations of dark hyphae to highly differentiated structures with a rind of smaller, dark thick-walled cells and a medulla of larger, colourless, thin-walled cells packed with food reserves (Willetts, 1971, 1972; Clémençon, 2004). Many of the sclerotial types described on pp. 18–20 are produced by Basidiomycota, e.g. the loose type (*Rhizoctonia solani*) and the strand type (*Sclerotium rolfsii*). Some sclerotia are massive, as in *Polyporus mylittae* where they may give rise to fruit bodies (carpogenic development; Figs. 18.13c,d).

Basidiocarps of many common agarics such as *Coprinus cinereus*, *Collybia tuberosa*, *Hygrophoropsis aurantiaca* and *Paxillus involutus* may develop from smaller sclerotia, and these are also characteristic of the clavarioid fungus *Typhula* (Corner, 1950). Sclerotial germination, especially of plant pathogenic fungi such as *Rhizoctonia*, is more usually by the outgrowth of mycelium (myceliogenic germination).

Pseudosclerotia with a similar function to sclerotia, but consisting of a compacted mass of intermixed substratum, soil, stones, etc., support the fruiting of certain polypores such as *Polyporus tuberaster* (the stone fungus, tuckahoe), and *Meripilus giganteus*.

18.8 | Asexual reproduction

Conidium formation is less commonly reported in the Basidiomycota than in the Ascomycota. Conidia may develop on monokaryotic or dikaryotic mycelia, sometimes on both. They may also form on basidiocarps. Conidia may have an asexual function in propagation and dispersal or may also fulfil a sexual role. We can only consider a few examples. In terms of structure and ontogeny, basidiomycete conidia are of three basic kinds which are summarized below (see Kendrick & Watling, 1979; Clémençon, 2004).

Fig 18.13 Hyphal aggregates of basidiomycetes. (a) White mycelial cords of *Megacollybia platyphylla* interconnecting a decaying log of wood serving as food base (far left) with two basidiocarps, exposed by removing the surface leaf litter. (b) Rhizomorphs of *Armillaria mellea* in their typical location between the bark (which has been stripped) and the wood of a dead tree. (c,d) The giant subterranean sclerotium of *Polyporus mylittae* which has produced an above-ground fruit body (c). When cut open, the sclerotium consists of a dark rind enclosing a medulla with a texture resembling compacted boiled rice (d). (c,d) reprinted from Fuhrer (2005), with permission by Bloomings Books Pty Ltd; original images kindly provided by B. Fuhrer.

18.8.1 Oidia (arthroconidia)

The development of arthroconidia in many ways resembles that found in ascomycetes (see p. 235). They are often termed oidia. They may be dry and dispersed by air currents, or wet and accumulating in slimy masses from which they are dispersed by insects, in water films or by rain splash. An example of dry oidia is seen in the agaric *Flammulina velutipes*, which fruits on dead tree stumps and logs in winter (see Plate 9d). Oidia may develop on monokaryotic and on clamped, dikaryotic mycelia, but the oidia formed on dikaryons are monokaryotic. They are formed by a process of de-dikaryotization in which each oidium comes to contain a single nucleus (Brodie, 1936; Ingold, 1980). Vacuoles appear at intervals in aerial hyphae, separating cylindrical lengths of uninucleate cytoplasm around which cell walls develop. The wall of the parent hypha then dissolves so that an irregular fragmented chain of oidia is formed (Fig. 18.14a). Oidia germinate by germ tubes formed at either or both ends of the spore. Dry oidia are not uncommon in members of the euagarics and polyporoid clades. Wet oidia are seen in *Coprinus cinereus* on monokaryons as well as dikaryons. They are formed on short, erect, branched or unbranched conidiophores (oidiophores) (Fig. 18.14b), the tips of which fragment into uninucleate, smooth-walled, cylindrical

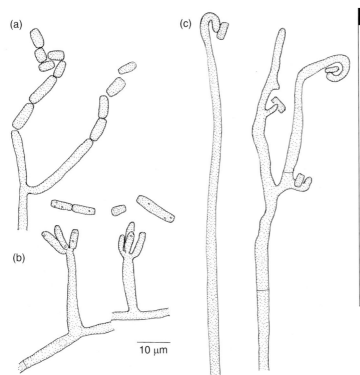

Fig 18.14 Arthroconidia in two Homobasidiomycetes. (a) Arthroconidia of *Flammulina velutipes* formed on a monokaryotic mycelium. When the chains of conidia disarticulate, the conidia remain dry and are dispersed by wind currents. (b) Arthroconidia (oidia) of *Coprinus cinereus* formed on a monokaryon. The sticky oidia accumulate in mucilage in globose heads and are dispersed by insects visiting the dung on which this fungus grows. (c) The homing reaction in *C. cinereus* as seen in an agar culture. The tips of lateral branches of monokaryotic hyphae have been stimulated chemotropically to grow or curve towards oidia of compatible mating type placed near them a few hours earlier. Plasmogamy between a hyphal tip and a compatible oidium is followed by transfer of a nucleus from the oidium into the monokaryon, converting it into a dikaryon.

10 μm

segments (Heinz & Niederpruem, 1970; Polak *et al.*, 1997). The oidia collect in mucilaginous globules from which they are dispersed by insects.

Oidia, whether wet or dry, can function as spermatizing agents. If an oidium is placed a little distance ahead of a monokaryotic hypha, the growing hypha changes direction, being attracted chemotropically towards the oidium. This phenomenon is termed the 'homing reaction' (Fig. 18.14c). The response has been detected over distances up to 75 μm. This is remarkable in view of the fact that the width of the approaching monokaryon hypha is about 2.5 μm and the growing zone of the hyphal tip is about 0.5 μm (Kemp, 1975a). The homing reaction is elicited not only between compatible oidium–hypha combinations but also between incompatible associations. It may even be triggered by oidia of different species. Where the oidium and approaching hypha are compatible, plasmogamy, i.e. fusion of the hyphal tip and the oidium, takes place, followed by nuclear

migration and the eventual establishment of a dikaryon. Plasmogamy may also occur between an oidium and an unrelated approaching hypha, i.e. one belonging to a different species. In this case the introduction of a nucleus from the oidium into a cell of the unrelated hypha results in a lethal response, involving the death of the receptor cell and possibly some adjacent cells. Kemp (1975a,b) has argued that the lethal response is important in maintaining interspecific barriers.

18.8.2 Blastic conidia

Blastic development involves the marked enlargement of a recognizable conidium initial before the conidium is delimited by a septum. This is usually achieved by a blowing-out of part of the conidiophore wall. There are many different ways in which blastic conidia can develop (Kendrick & Watling, 1979). For example, they may develop singly, synchronously in clusters, or in succession. The growth form

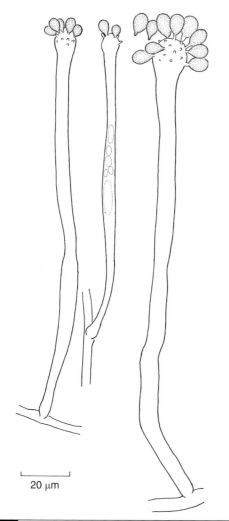

Fig 18.15 *Spiniger* state of *Heterobasidion annosum*. Conidiophores and conidia. Conidial development is blastic.

adopted by many basidiomycetous yeasts is an example of blastic development (see p. 659).

Heterobasidion annosum, a tree-pathogenic polypore, forms clusters of dry blastoconidia synchronously on the swollen tips of upright, club-shaped conidiophores (Fig. 18.15). Following detachment of the conidia, the surface of the swollen tip of the conidiophore bears spiny denticles. These conidiophores resemble the *Oedocephalum* type of conidia found in certain ascomycetes (see Fig. 14.4), but the anamorph name *Spiniger* has been given to them (Stalpers, 1974).

Auricularia auricula-judae, a parasite of elder (*Sambucus nigra*), develops successive blastic conidia from conidiophores on germinating basidiospores (Fig. 21.6c) or from monokaryotic hyphae. At discharge, each basidiospore is unicellular, but three transverse septa divide the basidiospore before germination and each of the resulting cells may form a short conidiophore which swells and curves at the tip to form a horseshoe-shaped (lunate) conidium, followed by further similar conidia.

Sistotrema hamatum (anamorph *Ingoldiella hamata*) is a subtropical aquatic basidiomycete with large, septate, branched conidia (Fig. 25.17; Nawawi & Webster, 1982). They consist of a main axis over 400 μm long with two to three tapering laterals. The tips of the branches are recurved. Conidia are formed on dikaryotic and on monokaryotic mycelia. Dikaryotic conidia are distinguished by the presence of clamp connections at the septa (Figs. 25.17a,b) which are absent from the otherwise similar monokaryotic conidia (Fig. 25.17c). Other species with aquatic branched conidia are recognizable as basidiomycetes either by the presence of clamp connections or dolipore septa within their conidia and mycelia (Nawawi, 1985; Webster, 1992).

18.8.3 Chlamydospores

The term chlamydospore is used here in a wide sense following the definition by Kirk *et al.* (2001) as 'an asexual one-celled spore (primarily for perennation, not dissemination) originating endogenously and singly within part of a pre-existing cell, by the contraction of the protoplast and possessing an inner secondary and often thickened hyaline or brown wall, usually impregnated with hydrophobic material.' A common example of a basidiomycete forming chlamydospores is *Laetiporus sulphureus*, a yellowish-orange bracket fungus parasitic on a range of tree hosts such as oak, willow and yew. In culture they are mostly formed terminally on aerial branched conidiophores which develop from a mycelium lacking clamp connections (Fig. 18.16a). In the mycelium within the substrate, intercalary chlamydospores are also formed. Adjacent to the

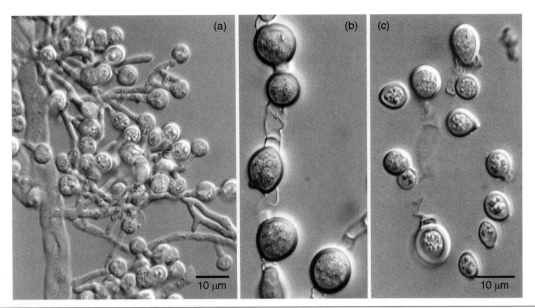

Fig 18.16 *Sporotrichum* state of *Laetiporus sulphureus*. (a) Branched aerial conidiophores with terminal chlamydospores. (b) Intercalary chlamydospores. (c) Detached chlamydospores. (a) and (b) to same scale.

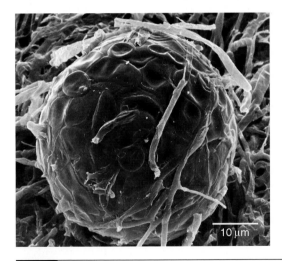

Fig 18.17 SEM image of a bulbil of *Minimedusa polyspora* produced in agar culture.

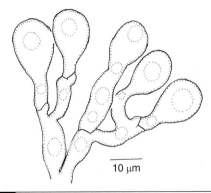

Fig 18.18 *Bulbillomyces farinosus*. Fragment of a large multicellular, branched propagule made up of inflated cells with clamp connections at their base. Air is trapped between the cells. From Abdullah (1980), with permission.

swollen segment packed with cytoplasm which becomes the chlamydospore, there are empty mycelial segments (Fig. 18.16b). The collapse of these empty cells brings about release of the chlamydospore (Fig. 18.16c) so that release is rhexolytic, as already seen in certain ascomycete conidia (see p. 235).

18.8.4 Bulbils

Certain basidiomycetes develop multicellular, pseudoparenchymatous propagules composed of thin-walled, undifferentiated, homogeneous cells in the form of raspberry-like bulbils. A classical example is the terrestrial basidiomycete *Minimedusa polyspora* (Fig. 18.17; Weresub & LeClair, 1971). Another good example of a multi-cellular propagule is seen in the semi-aquatic fungus *Bulbillomyces farinosus* (anamorph *Aegerita candida*). The anamorphic state (Fig. 18.18) appears on the surface of wet wood from freshwater streams as white clusters of clamped cells between which air is trapped, giving the propagule buoyancy. This is a typical feature of **aeroaquatic fungi** which form asexual propagules in air on the surface of leaves and branches of trees previously submerged in fresh water (see Section 25.3).

18.9 | Mating systems in basidiomycetes

18.9.1 Homothallic systems

About 10% of the Basidiomycota which have been tested are homothallic (Raper, 1966). Three types of homothallic behaviour may be distinguished, namely primary, secondary and unclassified homothallism.

Primary homothallism

In *Coprinus sterquilinus* a single basidiospore germinates to form a mycelium, which soon becomes organized into binucleate segments bearing clamp connections at the septa. There is no genetic distinction between the two nuclei in each cell, and this mycelium is capable of forming fruit bodies.

Secondary homothallism (pseudohomothallism)

In *Coprinus ephemerus* f. *bisporus* the basidia bear only two spores, but the spores are heterokaryotic. After meiosis two nuclei enter each spore and a mitotic division may follow. On germination, a single spore forms a dikaryotic mycelium with clamp connections, capable of fruiting. Occasional spores, on germination, give rise to non-clamped mycelia, and fruiting occurs only if these are paired in certain combinations, showing that the fungus is basically heterothallic. The cultivated mushroom, *Agaricus bisporus*, also has a mating system of this type. Although most basidia bear two spores, four-spored basidia do occur, and when monosporous cultures derived from basidiospores from four-spored

basidia are crossed, they produce fruiting mycelia in certain combinations. It has been suggested that a simple bipolar system (see below) is operating (Miller, 1971; Raper *et al.*, 1972). This situation occurs in a number of other two-spored basidiomycetes and is closely paralleled by that found in certain four-spored ascomycetes such as *Neurospora tetrasperma* (Raper, 1966). Fungi showing both secondary homothallism and heterothallic behaviour are said to be **amphithallic**.

Unclassified homothallism

The four-spored wild mushroom, *Agaricus campestris*, is homothallic in the sense that a mycelium derived from a single spore is capable of fruiting. There is nuclear fusion in the basidium, followed by two nuclear divisions, presumably meiotic. However, paired nuclei, conjugate nuclear divisions and clamp connections have not been observed.

18.9.2 Heterothallic systems

Amongst the remaining 90% of the Basidiomycota reported to be heterothallic, we can distinguish bipolar and tetrapolar conditions.

Bipolar

In species such as *Coprinus comatus* (the shaggy ink-cap) and *Piptoporus betulinus* (the birch polypore), when mycelia obtained from single spores from any one fruit body are mated together, dikaryons are formed in half the crosses. This can be explained on the basis of a single gene (or factor) with two alleles. Because only a single factor is involved, the genetic basis for the bipolar condition is described as unifactorial. Segregation of the two alleles at meiosis ensures that a single spore carries only one allele. Dikaryons are only formed between monokaryons carrying different alleles at the mating type locus. In fact, it is known that there may be numerous mating type alleles in a population of fruit bodies collected over a wide area (see below). About 25% of Basidiomycota examined have been shown to be bipolar. Most members of the Uredinales and Ustilaginales have mating systems of this type, although a few have more complex systems.

Tetrapolar

In the coprophilous ink-cap *Coprinus cinereus* or the wood-rotting *Schizophyllum commune*, fertile dikaryons result in one-quarter of the matings when primary mycelia derived from basidiospores from a single fruit body are intercrossed. The explanation originally proposed for this situation was that incompatibility is controlled by two genes (factors), with two alleles at each locus. Because two separate factors are involved, the genetic basis is termed bifactorial. Thus we can denote the two genes as A and B and their two alleles as A_1, A_2 and B_1, B_2, respectively. Consider the cross of a monokaryon bearing A_1B_1 with another bearing A_2B_2. This would result in a fertile dikaryon ($A_1B_1 + A_2B_2$). Such a dikaryon would form spores following meiosis and the spores would be of four kinds: A_1B_1, A_2B_2 (parentals), A_2B_1 and A_1B_2 (recombinants). In most cases studied, the proportions of the four kinds of spore are equal, showing that the A and B loci are unlinked, i.e. borne on different chromosomes.

Fertile dikaryons are only formed when the alleles present at each locus in the opposing monokaryons differ – e.g. in crosses of the type $A_1B_1 \times A_2B_2$ or $A_2B_1 \times A_1B_2$. Where there is an identical allele at either or both loci the cross is unsuccessful. Thus the success of inbreeding within the spores of any one fruit body is only 25% in tetrapolar species as compared with 50% in bipolar forms.

A species with tetrapolar heterothallism whose life cycle is unusual and difficult to interpret is *Armillaria mellea*. Most of the cells of the mycelium are monokaryotic, and there is no evidence of clamp connections in the mycelium or the rhizomorphs. Fruit body primordia arise from the monokaryotic rhizomorphs, and the cells of the young primordia are also monokaryotic. However, cells making up the gill tissue are dikaryotic, and these dikaryotic hyphae are associated with clamp connections, whilst the monokaryotic cells formed in the remaining tissue of the stem and cap have no clamps. Estimations of nuclear volume in monokaryotic and dikaryotic cells suggest that the nuclei of monokaryotic cells are diploid, whilst those of dikaryotic cells are haploid. It is presumed that the diploid nuclei undergo haploidization by an unknown mechanism during the formation of gill initials. Within the basidia, nuclear fusion and meiosis occur, and a single meiotic product enters each basidiospore. In the spore, the nucleus divides mitotically, and one daughter nucleus from each spore migrates back into the body of the basidium and degenerates (Korhonen & Hintikka, 1974; Tommerup & Broadbent, 1974; Ullrich & Anderson, 1978; Anderson & Ullrich, 1982).

Variations in the life cycle of *A. mellea* have been reported. In a form designated as 'Japanese *A. mellea*', Ota *et al.* (1998) presented evidence that four haploid nuclei appear after meiotic division of the diploid nucleus in the young basidium. These haploid nuclei fuse in pairs, resulting in two diploid nuclei which migrate into two of the developing basidiospores where they divide mitotically. One nucleus from each basidiospore returns to the basidium, leaving the spore containing one diploid nucleus. Occasionally nuclear migration fails to occur and the spore remains binucleate. Spores with diploid nuclei can complete the life-cycle by forming a mycelium competent to fruit. Ota *et al.* (1998) therefore concluded that the 'Japanese *A. mellea*' illustrates a kind of secondary homothallism.

18.9.3 Multiple alleles, complex loci

Although a single spore from one fruit body of *S. commune* or *C. cinereus* is compatible with only one-quarter of its fellow spores, crosses between spores from fruit bodies of different origin often result in 100% mating success, i.e. a spore from one fruit body can mate successfully with 100% of the spores from a different fruit body. The explanation of this phenomenon is that a large number of alleles is present in a population representing the species as a whole, instead of the single pair of alleles at each locus present in any one dikaryotic mycelium. Suppose that a second fruit body had the composition ($A_3B_3 + A_4B_4$), then all the four kinds of spore it produced,

A_3B_3, A_3B_4, A_4B_3 and A_4B_4 would be compatible with all the spores of the original fruit body, on the assumption that the essential requirement for fertility is that in any cross both alleles should differ at both loci. This high value for outbreeding success implies the existence of a large number of different mating type factors, and estimates based on isolates from worldwide collections of fruit bodies indicate that the number of mating type factors of certain species may be many thousands (Raper, 1966).

Our understanding of the structure and function of the mating type factors is derived from studies of three species, namely the two Homobasidiomycetes *S. commune* and *C. cinereus*, and the maize smut fungus *Ustilago maydis* (Ustilaginomycetes). Extensive literature is available on general aspects of this topic (see Kües & Casselton, 1992; Kämper *et al.*, 1994; Kothe, 1996; Kronstad & Staben, 1997; Brown & Casselton, 2001; Casselton, 2002) and on the individual species *C. cinereus* (Kües, 2000), *S. commune* (Stankis *et al.*, 1990; Ullrich *et al.*, 1991; Kothe, 1999) and *U. maydis* (Banuett, 1995).

18.9.4 Functions of the A and B loci

There is a close similarity between the functions and structure of the *A* and *B* mating type loci in *S. commune* and *C. cinereus* as described in Table 18.2. The fact that they influence many different functions indicates that their gene products are active as regulatory proteins. The *A* locus encodes two peptides which together make up a heterodimer transcription factor, i.e. the molecule is active only if its two halves are different from each other (see below and Fig. 23.8). The *B* locus of *S. commune* and *C. cinereus* directly encodes the peptide pheromone and a transmembrane receptor for pheromones of compatible strains. In this way it differs from the mating system of the ascomycete yeast *S. cerevisiae* in which the mating type loci encode regulatory genes whose products stimulate the transcription of pheromone and pheromone receptor genes located elsewhere in the genome (see Fig. 10.5).

18.9.5 Structure of the mating type factors

The results of crossing experiments between compatible monokaryons indicate that recombination may occur within the *A* and *B* loci to give novel mating types. This implies that both loci are complex. For *S. commune* it has been proposed that the *A* locus contains two sub-loci, *Aα* and *Aβ*. Similarly the *B* locus contains two sub-loci, *Bα* and *Bβ*. For each sub-locus, pairing tests revealed a number of 'alleles': 9 *Aα*, 32 *Aβ*, 9 *Bα* and 9 *Bβ* (Ullrich *et al.*, 1991). Recombination between the *Aα* and *Aβ* 'alleles' gives rise to 288 (i.e. 9 × 32) different *A* specificities. When these are multiplied by the 81 *B* specificities, over 20 000 possible mating types are generated. In *C. cinereus* there are an estimated 160 *A* specificities and 79 *B* specificities which together generate over 12 000 mating types (Casselton, 2002). The actual number of *Aα* and *Aβ* alleles is not known (Casselton & Olesnicky, 1998).

Table 18.2. | Functions of the *A* and *B* loci in *Schizophyllum commune* and *Coprinus cinereus*. The functions operate only if there are different specificities at the *A* and *B* loci.

Locus	Function
A-regulated	Pairing of nuclei in dikaryon
	Initiation of clamp cell formation
	Synchronized nuclear division
	Septation
B-regulated	Nuclear exchange between monokaryons
	Septal dissolution and nuclear migration
	Peg formation and clamp cell fusion
	Pheromone production

In both *S. commune* and *C. cinereus* it has been possible to obtain details of the structure of the sub-loci at a finer level of resolution. For example, the *Aα* locus of *C. cinereus* contains one gene pair (designated *a1* and *a2*) and the *Aβ* locus two gene pairs (*b1* and *b2*, *d1* and *d2*), each with a number of alleles. In the field, many strains have lost one or more of their maximum complement of six *A* genes (Fig. 18.19). The gene products are the subunits of the heterodimer transcription factor. The two different subunits (1 and 2) are encoded by compatible alleles. For example, a functional heterodimer can be formed from the product of an *a* allele at the *a1* position (e.g. *a1–1*) and a different one at the *a2* position (e.g. *a2–2*, *a2–3*, etc., but not *a2–1*). Likewise, heterodimers can be formed between the products of two compatible *b* or two *d* alleles, but there is a great deal of functional redundancy in the system in the sense that it is sufficient if only one of six possible heterodimers is formed. In other words, compatibility between two strains at the *A* locus is ensured if compatible alleles are present at the *a* or *b* or *d* genes. Casselton and Olesnicky (1998) have calculated that only 5–6 alleles would be required at each gene pair (e.g. 5 × 6 × 6) to account for the estimated 160 unique *A* gene combinations in *C. cinereus*.

The genetics of mating type behaviour in the maize smut fungus *Ustilago maydis* is, in many respects, similar to that described for *C. cinereus* and *S. commune*. This fungus is dimorphic, with a monokaryotic, saprotrophic yeast-like state which can be readily cultured, and a dikaryotic mycelial state which is parasitic on maize and requires living host cells for growth (see Fig. 23.1). *Ustilago maydis* is tetrapolar, unlike most smut fungi which are bipolar. Incompatibility is governed by two loci, designated *a* and *b*. Unfortunately the functions of the *b* mating type factor in *U. maydis* correspond to those of the *A* factor in *C. cinereus* and *S. commune*, and vice versa. There are two alleles at the *a* locus (a_1 and a_2) and about 25, possibly more, at the *b* locus. The *a* locus encodes a pheromone and a pheromone receptor; it controls the switch to filamentous growth but not pathogenicity. The *b* locus encodes the production of a heterodimeric DNA-binding protein involved in self/non-self recognition and also in pathogenicity; two different *b* alleles are required for pathogenic growth (see p. 643).

Much lower numbers of alleles have been estimated for the bird's nest fungi (Gasteromycetes; see p. 581). For instance, *Cyathus striatus* has 4 *A* and 5 *B* alleles, and *Crucibulum vulgare* 3 *A* and about 16 *B* alleles. It has been suggested that

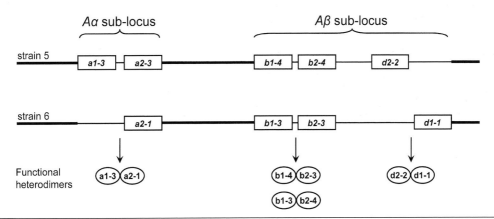

Fig 18.19 The *A* loci of two different field strains of *Coprinus cinereus*. Neither strain has the complete set of six genes, but none the less four different functional heterodimers can be formed from gene products of different (compatible) alleles at the *a*, *b* and *d* genes. Only one would be sufficient for overall compatibility at the *A* locus. Note that the position of the *d* genes is reversed relative to those of the *a* and *b* genes. Genetic recombination events are possible in homologous DNA regions (thick lines) but not within the regions covered by the gene pairs *a*, *b* or *d*. Redrawn and modified from Casselton and Olesnicky (1998).

the small number of incompatibility factors may be related to the specialized method of dispersal in which numerous basidiospores are packaged into a peridiole.

The molecular basis of bipolar (unifactorial) mating systems may be quite similar if one considers that here the A and B loci are simply linked to each other in close proximity on the same chromosome (see p. 637).

18.9.6 The Buller phenomenon

Buller (1931) discovered that a monokaryon of *C. cinereus* paired with a dikaryon of the same fungus could be converted to the dikaryotic state. The same phenomenon has been reported in some other bipolar and tetrapolar fungi. Conversion, i.e. dikaryotization, is brought about by nuclear migration from the dikaryon into the monokaryon. Different kinds of combination (di–mon matings) are possible (Raper, 1966).

There are two kinds of **legitimate combinations**. (1) In fully compatible combinations, a monokaryon is compatible with both nuclear components of the dikaryon, e.g. bipolar $(A_1 + A_2) \times A_3$ or tetrapolar $(A_1B_1 + A_2B_2) \times A_3B_3$. (2) In hemicompatible combinations, a monokaryon is compatible with only one of the nuclear components of the dikaryon, e.g. bipolar $(A_1 + A_2) \times A_2$ or tetrapolar $(A_1B_1 + A_2B_2) \times A_1B_1$. In **illegitimate** (incompatible) **combinations**, a monokaryon is compatible with neither nuclear component of the dikaryon, e.g. tetrapolar $(A_1B_1 + A_2B_2) \times A_1B_2$ or A_2B_1.

Surprising features were discovered in some such pairings. In compatible pairings using *Schizophyllum* it was found that the selection of a compatible nucleus from the dikaryon to dikaryotize the monokaryon was not a matter of chance. Consider the fully compatible di–mon mating $(A_1B_1 + A_2B_2) \times A_3B_3$. If conversion of the monokaryon by one of the nuclei from the dikaryon were entirely random, dikaryons $(A_1B_1 + A_3B_3)$ and $(A_2B_2 + A_3B_3)$ would be equally frequent. However, this is not the case; there is evidence of preferential selection of one mating type over the other, but the reasons for this selection are obscure. A second unexpected feature is the discovery that dikaryotization can occur in incompatible pairings. A possible reason for this phenomenon is that somatic recombination between the nuclei in the original dikaryon can occur to give rise to a nucleus compatible with that of the monokaryon (Raper, 1966).

An unusal mating phenomenon which is the equivalent of the Buller phenomenon has been discovered in *Armillaria mellea*. The vegetative phase of this fungus is mainly diploid, not dikaryotic. Its mating system is bifactorial, i.e. tetrapolar, controlled by A and B loci. When diploid and haploid mycelia are paired in certain combinations, mating occurs, i.e. the diploid mycelium is capable of dikaryotizing the haploid monokaryon (Anderson & Ullrich, 1982).

18.10 | Fungal individualism: vegetative incompatibility between dikaryons

When genetically distinct dikaryons belonging to the same species are paired together, they do not coalesce. Although their hyphae may fuse together, the cells of the resulting heteroplasmon die and often become darkly pigmented. This is the result of vegetative incompatibility. In contrast, when genetically identical dikaryons are paired, their mycelia intermingle. Vegetative incompatibility is readily demonstrated in culture (Fig. 18.20c) and is recognizable in the field as black bands of fungal cells at the interface between adjacent dikaryotic colonies in decaying tree stumps (Fig. 18.20a). The phenomenon was first discovered in the bracket fungus *Trametes versicolor*, the cause of white-rot in deciduous trees, but has since been found to be widespread amongst different ecological groups of basidiomycetes, including wood rotting, coprophilous, ectomycorrhizal and plant pathogenic species, and has led to the concept of fungal individualism (see Todd & Rayner, 1980; Rayner, 1991a,b). Pairing tests between dikaryotic isolates facilitate the determination of the limits and extent of an individual mycelium of a basidiomycete. In *T. versicolor*, tree trunks colonized by this fungus show a number of decay columns which can be traced through a series of transverse slices of wood extending the length of the trunk. In stumps of birch (*Betula*) these columns can be matched genetically with different clusters of fruit bodies emerging at the surface of the stump.

By making pairings of isolates from basidiocarps, decaying wood, and mycelial cords or rhizomorphs of *Megacollybia platyphylla* or *Armillaria bulbosa*, it has been shown that a single individual can extend over many hectares (Anderson *et al.*, 1979; Thompson & Rayner, 1982). Isolations from basidiocarps of the fairy ring fungus *Marasmius oreades* show that the same individual mycelium has spread in an annular fashion and estimates of the incremental growth rate indicate that the same individual may be several centuries old (Mallett & Harrison, 1988; Dix & Webster, 1995).

Vegetative incompatibility enables a fungus to distinguish between 'self' and 'non-self' and prevents the spread of genetic information in the form of nuclei and mitochondria and possibly also of fungal viruses from one mycelium to another of the same species. It thus helps to preserve the genetic integrity of an individual mycelium.

18.11 | Relationships

Basidiomycota are related to Ascomycota. Evidence for this view is based on similarities in composition and construction of the hyphal wall (e.g. presence of chitin; see Section 1.2.2), the molecular basis of mating type control (see p. 266), the production of similar conidial states, and molecular sequence data (see Bruns *et al.*, 1992; Tehler *et al.*, 2003). It has been

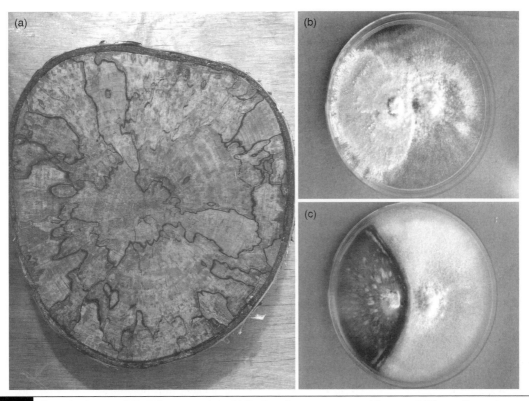

Fig 18.20 Vegetative incompatibility in *Trametes versicolor*. (a) Section of a tree stump containing several different dikaryotic colonies. At the interfaces between adjacent colonies double black lines indicate the incompatibility reaction. (b,c). Interactions between dikaryotic colonies in culture. (b) Reaction when two genetically identical colonies are inoculated near each other. The two mycelia intermingle. (c) Incompatible reaction when two genetically distinct dikaryons are paired.

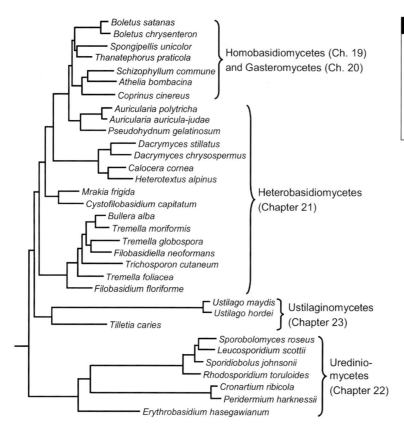

Fig 18.21 Phylogenetic tree of the Basidiomycota based on small nuclear (18S) rDNA gene analyses. The various taxonomic groups and the chapters covering them are indicated. Redrawn and modified from Nishida and Sugiyama (1994), with permission from *Mycoscience*.

postulated that, in evolutionary terms, basidiospores are the equivalent of ascospores whose development has become external instead of taking place endogenously. Clamp connections (Fig. 18.11), characteristic of dikaryotic hyphae of basidiomycetes, are seen as homologous to the croziers in ascogenous hyphae (see Fig. 8.10); both have the same function of re-distributing nuclei. Tehler *et al.* (2003) have indicated the strength of their belief that the two groups are closely related by classifying them together in the Dikaryomycota. They have suggested that this group is a sister group to the Glomeromycota (treated in this book within the Zygomycota as the order Glomales, see Section 7.6). Ascomycetes and basidiomycetes probably diverged from zygomycetes some 400–600 million years ago (Berbee & Taylor, 2001).

Basidiomycetes have a long history. Fossil records show that the characteristic feature of basidiomycete hyphae, the clamp connection, already existed some 300 million years ago in the Carboniferous period (Dennis, 1970). Clearly recognizable mushroom basidiocarps have been preserved in amber about 90–94 million years old (Hibbett *et al.*, 1997a). This also contained well-preserved basidiospores with prominent hilar appendices.

18.12 | Classification

The classification which we have chosen follows that proposed by McLaughlin *et al.* (2001). They have divided the phylum Basidiomycota into four classes:

1. Homobasidiomycetes. Fungi with holobasidia, e.g. agarics and polypores.
2. Heterobasidiomycetes. Fungi with heterobasidia, i.e. jelly fungi and their allies.
3. Urediniomycetes. Rust fungi.
4. Ustilaginomycetes. Smut fungi.

These taxonomic groups are broadly supported by phylogenetic analyses (Swann & Taylor, 1993). An example of the kind of arrangement currently proposed for the basidiomycetes is shown in Fig. 18.21. Because their basidia are arranged on a freely exposed hymenium, the Heterobasidiomycetes and Homobasidiomycetes are sometimes collectively termed **Hymenomycetes**. Various basidiomycetes in which the hymenium is not freely exposed and which do not discharge their basidiospores violently have adopted alternative methods of spore dispersal. They are related to a number of Homobasidiomycete groups, from which they have arisen in the course of evolution. Whilst recognizing that they do not form a natural group, they are here considered together as gasteromycetes because they share important biological features (see Chapter 20). Basidiomycetous yeasts are another artificial assemblage of fungi which have relationships with various taxonomic groups; these are also considered in a separate chapter (Chapter 24).

Homobasidiomycetes

19.1 | Introduction

Fungi included in the Homobasidiomycetes possess holobasidia, in contrast to the hetero-basidia (phragmobasidia) of the Heterobasidiomycetes, rusts and smuts (see Chapters 21–23). The traditional classification of the Homobasidiomycetes, founded by the nineteenth century Swedish mycologist Elias Fries, was based on a number of different arrangements of the hymenium on the hymenophore. The most common types, shown in Fig. 19.1, are (A) **agaricoid**, i.e. gill-bearing (lamellate); (B) **poroid**, i.e. bearing pores instead of gills; (C) **hydnoid**, i.e. with a toothed or spiny hymenium; (D) **clavate**, with a club-shaped or coralloid fruit body, the outside of which is covered by the hymenium; (E) **resupinate**, i.e. with a flattened (corticioid) hymenium appressed to the underside of solid surfaces; and (F) epigeous or (G) hypogeous **gasteroid** or **secotioid** (non-ballistosporic) hymenophores (see Chapter 20). It has long been suspected that the different hymenial arrangements have evolved separately in unrelated fungal groups, i.e. that they represent examples of convergent evolution. They can be interpreted as different ways of maximizing the hymenial area for a given amount of fungal tissue (Pöder, 1983; Pöder & Kirchmair, 1995). Examples of similar hymenophore arrangements in unrelated fungi are seen in the tubular hymenia characteristic of *Boletus* (Fig. 19.21) and *Trametes* (Fig. 19.26) or the toothed hymenia found in the homobasidiomycete *Hydnum* (Fig. 19.1c) and the heterobasidiomycete *Pseudohydnum* (Fig. 21.9b). Quite possibly, only one or a few genes are involved in the morphogenetic events resulting in these various hymenial structures, as has been shown for the transition of gill-bearing fungi to gasteromycetes (see pp. 578–580). There is much other evidence, e.g. in fruit body construction, ultrastructure, chemical reactions of basidio-carps, basidium cytology, spore colour and morphology, and molecular sequence data to support the view that gross morphological characters of hymenophores are unsatisfactory criteria on which to base a natural classification system.

Hibbett *et al.* (1997b) and Hibbett and Thorn (2001), in their preliminary attempts at a more natural classification, found the Homobasidiomycetes to be distributed amongst eight phylogenetic clades (Fig. 19.2), and this scheme has been confirmed and extended in subsequent work (e.g. Binder & Hibbett, 2002; Moncalvo *et al.*, 2002). The eight clades may ultimately be given the status of orders, and although it is tempting to use some of the existing order names in synonymy, e.g. Agaricales for the euagarics clade, the necessary emendations of orders have not yet been carried out. Therefore, we shall use the clade system for the time being. One serious limitation of it is that all large-scale attempts at Homobasidiomycete phylogeny undertaken to date have been based on nuclear and mitochon-drial ribosomal DNA, and confirmation by comparing other genes will be required before

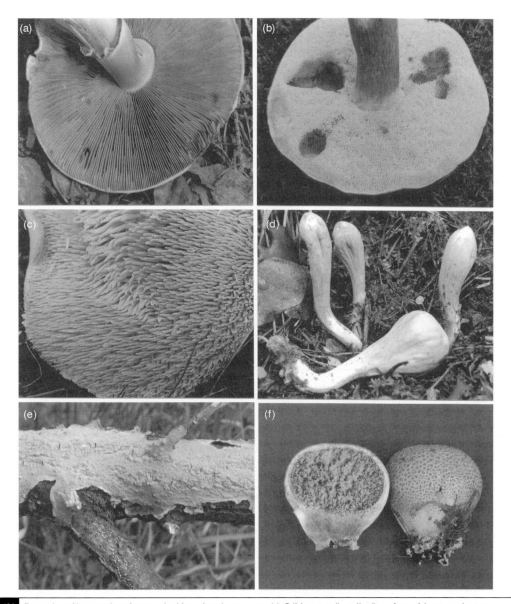

Fig 19.1 Examples of hymenial surfaces in the Homobasidiomycetes. (a) Gill-bearing (lamelloid) surface of *Agaricus silvaticus*. (b) Tubular (poroid) hymenium of *Boletus badius*. (c) Spiny (hydnoid) surface of *Hydnum repandum*. (d) Club-shaped (clavate) fruit body of *Clavariadelphus pistillaris*. The hymenium lines the surface of the fruit body. (e) Flattened (corticioid) hymenium of *Peniophora quercina* forming a crust on the underside of an oak twig. (f) Enclosed (gasteroid) fruit body of *Scleroderma citrinum*.

a more definitive taxonomic system can be put in place.

The eight clades, shown in Fig. 19.2, are as follows: (1) the polyporoid clade, including most members of the former order Polyporales; (2) the euagarics clade containing many members of the old order Agaricales, together with fungi from diverse other groups; (3) the boletoid clade; (4) a thelephoroid clade; (5) the russuloid clade including a particularly wide range of fruit body types; (6) the hymenochaetoid clade; (7) the cantharelloid clade; and (8) a gomphoid–phalloid clade. Table 19.1 indicates that each of these clades contains several of the hymenophore

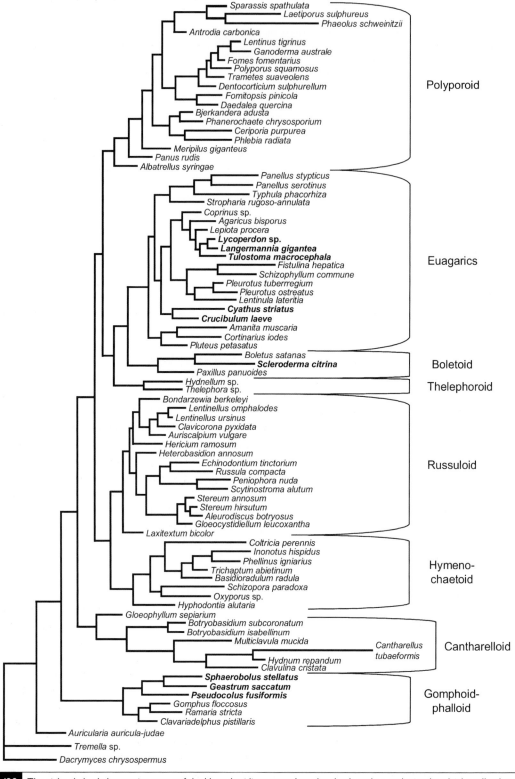

Fig 19.2 The eight-clade phylogenetic system of the Homobasidiomycetes based on both nuclear and mitochondrial small-subunit rDNA sequences. Gasteromycete genera are printed in bold. Redrawn from Hibbett *et al.* (1997b), with permission.
© 1997 National Academy of Sciences, U.S.A.

Table I9.I. Distribution of fruit body morphotypes among the major clades of Homobasidiomycetes shown in Fig. 19.2. From Hibbett and Thorn (2001).

Major clades	Hymenophore type						
	(A) Agaricoid	(B) Poroid	(C) Hydnoid	(D) Clavate	(E) Resupinate	(F) Epigeous gasteroid– secotioid	(G) Hypogeous gasteroid
(1) Polyporoid	+	+	+	+	+	+	
(2) Euagarics	+	+		+	+	+	+
(3) Boletoid	+	+	+		+	+	+
(4) Thelephoroid	+	+	+	+	+		
(5) Russuloid	+	+	+	+	+	+	+
(6) Hymenochaetoid	+	+	+	+	+		
(7) Cantharelloid	+		+	+	+		
(8) Gomphoid–phalloid	+		+	+	+	+	+

morphotypes shown in Fig. 19.1. This finding is perhaps best explained by the assumption that the first Homobasidiomycetes had a morphologically simple fruit body form, probably the flattened (resupinate) type, from which more complex fruit bodies arose as numerous independent evolutionary events (Hibbett & Binder, 2002). The class Gasteromycetes is an artificial taxon, comprising gasteroid and secotioid fruit body forms which have lost the capacity for violent basidiospore discharge and have adopted alternative mechanisms for spore dispersal (see Fig. 19.2). Because of unifying biological and morphological features, the gasteromycetes are described separately in Chapter 20, where we will refer to their affinity to the Homobasidiomycete clades.

19.2 | Structure and morphogenesis of basidiocarps

The fleshy basidiocarps of many fungi belonging to groups 2–5, 7 and 8 in Fig. 19.2 ('agarics') differ from the tougher fruit bodies found in the polyporoid and hymenochaetoid clades 1 and 6 ('polypores') in texture and construction. Agaric basidiocarps are often umbrella-shaped, with a centrally attached stalk (**stipe**) supporting the cap (**pileus**) beneath which are the gills or tubes lined by basidia. The polypores include bracket fungi with basidiocarps which often bear pores and are laterally attached to their substratum. These fruit bodies are usually firmer in texture, i.e. leathery, corky or woody. The differences in texture reflect fundamental principles of construction as shown by analysis of the hyphae composing the basidiocarp. There are also differences in development between agaric- and polypore-type fruit bodies.

19.2.1 Hyphal analysis

Corner (1932a,b, 1953) dissected the basidiocarp tissues of various polypores and showed that they were constructed of three distinctive types of hyphae – generative, binding and skeletal (see below). Later work by Corner and others has identified a further range of hyphal types making up basidiocarps (see Pegler, 1996; Kirk et al., 2001; Clémençon, 2004). This method of **hyphal analysis** provides valuable criteria which have greatly improved the taxonomy of the genera of agaric- and polypore-type fungi. However, it does not of itself provide the basis for a higher-level 'natural classification' of homobasidiomycetes.

There are three main hyphal types. (1) **Generative hyphae** are thin-walled near the margin of a basidiocarp but often thicker-walled behind, with or without clamp connections, usually with cytoplasmic contents. This kind of hypha is universally present in all basidiocarps at some stage of development. The generative hyphae produce basidia and other types of cell making up the hymenium, and they also give rise to the other kinds of hyphae from which the basidiocarp is constructed (Fig. 19.3a). (2) **Skeletal hyphae** are unbranched or sparsely branched, thick-walled hyphae with a narrow lumen. They arise as lateral branches of generative hyphae and form a rigid framework (Fig. 19.3c). (3) **Binding hyphae** (sometimes termed **ligative hyphae**) are much-branched, narrow, thick-walled hyphae of limited growth. These hyphae weave themselves between the other hyphae and bind them together (Fig. 19.3b).

Several other kinds of hypha have been described, some as intermediates between the above three principal systems. **Sarco-hyphae** are composed of long, greatly inflated, mostly unbranched cells $500-3000 \times 10-30\,\mu m$ with relatively narrow septa. They can be interpreted as skeletal, inflated generative hyphae. In *Amauroderma rugosum* (Ganodermataceae) **skeleto-ligative** hyphae resembling skeletal hyphae have thick-walled contorted branches and function in the same way as binding hyphae. **Arboriform** skeletal hyphae with terminal thick-walled branches are present in the basidiocarps of *Ganoderma* (see Figs. 19.23b,c). **Gloeoplerous** hyphae have dense oily contents (Fig. 19.27). The diverse hyphal types may be present in basidiocarps in different

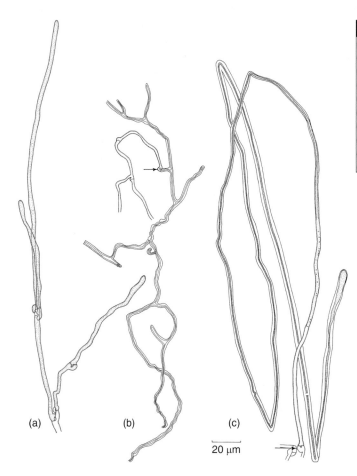

(a)   (b)   (c)

20 μm

Fig 19.3 Hyphal analysis of material dissected from the fruit body of a trimitic polypore, *Trametes versicolor*. (a) Generative hyphae characterized by thin walls, dense cytoplasmic contents and clamp connections. (b) Binding hyphae, branched, contorted and thick-walled. The arrow shows the origin from a generative hypha. (c) A skeletal hypha, unbranched and thick-walled, originating from a generative hypha (arrow).

combinations according to the **mitic system** of description (Gr. *mitos* = a thread of the warp).

Monomitic basidiocarps are made up of generative hyphae only. Most agaric fruit bodies are of this type, containing inflated generative hyphae with or without clamps. However, various modified cell types may also be present, e.g. the **lactifers** (laticifers) containing latex in *Lactarius* (Fig. 19.4). In *Lactarius* and *Russula* (russuloid clade), the flesh contains rosettes of globose, thin-walled **sphaerocysts** or sphaerocytes (see Figs. 19.4 and 19.9c) which give it a brittle texture. Basidiocarps of the polypore *Bjerkandera adusta* are also monomitic, but here the walls of the generative hyphae thicken with age. **Sarcomitic** construction is seen in the polypore-type basidiocarps of *Meripilus*, made up of inflated sarco-hyphae which function as skeletal elements.

Dimitic fruit bodies, produced by various members of the polyporoid and russuloid clades, may show several kinds of construction. Dimitic fruit bodies with binding hyphae (i.e. generative and binding hyphae) are found in the basidiocarp of *Laetiporus sulphureus* (see Plate 10b). The **dissepiments** (i.e. the blocks of flesh separating the tubes) are, however, monomitic. Dimitic fruit bodies with skeletal hyphae are found in *Heterobasidion annosum* (see Fig. 19.26b), whereas dimitic basidiocarps

with skeleto-ligative hyphae are found in *Lentinus* and *Ganoderma*.

Trimitic fruit bodies contain all three kinds of hypha shown in Fig. 19.3. A good example is *Trametes versicolor*, a common bracket fungus on hardwood stumps, trunks and branches (see Plate 10a). At the growing margin of the fruit body and in the dissepiments, construction is dimitic, composed of generative and skeletal hyphae. Binding hyphae develop in the adult flesh behind the growing margin. The term **sarcotrimitic** has been used to describe fruit body construction in *Trogia* where there are generative, binding and sarco-hyphae.

19.2.2 Development of basidiocarps

Basidiocarps begin their development from a hyphal knot, an aggregation of hyphae formed usually on the secondary (i.e. dikaryotic) mycelium. The surface of the hyphae forming the fruit body primordium is often non-wettable due to hydrophobin rodlets (Wessels, 1994, 2000). This property ensures that air-filled channels are present within basidiocarps, allowing efficient gas exchange to occur even under wet conditions (Lugones *et al.*, 1999). The first-formed hyphae making up the young fruit body are little differentiated from normal vegetative hyphae and are termed **protenchyma** but, as differentiation proceeds, the hyphae of the

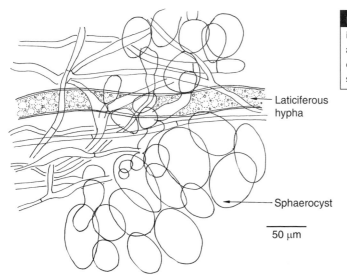

Laticiferous hypha

Sphaerocyst

50 μm

Fig 19.4 *Lactarius rufus.* Cells from the pileus, including thin-walled septate generative hyphae, a wider, thicker-walled laticiferous hypha which contains a milky latex, and clusters of globose sphaerocysts.

basiodiocarp become interwoven and often tightly packed, making up the **plectenchyma**. If the hyphae are less tightly packed, the resulting tissue is known as **pseudoparenchyma**. Although a range of cell types can be distinguished in the mature basidiocarp, they all develop from generative hyphae. Enlargement and differentiation of basidiocarps is associated with inflation of segments of these hyphae. Details of the morphogenesis of basidiocarps will be described later for selected genera.

Several kinds of development have been described, relating to whether or not the hymenophore is at all times exposed or is at first surrounded by other tissues. For example, young fruit bodies may be enveloped by a **universal veil** which is broken as the pileus expands, leaving a cup-like **volva** at the base of the stipe and broken scales on the cap as in *Amanita*. In some agarics the hymenophore is protected during development by a **partial veil** stretching from the edge of the cap to the stem. Where the partial veil is thin and cobweb-like as in *Cortinarius* it is termed the **cortina**, but where it is composed of firmer tissues it persists as a ring (**annulus**) on the stem. Agaric fruit bodies may thus have both universal and partial veils (Fig. 19.5), either, or neither. These different kinds of development have been distinguished by technical terms as follows (Reijnders, 1963, 1986; Moore, 1998).

Gymnocarpic hymenophores are naked from the time of their first appearance and are never enclosed by tissue. The pileus develops at the tip of the stipe and the hymenophore differentiates on the lower side. Gymnocarpic development is found in several unrelated genera, e.g. *Cantharellus, Boletus, Russula, Lactarius* and *Clitocybe* (Fig. 19.6a).

The **angiocarpic** hymenophore is enclosed by tissue during part of its development. There are two kinds of angiocarpic development. In **primary angiocarpy**, the pileus margin, hymenophore and sometimes also pileus and stipe differentiate beneath the surface of the primary tissue of the primordium (protenchyma). *Stropharia semiglobata* and *Amanita rubescens* are primarily angiocarpic (Figs. 19.6c,d). In a fruit body showing **secondary angiocarpy**, hyphae from an already differentiated surface grow out towards the exterior to enclose the primordium or part of it. The hyphae may extend from the margin of the pileus towards the stipe, or from the stipe to the pileus, or both. In *Lentinus tigrinus* hyphae from both the pileus margin and the stipe extend to enclose the developing gills (Fig. 19.6b).

The tissue of the monomitic agaric-type basidiocarp is plectenchymatous or pseudoparenchymatous. The fruit body expands due to inflation of the cells. Although differentiation into skeletal or binding hyphae is limited

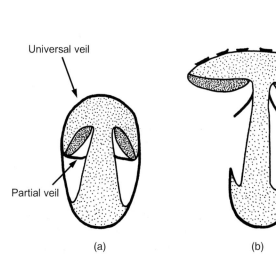

Universal veil

Partial veil

(a)

Cap scales (universal veil)

Annulus (partial veil)

Volva (universal veil)

(b)

Fig 19.5 Schematic drawing of an agaric-type fruit body showing both the universal and partial veils. (a) The button stage. (b) Fully expanded fruit body. Remnants of the universal veil are seen as cap scales and the volva, whereas the partial veil has formed a ring (annulus) around the stipe.

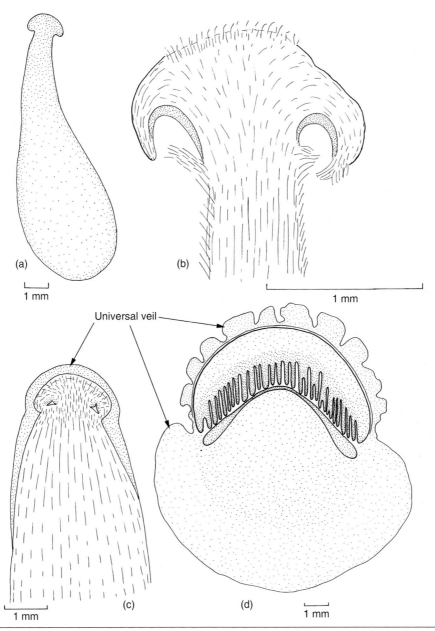

Universal veil

Fig 19.6 Basidiocarp development in some Agaricales illustrated by longitudinal sections (after Reijnders, 1963). (a) *Clitocybe clavipes*. Gymnocarpic development. (b) *Lentinus tigrinus*. Secondary angiocarpy resulting from extension of hyphae from pileus margin and stipe to enclose the previously differentiated hymenophore. (c) *Stropharia semiglobata*. Primary angiocarpy. Note the universal veil enclosing the upper part of the primordium. In mature fruit bodies this becomes gelatinous. The hymenophore is also enclosed by a partial veil. (d) *Amanita rubescens*. Tangential section. Note the break up of the universal veil to form scales on the surface of the pileus. The gill chamber is enclosed by a partial veil.

or absent, specialized tissues or cells may arise (see Fig. 19.4). In a study of the fine structure of the sporophore of *Agaricus campestris*, Manocha (1965) has shown that the stipe contains two kinds of cells — wide inflated cells and narrower thread-like cells. A similar differentiation is found in *Coprinus cinereus* (Moore, 1998). When portions of stipe tissue are placed on suitable agar media, only the thinner hyphae seem to give rise to vegetative growth (Borriss, 1934).

19.2.3 Types of lamellae

The gills of most lamellate fruit bodies are wedge-shaped in longitudinal section and are of the **aequi-hymenial** (aequi-hymeniiferous) type. This term refers to the fact that the hymenium develops in an equal manner all over the surface of the gill, i.e. basidial development is not localized at any one point on the gill. The wedge-shaped section may be an adaptation to minimize wastage of spores should the fruit body be tilted from the vertical. Buller (1909) calculated for the field mushroom *Agaricus campestris* that a displacement of 2°30′ from the vertical

would still allow all the spores to escape. Adjustments in the orientation of the stipe and, sometimes, of the gills themselves may further help to minimize wastage. The topic of **gravitropism** (gravimorphogenesis), i.e. the re-orientation by stipe bending or gill curvature in displaced basidiocarps to restore hymenia to a vertical position, is discussed on p. 546.

Gills of the **inaequi-hymenial** (inaequi-hymeniiferous) type are characteristic of the ink-caps (*Coprinus sensu lato*) where the gills are not wedge-shaped in section, but parallel-sided, and often held apart by **cystidia** (see Fig. 19.7). The term inaequi-hymenial refers to the fact that the hymenium develops in an *unequal* manner, with basidia ripening in zones. In *Coprinus* a wave

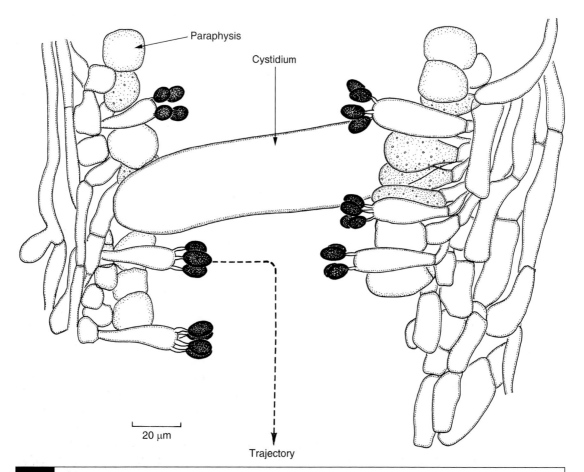

Fig 19.7 *Coprinus atramentarius.* Vertical section of the parallel-sided gills showing basidia, interspersed by globose paraphyses and a cystidium extending across the space between adjacent gills to make contact with the surface of the opposing gill. The dashed arrow indicates the trajectory (sporabola) taken by a projected spore.

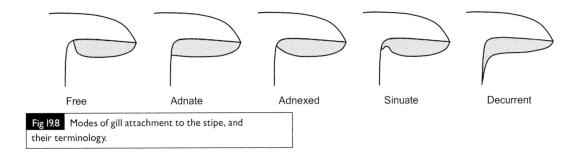

Free   Adnate   Adnexed   Sinuate   Decurrent

Fig 19.8 Modes of gill attachment to the stipe, and their terminology.

of gill maturation begins at the cap margin and passes slowly upwards and inwards. After the basidia at the lower edge of the gill have discharged their spores, the gill tissue undergoes autolysis (deliquescence) into an inky black liquid which drips away from the cap. The gravitropic gill curvature characteristic of aequihymenial types is absent, but the stipe may still curve to bring the gills into an approximately vertical position.

An important criterion for identification of gill-bearing agarics in the field is the way in which the lamellae are attached to the stipe (Fig. 19.8). Gills are said to be **free** if their blade does not touch the stipe. **Adnate** gills show attachment to the stipe with their entire base, whereas **adnexed** gills are attached only partially to the stipe. In **sinuate** gills, the gill margin shows an S-shaped curve near the point of junction with the stipe. **Decurrent** gills run down the surface of the stipe.

19.2.4 The hymenophoral trama

The hymenophore structures of some Homobasidiomycetes are shown in Fig. 19.9. In most hymenophores, there is a central group of hyphae running from the underside of the cap to the tip of the gill, pore or spine, and these hyphae are collectively called the **hymenophoral trama**. Various distinctive types have been recognized by Reijnders and Stalpers (1992), but these do not correlate well with the phylogeny of the Homobasidiomycetes as shown in Fig. 19.2.

1. In the trametoid type (e.g. in *Lentinus*, *Fistulina* and *Schizophyllum*), development begins with a bundle of parallel hyphae from which branches grow out in various directions and may become interwoven.

2. In the cantharelloid type, the developing hymenophore is at first smooth and covered by a palisade of hyphae which will form the hymenium. Later, locally increased activity in the subhymenium may result in irregular ridges (*Serpula*), more regular veins (*Cantharellus*, *Craterellus*) or gills (*Hygrophoropsis*).

3. The boletoid type is shown by *Boletus* and relatives (Fig. 19.9a). The first stage is made up of divergent hyphae which form a narrow central layer (mediostratum), followed by swelling and differentiation of the cells to form a lateral stratum. The diverging hyphae of the lateral strata curve sharply outwards to form the hymenium, and the term **divergent trama** is therefore used to describe this type of hymenophore. A subhymenium made up of narrow interwoven hyphae is also present.

4. The agaricoid type is by far the most common and includes the coprinoid, russuloid, agaricoid, pluteoid and amanitoid subtypes. In cross-section, gills usually appear differentiated into a central trama, a subhymenium and the hymenium (Fig. 19.9b). The hyphae in the trama often run parallel to each other, but in the pluteoid subtype (e.g. *Volvariella* and *Pluteus*), the wide hyphae making up the trama are arranged in a V-shaped pattern (Fig. 19.9d), and this is called the **inverted trama**. In the russuloid subtype (*Russula* and *Lactarius*), the mature trama is said to be intermixed because in addition to the narrow generative hyphae it contains swollen cells (sphaerocysts) of very different width (Fig. 19.9c). The amanitoid type shows a central mediostratum made up of narrow generative hyphae from which wider subhymenial elements diverge (Fig. 19.9e). This type of trama is described as **bilateral**.

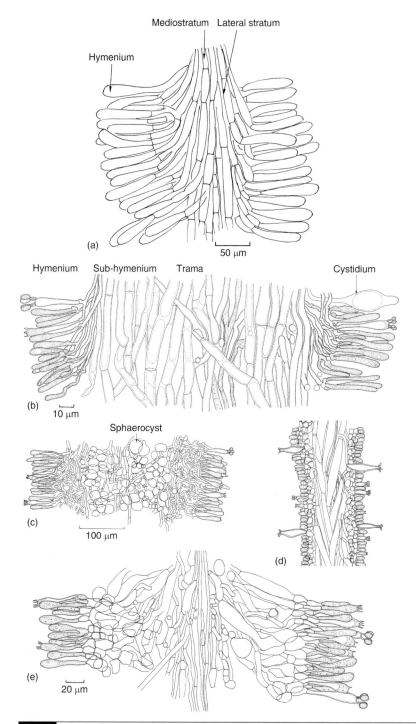

Mediostratum Lateral stratum

Hymenium

(a)

50 μm

Hymenium Sub-hymenium Trama Cystidium

(b)

10 μm

Sphaerocyst

(c)

100 μm

(d)

(e)

20 μm

Fig 19.9 Different types of hymenophoral trama. (a) The boletoid type shown by *Boletus edulis*. L.S. immature pore showing the divergent trama. (b–e) Vertical sections of gills of various agarics of the aequi-hymenial type showing different types of hymenophore. (b) The agaricoid type shown in *Flammulina velutipes*. (c) The russuloid subtype shown by *Russula cyanoxantha*. Note the globose sphaerocysts. (d) The pluteoid subtype shown by *Pluteus cervinus*. Note the converging (V-shaped) arrangement of the cells making up the trama (inverted trama) and the hooked cystidia. (e) The amanitoid subtype. Vertical section of gill of *Amanita rubescens*, showing the bilateral hymenophoral trama. (c) and (d) to same scale.

19.2.5 The hymenium

The ends of the tramal hyphae turn outwards to form a distinct layer of shorter cells, the subhymenium, which lies immediately beneath the hymenium of a palisade-like layer of ripe basidia, developing basidia (basidioles) and sometimes other structures such as cystidioles, cystidia and paraphyses. All these represent the terminal cells of hyphae making up the fruit body and are therefore homologous with basidia. Cystidioles are thin-walled, sterile elements of the hymenium, about the same diameter as the basidia, and usually protruding only slightly from the hymenial surface.

Cystidia are more varied. They are often enlarged conical or cylindrical cells which may arise in the hymenium along with the basidia (hymenial cystidia) or sometimes deeper, for example in the trama (tramal cystidia). Fusion of paired nuclei within cystidia has been reported and in some cases 2–4 basidiospores have been observed to develop on cystidia. In Coprinus and its allies, the cystidia on the face of the gill may stretch across the space between the gills (Fig. 19.7). The tip of a cystidium makes contact with a cell on the opposing hymenium, and this cell, termed a **cystesium** (from cystidium + Lat. haerare = to adhere), then differentiates by developing a granular, vacuolated cytoplasm and becomes cemented to the tip of the cystidium, so that the cystidium bridging the gill cavity becomes firmly attached to both hymenia (Horner & Moore, 1987; Moore, 1998). The role of the hymenial cystidia in Coprinus was initially thought to be that of buttresses helping to space the inaequi-hymenial gills apart. However, it is now believed that the cystidium–cystesium pairs function as tension elements, balancing the stretching forces generated by expansion growth of the cap and helping to straighten out the thin lamellae which are folded during early development (Chiu & Moore, 1990a). In Pluteus the tramal cystidia bear hook-like tips (Fig. 19.9d) whose function is not understood. The suggestion that they might deter animals such as slugs from eating the gill tissue is not supported by feeding experiments (Buller, 1924). In Volvariella bombycina, cystidia on the face of the gill have aqueous drops adhering to them and are thus believed to be secretory in function, releasing water vapour into the space surrounding the basidia (Chiu & Moore, 1990b). Various terms have been used to describe cystidia from different parts of the basidiocarp. Those on the gill face are termed facial cystidia or **pleurocystidia**, those on the gill margin **cheilocystidia**. Cystidia are not confined to the hymenium. Similar structures have been found on the surface of the pileus (**pileocystidia**) and the stipe (**caulocystidia**). We are still ignorant of the function of many of these structures. For a fuller discussion see Singer (1986) and Clémençon (2004).

Paraphyses are present in the hymenium of Coprinus. They develop on branches from the same hyphae which produce basidia (Fig. 19.7), arising after the young basidia are committed to meiosis. The tips of the paraphyses force themselves into the hymenium between developing basidia, grow to about two-thirds of the height of a basidium, become spherical and enlarge, thus playing an important role in the expansion of the maturing basidiocarp (Rosin & Moore, 1985; Moore, 1998).

19.3 | Importance of homobasidiomycetes

Homobasidiomycetes have a significant impact on our lives. The fruit bodies of thousands of species are potentially edible, although only some 40 species have been cultivated. According to worldwide production data for the year 1997 (Chang & Miles, 2004), the most popular cultivated mushrooms are Agaricus bisporus (2 000 000 t per annum), Lentinula edodes (1 500 000 t), Pleurotus spp. (875 600 t), Flammulina velutipes (284 000 t) and Volvariella volvacea (180 000 t). Mushroom production is expanding rapidly, especially in the Far East. However, some of the most prized edible mushrooms are mycorrhizal associates of trees which cannot yet be cultivated away from their host. Mycorrhizal associations involving Homobasidiomycetes are prominent in forest situations and are dealt with in more detail below and also on p. 581.

An equally important ecological role played by the Homobasidiomycetes in the global environment is that of saprotrophs involved in the decomposition of the two most abundant organic carbon sources, cellulose and lignin, thereby releasing nutrients locked up in wood and leaf litter. The degradation of wood is achieved in two different ways, white-rot and brown-rot, and these are described briefly on pp. 527–532. A few wood-rotting species are plant pathogens, e.g. *Armillaria mellea*, *Phellinus noxius* and *Crinipellis perniciosa*, whereas others, notably the dry rot fungus *Serpula lacrymans*, cause economic damage of timber built into houses.

19.3.1 Ectomycorrhiza

Many members of the Homobasidiomycetes (including gasteromycetes), as well as a few Ascomycota such as the truffles (*Tuber* spp.; see p. 423), form ectomycorrhizal associations with coniferous and broad-leaved trees. These are distinguished from the vesicular–arbuscular mycorrhiza between herbaceous plants and Zygomycota (see p. 217) in several key features which have been reviewed by Smith and Read (1997) and Peterson *et al.* (2004). The most immediately obvious is that the bulk of the fungal biomass is located *outside* the plant root, hence the term ectomycorrhiza. The colonization of a tree root by an ectomycorrhizal fungus has reciprocal morphogenetic effects. Lateral roots show stunted growth accompanied by increased branching which is often dichotomous (Fig. 19.10a) in response to fungal colonization, and they are covered by a thick sheath of hyphae, the **mantle** (Fig. 19.10b). There is also a limited colonization of the root cortex in the shape of the **Hartig net**, a system of unusually richly branched intercellular hyphae (Fig. 19.10b). The sequence of colonization events probably starts with hyphae being initially attracted to root tips by their exudates or possibly those of microorganisms associated with the rhizosphere, such as fluorescent pseudomonads (Garbaye, 1994; Smith & Read, 1997). Following contact with a root hair, hyphae grow alongside it until they meet the surface of the main root (Thomson *et al.*, 1989). There, morphogenetic changes are initiated, such as hyphal branching and anastomosis which lead to the establishment of the mantle (Massicotte *et al.*, 1987). These changes may result from the specific recognition of wall surface molecules between the root and fungus hypha (Giollant *et al.*, 1993; Lapeyrie & Mendgen, 1993). Outside the mantle, the mycelium may extend into the soil by a few centimetres, or much further if the fungus is capable of forming mycelial cords (see p. 581). The roots of different plants in complex forest ecosystems may be linked by common ectomycorrhizal fungi, and there may be a net transfer of carbon from sunlit plants to those growing in the shade (Leake *et al.*, 2004). Mineral nutrients, notably phosphate, as well as water are transported from the fungus to the plant. Carbohydrates travel the opposite way. Sucrose, the main transport carbohydrate in plants, is secreted into the apoplast and hydrolysed to give fructose and glucose. The latter is taken up by the fungus and converted to glycogen, trehalose or polyols (Smith & Read, 1997). Some 10% of the net photosynthetic assimilate may be allocated to mycorrhizal fungi which make up 20–30% of the microbial biomass in forest soils (Leake *et al.*, 2004). Fruit bodies are a major sink for translocated carbon. Colourless (achlorophyllous) plants may obtain their carbon by plugging into the mycorrhizal network, a strategy known as **mycoheterotrophy** (Smith & Read, 1997).

Fossil records of ectomycorrhizal associations date back some 50 million years (LePage *et al.*, 1997), although they are more likely to be around 200 million years old (Cairney, 2000). Mycorrhizae of this kind are particularly prominent in nutrient-poor or dry soils. Many ectomycorrhizal fungi have retained the capacity to produce hydrolytic enzymes and are capable of solubilizing, for example, phosphorus and nitrogen from complex sources (Perez-Moreno & Read, 2000). They are therefore often associated with humus layers in which saprotrophic species also grow. Ectomycorrhizal fungi may show a greater or lesser degree of host specificity; for instance, *Suillus grevillei* is associated almost exclusively with larch (*Larix* spp.), whereas *Amanita muscaria* (fly agaric), *Boletus edulis* (cep or penny bun) or *Cantharellus cibarius*

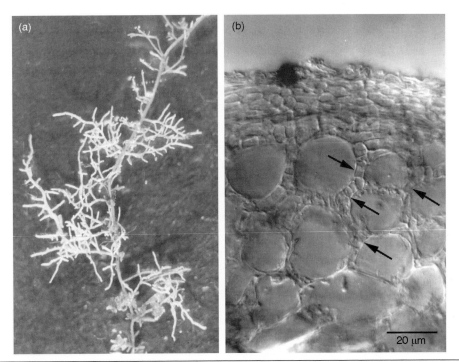

Fig 19.10 Ectomycorrhiza on beech roots collected from humus. (a) General view. Note the stunted growth and dichotomous branching of the lateral roots which are entirely covered by the mantle. (b) Transverse section through such a mycorrhizal root tip. A mantle several hyphae thick has formed a sheath around the epidermis. Branching hyphae grow between the outer cortical cells (arrows) to form the Hartig net. (a) kindly provided by A. E. Ashford.

(chanterelle) are found under both broad-leaved and coniferous trees.

The latter two species together with *Tricholoma matsutake* are the three most valuable mycorrhizal Homobasidiomycetes in commercial terms, with a combined annual crop value of over US$ 2 billion (Hall *et al.*, 2003). Considerable efforts have been made to develop methods for their cultivation, but success has been limited. Whilst it is possible to cultivate mycelium of these species in vitro on agar media and to inoculate seedlings of host trees, the crops of fruit bodies after outplanting have been disappointing or altogether absent (Wang & Hall, 2004). The situation is more encouraging for the ectomycorrhizal ascomycete truffles *Tuber melanosporum* and *T. magnatum* (see p. 423), but even with these species the establishment of commercial plantations has failed to reverse the decline in overall harvests due to overcollection and destruction of natural habitats.

Fortunately, there are several ectomycorrhizal species which, although inedible, can be used as reliable inoculum to promote the growth of trees in challenging environmental situations. Probably the most successful group is the gasteromycete genus *Pisolothus*, which is used extensively in afforestation programmes (see p. 581).

19.3.2 Degradation of wood

The complex architecture of wood is summarized in Fig. 19.11. The middle lamella separating adjacent cells consists of α-(1,4)-linked galacturonic acid polymers collectively called pectins, in which the carboxylic acid group may be derivatized by methylation or calcium salt formation. The primary wall consists chiefly of cellulose made up of β-(1,4)-linked glucose units. Hemicelluloses are also present; these comprise several heterogeneous polymers with a backbone of glucose, mannose or xylose in β-(1,4)-linkage,

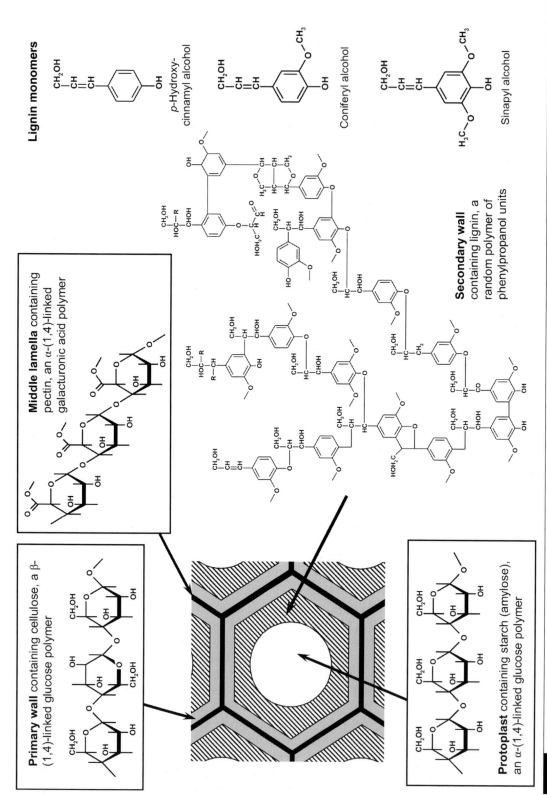

Lignin monomers

p-Hydroxy-cinnamyl alcohol

Coniferyl alcohol

Sinapyl alcohol

Secondary wall containing lignin, a random polymer of phenylpropanol units

Middle lamella containing pectin, an α-(1,4)-linked galacturonic acid polymer

Primary wall containing cellulose, a β-(1,4)-linked glucose polymer

Protoplast containing starch (amylose), an α-(1,4)-linked glucose polymer

Fig 19.11 Carbon-containing polymers and their localization in woody tissue. Lignin consists of a complex polymer (modified from Adler, 1977) of the three phenylpropanoid units shown, which may be linked in numerous different combinations.

with various sugars or uronic acids added as side chains. Since pectin, cellulose and hemicelluloses are all polymerized by enzymes in a regular fashion, they can also be degraded by hydrolytic

Fig 19.12 Wood rot symptoms. (a) White-rot of beech caused by *Trametes hirsuta* (left). Note the bleached appearance of the branch interior at the broken surface. Beech wood attacked by a brown-rot is also shown for comparison (right). (b) Trunk segments of *Picea abies* attacked by *Fomitopsis pinicola*. The brown-rot had caused a hollowing of the standing tree prior to felling. (c) Log of *Picea abies* colonized by *Fomitopsis pinicola*, showing brown rot symptoms. The wood has cracked into cube-like fragments which are easily ground into a powder.

enzymes, although accessibility problems may prevent this from happening *in situ* (see below). The architecture of secondary plant walls is radically different. Apart from hemicelluloses as a minor component, the principal building material is lignin, a polymer of aromatic alcohols which are cross-linked in a random fashion by free radical reactions. Lignin therefore has a complex, non-repetitive three-dimensional structure resistant to direct attack by enzymes. Only few basidiomycetes are able to mineralize lignin to H_2O and CO_2, and this is achieved by oxidative rather than hydrolytic enzymes. The result is that wood undergoes **white-rot** (see Plate 10a; Fig. 19.12a), i.e. it appears bleached because all its components are degraded more or less simultaneously, or lignin degradation precedes attack on cellulose. In contrast, many fungi which degrade cellulose leave behind the lignin component of wood which turns brown upon oxidation, and this type of decay is therefore called **brown-rot**. The removal of cellulose destroys the structural integrity of wood so that the lignin cracks into cubes (Fig. 19.12c) and ultimately breaks up into a powder which becomes incorporated into humus. Living trees may survive infection by brown-rot fungi if this is confined to the heartwood in the core of the trunk, leaving a sufficiently strong cylinder of sound wood (Fig. 19.12b). Prolonged or repeated attacks by such wood-rotting basidiomycetes are responsible for the hollowing of trunks observed in many historic trees.

Brown-rot

The hyphae of brown-rot fungi colonizing wood through its lignin-encased tubular cavities face the major problem of gaining access to degradable substrates. One point of attack is the middle lamella, which is exposed by simple or bordered pits in the secondary cell walls of adjacent cells, and polygalacturonases are indeed produced by many brown-rot fungi (Green & Clausen, 1999, 2003). A typical feature of brown-rots is the production of oxalic acid, which may be important in chelating calcium ions released at potentially toxic concentrations by the degradation of calcium pectate. Calcium oxalate crystals are common on the surfaces of many fungal hyphae. Oxalic acid may also chelate other ions such as Cu^{2+}, which is used as a preservative for the treatment of wood. The oxalic acid concentration may bring about strongly acidic conditions, sometimes as low as pH 1.7, which would be sufficient for acid hydrolysis of pectin and even cellulose (Green *et al.*, 1991). Both endo- and exo-enzymes of the cellulase complex are generally produced by brown-rot fungi (Hegarty *et al.*, 1987), and these may act in the usual synergistic way to convert cellulose to oligosaccharides and thence to glucose (Radford *et al.*, 1996). Hemicellulases are thought to act in a similar way. However, these enzymes may not gain access to their substrate where it is masked by the lignin-containing secondary wall. Although the detailed mechanism of brown-rot decay is still unknown, there is evidence that at least the initial attack on cellulose (and hemicellulose) is mediated by small molecules capable of penetrating the lignin layer. These may be the hydronium ion (H_3O^+) generated by oxalic acid in water, or another molecule such as the hydroxyl radical (HO·) released from hydrogen peroxide (H_2O_2) by the Fenton reaction ($Fe^{2+} + H_2O_2 \rightarrow Fe^{3+} + HO· + HO^-$). The mechanism of brown-rot decay is all the more mysterious because many brown-rot fungi do, in fact, produce enzymes capable of degrading lignin (Mtui & Nakamura, 2004), but these may have other functions, such as the detoxification of antimicrobial phenolics which are often present in wood at high concentrations (Rabinovich *et al.*, 2004).

White-rot

The suite of enzymes required to break down lignin has been most thoroughly examined in the white-rot fungus *Phanerochaete chrysosporium* (see de Jong *et al.*, 1994a; Heinzkill & Messner, 1997). There are numerous conflicting ideas about the enzymology of lignin degradation, and we can only generalize here. Initial attack on lignin is mediated by lignin peroxidases (LiP) and/or manganese peroxidases (MnP) which do not themselves enter the lignin layer. Instead, small diffusible molecules probably act as redox charge carriers between the enzymes and their substrate. A possible reaction scheme for LiP is shown in Fig. 19.13 (Heinzkill & Messner, 1997; ten Have & Teunissen, 2001). The enzyme consists of a single polypeptide chain and a protoporphyrin IX (haem) group which is buried deep inside the enzyme, accessible to small diffusible molecules through a narrow pore. LiP loses two electrons when it reduces H_2O_2 to water. This highly oxidized LiP I state is returned to the ground state in two steps, each associated with the one-electron oxidation of a reductant into its cation radical. Veratryl alcohol, produced in abundance by most white-rot fungi, is such a reductant. The two cation radicals leave the enzyme and either themselves attack the lignin structure, or pass on their charge to other small molecules. Either way, the extraction of one electron from an aromatic ring of lignin generates a structure reacting both as a cation and as a radical, leading to numerous possible degradation products. The conversion of LiP into LiP I requires H_2O_2 which is generated by extracellular enzymes such as glucose oxidase or aryl alcohol oxidase. The latter uses organohalogens such as 3-chloroanisylalcohol as substrates, and these may accumulate in the environment colonized by white-rot fungi (de Jong *et al.*, 1994b), even though organohalogens have traditionally been regarded as man-made (anthropogenic) environmental pollutants.

Manganese peroxidase (MnP) is closely related to LiP in its protein structure and its catalytic cycle, except that the co-factor is Mn^{2+} instead of veratryl alcohol. The oxidized Mn^{3+} may be stabilized by organic acids en route to the lignin substrate, where it catalyses a one-electron oxidation either directly or via charge transfer molecules, followed by recycling of Mn^{2+} back to the MnP enzyme. Copper-containing laccases are a third group of enzymes attacking lignin, and these are produced by a range of fungi much wider than that causing white-rot. By reducing O_2 to H_2O, laccases are capable of performing a four-electron oxidation either of a redox carrier or the final substrate itself. In this way, laccases may be able to degrade smaller lignin fragments, although they are probably incapable of a direct attack on intact lignin, due to steric problems. The precise role of laccases in lignin degradation

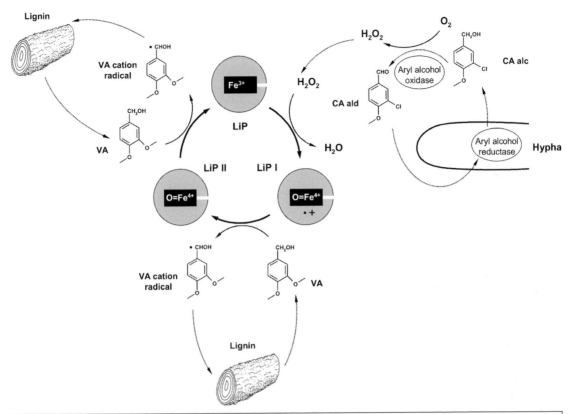

Fig 19.13 Reaction cycle of lignin peroxidase. The reduction of H_2O_2 to H_2O withdraws two electrons from the ground-state enzyme (LiP), one from the ferric ion to give the ferryl ion (Fe^{4+}), and the other from the protoporphyrin group itself which is converted to the porphyrin cation radical. This results in the highly oxidized LiP I state which catalyses two separate one-electron oxidations of a reductant, e.g. veratryl alcohol (VA), becoming reduced via the LiP II state to the ground state. The VA cation radical is regenerated when it withdraws one electron from lignin itself or from other molecules serving as redox charge carriers (not shown). H_2O_2 can be generated by several means, including the oxidation of 3-chloro-anisylalcohol (CA alc) to give its corresponding aldehyde (CA ald).

has not yet been fully characterized, although it is likely to be pivotal as most, if not all, white-rot fungi produce them, and in some white-rot species such as *Pycnoporus cinnabarinus* they may be the only kind of ligninolytic enzyme present (Eggert *et al.*, 1997). All three lignin-attacking enzymes are usually secreted as multiple isoforms which differ in their pH optima or catalytic properties (Kirk & Cullen, 1998). It should be noted that lignin degradation is always **co-metabolic**, i.e. white-rot fungi cannot utilize lignin as the sole source of carbon or energy, and they probably remove it for the purpose of gaining access to more easily degraded substrates such as cellulose. Although

white-rot fungi are generally considered to be more 'efficient' than brown-rots, the latter, in circumventing the effort of lignin degradation, probably obtain more glucose per unit metabolic effort.

The attack on lignin by means of charge transfer molecules being a non-specific reaction mechanism, ligninolytic enzymes will also degrade other substances, including man-made recalcitrant environmental pollutants such as polycyclic aromatic hydrocarbons, organohalogens, or the explosive TNT (see Gadd, 2001). White-rot fungi are therefore being investigated extensively for their potential in the **bioremediation** of contaminated environments. Another biotechnological application is the use of oxidative enzymes for the bleaching of wood pulp for paper production, or of textile dyes.

There is a further twist to the tale. Many fungal metabolites, including veratryl alcohol and 3-chloro-anisylalcohol (see Fig. 19.13), contain methoxy groups or methyl esters, whose methyl groups are added late during biosynthesis. Various molecules may act as donors of methyl groups, including chloromethane (CH$_3$Cl), a known greenhouse gas. A few wood-rotting fungi, notably *Phellinus* spp., have gained notoriety because they produce vastly more chloromethane than they require for their biosynthetic machinery. Watling and Harper (1998) estimated that atmospheric chloromethane, which is mainly of natural origin, accounts for 15–20% of the chlorine-mediated ozone destruction in the stratosphere, and that 86% of the total fungal chloromethane contribution can be attributed to the genus *Phellinus*.

19.4 | Euagarics clade

As shown in Table 19.1 and Fig. 19.2, the euagarics clade includes not only forms with lamellate (i.e. agaricoid) basidiocarps formerly classified in the Agaricales, but also forms with other hymenial configurations. This means that there are few, if any, reliable gross morphological characters by which members of the euagarics clade can be recognized. It is probable that agaricoid basidiocarp types have evolved repeatedly (Hibbett *et al.*, 1997b). Kirk *et al.* (2001) have recognized 26 families, 347 genera and about 10 000 species, but cautioned that there are difficulties in defining these families at present. In the account of some common genera which follows, the estimated numbers of species have been taken from Kirk *et al.* (2001) unless otherwise indicated.

19.4.1 Agaricaceae

The Agaricaceae are one of the most diverse families of the Agaricales, estimated to contain over 50 genera and some 900 species. The **spore print** (i.e. the accumulation of spores projected from a basidiocarp) may be white or coloured. There are also variations in the structure of the hymenophoral trama and the surface of the pileus. Partial and/or universal veils are usually present. Despite these variations, molecular evidence supports the view that the core genera of this family, including *Agaricus* and *Lepiota*, are monophyletic (Moncalvo *et al.*, 2000, 2002; Vellinga, 2004).

Agaricus (c. 200 spp.)

This is a large genus of fungi often synonymized with the term 'mushrooms', distributed mainly in temperate regions. *Agaricus* spp. are saprotrophs, growing in pastures and woodland litter. The mycelium is perennial and some species, e.g. *A. arvensis* (horse mushroom), *A. xanthodermus* (yellow stainer) and *A. tabularis* may fruit in rings. From measurements of the annual rate of increase in diameter, a ring of *A. tabularis* 60 m in diameter was estimated to be 250 years old. *Agaricus* basidiocarps are moderate to large in size, generally firm and white but sometimes changing colour upon bruising (e.g. in *A. xanthodermus*). There is a ring on the stem (two rings in *A. bitorquis*), but no volva. The gills are at first pink due to the colour of the cytoplasm of the spores, but, as the spores mature, their walls darken to a purple-brown, the colour of the spore print. Many species have edible basidiocarps and some are prized as food, e.g. *A. campestris* (field mushroom), *A. arvensis*, *A. macrosporus* and *A. silvaticus*. *Agaricus bisporus* (Fig. 19.14a), now sometimes called *A. brunnescens*, is the cultivated white button mushroom (see below) and is occasionally found in nature on manure heaps, garden waste and roadsides, and under *Cupressus* in coastal areas of California and France. The basidiocarps of *A. xanthodermus* may cause gastrointestinal upsets in some people, probably due to the presence of phenol which also gives the fruit bodies an unpleasant smell of carbolic acid (Gill & Strauch, 1984).

Cultivation of *Agaricus bisporus*

The white button mushroom has been cultivated for its edible fruit bodies for almost four centuries since collectors discovered that its spawn could be used to inoculate compost. The

Fig 19.14 Basidiocarps in the euagarics clade (I). (a) The commercial mushroom, *Agaricus bisporus*. (b) The parasol mushroom, *Macrolepiota procera*. In mature specimens the ring is moveable. (c) *Coprinus comatus*, the shaggy inkcap or lawyer's wig. (d) *Coprinus cinereus* fruiting in the laboratory on a dung/straw mixture. (e) *Pleurotus ostreatus*, the oyster mushroom.

widespread cultivars used in the mushroom industry originated in France, where it is called champignon de Paris or champignon de couche. There are probably about seven such ancestral lineages (Callac, 1995). Specimens growing in nature belong to two distinct groups, namely genetically diverse indigenous populations and more uniform 'cultivar-like' populations which probably represent escapes from cultivation (Xu *et al.*, 1997). Mushroom cultivation is now an important industry in temperate countries, and technical details have been described by van Griensven (1988) and Chang and Miles (2004). In 1996 global production exceeded

2 million tonnes (Moore & Chiu, 2001). In areas with a warmer climate, e.g. in Southern Europe, *A. bitorquis* is also grown.

Mushrooms were originally cultivated in caves, but today most are grown in specially constructed sheds. The basic substratum is a composted straw/manure mixture (Straatsma *et al.*, 1995). The heap of compost is kept outdoors where it heats up naturally to 80°C over a period of 2 weeks, with occasional turning. The heat produced during composting selects for the growth of specialized thermophilic microbes, the most important of which for mushroom production is the anamorphic mould *Scytalidium thermophilum* (syn. *Humicola insolens*) (Straatsma & Samson, 1993). The yield of mushrooms from a compost inoculated with *S. thermophilum* is doubled as compared with pasteurized controls (Straatsma *et al.*, 1994a,b) because the *A. bisporus* mycelium can grow partly on the residues associated with the activities of *S. thermophilum* and partly on its living mycelium and conidia (Bilay & Lelley, 1997). The heat produced during composting destroys many competing microbes, and further destruction is achieved by a pasteurization process in which the compost, now moved into a shed, is heated by steam to a temperature of 60°C for 8 h which minimizes diseases caused later by other fungi.

After cooling, the compost is inoculated with mushroom spawn pre-grown on sterilized cereal grains. The inoculated compost is incubated at 25°C for 2 weeks, and fruiting is induced by covering the colonized spawn with a 3–5 cm deep layer of casing soil or a special mixture of moist peat and chalk. The mycelium grows up through the casing layer and forms anastomosing mycelial cords from which 'pinheads' develop at the surface, some of which expand to form mature basidiocarps. In addition to any physical or chemical effect associated with the casing layer, there is also a biological component. This is shown by the fact that sterilized casing soil is far less effective than natural soil in encouraging fruiting. Re-inoculation of bacteria isolated from casing soil into sterile casing has shown that *Pseudomonas* spp. and especially *P. putida* are effective in inducing fruiting (Hayes *et al.*, 1969). The growth of *P. putida*,

in turn, is selectively stimulated by volatile substances such as ethanol emanating from the mushroom compost. Bacteria accumulate around *Agaricus* hyphae in a zone sometimes termed the **hyphosphere**.

Bacteria which are active in inducing basidiocarp formation might achieve their effect in two ways. Possibly, they induce starvation of *A. bisporus* hyphae in a manner similar to **fungistasis** in soil, where the depletion of nutrients by competing microbes inhibits fungal spore germination. An alternative idea is that bacteria in the casing layer absorb substances which retard the fruiting of *A. bisporus*. This idea is supported by observations that fruiting can be induced by replacing the casing layer with activated charcoal, believed to adsorb inhibitory substances (Long & Jacobs, 1974; Flegg & Wood, 1985; Noble *et al.*, 2003). The isolation of putative inhibitors is a difficult task, but one substance, 1-octen-3-ol, has been shown to inhibit primordium formation in plate culture.

Two weeks after application of the casing layer, fruiting is stimulated by forced air ventilation which lowers the temperature to 20°C and the CO_2 concentration to 300–1000 ppm. Cropping occurs after 10 days, followed by 2–3 more flushes at weekly intervals. The whole cycle from compost preparation to cropping is completed in 10 weeks.

Morphogenesis of *A. bisporus* basidiocarps

The pinheads which form at the surface of the casing soil attached to mycelial cords are basidiocarp primordia, consisting at first of loose aggregates of hyphae. Basidiocarp development has been followed by several workers (see Bonner *et al.*, 1956; Umar & van Griensven, 1997). The hyphal aggregates become compact and enlarge to form flattened 'buttons'. When these are about 2 mm high, there are signs of reorientation of hyphae in the region of the gills, and a gap develops beneath the gill region as a result of programmed cell death (apoptosis) to form the hymenial chamber, enclosed below by the partial veil. At a height of about 5 mm the stipe becomes recognizable by the parallel arrangement of hyphae and, by the time the primordium has grown to a height of 10 mm, the orientation of the upper stalk region is complete. Subsequent increase in height of the stipe is almost entirely due to cell expansion, although some nuclear and cell division also occur, especially in the upper region of the stipe where cell elongation is most marked (Craig *et al.*, 1977). The gills develop in radially arranged ridges made up of downward-growing hyphae terminating in a tightly packed palisade of cells which are the basidia. Although the hyphal segments making up the vegetative mycelium and the stipe are multinucleate, the number of nuclei in the basidia is reduced to two. As the pileus continues to expand, the partial veil is broken and persists as a ring attached to the stem (Fig. 19.14a). There is evidence that stipe elongation is promoted by a factor of unknown chemical identity, which is mainly produced by the ripening gills (Frazer, 1996). Konishi (1967) partially purified a substance which enhanced stipe elongation. The promoting substance included a mixture of amino acids, and it is not known if they function as nutrients or growth factors. It has also been claimed that a compound found chiefly in gill tissue of mushrooms, 10-oxo-*trans*-8-decenoic acid (ODA), stimulates mycelial growth and enhances elongation of the upper stipe (Mau *et al.*, 1992). Unfortunately, attempts to confirm these findings have been disappointing (Champavier *et al.*, 2000).

Biochemical changes occur during basidiocarp development (Hammond, 1985; de Groot *et al.*, 1998). Mannitol, glycogen and trehalose accumulate in basidiocarp primordia, lowering the water potential and thereby possibly causing uptake of water from the mycelium. This would lead to an increase in turgor pressure, causing cell enlargement and fruit body expansion (Hammond & Nichols, 1979). Rapid synthesis of chitin is correlated with stipe elongation (Craig *et al.*, 1979).

A hydrophobin-like protein is secreted by the vegetative mycelium of *A. bisporus* and an abundant hydrophobin (ABH1) forms hydrophobic rodlet layers on the surfaces of the cells in certain regions of the basidiocarp, especially those where there are air spaces, such as in the outer regions of the pileus and stipe, in the veil and in the core of the stipe, but not in the gills. The hydrophobic layer may be responsible for the non-wettability of the surface of the basidiocarp, preventing the inflow of water from the outside and possibly protecting against bacterial and fungal parasites (Lugones *et al.*, 1996).

Life cycle of *Agaricus bisporus*

The life cycle of *A. bisporus* is unusual (Miller, 1971). The majority of the spores formed on its two-spored basidia are at first binucleate but post-meiotic mitosis increases the number of nuclei to four. The spores are heterokaryotic for mating type, i.e. they contain non-sister nuclei with dissimilar *A* mating type idiomorphs (Evans, 1959; Elliott, 1985). The products of meiosis in the basidium are four haploid nuclei, two with mating type idiomorph A_1 and two with A_2. Since two nuclei enter each basidiospore, the spores may be homokaryotic ($A_1 + A_1$ or $A_2 + A_2$) or heterokaryotic ($A_1 + A_2$). Of the 12 possible pairings of the 4 haploid meiotic products, 4 are homokaryons and 8 are heterokaryons, i.e. in a ratio of $1:2$. Evans (1959) has claimed that the disproportionately high ratio of heterokaryotic spores results from the alignment of the nuclear spindles during meiosis in the basidia.

On germination the heterokaryotic spores give rise to a mycelium with multinucleate hyphal segments capable of forming basidiocarps. Mycelia from homokaryotic spores can only fruit following anastomosis with mycelia of opposite mating type. Basidia of normal cultivated *A. bisporus* may rarely bear three or four basidiospores, and most of these are initially uninucleate. Mycelium from such an aberrant spore will only fruit when mated with a mycelium of opposite mating type. Thus *A. bisporus* has two alternative kinds of mating behaviour, secondary homothallism or bipolar heterothallism. Such ambivalent behaviour is sometimes termed **amphithallic** (Lange, 1952) and is not confined to *A. bisporus*, being also found in some other basidiomycetes with two-spored basidia, e.g. certain species of *Coprinus* (euagarics clade) and the four-spored *Stereum sanguinolentum* (russuloid clade).

Breeding of *Agaricus bisporus*

The development of new strains of *A. bisporus* with superior qualities such as more efficient substrate conversion rates, rapid fruiting, higher yield, resistance to disease, ease of picking, extended shelf-life, better appearance, flavour or consistency is actively pursued by the mushroom industry (see Raper, 1985; Sonnenberg, 2000). This task is difficult but worthwhile in view of the high commercial value of the crop, and various techniques are being used:

1. Selection of strains collected in the field and arising during cultivation.

2. Hybridization. The small amount of variation in the gene pool of commercial stocks of *A. bisporus* creates little scope for recombination. However, a search for new genetic resources within wild *Agaricus* populations, the *Agaricus* Resource Program (ARP), has made available novel germ plasm which will prove useful in breeding (Kerrigan, 1996). Of particular interest is the discovery in the Sonoran desert (California) of tetrasporic populations, named *A. bisporus* var. *burnettii*, which are completely interfertile with commercial lines (Kerrigan *et al.*, 1994; Kerrigan, 1995). *Agaricus bisporus* var. *eurotetrasporus*, another four-spored variety, has been discovered in Europe. Whilst var. *burnettii* is amphithallic and predominantly heterothallic, var. *eurotetrasporus* is homothallic (Callac *et al.*, 2003).

3. Protoplast fusion. The fusion of protoplasts from different homokaryons is an alternative to the conventional hybridization technique involving the pairing of compatible homokaryotic mycelia.

4. Transformation (genetic modification) using *Agrobacterium* as a vector for the transforming DNA is also possible, but genetically modified food products are unpopular with consumers and certain political parties.

Macrolepiota (30 spp.)

The best-known species of *Macrolepiota* are the parasol *M. procera* (Fig. 19.14b) and the shaggy parasol *M. rhacodes*, both of which grow in parks, pastures and woodland. They are saprotrophs. The pale brown fruit bodies are large and generally considered good to eat, although those of *M. rhacodes* may cause gastric upsets.

A feature of both species is that the ring is detachable and is free to move up and down the stem. *Macrolepiota procera* and *M. rhacodes* have been separated from *Lepiota*, and Vellinga *et al.* (2003) have proposed that they should be separated from each other as well, with *M. procera* showing affinities with *Leucoagaricus* and *Leucocoprinus*, and *M. rhacodes* with *Agaricus*.

There are concerns about the ability of *M. procera* to concentrate mercury absorbed from the soil into the basidiocarp tissues, with values in the caps as high as $13\,\mu g\,g^{-1}$ reported from various sites in Poland. The mercury content of the caps and stalks was generally independent of the soil substrate concentration, suggesting a remarkable ability of the *M. procera* mycelium to bioconcentrate mercury (Gucia & Falandysz, 2003).

Coprinus

Molecular evidence indicates a relationship between lepiotoid fungi and two species of *Coprinus*, namely *C. comatus* and *C. sterquilinus* (Hopple & Vilgalys, 1999; Redhead *et al.*, 2001). Although it is now possible, with hindsight, to recognize other features shared by these species and other Agaricaceae, e.g. the shaggy surface of the cap or the moveable annulus (Fig. 19.14c), we will discuss them in the context of the many biological features uniting them with their former allies (see below).

19.4.2 Coprinaceae (now Psathyrellaceae)

Molecular studies have shown that the genus *Coprinus* is not monophyletic (Hopple & Vilgalys, 1999; Redhead, 2001; Redhead *et al.*, 2001), implying that the coprinoid character of deliquescent gills has evolved more than once. This finding has caused a taxonomic nightmare because one of the two species found to be related more closely to *Macrolepiota* than to *Coprinus* happened to be the type-species *C. comatus*. Its move from the Coprinaceae to the Agaricaceae meant that those numerous species remaining in the Coprinaceae were given new names, namely *Coprinopsis* (e.g. *C. atramentarius*, *C. cinereus*, *C. lagopus*, *C.*

psychromorbidus), *Coprinellus* (e.g. *C. bisporus*, *C. domesticus*, *C. heptemerus*, *C. micaceus*, *C. plagioporus*, *C. sassii*), and *Parasola* (e.g. *C. plicatilis*). Members of the genus *Coprinopsis* produce fruit bodies with hollow stipes and are closely related to *Psathyrella*, thereby presenting the opportunity to re-name Coprinaceae as Psathyrellaceae. However, because the old name *Coprinus* in a broad sense (*sensu lato*) continues to be used in most non-taxonomic studies, we follow that convention for the time being. It is also possible that there will eventually be an initiative to designate a new type-species of *Coprinus*, i.e. to change the name *C. comatus* rather than those of almost all the other species.

Coprinus sensu lato (c. 350 spp.)

This is the large group of 'ink-caps' with deliquescent inaequi-hymenial gills undergoing autodigestion at maturity (see Orton & Watling, 1979). Representatives of *Coprinus sensu lato* are cosmopolitan, fruiting on a great variety of substrata including soil, dung and wood (Orton & Watling, 1979). *Coprinus comatus* (the shaggy ink-cap or lawyer's wig; Fig. 19.14c) fruits on soil, especially on disturbed ground. It has a well-developed, moveable annulus. Another characteristic feature is the presence of an elastic strand of aggregated hyphae which extends through the lumen of the hollow stipe (Redhead, 2001). Basidiocarps of *C. atramentarius* (common ink-cap or pavement cracker) emerge at the soil surface from buried wood; those of *C. micaceus* (glistening ink-cap) are found in similar situations, or directly on broad-leaved tree stumps. *Coprinus domesticus* also fruits on dead wood of deciduous trees arising from a rust-coloured sporulating mat of mycelium, the *Ozonium* anamorphic state.

There are many coprophilous species, the most studied being *C. cinereus* (earlier erroneously named *C. lagopus*) on cattle and horse manure heaps and manured straw (Fig. 19.14d). This species grows and fruits well in the laboratory and has been the subject of much research, ably reviewed by Kües (2000), on the genetics of mating systems (Pukkila & Casselton, 1991; Casselton & Riquelme, 2004), cytology (Raju & Lu, 1970; L. Li *et al.*, 1999), morphogenesis (Moore, 1998), gravitropism (Moore *et al.*, 1996), stipe elongation (Kamada, 1994), nuclear migration, clamp formation, physiology of fruiting and other phenomena. Whilst most *Coprinus* species are saprotrophic, *C. psychromorbidus* (= 'diseased by cold') belongs to a group of low-temperature basidiomycetes which are plant pathogens. It is the cause of snow mould disease of grasses and other plants in Canada and also causes a post-harvest rot of apples in cold store. It can continue to grow actively under snow cover because it possesses special thermal hysteresis proteins with antifreeze properties (Sholberg & Gaudet, 1992; Hoshino *et al.*, 2003).

Basidiocarps of *Coprinus* have thin inaequi-hymenial gills with prominent pleurocystidia and cystesia (see p. 525). Basidiocarps range in size from a few millimetres in coprophilous forms to over 30 cm (*C. comatus*). Spore development and discharge occurs first at the base of the gill, followed by autodigestion of the discharged basidia and the hyphae which supported them. Chitinase and other hydrolytic enzymes play a major role in autodigestion (Iten & Matile, 1970; Miyake *et al.*, 1980). The digested tissue drips away from the base of the gills as a black fluid which can be used as writing ink. Meanwhile an upward wave of basidiospore discharge continues above the digested tissues. In certain species, the gills do not deliquesce, or do so only partially. In *C. curtus* some limited autodigestion occurs at the edge of the gill and the gills open, as in *C. plicatilis*, by a V-shaped groove which widens from above (Buller, 1909, 1931). Buller (1924) has shown that in many species of *Coprinus* the basidia are dimorphic, with long and short forms present in the same gill. The short forms are fully functional, i.e. the basidia ripen at different levels of the hymenium. This arrangement makes it possible to crowd a larger number of basidia into a given area without interference in spore release. Trimorphic and tetramorphic basidia are present in some species.

Sclerotia in Coprinus

Sclerotia are formed by several species. *Coprinus sterquilinus* develops its sclerotia at the surface of cattle dung and, under suitable conditions, basidiocarps develop from them (Buller, 1924).

In culture, the sclerotia of *C. cinereus* form on monokaryons as well as dikaryons growing aerially or submerged (Waters *et al.*, 1975a,b). They are quite small (100–1000 μm in diameter), dark brown to black and more or less spherical. These sclerotia do not develop basidiocarps. Three tissue layers are distinguishable in aerial sclerotia: an outer diffuse layer of apparently dead cells, a multi-layered rind of heavily pigmented, closely packed, thick-walled cells, and a medulla with predominantly thick-walled cells. These thick-walled cells at first accumulate rosettes of a glycogen-like polysaccharide in the cytoplasm, but this disappears as the cells mature, while a secondary cell wall layer increases in thickness. It is believed that the glycogen serves a temporary storage function but that the secondary wall represents the long-term storage compartment.

Stipe elongation in *Coprinus*

Shortly before spore discharge begins in *Coprinus*, the stipe undergoes rapid elongation, most of it taking place in its upper half. Although in *C. cinereus* the cells of the vegetative mycelium are dikaryotic, the cortical cells of the stipe undergo repeated nuclear division prior to rapid elongation so that they may contain between 32 and 156 nuclei. Stipe elongation is largely brought about by turgor-driven cell expansion, not cell division. The cells increase in length, but not in width, by about eightfold in the final 15 h of fruit body expansion. During that period, the osmotic pressure is actively maintained as water enters the cells. The solutes contributing to the maintenance of osmotic pressure have not been identified. The increase in stipe length is mirrored by an increase in the chitin content. Helically coiled chitin microfibrils are present in the walls of cortical cells, and it is believed that newly synthesized microfibrils are inserted between pre-existing ones, a process termed **diffuse extension growth** to distinguish it from the usual apical extension growth of hyphal tips (Kamada, 1994).

The force of the elongating stipe can be considerable. Buller (1931) placed weights on an expanding fruit body of *C. sterquilinus* which had a pileus height of 1.4 cm and diameter of 0.8 cm. This fruit body could lift a weight of 204 g. The cross-sectional area of the ring of tissue surrounding the hollow stipe was 29 mm^2 and calculations of the upward pressure of the stipe gave a value of two-thirds of an atmosphere ($0.7 \times 10^4\,\mathrm{N\,m^{-2}}$). This may explain why the expanding fruit bodies of *C. atramentarius* can crack asphalt and why other fungi can lift paving slabs.

Mating behaviour in *Coprinus*

Some *Coprinus* spp. are primarily homothallic, e.g. *C. heptemerus*. Species with two-spored basidia are often secondarily homothallic, e.g. *C. bisporus* (= *C. ephemerus* var. *bisporus*). Yet other species show bipolar (i.e. unifactorial) heterothallism (e.g. *C. comatus* and *C. ephemerus*), or tetrapolar (i.e. bifactorial) heterothallism (e.g. *C. cinereus*; see Section 18.9). M. Lange (1952) has proposed the term amphithallism (see p. 532) to describe the behaviour of species of *Coprinus* in which both homothallic and heterothallic mycelia can be raised from the same fruit body, giving as examples *C. sassii*, a bisporic species which is amphithallic-bipolar, and the amphithallic-tetrapolar *C. plagioporus*.

In *C. cinereus*, as in many other species, monokaryotic (rarely dikaryotic) mycelia form globose mucilaginous heads containing numerous oidia dispersed by insects. Plasmogamy preceding dikaryotization may be achieved by fusion between compatible monokaryotic hyphae or by fusion between a monokaryotic hypha and a compatible oidium. Monokaryons may also be dikaryotized following hyphal fusion with dikaryons (the Buller phenomenon; see pp. 508 and 566).

Edibility of *Coprinus* basidiocarps

Some species of *Coprinus* have basidiocarps which are edible when young. *Coprinus comatus* is an example. The fruit bodies of *C. atramentarius* are edible and harmless except if consumed with alcohol, which results in unpleasant symptoms of nausea and palpitations. The substance associated with this effect has been identified as coprine (Fig. 19.15a), and this is similar in its effects to the drug disulfuram (antabuse) which

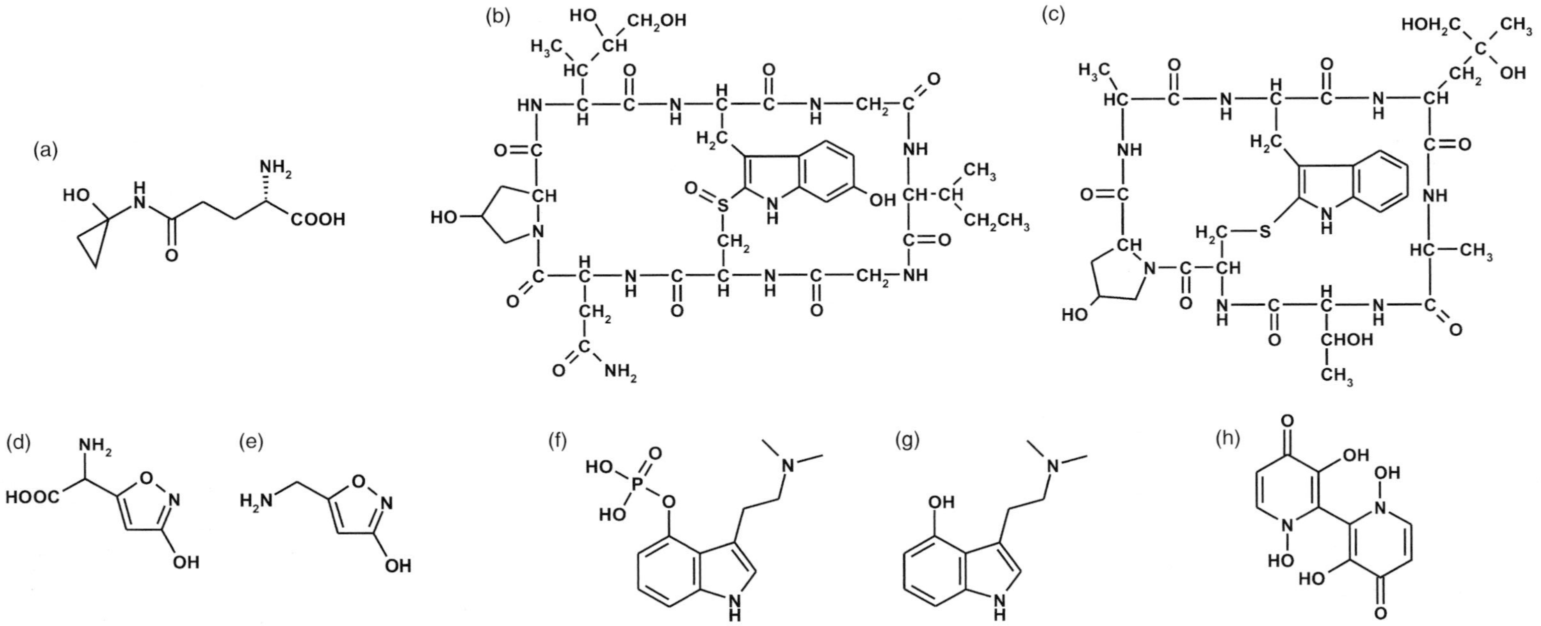

Fig 19.15 Secondary metabolites in the euagarics clade. Coprine (a) produced by *Coprinus atramentarius* is toxic only when consumed with alcohol. α-Amanitin (b) and phalloidin (c) are the toxic principles of death caps, especially *Amanita phalloides* and *A. virosa*. The alkaloids ibotenic acid (d) and its derivative muscimol (e) are found in the fly agaric, *Amanita muscaria*. They are hallucinogenic, probably mimicking neurotransmitters. Similarly, the indole alkaloids psilocybin (f) and its derivative psilocin (g), produced by *Psilocybe* spp., are hallucinogenic. Their structure is similar to the neurotransmitter serotonin. The bipyridyl toxin orellanine (h) is the cause of poisoning by *Cortinarius orellanus*.

is used in attempts to wean alcoholics from their addiction, although it is different chemically.

Interference competition involving *Coprinus*

Many species of *Coprinus* are coprophilous, typically with their basidiocarps appearing relatively late in the succession of fungi on dung (Dix & Webster, 1995). Certain species are known to suppress the fruiting of other fungi. A good example of this phenomenon is *C. heptemerus*, which inhibits the fruiting of many species in nature on rabbit dung and in culture. Its effect on the sensitive ascomycete *Ascobolus crenulatus* was tracked down to the moment when the hyphae of the two fungi make contact. Within minutes, the hyphal segment of *Ascobolus* touched by a hyphal tip of *Coprinus* is killed. There is a rapid loss of turgor, shown by the bulging of the septa of adjacent cells into the affected cell, which also loses the ability to undergo plasmolysis, whilst adjacent cells readily plasmolyse when bathed in hypertonic fluid. This form of competition has been termed **hyphal interference** or interference competition. It occurs in a range of genera of coprophilous basidiomycetes, but is not confined to this ecological group, having been demonstrated to be effective also in lignicolous fungi (Ikediugwu & Webster, 1970a,b; Ikediugwu *et al.*, 1970). Cell damage is very similar to the effects seen when self- and non-self hyphae of the same species confront each other. An oxidative burst is stimulated and hydrogen peroxide accumulates in cells of the sensitive partner. This is interpreted as a defence reaction (Silar, 2005).

19.4.3 Amanitaceae

Amanita (*c.* 500 spp.)

This is a large and important genus whose species form sheathing mycorrhiza with trees. *Amanita muscaria* ('fly agaric'; Plate 9a) is often associated with birch (*Betula*) but also grows in mycorrhizal association with *Abies*, *Pinus*, *Picea*, *Quercus* and other hosts. As is well known, the basidiocarps of some species are poisonous, especially those of *A. phalloides* (death cap), *A. virosa* (destroying angel), *A. pantherina* (panther cap) and *A. verna*, whilst those of *A. muscaria* are more hallucinogenic than poisonous. There are also species whose basidiocarps are excellent to eat, most notably *A. caesarea* (Caesar's mushroom; Plate 9b), which has been hunted enthusiastically in Southern European countries since Roman times. Emperor Claudius was an early connoisseur of *A. caesarea* and may have paid for his mycophagy with his life, probably falling victim to a poisoned mushroom dish manipulated by his wife Agrippina in AD 54 (Ramsbottom, 1953). Other edible species are *A. rubescens* (blusher), *A. vaginata* (grisette) and *A. fulva* (tawny grisette). In view of the possible confusion between edible and poisonous species, it is obviously best to avoid eating basidiocarps of any whose identity is uncertain.

The characteristic features of *Amanita* include a white spore print and the presence of a volva, i.e. the torn remnants of a universal veil. The volva persists as a cup at the base of the stipe and broken volva fragments may also adhere to the cap, as seen as the white scales on the red caps of *A. muscaria* (Plate 9a). Most species also have a ring (annulus) on the stem, the remnant of the partial veil which protected the gills during fruit body development, but there is no ring in some species formerly classified in *Amanitopsis* (e.g. *A. vaginata* and *A. fulva*).

Amanita poisoning

Symptoms after ingestion of fruit bodies of *A. phalloides* follow a characteristic time course over a period of 7 days (Faulstich & Zilker, 1994). After a mushroom meal and symptom-free interval (day 1), there is a period of emesis (vomiting), abdominal cramps and diarrhoea (day 2), followed by a period of remission (day 3) which is treacherous because it lures many patients into believing that they have overcome the poisoning. Meanwhile, severe liver damage is ongoing and symptoms resume with a vengeance with gastrointestinal bleeding (day 4), hepatic encephalopathy (brain damage, day 5), kidney failure (day 6) and death (day 7).

Amanita poisoning is caused by two toxins, namely the amatoxin α-amanitin and the phallotoxin phalloidin (Bresinsky & Besl, 1990; Wieland & Faulstich, 1991; Chilton, 1994, Wieland, 1996). Both are bicyclic oligopeptides (see Figs. 19.15b,c). Of these, the more damaging is α-amanitin which binds to hepatocytes (liver cells). It inhibits a nuclear polymerase responsible for transcribing DNA into mRNA, resulting in reduced protein synthesis at the ribosomes and ultimately the death of cells. Phalloidin acts by binding to G-actin in liver cells and brings about an efflux of potassium ions and lysosomal enzymes, which leads to cell destruction (Chilton, 1994).

The treatment of patients suffering from *A. phalloides* poisoning includes forced vomiting and other means of evacuating the stomach, orally applied activated charcoal to adsorb the toxins, external dialysis, infusion with silybinin, an extract from the milkthistle *Silybum marianum* which competes with α-amanitin for adsorption on to hepatocytes, and liver transplantation (Faulstich & Zilker, 1994).

The effects of ingesting basidiocarps of *A. muscaria* are less severe, being mainly hallucinogenic. Characteristic symptoms are drowsiness, followed by deep sleep in which vivid dreams occur. Recovery is generally complete, with no permanent or prolonged ill after-effects. The two toxins chiefly involved are the alkaloids ibotenic acid and muscimol (Figs. 19.15d,e; Michelot & Melendez-Howell, 2003), and the former is readily converted to the latter in the gut. The molecular structures of ibotenic acid and muscimol closely resemble those of two neurotransmitters, glutamic acid and γ-aminobutyric acid (GABA), respectively. Certain ethnic groups (e.g. in Siberia) have taken advantage of the hallucinogenic properties of *A. muscaria* to experience euphoria. Its use has extended to semi-religious practices in which shamans have induced themselves into trances in which they claim to have powers of revelation. Within 1 h of ingestion, ibotenic acid and muscimol are detectable in the urine of humans and also reindeer, and this may account for the tradition of drinking such urine in order to obtain a 'second-hand kick'. Such practices have now been discontinued in favour of alcoholic beverages, although *A. muscaria* is still occasionally taken as a recreational drug.

The name 'fly agaric' refers to the supposed insecticidal properties of *A. muscaria* caps. When soaked in milk the caps attract flies, possibly brought there in response to an attractant, diolein. The flies ingest the *Amanita* flesh, but it is likely that they are intoxicated rather than killed.

Amanita muscaria, easily recognized by its red cap adorned by white volva scales, is the best-known of all macro-fungi, featuring in countless illustrations, folk tales, religious and semi-religious ceremonies. It has been considered as the 'tree of life' or 'tree of knowledge' (Wasson, 1968). Michelot and Melendez-Howell (2003) have given an account of its chemistry, biology, toxicology and ethnomycology. It is widely distributed in Europe, North America and Asia, and has probably been introduced into other parts of the world as a mycorrhizal partner on the roots of imported trees. Molecular phylogeny studies suggest that it should be separated into at least three groups, corresponding to Eurasian, Eurasian subalpine and North American regions (Oda *et al.*, 2004).

19.4.4 Pluteaceae

Pluteus (c. 300 spp.)

Species of *Pluteus* are saprotrophs, growing mainly on rotting wood. *Pluteus cervinus* is common, fruiting on deciduous tree stumps, logs, fallen branches and on sawdust heaps. Characteristic features are the free gills and pink spore print. On the surface of the hymenium, tapering, thick-walled cystidia crowned by pointed prongs protrude (Fig. 19.9d). Their function is unknown.

The Pluteaceae are related to the Amanitaceae, and these two families appear as sister groups in phylogenetic analyses (Moncalvo *et al.*, 2002).

19.4.5 Pleurotaceae

This family includes only two genera, *Pleurotus* and *Hohenbuehelia* (Thorn *et al.*, 2000; Moncalvo *et al.*, 2002).

Pleurotus (20 spp.)

Pleurotus spp. (oyster mushrooms) have laterally attached, lamellate basidiocarps. Most cause white rot decay of wood. The basidiocarps are edible and several species, e.g. *P. ostreatus* and *P. sajor-caju*, are widely cultivated. Ligno-cellulosic waste products from a range of agricultural crops (e.g. wheat, rice, maize straw, banana leaves and dried water hyacinth) and industrial extraction processes can be used as substratum and cultivation can be practised in factories as well as on a smaller scale, e.g. in large cylindrical plastic bags in the context of a 'cottage industry' which can provide a valuable food supplement in developing countries (Poppe, 2000; Sánchez, 2004). Spent substratum can be used for further fungal fermentations or as animal feed. A problem with the cultivation of *Pleurotus* spp. is the massive amount of basidiospores released from an early developmental stage onwards, causing 'mushroom worker's lung'. In order to overcome this problem, sporeless mutants have been bred.

Pleurotus ostreatus, one of the best-known species, forms clusters of fan-shaped, bluish-grey basidiocarps at the base of deciduous tree stumps (Fig. 19.14e), especially beech (*Fagus sylvatica*). The gills are decurrent (running down the base of the stipe). Development in *P. ostreatus* is gymnocarpic and construction monomitic, but in some other species, e.g. *P. tuberregium*, skeletal hyphae are present. Some species develop thallic–arthric anamorphs, e.g. *P. cystidiosus* in which the conidiophores are synnematal. In addition to being able to decompose wood, *P. ostreatus* and some related species are nemato-phagous (Barron & Thorn, 1987; Hibbett & Thorn, 1994; Thorn *et al.*, 2000; see p. 679), and the nematode prey represents an important supplement of nitrogen which is often the growth-limiting nutrient in woody substrata. The ability to parasitize nematodes has been used as a criterion to support the classification of *P. tuberregium*, a tropical species whose basidiocarps arise from large subterranean sclerotia (Hibbett & Thorn, 1994).

Hohenbuehelia (50 spp.)

This genus includes terrestrial lignicolous species. The basidiocarps frequently contain gelatinized tissue. All species are nematopha-gous. Nematodes are trapped on hourglass-shaped (i.e. constricted) knobs covered with a toxic secretion, formed either directly on the clamped mycelium or at the tips of bent, tapering conidia assigned to the anamorph genus *Nematoctonus* (Barron, 1977; Barron & Dierkes, 1977; see Fig. 25.7).

19.4.6 Schizophyllaceae

Although the unique pattern of fruit body and gill morphogenesis seemed to set *Schizophyllum* clearly apart from the Agaricales (Donk, 1964), the euagarics clade is where we must now place it on the basis of DNA analyses (Moncalvo *et al.*, 2002). More confusingly still, the two fungi apparently closest to *Schizophyllum* are the polypore *Fistulina* and the gill-bearing genus *Volvariella*.

Schizophyllum

Schizophyllum commune has a worldwide distribution but is more common in warmer regions. It grows saprotrophically or parasitically on a wide range of woody substrata forming beige-coloured, fan-shaped, laterally attached fruit bodies with a furry upper surface. Since 1990 it has become common in Western Europe on plastic-wrapped hay silage bales from which clusters of basidiocarps burst out (Fig. 19.16a; Brady *et al.*, 2005). Its spores are wind-borne. James and Vilgalys (2001), working in the Caribbean, trapped spores sedimenting onto the surface of Petri dishes containing a pre-grown homokaryon culture. A successful trapping event was detected by the appearance of dikaryotic growth (with clamp connections). Deposition rates of 18 spores $m^{-2} h^{-1}$ indicate that there is ample inoculum to ensure the colonization of most available substrates as soon as they are available for decay.

The name *Schizophyllum* refers to the longitudinally 'split gills' which are a xeromorphic adaptation (Figs. 19.16b and 19.17). In dry weather the 'gills' curve inwards so that the hymenial surface is protected by a series of adjoining folds. The curvature is due to the shrinkage of thinner-walled hyphal layers on drying. Since the remaining tissue of the gill is composed of thick-walled clamped hyphae which do not shrink so readily, inward curvature follows. Dried basidiocarps can be stored for

Basidiocarps in the euagarics clade (2). (a,b) *Schizophyllum commune*. (a) Cluster of fruit bodies growing through the plastic covering of a silage bale. (b) Basidiocarp as seen from below. (c) *Volvariella speciosa*. Note the presence of a volva and the absence of a ring. (d) *Volvariella surrecta*, a mycoparasite fruiting on basidiocarps of *Clitocybe nebularis*. (e) *Panaeolus sphinctrinus*.

years and revive when placed in a moist environment. Water uptake occurs especially through the hairy upper surface, and the gills straighten out as the hymenium expands within 2–3 h. Spore discharge commences after 3–4 h, and basidia are readily seen in hand sections, rendering *S. commune* a suitable object for microscopic examination if classes have to be held out of season.

Schizophyllum commune is a 'model' fungus with a very extensive literature. It has been studied in detail by workers interested in its mating system, which is tetrapolar with complex *A* and *B* loci, each composed of two sub-loci, α and β. Several alleles have been identified at each sub-locus, creating over 20 000 mating type specificities (see p. 506). Other aspects of its biology which have been studied in depth

(a)

(b)

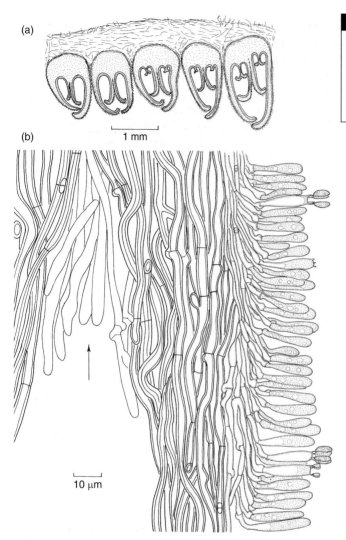

1 mm

10 μm

Fig 19.17 *Schizophyllum commune.* (a) V.S. portion of basidiocarp showing the divided inrolled 'gills'. (b) High-power drawing of part of the 'gill' in the region of the split (arrowed). Note that the hyphae in this region are thin-walled in contrast to the thicker-walled hyphae making up the rest of the flesh.

include genetics (Raper & Hoffman, 1974), nuclear migration (Snider, 1965) and morphogenesis (Wessels, 1993b, 1994). Dikaryotic mycelia fruit readily in culture. Monokaryotic fruiting may also occur (for references see Moore, 1998). The early stages of basidiocarp development have been studied by Raudaskoski and Viitanen (1982) and Raudaskoski and Vauras (1982). A short exposure to light and good aeration are necessary to induce sporulation; elevated CO_2 concentrations are inhibitory. Under dark conditions, mycelial growth is depressed beneath the surface of the medium under a layer of slimy material. Within 3 h after transfer of 3 day old dark-grown cultures to continuous light, there is abundant growth of aerial hyphae at the margin of the culture. After 9 h the emergent aerial hyphae become aggregated into a horseshoe-shaped structure which is the ventral surface of the fruit body. After 15 h basidial initials form the beginnings of the hymenium which develops over the entire lower surface of the basidiocarp. The outside of the fruit body consists of vertical and horizontal hyphal strands, and its outermost layer is formed by parallel backward growth of numerous hyphal tips towards the developing hymenium. By more rapid growth on one side, the fruit body may become fan-shaped. The split gills arise by marginal proliferation, and their number is increased by downgrowths from the flesh of the fruit body. The cytology of basidial development shows no unusual features.

Surface properties, especially the wettability of hyphal surfaces, play an important role in mycelial growth and fruit body development in *Schizophyllum*. Aerial hyphae, including those making up the basidiocarp, are strongly hydrophobic (i.e. non-wettable) because they are covered by parallel rodlets of special proteins, hydrophobins, which were first discovered during research with *S. commune*. Hydrophobins are amphipathic, i.e. they arrange themselves into sheets with one wettable (hydrophilic) face and a non-wettable (hydrophobic) face. At an air–water interface or over the surface of a hypha, the hydrophilic face attaches to the meniscus or to the hyphal surface whilst the hydrophobic layer faces outwards, giving it its non-wettable properties (Wessels, 1997, 2000). *Schizophyllum commune* has at least four hydrophobin genes. The gene *SC3* is active in both monokaryons and dikaryons, whereas *SC1*, *SC4* and *SC6* are active in dikaryons only. When grown experimentally in liquid culture, the hyphae of *S. commune* secrete the monomers of SC3p which accumulate at the air–liquid interface and also lower the surface tension of the culture liquid. This enables individual hyphae to penetrate the interface, and to grow into the air, simultaneously becoming coated with hydrophobin. The function of the product of the *SC4* gene, the hydrophobin SC4p, has been clarified. It covers the surface of the hyphae lining the numerous air channels which traverse the basidiocarp, preventing them from being clogged by water and thus permitting gas exchange to continue during respiration.

Volvariella

There are about 50 species of *Volvariella*, and their taxonomic position is still unsettled but may be close to *Schizophyllum* (Moncalvo *et al.*, 2002). The genus includes several species cultivated for their edible basidiocarps. The best known is *V. volvacea*, the paddy straw mushroom, widely grown in Asia on rice straw, cotton waste and other cellulose-rich agricultural waste products. In terms of world production it is one of the most important edible mushrooms. It is a warm-temperature fungus which can grow vegetatively at 32–34°C, with an optimum temperature for fruiting of 28–30°C. Under favourable conditions, the period between inoculation and harvest of fruit bodies is 8–10 d, the shortest in any cultivated fungus (Chang & Miles, 2004). The basidiocarps of *Volvariella* spp. are enclosed by a universal veil which persists as a prominent cup-like volva (Fig. 19.16c). There is no ring. The gills are free and the spore print is pink.

Volvariella volvacea has an unusual life cycle (Chang & Yau, 1971; Chiu, 1993). It is homothallic and its mycelium is haploid, being made up of multinucleate segments. There are no clamp connections and it is reported that nuclei can pass through the transverse septa which separate adjacent hyphal segments. Brown, thick-walled, multinucleate chlamydospores are borne on specialized branches of the aerial mycelium. They serve as asexual propagules under adverse conditions, germinating by hyphal growth. Young basidia contain two haploid nuclei which fuse to form a diploid nucleus. Meiosis gives rise to four haploid nuclei, one entering each of the four basidiospores. Basidiospores are therefore normally uninucleate (occasionally binucleate).

Volvariella bombycina, the silver silk straw mushroom, has a similar life cycle. It fruits readily in laboratory culture and its basidiocarp development has been studied by Chiu and Moore (1990b). Development is normally hemiangiocarpic, but angiocarpic (i.e. with the hymenium enclosed until a late stage) and gymnocarpic (hymenium exposed) development may also occur. Hemi-angiocarpic development has been arbitrarily divided into five phases: the bulb, button, egg, elongation and mature stages. During the button stage, a schizogenous cavity is formed, enclosed by the universal veil and basal bulb. The hymenophore develops within this cavity. Elongated cylindrical hyphae of uniform size develop on ridges, initiating the formation of the gills. These ridges do not extend to the future stipe, so the gills remain free. During the late egg stage, i.e. before rupture of the universal veil, local proliferation of the gills occurs, and at the site of proliferation the hymenophore is folded in a sinuous or labyrinth-like pattern in contrast with the regular folding seen in the mature stage. During the elongation stage, expansion of the stipe and cap causes the universal veil to rupture. In addition to the primary gills which run along the entire radius of the cap, secondary or tertiary gills, extending for shorter distances, develop either by successive bifurcation of an older gill near its inner, free edge or by the formation of a new hymenial layer at or near the root of an old gill. The hymenium includes basidia and larger, skittle-shaped facial cystidia which may extend across to the opposite gill face. Droplets observed on the cystidia suggest that the cystidia may have a secretory role.

Volvariella surrecta is mycoparasitic. Its basidiocarps grow on the caps of other agarics such as *Clitocybe nebularis* (Fig. 19.16d).

19.4.7 Bolbitiaceae
Panaeolus (25 spp.)
Most members are coprophilous but some grow in pastures. Common examples are *P. semi-ovatus* (sometimes classified in a separate genus *Anellaria*) and *P. sphinctrinus* (Fig. 19.16e). Both species fruit on cattle dung. The gills are wedge-shaped in section and aequi-hymeniiferous. The spores are black. *Panaeolus* spp. are commonly called mottle-gills, and this refers to the fact that the basidia ripen in patches, not uniformly, so that areas in which the spores are ripe appear darker than those in which the spores are still immature (Buller, 1922). Basidiocarps of some *Panaeolus* spp. contain psilocybin and are hallucinogenic (see p. 553).

19.4.8 Hygrophoraceae
Members of the Hygrophoraceae are sometimes included in Tricholomataceae but have been resolved in recent phylogenetic studies as a separate family (Moncalvo *et al.*, 2002). Representative genera are *Hygrophorus* (100 spp.) and *Hygrocybe* (150 spp.), also known as waxcaps because of the texture of their basidiocarps. Some species grow in woodland and may form ectomycorrhiza with trees, e.g. *Hygrophorus eburneus* from beech woods (*Fagus sylvatica*). However, the characteristic habitats of many

waxcaps are impoverished, non-fertilized pastures. Their basidiocarps are often slimy or gelatinous and brightly coloured, e.g. the blood-red *Hygrocybe coccinea* (scarlet waxcap; Plate 9c) or the greenish-yellow *Hygrocybe psittacina* (parrot waxcap). The basidiocarps of *Hygrocybe conica* (blackening waxcap) turn black when bruised.

19.4.9 Marasmiaceae
Although the Marasmiaceae, like most families in the euagarics clade, may well become substantially rearranged in the future, the work by Moncalvo *et al.* (2000) lends some support for the grouping together of key genera such as *Lentinula, Omphalotus, Marasmius, Crinipellis, Flammulina* and *Strobilurus*. Most fungi formerly known as *Collybia* are now also accommodated in the Marasmiaceae, assigned to the genera *Gymnopus* and *Rhodocollybia*. Several members of the Marasmiaceae are known to produce antifungal substances in pure culture, and one group, the strobilurins, has been developed into fungicides currently enjoying worldwide application (Sauter *et al.*, 1999; see p. 410). Some strobilurins and the related oudemansins have been shown to be secreted by their producers into colonized wood at concentrations which are toxic to potential competitors, thereby indicating their ecophysiological function as a means of resource capture and defence (Engler *et al.*, 1998). Whereas most members of the Marasmiaceae are saprotrophs on wood and humus, some are necrotrophic or, rarely, biotrophic pathogens of trees.

Marasmius (500 spp.)
Species of *Marasmius* often have rather tough, leathery basidiocarps which shrivel on drying but rapidly revive on wetting. The best known species is *M. oreades*, the fairy ring fungus whose basidiocarps are edible and may be dried (Fig. 19.18a). The mycelium of the fungus grows outwards from a central point and, as the mycelial front progresses radially, the older, trailing mycelium dies. This results in a ring of active mycelium, visible as a circle of green grass in lawns and pastures (Fig. 19.18b). Measurements of the radial extension of growing

Fig 19.18 Basidiocarps in the euagarics clade (3): Marasmiaceae. (a) *Marasmius oreades*. (b) Two fairy rings of *M. oreades* approaching each other on a lawn. The centres of the rings are to the top and bottom of the picture. (c) *Oudemansiella radicata* which grows attached to tree stumps and buried wood and has a long, tapering, underground base to the stem (pseudorhiza). (d) *Armillaria mellea*. (e) Flower cushion of cocoa tree infected by the monokaryotic mycelium of *Crinipellis perniciosa*. Both healthy flowers and swollen infected shoots are visible. (f) Basidiocarps of *C. perniciosa* on a dead cocoa twig. (e) and (f) from Griffith *et al.* (2003), *New Zealand Journal of Botany*, by copyright permission of the Royal Society of New Zealand; original images kindly provided by R. N. Birch, J. N. Hedger and G. W. Griffith.

rings show that they can extend outwards at a rate of about 1–3.5 cm per annum, and measurements of their diameter indicate that some, in permanent pastures, may be centuries old. The mycelium is intermingled with the grass roots usually near the soil surface at a depth of 8–10 cm, but it may extend to a depth of 30 cm,

depending on soil type. Closer inspection shows that there are three concentric rings at the soil surface: (1) an outer zone in which the grass is greener and taller than outside, and in which the fungus fruits; (2) a middle zone where the grass is dead and the ground bare, especially in dry seasons; and (3) an inner zone of stimulated growth often occupied by other plants which have colonized the previously bare ground. The stimulated growth of grass in the outer ring zone is associated with more rapid decomposition of soil organic matter and the concomitant release of nutrients. The amount of soil organic carbon may be lowered by about 50% as compared with a non-invaded turf. The death of grass in the bare central ring zone is probably due to a combination of factors, including parasitic attack by the fungus, drought resulting from impeded water percolation, and toxins of fungal origin (e.g. cyanide, HCN) which damage the grass root tips (see Dix & Webster, 1995).

Other common species of *Marasmius* include *M. androsaceus* with thin, black, horsehair-like rhizomorphs and small basidiocarps arising from pine needles, dead heather, etc., and *M. ramealis* which forms clusters of basidiocarps on dead twigs and herbaceous stems. *Marasmius rotula* grows in similar situations. Its gills are not attached directly to the stem but to a cylindrical collar.

Oudemansiella (10 spp.)

This is a genus containing temperate and tropical species. Some species, e.g. *O. mucida* (porcelain fungus), have a membranous ring on the stem, but in others, e.g. *O. radicata*, a ring is lacking. *Oudemansiella mucida* is parasitic on beech (*Fagus sylvatica*) and forms white basidiocarps with slimy caps on the branches of infected trees. *Oudemansiella radicata* is so called because the base of the stipe is prolonged into a tapering **pseudorhiza** which connects with buried woody branches extending for several cm beneath ground level (Fig. 19.18c; see also Buller, 1931).

Flammulina (10 spp.)

The best known species is *F. velutipes*, the velvet shank or winter fungus, which grows on deciduous tree stumps and branches and fruits in winter, forming golden yellow basidiocarps singly or in clusters (Plate 9d). Basidiocarps can survive being frozen and soon continue to discharge spores on thawing. In the past, most collections of *Flammulina* were named *F. velutipes*, but it is now believed that several taxa had

been included under this name (see Hughes *et al.*, 1999). The Latin and trivial names refer to the brown velvety hairs at the base of the stipe. The distribution of the fungus is throughout temperate regions of the Northern Hemisphere. An anamorph is produced as chains of dry arthroconidia which have already been described (see Fig. 18.14a). The basidiocarps of *F. velutipes* are edible and the fungus is cultivated for food in Japan as enoki-take (Chang & Miles, 2004). Spawn of the fungus develops on sawdust in plastic bags or on logs incubated at 21–24°C for 14–18 d and then subjected to a cold shock (4–10°C) for 3–5 d to induce fruiting, which follows within 5–8 d at a temperature of 10–16°C. Light is required for fruiting.

Graviperception and gravitropism in Flammulina

When normally erect basidiocarps of *F. velutipes* are displaced into a horizontal position, they respond within about 3 h by bending upwards to restore their original orientation (Fig. 19.19). Experimental studies, some conducted in orbit in a space laboratory, have helped elucidate the mechanism (Moore *et al.*, 1996; Kern *et al.*, 1998; Kern, 1999). Bending occurs by curvature of the stipe in a transitional zone (2–3 mm long) immediately beneath the cap, where the hyphae making up the stipe and cap intertwine. Although basidiocarps which develop on Earth are erect, those formed in space were randomly orientated, growing out in all directions from their substrate. Despite this, normal caps, basidia and spores were produced. This shows that gravitational force is needed for stipe orientation but not for other aspects of fruit body development. Two responses operate to control fruit body development, namely negative hydrotropism causing basidiocarps to grow away from their moist substrate into drier air, and negative gravitropism. Bending is caused by greater enlargement of the cells making up the lower flank of the stipe as compared with those of the upper flank. The cells of the lower flank are more vacuolate than those of the upper. In green plants with gravitropic responses, graviperception is correlated with the sedimentation of denser cytoplasmic particles, **statoliths**, which

Fig 19.19 Gravitropism of basidiocarps in *Flammulina velutipes* growing on a branch of gorse (*Ulex europaeus*). The branch was turned clockwise by 90° 4 h earlier, and the upper stipe regions are undergoing curvature (arrows) to bring the pilei back into the horizontal position.

congregate and are more abundant in the lower sides of cells in horizontally displaced organs. In *Flammulina*, Monzer (1996) considered that only the nuclei, with a density of $1.22\,\mathrm{g\,cm^{-3}}$, were valid candidates for gravity-related sedimentation to enable them to function as statoliths. In the transition zone there may be up to 10 nuclei per hyphal segment. These nuclei are enmeshed in filaments of F-actin, and Monzer (1995) has suggested that tension of the actin filaments associated with sedimentation of nuclei is transmitted to the plasma membrane and provides the trigger to initiate cellular changes involved in the gravitational response. Evidence supporting the involvement of actin filaments is that the gravitropic response is affected by the actin-depolymerizing drug cytochalasin D, but not by treatment with microtubule-inhibiting drugs.

Within 30 min of displacement of a basidiocarp from a vertical to a horizontal position, differences may be noted between the upper and lower flank cells. Microvesicles, most likely derived from the endoplasmic reticulum and Golgi cisternae, are more abundant in the lower flank cells. They fuse with the vacuoles, thereby contributing to the volume increase and turgor-driven enlargement of these cells. Vesicles containing wall precursor materials and enzymes also develop, providing for enlargement and stretching of the walls of lower flank cells. There are some 1.2 million hyphae present in the cross-section of a *Flammulina* stipe. They remain strictly parallel to each other but do not show anastomosis or direct contact with each other. It is believed that each separate hypha in the transition zone responds to the gravitational stimulus. Possibly a growth factor which inhibits the growth of upper flank cells is involved, but no such substance has yet been demonstrated or identified.

Detailed studies of gravitropism have also been made using stipes of the coprophilous inkcap *Coprinus cinereus* (Kher *et al.*, 1992). Although there are similarities to *Flammulina*, there are also differences. For example, the response time of *Coprinus* is much shorter than that of *Flammulina*, and curvature extends along the entire stipe rather than being restricted to the transition zone. These differences reflect the varied growth conditions of the two fungi (Moore *et al.*, 1996). *Flammulina velutipes* is lignicolous and its basidiocarps are relatively long-lived whilst *C. cinereus* is coprophilous and its basidiocarps are evanescent.

Armillaria (42 spp.)

Most species are root pathogens, especially of woody plants, and there is an extensive literature on their biology and pathogenicity (see Shaw & Kile, 1991; Holliday, 1998; Fox, 2000). In the past, most collections were identified as *A. mellea* (the 'honey fungus'), but this is now regarded as an aggregate of about 5–10 species, *A. mellea* agg. (Fig. 19.18d). Pegler (2000) has given a key to the European species. Basidiocarps may occur singly, but often grow in clumps of dozens or even hundreds on the stumps of dead trees. The fruit bodies are reported to cause gastro-intestinal upsets if eaten raw, but are edible if they are cooked and the cooking water is discarded. Most species are annulate (i.e. with a ring on the stem), but in A. tabescens a ring is lacking. Serious pathogens include A. luteobubalina and A. mellea sensu stricto on a very broad range of hosts, and A. ostoyae on conifers (Larix, Pinus, Pseudotsuga, Picea) and birch (Betula). Armillaria cepistipes and A. lutea are mainly saprotrophic.

In infected trees, sheets of white mycelium grow between the wood and bark, destroying the phloem and cambium. In plantations, infections arise when air-borne basidiospores colonize the cut ends of thinning stumps. Spread of infections is by root-to-root contact from diseased to healthy trees and by means of bootlace-like rhizomorphs (see and Fig. 18.13b). In this way adjacent trees in plantations become infected, resulting in group dying. A considerable area of forest may be affected by Armillaria, with an individual clone of A. bulbosa shown to extend over several hectares (Smith et al., 1992). The mycelium of Armillaria continues to grow saprotrophically after the death of an infected tree, and the zone of infected wood within the tree may be surrounded by dark, melanized, pseudosclerotial plates. Rhizomorphs may survive there for up to 40 years and continue to support fruiting over several years. Infected wood is often bioluminescent. Survival of the mycelium and its protection by pseudosclerotial plates makes the control of diseases caused by Armillaria difficult and expensive, although fungicidal treatment has been attempted (West, 2000). Alternative biological control measures using fungal antagonists such as Trichoderma spp. also hold promise. Integrated control based on a combination of fungicidal treatments and biological antagonists (fungi and nematodes) has had a limited measure of success (Raziq, 2000).

Although A. mellea agg. is generally regarded as a pathogen, it forms endotrophic mycorrhiza with several genera of chlorophyllous orchids in the tropics, and also with the colourless (achlorophyllous) orchid Gastrodia elata. As in other orchids (see p. 597) the seedlings of Gastrodia only become established following infection by the haploid basidiomycete Rhizoctonia. However, the Rhizoctonia infection is only a primary phase of limited duration, and secondary infection by Armillaria is essential for successful further growth of the orchid protocorm. The Armillaria mycelium within the protocorm is connected to mycelium growing parasitically on adjacent trees (Smith & Read, 1997).

Crinipellis (75 spp.)

Crinipellis perniciosa is the cause of a severe witches' broom disease of cocoa (Theobroma cacao) especially in South America (Purdy & Schmidt, 1996; Griffith et al., 2003). Young shoots are induced to proliferate and become swollen (Fig. 19.18e). There is considerable reduction in the yield of cocoa pods, and crop losses of up to 80% have been reported. Crinipellis perniciosa is hemibiotrophic; its monokaryotic mycelium is biotrophic but its dikaryotic mycelium is necrotrophic. It forms groups of small (pileus diameter: 5–15 mm) crimson to pink basidiocarps on dead cocoa twigs (Fig. 19.18f). Infection is by air-borne basidiospores which germinate under conditions of high humidity (e.g. dew) on meristematic tissues such as flowering cushions, young leaves and shoots. Germ tubes penetrating stomata or wounded tissue establish a monokaryotic mycelium which pervades the proliferating hypertrophied shoots. Infection is not systemic. The monokaryotic phase does not grow in agar culture, but growth can be induced in cocoa or potato callus tissue cultures. There are several biotypes of C. perniciosa infecting different host plants. The C (cacao) biotype is primarily homothallic. Infected host twigs and pods eventually become necrotic and death of host tissues is associated with a change in the nuclear condition of the mycelium to the dikaryotic state. Basidiocarps develop on the dikaryon in the dead twigs. The dikaryotic mycelium can be grown in culture, and fruit body development is also possible on agar or on a bran–vermiculite medium covered with a peat-based casing soil.

Control of witches' broom disease is difficult but crop losses can be reduced by sanitation, i.e. the regular removal of brooms and diseased pods. Other measures involve the use of *Trichoderma* isolates as agents of biological control. Some are antagonistic, but *T. stromaticum* is mycoparasitic on mycelium and basidiocarps (Samuels *et al.*, 2000; Sanogo *et al.*, 2002). Systemic fungicides are also used. Further, there is an ongoing search for hosts resistant to the pathogen which could be used for breeding purposes. The origin of the cocoa pathogen is thought to be native (wild) species of *Theobroma* growing in the forests, but other *C. perniciosa* strains attack members of the Solanaceae, *Bixa* spp., lianas (climbers) and woody debris on the forest floor.

A closely related fungus, *C. roreri* (*Myceliophthora roreri*), causes the damaging frosty pod rot disease of cacao. Infected pods are covered by white powdery masses of wind-dispersed spores, originally regarded as conidia. These spores are borne on a dikaryotic mycelium which can be grown in culture. Dikaryotic hyphae swell and branch to form sporophore initials. The nuclei in the dikaryon fuse and septa are laid down, resulting in chains of diploid cells which have been interpreted as the equivalent of probasidia. The diploid cells develop thick walls to become spores and their nuclei undergo meiosis, but division may be arrested at a binucleate state (after the first meiotic division) or proceed to the formation of nuclear tetrads. On germination, there are indications that a four-celled metabasidium may be produced, with sterigmata which function as infective hyphae. As in *C. perniciosa*, this monokaryophase is bio-trophic and can only be grown on living cocoa tissue. The life cycle of *C. roreri* is thus consider-ably modified, and a recognizable mushroom-like basidiocarp stage probably does not occur (Evans *et al.*, 2002, 2003; Griffith *et al.*, 2003).

19.4.10 Mycenaceae

Mycena (about 150 spp.)

This is a large polyphyletic genus of fungi, with most species clustering in phylogenetic analyses around the type-species, *Mycena galericulata*. This group has been called Mycenaceae by Moncalvo *et al.* (2002). *Mycena* spp. produce rather small, delicate basidiocarps which have long slender stipes and conical or bell-shaped caps. Fruit bodies may emerge singly or in clusters from wood, leaf litter and other debris such as twigs, pine cones and bracken petioles in woodland and pastures. Some species exude latex when the stipe is broken, e.g. *M. sanguinolenta* with blood red latex and *M. galopus* which produces a milk white exudate. Much is known about the autecology of *M. galopus*, which has a perennial mycelium when growing in coniferous litter and also fruits on a wide range of twiggy debris but does not grow in bulky wood masses, possibly because of restricted aeration there (Frankland, 1984; Dix & Webster, 1995). It is capable of growing on most of the constituents of leaf litter and its ability to break down lignin and cellulose enables it to function as a typical white-rot decay fungus. *Mycena galopus* is a key decomposer of oak leaves and, within two years of leaf fall, its mycelium may be present on 80% of fallen leaves. This species is regarded as a secondary colonizer, growing on plant material which is already well-colonized by other fungi. In this context, it is worth mentioning that many *Mycena* spp. produce antifungal metabolites, including strobilurins, and these may aid in the displacement of other wood-rotting fungi. Despite this, moribund fruit bodies of various *Mycena* spp., including strobilurin producers, may be parasitized by the zygomycete *Spinellus fusiger* (Plate 3e).

Basidiocarp development in *M. stylobates* growing on beech leaves has been described and well-illustrated by Walther *et al.* (2001). An irregular arrangement of interwoven hyphae within the leaf bursts through to form an ovoid structure at the surface, composed mainly of vertically arranged hyphae. Cells at the margin increase in diameter and enclose the early stages of the primordium entirely. Separation of this large-celled wrapping tissue from internal hyphae results in the formation of a ring-like groove at the base of the primordium and a layer of protective hyphae covering a central bulb. The cells of the outer layer of vertically arranged hyphae increase in diameter to form the stipe. Simultaneously, hyphae at the apex of the bulb form horizontal outgrowths, giving rise to the pileus. The development of the hymenophore starts with the formation of small alveoli on the lower surface of the pileus, near its margin.

19.4.11   Tricholomataceae

This family has not yet been clearly separated from related groups in recent phylogenetic analyses. Genera currently placed here include *Tricholoma*, *Lepista*, *Clitocybe*, *Termitomyces* and *Lyophyllum*. A few *Collybia* spp. (*C. tuberosa*, *C. cirrhata*, *C. cookei*) are also grouped here, whereas most of the genus *Collybia sensu lato* seems to belong to the Marasmiaceae (see p. 546). It is probable that many other genera included in the Tricholomataceae are also polyphyletic (Moncalvo *et al.*, 2000, 2002).

Clitocybe

This is a large genus with about 60 species in Britain. The basidiocarps are funnel-shaped with decurrent gills (Fig. 19.20a) and can be large, up to 20 cm in diameter in *C. geotropa*. *Clitocybe nebularis* may form fairy rings many metres in diameter in deciduous and coniferous woods. The fruit bodies of several species are edible and those of *C. odora* have a fragrant, aniseed-like flavour. *Clitocybe* spp. are non-mycorrhizal.

Lepista (50 spp.)

The basidiocarps of *Lepista* are generally known as 'blewits' and include several good edible species such as *L. nuda* and *L. saeva*. The gills are attached to the stem in a sinuate manner, i.e. with an S-shaped point of attachment (see Fig. 19.8). *Lepista* is distinguished from *Tricholoma* in having a pale pink spore print (white in *Tricholoma*). *Lepista* spp. are non-mycorrhizal, growing mainly among humus and leaf litter.

Tricholoma (200 spp.)

This genus of mostly ectomycorrhizal species is mainly North-temperate in distribution in woodlands and pastures. The basidiocarps have sinuate gills and produce a white spore-print.

The most important edible species is the matsutake mushroom, *T. matsutake*, which is highly valued, especially in the Far East. Unfortunately, basidiocarps still have to be collected from woodlands because all attempts at cultivating it for commercial production have failed (see Wang & Hall, 2004). However, Guerin-Languette *et al.* (2005) have succeeded in infecting mature pine roots with *T. matsutake* in the laboratory, and this may have laid the foundations for an inoculation protocol for forest trees in future. An edible European species is St George's mushroom (*T. gambosum* or *Calocybe gambosa*) which grows in pastures, often under trees, and fruits in spring around St George's Day (23 April). *Tricholoma sulphureum* is a common woodland species whose basidiocarps are readily recognized by their sulphureous colour and unpleasant smell.

19.4.12   *Laccaria*

The relatively small basidiocarps of *Laccaria* (25 spp.) are known as 'deceivers' because of the change of colour intensity of the cap surface between wet (intensely coloured) and dry (pale) conditions. *Laccaria* spp. are important ectomycorrhizal associates of forest trees and are used for artificial inoculation of trees to be planted out into challenging situations. They are usually among the early-stage colonizers of tree roots, being replaced by others as the succession proceeds (Last *et al.*, 1983, 1987; Kropp & Mueller, 1999). Among the best known species are *L. laccata*, growing with *Pinus* and a range of other hosts, *L. bicolor* which grows with *Abies*, *Pseudotsuga*, *Pinus* and *Picea*, and *L. amethystina* (amethyst deceiver), an associate of oak (*Quercus*) and beech (*Fagus*). All species of *Laccaria* have edible basidiocarps. The relationships of *Laccaria* are poorly resolved, and family assignment is uncertain at present.

Bertaux *et al.* (2003) have reported the presence of a bacterium (*Paenibacillus* sp.) within the hyphae of *L. bicolor*. Bacterial cells were visualized by fluorescence microscopy, and the escape of these bacteria in liquid culture conditions was shown to account for the sporadic bacterial 'contamination' of fermenter

Fig 19.20 Basidiocarps in the euagarics clade (4). (a) *Clitocybe nebularis*. (b) *Stropharia semiglobata*. (c) *Psilocybe semilanceata*, the magic mushroom. (d) *Hypholoma sublateritium*. (e) *Cortinarius purpureus*. Note the fibrillose cortina on the stipe of the young specimens (centre and right).

cultures. Intrahyphal bacteria appear to be rare in higher fungi, but they have been reported in certain Zygomycota, notably *Geosiphon* (p. 221) and *Rhizopus* (p. 184).

19.4.13 Strophariaceae

Against all expectations, species included among the Strophariaceae do seem to group together, albeit with weak statistical support, in recent phylogenetic analyses (Moncalvo *et al.*, 2002). Two groups of *Psilocybe* fall outside the core of the Strophariaceae, with the hallucinogenic species clustering separately from other *Psilocybe* spp.

Stropharia

Species of *Stropharia* fruit on soil, dung or wood. The developing gills are protected by a membranous veil which may persist as a solid annulus or a loose weft of fibrils. The spores are purplish-brown to black. *Stropharia semiglobata*, extremely common on several kinds of herbivore dung, has a viscid, yellowish, hemispherical cap and a ring of dark fibrils on the stem (Fig. 19.20b). *Stropharia aeruginosa* (verdigris agaric) has an attractive bluish-green viscid cap flecked with white scales and grows in woodland while *S. aurantiaca*, with bright orange caps, fruits on sawdust and on wood chippings.

Psilocybe (c. 300 spp.)

The basidiocarps of *Psilocybe* (Fig. 19.20c) are generally small and campanulate (bell-shaped), with attached gills and purple-brown spores. Some species have a fibrillose veil whilst others have a distinct annulus. The genus includes several species whose basidiocarps are hallucinogenic and are used as recreational drugs. *Psilocybe mexicana* (sacred mushroom, teonanácatl) has been used for centuries by Mexican Indians in religious ceremonies (Heim & Wasson, 1958), and *P. cubensis* (golden top or giggle mushroom) is cultivated to obtain hallucinogenic basidiocarps. In Europe, *P. semilanceata* (magic mushroom, liberty cap) fruits in late summer and autumn on the ground in sheep pasture and well-manured grassland. It is saprotrophic, its mycelium being associated with decaying grass roots. Fruit bodies are picked and dried, although their sale is illegal in many European countries. The main hallucinogens of *Psilocybe* are the alkaloids psilocybin and psilocin (Figs. 19.15f,g), which are *N*-methylated tryptamines. They operate on serotonergic systems of the brain and are similar in their effects to mescaline and LSD (Lincoff & Mitchel, 1977; Bresinsky & Besl, 1990). Psilocin is less stable than psilocybin, and when oxidized shows a blue discoloration. This may be the reason why the flesh of many of the hallucinogenic *Psilocybe* spp. turns blue when bruised or broken.

A readable account of the discovery of LSD and the elucidation of hallucinogenic principles in magic mushrooms has been written by the discoverer of both, Albert Hofmann. English translations of the original German text (Hofmann, 1979) ae readily available.

Hypholoma

The best-known species are *H. fasciculare* (sulphur tuft) and *H. sublateritium* (Fig. 19.20d), both growing on wood. They are sometimes classified in the genus *Naematoloma*. *Hypholoma fasciculare* is very common, forming clusters of yellow basidiocarps on many kinds of deciduous and coniferous tree stumps, whilst *H. sublateritium* fruits on deciduous tree stumps and has brick-red caps. There is a cottony veil on the stem and the spores are purplish-brown. *Hypholoma fasciculare* basidiocarps are inedible and bitter to taste. They cause gastro-intestinal irritation and there are occasional reports of death following ingestion (Lincoff & Mitchel, 1977). *Hypholoma fasciculare* is highly competitive against other wood-rotting fungi. It is capable of extending from a woody food base through the soil by means of mycelial cords in search of further woody substrata. It can also utilize leaf litter. As a wood-decaying fungus its strategy is **secondary resource capture**, i.e. the displacement of other fungi which have earlier colonized wood, killing their mycelia. It therefore tends to fruit late in succession (Rayner & Boddy, 1988; Boddy, 1993).

Pholiota (c. 150 spp.)

Most species of *Pholiota* grow on wood. *Pholiota squarrosa* is associated with soft, pale brown-rot (butt rot) of living trees of ash (*Fraxinus*), beech (*Fagus*) and poplar (*Populus*). Dense clusters of scaly brown basidiocarps are formed at the base of the trunk (Plate 9e). There is a prominent ring on the stem and the spores are smooth and rusty brown. It is best to avoid eating the basidiocarps of *Pholiota* spp. because serious ill-effects may follow, especially if they are consumed with alcohol. However, the nameko fungus (*P. nameko*) is cultivated commercially on sawdust-based substrate in Far Eastern countries for its edible basidiocarps (Chang & Miles, 2004). It is a moisture-loving fungus, growing in nature on dead trunks and stumps of deciduous trees at high altitudes in Japan and Taiwan. Its mating system is bipolar. Arthroconida develop on monokaryotic and on dikaryotic mycelia. Similarly, basidiocarps develop on monokaryons and on dikaryons.

19.4.14 Cortinariaceae

Cortinarius (c. 2000 spp.)

Because of its large number of species, this genus provides one of the toughest challenges to fungal taxonomists as well as field mycologists. In Britain alone, 230 species have been listed. The genus has been divided into several subgenera (e.g. *Cortinarius, Dermocybe, Leprocybe, Phlegmacium, Myxacium, Telamonia*). Species in the subgenus *Myxacium* have slimy caps and stems derived from a glutinous universal veil, whilst in the subgenus *Phlegmacium* only the cap is sticky whereas the stem is dry. The genus is distributed in temperate regions of the Northern Hemisphere. All species are mycorrhizal with trees, so that they fruit in woodlands and woodland margins. The fruit bodies are small to large, buff, clay-coloured, orange-brown or sometimes very colourful, e.g. violet in *C. violaceus* or blood-red in *C. sanguineus*, *C. purpureus* and related species. Young fruit bodies are enveloped in a filamentous or glutinous veil and developing gills are protected by a fibrillose **cortina** which may be evanescent or may persist, attached to the stem (Fig. 19.20e). The spores are mostly warty, and the spore print is cinnamon to rust-brown. It is unwise to attempt to eat *Cortinarius* basidiocarps because the edibility of many species is unknown and some are deadly poisonous. The most notorious among them is *C. orellanus*, which contains the bipyridyl toxin orellanine (Fig. 19.15h), the cause of delayed renal failure. Other toxins are cortinarins, which are cyclic peptides. The long delay of 2–20 days between ingestion of the mushroom and the onset of symptoms (nausea, vomiting, diarrhoea, gastric upset and abdominal pain) are characteristic features of *Cortinarius* poisoning. Death may ensue 2–6 months later (Bresinsky & Besl, 1990; Michelot & Tebbett, 1990).

19.5 | Boletoid clade

This clade includes not only the Boletales, but some other groups with basidiocarps which are dissimilar in appearance. Traditionally, the Boletales, typified by the genus *Boletus*, have included forms with fleshy, mushroom-like basidiocarps with tubular hymenophores. Later, based on morphological and chemical criteria, the concept was expanded to include gill-bearing forms such as *Paxillus* and *Hygrophoropsis*. Relationships were also suggested between poroid boletes and resupinate forms such as *Coniophora* or *Serpula*, and gasteroid genera such as *Scleroderma* or *Rhizopogon*. Molecular phylogenetic techniques have confirmed these relationships and a boletoid clade has been recognized to embrace this wider concept (Bruns *et al.*, 1998; Hibbett & Thorn, 2001). We shall study representatives of a gill-bearing group (Paxillaceae), poroid groups (Boletaceae, Suillaceae) and resupinate forms (Coniophoraceae).

19.5.1 Paxillaceae

Paxillus (15 spp.)

Most species of *Paxillus* are ectomycorrhizal (Wallander & Söderström, 1999), but *P. atrotomentosus* fruits on conifer stumps. The basidiocarps are soft and fleshy. A characteristic feature is that the hymenial tissue separates readily from the flesh of the cap. The roll-rim *Paxillus involutus* (Fig. 19.21a) has a brown, funnel-shaped cap with an inrolled margin, decurrent gills and brown spores. Upon bruising, an intense brown colour develops due to the accumulation of the pigment involutin (Fig. 19.22a), a member of the shikimic acid-derived pulvinic acid family typical of Boletales. Considering that it is a mycorrhizal species, *P. involutus* has an unusually wide host range, with 23 different tree species listed by Wallander and Söderström (1999). It is most commonly associated with birch (*Betula*) and oak (*Quercus*) in acid woodlands.

Fig 19.21 Basidiocarps in the boletoid clade. (a) *Paxillus involutus*. Note the inrolled cap margin and decurrent gills. The fruit bodies shown here have been attacked by the parasitic mould *Sepedonium chrysospermum* (white patches). (b) *Boletus* (*Xerocomus*) *chrysenteron*. (c) *Boletus edulis*, the cep or penny bun. (d) *Suillus grevillei*, an ectomycorrhizal associate of *Larix*. Note the ring on the stem. (e,f) *Serpula lacrymans*, the dry rot fungus. (e) Beam supporting the roof of a church showing the typical cracking transverse to the grain of the wood which also shows shrinkage. (f) Resupinate fruit body on a ceiling showing the shallow pores.

Spread through the soil to fresh young roots is by an effuse mycelium or by rhizomorphs. The fungus may survive in the soil by means of sclerotia. Some isolates are capable of saprotrophic growth and can fruit in the absence of a mycorrhizal host. Affinity with the Boletales has long been suspected because basidiocarps of *P. involutus* are commonly attacked by the bright yellow conidial state (*Sepedonium chrysospermum*) of the mycoparasitic ascomycete *Apiocrea chrysosperma*, which also attacks different boleti and gasteromycetes believed to be related to them (Fig. 19.21a; Plate 9h; p. 581).

The basidiocarps of *P. involutus*, although traditionally considered edible, are, in fact, poisonous. There are two symptoms. A heat-labile substance

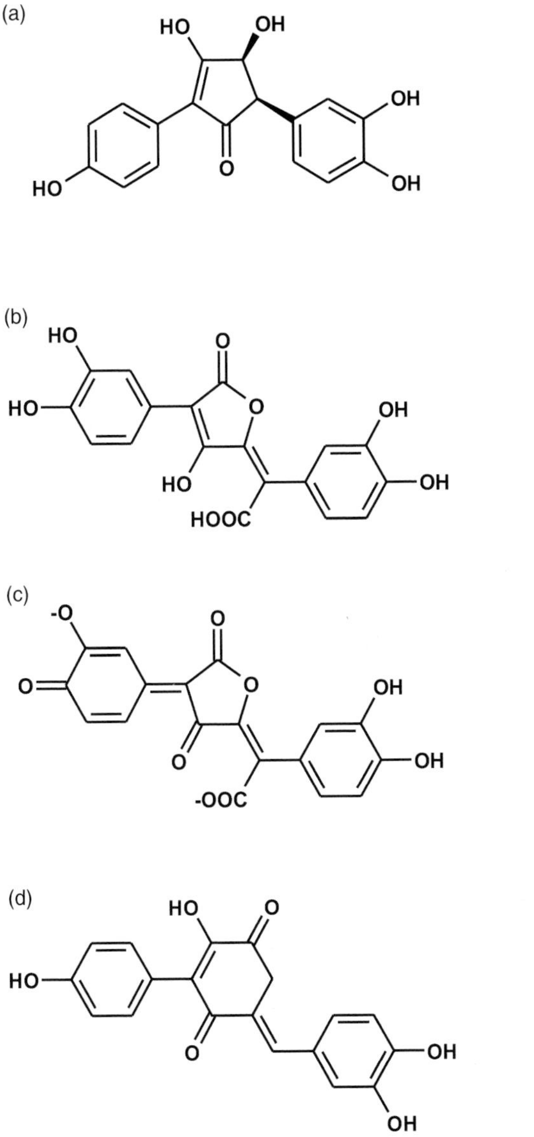

(a)

(b)

(c)

(d)

Fig 19.22 Pulvinic acid-type pigments typical of members of the boletoid clade. (a) Involutin, a brown pigment produced by *Paxillus involutus*. (b) Variegatic acid, a yellow pigment. (c) The dark blue oxidation product of variegatic acid upon bruising the fruit bodies of *Boletus erythropus* and certain other *Boletus* spp. (d) Grevillin B produced by a range of *Suillus* spp.

seems to be responsible for gastric upsets within a few hours of consuming raw or undercooked specimens, whereas haemolysis due to an allergic reaction can cause delayed liver failure and kidney damage only after repeated

consumption. The toxic principle(s) appear to be unknown as yet.

19.5.2 Boletaceae
Boletus (300 spp.)

Formerly comprising a broad range of pore-bearing fungi, the genus *Boletus* has been divided into a number of groups, often now recognized as separate genera. Species of *Boletus* are ecto-mycorrhizal. They have medium-sized, large or very large fleshy basidiocarps with a tubular hymenophore. The pores marking the openings of the hymenial tubes may be of the same yellowish-green colour as the tubes or may be coloured orange to blood-red as in *B. erythropus* (Plate 9f) and *B. satanas*. The pigments in *Boletus* are of the pulvinic acid group (Gill & Steglich, 1987), with variegatic acid (Fig. 19.22b) being the most common. Within seconds of bruising or cutting the basidiocarps of certain species such as *B. erythropus*, their flesh and pores become discoloured blue or bluish-black due to the enzyme-mediated oxidation of variegatic acid (Fig. 19.22c) and xerocomic acid.

One of the most common species, *B. chrysenteron* (Fig. 19.24b), sometimes classified in the separate genus *Xerocomus*, is easily recognized by the exposure of yellow or red flesh when the cap surface skin cracks. It is associated with broad-leaved trees. The stem of some species may be punctate, i.e. dotted with tiny warts (e.g. *B. erythropus*; Plate 9f), or veined (e.g. *B. edulis*). There are several species with delicious edible basidiocarps; amongst the best-known are *B. edulis* (cep or penny bun; Fig. 19.21c), *B. badius* (bay bolete; Plate 9h) and *B. appendiculatus*. Although it is possible to grow mycelium of *B. edulis* in pure culture, it is as yet impossible to induce it to form basidiocarps, and therefore this much-prized edible species continues to be collected in forests (Wang & Hall, 2004). Some *Boletus* spp. are safe to eat only after cooking, e.g. *B. luridus* and *B. erythropus*, whereas others have poisonous basidiocarps, notably *B. satanas* (devil's bolete) and *B. satanoides*.

Boletus parasiticus forms fruit bodies attached to the basidiocarps of the earth ball *Scleroderma*. Doubts have been expressed as to whether this fungus is actually parasitic, and quite possibly it is merely stimulated to fruit by the presence of the earth ball.

Leccinum (75 spp.)

Like *Boletus* spp., the genus *Leccinum* has large fleshy basidiocarps. A characteristic feature is that the stem is covered with scales composed of cystidia. *Leccinum scabrum* (brown birch bolete) and *L. versipelle* (orange birch bolete) are common ectomycorrhizal associates of birch (*Betula*). Basidiocarps of *Leccinum* are generally edible.

19.5.3 Suillaceae

Suillus (90–100 spp.)

Suillus species have medium-sized fleshy basidiocarps forming ectomycorrhizal associations with conifers. Their fruit bodies are usually of yellowish colours due to the abundance of pigments derived from the shikimic acid pathway, especially grevillins (Fig. 19.22d; Besl & Bresinsky, 1997). The cap may be dry and scaly but is more usually viscid or slimy. There are species with a ring on the stem (Fig. 19.21d), e.g. *S. grevillei* (larch bolete) and *S. luteus* (slippery jack), whilst in many others (e.g. *S. bovinus*, *S. granulatus*) there is no ring (Plate 9g). The spores are smooth and elongate, and pale brown to brown. The presence of bundles of cystidia in the hymenium is a feature which distinguishes this genus from other boletes, and the classification into a separate family is supported by phylogenetic studies (e.g. Grubisha *et al.*, 2001). Also included in the Suillaceae are the gasteromycete genus *Rhizopogon* (see p. 581) and the gill-bearing *Gomphidius*.

The distribution of species mirrors that of their mycorrhizal hosts, coniferous trees which are largely confined to North-temperate regions, only extending to other areas where introduced. There is a fairly high degree of host specificity. For example, in nature *S. grevillei* (Fig. 19.21d) is almost exclusively associated with *Larix* spp. (larch) whilst *S. bovinus* and *S. luteus* are associated with *Pinus* spp. This ecological specificity is probably associated with the effects of competition because in the laboratory, under aseptic conditions, a wider range of hosts has been

infected experimentally. Many species of *Suillus* have edible basidiocarps and some areas of planted pine (*P. radiata*) forests in South America are devoted to production of *S. luteus*, with astonishing annual productivity values of up to $1\,t$ dry weight ha^{-1} reported (Dahlberg & Finlay, 1999). *Suillus bovinus* and *S. variegatus* are common in Swedish pine plantations. Whereas *S. bovinus* is an early-stage mycorrhizal fungus, *S. variegatus* is often found in older stands. In open communities dominated by seedling pines, *S. bovinus* first develops a large number of **genets** (genetically defined mycelial individuals) derived from basidiospores. Once established, a colony extends by mycelial spread and by rhizomorphs to nearby host roots. The extent of a genet can be estimated by evidence of somatic incompatibility between mycelial isolates made from basidiocarps. As the original genets expand, they compete with each other. The number of genets decreases, but as the genets increase in size they may fragment so that parts of the same genet may become separated. The maximum rate of mycelial extension has been estimated to be about $20\,cm$ per annum and the age of the largest genet to be about 75 years (Dahlberg & Stenlid, 1990, 1994; Dahlberg, 1997).

19.5.4 Coniophoraceae

This group includes wood-rotting fungi such as *Serpula* and *Coniophora* which form spreading, crust-like resupinate fructifications. Morphological characteristics suggested that the Coniophoraceae have an affinity with Boletales (Pegler, 1991), and Hibbett and Thorn (2001) have included them in their bolete clade. This is supported by the presence of pulvinic acids in both *Serpula* and *Coniophora* (Gill & Steglich, 1987). Here we shall consider only *Serpula*.

Serpula

Serpula lacrymans (Figs. 19.21e,f) causes dry rot and is one of the most serious agents of timber decay in buildings. There is a very extensive literature (see Rayner & Boddy, 1988; Jennings & Bravery, 1991). Both hardwoods and softwoods are attacked but, because softwoods are more commonly used in building construction, it is on such timbers that the fungus is most frequently reported. There is evidence that *S. lacrymans* has become adapted to man-made habitats. In the wild, *S. lacrymans* has been collected on spruce logs in the Himalayas at an altitude of 8000–10 000 ft (Bagchee, 1954; Singh *et al.*, 1993), and it has also been reported from Northern Europe, Northern California and Siberia. It can continue to grow vegetatively at −2°C, and the optimum (23°C) and maximum (26°C) temperatures for growth are rather low.

Strains of *S. lacrymans* associated with houses are considered to belong to *S. lacrymans* var. *domesticus* and may have been spread by human activities in recent times. The Himalayas are often mentioned as a likely centre of origin, although Kauserud *et al.* (2004) put forward an alternative hypothesis featuring North America. Either way, possible vehicles for the dispersal of *S. lacrymans* var. *domesticus* may have been wooden sailing ships. Ramsbottom (1938), in his fascinating review of a wealth of original literature sources, concluded that wooden vessels were frequently infested by *S. lacrymans*, often being unsound even before being launched due to the careless use of non-seasoned timber combined with poor ventilation of the ship holds. Typical symptoms were described by a Commission of Inquiry, reporting to King James I in 1609 on the state of battleships in the Royal Navy (taken from Ramsbottom, 1938):

> In buylding and repaireing Shippes with greene Tymber, Planck and Trennels it is apparent both by demonstration to the Shippes danger and by heate of the Houlde meeting with the greenesse and sappines thereof doth immediately putrefie the same and drawes that Shippe to the Dock agayne for reparation within the space of six or seven yeares that would last twentie if it were seasoned as it ought and in all other partes of the world is accustomed. Adde hereunto experience at this day that many Shippes thus brought in to be repaired, subject to miscareinge upon employment, and besides they breed infection among the men that serve in them.

Only wood with a moisture content above about 20–25% of the oven-dry weight is susceptible to attack by the fungus. Well-dried and seasoned timber has a moisture content of 15–18%, and in a properly ventilated house this soon falls to 12–14% or lower. If woodwork becomes wet through contact with the soil, damp masonry, faulty construction or inadequate ventilation, then infection from air-borne basidiospores is likely to follow. Basidiospores germinate in the presence of free water on moist wood surfaces. The mycelium within the wood develops chiefly at the expense of the cellulose; lignin is not attacked, and the type of decay is a brown-rot. Well-rotted timber is shrunken with transverse cracks and has a dry crumbly texture. Water produced by the breakdown of cellulose (sometimes termed the water of metabolism) may be sufficient for further growth even if the air humidity is lowered below the point at which new basidiospore infections could arise. Up to 55.6% of the cellulose consumed may be available as metabolic water (see Bravery, 1991). As in many brown-rot fungi (see p. 527), oxalic acid is released into the environment, lowering the pH of the wood and mortar over which *S. lacrymans* is growing.

The epithet *lacrymans* (weeping) refers to the beads of moisture sometimes found on decaying timber, at the tips of hyphae and on mycelial cords. Sheets of mycelium may extend over the timber and adjacent brickwork, and the fungus is also capable of spreading several metres by means of mycelial cords up to 5 mm in diameter. The internal hyphae of the mycelial cords are exceptionally wide (up to 60 μm) and are modified for rapid conduction, enabling water and nutrients to be transported (Nuss *et al.*, 1991). Transport is by pressure-driven hydraulic flow (Jennings, 1987, 1991). Carbohydrate is transported mainly in the form of trehalose. The strands can penetrate mortar and stonework between walls and can spread throughout a building provided that there is enough wood as a food base. Strands remaining after removal of affected timber may still be able to initiate fresh infections.

Reproduction in *Serpula lacrymans*

This fungus is heterothallic and tetrapolar. Globally there are probably no more than four *A* and five *B* alleles (Schmidt & Moreth-Kebernik, 1991). Arthroconidia are formed on monokaryons but not on dikaryons. Basidiocarps develop from dikaryons as flat, fleshy resupinate structures (Nuss *et al.*, 1991). They may grow undetected under floorboards and in roof spaces for long periods and may reach a diameter of 1–2 m. The lower side is brown and corrugated into shallow pores supporting the hymenium (Fig. 19.21f). The folding of the hymenophore is the result of continuous thickening of the hymenium. The construction is at first monomitic, but becomes dimitic with the development of skeletal hyphae. Sporulation is continuous and it has been estimated that a basidiocarp measuring 100 cm^2 can produce 300 million spores h^{-1}. The immense numbers of rusty-brown basidiospores may form deposits visible to the naked eye on cobwebs, shelves, etc. In the basements of buildings containing fruit bodies, spore concentrations of around 80 000 m^{-3} air have been detected (see Hegarty, 1991), raising concern over respiratory allergy which may already have been referred to in the report to King James I (see above).

Control of dry rot
The economic consequences of failing to eradicate and control dry rot in a building can be severe. Modern methods aim to render the indoor environment hostile to *S. lacrymans* by eliminating all routes by which water could gain access to timber, accompanied by constantly high ventilation (Bravery, 1991; Palfreyman & White, 2003). A more traditional method is the removal of all infected timber and surrounding sound wood, and the treatment of any timber left in place with a recommended and approved fungicide. Replacement timbers should be selected for their durability properties and can be treated with pressure-impregnated fungicides such as copper/chromium/arsenic preservatives. Plaster can be treated with zinc oxychloride. Given the low temperature maximum of *S. lacrymans*, it is also sometimes possible to encase an entire building with a tent and subject it to thermal treatment (50°C). Since *S. lacrymans* is susceptible to attack by *Trichoderma* spp., biological control has been proposed but not yet put into practice (Palfreyman *et al.*, 1995). The most important control measure, however, is

proper construction to ensure that the moisture level of the timber remains below the point at which infection can be initiated.

19.6 | Polyporoid clade

Included in this clade are certain members of the Polyporales (= Aphyllophorales), a group comprising hymenomycetes in which (with a few exceptions) the hymenium is not borne on the surface of gills. It included bracket fungi (polypores), tooth fungi, coralloid fungi and forms with flattened or crust-like basidiocarps. However, morphological, anatomical and chemical studies have indicated that this was not a natural grouping, and molecular phylogenetic investigations have amply confirmed this view, with Hibbett and Thorn (2001) showing that the aphyllophoroid condition occurs in all eight clades of Homobasidiomycetes. Most aphyllophoralean fungi outside the polyporoid clade are now considered to belong to the russuloid clade (see Section 19.7). The taxonomic hierarchy within the polyporoid clade is too tentative to be adopted at present, and therefore we have desisted from using family names here.

Economically important wood-rotting bracket fungi are found in the polyporoid clade. As described in detail on pp. 519–522, two main types of wood decay caused by basidiomycetes have been recognized, namely brown-rot in which cellulose is destroyed whereas lignin is left essentially unchanged, and white-rot in which both lignin and cellulose are attacked. Both types of rot can be caused by members of the polyporoid clade, e.g. brown-rot by *Piptoporus betulinus* (Fig. 19.23d) and white-rot by *Trametes versicolor* (Plate 10a). However, both rots can also be caused by other groups of basidiomycetes.

The mycelium of members of the polyporoid clade is often perennial in large tree trunks and may give rise to a fresh crop of basidiocarps annually. In some species, e.g. *Fomes fomentarius* or *Ganoderma applanatum* (Figs. 19.23b,c), the fruit body itself may be perennial and new layers of hymenial tubes develop annually on the lower side of the basidiocarp. Typically

Fig 19.23 Basidiocarps in the polyporoid clade. (a) *Ganoderma lucidum*. Basidiocarps formed in culture from sawdust contained in plastic bags. (b,c) *Ganoderma applanatum*. (b) Two sporophores attached to a living beech tree. (c) Detached sporophore split vertically to show two layers of hymenial tubes. (d) *Piptoporus betulinus* on a dead birch trunk. (e) *Polyporus squamosus*, basidiocarp attached to a living sycamore tree. (f) *Polyporus brumalis* growing from a dead beech twig. (g) *Sparassis crispa* fruiting at the base of a living pine tree. a kindly provided by Y.-J. Yao.

polypore fruit bodies develop as fan-shaped brackets lacking stipes, but there are forms with lateral (eccentric) stipes, such as *Polyporus squamosus* (dryad's saddle; Fig. 19.23e), or even with centrally stalked fruit bodies, e.g. the winter polypore *Polyporus brumalis* (Fig. 19.23f). When the fruit bodies of polypores and other wood-rotting fungi develop on the underside of logs, they may be appressed to the surface of the wood and are then described as resupinate.

The hyphal construction (hyphal analysis) of polypore basidiocarps varies (see p. 517–519) and is useful in identification. In some, e.g. *Bjerkandera adusta*, construction is monomitic; the basidiocarps are composed entirely of generative hyphae. The basidiocarps of *Laetiporus sulphureus* (chicken of the woods; Plate 10b) are dimitic, whereas *Trametes versicolor* has trimitic basidiocarps. The distinction between the different kinds of construction is best appreciated by attempting to tear the fruit bodies of these fungi apart. *Trametes versicolor* basidiocarps tear with difficulty, in contrast to the cheese-like consistency of *L. sulphureus*. Various modifications to the different hyphal systems may occur with age. For example, in *L. sulphureus* the generative hyphae may become inflated. In *Polyporus squamosus* the binding hyphae arise relatively late following inflation of the generative hyphae, converting the sappy flesh of the fully grown fruit body to a drier and firmer texture. In *Piptoporus betulinus*, too, binding hyphae arise very late but ultimately replace the generative hyphae. The dissepiments (tissues between the pores) show a different construction, being dimitic with skeletal hyphae.

The polyporoid clade is a large group, probably containing about 70 genera and over 600 species. There is a very extensive literature. Notes on interesting features of some common polypores are given below.

Trametes

Trametes (Coriolus) versicolor (Plate 10a), colloquially called 'turkey tail', is a common saprotroph on various hardwood stumps and logs, causing white-rot. Both the mycelium and the fruit bodies are tolerant of desiccation. The annual fruit bodies have a zoned, multicoloured, velvety upper surface which readily absorbs rain. Details of the anatomy of the fruit body are shown in Fig. 19.24.

Basidiocarps of *T. versicolor* may come in a range of colours and shapes, and if such variations in fruit body appearance occur on a single log, they can be traced to distinct columns of decayed wood when the log is serially sectioned. Dark brown 'zone lines' separate the columns. All isolations from within a column yield an identical dikaryon, and the zone lines mark intraspecific antagonism between distinct dikaryons (see Fig. 18.20a). Therefore, within a log of wood the fungus does not behave as a single 'unit mycelium' but as a series of discrete individuals (Rayner & Todd, 1977, 1979). When monokaryons are inoculated experimentally into logs in the field, they quickly become converted into dikaryons by anastomosis with compatible monokaryotic colonies derived from air-borne basidiospores (Williams *et al.*, 1981). The individual dikaryotic colonies, once established, may persist and retain their integrity over several years, continuing to produce fresh crops of basidiocarps of the same genetic constitution each year. Antagonism, i.e. vegetative incompatibility between different dikaryons, can also be readily demonstrated in pure culture as shown in Fig. 18.20c. Monokaryotic mycelia form arthroconidia. This species is tetrapolar with multiple alleles.

Because of its prolific production of peroxidase-type enzymes, *T. versicolor* is used industrially in such processes as the bioremediation of textile dyes or the wood preservative pentachlorophenol (PCP), and in delignification and decolorization of Kraft woodpulp. Some *T. versicolor* strains used in industry are thermotolerant. In addition to enzymes, *T. versicolor* also produces polysaccharopeptides, and these are of commercial interest in anti-cancer therapy (reviewed by Cui & Chisti, 2003). A range of these substances are produced by different strains of the fungus. Their exact chemical composition is variable, with a branched sugar backbone consisting of β-(1,3) and α-(1,4) linkages and a protein content of about 30%. Polysaccharopeptides are extracted from mycelium grown in fermenters, purified, and administered orally. Although it is unclear how these large molecules can be taken up intact by the gut and how exactly they act to achieve the claimed results, several effects on the human body are suspected, with a general enhancement of the immune system being the most common. They are therefore considered useful as a complementation of other, more aggressive anti-cancer treatments.

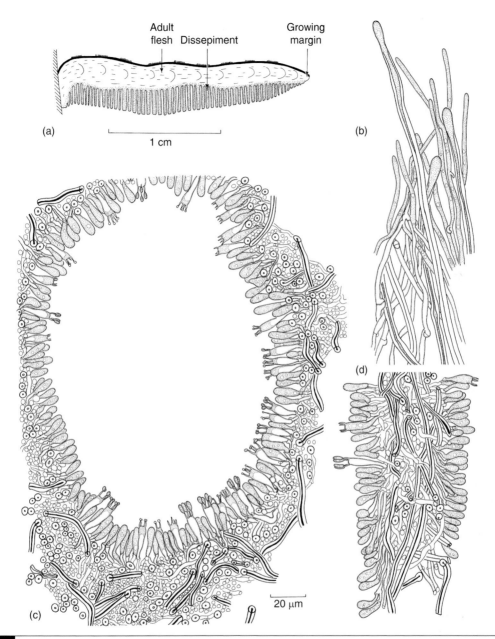

Adult
flesh Dissepiment

Growing
margin

(a)

1 cm

(b)

(d)

(c)

20 μm

Fig 19.24 *Trametes versicolor.* (a) Vertical section through a basidiocarp. (b) Group of hyphae teased from the growing margin. Only generative and skeletal hyphae are present here. In the adult flesh, binding hyphae are also present (see Fig. 19.3). (c) Transverse section across a pore showing the hymenium. (d) Longitudinal section of part of a dissepiment. The tissue contains only generative and skeletal hyphae. (b–d) to same scale.

Piptoporus

Piptoporus betulinus basidiocarps are a common sight on dead and dying birch trees (Fig. 19.23d). The fungus is probably a wound parasite, its basidiospores entering and germinating where branches have broken off. Infected trees show a brown-rot of the heart wood which first undergoes cubical cracking and later disintegrates as a powder. Although the brown-rot indicates cellulose decay, there is also evidence of peroxidase activity which suggests limited lignin breakdown. The fungus is bipolar with about 30 alleles at the mating type locus. Multiple infections of single birch trunks are comparatively rare, but where they do occur, transverse sections of trunks with multiple infections show a characteristic pattern of black 'zone lines' marking the interface between antagonistic dikaryons (Adams *et al.*, 1981). The dikaryons retain their integrity and persist over several years. A fresh crop of basidiocarps is produced annually on standing trees or on fallen trunks. The birch polypore was formerly known as the razor strop fungus in the days when the basidiocarps were used to hone 'cut throat' razors.

Laetiporus

The poroid *Laetiporus* and *Phaeolus*, together with *Sparassis* which produces lobed, cauliflower-like fruit bodies (Fig. 19.23g), form one of the few well-resolved branches within the polyporoid clade (Wang *et al.*, 2004). Biological features uniting these genera are that they cause wood decay of the brown-rot type often in living trees, and that their mating system is bipolar. The former species *L. sulphureus sensu lato* has now been subdivided into several new taxa which show a certain degree of host specificity (Burdsall & Banik, 2001). Infection is typically through wounds of living trees, causing an intense brown-rot in the heartwood of standing trees. The mycelium is clearly long-lived, with fresh crops of basidiocarps (Plate 10b) produced annually for several years in the living tree, and later from the dead trunk. *Laetiporus* is considered to be one of the main causes for the hollowing of old oak trees in parks. Fruit trees (especially apple) are also affected. In the Alps, broad-leaved trees are attacked at altitudes below 3000 ft whereas coniferous species are infected higher up in the mountains. In addition to basidiospores, *L. sulphureus* also produces a chlamydosporic conidial state called *Sporotrichum* (see Fig. 18.16).

Laetiporus sulphureus sensu lato has tradition-ally been considered an edible species, as its common name 'chicken of the woods' indicates.

However, consumption has been associated with gastrointestinal upsets (Jordan, 1995).

Polyporus

Polyporus squamosus (Fig. 19.23e) is a wound parasite of deciduous trees such as elm (*Ulmus*), beech (*Fagus*) and sycamore (*Acer pseudoplatanus*), producing an intensive white-rot. The mycelium persists on dead trunks, stumps and logs, and forms successive annual crops of basidiocarps during the early summer. The large, fan-shaped fruit bodies are creamy yellow, with brown scales. They are edible. Their texture is distinctly fleshy due to a dimitic structure with binding hyphae.

Ganoderma

There are more than 250 species of *Ganoderma*. They are white-rot fungi causing root and stem rots of hardwood and softwood hosts. A distinguishing feature is that the spore appears double-walled, with a dark-coloured inner layer bearing an ornamentation which pierces the hyaline outer one, so that the spore appears to have a spiny surface (Fig. 19.25). Mims and Seabury (1989) have interpreted the basidiospore wall of *Ganoderma lucidum* as comprising three

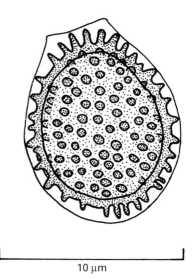

10 μm

Fig 19.25 *Ganoderma applanatum*. Basidiospore. The truncate portion is the apex of the spore. Spiny extensions of the darker inner spore wall penetrate the hyaline outer wall.

distinct components, an outermost primary wall, inter-wall pillars surrounded by electron-transparent regions, and an innermost secondary wall. Wall pillars, which initially develop immediately adjacent to the spore plasma membrane, eventually appear to fuse with both the primary and secondary walls.

Ganoderma applanatum and *G. adspersum* are two species with similar basidiocarps which are confused with each other but can be distinguished by their basidiospores which are larger in *G. adspersum*. Both are wound parasites of deciduous trees, producing large (up to 1 m), perennial, brown, woody brackets (conks) which form a fresh layer of hymenial tubes annually (Figs. 19.23b,c). Recently formed hymenial tubes of *G. applanatum* are chocolate-brown in colour in contrast to the paler appearance of older layers formed earlier. The older layers are less dense and have a lower nitrogen content than the current year's growth, suggesting that nitrogen is translocated from older, spore-depleted strata to the newly formed layers (Setliff, 1988). Nitrogen is often in limited supply in woody substrates, and its conservation is therefore important for wood-decaying fungi.

Ganoderma applanatum is less common than *G. adspersum* and is usually found on old trunks of beech (*Fagus sylvatica*), where it causes a white heart rot. The hyphal structure of the fruit body is trimitic. A characteristic feature is that the skeletal hyphae are of two types, namely arboriform, i.e. showing an unbranched basal part with a branched tapering end, and aciculiform, i.e. unbranched and usually with a sharp tip (Hansen, 1958; Furtado, 1965). The hymenial tubes of *G. applanatum* may be up to 2 cm in length and about 0.1 mm in diameter, i.e. 200 times as long as broad. The fall of the spores down this tube raises problems. The hard, rigid construction of the sporophore minimizes lateral disturbance to the vertical alignment of the tubes. Gregory (1957) has shown that the majority of the spores carry a positive electrostatic charge, but whether this charge has any relevance to the positioning of the spores during their fall by causing them to be repelled from the tube wall seems doubtful. It has been calculated that a large specimen may release as many as 20 million spores min^{-1} during the 5 or 6 months from May to September (Buller, 1922). Spore discharge can continue even during periods of drought, doubtless associated with uptake of water from the tree host (Ingold, 1954b, 1957). In addition to wind dispersal, basidiospores of *G. applanatum* have been shown to be dispersed by specialized mycophagous flies visiting the basidiocarps (Tuno, 1999). Basidiocarps of *G. applanatum* are parasitized by larvae of the mycophagous fly *Agathomyia wankowiczii*. The trama is stimulated to proliferate into conical or cylindrical gall-like outgrowths on the underside of the basidiocarp. When the larva has completed its development, it bores an exit hole through the tip of the gall and drops to the forest floor for pupation (Eisfelder & Herschel, 1966). Gall-forming insects rarely attack fungi, and even other *Ganoderma* spp. do not seem to be attacked by *A. wankowiczii*.

Spores placed on media suitable for germination may take 6–12 months to develop germ tubes. Pairings between monosporous mycelia show that the fungus is tetrapolar, with multiple alleles. The large output of spores may be related to the low probability of compatible spores infecting the same tree trunk.

As discussed above for *Trametes*, *Ganoderma* spp. have also been credited with having medicinal value against a wide range of ailments. *Ganoderma lucidum* (Fig. 19.23a), distinguished by its stalked, shiny (lacquered) basidiocarps, grows on the roots of deciduous trees. It is cultivated in China on sawdust in plastic bags for its basidiocarps from which polysaccharides and other pharmacologically active substances are extracted (Chang & Miles, 2004).

Phanerochaete

There are about 100 species to this genus of saprotrophic fungi forming resupinate, crust-like basidiocarps with smooth, wrinkled or spiny but non-poroid hymenial surfaces. By far the best-known species is *P. chrysosporium*, which has become a 'model' organism for the examination of lignin-degrading enzymes (see Fig. 19.13) and their potential biotechnological applications. There is a conidial state called *Sporotrichum pulverulentum* (Burdsall, 1981). Little is known about the ecological role of *P. chrysosporium* because this species is rarely found in nature.

19.7 | Russuloid clade

The russuloid clade is probably the most confusing group in the eight-clade system of Hibbett and Thorn (2001), containing the complete range of hymenophore types shown in Table 19.1 and Fig. 19.1. The phylogeny within this clade has been examined in detail by Larsson and Larsson (2003), who found that the main character uniting all members is the presence of **gloeocystidia**, i.e. cystidia filled with light-refractile (lipid-rich) contents (Fig. 19.29b). In *Stereum*, these are modified as laticiferous hyphae (Fig. 19.27). Other characters commonly used to identify russuloid fungi are diagnostic only at lower taxonomic ranks, e.g. the ornamented walls of basidiospores and their starch-positive (amyloid) staining with Melzer's iodine. We shall study representatives of agaricoid basidiocarps (Russulaceae), polyporoid fruit bodies (*Heterobasidion*), the spine-bearing *Auriscalpium* and corticioid forms (Stereaceae). The family arrangement is still tentative, and we give family names only where these have a fair chance of survival in future classification schemes.

19.7.1 Russulaceae
There are two important genera, *Lactarius* and *Russula*, and they share many features such as being ectomycorrhizal species, and having a characteristically brittle texture to the gills and general fruit bodies due to the presence of sphaerocysts embedded in the tissue (see Fig. 19.4). Basidiospores are typically amyloid and ornamented with warts or ridges.

Russula (*c.* 750 spp.)
The genus *Russula* presents a similar taxonomic nightmare to *Cortinarius*, and it is unlikely that this genus is monophyletic. Species are widely distributed and ectomycorrhizal. The basidiocarps (Fig. 19.26a) are moderate to large, often with a brightly coloured upper cap surface (white, yellow, green, red, purple or black). The gills are straight, arranged in a crowded but regular pattern, and vary from white to straw-coloured. The spores are ornamented with a network of branched ridges or plates. Some common species are *R. ochroleuca* (edible), *R. fellea* (bitter to the taste and inedible) and the sickener, *R. emetica* (poisonous). *Russula ochroleuca* fruits under coniferous and deciduous trees, *R. fellea* under beech (*Fagus sylvatica*) and *R. emetica* under pines.

Lactarius (*c.* 400 spp.)
The common name for *Lactarius* is milk cap, referring to the milky juice which exudes when the flesh of the cap is broken. The juice varies in colour from white to yellow, orange or violet and may change after exposure to the air. For example, the juice of *L. deliciosus* (saffron milk cap; Plate 10c) is carrot-coloured at first, but turns bright green upon prolonged exposure to air. The juice also varies in taste, being mild to faintly bitter in *L. quietus*, hot and acrid in *L. pyrogalus*, or first mild and then acrid in *L. rufus*. The juice is contained within broad laticiferous hyphae (see Fig. 19.4). As in *Russula*, the flesh also contains clusters of sphaerocysts. The gills are generally decurrent and the spores are ornamented like those of *Russula*. Some species of *Lactarius* have a narrow range of mycorrhizal partners, e.g. *L. torminosus* (woolly milk cap) and *L. turpis* (ugly milk cap) are associated with birch (*Betula*), and *L. deliciosus* and *L. deterrimus* with *Pinus* and *Picea*. There are several species with edible basidiocarps, but some are poisonous or their edibility is unknown. Fruit bodies of *L. deliciosus* are especially valued, and it is possible to produce them in plantations of artificially inoculated host trees.

19.7.2 Bondarzewiaceae
Heterobasidion, a seemingly typical polyporoid wood-degrading bracket fungus, has been placed in the russuloid clade in several phylogenetic analyses. Some authors have assigned it to

Fig 19.26 Basidiocarps in the russuloid clade. (a) *Russula atropurpurea*. (b) *Heterobasidion annosum*, the cause of butt rot of conifers. (c) *Stereum hirsutum* basidiocarps growing on an oak branch viewed from above. Note the white-rot at the broken end of the wood. (d) As (c) but viewed from below to show the smooth hymenium. (e) *Auriscalpium vulgare* basidiocarps growing from a buried pine cone. The hymenium is borne on spines on the lower side of the cap.

the family Bondarzewiaceae (Larsson & Larsson, 2003).

Biology of *Heterobasidion*

Heterobasidion annosum (Fig. 19.26b) is the cause of heart rot or butt rot of managed conifer plantations, and occasionally of deciduous trees such as birch in North-temperate regions (Korhonen & Stenlid, 1998; Asiegbu *et al.*, 2005). It is the most important pathogen of conifers in these areas. The disease is especially common on alkaline soils. The fungus is a necrotrophic parasite whose mycelium extends downwards into the root system and upwards into the stem, causing a stringy heart rot which damages the strength and affects the saleability of the wood. Root decay is often followed by wind throw. The fungus may survive in the root system, especially if it is resinous, for over 60 years. In Europe, three somatic incompatibility groups have been distinguished, an 'S' group (S for spruce) confined to Norway spruce (*Picea abies*), a 'P' group which

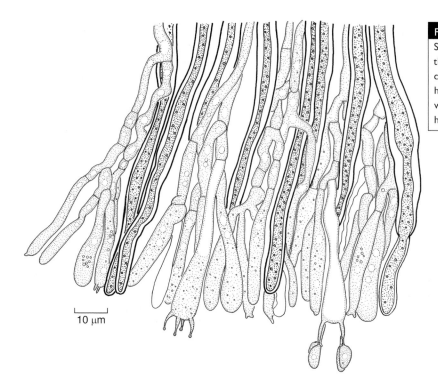

10 μm

Fig 19.27 *Stereum rugosum.* Section of hymenium. The thick-walled hyphae with dense contents are sanguinolentous hyphae which exude a red fluid when damaged, causing the hymenium to 'bleed'.

grows on pine, spruce, birch and other hosts, and an 'F' group growing on fir trees (*Abies* spp.) in Southern Europe. These groups, originally considered as infra-specific populations of *H. annosum*, have since been recognized as separate species (Niemelä & Korhonen, 1998). The name *H. annosum* is retained, but in a restricted sense for the 'P' group. The 'S' type is now named *H. parviporum* and the 'F' type *H. abietinum*. The first two groups also occur in North America.

The fruit bodies are perennial, at first thin and leathery, later hard and woody (the epithet *annosum* means aged). They are formed at the base of dead trees, stumps or beneath fallen logs and are readily identified whilst still actively growing by their orange-brown colour with a white margin. They are dimitic with skeletals. Several successive layers of hymenial tubes are formed on old basidiocarps, and basidiospores are produced throughout the year. The mating system is unifactorial (i.e. bipolar) with multiple alleles, probably over 100 (Korhonen & Stenlid, 1998). Basidiospores germinate to form a homokaryotic primary mycelium with multinucleate segments. Anastomosis with

a compatible homokaryon results in the formation of a heterokaryotic secondary mycelium also with multinucleate segments and clamp connections at some, but not all, septa. The conidial state (Fig. 18.15) has been assigned to the anamorph genus *Spiniger*. Conidiophores can develop on primary and secondary mycelium and the conidia may be uninucleate or multinucleate.

Freshly cut stump surfaces are hot-spots for infection by *H. annosum*, and the disease foci in plantations are often from thinning stumps which have become infected from air-borne spores. Insects are also possible vectors of conidia. Colonization of the roots of the stump is followed by spread to adjacent healthy trees by root-to-root contact and possibly by root-to-root graft because the fungus does not grow freely in the soil. Spread in this way may result in the development of genets (clones) many metres in diameter, in which several adjacent trees are infected by a secondary mycelium (heterokaryon) of identical genotype (Swedjemark & Stenlid, 1993). The pathogenic effect of *H. annosum* may be correlated with the secretion of fungal toxins, of which several have been described, predominantly benzofuran derivatives (Asiegbu et al., 1998).

Control of *Heterobasidion annosum*

Earlier methods of control were aimed at preventing stump infection by treatment with biocides such as creosote, sodium nitrite, urea, and boron-containing chemicals. The last two are still in use (Pratt et al., 1998; Asiegbu et al., 2005). An interesting alternative treatment, using biological control, is to inoculate freshly cut stumps with the conidia of the saprotrophic basidiomycete *Phlebiopsis gigantea* (= *Peniophora gigantea*; polyporoid clade) which competes with the parasite in the stumps and prevents its colonization, killing the cells of *Heterobasidion* by hyphal interference (Rishbeth, 1952, 1961; Ikediugwu et al., 1970). Conidial suspensions of *P. gigantea* are available commercially for use in biological control or, in conjunction with urea, in integrated control of stump infection in *Picea* plantations (Vasiliauskas et al., 2004, 2005). *Trichoderma harzianum* is another effective biological control agent (Holdenrieder & Greig, 1998).

19.7.3 Stereaceae

In this family the fruit body is flattened, appressed or resupinate, with a smooth (untextured) hymenium on the lower surface. Construction may be monomitic or dimitic with skeletals. Basidiospores are commonly smooth-walled and amyloid. Most members are lignicolous and saprotrophic but some are pathogens of trees. Hibbett and Thorn (2001) have placed the family in the russuloid clade, and Larsson and Larsson (2003) found that there are several groups of *Stereum*-like fungi in the russuloid clade. Forms with *Stereum*-like basidiocarps also occur in several other clades of basidiomycetes (Yoon et al., 2003).

Stereum sensu stricto

Basidiocarps of *Stereum* are common on decaying stumps and on attached and fallen tree branches. *Stereum hirsutum* (Figs. 19.26c,d) forms clusters of yellowish, fan-shaped leathery brackets with a hairy upper surface and a smooth hymenial surface on various angiospermous woody hosts, especially oak. It is a saprotrophic species and important as a cause of decay of sapwood of oak logs after felling. *Stereum gausapatum* is another common fungus on oak, which can grow parasitically on living trees, causing long narrow decay columns ('pipe rot') of the heartwood. When the hymenium is bruised it exudes a red latex. The same phenomenon is found in *S. rugosum*, common on coppice poles of hazel (*Corylus avellana*) and trunks of alder (*Alnus glutinosa*), and in *S. sanguinolentum*, a wound parasite or saprotroph on conifers. In most *Stereum* spp. there are specialized laticiferous (sanguinolentous) hyphae which extend through the flesh into the hymenium (Clémençon, 2004). These hyphae (Fig. 19.27) are thick-walled and are interpreted as modified skeletals homologous to the gloeocystidia of other members of the russuloid clade.

Mating in *Stereum*

The mating behaviour of *Stereum* spp. presents some unusual features. Germinating basidiospores form a multinucleate primary mycelium, often with whorls (verticils) of clamp connections at the septa, a condition described as **holocoenocytic** (Boidin, 1971). This feature led to the erroneous conclusion that *S. hirsutum* is homothallic, but in fact it shows unifactorial (i.e. bipolar) multi-allelic heterothallism (Coates et al., 1981). The mating type factor has been termed the 'C' factor. Compatible homokaryotic primary mycelia conjugate to form a heterokaryotic secondary mycelium which also produces whorled clamps. However, secondary mycelia can be distinguished by their yellowish pigmentation and often leathery surface, in contrast to the more delicate white primary mycelium. Some species of *Stereum* (*S. hirsutum*, *S. sanguinolentum*) include outcrossing and non-outcrossing populations (Ainsworth, 1987), whereas in others (e.g. *S. gausapatum* and *S. rugosum*) only outcrossing populations have been detected. In *S. hirsutum*, a feature of non-outcrossing homokaryons is their inability to accept non-self donor nuclei whilst they themselves can transfer nuclei to outcrossing strains (Ainsworth & Rayner, 1989; Ainsworth et al., 1990). In the short term, non-outcrossing strains may have the selective advantage in being able to exploit a particular habitat. They are also resistant to potential takeover or conversion to unfit or unstable genomic combinations.

Like many wood-rotting basidiomycetes, *Stereum* spp. show antagonistic reactions in the form of discoloured barrage zones where different strains confront each other, as seen in transverse sections of tree branches which contain more than one heterokaryon (Rayner & Boddy, 1988). Similar zones also develop on culture plates when dissimilar heterokaryons meet. Confrontations between heterokaryotic (secondary) and homokaryotic (primary) mycelia, equivalent to 'di–mon' mating in *Coprinus cinereus* and *Schizophyllum commune* (see p. 508), have been studied in *S. hirsutum* in search of evidence of the 'Buller phenomenon' (Coates & Rayner, 1985a). Such confrontations may be fully compatible, where both types of nuclei from the heterokaryon are potentially competent in bringing about dikaryotization of the homokaryon, or hemi-compatible, where only one kind of nucleus is competent. The Buller phenomenon does occur in *S. hirsutum*, with either one or two new heterokaryons formed per interaction. The genotypes of the new heterokaryons can be classified as composite (acceptor homokaryon plus heterokaryon component), parental (identical with the donor heterokaryon) and novel (different from all three parental and composite combinations). There is preferential selection of a non-sib-related component of the donor heterokaryon, i.e. a nucleus from an unrelated strain, not derived from the same fruit body as the recipient homokaryon.

The bow-tie reaction

Because *S. hirsutum* is bipolar, 50% of sib-matings (i.e. matings between primary mycelia from basidiospores of the same basidiocarp) are compatible, whilst non-sib matings are uniformly compatible because of multiple alleles at the mating type locus. In about one-third of incompatible sib-related matings a distinctive pattern of mycelial interaction occurs in agar cultures termed the 'bow-tie' reaction (see Fig. 19.28a; Coates *et al.*, 1981). A band of appressed sparse mycelium, shaped like a bow-tie in being widest at the edges, develops between the two homokaryons, and is bounded by narrow regions with exude watery droplets. Hyphae within the bow-tie band often burst, have granular contents and produce abundant irregular lateral branches (Figs. 19.28b,c). It is believed that the bow-tie region of a culture is occupied by a weakly growing heterokaryon. The bow-tie region often expands and may replace the mycelium of one of the component monokaryons. In some cases, a darkened zone of mutual antagonism develops between them. The development of bow-ties is controlled by a multi-allelic genetic factor, the B-factor, unlinked to the mating type C-factor.

Heterozygosity of the B-factor results in bow-tie formation which is only visible between incompatible homokaryons. The bow-tie phenomenon, although discovered in *S. hirsutum*, is not a feature peculiar to this fungus. Similar behaviour has been found in mating type compatible pairings of some other basidiomycetes, e.g. *S. gausapatum*, *Phanerochaete velutina*, *Mycena galopus* and *Coniophora puteana* (Coates & Rayner, 1985b).

Mating in *Stereum sanguinolentum*

The mating behaviour of *S. sanguinolentum* is also unusual (Calderoni *et al.*, 2003). Karyogamy, meiosis and post-meiotic mitosis may occur in the four-spored basidia or spore primordia. Most basidiospores contain two nuclei and are heterokaryotic, producing a mycelium capable of fruiting. *Stereum sanguinolentum* therefore shows secondary homothallism or pseudohomothallism. Pairings between single-basidiospore isolates from the same basidiocarp (intra-basidiome pairings) are somatically compatible but virtually all pairings between isolates from different basidiocarps (inter-basidiome pairings) are somatically incompatible, with demarcation lines developing between them. This indicates that in *S. sanguinolentum* there are numerous vegetative incompatibility groups (Calderoni *et al.* 2003).

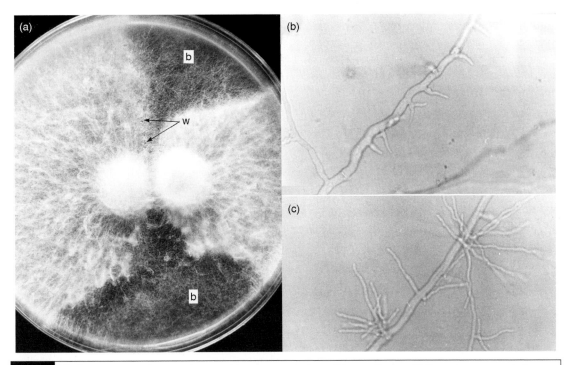

Fig 19.28 The bow-tie reaction between incompatible monospore isolates of *Stereum hirsutum*. (a) General appearance during early development; w, watery exudation; b, bow-tie region. (b,c) Abnormal branching of hyphae in the bow-tie region. From Coates *et al.* (1981), with permission from Elsevier. Photograph kindly provided by A. D. M. Rayner.

Chondrostereum purpureum

There are several species superficially resembling *Stereum* but falling outside the core group currently placed in the russuloid clade. One such fungus is *Chondrostereum purpureum* (formerly known as *Stereum purpureum*), with purplish leathery bracket-like or resupinate basidiocarps (Plate 10d). Its exact placement in the Homobasidiomycetes remains to be determined. It is a wound pathogen of numerous genera of deciduous trees but especially of members of Rosaceae, including plum and cherry trees in which it causes 'silver leaf' disease. The silver sheen on the leaves is due to the separation of the epidermis from the mesophyll induced by secretions of toxic secondary metabolites (Strunz *et al.*, 1997) and/or cellulase- and pectinase-type enzymes (Simpson *et al.*, 2001) from mycelium in the branches below the leaves. Although *C. purpureum* is of considerable economic importance as a pathogen of fruit trees, it has been proposed as a biocontrol agent in forest situations where coniferous trees are to be grown and broadleaved trees, such as alder or birch, occur as weeds (Shamoun, 2000).

Amylostereum

There are several species of *Amylostereum* related to each other, but of uncertain placement within the Homobasidiomycetes (Slippers *et al.*, 2003). They are necrotrophic wood-rotting wound pathogens of *Pinus* spp. and have entered a species-specific relationship with wood wasps of the genus *Sirex*, which distribute conidial (arthrosporic) inoculum in their internal glands and deposit it with mucilage surrounding their eggs, thereby providing the fungus with a suitable entry route into host trees. Decomposition of the wood by *Amylostereum* spp. facilitates feeding by the *Sirex* larvae, which take up inoculum prior to pupation

(Talbot, 1977). Severe infections of pine plantations are thought to be due to a synergism
between the fungus and its insect vector,
and are particularly common in the Southern
Hemisphere.

19.7.4 *Auriscalpium*

The fruit bodies of *Auriscalpium* have a toothed
hymenium, but molecular sequence studies have
shown that the genus is related to the gill-
bearing *Lentinellus* and the clavarioid fungus
Clavicorona (Pine *et al.*, 1999), both included in
the russuloid clade (Fig. 19.2). *Auriscalpium
vulgare*, the earpick fungus, is distributed in the
Northern Hemisphere. It grows on buried pine
cones, forming stalked, one-sided, brown, hairy
fruit bodies during autumn and winter
(Fig. 19.26e). The hyphal construction is dimitic
with skeletals. The hymenium is formed on
vertical, finger-like downgrowths from the
underside of the pileus. Interspersed amongst
the basidia are irregularly enlarged, thin-walled
hyphal tips with highly refractile contents,
gloeocystidia (Fig. 19.29). Basidiocarps show

rapid gravitropic readjustment to the vertical
position if displaced laterally. The mating
system of *Auriscalpium* is bifactorial, i.e.
tetrapolar (Petersen & Wu, 1992; Petersen &
Cifuentes, 1994).

19.8 | Thelephoroid clade

This includes the order Thelephorales, a small
group of predominantly ectomycorrhizal fungi
with variable basidiocarps. The most important
genus is *Thelephora* (*c.* 50 spp.). The earth fan
T. terrestris (Plate 10e) produces clusters of fan-
shaped basidiocarps which are chocolate-brown
in colour with a paler margin. They are
often formed around the stem of young trees,
seemingly 'choking' them. Basidiocarps of
T. terrestris superficially resemble those
of *Stereum* but are monomitic, composed of
clamped generative hyphae only. The basidiospores are brown and warty. *Thelephora terrestris*
fruits in association with coniferous trees
growing on light sandy soils and heaths. It is

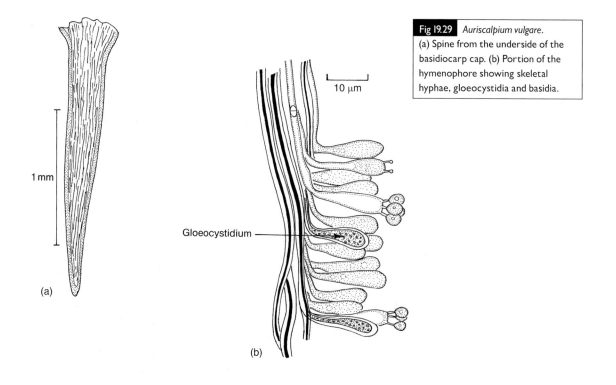

1 mm

(a)

Gloeocystidium

10 µm

(b)

Fig 19.29 *Auriscalpium vulgare.*
(a) Spine from the underside of the
basidiocarp cap. (b) Portion of the
hymenophore showing skeletal
hyphae, gloeocystidia and basidia.

one of a group of early-stage ectomycorrhizal associates of a variety of trees and also forms mycorrhiza with *Arbutus menziesii*, a member of the Ericaceae (Zak, 1976). According to Molina and Trappe (1984), *T. terrestris* represents the most abundant naturally present mycorrhizal fungus in bare-root nurseries.

19.9 | Hymenochaetoid clade

One feature that distinguishes the five Homobasidiomycete clades considered in the previous sections from the remaining three clades is the structure of the parenthesome, i.e. the membranous structure overarching the septal pore (see Fig. 18.10). In the five clades already described, the typical homobasidiomycete dolipore with a perforated parenthesome is found, whereas in the hymenochaetoid, cantharelloid and gomphoid–phalloid clades shown in Fig. 19.2, the parenthesome is generally imperforate (Hibbett & Thorn, 2001). Imperforate parenthesomes are also found in certain Heterobasidiomycetes, namely Dacrymycetales (Section 21.3) and Auriculariales (Section 21.4). This character has been discussed as being of phylogenetic significance, supporting the eight-clade system shown in Fig. 19.2. Hibbett and Thorn (2001) have estimated that the hymenochaetoid clade comprises about 630 spp. recruited from three families, namely the entire Hymenochaetaceae and parts of Corticiaceae and Polyporaceae.

19.9.1 Hymenochaetaceae

Most members of the hymenochaetoid clade are wood-decomposing fungi, exemplified by the white-rots *Inonotus* and *Phellinus* which often fruit on old living trees and continue to do so after the death of the host. In general terms, the basidiocarps are monomitic and annual in *Inonotus* but dimitic, hard and perennial in *Phellinus*. Generative hyphae typically lack clamp connections in both genera. Fruit body morphology is variable, with resupinate and bracket-like forms most commonly produced; the hymenium is usually poroid. There are

transitional forms between the genera *Inonotus* and *Phellinus*, and both have now been split up into several smaller units (Wagner & Fischer, 2001). Here we shall focus on *Phellinus sensu lato*.

Phellinus sensu lato (c. 180 spp.)

This genus is widespread and cosmopolitan, affecting both coniferous and broad-leaved trees. An interesting example is *P. weirii*, of which Hansen and Goheen (2000) have given a superb account. This species causes laminated root rot in native coniferous trees in the West Coast forests of North America, affecting especially Douglas fir (*Pseudotsuga menziesii*). Mycelium is able to penetrate intact roots and then spreads to adjacent trees by root-to-root contact. Trees of all ages are affected and are killed when extensive rot of the root system renders them unstable to storms, often several years or decades after initiation of infection. In this way, large genets of *P. weirii* slowly spread through the native forest, profoundly influencing the host population dynamics and shaping the forest landscape *en route*. Mycelium can also survive for decades in fallen logs. *Phellinus weirii* has a bipolar (unifactorial) mating-system although, according to Hansen and Goheen (2000), basidiospores are not an important part of the life cycle which relies mainly on clonal spread.

Phellinus noxius is the cause of root rot in numerous tree species in Central America and the Far East, Taiwan being particularly severely affected. Disease progression is unusually rapid for this group of pathogens, with infected trees sometimes dying within one growing season (Ann *et al.*, 2002). This pathogen may remain viable in dead colonized roots for several years if the soil is relatively dry, and the flooding of affected areas, where practicable, is an efficient control method.

There are also several species encountered in gardens, parks and forests in temperate climates, colonizing the trunks especially of mature and declining trees. Basidiocarps may be located several metres above ground. For example, *P. igniarius* (Plate 10f) forms basidiocarps predominantly on willow (*Salix*) and apple (*Malus*) trees,

P. pomaceus on plum trees (*Prunus*), and *P. robustus* on oak (*Quercus*).

Phellinus pomaceus has been extensively examined for the mechanism by which it produces the greenhouse gas chloromethane. This activity is indirectly associated with its lignin degradation system (see p. 527).

19.10 | Cantharelloid clade

This is a further group of Homobasidiomycetes with a wide range of morphologically different basidiocarps. Species likely to be encountered during forays belong to the Cantharellaceae, Hydnaceae and Clavulinaceae. Pine *et al.* (1999) have provided evidence of the relationship between these groups. Hibbett and Thorn (2001) have suggested that *Tulasnella* might be included in this group, but since members of this genus produce heterobasidia, we regard them as belonging to the Heterobasidiomycetes (see Fig. 21.1b).

19.10.1 Cantharellaceae

The genera *Cantharellus* and *Craterellus* belong to the most sought-after edible species. *Cantharellus* and related species are known as the chanterelles, with *C. cibarius* being the most readily recognized (Plate 10g). Another abundant, although less well-known, edible species is *C. tubaeformis* (Fig. 19.30a). The fruit bodies of most *Cantharellus* spp. appear in vibrant yellow or red colours due to carotenoid pigments, including canthaxanthin in the case of *C. cinnabarinus*. The fruit bodies are funnel-shaped, and the hymenium consists of shallow branching ridges which are strongly decurrent. The number of sterigmata and basidiospores on the basidium of *C. cibarius* seems to vary in the range of 4–7. In *Craterellus cornucopioides*, the horn of plenty or death's trumpet, the hymenium is smooth, and the fruit bodies are trumpet-shaped with a hollow stipe.

Members of this group are ectomycorrhizal with coniferous (*Picea*, *Pinus*) and broad-leaved (*Fagus*) trees, and it has been possible to grow *C. cibarius* on agar-based media and to

Fig 19.30 Basidiocarps in the cantharelloid clade. (a) *Cantharellus tubaeformis*. (b) *Hydnum rufescens*, the hedgehog fungus. (c) *Clavulina cristata*.

generate mycorrhizal associations under laboratory conditions (Danell, 1994). There is hope that a method may ultimately be developed towards commercial production of fruit bodies under controlled conditions (Wang & Hall, 2004).

19.10.2 Hydnaceae

Hydnum

The fruit bodies of the hedgehog fungi *Hydnum repandum* (see Fig. 19.1c) and *H. rufescens* (Fig. 19.30b) grow in deciduous and coniferous woodlands where they are ectomycorrhizal. They are more or less mushroom-shaped, with a cap and a central or lateral stipe. *Hydnum* spp. are good to eat and *H. repandum* is often collected for sale in mainland Europe. The basidiocarp construction is monomitic, with generative hyphae which become inflated, giving the fruit body a fleshy texture. The hymenium covers the tapering spines which develop from the lower side of the cap.

19.10.3 Clavulinaceae

Although the spindle-shaped fruit bodies of *Clavulina* are morphologically very similar to those of *Clavaria*, both genera have been placed in different clades. *Clavulina* spp. are saprotrophic fungi growing in the humus layer in forests and on lawns where they form coral- or spindle-shaped basidiocarps which are white, grey or pale yellow in colour. A common representative is *C. cristata*, which is a variable fungus with highly branched fructifications (Fig. 19.30c). A characteristic feature of the genus *Clavulina* is that the basidia are two-spored, narrowly cylindrical, and often undergo septation after spore discharge. The hymenium thickens with age. The fruit body construction is monomitic, with clamped inflated hyphae (Fig. 19.31).

19.11 | Gomphoid–phalloid clade

There are about 350 species in this group which has given rise to the most fascinating array of gasteromycetes, including the cannonball fungus, earth stars and stinkhorns. These are described in detail in Section 20.4. There are also some genera of actively spore-discharging basidiomycetes in the gomphoid–phalloid clade, especially with club-shaped and coralloid basidiocarps. These are grouped in the Clavariaceae which are briefly described below.

19.11.1 Clavariaceae

Hymenomycetes with smooth, branched or unbranched cylindrical or clavate fructifications were previously aggregated into this family, but microscopical and molecular phylogenetic analysis show that such an arrangement groups together unrelated forms, and it is clear that the clavarioid type of fructification has evolved

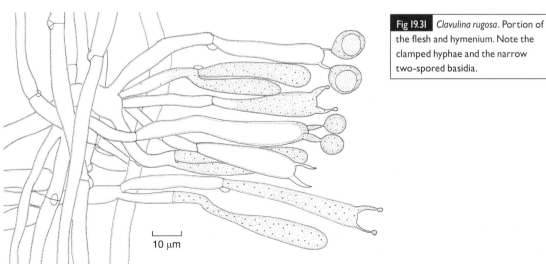

Fig 19.31 *Clavulina rugosa*. Portion of the flesh and hymenium. Note the clamped hyphae and the narrow two-spored basidia.

10 µm

independently in several unrelated basidio-mycete groups (Pine *et al.*, 1999). For example, *Clavulina* is now included in the cantharelloid clade (p. 575). Corner (1950) has monographed the clavarioid fungi and Petersen (1973) has given keys to genera.

Clavaria and its allies

This is a large genus of pasture and woodland fungi with cylindrical or club-shaped branched or unbranched fructifications. The flesh of *Clavaria* is made up of thin-walled hyphae which lack clamp connections and may become inflated and develop secondary septa. The hymenium which covers the whole surface of the fruit body usually consists of four-spored basidia with or without basal clamps, bearing colourless spores. A typical species is *C. vermicularis* which fruits in grassland, forming tufts of whitish, unbranched spindle-shaped basidiocarps. *Clavaria argillacea* forms yellow club-shaped fructifications on moors, heaths and peat bogs. In North America fructifications are consistently associated with ericaceous plants, including cultivated blueberries (*Vaccinium angustifolium* and *V. myrtilloides*), *Azalea* and *Erica* (Englander & Hull, 1980). In Australia a similar species is associated with cultivated *Azalea indica* (Seviour *et al.*, 1973). The close association suggests a mycorrhizal relationship, and although it has not been possible to synthesize mycorrhiza by inoculating aseptically grown ericaceous seedlings with mycelial cultures of *Clavaria*, evidence in support of a mycorrhizal partnership has been obtained by the demonstration of a two-way transfer of radioactive carbon and phosphorus between *Clavaria* and ericaceous plants, and by immunofluorescence studies in which hyphal coils were stained within root epidermal cells with conjugated antibodies raised against *Clavaria* basidiocarps (Mueller *et al.*, 1986). It is therefore likely that ericoid mycorrhizae develop following root infection by basidiomycetes as well as by ascomycetes such as *Hymenoscyphus ericae* (see p. 442), and both groups of fungi may infect the same root.

There are numerous other common representatives of the Clavariaceae. *Clavariadelphus pistillaris* (giant club) forms exceptionally large club-shaped fruit bodies ($7-30 \times 2-6$ cm) in beech woods on chalk (see Fig. 19.1d). The construction is monomitic, with clamps at the septa. As the fruit body matures, the hymenium becomes thicker by the development of further layers of basidia. Some of the more richly branched fairy clubs are placed in the genus *Ramaria*, distinguished by tougher flesh and pink, yellow or brown-coloured basidiospores, which are often rough. A particularly striking example is *R. botrytis* (Plate 10h). Most *Ramaria* spp. are ectomycorrhizal (Nouhra *et al.*, 2005); *R. stricta* is exceptional in growing on rotten wood.

Homobasidiomycetes: gasteromycetes

20.1 | Introduction

Gasteromycetes are an unnatural assemblage of basidiomycetes sharing the common negative character that the basidiospores are not discharged violently from their basidia. Instead of the ballistosporic basidiospores of other basidiomycetes which are asymmetric in side view (see Fig. 18.5), those of the gasteromycetes are usually symmetrically poised on their sterigmata or are sessile. Dring (1973) has termed such basidiospores **statismospores**. Commonly the basidia open into cavities within a fruit body, and the basidiospores are released into these cavities as the tissue between them breaks down or dries out. A recognizable fertile layer (hymenium) may be present or absent (see Reijnders, 2000). The internal production of basidiospores has given the gasteromycetes their name (Gr. *gaster* = stomach). The gasteromycete fruit body is termed the **gasterocarp**, and the spore mass enclosed by the gasterocarp wall (**peridium**) is the **gleba**. Sometimes, as in *Lycoperdon* or *Geastrum*, the gasterocarp opens by a pore through which the spores escape, but in forms with subterranean (hypogeous) fruit bodies there is no special opening, and it is possible that the spores are dispersed by rodents and other burrowing animals (Colgan & Claridge, 2002). In *Phallus* and its allies, the basidiospores are exhibited in a sticky mass attractive to insects, whilst in *Cyathus* and *Sphaerobolus* the spores are enclosed in separate glebal masses or **peridioles** which are dispersed as units. In spite of these variations in gasterocarp morphology, the life cycles of most gasteromycetes follow the general Homobasidiomycete pattern outlined in Fig. 18.4. Most species for which details are known appear to be heterothallic, with a basidiospore germinating to give a monokaryotic primary mycelium. Following fusion of compatible primary mycelia, a dikaryotic secondary mycelium is established, and this produces gasterocarps in which karyogamy and meiosis occur, and haploid basidiospores are formed. Dikaryotic asexual propagules are also known in some gasteromycetes.

Most members of the group are saprotrophic and grow on soil, rotting wood and other vegetation, or dung. Mycelial cords or rhizomorphs are often formed. *Rhizopogon* which produces hypogeous gasterocarps, and *Scleroderma* and *Pisolithus* with epigeous fruit bodies, are important ectomycorrhizal associates of forest trees. There are also two genera of aquatic gasteromycetes. *Nia vibrissa* grows on driftwood in the sea, forming globose, yellowish gasterocarps a few millimetres in diameter. Its basidiospores bear 4–5 radiating appendages (Jones & Jones, 1993). Such appendages are a typical adaptation to the aquatic habitat (see Section 25.2). *Limnoperdon* forms small, floating fruit bodies in freshwater swamps and marshes (Escobar *et al.*, 1976).

Because of the conspicuous shape and appearance of their gasterocarps, gasteromycetes have long attracted the attention of mycologists, and several keys and descriptions are available, including the books by Miller and Miller (1988),

Ellis and Ellis (1990) and Pegler *et al.* (1995). Species with hypogeous gasterocarps ('false truffles') have been described by Pegler *et al.* (1993).

20.2 | Evolution and phylogeny of gasteromycetes

In theory, the evolution of a gasteromycete from a hymenomycete ancestor requires only two morphogenetic changes, i.e. the production of a closed fruit body accompanied by the loss of the active spore discharge mechanism. The coincidence of these two features is shown by several examples of **secotioid** fruit bodies, i.e. basidiocarps in which the margin of the pileus fails to become detached from the stipe (Thiers, 1984). Watling (1971) described such aberrant development in the agaric *Psilocybe merdaria* growing in culture, where the failure of the fruit body to expose its hymenium coincided with morphological changes to the gills and basidia. Chiu *et al.* (1989) reported a similar case in *Volvariella bombycina*, and Hibbett *et al.* (1994) demonstrated that a naturally occurring recessive allele in a single gene is responsible for converting the lamellate fruit body of *Lentinus tigrinus* into a secotioid one. Except for the *L. tigrinus* mutant in which an existing hymenium is belatedly overgrown by a veil (see Fig. 19.6b), most secotioid forms seem to arise as a developmental defect causing incomplete differentiation of an agaric- or bolete-type basidiocarp (Thiers, 1984; Hibbett *et al.*, 1997b). Several species pairs are known in nature in which a secotioid form is closely related to a mushroom-type species, e.g. *Montagnea* and *Podaxis* (Fig. 20.1) related to *Coprinus comatus* (Fig. 19.14c), *Gastroboletus* related to *Boletus*, *Gastrosuillus* related to *Suillus*, *Hydnangium* related to *Laccaria*, or *Thaxterogaster* related to *Cortinarius* (Thiers, 1984; Mueller & Pine, 1994; Hopple & Vilgalys, 1999). Such evolutionary trends towards 'gasteromycetation' may be ongoing, and e.g. *Gastrosuillus laricinus* is thought to have arisen from *Suillus grevillei* as recently as

Fig 20.1 Fruit bodies of the secotioid mushroom *Podaxis pistillaris*, a close relative of the ink cap *Coprinus comatus* (see Fig. 19.14c). (a) Young fruit body. (b) Mature disintegrating fruit body. Original photographs kindly provided by A. E. Ashford.

70 years ago (Baura *et al.*, 1992). In contrast, the most ancient fossil gasteromycete found so far, an earth star resembling *Geastrum*, dates back to the Cretaceous period some 65–70 million years ago (Krassilov & Makulbekov, 2003).

One selective environmental pressure towards the secotioid and ultimately gasteromycete habit might be drought, since the very nature of the active basidiospore discharge mechanism by drop fusion (see p. 493) precludes its function at low humidity. It is perhaps no coincidence that secotioid fungi are particularly common in arid regions (Thiers, 1984). Secotioid fruit bodies are generally assumed to be the first step towards typical gasteromycete forms such as earth balls, puffballs and false truffles (Reijnders, 2000). However, mycologists are still at a loss to explain how the fantastically complicated fruit bodies, e.g. of the stinkhorns or bird's nest fungi, could have evolved from there.

Given the ease with which secotioid fruit bodies can arise, it is hardly surprising that gasteromycetes have evolved several times independently from hymenomycete ancestors, as indicated by numerous phylogenetic studies (see Fig. 19.2; Hibbett *et al.*, 1997b; Hibbett & Thorn, 2001). In subsequent sections of this chapter we shall consider the three most important groupings which are as follows (see Table 20.1):

1. Members of the euagarics clade (Section 19.4). The puffballs (Lycoperdaceae) and bird's nest fungi (Nidulariaceae) as well as a few smaller groups of gasteromycetes belong to this group. The Lycoperdaceae are close to *Macrolepiota* (Krüger *et al.*, 2001), whereas the Nidulariaceae cannot be placed accurately as yet but are likely to have arisen on a separate occasion. A further independent evolutionary event was that leading to the marine gasteromycete *Nia vibrissa* (Binder *et al.*, 2001).

2. Members of the boletoid clade (Section 19.5). Several gasteromycetes have their origin in the boletoid clade (Binder & Bresinsky, 2002). The most important group is the family Sclerodermataceae, i.e. the earth balls and their relatives (*Scleroderma*, *Pisolithus*, *Astraeus*) which

are closely related to *Gyrodon*. Another example is *Rhizopogon* which is close to *Suillus*. Like their actively spore-discharging relatives, these boletoid gasteromycetes are important ectomycorrhizal associates of trees.

3. The gomphoid–phalloid clade. This group contains the coral fungi and similar basidiomycetes with exposed hymenia and active basidiospore discharge (*Ramaria*, *Clavariadelphus*, *Gomphus*; see p. 575), as well as several important groups of gasteromycetes, namely the earth stars (*Geastrum* spp.), the cannonball fungus (*Sphaerobolus*), and the stinkhorns and their allies. The phylogeny of this grouping has been discussed by Humpert *et al.* (2001).

Although the artificial nature of the gasteromycetes as a taxonomic group has been known or suspected for many decades, it still comes as a shock to most mycologists to realize just how strongly convergent the evolution of these fungi has been. For instance, the implications from the results of phylogenetic studies (see Table 20.1) are that the earth stars (*Geastrum* spp.) have arisen independently of the barometer earth star (*Astraeus*), that the raindrop-mediated bellows mechanism of basidiospore release through an apical pore in puffballs and earth stars has evolved at least three times, and that the peridioles in the bird's nest fungi (*Cyathus*, *Crucibulum*), in *Sphaerobolus* and in *Pisolithus* are analogous rather than homologous structures. Referring to these and other findings made by molecular phylogeneticists, Reijnders (2000) concluded that 'if this key denotes real affinities, morphologists must be ashamed of their wrong conclusions'.

Ingold (1971) regarded the gasteromycetes as a biological group which, having lost the active spore discharge mechanism of their hymenomycete ancestors, have attempted a remarkable series of experiments in spore liberation. In order to explore this aspect of gasteromycete biology, we shall consider these fungi together in the present chapter, but drawing on the taxonomic framework as set out in Chapter 19.

Table 20.1. | Taxonomic affinities and spore release mechanisms of selected gasteromycetes.

Hymenomycete grouping	Gasteromycete genus	Gasterocarp type	Propagule	Dispersal [*]	Ecology
Euagarics clade	*Bovista, Calvatia*	Puffball (epigeous)	Basidiospore	A	Saprotrophic
	Lycoperdon	Puffball (epigeous)	Basidiospore	B	Saprotrophic
	Crucibulum, Cyathus	Bird's nest (epigeous)	Peridiole	C	Saprotrophic
	Nia	Gelatinous (epigeous)	Basidiospore	D	Saprotrophic
Boletoid clade	*Rhizopogon*	False truffle (hypogeous)	Basidiospore	E	Ectomycorrhizal
	Melanogaster	False truffle (hypogeous)	Basidiospore	E	Ectomycorrhizal
	Pisolithus	Earth ball (epigeous)	Peridiole or basidiospore	A	Ectomycorrhizal
	Scleroderma	Earth ball (epigeous)	Basidiospore	A	Ectomycorrhizal
	Astraeus	Earth star (epigeous)	Basidiospore	B	Ectomycorrhizal
	Calostoma	Puffball (epigeous)	Basidiospore	B	Saprotrophic
Gomphoid— phalloid clade	*Geastrum*	Earth star (epigeous)	Basidiospore	B	Saprotrophic
	Anthurus, Clathrus, Phallus, Mutinus	Stinkhorn, etc. (epigeous)	Basidiospore	F	Saprotrophic
	Sphaerobolus	Cannonball (epigeous)	Peridiole	G	Saprotrophic

*Dispersal mechanisms of propagules are as follows:

A. Disintegration of gasterocarp followed by release of propagules by wind or animal trampling.

B. Puffing through a pore after a raindrop hits the endoperidium ('bellows mechanism').

C. Splash cup dispersal following impact by a raindrop.

D. Passive release into water.

E. Distribution by burrowing animals and/or passive release into the soil following disintegration of the gasterocarp.

F. Insect dispersal following olfactory and visual attraction.

G. Active discharge of peridiole by tension-snap mechanism.

20.3 | Gasteromycetes in the euagarics clade

The euagarics clade contains some 10 000 fungi in 26 families (Hibbett & Thorn, 2001; Kirk *et al.*, 2001). Hymenia may be produced on the gills, pores and ridges of mushrooms and on the surface of coral-shaped fruit bodies, or basidia may be enclosed in gasterocarps. Among the gasteromycetes found within the euagarics, the two most important families are the Lycoperdaceae comprising puffballs and related forms, and the Nidulariaceae (bird's nest fungi).

20.3.1 Lycoperdaceae: puffballs

Puffballs such as *Lycoperdon*, *Vascellum* and *Calvatia* form a phylogenetically well-defined group which seems to be closely related to the genus *Macrolepiota* both on the basis of DNA sequence analyses (Krüger *et al.*, 2001) and because of similarities in the ontogeny and architecture of rhizomorphs (Agerer, 2002). The current family Lycoperdaceae contains 18 genera and 158 species of gasteromycetes with epigeous fruit bodies. The mature gasterocarp is thin-walled and either forms an apical pore (in *Lycoperdon*) or disintegrates from the apex downwards (e.g. in *Calvatia*, *Vascellum*, *Bovista*). Basidiospores are brown in colour and have warty or spiny walls, with the distal part of the basidial sterigma often remaining attached to mature spores (Portman *et al.*, 1997). Most species are saprotrophic on soil and humus.

Lycoperdon

About 50 species are known, producing fruit bodies which are pear-shaped or top-shaped. Most species grow on the ground. *Lycoperdon pyriforme* (Fig. 20.3) is unusual in growing directly on old stumps, rotting wood and sawdust heaps. It is not closely related to other *Lycoperdon* spp. and is now called *Morganella* by some authors. Gasterocarps of *Lycoperdon* spp. commonly arise on mycelial cords. The individual cells of the mycelium usually contain paired nuclei, but clamp connections are absent (Dowding & Bulmer, 1964). A longitudinal section of a young gasterocarp of *L. pyriforme* (Figs. 20.3a,b) shows that it is surrounded by a two-layered peridium, but as the fruit body expands the pseudoparenchymatous exoperidium may slough off or crack into numerous scales or warts (Fig. 20.2a) whilst the tougher endoperidium made up of both thick-walled and thin-walled hyphae remains unbroken, apart from a pore at the apex of the fruit body. The tissue within the peridium is differentiated into a non-sporing region or **sub-gleba** at the base of the gasterocarp, which extends as a columella into the sporulating region (gleba) in the upper part of the fruit body. The glebal tissue is sponge-like, containing numerous small cavities, and in the upper fertile part the cavities are lined by the hymenium. The tissue separating the hymenial chambers is made up of thick- and thin-walled hyphae. The thin-walled hyphae break down as the gasterocarp ripens, but the thick-walled hyphae persist to form the **capillitium** threads between which the spores are contained. The basidia lining the cavities of the gleba are rounded and bear one to four basidiospores symmetrically arranged on sterigmata of varying length (Fig. 20.3c). Young basidia are binucleate, and nuclear fusion and meiosis occur in the usual way. One nucleus migrates into each spore, and if fewer than four spores are produced, the spare nuclei degenerate in the basidium (Dowding & Bulmer, 1964). The basidiospores are not violently projected from the sterigmata. As the glebal tissue breaks down and dries, the spores are left as a brown dusty mass inside the fruit body. An apical pore develops by controlled lysis of the endoperidium (Fig. 20.2b). The thin upper layer of the endoperidium is elastic and acts as a bellows, and when rain drops impinge on this layer, small clouds of spores are puffed out (Gregory, 1949). Little is known of the mating behaviour of *Lycoperdon*.

Calvatia

Gasterocarps about the size of a rugby football are produced by *Calvatia* (*Langermannia*) *gigantea* (Fig. 20.2c) growing on grassland and on disturbed ground. There is no definite pore; the peridium breaks away to expose a brown spore mass. Buller (1909) estimated that the output of

Fig 20.2 Gasteromycetes in the euagarics clade. (a,b) *Lycoperdon perlatum*. Young gasterocarps (a) are ornamented by warts formed from the exoperidium. In older gasterocarps (b), the warts have sloughed off and the endoperidium has formed the apical pore. (c) *Calvatia gigantea*. Young fruit bodies growing with *Urtica dioica*. The coin is 2.5 cm in diameter. (d) *Cyathus stercoreus*. Gasterocarps (about 5 mm diameter) opening up to reveal the peridioles.

a specimen measuring $40 \times 28 \times 20$ cm was 7×10^{12} spores (see Table 18.1), and much larger specimens have been recorded. The spores are spherical with scattered warts. A more commonly encountered species is *C. excipuliformis*, which fruits on humus (Plate 11a).

When attempts are made to germinate the spores of this and other puffballs in the laboratory, the percentage of germination is extremely low, often less than 0.1%. Germination takes several weeks and is stimulated by the growth of yeasts (Bulmer, 1964; Wilson & Beneke, 1966).

20.3.2 Nidulariaceae: bird's nest fungi

Here the gasterocarps are funnel-shaped, and the gleba is differentiated into lens-shaped peridioles (glebal masses) which contain the basidiospores. Some 50 species in 4 genera are known, of which the most common examples are *Cyathus* and *Crucibulum*. Detailed and highly readable accounts of the biology of bird's nest fungi have been given by Brodie (1975, 1984). Members of this family are saprotrophic and are capable of degrading lignin (Wicklow *et al.*, 1984).

Cyathus

The fruit bodies of *C. olla* can be found in autumn growing amongst cereal stubble. *Cyathus striatus*, recognized by the furrowed inner wall of its cups, grows on old stumps and twigs whilst *C. stercoreus* grows on old dung patches. This last species can be made to fruit readily if mycelium

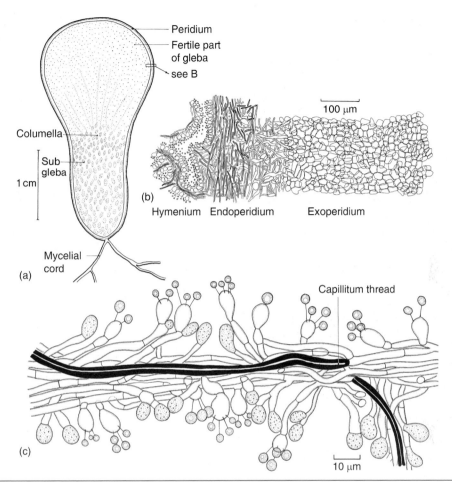

Peridium
Fertile part of gleba
see B

100 μm

Columella

Sub gleba

1 cm

(b)

Hymenium   Endoperidium   Exoperidium

Mycelial cord

(a)

Capillitum thread

(c)

10 μm

Fig 20.3 *Lycoperdon pyriforme.* (a) L.S. gasterocarp. (b) Portion of peridium and gleba. Note the pseudoparenchymatous exoperidium and the fibrous endoperidium. (c) Portion of gleba showing basidia, thin-walled hyphae and capillitium threads.

grown on a mixture of cow dung and straw is covered with a thin layer of casing soil and then left at room temperature for a few weeks (Fig. 20.2d; Webster & Weber, 1997). The fungus can also fruit on agar media (Lu, 1973). Light is essential for fruit body formation. Since the peridioles retain viability for many years if frozen, *C. stercoreus* provides a convenient example to study.

The first sign of fruit body development is the appearance of brown mycelial cords at the soil surface, on which knots of hyphae differentiate. In young gasterocarps, the mouth of the funnel is closed over by a thin papery **epiphragm** (Figs. 20.2d, 20.4a) which ruptures as the fruit body expands. Within the funnel, the **peridioles** develop. They are lens-shaped, slate-blue in

colour and attached to the peridium by a complex **funiculus**. In earlier stages of development in this and other species of *Cyathus*, the peridioles are separated by thin-walled hyphae which disappear at maturity (Walker, 1920). The peridiole wall consists of an outermost layer (**tunica**) made up of loosely interwoven hyphae, a dark **cortex**, and an inner layer of thick-walled but hyaline cells (Fig. 20.4g). The fertile centre of the peridiole is made up of thin-walled hyphae ('nurse hyphae') between which basidia develop. The basidia form 4–8 basidiospores and disappear soon afterwards, but the spores continue to enlarge and become thick-walled (Fig. 20.4g). Most of the nurse hyphae also break down, possibly providing nutrients for the enlarging spores.

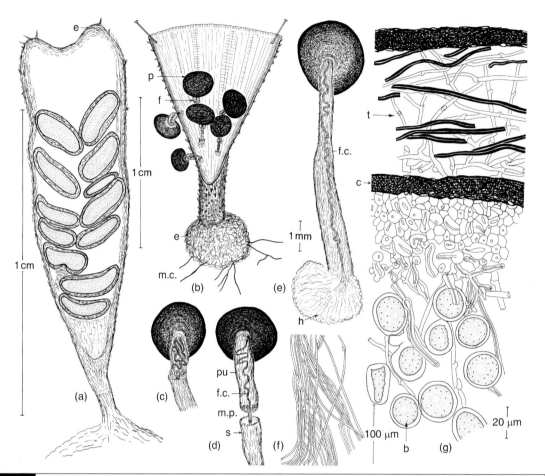

Fig 20.4 *Cyathus stercoreus.* (a) Section of immature gasterocarp showing peridioles. (b) Gasterocarp cut open and pinned back to show the attachment of the peridioles. (c–e) Details of structure of funiculus. (c) Condition of funiculus before stretching. (d) Stretched funiculus. Note the funicular cord within the purse. (e) Funicular cord extended after rupture of the purse. The base of the funicular cord is frayed out to form the hapteron. (f) Portion of funicular cord. Note the spirally coiled hyphae. The thickenings are modified clamp connections. (g) Detail of peridiole wall and contents (b, basidiospore; c, cortex; e, epiphragm; em, emplacement; f, funiculus; f.c., funicular cord; h, hapteron; m.c., mylial cords; m.p., middle piece; p, peridiole; pu, purse; s, sheath; t, tunica).

The gasterocarps of *Cyathus* and *Crucibulum* have been aptly termed splash cups because the peridioles are splashed out by the action of rain drops to distances of over 1 m. The key to understanding the mechanism of discharge lies in the structure of the funiculus. In *Cyathus* (Figs. 20.4d,e), the funiculus is made up of several characteristic structures (see Brodie, 1975). The **sheath** is a tubular network of hyphae attached to the inner surface of the gasterocarp. It terminates in the **middle piece** where the innermost hyphae of the sheath unite to form a short cord. The middle piece flares out at its top where its hyphae are attached to a cylindrical sac, the **purse**, which is firmly attached to the peridiole at a small depression. Folded up within the purse is a long strand of spirally coiled hyphae, the **funicular cord** (Fig. 20.4f). The free end of the funicular cord is composed of a tangled mass of adhesive hyphae, the **hapteron**.

Rain drops, which may be as much as 4 mm in diameter and have a terminal velocity of about $8 \, \text{m s}^{-1}$, fall into the cup. Drops of this size are most likely to drip from the woodland canopy (Savile & Hayhoe, 1978). The force creates a strong upward thrust which tears open the purse. The funicular cord, spirally coiled up within the purse, swells explosively, stretching to a length of about 2–3 mm in *C. stercoreus*, whilst in *C. striatus* it may be as long as 4–12 cm. As the peridiole is flicked away, the sticky hapteron at the base of the funicular cord helps to attach the peridiole to surrounding vegetation, and the momentum of the peridiole may cause the funicular cord to wrap around objects. Peridioles of *C. stercoreus* are presumably eaten by herbivorous animals and it is known that the basidiospores on release from the peridiole are stimulated to germinate by incubation around 37°C. Whether animals play a significant role in the dispersal of other bird's nest fungi is uncertain. The funiculus of *Crucibulum* is different from that of *Cyathus*, with a longer middle piece, a very short purse, and a funicular cord which is composed of relatively few hyphae only slightly coiled.

Both *Cyathus* and *Crucibulum* show tetrapolar heterothallism with relatively few alleles (generally not more than 15) at each locus, and this is the most usual condition within the Nidulariaceae (Burnett & Boulter, 1963; Lu, 1964).

20.4 | Gasteromycetes in the boletoid clade

The boletoid clade, as defined by molecular phylogeny (Hibbett & Thorn, 2001), contains fungi with a wide range of fruit body types, including lamellate (e.g. *Paxillus*), boletoid (e.g. *Boletus*, *Leccinum*, *Suillus*, *Xerocomus*) and resupinate forms (e.g. *Coniophora*, *Serpula*). These have been described previously (Section 19.5). Gasteromycete fungi have arisen from boletoid ancestors on several occasions, with the family Sclerodermataceae having an affinity with *Gyroporus* in a 'boletoid' branch, and the Rhizopogonaceae with *Suillus* ('suilloid' branch; Grubisha *et al.*, 2001; Binder & Bresinsky, 2002). Both families contain mainly ectomycorrhizal fungi (see Table 20.1).

Features of their association with tree roots are typical of members of the boletoid clade in that there is a large amount of fungal biomass extending from the mantle into the soil by means of mycelial cords or rhizomorphs which may be several metres long. This type of ectomycorrhiza appears to be particularly effective in exploiting a large volume of soil for nutrients, and it is also credited with improving the water status and thus the performance of the tree host under conditions of drought (Smith & Read, 1997; Agerer, 2001). The ability to form an extensive rhizomorph system may explain why mycorrhizal gasteromycetes belonging to the boletoid clade are particularly prominent in dry habitats. It is probable that long-distance transport processes in rhizomorphs are facilitated by peristaltic movement through a system of tubular vacuoles (Fig. 1.9; Ashford & Allaway, 2002). Certain ectomycorrhizal fungi such as *Rhizopogon*, *Scleroderma* and *Pisolithus* can be grown in pure culture, and basidiospore inoculum from their relatively large fruit bodies is also easily collected and stored. Hence, these species are suitable for laboratory-based research as well as inoculation of trees prior to outplanting into forestry situations.

In addition to the morphology of ectomycorrhiza, there are several further features betraying an affinity of the Sclerodermataceae and Rhizopogonaceae with the boletoid clade. For instance, pulvinic acid-type pigments typical of *Boletus*, *Suillus* and *Xerocomus* (see Fig. 19.22) are also found in their gasteromycete relatives, either in a pure form (e.g. variegatic acid, xerocomic acid) or as derivatives (Gill & Watling, 1986; Gill & Steglich, 1987; Winner *et al.*, 2004). Further, the mycoparasitic mould *Apiocrea chrysosperma* (anamorph *Sepedonium chrysospermum*), which frequently forms a golden yellow conidial crust on fruit bodies of *Boletus*, *Suillus*, *Xerocomus* and *Paxillus* (Plate 9h), also infects gasteromycetes such as *Scleroderma* and *Rhizopogon* (see Gill & Watling, 1986). Pathogens may be competent taxonomists!

20.4.1 Sclerodermataceae: earth balls and relatives

This family comprises some 50 species in 7 genera (Kirk *et al.*, 2001), including the earth ball *Scleroderma*, the stalked puffball *Calostoma*, the dye ball *Pisolithus*, and the barometer earth star *Astraeus*. Binder and Bresinsky (2002) have examined the phylogeny of the group and have recommended its partitioning into several families. All members seem to be ecto-mycorrhizal with trees. Here we shall consider the two most important genera, *Pisolithus* and *Scleroderma*.

Pisolithus

The best-known species is *P. tinctorius* (= *P. arhizus*), which is so called because its immature gasterocarps, when injured, produce an intense black dye (see Plate 11b). Because of the variability of gasterocarp appearance, the taxonomy of *Pisolithus* has been problematic, and initially *P. tinctorius* was thought to be of pan-global distribution, capable of associating with almost any ectomycorrhiza-forming tree species (Marx, 1977). *Pisolithus* is now known to consist of more than 10 species (Cairney, 2002; Martin *et al.*, 2002), with *P. tinctorius* distributed throughout the Northern Hemisphere and associated mainly with *Pinus* and *Quercus*. The centre of evolution of the genus is probably Australia, and *P. marmoratus* associated with *Eucalyptus* is regarded as the Southern Hemisphere equivalent of *P. tinctorius*. This and other species have been spread to South America, South East Asia and Africa, together with their host trees (*Eucalyptus*, *Acacia*) which are used in intensive forestry and in reforestation programmes (Dell *et al.*, 2002; Martin *et al.*, 2002). In addition to this anthropogenic disper-sal, there is also evidence that *Pisolithus* can travel long distances as air-borne basidiospores, e.g. from Australia to New Zealand (Moyersoen *et al.*, 2003).

Pisolithus spp. may be displaced by other ectomycorrhizal fungi in cool, wet situations (McAfee & Fortin, 1986) but are prominent in extreme environments, e.g. dry habitats with sandy soil, or areas polluted with heavy metals (Walker *et al.*, 1989; Smith & Read, 1997). In such situations, the growth of mycorrhizal trees can be increased several-fold relative to uninoculated controls. Benefits of *Pisolithus* infections to the host tree include enhanced provision of nutri-ents and water, detoxification of heavy metals, and protection against soil-borne plant pathogens. The considerable promise of *Pisolithus* is reflected by an immense body of literature which has been summarized admirably by Cairney and Chambers (1997) and Chambers and Cairney (1999).

Pisolithus has a tetrapolar mating system (Kope & Fortin, 1990), and although monokary-ons can infect tree roots, a full-scale ectomycor-rhizal association requires a dikaryotic mycelium. The establishment of a mycorrhiza proceeds in several steps. Chemotropic growth of *Pisolithus* hyphae towards host root tips is followed by the secretion of glycoprotein fibrils by the fungus during initial contact (Lei *et al.*, 1990). Dead or moribund cells in the root cap region are infected first; the mantle is then established within 48 h, and a Hartig net formed subsequently (Horan *et al.*, 1988; Lei *et al.*, 1990). Only root material grown after initial contact is colonized. Rhizomorphs radiate outwards for several metres, and these may partly account for the success of the *Pisolithus* mycorrhiza, especially in dry habitats. Another factor may be the formation of sclerotia which enable the fungus to survive adverse conditions in the soil (Grenville *et al.*, 1985). Eventually, fruit body initials are formed in the soil, with the maturing gasterocarps pushing through the surface. Basidiospores are produced inside numerous peridioles which disintegrate to release their spores passively (Plate 11b). The formation and maturation of peridioles proceeds from the tip to the base of the gasterocarp which gradually breaks up in the process. Gasterocarps can be sizeable, up to 20 cm tall.

Studies on *Pisolithus* mycelia in Australian eucalypt forests have revealed that genetically distinguishable individuals (genets) may be variable in size, ranging from less than 2 m^2 to 50 m^2 or more. Since these are interspersed, the smaller genets are interpreted as the result of recent re-colonization events from wind-dispersed basidiospore inoculum (Anderson *et al.*, 1998, 2001).

Scleroderma

There are about 25 species of *Scleroderma* (Sims *et al.*, 1995), three common temperate examples being *S. bovista*, *S. citrinum* (Fig. 20.5) and

Fig 20.5 *Scleroderma citrinum.* (a) Maturing gasterocarps, about 3–6 cm in diameter. One has been cut open to reveal the gleba containing purplish-black basidiospores. (b) Old gasterocarps. The peridium has cracked open, permitting passive dispersal of the black basidiospore mass.

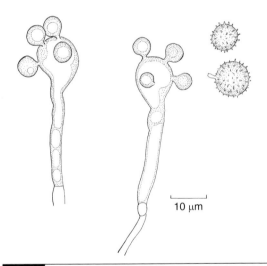

10 μm

Fig 20.6 *Scleroderma verrucosum.* Basidia and basidiospores. Note that the spores are almost sessile.

S. verrucosum (Pegler *et al.*, 1995). Earth balls are found in the autumn in acid woodlands and heaths under such trees as *Pinus*, *Betula*, *Quercus* and *Fagus* with which they form ectomycorrhizal associations. Mycorrhizal infection is easily reproduced under laboratory conditions, using aqueous spore suspensions (Parladé *et al.*, 1996). General aspects of mycorrhiza involving *Scleroderma* have been reviewed by Jeffries (1999).

In mature gasterocarps, the peridium is apparently a single, fairly thick layer. Although the glebal mass may be traversed by a system of sterile veins, there is no columella and no capillitium. The basidiospores are sessile (Fig. 20.6). When the gasterocarp is ripe, it cracks open irregularly and the dry spores escape (Fig. 20.5b). There is no well-developed bellows mechanism as in *Lycoperdon* (p. 579) or *Geastrum* (p. 588).

20.4.2 Rhizopogonaceae: beard truffles

This family comprises some 150 species in 4 genera. By far the most important genus is *Rhizopogon*, which is mycorrhizal mostly with coniferous trees. It originates from the Northern Hemisphere, with an unusually high diversity of species encountered in the Pacific Northwest where several host trees are native (Martín, 1996; Molina *et al.*, 1999). *Rhizopogon* is related to *Suillus* in the boletoid clade (Grubisha *et al.*, 2001).

Rhizopogon

Members of this genus form gasterocarps which resemble those of *Scleroderma* but are hypogeous, arising from mycelial cords (Plate 11c). The gasterocarps may be eaten by burrowing mammals, and basidiospores pass through their digestive tracts unharmed (Colgan & Claridge, 2002). *Rhizopogon luteolus* and *R. roseolus* are common and cosmopolitan under *Pinus* spp., whereas *R. vinicolor* and others are specific associates of Douglas fir (*Pseudotsuga menziesii*). Reviews of the genus have been written by Molina and Trappe (1994) and Molina *et al.* (1999).

Rhizopogon spp. provide similar benefits to their hosts as *Pisolithus* and are the subject of research and development activities in forestry. Inoculation is readily achieved by dusting seeds with basidiospores or immersing the roots of seedlings in spore suspensions (Parladé *et al.*, 1996). Although *Rhizopogon* spp. can be grown in pure culture, inoculation of host tree seedlings with mycelium is not generally efficient with this and related fungi, unless the hyphae are protected from mechanical damage, for example, by being grown inside porous or gel-like beads (see Smith & Read, 1997).

20.5 | Gasteromycetes in the gomphoid–phalloid clade

The gomphoid–phalloid clade contains some 350 species of morphologically diverse fungi (Hibbett & Thorn, 2001). Most of the species with active basidiospore discharge form coral- or club-shaped basidiocarps, e.g. *Ramaria* (see Plate 10h), *Clavariadelphus* and *Gomphus*. Several groups of well-known gasteromycetes also belong here (Hibbett *et al.*, 1997b; Humpert *et al.*, 2001), and these show the most spectacular spore dispersal mechanisms of all gasteromycetes, including the bellows mechanism, insect dispersal, and active discharge of peridioles. Details of the phylogeny or evolutionary history of these gasteromycetes still appear to be unknown, although some groups such as earth stars may be ancient (Krassilov & Makulbekov, 2003). Members of the gomphoid–phalloid clade are mostly saprotrophic on wood, other plant debris and humus, extending to the soil by means of mycelial cords.

20.5.1 Geastraceae: earth stars

There are some 50 species of earth stars. Descriptions of the common species may be found in the keys by Ellis and Ellis (1990) and Pegler *et al.* (1995). One of the most frequent and widespread species in temperate and subtropical forests is *Geastrum triplex* (Fig. 20.7a) which grows in the leaf litter of beech, sycamore and pine. The young fruit body is onion-shaped and develops at or just below the soil surface. The exoperidium is complex, consisting of a brown outer layer made of narrow hyphae mostly running longitudinally, and a paler pseudoparenchymatous inner layer. As the fruit body ripens, the whole of the exoperidium splits open from the tip in a stellate fashion and, due to swelling of the pseudoparenchyma cells of the exoperidium, the triangular flaps curve outwards and make contact with the soil, lifting the inner part of the fruit body into the air (Fig. 20.7a). The thin, papery endoperidium opens by an apical pore. Spores are puffed out by the bellows mechanism when rain drops strike the endoperidium (Ingold, 1971). The gleba contains a columella (sometimes termed a pseudocolumella) and capillitium, much as in the puffball *Lycoperdon* (Fig. 20.3) with which *Geastrum* is not related. Basidial development can only be observed in young unexpanded gasterocarps. The basidia are pear-shaped, with 4–6 (sometimes up to 8) spores borne on a knob-like extension of the pointed end.

20.5.2 *Sphaerobolus*

Sphaerobolus is unique among gasteromycetes in that it has developed an active discharge mechanism of peridioles (glebal masses), thereby reversing the loss of active basidiospore liberation. The precise taxonomic position of *Sphaerobolus* is still unclear at present; formerly grouped together with the bird's nest fungi (p. 578), it is now known to belong to the gomphoid–phalloid clade. Kirk *et al.* (2001) have included it in the Geastraceae.

Sphaerobolus stellatus forms globose orange gasterocarps about 2 mm in diameter. They are attached to rotten wood, rotting herbaceous stems, sacking and weathered dung of herbivores such as cow and sheep. Ripe fruit bodies open

Fig 20.7 Gasteromycetes belonging to the gomphoid–phalloid clade. (a) The earth star *Geastrum triplex*. (b) *Sphaerobolus stellatus*. Stages of maturation can be seen from left (immature gasterocarp) through the centre (two opened gasterocarps exhibiting glebal masses) to right (discharged gasterocarp with everted inner cup). (c) The veiled stinkhorn, *Phallus indusiatus*. This beautiful species is called 'queen of mushrooms' (kinoko no joou) in Japan. (d) The dog's stinkhorn, *Mutinus caninus*. (b) reproduced from Webster and Weber (1999), with permission from Elsevier. Original print of (c) kindly provided by N. Tuno.

to form a star-like arrangement of two cups fitting inside each other, attached only by the triangular tips of their teeth (Fig. 20.7b). Within the inner cup is a single brown peridiole or glebal mass about 1 mm in diameter. By sudden eversion of the inner cup, the peridiole is projected for a considerable distance. Buller (1933) has given a detailed account of peridiole discharge. He showed that the peridiole could be projected vertically for more than 2 m, and horizontally for over 4 m, with the record currently standing at 5.7 m (Ingold, 1971, 1972). The fungus can be cultivated if a peridiole is placed in a plate of oatmeal agar, and gasterocarps are produced after a few weeks' incubation in daylight on this medium or on chopped straw saturated with a nutrient solution (Flegler, 1984; Webster & Weber, 1999).

A section through an almost mature, but unopened, gasterocarp is shown in Fig. 20.8a. The peridiole is surrounded by a peridium in which six layers can be distinguished. Three of these layers form the structure of the outer cup. The three layers making up the inner cup consist of an outer layer of tangentially arranged interwoven hyphae, a central layer of radially elongated cells forming a kind of palisade, and a thin innermost layer of pseudoparenchyma whose cells undergo deliquescence before glebal discharge to form a liquid which bathes the gleba and lies in the bottom on the inner cup. Before the gasterocarp opens up, the cells of the palisade layer are rich in glycogen, but this is converted to glucose during ripening (Engel & Schneider, 1963). Intracellular accumulation of glucose causes the osmotic concentration of the cells to rise so that they absorb water and become more turgid. The swelling of the palisade layer is restrained by the tangentially arranged hyphae, and this sets up strains within the tissues of the inner cup which are only released by its turning inside out.

Light is necessary for development, and the opening of the fruit body is phototropic,

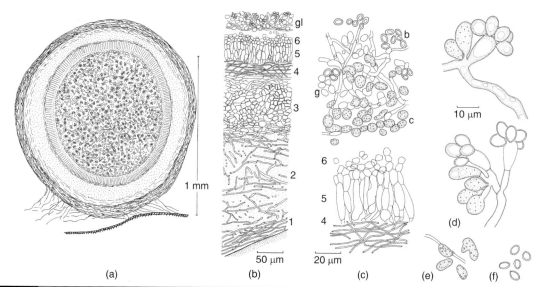

Fig 20.8 *Sphaerobolus stellatus.* (a) V.S. of nearly ripe gasterocarp showing the central glebal mass (peridiole) surrounded by a six-layered peridium. (b) Details of the peridial layers: I, outermost layer composed of interwoven hyphae; 2, layer in which the hyphae are separated by extensive mucilage; 3, pseudoparenchymatous layer; 4, fibrous layer; 5, palisade layer; 6, layer of lubricating cells; gl, outer layers of glebal mass. (c) Enlarged portion of layers 4—6 and portion of the glebal mass: c, cystidia; g, gemmae; b, basidiospores. (d) Clusters of basidia from unripe gasterocarps. There are usually 4—6 basidiospores. (e) Gemmae. (f) Basidiospores. (c, e) and (f) to same scale.

ensuring that the glebal mass is projected towards the light (Alasoadura, 1963). Peridiole discharge follows a diurnal rhythm, with release occurring during the light phase. In continuous light rhythmic discharge ceases, but in continuous darkness a culture previously exposed to alternating periods of 12 h of light and darkness continues to discharge peridioles rhythmically at times corresponding to the previous light periods, indicating an endogenous circadian rhythm.

The spherical peridiole (glebal mass) is surrounded by a dark brown sticky coat derived from the breakdown of the cells of the innermost peridial layer. Immediately within the brown outer coat of the glebal mass are layers of rounded cells sometimes termed cystidia (Fig. 20.8c). Apparently these cells are incapable of germination and their function is not known. The rest of the glebal mass consists of oval thick-walled haploid basidiospores and thinner-walled dikaryotic gemmae. About 4—8 basidiospores develop on the basidia some 2 days before discharge (Fig. 20.8d), but the basidia disappear as the glebal mass ripens. Gemmae arise either terminally or in an intercalary position on hyphae within the glebal mass. Oil-rich cells are also present. The sticky peridiole adheres readily to objects on which it is impacted, and after drying it is very difficult to dislodge even by a jet of water. Peridioles are viable for several years. Projectiles adhering to herbage may be eaten by animals, and this may explain the presence of fruit bodies on dung.

On germination the peridioles give rise to clamped hyphae which usually arise directly from the gemmae and not from the basidiospores. Most basidiospores, if they germinate, give rise to mycelia with simple septa. Pairings of monosporous mycelia have indicated that the fungus is usually heterothallic, although details of the mating system still appear unclear at present.

20.5.3 Phallaceae: stinkhorns

An original solution to the problem posed by the loss of active basidiospore discharge has been developed also by members of the Phallaceae which attract insects, especially cadaver-feeding flies such as bluebottles, to visit their gasterocarps. Attraction may be by the emission of a cadaverous smell or by colour, with gasterocarps of species like *Clathrus ruber* (Plate 11e) and *C. archeri* (Plate 11f) appearing dark red due to the accumulation of carotenoids, chiefly lycopene (Fiasson & Petersen, 1973; Gill & Steglich, 1987). There may be a synergism of attractions because species with brightly coloured gasterocarps tend to emit less evil smells than dull-coloured ones.

Common temperate examples of Phallaceae are the graphically named stinkhorn *Phallus impudicus* (Lat. *impudicus* = shameless; Plate 11d) and the dog's stinkhorn *Mutinus caninus* (Fig. 20.7d).

Phallus

In late summer and autumn, stinkhorns can be detected readily by their smell. They can be a common or even dominant component of the population of basidiocarps on the forest floor, as shown by Shorrocks and Charlesworth (1982) who estimated some 50 000–70 000 gasterocarps km^{-2} per season in a woodland. In such situations, stinkhorns and their primordia, the 'eggs', provide a major breeding ground for mycophagous flies. Eggs of *P. impudicus* are about 5 cm in diameter and develop from an extensive system of white mycelial cords which can be traced underground to a buried tree stump (see Fig. 1.12b; Grainger, 1962). A longitudinal section of an egg (Fig. 20.9a) shows a thin papery outer and inner peridium and a wider mass of jelly making up the middle peridium. The central part of the gasterocarp is differentiated into a cylindrical hollow stipe and a folded honeycomb-like **receptacle** which bears the fertile part of the gleba. Within the young gleba are cavities lined by basidia bearing up to nine spores (Fig. 20.9b), but as the glebal mass ripens the basidia disintegrate. Gasterocarps expand very rapidly: within a few hours the stipe may elongate from about 5 cm to a length of 15 cm or more, leaving behind the peridial remains as a volva at its base. A demonstration of gasterocarp erection in the laboratory by incubation of freshly collected ripe eggs in a moist chamber rarely fails to impress. The sudden expansion is probably at the expense of water stored within the jelly of the middle peridium. The mean weight of expanded stipes is more than twice that of unexpanded ones (Ingold,

1959). Expansion of the stipe of *P. impudicus* is accompanied by breakdown of glycogen and its conversion to sugar (Buller, 1933). A similar conversion has been reported in *P. indusiatus* (Fig. 20.7c) in which cells of the unexpanded stipe are folded but expand to almost 12 times their original volume during stipe elongation (Kinugawa, 1965).

At about the same time as the stipe of *P. impudicus* is elongating, the fertile glebal mass begins to release strong-smelling volatile substances which are attractive to flies, especially bluebottles (Plate 11d). Depending on the analytical methods used, the smell has been attributed to a range of substances including methylmercaptan and hydrogen sulphide (List & Freund, 1968) or dimethyl disulphide (see Fig. 15.4) and dimethyl trisulphide (Borg-Karlson *et al.*, 1994). Once a fly has located a gasterocarp, it is presented with the dark green glebal mass of basidiospores embedded in a liquid which contains sugars and also sweet-smelling substances such as phenylacetaldehyde and phenylethanol. Flies feed on the spore mass which is removed within a few hours, leaving behind the pale receptacle. The ingested basidiospores are defaecated, apparently unharmed, onto surrounding vegetation and the soil, often within a short time of ingestion. Tuno (1998) found that the gut of fruitflies (*Drosophila* spp.) feeding on *P. indusiatus* and *P. duplicatus* contained up to 240 000 basidiospores, and that of the larger muscid flies up to 1.7 million. Basidiospore germination was unaffected by passage through the gut. However, it is unknown how the mycelium from germinating basidiospores succeeds in reaching fresh tree stumps, and clonal spread by mycelial cords may be an important additional mode of reproduction.

There have been few studies of the nutrition and physiology of *Phallus*, but *P. ravenelii* has been shown to make good vegetative growth on a wide range of carbohydrates, and to require thiamine (Howard & Bigelow, 1969). The veiled stinkhorns *P. indusiatus* (Fig. 20.7c) and *P. duplicatus* are grown in China as a culinary speciality. Expanded fruit bodies are produced commercially from inoculated wood both in woodlands and indoors, and are marketed in a

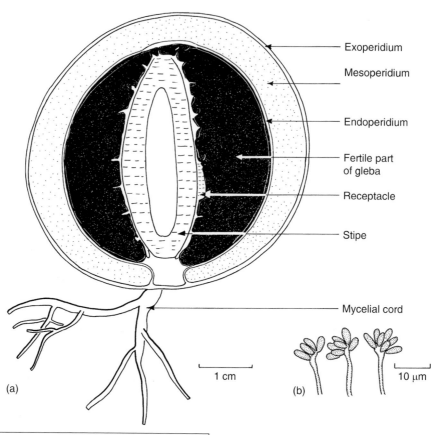

Exoperidium

Mesoperidium

Endoperidium

Fertile part of gleba

Receptacle

Stipe

Mycelial cord

1 cm

10 μm

(a)

(b)

Fig 20.9 *Phallus impudicus.* (a) L.S. 'egg' showing the unexpanded stipe. (b) Basidia.

dried form; sensibly, the volva and glebal mass are removed prior to marketing (Chang & Miles, 2004). Apparently, the egg stage is also eaten after it has been boiled. These species were formerly thought to belong to a separate genus, *Dictyophora*, on the basis of the presence of a veil. However, this character is no longer considered useful because even the common stinkhorn may occasionally produce a short veil and is then called *P. impudicus* var. *togatus* (see Kibby & Bingham, 2004).

Other genera

The general form of *Mutinus*, the dog's stinkhorn, is similar to *Phallus*, but the gasterocarps are smaller (Fig. 20.7d). The upper part of the stipe is orange in colour, and the smell is less overpowering. The receptacle bearing the glebal mass is not reticulate.

Clathrus ruber (Plate 11e) forms conspicuous, red cage-like gasterocarps in warm, dry habitats in the Mediterranean, occasionally extending to temperate climates. Other *Clathrus* spp. are tropical. *Clathrus archeri*, the squid fungus, was probably introduced to Europe from Australia some 100 years ago (Ramsbottom, 1953) and is now established in many localities. An even more bizarre Australian species is *Aseroe rubra*, the starfish fungus (Plate 11f).

Heterobasidiomycetes

21.1 | Introduction

The class Heterobasidiomycetes is approximately synonymous with the terms 'Phragmobasidio-mycetes' or 'jelly fungi' and contains fungi with the following characteristics.

1. The dolipore septum is complex, i.e. it is surrounded by a parenthesome. Parenthesomes are also found in the Homobasidiomycetes (Chapters 19 and 20), but not in the Urediniomycetes (Chapter 22) and Ustilaginomycetes (Chapter 23).

2. The basidia of Heterobasidiomycetes may be strongly lobed and often divided by transverse, oblique or longitudinal septa. Such basidia are loosely termed **heterobasidia**, especially if they arise directly from hyphae instead of teliospores as in most Urediniomycetes and Ustilaginomycetes. If the basidia are septate, they are also called **phragmobasidia**. The sterigma of the heterobasidium is unusually prominent and is often termed **epibasidium** (Martin, 1945). In contrast, the basidia of Homobasidiomycetes are club-shaped and always single-celled.

3. The fruit bodies of most Heterobasidiomycetes are simpler in architecture than those of Homobasidiomycetes, and the hymenium is not normally protected by a roof- or shelf-like architecture. In compensation, these simple fruit bodies are generally able to survive drying and rehydration, with fresh crops of basidiospores produced after each rehydration event. Fully hydrated basidiocarps are typically greatly swollen and gelatinous, hence the term 'jelly fungi' for the Heterobasidiomycetes.

4. The basidiospores of most species are capable of producing secondary spores which may be ballistoconidia, passively released conidia or yeast cells.

Species included in this class show considerable morphological diversity, and taxonomic concepts have been in a state of flux. The first workers to emphasize the importance of basidial morphology were Patouillard (1887) and Brefeld (1888). Most orders currently included were placed here by Martin (1945) and Bandoni (1984), and these are listed in Table 21.1. The inclusion of these groups is supported by DNA-based phylogenetic studies (Weiss & Oberwinkler, 2001). The life cycles of Heterobasidiomycetes, as far as they are known, show an alternation of monokaryotic and dikaryotic stages. The two broad subclasses, Heterobasidiomycetidae and Tremellomycetidae, can be distinguished by their monokaryotic phase being mycelial or yeast-like, respectively. All heterobasidiomycete yeasts discussed in Chapter 24 seem to belong to the Tremellomycetidae (Wells & Bandoni, 2001). Thorough and authoritative circumscriptions of the Heterobasidiomycetes have been written by Wells (1994) and Wells and Bandoni (2001).

Ecologically, Heterobasidiomycetes are associated with wood and other decaying plant matter, either as saprotrophs or as mycoparasites of saprotrophic fungi. Some species, especially in the Ceratobasidiales (Section. 21.2), play dual roles as necrotrophic pathogens of various plants and mycorrhizal associates of orchids. Heterobasidiomycetes are mainly terrestrial.

| Table 21.1. | Orders and selected genera currently included in the Heterobasidiomycetes. After Wells and Bandoni (2001). |

Subclass Heterobasidiomycetidae

1. Ceratobasidiales (see below)
   Ceratobasidium (anam. *Rhizoctonia* = *Ceratorhiza*).
   Thanatephorus (anam. *Rhizoctonia* = *Moniliopsis*).

2. Tulasnellales (see Fig. 21.1b)
   Tulasnella (anam. *Rhizoctonia*)

3. Dacrymycetales (p. 598)
   Calocera
   Dacrymyces
   Ditiola

4. Auriculariales (p. 601)
   Auricularia
   Exidia
   Pseudohydnum
   Sebacina

Subclass Tremellomycetidae

1. Tremellales (p. 604)
   Cryptococcus and *Bullera* (yeast forms; Table 24.1)
   Filobasidiella neoformans (see p. 660)
   Tremella

2. Christianseniales

3. Filobasidiales (yeast forms; see Table 24.1)

4. Cystofilobasidiales (see Table 24.1)
   Itersonilia (see p. 493)
   Phaffia, Xanthophyllomyces (see p. 665)

5. Trichosporonales (yeast forms; see Table 24.1)

21.2 | Ceratobasidiales

The most important members of this family belong to the anamorph genus *Rhizoctonia* which we shall consider in detail. Sneh *et al.* (1996) and Roberts (1999) have compiled important reference works on this group. The secondary hyphae of *Rhizoctonia* have conspicuous dolipore septa which are visible even with the light microscope (Tu *et al.*, 1977). Electron microscopy studies have revealed the parenthesomes to be perforated by several large pores (Müller *et al.*, 1998b). Clamp connections are not found in *Rhizoctonia* but may be present in other members of the order. The hyphae of *Rhizoctonia* are highly characteristic. Branches typically arise at a right angle to the leading hypha, with the branch point slightly constricted and a septum located a little way into the branch (Fig. 21.2a). Depending on the species, the compartments of vegetative hyphae are binucleate or multinucleate; uninucleate hyphae are uncommon. Teleomorphic states are rare and conidia are not produced, but sclerotia of the loose type (see Fig. 1.16a) are frequently seen. The form-genus *Rhizoctonia* has now been broken up into several taxa which correlate with different teleomorphs (Moore, 1987; Andersen & Stalpers, 1994). For instance, *Moniliopsis* has multinucleate hyphae and is referred to the teleomorph genus *Thanatephorus* whereas the hyphae of *Ceratorhiza* (teleomorph *Ceratobasidium*) are binucleate (Tu *et al.*, 1977; Vilgalys & Cubeta, 1994). Fungi resembling *Rhizoctonia* transgress the boundaries of orders, with some teleomorphs referable to the Tulasnellales. Indeed, a few *Rhizoctonia*-like fungi have even been assigned to the Ascomycota.

Whereas hyphae of *Rhizoctonia* are readily recognized as such, they offer few microscopic features for species identification, and a system based on **anastomosis groups**, i.e. the ability of a given isolate to undergo plasmogamy with hyphae of defined tester strains, has been developed (Sneh *et al.*, 1991). Such pairings may yield three different responses, with intermediate reactions also possible. (1) In genetically identical or closely related strains, anastomosis leads to perfect fusion. (2) In less closely related members of the same anastomosis group, anastomosis is followed by death of the fusion cell due to vegetative incompatibility. (3) No anastomosis occurs between members of different anastomosis groups. The best-studied taxon, *R. solani*, contains about a dozen anastomosis groups, some of which have been further divided into subgroups according to biochemical or other criteria. Although the individual anastomosis groups within *R. solani* and other taxa (e.g. *Ceratorhiza*) correlate with phylogenetic clusters obtained by DNA-based phylogeny (Kuninaga *et al.*, 1997; Gonzalez *et al.*, 2001) and also to a certain extent with the range of plant hosts affected (Sneh *et al.*, 1991), the question of the boundaries of biological species remains unanswerable at present.

The basidiocarps of Ceratobasidiales are thin, gelatinous and often resupinate. Their formation can sometimes be induced by covering an agar culture with soil (Warcup & Talbot, 1966). Hyphal tips due to develop into a probasidium are generally binucleate, and karyogamy is followed swiftly by meiosis. During the later stages of meiosis, four prominent sterigmata referred to as epibasidia are formed (Wells & Bandoni, 2001), but septa are not laid down, even in the mature basidium (Fig. 21.1a). The basidiospores are uninucleate and frequently germinate by repetition to produce uninucleate ballistoconidia. Eventually, hyphal germination occurs. Mating systems in the Ceratobasidiales are not well understood, and several species may be homothallic.

The basidium of the Tulasnellales differs in that the four epibasidia become separated from the metabasidium by septation after meiosis. Maturation of the four epibasidia may occur at different times (Fig. 21.1b).

21.2.1 *Rhizoctonia* in agriculture

The most important taxon is *R.* (*Moniliopsis*) *solani* (teleomorph *Thanatephorus cucumeris*), which causes a wide array of soil-borne necrotrophic diseases especially of herbaceous plants including all kinds of vegetables, rice, turfgrasses, and less frequently also woody tree hosts (Adam, 1988; Agrios, 2005). The most common disease is damping off of seedlings in which the infected hypocotyl region becomes water-soaked and no longer provides structural integrity (see p. 95), leading to pre- or post-emergence death of the seedling. In older plants, infection may not immediately cover the entire circumference of the plant stem, so that cankers and girdling of the stem may result. Roots are also frequently infected, as are aerial plant organs in contact with the soil or exposed to watersplash from the soil surface. The fungus can survive in the soil for several years as small sclerotia about 1–3 mm in diameter. These are often associated with the debris of host plants killed by the fungus. Sclerotia are occasionally seen as black scurf on the surface of potato tubers because they are not easily removed, even by assiduous washing. *Rhizoctonia* diseases are typically most severe at cool temperatures, presumably because the seedling stage of host plants is prolonged due to their slow growth. The infection process is similar to that of *Gaeumannomyces graminis*, i.e. hyphae branch and build up a cushion-like compound appressorium on the host plant surface, from which penetration is achieved.

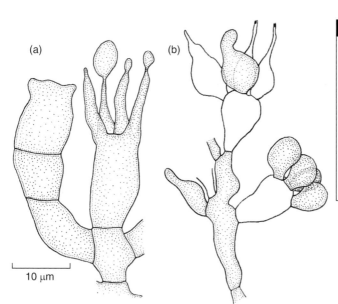

(a)

(b)

10 μm

Fig 21.1 (a) Basidium of *Thanatephorus cucumeris* (Ceratobasidiales). Note the four prominent epibasidia and the lack of septation within the basidium. (b) Basidium of *Gloeotulasnella cystidiophora* (Tulasnellales) showing various stages of maturity. The epibasidia are separated from the metabasidium by septation. Three epibasidia have already discharged their basidiospore, with one about to produce it. (a) redrawn from Warcup and Talbot (1962) with permission from Elsevier, (b) redrawn from Wells and Bandoni (2001) with kind permission of Springer Science and Business Media.

Simple appressoria may also be formed. Several lytic enzymes including cellulases, pectinases and cutinases are involved in colonizing and degrading plant tissue.

Another pathogenic species is *Rhizoctonia cerealis* (= *Ceratorhiza cerealis, C. ramicola*; teleomorph *Ceratobasidium cornigerum* = *C. cereale*), which causes sharp eyespot of cereals. As in *R. solani*, infection is favoured by cool conditions, and the disease is more common on winter cereals than spring-sown varieties. The fungus overwinters on stubble as mycelium and sclerotia, and infections give rise to spindle-shaped eyespot lesions on stems and leaves. These differ from true eyespot caused by *Tapesia yallundae* (p. 439) in being more sharply demarcated, with a reddish-brown margin enclosing an inner greyish region. Crop losses in cereals due to *R. cerealis* are not generally severe so that chemical control is not practised specifically against sharp eyespot (Parry, 1990). *Rhizoctonia cerealis* also causes yellow or brown patches in swards of turfgrass, as well as root and foliar infections in a range of other crop plants (Kataria & Hoffman, 1988).

Whereas chemical control of *Rhizoctonia* spp., like that of many other soil-borne pathogens, is difficult, biological control shows some promise. Several biocontrol organisms are effective under controlled laboratory and greenhouse conditions, including species of *Bacillus, Serratia* and *Streptomyces*, as well as 2,4-diacetylphloroglucinol-producing *Pseudomonas* spp. (see p. 385). The high efficacy of *Trichoderma* spp. against *Rhizoctonia* has been partially correlated with their secretion of cell wall-degrading enzymes, notably chitinases and β-(1,3)-glucanases (Innocenti *et al.*, 2003; Markovich & Kononova, 2003). As in the case of the cereal take-all pathogen *Gaeumannomyces graminis*, certain soils can acquire the capacity to suppress *Rhizoctonia* after several consecutive cultivation cycles with the same crop (Henis *et al.*, 1979; Mazzola, 2002), and this has been attributed to the build-up of biocontrol organisms, especially *Trichoderma* and *Pseudomonas* spp.

21.2.2 *Rhizoctonia* and orchid mycorrhiza

Stimulating accounts of this topic have been written by Arditti (1992), Smith and Read (1997),

Peterson *et al.* (1998) and Rasmussen (2002). All members of the plant family Orchidaceae (about 17 500 species) appear to be associated with mycorrhizal fungi at all stages of their life cycle in nature. In contrast to other types of mycorrhiza, there is a net flow of sugars from the fungal partner to the plant at least during the establishment of the orchid seedling. All colourless (non-photosynthetic) orchids continue to rely on this external supply throughout their lives, and even green orchids do not seem to share their photosynthetic products with the fungus (Alexander & Hadley, 1985). In orchid mycorrhiza, therefore, the plant parasitizes the fungus, and it is a curious fact that many of the fungi thus exploited are themselves serious plant pathogens, especially *Rhizoctonia* spp. (Roberts, 1999). Indeed, the very *Rhizoctonia* strains isolated as pathogens of other plants (e.g. *R. solani, R. cerealis*) can support the germination of orchid seeds (Figs. 21.2b–f). These as well as other species (e.g. *R. goodyerae-repentis, R. repens*) can also be isolated from mature orchid roots. Orchid mycorrhizal symbiosis therefore seems to be less specific than other forms of mycorrhiza (Masuhara *et al.*, 1993).

Orchid seeds are tiny and lack differentiated embryos or food reserves. In the absence of soluble external carbohydrates, they show only limited germination to form an intermediate stage called a protocorm. This may emit a few epidermal hairs before growth stalls (Fig. 21.2b). Further development of the protocorm (Fig. 21.2c) occurs only if a suitable soluble carbon source is added, or if a mycorrhizal fungus such as *Rhizoctonia* is allowed to grow from a food base (e.g. starch or cellulose) to the protocorm. Growth ensues even if the fungus is made to cross a barrier between the food base and the protocorms, thereby demonstrating net carbon translocation (Smith, 1966). This experiment is easily set up in the laboratory (Fig. 21.2d; Weber & Webster, 2001b). The main transport compound seems to be trehalose, and this may be hydrolysed to glucose and converted to sucrose by the plant (Smith, 1967; Smith & Read, 1997).

Infection of the orchid protocorm is initiated through the epidermal hairs (Fig. 21.2e) or through the suspensor tissue at the base of

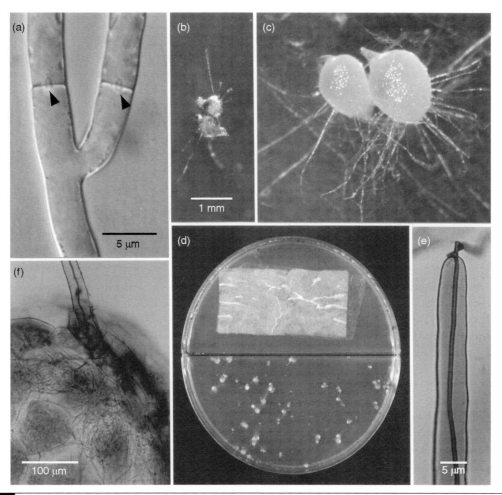

Fig 21.2 *Rhizoctonia cerealis* and its mycorrhiza with the heath spotted orchid (*Dactylorhiza maculata* ssp. *ericetorum*) in the laboratory. (a) Vegetative hypha showing typical branching and dolipore septa (arrowheads). (b) Seeds of *D. maculata* on agar after ten weeks without *Rhizoctonia*. Protocorms with a few epidermal hairs have formed. (c) Seeds after ten weeks but with *R. cerealis* spreading from a food base. The protocorms have grown and are differentiating shoot tips (arrows). Same scale as (b). (d) The split-plate experiment. *Rhizoctonia cerealis* has been inoculated onto a food base (tissue paper, top half) which is separated from the orchid seeds by a partition. The fungus has overgrown this barrier, and the orchid protocorms are using the translocated sugars derived from the degraded cellulose; 13 weeks after inoculation. (e) Penetration of an epidermal hair by *R. cerealis*. (f) Penetration of *R. cerealis* hyphae through an epidermal hair into the cortex of a protocorm where pelotons have formed. (b) and (d–f) reprinted from Weber and Webster (2001b), with permission from Elsevier.

the seedling. Penetration of the wall of a cell in the protocorm cortex invaginates the plasmalemma and results in the formation of a **peloton**, i.e. a dense mass of coiled hyphae (Fig. 21.2f). Initially, each hypha is ensheathed by the host plasmalemma, which is called the **perifungal membrane** and is functionally modified from the plasmalemma of uninfected regions (Peterson & Currah, 1990; Peterson *et al.*, 1996). An interfacial matrix of unknown composition is located between the perifungal membrane and the fungal cell wall. Within 24 h of formation, a peloton may begin to be degraded and is ultimately left behind as an amorphous clump of lysed hyphae surrounded by one continuous perifungal membrane (Hadley & Williamson, 1971; Peterson & Currah, 1990). Any one orchid cell can become repeatedly re-infected

(Uetake *et al.*, 1992), and the same protocorm, mature root or even individual cell can be colonized simultaneously by different fungi.

The unstable nature of the orchid mycorrhiza is indicated by the quick and repeated cycle of peloton formation and degradation, and the several different outcomes of the orchid–fungus interaction observed under laboratory conditions (Fig. 21.2d). A balanced mycorrhizal symbiosis will develop only in a proportion of protocorms, whereas other seeds of the same orchid species may be parasitized and killed by the fungus, or simply resist infection and stall in their development (Hadley, 1970; Smreciu & Currah, 1989; Beyrle *et al.*, 1995). There is also vevidence of a succession of mycorrhizal fungi during the development of an orchid in nature, and the mycorrhizal fungi isolated from adult plants may not support protocorm growth and vice versa (Xu & Mu, 1990; Zelmer *et al.*, 1996).

Mature orchids may be associated with *Rhizoctonia* spp. and/or a range of other Basidiomycota, including saprotrophic (e.g. *Mycena*), necrotrophic (e.g. *Armillaria*; see p. 546) or ectomycorrhizal species related to *Russula* and *Thelephora* (Rasmussen, 2002). Ectomycorrhizal fungi transport carbohydrates from their tree host to the orchid, thus allowing green orchids to grow even in densely shaded woodland conditions (Taylor & Bruns, 1997; McKendrick *et al.*, 2000; Bidartondo *et al.*, 2004). Interestingly, these fungi form typical ectomycorrhiza with their tree hosts but pelotons in infected orchid roots (Zelmer & Currah, 1995). The orchid therefore calls the shots in its symbiosis with basidiomycetes, and peloton formation and degradation is traditionally interpreted as a balanced defence reaction of the orchid against invasion attempts by the fungus.

21.3 | Dacrymycetales

This order is characterized by forked (**furcate**) basidia (Figs. 21.4 and 21.5), which are found in all species except *Dacrymyces unisporus*. The fruit bodies are coloured yellow or orange due to the presence of a wide range of carotenoids (for references, see Gill & Steglich, 1987). Fruit bodies are gelatinous and show a striking diversity of forms as exemplified by the cushion-like basidiocarps of *Dacrymyces stillatus* (Fig. 21.3) and the clavarioid ones of *Calocera viscosa* (Plate 11g).

Not much is known about the life cycle of the Dacrymycetales, but it is presumed that the usual basidiomycete pattern of alternating mono- and dikaryotic stages operates. There are no clamp connections. The dolipore-type septa

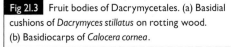

Fig 21.3 Fruit bodies of Dacrymycetales. (a) Basidial cushions of *Dacrymyces stillatus* on rotting wood. (b) Basidiocarps of *Calocera cornea*.

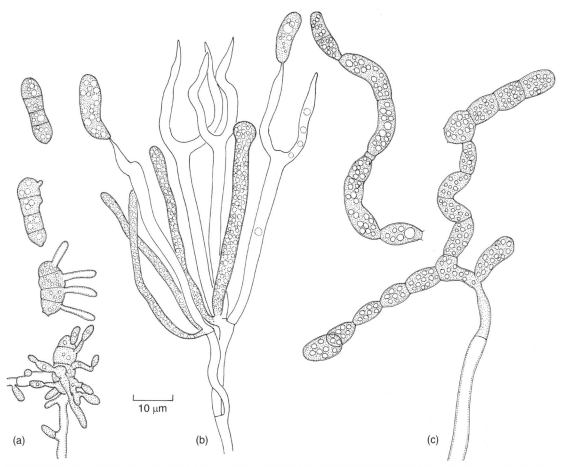

Fig 21.4 *Dacrymyces stillatus*. (a) Basidiospores showing germination by germ tubes or formation of conidia (bottom). (b) Basidia. Note that the attached basidiospores are unicellular. They become three-septate on germination. (c) Arthrospores from a conidial pustule.

are surrounded by parenthesomes without perforations (Wells, 1994). The probasidium arises from a dikaryotic hypha and is initially club-shaped. At this stage karyogamy occurs and is immediately followed by meiosis. Meanwhile the two epibasidia develop. Each of the two basidiospores seems to receive one nucleus, and the remaining two nuclei degenerate in the epibasidia (see Wells & Bandoni, 2001). Before the basidiospore germinates, it lays down one or more septa. Each spore segment can produce a haploid monokaryotic hypha or may give rise to conidia which in turn germinate by means of monokaryotic germ tubes (Ingold, 1983b). It is unclear in many species how and where dikaryotization occurs, but it is probably by fusion of monokaryotic hyphae. Mating systems, where known, are bifactorial (tetrapolar), i.e. with two mating type loci *A* and *B*.

Members of the Dacrymycetales are saprotrophic on wood and cause brown-rots, although some lignin degradation has also been observed (Seifert, 1983; Worrall *et al.*, 1997). The fruit bodies are common on decaying wood, including wood built into outdoor structures such as park benches or fences. There are about 70 species, and the order is monophyletic (Oberwinkler, 1993). Reid (1974) has given keys and descriptions of the common British and European species. The seminal features of the order have been summarized by Wells (1994) and Wells and Bandoni (2001).

21.3.1 *Dacrymyces*

Orange gelatinous cushions about 1–5 mm in diameter, so common on damp rotting wood, are the fructifications of *D. stillatus* (Fig. 21.3a). Close inspection with a hand lens reveals that the fruit bodies are of two kinds: soft, bright orange, hemispherical cushions and firmer, pale yellow, flatter structures. The bright orange cushions are conidial pustules which consist of hyphae whose tips are branched and fragment into numerous dikaryotic arthroconidia (Fig. 21.4c). The cells are packed with oil globules containing carotenoids. Such conidia are readily dispersed by rainsplash and are obviously similar in function to the splash-dispersed conidia of *Nectria cinnabarina* (see Plate 5d). The yellow flatter structures are basidial cushions which are attached centrally to the woody substratum. The surface layer is composed of clusters of forked basidia (Fig. 21.4b) which arise from dikaryotic hyphae. Each basidium forms two haploid basidiospores. After discharge, a basidiospore undergoes nuclear division and septation to give four cells. Depending on environmental conditions, each cell germinates by

means of a monokaryotic haploid germ tube or by a short conidiophore (denticle). Conidia may also arise on older hyphae. They germinate to give monokaryotic hyphae. Dikaryotization occurs when two compatible monokaryotic hyphae fuse. Mossebo and Amougou (2001) have shown that the parenthesome and dolipore complex dissolve in order to facilitate passage of nuclei through the septum in the course of dikaryotization. *Dacrymyces stillatus* is heterothallic with a bifactorial (tetrapolar) mating system. Cells of the secondary mycelium are usually but not unfailingly dikaryotic (Mossebo, 1998).

21.3.2 *Calocera*

At first sight the ubiquitous cylindrical orange outgrowths of *C. viscosa* from coniferous logs (Plate 11g), or the smaller *C. cornea* from hardwood logs (Fig. 21.3b), could be mistaken for species of *Clavaria*. However, the gelatinous texture and the characteristically forked basidia (Fig. 21.5) place them in the Dacrymycetales, and this placement has been confirmed by molecular phylogenetic studies (Weiss & Oberwinkler, 2001). Ingold (1983b) has carefully observed the fate

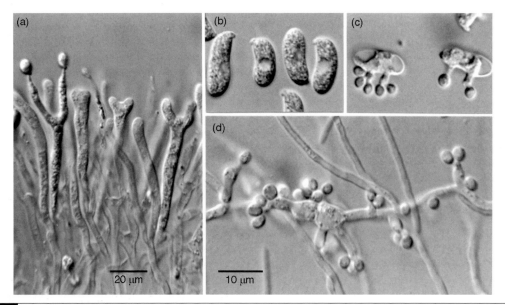

Fig 21.5 *Calocera viscosa*. (a) T.S. through hymenium with furcate basidia at different stages of development. (b) Freshly discharged basidiospores which are aseptate. The clear area in each spore shows the displacement of cytoplasmic contents by the single nucleus. (c) Two 24-hour-old basidiospores on tap-water agar. Each spore has produced two conidiophores bearing microconidia on denticles. (d) Direct germination of a basidiospore has given rise to monokaryotic hyphae which are forming microconidia. The septum dividing the spore into two is clearly visible. (b–d) to same scale.

of basidiospores in *C. viscosa*. Freshly discharged basidiospores are aseptate (Fig. 21.5b) but they soon develop one septum. Further development is by direct germination or formation of globose microconidia from basidiospore segments and from haploid monokaryotic hyphae (Figs. 21.5c,d), as described above for *D. stillatus*.

21.4 | Auriculariales

The order Auriculariales has been subject to numerous taxonomic rearrangements. As currently understood, its members produce both mono- and dikaryotic mycelia with dolipores and parenthesomes lacking perforations (see Fig. 18.10b). Basidia are septate (Wells, 1994; Wells & Bandoni, 2001). The Auriculariales are distinguishable from the Tremellales (Section 21.5) which have yeast-like monokaryotic stages, and from the Ceratobasidiales and Dacrymycetales which have aseptate basidia. Although taxonomic adjustments continue to be made, there is now little doubt that the order Auriculariales should contain both *Auricularia* with its transversely septate basidia, and *Exidia* and *Pseudohydnum*, which have basidia with longitudinal septa, the so-called tremelloid basidia (Weiss & Oberwinkler, 2001).

There are great variations in fruit body size and shape. Clamp connections may be present or absent, depending on species. Most species have a bifactorial mating system with multiple alleles at both loci (Wells, 1987, 1994; Wong & Wells, 1987; Wong, 1993). Depending on environmental conditions, basidiospores are typically able to germinate in several different ways, e.g. by repetition (ballistospore formation), as hyphae, or by forming sickle-shaped (**lunate**) microconidia. Members of the Auriculariales are saprotrophic on wood, causing intensive white-rots (Worrall *et al.*, 1997).

21.4.1 *Auricularia*

The Jew's ear fungus *A. auricula-judae* forms rubbery, ear-shaped fruit-bodies on branches of elder (*Sambucus*) (Plate 11h) and is a weak pathogen, growing on the wood and pith of living branches and on dead wood. A wide range of other hosts has been reported, on which *A. auricula-judae* causes a rapid white-rot similar to that produced by members of the polyporoid clade (see Plate 10a; Worrall *et al.*, 1997).

A section through the flesh of a fruit body shows a hairy upper surface, a central gelatinous layer containing narrow clamped hyphae, and a broad hymenium on the lower side (Fig. 21.6a). Details of basidiocarp anatomy are useful in classification (Lowy, 1952). The fruit body can dry to a hard brittle mass, but on wetting it quickly absorbs moisture and discharges spores within a few hours. The basidia are cylindrical and become divided into four cells by three transverse septa (Fig. 21.6b). Each cell of the basidium develops a long cylindrical epibasidium which extends to the surface of the hymenium and terminates in a conical sterigma bearing a basidiospore which is monokaryotic. At 20 µm or more in length, the *Auricularia* basidiospore is one of the largest objects propelled by the surface tension catapult mechanism (see Pringle *et al.*, 2005), and ballistospore discharge is easily observed with thin slices of basidiocarp material placed sideways on water agar (see Webster & Hard, 1998b). *Auricularia auricula-judae* is heterothallic with a bifactorial (tetrapolar) mating system, and there are indications of multiple alleles (see Wong, 1993).

Germination of the basidiospore can proceed in several different ways (Fig. 21.6c), and such a variability is common in the Auriculariales (Ingold, 1982a, 1984b). Washings from the hymenial surface of fruit bodies contain basidiospores undergoing repetitious germination by means of a sterigma which produces another ballistospore. Ingold (1982a) interpreted this as a second chance for the spore to get away from the fruit body. Basidiospores alighting on a nutrient-poor surface such as tap water agar lay down three transverse septa, and each of the resulting four cells may emit one or more extensions (denticles) which produce a cluster of lunate microconidia. Alternatively or additionally, germination may occur directly by means of a germ tube, and this mode of germination is found especially on slightly richer media such as cornmeal agar. Septa are laid down, the first one bulging

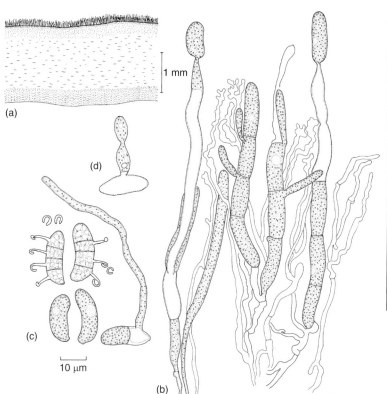

Fig 21.6 *Auricularia auricula-judae.*
(a) Section of fruit body. The hymenium is on the lower side.
(b) Squash preparation of the hymenium showing basidia. Note the transverse segmentation and the long epibasidia. The basidia are associated with branched hyphae.
(c) Basidiospores. Two are ungerminated; one has developed a septum and is germinating directly by means of a germ tube; and two basidiospores have become three-septate and are producing lunate conidia from short conidiophores (denticles).
(d) A basidiospore which had fallen back onto the hymenial surface and is germinating repetitiously to form a ballistoconidium. (b–d) to same scale.

backwards from the protoplast-containing germ tube towards the empty basidiospore. Such septa are interpreted as **retraction septa** (Ingold, 1982a). A richly branched monokaryotic mycelium of very fine hyphae develops, and after a while these hyphae form lateral or terminal denticles which produce clusters of lunate microconidia.

The lunate conidia are a feature found in many species of Auriculariales, but their significance in the life cycle is uncertain. Like basidiospores and the repetitious ballistoconidia produced from them, they are capable of germination to form monokaryotic hyphae, but they might also function directly as spermatia. The putative life cycle of *A. auricula-judae* is shown in Fig. 21.7.

Another common species is *A. mesenterica*, which forms thicker, hairy, fan-shaped fruit bodies on old stumps and logs of elm (*Ulmus*) and other trees. It, too, causes active wood decay and may occasionally be weakly pathogenic.

Auricularia as a cultivated mushroom

Although devoid of any distinctive taste, the fruit bodies of *Auricularia* are highly nutritious and possess a chewy, rubbery texture which renders them attractive ingredients for Far Eastern soups and stir fries. The main species cultivated for food is *A. polytricha* ('Mu-Erh'). The history of cultivation dates back to AD 600 in China (Cheng & Tu, 1978; Chang & Miles, 2004), making *Auricularia* the first cultivated mushroom for which we have historical records. Some 465 000 t of fresh fruit bodies are currently produced *per annum* (Pegler, 2001). Very conveniently, the fruit bodies can be stored dry for several months, and rehydrated when needed. Cultivation is traditionally performed by inoculating logs of suitable broadleaved trees with mycelial spawn. Infected logs can produce good crops for several years. The fungus is now also often cultivated in plastic bags filled with a mixture of sawdust and rice bran, allowing the fruit bodies to emerge through holes in the plastic.

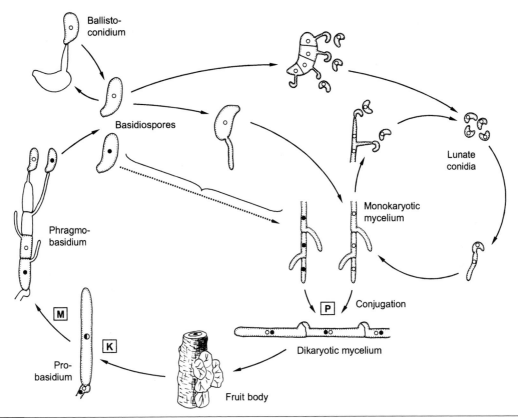

Ballisto-
conidium

Basidiospores

Lunate
conidia

Monokaryotic
mycelium

Phragmo-
basidium

M

K

Conjugation

P

Pro-
basidium

Dikaryotic mycelium

Fruit body

Fig 21.7 Life cycle of *Auricularia auricula-judae*. Depending on the substrate, a basidiospore has various options by which to germinate; these are equivalent for both mating types, but for clarity we show them here for only one of them (white nuclei). Basidiospores falling onto the fruit body hymenium may germinate by repetition to form another ballistospore. Depending on the nutrient status, basidiospores may germinate by formation of a monokaryotic mycelium or by producing lunate microconidia. The latter may also be produced by monokaryotic hyphae. Conjugation leads to the establishment of a dikaryotic mycelium which may form basidiocarps. Key events in the life cycle are plasmogamy (P), karyogamy (K) and meiosis (M). Haploid nuclei are drawn as empty or filled circles; the diploid nucleus is drawn larger and half-filled.

21.4.2 Other members of the Auriculariales

Exidia

Exidia glandulosa, sometimes called 'witches' butter', forms black rubbery fructifications on decaying branches of various woody hosts, especially lime (*Tilia*) and oak (*Quercus*) (Fig. 21.9a). The hymenium is borne on the lower side of the fruit body and in this species is studded with small black warty outgrowths. The basidia are formed deep within the hymenium and produce long epibasidia. They are divided by longitudinal instead of transverse septa (Fig. 21.8). Basidiospores of *E. glandulosa* show germination patterns identical to those described above for *A. auricula-judae*, and Ingold (1982b) found

that the only clear microscopic difference between these two species is the arrangement of septa in the basidium. Phylogenetic studies have confirmed the close relationship between *Auricularia* and *Exidia* (Weiss & Oberwinkler, 2001).

Pseudohydnum gelatinosum

This species grows on dead stumps and branches of coniferous trees. The fruit body is jelly-like in consistency and has a short eccentric stalk. The hymenium is on the lower side of the pileus and is arranged into numerous conical teeth (Fig. 21.9b) resembling those of *Hydnum*. The basidia are similar to those of *Exidia* in being longitudinally septate. As in *Exidia* and *Auricularia*, basidiospores failing to escape from

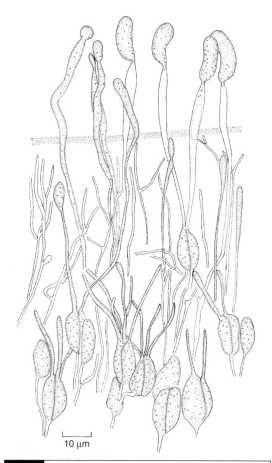

10 μm

Fig 21.8 *Exidia glandulosa.* Section of hymenium showing longitudinally divided basidia with long epibasidia extending to the surface.

the hymenium may germinate by repetition or, in other situations, by means of a germ tube, but lunate or other microconidia have not been observed (Ingold, 1985).

21.5 | Tremellales

Formerly understood as a broad taxon which included genera such as *Exidia* and *Pseudohydnum* (see p. 603), the order Tremellales is now restricted to fungi which possess a yeast-like haploid state, basidia divided by longitudinal septa (**tremelloid** basidia), and dikaryotic hyphae with a dolipore septum and a parenthesome which is sacculate, i.e. invaginated towards the septal pore (Fig. 21.10a; Berbee & Wells, 1988). Dikaryotic hyphae usually have clamp connections. The monokaryotic yeast state resembles heterobasidiomycete yeasts such as *Filobasidiella neoformans*, *Phaffia rhodozyma* and species of *Bullera* and *Cryptococcus*, all of which belong to the Tremellales and related orders within the Tremellomycetidae (Fell *et al.*, 2001). Heterobasidiomycetes growing predominantly in the yeast state are discussed in Chapter 24 (p. 660).

In the life cycle of the filamentous Tremellales considered here, the dikaryotic condition is the dominant phase and is re-established by

(a)　(b)

2 cm　　　1 cm

Fig 21.9 Fruit bodies of Auriculariales. (a) *Exidia glandulosa.* Fruit bodies on lime (*Tilia*). The hymenial surface bears black warts and is on one face of the fruit body. (b) *Pseudohydnum gelatinosum.* Fruit bodies seen from above (right) and below (left). The hymenium covers the surface of the spines.

conjugation of compatible yeast cells. The mating system has been described as 'modified tetrapolar' (Bandoni, 1963) because there are only two alleles at the *A* locus but multiple alleles at *B*. This system is uncommon in the Basidiomycota, which usually have multiple alleles at both loci, aiding outbreeding (see p. 507). The modified tetrapolar system of *Tremella* is, however, also found in *Ustilago maydis*, and the designation of loci is equivalent (p. 643). Thus, the *A* locus controls conjugation and the *B* locus growth of the resulting dikaryon. Both *A* alleles of *Tremella* encode peptide-type hormones (Sakagami *et al.*, 1981; Ishibashi *et al.*, 1984) and receptors for the hormone of the opposite mating type. In *T. mesenterica*, these hormones are linear peptides called tremerogen A-10 (12 amino acids) and tremerogen α-13 with 13 amino acids. Both peptides are derivatized with a farnesyl unit. Mating occurs by formation of conjugation tubes which requires the presence of the hormone of the opposite mating type; each hormone is produced constitutively, irrespective of the presence or absence of a compatible mating partner (Bandoni, 1965). Yeast cells with opposite alleles at *A* but like *B* alleles will conjugate but fail to initiate dikaryotic hyphal growth.

The fruit bodies of Tremellales are usually formed on wood, often in association with those of other fungi (Asco- and Basidiomycota) or with lichen thalli, which may be parasitized (Diederich, 1996; Chen, 1998). Parasitism is by intimate hyphal contact via haustorial branches (Figs. 21.10b,c; see below). The fruit bodies of Tremellales are highly variable in size, ranging from a limited hymenium on the mycelium of putative hosts to large structures (several centimetres) surrounding host basidiocarps or growing near them. Although often considered as saprotrophs, no special capacity to degrade wood seems to have been recorded.

Accounts of the Tremellales have been written by Bandoni (1987), Bandoni and Boekhout (1998) and Chen (1998).

21.5.1 *Tremella*

This is a large genus of some 80 species. Detailed descriptions of representatives of all species

groups have been given by Chen (1998). One of the commonest and most thoroughly examined species is *T. mesenterica*, whose yellow or orange gelatinous fruit bodies are readily seen on various woody hosts such as oak, willow, gorse and beech (Plate 11i). Variations in the intensity of fruit body coloration could be due to a stimulation of carotenoid synthesis by high light intensity because fruit bodies exposed to sunlight are often more deeply coloured than those in the shade (Wong *et al.*, 1985). The fruit bodies of *T. mesenterica* are usually associated with those of *Peniophora* in the field, and Zugmaier *et al.* (1994) have shown that hyphae of *Peniophora* spp. are parasitized in vivo and in vitro. Several *Tremella* spp. are more obviously mycoparasitic than *T. mesenterica*. For instance, *T. globospora* produces its fruit bodies within the perithecia of *Diaporthe*, and *T. encephala* overgrows the fructifications of its host, *Stereum*, which remain as a firm core in the *Tremella* basidiocarp. There are also several other genera within the Tremellales which parasitize other fungi (Bauer, 2004; Figs. 21.10b,c).

Parasitism is mediated by modified hyphae which Olive (1947) called '**haustorial branches**'. They consist of a swollen binucleate hyphal segment, delimited at its base by a clamp connection which puts forward one or several long thin filaments (Fig. 21.11a). The association of the swollen segment with the term 'haustorium' is unfortunate because the haustorial branch is formed outside the host cell. Where 'haustorial filaments' of *Tremella* or related fungi contact the hypha of a suitable host, the host wall is dissolved and a micropore is formed which establishes direct cytoplasmic contact between the filament and the host, apparently by fusion of the two plasma membranes (Figs. 21.10b,c).

The life cycle of *T. mesenterica* is complex and not yet fully understood (Fig. 21.12). Haploid basidiospores are discharged from long epibasidia by the surface tension catapult mechanism. Basidiospores failing to escape from the hymenium can germinate by repetition to form ballistoconidia (Ingold, 1982b). When landing on a suitable substrate, the basidiospore germinates by forming several buds which remain attached

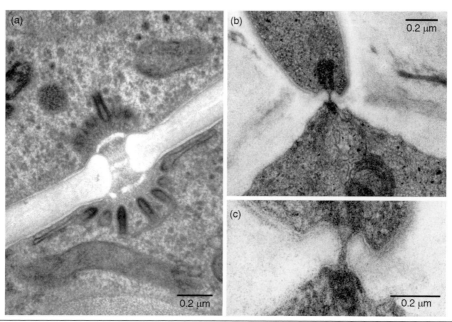

Fig 21.10 Ultrastructure of Tremellales. (a) Cupulate parenthesome and dolipore septum of *T. globospora*. (b,c) Tip of the haustorial filament of *Trimorphomyces papilionaceus* (top) which has made contact with a hypha of its host, *Arthrinium sphaerospermum* (bottom). Membrane continuity has been established in the micropore region (c). Original prints kindly provided by M. L. Berbee (a) and R. Bauer (b,c). a reprinted from Berbee and Wells (1988), with permission from *Mycologia*. ©The Mycological Society of America.

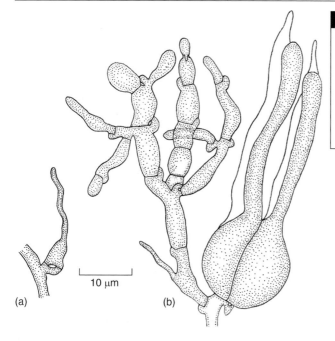

Fig 21.11 *Tremella mesenterica*. (a) Haustorial branch. The swollen dikaryotic branch arises from a clamp connection and emits filaments which may contact host hyphae. (b) Dikaryotic clamped hypha from the hymenium of a fruit body. One branch has produced a basidium whereas the other has formed dikaryotic conidia. Redrawn from Wells and Bandoni (2001) with kind permission of Springer Science and Business Media.

to the basidiospore. These buds act as conidiogenous cells by producing numerous minute blastoconidia, which in turn germinate by swelling and budding to give rise to the haploid yeast state. Fusion between compatible yeast cells re-establishes the dikaryotic mycelial phase. In addition to producing basidia, hyphae in the fruit body may also form a dikaryotic conidial

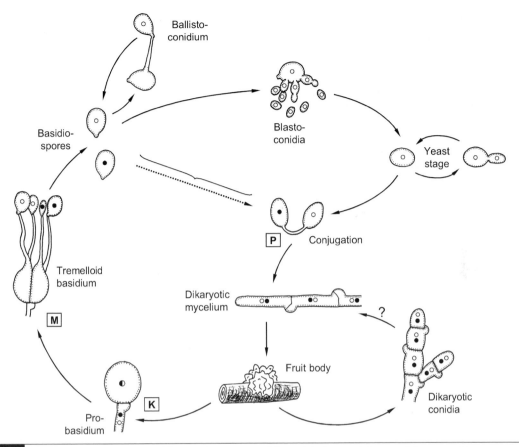

Fig 21.12 Life cycle of *Tremella mesenterica*. The monokaryotic part of the cycle is shown only for one of the two mating types (white nuclei). Basidiospores failing to leave the fruit body may gain a second chance by repetitious germination to form a ballistoconidium. On other substrates, basidiospores germinate by way of blastoconidia which give rise to yeast cells. Conjugation of two compatible yeast cells gives rise to the dikaryotic mycelial stage which forms fruit bodies producing dikaryotic conidia, haploid yeast cells by de-dikaryotization (not shown), and haploid basidiospores by meiosis. Key events in the life cycle are plasmogamy (P), karyogamy (K) and meiosis (M). Haploid nuclei are drawn as empty or filled circles; the diploid nucleus is drawn larger and half-filled.

state (Fig. 21.11b) with an uncertain role in the life cycle. Additionally, de-dikaryotization to give yeast cells has been observed in the fruit bodies of *T. mesenterica*.

Another common species is *T. foliacea* (sometimes synonymized with *T. frondosa*) with its flesh-coloured to pale brown, lobed or contorted fruit bodies on oak and beech stumps. In this species, basidiospores germinate on suitable substrates by giving rise directly to budding yeast cells (Fig. 21.13). This has also been observed in *T. encephala* (Ingold, 1985). There are no dikaryotic conidia in these two species. Most *Tremella* spp. are heterothallic with a modified tetrapolar mating system, but in *T. fuciformis* both homothallic

and heterothallic strains have been observed (Fox & Wong, 1990).

Cultivation of *Tremella*

The 'silver ear' fungus, *T. fuciformis*, has been cultivated in China on wood and sawdust for about 200 years (Chang & Miles, 2004). Although almost always associated with *Hypoxylon* spp. in nature (Chen, 1998), mycoparasitism does not seem to be obligate because cultivation is possible in monoculture. However, yields are greatly stimulated in the presence of a 'friend of the mycelium', i.e. the substrate is co-inoculated with *T. fuciformis* and *Hypoxylon archeri* or another suitable host species (Chang & Miles, 2004).

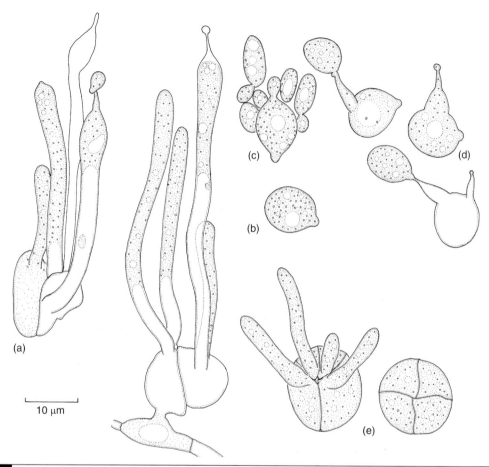

10 μm

Fig 21.13 *Tremella frondosa*. (a) Basidia showing epibasidia with various stages of spore development. (b) Freshly discharged basidiospore. (c) Basidiospore germinating on malt extract agar by budding to form yeast-like cells which also undergo budding. (d) Basidiospores germinating in water by repetition, i.e. by producing ballistospores. (e) Immature basidia seen from above, showing the division of the basidium into four cells by longitudinal septa.

The fruit bodies of *T. fuciformis* are consumed in a stewed form, especially as a dessert, and this species has traditionally also been viewed as a medicinal mushroom in the Far East. Life-prolonging, vitalizing and anti-cancer properties have been ascribed to *T. fuciformis*, and these effects seem to be due mainly to a stimulation of the immune system (Wasser, 2002). The biochemical basis is thought to be due to the production of exopolysaccharides by this and other *Tremella* spp., including *T. mesenterica* (Reshetnikov *et al.*, 2000). However, it is as yet unclear how such polysaccharides are assimilated by the human digestive system, and how they interact with human cells. Many Homobasidiomycetes produce exopolysaccharides that are said to possess similar medicinal properties, and these molecules are mainly β-(1,3)-glucans substituted with various sugar moieties. In contrast, in *Tremella* the biologically active molecules have a β-(1,3)-mannan backbone substituted with xylose and glucuronic acid (de Baets & Vandamme, 2001; Vinogradov *et al.*, 2004). A multitude of such mannans differing mainly in their substitution pattern is produced by the fruit bodies on solid substrata and by yeast cells in liquid culture. Yeast cells are encapsulated by these polysaccharides, as revealed by staining with Indian ink (see Fig. 24.1b). Structurally similar xyloglucuronomannans make up the capsule of the human pathogen *Cryptococcus neoformans* (see p. 661).

Urediniomycetes: Uredinales (rust fungi)

22.1 | Urediniomycetes

Following extensive re-arrangements, the class Urediniomycetes (about 8000 species) is now considered to be monophyletic, although the naming of orders and families is still proving difficult (Swann & Taylor, 1995; Kirk *et al.*, 2001; Swann *et al.*, 2001). The order Uredinales (rust fungi) is by far the largest (about 7000 species) and the most important. The order Microbotryales, although taxonomically part of the Urediniomycetes, is a group of fungi causing smut diseases and will be discussed in Chapter 23. Many Urediniomycetes belonging to several orders occur predominantly in the yeast state. An important group, the Sporidiales, contains the red yeasts *Sporidiobolus* and *Rhodosporidium* (anamorphs *Sporobolomyces* and *Rhodotorula*, respectively), and this order is considered in more detail on pp. 666–670.

General information on the groups included in the Urediniomycetes has been given by Swann *et al.* (2001). They have defined Urediniomycetes as fungi in which the processes of karyogamy and meiosis occur in distinct parts of the basidium, i.e. the probasidium and metabasidium, respectively. The metabasidium is typically transversely septate, with basidiospores produced laterally (see Fig. 22.2d). Another useful character is the structure of septa viewed by transmission electron microscopy. Urediniomycete septa are simple with a single pore which may be open or plugged, but they typically lack the dolipore arrangement found in other basidiomycetes (see Fig. 18.9). Clamp connections are absent.

22.2 | Uredinales: the rust fungi

Rust fungi (Uredinales) are a fascinating group of organisms. The life cycle of a typical rust species is among the most complex found anywhere in nature, consisting of five different spore stages on two plant hosts which are taxonomically entirely unrelated to each other. These pathogens infect most groups of vascular plants, including Pteridophytes (ferns), Gymnosperms, and Angiosperms (both monocots and dicots). Numerous fundamental questions about rust fungi remain to be answered, e.g. how a biotrophic organism manages to infect and parasitize two unrelated hosts using different mechanisms on either; how the five spore stages with their numerous different dispersal mechanisms could have evolved; how easily one or more of them can become aborted in derived (reduced) life cycles; how rust fungi survive in situations where one of their two hosts is unavailable; and how quickly new rust species or races spread to new habitats and then come to an equilibrium with their host plants. A protocol to generate stable transformants of rust fungi would greatly facilitate experimental work on them, but unfortunately this is not yet available.

The species concept in rust fungi is also challenging. Morphological species are readily recognized, but these can show considerable genetic adaptability by fragmenting into several forms which infect non-overlapping spectra of host species and thereby become reproductively isolated. Because of the identity of the host family as an additional taxonomic feature, species of rust fungi are relatively easily grouped into genera which are often monophyletic. At the higher taxonomic level, two large groups can be resolved weakly by phylogenetic analyses (Maier *et al.*, 2003; Wingfield *et al.*, 2004). They coincide with the Pucciniaceae and Melampsoraceae of earlier concepts (Dietel, 1928) in which these two families were distinguished by their teliospore or probasidium being stalked (Pucciniaceae) or unstalked (Melampsoraceae). Another point of distinction is that a particular spore-producing structure, the aecial stage (see below), is produced on angiosperm hosts by Pucciniaceae but on gymnosperms by Melampsoraceae (Wingfield *et al.*, 2004). Dietel's two broad groups are usually split up into about 13–14 smaller families (Kirk *et al.*, 2001; Cummins & Hiratsuka, 2003), but since the boundary lines are still being re-drawn from time to time, we prefer to adhere to the original concept for the purposes of this book.

The popular name 'rust fungi' refers to the reddish-brown colour of some of the spores which are produced in dense pustules on crop plants, giving them a 'rusted' appearance. This is especially true of the rusts on cereals, which have probably caused crop losses since the beginning of agriculture. Archaeological excavations have uncovered rusted cereal remains dating back to the Bronze Age (Kislev, 1982), and it is well known that the ancient Romans held a special festival, the Robigalia, to appease their rust gods. This took place on 25 April, i.e. at a time when the crop was particularly vulnerable to attack (Large, 1940). The Robigalia may be the origin of Rogation Sunday, a day of blessing of the crops which is still observed by some Christian churches every year in late April (Schuman, 1991).

Rusts can also cause serious economic damage on non-cereal crops, and examples are given in later sections of this chapter. Because of the immense economic importance of rust fungi, an enormous body of literature has been written,

and the wheat–*Puccinia graminis* system is probably the most thoroughly examined of all host–pathogen interactions involving fungi. None the less, no substantial integrated treatment of the rust fungi has appeared in the past two decades, the two-volume set on cereal rusts (Bushnell & Roelfs, 1984; Roelfs & Bushnell, 1985) having been the last major effort. However, good keys and species descriptions are available, e.g. in Grove (1913), Gäumann (1959), Wilson and Henderson (1966) and Cummins and Hiratsuka (2003).

22.2.1 The basic life cycle of rusts

The classical example of a rust fungus, *Puccinia graminis*, is the cause of black stem rust on wheat and other cereals (Fig. 22.1), which is described more fully in Section 22.3. The life cycle of rusts is homologous with that of the Homobasidiomycetes (Fig. 18.4) in being divided into stages of primary (homo- and mono-karyotic) and secondary (hetero- and di-karyotic) mycelium. The heterokaryotic phase is the main period in the life cycle of rusts and it is the only one in which some rusts can survive indefinitely under suitable conditions. The host plant species on which this stage is produced is therefore termed the **principal host**, with that bearing the homokaryotic mycelium called the **alternate host**.

On leaves of the principal host, *P. graminis* produces **urediniospores** in pustules called **uredinia**, and they rapidly spread the infection because they are capable of re-infecting the same host species. Urediniospores are produced on stalks from which they break off at maturity, being released and distributed passively by wind. Urediniospores of rust fungi have a relatively thick wall which is often pigmented and typically spiny. They are capable of surviving airborne for several weeks or months, which accounts for their long-distance dispersal sometimes over hundreds or even thousands of miles. Each urediniospore is binucleate, with the two nuclei being of opposite mating type. In temperate climates, uredinia of many rusts are gradually replaced by **telia** in autumn, especially on leaf sheaths and stems. **Teliospores** (= probasidia) of *Puccinia* are two-celled and thick-walled.

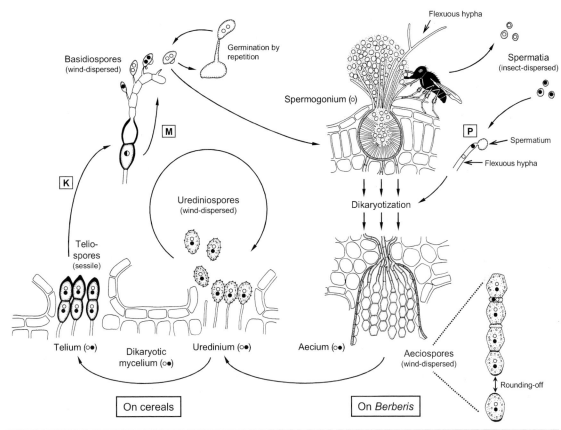

Fig 22.1 The life cycle of *Puccinia graminis*, with its heterokaryotic phase on cereals and the homokaryotic stage on barberry (*Berberis*). The different ways in which the five spore stages are released and dispersed are indicated. Teliospores overwinter after nuclear fusion, i.e. as diploid cells. During basidiosporogenesis, meiosis is followed by a mitotic division so that each basidiospore is a homokaryon containing two nuclei of the same mating type. Open and closed circles represent haploid nuclei of opposite mating type; diploid nuclei are larger and half-filled. Key events in the life cycle are plasmogamy (P), karyogamy (K) and meiosis (M).

Each cell contains two haploid nuclei of opposite mating-type which fuse to form a diploid nucleus (see Fig. 22.2c). The teliospore overwinters in the diploid state, firmly attached to the plant tissue on which it was produced. In spring, meiosis occurs while each teliospore cell emits a germ tube which becomes transversely septate and forms the promycelium or **metabasidium**. Each compartment of the metabasidium produces one basidiospore which initially contains one haploid nucleus. This commonly divides so that the mature basidiospore often contains two genetically identical haploid nuclei and is thus a dikaryotic homokaryon (Anikster, 1983). Basidiospores are actively liberated by the surface tension catapult mechanism involving Buller's drop as described before for other basidiomycetes (Section 18.5).

The basidiospores of *P. graminis* are unable to infect the cereal host but will infect the alternate host, barberry (*Berberis vulgaris*). The resulting primary mycelium is haploid, homo- and mono-karyotic. At the upper (adaxial) surface of the host leaf, the primary mycelium forms a flask-shaped fructification known as a **spermogonium**. Within the mesophyll, knots of hyphae form a **proto-aecium** but do not develop further at this stage. Minute uninucleate **spermatia** are produced from annellide-like structures (Littlefield & Heath, 1979) within the main body of the spermogonium, and they aggregate within a rim of hairs (periphyses) at the spermogonial

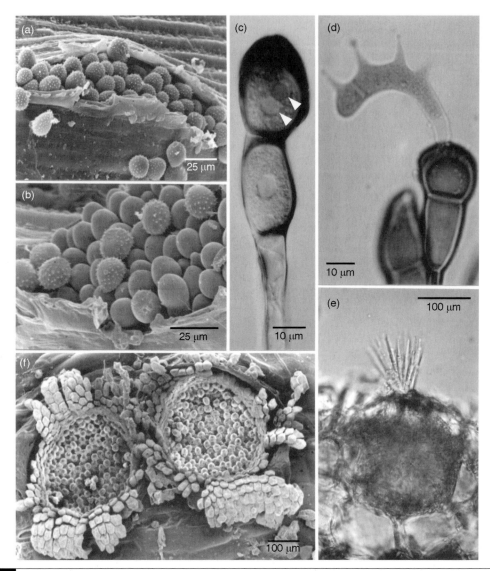

Fig 22.2 Examples of spore stages in rust fungi. (a) Uredinium of *Puccinia obscura* on its principal host, the wood-rush *Luzula campestris*. Note the spiny urediniospores. (b) Development of a telium within an old uredinium of *P. obscura*. Some urediniospores are still seen on top of the smooth-walled teliospores pushing through. (c) Karyogamy in a teliospore of *Puccinia lagenophorae*, a demicyclic rust on *Senecio vulgaris*. The terminal cell still contains two paired nuclei (arrowheads) whereas in the basal cell the two nuclei have already fused to give one larger (diploid) nucleus. (d) Production of a promycelium (metabasidium) by *Puccinia graminis*. Four sterigmata have formed, each about to produce one basidiospore. (e) Spermogonium of *P. obscura* on its alternate host, *Bellis perennis*. Note the flask-shaped body and the ring of stiff periphyses which protrude beyond the upper host epidermis. (f) Aecia of *P. lagenophorae* on *Senecio vulgaris*.

ostiole in a sugary liquid which attracts insects. These carry spermatia around, and fertilization occurs if a spermatium makes contact with a modified periphysis called **receptive hypha** of a spermogonium of opposite mating type. Spermatia are incapable of independent germination to establish new infections but perform solely a sexual role in fertilization. Most rust fungi are heterothallic with a bipolar mating system, although Narisawa *et al.* (1994) have claimed that *P. coronata* has a tetrapolar mating system with two mating type loci.

Homothallic species are also known, and these generally do not produce spermogonia. A generalized account of rust life cycles has been given by Buller (1950).

Following fertilization, the nucleus from the donor spermatium divides repeatedly and migrates down the receptive hypha into the receiving primary mycelium which thereby undergoes dikaryotization. When compatible nuclei reach the proto-aecium, this becomes converted into an **aecium** which breaches the lower epidermis. Heterokaryotic **aeciospores** develop and are released by a sudden rounding-off of the flattened wall separating adjacent spores, thus flicking the spores into the air. Aeciospores are relatively thin-walled and bear warty rather than spiny surface ornamentations. They are often brightly coloured due to the abundance of carotenoids accumulating in lipid droplets (Plate 12a). Aeciospores of *P. graminis* are unable to re-infect *Berberis*, but are infective on the principal grass host. The resulting secondary mycelium produces urediniospores, thus completing the life cycle.

Examples of the different spore types and the pustules (sori) producing them are shown in Fig. 22.2. Spores and sori have been given various names, and we follow the naming used in the *Dictionary of Fungi* (Kirk *et al.*, 2001). It is customary to assign Roman numerals 0, I, II, III or IV to the distinct spore types, and these numbers provide a convenient shorthand for describing the range of spores found in a given rust. The most important terms are summarized in Table 22.1.

The various spore stages of rust fungi differ greatly in their length of survival. Although details are dependent on species and conditions of storage (temperature, state of hydration, light), it may be generalized that the maximum survival period in the field is in the order of days (spermatia and basidiospores), weeks (aeciospores), a few months (urediniospores) and several months to more than a year (teliospores).

22.2.2 Derived life cycles and Tranzschel's Law

Rust fungi which must alternate between two different host plants in order to complete their life cycle are called **heteroecious**. In contrast, **autoecious** species are confined to one host plant. Species whose life cycle contains all five possible spore stages are called **macrocyclic**. In this terminology, *P. graminis* is a macrocyclic heteroecious rust whereas *P. menthae*, which produces all five spore stages on one host (mint), is macrocyclic but autoecious. Many rusts have derived life cycles in which one or more spore stages have been omitted. One common variation is the absence of uredinia in heterocyclic rusts. For instance, *Gymnosporangium fuscum* (see p. 629) produces spermogonia and aecia on the leaves of pear trees and has *Juniperus* spp. as its principal host for production of telia. Such rusts which lack uredinia are called **demicyclic** or **-opsis forms**.

An autoecious version of the demicyclic theme is presented by *P. lagenophorae* which produces spermogonia, aecia and telia on *Senecio* spp. (see Figs. 22.2c–f). In functional terms, the aecia could be considered uredinia because the aeciospores infect the same host species on which they were produced, and because in old aecia the

Table 22.1.	Generally accepted terminology of the sori and spore states of rust fungi.				
	0	I	II	III	IV
Sorus	**Spermogonium** Pycnium	**Aecium** Aecidiosorus Aecidium	**Uredinium** Uredosorus Uredium	**Telium** Teleutosorus	**Basidium** Metabasidium Promycelium
Spore	**Spermatium** Pycniospore	**Aeciospore** Aecidiospore	**Urediniospore** Uredospore Urediospore	**Teliospore** Teleutospore	**Basidiospore** Sporidium

aeciospores can be displaced by teliospores. Such aecia are sometimes termed aecidioid uredinia.

Another common variation is the **microcyclic** one in which both aecia and uredinia are lacking. Such rusts are, of course, autoecious. Many microcyclic rusts do not undergo sexual reproduction in the sense of sexual recombination, i.e. spermogonia are absent and the only reproductive unit is the teliospore with or without basidiospores. Such species are either homothallic or asexual. The wide array of possible nuclear cycles has been summarized by Ono (2002).

An example of a microcyclic rust fungus is *P. mesnieriana*, which produces telia on *Rhamnus catharticus* (buckthorn). The teliospores germinate to produce metabasidia, but only two basidiospores are formed per basidium (Anikster & Wahl, 1985). This species is homothallic because a single basidiospore can infect *R. catharticus*, giving rise to a telium-producing infection. The teliospores are highly characteristic because they carry a 'crown' of spiny outgrowths. Very similar teliospores (see Fig. 22.13a) are produced by the macrocyclic 'crown rust', *Puccinia coronata*, which also colonizes *Rhamnus* and related plants as alternate hosts, but infects grasses and cereals, especially oat (*Avena*), as the principal (i.e. uredinial and telial) host. Working in the late nineteenth and early twentieth century when the elucidation of rust life cycles was very much in vogue, the Russian mycologist Vladimir Tranzschel generalized that an unknown aecial stage of a macrocyclic uredinial/telial rust should be sought on plant species infected by microcyclic rusts with morphologically similar teliospores. This rule, known as **Tranzschel's Law**, has been applied several times, and DNA-based studies have confirmed the close relationship, for example between *P. mesnieriana* and *P. coronata* (Zambino & Szabo, 1993; Shattock & Preece, 2000). In other words, the telia of the microcyclic species mimic the aecia of the macrocyclic ancestor. These two species are then said to be **correlated**. The specialization of a recently evolved microcyclic rust onto the alternate host of the ancestor rust species may have something to do with the observation that the alternate hosts are almost always perennial, whereas principal hosts may be annual (Shattock & Preece, 2000). Examples of the reverse case, i.e. the evolution of the microcyclic species on the principal host of the macrocyclic ancestral species, do not seem to be known.

22.2.3 The infection process

Concise reviews of this vast topic have been written by Heath and Skalamera (1997), Mendgen (1997) and Hahn (2000). One of the differences between germ tubes arising from basidiospores and those arising from heterokaryotic spores (aeciospores or urediniospores) is that the former usually penetrate the cuticle directly without an appressorium, whereas the latter usually form an appressorium and preferentially penetrate through stomata (Figs. 22.4b, 22.5; Mendgen, 1997). An exception to this generalization is presented by germinating urediniospores of *Phakopsora* spp., which show appressorium-mediated direct penetration of the cuticle (Adendorff & Rijkenberg, 2000). There are also differences in the carbohydrate polymers making up the walls of monokaryotic and dikaryotic stages of the same species of rust fungus, and at different steps of the infection process (Freytag & Mendgen, 1991). Relatively little is known about infection from basidiospores, not least because these are difficult to obtain in the quantity required for experiments (see Gold & Mendgen, 1991). We shall therefore focus on the infection process arising from germinating urediniospores, since these are the most thoroughly researched system and have a major impact on agriculture as carriers of the repeated infection cycle.

Before urediniospores can germinate, germination autoinhibitors such as methyl-*cis*-3, 4-dimethoxycinnamate (e.g. in *Uromyces appendiculatus*) must be diluted out or degraded. This substance is active at concentrations in the 10^{-11} M range (Macko *et al.*, 1970; Staples, 2000), i.e. at a similarly low concentration as the sex hormones of *Achlya* (see p. 86). This makes it one of the most potent biological molecules. Wolf (1982) has suggested that the autoinhibitor acts by blocking the lysis of the germ pore. Lysis must

(a)

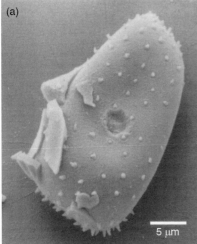

5 μm

(b)

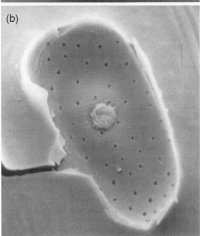

Fig 22.3 Attachment of a hydrated urediniospore of *Uromyces viciae-fabae* to the surface of a broad-bean leaf I h after contact. (a) Spore removed with sticky tape. The germ pore has partially lysed. (b) The adhesion pad on the leaf from which the spore was removed. Some of the germ pore material has become incorporated into the pad which has made firm contact with the host cuticle. From Clement *et al.* (1997), with permission from Elsevier. Original image kindly provided by J. A. Clement.

occur before the germ tube can emerge (see Fig. 22.3).

Attachment of urediniospores to surfaces is a multi-step process. Initial attachment is probably purely physical and based on hydrophobic interactions since it is stronger on hydrophobic than hydrophilic surfaces (Terhune & Hoch, 1993). As soon as the urediniospore becomes fully hydrated upon contact with water, there is

evidence of the formation of an **adhesion pad** of unknown composition (Fig. 22.3; Clement *et al.*, 1997). Further, cutinases and other esterases are released, and their activity is thought to modify the surface properties of the host cuticle, cementing the attachment pad to the host surface (Deising *et al.*, 1992). The germ tube is also tightly attached to the host surface by means of a glue which probably consists of glucans and proteins (Epstein *et al.*, 1985; Chaubal *et al.*, 1991).

The process of appressorium differentiation by dikaryotic stages of rusts may be unique among fungi in being triggered by **thigmotropism**. A ridge about 0.5 μm high is required and sufficient to induce appressorium differentiation even on chemically inert surfaces (Fig. 22.4a; Hoch *et al.*, 1987; Allen *et al.*, 1991). In nature, the relevant topographic feature is the stomatal lip, i.e. the point where the cuticle broke during the developmental expansion and opening of the stoma (Terhune *et al.*, 1991). Firm attachment of the germ tube to the surface is required for the perception of the signal for appressorium induction in urediniospore germ tubes of *U. appendiculatus*, and there is evidence that the reception of the physical signal involves microtubules and integrin-like molecules (Corrêa *et al.*, 1996) and is transmitted via stretch-activated Ca^{2+} channels (Zhou *et al.*, 1991). We have already come across stretch-activated Ca^{2+} channels and integrin as possible regulators of the rate of hyphal tip extension (see p. 8). Once the signal has been perceived, an appressorium differentiates in as little as 60 min.

Following the differentiation of an appressorium over a stoma, a thin penetration hypha develops which swells beneath the guard cells to form the **substomatal vesicle** (Fig. 22.5). From this, intercellular hyphae grow and form appressorium-like structures called **haustorial mother cells** on the surface of leaf mesophyll cells. The haustorial mother cell co-ordinates penetration of the plant cell, leading to the formation of a **haustorium**. Morphologically recognizable haustoria are not normally formed by monokaryotic rust stages; instead, the hyphae appear to grow through mesophyll cells whose plasmalemma invaginates around them

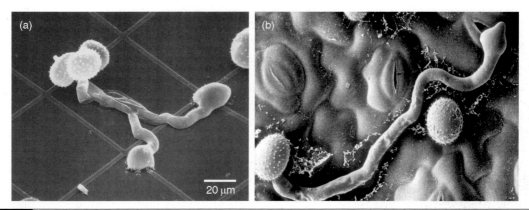

Fig 22.4 Appressorium formation by urediniospore germ tubes. (a) The runner bean rust *Uromyces appendiculatus*. Urediniospores inoculated onto a polystyrene surface with artificial ridges (0.5 μm high) have germinated and differentiated appressoria upon reaching the ridge. (b) The broad bean rust *Uromyces viciae-fabae*. The germ tube has followed topographical features of the host surface until it has reached a stoma, over which an appressorium has formed. (a) Reprinted from Allen *et al.* (1991) with permission by APS Press. Original print kindly provided by H.C. Hoch. (b) reprinted from Mendgen (1997), with kind permission of Springer Science and Business Media. Original print kindly provided by K. Mendgen. Both images approximately to same scale.

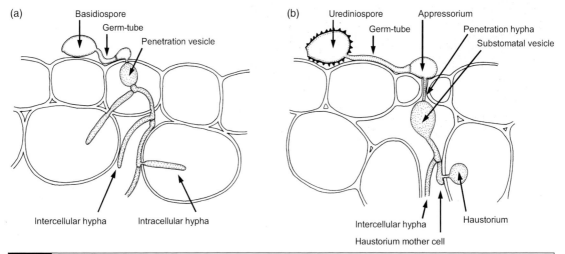

Fig 22.5 Schematic comparison of infection processes in *Uromyces*. (a) Germinating basidiospore. The monokaryotic germ tube infects directly through the epidermal wall without forming a pronounced appressorium. Growth is both inter- and intracellular. Differentiated haustoria are not formed, and intracellular hyphae are regarded as M-haustoria. (b) Germinating urediniospore. The dikaryotic germ tube forms an appressorium above a stoma and infects through the stomatal opening. Growth *in planta* is mainly intercellular. Stalked globose haustoria (D-haustoria) arising from haustorial mother cells are the only intracellular organs of note. Both M-haustoria and D-haustoria are surrounded by a matrix (not drawn). Based on Mendgen (1997) and Mendgen and Hahn (2002).

(Gold & Mendgen, 1991). These are sometimes called M-haustoria. In contrast, differentiated haustoria are formed by dikaryotic hyphae (D-haustoria), and valuable physiological work has been performed on them. This is summarized in the following section.

22.2.4 The physiology of biotrophy in rust fungi

The dikaryotic haustorium of rust fungi (Fig. 22.7) is functionally very similar to that of the Erysiphales (see Fig. 13.5). The haustorial mother cell emits a narrow penetration tube

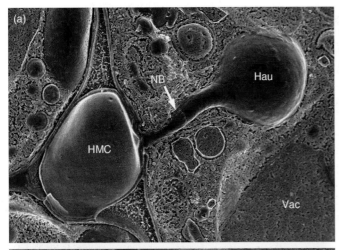

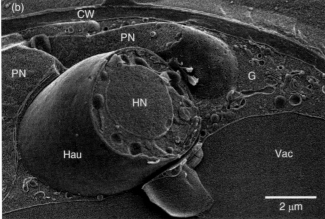

Fig 22.6 Cryo SEM of freeze-fractured haustoria of *Uromyces viciae-fabae*. (a) Intact haustorium (Hau) connected to the intercellular haustorial mother cell (HMC) by a thin penetration tube. The neck-band (NB) is visible in the penetration tube. The vacuole (Vac) of the infected plant cell is also obvious. (b) Fracture through a haustorium, revealing one of its two nuclei (HN). The nucleus of the infected plant cell (PN) is closely associated with the haustorium, a feature frequently observed in rust infections. The host cell wall (CW), vacuole (Vac) and a Golgi stack (G) are also visible. Both images to same scale. Previously unpublished images very kindly provided by E. Kemen and K. Mendgen.

which swells inside the host cell to form the haustorial body. This contains a full complement of organelles, including two nuclei. In some rusts, especially the cereal-infecting species, these two nuclei sometimes fuse into one diploid nucleus (Harder & Chong, 1984). From the inside outwards, the haustorial cytoplasm is surrounded by the haustorial membrane, the haustorial wall, the extrahaustorial matrix and the extrahaustorial membrane (i.e. the modified plant plasmalemma). The haustorial matrix is sealed against the apoplast outside the infected plant cell by means of a neckband (Fig. 22.6a). The host nucleus is often closely associated with the extrahaustorial membrane (Fig. 22.6b).

As in the Erysiphales (see p. 398), the extra-haustorial membrane surrounding D-haustoria seems to lack ATPase activity (Baka *et al.*, 1995),

thus indicating that the infected plant cell has no effective means to restrict the efflux of metabolites into the haustorial matrix. In contrast, ATPase activity in the fungal haustorial membrane may actually be increased relative to normal hyphae (Struck *et al.*, 1996). Not surprisingly, proton-driven hexose and amino acid uptake appears to occur from the matrix across the haustorial membrane into the haustorium (Fig. 22.7; Voegele & Mendgen, 2003). The uptake mechanism is thus equivalent to the uptake of solutes into growing hyphae (see Fig. 1.11). Whereas powdery mildews appear to rely on the host plant for the hydrolysis of the transport disaccharide sucrose into the hexoses fructose and glucose prior to uptake into the haustorium (p. 398), Voegele and Mendgen (2003) have suggested that the rust haustorium secretes

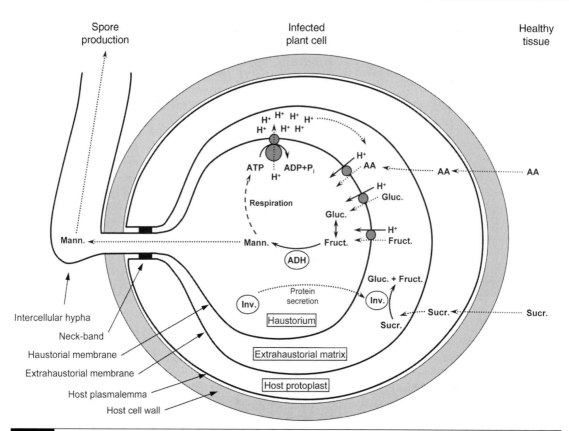

Fig 22.7 Diagram showing nutrient uptake mechanisms in a D-haustorium. Enzymatic reactions and proton-driven pumping are indicated by solid arrows; dotted arrows indicate diffusion or translocation processes. Sucrose (Sucr.) is hydrolysed to fructose (Fruct.) and glucose (Gluc.) by invertase (Inv.) secreted into the extrahaustorial matrix. These monosaccharides as well as amino acids (AA) are taken up across the haustorial membrane by specialized porter proteins fuelled by a transmembrane H^+ gradient. Fructose is converted to the fungal transport compound mannitol (Mann.) by alcohol dehydrogenase within the haustorium. Modified and redrawn from Voegele and Mendgen (2003).

invertase into the matrix to perform the necessary hydrolysis (Fig. 22.7).

All rusts are obligately biotrophic in nature, i.e. they need to parasitize living host plants in order to complete their life cycle. The fact that many of them can now be cultivated on agar media in the laboratory does not alter their status as obligate biotrophs in nature (P. G. Williams, 1984). The initial report of the cultivation of *P. graminis* on a simple medium (Williams *et al.*, 1966) raised high hopes for a breakthrough in research on rusts, but interest has waned in recent years. Excellent accounts recalling the excitement of discovery have been written by Maclean (1982) and P. G. Williams (1984). It is now possible to cultivate rust fungi on relatively

simple agar media based on minerals (e.g. Czapek-Dox agar) with sucrose or glucose as carbon source and yeast extract, peptone or certain amino acids as a nitrogen source.

22.2.5 Host resistance

Resistance against rust infections can manifest itself at different stages of the infection process. Most commonly it becomes evident as a hypersensitive response when the infectious hypha attempts to breach the first host cell wall (Heath & Skalamera, 1997). With monokaryotic stages of rust fungi, this is the direct penetration through the wall of the epidermal cell, whereas the resistance response to an incompatible dikaryotic strain occurs later, during or after the formation of the first haustorium (Heath, 1982, 2002). Either way, the hypersensitive response triggers biochemical events involved in local and systemic defence against the pathogen (Heath, 2000). One point of difference between penetration by rust fungi and powdery mildews is that the host plant mounts some resistance response against the latter even if the interaction ultimately turns out to be compatible, whereas no such recognition occurs if a compatible rust germ tube penetrates (Heath, 2002). The ultrastructural consequence is that powdery mildew haustoria commonly have a callose ring around their neck, whereas this is absent or reduced in the case of rust haustoria (Heath & Skalamera, 1997).

The **gene-for-gene concept** was proposed to explain the specific interactions between the rust *Melampsora lini* and its host, *Linum usitatissimum* (flax). It postulates that for every resistance gene of the host plant there is a matching virulence gene in the pathogen (see Flor, 1971). Resistance is usually dominant (*R*) whereas virulence is recessive (*a*). An incompatible interaction results if a rust fungus with an avirulence allele (*A*) attempts to infect a host carrying the matching resistance (*R*) allele. The identity of most molecules interacting in recognition is still unknown. However, Catanzariti *et al.* (2006) have demonstrated that the protein products of several avirulence genes are secreted by developing haustoria of *M. lini*, and that these seem to enter the cytoplasm of infected host cells where recognition leading to hypersensitivity occurs. The localization of proteins of haustorial origin in host cells, including the host nucleus, was also demonstrated by Kemen *et al.* (2005) for *Uromyces* spp. on broad bean (*Vicia faba*). These authors suggested that fungal avirulence genes might encode transcription factors involved in manipulating the host's metabolism during the biotrophic interaction. In this theory, avirulence would result if the host cell managed to detect the fungal transcription factor as foreign. It is not yet known how the avirulence proteins are translocated into the plant cell. The molecular basis of gene-for-gene interactions involving rust fungi is therefore fundamentally different from that, for example, in

Cladosporium fulvum infecting tomato, where recognition events occur at the host plasma membrane (see p. 482).

Rusts, like most other fungi, possess an uncanny ability to overcome **major gene resistance** based on gene-for-gene interactions, especially if a single resistance gene is involved. Genetic variation is enhanced by the ability to reproduce sexually in the field if the alternate host is available. Even in the absence of the alternate host, however, rust fungi can still undergo genetic recombination by anastomosis and nuclear exchange in various other ways (see p. 625). Different **races** of a given pathogen can be distinguished by their ability to infect any host in a set of cultivars containing defined resistance genes alone or in combination. Such tests are routinely employed by plant pathologists for the identification of races, and for monitoring their spread (see p. 626).

Major gene resistance against rust fungi was recognized early in the twentieth century by Biffen (1905) and others, and extensive breeding programmes were initiated. Typically a new cultivar produces excellent results for a few cropping seasons until resistant races develop and spread in the field, thereby rendering the breeders' efforts futile. This is the 'boom-and-bust' cycle. The 'bust' of a cultivar can be delayed if it contains a 'pyramid' of several resistance genes which the pathogen may be unable to overcome. Sometimes the breeding for resistance against one pathogen can lead to susceptibility to another, as in the case of the Victoria oat cultivar which showed good resistance against *P. coronata* f. sp. *avenae* but was devastated by *Cochliobolus victoriae* (see p. 471). In addition to major gene resistance which is commonly associated with the hypersensitive response, there are other types of resistance. Some plant genes do not afford total resistance but give a moderate degree of resistance. If a combination of several genes is involved, the resistance may well be more durable in the field than single-gene resistance. An example of such 'field' or 'partial' resistance is the reduced formation of appressoria over stomata of grasses covered by a particularly thick wax layer, which seems to mask the topographic features of the stomatal surface acting as the thigmotropic signal (Rubiales & Niks, 1996). There is also adult-plant resistance, usually an interaction between various genes which retard the progress of infection to such a degree that the losses are reduced to economically acceptable levels.

22.3 | *Puccinia graminis*, the cause of black stem rust

A connection between the barberry bush (*Berberis vulgaris*) and rust disease on wheat and other cereals had been suspected for centuries by farmers who noticed the frequent occurrence of crop damage around or downwind from barberry bushes. Further, an altogether different-looking fungus known as *Aecidium berberidis* was known to infect barberry leaves. The formal proof that the barberry and wheat pathogens were different stages of the same species was made by Anton de Bary who, in 1865, performed cross-inoculation experiments to show that basidiospores derived from germinating teliospores from wheat were

able to cause spermogonial and aecial infections on the barberry. This story has been recounted many times, but nobody has told it better than Large (1940).

Black stem rust affects especially wheat but also most other cereals and a range of wild grasses. Crop losses can be severe as a reduction in quantity as well as quality of the grain yield. The host epidermis can become ruptured over much of its surface in severe infections, thereby debilitating the plants (see Fig. 22.8d). Serious epidemics have occurred in the past, e.g. in the USA in 1904 and especially during the war year 1916. In fact, losses were so severe (up to 50% in the Great Plains; Eversmeyer & Kramer, 2000) that *P. graminis* was seriously considered and developed as a biological weapon by the US Government during the 1960s (Line & Griffith, 2001). On susceptible wheat cultivars and without chemical protection, *P. graminis* can cause total crop failure.

Although it is an ecologically obligate biotroph, *Puccinia graminis* is able to infect an astonishingly wide range of grass and cereal hosts. Gäumann (1959) listed 365 host species in

Fig 22.8 *Puccinia graminis.* (a) Spermogonial pustules on the upper surface of a leaf of *Berberis vulgaris*. Note the drops of nectar. (b) Aecia on the underside of a *Berberis* leaf. The outer frilly layer is the white peridium, within which is a mass of orange-coloured aeciospores. (c) Wheat leaf showing uredinia which appear as reddish-brown powdery masses. (d) Wheat straw showing telia as black raised pustules.

54 genera, and this list is still expanding. However, no single isolate of *P. graminis* is able to infect all these host species. Instead, the species *P. graminis* can be separated into several specialized forms which have become adapted to one or a few principal host species. These are termed **formae speciales** (sing. *forma specialis*, abbreviated as 'f. sp.') because they cannot be distinguished reliably from each other on morphological criteria. A morphologically distinct form of the same species which can be clearly identified by microscopy or other means would be called *varietas* (abbreviated as 'var.').

The most important *formae speciales* are *P. graminis* f. sp. *tritici* (on wheat), f. sp. *avenae* (on oat) and f. sp. *secalis* (on rye). In addition, there are several forms on wild grasses, e.g. f. sp. *phleipratensis*, f. sp. *lolii* and f. sp. *agrostidis* (Wilson & Henderson, 1966; Anikster, 1984). It is possible to produce hybrids between some of these *formae speciales*, e.g. between f. sp. *tritici* and f. sp. *secalis* (Green, 1971). The hybrid aeciospores are not very virulent on either principal host, and Green (1971) has argued that the hybrids resemble a more primitive form of *Puccinia graminis* with low virulence and a wide host range, and that evolution in stem rust of cereals is progressing from low virulence and a wide host range to high virulence and a narrowed host range.

Each *forma specialis* on cereals in turn forms hundreds of races distinguished by the infection responses of differential host cultivars, with new races continually evolving (see p. 626). This feature highlights the remarkable genetic and physiological flexibility of rust fungi.

22.3.1 *Puccinia graminis* on barberry

The basidiospores are released in spring from overwintered cereal stubble at about the time when fresh barberry leaves unfold. Basidiospores are able to germinate by repetition if they do not land on a suitable host surface (Fig. 22.10e). Infection gives rise to a haploid monokaryotic mycelium which shows inter- and intra-cellular growth and colonizes the host tissue extensively. Generally, monokaryotic stages of rust fungi show more widespread colonization of host tissue than their dikaryotic counterparts.

Viewed from the surface, the colonized barberry leaf area appears as a yellowish circular lesion. On the upper surface of this lesion, several flask-shaped spermogonia develop whose necks protrude beyond the epidermal layer. Among the orange-coloured tapering periphyses surrounding the opening of each spermogonium are several thinner, hyaline branched hyphae, the flexuous (or receptive) hyphae. Lining the inside surface of the spermogonium are tapering annellides which give rise to small uninucleate spermatia. These ooze out through the mouth of the spermogonium and are held by the periphyses in a drop of sticky sweet-smelling liquid (Figs. 22.8a, 22.9a). Within the mesophyll of the barberry leaf, the haploid mycelium gives rise to several spherical structures called proto-aecia. These are mostly made up of large-celled pseudoparenchyma, but in the upper region is a cap of smaller, denser cells (Fig. 22.9b).

Single haploid lesions are incapable of further development unless cross-fertilization occurs. The sweet-smelling spermatial exudate contains fructose and several volatile substances (see p. 629) which attract insects feeding on the nectar and carrying the spermatia around by visiting several distinct pustules. The haploid pustules are of either of the two mating types, (+) or (−), and if a (+) spermatium is brought close to a flexuous hypha of opposite mating type, it produces a short germ tube which anastomoses with the flexuous hypha (Craigie, 1927; Buller, 1950). Nuclear transfer is followed by repeated division and migration of the introduced nucleus towards the proto-aecium (Craigie & Green, 1962). This results in the dikaryotization of the haploid mycelium until binucleate cells become visible in the cap region of the proto-aecium after about 3 days. The binucleate cells now start to give rise to chains of alternating long and short cells which are also binucleate. The longer cells enlarge and become aeciospores, but the shorter cells disintegrate as the spore chains develop (Fig. 22.9c). During the development of the spore chains, the large pseudoparenchymatous cells of the proto-aecium are also crushed and pushed aside. Surrounding the chains of spores is a specially differentiated layer of cells homologous with the

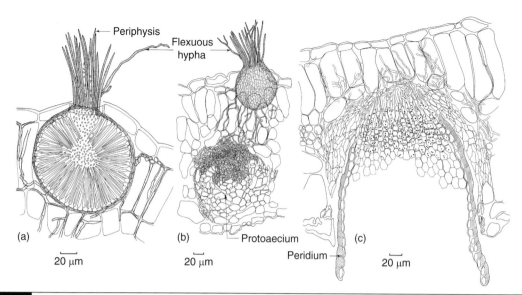

Periphysis
Flexuous hypha
(a)
20 μm
(b)
20 μm
Protoaecium
Peridium
(c)
20 μm

Fig 22.9 *Puccinia graminis*. (a) T.S. spermogonium on a leaf of *Berberis vulgaris*. The spermogonium has penetrated the upper epidermis. The wall of the spermogonium is lined by tapering annellides which give rise to spermatia. (b) T.S. leaf of *B. vulgaris* showing a spermogonium and a proto-aecium. (c) T.S. leaf of *B. vulgaris* showing an aecium in section. The aecium has burst through the lower epidermis of the host leaf. Note the columns consisting of alternating large and small cells. The large cells are the aeciospores.

spore chains, whose outer walls are thick and fibrous. This layer forms a clearly defined border or **peridium** surrounding the spores. Eventually the peridium and spore chains burst through the lower epidermis. The peridium ruptures and aeciospores are now visible as orange-coloured cells enclosed by the white cup-like peridium (Fig. 22.8b). Since the aecia are commonly seen clustered together beneath a spermogonial lesion, this stage is popularly known as the cluster-cup stage. In a section through the centre of a group of young aecia, it is usually possible to find the spermogonia penetrating the upper epidermis and the aecia penetrating the lower. The aeciospores are violently projected from the end of the spore chain by rounding off of the flattened interface between adjacent spores (Ingold, 1971). Aeciospores are unable to re-infect barberry but readily infect wheat or other grasses.

It is relatively easy to establish and maintain *P. graminis* in garden situations by planting *B. vulgaris* and allowing *Agropyron repens* to grow underneath (Webster *et al.*, 1999). The form most likely to grow is *P. graminis* f. sp. *secalis* which readily alternates between *Berberis* and grasses in Britain (Wilson & Henderson, 1966). In other countries the connection between *Berberis* and *P. graminis* f. sp. *tritici* is strong, and the first barberry eradication laws were implemented long before de Bary had formally proven the connection. The first recorded cases of such laws were in Rouen in 1660 and Massachusetts in 1755. The greatest effort by far was the barberry eradication campaign undertaken as a consequence of the 1916 epidemic in the United States. Widespread eradication of susceptible *Berberis* spp. started in 1918 and continued on a massive scale well into the 1930s, and locally for several decades afterwards. Overall, this campaign is considered to have been successful, with wheat rust epidemics much reduced in severity, especially on a local scale. Because aeciospores are not particularly long-lived and are not produced in vast numbers, the direct impact of barberry bushes on wheat crops is limited to less than 2 miles. Perhaps more importantly, barberry eradication also retarded the evolution of new races of *P. graminis* due to the reduced ability of the pathogen to reproduce sexually. Campbell and Long (2001) have provided a highly readable account of the eradication campaign in the USA.

22.3.2   *Puccinia graminis* on cereals

A symptom of infection on wheat leaves and stems is the appearance of brick-red pustules (uredinia) between the veins. Uredinia contain stalked, one-celled dikaryotic urediniospores which burst through the epidermis (Figs. 22.8c and 22.10a,b). These have a spiny wall which has four thinner areas (germ pores) near the middle of the spore. The urediniospores are detached by wind and blown to fresh wheat leaves upon which they germinate by extruding a germ tube from one of the germ pores. Germination requires free water (e.g. night-time dew) and proceeds optimally at about 20°C. Infection

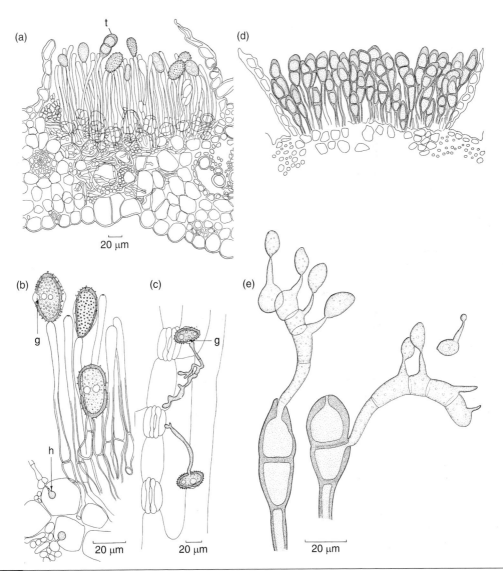

Fig 22.10   *Puccinia graminis* f. sp. *secalis*. (a) T.S. stem through a uredinium. The stalked unicellular urediniospores are protruding through the ruptured host epidermis. A teliospore (t) has also been formed. (b) Higher-power detail of urediniospores. Note the germ pores (g) and the haustoria (h) in the host cells. (c) Germination of urediniospores on host leaf. Note the directional growth of the germ tubes perpendicular to the long axes of the epidermal cells, towards the stomata. (d) T.S. leaf sheath through a telium. The stalked teliospores are projecting through the ruptured epidermis. Drawing to same scale as (a). (e) Germination of teliospores to form metabasidia bearing sterigmata and basidiospores. One basidiospore is giving rise to a secondary spore.

follows the usual pattern for dikaryotic propagules, i.e. via an appressorium through a stoma as shown in Fig. 22.5b. Within about 7–21 days of infection, a new crop of urediniospores is formed so that inoculum can build up rapidly within a crop. A single uredinium may produce between 50 000 to 400 000 spores, and there may be 4–5 generations of urediniospores in the growing period of a wheat crop. Disease development occurs best at daytime temperatures around 30°C. This may explain why stem rust of wheat is particularly prevalent in areas with a continental climate, whereas it is not a serious disease, for example, in Britain.

Puccinia graminis f. sp. *tritici* can survive the winter in the mild climates, e.g. of Mexico and the Southern USA (see Fig. 22.11) as uredinial infections on its principal host, i.e. wheat crops and volunteer plants. Despite the almost complete elimination of the barberry, infections with *P. graminis* f. sp. *tritici* still occur every year in North America from uredinial lesions overwintering in the South. Urediniospores are longer-lived than aeciospores, and waves of inoculum can move northwards with the prevailing southerly winds. The track taken is called the '*Puccinia* pathway' (Agrios, 2005). Using small aeroplanes, urediniospores have been detected at altitudes as high as 5000 feet. The total distance travelled each year from the Southern USA to Canada can be in excess of 2000 miles (Stakman & Christensen, 1946), but in most cases migration occurs in several shorter intervals, with new crops of uredinia being produced *en route* (Eversmeyer & Kramer, 2000). Consequently, infections start at successively later dates the further north the epidemic moves (Fig. 22.11). Attempts have been made to interrupt this migration by planting cultivars with different resistance genes in different regions of the *Puccinia* pathway (Frey *et al.*, 1973). Towards the end of the season there may be a reversal of the flow of air, so that wheat

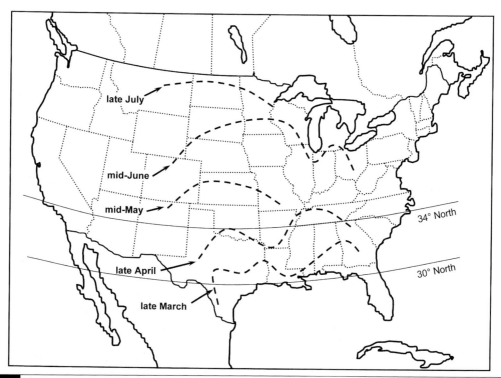

Fig 22.11 The '*Puccinia* pathway' in North America, with dates indicating the average annual arrival of *P. graminis* f. sp. *tritici* in the wheat fields. The fungus regularly overwinters in the uredinial state south of 30° Northern latitude, and occasionally as far north as 34°. Redrawn from Roelfs (1986) and several other sources.

in Texas and Mexico may be infected from urediniospores derived from the northern areas (Pady & Johnston, 1955). Similar long-distance transport of urediniospores has been reported from Europe and India (Nagarajan & Singh, 1990).

Late in the season, uredinial lesions gradually turn to producing teliospores instead of urediniospores. Telial pustules appear as black raised streaks along leaf-sheaths and stems of infected plants (Fig. 22.8d). Teliospores are initially binucleate, but soon the two nuclei fuse and the spores survive the winter in the diploid state. A period of maturation corresponding to winter dormancy is required before teliospores are competent to germinate. When they do germinate, each of the two cells emits a four-celled promycelium or metabasidium. Meiosis gives rise to four nuclei which migrate into the four basidiospores and then perform one further mitotic division. Each cell of the metabasidium bears one basidiospore containing two nuclei of the same mating type. The basidiospores are short-lived and do not travel far, and the teliospore state is of no significance in an agricultural context if no barberry bushes are in the vicinity.

22.3.3 Resistance breeding and physiological races of *P. graminis* f. sp. *tritici*

If the spores of a given isolate of *P. graminis* f. sp. *tritici* are inoculated onto a range of wheat cultivars, these hosts will differ in their response. Some may prove to be resistant, others highly susceptible, whilst yet others may be intermediate in their reaction. Spores from a second source may give an entirely different pattern of response. Using the reactions of different wheat cultivars, it has proven possible to classify *P. graminis* f. sp. *tritici* into over 300 physiological races. The existence of such a large number of races may be due to several factors.

1. Sexual recombination. New races of rust are often found adjacent to barberry bushes. Further, following inoculation with a single race, a number of variants may be found among the aeciospore progeny.

2. Anastomosis on the principal host. If two uredinial mycelia growing on the same host leaf come into hyphal contact, anastomosis may result. This can lead to the exchange of genetic material known as **somatic hybridization**. The simplest way is the exchange of entire nuclei which will lead to the formation of new heterokaryons and their dissemination as urediniospores (Ellingboe, 1961). However, when two different races of *P. graminis* were inoculated simultaneously onto a susceptible wheat cultivar, at least 15 different races were identified among the urediniospore progeny, which is more than would be expected from simple nuclear rearrangement (Watson & Luig, 1958; Bridgmon, 1959). *Puccinia graminis* contains a haploid set of 18 chromosomes (Boehm *et al.*, 1992), and the exchange of any of them in the dikaryotic mycelium during synchronous nuclear division is conceivable (Hartley & Williams, 1971). Additionally, parasexual recombination between chromosomes could occur. Park *et al.* (1999) have provided evidence of the occurrence of somatic hybridization in *P. triticina* in the field.

3. New races may arise by mutation, as shown by Park *et al.* (1995) for the brown leaf rust of wheat, *P. triticina*.

The existence of this large range of rust races complicates the task of the plant breeder, but fortunately all these races are not prevalent in an area at any one time. Therefore the breeding of resistant cultivars remains practicable and profitable (Johnson, 1953; Dyck & Kerber, 1985). The frequency of different rust races varies over the years, reflecting the changes in the wheat cultivars planted (Fig. 22.12). A new race may suddenly build up in frequency, as shown by the appearance of Race 15 (actually a sub-race or biotype referred to as Race 15B) in Canada. The appearance of new races presents a problem to plant breeders, and it is clear that the breeding and release of resistant cultivars has to be carefully co-ordinated.

The original scheme for using host differentials was devised by Stakman and Levine (1922) who used 12 different wheat cultivars ('standard differentials') to discriminate between the physiological races ('standard races'). The standard differentials frequently failed to reveal important

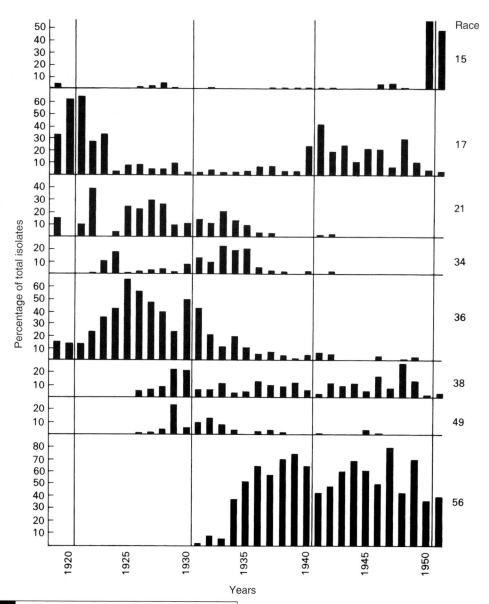

Fig 22.12 Diagrammatic representation of the distribution in Canada of eight physiological races of wheat stem rust during the period 1919–1951 (after Johnson, 1953).

changes in the virulence of the rust population, and this led to the use of supplementary differentials to differentiate between further physiological races. Gradually the nomenclature for numbering rust races became quite complicated. In the wake of the acceptance of the gene-for-gene concept (see Flor, 1971), a more rational way of identifying the physiological races was to base their classification on the resistance genes of the host plant. Using the wheat cultivar Marquis, several 'single gene lines', i.e. wheat cultivars carrying single resistance genes, have been generated. The resistance genes were given the symbols $Sr1$, $Sr2$, etc. (Sr for 'stem rust'). Some 40 resistance genes are known (Roelfs, 1985). Efforts at characterizing pathogen races are now standardized by means of an international nomenclatural system (Roelfs & Martens, 1988). An authoritative account of the history and importance of rust race characterization has been written by Roelfs (1984).

The serious race 15B epidemics of the early 1950s (see Fig. 22.12) resulted in the replacement of the then prevalent cultivars by others such as 'Selkirk', which carried several resistance genes and remained resistant to *P. graminis* f. sp. *tritici* in North America for several decades (Roelfs, 1985). Several wheat varieties more recently released have also retained their resistance for many years in the field (Roelfs, 1984; Dyck & Kerber, 1985), so that no major stem rust epidemic has occurred in the USA during the past 50 years. This success is probably due to two main reasons, namely the combination of several resistance genes in one cultivar which the rust fungus is apparently unable to overcome, and the stabilization of rust races by the elimination of sexual reproduction as a consequence of the barberry eradication campaign.

22.4 | Other cereal rusts

Although potentially the most damaging cereal rust, *P. graminis* has lost much of its menace for reasons discussed in the preceding section. Several other rust species now cause more serious crop losses than *P. graminis*. These are briefly discussed below. Cereal rusts are primarily controlled by breeding of resistant cereal cultivars, and this is an ongoing battle against new races which, once arisen, are capable of spreading rapidly and on a global scale (Chen, 2005; Kolmer, 2005). Chemical control by the application of fungicides is practised to protect crops grown for seed production, and to control severe outbreaks of rust if resistance fails. Fungicides used include the systemic strobilurin-type compounds and the ergosterol biosynthesis-inhibiting triazoles (see Fig. 13.15), as well as various protectant compounds.

22.4.1 *Puccinia triticina* (brown leaf rust of wheat)

Puccinia triticina was formerly named *P. recondita* f. sp. *tritici*, but Zambino and Szabo (1993) and Anikster *et al.* (1997) have shown that it is not closely related to *P. recondita*, which infects rye but is only of minor significance. *Puccinia triticina* has displaced *P. graminis* as the economically most important rust of wheat especially in North America and Eastern Europe (Samborski, 1985). This species is macrocyclic, with *Thalictrum speciosissimum* (= *T. flavum*; Ranunculaceae) serving as the alternate host, although it seems to survive mainly in the uredinial state. Its epidemiology is therefore similar to that of *P. graminis* f. sp. *tritici* (Eversmeyer & Kramer, 2000). The temperature optima for urediniospore germination and infection (about 18°C) and sporulation (25°C) are lower than those of *P. graminis* (Singh *et al.*, 2002). Teliospores are not readily formed, and the disease symptoms are easily recognized by the chocolate-brown uredinial lesions on wheat leaves. There is no infection of the stem, in contrast to *P. graminis*. Triticale, a hybrid between wheat (*Triticum*) and rye (*Secale*), is also affected. About 50 *Lr* ('leaf rust') resistance genes are known, and wheat leaf rust is currently controlled mainly by using cultivars containing a pyramid of several of these *Lr* genes.

22.4.2 *Puccinia striiformis* (stripe rust or yellow rust on wheat and barley)

The alternate host of this rust species has not yet been found, and it may no longer have one. Therefore, its life cycle is probably confined to the grass or cereal hosts on which uredinia and telia are produced. This rust can be distinguished from *P. triticina* by its yellow uredinia which are arranged in stripe-like rows following the leaf veins. There are several *formae speciales*, but their host ranges overlap to a greater extent than in the other cereal rusts and they are therefore less clearly delimited. The most important forms are those on wheat (f. sp. *tritici*) and barley (f. sp. *hordei*). The temperature optima of *P. striiformis* are the lowest of any of the common cereal rusts, with urediniospore germination and penetration around 10°C, and urediniospore production at 12–15°C (Singh *et al.*, 2002). This species overwinters by using winter crops and volunteers as a 'green bridge', and can cause early epidemics in spring. Zadoks and Bouwman (1985) have estimated that a single uredinium per hectare is sufficient to spark an epidemic. Even if no lesions are seen on winter wheat, the rust may still be present, surviving as latent infections. Crop losses can be severe in the cooler climates of North-Western Europe and North America where winter cereals, especially winter wheat, are widely grown (Stubbs, 1985). Like most other rusts, *P. striiformis* is controlled mainly by resistance breeding (Line, 2002). Chen (2005) has given a detailed account of recent epidemics and migrations of *P. striiformis*.

22.4.3 *Puccinia hordei* (barley leaf rust or 'dwarf rust')

This macrocyclic rust alternates between *Ornithogalum* spp. (Hyacinthaceae) as the alternate host and cultivated and wild barley (*Hordeum*) species as principal host. Clifford (1985) considered this species to be the most important rust on barley. The uredinial lesions are brownish in colour and are smaller than those of other rusts, hence the name 'dwarf rust'. In agricultural situations, the alternate host is probably of little importance as the fungus can overwinter in the uredinial state on volunteer plants or winter barley. Its optimum temperatures for infection (15°C) and sporulation (10−20°C) are almost as low as those for *P. striiformis*. In Europe in the 1970s, *P. hordei* caused major crop losses because the cultivars sown were highly susceptible. However, the situation seems to have improved in recent years, and several cultivars showing good partial resistance have been bred (Niks *et al.*, 2000).

22.4.4 *Puccinia coronata* (crown rust of oats and forage grasses)

This rust is so named because of the spiny extensions at the apex of its teliospore (see Fig. 22.13a). It is the most damaging disease on oats and has probably been a pathogen since oats were first cultivated (Simons, 1985). In addition to oats, numerous other grasses are infected, including important forage grasses such as the perennial ryegrass, *Lolium perenne*, for which separate resistance breeding programmes have been initiated (Kimbeng, 1999). Heavily infected meadows can acquire a distinctly orange hue

due to the urediniospores being produced in great abundance. There are no *formae speciales* other than f. sp. *avenae* on oats because the delimitation between strains on different grasses is not clear cut. *Puccinia coronata* is macrocyclic, alternating with *Rhamnus* (buckthorn) shrubs. As with *P. graminis* and *P. triticina*, a standardized nomenclature for identifying races of *P. coronata* f. sp. *avenae* has been proposed. This is based on a differential comprising 16 oat cultivars carrying single resistance (*Pc*) genes (Chong *et al.*, 2000). In total, about 100 resistance genes are known, and resistance breeding is the main strategy to control oat rust. Heavy infections of susceptible cultivars can result in total crop failure.

Even well-known rust species can harbour surprises upon close inspection, and one of these was the description of a form of *P. coronata* infecting *Bromus inermis* as its principal host. Anikster *et al.* (2003) described that germinating teliospores of this form produced only two basidiospores, each of which had four instead of the usual two nuclei and was self-fertile, i.e. infection on *Rhamnus cathartica* gave rise to aecia without spermogonia. This fungus is therefore homothallic, like the correlated microcyclic *P. mesnieriana* (Anikster & Wahl, 1985) and probably the barley form of *P. coronata*. In contrast, *P. coronata* f. sp. *avenae* is heterothallic with a tetrapolar mating system, i.e. two mating type loci with two alleles each, instead of the usual bipolar system (Narisawa *et al.*, 1994).

22.4.5 The origin of cereal rusts

It is generally assumed that the cereal rusts evolved at the centre of origin of the wild grasses which were the progeny of cultivated cereals. The argument is strengthened if the alternate host is native to the same area. This topic has been discussed extensively by Anikster and Wahl (1979) and Wahl *et al.* (1984). *Puccinia graminis* is thought to have evolved with *Berberis vulgaris*, which is of Asian origin, moving westwards to an area from Transcaucasia to the Western Mediterranean, which is the centre of origin of wheat and rye. Westward migration must have occurred early in the history of agriculture because 3300-year-old *P. graminis*-infected cereal

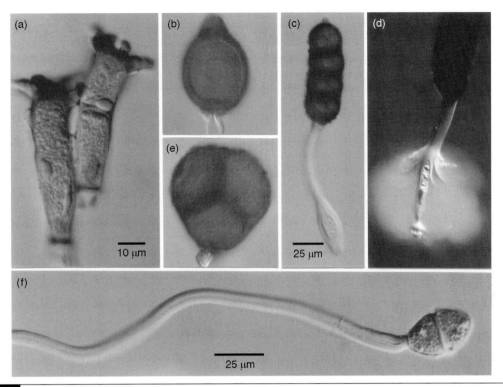

Fig 22.13 Teliospores of rust fungi belonging to the Pucciniaceae *sensu* Dietel. (a) *Puccinia coronata*. Note the 'crown' of spine-like wall extensions at the teliospore apex. (b) *Uromyces appendiculatus*. The teliospore is one-celled. (c,d) *Phragmidium mucronatum*. (c) Teliospore freshly mounted in water. (d) Spore after about 15 min in water. The base of the pedicel has broken, exuding a large amount of mucilage which has been made visible by replacing the water with dilute Indian ink. (e) *Triphragmium ulmariae*. The teliospore is three-celled. (f) *Gymnosporangium fuscum*. Note the long teliospore stalk. (a,b,e) to same scale; (c,d) to same scale.

remains have been found in Israel (Kislev, 1982). The centre of origin of *P. triticina* and its aecial host, *Thalictrum*, is probably also in the Western Mediterranean, as is that of *Puccinia striiformis*. *Rhamnus* and wild oats are common in Israel, as is *P. mesnieriana*, suggesting that this is where *P. coronata* may have its origin. Wild barley species, especially *Hordeum spontaneum*, are also very common in Israel and adjacent countries where barley has been cultivated since the dawn of agriculture. Since the alternate host of *P. hordei*, *Ornithogalum*, is also found there and is profusely infected, this region may be where *P. hordei* evolved.

22.5 | *Puccinia* and *Uromyces*

The genus *Puccinia* is by far the largest among the Uredinales, comprising about 4000 species (Kirk *et al.*, 2001). The most readily recognized characteristic is the teliospore which usually contains two darkly pigmented cells borne on a thin hyaline stalk. *Puccinia* spp. are pathogenic mainly on Angiosperms, and especially on grasses as principal hosts. The genus *Uromyces* (about 800 spp.) is similar in appearance to *Puccinia*, differing mainly in its teliospore, which is commonly one-celled (Fig. 22.13b). Several *Puccinia* spp. also produce a certain proportion of one-celled teliospores (called mesospores) in their telia, and it is therefore easy to imagine how the one-celled teliospore of *Uromyces* could have evolved. DNA sequencing data support a close relationship between the genera *Puccinia* and *Uromyces* (Maier *et al.*, 2003).

22.5.1 Other common *Puccinia* spp.

Puccinia sorghi was so named because Schweinitz, when he first described it, erroneously believed that he was dealing with infected *Sorghum* material. It is, however, a pathogen of maize and can be encountered wherever that crop is grown. Since maize is native to Central America, the pathogen probably also originated there. In Europe, crop losses are not generally sufficiently severe to warrant control measures (Smith *et al.*, 1988), but elsewhere breeding programmes are in operation and fungicides are also occasionally used (Hooker, 1985). The fungus produces numerous chocolate-brown uredinia on the leaves of grain and fodder maize. It is macrocyclic and heteroecious, alternating with *Oxalis* spp.

Several macrocyclic *Puccinia* spp. not causing agriculturally significant disease are frequently encountered in the field. An example is *P. caricina* which has sedges (*Carex* spp.) as its principal host and produces spermogonia and aecia on *Urtica* (stinging nettle). The latter are associated with growth deformations (Plate 12b). Another ubiquitous species is *P. poarum*, with grasses of the genus *Poa* as principal and *Tussilago farfara* (coltsfoot) as alternate hosts. This species is unusual in completing two life cycles per growing season, with aecia appearing in May and August (Wilson & Henderson, 1966). Mint rust (*P. menthae*) is autoecious and macrocyclic. All the above species are abundant in Europe and on most other continents.

Puccinia punctiformis is a systemic autoecious rust attacking *Cirsium arvense* (creeping thistle). In spring, infected plants are clearly distinguishable by their yellowish appearance and appressed leaves, and by the strong sweet smell associated with numerous spermogonia which develop all over the infected shoots. These aromatic substances include common fragrance molecules such as benzaldehyde, 2-phenylethanol, indole and phenylacetaldehyde (Connick & French, 1991), and it is possible to smell infected thistles from a distance of several metres. All these substances are produced by a wide range of fungi and, with the exception of indole, also by the host plant during flowering. The infections develop from a dikaryotic mycelium overwintering systemically in the rootstock. The haploid condition of the spermogonial tissue is probably established by

de-dikaryotization as new shoots are infected in spring. Transfer of spermatia to compatible flexuous hyphae results in dikaryotization, and the next structures to form resemble uredinia with chocolate-brown spores. They are sometimes called uredinoid aecia. The urediniospores can infect healthy thistles and normal uredinia develop on these. Later in the season, teliospores develop (Buller, 1950).

A yet more specialized way to attract insect pollinators has been developed by a complex of rusts attacking Brassicaceae (*Arabis* spp.). Infected plants produce the same fragrances as *Cirsium* infected by *P. punctiformis*, plus many more (Raguso & Roy, 1998). Further, infected plant organs are morphologically modified to form flower-like structures called pseudoflowers. Intriguingly, these do not resemble the flowers of the host, but those of taxonomically unrelated plant species such as buttercups (*Ranunculus* spp.), which often grow in the same habitats (Roy, 1994). The combination of scent, shape, colour and nectar in this floral mimicry is necessary to attract a wide range of insects (Roy, 1994; Roy & Raguso, 1997). Several rust species can stimulate the production of such pseudoflowers on *Arabis* spp., including the macrocyclic heteroecious *P. monoica* which alternates with grasses, and correlated autoecious and microcyclic species (Roy *et al.*, 1998).

22.5.2 *Uromyces*

Common species of *Uromyces* are *U. ficariae*, a microcyclic species forming brown telia on *Ranunculus ficaria*, and *U. dactylidis* with urediniospores and teliospores on grasses (*Dactylis*, *Festuca* and *Poa*) and aecia on *Ranunculus* spp. *Uromyces dianthi* produces uredinia and telia on *Dianthus* (carnation) and other ornamental flowers, with *Euphorbia* spp. as the alternate host. However, the most important species of *Uromyces* are those infecting legumes (Fabaceae), and because of the ease with which legumes such as *Pisum*, *Phaseolus* and *Medicago* can be cultivated in the laboratory, their rusts have been used for numerous studies of fundamental physiological aspects such as differentiation of infection structures and haustorial functioning, which have been mentioned at the beginning of this chapter.

Uromyces appendiculatus is a macrocyclic autoecious rust on French bean (*Phaseolus vulgaris*) and many other leguminous plants. Several varieties have been distinguished. The fungus survives the winter as teliospores on plant debris or, in milder conditions, as urediniospores (McMillan *et al.*, 2003). As one would expect of a situation in which a rust fungus undergoing sexual reproduction is being controlled by the breeding of resistant cultivars, numerous (more than 200) physiological races of *U. appendiculatus* are known. Control is by means of resistant cultivars or, if these fail, with protectant fungicides (maneb and chlorothalonil). Strobilurin-type compounds are also now used against *U. appendiculatus*. Since beans are a higher-value crop than cereals, fungicide applications are generally more profitable. *Uromyces viciae-fabae* has a similar life cycle to that of *U. appendiculatus*, but it infects broad bean (*Vicia faba*) and numerous other plants. It is a cosmopolitan species but seems to cause lesser damage than *U. appendiculatus*.

Uromyces pisi is heteroecious, producing spermogonia and aecia on *Euphorbia cyparissias* and uredinia and telia on various leguminous hosts, especially pea (*Pisum sativum*). Infections on the alternate host are of biological interest because the fungus infects systemically and causes its alternate host to produce pseudoflowers, at the expense of its real flowers (Pfunder & Roy, 2000). The name *U. pisi* is now known to cover a complex of several biologically distinct species which are united by their alternate host but infect different members of the Fabaceae as principal hosts (Pfunder *et al.*, 2001).

22.6 | Other members of the Pucciniaceae

22.6.1 *Phragmidium*

All *Phragmidium* spp. are autoecious and confined to the Rosaceae. The two best-known species, both extremely common in Europe, are *P. violaceum* on leaves of the *Rubus fruticosus* species aggregate (bramble or blackberry; Plate 12c) and

P. mucronatum on roses. Both are macrocyclic. The uredinial and telial stages are readily recognized from above as chlorotic or purple lesions, and the spores are formed on the underside of the host leaves. They are yellowish-orange (urediniospores) or violet to black (teliospores). The teliospores contain mostly 4 (*P. violaceum*) or 6–8 (*P. mucronatum*) cells with very dark and rough walls, and are borne on a long stalk (pedicel) which becomes easily detached from the lesions. When teliospores are mounted in water, the base of the pedicel breaks and exudes a large amount of mucilage (Figs. 22.13c,d). In nature, this may facilitate the attachment of the teliospore to adjacent surfaces during overwintering (Ingold *et al.*, 1981). The teliospores of *P. mucronatum* were among the first fungus spores illustrated by Robert Hooke in 1667 soon after the invention of the microscope (see Large, 1940).

Phragmidium violaceum shows promise as a biocontrol agent against the spread of brambles which were introduced into Australia and are spreading there in their typically uncontrollable manner (Mahr & Bruzzese, 1998). *Phragmidium mucronatum* is not tolerated by rose breeders or hobby gardeners; it is controlled by fungicide applications. Removal of fallen leaves in autumn can also help to control the disease.

22.6.2 *Gymnosporangium*

There are about 60 species of *Gymnosporangium* which produce their spermogonia and aecia on members of the Rosaceae, and telia on Cupressaceae. Uredinia are not normally formed. One of the most commonly encountered species is *G. fuscum* (= *G. sabinae*), the pear trellis rust, which alternates between *Juniperus* spp. of the *sabina* group (principal host) and pear trees. The dikaryotic mycelium can survive for many years in the principal host, and repeated production of telia is associated with a spindle-shaped swelling or canker of the infected trunk or twig. Spore-bearing telia are produced in the spring as horn-like outgrowths which greatly expand during rainfall due to swelling of the long teliospore stalks (Plate 12d; Fig. 22.13f). Teliospores at the surface of a swollen telial horn germinate to release basidiospores at the time of leaf bud of the alternate host. Between late spring and autumn, infected pear leaves show bright orange-coloured lesions up to 1 cm in diameter (Plate 12e). These give rise to spermogonia at the upper leaf surface, and later to aecia at the lower surface. The aecia are greatly swollen and elongated. Eventually, they rupture at their sides, leaving a cap joined to the aecium by trellis-like threads (Plate 12f). The characteristic tube-like aecia of *Gymnosporangium* are referred to as **roestelioid**, after the generic name *Roestelia* previously given to such aecial stages. Aeciospores infect susceptible *Juniperus* spp. in the autumn (Borno & van der Kamp, 1975).

Gymnosporangium fuscum can cause serious damage to pear trees because heavy infections reduce the photosynthetic capacity of the trees, and yields and ultimately the trees themselves decline after severe successive infections. This species is very much on the increase in Europe and elsewhere, especially because of the planting of ornamental *Juniperus* spp. (e.g. Chinese juniper) and pear trees side-by-side in gardens. Whereas the pathogen needs to infect the alternate host afresh each year, it can survive almost indefinitely in its principal host once this has become infected. Several communities and countries have therefore launched eradication schemes to remove infected *Juniperus* trees. Since the basidiospores do not normally travel distances longer than about a mile, this approach can reduce the infection pressure to acceptable limits. Chemical control of *G. fuscum* on pear trees is possible but obviously not desirable (Ormrod *et al.*, 1984).

Several related species of *Gymnosporangium* are also commonly encountered, e.g. *G. clavariiforme*, which alternates between *Juniperus communis* (a plant not infected by *G. fuscum*) and *Crataegus* spp., or *G. cornutum* on *J. communis* and *Sorbus aucuparia* (mountain ash, rowan). *Gymnosporangium juniperi-virginianae* is the cause of North American cedar-apple rust, alternating between the 'Eastern red cedar' (*Juniperus virginiana*) and apple (*Malus* spp.). On the principal host, telia often develop from gall-like deformations called 'cedar-apples'. This species causes considerable economic damage in North America where it is native. It does not seem to be present in Europe as

yet (Smith *et al.*, 1988). It can be controlled by fungicide applications similar to those effective against apple scab caused by *Venturia inaequalis* (see p. 478).

22.6.3 *Hemileia vastatrix*

The phylogenetic position of *H. vastatrix*, the cause of coffee leaf rust, has not yet been resolved. It appears to be an ancient rust lineage possibly pre-dating the separation between Pucciniaceae and Melampsoraceae *sensu* Dietel (Wingfield *et al.*, 2004). We assume that it belongs to the Pucciniaceae *sensu* Dietel because its teliospores are stalked. Coffee is one of the most valuable commodities on the world market, and *H. vastatrix* causes the most important disease of it. So far, only urediniospore, teliospore and basidiospore stages have been reported, and an alternate host has not been found. Therefore, it is likely that the disease is spread exclusively by urediniospores. *Hemileia* was so named because its urediniospores are unusual in being half smooth, with the other half of their surface spiny like the urediniospores of other rusts. Urediniospores of *H. vastatrix* infect coffee leaves in the typical rust fashion, i.e. they require free water for germination and form appressoria over stomata. Resistance is expressed as a hypersensitive response after the formation of the first haustorium (Silva *et al.*, 2002). Unusual features include details of appressorium formation (Coutinho *et al.*, 1993), the fact that infection occurs exclusively through stomata located on the lower (abaxial) leaf surface, and the formation of urediniospores and teliospores through stomata (Ward, 1882), i.e. typical rust pustules rupturing the epidermis are not formed. Another remarkable feature is that removal of the coffee berries, which are not themselves infected by *H. vastatrix*, drastically reduces the severity of the disease (Monaco, 1977).

The story of coffee rust is fascinating and its first part has been told eloquently by Large (1940). The disease was discovered on cultivated coffee in Ceylon (now Sri Lanka) in about 1869. The origin of *H. vastatrix* seems obscure. Ceylon was then the major coffee-growing region for the

British Empire in which coffee drinking was fashionable. Initial outbreaks of the rust were not taken seriously because severe infections, which led to the defoliation of trees, showed their effect only in subsequent seasons in the shape of a gradual debilitation of the coffee plants (see Brown *et al.*, 1995). After the failure of several successive crops, H. Marshall Ward was sent to Ceylon in 1879 to investigate the problem, and he succeeded in elucidating the life cycle of *H. vastatrix* and the details of the infection process (Ward, 1882). Unfortunately, by that time it was too late to save the coffee production of Ceylon and other South East Asian countries to which the rust had spread in the meantime. As a consequence of coffee rust, tea became the main crop of Ceylon, and tea drinking was promoted throughout the Empire (Schuman, 1991).

The spread of coffee rust to all major coffee-growing regions of the world (except Hawaii) is charted in Fig. 22.14. It is probably the combined result of human travel and natural long-distance transport. The much-dreaded jump from West Africa, where the disease arrived in the 1960s, to Brazil probably occurred during a period between January and April in the late 1960s as a one-off transport of urediniospores by wind (Bowden *et al.*, 1971). A similar chance event of wind-borne intercontinental rust spore traffic may have happened in June 1978 when the sugarcane rust (*Puccinia melanocephala*) arrived in the Dominican Republic, probably from Cameroon (Purdy *et al.*, 1985; Brown & Hovmøller, 2002). Within coffee plantations, urediniospores of *H. vastatrix* may be spread both by wind and by rain splash.

Coffee rust can be controlled by fungicide applications, with copper-containing compounds being the most useful even today (Bock, 1962; Kushalappa & Eskes, 1989). Since the timing of fungicide application is crucial, disease forecast models to optimize fungicide applications are being developed. Resistance breeding is also promising. The main cultivated coffee plant, *Coffea arabica*, was clonally propagated by the Dutch in the late seventeenth century and is therefore genetically quite uniform across many coffee-growing areas, but the introduction of

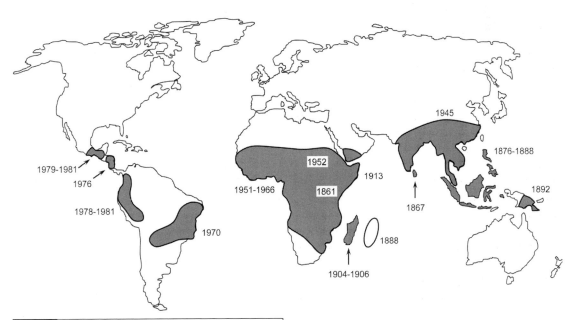

Fig 22.14 The journey of coffee rust (*Hemileia vastatrix*) around the world. Time points of first records are indicated. Redrawn from Schuman (1991), with supplementary data from Monaco (1977) and Schieber and Zentmyer (1984).

resistance genes from wild coffee species can give rise to cultivars with stable partial resistance. Altogether 9 major resistance genes and about 30 relatively stable races of *H. vastatrix* are known (Monaco, 1977; Kushalappa & Eskes, 1989).

22.7 | Melampsoraceae

22.7.1 *Cronartium*

About 20 species of *Cronartium* are known. They are macrocyclic and heteroecious, alternating between various dicotyledonous plants as principal hosts and *Pinus* spp. as alternate hosts. The one-celled teliospores of *Cronartium* are produced in long columns projecting out of the telia. The teliospores are attached to each other but not to their host tissue, and they are unstalked. In general, the mycelium on the alternate host is perennial whereas the principal host needs to be re-infected afresh each growing season. Reduced forms with aecia on *Pinus* spp. are named *Peridermium* (or *Endocronartium*). Their ability to re-infect the *Pinus* host by means of aeciospores produced from aecia indicates that these are aecidioid uredinia (Ono, 2002). *Cronartium* and *Peridermium* spp. form a well-resolved monophyletic group of rust species also clustering together with their telial hosts (Vogler & Bruns, 1998). Economic damage is generally caused mainly by the infection of *Pinus* spp. The symptoms are similar between the various species considered below.

Cronartium ribicola causes white pine blister rust, which alternates between *Ribes* spp. (currants, gooseberry etc.) and those *Pinus* spp. which produce their needles in groups of five (e.g. *P. strobus*). Infection of the alternate host proceeds by basidiospores germinating on the needles of *Pinus* spp. in autumn. During the following spring, infection spreads to the needle base and ultimately to the branches or main stem where lesions become visible as cankers. These produce spermogonia and aecia (Smith *et al.*, 1988). Large cankers can girdle the entire trunk of trees, causing host death and widespread economic and ecological damage to *Pinus* forests especially in North America and Europe. In summer, aeciospores infect *Ribes* spp., where uredinial and telial lesions arise on the leaves. Severe infections can cause premature leaf abscission, but this stage of the disease is not economically as serious as that on *Pinus*. It is at present impossible to control *C. ribicola*. Since its basidiospores cannot travel long distances, a massive attempt at eradicating the principal host was launched in North America. However, this campaign failed because *Ribes* spp. are too abundant and re-grow easily from roots or seeds remaining in the soil (Maloy, 1997). Fungicide treatment of infected trees was attempted by aerial sprays with cycloheximide (actidione), but this was ineffective (Dimond, 1966; Maloy, 1997). More recently, attention has turned towards the breeding of resistant cultivars. Although *C. ribicola* is probably of Asian origin and was introduced into Europe in the eighteenth century and to North America around 1900, major gene resistance does exist among individual plants of *Pinus* spp. previously unexposed to the rust (Kinloch & Dupper, 2002). Resistance is often based on a gene-for-gene interaction whereby incompatibility commonly occurs as a hypersensitive response at the stage of needle infection (Jurgens *et al.*, 2003). Resistance breeding holds promise because the populations of *C. ribicola* in Europe and North America are genetically quite uniform. The alternative to resistance breeding is to give up *P. strobus* as a forestry tree, but the ecological damage to forests which regenerate by natural rejuvenation remains immense (Kinloch, 2003).

Cronartium quercuum is the cause of fusiform rust especially on *Pinus taeda* and *P. elliottii*. These species were planted in the South-Eastern USA where *C. quercuum* originated on native, relatively resistant *Pinus* spp. The principal hosts are *Quercus* spp. The life cycle is similar to that of *C. ribicola*, as is the economic damage, with the exception that this species is still confined to the USA. The disease is now kept under control by the planting of rust-resistant cultivars (Powers & Kuhlman, 1997; Schmidt, 2003).

Cronartium flaccidum causes resin top disease on Scots pine (*Pinus sylvestris*), *P. nigra*, *P. pinaster* and other species in Europe (Smith *et al.*, 1988). The principal hosts are various herbaceous

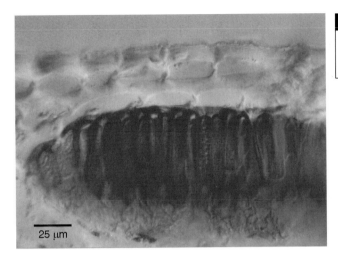

25 μm

Fig 22.15 Section through a telial lesion of *Melampsora euphorbiae* on *Euphorbia peplus*. The teliospores are formed in a crust-like layer beneath the epidermis.

plants. A similar disease is caused by the purely aecial (anamorphic) *Peridermium pini*. Both species are genetically identical, and Hantula *et al.* (2002) have recommended that they be considered as one species.

22.7.2 *Melampsora* and *Melampsoridium*

In *Melampsora* spp. the unicellular teliospores are sessile and often form a subepidermal crust (Fig. 22.15). Germination is by an external metabasidium of the usual type. The aecia lack peridia so that they are diffuse instead of cup-shaped. Such diffuse aecia are called **caeomata** (singular caeoma). *Melampsora lini* var. *lini* is an autoecious rust common on *Linum catharticum*, and *M. lini* var. *liniperda* infects cultivated flax. It was this fungus which Flor (1955, 1971) used for his pioneering work on the gene-for-gene hypothesis. Several *Melampsora* spp. (*M. populnea*, *M. larici-populina*, *M. medusae*, *M. allii-populina*) produce brightly coloured and extremely abundant uredinia and, later in the season, telia on *Populus* (poplar) trees (Plate 12g). Their alternate hosts are *Larix*, *Abies*, *Picea*, *Pinus* and *Allium* spp.

(Smith *et al.*, 1988). Some of these rusts seem to differ very little other than in their aecial host, and much more work is required before an acceptable species concept is in place. Further, hybridization between different rusts can occur (Spiers & Hopcroft, 1994). There is an even more bewildering complex of *Melampsora* rusts on willows (*Salix* spp.), alternating with *Allium*, *Larix*, *Ribes* or orchid species (Smith *et al.*, 1988; Pei *et al.*, 1993). The most important species is *M. epitea*, which alternates with *Larix* and exists as numerous races. Since willow and poplar trees can be important crops in certain forest situations, efforts at resistance breeding are being made (Ramstedt, 1999).

Melampsoridium betulinum causes a rust on birch as its uredinial and telial host, on which it is extremely common everywhere. *Larix* is the alternate host. In contrast to *Melampsora* spp., aecia of *Melampsoridium* are surrounded by a peridium, and phylogenetic studies have also shown that the genera *Melampsora* and *Melampsoridium* are not particularly closely related (Maier *et al.*, 2003).

Ustilaginomycetes: smut fungi and their allies

23.1 | Ustilaginomycetes

The Ustilaginomycetes are one of the four main classes of Basidiomycota and contain about 1500 species (Kirk *et al.*, 2001). In its present form as circumscribed by Begerow *et al.* (1997) and Bauer *et al.* (1997, 2001), this group is monophyletic. Hypha-producing Ustilaginomycetes are united by their lifestyle as ecologically obligate plant pathogens, often with an additional free-living (saprotrophic) yeast phase. They can be distinguished from the rust fungi in that haustoria are either altogether absent or, where present, take the shape of simple intracellular hyphae or hyphal extensions which invaginate the host plasmalemma but are not differentiated into a narrow neck and a wider haustorial body. Further, intracellular hyphae of Ustilaginomycetes usually secrete a thick sheath which is readily visible by transmission electron microscopy (see Figs. 23.6 and 23.17). The septa either lack perforations or contain simple pores or dolipores which are similar to those in the Urediniomycetes in lacking parenthesomes. True clamp connections are not usually found. The basidia of smut fungi produce numerous basidiospores whereas those of rust fungi usually produce only four.

The class Ustilaginomycetes has been divided into three subclasses by Begerow *et al.* (1997, 2000). We shall consider representatives of two of these. The Ustilaginomycetidae (Section 23.2) are the most important plant-pathogenic Ustilaginomycetes, causing smut-like symptoms.

Typical members of the Exobasidiomycetidae (Section 23.4) cause other biotrophic diseases and are distinguished from the former by producing basidia directly from parasitic mycelium, not from teliospores. An exception is *Tilletia*, which is so similar to smuts in the Ustilaginomycetidae that we discuss it alongside them (p. 650). The third group, Entorrhizomycetidae, contains fungi which produce galls on the roots of Cyperaceae and Juncaceae (Vánky, 1994). Numerous Ustilaginomycetes live exclusively or predominantly as yeasts, and an example of this lifestyle is the order Malasseziales described on p. 670.

Fungi causing smut-like disease symptoms (Fig. 23.7, Plate 12h) have arisen at least twice within the Basidiomycota, namely in the Ustilaginomycetes and the Urediniomycetes. Urediniomycetous smuts are found in the Microbotryales, and since these share many biological features with the 'true' smuts in the Ustilaginomycetes and have traditionally been studied by smut specialists, we shall consider the Microbotryales in this chapter (Section 23.3). Therefore, in our usage the term 'smut fungus' has a biological rather than a taxonomic meaning.

23.2 | The 'true' smut fungi (Ustilaginomycetes)

The word 'smut' describes the causal fungus or the symptoms of a particular group of plant diseases in which loose masses (**sori**) of dark spores are produced in infected plant organs. Leaves, stems, flowers and seeds of grasses and other herbaceous plants are particularly frequently attacked. The term 'bunt' is sometimes used for smut fungi infecting the ovaries of their hosts, the seed becoming filled with teliospores in place of the embryo. Host tissues from which the spores are released often appear as if burnt or scorched, and this is why de Bary (1853) used the term 'Brandpilze' to describe the smut fungi. Various names have been given to these spores, e.g. teliospore, chlamydospore, brand spore, melanospore, ustospore or ustilospore. We prefer to call them **teliospores** because their function in the life cycle of smut fungi is equivalent to that of teliospores in the rust life cycle, i.e. they are the site of nuclear fusion and, on germination, give rise to the promycelium in which meiosis occurs.

Several good monographs have been written about smut fungi, especially by Vánky (1987, 1994). The surface ornamentation of teliospores (Figs. 23.2, 23.9) and their method of germination (Figs. 23.3, 23.4, 23.5, 23.10, 23.12) as well as the type of symptoms and host range are important characters in identification.

23.2.1 The life cycle of smut fungi

The life cycle of smut fungi can be divided into two phases, a yeast-like monokaryotic (homokaryotic) form which grows saprotrophically but is unable to infect plants, and a predominantly dikaryotic (heterokaryotic) mycelial phase which is infectious on host plants but cannot grow saprotrophically (Fig. 23.1). The infectious dikaryotic mycelium is often systemic, showing inter- or intra-cellular growth. Meristematic host tissues are particularly densely colonized. Usually there are no specialized haustoria, but hyphae may enter or even grow through host cells, invaginating the host plasmalemma. Intracellular hyphae are surrounded by a partial or complete sheath (Fig. 23.6) which is formed partly by the secretory activity of the pathogen and partly by the contributions from the infected host cell (Bauer *et al.*, 1997). During sorus development, the dikaryotic hyphae proliferate and mass together in intercellular spaces, often

destroying the softer internal host tissues but remaining enclosed by the host epidermis. There are no phialides or other specialized spore-producing structures, but most hyphal segments can become converted to spores. The sporogenous hyphae are composed of binucleate cells. After nuclear fusion, the walls thicken and gelatinize, and the cells fragment. They then enlarge and become globose, and finally a thick wall is laid down (Snetselaar & Mims, 1994; Banuett & Herskowitz, 1996). The gelatinous matrix disappears at maturity. The ripe, uninucleate, diploid teliospores have thick, usually dark walls which may be smooth or ornamented by spines or reticulations (Vánky, 1987, 1994; Piepenbring *et al.*, 1998a,b). In some genera, e.g. *Urocystis*, a central fertile cell is surrounded by a group of sterile cells.

The teliospores are commonly dispersed by wind or become attached to the surface of seeds. Teliospores of many smut fungi germinate in a similar way to those of rust fungi by forming a septate promycelium. The teliospore is therefore the probasidium, with the promycelium being the metabasidium. If this is septate, it is a **phragmobasidium**. Teliospore germination can proceed in several different ways which are shown in subsequent sections. Typically, numerous haploid basidiospores (often called **sporidia**) are produced by direct budding from segments of the promycelium, and these are unable to re-infect the host. Instead, they germinate by further budding to form elongated yeast cells which are capable of prolonged saprotrophic growth. Compatible yeast cells meeting on the surface of the host plant conjugate, and this initiates the dikaryotic hyphal stage which is able to infect the host plant.

23.2.2 Mating systems

Many smut fungi are heterothallic with a unifactorial (bipolar) mating system (i.e. one mating type locus with two alleles), but *Ustilago maydis* and a few other species are tetrapolar, with two mating type loci located on different chromosomes. The *a* locus has two idiomorphs and controls fusion of sporidia which is driven by the exchange of mating pheromones between

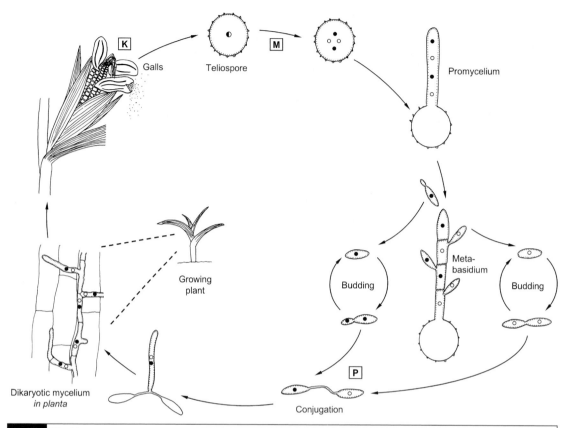

Fig 23.1 The life cycle of *Ustilago maydis*, diagrammatic and not to scale. Teliospores are formed in galls developing from dikaryotic hyphae. Karyogamy (K) occurs in the teliospore, followed by meiosis (M) and germination to form a three-septate promycelium. Each segment buds off numerous haploid sporidia. Sporidia bud in a yeast-like manner and are capable of saprotrophic growth but unable to infect maize plants. When yeast cells or sporidia of compatible mating type meet, conjugation tubes are formed, and their fusion (plasmogamy, P) gives rise to a dikaryotic hypha which is infectious on maize. Open and closed circles represent haploid nuclei of opposite mating type; the diploid nucleus is drawn larger and half-filled.

compatible cells. The multiallelic *b* locus controls growth of the dikaryon, as well as its ability to infect the host and to complete sexual development (see p. 643). In the bipolar species *U. hordei*, the genetic control of fusion and dikaryon growth is very similar to that in *U. maydis*, with the important differences that there are only two instead of multiple *b* alleles and that the *a* and *b* loci are tightly linked on the same chromosome (Lee *et al.*, 1999). Genetic recombination during meiosis is suppressed in the region containing the *a* and *b* loci. This region is substantially larger than the mating type genes themselves, making up about one-sixth of the chromosome on which it resides. Hence, this chromosome is regarded as a primitive form of

the sex chromosome as found, for example, in mammalian organisms (Fraser & Heitman, 2004). Mammalian sex chromosomes are recognizably different when they are condensed at mitosis, the X chromosome being very much larger than the Y chromosome. Intriguingly, measurable size differences have also been found between the two sex chromosomes of the urediniomycetous smut, *Microbotryum violaceum* (Hood, 2002).

The smut fungi are **dimorphic**, producing both yeast cells and true hyphae. This feature has aroused the interest of cell biologists, and especially *Ustilago maydis* is being used extensively as a tool to examine fundamental aspects of eukaryotic biology (see Fig. 23.8).

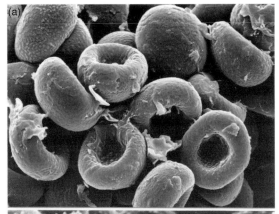

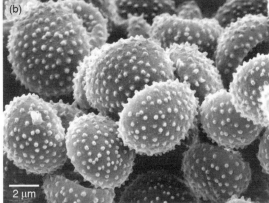

Fig 23.2 Teliospore surfaces of *Ustilago* spp. (a) Smooth surface in *U. hordei*. (b) Spiny surface in *U. nuda*. From Vánky (1994); original prints kindly provided by K. Vánky.

23.2.3 Teliospore release and dispersal

Piepenbring *et al.* (1998c) have written an excellent review of the teliospore as a dispersal unit in a wide range of smut fungi, and we can only summarize a few salient features here. Teliospores of most species are produced in a sorus enclosed by a thin layer of host tissue. They are generally released dry. Spore release is often by the simple rupture of the host epidermis surrounding the sorus, possibly as a result of pressure exerted by the expanding teliospores. The host epidermis may aid in the release of teliospores, e.g. if it is hit by water drops. The mechanism in this case is similar to that found in puffballs (see p. 578). Gusts of wind which shake infected host organs may also be effective in releasing smut spores over time. Alternatively,

entire infected host organs or galls may be dispersed, and teliospores or mycelium of smut fungi may be borne on or in seeds.

The most common dispersal route in the smut fungi is by wind-blown teliospores, and the transatlantic wind-borne passage of African sugarcane smut, *Sporisorium scitamineum*, has been proposed as the most likely route of entry into the Caribbean. Teliospores of many smuts may adhere to the seeds of their host plants and germinate with them. Certain species causing loose smuts of cereals (see p. 639) systemically infect the living embryo of the cereal grain. Grains contaminated by external spore dusts or systemic infections can be spread by human transport. Insect dispersal is important in the anther smut, *Microbotryum violaceum*, which is related to the rust fungi (see p. 652).

23.2.4 *Ustilago* species on monocotyledonous hosts

Several *Ustilago* spp. cause diseases on grasses and cereals. Sori of the agriculturally most important species, *U. hordei*, *U. nuda*, *U. tritici* and *U. avenae*, are produced in place of the developing seeds. Other *Ustilago* spp. affect the leaves of their hosts, e.g. *U. filiformis* (formerly *U. longissima*), which causes leaf stripe smut on *Glyceria* spp. *Ustilago maydis* infects both vegetative and reproductive organs of its host (Plate 12h), causing gall-like deformations. This species is considered in detail on pp. 643–647. Many *Ustilago* spp. have also been described from dicotyledonous hosts, but it now seems that all of them belong to *Microbotryum* (Vánky, 1998, 1999; Almaraz *et al.*, 2002; see p. 652).

Phylogenetic studies have shown *U. hordei*, *U. nuda*, *U. tritici* and *U. avenae* to be closely related to each other, to the point where it becomes difficult to distinguish individual species by DNA sequences and microscopy. They are regarded as the core species of *Ustilago* (Stoll *et al.*, 2003). Since these species can hybridize with each other, it has been proposed that they should be merged into one taxon, *U. segetum* (see Bakkeren *et al.*, 2000). However, because the existing names are so well-established and because well-known plant diseases are caused by

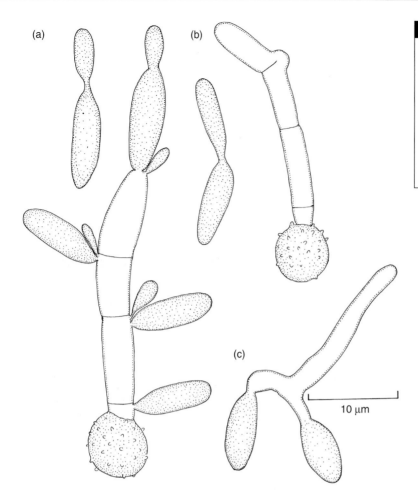

(a)

(b)

(c)

Fig 23.3 Teliospore germination in *Ustilago avenae*. (a) Germinating teliospore showing the four-celled promycelium, each cell of which is producing sporidia. Budding of detached sporidia is also shown. (b) Germinating teliospore showing fusion of the two terminal cells to initiate a dikaryon. (c) Fusion of conjugation tubes from two basidiospores to initiate a dikaryon.

10 μm

these cereal smuts, we prefer to keep their original names for the time being. The corn smut pathogen, *U. maydis*, is not closely related to the core *Ustilago* spp. and instead occupies an intermediate position between *Ustilago* and *Sporisorium* (Stoll *et al.*, 2003). In total, there are about 230 *Ustilago* spp. and 190 *Sporisorium* spp. (Kirk *et al.*, 2001), although such numbers obviously vary with the species concept adopted.

The surface of teliospores is an important aid in identification. *Ustilago hordei* has a smooth teliospore surface whereas most other cereal smuts have spiny surfaces (Figs. 23.2a,b). Huang and Nielsen (1984) have shown that this difference between smooth and spiny surfaces is due to only two genes. Many of those *Ustilago* spp. now grouped with *Microbotryum* have teliospore surfaces with conspicuous reticulations or

striations. There are also variations in the processes of teliospore germination in *Ustilago* (Ingold, 1983c; Vánky, 1994). The classical pattern is shown by *U. avenae* (Fig. 23.3), *U. hordei* and *U. maydis* in which the promycelium is three-septate. All four cells give rise to sporidia, and compatible sporidia fuse following pheromone stimulation (see p. 643). When attached to a host plant surface, teliospores of *U. avenae* germinate in a different way; instead of producing sporidia, adjacent compatible cells of the promycelium fuse directly (Fig. 23.3b; Vánky, 1994). In *U. nuda* (Fig. 23.4), direct fusion of promycelium cells occurs both on agar and in nature. However, following synchronous division of the two nuclei in the fusion cell, septa are laid down such that a mosaic of both mono- and dikaryotic cells results. Compatible monokaryotic cells may fuse

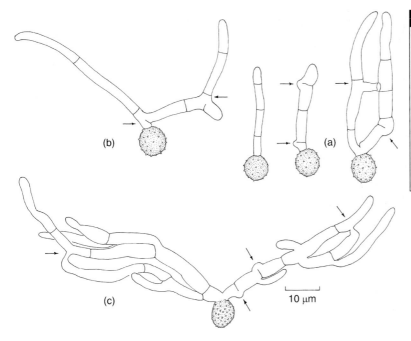

Fig 23.4 Teliospore germination in *Ustilago nuda*. (a) Three teliospores 20 h after germination, showing various stages of development of the promycelium. The arrows indicate points where cell fusion has occurred. (b) A later stage showing the extension of mycelium from the fusion cells. (c) Two-day-old germinating teliospore showing repeated cell fusions (arrows).

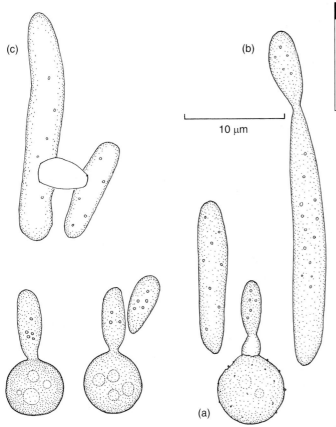

Fig 23.5 Teliospore germination in *Ustilago filiformis*. (a) Teliospore germination by successive development of sporidia. Note the absence of an extended promycelium. The first-formed sporidia are short. (b) Sporidium showing budding. (c) Sporidia conjugating. A dikaryotic mycelium arises following conjugation.

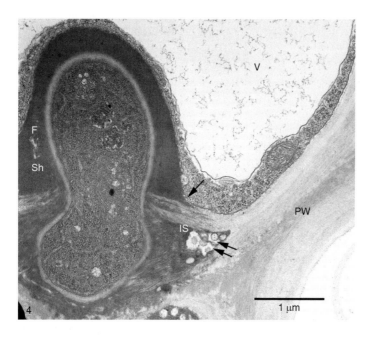

Fig 23.6 TEM of *Ustilago hordei* infecting a compatible barley cultivar. The host cell consists of a large vacuole (V) surrounded by a thin layer of cortical cytoplasm. The intracellular fungal hypha (F) has penetrated the plant cell wall (PW) and invaginated the host plasmalemma (single arrow); note the thick electron-dense sheath (Sh) between the hyphal wall and the host plasmalemma. The double arrow points to vesicles in the intercellular space (IS). Photomicrograph taken by G. Hu; reprinted from Hu *et al.* (2003), with permission from Elsevier.

Fig 23.7 Loose smuts. (a) Infection of *Ustilago avenae* on oats. All seeds on the affected inflorescence have been replaced by teliospore sori. (b) *Ustilago tritici* causing loose smut on wheat. A healthy wheat ear is shown on the left.

dikaryophase is established by fusion of these germ tubes (Malik, 1974). *Ustilago filiformis* has teliospores which, on germination, do not produce an obvious promycelium, but merely a short tube from which sporidia are budded off successively (Fig. 23.5). All of the above fusion events ultimately result in a dikaryotic hypha which infects the host plant.

In contrast to *U. maydis*, many small-grain cereal smuts entertain a gene-for-gene relationship with their hosts. In incompatible interactions, the infection hypha is encased in an unusually thick sheath as soon as it penetrates the first epidermal cell. This is closely followed by the death of the infected cell through a hypersensitive response (Hu *et al.*, 2003). The breeding of resistant cultivars can be an effective means to control these pathogens (Smith *et al.*, 1988). As we have already seen in the rust fungi (p. 625), the evolution of numerous races of smut fungi causes difficulties in cultivating cereals carrying major resistance genes. The spread of these races can be monitored in time and space by screening field isolates against a differential of cereal cultivars (Menzies *et al.*, 2003). Distinct smut races show a high tendency to hybridize under natural conditions, so that new races can arise rapidly (Thomas, 1984).

with each other again at a later stage (Nielsen, 1988). In *U. tritici* there are also no sporidia, but here each uninucleate cell of the septate promycelium gives rise to a germ tube, and the

Ustilago hordei

This is the type species of *Ustilago*, causing **covered smut** of barley and oats. Crop losses are usually less than 1% in well-managed agrosystems (Thomas & Menzies, 1997). The infection cycle has been described in detail by Hu *et al.* (2002). The term 'covered smut' implies that the teliospores produced in the cereal grains replace the internal tissues but remain covered by the outer layer (pericarp) of the grains and are released only during threshing. Teliospores attached to the outer surface of seeds will germinate at the time of host germination. The fusion of compatible monokaryotic yeast cells results in a dikaryotic hypha which infects the young seedling. The epidermal cells are penetrated directly from slightly swollen hyphal tips, which may be regarded as rudimentary appressoria. There then follows an extended phase of systemic growth without outward symptoms of infection. Colonization of the host is chiefly intracellular but also intercellular. Differentiated haustoria are not produced. Instead, hyphae grow through host cells, invaginating the plasmalemma in the process. A thick sheath is deposited between the plasma membranes of the host and the pathogen (Fig. 23.6). By the time the mycelium reaches the apical meristem (about 40–55 days post infection), the host has usually formed the inflorescence initials, and massive proliferation of the mycelium occurs from that stage onwards. The developing spike tissue becomes filled with hyphae which then branch profusely and form teliospores. This infection sequence of a prolonged symptomless colonization followed by symptom development in the reproductive structures of the host is typical of the small-grain cereal smuts but differs from the infection of maize by *U. maydis* (see below).

Ustilago avenae, U. nuda and U. tritici

These species cause **loose smut** of their cereal hosts, i.e. the kernels are replaced by sori producing teliospores, with entire glumes usually destroyed. *Ustilago avenae* infects oats (Fig. 23.7a) and false oat-grass, *Arrhenatherum elatius*. There are numerous races which are specific to different oat cultivars. The disease cycle is complex. Teliospores have been shown to survive for 13 years and thus the fungus can infect seedlings from teliospores dusted onto the outer surface of seeds. Additionally, it is capable of limited systemic infection of the seed pericarp which is initiated during flowering (Neergaard, 1977), and such infections are genuinely seed-borne. Either way, the fungus systemically colonizes the growing host plant and proliferates during flowering to produce a crop of teliospores. Systemically infected hosts flower slightly earlier than healthy plants, and they release their teliospores at the time when healthy plants flower. Teliospores released at that point can germinate on healthy flowers (Mills, 1967), leading to systemic infections which may be carried over to the next growing season within viable seeds.

Ustilago nuda and *U. tritici* cause loose smut on barley and wheat, respectively (Fig. 23.7b). Their disease cycles are very similar to each other. Systemic infection of the host plant gives rise to smutted heads and, as in the case of *U. avenae*, flowering of smutted plants occurs slightly earlier than that of healthy plants, so that the disease can spread to uninfected flowers by means of teliospores. The normal entry point of dikaryotic mycelium was formerly thought to be the stigma of healthy flowers, but Batts (1955) showed that it is, in fact, the young tissue at the base of the ovary. The mycelium survives systemically in infected embryos. In spring, systemic infections spread when these embryos germinate. In contrast to *U. avenae*, teliospores of *U. nuda* and *U. tritici* are short-lived, rarely surviving for more than a few days under normal conditions. Hence, infection from teliospore dust on the surface of seeds does not seem to be an important infection route.

23.2.5 *Ustilago maydis*

This species is the cause of corn smut and is by far the most thoroughly researched member of the Ustilaginomycetes (reviewed by Kahmann *et al.*, 2000). One of the early research highlights using *U. maydis* was on mitotic recombination, leading to the development of the 'Holliday model' to explain the exchange of DNA strands at 'Holliday junctions' between paired DNA double helices (see p. 317; Holliday, 1962, 1964). Areas of ongoing research on *U. maydis* are described in subsequent sections. Work on *U. maydis* has greatly benefited from the ease with which the haploid phase of this fungus can be grown in culture and transformed by molecular biology tools. The complete genome of this fungus has now been sequenced, and this should stimulate further research.

The teliospores of *U. maydis* are long-lived and can survive in the environment (e.g. the soil) for several years. Dikaryotic hyphae resulting from the fusion of haploid sporidia can infect any above-ground tissue of its host (wild and cultivated *Zea mays*) by entry through stomata or direct, appressorium-mediated penetration of the epidermis (Snetselaar & Mims, 1992; Banuett & Herskowitz, 1996). Infected host organs become strongly hypertrophied to form galls. This is in contrast to the small-grain cereal smuts described above, in which symptom development is confined to the developing seeds of the host, and infection usually occurs at the seedling stage followed by a prolonged symptomless phase. In *U. maydis*, the entire process from dikaryon formation on the plant surface to the release of mature teliospores may take as little as 2 weeks (Banuett & Herskowitz, 1996). In the field, infections by *U. maydis* are most commonly observed on the cobs presumably because the stigmata with their thin epidermal layer are most readily penetrated by the fungus (Snetselaar & Mims, 1993). As a result of infection, the developing seeds on the corn cob become replaced by gall-like outgrowths ('tumours') which measure about 1–5 cm in diameter (Plate 12h). Although not generally welcomed by farmers, such infected cobs are prized as a delicacy in the Mexican cuisine (Pataky & Chandler, 2003).

Monokaryotic yeast cells of *U. maydis* synthesize auxins, especially indole-3-acetic acid, in pure culture (see Basse *et al.*, 1996), and infected hypertrophied host tissue also shows elevated concentrations of this plant growth hormone. It therefore seems likely that the production of growth hormones by *U. maydis in planta* contributes to the development of the striking disease symptoms, although this has not yet been

formally proven. The mycelium of *U. maydis* ramifies in these hypertrophied tissues, followed by hyphal fragmentation and production of teliospores. Although the dikaryotic phase is obligately biotrophic in nature, *U. maydis* can be stimulated to complete its life cycle in artificial conditions if it is grown on living cell cultures of *Zea mays* separated from the dikaryotic mycelium by a membrane permitting the diffusion of metabolites (Ruiz-Herrera *et al.*, 1999). Deviations from the life cycle as shown in Fig. 23.1 have been described by Kahmann *et al.* (2000).

Mating and dikaryon establishment

Ustilago maydis is heterothallic and has two mating type loci. A given dikaryon is fully pathogenic only if it contains different idiomorphs or alleles at both loci. Since it will produce haploid progeny of four genetically distinct types, the mating system is said to be tetrapolar. Locus *a* has two idiomorphs, *a1* and *a2*, each of which encodes a mating pheromone and the receptor for the pheromone encoded by the opposite idiomorph. Two haploid cells will mate if they contain opposite idiomorphs at locus *a*, irrespective of their *b* alleles. Conjugation can be induced experimentally even between cells containing the same idiomorph if the matching pheromones are added. Mating is therefore purely driven by the pheromones which are peptides containing 13 (peptide a1) or 9 (peptide a2) amino acids derivatized with a lipid (farnesyl) side-chain (Spellig *et al.*, 1994). A cell whose receptor has bound the pheromone of the opposite mating type will arrest its cell cycle in the G2 position and undergo a morphogenetic change to produce a thin flexible conjugation tube which grows chemotropically towards the pheromone source (Snetselaar *et al.*, 1996).

The ability to infect maize plants and to complete the life cycle is tightly linked to the ability of the fusion cell to form dikaryotic hyphae. This is controlled by the *b* locus after conjugation of compatible cells. In contrast to the *a* locus which has two idiomorphs with low sequence homology, the *b* locus has about 25 alleles which are genetically relatively similar to each other. Each allele encodes two proteins

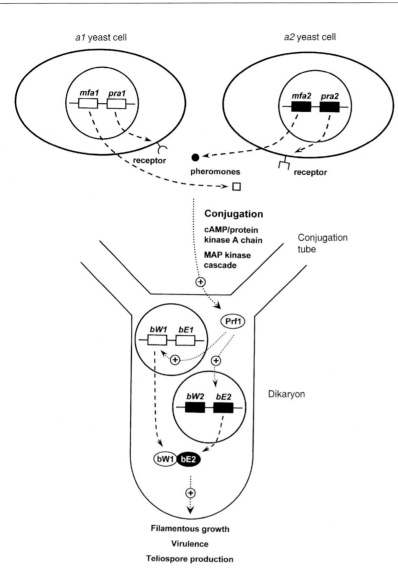

Fig 23.8 The roles of the *a* and *b* loci in the conjugation between monokaryotic yeast cells and the establishment of dikaryotic filamentous growth in *Ustilago maydis*. Signalling events are represented by dotted lines; diffusion or transport of signalling molecules is indicated by dashed lines. Based on Bölker (2001).

called bE and bW (Fig. 23.8), and the dikaryotic phenotype depends on the formation of a heterodimer of proteins transcribed from different alleles. For instance, bE1 and bW2 can dimerize but bE1 and bW1 cannot. The heterodimer is a transcription factor which triggers signalling cascades involved in maintaining dikaryon stability, hyphal growth and virulence on maize. At the same time, further mating reactions are suppressed.

Signalling events involved in conjugation and the subsequent dimorphic switch from yeast-like monokaryotic to hyphal dikaryotic growth are being extensively examined and have been reviewed, for example, by Banuett (1995), Bölker (2001), Martínez-Espinoza *et al.* (2002) and Feldbrügge *et al.* (2004). Although it is impossible to discuss the details here, we can summarize that a cAMP and protein kinase A signalling chain is involved in maintaining growth by budding, whereas a MAP kinase cascade is crucial for the switch to hyphal growth and pathogenicity. These two branches seem to converge on a transcription factor called Prf1 which is subject to phosphorylation at different sites by protein kinase A and a MAP kinase (see Fig. 23.8). Of course, the cAMP/protein kinase A and MAP kinase signalling chains fulfil many other functions at subsequent developmental stages such as gall formation and teliospore production (Gold *et al.*, 1997). For instance, the *fuz7* gene product which encodes a MAP kinase–kinase involved in the signalling chain shown in Fig. 23.8 is also involved in the process of hyphal fragmentation leading to teliospore production (Banuett & Herskowitz, 1996). Further, there is extensive cross-talk between the cAMP/ protein kinase A and MAP kinase chains at different time-points in development.

The cytoskeleton

Much valuable work has been performed on *U. maydis* to assign various cellular transport phenomena to particular elements of the cytoskeleton and their associated motor proteins. Thus, microtubules have been implicated in the transport of nuclei, mitochondria, vacuoles and the endoplasmic reticulum, as well as secretory vesicles in exocytosis and endosomes in endocytosis (Steinberg, 2000; Basse & Steinberg, 2004; Steinberg & Fuchs, 2004). Bidirectional movement can be achieved by kinesin motors which move towards the polymerizing end (plus end) of microtubules, and dynein which moves towards the minus end. Actin cables and their myosin motors are also involved in morphogenetic events and transport processes. Since the genome of *U. maydis* has been completely sequenced, the number of genes encoding myosin (3), dynein (1) and kinesin (10) is known (Basse & Steinberg, 2004) and further rapid progress on the role of the cytoskeleton in morphogenesis and cellular transport can be expected with *U. maydis* as an experimental organism.

Mycoviruses and killer toxins in *U. maydis*

Although mycoviruses are not uncommon in fungi, few of them have been investigated in detail. Two examples we have encountered in earlier chapters of this book are the virus-like particles in *S. cerevisiae* (p. 273) and related yeasts which contain double-stranded RNA (dsRNA) encoding killer toxins, and the hypovirulence-causing dsRNA viruses of *Cryphonectria parasitica* (p. 375) and *Ophiostoma novo-ulmi* (p. 366). Mycoviruses infecting *Ustilago maydis* are similar to those of *S. cerevisiae* in that they encode killer toxins. The first evidence of them was found when certain *U. maydis* strains killed sexually compatible strains upon anastomosis in mating assays. The cytoplasmic inheritance of the killing trait, the proteinaceous nature of the toxin and the presence of virus particles in killer strains were quickly established (Hankin & Puhalla, 1971; Wood & Bozarth, 1973). There are three types of virus (P1, P4 and P6) each encoding its own killer protein (KP) toxin. Day (1981) showed that virus-infected *U. maydis* strains are common in field populations.

A great deal is now known about viruses of *U. maydis* (see Magliani *et al.*, 1997; Martínez-Espinoza *et al.*, 2002). Their genomes are fragmented into three size classes of dsRNA, whereby each size class can have several members. In total, there are six dsRNA fragments in P1, seven in P4 and five in P6, with one capsid able to accommodate either one H or one to several M chains (Bozarth *et al.*, 1981). In all three viruses, a heavy (H) segment encodes the capsid protein and the replication machinery, and H segments are also essential for the maintenance of the medium-sized (M) and light (L) segments. The toxins are encoded by the M fragments, and their synthesis as prepropolypeptide chains followed by proteolytic cleavage and secretion of the mature toxins is similar to that of the *S. cerevisiae* killer toxins (p. 273). The function of the L segments is unknown at present.

The modes of action of the three toxins seem to be diverse. The best-characterized is KP4 which is active as a monomer blocking certain types of Ca^{2+} uptake channel. This activity can be observed in susceptible *U. maydis* strains as well as in mammalian cells, where it acts in a similar way to the black mamba snake venom, calciseptine (Gage *et al.*, 2001, 2002). The KP1 and KP6 toxins are released as two separate polypeptides after proteolytic cleavage, but in contrast to the yeast killer toxins these do not re-associate with each other by covalent (disulphide) bonds. Whilst the mode of action of KP1 is unknown, KP6 may form membrane pores which disrupt the ionic balance of the target cells (N. Li *et al.*, 1999). The toxin-producing cell must obviously be resistant to its own toxin. In contrast to the *S. cerevisiae* system where the unprocessed toxin precursors bestow resistance, in toxin-producing *U. maydis* strains resistance is encoded by nuclear genes. Resistance is specific, i.e. a KP1-producing strain is sensitive to KP4 and KP6. All three toxins are also effective against other members of the Ustilaginales (Koltin & Day, 1975).

Attempts have been made to transform wheat plants with the KP4 toxin gene of the *U. maydis* virus P4 in order to engineer cultivars with resistance against *Tilletia caries* (Clausen *et al.*, 2000), but no recent reports seem to have been published on this subject. Even if cultivars with good resistance in the field can be produced, it is unlikely that a transgenic crop plant expressing a calciseptine-like toxin will gain acceptance with regulatory authorities or the general public.

23.2.6   *Tilletia*

There are about 125 species of *Tilletia*, all of which parasitize grasses (Poaceae). In economic terms, the most important pathogens are *T. caries* (= *T. tritici*) and *T. indica* on wheat, *T. controversa* on wheat and other cereals, and *T. horrida* on rice. Many other *Tilletia* spp. infect wild grasses (Vánky, 1994). Teliospores of *Tilletia* spp. typically bear reticulate surface ornamentations (Fig. 23.9). They are long-lived in the soil and on the exterior of seeds. Infections are systemic and result in covered smut symptoms in the seeds.

Tilletia caries

This species is cosmopolitan and causes 'common bunt', the best-known covered smut disease of wheat. The entire interior of the infected grain becomes converted to a greenish-brown teliospore sorus surrounded by the pericarp. Such sori are called 'bunt balls'. The teliospores are not released until threshing, when they are dusted onto the surface of healthy grains. If heavily contaminated crops are processed, the teliospore concentrations in the air can be sufficiently high to cause dust explosions in mills or storage facilities. Respiratory allergies among millers caused by *T. caries* were also common in the past, but are much less frequent now because the incidence of common bunt has declined.

Crushed sori have a fishy smell caused by the presence of trimethylamine. For this reason, the disease is also known as 'stinking smut', and flour made from contaminated grain is unfit for human consumption. Since teliospores are not readily released in the field, Piepenbring *et al.* (1998c) have suggested that *T. caries* may be an example of a fungus that has adapted to humans as a dispersal vector.

This fungus was a major pathogen in the past and occupies a special place in the history of plant pathology (Large, 1940; Ainsworth, 1981). In 1752, Mathieu Tillet, by careful experimentation, demonstrated that the common bunt disease was associated with dusting the seeds with bunt spores prior to sowing. He also reported that incidence of the disease was somewhat reduced by steeping the seeds in sea water and lime prior to sowing. In 1807, Bénédict Prévost observed the process of teliospore germination with his microscope, and he proposed that the bunt disease was caused by a living organism. Further, he discovered that teliospore germination was inhibited by copper salts, and that the treatment of wheat seeds with dilute copper sulphate solution was effective against infections of *T. caries*. The combination of copper and lime became known as the famous Bordeaux mixture only much later, in about

Fig 23.9 SEM of teliospores of *Tilletia caries*. From Vánky (1994); original prints kindly provided by K. Vánky.

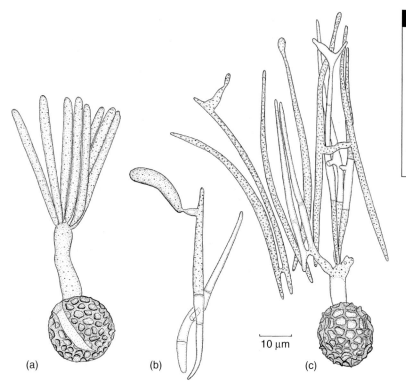

Fig 23.10 *Tilletia caries.*
(a) Germinating teliospore showing a non-septate promycelium and a crown of primary sporidia.
(b) Two detached primary sporidia showing conjugation. A secondary sporidium has developed from one of the primary sporidia.
(c) Primary sporidia attached to the promycelium showing conjugation.

10 μm

(a)   (b)   (c)

1882, when Alexis Millardet discovered its ability to control downy mildew of vines (see p. 119).

The process of germination of *T. caries* teliospores is shown in Fig. 23.10. The diploid nucleus in the mature teliospore divides meiotically, and one or more mitotic divisions follow so that 8 or 16 nuclei are formed. The promycelium is often but not invariably aseptate, and from its tip narrow curved uninucleate primary sporidia arise, corresponding in number to the number of nuclei in the young promycelium (Fig. 23.10a). The primary sporidia mate in pairs by means of short conjugation tubes, often whilst still attached to the tip of the promycelium (Fig. 23.10c). Detached primary sporidia may also conjugate. During conjugation, a nucleus from one primary sporidium passes into the other sporidium which therefore becomes binucleate. Each H-shaped pair of primary sporidia develops a single lateral sterigma on which a curved binucleate spore develops (Fig. 23.10b). This spore is projected actively from the sterigma using the surface-tension catapult mechanism, and it is sometimes referred to as the **secondary sporidium**. Because of its characteristic method of discharge, Buller and Vanterpool (1933) have interpreted this spore as a basidiospore. The secondary sporidium brings about infection of the host.

Teliospores of *T. caries* (Fig. 23.9) are viable for up to 15 years and germinate along with the seeds if contaminated grain is sown. Secondary sporidia produce germ tubes which infect the coleoptiles of the seedlings. Infection is systemic, the mycelium growing through the tissues of the shoot, and by suitable techniques it is possible to isolate the dikaryotic mycelium from infected host tissues (Trione, 1972). Although infected plants may grow less vigorously than uninfected ones, they show no outward sign of infection until the ears are almost ripe.

Tilletia controversa

This species is closely related to *T. caries* and is most damaging on wheat, in addition to infecting other cereals and grasses. The crucial difference is that teliospores of *T. controversa* germinate at much lower temperatures than those of *T. caries*. In consequence, *T. controversa* causes bunt mainly on winter wheat, infecting overwintering plants under snow cover from soil-borne teliospores (Purdy et al., 1963). In contrast to other Tilletia spp., spores dusted onto the seed surface are unimportant as inoculum because germination of winter wheat seeds precedes teliospore germination by a few months. The disease caused by T. controversa is called 'dwarf bunt' because infected plants show stunted growth. Like T. caries, T. controversa has a gene-for-gene relationship with its host, and breeding for resistance is an efficient control method (Fuentes-Dávila et al., 2002). The life cycle of T. controversa is similar to that of T. caries, except that T. controversa has a bipolar multiallelic mating system whereas that of T. caries is bipolar with only two mating type alleles.

Tilletia controversa is cosmopolitan but is not in itself a serious pathogen. However, it has acquired notoriety by being used as the reason for establishing import restrictions on wheat imports from the United States by China during the 1970s to 1990s (Mathre, 1996).

Tilletia indica

The disease caused by T. indica is called Karnal bunt, named after the city in India where it was first described (Mitra, 1931). It is also called partial bunt because the teliospore sori often fill only part of the infected wheat grain, leaving the embryo unaffected. In contrast to T. caries and T. controversa, T. indica does not appear to grow systemically. Instead, teliospores germinate by producing numerous (up to 180) haploid primary needle-shaped sporidia from the tip of the aseptate promycelium. The primary sporidia germinate by monokaryotic haploid hyphae, and these in turn produce a further crop of haploid monokaryotic secondary sporidia. Two types may be produced, namely a repetition of the needle-shaped form or a sausage-shaped sporidium liberated actively by the surface-tension catapult mechanism. In contrast to T. caries and T. controversa, H-shaped fusion cells are not formed and both types of secondary sporidium are monokaryotic and haploid. Secondary sporidia can germinate to produce a mycelium giving rise to further sporidia of either type (Dhaliwal, 1989). The actively liberated form enables the fungus to work its way up on the outside of the leaves of growing wheat plants until it reaches the flag leaf. There, fusion of compatible monokaryotic haploid hyphal segments may occur, with the resulting hetero-karyotic mycelium causing infections of individual florets of the immature wheat ear (Goates, 1988; Nagarajan et al., 1997).

Occurrence and severity of Karnal bunt have increased in high-yielding crop systems, but the disease is not in itself serious, causing crop losses less than 1% per annum even in severe epidemics (Nagarajan et al., 1997). In addition, several resistant cultivars are available. Although bunt balls contain trimethylamine, flour made from crops with up to 4% infected grains is still fit for human consumption (Fuentes-Dávila et al, 2002). As in the case of T. controversa, the major threat posed by T. indica is a legal one. Countries as yet free from the disease may ban wheat imports from those affected by T. indica. For example, the United States imposed quarantine regulations to prevent the spread of Karnal bunt to North America, only to find bans imposed on US exports by other countries when T. indica was eventually discovered in Arizona in 1996 (Palm, 1999). Rush et al. (2005) have given a fascinating account of the complex interactions between agriculture, politics, international trade and research in dealing with Karnal bunt in the United States, concluding that the threat posed by this disease was initially overstated.

23.2.7 Urocystis

The distinguishing feature of the genus Urocystis is that the teliospore consists of one or more melanized fertile cells surrounded by several sterile cells (Figs. 23.11, 23.12). This type of compound teliospore is called a spore ball. There are about 140 species (Vánky, 1994). An important species is U. tritici which causes leaf stripe-smut or flag smut on wheat in warm climates. This was formerly called U. agropyri, a name now applied in a more restricted sense to forms on wild grasses such as Agropyron and Elymus spp. (Fig. 23.11). The fertile cell of a spore ball of Urocystis germinates by producing an aseptate promycelium which gives rise to about four primary sporidia (Fig. 23.12). These sporidia

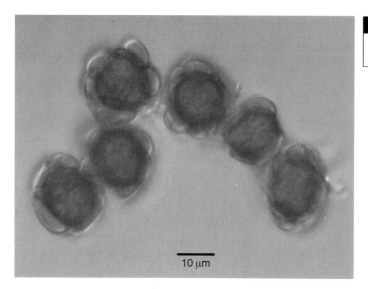

Fig 23.11 Teliospores of *Urocystis agropyri*. A dark-walled fertile cell is surrounded by several flattened hyaline sterile cells.

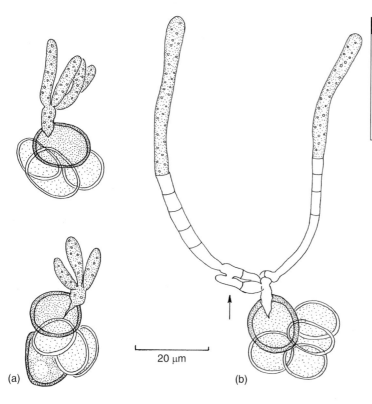

Fig 23.12 *Urocystis anemones*. (a) Spore balls showing germination after 24 h incubation. A promycelium develops from a fertile cell, bearing a crown of three or four sporidia. (b) Development after 48 h. The sporidia have fused (arrow) and a septate infection hypha develops from the fusion cell.

(a)   (b)

conjugate in pairs whilst still attached to the promycelium. The two haploid nuclei then migrate into an infection hypha which develops from the conjugation tube (Thirumalachar & Dickson, 1949).

Urocystis tritici infects its wheat host by appressorium-mediated penetration of the infection hypha through the host epidermis (Nelson & Durán, 1984). Infection occurs at the early seedling stage and is systemic. The outbreak of symptoms is delayed, with the typical stripe-like sori breaking through the epidermis of several leaves after flowering of the host plant. Heavily infected plants may abort their tillers.

Teliospores of *U. tritici* survive in the soil and attached to seeds (Fuentes-Dávila *et al.*, 2002). *Urocystis agropyri* is common on wild grasses, and another common species is *U. anemones* which causes a leaf smut on *Anemone* and *Ranunculus* leaves.

23.2.8 Control of smut diseases

Control of loose and covered smuts presents very different problems. Whilst the surface of the grain is merely *contaminated* with the spores of covered smuts, in the case of loose smuts the ripe grain is already *infected* by a mycelium within the embryo. The control of covered smuts by means of fungicidal seed dressings is therefore simple, and it is standard practice for seed grains to be treated by seed merchants in this way. The first effective seed treatment was the steeping of seed grains in a dilute copper sulphate solution prior to sowing (Large, 1940). In the middle of the twentieth century, seed dressings containing organic mercury compounds were widely used, but these are now banned due to their high general toxicity. Instead, cocktails of fungicides are employed. In most countries with a well-developed agriculture, bunt of wheat is now a rare disease. For example, the incidence of bunt balls in seed samples sent to the Official Seed Testing Station at Cambridge fell from 12−33% in 1921−1925 to 0.2−0.3% in 1955−1957 (Marshall, 1960).

Control of the loose cereal smuts was impossible until the Danish plant pathologist Jens Ludwig Jensen invented the hot water treatment method in the 1880s (Large, 1940; Ainsworth, 1981). This was based on the observation that *Ustilago* spp. are less heat-tolerant than cereals, and infected seeds could be effectively disinfected by soaking them first in cold water for 5 h followed by a dip in hot water; 10 min at 54°C for wheat and 15 min at 52°C for barley (Fischer & Holton, 1957; Ainsworth, 1981). Today, effective fungicidal seed dressings are available and these usually combine systemic fungicides such as carboxin (Fig. 23.13) or its derivatives with protectant fungicides such as captan, maneb or pentachloronitrobenzene (Kulka & von Schmeling, 1995). Such seed dressings control loose and covered smuts, in addition to numerous other fungal diseases.

Since infection of next season's grain with loose smuts occurs at flowering, one obvious method of control is to inspect crops grown for seed at flowering time and to assess the incidence of smutted heads. In these so-called seed certification schemes, only crops which contain fewer than a defined limit, e.g. one smutted ear in 10 000 ears, are approved for use as seed stocks (Doling, 1966). It is also possible to detect the presence of loose smut mycelium within the embryos by microscopic examination (Morton, 1961) or PCR-based methods (Pearce, 1998). The latter can also detect the spores of any other fungus of interest on the seed coat, including covered smuts such as *T. caries*, *T. controversa* and *T. indica*. PCR-based methods are being developed rapidly, partly because of the need for rapid testing of exported or imported agricultural produce for contamination, and partly for the purpose of thwarting bioterrorist attacks (Schaad *et al.*, 2003).

If all else fails, many systemic fungicides, e.g. sterol demethylation inhibitors (see p. 410), are effective against the systemic smuts when sprayed onto the growing crop. In practice, these fungicides are often applied to control infections caused by non-smut cereal pathogens, with the suppression of *Ustilago* spp. as an additional bonus which often goes unnoticed by the farmer (Jones, 1999).

Although many crop plants and their wild relatives possess resistance genes against smut fungi, smut-resistant cultivars are less important in an agricultural context than those with resistance against other biotrophic pathogens, e.g. rusts (see pp. 625 and 627) or powdery

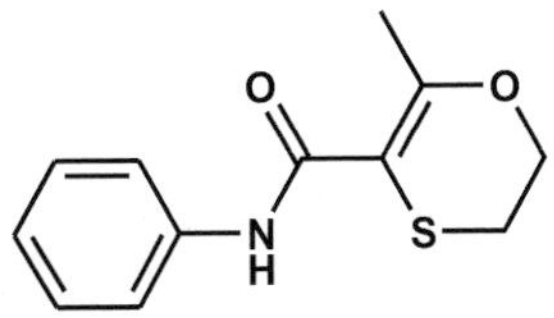

Fig 23.13 Molecular structure of carboxin, a systemic fungicide commonly applied as a seed dressing. Carboxin acts by inhibiting the enzyme succinate dehydrogenase (complex II) in fungal mitochondrial respiration (see Uesugi, 1998).

mildews (p. 408). Avirulence genes involved in gene-for-gene interactions have been desribed for several cereal smuts (Sidhu & Person, 1972; Eckstein *et al.*, 2002; Hu *et al.*, 2003). The hypersensitive response in incompatible interactions is often microscopically small (Hu *et al.*, 2003). Field resistance against smut fungi may also occur as a combination of several genes (Nelson *et al.*, 1998). An interesting case of single-gene resistance has been reported by Wilson and Hanna (1998) as trichome-less mutants of pearl millet (*Pennisetum glaucum*) showing a 50% reduction in disease severity caused by the smut *Moesziomyces penicillariae*. Curiously little is known about resistance mechanisms in *Ustilago maydis* despite its status as a well-examined 'model organism' and the availability of its genome sequence.

23.3 | Microbotryales (Urediniomycetes)

The order Microbotryales contains smut fungi with features almost indistinguishable from those described in the preceding section. This includes their general life cycle of teliospores germinating by means of a promycelium which produces haploid yeast-like sporidia, and the infection of host plants following establishment of a dikaryotic infection hypha. Intriguingly, however, phylogenetic studies have clearly shown the Microbotryales to belong to the Urediniomycetes (Blanz & Gottschalk, 1984; Begerow *et al.*, 1997; Swann *et al.*, 1999). There are certain microscopic details which distinguish them from Ustilaginomycetes, notably that growth by *Microbotryum* and allied species *in planta* is exclusively intercellular (Bauer *et al.*, 1997) and that the teliospores have a violet-purplish rather than brown pigmentation.

The most important genus in the Microbotryales is *Microbotryum* which currently contains about 75 species (Vánky, 1998, 1999). All of them parasitize dicotyledonous hosts, and they can be resolved by DNA sequence analysis into two groups (Almaraz *et al.*, 2002). The original

genus *Microbotryum* parasitizes the anthers of plants belonging to the Caryophyllaceae, whereas species of the second group cause smuts on various organs of other host plants. Until recently, members of the second group were considered to belong to *Ustilago*. The eventful taxonomic history of these 'dicot *Ustilago*' spp. has been recounted by Vánky (1998, 1999); based on the findings of Almaraz *et al.* (2002) they will probably have to be given a new generic name, adding at least one further twist to the story.

Good descriptions of many 'dicot *Ustilago*' spp. have been given by Vánky (1994, 1998). We shall devote our attention to *Microbotryum sensu stricto*, cause of anther smut on Caryophyllaceae.

23.3.1 *Microbotryum violaceum*

Anther smut of Caryophyllaceae has long fascinated biologists with a wide range of research interests. Although infection is systemic throughout the host plant, the disease symptoms are confined to the anthers in which pollen is replaced by purple teliospores. Several members of the Caryophyllaceae are infected, and there is evidence that *Microbotryum violaceum* (formerly called *Ustilago violacea*), the causal fungus, is undergoing speciation in diverse host plants (Antonovics *et al.*, 2002), so that species delimitation is difficult at present (see Vánky, 1994). In dioecious hosts such as red or white campion (*Silene dioica* and *S. alba*), the flowers of infected female plants are stimulated to produce anthers, i.e. their morphology changes from female to male. A thorough review of the general biology of *M. violaceum* has been written by Day and Garber (1988).

Life cycle and genetic recombination
The diploid teliospores of *M. violaceum* are transported from diseased to healthy flowers by pollinating insects. They germinate by forming a promycelium (Fig. 23.14) which often separates into an exterior three-celled section and a one-celled fragment remaining in the teliospore. Each promycelium segment produces sporidia. The three-celled segment becomes readily detached from the teliospore and may continue to develop sporidia after separation.

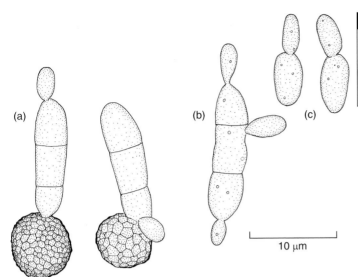

Fig 23.14 *Microbotryum violaceum.*
(a) Germinating teliospore showing three cells of the promycelium. (b) Detached three-celled promycelium segment producing sporidia. (c) Detached sporidia reproducing as yeast cells.

(a)

(b)

(c)

10 μm

In nutrient-rich conditions and at temperatures above 25°C, the haploid sporidia reproduce by yeast-like budding, whereas mating occurs at lower temperatures in nutrient-poor media such as tapwater agar. *Microbotryum violaceum* is heterothallic and bipolar, and it was the first smut fungus for which such an incompatibility system was demonstrated (Kniep, 1919). During mating, an *a2* sporidium or yeast cell produces a conjugation tube which grows towards the *a1* cell. Provided that α-tocopherol (vitamin E) is present, fusion is followed by the production of a dikaryotic hypha which infects the flower of the host plant (Castle & Day, 1984).

In the absence of α-tocopherol or on nutrient-rich media, de-dikaryotization occurs and haploid or aneuploid yeast cells are formed by the loss of chromosomes. Several other complications exist in the life cycle of *M. violaceum*, illustrating the wealth of non-meiotic mechanisms of recombination encountered in fungi (Day & Garber, 1988; Day, 1998). For instance, nuclear fusion not associated with meiosis is readily inducible in *M. violaceum*, and this generates somatic diploids. Mitotic crossing-over occurs at a high frequency in somatic diploids. As already mentioned, *M. violaceum* has dimorphic mating type chromosomes (Hood, 2002), and whilst the mating type locus is a hot spot for mitotic recombination, meiotic recombination does not occur in this region of the genome.

Somatic diploids can be mated if they are homozygous for mating type, so that triploid or tetraploid strains can be generated. In nutrient-rich media, these strains grow as yeast cells but are unstable due to the gradual loss of chromosomes. The size of the yeast cells is correlated with their ploidy level. There is also evidence for the activity of transposons in enhancing genetic recombination in *M. violaceum* (Garber & Ruddat, 2002). The genetic flexibility of *M. violaceum* may explain the existence of numerous strains in nature.

Fimbriae and conjugation

The process of sporidial conjugation has been extensively studied in *M. violaceum*. Within 2 h of placing sporidia of opposite mating type on a suitable medium, a series of events occurs which culminates in the formation of conjugation tubes and plasmogamy. Poon and Day (1974) observed an early adhesion between the two sporidia before any wall-to-wall contact was established, and they discovered that this initial contact was mediated by fimbriae (Fig. 23.15) whose existence in fungi had not been appreciated previously. Subsequent investigations revealed that fungi from most major groups (Zygomycota, Ascomycota, Basidiomycota) possess such

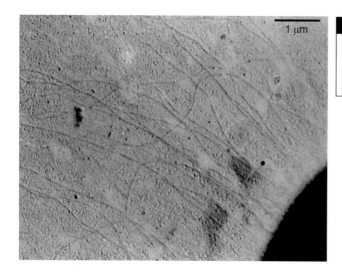

Fig 23.15 Fimbriae of *M. violaceum*. Shadowing-TEM view. Reprinted from Gardiner and Day (1988) by copyright permission of the National Research Council of Canada; original print kindly provided by A. W. Day.

fimbriae, and that these have a similar biochemical composition. Fimbriae can be of variable length (up to 20 µm) but have a uniform diameter of 7 nm. Most of the biochemical work on fungal fimbriae was done with *M. violaceum*, and this subject has been well reviewed by Celerin and Day (1998).

Most fungal fimbriae consist of three structural components. The main component is proteinaceous, with strong homology to collagen which had been thought previously to exist only in animals (Celerin *et al.*, 1996). The second component consists of polysaccharide glycosylation chains on the collagen protein. Whereas deglycosylated fungal collagen can still polymerize to form fimbriae, these become unstable in extreme conditions. Further, the glycosylation chains are involved in the specificity of mating recognition in *M. violaceum*, with the fimbrial protein component of *a1* sporidia binding to the mannose residues on the proteins of the *a2* fimbriae, i.e. acting as a lectin (Castle *et al.*, 1996). The third component of fungal fimbriae is a short (30 nucleotides) single-stranded fimbrial RNA (fRNA), the function of which is as yet unknown (Celerin *et al.*, 1994). The fimbriae of *M. violaceum* do not traverse the cell wall, but are anchored to it by means of a protein which may be functionally analogous to mammalian fibrinogen.

Whereas *M. violaceum* produces only one type of fimbria, other fungi have several morphologically distinguishable types (Gardiner & Day, 1988). In *Candida albicans* (p. 277), fimbriae play an important part in adhesion to human tissue and initiation of infection, and the fimbriae involved in this process are not made up of collagen (Yu *et al.*, 1994).

Host–pathogen interactions

Initial infection of the host plants by *M. violaceum* occurs if a pollinating insect carries teliospores from a diseased to a healthy flower (Jennersten, 1988). Anther smut is thus a sexually transmitted disease. The entire flower is colonized by intercellular dikaryotic mycelium. Infection results in reproductive sterility of the host; in male flowers, the anthers become converted to teliospore sori whereas in female flowers, the female traits are suppressed and anthers are formed which carry teliospores. This development has been described in detail for infections of *Silene latifolia* (Uchida *et al.*, 2003). Diploid female plants carry two X chromosomes in addition to 22 autosomes, whereas male plants contain 22 autosomes plus one X and one Y chromosome. The Y chromosome encodes genes that suppress the development of female traits and promote stamen development, thereby making the flower male. *Microbotryum violaceum* must therefore produce a signal called male-sterility restoration factor that substitutes for the Y chromosome in female flowers (Uchida *et al.*, 2003). Its identity is as yet unknown, but it must be a signal molecule or transcription factor acting on genes encoded on the autosomes or the X chromosome, since the expression patterns of some male-specific genes in healthy male and infected female flowers are similar (Scutt *et al.*, 1997).

In perennial host species, *M. violaceum* can colonize the entire host in the course of 1–2 years, overwintering systemically in the root system (Alexander & Antonovics, 1988). Infected plants acquire lifelong sterility, i.e. they will continue to produce flowers but these will all be male, developing anthers in which pollen grains are replaced by teliospores.

The ecology of *M. violaceum*

The ecological effects of *M. violaceum* infections have been considered by several workers. Systemically infected host plants flower earlier than healthy plants, and this may force pollinators such as bumblebees to visit infected flowers early in the season. Consequently, teliospore dispersal peaks earlier than pollen dispersal (Jennersten, 1988). In direct comparison, diseased flowers are less attractive to bumblebees (Shykoff & Bucheli, 1995), possibly because they contain less nectar and are asymmetric and smaller than healthy flowers (Shykoff & Kaltz, 1998). A pollinator preference for healthy flowers may be advantageous to the pathogen because it enhances the opportunity for spreading the infection following an accidental visit to a diseased plant (Shykoff & Bucheli, 1995). Although systemically infected host plants produce an increased number of flowers which also remain open for longer, the drawback is that the root biomass is decreased, thereby potentially affecting the chance of the pathogen to overwinter systemically. Since healthy female flowers remain open for much longer than healthy male flowers, they are at greater risk of infection (Kaltz & Shykoff, 2001). As in many venereal diseases, a more attractive display draws a greater number of visitors, and one possible adaptation to *M. violaceum* is predicted to be a reduction in flower size. Another might be the change from a perennial to an annual habit, as this would prevent overwintering of the pathogen (Kaltz & Shykoff, 2001). In situations where perennial hosts are forced into an annual

habit, e.g. near field borders or railway lines subject to regular clearing, infections by *M. violaceum* are rare or absent (Alexander & Antonovics, 1988). The interaction between *M. violaceum* and its hosts may therefore have different outcomes, ranging from local extinction of the host to that of the pathogen.

23.4 | Exobasidiales (Ustilaginomycetes)

The order Exobasidiales (subclass Exobasidiomycetidae) is a well-defined group (Begerow *et al.*, 1997, 2002) of ecologically obligate, biotrophic plant pathogens containing less than 100 species. Most species cause systemic or limited infections of shoots or leaves, and these are often accompanied by hypertrophy of the infected tissue. The most important genus is *Exobasidium*. Fungi grouped in the Exobasidiales firmly belong to the Ustilaginomycetes and possess a haploid saprotrophic yeast-like phase and a dikaryotic biotrophic phase. However, they differ from the 'true smuts' (Ustilaginomycetidae) in several aspects:

1. They do not produce teliospores, but basidia are formed directly on the surface of the infected host.

2. The basidium appears similar to that of the Homobasidiomycetes, i.e. it looks like a holobasidium, not divided into a pro- and metabasidium as in smut fungi (but see below).

3. Basidiospores are violently discharged using the surface-tension catapult mechanism, and they often become septate after discharge.

4. The mycelium is usually intercellular, with haustoria often present. These differ from the intracellular hyphae of smut fungi in not being completely surrounded by a sheath. Instead, a complex interaction apparatus is formed at the haustorial apex (Fig. 23.17).

23.4.1 *Exobasidium*

About 50 *Exobasidium* spp. are known. All of them cause either local infections of individual leaves or more widespread, sometimes systemic infections of whole shoots or shoot tips

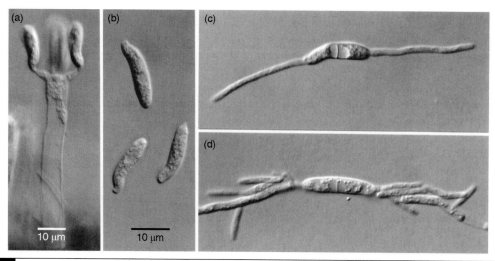

Fig 23.16 *Exobasidium japonicum.* (a) Basidium on the lower surface of an evergreen azalea. (b) Freshly discharged aseptate basidiospores. (c) Basidiospore after a few hours' incubation on potato dextrose agar. The spore has formed a transverse septum and has germinated. (d) Basidiospore after 12 h. The germ tubes have branched and are budding off elongated yeast cells.

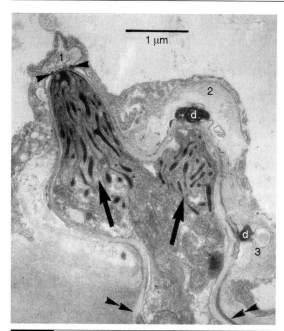

Fig 23.17 Lobed haustorium of *Exobasidium* sp. Three interaction sites with the host cytoplasm are numbered. Site 1 is shown in median section, and the interaction ring is visible (arrowheads). Sites 2 and 3 are in tangential section, showing electron-dense deposits (d). Tubulovesicular cytoplasmic structures involved in secretion of the electron-dense matter are also visible (arrows). The penetration site of the host wall is at the bottom of the picture (double arrowheads). Reprinted from Begerow et al. (2002), *Mycological Progress*, with permission of IHW-Verlag; original print kindly provided by D. Begerow.

(Plates 12i,j). Infection commonly gives rise to hypertrophied tissue. Keys and descriptions have been given by Nannfeldt (1981) and Ing (1998). Frequently infected host plants include members of the Ericaceae, e.g. *Vaccinium* (bilberry, blueberry) in heaths, moors and forests, and ornamental azaleas and *Rhododendron* spp. in gardens. Cosmopolitan examples are *E. japonicum* on evergreen azaleas (Plate 12i; Fig. 23.16) and *E. vaccinii* on *Vaccinium* spp. (Plate 12j). A second group of host plants is the Theaceae family, with the tea plant (*Camellia sinensis*) a prominent casualty of *E. vexans* which adversely affects the quantity and quality of the tea harvest (Gulati *et al.*, 1999).

Infections by *Exobasidium* spp. are easily confused with symptoms caused by *Taphrina* (Plate 4a,c), and these two genera, although phylogenetically unrelated, have much in common. Plant tissues infected by *Exobasidium* are often swollen and show a pale or reddish discoloration. There is an obvious hormonal imbalance in the infected host tissue, although no detailed examinations have been carried out. Infected shoots are often affected in their reproductive development (Wolfe & Rissler, 1999). A whitish superficial layer of basidia is eventually produced on the surface of infected host tissue, and these carry about 2–8 sterigmata

(Fig. 23.16a). Basidiospores are initially aseptate (Fig. 23.16b), but in many species they become septate after their discharge, and both cells can germinate by emitting germ tubes (Fig. 23.16c). Elongated haploid yeast cells are produced from these germ tubes or directly from the basidiospore cell (Fig. 23.16d). Several *Exobasidium* spp. grow as yeast cells in culture. Infection of host plants may occur from basidiospores or yeast cells and gives rise to a dikaryotic intercellular mycelium (Mims & Nickerson, 1986). *Exobasidium* spp. may overwinter systemically in their host, or as spores on the outside, e.g. in bud scales or bark.

Haustoria are formed by the intercellular hyphae, and these show unique characteristics in being lobed and producing an electron-dense apical ring at the localized point of contact with the host plasmalemma. This then becomes elaborated into an apical cap which is associated with the host wall (Bauer *et al.*, 1997) or directly with the host plasmalemma (Mims, 1982). The ring and cap material is secreted by the haustorium from an elaborate tubular membrane system (Fig. 23.17). This cap may be homologous to the thick sheath which surrounds the intracellular hyphae of smut fungi (see Fig. 23.6).

Although the basidia of *Exobasidium* look like typical club-shaped holobasidia, their development differs from that in the homobasidiomycetes (Mims *et al.*, 1987). Nuclear fusion occurs in a subterminal cell (strictly speaking the probasidium), and then the nucleus migrates into the basidium proper where meiosis is completed. The basidium of *Exobasidium* is thus a metabasidium in disguise, and the pattern of basidial development is equivalent to that in the rust and smut fungi. A further unusual feature is that the hilar appendices of basidiospores on the basidium point outwards (Fig. 23.16a), not inwards as in Homobasidiomycetes.

Basidiomycete yeasts

24.1 | Introduction

Fungi living predominantly or exclusively as yeasts are encountered in three classes of Basidiomycota, namely the Heterobasidiomycetes (Chapter 21), Urediniomycetes (Chapter 22) and Ustilaginomycetes (Chapter 23). We shall discuss basidiomycete yeasts together in the present chapter because these organisms, although taxonomically diverse, are unified by many biological features.

24.1.1 Ecology

Little is known about the ecology of basidiomycete yeasts. They occur in marine and freshwater habitats, the soil and the plant rhizosphere, and especially on above-ground plant surfaces such as tree bark, leaves, flowers and fruits. A certain degree of specificity of yeast species relative to plant species or organs of a given plant host has been observed (Phaff, 1990). Basidiomycete yeasts may be found in all climatic zones from the arctic to the tropics. They generally exist as saprotrophic phylloplane organisms. When nutrients become available, there may be a steep increase in the population density of these yeasts. Many yeasts isolated from soil have their origin in vegetation which becomes incorporated into humus after leaf fall. Basidiomycete yeasts are not noted as plant pathogens, but some species can infect animals. *Cryptococcus neoformans* is one of the most serious fungal pathogens of humans (pp. 661–665). Members of another group (*Malassezia* spp.) live commensally on the skin of humans and other mammals, causing superficial dermatomycoses under suitable conditions (p. 671).

Basidiomycete yeasts therefore share many ecological features with their ascomycete counterparts, and they are easily isolated following similar procedures, i.e. by plating out soil suspensions, leaf washings or filters bearing water samples onto standard agar media augmented with antibacterial antibiotics (p. 262). One good way to isolate ballistoconidium-forming yeasts is to attach a piece of vegetation to the underside of a Petri dish lid, permitting ballistosporic yeasts to shower their spores onto the agar medium. Within 2–3 days, yeasts such as *Sporobolomyces* spp. will grow and can be isolated in pure culture. Yeasts are preserved in an active state on agar slopes at 4°C. Lyophilized preparations can also be made from vegetative cells of many species, and these remain viable for several years.

Many basidiomycete yeasts, the so-called 'red yeasts', are coloured yellow, orange, pink or red due to the presence of carotenoids (see Fig. 24.8). Red yeasts are found in all three classes containing basidiomycete yeasts, but they are rare among ascomycetes (see p. 253 for the *Lalaria* state of *Taphrina*). Carotenoid production can be of commercial value, as in *Phaffia rhodozyma* which produces astaxanthin, an important food and feed pigment (p. 665). Other commercial applications of basidiomycete yeasts are as potential producers of lipid, e.g. as cocoa butter substitute (Ratledge, 1997), and as biocontrol agents against storage rots of fruits

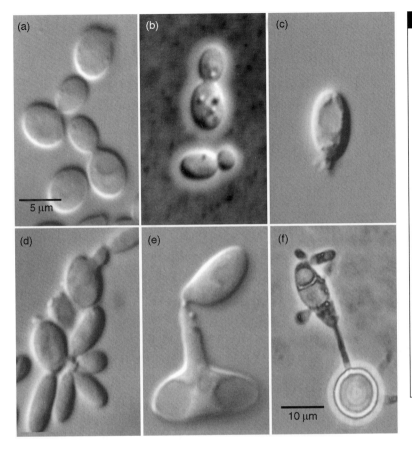

Fig 24.1 Basidiomycete yeasts as seen by light microscopy. (a) Budding in *Dioszegia* sp. (Heterobasidiomycetes). The yeast cells are isodiametric with a slightly pointed bud region. (b) Budding cells of *Rhodotorula glutinis* (Urediniomycetes). This species produces a prominent polysaccharide capsule which has been revealed by mounting yeast cells in Indian ink. (c) Collarette at the site of repeated bud formation in *Sakaguchia dacryoidea* (Urediniomycetes). (d) Annellidic structure formed by a member of the *Dioszegia hungarica* group (Heterobasidiomycetes). (e) Ballistoconidium formation in *Sporobolomyces roseus* (Urediniomycetes). (f) Septate stalked basidium arising from a teliospore of *Sporidiobolus ruineniae* (Urediniomycetes). (a–e) to same scale. Original print of f kindly provided by J. W. Fell.

caused by filamentous fungi (Janisiewicz & Korsten, 2002).

Detailed descriptions of basidiomycete yeasts along with their ascomycete counterparts may be found in Kurtzman and Fell (1998) and Barnett *et al.* (2000). A useful general introduction to the topic is that by Fell *et al.* (2001).

24.1.2 Morphology and life cycles

Asexual reproduction is mainly by budding. In contrast to ascomycete yeasts, the budding sites are confined to either or both poles of the vegetative cells which are usually ellipsoid or elongated (Figs. 24.1b,d) but may be isodiametric (Fig. 24.1a). Yeast cells may be coated by a prominent polysaccharide capsule (Fig. 24.1b), giving the culture a slimy appearance. Repeated budding at the same site may lead to the formation of collarettes (Fig. 24.1c) or even annellide-like structures (Fig. 24.1d). The formation of daughter cells is **enteroblastic**, i.e. only the inner wall of the mother cell extends to form the wall of the daughter cell. The whole budding yeast cell can thus be viewed as a phialide. This is in contrast to most ascomycete yeasts which bud in a holoblastic manner (i.e. the daughter wall is continuous with the entire wall of the mother cell). Given a little practice, it is often possible with the light microscope to recognize a basidiomycete yeast as such. Transmission electron microscopy has revealed differences between the two-layered cell walls of ascomycetes and the multi-layered lamellate walls of basidiomycete yeasts (Kreger-van Rij & Veenhuis, 1971). These differences have been correlated with the reaction of basidiomycete but not ascomycete yeasts with the diazonium blue B stain (Simmons & Ahern, 1987).

Apart from budding, several genera of basidiomycete yeasts such as *Sporobolomyces* and its teleomorph *Sporidiobolus* (Urediniomycetes) or *Bullera* and its teleomorph *Bulleromyces*

(Heterobasidiomycetes) also produce conidia which are actively liberated in the manner of basidiospores, i.e. by means of the surface-tension catapult mechanism involving Buller's drop (Fig. 24.1e). Since these are asexual propagules, they are called **ballistoconidia**. Their existence provided one of the first clues that the yeasts producing them belong to the Basidiomycota (Kluyver & van Niel, 1927).

Sexual reproduction is relatively rare in basidiomycete yeasts except, of course, for those species which have dominant mycelial diploid or dikaryotic stages such as the smut fungi discussed in Chapter 23, or jelly fungi and allies (Chapter 21). Following mating between compatible yeast cells, a limited dikaryotic mycelium often bearing clamp connections may arise, and this produces basidia either directly or, more commonly, via thick-walled resting cells called teliospores. There, karyogamy occurs, i.e. the teliospores function as probasidia. Teliospores germinate in a manner described in detail for rust and smut fungi, namely by the production of a promycelium (= metabasidium) which may or may not undergo transverse septation. The basidiospores thus produced germinate by budding as yeast cells (Fig. 24.1f).

Where present, septa can be examined by transmission electron microscopy for their ultrastructure, and this is an important feature in classification. In general, the septa of yeasts belonging to Urediniomycetes and Ustilaginomycetes contain simple pores whereas those of heterobasidiomycete yeasts have dolipores, often with a parenthesome (Fell *et al.*, 2001).

24.1.3 Phylogeny of basidiomycete yeasts

In addition to examining morphological features as outlined above, several biochemical tests can be performed to characterize yeasts. Such tests, e.g. carbon source utilization, the identity of coenzyme Q or the spectrum of killer toxins produced, have been employed extensively in the past to identify yeasts and are still relevant today (Yarrow, 1998; Fell *et al.*, 2001). However, the results are rarely clear-cut, and taxonomic confusion has resulted due to the extensive overlap of features between members of different taxa.

Hence, the names of many basidiomycete yeasts are of descriptive rather than taxonomic value and may be found in several phylogenetically distinct clades (Table 24.1). For instance, the main feature to distinguish *Sporobolomyces* from *Rhodotorula* is the presence or absence (respectively) of ballistoconidia, and it is now known that these character states, and hence the two generic names, are of little taxonomic relevance.

Major work is currently being carried out in order to establish phylogenetically coherent groups of yeasts, using especially ribosomal DNA but also increasingly other gene sequences. Valuable recent contributions to phylogeny at higher taxonomic levels are those by Fell *et al.* (2000, 2001) and Scorzetti *et al.* (2002) in which the heterobasidiomycete, urediniomycete and ustilaginomycete clades of yeasts have been circumscribed (Table 24.1). The integration of some of these clades into the taxonomy of filamentous basidiomycetes has yet to be accomplished. Additionally, numerous publications have dealt with the analysis of taxa at the level of genus or species. Hence, although it is generally estimated that only 1–5% of all basidiomycete yeasts have been discovered as yet, a large database of DNA sequences is already available. A convenient side effect of this work is that the identification of new isolates or at least their assignment to a given family or genus is a relatively straightforward matter if their rDNA sequences can be obtained. Biochemical tests and microscopy can then be used to verify and extend the identification.

24.2 | Heterobasidiomycete yeasts

This is a large and morphologically diverse group of yeasts. We shall focus on two species which have been particularly well examined, the serious human pathogen *Filobasidiella neoformans* and the astaxanthin producer *Phaffia rhodozyma*. *Trichosporon* spp., which are occasional opportunistic human pathogens, are not further discussed. They show a diversity of growth forms, including hyphae, pseudohyphae, yeast cells, blastoconidia and arthroconidia.

24.2.1 *Filobasidiella* (*Cryptococcus*) *neoformans*

The anamorph genus *Cryptococcus* comprises some 30–40 species which are scattered throughout all 4 yeast-containing orders of the Heterobasidiomycetes, and with a range of different teleomorphs (Table 24.1). Clearly, therefore, the genus is polyphyletic. *Cryptococcus* spp. are ubiquitous, being found in all climatic zones on plant material and in the soil. One species, *C. neoformans* (teleomorph *Filobasidiella neoformans*), is a human pathogen and among the most serious causes of mycosis in man. Both individuals with a healthy immune system and, more frequently, immunocompromised patients such as AIDS sufferers or organ transplant patients, can be attacked. There is a vast amount of literature on *C. neoformans*, including an authoritative monographic treatment (Casadevall & Perfect, 1998) and several good reviews (e.g. Kwon-Chung & Bennett, 1992; Mitchell & Perfect, 1995; Buchanan & Murphy, 1998).

Varieties of *C. neoformans* and their habitats

One important feature of *C. neoformans* is the coat of mucilage which surrounds actively growing yeast cells. There are four mucilage serotypes, A–D, and these have been correlated with two varieties of *C. neoformans*, i.e. var. *gattii* (serotypes B and C) and var. *neoformans* (A and D). Franzot *et al.* (1999) have put forward evidence that the latter should be separated into var. *neoformans* (D) and var. *grubii* (A). Some serotypes cannot be assigned clearly to any one variety (e.g. serotype AD). Although the three varieties probably diverged some 18–37 million years ago, they still show occasional hybridization in nature and the ability to mate in the laboratory. Many of the hybrids isolated from nature are of recent origin, and there is evidence that human activity has brought together strains which were previously living in geographic isolation. Thus, the species *C. neoformans* may be undergoing de-diversification at present (Xu *et al.*, 2000). The anamorph–teleomorph connection is such that *C. neoformans* var. *neoformans* and var. *grubii* correspond to *F. neoformans* var. *neoformans*, whereas the teleomorph of *C. neoformans* var. *gattii* is *F. neoformans* var. *bacillispora* which differs in that its basidiospores are rod-shaped rather than spherical or ellipsoid.

As with most if not all fungi pathogenic to man, the ability of *C. neoformans* to cause disease is serendipitous, and the human body represents a dead-end in the life cycle of the pathogen. Disease outbreaks are preceded by contact of humans with the fungus in its natural habitat, but the precise identity of this has been difficult to track down. Ellis and Pfeiffer (1990a) noted the geographic co-occurrence of human cases caused by *C. neoformans* var. *gattii* and the distribution of *Eucalyptus camaldulensis* trees in Australia, and were able to show that the fungus is associated with the flowers, bark and litter of *Eucalyptus*. In other countries, different plants may also act as hosts. Further, clinical and environmental isolates of *C. neoformans* var. *gattii* in Australia have been shown to be genetically identical (Sorrell *et al.*, 1996). Thirdly, the confinement of *C. neoformans* var. *gattii* to rural areas may explain why its abundance has barely increased in the wake of the AIDS epidemic, in marked contrast to var. *neoformans* and especially var. *grubii* which accounts for 99% of infections in AIDS patients (Mitchell & Perfect, 1995). Both these latter varieties are common in urban habitats, being associated with the nests and droppings of birds, especially pigeons (Emmons, 1955) but also caged birds. Although birds certainly spread the fungus, they are more likely to be vectors than primary hosts in nature because they are not themselves affected by cryptococcosis. As yet unidentified plants are probably the primary substratum, and Lazéra *et al.* (1996) have shown that *C. neoformans* var. *neoformans* (presumably including var. *grubii*) occurs in the hollows of living tree trunks.

Whereas *C. neoformans* var. *gattii*, like *Eucalyptus camaldulensis*, is confined to tropical and subtropical areas, vars. *neoformans* and *grubii* are cosmopolitan.

Life cycle

The life cycle of *F. neoformans* is unusual in several respects (Kwon-Chung, 1998) as shown in Fig. 24.2. The fungus is heterothallic with a bipolar mating system (two mating types, *a* and *α*). The yeast cells are haploid and uninucleate,

and they reproduce by budding. Conjugation between two cells of compatible mating types gives rise to a limited dikaryotic mycelium with septa bearing clamp connections. From this, elongated metabasidia arise and terminate in a swollen tip. Nuclear fusion occurs in the basidial stalk, and meiosis followed by repeated mitotic divisions in the swollen apex. This produces four patches which bud off a chain of basidiospores, each of which contains a single haploid nucleus. Meanwhile, mitosis continues in the basidial cytoplasm. Each patch can produce numerous basidiospores so that a column consisting of four chains is formed. The migration of nuclei into the basidiospores is random so that each chain may contain basidiospores of both mating types. This kind of basidium is unique to *F. neoformans*.

'Self-fertility' is a frequently observed phenomenon in *α* mating type strains of *F. neoformans* and involves the production of a monokaryotic mycelium with incomplete clamp connections, giving rise to basidia and basidiospores presumably without meiosis (Wickes *et al.*, 1996). These should therefore be called conidia. Of all varieties of *C. neoformans*, strains containing mating type *α* are isolated much more frequently than *a* strains from natural situations as well as from patients (Kwon-Chung & Bennett, 1978), and the ability of *α* but not *a* strains to form dry wind-dispersed conidia may explain their greater abundance in nature. Kwon-Chung *et al.* (1992), working on strains of *C. neoformans* var. *neoformans* which were genetically identical to each other except for their mating type, found that the *α* strain was also more virulent than the *a* strain. The conclusion was that the *α* idiomorph might encode virulence factors.

Both *α* and *a* mating type idiomorphs of *C. neoformans* var. *neoformans* and var. *grubii* have now been sequenced (Lengeler *et al.*, 2002) and were found to be unusually large (105–130 kb). Approximately 20 genes are encoded by the entire sequence, including pheromones, receptors, transcription factors and components of signalling cascades. Several genes are unique to either *α* or *a*, and even those common to both are arranged in different positions, making recombination within the mating type region impossible. Lengeler *et al.* (2002) have likened the

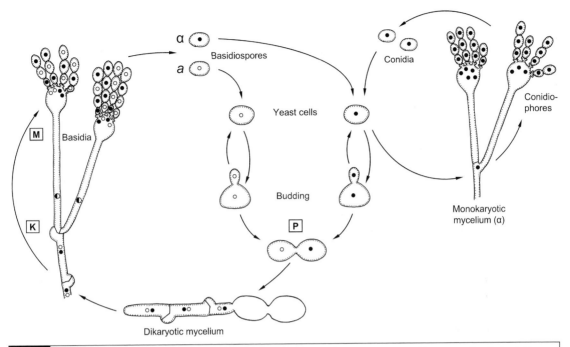

Fig 24.2 Life cycle of *Filobasidiella neoformans*. Note that only the α but not the *a* mating type is able to reproduce by haploid conidia. Based partly on Kwon-Chung and Bennett (1992) and Kwon-Chung (1998). Open and closed circles represent haploid nuclei of opposite mating type; diploid nuclei are larger and half-filled. Key events in the life cycle are plasmogamy (P), karyogamy (K) and meiosis (M).

mating type idiomorph of *C. neoformans* to the sex chromosomes of other organisms. This situation is most unusual for Basidiomycota which typically have two unlinked mating type loci (e.g. *A* and *B* in *Coprinus cinereus* or *Schizophyllum commune*; see p. 508) encoding a receptor/pheromone system and a transcription factor. Another, though structurally different, example of a sex chromosome may be found in *Ustilago hordei* (see p. 637).

Infection and therapy

The available evidence suggests that infection with *C. neoformans* is initiated by inhalation, and that the disease is not spread between humans. This is probably because the infectious particles must be very small (<3 μm) in order to reach the alveoli, and basidiospores or conidia, which are about 2 μm in diameter in *C. neoformans* var. *neoformans* and var. *grubii*, are probably the main agents of infection. They germinate by producing yeast cells which are larger and are additionally surrounded by a thick polysaccharide capsule.

Only the yeast stage is found in the human host (Ellis & Pfeiffer, 1990b).

Immunocompetent hosts may eliminate infections with their immune system, or may restrict the fungus to a latent stage in the lungs, in which case treatment may not be necessary. Cryptococcal pneumonia may develop, but much more dangerous is the rapid dissemination in the blood stream which happens especially in immunocompromised hosts, leading to infections of a range of organs such as bones, the skin or the eye. By far the most common site of such disseminated cryptococcosis is the brain where a condition with symptoms resembling meningitis is established. This is properly called meningoencephalitis because *C. neoformans* infects not just the meninges (i.e. membranes surrounding the brain), but also the brain itself where inflammations, lesions or accumulation of fluid may develop (Casadevall & Perfect, 1998). Untreated meningoencephalitis is always fatal even in immunocompetent patients, but treatment by daily doses of amphotericin B and

5-fluorocytosine over several weeks is effective if the disease has been diagnosed sufficiently early (Mitchell & Perfect, 1995). For AIDS sufferers, 'the therapeutic goal is to ablate the symptoms and signs of cryptococcosis until the patient dies of other causes' (Kwon-Chung & Bennett, 1992). Antifungal drugs, especially fluconazole, may have to be administered for the rest of the patient's life because of the high risk of relapses. Cryptococcosis, like certain other fungal diseases, is an AIDS-defining illness, i.e. patients diagnosed as HIV-positive and suffering from cryptococcosis are considered to have developed AIDS.

Virulence factors

Much work is being done to identify the properties which enable *C. neoformans* to cause disease (reviewed by Buchanan & Murphy, 1998; Casadevall & Perfect, 1998; Perfect, 2004). The most important and obvious virulence factor is the capsule which surrounds actively growing yeast cells of *C. neoformans* and protects them from adverse environmental effects. It also seems to inhibit the phagocytosis of yeast cells by macrophages. The capsule can be visualized by light microscopy of yeast cells mounted in India ink and can be rather more substantial than that shown for *Rhodotorula* in Fig. 24.1b. The major capsule polysaccharide of *C. neoformans* is a linear α-(1,3)-mannan chain, of which roughly every third mannose moiety is substituted with a single β-(1,2)-glucuronic acid unit. The presence of xylose determines the antigenic properties of the capsule; one (serotype D), two (A), three (B) or four (C) xylose residues may be present for every three mannose moieties (Cherniak & Sundstrom, 1994). The capsule polysaccharides are produced in such profusion that they can be detected in the blood serum and other body fluids of infected patients, and this is an important diagnostic tool (Kwon-Chung & Bennett, 1992). Mutants unable to produce a capsule in the host are apathogenic.

Both *C. neoformans* var. *gattii* and vars. *grubii/neoformans* have been shown to undergo switches in colony phenotype from smooth to mucoid, wrinked or pseudohyphal. Switching between smooth and mucoid appears to be readily reversible in the mammalian host in var. *gattii*, but less so in the latter two. The mucoid type is characterized by an increase in the thickness of the polysaccharide capsule. In *C. neoformans* var. *neoformans* and var. *grubii* a thick capsule coincides with an enhanced resistance to antimycotics, and mucoid strains may be selected by prolonged chemotherapy (Guerrero *et al.*, 2006). In mice infected with *C. neoformans* var. *gattii*, Jain *et al.* (2006) have shown that mucoid forms are associated preferentially with pulmonary infections, presumably because of an enhanced resistance to intracellular digestion by phagocytes. Smooth cells with their thinner coat were preferentially isolated from infected brain tissue. In contrast to morphotype switching in *Candida albicans* (p. 277), no sexual function has been suggested for switching in *C. neoformans*.

A second important virulence factor is melanin. The ability of *C. neoformans* to synthesize melanin distinguishes this species from other members of the genus, and the melanized cell wall has been suggested to protect the cell against oxidative stress such as that encountered during the oxidative burst after ingestion by macrophages (Wang *et al.*, 1995). The pathway of melanin biosynthesis in *C. neoformans* is different from the dihydroxynaphthalene (DHN) route found in most fungi (see Fig. 12.46). The precursor molecule is 3,4-dihydroxyphenylalanine (DOPA) which cannot be synthesized by *C. neoformans* but, if present, can be oxidized to quinones by a laccase-type enzyme, and these quinones spontaneously polymerize to melanin (Salas *et al.*, 1996). Catecholamines such as DOPA are present at high levels in the brain, and this may explain the preferential accumulation of *C. neoformans* in brain tissue (Polachek *et al.*, 1990).

A third important factor is, of course, the ability of *C. neoformans* to grow at 37–39°C, which is unique among *Cryptococcus* spp. Several gene products are required for growth at 37°C, and prominent among them is calcineurin, a Ca^{2+}-regulated serine/threonine phosphatase involved in eukaryotic cellular signalling (Odom *et al.*, 1997). Calcineurin is the target of a complex formed between a cyclophilin protein involved in protein folding, and cyclosporin A (see Fig. 12.24a), an immunosuppressive drug widely used to prevent the graft rejection reaction after organ transplantations. In vitro, growth of *C. neoformans* is inhibited by cyclosporin A at 37°C but not at lower temperatures. The use of cyclosporin A for human antifungal therapy is impossible because its immunosuppressive activity outweighs the antifungal effect. However, Cruz *et al.* (2000) have reported cyclosporin derivatives which are antifungal but do not have immunosuppressive properties, i.e. they appear to interfere with fungal calcineurin signalling but not with the equivalent human signalling chain involved in the immune response. It remains to be seen whether these substances are sufficiently specific for use as new antifungal drugs.

24.2.2 *Phaffia* and *Xanthophyllomyces*

Phaffia rhodozyma and its putative teleomorph *Xanthophyllomyces dendrorhous* have aroused considerable interest because they are among very few fungi producing the commercially valuable carotenoid pigment astaxanthin (see Fig. 24.8), and probably the only ones to do so in pure culture. Astaxanthin is of importance in the fish farming industry because salmonid fish (salmon and trout) require a minimum level of astaxanthin in their food for healthy growth. One reason for this may be that these fish contain high levels of polyunsaturated fatty acids which are susceptible to peroxidation by reactive oxygen species such as the hydroxyl radical (HO·) or superoxide radical ($O_2 \cdot^-$). Astaxanthin and the related pigment canthaxanthin are also responsible for the orange pigmentation of salmon steak. In nature, astaxanthin travels the food chain phytoplankton → zooplankton → larger crustaceans → salmon. Simple methods to extract and analyse astaxanthin from *Phaffia* and salmon have been described by Weber and Davoli (2003). In view of the correlation between oxidative processes and cancer or degenerative diseases related to ageing, astaxanthin is also gaining popularity as a 'nutraceutical', i.e. an additive to the human diet.

Astaxanthin is now produced commercially by total chemical synthesis, by extracting it from the exoskeletons of shellfish, or by microbial fermentation using *Phaffia* or the alga *Haematococcus pluvialis*. Whereas wild-type strains of *Phaffia* synthesize a limited amount of astaxanthin (typically less than $300\,\mu\mathrm{g\,g^{-1}}$ dry weight), the astaxanthin levels in strains used for industrial production are at least 10-fold higher. In-depth reviews of biotechnological aspects of astaxanthin production have been written by Johnson and An (1991) and Johnson and Schroeder (1995a,b).

It is unclear why, among the numerous red yeasts, *Phaffia* and *Xanthophyllomyces* are the only ones as yet known to synthesize astaxanthin. Part of the answer may lie in their unusual habitat, all strains known to date having been isolated from the slime fluxes of broad-leaved trees, especially birch (*Betula* spp.) in Alaska, Japan, Scandinavia and Russia (Phaff, 1990). Schroeder and Johnson (1995) have presented evidence that birch sap contains a photosensitizer, i.e. a substance that becomes energized by UV light. When this passes on its excitation energy, it can transform ground-state oxygen (triplet oxygen, $^3O_2$) to the reactive singlet oxygen ($^1O_2$) state. This can be returned to its ground state under dissipation of the excess energy as heat by astaxanthin and other carotenoids. Astaxanthin appears to be necessary for the survival of *Phaffia* under the oxidizing conditions prevalent in birch sap. In fact, cultivation and mutant studies have shown that astaxanthin is able to protect its producing organism against a wide range of oxidative stresses caused by H_2O_2, singlet oxygen and the hydroxyl and superoxide radicals, and that astaxanthin biosynthesis is increased under such conditions (Schroeder & Johnson, 1993). In contrast, β-carotene is the main pigment at low oxygen partial pressure, i.e. *Phaffia* cultures grown under reduced aeration appear yellow instead of orange (see Fig. 24.8).

The original isolate of *Phaffia* is a purely asexual strain, but other isolates show sexual reproduction under suitable conditions characterized by nitrogen starvation and the presence of polyols (Kucsera *et al.*, 1998). This perfect state was named *Xanthophyllomyces dendrorhous*, and there is still disagreement as to whether *Phaffia* and *Xanthophyllomyces* represent the same or closely related species (Fell & Blatt, 1999; Kucsera *et al.*, 2000). *Xanthophyllomyces* appears to be homothallic, vegetative cells being diploid. Mating can occur between a mother cell and its bud, giving rise to a tetraploid zygote (Kucsera *et al.*, 1998). A long thin aseptate metabasidium is formed, and the fusion nucleus migrates to the tip of it, undergoing meiosis in the course of the journey. About 2–7 diploid basidiospores are produced and released passively (Fig. 24.4c). Vegetative reproduction is by budding (Fig. 24.4a), but the nuclear events are unusual in that the mother nucleus migrates into the bud and divides there, followed by the return of one of the daughter nuclei to the mother cell (Slaninova *et al.*, 1999). Thick-walled chlamydospores (Fig. 24.4b) are occasionally formed, and these germinate by mitotic budding, i.e. they are not equivalent to the teliospores of other basidiomycete yeasts. The life cycle of *Xanthophyllomyces rhodozyma* is summarized in Fig. 24.3. In general, there is a considerable variation between different isolates; for instance, the number of chromosomes ranges from 7 to 17 (Kucsera *et al.*, 2000).

24.3 | Urediniomycete yeasts

Yeasts belonging to the Urediniomycetes appear to fall into four phylogenetic groups (see Table 24.1), some of which still need to be integrated into higher taxa (families and orders). Many species contain carotenoids, i.e. they are red yeasts. Some of them, especially species belonging to the *Sporodiobolus* clade (provisional order Sporidiales), are extremely common in the environment. For this reason, we shall consider them here. There are two important genera in the Sporidiales, *Sporidiobolus* (anamorph *Sporobolomyces*) and *Rhodosporidium* (anamorph *Rhodotorula*).

24.3.1 *Sporidiales*

In addition to budding, *Sporobolomyces* spp. also form asexual spores which are ejected from a sterigma into the air by the surface-tension catapult mechanism and are therefore called ballistoconidia. If an agar culture is incubated upside down, a mirror image of the colony will be deposited as spores on the Petri dish lid. For this reason, *Sporobolomyces* is known as a mirror yeast. One of the most frequently encountered species, *S. roseus*, is now known to represent a

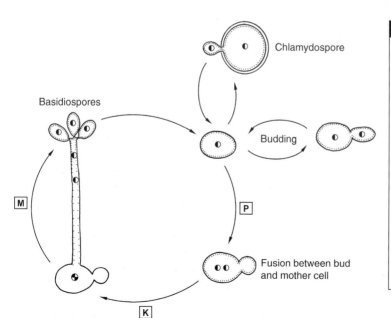

Basidiospores

Chlamydospore

Budding

Fusion between bud and mother cell

M

P

K

Fig 24.3 Life cycle of *Xanthophyllomyces dendrorhous*. Vegetative cells are diploid and reproduce by budding. The fungus is homothallic, and fusion between mother and daughter cell establishes a tetraploid zygote which forms an elongate aseptate basidium in which meiosis occurs. The diploid basidiospores germinate by budding. *Xanthophyllomyces* can survive unfavourable conditions by means of thick-walled chlamydospores. Half-filled circles represent diploid nuclei; tetraploid nuclei are larger and divided into four segments. Key events in the life cycle are plasmogamy (P), karyogamy (K) and meiosis (M).

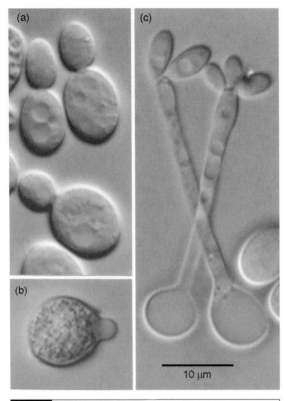

Fig 24.4 Microscopic features of *Phaffia rhodozyma* (a) and *Xanthophyllomyces* (b,c), a closely related teleomorph. (a) Vegetatively dividing yeast cells. (b) Thick-walled chlamydospore in the process of germination by budding about 3 h after transfer to a fresh medium. (c) Aseptate elongate basidia of *Xanthophyllomyces* sp. producing basidiospores at their tips. All images to same scale.

complex of several species which are difficult to distinguish by any means other than DNA sequencing (Bai *et al.*, 2002). *Rhodotorula* spp. also reproduce vegetatively by budding but are unable to produce ballistoconidia. Apart from this difference, the two genera are very close to each other, overlapping in phylogenetic trees and also possessing similar life cycles.

Life cycle

The budding phase of *Sporobolomyces roseus* is uninucleate, and nuclear division occurs at the time of bud formation (Buller, 1933). If ballistoconidia are to be formed, a conical sterigma develops vertically and bears an asymmetric spore which strongly resembles a basidiospore (Figs. 24.1e and 24.5). A daughter nucleus passes

into this spore. Following the fusion of Buller's drop with the adaxial blob, the spore is flicked away for a distance of about 0.1 mm. Colonies of *S. roseus* begin to form ballistoconidia after 2–3 days on most agar media, and they are recognized by the presence of numerous satellite colonies outside of the margin of the parent colony. A single sterigma may produce a second or even a third spore, and occasionally two or three sterigmata arise from one cell.

Although *S. roseus* can form pseudohyphae and true hyphae, no sexual reproduction has been observed. However, a related species, *Sporidiobolus* (*Sporobolomyces, Aessosporon*) *salmonicolor* (formerly also called *S. odorus*), does undergo the full sexual cycle (Fig. 24.7). This species is heterothallic with a bipolar (unifactorial) mating system (Fell & Statzell-Tallman, 1981). Following conjugation between compatible haploid uninucleate cells, a dikaryotic mycelium with clamp connections develops and eventually produces globose binucleate teliospores (Fig. 24.6; Bandoni *et al.*, 1971). Karyogamy follows. Meiosis occurs during teliospore germination, which gives rise to an aseptate metabasidium. This buds off haploid monokaryotic basidiospores at its tip. Life cycles of this kind have also been described in some other members of the Sporidiales, with minor variations such as the presence of transverse septa in the metabasidia, or the absence of ballistoconidia (see Fig. 24.1f).

There is, however, a complication in the life cycle of *S. salmonicolor* because, according to van der Walt (1970), yeast cells may be haploid or diploid. The diploid cells are larger than the haploid ones. Both types are capable of reproducing by budding and by producing ballistoconidia, but only the diploid cells can additionally develop directly into thick-walled teliospores. These germinate by means of a short aseptate promycelium producing yeast-like basidiospores at its tip (Fig. 24.7).

According to Fell and Statzell-Tallman (1998), a similar life cycle is found in *Rhodosporidium toruloides* (anamorph *Rhodotorula glutinis*). In this species, diploid yeast cells are thought to arise if there is a failure of meiosis during teliospore germination. In contrast to the observations by

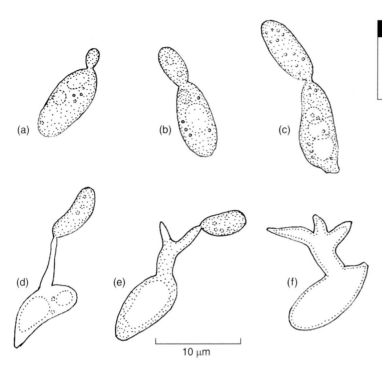

(a)

(b)

(c)

(d)

(e)

(f)

10 μm

van der Walt (1970) on *Sporidiobolus*, the diploid yeast cells of *R. toruloides* do not develop into teliospores directly, but first produce a mycelium which in turn forms teliospores. Conjugation between haploid cells of compatible mating type is mediated by peptide hormones. One of them, rhodotorucine A, has been identified as an undecapeptide with a farnesyl side chain (Kamiya *et al.*, 1979).

Some species of *Rhodosporidium* and *Sporidiobolus* are entirely homothallic, and here all yeast cells seem to be diploid (Fell & Statzell-Tallman, 1998; Statzell-Tallman & Fell, 1998). Meiosis in a diploid yeast cell establishes a clamped dikaryotic mycelium which produces teliospores. Germinating teliospores give rise to metabasidia which produce diploid basidiospores.

Ecology

Sporobolomyces, *Rhodotorula* and other yeasts belonging to the Urediniomycetes are found in diverse habitats such as the soil (Sláviková & Vadkertiová, 2003) and the sea, including deep sea locations (Nagahama *et al.*, 2001). However, their most prominent habitat is healthy and moribund vegetation (Nakase, 2000) from which

they can be isolated throughout the year. Together with the black yeast *Aureobasidium* (Ascomycota; see p. 484), they form a major component of the phylloplane yeast population. Breeze and Dix (1981) have estimated that yeasts (including ascomycetes) produce up to 50 times more biomass than hyphal fungi on *Acer* leaves throughout the growing season. *Sporobolomyces roseus* is the most abundant phylloplane yeast, comprising for example 76% of the yeast population on grapes (de la Torre *et al.*, 1999). Scanning electron microscopy studies have shown that cells of *S. roseus* may form sheets of mucilage by which they adhere to the leaf surface. There is no evidence that the growth of the yeasts causes corrosion of the cuticle (Bashi & Fokkema, 1976). There is interest in the suggestion that *Sporobolomyces* or *Rhodotorula* on leaf surfaces may compete with foliar pathogens, and that biological control of the pathogens might be possible (Fokkema & van der Meulen, 1976). Such an approach is especially promising in the control of post-harvest diseases because the incubation conditions can be more precisely controlled (Janisiewicz & Bors, 1995; Janisiewicz & Korsten, 2002). Many *Sporobolomyces*

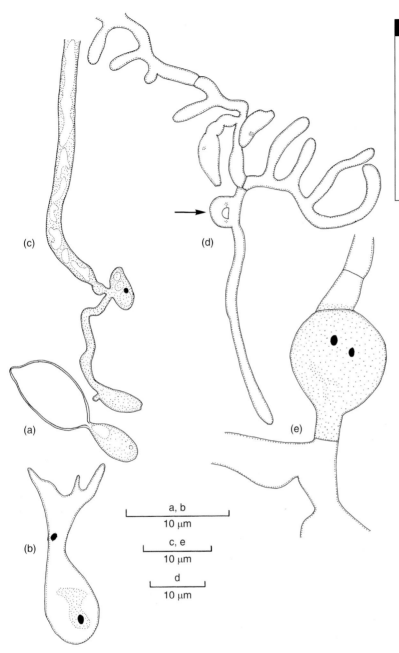

Fig 24.6 *Sporidiobolus salmonicolor.*
(a) Budding cell. (b) Cell with
sterigmata. (c) Conjugation of
haploid cells and initiation of a
dikaryotic hypha. (d) Young
dikaryotic hyphae, the original
conjugants obscured. The arrow
indicates the first clamp
connection. (e) Late stage of
chlamydospore development;
the paired nuclei are visible.
After Bandoni *et al.* (1971).

(c)
(d)
(a)
(b)
(e)

a, b
10 μm

c, e
10 μm

d
10 μm

and *Rhodotorula* spp. appear to be associated with lesions caused by plant pathogens such as rust fungi, or by parasites such as nematodes.

Ballistoconidia of *Sporobolomyces* are frequent in the air, especially during warm summer nights, and concentrations of these spores may reach values of up to $10^6 \, \mathrm{m}^{-3}$ (Gregory & Sreeramulu, 1958). Such high concentrations are a cause for concern because *Sporobolomyces* has

been shown to be a respiratory allergen (Evans, 1965).

Phylloplane yeast populations occupy an exposed habitat and are therefore susceptible to environmental changes. *Sporobolomyces roseus* has been proposed as an indicator of air quality, low colony counts correlating with heavy air pollution (Dowding & Richardson, 1990). The results can be superimposed on those obtainable

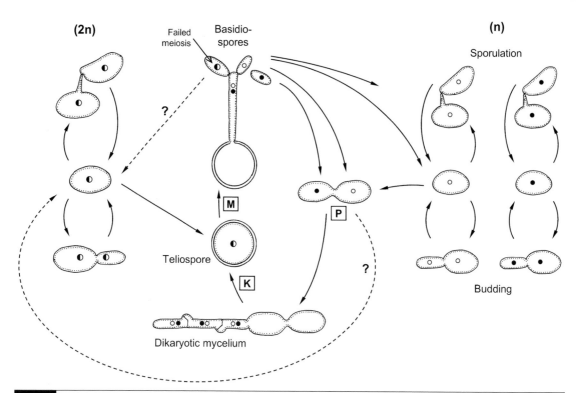

(2n)

Failed meiosis

Basidio-spores

(n)

Sporulation

?

M

Teliospore

P

K

?

Budding

Dikaryotic mycelium

Fig 24.7 Possible life cycle of *Sporidiobolus salmonicolor*. Both diploid stages (large split nuclei) and haploid stages (small black or white nuclei) exist and can reproduce by budding and ballistoconidium formation. Plasmogamy of two yeast cells of compatible mating types or conversion of a diploid yeast cell leads to the formation of a teliospore which undergoes meiosis upon germination. Failure of meiosis in the metabasidium or conjugation between two haploid yeast cells could give rise to diploid yeast cells. Key events in the life cycle are plasmogamy (P), karyogamy (K) and meiosis (M).

with the biomonitoring of lichens (see p. 454), and both groups of organisms are susceptible to SO_2.

Phylloplane yeasts including *S. roseus* are also sensitive to UV radiation, and enhanced irradiation leads to a decline in their abundance on leaves (Newsham *et al.*, 1997). However, some protection may be afforded by the carotenoids produced by red yeasts, since carotenoid-deficient mutants are more sensitive to UV light than the red wild-type strains. Pure-culture studies with *Sporobolomyces* and *Rhodotorula* spp. often show enhanced production of carotenoids under oxidative stress such as high aeration or the addition of radical generators. Further, the carotenoid spectrum may also change; if it does, the shift is commonly from γ- and/or β-carotene under reduced oxygen pressure to torulene and oxidized carotenoids (e.g. torularhodin) at oxidative stress (Fig. 24.8; Sakaki *et al.*, 2002;

Davoli *et al.*, 2004). This may represent an adaptive strategy because many of the oxygen-containing carotenoids (i.e. xanthophylls) have superior anti-oxidant properties as compared to β-carotene (Martin *et al.*, 1999). Other metabolites (e.g. vitamin E) or enzymes (superoxide dismutase, catalase) with anti-oxidant activities may also be produced by red yeasts so that the correlation between carotenoid production and anti-oxidant protection is not always absolute.

24.4 | Ustilaginomycete yeasts

Most yeasts belonging to the Ustilaginomycetes are the monokaryotic stages of smut fungi (Section. 23.2) or of *Exobasidium* (Section 23.4) and live saprotrophically on plant surfaces.

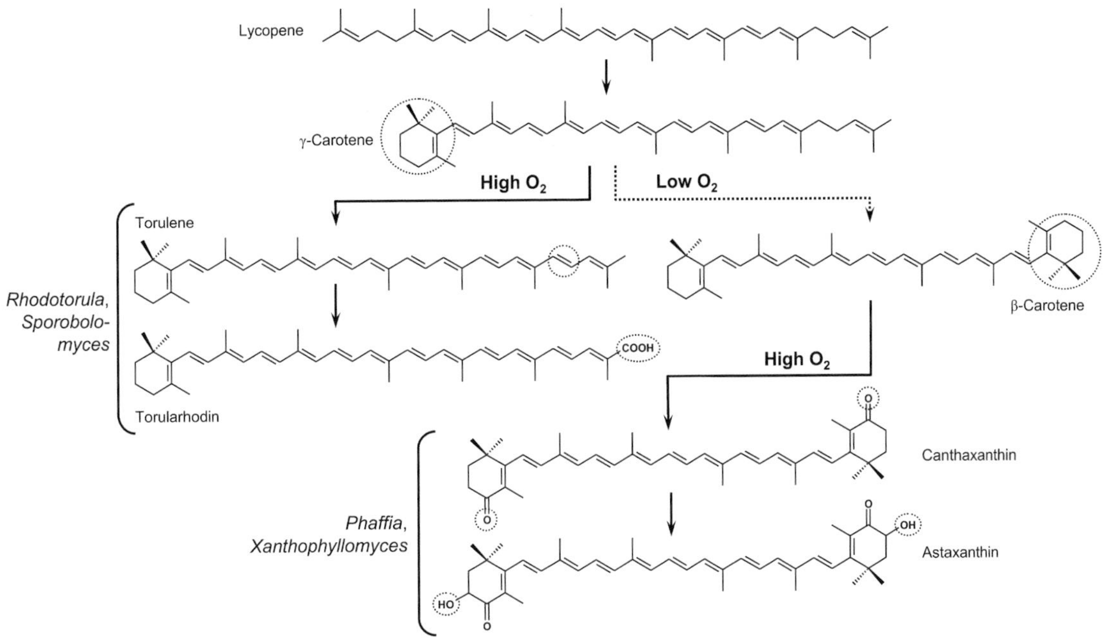

Fig 24.8 Examples of the effect of aeration on carotenoid production in red yeasts belonging to the Basidiomycota. In most red yeasts, carotenoid biosynthesis progresses via lycopene at least to the first end-group cyclization to give γ-carotene. Several yeasts adjust their carotenoid spectrum in response to the level of oxidative stress in their environment. In *Sporobolomyces* and *Rhodotorula* spp. (Urediniomycetes), γ-carotene and/or the bicyclic β-carotene accumulate under microaerophilic conditions (dotted arrow), whereas torulene and oxygen-containing carotenoids (xanthophylls) such as torularhodin are produced under oxidative stress (solid arrows). In contrast, in *Phaffia rhodozyma* (Heterobasidiomycetes) the β-carotene produced at low partial oxygen pressure is directly oxidized to bicyclic xanthophylls (notably astaxanthin) under oxidative stress.

Here we shall briefly consider a third group of Ustilaginomycetes, the Malasseziales, which colonize the skin of warm-blooded animals, i.e. mammals and birds. Good accounts of this order have been written by Ashbee and Evans (2002), Crespo Erchiga and Delgado Florencio (2002), Inamadar and Palit (2003) and Batra *et al.* (2005).

24.4.1 Malasseziales

This group has been named after the French pathologist Louis C. Malassez who was one of the pioneers in this field. Following an eventful taxonomic history, there is now only one genus, *Malassezia* (formerly *Pityrosporum*) with seven species (Guillot & Guého, 1995; Guého *et al.*, 1996). All of them are haploid without known sexual stages.

The yeast cells show several unusual characteristics by which they can be recognized. Budding is enteroblastic, with the daughter cell arising in a unipolar fashion from an exceptionally broad ring-like bud scar of the mother cell (Guého & Meyer, 1989). The ultrastructure of *Malassezia* is most unusual. The cell wall is thick, with its inner surface sculptured into spiral ridges (Fig. 24.9) which cast the plasma membrane into corresponding grooves (David *et al.*, 2003). There is considerable morphological plasticity in some *Malassezia* spp., notably *M. furfur*, in that the shape of yeast cells may change from globose to elongated after repeated subculturing. Hyphae may also be formed under certain conditions.

All but one species (*M. pachydermatis*) are obligately lipophilic and are isolated from skin samples if standard agar media are overlaid with a thin film of olive oil. It is possible to isolate most *Malassezia* spp. from the skin of humans and other animals, with *M. pachydermatis*

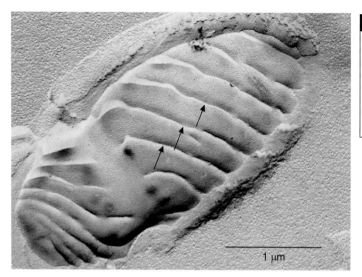

1 μm

Fig 24.9 *Malassezia pachydermatis.* Freeze—fracture image of a budding cell showing the inner surface of the wall. The spiral ridges are not seen in the collar region. From David *et al.* (2003), by copyright permission of *Scripta Media*, Brno. Original image kindly provided by M. David, M. Gabriel and M. Kopecká.

showing a predilection for dogs and cats where it can cause skin infections (Guillot & Bond, 1999). Most healthy humans carry *Malassezia* spp., but the species composition and density of colonization are altered in patients affected by dermatitis (Sugita *et al.*, 2001; Crespo Erchiga & Delgado Florencio, 2002). The pattern of colonization is dependent upon several factors, e.g. the degree of sweating and the amount of lipid produced. *Malassezia* spp. are often most abundant on people in early adulthood because at that age the lipid-producing sebaceous glands at the base of the hair shaft are most active. A clear role in pathogenesis has been demonstrated only for *M. globosa* which causes a superficial mycosis known as *pityriasis versicolor* in adults below middle-age (Gupta *et al.*, 2002). The disease is more common in the tropics than in cooler climates. Infected skin areas differ in pigmentation from the normal skin. Infections are associated mainly with the hyphal growth form of *M. globosa* whereas commensal growth is yeast-like (Crespo Erchiga & Delgado Florencio, 2002). Occasionally, *Malassezia* spp. also cause contaminations of catheters.

The association of *Malassezia* spp. with dandruff has been suggested but is not yet proven (Piérard-Franchimond *et al.*, 2000). Evidence in favour of the argument is that dandruff, like pityriasis versicolor and other superficial skin infections caused by *Malassezia* spp., disappears upon treatment with shampoos or lotions containing selenium sulphide or other topical antifungal agents (Kwon-Chung & Bennett, 1992). Further, some *Malassezia* infections are more common in immunocompromised patients than in healthy subjects, and there is evidence of both humoral and cell-mediated immune responses against *Malassezia* in immunocompetent humans (Ashbee & Evans, 2002).

25

Anamorphic fungi (nematophagous and aquatic forms)

Throughout this book we have attempted to consider fungi showing predominantly or purely asexual reproduction together with their known or suspected teleomorphs. However, certain groups of taxonomically diverse fungi colonizing the same specialized habitats or substrates are best understood in their ecological context, especially if they show strikingly similar adaptations and morphology despite their different evolutionary histories. Two cases illustrating such convergent evolution among anamorphic fungi are the nematophagous habit and the aquatic habitat, which we shall consider in turn in this chapter.

25.1 | Nematophagous fungi

Nematodes are a very varied group of invertebrates. They are particularly common as free-living saprotrophic species in the soil, around plant roots, on dung and in all kinds of decomposing plant matter, as well as in freshwater and marine habitats. Most saprotrophic nematodes feed on bacteria, although fungal hyphae may also be consumed. Other species parasitize animals, releasing their eggs or motile stages into the environment when their hosts defaecate. Plant-parasitic species chiefly attack roots as free-living or sedentary organisms. Sedentary species form adult stages inside plant root tissues where

they cause the economically important root knot diseases (*Meloidogyne* spp.) or root cyst diseases (*Heterodera* spp. and *Globodera* spp.). Plant-parasitic nematodes are readily recognized because their mouth parts are modified as stylets with which they penetrate plant tissues. Gravid females of cyst nematodes enlarge, and their bodies become converted into a hardened cyst containing the eggs.

Fungi have evolved a range of mechanisms to attack nematodes, which can be grouped into three broad categories described below. A summary of genera and their taxonomic relationships is presented in Table 25.1. Nematophagous fungi are common in most types of soil which are rich in organic matter, and they have been found in arctic, temperate and tropical climates (see Dix & Webster, 1995). There is no strong evidence of host selectivity.

Predatory nematophagous fungi produce a sizeable mycelium in the soil, with trapping devices formed at intervals along the length of the hyphae. The varied and occasionally spectacular trapping mechanisms have captured the imagination of generations of mycology students. The most common traps are sticky knobs, adhesive networks, non-constricting rings or constricting rings (see pp. 675–680). Penetration of captured nematodes is followed by the growth of trophic hyphae throughout the nematode body, and digestion of its contents. Because most predatory nematophagous fungi

Table 25.1. Examples of the diversity of nematophagous fungi.

Genus	Taxonomic affinity	Mode of parasitism
Predatory fungi		
Acaulopage, Stylopage	Zygomycota	Adhesive hyphae
Gamsylella, Dactylellina, Arthrobotrys (incl. *Dactylella, Dactylaria, Monacrosporium, Duddingtonia*)	Orbiliaceae (Ascomycota)	Adhesive knobs, columns, nets, non-constricting rings
Drechslerella (incl. *Arthrobotrys, Dactylella, Monacrosporium*)	Orbiliaceae (Ascomycota)	Constricting rings
Nematoctonus (*Hohenbuehelia*), *Pleurotus*	Pleurotaceae (euagarics clade, Basidiomycota)	Adhesive or poisonous knobs
Endoparasitic fungi		
Haptoglossa	Plasmodiophoromycota or Oomycota	Gun cells (see Fig. 3.9)
Myzocytium, Nematophthora	Oomycota	Encysting zoospores
Catenaria	Chytridiomycota	Encysting zoospores
Harposporium, Drechmeria, Verticillium, Hirsutella	Clavicipitaceae (Pyrenomycetes, Ascomycota)	Ingestion or attachment of conidia
Egg and cyst parasites		
Rhopalomyces	Zygomycota	Hyphal colonization of eggs
Pochonia chlamydosporia	Clavicipitaceae (Pyrenomycetes, Ascomycota)	Hyphal colonization of cysts
Paecilomyces lilacinus	Ascomycota (*incertae sedis*)	Hyphal colonization of eggs

have a high saprotrophic potential, including the ability to degrade cellulose, nematodes are probably utilized mainly as a nitrogen supplement (Barron, 1992).

In the **endoparasitic** nematophagous fungi (see p. 680) there is no extensive mycelial development outside the nematode host, and these species must therefore be regarded as obligate parasites in ecological terms. They exist in the soil as spores which may either become attached to the body of the host, or become ingested. The spores then germinate and penetrate the animal, developing a mycelium within the body of the nematode. Only reproductive hyphae (conidiophores) penetrate to the outside of the dead colonized nematode.

Parasites of eggs and cysts (p. 684) are found in several different taxonomic groups. Typically these fungi can be isolated from the soil or rhizosphere as well as from nematode eggs or cysts, and are therefore viewed as opportunistic saprotrophs. The most thoroughly studied representatives are *Paecilomyces lilacinus* and *Pochonia chlamydosporia* (formerly *Verticillium chlamydosporium*).

We owe much of our knowledge of nemato-phagous fungi to the numerous publications by Charles Drechsler. These are cited in many of the general accounts given by Duddington (1955, 1957), Barron (1977, 1981), Dowe (1987), Gray (1987), Nordbring-Hertz (1988) and Dix and Webster (1995). A superb film has been produced by Nordbring-Hertz *et al.* (1995). Cooke and Godfrey (1964) have provided a key to identification.

25.1.1 Predatory fungi belonging to Ascomycota

Predatory nematophagous fungi are easy to study by means of the sprinkle-plate technique. A small pinch of soil is added to a Petri dish containing tapwater agar or dilute cornmeal agar, and the dish is incubated for a few weeks at room temperature. Saprotrophic nematodes present in the soil will crawl out over the agar surface, feeding on bacteria. If predatory fungi are present in the soil, they develop structures for trapping nematodes, and the trapped dead or dying animals are easy to see with a dissecting microscope. Conidia will be produced around colonized nematodes, and if these are trans-ferred to standard media such as cornmeal agar, mycelial colonies will grow. Whilst a few zygomycetes belonging to the order Zoopagales trap nematodes by secreting an adhesive from undifferentiated hyphae or short hyphal branches upon contact with nematodes (Drechsler, 1962; Saikawa & Morikawa, 1985), and two basidiomycete genera also trap nema-todes (see Fig. 25.7), the most striking predatory species are conidial forms of Ascomycota. Detailed species descriptions may be found in the works by van Oorschot (1985) and Rubner (1996).

The taxonomy of predatory ascomycetes has traditionally been based on the morphology of conidia, especially the number of septa, and the degree of clustering of conidia on the conidio-phore. However, DNA-based phylogenetic approaches have confirmed long-held suspicions that these criteria are artificial, and that more natural taxa can be obtained by grouping predatory ascomycetes according to the type of trap they form. These results have been summa-rized and discussed by Scholler *et al.* (1999). All known trap-forming ascomycetes described so far belong to the family Orbiliaceae (Hagedorn & Scholler, 1999), with some species shown to produce an apothecium referrable to *Orbilia* itself (Pfister, 1997). This family is still of uncertain affinity (*incertae sedis*) within the Ascomycota but may belong to the Pezizales. The following trapping structures have been described.

Adhesive knobs and lateral branches. Single-celled globose knobs, covered by a sticky secretion and spaced at intervals along a hypha, form the morphologically simplest trapping organs. These knobs are borne directly on the hypha or on short lateral branches in such a way that a nematode may become attached to several knobs (Fig. 25.1). A different type of sticky knob is produced on thin stalks (see Fig. 25.5b). Sometimes a nematode attached to an adhesive knob may pull it off the subtending hypha by its violent movement, but respite is only temporary because penetration of the host may still occur from such detached knobs. Barron (1977) has suggested that detachable knobs may provide an effective means of dispersal. In some species, lateral knob-bearing branches may develop in sufficient proximity to each other for anasto-mosis to take place, and this results in the formation of a primitive two-dimensional sticky network. Scholler *et al.* (1999) have proposed two genera to accommodate fungi with knob-like traps, namely *Dactylellina* (formerly *Monacrosporium*, *Arthrobotrys* and *Dactylella* spp.) for species with stalked knobs, and *Gamsylella* (formerly *Dactylella* spp.) for forms with unstalked knobs and/or primitive nets.

Adhesive nets. One of the most common types of trap is a three-dimensional adhesive network formed by anastomosis of the recurved hyphal tips of a lateral branch system. The network is lifted above the general level of the mycelium. The entire surface of the network is covered by an adhesive, as shown by scanning electron microscopy (Fig. 25.2; Nordbring-Hertz, 1972), and a nematode which thrusts its body into the network is quickly immobilized. Scholler *et al.* (1999) have grouped fungi

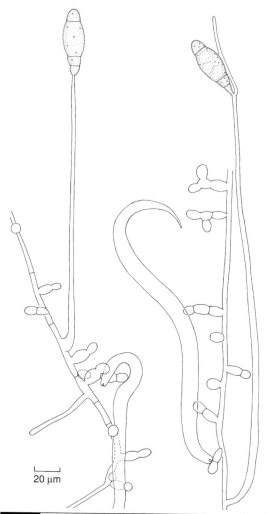

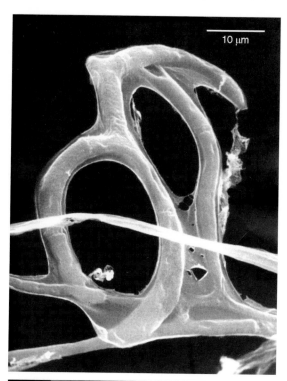

Fig 25.1 *Gamsylella* sp. Hyphae showing short lateral branches modified as sticky knob traps, with nematodes attached at several points. This species would be called *Monacrosporium* sp. in traditional nomenclature because of the single conidium produced at the end of the conidiophore.

producing traps of this kind in the genus *Arthrobotrys*. Adhesive nets are illustrated for *A eudermata* (syn. *Monacrosporium eudermatum*, *Duddingtonia flagrans*) in Fig. 25.3 and for *A. robusta* (Fig. 25.4). *Arthrobotrys oligospora* (Fig. 25.2) is by far the most thoroughly investigated member of the genus.

Non-constricting rings. A number of predatory fungi ensnare their prey by three-celled rings which are formed by recurvature of the tip of a lateral branch, followed by its anastomosis with itself. Non-constricting rings have a sticky inner surface. A nematode thrusting its body into such a loop may become tightly wedged inside it, and may find it impossible to retract. The point of junction of the loop to the subtending hypha is often weak, and the struggling nematode may detach the loop. Occasionally, a single nematode bearing several loops may be seen. The detached loops are still capable of penetrating and killing the nematode. The action of this trap is passive, i.e. there is no inflation of the ring. Fungi producing non-constricting rings also form stalked adhesive knobs and are included in *Dactylellina* (Scholler *et al.*, 1999). An example is *D. haptotyla* (syn. *Dactylaria candida*), shown in Fig. 25.5.

Constricting rings. The most dramatic type of trap is the constricting ring trap, which develops

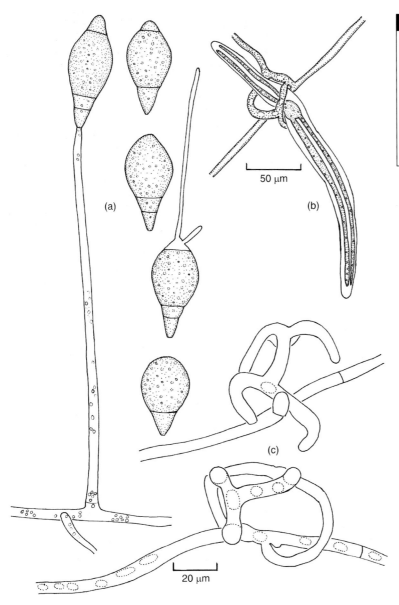

Fig 25.3 *Arthrobotrys eudermata* (syn. *Monacrosporium eudermatum*, *Duddingtonia flagrans*).
(a) Conidiophore with its single attached conidium. Several detached conidia are also shown, one germinating. (b) Trapped nematode showing infection bulb and assimilative hyphae. (c) Adhesive trapping networks.

(a)

(b)

50 μm

(c)

20 μm

in the same way as the non-constricting ring but differs from it in that the three ring cells are able to inflate rapidly following stimulation by mechanical contact with the inner ring surface. Inflation occludes the lumen of the ring, severely constricting any nematode trapped in it. The surface of constricting rings does not carry any adhesive. Fungi producing constricting rings have been assigned to the genus *Drechslerella* by Scholler *et al.* (1999) and were formerly distributed in *Arthrobotrys* and *Dactylella*. This type of trap is illustrated in Fig. 25.6.

The mechanism of ring closure is of interest. The insertion of a fine glass rod into a ring by means of a micromanipulator, followed by gentle friction of one of the cells, can trigger off the closure of the trap. Other stimuli, such as heat or a stream of dry air, are also effective. The enlargement of the cells is accompanied by vacuolation of their contents, and by elastic stretching of the *inside* wall of the ring, whilst the outer wall does not change shape (Estey & Tzean, 1976). Ring closure is complete within 0.1 s. Enlargement of the three cells

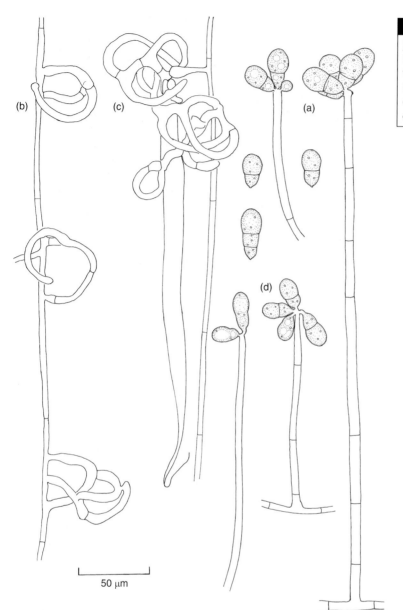

Fig 25.4 (a–c) *Arthrobotrys robusta*.
(a) Conidiophores and conidia.
(b) Mycelium with anastomosing
traps. (c) Nematode caught in trap.
(d) *Arthrobotrys oligospora*. Two
conidiophores showing the sequence
of conidial development.

50 μm

making up the ring is not simultaneous, but one cell inflates a fraction of a second before the others. By immersing rings in 0.3–0.5 M sucrose and inducing trap closure by heat, Muller (1958) succeeded in slowing down the rate of ring closure by a factor of 100, so that the process took about 10 s. Estimations of the volume change of the cells during constriction showed a threefold increase.

Several physiological changes must take place during closure. Chen *et al.* (2001) have provided evidence of a signalling chain in which a physical stimulus at the inner ring surface activates a trimeric G protein, and transduction of this signal via inositol trisphosphate, Ca^{2+} and calmodulin (see Fig. 12.48) leads to the release of osmotically active molecules within the ring cells, coupled with the opening of water channels at the plasma membrane surface. Muller (1958) estimated that there must be an uptake of $18\,000\,\mu m^3$ of water in 0.1 s, and water uptake may therefore take place over much of the

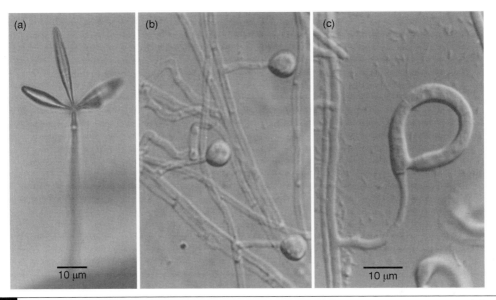

Fig 25.5 *Dactylellina haptotyla* (syn. *Dactylaria candida*). (a) Conidiophore arising from the substrate, producing a terminal cluster of septate spindle-shaped conidia. (b) Stalked unicellular knob traps. (c) Non-constricting ring traps. (b,c) to same scale.

surface of the ring cells. Since the inner part of the ring wall rapidly changes shape, it is likely that there is a slippage of microfibrils making up the wall. Dowsett *et al.* (1977) showed that the inner (luminal) wall of the ring cells in *Drechslerella* (*Dactylella*) *brochopaga* is thicker than the outer wall. The luminal wall is initially four-layered, but the two external wall layers rupture during expansion.

The threefold increase in cell volume and the commensurate increase in surface area, together with the process of vacuolation, clearly necessitate a rapid rearrangement of membrane material within the cell. Heintz and Pramer (1972) discovered a labyrinth of membrane-bound material close to the plasma membrane on the inside of unexpanded ring cells in *Drechslerella* (*Arthrobotrys*) *dactyloides*. In expanded rings, a more usual type of plasma membrane organization was found, suggesting that the membrane-bound inclusions had contributed to the formation of the enlarged plasma membrane (see also Dowsett *et al.*, 1977).

25.1.2 Predatory fungi belonging to Basidiomycota

Nematode-destroying basidiomycetes seem to be confined to two closely related genera,

Hohenbuehelia and *Pleurotus*, both belonging to the Pleurotaceae in the euagarics clade (p. 541; Thorn *et al.*, 2000). Both genera contain wood-rotting species forming clamped hyphae. *Hohenbuehelia* produces a *Nematoctonus* anamorph which is of relevance in nematode capture (Fig. 25.7).

The unusual killing mechanism of *Pleurotus* spp. has been described by Barron and Thorn (1987). Hyphae produce minute unicellular lollipop-shaped spathulate cells which secrete droplets of a toxin-containing liquid. Various toxins have been identified (p. 682). Nematodes touching such a drop show dramatic symptoms of a shrinkage of the head region and deformation of the oesophagus. Onset of symptoms is within one to several minutes and is quickly followed by lethargy, so that the affected nematode becomes immobilized close by. Hyphae of *Pleurotus* show rapid tropic growth towards orifices of the paralysed nematode, and colonization ensues within one to several hours (Fig. 25.7a).

Hohenbuehelia possesses two modes of parasitism. Hourglass-shaped unicellular mycelial branches produce a large drop of non-toxic glue by which nematodes are trapped in the usual predatory way (Fig. 25.7b). In addition, both

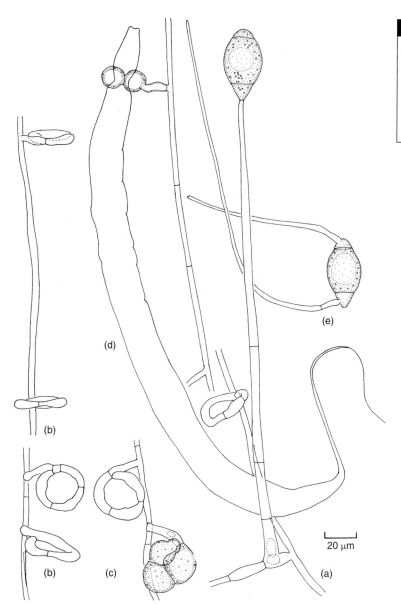

Fig 25.6 *Drechslerella* sp.
(a) Conidiophore with a single terminal conidium typical of the former genus *Monacrosporium*.
(b) Three-celled constricting ring traps. (c) Two traps, one inflated.
(d) Nematode caught in constricting ring trap. (e) Germinating conidium.

20 μm

Hohenbuehelia basidiospores and *Nematoctonus*-type conidia produce sticky drops (Barron & Dierkes, 1977). Drops are formed at the tapering ends of the conidia which are sometimes bent at an angle, attaching them to the cuticle of a nematode brushing them (Figs. 25.7c,d). *Hohenbuehelia* therefore shows a transition of predatory and endoparasitic features (Poloczek & Webster, 1994). There are also transitions between *Hohenbuehelia* and *Pleurotus* because Thorn *et al.* (2000) have described a

Hohenbuehelia sp. producing both sticky knobs and toxin-secreting lollipop-shaped branches. Unidentified toxins have also been shown to be present in the sticky drops on *Nematoctonus*-type conidia (Giuma *et al.*, 1973).

25.1.3 Endoparasitic nematophagous fungi
Several genera of anamorphic fungi include forms which are endoparasitic. Although some of these fungi can be grown in pure culture in the laboratory, in nature probably most of them

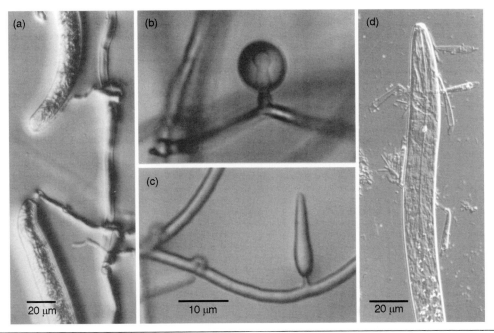

Fig 25.7 Nematode-trapping Basidiomycota. (a) Growth of *Pleurotus* hyphae towards buccal cavities of two nematodes killed at za distance by secreted toxins. (b–d) *Nematoctonus* state of *Hohenbuehelia*. (b) Hourglass-shaped cell producing an adhesive. (c) Production of an awl-shaped conidium from a hypha bearing clamp connections. (d) Conidia attached to the cuticle of a dead nematode by a sticky knob at the tip of a bent neck. Infection has given rise to trophic hyphae. (b,c) to same scale.

are obligate parasites without a free-living saprotrophic phase. In some species, the conidia attach themselves to the cuticle of the host and, on germination, penetrate into the body cavity, eventually filling it with hyphae. *Drechmeria coniospora* (Figs. 25.8a–c) has conical conidia which, at maturity, bear a globose, adhesive knob at the narrow end. Such conidia readily adhere to the cuticle of a nematode which brushes past them, and they become preferentially attached to the buccal end of nematodes (Jansson *et al.*, 1985). In some other forms, infection follows ingestion of a conidium. When the crescent-shaped conidia of *Harposporium anguillulae* are ingested, the pointed end of the spore becomes lodged in the wall of the oesophagus and, upon germination, penetrates the body cavity from within (Aschner & Kohn, 1958). The conidiophores emerging from the dead host bear sessile, subglobose phialides (Figs. 25.8d,e). The mycelium within the host may form chlamydospores which presumably survive in the soil when the body of the nematode decays.

Little is known about the taxonomic position of endoparasitic fungi. Several genera contain members parasitizing soil invertebrates other than nematodes. For instance, the teleomorph of *Harposporium*, *Podocrella* (syn. *Atricordyceps*), is related to the insect-parasitic genus *Cordyceps* in the Clavicipitaceae (see p. 360). There are also numerous examples of lower fungi adopting an endoparasitic mode of life, such as *Catenaria anguillulae* (Chytridiomycota) and *Myzocytium* spp. (Oomycota) which infect nematodes by means of zoospores, or *Haptoglossa* which produces the spectacular gun cells shown in Fig. 3.9.

Although endoparasitic fungi can sometimes be observed on soil-sprinkle plates, a more efficient technique for their detection is the Baermann funnel. A conical funnel is fitted with rubber tubing at its base, and the opening is sealed with a clamp. The funnel is lined with tissue paper or filter paper; the soil sample is filled into the funnel and submerged in water. Nematodes will migrate through the filter paper

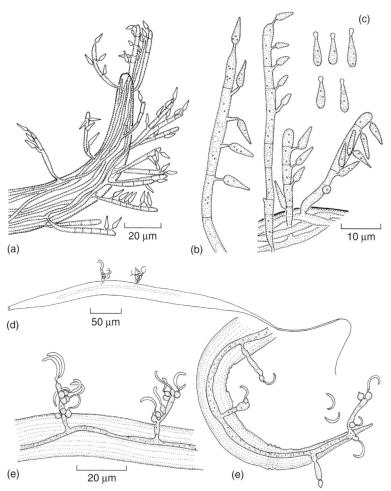

(a)

(b)

(c)

(d)

(e)

(e)

20 μm

10 μm

50 μm

20 μm

Fig 25.8 Endoparasitic nematophagous fungi. (a–c) *Drechmeria coniospora*. (a) Dead nematode containing hyphae, and bearing emergent conidiophores. (b) Conidiophores. (c) Mature conidia showing terminal adhesive bulbs. (d,e) *Harposporium anguillulae*. (d) Dead nematode showing the extent of internal mycelium, and external conidiophores. (e) Enlargement to show details of conidiophores which bear subglobose phialides and sickle-shaped phialoconidia.

and congregate at the bottom of the tube, from which they can be collected onto tap-water agar plates and observed by microscopy. After a few hours' incubation, non-motile nematodes may be found, some of them infected with endoparasitic fungi. These may develop reproductive structures. Bailey and Gray (1989) have discussed various isolation techniques for nematophagous fungi.

25.1.4 Nematode–fungus interactions
Stimulation of trap formation
Many nematophagous fungi do not produce trapping organs in pure culture, but the addition of nematodes or nematode extracts will induce their development within 24–48 h. The inducing molecules include small oligopeptides containing non-polar and aromatic amino acids, such as the phenylalanyl-valine dipeptide (Friman *et al.*, 1985). These are present in exudates from nematodes (Nordbring-Hertz, 1977). Other substances, such as horse serum or yeast extract, are also effective in inducing trap formation, and in natural substrates such as soil or dung conidia may germinate directly by formation of a trap (Persmark & Nordbring-Hertz, 1997).

In general terms, the readiness to form traps has been found to differ between various genera and is inversely correlated with their saprotrophic capacity (Cooke, 1964; Barron, 1992). Nordbring-Hertz and Jansson (1984) have categorized nematophagous fungi into three groups. Thus, the net-forming genus *Arthrobotrys* (group 1), credited with the highest competitiveness in soil, forms its traps only upon induction by the presence of nematodes, whereas producers of knobs and rings (*Dactylellina*, *Gamsylella* and *Drechslerella*; group 2) form traps more readily but have a more limited capacity to survive saprotrophically in soil. The obligate end of the spectrum from saprotrophy to parasitism among nematophagous fungi is occupied by some of the endoparasitic fungi (group 3). The conidia of these species form sticky drops constitutively and have no saprotrophic ability.

Mycelium with traps – whether formed constitutively or induced by the presence of nematodes – has been shown to attract nematodes significantly more strongly than mycelium without traps (Field & Webster, 1977; Jansson & Nordbring-Hertz, 1980). The strength of attraction can be correlated with the saprotrophic capacity of the fungus in question, i.e. group 3 fungi exert the highest attraction to nematodes, with group 2 fungi being intermediate and group 1 fungi attracting nematodes to a lesser extent even if trap formation has been induced (Jansson & Nordbring-Hertz, 1979). The identity of the nematode-attracting substances is as yet unknown.

Adhesion

The strength of adhesion of a trap or conidium to the nematode prey is remarkable, and much work has been carried out to characterize the glue involved. This has been summarized by Nordbring-Hertz (1988) and Tunlid et al. (1992), who presented evidence of the involvement of lectins, i.e. proteins that bind to specific carbohydrate residues (receptors). Different predatory and endoparasitic nematophagous fungi differ in the kind of lectin they produce, and this may partially account for the specificity of binding observed in some cases, e.g. in *Drechmeria coniospora* whose conidia attach mainly to the buccal end of its prey. It seems that the lectins are located within the glue on the trap or conidium, and that the carbohydrate-based receptors recognized by the lectins are part of the glycosylation chains of proteins on the nematode surface.

Toxins

The capture of a nematode by a predatory fungus is soon followed by its death, sometimes after a quick but violent struggle. In the case of constricting ring traps, the stricture of the body may well be a contributory cause of death, but there is evidence that toxins are also produced by certain fungi. Stadler et al. (1993) isolated the common fatty acid linoleic acid as a nematicidal principle of *Arthrobotrys oligospora* and other species. Linoleic acid was produced at higher amounts by trap-forming cultures than by uninduced ones, and it was highly toxic against nematodes in vitro, even reproducing the typical symptoms of hyperactivity followed by paralysis. In the case of *Pleurotus* spp., two toxins have been isolated, namely *trans*-2-decenedioic acid (Kwok et al., 1992) and linoleic acid (see Anke et al., 1995). The production of antibacterial antibiotics by numerous nematophagous fungi has been interpreted as a substrate defence strategy against bacterial competitors during the colonization of a killed nematode (Anke et al., 1995).

The infection process

Infection of nematodes often begins before the animal is dead. Enzymatic and turgor pressure-driven mechanisms have been implicated (Veenhuis et al., 1985; Dijksterhuis et al., 1990). Since the nematode cuticle consists mainly of collagen-type proteins and because serine proteases of the subtilisin type are known from a wide diversity of nematophagous fungi, these are generally assumed to play an important role in infection of the host and its subsequent degradation (Åhman et al., 2002; Morton et al., 2004), and their transcription is enhanced by the presence of nematode cuticle (Åhman et al., 1996). A subtilisin from *A. oligospora* has even been found to possess nematotoxic properties, hinting at several roles which these proteases may play during infection (Åhman et al., 2002).

Details of the infection process differ between predatory and endoparasitic nematodes. In predatory species, penetration of the cuticle is by means of an appressorium or a hypha emitted by that part of the trap which is in contact with the cuticle. Immediately within the nematode body, the penetration hypha swells to form a globose vesicle, the **infection bulb** (see Fig. 25.3b), from which assimilative hyphae radiate throughout the animal, now dead. The cytoplasm of traps of most predatory fungi contains an unusual abundance of dense bodies (microbodies) identified as peroxisomes. Their function is currently unknown but may be related to storage (Veenhuis et al., 1989). Similar organelles, though of different cytological origin, are seen in the assimilative hyphae within the nematode. It is likely that they are involved in amino acid assimilation and/or the degradation of lipids during the digestion of the nematode contents (Dijksterhuis et al., 1993).

The sequence of infection-related development in the endoparasitic conidial species *Drechmeria coniospora* and *Verticillium balanoides* has been described by Dijksterhuis et al. (1990, 1991) and Sjollema et al. (1993). A hypha growing out through the adhesive pad at the conidial apex forms an appressorium on the nematode cuticle; this mediates penetration. There is no infection bulb, and the nematode may remain alive while the trophic hyphae proliferate within its body. As in predatory species, numerous microbodies and lipid droplets are seen within the trophic hyphae. The extent of mycelium produced by endoparasitic nematophagous species in nematodes is often limited, with most of the captured biomass being converted to conidia.

25.1.5 Opportunistic parasites of eggs and cysts

Numerous saprotrophic soil fungi have been shown to be associated with the eggs and cysts of nematodes, especially sedentary species parasitizing plant roots (Stiles & Glawe, 1989). Even though some host specialization may occur, all fungi colonizing nematode eggs and cysts are currently considered opportunistic parasites (Siddiqui & Mahmood, 1996). They reproduce as conidia and/or chlamydospores, like most soil fungi. Infection is by means of hyphal tips, and no specialized infection structures are formed

apart from appressoria in some species. Nematode eggs contain chitin in addition to collagen, and chitinases as well as serine proteases have been demonstrated in fungal egg parasites (Morton et al., 2004). The most important species as potential biological control agents are *Pochonia chlamydosporia* which produces conidia and multicellular melanized chlamydospores, and *Paecilomyces lilacinus* (Siddiqui & Mahmood, 1996; Kerry & Jaffee, 1997).

25.1.6 Biological control of nematodes by parasitic fungi

Some nematodes are serious parasites of plants and animals, and attempts have been made to use all three ecological groups of nematophagous fungi described above – predatory, endoparasitic and egg- or cyst-colonizing forms – for biological control. Although the promise held by nematophagous fungi is high and some success has been reported, no commercial breakthrough has as yet been achieved.

Predatory nematophagous fungi tend to show only limited competitiveness in non-native soils (Siddiqui & Mahmood, 1996). Although this may preclude their use in the biological control of plant-parasitic nematodes, the situation is different with animal parasites. Many trap-forming fungi occur on dung along with the larval stages of nematodes parasitizing herbivorous animals. Biological control should therefore be feasible, but a major obstacle is the requirement for the spores of such potential biocontrol agents to survive the stringent passage through the herbivore gut. This ability has been demonstrated for *Arthrobotrys eudermata* (syn. *Duddingtonia flagrans*), which is unusual among predatory fungi in forming thick-walled chlamydospores in addition to conidia. There is considerable current interest in biological control based on feed supplemented with chlamydospores of *A. eudermata* (reviewed by Larsen, 2000).

There are obvious problems in using endoparasitic nematophagous fungi for biological control due to practical difficulties in producing sufficient inoculum of many species, and the fundamental problem of their low saprotrophic capacity. The latter feature may be responsible for observations of a tight coupling between the population dynamics of root cyst nematodes and those of their fungal endoparasite, *Hirsutella rhossiliensis* (Jaffee, 1992). This species can be grown in pure culture, and although conidia generated in this way are not infectious, hyphal inoculum might be useful for biological control in the future (Kerry & Jaffee, 1997).

Opportunistic egg and cyst parasites, notably *Pochonia chlamydosporia* and *Paecilomyces lilacinus*, possess a high ability to colonize plant roots in agricultural soils, and are therefore potentially useful in the biological control of plant-parasitic nematodes. Both have suppressed root knot and cyst nematodes in experiments under controlled conditions (Siddiqui & Mahmood, 1996), although a major problem is that egg masses or cysts embedded in root tissue cannot be attacked by either fungus (Kerry, 2000). *Pochonia chlamydosporia* has been associated with the suppressive properties of agricultural soils against the cereal cyst nematode (Kerry *et al.*, 1982).

25.2 | Aquatic hyphomycetes (Ingoldian fungi)

If a sample of foam from a rapid stream flowing through deciduous woodland is examined microscopically, especially in autumn after leaf fall, it will be found to contain a rich variety of conidia of unusual shape (Fig. 25.9). Many are quite large, spanning 100 μm or more. These conidia belong to aquatic hyphomycetes which grow on decaying leaves and twigs. They are also referred to as Ingoldian fungi in honour of C.T. Ingold who pioneered their study. If decaying leaves are collected from the stream and incubated in a shallow layer of water at a temperature of 10–20°C, numerous conidiophores of these fungi will develop in a few hours. Ingoldian fungi have a worldwide

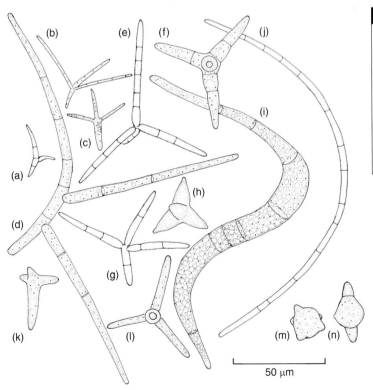

Fig 25.9 Conidia of aquatic hyphomycetes from river foam.
(a) *Volucrispora* sp. (b) *Alatospora acuminata*.
(c) *Clavatospora longibrachiata*. (d) *Tricladium splendens*. (e) *Lemonniera aquatica*.
(f) *Lemonniera terrestris*. (g) *Articulospora tetracladia*. (h) *Clavatospora stellata*.
(i) *Anguillospora crassa*. (j) *Anguillospora* sp.
(k) *Heliscus lugdunensis*. (l) Unidentified.
(m) *Margaritispora aquatica*.
(n) *Tumularia aquatica*.

50 μm

distribution. An excellent guide to the group has been written by Ingold (1975) himself, and Webster and Descals (1981) have provided keys to the then-known species. Over 300 species have been described so far. Two common spore shapes can be recognized – the tetraradiate or branched conidium, and the sigmoid or steep-helix type. Both types of spore may develop in a variety of different ways, and it is clear that these spore shapes represent a number of separate lines of convergent evolution. Evidence for the view that tetraradiate and sigmoid propagules have evolved independently several times is provided by (1) developmental studies, (2) anamorph–teleomorph connections, and (3) observations of the occurrence of tetraradiate propagules in unrelated aquatic organisms.

Table 25.2 lists some of the connections between aquatic hyphomycetes and different groups of ascomycetes and basidiomycetes. It is obvious that some anamorph genera, e.g. *Tricladium* or *Anguillospora*, are unnatural (poly-phyletic), i.e. they are not made up of related species. The teleomorphs for which connections have been established develop more readily on submerged or partially submerged wood than on leaves, although a few leaf-borne teleomorphic states have been described. The pleomorphic nature of Ingoldian fungi, as in other fungal groups, presents problems of nomenclature. Strictly, the name adopted should be that of

Table 25.2. Examples of the taxonomic diversity of Ingoldian fungi (based on Webster, 1992 and Descals et al., 1998).

Anamorph genus	Teleomorph genus	Taxonomic affinity
Branched conidial Ascomycota		
Actinosporella	*Miladina*	Pezizales
Anavirga	*Vibrissea*	Helotiales
Articulospora	*Hymenoscyphus*	Helotiales
Casaresia	*Mollisia*	Helotiales
Clavariopsis	*Massarina*	Dothideales (Loculoascomycetes)
Dwayaangam	*Orbilia*	Orbiliaceae
Geniculospora	*Hymenoscyphus*	Helotiales
Tricladium	*Hymenoscyphus,* *Hydrocina, Cudoniella*	Helotiales
Varicosporium	*Hymenoscyphus*	Helotiales
Branched conidial Basidiomycota		
Taeniomyces	*Fibulomyces*	Polyporoid clade (Homobasidiomycetes)
Ingoldiella	*Sistotrema*	Polyporoid clade (Homobasidiomycetes)
Crucella	*Camptobasidium*	Atractiellales (Urediniomycetes)
Sigmoid conidial Ascomycota		
Anguillospora	*Mollisia, Pezoloma,* *Hymenoscyphus, Loramyces*	Helotiales
Anguillospora	*Orbilia*	Orbiliaceae
Anguillospora	*Massarina*	Dothideales (Loculoascomycetes)
Flagellospora	*Nectria*	Hypocreales (Pyrenomycetes)
Conidial Ascomycota of other shapes		
Heliscus, Cylindrocarpon	*Nectria*	Hypocreales (Pyrenomycetes)
Dimorphospora	*Hymenoscyphus*	Helotiales
Tumularia	*Massarina*	Dothideales (Loculoascomycetes)

the teleomorph, but in practice, because the teleomorphs are less frequently encountered than the anamorphs, it is the name of the latter which is generally used.

25.2.1 Tetraradiate and other branched conidia

A few examples of the development of branched conidia illustrate great variation which can be followed by making observations of spore development over a period of a few hours on leaf fragments bearing conidiophores or from pieces of agar culture incubated in water. In some cases development is aided by placing culture pieces in special flow cells which permit microscopic observations to be made under continuous water flow over time.

Phialidic tetraradiate conidia

A good example is *Lemonniera aquatica* (Fig. 25.10a), probably the most common of the six species in this genus. Conidiophores develop from mycelium embedded in the leaf tissues or from chlamydospores or sclerotia, and terminate in 1–3 phialides. From the tip of the phialide a tetrahedral conidium primordium develops, and the four corners of the tetrahedron extend simultaneously to form cylindrical arms which may become septate. The mature conidium is thus attached centrally to the phialide at the point of divergence of the arms. When the first-formed conidium is detached to be carried away by water currents, a second conidium develops, and others follow.

Phialides of *Alatospora* (Fig. 25.10b) develop singly at the tips of short, inconspicuous conidiophores. Midway along the length of an elongated spore initial, two divergent lateral arms arise and extend simultaneously.

Heliscus lugdunensis (Fig. 25.10c) is a common early colonizer of the bark of twigs which have fallen into streams. Its conidiophores develop on sporodochium-like pustules and branch repeatedly, terminating in phialides. Conidia which develop underwater are clove-shaped with short, conical projections at the upper end, whereas more cylindrical conidia are produced under aerial conditions, e.g. on a twig incubated in a moist chamber or on agar culture. The

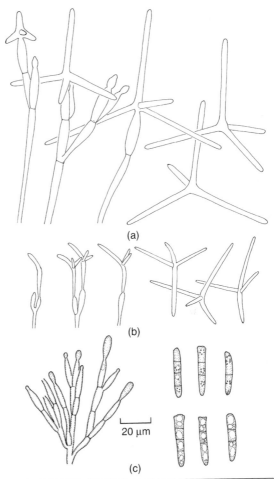

Fig 25.10 Three phialidic aquatic hyphomycetes. (a) *Lemonniera aquatica*. (b) *Alatospora acuminata*. (c) *Heliscus lugdunensis*.

teleomorph is *Nectria lugdunensis* which forms bright red perithecia on half-submerged twigs.

Blastic tetraradiate conidia

There are numerous examples of blastic conidial development in aquatic hyphomycetes. *Articulospora tetracladia* (Fig. 25.11a) forms short conidiophores extending from mycelium within a leaf. At the tip of the conidiophore the first arm develops as a cylindrical bud. At the apex of this first arm, three further cylindrical buds develop in turn. A narrow constriction or joint marks the point of attachment of these later-formed arms to the first (hence the name *Articulospora*, a jointed spore). The mycelium and

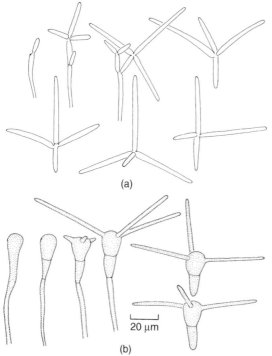

(a)

(b)

20 µm

Fig 25.11 Two blastic aquatic hyphomycetes. (a) *Articulospora tetracladia*. The arms of the conidia develop successively. (b) *Clavariopsis aquatica*. The top-shaped body of the conidium develops first, followed by simultaneous development of the three thinner arms.

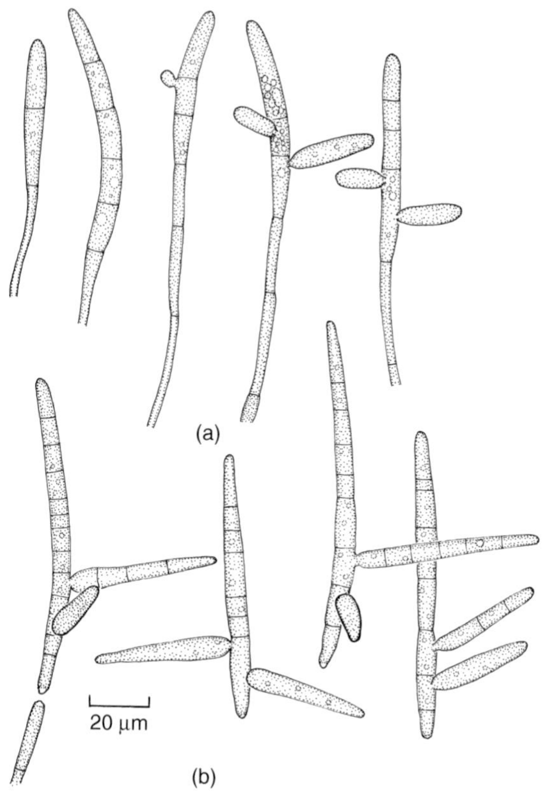

(a)

20 µm

(b)

Fig 25.12 *Tricladium splendens*. (a) Stages in blastic development of conidia. A club-shaped main axis develops lateral arms successively from different points along its length. (b) Mature detached conidia.

conidia of *Articulospora* are hyaline. The teleomorph is an inoperculate discomycete, *Hymenoscyphus tetracladius*.

Clavariopsis aquatica (Fig. 25.11b) has dark mycelium and conidia. The conidia have a broad, obconical body with a rounded tip bearing three cylindrical arms which develop simultaneously. The mature conidium usually has a single septum in the central body. In culture, a spermogonial state has been found, a dark-coloured pycnidium containing minute, colourless spermatia. The teleomorph belongs to the genus *Massarina* (Loculoascomycetes).

Tricladium splendens (Fig. 25.12) also has dark-coloured mycelium and conidia. The apex of the conidiophore develops a club-shaped swelling which becomes septate and forms the main axis of the conidium. A bud develops at one point on the main axis, to be followed by a second, at a different point. The arms taper and are constricted where they join the main axis. Its teleomorph is *Hymenoscyphus splendens*.

Tetrachaetum elegans (Fig. 25.13) has a hyaline mycelium and conidia which are relatively large, spanning up to 200 µm. The conidium develops by the curvature of the main axis which is narrowly cylindrical. Two laterals arise at a common point about halfway along the main axis, and develop simultaneously. *Varicosporium elodeae* (Fig. 25.14) and *Dendrospora erecta* (Fig. 25.15) bear blastoconidia which are more highly branched, with further branches developing from the primary laterals.

Branched conidia with clamp connections and dolipore septa

A number of branched conidia found in water or foam have clamp connections at their septa, showing that they are basidiomycetes. They

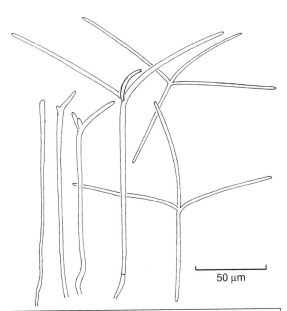

Fig 25.13 *Tetrachaetum elegans*. The main axis of the conidium bends and, at the point of curvature, two lateral arms arise simultaneously.

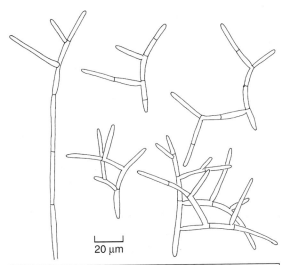

Fig 25.14 *Varicosporium elodeae*. Branched blastoconidia formed by repeated branching of the lateral arms which develop mostly from one side of the main axis.

include *Taeniomyces* (Fig. 25.16) which has a conidium somewhat resembling a *Tricladium*, but with a single clamp connection at the septum lying between the two arms. The conidium is dikaryotic, and the basidial state has been

obtained in cultures derived from a conidium. The teleomorph is a species of *Fibulomyces* probably belonging to the polyporoid clade.

Ingoldiella hamata (Fig. 25.17), a tropical aquatic fungus, has large, dikaryotic conidia with numerous clamp connections. The basidial state is *Sistotrema hamatum* (polyporoid clade) with eight-spored basidia. Single basidiospores germinate to form monokaryotic mycelia on which monokaryotic conidia develop. These closely resemble dikaryotic conidia but lack clamp connections.

There are other aquatic hyphomycetes with basidiomycetous affinities indicated by the possession of dolipore septa within the mycelium and the conidium. Examples include *Dendrosporomyces prolifer* and *D. splendens* which have conidia with a strong morphological resemblance to *Dendrospora*, and *Tricladiomyces malaysianum* with spores resembling those of a *Tricladium* but with dolipore septa (Nawawi, 1985).

25.2.2 Sigmoid conidia

A similar range of different types of conidial ontogeny can be demonstrated for sigmoid conidia.

Phialoconidia

Flagellospora curvula (Fig. 25.18a) has narrow, sigmoid phialoconidia developing from phialides on a sparsely branched conidiophore. A more richly branched, penicillate arrangement of phialides is found in *F. penicillioides* (Fig. 25.18b), which has a *Nectria* teleomorph.

Blastoconidia

Anguillospora has blastic, sigmoid conidia, but evidence from the known teleomorphs indicates that this form-genus is very heterogeneous, including species from unrelated groups of Ascomycota (see Table 25.2). There are also differences in the mechanism of conidial separation. *Anguillospora longissima* (Fig. 25.19) has dark mycelium and conidia. The conidia develop as club-shaped swellings from the apices of conidiophores. The conidium becomes septate and helically curved. Conidial separation is rhexolytic, brought about by the collapse

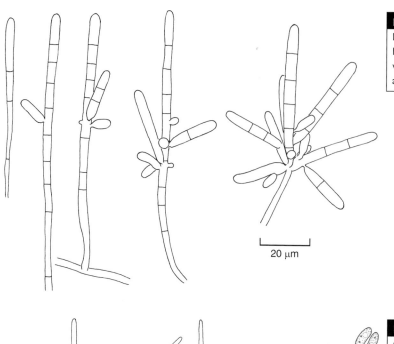

Fig 25.15 *Dendrospora erecta*. Much-branched blastoconidia formed by repeated branching of lateral arms which develop on all sides of the main axis of the conidium.

20 μm

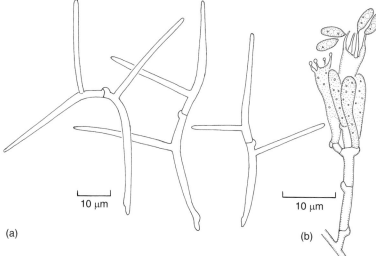

Fig 25.16 *Taeniomyces gracilis*. (a) Mature detached conidia. Note the clamp connection between the lateral arms. (b) Basidial state (*Fibulomyces* sp.).

10 μm

(a)

10 μm

(b)

of a special separating cell at the base of the conidium. The contents of the separating cell disintegrate, and the cell wall breaks down at a line of weakness near the middle. When the conidium separates, it carries at its base a little collar which represents half of the empty separating cell. Similarly, the apex of the conidiophore bears a collar after detachment of the first conidium. The conidiophore may develop a second conidium by percurrent extension through the remnants of the first separating cell and, after several conidia have been formed, a

succession of collars may be found at the tip of the conidiophore. The teleomorph of *A. longissima* is a pseudothecium-forming species of *Massarina* found on twigs in streams.

A second species, *A. furtiva*, has conidia which closely resemble those of *A. longissima*, but cultures derived from such conidia have developed apothecia of an inoperculate discomycete, *Pezoloma* sp. In *A. furtiva* there is no separating cell and conidium separation is schizolytic, i.e. by dissolution of a septum at its base (Descals *et al.*, 1998). In *A. crassa*, which has fatter

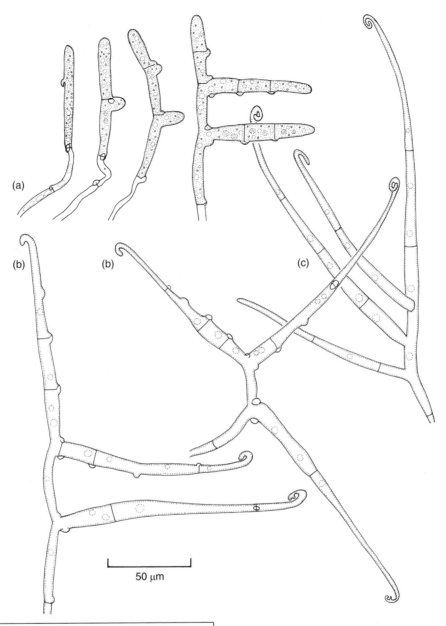

Fig 25.17 *Ingoldiella hamata*. (a) Developing dikaryotic conidia. Note that the septa bear clamp connections. (b) Mature dikaryotic conidia. (c) Monokaryotic conidium lacking clamp connections.

conidia, separation is also brought about by septum dissolution, and this species also has an inoperculate discomycete as its teleomorph, *Mollisia uda*.

Lunulospora curvula (Fig. 25.18c) has crescent-shaped blastoconidia which develop from specialized conidiogenous cells at the apex of dark conidiophores. This fungus is more common in warmer countries than in temperate regions, and in Britain its season of maximum abundance is in late summer and autumn.

25.2.3 Other types of spore

Not all aquatic hyphomycetes have branched or sigmoid conidia. *Margaritispora aquatica* forms hyaline, globose phialoconidia bearing a few conical protrusions (Fig. 25.20a), whilst

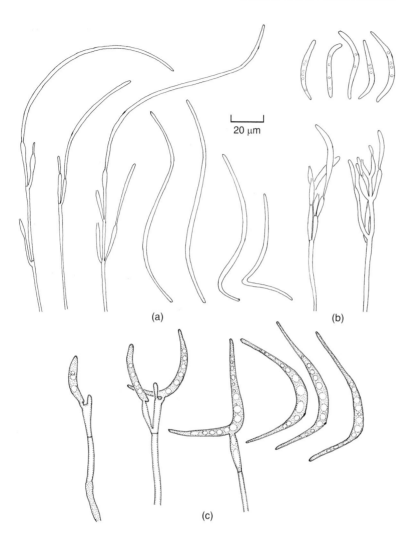

Fig 25.18 Three aquatic hyphomycetes with sigmoid conidia. (a) *Flagellospora curvuula* conidiophores with phialides and sigmoid phialoconidia. (b) *Flagellospora penicillioides* conidiophores with phialides and phialoconidia. (c) *Lunulospora curvula*, showing blastic development of crescent-shaped conidia.

20 µm

(a)

(b)

(c)

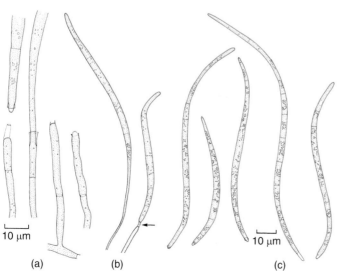

Fig 25.19 *Anguillospora longissima*, an aquatic hyphomycete with blastic sigmoid conidia. (a) Detachment of conidium showing the remnants of the separating cell, and percurrent proliferation. (b) Developing conidia. The arrow marks a separating cell. (c) Mature detached conidia.

10 µm

10 µm

(a)

(b)

(c)

Tumularia aquatica (Fig. 25.20b) has pear-shaped or broadly fusiform blastoconidia which separate by septal dissolution. The teleomorph of this fungus is *Massarina aquatica* which forms pseudothecia on submerged wood in streams.

25.2.4 The significance of spore shape in aquatic fungi

We have seen that tetraradiate conidia may develop in a variety of ways and are produced both by Ascomycota and Basidiomycota. Tetraradiate propagules have also been found as secondary conidia of Entomophthoraceae attacking aquatic insects (see Fig. 7.42; Descals *et al.*, 1981), and as basidiospores in the marine fungi *Digitatospora marina* (Homobasidiomycetes) and *Nia vibrissa* (gasteromycetes) (Doguet, 1962, 1967). The brown alga *Sphacelaria* also forms tetraradiate propagules. There is thus ample evidence for the view that this type of structure has evolved repeatedly in aquatic environments.

The functional significance of convergent evolution of tetraradiate and sigmoid spore shapes has been discussed by Webster (1987). Experimental studies have shown that tetraradiate propagules are more effectively trapped by impaction onto underwater objects than spores of more conventional shape. This is because a tetraradiate propagule making contact with a surface will achieve a three-point landing, a very stable form of attachment. Attachment is aided by secretion of mucilage at the tips of the arms which quickly develop 'appressoria' and germ tubes, but the fourth arm not in contact with a surface fails to differentiate in this way (Read *et al.*, 1991, 1992a,b; Jones, 1994). Once appressoria have been formed, the propagules are very resistant to detachment.

Trapping efficiency may not be the only advantage of a tetraradiate propagule because these spores may also be found among leaves in terrestrial habitats (Bandoni, 1972; Park, 1974).

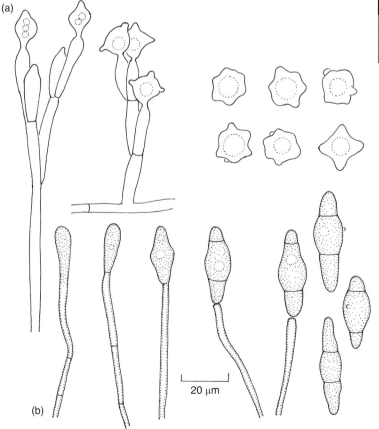

Fig 25.20 (a) *Margaritispora aquatica*, conidiophores, phialides and phialoconidia. (b) *Tumularia aquatica*, conidiophores and blastoconidia.

20 μm

Bandoni (1974) has advanced the idea that tetraradiate spores may be adapted to movement in surface films of water.

Although sigmoid spores are less efficiently trapped than tetraradiate spores, they, too, can develop in different ways and in unrelated groups of fungi which have adopted an aquatic habit, suggesting that their shape has selective value. Observations on sigmoid spores moving with the current flow in flat capillary tubes show that as they approach a surface they tumble end over end and come to rest with one spore tip in contact with the surface. Immediately after arrest, the spore swings parallel to the current, thus minimizing the shear forces acting to detach the spore (Webster & Davey, 1984). As in tetraradiate spores, the tips of the arms of sigmoid spores secrete mucilage, possibly in response to a thigmotropic stimulus associated with their tumbling movements. Mucilage is also present on the outside of the sigmoid spore of *Mycocentrospora filiformis* prior to it making contact with a surface (Au *et al.*, 1996). Current flow forces the spore into contact with the surface at a second point along its length. Germination occurs by the development of a germ tube from that end of the spore which is in contact with the surface. Thus, in contrast to the three-point contact associated with tetraradiate spore shape, sigmoid spores make two points of contact with surfaces to which they adhere.

25.2.5 Spores in stream foam

Foam is an effective trap for both tetraradiate and sigmoid spores (see Fig. 25.9). Spores suspended in water or caught in stream foam rarely germinate and many studies on the distribution and seasonal abundance of aquatic hyphomycetes using preserved foam samples have been made. In experiments in which air bubbles were passed through concentrated suspensions of conidia, the concentration of suspended spores fell very rapidly. Tetraradiate conidia were removed more readily than conidia of sigmoid or other shape (Iqbal & Webster, 1973a). Comparisons of spores collected in foam or by Millipore filtration from the same stream

indicates that the spore content of foam over-represents the tetraradiate type of conidium in relation to other spore types known to be present. It cannot be assumed that all propagules found in stream foam originate from within the stream. Some come from fungi growing on the living leaves of riparian trees and the spores are brought into the stream in raindrops or from rainwater draining down the tree trunks. Such fungi have been distinguished as terrestrial–aquatic hyphomycetes (Ando & Tubaki, 1984a,b).

25.2.6 Adaptations of Ingoldian fungi to the aquatic habitat

Ingoldian fungi represent an ecological group of fungi sharing a common habitat, typically leaves and twigs in rapidly flowing streams. It is believed that they have been derived from terrestrial ancestors and are able to colonize their habitat by virtue of a number of adaptations. These include effective spore attachment mechanisms associated with tetraradiate and sigmoid spore shape, rapid germination, and rapid growth and sporulation enhanced by turbulence and rapid water flow (Webster & Towfik, 1972; Sanders & Webster, 1980). Physiological adaptations include the possession of a range of enzymes enabling them to degrade their substrata (see below), and an ability to grow at low temperatures, sometimes approaching 0°C, so that they can continue growth and sporulation on submerged deciduous tree leaves after the autumn pulse of leaf fall (Koske & Duncan, 1974). Despite their typical environment, many aquatic hyphomycetes can survive for several weeks on dried leaves previously colonized in streams and brought out by flooding or by falling water levels (Sanders & Webster, 1978). It is possible that overland dispersal is achieved if colonized dried leaves are blown about by wind and re-deposited in streams.

25.2.7 Ecophysiological studies

There have been extensive studies on the ecology and physiology of Ingoldian hyphomycetes stimulated by the discovery that they play an essential role in rendering leaves which they colonize in streams more palatable and nutritious to aquatic invertebrates feeding on them (Bärlocher, 1992; Dix & Webster, 1995; see below). Ecological studies have been facilitated by the fact that these fungi can, in general, be recognized to a high degree of certainty from their spores, which is not the case for most other groups of fungi.

Distribution

Ingoldian hyphomycetes have a worldwide distribution from the equator to the Arctic. They are found most frequently in babbling brooks overhung by deciduous trees and are less abundant in wider rivers or where streams flow through afforested regions in which trees have been clearfelled on both sides (Metwalli & Shearer, 1989). They are relatively infrequent in streams in mountainous or moorland areas devoid of riparian trees, but can grow there on plants such as rushes (*Juncus* spp.). As rivers flow towards the sea and the brackish condition is encountered in the estuaries, Ingoldian hyphomycetes decline in frequency. A few species grow in lakes, but the lotic (flowing) habitat is preferred to the lentic (smooth).

Substrates

The main substrates for aquatic hyphomycetes are leaves of deciduous trees. In general these fungi show little host specificity but certain tree leaves support a richer fungal population with more abundant sporulation than others (Gulis, 2001). A particularly rich mycota is associated with the leaves of alder (*Alnus glutinosa*), a common riparian tree, and this is possibly correlated with their relatively high nitrogen content due to the fact that *Alnus* can fix gaseous N_2. In contrast, the leaves of beech (*Fagus sylvatica*) are a poor substrate. Needles of conifers are resistant to colonization by aquatic hyphomycetes, related partly to their thick cuticles and also to the presence of compounds inhibitory to mycelial growth (Bärlocher & Oertli, 1978a,b). Wood which has fallen into streams is an important substratum because it is more enduring than leaves, which decompose more rapidly and are consumed by animals or scoured by

water currents, so that they may not survive in great quantity to provide inoculum for the next season's autumnal input. Additionally, colonized wood is a substratum for the development of teleomorphs (Shearer, 1992). Living riparian tree roots also support a population of aquatic hyphomycetes (Fisher *et al.*, 1991; Sridhar & Bärlocher, 1992a,b), with roots of alder and willow (*Salix*) particularly important because they extend into streams. Aquatic hyphomycetes have even been reported from beech roots growing in woodland soil (Waid, 1954). In lowland streams and rivers a few species may colonize living and decaying leaves of in-stream macrophytes (Kirby *et al.*, 1990). Treeholes, cavities formed within tree trunks where the stumps of fallen branches have decayed, intermittently fill with rain water. Samples of the water and the leaf and other debris which accumulates in treeholes reveal the frequent presence of Ingoldian fungi such as *Alatospora acuminata* (Gönczöl & Révay, 2003).

Spore concentrations in streams

Concentrations of aquatic hyphomycete spores in streams can be readily estimated by the filtration of water samples through a Millipore filter (preferably with 8 μm pore size, to facilitate rapid filtration) followed by treatment which stains the spores and renders the filter transparent. Concentrations reach a peak soon after the main period of autumnal deciduous leaf fall in temperate countries. Spore counts as high as 2–$3 \times 10^4 l^{-1}$ have been made in October and November in Britain and elsewhere (Iqbal & Webster, 1973b). As the leaves are decomposed, consumed or swept downstream, the concentration of spores in suspension may fall to undetectable levels. This points to the significance of colonization of more enduring woody substrata and also growth in the roots of riparian trees which extend into the water.

Feeding of invertebrates on aquatic hyphomycetes

It is now becoming clear that aquatic hyphomycetes play an important role in the cycling of nutrients in streams (e.g. Kaushik & Hynes, 1968, 1971; Bärlocher & Kendrick, 1973a,b; Suberkropp

& Klug, 1976, 1980). Streams bordered by trees receive the bulk of their fixed carbon input not from attached macrophytes or algae, but from leaves, twigs and other debris shed by trees and other plants. Such material is relatively poor in nitrogen or is otherwise unpalatable to the invertebrate animal population. The colonization of leaf litter by aquatic fungi and bacteria is an important part of the 'processing' which makes it a more attractive substrate to animals (Berrie, 1975). There are two main reasons for this, namely pre-digestion and enrichment in organic nitrogen.

First, the activities of the fungi soften the leaf tissues. Aquatic hyphomycetes possess a range of pectolytic, cellulolytic, proteolytic and ligninolytic enzymes capable of degrading leaf tissues (Suberkropp & Klug, 1980; Chamier, 1985; Zemek et al., 1985; Gessner et al., 1997). Colonized softened leaves are more easily grazed and shredded than uncolonized leaves by aquatic invertebrates such as Gammarus and Asellus and by the larvae of aquatic insects feeding directly on leaf tissue or on particulate organic matter released into streams as leaves decay. In general, aquatic animals do not possess enzymes capable of degrading the cell walls of leaf tissue, but the enzymes present within ingested leaf fragments may continue to be active within the animals' guts. When presented with a choice of uncolonized or colonized leaf tissue, either as separate discs or as patches on the same leaf, caddis fly larvae feed preferentially on colonized tissue (Arsuffi & Suberkropp, 1985).

Second, the protein content of leaf material is enhanced by microbial colonization. Aquatic fungi can concentrate inorganic nitrogen present in solution in the water at low concentrations but in large total amounts and, making use of the organic matter in the leaves, manufacture microbial protein. Fungi can make up over 90% of the microbial biomass which develops on decomposing leaves in streams. Aquatic animals may feed directly on the fungal mycelium and on fungal spores. Detritivorous animals such as Gammarus pulex and Asellus aquaticus fed on a fungus diet make a much greater weight increase than those fed solely on a diet of uncolonized leaves. They are also more fecund, i.e. produce more eggs per brood (Graca et al., 1993). In order to sustain their restricted growth rate on leaf diets, the animals consume about 10 times more leaf material (by dry weight) than individuals fed on fungus diets. Thus aquatic hyphomycetes play a very important role as intermediaries in the diet of aquatic invertebrates and their major food source, the leaves of riparian trees. Since aquatic invertebrates in turn provide the food source of other animals, including fish, the activities of aquatic hyphomycetes are vital to the food chain in maintaining stream productivity.

25.3 | Aero-aquatic fungi

If leaves and twigs from the mud surface of stagnant pools or slow-running ditches are rinsed and incubated at room temperature in a humid environment (e.g. a Petri dish or plastic box lined with wet blotting paper), fungi with very characteristic large conidia usually develop within a few days. The common feature of the conidia of these fungi is that they trap air as they develop, which assists in floating off the conidia if the substratum is submerged in water. Conidia of this type have been termed **bubble-trap propagules** (Michaelides & Kendrick, 1982). Such fungi grow vegetatively on leaves and twigs, often in water with quite low amounts of dissolved oxygen. Under submerged conditions these fungi do not sporulate, but do so only after incubation under aerial conditions in which a moist interface between air and water is provided, as might happen at a pond margin as the water dries up and previously submerged twigs or leaves become exposed to air. They have therefore been termed aero-aquatic fungi.

Aero-aquatic fungi are an ecological group of organisms without phylogenetic coherence, as shown in Table 25.3. Although the taxonomy of the group is fairly well known (see e.g. Linder, 1929; Moore, 1955; Webster & Descals, 1981; Goos, 1987; Voglmayr, 2000), it is likely that many more species remain to be discovered. Careful studies should also reveal new ascomycetous teleomorphs because several species

Table 25.3. | Examples of the taxonomic diversity of aero-aquatic fungi.

Anamorph genus	Teleomorph genus	Taxonomic affinity
Oomycota		
	Medusoides (oogonium)	Pythiogetonaceae (Pythiales)
Ascomycota		
Clathrosphaerina	*Hyaloscypha*	Helotiales
Helicoon	*Orbilia*	Orbiliaceae
Helicodendron	*Mollisia, Hymenoscyphus*	Helotiales
Helicodendron	*Lambertella, Herpotrichiella*	Chaetothyriales (see p. 484)
Helicodendron	*Tyrannosorus*	Dothideales (Loculoascomycetes)
Pseudaegerita	*Hyaloscypha*	Helotiales
Basidiomycota		
Aegerita	*Bulbillomyces*	Polyporoid clade
Aegeritina	*Subulicystidium*	Polyporoid clade
(Unknown)	*Limnoperdon* (basidiocarp)	Gasteromycetes
Akenomyces (Sclerotium)	(Unknown)	*Incertae sedis*

of aero-aquatic fungi form microconidial syna-namorphs in culture, the spores of which fail to germinate. These are probably spermatia.

25.3.1 Development of propagules

There are various ways in which conidia of aero-aquatic hyphomycetes may develop. In *Helicoon* (Fig. 25.21a) they develop as cylindrical or barrel-shaped spirals. The conidia vary in colour from hyaline to black. The direction of coiling of the spirals (looking upwards from the apex of the conidiophore) is clockwise in *H. richonis* whereas in some other helicosporous fungi the direction of coiling is counter-clockwise. The direction appears to be constant for a given species. In *Helicoon* the conidia themselves do not branch, but in *Helicodendron* (*Hd.*) which is clearly a polyphyletic anamorph genus, the conidia may bear further conidia as lateral branches (Figs. 25.21b,c; 25.22a). *Beverwijkella pulmonaria*, probably an anamorphic ascomycete (Fig. 25.23a), forms uni- or bi-lobed, balloon-like structures. At the surface, aggregates of dark, thick-walled, tightly packed cells develop, with air trapped inside the cavity which they enclose (Michaelides & Kendrick, 1982). *Spirosphaera*, another anamorphic ascomycete, achieves the same end by the formation of globose propagules made up of richly branched, incurved hyphae (Fig. 25.23b). Yet another way of entrapping air within the propagule is shown by *Clathrosphaerina zalewskii* which forms hollow, spherical propagules with a lattice wall, resembling practice golf balls. These clathrate structures are formed by the repeated dichotomy of the arms of the developing conidium, which then curve inwards and join firmly where the tips of the arms touch (Figs. 25.22b, 25.24). This fungus has a minute inoperculate dis-comycete teleomorph, a species of *Hyaloscypha* whose apothecia develop in air on twigs or pieces of wood which have previously been submerged.

Several bubble-trap propagules are also known among Basidiomycota. An example is *Aegerita candida* (teleomorph *Bulbillomyces farinosus*), which forms its conidia (bulbils) and basidia on the surface of wet, previously submerged wood. The propagule is made up of tightly clustered aggregates of inflated, dikaryotic clamped cells between which air is entrapped (see Figs. 18.8 and 25.22c). The propagules of *Aegeritina tortuosa* (teleomorph *Subulicystidium longisporum*) which also grows on wet wood resemble those of *A. candida* but are not clamped.

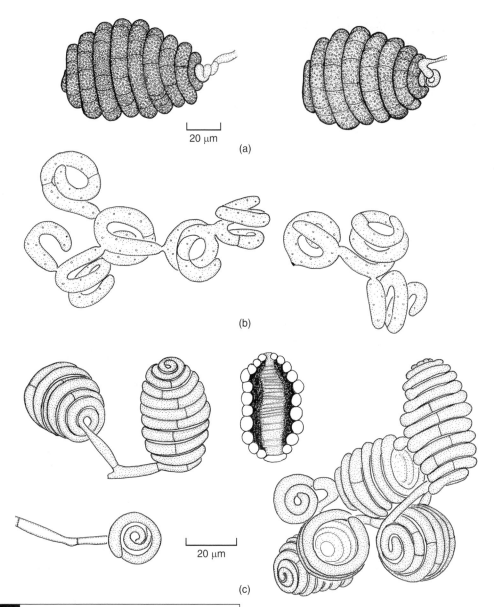

20 μm

(a)

(b)

20 μm

(c)

Fig 25.21 Some aero-aquatic helicosporous fungi. (a) *Helicoon richonis*. (b) *Helicodendron triglitziense*. (c) *Helicodendron conglomeratum*. The central spore is drawn in optical section to show the trapped air bubble.

25.3.2 Ecophysiological studies

Although aero-aquatic fungi are taxonomically diverse, they share several common physiological features which help in understanding their ecology (for references see Webster & Descals, 1981; Dix & Webster, 1995; Voglmayr, 2000).

Simple techniques have aided studies of their ecology. Quantitative studies on colonization and survival have been made using small discs of leaves of beech, *Fagus sylvatica*. Sterile or artificially inoculated discs can be submerged among the accumulated leaf detritus in a pond and recovered at intervals in order to monitor the development of conidia for quantifying colonization.

The bubble-trap propagules of aero-aquatic fungi are hydrophobic and float ungerminated at the water surface of stagnant ponds. Autumn-

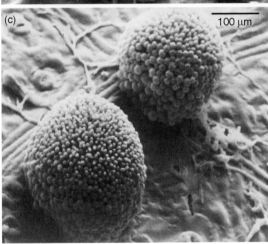

Fig 25.22 Propagules of aero-aquatic fungi which have developed at the surface of moist leaves incubated in air. (a) *Helicodendron giganteum*. Note that secondary conidia can develop as branches from the first-formed conidia. (b) *Clathrosphaerina zalewskii*. (c) *Aegerita candida*. All images to same scale. Reprinted from Dix and Webster (1995); original micrographs kindly provided by P. J. Fisher.

shed leaves which fall onto the water are rapidly colonized (Premdas & Kendrick, 1991). The leaves eventually sink to the bottom of the pond and growth of the fungi continues so long as dissolved oxygen is available in the water. Underwater colonization may also take place

by leaf-to-leaf contact, shown by the fact that sterilized beech leaf discs submerged among detritus develop conidia of aero-aquatic fungi when later incubated under suitable conditions out of water. If there is a substantial accumulation of fallen leaves, the metabolic activity of decomposer organisms will lead to anaerobic conditions and the evolution of hydrogen sulphide (H_2S) which is toxic at low concentrations to eukaryotic organisms. Anaerobic conditions are evidenced by the sulphurous smell of disturbed mud and by the black colour of the silt caused by the accumulation of metallic sulphides, especially iron sulphide. Although aero-aquatic hyphomycetes grow best at atmospheric oxygen levels, their growth in anoxic conditions is still superior to that of other fungi (Fisher & Webster, 1979). Under strictly anaerobic conditions, five species of *Helicodendron* showed almost 100% survival for 6 months, and substantial survival even after 12 months (Field & Webster, 1983). Survival in most cases appears to be by thick-walled hyphae because chlamydospores and sclerotia are rarely found. Similar comparative studies of the survival of aero-aquatic fungi and aquatic hyphomycetes under anaerobic conditions in the presence of low concentrations of H_2S showed better survival of aero-aquatic fungi than aquatic hyphomycetes (Field & Webster, 1985).

Following a prolonged period of submersion under the anaerobic or near-anaerobic conditions of the bottom silt of a pond, it takes several days for sporulation of aero-aquatic fungi to commence in air. Incubation in well-aerated water before exposure to air improves the recovery, suggesting that a period of aerobic growth is a stimulus to sporulation. For most species studied, exposure to light also enhances sporulation (Fisher & Webster, 1978).

Aero-aquatic fungi (*Helicodendron* spp.) grown on a homogenized beech leaf mash can survive on the soil surface for several months if air-dried (Fisher, 1978), suggesting a capacity for vegetative existence out of water, a conclusion confirmed by reports of some species from soil (Abdullah & Webster, 1980). Dispersal of colonized wind-blown leaves from one body of water to another is clearly a possibility. Other possible

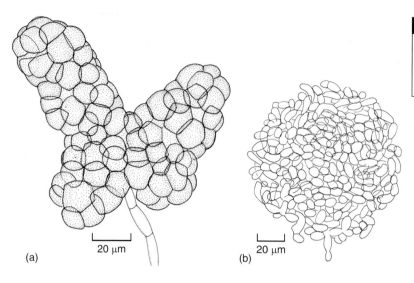

(a)

(b)

Fig 25.23 Air-trapping multicellular propagules of two aero-aquatic fungi. (a) *Beverwijkella pulmonaria*. (b) *Spirosphaera* sp.

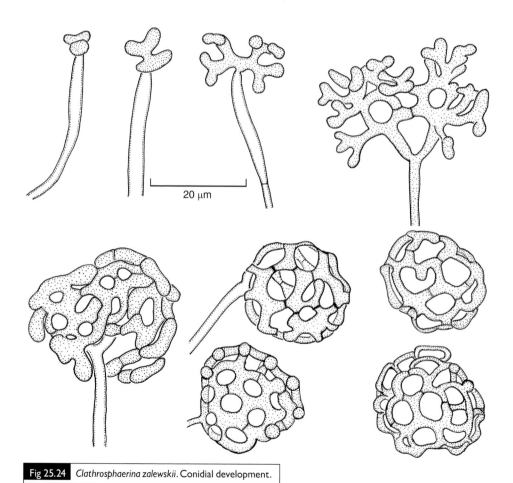

Fig 25.24 *Clathrosphaerina zalewskii*. Conidial development.

means of dispersal are the carriage of wind-borne ascospores and basidiospores of the teleomorphs, passive dispersal of conidia by waterfowl, and conidium dispersal at the water surface during flooding.

Aero-aquatic fungi show some host specificity. Leaves and twigs of broad-leaved trees are colonized by a wide range of species, but leaves of coniferous needles such as *Pinus* have a more restricted mycota including *Hd. fractum* and *Hd. hyalinum*. Monocotyledonous hosts such as grasses, rushes and sedges support few species, but exceptions are *Hd. praetermissum*, *Spirosphaera carici-graminis* and *S. minuta*. Fruiting of *Aegerita* and *Aegeritina* is supported on woody substrates rather than leaves, whereas *Pseudaegerita* spp. fruit both on rotten wood and on leaves. There are also differences in frequency in relation to water chemistry (Voglmayr, 2000). For example *B. pulmonaria*, *Hd. conglomeratum*, *Hd. tubulosum* and *Helicoon fuscosporum* are found in eutrophic waters, *Spirosphaera minuta* is characteristic of oligotrophic conditions, and *Candelabrum desmidiaceum* is mostly found in dystrophic conditions. Other species have a wider tolerance. *Helicodendron triglitziense* and *S. floriformis* are two very common species found in dystrophic to eutrophic waters.

Like so many fungi described in this book, aero-aquatic species have adopted, by convergent evolution, a suite of flexible characters which has enabled them to explore a habitat of extremes, in this case the amphibious environment involving changes between anaerobic and aerobic conditions, and between aquatic and terrestrial lifestyles.

References

Abbott, S. P. & Currah, R. S. (1997). The Helvellaceae: systematic revision and occurrence in northwestern North America. *Mycotaxon*, **62**, 1–125.

Abbott, S. P. & Sigler, L. (2001). Heterothallism in the Microascaceae demonstrated by three species in the *Scopulariopsis brevicaulis* series. *Mycologia*, **93**, 1211–1220.

Abdullah, S. K. (1980). British aero-aquatic fungi Ph.D. thesis, University of Exeter, UK.

Abdullah, S. K. & Webster, J. (1980). Occurrence of aero-aquatic fungi in soil. *Transactions of the British Mycological Society*, **75**, 511–514.

Adam, G. C. (1988). *Thanatephorus cucumeris* (*Rhizoctonia solani*), a species of wide host range. *Advances in Plant Pathology*, **6**, 535–552.

Adams, M. J. (1991). Transmission of plant viruses by fungi. *Annals of Applied Biology*, **118**, 479–492.

Adams, T. H. & Yu, J.-H. (1998). Coordinate control of secondary metabolite production and asexual sporulation in *Aspergillus nidulans*. *Current Opinion in Microbiology*, **1**, 674–677.

Adams, T. H., Wieser, J. K. & Yu, J.-H. (1998). Asexual sporulation in *Aspergillus nidulans*. *Microbiology and Molecular Biology Reviews*, **62**, 35–54.

Adams, T. J. H., Todd, N. K. & Rayner, A. D. M. (1981). Antagonism between dikaryons of *Piptoporus betulinus*. *Transactions of the British Mycological Society*, **76**, 510–513.

Adendorff, R. & Rijkenberg, F. H. J. (2000). Scanning electron microscopy of direct host leaf penetration by urediospore-derived infection structures of *Phakopsora apoda*. *Mycological Research*, **104**, 317–324.

Adler, E. (1977). Lignin chemistry. Past, present and future. *Wood Science and Technology*, **11**, 169–218.

Agerer, R. (2001). Exploration types of ectomycorrhizae. A proposal to classify ectomycorrhizal mycelial systems according to their patterns of differentiation and putative ecological importance. *Mycorrhiza*, **11**, 107–114.

Agerer, R. (2002). Rhizomorph structures confirm the relationship between Lycoperdales and Agaricaceae (Hymenomycetes, Basidiomycota). *Nova Hedwigia*, **75**, 367–385.

Agrios, G. N. (2005). *Plant Pathology*, 5th edn. Amsterdam: Elsevier.

Ahmadjian, V. (1993). *The Lichen Symbiosis*. New York: John Wiley.

Ahmadjian, V. & Jacobs, J. B. (1981). Relationship between fungus and alga in the lichen *Cladonia cristatella* Tuck. *Nature*, **289**, 169–172.

Åhman, J., Ek, B., Rask, L. & Tunlid, A. (1996). Sequence analysis and regulation of a cuticle-degrading serine protease from the nematophagous fungus *Arthrobotrys oligospora*. *Microbiology*, **142**, 1605–1616.

Åhman, J., Johansson, T., Olsson, M., *et al.* (2002). Improving the pathogenicity of a nematode-trapping fungus by genetic engineering of a subtilisin with nematotoxic activity. *Applied and Environmental Microbiology*, **68**, 3408–3415.

Ahmed, S. I. & Cain, R. F. (1972). Revision of the genera *Sporormia* and *Sporormiella*. *Canadian Journal of Botany*, **50**, 419–477.

Aidoo, K. E., Smith, J. E. & Wood, B. J. B. (1994). Industrial aspects of soy sauce fermentations using *Aspergillus*. In *The Genus Aspergillus. From Taxonomy and Genetics to Industrial Application*, ed. K. A. Powell, A. Renwick & J. F. Peberdy. New York: Plenum Press, pp. 155–169.

Ainsworth, A. M. (1987). Occurrence and interactions of outcrossing and non-outcrossing populations of *Stereum, Phanerochaete* and *Coniophora*. In *Evolutionary Biology of Fungi*, ed. A. D. M. Rayner, C. M. Brasier & D. Moore. Cambridge: Cambridge University Press, pp. 285–299.

Ainsworth, A. M. & Rayner, A. D. M. (1989). Hyphal and mycelial responses associated with genetic exchange within and between species of the basidiomycete genus *Stereum*. *Journal of General Microbiology*, **135**, 1643–1659.

Ainsworth, A. M., Rayner, A. D. M., Broxholme, S. J. & Beeching, J. R. (1990). Occurrence of unilateral genetic transfer and genomic replacement between strains of *Stereum hirsutum* from non-outcrossing and outcrossing populations. *New Phytologist*, **115**, 119–128.

Ainsworth, G. C. (1973). Introduction and keys to higher taxa. In *The Fungi. An Advanced Treatise* **IVB**: *A Taxonomic Review with Keys*, ed. G. C. Ainsworth, F. K. Sparrow & A. S. Sussman. New York: Academic Press, pp. 1–7.

Ainsworth, G. C. (1976). *Introduction to the History of Mycology*. Cambridge: Cambridge University Press.

Ainsworth, G. C. (1981). *Introduction to the History of Plant Pathology*. Cambridge: Cambridge University Press.

Aist, J. R. & Israel, H. W. (1977). Timing and significance of papilla formation during host penetration by *Olpidium brassicae*. *Phytopathology*, **67**, 187–194.

Aist, J. R. & Williams, P. H. (1971). The cytology and kinetics of cabbage root hair penetration by *Plasmodiophora brassicae*. *Canadian Journal of Botany*, **49**, 2023–2034.

Akashi, T., Kanbe, T. & Tanaka, K. (1997). Localized accumulation of cell wall mannoproteins and endomembranes induced by brefeldin A in hyphal cells of the dimorphic yeast *Candida albicans*. *Protoplasma*, **197**, 45–56.

Akins, R. A. (2005). An update on antifungal targets and mechanisms of resistance in *Candida*. *Medical Mycology*, **43**, 285–318.

Alasoadura, S. O. (1963). Fruiting in *Sphaerobolus* with special reference to light. *Annals of Botany* N. S., **27**, 123–145.

Alberts, B., Johnson, A., Lewis, J., *et al.* (2002). *Molecular Biology of the Cell*, 4th edn. New York: Garland Science.

Alcorn, J. L. (1981). Ascus structure and function in *Cochliobolus* species. *Mycotaxon*, **13**, 349–360.

Alcorn, J. L. (1988). The taxonomy of 'Helminthosporium' species. *Annual Review of Plant Pathology*, **26**, 37–56.

Alderman, D. J. & Polglase, J. L. (1986). *Aphanomyces astaci*: isolation and culture. *Journal of Fish Diseases*, **9**, 367–379.

Alderman, D. J., Holdich, D. & Reeve, I. (1990). Signal crayfish as vectors in crayfish plague in Britain. *Aquaculture*, **86**, 3–6.

Alderman, S. C. (2003). Diversity and speciation in *Claviceps*. In *Clavicipitalean Fungi*, ed. J. F. White, C. W. Bacon, N. L. Hywel-Jones & J. W. Spatafora. New York: Marcel Dekker Inc., pp. 190–245.

Alexander, C. & Hadley, G. (1985). Carbon movement between host and mycorrhizal endophyte during development of the orchid *Goodyera repens* Br. *New Phytologist*, **101**, 657–665.

Alexander, H. M. & Antonovics, J. (1988). Disease spread and population dynamics of anther-smut infection of *Silene alba* caused by the fungus *Ustilago violacea*. *Journal of Ecology*, **76**, 91–104.

Alexopoulos, C. J., Mims, C. W. & Blackwell, M. (1996). *Introductory Mycology*, 4th edn. New York: Wiley.

Allen, E. A., Hoch, H. C., Stavely, J. R. & Steadman, J. R. (1991). Uniformity among races of *Uromyces appendiculatus* in response to topographic signaling for appressorium formation. *Phytopathology*, **81**, 883–887.

Allen, M. F. (1991). *The Ecology of Mycorrhizae*. Cambridge: Cambridge University Press.

Allen, M. F., Klironomos, J. N. & Harney, S. (1997). The epidemiology of mycorrhizal fungal infection during succession. In *The Mycota VB: Plant Relationships*, ed. G. C. Carroll & P. Tudzynski. Berlin: Springer-Verlag, pp. 169–83.

Almaraz, T., Roux, C., Maumont, S. & Durrieu, G. (2002). Phylogenetic relationships among smut fungi parasitizing dicotyledons based on ITS sequence analysis. *Mycological Research*, **106**, 541–548.

Amagai, A. (1992). Induction of heterothallic and homothallic zygotes in *Dictyostelium discoideum* by ethylene. *Development Growth and Differentiation*, **34**, 293–299.

Amano, K. (1986). *Host Range and Geographical Distribution of the Powdery Mildew Fungi*. Tokyo: Japan Scientific Societies Press.

Amon, J. P. & French, K. H. (2004). Photoresponses of the marine protist *Ulkenia* sp. zoospores to ambient, artificial and bioluminescent light. *Mycologia*, **96**, 463–469.

Amon, J. P. & Perkins, F. O. (1968). Structure of *Labyrinthula* sp. zoospores. *Journal of Protozoology*, **15**, 543–546.

Anagnostakis, S. L. (1987). Chestnut blight: the classical problem of an introduced pathogen. *Mycologia*, **79**, 23–37.

Anagnostakis, S. L. & Day, P. R. (1979). Hypovirulence conversion in *Endothia parasitica*. *Phytopathology*, **69**, 1226–1229.

Andersen, R. A., Barr, D. J. S., Lynn, D. H., *et al.* (1991). Terminology and nomenclature of the cytoskeletal elements associated with the flagellar/ciliary apparatus in protists. *Protoplasma*, **164**, 1–8.

Andersen, T. F. & Stalpers, J. A. (1994). A check-list of *Rhizoctonia* epithets. *Mycotaxon*, **51**, 437–457.

Anderson, D. L., Giacon, H. & Gibson, N. L. (1997). Detection and thermal destruction of the chalkbrood fungus (*Ascosphaera apis*) in honey. *Journal of Apicultural Research*, **36**, 163–168.

Anderson, D. L., Gibbs, A. J. & Gibson, N. L. (1998). Identification and phylogeny of spore–cyst fungi (*Ascosphaera* spp.) using ribosomal DNA sequences. *Mycological Research*, **102**, 541–547.

Anderson, I. C., Chambers, S. M. & Cairney, J. W. G. (1998). Use of molecular methods to estimate the size and distribution of mycelial individuals of the ectomycorrhizal basidiomycete *Pisolithus tinctorius*. *Mycological Research*, **102**, 295–300.

Anderson, I. C., Chambers, S. M. & Cairney, J. W. G. (2001). Distribution and persistence of Australian *Pisolithus* species genets at native sclerophyll forest field sites. *Mycological Research*, **105**, 971–976.

Anderson, J.B. & Ullrich, R.C. (1982). Diploids of *Armillaria mellea*: synthesis, stability, and mating behavior. *Canadian Journal of Botany*, **60**, 432–439.

Anderson, J.B., Ullrich, R.C., Roth, L.F. & Filip, G.M. (1979). Genetic identification of clones of *Armillaria mellea* in coniferous forests in Washington. *Phytopathology*, **69**, 1109–1111.

Ando, K. & Tubaki, K. (1984a). Some undescribed hyphomycetes in the raindrops from intact leaf surfaces. *Transactions of the Mycological Society of Japan*, **25**, 21–37.

Ando, K. & Tubaki, K. (1984b). Some undescribed hyphomycetes in rainwater draining from intact trees. *Transactions of the Mycological Society of Japan*, **25**, 39–47.

Andrianopoulos, A. (2002). Control of morphogenesis in the human fungal pathogen *Penicillium marneffei*. *International Journal of Medical Microbiology*, **292**, 331–347.

Andrivon, D. (1995). Biology, ecology, and epidemiology of the potato late blight pathogen *Phytophthora infestans* in soil. *Phytopathology*, **85**, 1053–1056.

Anikster, Y. (1983). Binucleate basidiospores – a general rule in rust fungi. *Transactions of the British Mycological Society*, **81**, 624–626.

Anikster, Y. (1984). The *formae speciales*. In *The Cereal Rusts 1: Origins, Specificity, Structure, and Physiology*, ed. W.R. Bushnell & A.P. Roelfs. Orlando: Academic Press, pp. 115–130.

Anikster, Y. & Wahl, I. (1979). Coevolution of the rust fungi on Gramineae and Liliaceae and their hosts. *Annual Review of Phytopathology*, **17**, 367–403.

Anikster, Y. & Wahl, I. (1985). Basidiospore formation and self-fertility in *Puccinia mesnieriana*. *Transactions of the British Mycological Society*, **84**, 164–167.

Anikster, Y., Bushnell, W.R., Eilam, T., Manisterski, J. & Roelfs, A.P. (1997). *Puccinia recondita* causing leaf rust on cultivated wheats, wild wheats, and rye. *Canadian Journal of Botany*, **75**, 2082–2096.

Anikster, Y., Eilam, T., Manisterski, J. & Leonard, K.J. (2003). Self-fertility and other distinguishing characteristics of a new morphotype of *Puccinia coronata* pathogenic on smooth brome grass. *Mycologia*, **95**, 87–97.

Anke, H., Stadler, M., Mayer, A. & Sterner, O. (1995). Secondary metabolites with nematicidal and anti-microbial activity from nematophagous fungi and Ascomycetes. *Canadian Journal of Botany*, **73**, S932–S939.

Anke, T. (1997). Strobilurins. In *Fungal Biotechnology*, ed. T. Anke. Weinheim, Germany: Chapman & Hall, pp. 206–212.

Anke, T., Oberwinkler, F., Steglich, W. & Schramm, G. (1977). The strobilurins – new antifungal antibiotics from the basidiomycete *Strobilurus tenacellus* (Pers. Ex Fr.) Sing. *Journal of Antibiotics*, **30**, 806–810.

Ann, P.-J., Chang, T.-T. & Ko, W.-H. (2002). *Phellinus noxius* brown root rot of fruit and ornamental trees in Taiwan. *Plant Disease*, **86**, 820–826.

Ansell, P.J. & Young, T.W.K. (1982). Light and electron-microscopic study of *Mortierella* species. *Nova Hedwigia*, **36**, 309–339.

Ansell, P.J. & Young, T.W.K. (1983). Light and electron microscopy of *Mortierella indohi* zygospores. *Mycologia*, **75**, 64–69.

Ansell, P.J. & Young, T.W.K. (1988). The zygospore of *Mortierella indohi*. *Transactions of the British Mycological Society*, **91**, 221–226.

Antelo, L., Cosio, E.G., Hertkorn, N. & Ebel, J. (1998). Partial purification of a GTP-insensitive $(1\rightarrow 3)$-β-glucan synthase from *Phytophthora sojae*. *FEBS Letters*, **433**, 191–195.

Antonovics, J., Hood, M. & Partain, J. (2002). The ecology and genetics of a host shift: *Microbotryum* as a model system. *American Naturalist*, **160**, S40–S53.

Arabi, M.I.E. & Jawhar, M. (2002). The ability of barley powdery mildew to grow *in vitro*. *Journal of Phytopathology*, **150**, 305–307.

Archibald, J.M. & Keeling, P.J. (2004). Actin and ubiquitin protein sequences support a cercozoan/foraminiferan ancestry for the plasmodiophorid plant pathogens. *Journal of Eukaryotic Microbiology*, **51**, 113–118.

Arditti, J.H. (1992). *Fundamentals of Orchid Biology*. New York: John Wiley & Sons.

Arellano, M., Cartagena-Lirola, H., Hajibagheri, M.A.N., Durán, A. & Valdivieso, M.H. (2000). Proper ascospore maturation requires the chs^+ chitin synthase gene in *Schizosaccharomyces pombe*. *Molecular Microbiology*, **35**, 79–89.

Arenal, F., Platas, G., Monte, E. & Peláez, F. (2000). ITS sequencing support for *Epicoccum nigrum* and *Phoma epicoccina* being the same biological species. *Mycological Research*, **104**, 301–303.

Aristizabal, B.H., Clemons, K.V., Stevens, D.A. & Restrepo, A. (1998). Morphological transition of *Paracoccidioides brasiliensis* conidia to yeast cells: in vivo inhibition in females. *Infection and Immunity*, **66**, 5587–5591.

Arnold, C.A. (1928). The development of the perithecium and spermagonium of *Sporormia leporina*. *American Journal of Botany*, **15**, 241–245.

Arnold, D. L., Blakesley, D. & Clarkson, J. M. (1996). Evidence for the growth of *Plasmodiophora brassicae in vitro*. *Mycological Research*, **100**, 535–540.

Arroyo, M., de la Mata, I., Acebal, C. & Castillón, M. P. (2003). Biotechnological applications of penicillin acylases: state-of-the-art. *Applied Microbiology and Biotechnology*, **60**, 507–514.

Arsuffi, T. L. & Suberkropp, K. (1985). Selective feeding by stream caddisfly (Trichoptera) detritivores on leaves with fungal-colonized patches. *Oikos*, **45**, 50–58.

Aschner, M. & Kohn, S. (1958). The biology of *Harposporium anguillulae*. *Journal of General Microbiology*, **19**, 182–189.

Ashbee, H. R. & Evans, E. G. V. (2002). Immunology of diseases associated with *Malassezia* species. *Clinical Microbiology Reviews*, **15**, 21–57.

Ashford, A. E. & Allaway, W. G. (2002). The role of the motile tubular vacuole system in mycorrhizal fungi. *Plant and Soil*, **244**, 177–187.

Ashford, A. E., Cole, L. & Hyde, G. J. (2001). Motile tubular vacuole systems. In *The Mycota VIII: Biology of the Fungal Cell*, ed. R. J. Howard & N. A. R. Gow. Berlin: Springer-Verlag, pp. 243–265.

Asiegbu, F. O., Adomas, A. & Stenlid, J. (2005). Conifer root and butt rot caused by *Heterobasidion annosum* (Fr.) Bref. *s.l. Molecular Plant Pathology*, **6**, 395–409.

Asiegbu, F. O., Johansson, M., Woodward, S. & Hütterman, A. (1998). Biochemistry of the host–parasite interaction. In *Heterobasidion annosum. Biology, Ecology, Impact and Control*, ed. S. Woodward, J. Stenlid, R. Karjalainen & A. Hütterman. Wallingford, UK: CAB International, pp. 167–193.

Au, D. W. T., Jones, E. B. G., Moss, S. T. & Hodgkiss, I. J. (1996). The role of mucilage in the attachment of conidia, germ tubes and appressoria in the saprobic aquatic hyphomycetes *Lemonniera aquatica* and *Mycocentrospora filiformis*. *Canadian Journal of Botany*, **74**, 1789–1800.

Aust, H.-J. (1981). Über den Verlauf von Mehltauepidemien innerhalb des Agrar-Ökosystems Gerstenfeld. *Acta Phytomedica*, **7**, 1–76.

Austin, D. J., Bu'lock, J. D. & Gooday, G. W. (1969). Trisporic acids: sexual hormones for *Mucor mucedo* and *Blakeslea trispora*. *Nature*, **223**, 1178–1179.

Austwick, P. C. K. (1976). Environmental aspects of *Mortierella wolfii* infection in cattle. *New Zealand Journal of Agricultural Research*, **19**, 25–33.

Avis, T. J. & Bélanger, R. R. (2002). Mechanisms and means of detection of biocontrol activity of *Pseudozyma* yeasts against plant-pathogenic fungi. *FEMS Yeast Research*, **2**, 5–8.

Ayers, W. A. & Lumsden, R. D. (1977). Mycoparasitism of oospores of *Pythium* and *Aphanomyces* species by *Hyphochytrium catenoides*. *Canadian Journal of Microbiology*, **23**, 38–44.

Aylmore, R. C., Wakley, G. E. & Todd, N. K. (1984). Septal sealing in the basidiomycete *Coriolus versicolor*. *Journal of General Microbiology*, **130**, 2975–2982.

Baba, M. & Osumi, M. (1987). Transmission and scanning electron microscopic examination of intracellular organelles in freeze-substituted *Kloeckera* and *Saccharomyces cerevisiae* yeast cells. *Journal of Electron Microscopy Technique*, **5**, 249–261.

Backhouse, D. & Willetts, H. J. (1985). Histochemical changes during conidiogenic germination of sclerotia of *Botrytis cinerea*. *Canadian Journal of Microbiology*, **31**, 282–286.

Bacon, C. W. & Hinton, D. M. (1988). Ascospore iterative germination in *Epichloe typhina*. *Transactions of the British Mycological Society*, **90**, 563–569.

Bacon, C. W. & Hinton, D. M. (1991). Microcyclic conidiation cycles in *Epichloe typhina*. *Mycologia*, **83**, 743–751.

Bacon, C. W., Porter, J. K., Robbins, J. D. & Luttrell, E. S. (1977). *Epichloe typhina* from toxic tall fescue grasses. *Applied and Environmental Microbiology*, **34**, 576–581.

Bacon, C. W., Richardson, M. D. & White, J. F. (1997). Modification and uses of endophyte-enhanced turf-grasses. *Crop Science*, **37**, 1415–1425.

Bagchee, K. (1954). *Merulius lacrymans* (Wulf.) Fr. in India. *Sydowia*, **8**, 80–85.

Bähler, J., Steever, A. B., Wheatley, S., *et al.* (1998). Role of Polo kinase and Mid1p in determining the site of cell division in fission yeast. *Journal of Cell Biology*, **143**, 1603–1616.

Bai, F.-Y., Zhao, J.-H., Takashima, M., *et al.* (2002). Reclassification of the *Sporobolomyces roseus* and the *Sporidiobolus pararoseus* complexes, with the description of *Sporobolomyces phaffii* sp. nov. *International Journal of Systematic and Evolutionary Microbiology*, **52**, 2309–2314.

Bailey, F. & Gray, N. F. (1989). The comparison of isolation techniques for nematophagous fungi from soil. *Annals of Applied Biology*, **114**, 125–132.

Bailey, J. A. & Jeger, M. J., eds. (1992). *Colletotrichum: Biology, Pathology and Control*. Wallingford, UK: CAB International.

Bailey, J. A., O'Connell, R. J., Pring, R. J. & Nash, C. (1992). Infection strategies of *Colletotrichum* species. In *Colletotrichum: Biology, Pathology and Control*, ed. J. A. Bailey & M. J. Jeger. Wallingford, UK: CAB International, pp. 88–120.

Baka, Z. A., Larous, L. & Lösel, D. M. (1995). Distribution of ATPase activity at the host–pathogen interfaces of rust infections. *Physiological and Molecular Plant Pathology*, **47**, 67–82.

Bakkeren, G., Kronstad, J. W. & Lévesque, C. A. (2000). Comparison of AFLP fingerprints and ITS sequences as phylogenetic markers in Ustilaginomycetes. *Mycologia*, **92**, 510–521.

Balazy, S. (1984). On rhizoids of *Entomophthora muscae* (Cohn) Fresenius (Entomophthorales: Entomophthoraceae). *Mycotaxon*, **19**, 397–407.

Baldauf, S. L. (1999). A search for the origins of animals and fungi: comparing and combining molecular data. *American Naturalist*, **154**, S178–S188.

Baldauf, S. L., Roger, A. J., Wenk-Siefert, I. & Doolittle, W. F. (2000). A kingdom-level phylogeny of eukaryotes based on combined protein data. *Science*, **290**, 972–977.

Bandoni, R. J. (1963). Conjugation in *Tremella mesenterica*. *Canadian Journal of Botany*, **41**, 467–474.

Bandoni, R. J. (1965). Secondary control of conjugation in *Tremella mesenterica*. *Canadian Journal of Botany*, **43**, 627–630.

Bandoni, R. J. (1972). Terrestrial occurrence of some aquatic hyphomycetes. *Canadian Journal of Botany*, **50**, 2283–2288.

Bandoni, R. J. (1974). Mycological observations on the aqueous films covering decaying leaves and other litter. *Transactions of the Mycological Society of Japan*, **15**, 309–315.

Bandoni, R. J. (1984). The Tremellales and Auriculariales: an alternative classification. *Transactions of the Mycological Society of Japan*, **25**, 489–530.

Bandoni, R. J. (1987). Taxonomic overview of the Tremellales. *Studies in Mycology*, **30**, 87–110.

Bandoni, R. J. & Boekhout, T. (1998). Tremelloid genera with yeast phases. In *The Yeasts. A Taxonomic Study*, 4th edn, ed. C. P. Kurtzman & J. W. Fell. Amsterdam: Elsevier, pp. 705–717.

Bandoni, R. J., Lobo, K. J. & Brezden, S. A. (1971). Conjugation and chlamydospores in *Sporobolomyces odorus*. *Canadian Journal of Botany*, **49**, 683–686.

Banuett, F. (1995). Genetics of *Ustilago maydis*, a fungal pathogen that induces tumours in maize. *Annual Review of Genetics*, **29**, 179–208.

Banuett, F. & Herskowitz, I. (1996). Discrete developmental stages during teliospore formation in the corn smut fungus, *Ustilago maydis*. *Development*, **122**, 2965–2976.

Bao, J. R. & Lasarovits, G. (2001). Differential colonization of tomato roots by nonpathogenic and pathogenic *Fusarium oxysporum* strains may influence fusarium wilt control. *Phytopathology*, **91**, 449–456.

Bardwell, L. (2004). A walk-through of the yeast mating pheromone response pathway. *Peptides*, **25**, 1465–1476.

Barksdale, A. W. (1960). Inter-thallic sexual reactions in *Achlya*, a genus of the aquatic fungi. *American Journal of Botany*, **47**, 14–23.

Barksdale, A. W. (1962). Effect of nutritional deficiency on growth and sexual reproduction of *Achlya ambisexualis*. *American Journal of Botany*, **49**, 633–638.

Barksdale, A. W. (1963). The uptake of exogenous hormone A by certain strains of *Achlya*. *Mycologia*, **55**, 164–171.

Barksdale, A. W. (1965). *Achlya ambisexualis* and a new cross-conjugating species of *Achlya*. *Mycologia*, **57**, 493–501.

Barksdale, A. W. (1966). Segregation of sex in the progeny of a selfed heterozygote of *Achlya ambisexualis*. *Mycologia*, **58**, 802–804.

Barksdale, A. W. (1967). The sexual hormones of the fungus *Achlya*. *Annals of the New York Academy of Sciences*, **144**, 313–319.

Barksdale, A. W. (1969). Sexual hormones of *Achlya* and other fungi. *Science*, **166**, 831–837.

Barksdale, A. W. (1970). Nutrition and antheridiol-induced branching in *Achlya ambisexualis*. *Mycologia*, **62**, 411–420.

Bärlocher, F., ed. (1992). *The Ecology of Aquatic Hyphomycetes. Ecological Studies* **94**. Berlin: Springer-Verlag.

Bärlocher, F. & Kendrick, W. B. (1973a). Fungi in the diet of *Gammarus pseudolimnaeus* (Amphipoda). *Oikos*, **24**, 295–300.

Bärlocher, F. & Kendrick, W. B. (1973b). Fungi and food preference of *Gammarus pseudolimnaeus*. *Archiv für Hydrobiologie*, **72**, 501–516.

Bärlocher, F. & Oertli, J. J. (1978a). Colonization of conifer needles by aquatic hyphomycetes. *Canadian Journal of Botany*, **56**, 57–62.

Bärlocher, F. & Oertli, J. J. (1978b). Inhibitors of aquatic hyphomycetes in dead conifer needles. *Mycologia*, **70**, 964–974.

Barnett, J. A., Payne, R. W. & Yarrow, D. (2000). *Yeasts: Characteristics and Identification*, 3rd edn. Cambridge: Cambridge University Press.

Barr, D. J. S. (1980). An outline for the reclassification of the Chytridiales, and for a new order, the Spizellomycetales. *Canadian Journal of Botany*, **58**, 2380–2394.

Barr, D. J. S. (1987). *Rhizophlyctis rosea*. In *Zoosporic Fungi in Teaching and Research*, ed. M. S. Fuller & A. Jaworski. Athens, USA: Southeastern Publishing Corporation, pp. 30–11.

Barr, D. J. S. (1988). Zoosporic plant parasites as fungal vectors of viruses: taxonomy and life cycles of species involved. In *Viruses with Fungal Vectors*, ed. J. I. Cooper & M. J. C. Asher. Sudbury, UK: Lavenham Press Ltd, pp. 123–137.

Barr, D. J. S. (1990). Phylum Chytridiomycota. In *Handbook of Protoctista*, ed. L. Margulis, J. O. Corliss, M. Melkonian & D. J. Chapman. Boston: Jones & Bartlett, pp. 454–466.

Barr, D. J. S. (1992). Evolution and kingdoms of organisms from the perspective of a mycologist. *Mycologia*, **84**, 1–11.

Barr, D. J. S. (2001). Chytridiomycota. In *The Mycota* **VIIA**: *Systematics and Evolution*, ed. D. J. McLaughlin, E. G. McLaughlin & P. A. Lemke. Berlin: Springer-Verlag, pp. 93–112.

Barr, D. J. S. & Allan, P. M. E. (1985). A comparison of the flagellar apparatus in *Phytophthora*, *Saprolegnia*, *Thraustochytrium*, and *Rhizidiomyces*. *Canadian Journal of Botany*, **63**, 138–154.

Barr, D. J. S. & Désaulniers, N. L. (1988). Precise configuration of the chytrid zoospore. *Canadian Journal of Botany*, **66**, 869–876.

Barr, D. J. S. & Hartmann, V. E. (1977). Zoospore ultrastructure of *Olpidium brassicae* and *Rhizophlyctis rosea*. *Canadian Journal of Botany*, **55**, 1221–1235.

Barr, D. J. S., Kudo, H., Jakober, K. D. & Cheng, K. J. (1989). Morphology and development of rumen fungi: *Neocallimastix* sp., *Piromyces communis*, and *Orpinomyces bovis* gen. nov., sp. nov. *Canadian Journal of Botany*, **67**, 2815–2824.

Barr, M. E. (1958). Life history studies of *Mycosphaerella tassiana* and *M. typhae*. *Mycologia*, **50**, 501–513.

Barr, M. E. (1983). The ascomycete connection. *Mycologia*, **75**, 1–13.

Barr, M. E. (2001). Ascomycota. In *The Mycota* **VIIA**: *Systematics and Evolution*, ed. D. J. McLaughlin, E. G. McLaughlin & P. A. Lemke. Berlin: Springer-Verlag, pp. 161–177.

Barr, M. E. & Huhndorf, S. M. (2001). Loculoascomycetes. In *The Mycota* **VIIA**: *Systematics and Evolution*, ed. D. J. McLaughlin, E. G. McLaughlin & P. A. Lemke. Berlin: Springer-Verlag, pp. 283–305.

Barron, G. L. (1977). *The Nematode-Destroying Fungi*. Guelph: Canadian Biological Publications.

Barron, G. L. (1981). Predators and parasites of microscopic animals. In *The Biology of Conidial Fungi* II, ed. G. T. Cole & B. Kendrick. New York: Academic Press, pp. 167–200.

Barron, G. L. (1992). Lignolytic and cellulolytic fungi as predators and parasites. In *The Fungal Community. Its Organization and Role in the Ecosystem*, 2nd edn, ed. G. C. Carroll & D. T. Wicklow. New York: Marcel Dekker, pp. 311–326.

Barron, G. L. & Dierkes, Y. (1977). Nematophagous fungi: *Hohenbuehelia*, the perfect state of *Nematoctonus*. *Canadian Journal of Botany*, **55**, 3054–3062.

Barron, G. L. & Szijarto, E. (1986). A new species of *Olpidium* parasitic in nematode eggs. *Mycologia*, **78**, 972–975.

Barron, G. L. & Thorn, R. G. (1987). Destruction of nematodes by species of *Pleurotus*. *Canadian Journal of Botany*, **65**, 774–778.

Barron, G. L., Cain, R. F. & Gilman, J. C. (1961). The genus *Microascus*. *Canadian Journal of Botany*, **39**, 1609–1631.

Barron, J. L. & Hill, E. P. (1974). Ultrastructure of zoosporogenesis in *Allomyces macrogynus*. *Journal of General Microbiology*, **80**, 319–327.

Barry, D., Callot, G., Janex-Favre, M. C. & Parguey-Leduc, A. (1993). Morphologie et structure des hyphes externes observées sur le péridium des *Tuber* à écailles: evolution au cours du développement de l'ascocarpe. *Canadian Journal of Botany*, **71**, 609–619.

Barry, D., Staunton, S. & Callot, G. (1994). Mode of absorption of water and nutrients by ascocarps of *Tuber melanosporum* and *Tuber aestivum*. *Canadian Journal of Botany*, **72**, 317–322.

Barry, D., Jaillard, B., Staunton, S. & Callot, G. (1995). Translocation and metabolism of phosphate following absorption by ascocarps of *Tuber melanosporum* and *Tuber aestivum*. *Mycological Research*, **99**, 167–172.

Barstow, W. E. (1987) *Catenaria anguillulae*. In *Zoosporic Fungi in Teaching and Research*, ed. M. S. Fuller & A. Jaworski. Athens, USA: Southeastern Publishing Corporation, pp. 34–35.

Barstow, W. E. & Pommerville, J. (1980). The ultrastructure of cell wall formation and of gamma particles during encystment of *Allomyces macrogynus* zoospores. *Archiv für Mikrobiologie*, **128**, 179–189.

Bartnicki-Garcia, S. (1968). Cell wall chemistry, morphogenesis, and taxonomy of fungi. *Annual Review of Microbiology*, **42**, 57–69.

Bartnicki-Garcia, S. (1987). The cell wall in fungal evolution. In *Evolutionary Biology of the Fungi*, ed. A. D. M. Rayner, C. M. Brasier & D. Moore. Cambridge: Cambridge University Press, pp. 389–403.

Bartnicki-Garcia, S. (1996). The hypha: unifying thread of the fungal kingdom. In *A Century of Mycology*, ed. B. C. Sutton. Cambridge: Cambridge University Press, pp. 105–133.

Bartnicki-Garcia, S. (2002). Hyphal tip growth: outstanding questions. In *Molecular Biology of Fungal Development*, ed. H.D. Osiewacz. New York: Marcel Dekker, pp. 29–58.

Bartnicki-Garcia, S. & Wang, M.C. (1983). Biochemical aspects of morphogenesis in *Phytophthora*. In *Phytophthora. Its Biology, Taxonomy, Ecology, and Pathology*, ed. D.C. Erwin, S. Bartnicki-Garcia & P.H. Tsao. St. Paul: APS Press, pp. 121–137.

Bartnicki-Garcia, S., Ruiz-Herrera, J. & Bracker, C.E. (1979). Chitosomes and chitin synthesis. In *Fungal Walls and Hyphal Growth*, ed. J.H. Burnett & A.P.J. Trinci. Cambridge: Cambridge University Press, pp. 149–168.

Barve, M.P., Arie, T., Salimath, S.S., Muehlbauer, F.J. & Peever, T.L. (2003). Cloning and characterization of the mating type (*MAT*) locus from *Ascochyta rabiei* (teleomorph: *Didymella rabiei*) and a *MAT* phylogeny of legume-associated *Ascochyta* spp. *Fungal Genetics and Biology*, **39**, 151–167.

Bashi, E. & Fokkema, N.J. (1976). Scanning electron microscopy of *Sporobolomyces roseus* on wheat leaves. *Transactions of the British Mycological Society*, **67**, 500–505.

Basith, M. & Madelin, M.F. (1968). Studies on the production of perithecial stromata by *Cordyceps militaris* in artificial culture. *Canadian Journal of Botany*, **46**, 473–480.

Basse, C. & Steinberg, G. (2004). *Ustilago maydis*, model system for analysis of the molecular basis of fungal pathogenicity. *Molecular Plant Pathology*, **5**, 83–92.

Basse, C.W., Lottspeich, F., Steglich, W. & Kahmann, R. (1996). Two potential indole-3-acetaldehyde dehydrogenases in the phytopathogenic fungus *Ustilago maydis*. *European Journal of Biochemistry*, **242**, 648–656.

Batra, L.R. (1973). Nematosporaceae (Hemiascomycetidae): taxonomy, pathogenicity, distribution and vector relations. *ARS Technical Bulletin*, **1469**, 1–71. Washington DC: US Department of Agriculture.

Batra, R., Boekhout, T., Guého, E., *et al.* (2005). *Malassezia* Baillon, emerging clinical yeasts. *FEMS Yeast Research*, **5**, 1101–1113.

Batts, C.C.V. (1955). Observations on the infection of wheat by loose smut (*Ustilago tritici* (Pers.) Rostr.). *Transactions of the British Mycological Society*, **38**, 465–475.

Bauer, R. (2004). Basidiomycetous interfungal cellular interactions – a synopsis. In *Frontiers in Basidiomycete Mycology*, ed. R. Agerer, M. Piepenbring & P. Blanz. Eching, Germany: IHW-Verlag, pp. 325–337.

Bauer, R., Oberwinkler, F. & Vánky, K. (1997). Ultrastructural markers and systematics in smut fungi and allied taxa. *Canadian Journal of Botany*, **75**, 1273–1314.

Bauer, R., Begerow, D., Oberwinkler, F., Piepenbring, M. & Berbee, M.L. (2001). Ustilaginomycetes. In *The Mycota* **VIIB**: *Systematics and Evolution*, ed. D.J. McLaughlin, E.G. McLaughlin & P.A. Lemke. Berlin: Springer-Verlag, pp. 57–83.

Baura, G., Szaro, T.M. & Bruns, T.D. (1992). *Gastrosuillus laricinus* is a recent derivative of *Suillus grevillei*: molecular data. *Mycologia*, **84**, 592–597.

Bayman, P. & Cotty, J.P. (1993). Genetic diversity in *Aspergillus flavus*: association with aflatoxin production and morphology. *Canadian Journal of Botany*, **71**, 23–31.

Bayne, H.G. & Michener, H.D. (1979). Heat resistance of *Byssochlamys* ascospores. *Applied and Environmental Microbiology*, **37**, 449–453.

Beadle, G.W. & Tatum, E.L. (1941). Genetic control of biochemical reactions in *Neurospora*. *Proceedings of the National Academy of Sciences of the USA*, **27**, 499–506.

Beakes, G.W. (1981). Ultrastructural aspects of oospore differentiation. In *The Fungal Spore: Morphogenetic Controls*, ed. H.R. Hohl & G. Turian. London: Academic Press, pp. 71–94.

Beakes, G.W. (1983). A comparative account of cyst coat ontogeny in saprophytic and fish-lesion (pathogenic) isolates of the *Saprolegnia diclina-parasitica* complex. *Canadian Journal of Botany*, **61**, 603–625.

Beakes, G.W. (1987). Oomycete phylogeny: ultrastructural perspectives. In *Evolutionary Biology of the Fungi*, ed. A.D.M. Rayner, C.M. Brasier & D. Moore. Cambridge: Cambridge University Press, pp. 405–421.

Beakes, G.W. (1998). Relationships between lower fungi and protozoa. In *Evolutionary Relationships among Protozoa*, ed. G.H. Coombs, K. Vickerman, M.A. Sleigh & A. Warren. London: Chapman & Hall, pp. 351–373.

Beakes, G.W. & Gay, J.L. (1978a). Light and electron microscopy of oospore maturation in *Saprolegnia furcata*. 1. Cytoplasmic changes. *Transactions of the British Mycological Society*, **71**, 11–24.

Beakes, G.W. & Gay, J.L. (1978b). Light and electron microscopy of oospore maturation in *Saprolegnia furcata*. 2. Wall development. *Transactions of the British Mycological Society*, **71**, 25–35.

Beakes, G.W. & Glockling, S.L. (1998). Injection tube differentiation in gun cells of a *Haptoglossa* species which infects nematodes. *Fungal Genetics and Biology*, **24**, 45–68.

Beakes, G.W. & Glockling, S.L. (2000). An ultrastructural analysis of organelle arrangement during gun (infection) cell differentiation in the nematode parasite *Haptoglossa dickii. Mycological Research*, **104**, 1258–1269.

Beakes, G.W., Canter, H.M. & Jaworski, G.H.M. (1993). Sporangium differentiation and zoospore fine-structure of the chytrid *Rhizophydium planktonicum*, a fungal parasite of *Asterionella formosa. Mycological Research*, **97**, 1059–1074.

Beaumont, A. (1947). Dependence on the weather of the date of potato blight epidemics. *Transactions of the British Mycological Society*, **31**, 45–53.

Bécard, G. & Fortin, J.A. (1988). Early events of vesicular–arbuscular mycorrhiza formation in Ri T-DNA transformed roots. *New Phytologist*, **108**, 211–218.

Beckett, A. (1976a). Ultrastructural studies on exogenously dormant ascospores of *Daldinia concentrica. Canadian Journal of Botany*, **54**, 689–697.

Beckett, A. (1976b). Ultrastructural studies on germinating ascospores of *Daldinia concentrica. Canadian Journal of Botany*, **54**, 698–705.

Beckett, A. (1979a). Ultrastructure and development of the ascospore germ slit in *Xylaria longipes. Transactions of the British Mycological Society*, **72**, 269–276.

Beckett, A. (1979b). Comparative ultrastructure of the ascospore germ slit in *Xylaria* and *Daldinia. Transactions of the British Mycological Society*, **72**, 320–322.

Beckett, A. (1981). The ascus with an apical pore: development, composition and function. In *Ascomycete Systematics, the Luttrellian Concept*, ed. D.R. Reynolds. New York: Springer-Verlag, pp. 7–26.

Beckett, A. & Crawford, R.M. (1973). The development and fine structure of the ascus apex and its role during spore discharge in *Xylaria longipes. New Phytologist*, **72**, 357–369.

Beckett, A. & Wilson, I.M. (1968). Ascus cytology of *Podospora anserina. Journal of General Microbiology*, **53**, 81–87.

Beckett, A., Barton, R. & Wilson, I.M. (1968). Fine structure of the wall and appendage formation in ascospores of *Podospora anserina. Journal of General Microbiology*, **53**, 89–94.

Beckett, A., Heath, I.B. & McLaughlin, D.J. (1974). *An Atlas of Fungal Ultrastructure*. London: Longman.

Beckman, C.H. (1987). *The Nature of Wilt Diseases of Plants*. St. Paul, USA: APS Press.

Beever, R.E. & Weeds, P.L. (2004). Taxonomy and genetic variation of *Botrytis* and *Botryotinia*. In *Botrytis: Biology, Pathology and Control*, ed. Y. Elad, B. Williamson, P. Tudzynski & N. Delen. Dordrecht: Kluwer, pp. 29–52.

Beever, R.E., Redgewell, R.J. & Dempsey, G.P. (1978). Function of rodlets on the surface of fungal spores. *Nature*, **272**, 608–610.

Begerow, D., Bauer, R. & Oberwinkler, F. (1997). Phylogenetic studies on large subunit ribosomal DNA sequences of smut fungi and related taxa. *Canadian Journal of Botany*, **75**, 2045–2056.

Begerow, D., Bauer, R. & Boekhout, T. (2000). Phylogenetic placements of ustilaginomycetous anamorphs as deduced from nuclear LSU rDNA sequences. *Mycological Research*, **104**, 53–60.

Begerow, D., Bauer, R. & Oberwinkler, F. (2002). The Exobasidiales: an evolutionary hypothesis. *Mycological Progress*, **1**, 187–199.

Bélanger, R.R. & Labbé, C. (2002). Control of powdery mildews without chemicals: prophylactic and biological alternatives for horticultural crops. In *The Powdery Mildews. A Comprehensive Treatise*, ed. R.R. Bélanger, W.R. Bushnell, A.J. Dik & T.L.W. Carver. St. Paul, USA: APS Press, pp. 256–267.

Bell, A. & Mahoney, D.P. (1995). Coprophilous fungi in New Zealand. I. *Podospora* species with swollen agglutinated perithecial hairs. *Mycologia*, **87**, 375–396.

Bell, A. & Mahoney, D.P. (1996). Perithecium development in *Podospora tetraspora* and *Podospora vesticola. Mycologia*, **88**, 163–170.

Bell, A. & Mahoney, D.P. (1997). Coprophilous fungi in New Zealand. II. *Podospora* species with coriaceous perithecia. *Mycologia*, **89**, 908–915.

Bell, A.A. & Wheeler, M.H. (1986). Biosynthesis and functions of fungal melanins. *Annual Review of Phytopathology*, **24**, 411–451.

Bellemère, A. (1994). Asci and ascospores in ascomycete systematics. In *Ascomycete Systematics. Problems and Perspectives in the Nineties*, ed. D.L. Hawksworth. New York: Plenum Press, pp. 111–126.

Benjamin, R.K. (1962). A new *Basidiobolus* that forms microspores. *Aliso*, **5**, 223–233.

Benjamin, R.K. (1966). The merosporangium. *Mycologia*, **58**, 1–42.

Benjamin, R.K. (1979). Zygomycetes and their spores. In *The Whole Fungus* 2, ed. B. Kendrick. Ottawa: National Museum of Natural Sciences; Kananaskis Foundation, pp. 573–616.

Benjamin, R.K. & Mehrotra, B.S. (1963). Obligate azygospore formation in two species of *Mucor* (Mucorales). *Aliso*, **5**, 235–245.

Bennett, R.J. & Johnson, A.D. (2005). Mating in *Candida albicans* and the search for a sexual cycle. *Annual Review of Microbiology*, **59**, 233–255.

Benny, G. L. (1992). Observations on the Thamnidiaceae (Mucorales). V. *Thamnidium*. *Mycologia*, **84**, 834–842.

Benny, G. L. (2001). Zygomycota: Trichomycetes. In *The Mycota VIIA: Systematics and Evolution*, ed. D. J. McLaughlin, E. G. McLaughlin & P. A. Lemke. Berlin: Springer-Verlag, pp. 147–160.

Benny, G. L. & Benjamin, R. K. (1976). Observations on Thamnidiaceae (Mucorales). II. *Chaetocladium, Cokeromyces, Mycotypha* and *Phascolomyces*. *Aliso*, **8**, 391–424.

Benny, G. L. & Benjamin, R. K. (1993). Observations on Thamnidiaceae. 6. Two new species of *Dichotomocladium* and the zygospores of *D. hesseltinei* (Chaetocladiaceae). *Mycologia*, **85**, 660–671.

Benny, G. L., Humber, R. A., & Morton, J. B. (2001). Zygomycota: Zygomycetes. In *The Mycota VIIA: Systematics and Evolution*, ed. D. J. McLaughlin, E. G. McLaughlin & P. A. Lemke. Berlin: Springer-Verlag, pp. 113–146.

Berbee, M. L. (1996). Loculoascomycete origins and evolution of filamentous ascomycete morphology based on 18S rRNA gene sequence data. *Molecular Biology and Evolution*, **13**, 462–470.

Berbee, M. L. (2001). The phylogeny of plant and animal pathogens in the Ascomycota. *Physiological and Molecular Plant Pathology*, **59**, 165–187.

Berbee, M. L. & Taylor, J. W. (1992a). Two ascomycete classes based on fruiting-body characters and ribosomal DNA sequence. *Molecular Biology and Evolution*, **9**, 278–284.

Berbee, M. L. & Taylor, J. W. (1992b). 18S ribosomal RNA gene sequence characters place the human pathogen *Sporothrix schenckii* in the genus *Ophiostoma*. *Experimental Mycology*, **16**, 87–91.

Berbee, M. L. & Taylor, J. W. (1993). Dating the evolutionary radiations of the true fungi. *Canadian Journal of Botany*, **71**, 1114–1127.

Berbee, M. L. & Taylor, J. W. (1999). Fungal phylogeny. In *Molecular Fungal Biology*, ed. R. P. Oliver & M. Schweizer. Cambridge: Cambridge University Press, pp. 21–77.

Berbee, M. L. & Taylor, J. W. (2001). Fungal molecular evolution: gene trees and geologic time. In *The Mycota VIIB: Systematics and Evolution*, ed. D. J. McLaughlin, E. G. McLaughlin & P. A. Lemke. Berlin: Springer-Verlag, pp. 229–245.

Berbee, M. L. & Wells, K. (1988). Ultrastructural studies of mitosis and the septal pore apparatus in *Tremella globospora*. *Mycologia*, **80**, 479–492.

Berbee, M. L. & Wells, K. (1989). Light and electron microscopic studies of meiosis and basidium ontogeny in *Clavicorona pyxidata*. *Mycologia*, **81**, 20–41.

Berbee, M. L., Yoshimura, A., Sugiyama, J. & Taylor, J. W. (1995). Is *Penicillium* monophyletic? An evaluation of phylogeny in the family Trichocomaceae from 18S, 5.8S and ITS ribosomal DNA sequence data. *Mycologia*, **87**, 210–222.

Berbee, M. L., Pirseyedi, M. & Hubbard, S. (1999). *Cochliobolus* phylogenetics and the origin of known, highly virulent pathogens, inferred from ITS and glyceraldehyde-3-phosphate dehydrogenase gene sequences. *Mycologia*, **91**, 964–977.

Berbee, M. L., Payne, B. P., Zhang, G., Roberts, R. G. & Turgeon, B. G. (2003). Shared ITS DNA substitutions in isolates of opposite mating type reveal a recombining history for three presumed asexual species in the filamentous ascomycetous genus *Alternaria*. *Mycological Research*, **107**, 169–182.

Berch, S. M., Allen, T. R. & Berbee, M. L. (2002). Molecular detection, community structure and phylogeny of ericoid mycorrhizal fungi. *Plant and Soil*, **244**, 55–66.

Bérdy, J. (1986). Further antibiotics with practical application. In *Biotechnology 4: Microbial Products II*, ed. H. Pape & H.-J. Rehm. Weinheim, Germany: VCH, pp. 465–507.

Bergman, K., Burke, P. V., Cerdá-Olmedo, E., *et al.* (1969). *Phycomyces. Bacteriological Reviews*, **33**, 99–157.

Berrie, A. D. (1975). Detritus, micro-organisms and animals in fresh water. In *The Role of Terrestrial and Aquatic Organisms in Decomposition Processes*, ed. J. M. Anderson & A. MacFadyen. Oxford: Blackwell Scientific, pp. 323–338.

Berta, G. & Fusconi, A. (1983). Ascosporogenesis in *Tuber magnatum*. *Transactions of the British Mycological Society*, **80**, 201–207.

Bertaux, J., Schmid, M., Prevost-Bourne, N. C., *et al.* (2003). In situ identification of intracellular bacteria related to *Paenibacillus* spp. in the mycelium of the ectomycorrhizal fungus *Laccaria bicolor* S238N. *Applied and Environmental Microbiology*, **69**, 4243–4248.

Bertrand, H. (2000). Role of mitochondrial DNA in the senescence and hypovirulence of fungi and potential for plant disease control. *Annual Review of Phytopathology*, **38**, 397–422.

Besl, H. & Bresinsky, A. (1997). Chemosystematics of *Suillaceae* and *Gomphidiaceae* (suborder *Suillineae*). *Plant Systematics and Evolution*, **206**, 223–242.

Bettiol, W. (1999). Effectiveness of cow's milk against zucchini squash powdery mildew (*Sphaerotheca fuliginea*) in greenhouse conditions. *Crop Protection*, **18**, 489–492.

Beuchat, L. R. (1995). Indigenous fermented foods. In *Biotechnology* **9**: *Enzymes, Biomass, Food and Feed*, ed. G. Reed & T. W. Nagodawithana. Weinheim, Germany: VCH, pp. 505–559.

Beuchat, L. R. & Toledo, R. T. (1977). Behaviour of *Byssochlamys nivea* ascospores in fruit syrups. *Transactions of the British Mycological Society*, **68**, 65–71.

Beyrle, H. F., Smith, S. E., Peterson, R. L. & Franco, C. M. M. (1995). Colonization of *Orchis morio* protocorms by a mycorrhizal fungus: effects of nitrogen nutrition and glyphosate in modifying the responses. *Canadian Journal of Botany*, **73**, 1128–1140.

Bhatnagar, D., Yu, J. & Ehrlich, K. C. (2002). Toxins of filamentous fungi. *Chemical Immunology*, **81**, 167–206.

Bidartondo, M. I., Burghardt, B., Gebauer, G., Bruns, T. D. & Read, D. J. (2004). Changing partners in the dark: isotopic and molecular evidence of ectomycorrhizal liaisons between forest orchids and trees. *Proceedings of the Royal Society*, **B 271**, 1799–1806.

Biffen, R. H. (1905). Mendel's laws of inheritance and wheat breeding. *Journal of Agricultural Science*, **1**, 4–48.

Bigelow, D. M., Olsen, M. W. & Gilbertson, R. L. (2005). *Labyrinthula terrestris* sp. nov., a new pathogen of turf grass. *Mycologia*, **97**, 185–190.

Bilay, V. T. & Lelley, J. I. (1997). Growth of mycelium of *Agaricus bisporus* on biomass and conidium of *Humicola insolens*. *Journal of Applied Botany — Angewandte Botanik*, **71**, 21–23.

Bimbó, A., Liu, J. & Balasubramanian, M. K. (2005). Roles of Pdk1p, a fission yeast protein related to phosphoinositide-dependent protein kinase, in the regulation of mitosis and cytokinesis. *Molecular Biology of the Cell*, **16**, 3162–3175.

Binder, M. & Bresinsky, A. (2002). Derivation of a polymorphic lineage of Gasteromycetes from boletoid ancestors. *Mycologia*, **94**, 85–98.

Binder, M. & Hibbett, D. S. (2002). Higher-level phylogenetic relationships of Homobasidiomycetes (mushroom-forming fungi) inferred from four rDNA regions. *Molecular Phylogenetics and Evolution*, **22**, 76–90.

Binder, M., Hibbett, D. S. & Molitoris, H. P. (2001). Phylogenetic relationships of the marine gasteromycete *Nia vibrissa*. *Mycologia*, **93**, 679–688.

Bishop, C. D. & Cooper, R. M. (1983). An ultrastructural study of root invasion in three vascular wilt diseases. *Physiological Plant Pathology*, **22**, 15–27.

Bistis, G. (1956). Sexuality in *Ascobolus stercorarius*. I. Morphology of the ascogonium; plasmogamy; evidence for a sexual hormonal mechanism. *American Journal of Botany*, **43**, 389–394.

Bistis, G. (1957). Sexuality in *Ascobolus stercorarius*. II. Preliminary experiments on various aspects of the sexual process. *American Journal of Botany*, **44**, 436–443.

Bistis, G. N. (1983). Evidence for diffusible mating-type specific trichogyne attractants in *Neurospora crassa*. *Experimental Mycology*, **7**, 292–295.

Bistis, G. N. (1998). Physiological heterothallism and sexuality in euascomycetes: a partial history. *Fungal Genetics and Biology*, **23**, 213–222.

Bistis, G. & Raper, J. R. (1963). Heterothallism and sexuality in *Ascobolus stercorarius*. *American Journal of Botany*, **50**, 880–891.

Black, W. (1952). A genetical basis for the classification of strains of *Phytophthora infestans*. *Proceedings of the Royal Society of Edinburgh*, **B 65**, 36–51.

Blackwell, E. M. (1943). The life history of *Phytophthora cactorum* (Leb. & Cohn) Schroet. *Transactions of the British Mycological Society*, **26**, 71–89.

Blackwell, M. (1994). Minute mycological mysteries: the influence of arthropods on the lives of fungi. *Mycologia*, **86**, 1–17.

Blackwell, M. & Malloch, D. (1989). Similarity of *Amphoromorpha* and secondary capilliconidia of *Basidiobolus*. *Mycologia*, **81**, 735–741.

Blackwell, W. H. & Powell, M. J. (1999). The nomenclatural propriety of *Rhizophlyctis rosea*. *Mycotaxon*, **70**, 213–217.

Blakeslee, A. F. (1906). Zygospore germinations in the Mucorineae. *Annales Mycologici*, **4**, 1–28.

Bland, C. E. & Charles, T. M. (1972). Fine structure of *Pilobolus*: surface and wall structure. *Mycologia*, **64**, 774–785.

Blanton, R. L. (1990). Phylum Acrasea. In *Handbook of Protoctista*, ed. L. Margulis, J. O. Corliss, M. Melkonian & D. J. Chapman. Boston: Jones & Bartlett, pp. 75–87.

Blanz, P. A. & Gottschalk, M. (1984). A comparison of 5S ribosomal RNA nucleotide sequences from smut fungi. *Systematic and Applied Microbiology*, **5**, 518–526.

Blodgett, D. J. (2001). Fescue toxicosis. *The Veterinary Clinics of North America. Equine Practice*, **17**, 567–577.

Bluemke, W. & Schrader, J. (2001). Integrated bioprocess for enhanced production of natural flavors and fragrances by *Ceratocystis moniliformis*. *Biomolecular Engineering*, **17**, 137–142.

Bly, J. E., Lawson, L. A., Dale, D. J., *et al.* (1992). Winter saprolegniosis in channel catfish. *Diseases of Aquatic Organisms*, **13**, 155–164.

Bobbitt, T. F. & Nordin, J. H. (1982). Production and composition of an exocellular nigeran-protein complex isolated from cultures of *Aspergillus awamori*. *Journal of Bacteriology*, **150**, 365–376.

Bock, K. R. (1962). Control of coffee leaf rust in Kenya Colony. *Transactions of the British Mycological Society*, **45**, 301–313.

Boddy, L. (1993). Saprotrophic cord-forming fungi: warfare strategies and other ecological aspects. *Mycological Research*, **97**, 641–655.

Boddy, L. & Rayner, A. D. M. (1983). Origins of decay in living deciduous trees: the role of moisture content and a re-appraisal of the expanded concept of tree decay. *New Phytologist*, **94**, 623–641.

Boddy, L., Gibbon, O. M. & Grundy, M. A. (1985). Ecology of *Daldinia concentrica*: effect of abiotic variables on mycelial extension and interspecific interactions. *Transactions of the British Mycological Society*, **85**, 201–211.

Boehm, E. W. A., Wenstrom, J. C., McLaughlin, D. J., *et al.* (1992). An ultrastructural pachytene karyotype for *Puccinia graminis* f. sp. *tritici*. *Canadian Journal of Botany*, **70**, 401–413.

Boerema, G. H., de Gruyter, J., Noordeloos, M. E. & Hamers, M. E. C. (2004). *Phoma Identification Manual*. Wallingford, UK: CABI Publishing.

Bogus, M. I. & Scheller, K. (2002). Extraction of an insecticidal protein fraction from the parasitic fungus *Conidiobolus coronatus* (Entomophthorales). *Acta Parasitologica*, **47**, 66–72.

Bogus, M. I. & Szczepanik, M. (2000). Histopathology of *Conidiobolus coronatus* (Entomophthorales) infection in *Galleria mellonella* (Lepidoptera) larvae. *Acta Parasitologica*, **45**, 48–54.

Boidin, J. (1971). Nuclear behaviour in the mycelium and the evolution of the Basidiomycetes. In *Evolution in the Higher Basidiomycetes*, ed. R. H. Petersen. Knoxville: University of Tennessee Press, pp. 129–148.

Boland, G. J. & Hall, R. (1994). Index of plant hosts of *Sclerotinia sclerotiorum*. *Canadian Journal of Plant Pathology*, **16**, 93–108.

Bölker, M. (2001). *Ustilago maydis* – a valuable model system for the study of fungal dimorphism and virulence. *Microbiology*, **147**, 1395–1401.

Bolton, M. D., Thomma, B. P. H. J. & Nelson, B. D. (2006). *Sclerotinia sclerotiorum* (Lib.) de Bary: biology and molecular traits of a cosmopolitan pathogen. *Molecular Plant Pathology*, **7**, 1–16.

Bonner, J. T. (1944). A descriptive study of the development of the slime mold *Dictyostelium discoideum*. *American Journal of Botany*, **31**, 175–182.

Bonner, J. T. (1999). The history of the cellular slime moulds as a 'model system' for developmental biology. *Journal of Biosciences*, **24**, 7–12.

Bonner, J. T. (2001). *First Signals: The Evolution of Multicellular Development*. Princeton and Oxford: Princeton University Press.

Bonner, J. T., Kane, K. K. & Levey, R. H. (1956). Studies on the mechanics of growth in the common mushroom. *Mycologia*, **48**, 13–19.

Boone, C., Bussey, H., Greene, D., Thomas, D. Y. & Vernet, T. (1986). Yeast killer toxin: site-directed mutations implicate the precursor protein as the immunity component. *Cell*, **46**, 105–113.

Booth, C. (1971). *The Genus Fusarium*. Kew: Commonwealth Mycological Institute.

Borel, J. F. (1986). Cyclosporin and its future. *Progress in Allergy*, **38**, 9–18.

Borges-Walmsley, M. I., Chen, D., Shu, X. & Walmsley, A. R. (2002). The pathobiology of *Paracoccidioides brasiliensis*. *Trends in Microbiology*, **10**, 80–87.

Borg-Karlson, A.-K., Englund, F. O. & Unelius, C. R. (1994). Dimethyl oligosulphides, major volatiles released from *Sauromatum guttatum* and *Phallus impudicus*. *Phytochemistry*, **35**, 321–323.

Borneman, A. R., Hynes, M. J. & Andrianopoulos, A. (2000). The *abaA* homologue of *Penicillium marneffei* participates in two developmental programmes: conidiation and dimorphic growth. *Molecular Microbiology*, **38**, 1034–1047.

Borno, C. & van der Kamp, B. J. (1975). Timing of infection and development of *Gymnosporangium fuscum* on *Juniperus*. *Canadian Journal of Botany*, **53**, 1266–1269.

Borriss, H. (1934). Beiträge zur Wachstums- und Entwicklungsphysiologie der Fruchtkörper von *Coprinus lagopus*. *Planta*, **22**, 28–69.

Bos, C. J. & Swart, K. (1995). Genetics of *Aspergillus*. In *The Mycota II: Genetics and Biotechnology*, ed. U. Kück. Berlin: Springer-Verlag, pp. 19–33.

Botella, M. A., Parker, J. E., Frost, L. N., *et al.* (1998). Three genes of the *Arabidopsis RPP1* complex resistance locus recognize distinct *Peronospora parasitica* avirulence determinants. *Plant Cell*, **10**, 1847–1860.

Boucias, D. G. & Pendland, J. C. (1998). *Principles of Insect Pathology*. Boston: Kluwer Academic Publishers.

Bourett, T. M. & Howard, R. J. (1990). *In vitro* development of penetration structures in the rice blast fungus, *Magnaporthe grisea*. *Canadian Journal of Botany*, **68**, 329–342.

Bourett, T. M. & Howard, R. J. (1992). Actin in penetration pegs of the fungal rice blast pathogen, *Magnaporthe grisea*. *Protoplasma*, **168**, 20–26.

Bourett, T. M., Czymmek, K. J. & Howard, R. J. (1998). An improved method for affinity probe localization in whole cells of filamentous fungi. *Fungal Genetics and Biology*, **24**, 3–13.

Bourke, A. (1991). Potato blight in Europe in 1845: the scientific controversy. In *Phytophthora*, ed. J. A. Lucas, R. C. Shattock, D. S. Shaw & L. R. Cooke. Cambridge: Cambridge University Press, pp. 12–24.

Bowden, J., Gregory, P. H. & Johnson, C. G. (1971). Possible wind transport of coffee leaf rust across the Atlantic Ocean. *Nature*, **229**, 500–501.

Bowman, B. H., Taylor, J. W., Brownlee, A. G., *et al.* (1992). Molecular evolution of the fungi: relationship of the Basidiomycetes, Ascomycetes and Chytridiomycetes. *Molecular Biology and Evolution*, **9**, 285–296.

Boxma, B., Voncken, F., Jannink, S., *et al.* (2004). The anaerobic chytridiomycete fungus *Piromyces* sp. E2 produces ethanol via pyruvate : formate lyase and an alcohol dehydrogenase E. *Molecular Microbiology*, **51**, 1389–1399.

Bozarth, R. F., Koltin, Y., Weissman, M. B., *et al.* (1981). The molecular weight and packaging of dsRNAs in the mycovirus from *Ustilago maydis* killer strains. *Virology*, **113**, 492–502.

Boze, H., Moulin, G. & Galzy, P. (1995). Production of microbial biomass. In *Biotechnology* **9***: Enzymes, Biomass, Food and Feed*, ed. G. Reed & T. W. Nagodawithana. Weinheim, Germany: VCH, pp. 167–220.

Bracker, C. E. (1968a). The ultrastructure and development of sporangia of *Gilbertella persicaria*. *Mycologia*, **60**, 1016–1067.

Bracker, C. E. (1968b). Ultrastructure of the haustorial apparatus of *Erysiphe graminis* and its relationship to the epidermal cell of barley. *Phytopathology*, **58**, 12–30.

Bradbury, E. M., Inglis, R. J., Matthews, H. R. & Langan, T. A. (1974). Molecular basis of control of mitotic cell division in eukaryotes. *Nature*, **249**, 553–556.

Bradley, W. H. (1967). Two aquatic fungi (Chytridiales) of Eocene age from the Green River Formation of Wyoming. *American Journal of Botany*, **54**, 577–582.

Bradshaw, A. D. (1959). Population differentiation in *Agrostis tenuis* Sibth. II. The incidence and significance of infection by *Epichloe typhina*. *New Phytologist*, **58**, 310–315.

Bradshaw, J. E., Gemmell, D. J. & Wilson, R. N. (1997). Transfer of resistance to clubroot (*Plasmodiophora brassicae*) to swedes (*Brassica napus* L. var. *napobrassica* Peterm.) from *B. rapa*. *Annals of Applied Biology*, **130**, 337–348.

Bradshaw, R. E., Bennett, J. W. & Peberdy, J. F. (1983). Parasexual analysis of *Aspergillus parasiticus*. *Journal of General Microbiology*, **129**, 2117–2123.

Brady, K. C., O'Kiely, P., Forristal, P. D. & Fuller, H. (2005). *Schizophyllum commune* on big-bale grass silage in Ireland. *Mycologist*, **19**, 30–35.

Brandhorst, T. T., Rooney, P. J., Sullivan, T. D. & Klein, B. (2002). Molecular genetic analysis of *Blastomyces dermatitidis* reveals new insights about pathogenic mechanisms. *International Journal of Medical Microbiology*, **292**, 363–371.

Braselton, J. P. (1995). Current status of the Plasmodiophorids. *Critical Reviews in Microbiology*, **21**, 263–275.

Braselton, J. P. (2001). Plasmodiophoromycota. In *The Mycota* **VIIA***: Systematics and Evolution*, ed. D. J. McLaughlin, E. G. McLaughlin & P. A. Lemke. Berlin: Springer-Verlag, pp. 81–91.

Brasier, C. M. (1975a). Stimulation of sex organ formation in *Phytophthora* by antagonistic species of *Trichoderma*. I. The effect *in vitro*. *New Phytologist*, **74**, 183–194.

Brasier, C. M. (1975b). Stimulation of sex organ formation in *Phytophthora* by antagonistic species of *Trichoderma*. II. Ecological implications. *New Phytologist*, **74**, 195–198.

Brasier, C. M. (1991a). Current questions in *Phytophthora* systematics: the role of the population approach. In *Phytophthora*, ed. J. A. Lucas, R. C. Shattock, D. S. Shaw & L. R. Cooke. Cambridge: Cambridge University Press, pp. 104–128.

Brasier, C. M. (1991b). *Ophiostoma novo-ulmi* sp. nov., causative agent of the current Dutch elm disease pandemics. *Mycopathologia*, **115**, 151–161.

Brasier, C. M. (1993). The genetic system as a fungal taxonomic tool: gene flow, molecular variation and sibling species in the *Ophiostoma piceae*–*Ophiostoma ulmi* complex and its taxonomic and ecological significance. In *Ceratocystis and Ophiostoma. Taxonomy, Ecology and Pathogenicity*, ed. M. J. Wingfield, K. A. Seifert & J. F. Webber. St. Paul, USA: APS Press, pp. 77–92.

Brasier, C. M. (2001). Rapid evolution of introduced plant pathogens via interspecific hybridization. *BioScience*, **51**, 123–133.

Brasier, C. M. & Kirk, S. A. (2001). Designation of the EAN and NAN races of *Ophiostoma novo-ulmi* as subspecies. *Mycological Research*, **105**, 547–554.

Brasier, C. M. & Mehrotra, M. D. (1995). *Ophiostoma himalulmi* sp. nov., a new species of Dutch elm disease fungus endemic in the Himalayas. *Mycological Research*, **99**, 205–215.

Brasier, C. M., Kirk, S. A., Pipe, N. D. & Buck, K. W. (1998). Rare interspecific hybrids in natural populations of the Dutch elm disease pathogens *Opiostoma ulmi* and *O. novo-ulmi*. *Mycological Research*, **102**, 45–57.

Brasier, C. M., Kirk, S. A., Delcan, J., *et al.* (2004). *Phytophthora alni* sp. nov. and its variants: designation of emerging heteroploid hybrid pathogens spreading on *Alnus* trees. *Mycological Research*, **108**, 1172–1184.

Braun, U. (1987). A monograph of the Erysiphales (powdery mildews). *Nova Hedwigia Beihefte*, **89**, 1–700.

Braun, U. (1995). *The Powdery Mildews (Erysiphales) of Europe*. Jena: G. Fischer Verlag.

Braun, U., Cook, R. T. A., Inman, A. J. & Shin, H.-D. (2002). The taxonomy of the powdery mildews. In *The Powdery Mildews. A Comprehensive Treatise*, ed. R. R. Bélanger, W. R. Bushnell, A. J. Dik, & T. L. W. Carver. St. Paul, USA: APS Press, pp. 13–55.

Bravery, A. F. (1991). The strategy for the eradication of *Serpula lacrymans*. In *Serpula lacrymans. Fundamental Biology and Control Strategies*, ed. D. H. Jennings & A. F. Bravery. Chichester: John Wiley & Sons, pp. 117–130.

Breeze, E. M. & Dix, N. J. (1981). Seasonal analysis of the fungal community on *Acer platanoides* leaves. *Transactions of the British Mycological Society*, **77**, 321–328.

Brefeld, O. (1876). Ueber die Zygosporenbildung bei *Mortierella rostafinskii* nebst Bemerkungen über die Systematik der Zygomyceten. *Sitzungsberichte der Gesellschaft Naturforschender Freunde zu Berlin*, 91.

Brefeld, O. (1881). Botanische Untersuchungen über Schimmelpilze. 4. *Pilobolus. Untersuchungen aus dem Gesammtgebiete der Mykologie IV*. Lepizig: Arthur Felix.

Brefeld, O. (1888). Basidiomyceten II. Protobasidiomyceten. *Untersuchungen aus dem Gesammtgebiete der Mykologie VII*. Leipzig: Arthur Felix.

Breitenbach, J. & Kränzlin, F. (1984). *Fungi of Switzerland. 1. Ascomycetes*. Lucerne: Verlag Mykologia.

Breitenbach, M. & Simon-Nobbe, B. (2002). The allergens of *Cladosporium herbarum* and *Alternaria alternata*. *Chemical Immunology*, **81**, 48–72.

Bresinsky, A. & Besl, H. (1990). *A Colour Atlas of Poisonous Fungi*. London: Manson Publishing.

Brewer, J. G. & Boerema, G. H. (1965). Electron microscope observations on the development of pycnidiospores in *Phoma* and *Ascochyta* spp. *Proceedings of the Academy of Sciences, Amsterdam*, **C 68**, 86–97.

Brian, P. W. (1960). Griseofulvin. *Transactions of the British Mycological Society*, **43**, 1–13.

Bridgmon, G. H. (1959). Production of new races of *Puccinia graminis* var. *tritici* by vegetative fusion. *Phytopathology*, **49**, 386–388.

Brierley, W. B. (1918). The microconidia of *Botrytis cinerea*. *Kew Bulletin*, 129–146.

Brobyn, P. J. & Wilding, N. (1977). Invasive and developmental processes in *Entomophthora* species infecting aphids. *Transactions of the British Mycological Society*, **69**, 349–366.

Brobyn, P. J. & Wilding, N. (1983). Invasive and developmental processes in *Entomophthora muscae* infecting houseflies. *Transactions of the British Mycological Society*, **80**, 1–8.

Brodie, H. J. (1936). The occurrence and function of oidia in the Hymenomycetes. *American Journal of Botany*, **23**, 309–327.

Brodie, H. J. (1975). *The Bird's Nest Fungi*. Toronto: Toronto University Press.

Brodie, H. J. (1984). More bird's nest fungi. A supplement to 'The Bird's Nest Fungi' (1975). *Lejeunia N. S.*, **112**, 1–69.

Brodie, H. J. & Gregory, P. H. (1953). The action of wind in the dispersal of spores from cup-shaped plant structures. *Canadian Journal of Botany*, **31**, 402–410.

Brodo, I. M. & Richardson, D. H. S. (1978). Chimeroid associations in the genus *Peltigera*. *Lichenologist*, **10**, 157–170.

Brodo, I. M., Sharnoff, S. D. & Sharnoff, S. (2001). *Lichens of North America*. New Haven: Yale University Press.

Brooke, A. G. (1994). Industrial fermentation and *Aspergillus* citric acid. In *The Genus Aspergillus. From Taxonomy and Genetics to Industrial Application*, ed. K. A. Powell, A. Renwick & J. F. Peberdy. New York: Plenum Press, pp. 129–134.

Brown, A. J. & Casselton, L. A. (2001). Mating in mushrooms: increasing the chances but prolonging the affair. *Trends in Genetics*, **17**, 393–400.

Brown, A. J. P. & Gow, N. A. R. (2002). Signal transduction and morphogenesis in *Candida albicans*. In *The Mycota VIII: Biology of the Fungal Cell*, ed. R. J. Howard & N. A. R. Gow. Berlin: Springer-Verlag, pp. 55–71.

Brown, J. K. M. (2002). Comparative genetics of avirulence and fungicide resistance in the powdery mildew fungi. In *The Powdery Mildews. A Comprehensive Treatise*, ed. R. R. Bélanger, W. R. Bushnell, A. J. Dik & T. L. W. Carver. St. Paul, USA: APS Press, pp. 56–65.

Brown, J. K. M. & Hovmøller, M. S. (2002). Aerial dispersal of pathogens on the global and continental scales and its impact on plant disease. *Science*, **297**, 537–541.

Brown, J. K. M., Simpson, C. G. & Wolfe, M. S. (1993). Adaptation of barley powdery mildew populations in England to varieties with two resistance genes. *Plant Pathology*, **42**, 108–115.

Brown, J.S., Whan, J.H., Kenny, M.K. & Merriman, P.R. (1995). The effect of coffee leaf rust on foliation and yield of coffee in Papua New Guinea. *Crop Protection*, **14**, 589–592.

Bruggink, A., Straathof, A.J. & van der Wielen, L.A. (2003). A 'fine' chemical industry for life science products: green solutions to chemical challenges. *Advances in Biochemical Engineering/Biotechnology*, **80**, 69–113.

Bruin, G.C.A. & Edgington, L.V. (1983). The chemical control of diseases caused by zoosporic fungi. A many-sided problem. In *Zoosporic Plant Pathogens*, ed. S.T. Buczacki. London: Academic Press, pp. 193–232.

Bruning, K. (1991a). Infection of the diatom *Asterionella* by a chytrid. I. Effects of light on reproduction and infectivity of the parasite. *Journal of Plankton Research*, **13**, 103–117.

Bruning, K. (1991b). Effects of phosphorus limitation on the epidemiology of a chytrid phytoplankton parasite. *Freshwater Biology*, **25**, 409–417.

Bruning, K. (1991c). Effects of temperature and light on the population dynamics of the *Asterionella–Rhizophydium* association. *Journal of Plankton Research*, **13**, 707–719.

Brunner, D. & Nurse, P. (2000). New concepts in fission yeast morphogenesis. *Philosophical Transactions of the Royal Society*, B **355**, 873–877.

Bruns, T.D., White, T.J. & Taylor, J.W. (1991). Fungal molecular systematics. *Annual Review of Ecology and Systematics*, **22**, 525–564.

Bruns, T.D., Vilgalys, R., Barns, S.M., *et al.* (1992). Evolutionary relationships within the fungi: analyses of nuclear small subunit rRNA sequences. *Molecular Phylogenetics and Evolution*, **1**, 231–241.

Bruns, T.D., Szaro, T.M., Gardes, M., *et al.* (1998). A sequence database for the identification of ectomycorrhizal basidiomycetes by phylogenetic analysis. *Molecular Ecology*, **7**, 257–272.

Bryan, G.T., Daniels, M.J. & Osbourn, A.E. (1995). Comparison of fungi within the *Gaeumannomyces–Phialophora* complex by analysis of ribosomal DNA sequences. *Applied and Environmental Microbiology*, **61**, 681–689.

Buchanan, K.L. & Murphy, J.W. (1998). What makes *Cryptococcus neoformans* a pathogen? *Emerging Infectious Diseases*, **4**, 71–83.

Buchta, M. & Cvak, L. (1999). Ergot alkaloids and other metabolites of the genus *Claviceps*. In *Ergot. The Genus Claviceps*, ed. V. Křen & L. Cvak. Amsterdam: Harwood Academic Publishers, pp. 173–200.

Buczacki, S.T. (1983). *Plasmodiophora*, an interrelationship between biological and practical problems. In *Zoosporic Plant Pathogens: A Modern Perspective*, ed. S.T. Buczacki. London: Academic Press, pp. 161–191.

Buczacki, S.T., Toxopeus, H., Mattusch, P., *et al.* (1975). Study of physiologic specialization in *Plasmodiophora brassicae*: proposals for an attempted rationalization through an international approach. *Transactions of the British Mycological Society*, **65**, 295–303.

Büdel, B. & Scheidegger, C. (1996). Thallus morphology and anatomy. In *Lichen Biology*, ed. T.H. Nash. Cambridge: Cambridge University Press, pp. 37–64.

Budge, S.P. & Whipps, J.M. (2001). Potential for integrated control of *Sclerotinia sclerotiorum* in glasshouse lettuce using *Coniothyrium minitans* and reduced fungicide application. *Phytopathology*, **91**, 221–227.

Buller, A.H.R. (1909). *Researches on Fungi*, **1**. London: Longmans Green & Co.

Buller, A.H.R. (1922). *Researches on Fungi*, **2**. London: Longmans Green & Co.

Buller, A.H.R. (1924). *Researches on Fungi*, **3**. London: Longmans Green & Co.

Buller, A.H.R. (1931). *Researches on Fungi*, **4**. London: Longmans, Green & Co.

Buller, A.H.R. (1933). *Researches on Fungi*, **5**. London: Longmans, Green & Co.

Buller, A.H.R. (1934). *Researches on Fungi*, **6**. London: Longmans, Green & Co.

Buller, A.H.R. (1950). *Researches on Fungi*, **7**. Toronto: Royal Society of Canada; Toronto University Press.

Buller, A.H.R. & Vanterpool, T.C. (1933). The violent discharge of the basidiospores (secondary conidia) of *Tilletia tritici*. In A.H.R. Buller, *Researches on Fungi*, **5**, pp. 207–278. London: Longmans, Green & Co.

Bullerwell, C.E. & Lang, B.F. (2005). Fungal evolution: the case of the vanishing mitochondrion. *Current Opinion in Microbiology*, **8**, 362–369.

Bulmer, G.S. (1964). Spore germination of forty-two species of puffballs. *Mycologia*, **56**, 630–632.

Bultman, T.L., White, J.F., Bodish, T.I. & Welch, A.M. (1998). A new kind of mutualism by insects and fungi. *Mycological Research*, **102**, 235–238.

Bunyard, B.A., Nicholson, M.S. & Royse, D.F. (1995). Phylogenetic resolution of *Morchella*, *Verpa*, and *Disciotis* (Pezizales: Morchellaceae) based on restriction enzyme analysis of the 28S ribosomal RNA gene. *Experimental Mycology*, **19**, 223–233.

Burdsall, H.H. (1981). The taxonomy of *Sporotrichum pruinosum* and *Sporotrichum pulverulentum/ Phanerochaete chrysosporium*. *Mycologia*, **73**, 675–680.

Burdsall, H. H. & Banik, M. T. (2001). The genus *Laetiporus* in North America. *Harvard Papers in Botany*, **6**, 43–55.

Burgeff, H. (1920). Über den Parasitismus des *Chaetocladium* und die heterokaroyotische Natur der von ihm auf Mucorineenerzeugten Gallen. *Zeitschrift für Botanik*, **12**, 1–35.

Burgeff, H. (1924). Untersuchungen über Sexualität und Parasitismus bei Mucorineen. I. *Botanische Abhandlungen*, **4**, 1–135.

Burnett, J. H. (1965). The natural history of recombination systems. In *Incompatibilty in Fungi*, ed. K. Esser & J. R. Raper. Berlin: Springer-Verlag, pp. 98–113.

Burnett, J. H. (1976). *Fundamentals of Mycology*, 2nd edn. London: Edward Arnold.

Burnett, J. H. & Boulter, M. E. (1963). The mating systems of fungi. II. Mating systems of the Gasteromycetes *Mycocalia denudata* and *M. duriaeana*. *New Phytologist*, **62**, 217–236.

Burney, D. A., Robinson, G. S. & Pigott Burney, L. (2003). *Sporormiella* and the late Holocene extinctions in Madagascar. *Proceedings of the National Academy of Sciences of the USA*, **100**, 10 800–10 8005.

Burr, A. W. & Beakes, G. W. (1994). Characterization of zoospore and cyst surface structure in saprophytic and fish pathogenic *Saprolegnia* species (oomycete fungal protists). *Protoplasma*, **181**, 142–163.

Burt, A., Carter, D. A., Koenig, G. L., White, T. J. & Taylor, J. W. (1996). Molecular markers reveal cryptic sex in the human pathogen *Coccidioides immitis*. *Proceedings of the National Academy of Sciences of the USA*, **93**, 770–773.

Buscot, F. (1989). Field observations on growth and development of *Morchella rotunda* and *Mitrophora semi-libera* in relation to forest soil temperature. *Canadian Journal of Botany*, **67**, 589–593.

Buscot, F. (1994). Ectomycorrhizal types and endo-bacteria associated with ectomycorrhizas of *Morchella elata* (Fr.) Boudier with *Picea abies* (L.) Karst. *Mycorrhiza*, **4**, 223–232.

Buscot, F. & Kottke, I. (1990). Histocytological study of the association between *Morchella esculenta* and the root system of *Picea abies*. *New Phytologist*, **116**, 425–430.

Buscot, F. & Roux, J. (1987). Association between living roots and ascocarps of *Morchella rotunda* (Pers.) Boudier. *Transactions of the British Mycological Society*, **89**, 249–252.

Bushnell, W. R. & Gay, J. (1978). Accumulation of solutes in relation to the structure and function of haustoria in powdery mildews. In *The Powdery Mildews*, ed. D. M. Spencer. London, New York and San Francisco: Academic Press, pp. 183–235.

Bushnell, W. R. & Roelfs, A. P., eds. (1984). *The Cereal Rusts* **1**: *Origins, Specificity, Structure, and Physiology*. Orlando: Academic Press.

Butler, E. J. & Jones, S. G. (1949). *Plant Pathology*. London: Macmillan.

Butler, G. M. (1966). Vegetative structure. In *The Fungi. An Advanced Treatise* **II**: *The Fungal Organism*. New York and London: Academic Press, pp. 83–112.

Butler, M. J. & Day, A. W. (1998). Fungal melanins: a review. *Canadian Journal of Microbiology*, **44**, 1115–1136.

Butt, T. M., Beckett, A. & Wilding, N. (1981). Protoplasts of the *in vivo* life cycle of *Erynia neoaphidis*. *Journal of General Microbiology*, **127**, 417–421.

Butt, T. M., Beckett, A. & Wilding, N. (1990). A histological study of the invasive and developmental process of the aphid pathogen *Erynia neoaphidis* (Zygomycotina: Entomophthorales) in the pea aphid *Acyrthosiphon pisum*. *Canadian Journal of Botany*, **68**, 2153–2163.

Butterfield, N. J. (2005). Probable Proterozoic fungi. *Paleobiology*, **31**, 165–182.

Byrde, R. J. W. & Willetts, H. J. (1977). *The Brown Rot Fungi of Fruit. Their Biology and Control*. Oxford: Pergamon Press.

Cai, L., Jeewon, R. & Hyde, K. D. (2005). Phylogenetic evaluation and taxonomic revision of *Schizothecium* based on ribosomal DNA and protein coding genes. *Fungal Diversity*, **19**, 1–21.

Cairney, J. W. G. (1991). Rhizomorphs: organs of exploration or exploitation? *Mycologist*, **5**, 5–10.

Cairney, J. W. G. (1992). Translocation of solutes in ectomycorrhizal and saprotrophic rhizomorphs. *Mycological Research*, **96**, 135–141.

Cairney, J. W. G. (2000). Evolution of mycorrhiza systems. *Naturwissenschaften*, **76**, 467–475.

Cairney, J. W. G. (2002). *Pisolithus* – death of the pan-global super fungus. *New Phytologist*, **153**, 199–201.

Cairney, J. W. G. & Chambers, S. M. (1997). Interactions between *Pisolithus tinctorius* and its hosts: a review of current knowledge. *Mycorrhiza*, **7**, 117–131.

Çakar, Z. P., Sauer, U., Bailey, J. E., *et al.* (2000). Vacuolar morphology and cell cycle distribution are modified by leucine limitation in auxotrophic *Saccharomyces cerevisiae*. *Biologie Cellulaire*, **92**, 629–637.

Calderoni, M., Sieber, T. N. & Holdenrieder, O. (2003). *Stereum sanguinolentum*: is it an amphithallic basidiomycete? *Mycologia*, **95**, 232–238.

Calduch, M., Gené, J., Cano, J., Stchigel, A. M. & Guarro, J. (2004). Three new species of *Oidiodendron* Robak from Spain. *Studies in Mycology*, **50**, 159–170.

Callac, P. (1995). Breeding of edible fungi with emphasis among the French genetic resources of *Agaricus bisporus*. *Canadian Journal of Botany*, **73**, S980–S986.

Callac, P., Jacobé de Haut, I., Imbernon, M., *et al.* (2003). A novel homothallic variety of *Agaricus bisporus* comprises rare tetrasporic isolates from Europe. *Mycologia*, **95**, 222–231.

Callaghan, A.A. (1969a). Light and spore discharge in Entomophthorales. *Transactions of the British Mycological Society*, **53**, 87–97.

Callaghan, A.A. (1969b). Morphogenesis in *Basidiobolus ranarum*. *Transactions of the British Mycological Society*, **53**, 99–108.

Callaghan, A.A. (2004). Teaching techniques for mycology: 22. Disclosure of *Conidiobolus* and *Basidiobolus* from soils and leaf litters. *Mycologist*, **18**, 29–32.

Callan, B.E. (1993). Rhizina root rot of conifers. *Forest Pest Leaflet*, **56**. Victoria, B.C.: Pacific Forestry Centre.

Callot, G. (1999). *La Truffe, la Terre, et la Vie*. Paris: INRA.

Calvo-Mendez, C. & Ruiz-Herrera, J. (1987). Biosynthesis of chitosan in membrane fractions from *Mucor rouxii* by the concerted action of chitin synthetase and a particulate deacetylase. *Experimental Mycology*, **11**, 128–140.

Câmara, M.P.S., Palm, M.E., van Berkum, P. & O'Neill, N.R. (2002). Molecular phylogeny of *Leptosphaeria* and *Phaeosphaeria*. *Mycologia*, **94**, 630–640.

Campbell, A.H. (1933). Zone lines in plant tissues I. The black lines formed by *Xylaria polymorpha* (Pers.) Grey in hardwoods. *Annals of Applied Biology*, **20**, 123–145.

Campbell, C.L. & Long, D.L. (2001). The campaign to eradicate the common barberry in the United States. In *Stem Rust of Wheat. From Ancient Enemy to Modern Foe*, ed. P.D. Peterson. St. Paul, USA: APS Press, pp. 16–50.

Campbell, G.F. & Crous, P.W. (2003). Genetic stability of net × spot hybrid progeny of the barley pathogen *Pyrenophora teres*. *Australasian Plant Pathology*, **32**, 283–287.

Campbell, R. (1968). An electron microscope study of spore structure and development in *Alternaria brassicicola*. *Journal of General Microbiology*, **54**, 381–392.

Campbell, R.N. (1985). Longevity of *Olpidium brassicae* in air-dry soil and the persistence of the lettuce big-vein agent. *Canadian Journal of Botany*, **63**, 2288–2289.

Campbell, R.N. (1996). Fungal transmission of plant viruses. *Annual Review of Phytopathology*, **34**, 87–108.

Campbell, R.N. & Fry, P.R. (1966). The nature of the associations between *Olpidium brassicae* and lettuce big vein and tobacco necrosis viruses. *Virology*, **29**, 222–233.

Canham, S.C. (1969). Taxonomy and morphology of *Hypocrea citrina*. *Mycologia*, **61**, 315–331.

Cannon, P.F. & Minter, D.W. (1986). The Rhytismataceae of the Indian subcontinent. *Mycological Papers*, **155**, 1–123.

Canter, H.M. (1950). Studies on British chytrids. IX. *Anisolpidium stigeoclonii* (de Wildeman) n. comb. *Transactions of the British Mycological Society*, **33**, 335–344.

Canter, H.M. & Jaworski, G.H.M. (1978). The isolation, maintenance and host range studies of the chytrid *Rhizophydium planktonicum* Canter emend., parasitic on *Asterionella formosa* Hassall. *Annals of Botany* N. S., **42**, 967–979.

Canter, H.M. & Jaworski, G.H.M. (1979). The occurrence of a hypersensitive reaction in the planktonic diatom *Asterionella formosa* Hassall parasitized by the chytrid *Rhizophydium planktonicum* Canter emend., in culture. *New Phytologist*, **82**, 187–206.

Canter, H.M. & Jaworski, G.H.M. (1981). The effect of light and darkness upon infection of *Asterionella formosa* Hassall by the chytrid *Rhizophydium planktonicum* Canter emend. *Annals of Botany* N. S., **47**, 13–30.

Canter, H.M. & Lund, J.W.G. (1948). Studies on plankton parasites. I. Fluctuations in the numbers of *Asterionella formosa* Hass., in relation to fungal epidemics. *New Phytologist*, **47**, 238–261.

Canter, H.M. & Lund, J.W.G. (1953). The parasitism of diatoms with special reference to lakes in the English Lake District. *Transactions of the British Mycological Society*, **36**, 13–37.

Canter, H.M. & Willoughby, L.G. (1964). A parasitic *Blastocladiella* from Windermere plankton. *Journal of the Royal Microscopical Society*, **83**, 365–372.

Cantino, E.C. (1955). Physiology and phylogeny in the water-molds – a re-evaluation. *Quarterly Review of Biology*, **25**, 269–277.

Cantino, E.C., Lovett, J.S., Leak, L.V. & Lythgoe, J. (1963). The single mitochondrion, fine structure, and germination of the spore of *Blastocladiella emersonii*. *Journal of General Microbiology*, **31**, 393–404.

Cantino, E.C., Truesdell, L.C. & Shaw, D.S. (1968). Life history of the motile spore of *Blastocladiella emersonii*: a study in cell differentiation. *Journal of the Elisha Mitchell Scientific Society*, **84**, 125–146.

Cao, H., Li, X. & Dong, X. (1998). Generation of broad-spectrum disease resistance by overexpression of an essential regulatory gene in systemic acquired resistance. *Proceedings of the National Academy of Sciences of the USA*, **95**, 6531–6536.

Carbone, I. & Kohn, L. M. (1993). Ribosomal DNA sequence divergence within internal transcribed spacer 1 of the Sclerotiniaceae. *Mycologia*, **85**, 415–427.

Carisse, O., Philion, V., Rolland, D. & Bernier, J. (2000). Effect of fall application of fungal antagonists on spring ascospore production of the apple scap pathogen, *Venturia inaequalis*. *Phytopathology*, **90**, 31–37.

Carlile, M. J. (1971). Myxomycetes and other slime moulds. In *Methods in Microbiology*, ed. C. Booth. New York: Academic Press, pp. 237–265.

Carlile, M. J. (1983). Motility, taxis, and tropism in *Phytophthora*. In *Phytophthora. Its Biology, Taxonomy, Ecology, and Pathology*, ed. D. C. Erwin, S. Bartnicki-Garcia & P. H. Tsao. St. Paul: APS Press, pp. 95–107.

Carlile, M. J. (1995). The success of the hypha and mycelium. In *The Growing Fungus*, ed. N. A. R. Gow & G. M. Gadd. London: Chapman & Hall, pp. 3–19.

Carlile, M. J. (1996a). The discovery of fungal sex hormones: I. Sirenin. *Mycologist*, **10**, 3–6.

Carlile, M. J. (1996b). The discovery of fungal sex hormones: II. Antheridiol. *Mycologist*, **10**, 113–117.

Carlile, M. J. & Dee, J. (1967). Plasmodial and lethal interaction between strains in a myxomycete. *Nature*, **215**, 832–834.

Caron, C. (1995). Commercial production of baker's yeast and wine yeast. In *Biotechnology* **9**: *Enzymes, Biomass, Food and Feed*, ed. G. Reed & T. W. Nagodawithana. Weinheim, Germany: VCH, pp. 321–351.

Carpenter, S. E. & Trappe, J. M. (1985). Phoenicoid fungi: a proposed term for fungi that fruit after heat treatment of substrates. *Mycotaxon*, **23**, 203–206.

Carreiro, M. M. & Koske, R. E. (1992). Room temperature isolations can bias against selection of low temperature microfungi in temperate forest soils. *Mycologia*, **84**, 886–900.

Carroll, F. E. & Carroll, G. C. (1971). Fine structural studies on 'poroconidium' formation in *Stemphylium botryosum*. In *Taxonomy of Fungi Imperfecti*, ed. W. B. Kendrick. Toronto: Toronto University Press, pp. 75–91.

Carruthers, R. I., Haynes, D. L. & MacLeod, D. M. (1985). *Entomophthora muscae* (Entomophthorales: Entomophthoraceae) mycosis in the onion fly *Delia antiqua* (Diptera: Anthomyiidae). *Journal of Invertebrate Pathology*, **45**, 81–93.

Carver, T. L. W. & Bushnell, W. R. (1983). The probable role of primary germ tubes in water uptake before infection by *Erysiphe graminis*. *Physiological and Molecular Plant Pathology*, **31**, 237–250.

Carver, T. L. W. & Ingerson, S. M. (1987). Responses of *Erysiphe graminis* germlings to contact with artificial and host surfaces. *Physiological and Molecular Plant Pathology*, **30**, 359–372.

Carver, T. L. W., Ingerson-Morris, S. M., Thomas, B. J. & Zeyen, R. J. (1995). Early interactions during powdery mildew infection. *Canadian Journal of Botany*, **73**, S632–S639.

Carver, T. L. W., Kunoh, H., Thomas, B. J. & Nicholson, R. L. (1999). Release and visualization of the extra-cellular matrix of conidia of *Blumeria graminis*. *Mycological Research*, **103**, 547–560.

Casadevall, A. & Perfect, J. R. (1998). *Cryptococcus neoformans*. Washington, DC: ASM Press.

Casselton, L. A. (2002). Mate recognition in fungi. *Heredity*, **88**, 142–147.

Casselton, L. A. & Economou, A. (1985). Dikaryon formation. In *Developmental Biology of Higher Fungi*, ed. D. Moore, L. A. Casselton, D. A. Wood, & J. C. Frankland. Cambridge: Cambridge University Press, pp. 213–229.

Casselton, L. A. & Olesnicky, N. S. (1998). Molecular genetics of mating recognition in basidiomycete fungi. *Microbiological and Molecular Biology Reviews*, **62**, 55–70.

Casselton, L. A. & Riquelme, M. (2004). Genetics of *Coprinus*. In *The Mycota* **II**: *Genetics and Biotechnology*, 2nd edn, ed. U. Kück. Berlin: Springer-Verlag, pp. 37–52.

Castle, A., Speranzini, D., Rghei, N., *et al.* (1998). Morphological and molecular identification of *Trichoderma* isolates on North American mushroom farms. *Applied and Environmental Microbiology*, **64**, 133–137.

Castle, A. J. & Day, A. W. (1984). Isolation and identification of α-tocopherol as an inducer of the parasitic phase of *Ustilago violacea*. *Phytopathology*, **74**, 1194–1200.

Castle, A. J., Stocco, N. & Boulianne, R. (1996). Fimbrial-dependent mating in *Microbotryum violaceum* involves a mannose–lectin interaction. *Canadian Journal of Microbiology*, **42**, 461–466.

Castle, E. S. (1942). Spiral growth and reversal of spiraling in *Phycomyces*, and their bearing on primary wall structure. *American Journal of Botany*, **29**, 664–672.

Castlebury, L. A. & Domier, L. L. (1998). Small subunit ribosomal RNA gene phylogeny of *Plasmodiophora brassicae*. *Mycologia*, **90**, 102–107.

Castlebury, L. A. & Glawe, D. A. (1993). A comparison of three techniques for inoculating Chinese cabbage with *Plasmodiophora brassicae*. *Mycologia*, **85**, 866–867.

Castlebury, L. A., Rossman, A. Y., Jaklitsch, W. J. & Vasilyeva, L. N. (2002). A preliminary overview of the Diaporthales based on large subunit nuclear ribosomal DNA sequences. *Mycologia*, **94**, 1017–1031.

Castro, C. D., Koretsky, A. P. & Domach, M. M. (1999). NMR-observed phosphate trafficking and polyphosphate dynamics in wild-type and *vph1–1* mutant *Saccharomyces cerevisiae* in response to stresses. *Biotechnology Progress*, **15**, 65–73.

Catanzariti, A.-M., Dodds, P. N., Lawrence, G. J., Ayliffe, M. A. & Ellis, J. G. (2006). Haustorially expressed secreted proteins from flax rust are highly enriched for avirulence elicitors. *Plant Cell*, **18**, 243–256.

Caten, C. E. & Jinks, J. E. (1966). Heterokaryosis: its significance in wild homothallic ascomycetes and Fungi Imperfecti. *Transactions of the British Mycological Society*, **49**, 81–93.

Cavalier-Smith, T. (1981). Eukaryote kingdoms: seven or nine? *BioSystems*, **14**, 461–481.

Cavalier-Smith, T. (1986). The kingdom Chromista: origin and systematics. In *Progress in Phycological Research*, ed. F. E. Round & D. J. Chapman. Bristol: Biopress, pp. 309–47.

Cavalier-Smith, T. (1987). The origin of fungi and pseudofungi. In *Evolutionary Biology of Fungi*, ed. A. D. M. Rayner, C. M. Brasier & D. Moore. Cambridge: Cambridge University Press, pp. 339–53.

Cavalier-Smith, T. (1998). A revised six-kingdom system for life. *Biological Reviews*, **73**, 203–266.

Cavalier-Smith, T. (2001). What are fungi? In *The Mycota VIIA: Systematics and Evolution*, ed. D. J. McLaughlin, E. G. McLaughlin & P. A. Lemke. Berlin: Springer-Verlag, pp. 3–37.

Cavender, J. C. (1990). Phylum Dictyostelida. In *Handbook of Protoctista*, ed. L. Margulis, J. O. Corliss, M. Melkonian & D. J. Chapman. Boston: Jones & Bartlett, pp. 88–101.

Ceccato-Antonini, S. R. & Sudbery, P. E. (2004). Filamentous growth in *Saccharomyces cerevisiae*. *Brazilian Journal of Microbiology*, **35**, 173–181.

Celerin, M. & Day, A. W. (1998). Sex, smut, and RNA: the complexity of fungal fimbriae. *International Journal of Plant Sciences*, **159**, 175–184.

Celerin, M., Laudenbach, D. E., Bancroft, J. B. & Day, A. W. (1994). Evidence that fimbriae of the smut fungus *Microbotryum violaceum* contain RNA. *Microbiology*, **140**, 2699–2704.

Celerin, M., Ray, J. M., Schisler, N. J., *et al.* (1996). Fungal fimbriae are composed of collagen. *EMBO Journal*, **15**, 4445–4453.

Cerdá-Olmedo, E. (1975). The genetics of *Phycomyces blakesleeanus*. *Genetical Research*, **25**, 285–296.

Cerdá-Olmedo, E. (2001). *Phycomyces* and the biology of light and color. *FEMS Microbiology Reviews*, **25**, 503–512.

Cerdá-Olmedo, E. & Lipson, E. D., eds. (1987). *Phycomyces*. New York: Cold Spring Harbor Laboratory Press.

Cerenius, L., Söderhäll, K., Persson, M. & Ajaxon, R. (1988). The crayfish plague fungus, *Aphanomyces astaci* – diagnosis, isolation and pathology. *Freshwater Crayfish*, **7**, 131–144.

Certik, M. & Shimizu, S. (1999). Biosynthesis and regulation of microbial polyunsaturated fatty acid production. *Journal of Bioscience and Bioengineering*, **87**, 1–14.

Cessna, S. G., Sears, V. E., Dickman, M. B. & Low, P. S. (2000). Oxalic acid, a pathogenicity factor for *Sclerotinia sclerotiorum*, suppresses the oxidative burst of the host plant. *Plant Cell*, **12**, 2192–2199.

Chadefaud, M. (1942). Études d'asques. II. Structure et anatomie comparée de l'appareil apical des asques chez divers Discomycètes et Pyrenomycètes. *Revue de Mycologie*, **7**, 57–88.

Chalutz, E. & Wilson, C. L. (1990). Postharvest biocontrol of green and blue mold and sour rot of citrus fruit by *Debaryomyces hansenii*. *Plant Disease*, **74**, 134–137.

Chambers, S. M. & Cairney, J. W. G. (1999). *Pisolithus*. In *Ectomycorrhizal Fungi: Key Genera in Profile*, ed. J. W. G. Cairney & S. M. Chambers. Berlin: Springer-Verlag, pp. 1–31.

Chambers, T. C. & Willoughby, L. G. (1964). The fine structure of *Rhizophlyctis rosea*, a soil Phycomycete. *Journal of the Royal Microscopical Society*, **83**, 355–364.

Chambers, T. C., Markus, K. & Willoughby, L. G. (1967). The fine structure of the mature zoosporangium of *Nowakowskiella profusa*. *Journal of General Microbiology*, **46**, 135–141.

Chamier, A. C. (1985). Cell-wall degrading enzymes of aquatic hyphomycetes: a review. *Botanical Journal of the Linnean Society*, **91**, 67–81.

Champavier, Y., Pommier, M.-T., Aprin, N., Voiland, A. & Pellon, G. (2000). 10-oxo-*trans*-8-decenoic acid (ODA); production, biological activities, and comparison of other hormone-like substances in *Agaricus bisporus*. *Enzyme and Microbial Technology*, **26**, 243–251.

Chang, C. W., Yang, H. C. & Leu, L. S. (1984). Zygospore formation of *Choanephora cucurbitarum*. *Transactions of the Mycological Society of Japan*, **25**, 67–74.

Chang, S. T. & Miles, P. G. (2004). *Mushrooms: Cultivation, Nutritional Value, Medicinal Effect and Environmental Impact*, 2nd edn. Boca Raton: CRC Press.

Chang, S. T. & Yau, C. K. (1971). *Volvariella volvacea* and its life history. *American Journal of Botany*, **58**, 552–561.

Chapela, I. H. (1989). Fungi in healthy stems and branches of American beech and aspen: a comparative study. *New Phytologist*, **113**, 65–75.

Chapela, I. H. & Boddy, L. (1988a). Fungal colonization of attached beech branches. I. Early stages of development of fungal communities. *New Phytologist*, **110**, 39–45.

Chapela, I. H. & Boddy, L. (1988b). Fungal colonization of attached beech branches. II. Spatial and temporal organization of communities arising from latent invaders in bark and functional sapwood under different moisture regimes. *New Phytologist*, **110**, 47–57.

Chapela, I. H., Petrini, O. & Petrini, L. E. (1990). Unusual ascospore germination in *Hypoxylon fragiforme*: first steps in the establishment of an endophytic symbiosis. *Canadian Journal of Botany*, **68**, 2571–2575.

Chapela, I. H., Petrini, O. & Hagmann, L. (1991). Monolignol glucosides as specific recognition messengers in fungal/plant symbioses. *Physiological and Molecular Plant Pathology*, **39**, 289–298.

Chapela, I. H., Petrini, O. & Bielser, G. (1993). The physiology of ascospore eclosion in *Hypoxylon fragiforme*: mechanisms in the early recognition and establishment of an endophytic symbiosis. *Mycological Research*, **97**, 157–162.

Charnley, A. K. & St. Leger, R. J. (1991). The role of cuticle-degrading enzymes in fungal pathogenesis in insects. In *The Fungal Spore and Disease Initiation in Plants and Animals*, ed. G. T. Cole & H. C. Hoch. New York: Plenum Press, pp. 267–286.

Chaubal, R., Wilmot, V. A. & Wynn, W. K. (1991). Visualization, adhesiveness, and cytochemistry of the extracellular matrix produced by germ tubes of *Puccinia sorghi*. *Canadian Journal of Botany*, **69**, 2044–2054.

Chaverri, P. & Samuels, G. J. (2003). *Hypocrea/Trichoderma* (*Ascomycota, Hypocreales, Hypocreaceae*): species with green ascospores. *Studies in Mycology*, **48**, 1–116.

Chaverri, P., Samuels, G. J. & Stewart, E. L. (2001). *Hypocrea virens*, the teleomorph of *Trichoderma virens*. *Mycologia*, **93**, 1113–1124.

Chazalet, V., Debeaupuis, J.-P., Sarfati, J., *et al.* (1998). Molecular typing of environmental and patient isolates of *Aspergillus fumigatus* from various hospital settings. *Journal of Clinical Microbiology*, **36**, 1494–1500.

Chelkowski, J. & Visconti, A., eds. (1992). *Alternaria: Biology, Plant Diseases and Metabolites*. Amsterdam: Elsevier.

Chen, B., Gao, S., Choi, G. H. & Nuss, D. L. (1996). Extensive alteration of fungal gene transcript accumulation and elevation of G-protein-regulated cAMP levels by a virulence-attenuating hypovirus. *Proceedings of the National Academy of Sciences of the USA*, **93**, 7996–8000.

Chen, C.-J. (1998). Morphological and molecular studies in the genus *Tremella*. *Bibliotheca Mycologica*, **174**, 1–225.

Chen, T.-H., Hsu, C.-S., Tsai, P.-J., Ho, Y.-F. & Lin, N.-S. (2001). Heterotrimeric G-protein and signal transduction in the nematode-trapping fungus *Arthrobotrys dactyloides*. *Planta*, **212**, 858–863.

Chen, X. M. (2005). Epidemiology and control of stripe rust (*Puccinia striiformis* f. sp. *tritici*) on wheat. *Canadian Journal of Plant Pathology*, **27**, 314–337.

Cheng, S. & Tu, C. C. (1978). *Auricularia* spp. In *The Biology and Cultivation of Edible Mushrooms*, ed. S. T. Chang & W. A. Hayes. New York: Academic Press, pp. 605–625.

Cherniak, R. & Sundstrom, J. B. (1994). Polysaccharide antigens of the capsule of *Cryptococcus neoformans*. *Infection and Immunity*, **62**, 1507–1512.

Cherry, J. M., Ball, C., Weng, S., *et al.* (1997). Genetic and physical maps of *Saccharomyces cerevisiae*. *Nature*, **387**, 67–73.

Chesters, C. G. C. & Greenhalgh, G. N. (1964). *Geniculosporium serpens* gen. et sp. nov., the imperfect state of *Hypoxylon serpens*. *Transactions of the British Mycological Society*, **47**, 393–401.

Chet, I. (1987). Trichoderma – application, mode of action, and potential as a biocontrol agent of soilborne plant pathogenic fungi. In *Innovative Approaches to Plant Disease Control*, ed. I. Chet. New York: John Wiley & Sons, pp. 137–160.

Chet, I., Henis, Y. & Kislev, N. (1969). Ultrastructure of sclerotia and hyphae of *Sclerotium rolfsii* Sacc. *Journal of General Microbiology*, **57**, 143–147.

Chilton, W. S. (1994). The chemistry and mode of action of mushroom toxins. In *Handbook of Mushroom Poisoning. Diagnosis and Treatment*, 2nd edn, ed. D. G. Spoerke & B. H. Rumack. Boca Raton: CRC Press, pp. 165–231.

Chilvers, G. A., Lapeyrie, F. F. & Douglass, P. A. (1985). A contrast between Oomycetes and other taxa of mycelial fungi in regard to the metachromatic granule formation. *New Phytologist*, **99**, 203–210.

Chiu, S. W. (1993). Evidence for a haploid life-cycle in *Volvaria volvacea* from microspectrophotometric measurements and observations of nuclear behaviour. *Mycological Research*, **97**, 1481–1485.

Chiu, S.W. & Moore, D. (1990a). A mechanism for gill pattern formation in *Coprinus cinereus*. *Mycological Research*, **94**, 320–326.

Chiu, S.W. & Moore, D. (1990b). Development of the basidiome of *Volvariella bombycina*. *Mycological Research*, **94**, 327–337.

Chiu, S.W., Moore, D. & Chang, S.T. (1989). Basidiome polymorphism in *Volvariella bombycina*. *Mycological Research*, **92**, 69–77.

Chong, J., Leonard, K.J. & Salmeron, J.J. (2000). A North American system of nomenclature for *Puccinia coronata* f. sp. *avenae*. *Plant Disease*, **84**, 580–585.

Chou, H.H. & Wu, W.S. (2002). Phylogenetic analysis of internal transcribed spacer regions of the genus *Alternaria*, and the significance of filament-beaked conidia. *Mycological Research*, **106**, 164–169.

Christensen, M.J. (1996). Antifungal activity in grasses infected with *Acremonium* and *Epichloe* endophytes. *Australasian Plant Pathology*, **25**, 186–191.

Christensen, M.J., Bennett, R.J. & Schmid, J. (2001). Vascular bundle colonization by *Neotyphodium* endophytes in natural and novel associations with grasses. *Mycological Research*, **105**, 1239–1245.

Christensen, M.J., Bennett, R.J. & Schmid, J. (2002). Growth of *Epichloe/Neotyphodium* and p-endophytes in leaves of *Lolium* and *Festuca* grasses. *Mycological Research*, **106**, 93–106.

Christiansen, S.K., Wirsel, S., Yun, S.-H., Yoder, O.C. & Turgeon, B.G. (1998). The two *Cochliobolus* mating type genes are conserved among species but one of them is missing in *C. victoriae*. *Mycological Research*, **102**, 919–929.

Chung, K.-R. & Schardl, C.L. (1997a). Sexual cycle and horizontal transmission of the grass symbiont, *Epichloe typhina*. *Mycological Research*, **101**, 295–301.

Chung, K.-R. & Schardl, C.L. (1997b). Vegetative compatibility between and within *Epichloe* species. *Mycologia*, **89**, 558–565.

Cisar, C.R. & TeBeest, D.O. (1999). Mating system of the filamentous ascomycete, *Glomerella cingulata*. *Current Genetics*, **35**, 127–133.

Clark, J., Haskins, E.F. & Stephenson, S.L. (2004). Culture and reproductive systems of 11 species of Mycetozoans. *Mycologia*, **96**, 36–40.

Clark, J.S.C. & Spencer-Phillips, P.T.N. (1993). Accumulation of photoassimilate by *Peronospora viciae* (Berk.) Casp. and leaves of *Pisum sativum* L.: evidence for nutrient uptake via intercellular hyphae. *New Phytologist*, **124**, 107–119.

Clarke, D.D. & Akhkha, A. (2002). Population genetics of powdery mildew-natural plant pathosystems. In *The Powdery Mildews. A Comprehensive Treatise*, ed. R.R. Bélanger, W.R. Bushnell, A.J. Dik & T.L.W. Carver. St. Paul, USA: APS Press, pp. 200–218.

Claus, R., Hoppen, H.O. & Karg, H. (1981). The secret of truffles: a steroidal pheromone? *Experientia*, **37**, 1178–1179.

Clausen, M., Kräuter, R., Schachermayr, G., Potrykus, I. & Sautter, C. (2000). Antifungal activity of a virally encoded gene in transgenic wheat. *Nature Biotechnology*, **18**, 446–449.

Claxton, J.R., Potter, U.J., Blakesley, D. & Clarkson, J.M. (1996). An ultrastructural study of the interaction between *Spongospora subterranea* f. sp. *nasturtii* and watercress roots. *Mycological Research*, **100**, 1431–1439.

Clay, C.M. & Walsh, J.A. (1997). *Spongospora subterranea* f. sp. *nasturtii*, ultrastructure of the plasmodial-host interface, food vacuoles, flagellar apparatus and exit pores. *Mycological Research*, **101**, 737–744.

Clay, K. (1988). Fungal endophytes of grasses: a defensive mutualism between plants and fungi. *Ecology*, **69**, 10–16.

Clay, K. & Schardl, C. (2002). Evolutionary origins and ecological consequences of endophyte symbiosis with grasses. *American Naturalist*, **160**, S99–S127.

Clémençon, H. (2004). *Cytology and Plectology of the Hymenomycetes*. Berlin and Stuttgart: J. Cramer.

Clement, J.A., Porter, R., Butt, T.M. & Beckett, A. (1997). Characteristics of adhesion pads formed during imbibition and germination of urediniospores of *Uromyces viciae-fabae* on host and synthetic surfaces. *Mycological Research*, **101**, 1445–1458.

Clifford, B.C. (1985). Barley leaf rust. In *The Cereal Rusts 2*: *Diseases, Distribution, Epidemiology, and Control*, ed. A.P. Roelfs & R. Bushnell. Orlando: Academic Press, pp. 173–205.

Clutterbuck, A.J. (1995). Genetics of fungi. In *The Growing Fungus*, ed. N.A.R. Gow & G.M. Gadd. London: Chapman & Hall, pp. 239–253.

Coates, D. & Rayner, A.D.M. (1985a). Heterokaryon–homokaryon interactions in *Stereum hirsutum*. *Transactions of the British Mycological Society*, **84**, 637–645.

Coates, D. & Rayner, A.D.M. (1985b). Genetic control and variation in expression of the 'bow-tie' reactions between homokaryons of *Stereum hirsutum*. *Transactions of the British Mycological Society*, **84**, 191–205.

Coates, D., Rayner, A. D. M. & Todd, N. K. (1981). Mating behaviour, mycelial antagonism and the establishment of individuals in *Stereum hirsutum*. *Transactions of the British Mycological Society*, **76**, 41–51.

Cocchietto, M., Skert, N., Nimis, P. L. & Sava, G. (2002). A review on usnic acid, an interesting natural compound. *Naturwissenschaften*, **89**, 137–146.

Cochrane, B. J., Brown, J., Wain, R. P., Yangco, B. G. & Testrake, D. (1989). Genetic studies in the genus *Basidiobolus* I. Isozyme variation among isolates of human and natural populations. *Mycologia*, **81**, 504–513.

Coffey, M. D. (1975). Ultrastructural features of the haustorial apparatus of the white blister fungus *Albugo candida*. *Canadian Journal of Botany*, **53**, 1285–1299.

Coffey, M. D. & Gees, R. (1991). The cytology of development. In *Phytophthora infestans, the Cause of Late Blight of Potato*, ed. D. S. Ingram & P. H. Williams. London: Academic Press, pp. 31–51.

Coffey, M. D. & Wilson, U. E. (1983). An ultrastructural study of the late-blight fungus *Phytophthora infestans* and its interaction with the foliage of two potato cultivars possessing different levels of general (field) resistance. *Canadian Journal of Botany*, **61**, 2669–2685.

Cole, G. T. (1975). The thallic mode of conidiogenesis in the Fungi Imperfecti. *Canadian Journal of Botany*, **53**, 2983–3001.

Cole, G. T. (1986). Models of cell differentiation in conidial fungi. *Microbiological Reviews*, **50**, 95–132.

Cole, G. T. & Kendrick, W. B. (1969a). Conidium ontogeny in hyphomycetes. The annellophores of *Scopulariopsis brevicaulis*. *Canadian Journal of Botany*, **47**, 925–929.

Cole, G. T. & Kendrick, W. B. (1969b). Conidium ontogeny in hyphomycetes. The arthrospores of *Oidiodendron* and *Geotrichum*, and the endoarthrospores of *Sporendonema*. *Canadian Journal of Botany*, **47**, 1773–1780.

Cole, G. T. & Samson, R. A. (1979). *Patterns of Development in Conidial Fungi*. London: Pitman.

Cole, L., Orlovich, D. A. & Ashford, A. E. (1998). Structure, function, and motility of vacuoles in filamentous fungi. *Fungal Genetics and Biology*, **24**, 86–100.

Coley-Smith, J. R. (1959). Studies of the biology of *Sclerotium cepivorum* Berk. III. Host range; persistence and viability of sclerotia. *Annals of Applied Biology*, **47**, 511–518.

Coley-Smith, J. R. (1986). A comparison of flavor and odor compounds of onion, leek, garlic and *Allium fistulosum* in relation to germination of sclerotia of *Sclerotium cepivorum*. *Plant Pathology*, **35**, 370–376.

Coley-Smith, J. R. & Cooke, R. C. (1971). Survival and germination of fungal sclerotia. *Annual Review of Phytopathology*, **9**, 65–92.

Coley-Smith, J. R. & King, J. E. (1969). The production by species of *Allium* of alkyl sulphides and their effect on germination of sclerotia of *Sclerotium cepivorum* Berk. *Annals of Applied Biology*, **64**, 289–301.

Coley-Smith, J. R. & Parfitt, D. (1986). Some effects of diallyl disulphide on sclerotia of *Sclerotium cepivorum*: possible novel control method for white rot disease of onions. *Pesticide Science*, **37**, 587–594.

Coley-Smith, J. R., Verhoeff, K. & Jarvis, W. R., eds. (1980). *The Biology of Botrytis*. London: Academic Press.

Colgan, W. & Claridge, A. W. (2002). Mycorrhizal effectiveness of *Rhizopogon* spores recovered from faecal pellets of small forest-dwelling mammals. *Mycological Research*, **106**, 314–320.

Collinge, A. J. & Trinci, A. P. J. (1974). Hyphal tips of wild type and spreading colonial mutants of *Neurospora crassa*. *Archives of Microbiology*, **99**, 353–368.

Collinge, D. B., Gregersen, P. L. & Thordal-Christensen, H. (2002). The nature and role of defense response genes in cereals. In *The Powdery Mildews. A Comprehensive Treatise*, ed. R. R. Bélanger, W. R. Bushnell, A. J. Dik & T. L. W. Carver. St. Paul, USA: APS Press, pp. 146–160.

Collins, N. C., Sadanandom, A. & Schulze-Lefert, P. (2002). Genes and molecular mechanisms controlling powdery mildew resistance in barley. In *The Powdery Mildews. A Comprehensive Treatise*, ed. R. R. Bélanger, W. R. Bushnell, A. J. Dik & T. L. W. Carver. St. Paul, USA: APS Press, pp. 134–145.

Colot, H. V., Loros, J. J. & Dunlap, J. C. (2005). Temperature-modulated alternative splicing and promoter use in the circadian clock gene *frequency*. *Molecular Biology of the Cell*, **16**, 5563–5571.

Comménil, P., Belingheri, L. & Dehorter, B. (1998). Antilipase antibodies prevent infection of tomato leaves by *Botrytis cinerea*. *Physiological and Molecular Plant Pathology*, **52**, 1–14.

Connick, W. J. & French, R. C. (1991). Volatiles emitted during the sexual stage of the Canada thistle rust fungus and by thistle flowers. *Journal of Agricultural and Food Chemistry*, **39**, 185–188.

Cook, R. T. A., Inman, A. J. & Billings, C. (1997). Identification and classification of powdery mildew anamorphs using light and scanning electron microscopy and host data. *Mycological Research*, **101**, 975–1002.

Cooke, D. E. L., Drenth, A., Duncan, J. M., Wagels, G. & Brasier, C. M. (2000). A molecular phylogeny of *Phytophthora* and related oomycetes. *Fungal Genetics and Biology*, **30**, 17–32.

Cooke, L. R. & Little, G. (2002). The effect of foliar application of phosphonate formulations on the susceptibility of potato tubers to late blight. *Pest Management Science*, **58**, 17–25.

Cooke, R. C. (1964). Ecological characteristics of nematode-trapping hyphomycetes. II. Germination of conidia in soil. *Annals of Applied Biology*, **54**, 375–379.

Cooke, R. C. & Godfrey, B. E. S. (1964). A key to the nematode-destroying fungi. *Transactions of the British Mycological Society*, **47**, 61–74.

Cooke, R. C. & Mitchell, D. T. (1970). Carbohydrate physiology of sclerotia of *Claviceps purpurea* during dormancy and germination. *Transactions of the British Mycological Society*, **54**, 93–99.

Cooke, R. C. & Rayner, A. D. M. (1984). *Ecology of Saprotrophic Fungi.* London: Longman.

Cooney, D. G. & Emerson, R. (1964). *Thermophilic Fungi, an Account of their Biology, Activities and Classification.* San Francisco: Freeman.

Cooney, E. W., Barr, D. J. S. & Barstow, W. E. (1985). The ultrastructure of the zoospore of *Hyphochytrium catenoides*. *Canadian Journal of Botany*, **63**, 497–505.

Coppin, E., Debuchy, R., Arnaise, S. & Pickard, M. (1997). Mating types and sexual development in filamentous ascomycetes. *Microbiology and Molecular Biology Reviews*, **61**, 411–428.

Coremans-Pelseneer, J. (1973). Isolation of *Basidiobolus* meristospores from natural sources. *Mycopathologia et Mycologia Applicata*, **49**, 173–176.

Corlett, M. (1966). Developmental studies in the Microascaceae. *Canadian Journal of Botany*, **44**, 79–88.

Corlett, M. (1973). Observations and comments on the *Pleospora* centrum type. *Nova Hedwigia*, **24**, 347–360.

Corlett, M. (1991). An annotated list of the published names in *Mycosphaerella* and *Sphaerella*. *Mycologia Memoir*, **18**, 1–328.

Cormack, B. P., Ghori, N. & Falkow, S. (1999). An adhesin of the yeast pathogen *Candida glabrata* mediating adherence to human epithelial cells. *Science*, **285**, 578–582.

Cornelius, G. & Nakashima, H. (1987). Vacuoles play a decisive role in calcium homeostasis in *Neurospora crassa*. *Journal of General Microbiology*, **133**, 2341–2347.

Corner, E. J. H. (1929). Studies in the morphology of discomycetes. II. The structure and development of the ascocarp. *Transactions of the British Mycological Society*, **14**, 275–291.

Corner, E. J. H. (1932a). The fruit body of *Polystictus xanthopus*. *Annals of Botany*, **46**, 71–111.

Corner, E. J. H. (1932b). A *Fomes* with two systems of hyphae. *Transactions of the British Mycological Society*, **17**, 51–81.

Corner, E. J. H. (1948). Studies in the basidium. I. The ampoule effect, with a note on nomenclature. *New Phytologist*, **47**, 22–51.

Corner, E. J. H. (1950). *A Monograph of Clavaria and Allied Genera.* London: Oxford University Press.

Corner, E. J. H. (1953). The construction of polypores. 1. Introduction to *Polyporus sulphureus, P. squamosus, P. betulinus* and *Polystictus microcyclus*. *Phytomorphology*, **3**, 152–169.

Corrêa, A., Staples, R. C. & Hoch, H. C. (1996). Inhibition of thigmostimulated cell differentiation with RGD-peptides in *Uromyces* germlings. *Protoplasma*, **194**, 91–102.

Cortés, J. C. G., Ishiguro, J., Durán, A. & Ribas, C. (2002). Localization of the (1,3)β-D-glucan synthase catalytic subunit homologue Bgs1p/Cps1p from fission yeast suggests that it is involved in septation, polarized growth, mating, spore wall formation and spore germination. *Journal of Cell Science*, **115**, 4081–4096.

Cortesi, P., Gadoury, D. M., Seem, R. C. & Pearson, R. C. (1995). Distribution and retention of cleistothecia of *Uncinula necator* on the bark of grapevines. *Plant Disease*, **79**, 15–19.

Cortesi, P., McCulloch, C. E., Song, H., Lin, H. & Milgroom, M. G. (2001). Genetic control of horizontal virus transmission in the chestnut blight fungus, *Cryphonectria parasitica*. *Genetics*, **159**, 107–118.

Costantin, J. (1936). La culture de la morille et sa forme conidienne. *Annales des Sciences Naturelles Botaniques. Série 10*, **18**, 111–129.

Cotter, D. A., Sands, T. W., Virdy, K. J., *et al.* (1992). Patterning of development in *Dictyostelium discoideum*: factors regulating growth, differentiation, spore dormancy, and germination. *Biochemistry and Cell Biology*, **70**, 892–919.

Cotty, P. J., Bayman, P., Egel, D. S. & Elias, K. S. (1994). Agriculture, aflatoxins and *Aspergillus*. In *The Genus Aspergillus. From Taxonomy and Genetics to Industrial Application*, ed. K. A. Powell, A. Renwick & J. F. Peberdy. New York: Plenum Press, pp. 1–27.

Couch, B. C., Fudal, I., Lebrun, M.-H., *et al.* (2005). Origins of host-specific populations of the blast pathogen *Magnaporthe oryzae* in crop domestication with subsequent expansion of pandemic clones on rice and weeds of rice. *Genetics*, **170**, 613–630.

Couch, J. N. (1939). Heterothallism in the Chytridiales. *Journal of the Elisha Mitchell Scientific Society*, **55**, 409–414.

Couch, J. N. (1945). Observations on the fungus *Catenaria*. *Mycologia*, **37**, 163–193.

Couch, J. N. & Bland, C. E., eds. (1985). *The Genus Coelomomyces.* Orlando: Academic Press.

Coustou, V., Deleu, C., Saupe, S. & Begueret, J. (1997). The protein product of the *het-s* heterokaryon incompatibility gene of the fungus *Podospora anserina* behaves as a prion analog. *Proceedings of the National Academy of Sciences of the USA*, **94**, 9773–9778.

Coustou-Linares, V., Maddelein, M.-L., Bégueret, J. & Saupe, S. (2001). *In vivo* aggregation of the HET-s prion protein of the fungus *Podospora anserina*. *Molecular Microbiology*, **42**, 1325–1335.

Coutinho, T. A., Rijkenberg, F. H. J. & van Asch, M. A. J. (1993). Appressorium formation by *Hemileia vastatrix*. *Mycological Research*, **97**, 951–956.

Cowen, L. E., Anderson, J. B. & Kohn, L. M. (2002). Evolution of drug resistance in *Candida albicans*. *Annual Review of Microbiology*, **56**, 139–165.

Cox, A. E. & Large, E. C. (1960). Potato blight epidemics throughout the world. *U. S. Department of Agriculture Handbook*, 174. Washington: USDA.

Cox, R. A. & Magee, D. M. (2004). Coccidioidomycosis: host response and vaccine development. *Clinical Microbiology Reviews*, **17**, 804–839.

Craft, C. M. & Nelson, E. B. (1996). Microbial properties of composts that suppress damping-off and root rot of creeping bentgrass caused by *Pythium graminicola*. *Applied and Environmental Microbiology*, **62**, 1550–1557.

Craig, G. D., Gull, K. & Wood, D. A. (1977). Stipe elongation in *Agaricus bisporus*. *Journal of General Microbiology*, **102**, 337–347.

Craig, G. D., Newsam, R. J., Gull, K. & Wood, D. A. (1979). An ultrastructural and autoradiographic study of stipe elongation in *Agaricus bisporus*. *Protoplasma*, **98**, 15–29.

Craigie, J. H. (1927). Discovery of the function of the pycnia of the rust fungi. *Nature*, **120**, 765–767.

Craigie, J. H. & Green, G. J. (1962). Nuclear behaviour leading to conjugate association in haploid infections of *Puccinia graminis*. *Canadian Journal of Botany*, **40**, 163–178.

Cramer, C. L. & Davis, R. H. (1984). Polyphosphate–cation interaction in the amino acid-containing vacuole of *Neurospora crassa*. *Journal of Biological Chemistry*, **259**, 5152–5157.

Creppy, E. E. (2002). Update of survey, regulation and toxic effects of mycotoxins in Europe. *Toxicology Letters*, **127**, 19–28.

Crespo Erchiga, V. & Delgado Florencio, V. (2002). *Malassezia* species in skin diseases. *Current Opinion in Infectious Diseases*, **15**, 133–142.

Crittenden, P. D., David, J. C., Hawksworth, D. L. & Campbell, F. S. (1995). Attempted isolation and success in the culturing of a broad spectrum of lichen-forming and lichenicolous fungi. *New Phytologist*, **130**, 267–297.

Crous, P. W., Aptroot, A., Kang, J. C., Braun, U. & Wingfield, M. J. (2000). The genus *Mycosphaerella* and its anamorphs. *Studies in Mycology*, **45**, 107–121.

Crous, P. W., Kang, J.-C. & Braun, U. (2001). A phylogenetic redefinition of anamorph genera in *Mycosphaerella* based on ITS rDNA sequence and morphology. *Mycologia*, **93**, 1081–1101.

Crowe, F. J., Hall, D. H., Greathead, A. S. & Baghott, K. G. (1980). Inoculum density of *Sclerotium cepivorum* and the incidence of white rot of onion and garlic. *Phytopathology*, **70**, 64–69.

Croxall, H. E., Collingwood, C. A. & Jenkins, J. E. E. (1951). Observations on brown rot (*Sclerotinia fructigena*) of apple in relation to injury by earwigs (*Forficula auricularia*). *Annals of Applied Biology*, **38**, 833–843.

Crute, I. R. (1984). The integrated use of genetic and chemical methods for the control of lettuce downy mildew (*Bremia lactucae*). *Crop Protection*, **3**, 223–242.

Crute, I. R. & Dixon, G. R. (1981). Downy mildew diseases caused by the genus *Bremia*. In *The Downy Mildews*, ed. D. M. Spencer. London: Academic Press, pp. 421–460.

Cruz, M. C., del Poeta, M., Wang, P., *et al.* (2000). Immunosuppressive and nonimmunosuppressive cyclosporine analogs are toxic to the opportunistic fungal pathogen *Cryptococcus neoformans* via cyclophilin-dependent inhibition of calcineurin. *Antimicrobial Agents and Chemotherapy*, **44**, 143–149.

Cuda, J. P., Hornby, J. A., Cotterill, B. & Cattell, M. (1997). Evaluation of *Lagenidium giganteum* for biological control of *Mansonia* mosquitoes in Florida (Diptera: Culicidae). *Biological Control*, **8**, 124–130.

Cuenca-Estrella, M., Gomez-Lopez, A., Mellado, E., *et al.* (2003). *Scopulariopsis brevicaulis*, a fungal pathogen resistant to broad-spectrum antifungal agents. *Antimicrobial Agents and Chemotherapy*, **47**, 2339–2341.

Cui, J. & Chisti, Y. (2003). Polysaccharopeptides of *Coriolus versicolor*: physiological activity, uses, and production. *Biotechnology Advances*, **21**, 109–122.

Culberson, C. F. & Ahmadjian, V. (1980). Artificial reestablishment of lichens. II: Secondary products of resynthesised *Cladonia cristatella* and *Lecanora chrysoleuca*. *Mycologia*, **72**, 90–109.

Cullum, F. J. & Webster, J. (1977). Cleistocarp dehiscence in *Phyllactinia*. *Transactions of the British Mycological Society*, **68**, 316–320.

Culvenor, C. C., Beck, A. B., Clarke, M., *et al.* (1977). Isolation of toxic metabolites of *Phomopsis leptostromiformis* responsible for lupinosis. *Australian Journal of Biological Science*, **30**, 269–277.

Cummins, G. B. & Hiratsuka, Y. (2003). *Illustrated Genera of Rust Fungi*, 3rd edn. St. Paul, USA: APS Press.

Cunfer, B. M. (2000). *Stagonospora* and *Septoria* diseases of barley, oat, and rye. *Canadian Journal of Plant Pathology*, **22**, 332–348.

Cunfer, B. M. & Seckinger, A. (1977). Survival of *Claviceps purpurea* and *C. paspali* sclerotia. *Mycologia*, **69**, 1142–1148.

Cunfer, B. M. & Ueng, P. P. (1999). Taxonomy and identification of *Septoria* and *Stagonospora* species on small-grain cereals. *Annual Review of Phytopathology*, **37**, 267–284.

Currah, R. S. (1985). Taxonomy of the Onygenales: Arthrodermataceae, Gymnoascaceae, Myxotrichaceae and Onygenaceae. *Mycotaxon*, **24**, 1–216.

Currah, R. S. (1994). Peridial morphology and evolution in the prototunicate ascomycetes. In *Ascomycete Systematics. Problems and Perspectives in the Nineties*, ed. D. L. Hawksworth. New York: Academic Press, pp. 281–293.

Currah, R. S., Smreciu, E. A., Lehesvirta, T., Niemi, M. & Larsen, K. W. (2000). Fungi in the winter diets of northern flying squirrels and red squirrels in the boreal mixedwood forest of northeastern Alberta. *Canadian Journal of Botany*, **78**, 1514–1520.

Curtis, F. C., Evans, G. H., Lillis, V., Lewis, D. H. & Cooke, R. C. (1978). Studies on Mucoralean parasites. I. Some effects of *Piptocephalis* species on host growth. *New Phytologist*, **80**, 157–165.

Curtis, K. M. (1921). The life history and cytology of *Synchytrium endobioticum* (Schilb.) Perc., the cause of wart disease in potato. *Philosophical Transactions of the Royal Society*, **B 210**, 409–478.

Curtis, M. J. & Wolpert, T. J. (2002). The oat mitochondrial permeability transition and its implication in victorin binding and induced cell death. *Plant Journal*, **29**, 295–312.

Cushion, M. T. (2004). *Pneumocystis*: unraveling the cloak of obscurity. *Trends in Microbiology*, **12**, 243–249.

Czeczuga, B. (1983). Mutatoxanthin, the dominant carotenoid in lichens of the *Xanthoria* genus. *Biochemical Systematics and Ecology*, **11**, 329–331.

Czymmek, K. J., Bourett, T. M. & Howard, R. J. (1996). Immunolocalization of tubulin and actin in thick-sectioned fungal hyphae and methacrylate de-embedment. *Journal of Microscopy*, **181**, 153–161.

Daboussi, M.-J. & Capy, P. (2003). Transposable elements in filamentous fungi. *Annual Review of Microbiology*, **57**, 275–299.

Dahlberg, A. (1997). Population ecology of *Suillus variegatus* in old Swedish Scots pine forests. *Mycological Research*, **101**, 47–54.

Dahlberg, A. & Finlay, R. D. (1999). *Suillus*. In *Ectomycorrhizal Fungi. Key Genera in Profile*, ed. J. W. G. Cairney & S. M. Chambers. Berlin: Springer-Verlag, pp. 33–64.

Dahlberg, A. & Stenlid, J. (1990). Population structure and dynamics in *Suillus bovinus* as indicated by spatial distribution of fungal clones. *New Phytologist*, **115**, 487–493.

Dahlberg, A. & Stenlid, J. (1994). Size, distribution and biomass of genets in populations of *Suillus bovinus* (L.: Fr.) Roussel revealed by somatic incompatibility. *New Phytologist*, **128**, 225–234.

Dahlstrom, J. L., Smith, J. E. & Weber, N. S. (2000). Mycorrhiza-like interaction by *Morchella* species with species of the Pinaceae in pure culture synthesis. *Mycorrhiza*, **9**, 279–285.

Daly, R. & Hearn, M. T. W. (2005). Expression of heterologous proteins in *Pichia pastoris*: a useful experimental tool in protein engineering and production. *Journal of Molecular Recognition*, **18**, 119–138.

Danell, E. (1994). Formation and growth of the ectomycorrhiza of *Cantharellus cibarius*. *Mycorrhiza*, **5**, 89–97.

Daniels, A., Lucas, J. A. & Peberdy, J. F. (1991). Morphology and ultrastructure of W and R pathotypes of *Pseudocercosporella herpotrichoides* on wheat seedlings. *Mycological Research*, **95**, 385–397.

Daniels, A., Papaikonomou, M., Dyer, P. S. & Lucas, J. A. (1995). Infection of wheat seedlings by ascospores of *Tapesia yallundae*: morphology of the infection process and evidence for recombination. *Phytopathology*, **85**, 918–927.

Daniels, J. A. (1961). *Chaetomium piluliferum* sp. nov., the perfect state of *Botryotrichum piluliferum*. *Transactions of the British Mycological Society*, **44**, 79–86.

Dao, D. N., Kessin, R. H. & Ennis, H. L. (2000). Developmental cheating and the evolutionary biology of *Dictyostelium* and *Myxococcus*. *Microbiology*, **146**, 1505–1512.

Datema, R., van den Ende, H. & Wessels, J. G. H. (1977). The hyphal wall of *Mucor mucedo*. 1. Polyanionic polymers. *European Journal of Biochemistry*, **80**, 611–619.

Daub, M. E. & Ehrenshaft, M. (2000). The photoactivated *Cercospora* toxin cercosporin: contributions to plant disease and fundamental biology. *Annual Review of Phytopathology*, **38**, 461–490.

Davey, C. B. & Papavizas, G. C. (1962). Growth and sexual reproduction of *Aphanomyces euteiches* as affected by the oxidation state of sulfur. *American Journal of Botany*, **49**, 400–404.

Davey, J. (1992). Mating pheromones of the fission yeast *Schizosaccharomyces pombe*: purification and structural characterization of M-factor and isolation and analysis of two genes encoding the pheromone. *EMBO Journal*, **11**, 951–960.

Davey, J. (1998). Fusion of a fission yeast. *Yeast*, **14**, 1529–1566.

David, M., Gabriel, M. & Kopecká, M. (2003). Unusual ultrastructural characteristics of the yeast *Malassezia pachydermatis*. *Scripta Medica*, **76**, 173–186.

Davidse, L. C. & Ishii, H. (1995). Biochemical and molecular aspects of the mechanisms of action of benzimidazoles, N-phenylcarbamates and N-phenylformamidoximes and the mechanisms of resistance to these compounds in fungi. In *Modern Selective Fungicides. Properties, Applications, Mechanisms of Action*, 2nd edn, ed. H. Lyr. New York: Gustav Fischer Verlag, pp. 305–322.

Davidse, L. C., Hofman, A. E. & Velthuis, G. C. M. (1983). Specific interference of metalaxyl with endogenous RNA polymerase activity in isolated nuclei from *Phytophthora megasperma* f. sp. *medicaginis*. *Experimental Mycology*, **7**, 344–361.

Davidse, L. C., van den Berg-Velthuis, G. C. M., Mantel, B. C. & Jespers, A. B. K. (1991). Phenylamides and *Phytophthora*. In *Phytophthora*, ed. J. A. Lucas, R. C. Shattock, D. S. Shaw & L. R. Cooke. Cambridge: Cambridge University Press, pp. 349–360.

Davis, E. E. (1967). Zygospore formation in *Syzygites megalocarpus*. *Canadian Journal of Botany*, **45**, 531–532.

Davis, R. H. (1995). The genetics of *Neurospora*. In *The Mycota II: Genetics and Biotechnology*, ed. W. Kück. Berlin: Springer-Verlag, pp. 3–18.

Davoli, P., Mierau, V. & Weber, R. W. S. (2004). Carotenoids and fatty acids in red yeasts *Sporobolomyces roseus* and *Rhodotorula glutinis*. *Applied Biochemistry and Microbiology*, **40**, 460–465.

Dawe, A. L. & Nuss, D. L. (2001). Hypoviruses and chestnut blight: exploiting viruses to understand and modulate fungal pathogenesis. *Annual Review of Genetics*, **35**, 1–29.

Day, A. R., Poon, N. H. & Stewart, G. G. (1975). Fungal fimbriae. III. The effect on flocculation in *Saccharomyces*. *Canadian Journal of Microbiology*, **21**, 558–564.

Day, A. W. (1998). Nonmeiotic mechanisms of recombination in the anther smut *Microbotryum violaceum*. *International Journal of Plant Sciences*, **159**, 185–191.

Day, A. W. & Garber, E. D. (1988). *Ustilago violacea*, anther smut of the Caryophyllaceae. *Advances in Plant Pathology*, **6**, 457–482.

Day, P. R. (1974). *Genetics of Host–Parasite Interaction*. San Francisco: W. H. Freeman & Co.

Day, P. R. (1981). Fungal virus populations in corn smut from Connecticut. *Mycologia*, **73**, 379–391.

Deacon, J. W. & Donaldson, S. P. (1993). Molecular recognition in the homing responses of zoosporic fungi, with special reference to *Pythium* and *Phytophthora*. *Mycological Research*, **97**, 1153–1171.

Deacon, J. W., Laing, S. A. K. & Berry, L. A. (1990). *Pythium mycoparasiticum* sp. nov., an aggressive mycoparasite from British soils. *Mycotaxon*, **42**, 1–8.

Dean, R. A. (1997). Signal pathways and appressorium morphogenesis. *Annual Review of Phytopathology*, **35**, 211–234.

de Baets, S. & Vandamme, E. J. (2001). Extracellular *Tremella* polysaccharides: structure, properties and applications. *Biotechnology Letters*, **23**, 1361–1366.

de Bary, H. A. (1853) *Untersuchungen über die Brandpilze und die durch sie verursachten Krankheiten der Pflanzen mit Rücksicht auf das Getreide und andere Nutzpflanzen*. Berlin: Müller.

de Bary, A. (1866). *Morphologie und Physiologie der Pilze, Flechten und Myxomyceten*. Leipzig: Wilhelm Engelmann.

de Bary, A. (1887). *Comparative Morphology and Biology of the Fungi, Mycetozoa and Bacteria*, 2nd edn (English translation). Oxford: Clarendon Press.

Debuchy, R. & Coppin, E. (1992). The mating types of *Podospora anserina*: functional analysis and sequence of the fertilization domains. *Molecular and General Genetics*, **233**, 113–121.

Deckert, R. J., Melville, L. H. & Peterson, R. L. (2001). Structural features of a *Lophodermium* endophyte during the cryptic life-cycle phase in the foliage of *Pinus strobus*. *Mycological Research*, **105**, 991–997.

de Cock, A. W. A. M., Mendoza, L., Padhye, A. A., Ajello, L. & Prell, H. H. (1987). *Pythium insidiosum* sp. nov., the etiologic agent of pythiosis. *Journal of Clinical Microbiology*, **25**, 344–349.

Decottignies, A., Sanchez-Perez, I. & Nurse, P. (2003). *Schizosaccharomyces pombe* essential genes: a pilot study. *Genome Research*, **13**, 399–406.

Degawa, Y. & Tokamasu, S. (1997). Zygospore formation in *Mortierella capitata*. *Mycoscience*, **38**, 384–394.

Degawa, Y. & Tokamasu, S. (1998). Zygospore formation in *Mortierella umbellata*. *Mycological Research*, **102**, 593–598.

Degousée, N., Gupta, G. D., Lew, R. R. & Heath, I. B. (2000). A putative spectrin-containing membrane skeleton in hyphal tips of *Neurospora crassa*. *Fungal Genetics and Biology*, **30**, 33–44.

De Groot, P. W. J., Visser, J., Van Griensven, L. J. L. D. & Schaap, P. J. (1998). Biochemical and molecular aspects of growth and fruiting of the edible mushroom *Agaricus bisporus*. *Mycological Research*, **102**, 1297–1308.

de Hoog, G. S. (1974). The genera *Blastobotrys*, *Sporothrix*, *Calcarisporium* and *Calcarisporiella* gen. nov. *Studies in Mycology*, **7**, 1–84.

de Hoog, G. S. (1977). *Rhinocladiella* and allied genera. *Studies in Mycology*, **15**, 1–140.

de Hoog, G. S. & Scheffer, R. J. (1984). *Ceratocystis* versus *Ophiostoma*: a reappraisal. *Mycologia*, **76**, 292–299.

de Hoog, G. S. & Smith, M. T. (2004). Ribosomal gene phylogeny and species delimitation in *Geotrichum* and its teleomorphs. *Studies in Mycology*, **50**, 489–515.

de Hoog, G. S., Kurtzman, C. P., Phaff, H. J. & Miller, M. W. (1998). *Eremothecium* Borzi emend. Kurtzman. In *The Yeasts: A Taxonomic Study*, 4th edn, ed. C. P. Kurtzman & J. W. Fell. Elsevier: Amsterdam, pp. 201–208.

de Hoog, G. S., Guarro, J., Gené, J. & Figueras, M. J. (2000a). *Atlas of Clinical Fungi*, 2nd edn. Baarn, Netherlands: Centraalbureau voor Schimmelcultures; Reus, Spain: Universitat Rovira i Virgili.

de Hoog, G. S., Queiroz-Telles, F., Haase, G., *et al.* (2000b). Black fungi: clinical and pathogenic approaches. *Medical Mycology*, **38**, Supplement 1, 243–250.

Deising, H. B., Nicholson, R. B., Haug, M., Howard, R. J. & Mendgen, K. (1992). Adhesion pad formation and the involvement of cutinase and esterases in the attachment of uredospores to the host cuticle. *Plant Cell*, **4**, 1101–1111.

de Jong, E., Field, J. A. & de Bont, J. A. M. (1994a). Aryl alcohols in the physiology of ligninolytic fungi. *FEMS Microbiology Reviews*, **13**, 153–187.

de Jong, E., Field, J. A., Spinnler, H. E., Wijnberg, J. B. P. A. & de Bont, J. A. M. (1994b). Significant biogenesis of chlorinated aromatics by fungi in natural environments. *Applied and Environmental Microbiology*, **60**, 264–270.

de Jong, J. C., McCormack, B. J., Smirnoff, N. & Talbot, N. J. (1997). Glycerol generates turgor in rice blast. *Nature*, **389**, 244–245.

Dekhuijzen, H. M. (1980). The occurrence of free and bound cytokinins in clubroots and *Plasmodiophora brassicae* infected turnip tissue cultures. *Physiologia Plantarum*, **49**, 169–176.

de la Torre, M. J., Millan, M. C., Perez-Juan, P., Morales, J. & Ortega, J. M. (1999). Indigenous yeasts associated with two *Vitis vinifera* grape varieties cultured in southern Spain. *Microbios*, **100**, 27–40.

Dell, B., Malajczuk, N. & Dunstan, W. A. (2002). Persistence of some Australian *Pisolithus* species introduced into eucalypt plantations in China. *Forest Ecology and Management*, **169**, 271–281.

Delmas, J. (1978). *Tuber* species. In *The Biology and Cultivation of Edible Mushrooms*, ed. S. T. Chang & W. A. Hayes. New York: Academic Press, pp. 645–681.

Demain, A. L. & Elander, R. P. (1999). The β-lactam antibiotics: past, present, and future. *Antonie van Leeuwenhoek*, **75**, 5–19.

Dengis, P. B. & Rouxhet, P. G. (1997). Surface properties of top- and bottom-fermenting yeast. *Yeast*, **13**, 931–943.

Dennis, R. L. (1970). A middle Pennsylvanian basidiomycete mycelium with clamp connections. *Mycologia*, **62**, 578–584.

Dennis, R. W. G. (1960). *British Cup Fungi and Their Allies*. London: Ray Society.

Dennis, R. W. G. (1981). *British Ascomycetes*. Vaduz: J. Cramer.

de Nobel, H. & Barnett, J. A. (1991). Passage of molecules through yeast cell walls: a brief essay-review. *Yeast*, **7**, 313–323.

de Nobel, H., Sietsma, J. H., van den Ende, H. & Klis, F. M. (2001). Molecular organization and construction of the fungal cell wall. In *The Mycota* **VIII**: *Biology of the Fungal Cell*, ed. R. J. Howard & N. A. R. Gow. Berlin: Springer-Verlag, pp. 181–200.

Descals, E. & Webster, J. (1984). Branched aquatic conidia in *Erynia* and *Entomophthora sensu lato*. *Transactions of the British Mycological Society*, **83**, 669–682.

Descals, E., Webster, J., Ladle, M. & Bass, J. A. B. (1981). Variations in asexual reproduction in species of *Entomophthora* on aquatic insects. *Transactions of the British Mycological Society*, **77**, 85–102.

Descals, E., Marvanová, L. & Webster, J. (1998). New taxa and combinations in aquatic hyphomycetes. *Canadian Journal of Botany*, **76**, 1647–1659.

Deshpande, M. S., Rale, V. B. & Lynch, J. M. (1992). *Aureobasidium pullulans* in applied microbiology: a status report. *Enzyme and Microbial Technology*, **14**, 514–527.

de Silva, L., Youatt, J., Gooday, G. R. & Gow, N. A. R. (1992). Inwardly directed ionic currents of *Allomyces macrogynus* and other water moulds indicate sites of proton-driven nutrient transport but are incidental to tip growth. *Mycological Research*, **96**, 925–931.

de Vrije, T., Antoine, N., Buitelaar, R.M., *et al.* (2001). The fungal biocontrol agent *Coniothyrium minitans*: production by solid-state fermentation, application and marketing. *Applied Microbiology and Biotechnology*, **56**, 58–68.

De Wolf, E.D., Effertz, R.J., Ali, S. & Francl, L.J. (1998). Vistas of tan spot research. *Canadian Journal of Plant Pathology*, **20**, 349–370.

deZwaan, T.M., Carroll, A.M., Valent, B. & Sweigard, J.A. (1999). *Magnaporthe grisea* Pth11p is a novel plasma membrane protein that mediates appressorium differentiation in response to inductive substrate cues. *Plant Cell*, **11**, 2013–2030.

Dhaliwal, H.S. (1989). Multiplication of secondary sporidia of *Tilletia indica* on soil and wheat leaves and spikes and incidence of Karnal bunt. *Canadian Journal of Botany*, **67**, 2387–2390.

Dhingra, O.D., Mizubuti, E.S.G. & Santana, F.M. (2003). *Chaetomium globosum* for reducing primary inoculum of *Diaporthe phaseolorum* f. sp. *meridionalis* in soil-surface soybean stubble in field conditions. *Biological Control*, **26**, 302–310.

Dick, M.W. (1970). Saprolegniaceae on insect exuviae. *Transactions of the British Mycological Society*, **55**, 449–459.

Dick, M.W. (1983). Validation of the class name Hyphochytriomycetes. In *Zoosporic Plant Pathogens*, ed. S.T. Buczacki. London: Academic Press, p. 285.

Dick, M.W. (1990a). Phylum Oomycota. In *Handbook of Protoctista*, ed. L. Margulis, J.O. Corliss, M. Melkonian & D.J. Chapman. Boston: Jones & Bartlett, pp. 661–685.

Dick, M.W. (1990b). *Keys to Pythium*. Reading, UK: University of Reading.

Dick, M.W. (1995). Sexual reproduction in the Peronosporomycetes (chromistan fungi). *Canadian Journal of Botany*, **73**, S712–S724.

Dick, M.W. (1997). Fungi, flagella and phylogeny. *Mycological Research*, **101**, 385–394.

Dick, M.W. (1998). The species and systematic position of *Crypticola* in the Peronosporomycetes, and new names for the genus *Halocrusticidia* and species therein. *Mycological Research*, **102**, 1062–1066.

Dick, M.W. (2001a). *Straminipilous Fungi*. Dordrecht: Kluwer Academic Publishers.

Dick, M.W. (2001b). The Peronosporomycetes. In *The Mycota **VIIA**: Systematics and Evolution*, ed. D.J. McLaughlin, E.G. McLaughlin & P.A. Lemke. Berlin: Springer-Verlag, pp. 39–72.

Dick, M.W. (2002). Towards an understanding of the evolution of the downy mildews. In *Advances in Downy Mildew Research*, ed. P.T.N. Spencer-Phillips, U. Gisi & A. Labeda. Dordrecht: Kluwer Academic Publishers, pp. 1–57.

Dick, M.W. (2003). The Peronosporomycetes and other flagellate fungi. In *Pathogenic Fungi in Humans and Animals*, 2nd edn, ed. D.H. Howard. New York: Marcel Dekker, pp. 17–66.

Dick, M.W., Vick, M.C., Gibbings, J.G., Hedderson, T.A. & Lopez Lastra, C.C. (1999). 18S rDNA for species of *Leptolegnia* and other Peronosporomycetes: justification for the subclass taxa Saprolegniomycetidae and Peronosporomycetidae and division of the Saproleniaceae *sensu lato* into the families Leptolegniaceae and Saprolegniaceae. *Mycological Research*, **103**, 1119–1125.

Dickinson, C.H. & Greenhalgh, J.R. (1977). Host range and taxonomy of *Peronospora* on crucifers. *Transactions of the British Mycological Society*, **69**, 111–116.

Diederich, P. (1996). The lichenicolous Heterobasidiomycetes. *Bibliotheca Mycologica*, **61**, 1–198. Stuttgart: Gebr. Borntraeges.

Diéguez-Uribeondo, J., Gierz, G. & Bartnicki-García, S. (2004). Image analysis of hyphal morphogenesis in Saprolegniaceae (Oomycetes). *Fungal Genetics and Biology*, **41**, 293–307.

Dietel, P. (1928). Reihe Uredinales. In *Die Natürlichen Pflanzenfamilien 6*, 2nd edn, ed. A. Engler. Leipzig: W. Engelmann, pp. 24–98.

Dijksterhuis, J., Veenhuis, M. & Harder, W. (1990). Ultrastructural study of adhesion and initial stages of infection of nematodes by conidia of *Drechmeria coniospora*. *Mycological Research*, **94**, 1–8.

Dijksterhuis, J., Harder, W., Wyss, U. & Veenhuis, M. (1991). Colonization and digestion of nematodes by the endoparasitic nematophagous fungus *Drechmeria coniospora*. *Mycological Research*, **95**, 873–878.

Dijksterhuis, J., Harder, W. & Veenhuis, M. (1993). Proliferation and function of microbodies in the nematophagous fungus *Arthrobotrys oligospora* during growth on oleic acid or D-alanine as the sole carbon source. *FEMS Microbiology Letters*, **112**, 125–130.

Dimond, A.E. (1966). Effectiveness of antibiotics against forest tree rusts: a summary of present status. *Journal of Forestry*, **64**, 379–382.

Di Pietro, A., Madrid, M.P., Caracuel, Z., Delgado-Jarana, J. & Roncero, M.I.G. (2003). *Fusarium oxysporum*: exploring the molecular arsenal of a vascular wilt fungus. *Molecular Plant Pathology*, **4**, 315–325.

Dissing, H. (1986). The genus *Helvella* in Europe with special emphasis on the species found in Norden. *Dansk Botanisk Arkiv*, **25**, 1–172.

Dissing, H., Eckblad, F.-E. & Lange, M. (2000). Pezizales. In *Nordic Macromycetes*. **1**. *Ascomycetes*, ed. L. Hansen & H. Knudsen. Copenhagen: Nordswamp, pp. 55–127.

Dittrich, H. H. (1995). Wine and brandy. In *Biotechnology* **9**: *Enzymes, Biomass, Food and Feed*, ed. G. Reed & T. W. Nagodawithana. Weinheim, Germany: VCH, pp. 463–504.

Dix, N. J. & Webster, J. (1995). *Fungal Ecology*. London: Chapman & Hall.

Dixon, D. M. (2001). *Coccidioides immitis* as a select agent of bioterrorism. *Journal of Applied Microbiology*, **91**, 602–605.

Dixon, K. P., Xu, J.-R., Smirnoff, N. & Talbot, N. J. (1999). Independent signaling pathways regulate cellular turgor during hyperosmotic stress and appressorium-mediated plant infection by *Magnaporthe grisea*. *Plant Cell*, **11**, 2045–2058.

Dodd, J. L. & McCracken, D. A. (1972). Starch in fungi. Its molecular structure in three genera and an hypothesis concerning its physiological role. *Mycologia*, **64**, 1341–1343.

Dodge, B. O. (1937). The perithecial cavity formation in a *Leptosphaeria* on *Opuntia*. *Mycologia*, **29**, 707–716.

Doggett, M. S. & Porter, D. (1996). Sexual reproduction in the fungal parasite, *Zygorhizidium planktonicum*. *Mycologia*, **88**, 720–732.

Doguet, G. (1962). *Digitatospora marina* n.g., n.sp., basidiomycète marin. *Comptes Rendus Hebdomadaires des Séances de l'Académie des Sciences*, **D 254**, 4336–4338.

Doguet, G. (1967). *Nia vibrissa* Moore et Myers, remarquable basidiomycète marin. *Comptes Rendus Hebdomadaires des Séances de l'Académie des Sciences*, **D 265**, 1780–1783.

Doling, D. A. (1966). Loose smut in wheat and barley. *Agriculture*, **73**, 523–527.

Domnas, A., Jaronski, S. & Hanton, W. K. (1986). The zoospore and flagellar mastigonemes of *Lagenidium giganteum* (Oomycetes, Lagenidiales). *Mycologia*, **78**, 810–817.

Domsch, K. H., Gams, W. & Anderson, T.-H. (1980). *Compendium of Soil Fungi*. London: Academic Press.

Donaldson, S. P. & Deacon, J. W. (1993). Changes in motility of *Pythium* zoospores induced by calcium and calcium-modulating drugs. *Mycological Research*, **97**, 877–883.

Dong, J. O., Chen, W. & Crane, J. L. (1998). Phylogenetic studies of the Leptosphaeriaceae, Pleosporaceae and some other Loculoascomycetes based on nuclear ribosomal DNA sequences. *Mycological Research*, **102**, 151–156.

Donk, M. A. (1964). A conspectus of the families of the Aphyllophorales. *Persoonia*, **3**, 199–324.

Donofrio, N. M. & Delaney, T. P. (2001). Abnormal callose response phenotype and hypersusceptibility to *Peronospora parasitica* in defence-compromised *Arabidopsis nim1–1* and salicylate hydroxylase-expressing plants. *Molecular Plant–Microbe Interactions*, **14**, 439–450.

Doss, R. P. (1999). Composition and enzymatic activity of the extracellular matrix secreted by germlings of *Botrytis cinerea*. *Applied and Environmental Microbiology*, **65**, 404–408.

Doss, R. P., Potter, S. W., Soeldner, A. H., Christian, J. K. & Fukunaga, L. E. (1995). Adhesion of germlings of *Botrytis cinerea*. *Applied and Environmental Microbiology*, **61**, 260–265.

Doss, R. P., Potter, S. W., Christian, J. K., Soeldner, A. H. & Chastagner, G. A. (1997). The conidial surface of *Botrytis cinerea* and several other *Botrytis* species. *Canadian Journal of Botany*, **75**, 612–617.

Doster, M. A. & Fry, W. E. (1991). Evaluation by computer simulation of strategies to time metalaxyl applications for improved control of potato late blight. *Crop Protection*, **10**, 209–214.

Dowding, E. S. & Bulmer, G. S. (1964). Notes on the cytology and sexuality of puffballs. *Canadian Journal of Microbiology*, **10**, 783–789.

Dowding, P. & Richardson, D. H. S. (1990). Leafyeasts as indicators of air quality in Europe. *Environmental Pollution*, **66**, 223–235.

Dowe, A. (1987). *Räuberische Pilze*. Wittenberg: A. Ziemsen Verlag.

Dowley, L. J., Bannon, E., Cooke, L. R., Keane, T. & O'Sullivan, E., eds. (1995). *Phytophthora infestans 150*. Dublin: Boole Press.

Dowsett, J. A., Reid, J. & van Caeseele, L. (1977). Transmission and scanning electron microscope observations on the trapping of nematodes by *Dactylaria brochopaga*. *Canadian Journal of Botany*, **55**, 2945–2955.

Drayton, F. L. & Groves, J. W. (1952). *Stromatinia narcissi*, a new sexually dimorphic discomycete. *Mycologia*, **48**, 655–676.

Drechsler, C. (1956). Supplementary developmental stages of *Basidiobolus ranarum* and *Basidiobolus haptosporus*. *Mycologia*, **48**, 655–676.

Drechsler, C. (1960). Two root rot fungi closely related to *Pythium ultimum*. *Sydowia*, **14**, 107–115.

Drechsler, C. (1962). A nematode-capturing phycomycete with distally adhesive branches and proximally imbedded fusiform conidia. *American Journal of Botany*, **49**, 1089–1095.

Dreyfuss, M., Härri, E., Hofmann, H., *et al.* (1976). Cyclosporin A and C; new metabolites from *Trichoderma polysporum* (Link ex Pers.) Rifai. *European Journal of Applied Microbiology*, 3, 125–133.

Dring, D. M. (1973). Gasteromycetes. In *The Fungi, an Advanced Treatise IVB: A Taxonomic Review with Keys*, ed. G. C. Ainsworth, F. K. Sparrow & A. S. Sussman. New York: Academic Press, pp. 451–478.

Drinkard, L. C., Nelson, G. E. & Sutter, R. P. (1982). Growth arrest: a prerequisite for sexual development in *Phycomyces blakesleeanus*. *Experimental Mycology*, **6**, 52–59.

Dubourdieu, D., Pucheu-Planté, B., Mercier, M. & Ribéreau-Gayon, P. (1978a). Structure, rôle et localisation du glucane exo-cellulaire sécrété par *Botrytis cinerea* dans la baie de Raisin. *Comptes Rendus Hebdomadaires des Séances de l'Académie des Sciences*, **D 287**, 571–573.

Dubourdieu, D., Fournet, B., Bertrand, A. & Ribéreau-Gayon, P. (1978b). Identification du glucane sécrété dans la baie de Raisin par *Botrytis cinerea*. *Comptes Rendus Hebdomadaires des Séances de l'Académie des Sciences*, **D 286**, 229–231.

Duddington, C. L. (1955). Fungi that attack microscopic animals. *Biological Reviews*, **21**, 377–439.

Duddington, C. L. (1957). *The Friendly Fungi. A New Approach to the Eelworm Problem*. London: Faber and Faber.

Dufour, E., Boulay, J., Rincheval, V. & Sainsard-Chanet, A. (2000). A causal link between respiration and senescence in *Podospora anserina*. *Proceedings of the National Academy of Sciences of the USA*, **97**, 4138–4143.

Dughi, R. (1956). La signification des appareils apicaux des asques de *Gymnophysma* et de *Chlamydophysma*. *Comptes Rendus Hebdomadaires des Séances de l'Académie des Sciences de Paris*, **243**, 750–752.

Duncan, E. G. & Galbraith, M. H. (1972). Post-meiotic events in Homobasidiomycetidae. *Transactions of the British Mycological Society*, **58**, 387–392.

Duncan, K. E. & Howard, R. J. (2000). Cytological analysis of wheat infection by the leaf blotch pathogen *Mycosphaerella graminicola*. *Mycological Research*, **104**, 1074–1082.

Duncan, R. A., Sullivan, R., Alderman, S. C., Spatafora, J. W. & White, J. F. (2002). An ergot adapted to the aquatic environment. *Mycotaxon*, **81**, 11–25.

Dunlap, J. C. (1999). Molecular bases for circadian clocks. *Cell*, **96**, 271–290.

Dunlap, J. C. & Loros, J. J. (2004). The *Neurospora* circadian system. *Journal of Biological Rhythms*, **19**, 414–424.

Dunphy, G. B. & Nolan, R. A. (1982). Cellular immune response of spruce budworm larvae to *Entomophthora aulicae* and other test particles. *Invertebrate Pathology*, **39**, 81–92.

Dupont, B., Crewe Brown, H. H., Westermann, K., *et al.* (2000). Mycoses in AIDS. *Medical Mycology*, **38**, Supplement 1, 259–267.

Durand, H., Clanet, M. & Tiraby, G. (1988). Genetic improvement of *Trichoderma reesei* for large scale cellulase production. *Enzyme and Microbial Technology*, **10**, 341–346.

Dyal, S. D. & Narine, S. S. (2005). Implications for the use of *Mortierella* fungi in the industrial production of essential fatty acids. *Food Research International*, **38**, 445–467.

Dyck, P. L. & Kerber, E. R. (1985). Resistance of the race-specific type. In *The Cereal Rusts 2: Diseases, Distribution, Epidemiology, and Control*, ed. A. P. Roelfs & W. R. Bushnell. Orlando: Academic Press, pp. 469–500.

Dykstra, M. J. (1994). Ballistosporic conidia in *Basidiobolus ranarum*: the influence of light and nutrition on the production of conidia and endospores (sporangiospores). *Mycologia*, **86**, 494–501.

Dykstra, M. J. & Bradley-Kerr, B. (1994). The adhesive droplet of capilliconidia of *Basidiobolus ranarum* exhibits unique ultrastructural features. *Mycologia*, **86**, 336–342.

Dylewski, D. P. (1990). Phylum Plasmodiophoromycota. In *Handbook of Protoctista*, ed. L. Margulis, J. O. Corliss, M. Melkonian & D. J. Chapman. Boston: Jones & Bartlett, pp. 399–416.

Eadie, M. J. (2004). Ergot of rye – the first specific for migraine. *Journal of Clinical Neuroscience*, **11**, 4–7.

Eaton, R. A. & Hale, M. D. C. (1993). *Wood, Decay, Pests and Prevention*. London: Chapman & Hall.

Eckstein, P. E., Krasichynska, N., Voth, D., *et al.* (2002). Development of PCR-based markers for a gene (*Un8*) conferring true loose smut resistance in barley. *Canadian Journal of Plant Pathology*, **24**, 46–53.

Edelmann, R. E. & Klomparens, K. L. (1994). The ultrastructural development of sporangiospores in multispored sporangia of *Zygorhynchus heterogamus* with a hypothesis for sporangial wall dissolution. *Mycologia*, **86**, 57–71.

Edgar, J.A., Frahn, J.L., Cockrum, P.A. & Culvenor, C.C.J. (1986). Lupinosis. The chemistry and biochemistry of the phomopsins. In *Mycotoxins and Phytotoxins*, ed. P.S. Stern & R. Vleggaar. Amsterdam: Elsevier, pp. 169–86.

Edlind, T.D. & Katiyar, S.K. (2004). The echinocandin 'target' identified by cross-linking is a homolog of Pil1 and Lsp1, sphingolipid-dependent regulators of cell wall integrity signaling. *Antimicrobial Agents and Chemotherapy*, **48**, 4491.

Edwards, H.H. & Allen, P.J. (1970). A fine-structure study of the primary infection process of barley infected with *Erysiphe graminis* f. sp. *hordei*. *Phytopathology*, **60**, 1504–1509.

Eggert, C., Temp, U. & Eriksson, K.-E.L. (1997). Laccase is essential for lignin degradation by the white-rot fungus *Pycnoporus cinnabarinus*. *FEBS Letters*, **407**, 89–92.

Eisfelder, I. & Herschel, K. (1966). *Agathomyia wankowiczi* Schnabl, die 'Zitzengallenfliege' aus *Ganoderma applanatum*. *Westfälische Pilzbriefe*, **6**, 5–10.

El-Abyad, M.S.H. & Webster, J. (1968a). Studies on pyrophilous Discomycetes. I. Comparative physiological studies. *Transactions of the British Mycological Society*, **51**, 353–367.

El-Abyad, M.S.H. & Webster, J. (1968b). Studies on pyrophilous Discomycetes. II. Competition. *Transactions of the British Mycological Society*, **51**, 369–375.

Elad, Y., Lifshitz, R. & Baker, R. (1985). Enzymatic activity of the mycoparasite *Pythium nunn* during interaction with host and non-host fungi. *Physiological Plant Pathology*, **27**, 131–148.

Elad, Y., Williamson, B., Tudzynski, P. & Delen, N., eds. (2004). *Botrytis: Biology, Pathology and Control*. Dordrecht: Kluwer.

El-Ani, A.S. (1971). Chromosome numbers in the Hypocreales. II. Ascus development in *Nectria cinnabarina*. *American Journal of Botany*, **58**, 56–60.

Elix, J.A. (1996). Biochemistry and secondary metabolites. In *Lichen Biology*, ed. T.H. Nash. Cambridge: Cambridge University Press, pp. 154–180.

Ellingboe, A.H. (1961). Somatic recombination in *Puccinia graminis* var. *tritici*. *Phytopathology*, **51**, 13–15.

Elliott, C.G. (1994). *Reproduction in Fungi. Genetic and Physiological Aspects*. London: Chapman & Hall.

Elliott, E.W. (1949). The swarm cells of myxomycetes. *Mycologia*, **41**, 141–170.

Elliott, T.J. (1985). The general biology of the mushroom. In *The Biology and Technology of the Cultivated Mushroom*, ed. P.B. Flegg, D.M. Spencer & D.A. Wood. Chichester: John Wiley & Sons, pp. 9–22.

Ellis, D.H. & Pfeiffer, T.J. (1990a). Natural habitat of *Cryptococcus neoformans* var. *gattii*. *Journal of Clinical Microbiology*, **28**, 1642–1644.

Ellis, D.H. & Pfeiffer, T.J. (1990b). Ecology, life cycle, and infectious propagule of *Cryptococcus neoformans*. *The Lancet*, **336**, 923–925.

Ellis, M.B. (1971a). *Dematiaceous Hyphomycetes*. Kew: Commonwealth Mycological Institute.

Ellis, M.B. (1971b). Porospores. In *Taxonomy of Fungi Imperfecti*, ed. W.B. Kendrick. Toronto: University of Toronto Press, pp. 71–4.

Ellis, M.B. (1976). *More Dematiaceous Hyphomycetes*. Kew: Commonwealth Mycological Institute.

Ellis, M.B. & Ellis, J.P. (1990). *Fungi Without Gills (Hymenomycetes and Gasteromycetes)*. London: Chapman & Hall.

Ellis, M.B. & Ellis, J.P. (1998). *Microfungi on Miscellaneous Substrates*. Slough: Richmond Publishing.

El-Shafie, A.K. & Webster, J. (1980). Ascospore liberation in *Cochliobolus cymbopogonis*. *Transactions of the British Mycological Society*, **75**, 141–146.

Embley, T.M., van der Giezen, M., Horner, D.S., Dyal, P.L. & Foster, P. (2002). Mitochondria and hydrogenosomes are two forms of the same fundamental organelle. *Philosophical Transactions of the Royal Society*, **B 358**, 191–203.

Emerson, R. (1941). An experimental study of the life cycles and taxonomy of *Allomyces*. *Lloydia*, **4**, 77–144.

Emerson, R. (1958). Mycological organization. *Mycologia*, **50**, 589–621.

Emerson, R. & Natvig, D.O. (1981). Adaptation of fungi to stagnant waters. In *The Fungal Community. Its Organization and Role in the Ecosystem*, ed. D.T. Wicklow & G.C. Carroll. New York: Marcel Dekker, pp. 109–128.

Emerson, R. & Robertson, J.A. (1974). Two new members of the Blastocladiaceae. I. Taxonomy, with evaluation of genera and inter-relationships in the family. *American Journal of Botany*, **61**, 303–317.

Emerson, R. & Weston, W.H. (1967). *Aqualinderella fermentans* gen. et sp. nov., a phycomycete adapted to stagnant waters. I. Morphology and occurrence in nature. *American Journal of Botany*, **54**, 702–719.

Emerson, R. & Wilson, C.M. (1954). Interspecific hybrids and the cytogenetics and cytotaxonomy of *Eu-Allomyces*. *Mycologia*, **46**, 393–434.

Emmons, C.W. (1955). Saprophytic sources of *Cryptococcus neoformans* associated with the pigeon (*Columba livia*). *American Journal of Hygiene*, **62**, 227–232.

Engel, H. & Schneider, J.C. (1963). Die Umwandlung von Glykogen in Zucker in den Fruchtkörpern von *Sphaerobolus stellatus* (Tode) Pers., vor ihrem Abschuss. *Berichte der Deutschen Botanischen Gesellschaft*, **75**, 397–400.

Englander, L. & Hull, R.J. (1980). Reciprocal transfer of nutrients between ericaceous plants and a *Clavaria* sp. *New Phytologist*, **84**, 661–667.

Engler, M., Anke, T. & Sterner, O. (1998). Production of antibiotics by *Collybia nivalis*, *Omphalotus olearius*, a *Favolaschia* and a *Pterula* species on natural substrates. *Zeitschrift für Naturforschung*, **53c**, 318–324.

Epstein, L., Laccetti, L., Staples, R.C., Hoch, H.C. & Hoose, W.A. (1985). Extracellular proteins associated with induction of differentiation in bean rust uredospore germlings. *Phytopathology*, **75**, 1073–1076.

Eriksson, O.E. (1981). The families of bitunicate ascomycetes. *Opera Botanica*, **60**, 1–220.

Eriksson, O.E., Baral, H.-O., Currah, R.S., *et al.* (2003). Outline of Ascomycota. *Myconet*, **9**, 1–89.

Erwin, D.C. & Ribeiro, O.K. (1996). *Phytophthora Diseases Worldwide*. St. Paul: APS Press.

Erwin, D.C., Bartnicki-Garcia, S. & Tsao, P.H., eds. (1983). *Phytophthora. Its Biology, Taxonomy, Ecology, and Pathology*. St. Paul: APS Press.

Escobar, G., McCabe, D.E. & Harpel, C.W. (1976). *Limnoperdon*, a floating Gasteromycete isolated from marshes. *Mycologia*, **68**, 874–880.

Eslava, A.P. & Alvarez, M.I. (1987). Crosses. In *Phycomyces*, ed. E. Cerdá-Olmedo & E.D. Lipson. Cold Spring Harbor, New York: Cold Spring Harbor Laboratory, pp. 361–365.

Eslava, A.P., Alvarez, M.I. & Delbrück, M. (1975a). Meiosis in *Phycomyces*. *Proceedings of the National Academy of Sciences of the USA*, **72**, 4076–4080.

Eslava, A.P., Alvarez, M.I., Burke, P.V. & Delbrück, M. (1975b). Genetic recombination in sexual crosses in *Phycomyces*. *Genetics*, **80**, 445–462.

Esser, K. (1974). *Podospora anserina*. In *Handbook of Genetics*, ed. R.C. King. New York: Plenum Press, pp. 531–551.

Esser, K. & Blaich, R. (1994). Heterogenic incompatibility in fungi. In *The Mycota I: Growth, Differentiation and Sexuality*, ed. J.G.H. Wessels & F. Meinhardt. Berlin: Springer-Verlag, pp. 211–232.

Estey, R.H. & Tzean, S.S. (1976). Scanning electron microscopy of fungal nematode-trapping devices. *Transactions of the British Mycological Society*, **66**, 520–522.

Eucker, J., Sezer, O., Graf, B. & Possinger, K. (2001). Mucormycoses. *Mycoses*, **44**, 253–260.

Evans, G.H. & Cooke, R.C. (1982). Studies on Mucoralean mycoparasites. 3. Diffusible factors from *Mortierella vinacea* Dixon-Stewart that direct germ tube growth in *Piptocephalis fimbriata* Richardson & Leadbeater. *New Phytologist*, **91**, 245–253.

Evans, G.H., Lewis, D.H. & Cooke, R.C. (1978). Studies on Mucoralean mycoparasites. 2. Persistent yeast-phase growth of *Mycotypha microspora* Fenner when infected by *Piptocephalis fimbriata* Richardson & Leadbeater. *New Phytologist*, **81**, 629–635.

Evans, H.C. (1989). Mycopathogens of insects of epigeal and aerial habitats. In *Insect–Fungus Interactions*, ed. N. Wilding, N.M. Collins, P.M. Hammond & J.F. Webber. London: Academic Press, pp. 205–238.

Evans, H.C. (2003). Use of Clavicipitalean fungi for the biological control of arthropod pests. In *Clavicipitalean Fungi*, ed. F. White, C.W. Bacon, N.L. Hywel-Jones & J.W. Spatafora. New York: Marcel Dekker Inc., pp. 517–548.

Evans, H.C. & Samson, R.A. (1982). *Cordyceps* species and their anamorphs pathogenic on ants (Formicidae) in tropical forest ecosystems. I. The *Cephalotes* (Myrmicinae) complex. *Transactions of the British Mycological Society*, **79**, 431–453.

Evans, H.C. & Samson, R.A. (1984). *Cordyceps* species and their anamorphs pathogenic on ants (Formicidae) in tropical forest ecosystems. II. The *Campanotus* (Formicinae) complex. *Transactions of the British Mycological Society*, **82**, 127–150.

Evans, H.C., Holmes, K.A., Phillips, W. & Wilkinson, M.J. (2002). What's in a name: *Crinipellis*, the final resting place for the frosty pod rot pathogen of cocoa. *Mycologist*, **16**, 148–152.

Evans, H.C., Holmes, K.A. & Reid, A.P. (2003). Phylogeny of the frosty pod rot pathogen of cocoa. *Plant Pathology*, **52**, 476–485.

Evans, H.J. (1959). Nuclear behaviour in the cultivated mushroom. *Chromosoma*, **10**, 411–419.

Evans, R.G. (1965). *Sporobolomyces* as a cause of respiratory allergy. *Acta Allergologica*, **20**, 197–205.

Eversmeyer, M.G. & Kramer, C.L. (2000). Epidemiology of wheat leaf and stem rust in the Central Great Plains of the USA. *Annual Review of Phytopathology*, **38**, 491–513.

Eyal, Z. (1999). The *Septoria tritici* and *Stagonospora nodorum* blotch diseases of wheat. *European Journal of Plant Pathology*, **105**, 629–641.

Fahselt, D. (1994). Secondary biochemistry of lichens. *Symbiosis*, **16**, 117–165.

Fahselt, D. (1996). Individuals, populations and population ecology. In *Lichen Biology*, ed. T.H. Nash. Cambridge: Cambridge University Press, pp. 181–98.

Falk, S.P., Gadoury, D.M., Cortesi, P., Pearson, R.C. & Seem, R.C. (1995). Parasitism of *Uncinula necator* cleistothecia by the mycoparasite *Ampelomyces quisqualis*. *Phytopathology*, **85**, 794–800.

Fan, K.W., Vrijmoed, L.L.P. & Jones, E.B.G. (2002). Zoospore chemotaxis of mangrove thraustochytrids from Hong Kong. *Mycologia*, **94**, 569–578.

Fang, J.G. & Tsao, P.H. (1995). Evaluation of *Pythium nunn* as a potential biocontrol agent against *Phytophthora* root rot of azalea and sweet orange. *Phytopathology*, **85**, 29–36.

Faretra, F., Antonacci, E. & Pollastro, S. (1988). Sexual behaviour and mating system of *Botryotinia fuckeliana*, teleomorph of *Botrytis cinerea*. *Journal of General Microbiology*, **134**, 2543–2550.

Farr, D.F., Bills, G.F., Chamuris, G.P. & Rossman, A.Y. (1989). *Fungi on Plants and Plant Products in the United States*. St. Paul: APS Press.

Fasolo-Bonfante, P. & Brunel, A. (1972). Caryological features in a mycorrhizal fungus: *Tuber*. *Allionia*, **18**, 5–11.

Fassi, B., Fontana, A. & Trappe, J.M. (1969). Ectomycorrhizae formed by *Endogone lactiflua* with species of *Pinus* and *Pseudotsuga*. *Mycologia*, **61**, 412–414.

Fatehi, J., Bridge, P.D. & Punithalingam, E. (2003). Molecular relatedness within the 'Ascochyta pinodes-complex'. *Mycopathologia*, **156**, 317–327.

Faulstich, H. & Zilker, T.R. (1994). Amatoxins. In *Handbook of Mushroom Poisoning. Diagnosis and Treatment*, 2nd edn, ed. D.G. Spoerke & B.H. Rumach. Boca Raton: CRC Press, pp. 233–264.

Federici, B.A. (1977). Differential pigmentation in the sexual phase of *Coelomomyces*. *Nature*, **267**, 514–515.

Feldbrügge, M., Kämper, J., Steinberg, G. & Kahmann, R. (2004). Regulation of mating and pathogenic development in *Ustilago maydis*. *Current Opinion in Microbiology*, **7**, 666–672.

Fell, J.W. & Blatt, G.M. (1999). Separation of strains of the yeasts *Xanthophyllomyces dendrorhous* and *Phaffia rhodozyma* based on rDNA IGS and ITS sequence analysis. *Journal of Industrial Microbiology and Biotechnology*, **23**, 677–681.

Fell, J.W. & Statzell-Tallman, A. (1981). Heterothallism in the basidiomycetous genus *Sporidiobolus*. *Current Microbiology*, **5**, 77–82.

Fell, J.W. & Statzell-Tallman, A. (1998). *Rhodosporidium* Banno. In *The Yeasts. A Taxonomic Study*, 4th edn, ed. C.P. Kurtzman & J.W. Fell. Amsterdam: Elsevier, pp. 678–692.

Fell, J.W., Boekhout, T., Fonseca, A., Scorzetti, G. & Statzell-Tallman, A. (2000). Biodiversity and systematics of basidiomycetous yeasts as determined by large-subunit rDNA D1/D2 domain sequence analysis. *International Journal of Systematic and Evolutionary Microbiology*, **50**, 1351–1371.

Fell, J.W., Boekhout, T., Fonseca, A. & Sampaio, J.P. (2001). Basidiomycetous yeasts. In *The Mycota* **VIIB**: *Systematics and Evolution*, ed. D.J. McLaughlin, E.G. McLaughlin & P.A. Lemke. Berlin: Springer-Verlag, pp. 3–35.

Fiasson, J.L. & Petersen, R.H. (1973). Carotenes in the fungus *Clathrus ruber* (Gasteromycetes). *Mycologia*, **65**, 201–203.

Fiddy, C. & Trinci, A.P.J. (1976). Mitosis, septation, branching and the duplication cycle in *Aspergillus nidulans*. *Journal of General Microbiology*, **97**, 169–184.

Field, J.I. & Webster, J. (1977). Traps of predacious fungi attract nematodes. *Transactions of the British Mycological Society*, **68**, 467–470.

Field, J.I. & Webster, J. (1983). Anaerobic survival of aquatic fungi. *Transactions of the British Mycological Society*, **81**, 365–369.

Field, J.I. & Webster, J. (1985). Effects of sulphide on survival of aero-aquatic and aquatic hyphomycetes. *Transactions of the British Mycological Society*, **85**, 193–199.

Filtenborg, O., Frisvad, J.C. & Samson, R.A. (2002). Specific association of fungi to foods and influence of physical environmental factors. In *Introduction to Food- and Airborne Fungi*, 6th edn, ed. R.A. Samson, E.S. Hoekstra, J.C. Frisvad & O. Filtenborg. Utrecht: Centraalbureau voor Schimmelcultures, pp. 306–20.

Fincham, J.R.S. & Day, P.R. (1971). *Fungal Genetics*, 3rd edn. Oxford: Blackwell Scientific.

Finger, F.P. & Novick, P. (1998). Spatial regulation of exocytosis: lessons from yeast. *Journal of Cell Biology*, **142**, 609–612.

Fischer, A. (1892). Die Pilze. IV Abt. Phycomycetes. In *Kryptogamenflora von Deutschland, Oesterreich und der Schweiz*, ed. L. Rabenhorst. Leipzig: Eduard Kummer.

Fischer, F.G. & Werner, G. (1958). Die Chemotaxis der Schwärmsporen von Wasserpilzen (Saprolegniaceen). *Hoppe-Seyler's Zeitschrift für Physiologische Chemie*, **310**, 92–96.

Fischer, G.W. & Holton, C.S. (1957). *Biology and Control of the Smut Fungi*. New York: Ronald Press Company.

Fischer, M., Cox, J., Davis, D.J., *et al.* (2004). New information on the mechanism of forcible ascospore discharge from *Ascobolus immersus*. *Fungal Genetics and Biology*, **41**, 698–707.

Fischer-Parton, S., Parton, R. M., Hickey, P. C., *et al.* (2000). Confocal microscopy of FM4−64 as a tool for analysing endocytosis and vesicle trafficking in living fungal hyphae. *Journal of Microscopy*, **198**, 246−259.

Fisher, K. E., Lowry, D. S. & Roberson, R. W. (2000). Cytoplasmic cleavage in living sporangia of *Allomyces macrogynus*. *Journal of Microscopy*, **198**, 260−269.

Fisher, M. C., Koenig, G. L., White, T. J. & Taylor, J. T. (2002). Molecular and phenotypic description of *Coccidioides posadasii* sp. nov., previously recognized as the non-California population of *Coccidioides immitis*. *Mycologia*, **94**, 73−84.

Fisher, P. J. (1978). Survival of aero-aquatic fungi on land. *Transactions of the British Mycological Society*, **71**, 419−423.

Fisher, P. J. & Webster, J. (1978). Sporulation of aero-aquatic fungi under different gas regimes in light and darkness. *Transactions of the British Mycological Society*, **71**, 465−468.

Fisher, P. J. & Webster, J. (1979). Effect of oxygen and carbon dioxide on growth of four aero-aquatic hyphomycetes. *Transactions of the British Mycological Society*, **72**, 57−61.

Fisher, P. J., Webster, J. & Petrini, O. (1991). Aquatic hyphomycetes and other fungi in living aquatic and terrestrial roots of *Alnus glutinosa*. *Mycological Research*, **95**, 543−547.

Fitt, B. D. L., Goulds, A. & Polley, R. W. (1988). Eyespot (*Pseudocercosporella herpotrichoides*) epidemiology in relation to prediction of disease severity and yield loss in winter wheat: a review. *Plant Pathology*, **37**, 311−328.

Flaishman, M. A. & Kollatukudy, P. E. (1994). Timing of fungal invasion using host's ripening hormone as a signal. *Proceedings of the National Academy of Sciences of the USA*, **91**, 6579−6583.

Fleet, G. H. (1990). Food spoilage yeasts. In *Yeast Technology*, ed. J. F. T. Spencer & D. M. Spencer. Berlin: Springer-Verlag, pp. 124−166.

Flegg, P. B. & Wood, D. A. (1985). Growth and fruiting. In *The Biology and Technology of the Cultivated Mushroom*, ed. P. B. Flegg, D. M. Spencer & D. A. Wood. Chichester: John Wiley & Sons, pp. 141−177.

Flegler, S. L. (1984). An improved method for production of *Sphaerobolus* fruit bodies in culture. *Mycologia*, **76**, 944−946.

Fleming, A. (1929). On the antibacterial action of cultures of *Penicillium*, with a special reference to their use in the isolation of *B. influenzae*. *British Journal of Experimental Pathology*, **10**, 226−236.

Fleming, A. (1944). The discovery of penicillin. *British Medical Bulletin*, **2**, 4−5.

Fletcher, H. J. (1969). The development and tropisms of *Pilaira anomala*. *Transactions of the British Mycological Society*, **53**, 130−132.

Fletcher, H. J. (1973). The sporangiophore of *Pilaira* species. *Transactions of the British Mycological Society*, **61**, 553−568.

Fletcher, J. (1972). Fine structure of developing merosporangia and sporangiospores of *Syncephalastrum racemosum*. *Archiv für Mikrobiologie*, **87**, 269−284.

Fletcher, J. (1973a). The distribution of cytoplasmic vesicles, multivesicular bodies and paramural bodies in elongating sporangiophores and swelling sporangia of *Thamnidium elegans* Link. *Annals of Botany*, **37**, 955−961.

Fletcher, J. (1973b). Ultrastructural changes associated with spore formation in sporangia and sporangiola of *Thamnidium elegans* Link. *Annals of Botany*, **37**, 963−972.

Fletcher, L. R. & Harvey, I. C. (1981). An association of a *Lolium* endophyte with ryegrass staggers. *New Zealand Veterinary Journal*, **29**, 185−186.

Flieger, M., Wurst, M. & Shelby, R. (1997). Ergot alkaloids − sources, structures and analytical methods. *Folia Microbiologica*, **42**, 3−30.

Flor, H. H. (1955). Host−parasite interaction in flax rust − its genetics and other implications. *Phytopathology*, **45**, 680−685.

Flor, H. H. (1971). Current status of the gene-for-gene concept. *Annual Review of Phytopathology*, **9**, 275−296.

Flores, R., Dederichs, A., Cerdá-Olmedo, E. & Hertel, R. (1999). Flavin-binding sites in *Phycomyces*. *Plant Biology*, **1**, 645−655.

Fokkema, N. J. & van der Meulen, F. (1976). Antagonism of yeastlike phyllosphere fungi against *Septoria nodorum* on wheat leaves. *Netherlands Journal of Plant Pathology*, **82**, 13−16.

Foley, M. F. & Deacon, J. W. (1985). Isolation of *Pythium oligandrum* and other necrotrophic mycoparasites from soil. *Transactions of the British Mycological Society*, **85**, 631−639.

Foley, M. F. & Deacon, J. W. (1986a). Physiological differences between mycoparasitic and plant-pathogenic *Pythium* spp. *Transactions of the British Mycological Society*, **86**, 225−231.

Foley, M. F. & Deacon, J. W. (1986b). Susceptibility of *Pythium* spp. and other soil fungi to antagonism by the mycoparasite *Pythium oligandrum*. *Soil Biology and Biochemistry*, **18**, 91−95.

Fontaine, T., Hartland, R. P., Beauvais, A., Diaquin, M. & Latgé, J.-P. (1997). Purification and characterization of an endo-1,3-β-glucanase from *Aspergillus fumigatus*. *European Journal of Biochemistry*, **243**, 315−321.

Foster, S. J. & Fitt, B. D. L. (2004). Isolation and characterisation of the mating-type (*MAT*) locus from *Rhynchosporium secalis*. *Current Genetics*, **44**, 277–286.

Fox, R. D. & Wong, G. J. (1990). Homothallism and heterothallism in *Tremella fuciformis*. *Canadian Journal of Botany*, **68**, 107–111.

Fox, R. T. V., ed. (2000). *Armillaria Root Rot: Biology and Control of Honey Fungus*. Andover, UK: Intercept Ltd.

Franke, J. & Kessin, R. (1977). A defined minimal medium for axenic strains of *Dictyostelium discoideum*. *Proceedings of the National Academy of Sciences of the USA*, **74**, 2157–2161.

Frankland, J. C. (1984). Autecology and the mycelium of a woodland litter decomposer. In *The Ecology and Physiology of the Fungal Mycelium*, ed. D. H. Jennings & A. D. M. Rayner. Cambridge: Cambridge University Press, pp. 241–260.

Franzot, S. P., Salkin, I. F. & Casadevall, A. (1999). *Cryptococcus neoformans* var. *grubii*: separate varietal status for *Cryptococcus neoformans* serotype A isolates. *Journal of Clinical Microbiology*, **37**, 838–840.

Fraser, J. A. & Heitman, J. (2004). Evolution of fungal sex chromosomes. *Molecular Microbiology*, **51**, 299–306.

Fraymouth, J. (1956). Haustoria of the Peronosporales. *Transactions of the British Mycological Society*, **39**, 79–107.

Frazer, L. N. (1996). Control of growth and patterning in the fungal fruiting structure. A case for the involvement of hormones. In *Patterns in Fungal Development*, ed. S.-W. Chiu & D. Moore. Cambridge: Cambridge University Press, pp. 156–181.

Frederick, L. (1990). Phylum plasmodial slime molds, class Myxomycota. In *Handbook of Protoctista*, ed. L. Margulis, J. O. Corliss, M. Melkonian & D. J. Chapman. Boston: Jones & Bartlett, pp. 467–483.

Frederick, L., Uecker, F. A. & Benjamin, C. R. (1969). A new species of *Neurospora* from the soil of West Pakistan. *Mycologia*, **61**, 1077–1084.

Frederickson, D. E. & Mantle, P. G. (1989). Secondary conidition of *Sphacelia sorghi* on sorghum, a novel factor in the epidemiology of the ergot disease. *Mycological Research*, **93**, 497–502.

Frederickson, D. E., Mantle, P. G. & De Milliano, W. A. J. (1991). *Claviceps africana* sp. nov.: the distinctive ergot pathogen of sorghum in Africa. *Mycological Research*, **95**, 1101–1107.

Freeman, S. & Rodriguez, R. J. (1993). Genetic conversion of a fungal pathogen to a non-pathogenic, endophytic mutualist. *Science*, **260**, 75–78.

Frey, K. J., Browning, J. A. & Simons, M. D. (1973). Management of host resistance genes to control diseases. *Zeitschrift für Pflanzenkrankheiten und Pflanzenschutz*, **80**, 160–180.

Freytag, S. & Mendgen, K. (1991). Carbohydrates on the surface of urediniospore- and basidiospore-derived infection structures of heteroecious and autoecious rust fungi. *New Phytologist*, **119**, 527–534.

Friend, J. (1991). The biochemistry and cell biology of interaction. In *Phytophthora infestans, the Cause of Late Blight of Potato*, ed. D. S. Ingram & P. H. Williams. London: Academic Press, pp. 85–129.

Friman, E., Olsson, S. & Nordbring-Hertz, B. (1985). Heavy trap formation by *Arthrobotrys oligospora* in liquid culture. *FEMS Microbiology Ecology*, **31**, 17–21.

Frisvad, J. C. & Samson, R. A. (2004). Polyphasic taxonomy of *Penicillium* subgenus *Penicillium*. A guide to identification of food and air-borne terverticillate Penicillia and their mycotoxins. *Studies in Mycology*, **49**, 1–173.

Frisvad, J. C., Smedsgaard, J., Larsen, T. O. & Samson, R. A. (2004). Mycotoxins, drugs and other extrolites produced by species in *Penicillium* subgenus *Penicillium*. *Studies in Mycology*, **49**, 201–242.

Fry, W. E. & Goodwin, S. B. (1997). Resurgence of the Irish potato famine fungus. *BioScience*, **47**, 363–371.

Fry, W. E., Goodwin, S. B., Dyer, A. T., *et al.* (1993). Historical and recent migrations of *Phytophthora infestans*: chronology, pathways, and implications. *Plant Disease*, **77**, 653–661.

Fuentes-Dávila, G., Goates, B. J., Thomas, P., Nielsen, J. & Ballantyne, B. (2002). Smut diseases. In *Bread Wheat. Improvement and Production*, ed. B. C. Curtis, S. Rajaram & H. Gómez Macpherson. Rome: Food and Agriculture Organization of the United Nations, pp. 251–71.

Fuhrer, B. (2005). *A Field Guide to Australian Fungi*. Melbourne: Bloomings Books Pty Ltd.

Fukshansky, L. (1993). Intracellular processing of a spatially non-uniform stimulus: case study of phototropism in *Phycomyces*. *Journal of Photochemistry and Photobiology B: Biology*, **19**, 161–186.

Fuller, J. C. (1969). *The Day of St Anthony's Fire*. New York: Macmillan.

Fuller, M. S. (1976). The zoospore, hallmark of aquatic fungi. *Mycologia*, **69**, 1–20.

Fuller, M. S. (1990). Phylum Hyphochytriomycota. In *Handbook of Protoctista*, ed. L. Margulis, J. O. Corliss, M. Melkonian & D. J. Chapman. Boston: Jones & Bartlett, pp. 380–7.

Fuller, M. S. (2001). Hyphochytriomycota. In *The Mycota VIIA: Systematics and Evolution*, ed. D. J. McLaughlin, E. G. McLaughlin & P. A. Lemke. Berlin: Springer-Verlag, pp. 73–80.

Fuller, M. S. & Clay, R. P. (1993). Observations on *Gonapodya* in pure culture: growth, development and cell wall characterization. *Mycologia*, **85**, 38–45.

Fuller, M. S. & Olson, L. W. (1971). The zoospore of *Allomyces. Journal of General Microbiology*, **66**, 171–183.

Fuller, M. S. & Rakatansky, R. M. (1966). A preliminary study of the carotenoids in *Acrasis rosea. Canadian Journal of Botany*, **44**, 269–274.

Furch, B. & Gooday, G. W. (1978). Sporopollenin in *Phycomyces blakesleeanus. Transactions of the British Mycological Society*, **70**, 307–309.

Furtado, S. (1965). Relation of microstructures to the taxonomy of the Ganodermatoideae (Polyporaceae). *Mycologia*, **57**, 588–611.

Gadd, G. M., ed. (2001). *Fungi in Bioremediation*. Cambridge: Cambridge University Press.

Gadoury, D. M. & Pearson, R. C. (1988). Initiation, development, dispersal, and survival of cleistothecia of *Uncinula necator* in New York vineyards. *Phytopathology*, **78**, 1413–1421.

Gage, M. J., Bruenn, J., Fisher, M., Sanders, D. & Smith, T. J. (2001). KP4 fungal toxin inhibits growth in *Ustilago maydis* by blocking calcium uptake. *Molecular Microbiology*, **41**, 775–785.

Gage, M. J., Rane, S. G., Hockerman, G. H. & Smith, T. J. (2002). The virally encoded fungal toxin KP4 specifically blocks L-type voltage-gated calcium channels. *Molecular Pharmacology*, **61**, 936–944.

Gall, A. M. & Elliot, C. G. (1985). Control of sexual reproduction in *Pythium sylvaticum. Transactions of the British Mycological Society*, **84**, 629–636.

Galland, P. & Tölle, N. (2003). Light-induced fluorescence changes in *Phycomyces*: evidence for blue light-receptor associated flavo-semiquinones. *Planta*, **217**, 971–982.

Gammie, A. E., Brizzio, V. & Rose, M. D. (1998). Distinct morphological phenotypes of cell fusion mutants. *Molecular Biology of the Cell*, **9**, 1395–1410.

Gamow, R. I., Ruiz-Herrera, J. & Fischer, E. P. (1987). The cell wall of *Phycomyces*. In *Phycomyces*, ed. E. Cerdá-Olmedo & E. D. Lipson. Cold Spring Harbor, New York: Cold Spring Harbor Laboratory, pp. 223–246.

Gams, W. (1977). A key to the species of *Mortierella. Persoonia*, **9**, 381–391.

Gams, W. & Bissett, J. (1998). Morphology and identification. In *Trichoderma and Gliocladium, I: Basic Biology, Taxonomy and Genetics*, ed. C. P. Kubicek & G. F. Harman. London: Taylor & Francis, pp. 3–34.

Gams, W. & Zare, R. (2003). A taxonomic review of the Clavicipitaceous anamorphs parasitizing nematodes and other microinvertebrates. In *Clavicipitalean Fungi*, ed. J. F. White, C. W. Bacon, N. L. Hywel-Jones & J. W. Spatafora. New York: Marcel Dekker Inc., pp. 17–73.

Gamundí, I. J. (1991). Review of recent advances in the knowledge of the Cyttariales. *Systema Ascomycetum*, **10**, 69–77.

Garbaye, G. (1994). Helper bacteria: a new dimension to the mycorrhizal symbiosis. *New Phytologist*, **128**, 197–210.

Garber, E. D. & Ruddat, M. (2002). Transmission genetics of *Microbotryum violaceum* (*Ustilago violacea*): a case history. *Advances in Applied Microbiology*, **51**, 107–127.

Garber, R. C. & Aist, J. R. (1979). The ultrastructure of meiosis in *Plasmodiophora brassicae* (Plasmodiophorales). *Canadian Journal of Botany*, **57**, 2509–2518.

García-Pajón, C. M. & Collado, I. G. (2003). Secondary metabolites isolated from *Colletotrichum* species. *Natural Product Reports*, **20**, 426–431.

Gardiner, R. B. & Day, A. W. (1988). Surface proteinaceous fibrils (fimbriae) on filamentous fungi. *Canadian Journal of Botany*, **66**, 2474–2484.

Gargas, A. & Taylor, J. W. (1995). Phylogeny of discomycetes and early radiations of the apothecial Ascomycotina inferred from SSU rDNA sequence data. *Experimental Mycology*, **19**, 7–15.

Gargas, A., DePriest, P. T., Grube, M. & Tehler, A. (1995). Multiple origins of lichen symbioses in fungi suggested by SSU rDNA phylogeny. *Science*, **268**, 1492–1495.

Garrett, S. D. (1953). Rhizomorph behaviour in *Armillaria mellea* (Vahl) Quél. I. Factors controlling rhizomorph initiation by *Armillaria mellea* in pure culture. *Annals of Botany* N.S., **17**, 63–79.

Garrett, S. D. (1970). *Pathogenic Root-Infecting Fungi*. Cambridge: Cambridge University Press.

Garrett, R. G. & Tomlinson, J. A. (1967). Isolate differences in *Olpidium brassicae. Transactions of the British Mycological Society*, **50**, 429–435.

Garrill, A. (1995). Transport. In *The Growing Fungus*, ed. N. A. R. Gow & G. M. Gadd. London: Chapman & Hall, pp. 163–181.

Garrill, A., Jackson, S. L., Lew, R. R. & Heath, I. B. (1993). Ion channel activity and tip growth: tip localised stretch-activated channels generate an essential Ca^{2+} gradient in the oomycete *Saprolegnia ferax. European Journal of Cell Biology*, **60**, 358–365.

Garrison, R. G. & Boyd, K. S. (1973). Dimorphism of *Penicillium marneffei* as observed by electron microscopy. *Canadian Journal of Microbiology*, **19**, 1305–1309.

Gauger, W. L. (1961). The germination of zygospores of *Rhizopus stolonifer. American Journal of Botany*, **48**, 427–429.

Gauger, W. L. (1966). Sexuality in an azygosporic strain of *Mucor hiemalis*. I. Breakdown of the azygosporic component. *American Journal of Botany*, **53**, 751–755.

Gauger, W. L. (1975). Further studies on sexuality in azygosporic strains of *Mucor hiemalis*. *Transactions of the British Mycological Society*, **64**, 113–118.

Gäumann, E. (1959). *Die Rostpilze Mitteleuropas*. Bern: Büchler & Co.

Gauriloff, L. P. & Fuller, M. S. (1987). *Rhizophydium sphaerocarpum*. In *Zoosporic Fungi in Teaching and Research*, ed. M. S. Fuller & A. Jaworski. Athens, USA: Southeastern Publishing Corporation, pp. 18–19.

Gay, J. L., Salzberg, A. & Woods, A. M. (1987). Dynamic experimental evidence for the plasmamembrane ATPase domain hypothesis of haustorial transport and for ionic coupling of the haustorium of *Erysiphe graminis* to the host cell (*Hordeum vulgare*). *New Phytologist*, **107**, 541–548.

Gay, L., Chanzy, H., Bulone, V., Girard, V. & Fèvre, M. (1993). Synthesis *in vitro* of crystalline chitin by a solubilized enzyme from the cellulosic fungus *Saprolegnia monoica*. *Journal of General Microbiology*, **139**, 2117–2122.

Gehrig, H., Schüssler, A. & Kluge, M. (1996). *Geosiphon pyriforme*, a fungus forming endocytobiosis with *Nostoc* (Cyanobacteria) is an ancestral member of the Glomales: evidence by SSU rRNA analysis. *Journal of Molecular Evolution*, **43**, 71–81.

Geiser, D. M. & LoBuglio, K. F. (2001). The monophyletic Plectomycetes: Ascosphaerales, Onygenales, Eurotiales. In *The Mycota VIIA: Systematics and Evolution*, ed. D. J. McLaughlin, E. G. McLaughlin & P. A. Lemke. Berlin: Springer-Verlag, pp. 201–19.

Geiser, D. M., Timberlake, W. E. & Arnold, M. L. (1996). Loss of meiosis in *Aspergillus*. *Molecular Biology and Evolution*, **13**, 809–817.

Geiser, D. M., Dorner, J. W., Horn, B. W. & Taylor, J. W. (2000). The phylogenetics of mycotoxin and sclerotium production in *Aspergillus flavus* and *Aspergillus oryzae*. *Fungal Genetics and Biology*, **31**, 169–179.

Geiser, D. M., Jiménez-Gasco, M. M., Kang, S., *et al.* (2004). FUSARIUM-ID v. 1.0: a DNA sequence database for identifying *Fusarium*. *European Journal of Plant Pathology*, **110**, 473–479.

Gemmill, T. R. & Trimble, R. B. (1999). Overview of N- and O-linked oligosaccharide structures found in various yeast species. *Biochimica et Biophysica Acta*, **1426**, 227–237.

Georgiou, C. D. & Petropoulou, K. P. (2001). Role of erythroascorbate and ascorbate in sclerotial differentiation in *Sclerotinia sclerotiorum*. *Mycological Research*, **105**, 1364–1370.

Georgiou, C. D., Tairis, N. & Polycratis, A. (2001). Production of β-carotene by *Sclerotinia sclerotiorum* and its role in sclerotium differentiation. *Mycological Research*, **105**, 1100–1115.

Georgopapadakou, N. H. (1998). Antifungals: mechanism of action and resistance, established and novel drugs. *Current Opinion in Microbiology*, **1**, 547–557.

Gerdemann, J. W. & Nicolson, T. H. (1963). Spores of mycorrhizal *Endogone* species extracted by wet sieving and decanting. *Transactions of the British Mycological Society*, **46**, 235–244.

Gerlach, W. & Nirenberg, H. I. (1982). *The Genus Fusarium – a Pictorial Atlas*. Berlin-Dahlem: Biologische Bundesanstalt für Land- und Forstwirtschaft.

Gerlagh, M., Goossen-van de Geijn, H. M., Hoogland, A. E. & Vereijken, P. F. G. (2003). Quantitative aspects of infection of *Sclerotinia sclerotiorum* sclerotia by *Coniothyrium minitans* – timing of application, concentration and quality of conidial suspension of the mycoparasite. *European Journal of Plant Pathology*, **109**, 489–502.

Gernandt, D. S. & Stone, J. K. (1999). Phylogenetic analysis of nuclear ribosomal DNA places the nematode parasite, *Drechmeria coniospora*, in Clavicipitaceae. *Mycologia*, **91**, 993–1000.

Gernandt, D. S., Platt, J. L., Stone, J. K., *et al.* (2001). Phylogenetics of Helotiales and Rhytismatales based on partial small subunit nuclear ribosomal DNA sequences. *Mycologia*, **93**, 915–933.

Gessner, M. O., Suberkropp, K. & Chauvet, E. (1997). Decomposition of plant litter by fungi in marine and freshwater ecosystems. In *The Mycota IV: Environmental and Microbial Relationships*, ed. D. T. Wicklow & B. Söderström. Berlin: Springer-Verlag, pp. 302–332.

Gessner, R. V. (1995). Genetics and systematics of North American populations of *Morchella*. *Canadian Journal of Botany*, **73**, S967–S972.

Gessner, R. V., Romano, M. A. & Schulz, R. W. (1987). Allelic variation and segregation in *Morchella deliciosa* and *M. esculenta*. *Mycologia*, **79**, 683–687.

Ghannoum, M. A. & Rice, L. B. (1999). Antifungal agents: mode of action, mechanisms of resistance, and correlation of these mechanisms with bacterial resistance. *Clinical Microbiology Reviews*, **12**, 501–517.

Gibbs, J. N. (1993). The biology of ophiostomoid fungi causing sapstain in trees and freshly cut logs. In *Ceratocystis and Ophiostoma. Taxonomy, Ecology and Pathogenicity*, ed. M. J. Wingfield, K. A. Seifert & J. F. Webber. St. Paul, USA: APS Press, pp. 153–160.

Giese, H., Hippe-Sanwald, S., Somerville, S. & Weller, J. (1997). *Erysiphe graminis*. In *The Mycota VB: Plant Relationships*, ed. G. C. Carroll & P. Tudzynski. Berlin: Springer-Verlag, pp. 55–78.

Giesy, R. M. & Day, P. R. (1965). The septal pores of *Coprinus lagopus* (Fr.) *sensu* Buller in relation to nuclear migration. *American Journal of Botany*, **52**, 287–293.

Gil, F. & Gay, J. L. (1977). Ultrastructural and physiological properties of the host interfacial components of haustoria of *Erysiphe pisi in vivo* and *in vitro*. *Physiological Plant Pathology*, **10**, 1–10.

Gil-ad, N. L., Bar-Nun, N. & Mayer, A. M. (2001). The possible function of the glucan sheath of *Botrytis cinerea*: effects on the distribution of enzyme activities. *FEMS Microbiology Letters*, **199**, 109–113.

Gilijamse, E., Frinking, H. D. & Jeger, M. J. (1997). Occurrence and epidemiology of pearl millet downy mildew, *Sclerospora graminicola*, in southwest Niger. *International Journal of Pest Management*, **43**, 279–283.

Gill, M. & Steglich, W. (1987). *Pigments of Fungi (Macromycetes)*. Vienna: Springer-Verlag.

Gill, M. & Strauch, R. J. (1984). Constituents of *Agaricus xanthodermus* Genevier: the first naturally endogenous azo compound and toxic phenolic metabolites. *Zeitschrift für Naturforschung*, **39c**, 1027–1029.

Gill, M. & Watling, R. (1986). The relationships of *Pisolithus* (Sclerodermataceae) to other fleshy fungi with particular reference to the occurrence and taxonomic significance of hydroxylated pulvinic acids. *Plant Systematics and Evolution*, **154**, 225–236.

Gilles, T., Ashby, A., Fitt, B. D. L. & Cole, T. (2001). Development of *Pyrenopeziza brassicae* apothecia on agar and oilseed rape debris. *Mycological Research*, **105**, 705–714.

Gillis, A. M. (1993). The magnificent devastator gets around. *Bioscience*, **43**, 368–371.

Gimeno, C. J., Ljungdahl, P. O., Styles, C. A. & Fink, G. R. (1992). Unipolar cell divisions in the yeast *S. cerevisiae* lead to filamentous growth: regulation by starvation and *RAS*. *Cell*, **68**, 1077–1090.

Gindin, G. & Ben Ze'ev, I. S. (1994). Natural occurrence of and inoculation experiments with *Conidiobolus coronatus* and *Conidiobolus* sp. in glasshouse populations of *Bemisia tabaci*. *Phytoparasitica*, **22**, 187–208.

Ginman, A. & Young, T. W. K. (1989). Azygospore morphology in *Mucor azygospora* and *M. bainieri*. *Mycological Research*, **93**, 314–320.

Giollant, M., Guillot, J., Damez, M., *et al.* (1993). Characterization of a lectin from *Lactarius deterrimus*: research on the possible involvement of the fungal lectin in recognition between mushroom and spruce during the early stages of mycorrhizae formation. *Plant Physiology*, **101**, 513–522.

Giovannetti, G., Roth-Bejerano, N., Zanini, E. & Kagan-Zur, V. (1994). Truffles and their cultivation. *Horticultural Reviews*, **16**, 71–107.

Giraud, T., Fortini, D., Levis, C., *et al.* (1999). Two sibling species of the *Botrytis cinerea* complex, *transposa* and *vacuma*, are found in sympatry on numerous host plants. *Phytopathology*, **89**, 967–973.

Girbardt, M. (1968). The ultrastructure and dynamics of the moving nucleus. *Symposia of the Society for Experimental Biology*, **22**, 249–259.

Girbardt, M. (1969). Die Ultrastruktur der Apikalregion von Pilzhyphen. *Protoplasma*, **67**, 413–441.

Gisi, U. (1983). Biophysical aspects of the development of *Phytophthora*. In *Phytophthora. Its Biology, Taxonomy, Ecology, and Pathology*, ed. D. C. Erwin, S. Bartnicki-Garcia & P. H. Tsao. St. Paul: APS Press, pp. 109–119.

Giuma, A. Y., Hackett, A. M. & Cooke, R. C. (1973). Thermostable nematotoxins produced by germinating conidia of some endozoic fungi. *Transactions of the British Mycological Society*, **60**, 49–56.

Glass, N. L. & Lorimer, A. J. (1991). Ascomycete mating type genes. In *More Gene Manipulations in Fungi*, ed. J. W. Bennett & L. L. Lasure. San Diego: Academic Press, pp. 194–216.

Glass, N. L., Vollmer, S. J., Staben, C., *et al.* (1988). DNAs of the two mating-type alleles of *Neurospora crassa* are highly dissimilar. *Science*, **241**, 570–573.

Glass, N. L., Grötelueschen, J. & Metzenberg, R. L. (1990). *Neurospora crassa* A mating-type region. *Proceedings of the National Academy of Sciences of the USA*, **87**, 4912–4916.

Glass, N. L., Jacobson, D. J. & Shiu, P. K. T. (2000). The genetics of hyphal fusion and vegetative incompatibility in filamentous ascomycete fungi. *Annual Review of Genetics*, **34**, 165–186.

Gleason, F. H. (1976). The physiology of the lower freshwater fungi. In *Recent Advances in Aquatic Mycology*, ed. E. B. Gareth-Jones. London: Elek Science, pp. 543–572.

Gleason, F. H., Letcher, P. M. & McGee, P. A. (2004). Some *Chytridiomycota* in soil recover from drying and high temperatures. *Mycological Research*, **108**, 583–589.

Glenn, A. E., Bacon, C. W., Price, R. & Hanlin, R. T. (1996). Molecular phylogeny of *Acremonium* and its taxonomic implications. *Mycologia*, **88**, 369–383.

Glockling, S. L. (1998). Isolation of a new species of rotifer-attacking *Olpidium*. *Mycological Research*, **102**, 206–208.

Glockling, S. L. & Beakes, G. W. (2000a). Video microscopy of spore development in *Haptoglossa heteromorpha*, a new species from cow dung. *Mycologia*, **92**, 747–753.

Glockling, S. L. & Beakes, G. W. (2000b). An ultrastructural study of sporidium formation during infection of a rhabditid nematode by large gun cells of

Haptoglossa heteromorpha. Journal of Invertebrate Pathology, **76**, 208–215.

Goates, B. J. (1988). Histology of infection of wheat by *Tilletia indica*, the Karnal bunt pathogen. *Phytopathology*, **78**, 1434–1441.

Godoy, G., Steadman, J. R., Dickman, M. B. & Dam, R. (1990). Use of mutants to demonstrate the role of oxalic acid in pathogenicity of *Sclerotinia sclerotiorum* on *Phaseolus vulgaris*. *Physiological and Molecular Plant Pathology*, **37**, 179–191.

Gold, R. E. & Mendgen, K. (1991). Rust basidiospore germlings and disease initiation. In *The Fungal Spore and Disease Initiation in Plants and Animals*, ed. G. T. Cole & H. C. Hoch. New York: Plenum Press, pp. 67–99.

Gold, S. E., Brogdon, S. M., Mayorga, M. E. & Kronstad, J. W. (1997). The *Ustilago maydis* regulatory subunit of a cAMP-dependent protein kinase is required for gall formation in maize. *Plant Cell*, **9**, 1585–1594.

Goldstein, B. (1923). Resting spores of *Empusa muscae*. *Bulletin of the Torrey Botanical Club*, **50**, 317–327.

Goldstein, S. (1960). Factors affecting the growth and pigmentation of *Cladochytrium replicatum*. *Mycologia*, **52**, 490–498.

Goldstein, S. (1961). Studies of two polycentric chytrids in pure culture. *American Journal of Botany*, **48**, 294–298.

Gönczöl, J. & Révay, A. (2003). Treehole fungal communities: aquatic, aero-aquatic and dematiaceous hyphomycetes. *Fungal Diversity*, **12**, 19–34.

Gonzalez, D., Carling, D. E., Kuninaga, S., Vilgalys, R. & Cubeta, M. A. (2001). Ribosomal DNA systematics of *Ceratobasidium* and *Thanatephorus* with *Rhizoctonia* anamorphs. *Mycologia*, **93**, 1138–1150.

Gooday, G. W. (1994). Hormones in mycelial fungi. In *The Mycota I: Growth, Differentiation and Sexuality*, ed. J. G. H. Wessels & F. Meinhardt. Berlin: Springer-Verlag, pp. 401–12.

Gooday, G. W. (1995). Cell walls. In *The Growing Fungus*, ed. N. A. R. Gow & G. M. Gadd. London: Chapman & Hall, pp. 45–62.

Gooday, G. W. & Adams, D. J. (1992). Sex hormones and fungi. *Advances in Microbial Physiology*, **34**, 69–145.

Gooday, G. W. & Carlile, M. J. (1997). The discovery of fungal sex hormones: III. Trisporic acid and its precursors. *Mycologist*, **11**, 126–130.

Gooday, G. W., Fawcett, P., Green, D. & Shaw, D. (1973). The formation of sporopollenin in the zygospore wall of *Mucor mucedo*: a role for the sexual carotenogenesis in the Mucorales. *Journal of General Microbiology*, **74**, 233–239.

Goodwin, S. B. (2002). The barley scald pathogen *Rhynchosporium secalis* is closely related to the discomycetes *Tapesia* and *Pyrenopeziza*. *Mycological Research*, **106**, 645–654.

Goodwin, S. B. & Zismann, V. L. (2001). Phylogenetic analyses of the ITS region of ribosomal DNA reveal that *Septoria passerinii* from barley is closely related to the wheat pathogen *Mycosphaerella graminicola*. *Mycologia*, **93**, 934–946.

Goodwin, S. B., Cohen, B. A. & Fry, W. E. (1994a). Panglobal distribution of a single clonal lineage of the Irish potato famine fungus. *Proceedings of the National Academy of Sciences of the USA*, **91**, 11 591–595.

Goodwin, S. B., Cohen, B. A., Deahl, K. L. & Fry, W. E. (1994b). Migration from northern Mexico as the probable cause of recent genetic changes in populations of *Phytophthora infestans* in the United States and Canada. *Phytopathology*, **84**, 553–558.

Goodwin, S. B., Sujkowski, L. S., Dyer, A. T., Fry, B. A. & Fry, W. E. (1995). Direct detection of gene flow and probable sexual reproduction of *Phytophthora infestans* in northern North America. *Phytopathology*, **85**, 473–479.

Goodwin, S. B., Dunkle, L. & Zismann, V. L. (2001). Phylogenetic analysis of *Cercospora* and *Mycosphaerella* based on the internal transcribed spacer region of ribosomal DNA. *Phytopathology*, **91**, 648–658.

Goos, R. D. (1987). Fungi with a twist. The helicosporous hyphomycetes. *Mycologia*, **79**, 1–22.

Gordon, C. C. (1966). A re-interpretation of the ontogeny of the ascocarp of species of the Erysiphaceae. *American Journal of Botany*, **53**, 652–662.

Gordon, T. R. (1993). Genetic variation and adaptive potential in an asexual soil-borne fungus. In *The Fungal Holomorph. Mitotic, Meiotic and Pleomorphic Speciation in Fungal Systematics*, ed. D. R. Reynolds & J. W. Taylor. Wallingford, UK: CAB International, pp. 217–224.

Gordon, T. R. & Martyn, R. D. (1997). The evolutionary biology of *Fusarium oxysporum*. *Annual Review of Phytopathology*, **35**, 112–128.

Gottlieb, A. M. & Lichtwardt, R. W. (2001). Molecular variation within and among species of Harpellales. *Mycologia*, **93**, 66–81.

Govindan, B., Bowser, R. & Novick, P. (1995). The role of Myo2, a yeast class V myosin, in vesicular transport. *Journal of Cell Biology*, **128**, 1055–1069.

Govrin, E. M. & Levine, A. (2000). The hypersensitive response facilitates plant infection by the necrotrophic pathogen *Botrytis cinerea*. *Current Biology*, **10**, 751–757.

Gow, N. A. R. (1995). Yeast—hyphal dimorphism. In *The Growing Fungus*, ed. N. A. R. Gow & G. M. Gadd. London: Chapman & Hall, pp. 403—422.

Gow, N. A. R. (2002). *Candida albicans* switches mates. *Molecular Cell*, **10**, 217—218.

Gow, N. A. R. & Gooday, G. W. (1987). Effects of antheridiol on growth, branching and electrical currents of hyphae of *Achlya ambisexualis*. *Journal of General Microbiology*, **133**, 3531—3535.

Gow, N. A. R., Bates, S., Brown, A. J., *et al.* (1999). *Candida* cell wall mannosylation: importance in host—fungus interaction and potential as a target for the development of antifungal drugs. *Biochemical Society Transactions*, **27**, 512—516.

Graca, M. A. S., Maltby, L. & Calow, P. (1993). Importance of fungi in the diet of *Gammarus pulex* and *Asellus aquaticus*. II. Effects on growth, reproduction and physiology. *Oecologia*, **96**, 304—309.

Gradmann, D., Hansen, U.-P., Long, W. S., Slayman, C. L. & Warncke, J. (1978). Current—voltage relationships for the plasma membrane and its principal electrogenic pump in *Neurospora crassa*. I. Steady-state conditions. *Journal of Membrane Biology*, **39**, 333—367.

Graham, T. R. & Emr, S. D. (1991). Compartmental organization of Golgi-specific protein modification and vacuolar protein sorting events defined in a yeast *sec18* (*NSF*) mutant. *Journal of Cell Biology*, **114**, 207—218.

Grainger, J. (1962). Vegetative and fructifying growth in *Phallus impudicus*. *Transactions of the British Mycological Society*, **45**, 147—155.

Gräser, Y., el Fari, M., Vilgalys, R., *et al.* (1999). Phylogeny and taxonomy of the family Arthrodermataceae (dermatophytes) using sequence analysis of the ribosomal ITS region. *Medical Mycology*, **37**, 105—114.

Gray, N. F. (1987). Nematophagous fungi with particular reference to their ecology. *Biological Reviews*, **62**, 245—304.

Green, F. & Clausen, C. A. (1999). Production of polygalacturonase and increase of longitudinal gas permeability in Southern Pine by brown-rot and white-rot fungi. *Holzforschung*, **53**, 563—568.

Green, F. & Clausen, C. A. (2003). Copper tolerance of brown-rot fungi: time course of oxalic acid production. *International Biodeterioration and Biodegradation*, **51**, 145—149.

Green, F., Larsen, M. J., Winandy, J. E. & Highley, T. L. (1991). Role of oxalic acid in incipient brown-rot decay. *Material und Organismen*, **26**, 191—213.

Green, G. J. (1971). Hybridization between *Puccinia graminis tritici* and *Puccinia graminis secalis* and its evolutionary implications. *Canadian Journal of Botany*, **49**, 2089—2095.

Green, J. R., Carver, T. L. W. & Gurr, S. J. (2002). The formation and function of infection and feeding structures. In *The Powdery Mildews. A Comprehensive Treatise*, ed. R. R. Bélanger, W. R. Bushnell, A. J. Dik & T. L. W. Carver. St. Paul, USA: APS Press, pp. 66—82.

Greenhalgh, G. N. & Bevan, R. J. (1978). Response of *Rhytisma acerinum* to air pollution. *Transactions of the British Mycological Society*, **71**, 491—523.

Greenhalgh, G. N. & Evans, L. V. (1967). The structure of the ascus apex in *Hypoxylon fragiforme* with reference to ascospore release in this and related species. *Transactions of the British Mycological Society*, **50**, 183—188.

Gregory, P. H. (1949). The operation of the puff-ball mechanism of *Lycoperdon perlatum* by raindrops shown by ultra-high-speed Schlieren cinematography. *Transactions of the British Mycological Society*, **32**, 11—15.

Gregory, P. H. (1957). Electrostatic charges on spores of fungi in air. *Nature*, **180**, 330.

Gregory, P. H. (1966). The fungus spore: what it is and what it does. In *The Fungal Spore*, ed. M. F. Madelin. London: Butterworths, pp. 1—14.

Gregory, P. H. (1973). *Microbiology of the Atmosphere*, 2nd edn. Aylesbury, Bucks: Leonard Hill.

Gregory, P. H. & Sreeramulu, T. (1958). Air spora of an estuary. *Transactions of the British Mycological Society*, **41**, 145—156.

Greif, M. D. & Currah, R. S. (2003). A functional interpretation of the role of the reticuloperidium in whole-ascoma dispersal by arthropods. *Mycological Research*, **107**, 77—81.

Grente, J. & Sauret, S. (1969). L'hypovirulence exclusive, phenomène original en pathologie végétale. *Comptes Rendus Hebdomadaires des Séances de l'Académie des Sciences*, **D 286**, 2347—2350.

Grenville, D. J., Peterson, R. L. & Piché, Y. (1985). The development, structure, and histochemistry of sclerotia of ectomycorrhizal fungi. I. *Pisolithus tinctorius*. *Canadian Journal of Botany*, **63**, 1402—1411.

Gries, C. (1996). Lichens as indicators of air pollution. In *Lichen Biology*, ed. T. H. Nash. Cambridge: Cambridge University Press, pp. 240—254.

Griffin, D. H. (1994). *Fungal Physiology*, 2nd edn. New York: Wiley-Liss.

Griffith, G. W., Nicholson, J., Nenninger, A., Birch, R. N. & Hedger, J. N. (2003). Witches' brooms and frosty pods: two major pathogens of cacao. *New Zealand Journal of Botany*, **41**, 423—435.

Griffith, J. M., Davis, A. J. & Grant, B. R. (1992). Target sites of fungicides to control Oomycetes. In *Target Sites of Fungicide Action*, ed. W. Köller. Boca Raton: CRC Press, pp. 69–100.

Griffiths, A. J. F. (1992). Fungal senescence. *Annual Review of Genetics*, **26**, 351–372.

Griffiths, D. A. (1974). The origin, structure and function of chlamydospores in fungi. *Nova Hedwigia*, **25**, 503–547.

Grime, J. P., Mackey, J. M. L., Hillier, S. H. & Read, S. J. (1987). Floristic diversity in a model system using experimental microcosms. *Nature*, **328**, 420–422.

Groff, J. M., Mughannam, A., McDowell, T. S., *et al.* (1991). An epizootic of cutaneous zygomycosis in cultured dwarf African clawed frogs (*Hymenochirus curtipes*) due to *Basidiobolus ranarum*. *Journal of Medical and Veterinary Mycology*, **29**, 215–223.

Grogan, R. G., Zink, F. W., Hewitt, W. B. & Kimble, K. A. (1958). The association of *Olpidium* with the big-vein disease of lettuce. *Phytopathology*, **48**, 292–297.

Gross, G. (1987). Zu europäischen Sippen der Gattung *Tuber*. In *Atlas der Pilze des Saarlandes 2*, ed. H. Derbsch & J. A. Schmitt. Saarbrücken: Verlag der Derlattinia, pp. 79–100.

Grove, S. N., Bracker, C. E. & Morré, D. J. (1968). Cytomembrane differentiation in the endoplasmic reticulum–Golgi apparatus–vesicle complex. *Science*, **161**, 171–173.

Grove, W. B. (1913). *The British Rust Fungi (Uredinales). Their Biology and Classification*. Cambridge: Cambridge University Press.

Grove, W. B. (1934). A systematic account and arrangement of the Pilobolidae. In Buller, A. H. R.: *Researches on Fungi*, 6, pp. 190–224. London: Longmans, Green & Co.

Grove, W. B. (1935). *British Stem- and Leaf-Fungi (Coelomycetes) I*. Cambridge: Cambridge University Press.

Grsic, S., Kirchheim, B., Pieper, K., *et al.* (1999). Induction of auxin biosynthetic enzymes by jasmonic acid and in clubroot diseased Chinese cabbage plants. *Physiologia Plantarum*, **105**, 521–531.

Grube, M. & Winka, K. (2002). Progress in understanding the evolution and classification of lichenized ascomycetes. *Mycologist*, **16**, 67–76.

Grubisha, L. C., Trappe, J. M., Molina, R. & Spatafora, J. W. (2001). Biology of the ectomycorrhizal genus *Rhizopogon*. V. Phylogenetic relationships in the Boletales inferred from LSU rDNA sequences. *Mycologia*, **93**, 82–89.

Grylls, B. T. & Seifert, K. A. (1993). A synoptic key to species of *Ophiostoma*, *Ceratocystis* and *Ceratocystiopsis*. In *Ceratocystis and Ophiostoma. Taxonomy, Ecology and Pathogenicity*, ed. M. J. Wingfield, K. A. Seifert & J. F. Webber. St. Paul, USA: APS Press, pp. 261–268.

Gu, Y. H. & Ko, W. H. (1998). Occurrence of a parasexual cycle following the transfer of isolated nuclei into protoplasts of *Phytophthora parasitica*. *Current Genetics*, **34**, 120–123.

Guarro, J. & Figueras, M. J. (1989). Morphological studies in *Chaetomium*: the peridium, the ostiolar apparatus and some considerations about the centrum. *Cryptogamic Botany*, **1**, 97–140.

Guarro, J. & von Arx, J. A. (1987). The ascomycete genus *Sordaria*. *Persoonia*, **13**, 301–313.

Gubler, F. & Hardham, A. R. (1988). Secretion of adhesive material during encystment of *Phytophthora cinnamomi* zoospores, characterized by immunogold labeling with monoclonal antibodies to components of peripheral vesicles. *Journal of Cell Science*, **90**, 225–235.

Gubler, F. & Hardham, A. R. (1990). Protein storage in large peripheral vesicles in *Phytophthora* zoospores and its breakdown after cyst germination. *Experimental Mycology*, **14**, 393–404.

Gucia, M. & Falandysz, J. (2003). Total mercury content in parasol mushroom *Macrolepiota procera* from various sites in Poland. *Journal de Physique IV*, **107**, 581–584.

Guého, E. & Meyer, S. A. (1989). A reevaluation of the genus *Malassezia* by means of genome comparison. *Antonie van Leeuwenhoek*, **55**, 245–251.

Guého, E., Midgley, G. & Guillot, J. (1996). The genus *Malassezia* with description of four new species. *Antonie van Leeuwenhoek*, **69**, 337–355.

Guého, E., Leclerc, M. C., de Hoog, G. S. & Dupont, B. (1997). Molecular taxonomy and epidemiology of *Blastomyces* and *Histoplasma* species. *Mycoses*, **40**, 69–81.

Guerin-Languette, A., Matsushita, N., Lapeyrie, F., Shindo, K. & Suzuki, K. (2005). Successful inoculation of mature pine with *Tricholoma matsutake*. *Mycorrhiza*, **15**, 301–305.

Guerrero, A., Jain, N., Goldman, D. L. & Fries, B. C. (2006). Phenotypic switching in *Cryptococcus neoformans*. *Microbiology*, **152**, 3–9.

Gugnani, H. C. (1992). Entomophthoromycosis due to *Conidiobolus*. *European Journal of Epidemiology*, **8**, 391–396.

Gugnani, H. C. (1999). A review of zygomycosis due to *Basidiobolus ranarum*. *European Journal of Epidemiology*, **15**, 923–929.

Guillot, J. & Bond, R. (1999). *Malassezia pachydermatis*: a review. *Medical Mycology*, **37**, 295–306.

Guillot, J. & Guého, E. (1995). The diversity of *Malassezia* yeasts confirmed by rRNA sequence and nuclear DNA comparisons. *Antonie van Leeuwenhoek*, **67**, 297–314.

Gulati, A., Gulati, A., Ravindranath, S. D. & Gupta, A. K. (1999). Variation in chemical composition and quality of tea (*Camellia sinensis*) with increasing blister blight (*Exobasidium vexans*) severity. *Mycological Research*, **103**, 1380–1384.

Gulis, V. (2001). Are there any substrate preferences in aquatic hyphomycetes? *Mycological Research*, **105**, 1088–1093.

Gull, K. (1978). Form and function of septa in filamentous fungi. In *The Filamentous Fungi, 3. Developmental Mycology*, ed. J. E. Smith & D. R. Berry. London: Edward Arnold, pp. 78–93.

Guo, L.-Y. & Michailides, T. J. (1998). Factors affecting the rate and mode of germination of *Mucor piriformis* zygospores. *Mycological Research*, **102**, 815–819.

Gupta, A. K., Einarson, T. R., Summerbell, R. C. & Shear, N. H. (1998). An overview of topical antifungal therapy in dermatomycoses. A North American perspective. *Drugs*, **55**, 645–674.

Gupta, A. K., Bluhm, R. & Summerbell, R. (2002). *Pityriasis versicolor. Journal of the European Academy of Dermatology and Venereology*, **16**, 19–33.

Guttenberger, M. (2000). Arbuscules of vesicular–arbuscular fungi inhabit an acidic compartment within plant roots. *Planta*, **211**, 299–304.

Gwynne, D. I. (1992). Foreign proteins. In *Applied Molecular Genetics of Filamentous Fungi*, ed. J. R. Kinghorn & G. Turner. Glasgow: Blackie, pp. 132–51.

Haase, G., Sonntag, L., Melzer-Krick, B. & de Hoog, G. S. (1999). Phylogenetic inference by SSU-gene analysis of members of the *Herpotrichiellaceae* with special reference to human pathogenic species. *Studies in Mycology*, **43**, 80–97.

Haber, J. E. (1998). Mating-type gene switching in *Saccharomyces cerevisiae. Annual Review of Genetics*, **32**, 561–599.

Hadley, G. (1968). Development of stromata in *Claviceps purpurea. Transactions of the British Mycological Society*, **51**, 763–769.

Hadley, G. (1970). Non-specificity of symbiotic infection in orchid mycorrhiza. *New Phytologist*, **69**, 1015–1023.

Hadley, G. & Williamson, B. (1971). Analysis of the post-infection growth stimulus in orchid mycorrhiza. *New Phytologist*, **70**, 445–455.

Hagan, I. M. (1998). The fission yeast microtubule cytoskeleton. *Journal of Cell Science*, **111**, 1603–1612.

Hagedorn, G. & Scholler, M. (1999). A reevaluation of predatory orbiliaceous fungi. I. Phylogenetic analysis using rDNA sequence data. *Sydowia*, **51**, 27–48.

Hahn, M. (2000). The rust fungi. In *Fungal Pathology*, ed. J. W. Kronstad. Dordrecht: Kluwer, pp. 267–306.

Hajek, A. E., Wraight, S. P. & Vandenberg, J. D. (2001). Control of arthropods using pathogenic fungi. In *Bio-Exploitation of Filamentous Fungi*, ed. S. B. Pointing & K. D. Hyde. Hong Kong: Fungal Diversity Press, pp. 309–347.

Hale, M. E. (1983). *The Biology of Lichens*, 3rd edn. Baltimore: Edward Arnold.

Hall, I. R. (1984). Taxonomy of VA mycorrhizal fungi. In *VA Mycorrhiza*, ed. C. L. Powell & D. J. Bagyaraj. Boca Raton: CRC Press, pp. 57–94.

Hall, I. R., Brown, G. & Byars, J. (1994). *The Black Truffle, its History, Uses and Cultivation*. Lincoln, New Zealand: Crop and Food Research.

Hall, I. R., Yun, W. & Amicucci, A. (2003). Cultivation of edible ectomycorrhizal mushrooms. *Trends in Biotechnology*, **21**, 433–438.

Hallett, I. C. & Dick, M. W. (1986). Fine structure of zoospore cyst ornamentation in the Saprolegniaceae and Pythiaceae. *Transactions of the British Mycological Society*, **86**, 457–463.

Hamer, J. E., Howard, R. J., Chumley, F. G. & Valent, B. (1988). A mechanism for surface attachment in spores of a plant pathogenic fungus. *Science*, **239**, 288–290.

Hammett, K. R. W. & Manners, J. G. (1971). Conidium liberation in *Erysiphe graminis*. I. Visual and statistical analysis of spore trap records. *Transactions of the British Mycological Society*, **56**, 387–401.

Hammett, K. R. W. & Manners, J. G. (1974). Conidium liberation in *Erysiphe graminis*. III. Wind tunnel studies. *Transactions of the British Mycological Society*, **62**, 267–282.

Hammill, T. M. (1981). Mucoralean sporangiosporogenesis. In *The Fungal Spore: Morphogenetic Controls*, ed. G. Turian & H. R. Hohl. New York: Academic Press, pp. 173–194.

Hammill, T. M. & Secor, D. L. (1983). The number of nuclei in sporangiospores of *Mucor mucedo. Mycologia*, **75**, 648–655.

Hammond, J. B. W. (1985). The biochemistry of *Agaricus* fructification. In *Developmental Biology of Higher Fungi*, ed. D. Moore, L. A. Casselton, D. A. Wood & J. C. Frankland. Cambridge: Cambridge University Press, pp. 339–401.

Hammond, J. B. W. & Nichols, R. (1979). Carbohydrate metabolism in *Agaricus bisporus*: changes in non-structural carbohydrates during periodic fruiting (flushing). *New Phytologist*, **83**, 723–730.

Hammond, J. R. (1995). Genetically-modified brewing yeasts for the 21st century. Progress to date. *Yeast*, **11**, 1613–1627.

Hampson, M.C. (1988). Control of potato wart disease through the application of chemical soil treatments: an historical review of earlier studies (1909–1928). *EPPO Bulletin*, **18**, 153–161.

Hampson, M.C. & Coombes, J.W. (1991). Use of crabshell meal to control potato wart in Newfoundland. *Canadian Journal of Plant Pathology*, **13**, 97–105.

Hampson, M.C., Yang, A.F. & Bal, A.K. (1994). Ultrastructure of *Synchytrium endobioticum* and enhancement of germination using snails. *Mycologia*, **86**, 733–740.

Hankin, L. & Puhalla, J.E. (1971). Nature of a factor causing interstrain lethality in *Ustilago maydis*. *Phytopathology*, **61**, 50–53.

Hanlin, R.T. (1971). Morphology of *Nectria haematococca*. *American Journal of Botany*, **58**, 105–116.

Hansen, E.M. & Goheem, E.M. (2000). *Phellinus weirii* and other native root pathogens as determinants of forest structure and process in Western North America. *Annual Review of Phytopathology*, **38**, 515–539.

Hansen, K., Laessøe, T. & Pfister, D.H. (2001). Phylogenetics of the Pezizaceae, with an emphasis on *Peziza*. *Mycologia*, **93**, 958–990.

Hansen, K., LoBuglio, K.F. & Pfister, D.H. (2005). Evolutionary relationships of the cup-fungus genus *Peziza* and Pezizaceae inferred from multiple nuclear genes: RPB2, β-tubulin, and LSU rDNA. *Molecular Phylogenetics and Evolution*, **36**, 1–23.

Hansen, L. (1958). On the anatomy of the Danish species of *Ganoderma*. *Botanisk Tidskrift*, **54**, 333–352.

Hansen, L. & Knudsen, H., eds. (2000). *Nordic Macromycetes* **1**. *Ascomycetes*. Copenhagen: Nordsvamp.

Hanssen, H.-P. (1993). Volatile metabolites produced by species of *Ophiostoma* and *Ceratocystis*. In *Ceratocystis and Ophiostoma. Taxonomy, Ecology and Pathogenicity*, ed. M.J. Wingfield, K.A. Seifert & J.F. Webber. St. Paul, USA: APS Press, pp. 117–125.

Hantula, J., Kasanen, R., Kaitera, J. & Moricca, S. (2002). Analyses of genetic variation suggest that pine rusts *Cronartium flaccidum* and *Peridermium pini* belong to the same species. *Mycological Research*, **106**, 203–209.

Harder, D.E. & Chong, J. (1984). Structure and physiology of haustoria. In *The Cereal Rusts* **1**: *Origins, Specificity, Structure, and Physiology*, ed. W.R. Bushnell & A.P. Roelfs. Orlando: Academic Press, pp. 431–476.

Hardham, A.R. (1995). Polarity of vesicle distribution in oomycete zoospores: development of polarity and importance for infection. *Canadian Journal of Botany*, **73**, S400–S407.

Hardham, A.R. & Gubler, F. (1990). Polarity of attachment of zoospores of a root pathogen and prealignment of the emerging germ tube. *Cell Biology International Reports*, **14**, 947–956.

Hardham, A.R. & Hyde, G.J. (1997). Asexual sporulation in the oomycetes. *Advances in Botanical Research*, **24**, 353–398.

Hardham, A.R., Gubler, F. & Duniec, J. (1991). Ultrastructural and immunological studies of zoospores of *Phytophthora*. In *Phytophthora*, ed. J.A. Lucas, R.C. Shattock, D.S. Shaw & L.R. Cooke. Cambridge: Cambridge University Press, pp. 50–69.

Harold, F.M. (1994). Ionic and electric dimensions of hyphal growth. In *The Mycota* **I**: *Growth, Differentiation and Sexuality*, ed. J.G.H. Wessels & F. Meinhardt. Berlin: Springer-Verlag, pp. 489–509.

Harrington, F.A., Pfister, D.H., Potter, D. & Donoghue, M.J. (1999). Phylogenetic studies within the Pezizales. I. 18S rDNA sequence data and classification. *Mycologia*, **91**, 41–50.

Harrington, T.C. & McNew, D.L. (1997). Self-fertility and uni-directional mating-type switching in *Ceratocystis coerulescens*, a filamentous ascomycete. *Current Genetics*, **32**, 52–59.

Harrington, T.C. & McNew, D.L. (2003). Phylogenetic analysis places the *Phialophora*-like anamorph genus *Cadophora* in the Helotiales. *Mycotaxon*, **87**, 141–151.

Harrington, T.C. & Wingfield, M.J. (1998). The *Ceratocystis* species on conifers. *Canadian Journal of Botany*, **76**, 1446–1457.

Harris, J.S. & Taber, W.A. (1973). Ultrastructure and morphogenesis of the synnema of *Ceratocystis ulmi*. *Canadian Journal of Botany*, **51**, 1565–1571.

Harrison, J.G., Searle, R.J. & Williams, N.A. (1997). Powdery scab of potato – a review. *Plant Pathology*, **46**, 1–25.

Harrison, J.S. (1993). Food and fodder yeasts. In *The Yeasts* **5**: *Yeast Technology*, ed. A.H. Rose & J.S. Harrison. London: Academic Press, pp. 399–433.

Harrison, T.S. & Levitz, S.M. (1996). Infections due to the dimorphic fungi. In *The Mycota* **VI**: *Human and Animal Relationships*, ed. D.H. Howard & J.D. Miller. Berlin: Springer-Verlag, pp. 125–146.

Harrold, C.E. (1950). Studies in the genus *Eremascus*. I. The re-discovery of *Eremascus albus* Eidam and some new observations concerning its life history and cytology. *Annals of Botany* N.S., **14**, 127–148.

Harrower, K.M. (1976). The micropycnidiospores of *Leptosphaeria nodorum*. *Transactions of the British Mycological Society*, **76**, 335–336.

Hartley, M.J. & Williams, P.G. (1971). Genotypic variation within a phenotype as a possible basis for somatic hybridization in rust fungi. *Canadian Journal of Botany*, **49**, 1085–1087.

Hartmeier, W. & Reiss, M. (2002). Production of beer and wine. In *The Mycota* X: *Industrial Applications*, ed. H.D. Osiewacz. Berlin: Springer-Verlag, pp. 49–65.

Hashmi, M.H., Morgan-Jones, G. & Kendrick, B. (1972). Conidium ontogeny in hyphomycetes. *Monilia* state of *Neurospora sitophila* and *Sclerotinia laxa*. *Canadian Journal of Botany*, **500**, 1461–1463.

Hashmi, M.H., Morgan-Jones, G. & Kendrick, B. (1973). Conidium ontogeny in hyphomycetes. The blastoconidia of *Cladosporium herbarum* and *Torula herbarum*. *Canadian Journal of Botany*, **51**, 1089–1091.

Hause, B. & Fester, T. (2005). Molecular and cell biology of arbuscular mycorrhizal symbiosis. *Planta*, **221**, 184–196.

Hausner, G., Reid, J. & Klassen, G.R. (1993). On the phylogeny of *Ophiostoma*, *Ceratocystis* s.s., and *Microascus*, and relationships within *Ophiostoma* based on partial ribosomal DNA sequences. *Canadian Journal of Botany*, **71**, 1249–1265.

Hausner, G., Belkhiri, A. & Klassen, G.R. (2000). Phylogenetic analysis of the small subunit ribosomal RNA gene of the hyphochytrid *Rhizidiomyces apophysatus*. *Canadian Journal of Botany*, **78**, 124–128.

Hawes, C.R. & Beckett, A. (1977). Conidium ontogeny in *Thielaviopsis basicola*. *Transactions of the British Mycological Society*, **68**, 304–307.

Hawker, L.E. & Beckett, A. (1971). Fine structure and development of the zygospore of *Rhizopus sexualis* (Smith) Callen. *Philosophical Transactions of the Royal Society*, **B 263**, 71–100.

Hawker, L.E., Thomas, B. & Beckett, A. (1970). An electron microscope study of structure and germination of conidia of *Cunninghamella elegans*. *Journal of General Microbiology*, **60**, 181–189.

Hawksworth, D.L. (2001). The magnitude of fungal diversity: the 1.5 million species estimate revisited. *Mycological Research*, **105**, 1422–1432.

Hawksworth, D.L. & Rose, F. (1970). Qualitative scale for estimating sulphur dioxide air pollution in England and Wales using epiphytic lichens. *Nature*, **227**, 145–148.

Hayashi, K., Schoonbek, H.-J. & de Waard, M.A. (2002). *Bcmfs1*, a novel major facilitator superfamily transporter from *Botrytis cinerea*, provides tolerance towards the natural toxic compounds camptothecin and cercosporin and towards fungicides. *Applied and Environmental Microbiology*, **68**, 4996–5004.

Hayes, W.A., Randle, P.E. & Last, F.T. (1969). The nature of the microbial stimulus affecting sporophore production in *Agaricus bisporus* (Lange) Singer. *Annals of Applied Biology*, **64**, 177–187.

Hayles, J. & Nurse, P. (2001). A journey into space. *Nature Reviews in Molecular Cell Biology*, **2**, 647–656.

Heath, I.B. (1987). Preservation of a labile cortical array of actin filaments in growing hyphal tips of the fungus *Saprolegnia ferax*. *European Journal of Cell Biology*, **44**, 10–16.

Heath, I.B. (1994). The cytoskeleton in hyphal growth, organelle movements, and mitosis. In *The Mycota* I: *Growth, Differentiation and Sexuality*, ed. J.G.H. Wessels & F. Meinhardt. Berlin: Springer-Verlag, pp. 43–65.

Heath, I.B. (1995a). The cytoskeleton. In *The Growing Fungus*, ed. N.A.R. Gow & G.M. Gadd. London: Chapman & Hall, pp. 99–134.

Heath, I.B. (1995b). Integration and regulation of hyphal tip growth. *Canadian Journal of Botany*, **73**, S131–S139.

Heath, I.B. (2001). Bridging the divide: cytoskeleton–-plasma membrane–cell wall interactions in growth and development. In *The Mycota* VIII: *Biology of the Fungal Cell*, ed. R.J. Howard & N.A.R. Gow. Berlin: Springer-Verlag, pp. 201–23.

Heath, I.B. & Harold, R.L. (1992). Actin has multiple roles in the formation and architecture of zoospores of the oomycetes, *Saprolegnia ferax* and *Achlya bisexualis*. *Journal of Cell Science*, **102**, 611–627.

Heath, I.B. & Steinberg, G. (1999). Mechanisms of hyphal tip growth: tube dwelling amoebae revisited. *Fungal Genetics and Biology*, **28**, 79–93.

Heath, I.B., Greenwood, A.D. & Griffiths, H.B. (1970). The origin of flimmer in *Saprolegnia*, *Dictyuchus*, *Synura* and *Cryptomonas*. *Journal of Cell Science*, **7**, 445–461.

Heath, I.B., Bauchop, T. & Skipp, R.K. (1983). Assignment of the rumen anaerobe *Neocallimastix frontalis* to the Spizellomycetales (Chytridiomycetes) on the basis of its polyflagellate zoospore ultrastructure. *Canadian Journal of Botany*, **61**, 295–307.

Heath, M.C. (1982). Host defense mechanisms against infection by rust fungi. In *The Rust Fungi*, ed. K.J. Scott & A.K. Chakravorty. London: Academic Press, pp. 223–245.

Heath, M.C. (2000). Hypersensitive response-related death. *Plant Molecular Biology*, **44**, 321–334.

Heath, M.C. (2002). Cellular interactions between biotrophic fungal pathogens and host or nonhost plants. *Canadian Journal of Plant Pathology*, **24**, 259–264.

Heath, M.C. & Skalamera, D. (1997). Cellular interactions between plants and biotrophic fungal parasites. *Advances in Botanical Research*, **24**, 195–225.

Heath, M.C., Howard, R.J., Valent, B. & Chumley, F.G. (1992). Ultrastructural interactions of one strain of *Magnaporthe grisea* with goosegrass and weeping lovegrass. *Canadian Journal of Botany*, **70**, 779–787.

Hedge, Y. & Kolattukudy, P.E. (1997). Cuticular waxes relieve self-inhibition of germination and appressorium formation by the conidia of *Magnaporthe grisea*. *Physiological and Molecular Plant Pathology*, **51**, 75–84.

Hedger, J.N., Lewis, P. & Gitay, H. (1993). Litter-trapping by fungi in moist tropical forest. In *Aspects of Tropical Mycology*, ed. S. Isaac, R. Watling, A.J.S. Whalley & J.C. Frankland. Cambridge: Cambridge University Press, pp. 15–35.

Hegarty, B. (1991). Factors affecting the fruiting of the dry rot fungus *Serpula lacrymans*. In *Serpula lacrymans. Fundamental Biology and Control Strategies*, ed. D.H. Jennings & A.F. Bravery. Chichester: John Wiley & Sons, pp. 39–53.

Hegarty, B., Steinfurth, A., Liese, W. & Schmidt, O. (1987). Comparative investigations on wood decay and cellulolytic and xylanolytic activity of some basidiomycete fungi. *Holzforschung*, **41**, 265–269.

Hegedus, D.D. & Rimmer, S.R. (2005). *Sclerotinia sclerotiorum*: when 'to be or not to be' a pathogen? *FEMS Microbiology Letters*, **251**, 177–184.

Heim, P. (1960). Évolution du *Spongospora* parasite des racines du Cresson. *Revue de Mycologie*, **25**, 3–12.

Heim, R. & Wasson, R.G. (1958). *Les Champignons Hallucinogènes du Mexique*. Paris: Musée Nationale d'Histoire Naturelle.

Heiniger, U. & Rigling, D. (1994). Biological control of chestnut blight in Europe. *Annual Review of Phytopathology*, **32**, 581–599.

Heintz, C.E. & Niederpruem, D.J. (1970). Ultrastructure and respiration of oidia and basidiospores of *Coprinus lagopus* (sensu Buller). *Canadian Journal of Microbiology*, **16**, 481–484.

Heintz, C.E. & Pramer, D. (1972). Ultrastructure of nematode-trapping fungi. *Journal of Bacteriology*, **110**, 1163–1170.

Heinzkill, M. & Messner, K. (1997). The ligninolytic system of fungi. In *Fungal Biotechnology*, ed. T. Anke. London: Chapman & Hall, pp. 213–27.

Heller, A. & Gierth, K. (2001). Cytological observations of the infection process by *Phomopsis helianthi* (Munt.-Cvet) in leaves of sunflower. *Journal of Phytopathology*, **149**, 347–357.

Hemmati, F., Pell, J.K., McCartney, H.A. & Deadman, M.L. (2001). Airborne concentration of conidia of *Erynia neoaphidis* above cereal fields. *Mycological Research*, **105**, 485–489.

Hemmati, F., Pell, J.K., McCartney, H.A. & Deadman, M.L. (2002). Aerodynamic diameter of conidia of *Erynia neoaphidis* and other entomophthoralean fungi. *Mycological Research*, **106**, 233–238.

Hemmes, D.E. (1983). Cytology of *Phytophthora*. In *Phytophthora. Its Biology, Taxonomy, Ecology, and Pathology*, ed. D.C. Erwin, S. Bartnicki-Garcia & P.H. Tsao. St. Paul: APS Press, pp. 9–40.

Hendrix, F.E. & Campbell, W.A. (1973). Pythiums as plant pathogens. *Annual Review of Phytopathology*, **11**, 77–98.

Henis, Y., Ghaffar, A. & Baker, R. (1979). Factors affecting suppressiveness to *Rhizoctonia solani* in soil. *Phytopathology*, **69**, 1164–1169.

Hennebert, G.L. & Weresub, L.K. (1977). Terms for states of fungi, their names and types. *Mycotaxon*, **6**, 207–211.

Henricot, B. & Prior, C. (2004). *Phytophthora ramorum*, the cause of sudden oak death or ramorum leaf blight and dieback. *Mycologist*, **18**, 151–156.

Hepper, C.M. (1984). Isolation and culture of VA mycorrhizal (VAM) fungi. In *VA Mycorrhiza*, ed. C.L. Powell & D.J. Bagyaraj. Boca Raton: CRC Press, pp. 95–112.

Hermansen, J.E., Torp, U. & Prahm, L.P. (1978). Studies of transport of cereal mildew and rust fungi across the North Sea. *Grana*, **17**, 41–46.

Herrera-Estrella, L. & Ruiz-Herrera, J. (1983). Light responses in *Phycomyces blakesleeanus*. Evidence for the roles of chitin biosynthesis and breakdown. *Experimental Mycology*, **7**, 362–369.

Herrera-Estrella, L., Chavez, B. & Ruiz-Herrera, J. (1982). Presence of chitosomes in the cytoplasm of *Phycomyces blakesleeanus* and the synthesis of chitin microfibrils. *Experimental Mycology*, **6**, 385–388.

Hervey, A.G., Bistis, G. & Leong, I. (1978). Culture studies of single ascospore isolates of *Morchella esculenta*. *Mycologia*, **70**, 1269–1274.

Hesseltine, C.W. (1957). The genus *Syzygites* (Mucoraceae). *Lloydia*, **20**, 228–237.

Hesseltine, C.W. (1991). Zygomycetes in food fermentations. *Mycologist*, **5**, 162–169.

Hesseltine, C.W. & Anderson, P. (1956). The genus *Thamnidium* and a study of the formation of its zygospores. *American Journal of Botany*, **43**, 696–703.

Hesseltine, C.W. & Rogers, R. (1987). Dark-period induction of zygospores in *Mucor*. *Mycologia*, **79**, 289–297.

Hesseltine, C.W., Whitehill, A.R., Pidacks, C., et al. (1953). Coprogen, a new growth factor present in dung required by *Pilobolus*. *Mycologia*, **45**, 7–19.

Hesseltine, C.W., Benjamin, R.K. & Mehrotra, B.S. (1959). The genus *Zygorhynchus*. *Mycologia*, **51**, 173–194.

Hibbett, D.S. & Binder, M. (2002). Evolution of complex fruiting-body morphologies in homobasidiomycetes. *Proceedings of the Royal Society B*, **269**, 1963–1969.

Hibbett, D.S. & Thorn, R.G. (1994). Nematode-trapping in *Pleurotus tuberregium*. *Mycologia*, **86**, 696–699.

Hibbett, D.S. & Thorn, R.G. (2001). Basidiomycota. Homobasidiomycetes. In *The Mycota VIIB: Systematics and Evolution*, ed. D.J. McLaughlin, E.G. McLaughlin & P.A. Lemke. Berlin: Springer-Verlag, pp. 121–68.

Hibbett, D.S., Tsuneda, A. & Murakami, S. (1994). The secotioid form of *Lentinus tigrinus*: genetics and development of a fungal morphological innovation. *American Journal of Botany*, **81**, 466–478.

Hibbett, D.S., Grimaldi, D. & Donoghue, M.J. (1997a). Fossil mushrooms from Miocene and Cretaceous ambers and the evolution of homobasidiomycetes. *American Journal of Botany*, **84**, 981–991.

Hibbett, D.S., Pine, E.M., Langer, E., Langer, G. & Donoghue, M.J. (1997b). Evolution of gilled mushrooms and puffballs inferred from ribosomal DNA sequences. *Proceedings of the National Academy of Sciences of the USA*, **94**, 12 002–12 006.

Hicks, J.B., Strathern, J.N. & Herskowitz, I. (1977). The cassette model of mating-type interconversion. In *DNA Insertion Elements, Plasmids and Episomes*, ed. A.I. Bukhari, J.A. Shapiro & S.L. Adhya. Cold Spring Harbor: Cold Spring Harbor Laboratory Press, pp. 457–462.

Higham, M.T. & Cole, K.M. (1982). Fine structure of sporangiole development in *Choanephora cucurbitarum* (Mucorales). *Canadian Journal of Botany*, **60**, 2313–2334.

Hill, D.J. & Smith, D.C. (1972). Lichen physiology. XII. The 'inhibition technique'. *New Phytologist*, **71**, 15–30.

Hillman, B.I. & Suzuki, N. (2004). Viruses of the chestnut blight fungus, *Cryphonectria parasitica*. *Advances in Virus Research*, **63**, 423–472.

Hintikka, V. (1973). A note on the polarity of *Armillariella mellea*. *Karstenia*, **13**, 32–39.

Hinton, D.M. & Bacon, C.W. (1985). The distribution and ultrastructure of the endophyte of toxic tall fescue. *Canadian Journal of Botany*, **63**, 36–42.

Hippe-Sanwald, S., Hermanns, M. & Somerville, S.C. (1992). Ultrastructural comparison of incompatible and compatible interactions in the barley powdery mildew disease. *Protoplasma*, **168**, 27–40.

Hirai, A., Kano, R., Nakamura, Y., Wantanabe, S. & Hasegawa, A. (2003). Molecular taxonomy of dermatophytes and related fungi by chitin synthase 1 (CHS1) gene sequences. *Antonie van Leeuwenhoek*, **83**, 11–20.

Hirst, J.M. (1953). Changes in atmospheric spore content: diurnal periodicity and the effects of weather. *Transactions of the British Mycological Society*, **36**, 375–393.

Hirst, J.M. & Stedman, O.J. (1960). The epidemiology of *Phytophthora infestans*. II. The source of inoculum. *Annals of Applied Biology*, **48**, 489–517.

Hiruki, C. (1994). Multiple transmission of plant viruses by *Olpidium brassicae*. *Canadian Journal of Plant Pathology*, **16**, 261–265.

Hiura, U. (1978). Genetic basis of *formae speciales* in *Erysiphe graminis* DC. In *The Powdery Mildews*, ed. D.M. Spencer. London: Academic Press, pp. 101–128.

Ho, H.-M. & Chen, Z.-C. (1998). Ultrastructural study of wall ontogeny during zygosporogenesis in *Rhizopus stolonifer* (Mucoraceae), an amended model. *Botanical Bulletin of Academia Sinica*, **39**, 269–277.

Ho, M.H.-M., Castañeda, R.F., Dugan, F.M. & Jong, S.C. (1999). *Cladosporium* and *Cladophialophora* in culture: descriptions and an expanded key. *Mycotaxon*, **72**, 115–157.

Hobbie, E.A., Weber, N.S. & Trappe, J.M. (2001). Mycorrhizal vs. saprotrophic status of fungi: the isotopic evidence. *New Phytologist*, **150**, 601–610.

Hoch, H.C., Staples, R.C., Whitehead, B., Comeau, J. & Wolf, E.D. (1987). Signaling for growth orientation and cell differentiation by surface topography in *Uromyces*. *Science*, **235**, 1659–1662.

Hock, B., Bahr, M., Walk, R.-A. & Nitschke, U. (1978). The control of fruiting body formation in the ascomycete *Sordaria macrospora* Auersw. by regulation of hyphal development. *Planta*, **141**, 93–103.

Hocking, D. (1963). β-Carotene and sexuality in the Mucoraceae. *Nature*, **197**, 404.

Hodge, K.T. (2003). Clavicipitaceous anamorphs. In *Clavicipitalean Fungi*, ed. J.F. White, C.W. Bacon, N.L. Hywel-Jones & J.W. Spatafora. New York: Marcel Dekker Inc., pp. 75–123.

Hodge, K.T., Krasnoff, S.B. & Humber, R.A. (1996). *Tolypocladium inflatum* is the anamorph of *Cordyceps subsessilis*. *Mycologia*, **88**, 715–719.

Hofmann, A. (1979). *LSD – Mein Sorgenkind*. Stuttgart: J.G. Cotta.

Hogan, L.H., Klein, B.S. & Levitz, S.M. (1996). Virulence factors of medically important fungi. *Clinical Microbiology Reviews*, **9**, 469–488.

Hohl, H. R. & Iselin, K. (1984). Strains of *Phytophthora infestans* from Switzerland with A$_2$ mating type behaviour. *Transactions of the British Mycological Society*, **83**, 529–530.

Hohmeyer, H. (1986). Ein Schlüssel zu den europäischen Arten der Gattung *Peziza* L. *Zeitschrift für Mykologie*, **52**, 161–188.

Hohn, T. M., Lovett, J. S. & Bracker, C. E. (1984). Characterization of the major proteins in gamma particles, cytoplasmic organelles in *Blastocladiella emersonii* zoospores. *Journal of Bacteriology*, **158**, 253–263.

Holdenrieder, O. & Greig, B. J. W. (1998). Biological methods of control. In *Heterobasidion annosum. Biology, Ecology, Impact and Control*, ed. S. Woodward, J. Stenlid, R. Karjalainan & A. Hüttermann. Wallingford, UK: CAB International, pp. 235–258.

Holfeld, H. (1998). Fungal infections of the phytoplankton: seasonality, minimal host density, and specificity in a mesotrophic lake. *New Phytologist*, **138**, 507–517.

Hölker, U., Érsek, T. & Hofer, M. (1993). Changes in ion fluxes and the energy demand during zoospore development in *Phytophthora infestans* zoospores. *Folia Microbiologica*, **38**, 193–200.

Holliday, P. (1998). *A Dictionary of Plant Pathology*, 2nd edn. Cambridge: Cambridge University Press.

Holliday, R. (1962). Mutation and replication in *Ustilago maydis*. *Genetical Research*, **3**, 472–486.

Holliday, R. (1964). A mechanism for gene conversion in fungi. *Genetical Research*, **5**, 282–304.

Hollomon, D. W. & Schmidt, H.-H. (1995). 2-Aminopyrimidine fungicides. In *Modern Selective Fungicides. Properties, Applications, Mechanisms of Action*, 2nd edn, ed. H. Lyr. New York: Gustav Fischer Verlag, pp. 355–371.

Hollomon, D. W. & Wheeler, I. E. (2002). Controlling powdery mildews with chemistry. In *The Powdery Mildews. A Comprehensive Treatise*, ed. R. R. Bélanger, W. R. Bushnell, A. J. Dik & T. L. W. Carver. St. Paul, USA: APS Press, pp. 249–255.

Holloway, S. A. & Heath, I. B. (1977a). Morphogenesis and the role of microtubules in synchronous populations of *Saprolegnia* zoospores. *Experimental Mycology*, **1**, 9–29.

Holloway, S. A. & Heath, I. B. (1977b). An ultrastructural analysis of changes in organelle arrangement and structure between the various spore types of *Saprolegnia*. *Canadian Journal of Botany*, **55**, 1328–1339.

Holm, L. (1959). Some comments on the ascocarps of the Pyrenomycetes. *Mycologia*, **50**, 777–788.

Holst-Jensen, A., Kohn, L. M. & Schumacher, T. (1997). Nuclear rDNA phylogeny of the Sclerotiniaceae. *Mycologia*, **89**, 885–899.

Holtorf, S., Ludwig-Müller, J., Apel, & Bohlmann, H. (1998). High level expression of a viscotoxin in *Arabidopsis thaliana* gives enhanced resistance against *Plasmodiophora brassicae*. *Plant Molecular Biology*, **36**, 673–680.

Holtz, B. A., Michailides, T. J. & Hong, C. (1998). Development of apothecia from stone fruit infected and stromatized by *Monilinia fructicola* in California. *Plant Disease*, **82**, 1375–1380.

Holub, E. B., Brose, E., Tor, M., Clay, C., Crute, I. R. & Beynon, J. L. (1995). Phenotypic and genotypic variation in the interaction between *Arabidopsis thaliana* and *Albugo candida*. *Molecular Plant–Microbe Interactions*, **8**, 916–928.

Honegger, R. (1978). The ascus apex in lichenized fungi I. The *Lecanora-, Peltigera- and Teleoschistes*-types. *Lichenologist*, **10**, 47–67.

Honegger, R. (1986). Ultrastructural studies in lichens. I. Haustorial types and their frequencies in a range of lichens with trebouxioid phycobionts. *New Phytologist*, **103**, 785–795.

Honegger, R. (1993). Developmental biology of lichens. *New Phytologist*, **125**, 659–677.

Honegger, R. (1997). Metabolic interactions at the mycobiont-photobiont interface in lichens. In *The Mycota VA: Plant Relationships*, ed. G. C. Carroll & P. Tudzynski. Berlin: Springer-Verlag, pp. 209–221.

Honegger, R. (2001). The symbiotic phenotype of lichen-forming ascomycetes. In *The Mycota IX: Fungal Associations*, ed. B. Hock. Berlin: Springer-Verlag, pp. 165–188.

Hood, M. E. (2002). Dimorphic mating-type chromosomes in the fungus *Microbotryum violaceum*. *Genetics*, **160**, 457–461.

Hooker, A. L. (1985). Corn and sorghum rusts. In *The Cereal Rusts 2: Diseases, Distribution, Epidemiology, and Control*, ed. A. P. Roelfs & W. R. Bushnell. Orlando: Academic Press, pp. 207–236.

Hopple, J. S. & Vilgalys, R. (1999). Phylogenetic relationships in the mushroom genus *Coprinus* and dark-spored allies based on sequence data from the nuclear gene coding for the large ribosomal subunit RNA: divergent domains, outgroups and monophyly. *Molecular Phylogenetics and Evolution*, **13**, 1–19.

Horák, J. (2003). The role of ubiquitin in down-regulation and intracellular sorting of membrane proteins: insights from yeast. *Biochimica et Biophysica Acta*, **1614**, 139–155.

Horan, D.P., Chilvers, G.A. & Lapeyrie, F.F. (1988). Time sequence of the infection process in eucalypt ectomycorrhizas. *New Phytologist*, **109**, 451–458.

Horgen, P.A. & Griffin, D.H. (1969). Structure and germination of *Blastocladiella emersonii* resistant sporangia. *American Journal of Botany*, **56**, 22–25.

Horgen, P.A., Meyer, R.J., Franklin, A.L., Anderson, J.B. & Filion, W.G. (1985). Motile spores from resistant sporangia of *Blastocladiella emersonii* possess one-half the DNA of spores from ordinary colourless sporangia. *Experimental Mycology*, **9**, 70–73.

Horio, T. & Oakley, B.R. (2005). The role of microtubules in rapid hyphal tip growth of *Aspergillus nidulans*. *Molecular Biology of the Cell*, **16**, 918–926.

Horn, B.W. (1989a). Ultrastructural changes in trichospores of *Smittium culisetae* and *S. culicis* during *in vitro* sporangiospore extrusion and holdfast formation. *Mycologia*, **81**, 742–753.

Horn, B.W. (1989b). Requirement for potassium and pH shift in host-mediated sporangiospore extrusion from trichospores of *Smittium culisetae* and other *Smittium* species. *Mycological Research*, **93**, 303–313.

Horn, B.W. (1990). Physiological changes associated with sporangiospore extrusion from trichospores of *Smittium culisetae*. *Experimental Mycology*, **14**, 113–123.

Horner, J. & Moore, D. (1987). Cystidial morphogenetic field in the hymenium of *Coprinus cinereus*. *Transactions of the British Mycological Society*, **88**, 479–488.

Hornsey, I.S. (2003). *A History of Beer and Brewing*. Cambridge: Royal Society of Chemistry.

Horré, R. & de Hoog, G.S. (1999). Primary cerebral infections caused by melanized fungi: a review. *Studies in Mycology*, **43**, 176–193.

Horsch, M., Mayer, C., Sennhauser, U. & Rast, D.M. (1997). β-N-acetylhexosaminidase: a target for the design of antifungal agents. *Pharmacology Therapy*, **76**, 187–218.

Hoshino, T., Kiriaki, M., Ohgiya, S., *et al.* (2003). Antifreeze proteins from snow mold fungi. *Canadian Journal of Botany*, **81**, 1175–1181.

Hostinová, E. (2002). Amylolytic enzymes produced by the yeast *Saccharomycopsis fibuligera*. *Biologia*, **57**, Supplement 11, 247–251.

Howard, D.H., Weitzman, I. & Padhye, A.A. (2003). Onygenales: Arthrodermataceae. In *Pathogenic Fungi in Humans and Animals* 2nd edn, ed. D.H. Howard. New York: Marcel Dekker, pp. 141–194.

Howard, K.L. (1971). Oospore types in the Saprolegniaceae. *Mycologia*, **63**, 679–686.

Howard, K.L. & Bigelow, H.E. (1969). Nutritional studies on two Gasteromycetes: *Phallus ravenelii* and *Crucibulum levis*. *Mycologia*, **61**, 606–613.

Howard, R.J. (1981). Ultrastructural analysis of hyphal tip cell growth in fungi: Spitzenkörper, cytoskeleton and endomembranes after freeze-substitution. *Journal of Cell Science*, **48**, 89–103.

Howard, R.J. (1994). Cell biology of pathogenesis. In *Rice Blast Disease*, ed. R.S. Zeiger, S. Leong & P.S. Teng. Wallingford, UK: CAB International, pp. 3–22.

Howard, R.J. (1997). Breaching the outer barriers – cuticle and cell wall penetration. In *The Mycota VA: Plant Relationships*, ed. G.C. Carroll & P. Tudzynski. Berlin: Springer-Verlag, pp. 43–60.

Howard, R.J. & Aist, J.R. (1977). Effects of MBC on hyphal tip organization, growth and mitosis of *Fusarium acuminatum*, and their antagonism by D_2O. *Protoplasma*, **92**, 195–210.

Howard, R.J. & Aist, J.R. (1980). Cytoplasmic microtubules and fungal morphogenesis: ultrastructural effects of methyl benzimidazole-2-ylcarbamate determined by freeze-substitution of hyphal tip cells. *Journal of Cell Biology*, **87**, 55–64.

Howard, R.J. & Valent, B. (1996). Breaking and entering: host penetration by the fungal rice blast pathogen *Magnaporthe grisea*. *Annual Review of Microbiology*, **50**, 491–512.

Howard, R.J., Ferrari, M.A., Roach, D.H. & Money, N.P. (1991). Penetration of hard substrates by a fungus employing enormous turgor pressures. *Proceedings of the National Academy of Sciences of the USA*, **88**, 11 281–284.

Howlett, B.J., Idnurm, A. & Pedras, M.S.C. (2001). *Leptosphaeria maculans*, the causal agent of blackleg disease of Brassicas. *Fungal Genetics and Biology*, **33**, 1–14.

Hsam, S.L.K. & Zeller, F.J. (2002). Breeding for powdery mildew resistance in common wheat (*Triticum aestivum* L.). In *The Powdery Mildews. A Comprehensive Treatise*, ed. R.R. Bélanger, W.R. Bushnell, A.J. Dik & T.L.W. Carver. St. Paul, USA: APS Press, pp. 219–38.

Hu, F.-M., Zheng, R.-Y. & Chen, G.-Q. (1989). A redelimitation of the species of *Pilobolus*. *Mycosystema*, **2**, 111–133.

Hu, G.G., Linning, R. & Bakkeren, G. (2002). Sporidial mating and infection process of the smut fungus, *Ustilago hordei*, in susceptible barley. *Canadian Journal of Botany*, **80**, 1103–1114.

Hu, G.G., Linning, R. & Bakkeren, G. (2003). Ultrastructural comparison of a compatible and incompatible interaction triggered by the presence of an avirulence gene during early infection of the smut fungus, *Ustilago hordei*, in barley. *Physiological and Molecular Plant Pathology*, **62**, 155–166.

Huang, B., Li, Z.G., Fan, M.Z. & Li, Z.Z. (2002). Molecular identification of the teleomorph of *Beauveria bassiana*. *Mycotaxon*, **81**, 229–236.

Huang, H.-Q. & Nielsen, J. (1984). Hybridization of the seedling-infecting *Ustilago* spp. pathogenic on barley and oats, and a study of the genotypes conditioning the morphology of their spore walls. *Canadian Journal of Botany*, **62**, 603–608.

Hua-Van, A., Langin, T. & Daboussi, M.-J. (2001). Evolutionary history of the *impala* transposon in *Fusarium oxysporum*. *Molecular Biology and Evolution*, **18**, 1959–1969.

Hubalek, Z. (2000). Keratinolytic fungi associated with free-living mammals and birds. In *Biology of Dermatophytes and Other Keratinolytic Fungi*, ed. R.K.S. Kushwara & J. Guarro. Bilbao: Revista Iberoamericana de Micología, pp. 93–103.

Hube, B. & Naglik, J. (2001). *Candida albicans* proteinases: resolving the mystery of a gene family. *Microbiology*, **147**, 1997–2005.

Hudson, R.E., Aukema, J.E., Rispe, C. & Roze, D. (2002). Altruism, cheating, and anticheater adaptations in cellular slime molds. *American Naturalist*, **160**, 31–43.

Hudspeth, D.S.S., Nadler, S.A. & Hudspeth, M.E.S. (2000). A *COX2* molecular phylogeny of the Peronosporomycetes. *Mycologia*, **92**, 674–684.

Hughes, K.W., McGhee, L.L., Methven, A.S., Johnson, J.E. & Petersen, R.H. (1999). Patterns of geographic speciation in the genus *Flammulina* based on sequences of the ribosomal ITS–5.8S–ITS2 area. *Mycologia*, **91**, 978–986.

Hughes, S.J. (1953). Conidiophores, conidia and classification. *Canadian Journal of Botany*, **31**, 577–659.

Hughes, S.J. (1985). The term chlamydospore. In *Filamentous Micro-Organisms. Biomedical Aspects*, ed. T. Arai. Tokyo: Japanese Scientific Societies Press, pp. 1–20.

Huhndorf, S.M., Miller, A.N. & Fernández, F.A. (2004). Molecular systematics of the Sordariales: the order and the family Lasiosphaeriaceae redefined. *Mycologia*, **96**, 368–387.

Humber, R.A. (1989). Synopsis of a revised classification for the Entomophthorales (Zygomycotina). *Mycotaxon*, **34**, 441–460.

Humber, R.A. (1997). Fungi: Identification. In *Manual of Techniques in Insect Pathology*, ed. L.A. Lacey. San Diego: Academic Press, pp. 153–185.

Humber, R.A. (2000). Fungal pathogens and parasites of insects. In *Applied Microbial Systematics*, ed. F. Priest & M. Goodfellow. London: Chapman & Hall, pp. 199–230.

Humpert, A.J., Muench, E.L., Giachini, A.J., Castellano, M.A. & Spatafora, J.W. (2001). Molecular phylogenetics of *Ramaria* and related genera: evidence from nuclear large subunit and mitochondrial small subunit rDNA sequences. *Mycologia*, **93**, 465–477.

Hung, C.-Y. (1977). Ultrastructural studies of ascospore liberation in *Pyronema domesticum*. *Canadian Journal of Botany*, **55**, 2544–2549.

Hung, C.-Y. & Wells, K. (1971). Light and electron microscopic studies of crozier development in *Pyronema domesticum*. *Journal of General Microbiology*, **66**, 15–27.

Hunsley, D. & Burnett, J.H. (1970). The ultrastructural architecture of the walls of some hyphal fungi. *Journal of General Microbiology*, **62**, 203–218.

Hutchison, L.J. & Kawchuk, L.M. (1998). *Spongospora subterranea* f. sp. *subterranea*. *Canadian Journal of Plant Pathology*, **20**, 118–119.

Hyde, G.J. & Hardham, A.R. (1992). Confocal microscopy of microtubule arrays in cryosectioned sporangia of *Phytophthora cinnamomi*. *Experimental Mycology*, **16**, 207–218.

Hyde, G.J. & Hardham, A.R. (1993). Microtubules regulate the generation of polarity in zoospores of *Phytophthora cinnamomi*. *European Journal of Cell Biology*, **62**, 75–85.

Hyde, G.J. & Heath, I.B. (1997). Ca²⁺ gradients in hyphae and branches of *Saprolegnia ferax*. *Fungal Genetics and Biology*, **21**, 238–251.

Hyde, G.J., Gubler, F. & Hardham, A.R. (1991). Ultrastructure of zoosporogenesis in *Phytophthora cinnamomi*. *Mycological Research*, **95**, 577–591.

Hyde, K.D. & Jones, E.B.G. (1989). Observations on ascospore morphology in marine fungi and their attachment to surfaces. *Botanica Marina*, **32**, 205–208.

Hyde, K.D., Moss, S.T. & Jones, E.B.G. (1989). Attachment studies in marine fungi. *Biofouling*, **1**, 287–298.

Hywel-Jones, N.L. (2002). Multiples of eight in *Cordyceps* ascospores. *Mycological Research*, **106**, 2–3.

Hywel-Jones, N. & Webster, J. (1986a). Scanning electron microscope study of external development of *Erynia conica* on *Simulium*. *Transactions of the British Mycological Society*, **86**, 393–399.

Hywel-Jones, N. & Webster, J. (1986b). Mode of infection of *Simulium* by *Erynia conica*. *Transactions of the British Mycological Society*, **87**, 381–387.

Ikediugwu, F.E.O. & Webster, J. (1970a). Antagonism between *Coprinus heptemerus* and other coprophilous fungi. *Transactions of the British Mycological Society*, **54**, 181–204.

Ikediugwu, F.E.O. & Webster, J. (1970b). Hyphal interference in a range of coprophilous fungi. *Transactions of the British Mycological Society*, **54**, 205–210.

Ikediugwu, F.E.O., Dennis, C. & Webster, J. (1970). Hyphal interference by *Peniophora gigantea* against *Heterobasidion annosum*. *Transactions of the British Mycological Society*, **54**, 307–309.

Imai, Y. & Yamamoto, M. (1994). The fission yeast mating pheromone P-factor: its molecular structure, gene structure, and ability to induce gene expression and G1 arrest in the mating partner. *Genes and Development*, **8**, 328–338.

Inácio, J., Rodrigues, M.G., Sobral, P. & Fonseca, Á. (2004). Characterisation and classification of phylloplane yeasts from Portugal related to the genus *Taphrina* and description of five novel *Lalaria* species. *FEMS Yeast Research*, **4**, 541–555.

Inamadar, A.C. & Palit, A. (2003). The genus *Malassezia* and human disease. *Indian Journal of Dermatology, Venereology and Leprology*, **69**, 265–270.

Ing, B. (1998). *Exobasidium* in the British Isles. *Mycologist*, **12**, 80–82.

Ing, B. (1999). *The Myxomycetes of Britain and Ireland*. Slough: Richmond Publishing Company.

Inglis, G.D., Yanke, L.J., Kawchuk, L.M. & McAllister, T.A. (1999). The influence of bacterial inoculants on the microbial ecology of aerobic spoilage of barley silage. *Canadian Journal of Microbiology*, **45**, 77–87.

Inglis, R.J., Langan, T.A., Matthews, H.R., Hardie, D.G. & Bradbury, E.M. (1976). Advance of mitosis by histone phosphokinase. *Experimental Cell Research*, **97**, 418–425.

Ingold, C.T. (1939). *Spore Discharge in Land Plants*. Oxford: Clarendon Press.

Ingold, C.T. (1946). Spore discharge in *Daldinia concentrica*. *Transactions of the British Mycological Society*, **29**, 43–51.

Ingold, C.T. (1953). *Dispersal in Fungi*. Oxford: Clarendon Press.

Ingold, C.T. (1954a). Ascospore form. *Transactions of the British Mycological Society*, **37**, 19–21.

Ingold, C.T. (1954b). The ascogenous hyphae in *Daldinia concentrica*. *Transactions of the British Mycological Society*, **37**, 108–110.

Ingold, C.T. (1956). A gas phase in viable fungal spores. *Nature*, **177**, 1242–1243.

Ingold, C.T. (1957). Spore liberation in higher fungi. *Endeavour*, **16**, 78–83.

Ingold, C.T. (1959). Jelly as a water reserve in fungi. *Transactions of the British Mycological Society*, **42**, 475–478.

Ingold, C.T. (1966). Aspects of spore liberation: violent discharge. In *The Fungus Spore*, ed. M.F. Madelin. London: Butterworths, pp. 113–32.

Ingold, C.T. (1971). *Fungal Spores. Their Liberation and Dispersal*. Oxford: Clarendon Press.

Ingold, C.T. (1972). *Sphaerobolus*: the story of a fungus. *Transactions of the British Mycological Society*, **58**, 179–195.

Ingold, C.T. (1975). *An Illustrated Guide to Aquatic and Water-Borne Hyphomycetes (Fungi Imperfecti) with Notes on their Biology*. Ambleside, UK: Freshwater Biological Association.

Ingold, C.T. (1980). Mycelia, oidia and sporophore initials in *Flammulina velutipes*. *Transactions of the British Mycological Society*, **75**, 107–116.

Ingold, C.T. (1981). The first-formed phialoconidium of *Thielaviopsis basicola*. *Transactions of the British Mycological Society*, **76**, 517–519.

Ingold, C.T. (1982a). Basidiospore germination and conidium development in *Auricularia*. *Transactions of the British Mycological Society*, **78**, 161–166.

Ingold, C.T. (1982b). Basidiospore germination and conidium formation in *Exidia glandulosa* and *Tremella mesenterica*. *Transactions of the British Mycological Society*, **79**, 370–373.

Ingold, C.T. (1983a). Structure and development in an isolate of *Itersonilia perplexans*. *Transactions of the British Mycological Society*, **80**, 365–368.

Ingold, C.T. (1983b). Basidiospore germination and conidium development in Dacrymycetales. *Transactions of the British Mycological Society*, **81**, 563–571.

Ingold, C.T. (1983c). The basidium in *Ustilago*. *Transactions of the British Mycological Society*, **81**, 573–584.

Ingold, C.T. (1984a). Further observations on *Itersonilia*. *Transactions of the British Mycological Society*, **83**, 166–174.

Ingold, C.T. (1984b). Patterns of ballistospore germination in *Tilletiopsis*, *Auricularia* and *Tulasnella*. *Transactions of the British Mycological Society*, **83**, 583–591.

Ingold, C.T. (1985). Observations on spores and their germination in certain Heterobasidiomycetes. *Transactions of the British Mycological Society*, **85**, 417–423.

Ingold, C.T. & Cox, V.J. (1955). Periodicity of spore discharge in *Daldinia concentrica*. *Annals of Botany* N. S., **21**, 201–209.

Ingold, C.T. & Hadland, S.A. (1959). The ballistics of *Sordaria*. *New Phytologist*, **58**, 46–57.

Ingold, C. T. & Zoberi, M. H. (1963). The asexual apparatus of Mucorales in relation to spore liberation. *Transactions of the British Mycological Society*, **46**, 115–134.

Ingold, C. T., Davey, R. A. & Wakley, G. (1981). The teliospore pedicel of *Phragmidium mucronatum*. *Transactions of the British Mycological Society*, **77**, 439–442.

Ingram, D. S. & Williams, P. H., eds. (1991). *Phytophthora infestans, the Cause of Late Blight of Potato*. London: Academic Press.

Innes, J. L. (1988). The use of lichens in dating. In *CRC Handbook of Lichenology III*, ed. M. Galun. Boca Raton: CRC Press, pp. 75–91.

Innocenti, G., Roberti, R., Montanari, M. & Zakrisson, E. (2003). Efficacy of microorganisms antagonistic to *Rhizoctonia cerealis* and their cell wall degrading enzymatic activities. *Mycological Research*, **107**, 421–427.

Iqbal, S. H. & Webster, J. (1973a). The trapping of aquatic hyphomycete spores by air bubbles. *Transactions of the British Mycological Society*, **60**, 37–48.

Iqbal, S. H. & Webster, J. (1973b). Aquatic hyphomycete spora of the River Exe and its tributaries. *Transactions of the British Mycological Society*, **61**, 331–346.

Isaka, M., Kittakoop, P. & Thebtaranonth, Y. (2003). Secondary metabolites of Clavicipitalean fungi. In *Clavicipitalean Fungi*, ed. J. F. White, C. W. Bacon, N. L. Hywel-Jones & J. W. Spatafora. New York: Marcel Dekker Inc., pp. 355–397.

Ishibashi, Y., Sakagami, Y., Isogai, A. & Suzuki, A. (1984). Structures of tremerogens A-9291–I and A-9291–VIII: peptidyl sex hormones of *Tremella brasiliensis*. *Biochemistry*, **23**, 1399–1404.

Issakainen, J., Jalava, J., Hyvönen, J., et al. (2003). Relationships of *Scopulariopsis* based on LSU rDNA sequences. *Medical Mycology*, **41**, 31–42.

Iten, W. & Matile, P. (1970). Role of chitinase and other lysosomal enzymes of *Coprinus lagopus* in the autolysis of fruiting bodies. *Journal of General Microbiology*, **61**, 301–309.

Ito, H., Hanyaku, H., Harada, T., Mochizuki, T. & Tanaka, S. (1998). Ultrastructure of the ascospore formation of *Arthroderma simii*. *Mycoses*, **41**, 133–137.

Jackson, G. V. H. & Gay, J. L. (1976). Perennation of *Sphaerotheca mors-uvae* as cleistothecia with particular reference to microbial activity. *Transactions of the British Mycological Society*, **66**, 463–471.

Jaffee, B. A. (1992). Population biology and biological control of nematodes. *Canadian Journal of Microbiology*, **38**, 359–364.

Jahn, M., Munger, H. M. & McCreight, J. D. (2002). Breeding cucurbit crops for powdery mildew resistance. In *The Powdery Mildews. A Comprehensive Treatise*, ed. R. R. Bélanger, W. R. Bushnell, A. J. Dik & T. L. W. Carver. St. Paul, USA: APS Press, pp. 239–248.

Jailloux, F., Thind, T. & Clerjeau, M. (1998). Release, germination, and pathogenicity of ascospores of *Uncinula necator* under controlled conditions. *Canadian Journal of Botany*, **76**, 777–781.

Jailloux, F., Willocquet, L., Chapuis, L. & Froidefond, G. (1999). Effect of weather factors on the release of ascospores of *Uncinula necator*, the cause of grape powdery mildew, in the Bordeaux region. *Canadian Journal of Botany*, **77**, 1044–1051.

Jain, N., Li, L., McFadden, D. C., et al. (2006). Phenotypic switching in a *Cryptococcus neoformans* variety *gattii* strain is associated with changes in virulence and promotes dissemination to the central nervous system. *Infection and Immunity*, **74**, 896–903.

Jakobsen, I. & Rosendahl, L. (1990). Carbon flow into soil and external hyphae from roots of mycorrhizal cucumber plants. *New Phytologist*, **115**, 77–83.

Jakobsen, M., Cantor, M. D. & Jespersen, L. (2002). Production of bread, cheese and meat. In *The Mycota X: Industrial Applications*, ed. H. D. Osiewacz. Berlin: Springer-Verlag, pp. 3–22.

James, T. Y. & Vilgalys, R. (2001). Abundance and diversity of *Schizophyllum commune* spore clouds in the Caribbean detected by selective sampling. *Molecular Ecology*, **10**, 471–479.

James, T. Y., Porter, D., Leander, C. A., Vilgalys, R. & Longcore, J. E. (2000). Molecular phylogenetics of the Chytridiomycota supports the utility of ultrastructural data in chytrid systematics. *Canadian Journal of Botany*, **78**, 336–350.

Janex-Favre, M. C. & Parguey-Leduc, A. (2002). Particularités des ascospores et de l'hyménium des truffes (Ascomycètes) I. Développement et structure des ascocarpes. *Cryptogamie Mycologie*, **23**, 103–128.

Janex-Favre, M.-C., Parguey-Leduc, A. & Bruxelles, G. (1998). L'hyménium de *Morchella deliciosa* Fr. (Ascomycètes, Discomycètes). *Cryptogamie Bryologie–Lichénologie*, **19**, 293–304.

Janisiewicz, W. J. & Bors, B. (1995). Development of a microbial community of bacterial and yeast antagonists to control wound-invading postharvest pathogens of fruits. *Applied and Environmental Microbiology*, **61**, 3261–3267.

Janisiewicz, W. J. & Korsten, L. (2002). Biological control of postharvest diseases of fruits. *Annual Review of Phytopathology*, **40**, 411–441.

Jansson, H.-B. & Nordbring-Hertz, B. (1979). Attraction of nematodes to living mycelium of nematophagous fungi. *Journal of General Microbiology*, **112**, 89–93.

Jansson, H.-B. & Nordbring-Hertz, B. (1980). Interactions between nematophagous fungi and plant-parasitic nematodes: attraction, induction of trap formation and capture. *Nematologica*, **26**, 383–389.

Jansson, H.-B., Jeyaprakash, A. & Zuckerman, B.M. (1985). Differential adhesion and infection of nematodes by the endoparasitic fungus *Meria coniospora* (Deuteromycetes). *Applied and Environmental Microbiology*, **49**, 552–555.

Jarl, K. (1969). Symba yeast process. *Food Technology*, **23**, 23–26.

Jarvis, W.R., Gubler, W.D. & Grove, G.G. (2002). Epidemiology of powdery mildews in agricultural pathosystems. In *The Powdery Mildews. A Comprehensive Treatise*, ed. R.R. Bélanger, W.R. Bushnell, A.J. Dik & T.L.W. Carver. St. Paul, USA: APS Press, pp. 169–199.

Javadekar, V.S., Sivaraman, H. & Gokhale, D.V. (1995). Industrial yeast strain improvement: construction of a highly flocculent yeast with a killer character by protoplast fusion. *Journal of Industrial Microbiology*, **15**, 94–102.

Jazwinski, S.M. (2002). Growing old: metabolic control and yeast aging. *Annual Review of Microbiology*, **56**, 769–792.

Jee, H.J., Ho, H.H. & Cho, W.D. (2000). *Pythiogeton zeae* sp. nov. causing root and basal stalk rot of corn in Korea. *Mycologia*, **92**, 522–527.

Jeffries, P. (1999). *Scleroderma*. In *Ectomycorrhizal Fungi: Key Genera in Profile*, ed. J.W.G. Cairney & S.M. Chambers. Berlin: Springer-Verlag, pp. 187–200.

Jeffries, P. & Young, T.W.K. (1975). Ultrastructure of the sporangiospores of *Piptocephalis unispora*. *Archiv für Mikrobiologie*, **105**, 329–333.

Jeffries, P. & Young, T.W.K. (1976). Ultrastructure of infection of *Cokeromyces recurvatus* by *Piptocephalis unispora* (Mucorales). *Archives of Microbiology*, **109**, 277–288.

Jeffries, P. & Young, T.W.K. (1994). *Interfungal Parasitic Relationships*. Wallingford, UK: CAB International.

Jeffries, T.W. & Kurtzman, C.P. (1994). Strain selection, taxonomy, and genetics of xylose-fermenting yeasts. *Enzyme and Microbial Technology*, **16**, 922–932.

Jeger, M.J., Gilijamse, E., Bock, C.H. & Frinking, H.D. (1998). The epidemiology, variability and control of the downy mildews of pearl millet and sorghum, with particular reference to Africa. *Plant Pathology*, **47**, 544–569.

Jeng, R.S. & Hubbes, M. (1980). Ultrastructure of *Ceratocystis ulmi*. II. Ascogenous system and ascosporogenesis. *European Journal of Forest Pathology*, **10**, 104–116.

Jenkinson, P. & Parry, D.W. (1994). Splash dispersal of conidia of *Fusarium culmorum* and *Fusarium avenaceum*. *Mycological Research*, **98**, 506–510.

Jenkyn, J.F. & Bainbridge, A. (1978). Biology and pathology of cereal powdery mildews. In *The Powdery Mildews*, ed. D.M. Spencer. London: Academic Press, pp. 283–321.

Jennersten, O. (1988). Insect dispersal of fungal disease: effects of *Ustilago* infection on pollinator attraction in *Viscaria vulgaris*. *Oikos*, **51**, 163–170.

Jennings, D.H. (1987). Translocation of solutes in fungi. *Biological Reviews*, **62**, 215–243.

Jennings, D.H. (1991). The physiology and biochemistry of the vegetative mycelium. In *Serpula lacrymans. Fundamental Biology and Control Strategies*, ed. D.H. Jennings & A.F. Bravery. Chichester: John Wiley & Sons, pp. 55–79.

Jennings, D.H. (1995). *The Physiology of Fungal Nutrition*. Cambridge: Cambridge University Press.

Jennings, D.H. & Bravery, A.F., eds. (1991). *Serpula lacrymans. Fundamental Biology and Control Strategies*. Chichester: John Wiley & Sons.

Jennings, D.H. & Watkinson, S.C. (1982). Structure and development of mycelial strands in *Serpula lacrimans*. *Transactions of the British Mycological Society*, **78**, 465–474.

Jensen, A.B., Gargas, A., Eilenberg, J. & Rosendahl, S. (1998). Relationships of the insect-pathogenic order Entomophthorales (Zygomycota, Fungi) based on phylogenetic analyses of nuclear small subunit ribosomal DNA sequences (SSU rDNA). *Fungal Genetics and Biology*, **24**, 325–334.

Jensen, J.D. (1983). The development of *Diaporthe phaseolorum* variety *sojae* in culture. *Mycologia*, **75**, 1074–1091.

Jensen, R.E., Hobbs, A.E.A., Cerveny, K.L. & Sesaki, H. (2000). Yeast mitochondrial dynamics: fusion, division, segregation, and shape. *Microscopy Research and Technique*, **51**, 573–583.

Jenson, I. (1998). Bread and baker's yeast. In *Microbiology of Fermented Foods*, 2nd edn, ed. B.J.B. Wood. London: Blackie Academic, pp. 172–95.

Jesenská, Z., Piecková, E. & Bernát, D. (1993). Heat resistance of fungi from soil. *International Journal of Food Microbiology*, **19**, 187–192.

Ji, T. & Dayal, R. (1971). Studies in the life cycle of *Allomyces javanicus* Kniep. *Hydrobiologia*, **37**, 245–251.

Jiang, J., Stephenson, L. W., Erwin, D. C. & Leary, J. V. (1989). Nuclear changes in *Phytophthora* during oospore maturation and germination. *Mycological Research*, **92**, 463–469.

Jiang, Y. & Yao, Y.-J. (2002). Names related to *Cordyceps sinensis* anamorph. *Mycotaxon*, **84**, 245–254.

Joffe, A. Z. (1986). *Fusarium Species: Their Biology and Toxicology*. New York: John Wiley & Sons.

Johannesson, H., Læssøe, T. & Stenlid, J. (2000). Molecular and morphological investigation of *Daldinia* in northern Europe. *Mycological Research*, **104**, 275–280.

Johannesson, H., Gustafsson, M. & Stenlid, J. (2001). Local population structure of the wood decay ascomycete *Daldinia loculata*. *Mycologia*, **93**, 440–446.

Johnson, B. F., Sowden, L. C., Walker, T., Yoo, B. Y. & Calleja, G. B. (1989). Use of electron microscopy to characterize the surfaces of flocculent and nonflocculent yeast cells. *Canadian Journal of Microbiology*, **35**, 1081–1086.

Johnson, E. A. & An, G.-H. (1991). Astaxanthin from microbial sources. *Critical Reviews in Biotechnology*, **11**, 297–326.

Johnson, E. A. & Schroeder, W. A. (1995a). Microbial carotenoids. *Advances in Biochemical Engineering/Biotechnology*, **53**, 119–178.

Johnson, E. A. & Schroeder, W. (1995b). Astaxanthin from the yeast *Phaffia rhodozyma*. *Studies in Mycology*, **38**, 81–90.

Johnson, E. M. & Valleau, W. D. (1949). Synonymy in some common species of *Cercospora*. *Phytopathology*, **39**, 763–770.

Johnson, T. (1953). Variation in the rusts of cereals. *Biological Reviews*, **28**, 105–157.

Johnson, T. W. (1969). The aquatic fungi of Iceland: *Olpidium* (Braun) Rabenhorst. *Archiv für Mikrobiologie*, **69**, 1–11.

Johnson, T. W. (1973). Aquatic fungi from Iceland: some polycentric species. *Mycologia*, **65**, 1337–1355.

Johnson, T. W. (1977). Resting spore germination in three chytrids. *Mycologia*, **69**, 34–45.

Johnston, P. R. (1997). Tropical Rhytismatales. In *Biodiversity of Tropical Microfungi*, ed. K. D. Hyde. Hong Kong: Hong Kong University Press, pp. 241–254.

Joly, P. (1964). *Le Genre Alternaria*. Paris: Editions Paul Chevalier.

Jones, A. M. & Jones, E. B. G. (1993). Observations on the marine gasteromycete *Nia vibrissa*. *Mycological Research*, **97**, 1–6.

Jones, B. E. & Gooday, G. W. (1977). Lectin binding to sexual cells in fungi. *Biochemical Society Transactions*, **5**, 717–719.

Jones, D., McHardy, W. J. & Wilson, M. J. (1976). Ultrastructure and chemical composition of spines in Mucorales. *Transactions of the British Mycological Society*, **66**, 153–157.

Jones, E. B. G. (1994). Fungal adhesion. *Mycological Research*, **98**, 961–981.

Jones, P. (1999). Control of loose smut (*Ustilago nuda* and *U. tritici*) infections in barley and wheat by foliar applications of systemic fungicides. *European Journal of Plant Pathology*, **105**, 729–732.

Jones, S. G. (1925). Life-history and cytology of *Rhytisma acerinum* (Pers.) Fries. *Annals of Botany*, **39**, 41–73.

Jones, S. W., Donaldson, S. P. & Deacon, J. W. (1991). Behaviour of zoospores and zoospore cysts in relation to root infection by *Pythium aphanidermatum*. *New Phytologist*, **117**, 289–301.

Joosten, M. H. A. J., Hendrickx, L. J. M. & de Wit, P. G. J. M. (1990). Carbohydrate composition of apoplastic fluids isolated from tomato leaves inoculated with virulent or avirulent races of *Cladosporium fulvum* (syn. *Fulvia fulva*). *Netherlands Journal of Plant Pathology*, **96**, 103–112.

Joosten, M. H. A. J., Honée, G., van Kan, J. A. L. & de Wit, P. J. G. M. (1997). The gene-for-gene concept in plant-pathogen interactions: tomato–*Cladosporium fulvum*. In *The Mycota VB: Plant Relationships*, ed. G. C. Carroll & P. Tudzynski. Berlin: Springer-Verlag, pp. 3–16.

Jordan, M. (1995). Evidence of severe allergic reactions to *Laetiporus sulphureus*. *Mycologist*, **9**, 157–158.

Jørgensen, J. H. (1992). Discovery, characterization and exploitation of *Mlo* powdery mildew resistance genes in barley. *Euphytica*, **63**, 141–152.

Jørgensen, J. H. (1994). Genetics of powdery mildew resistance in barley. *Critical Reviews in Plant Science*, **13**, 97–119.

Ju, Y.-M. & Rogers, J. D. (1996). *A Revision of the Genus Hypoxylon*. St. Paul, USA: APS Press.

Ju, Y.-M., Rogers, J. D. & San Martin, F. (1997). A revision of the genus *Daldinia*. *Mycotaxon*, **61**, 243–293.

Jurgens, J. A., Blanchette, R. A., Zambino, P. J. & David, A. (2003). Histology of white pine blister rust in needles of resistant and susceptible Eastern white pine. *Plant Disease*, **87**, 1026–1030.

Kahmann, R., Steinberg, G., Basse, C., Feldbrügge, M. & Kämper, J. (2000). *Ustilago maydis*, the causative agent of corn smut disease. In *Fungal Pathology*, ed. J. W. Kronstad. Dordrecht: Kluwer Academic Publishers, pp. 347–371.

Kakani, K., Robbins, M. & Rochon, D. (2003). Evidence that binding of cucumber necrosis virus to vector zoospores involves recognition of oligosaccharides. *Journal of Virology*, **77**, 3922–3928.

Kaltz, O. & Shykoff, J. A. (2001). Male and female *Silene latifolia* plants differ in per-contact risk of infection by a sexually transmitted disease. *Journal of Ecology*, **89**, 99–109.

Kam, A. P. & Xu, J. (2002). Diversity of commensal yeasts within and among healthy hosts. *Diagnostic Microbiology and Infectious Disease*, **43**, 19–28.

Kamada, T. (1994). Stipe elongation in fruit bodies. In *The Mycota I: Growth, Differentiation and Sexuality*, ed. J. G. H. Wessels & F. Meinhardt. Berlin: Springer-Verlag, pp. 367–379.

Kamada, T., Hirai, K. & Fujii, M. (1993). The role of the cytoskeleton in the pairing and positioning of the two nuclei in the apical cells of the dikaryon of the basidiomycete *Coprinus cinereus*. *Experimental Mycology*, **17**, 338–344.

Kaminskyj, S. G. W. & Hamer, J. E. (1998). *hyp* loci control cell pattern formation in the vegetative mycelium of *Aspergillus nidulans*. *Genetics*, **148**, 669–680.

Kaminskyj, S. G. W. & Heath, I. B. (1996). Studies on *Saprolegnia ferax* suggest the general importance of the cytoplasm in determining hyphal morphology. *Mycologia*, **88**, 20–37.

Kamiya, Y., Sakurai, A., Tamura, S., *et al.* (1979). Structure of rhodotorucine A, a peptidyl factor, inducing mating tube formation in *Rhodosporidium toruloides*. *Agricultural and Biological Chemistry*, **43**, 363–369.

Kämper, J., Bölker, M. & Kahmann, R. (1994). Mating-type genes in Heterobasidiomycetes. In *The Mycota I: Growth, Differentiation and Sexuality*, ed. J. G. H. Wessels & F. Meinhardt. Berlin: Springer-Verlag, pp. 323–332.

Kapat, A., Zimand, G. & Elad, Y. (1998). Biosynthesis of pathogenicity hydrolytic enzymes by *Botrytis cinerea* during infection of bean leaves and *in vitro*. *Mycological Research*, **102**, 1017–1024.

Kaplan, J. D. & Goos, R. D. (1982). The effect of water potential on zygospore formation in *Syzygites megalocarpus*. *Mycologia*, **74**, 684–686.

Kapteyn, J. C., van den Ende, H. & Klis, F. M. (1999). The contribution of yeast wall proteins to the organization of the yeast cell wall. *Biochimica et Biophysica Acta*, **1426**, 373–383.

Karaffa, L. & Kubicek, C. P. (2003). *Aspergillus niger* citric acid accumulation: do we understand this well working black box? *Applied Microbiology and Biotechnology*, **61**, 189–196.

Karjalainen, R. & Lounatmaa, K. (1986). Ultrastructure of penetration and colonization of wheat leaves by *Septoria nodorum*. *Physiological and Molecular Plant Pathology*, **29**, 263–270.

Karling, J. S. (1950). The genus *Physoderma*. *Lloydia*, **13**, 1–71.

Karling, J. S. (1964). *Synchytrium*. New York and London: Academic Press.

Karling, J. S. (1968). *The Plasmodiophorales*. New York and London: Hafner Publishing Company.

Karling, J. S. (1969). Zoosporic fungi in Oceania. VII. Fusions in *Rhizophlyctis*. *American Journal of Botany*, **56**, 211–221.

Karling, J. S. (1973). A note on *Blastocladiella* (Blastocladiaceae). *Mycopathologia et Mycologia Applicata*, **49**, 169–172.

Karling, J. S. (1977). *Chytridiomycetarum Iconographia*. Monticello and New York: Lubrecht & Cramer.

Kars, I. & van Kan, J. A. L. (2004). Extracellular enzymes and metabolites involved in pathogenesis of *Botrytis*. In *Botrytis: Biology, Pathology and Control*, ed. Y. Elad, B. Williamson, P. Tudzynski & N. Delen. Dordrecht: Kluwer, pp. 99–118.

Kataria, H. R. & Hoffmann, G. M. (1988). A critical review of plant pathogenic species of *Ceratobasidium* Rogers. *Zeitschrift für Pflanzenkrankheiten und Pflanzenschutz*, **95**, 81–107.

Kauppi, M. (1979). The exploitation of *Cladonia stellaris* in Finland. *Lichenologist*, **11**, 85–89.

Kauserud, H., Högberg, N., Knudsen, H., Elborne, S. A. & Schumacher, T. (2004). Molecular phylogenetics suggest a North American link between the anthropogenic dry rot fungus *Serpula lacrymans* and its wild relative *S. himantioides*. *Molecular Ecology*, **13**, 3137–3146.

Kaushik, N. K. & Hynes, H. B. N. (1968). Experimental study on the role of autumn-shed leaves in aquatic environments. *Journal of Ecology*, **56**, 229–243.

Kaushik, N. K. & Hynes, H. B. N. (1971). The fate of dead leaves that fall into streams. *Archiv für Hydrobiologie*, **68**, 465–515.

Keenan, K. A. & Weiss, R. L. (1997). Characterization of vacuolar arginine uptake and amino acid efflux in *Neurospora crassa* using cupric ion to permeabilize the plasma membrane. *Fungal Genetics and Biology*, **22**, 177–190.

Kemen, E., Kemen, A. C., Rafiqi, M., *et al.* (2005). Identification of a protein from rust fungi transferred from haustoria into infected plant cells. *Molecular Plant–Microbe Interactions*, **18**, 1130–1139.

Kemp, R. F. O. (1975a). Breeding biology of *Coprinus* species in the section *Lanatuli*. *Transactions of the British Mycological Society*, **65**, 375–388.

Kemp, R. F. O. (1975b). Oidia, plasmogamy and speciation in basidiomycetes. *Biological Journal of the Linnean Society*, **7**, S57–S69.

Kendrick, B., ed. (1971). *Taxonomy of Fungi Imperfecti.* Toronto and Buffalo: Toronto University Press.

Kendrick, W. B. & Cole, G. T. (1968). Conidium ontogeny in Hyphomycetes. The sympodulae of *Beauveria* and *Curvularia. Canadian Journal of Botany*, **46**, 1297–1301.

Kendrick, B. & Watling, R. (1979). Mitospores in basidiomycetes. In *The Whole Fungus 2*, ed. B. Kendrick. Ottawa: National Museum of Natural Sciences, National Museums of Canada, Kananaskis Foundation, pp. 473–545.

Kerkenaar, A. (1995). Mechanism of action of cyclic amine fungicides: morpholines and piperidines. In *Modern Selective Fungicides. Properties, Applications, Mechanisms of Action*, 2nd edn, ed. H. Lyr. New York: Gustav Fischer Verlag, pp. 185–204.

Kern, H. & Naef-Roth, S. (1975). Zur Bildung von Auxinen und Cytokininen durch *Taphrina*-Arten. *Phytopathologische Zeitschrift*, **83**, 103–108.

Kern, V. D. (1999). Gravitropism of basidiomycetous fungi. *Life Sciences: Microgravity Research*, **24**, 697–706.

Kern, V. D., Rehm, A. & Hock, B. (1998). Gravitropic bending of fruit bodies – a model based on hyphal gravisensing and cooperativity. *Life Sciences: Microgravity Research*, **21**, 1173–1178.

Kerrigan, R. W. (1995). Global genetic resources for *Agaricus* breeding and cultivation. *Canadian Journal of Botany*, **73**, S973–S979.

Kerrigan, R. W. (1996). Characteristics of a large collection of wild edible mushroom germplasm: the *Agaricus* Resource Program. In *Culture Collections to Improve the Quality of Life*, ed. R. A. Samson, J. A. Stalpers, O. van der Mei & A. H. Stouthamer. Baarn: Centraalbureau voor Schimmelcultures, pp. 302–308.

Kerrigan, R. W., Imbernon, M., Callac, P., Billette, C. & Oliver, J. M. (1994). The heterothallic life cycle of *Agaricus bisporus* var *burnettii* and the inheritance of the tetrasporic trait. *Experimental Mycology*, **18**, 193–210.

Kerry, B. R. (2000). Rhizosphere interactions and the exploitation of microbial agents for the biological control of plant-parasitic nematodes. *Annual Review of Phytopathology*, **38**, 423–441.

Kerry, B. R. & Jaffee, B. A. (1997). Fungi as biological control agents for plant parasitic nematodes. In *The Mycota IV: Environmental and Microbial Relationships*, ed. D. T. Wicklow & B. Söderström. Berlin: Springer-Verlag, pp. 203–18.

Kerry, B. R., Crump, D. H. & Mullen, L. A. (1982). Studies of the cereal cyst-nematode, *Heterodera avenae* under continuous cereals, 1975–1978. II. Fungal parasitism of nematode females and eggs. *Annals of Applied Biology*, **100**, 489–499.

Kerwin, J. L. & Washino, R. K. (1986). Oosporogenesis by *Lagenidium giganteum*: induction and maturation are regulated by calcium and calmodulin. *Canadian Journal of Microbiology*, **32**, 663–672.

Kerwin, J. L., Johnson, L. M., Whisler, H. C. & Tuininga, A. R. (1992). Infection and morphogenesis of *Pythium marinum* in species of *Porphyra* and other red algae. *Canadian Journal of Botany*, **70**, 1017–1024.

Keskin, B. & Fuchs, W. H. (1969). Der Infektionsvorgang bei *Polymyxa betae. Archiv für Mikrobiologie*, **68**, 218–226.

Kessin, R. H. (2001). *Dictyostelium. Evolution, Cell Biology, and the Development of Multicellularity.* Cambridge: Cambridge University Press.

Khairi, S. M. & Preece, T. F. (1978). Hawthorn powdery mildew: overwintering mycelium in buds and the effect of clipping hedges on disease epidemiology. *Transactions of the British Mycological Society*, **71**, 399–404.

Khan, S. R. & Kimbrough, J. W. (1982). A reevaluation of the basidiomycetes based upon septal and basidial structures. *Mycotaxon*, **15**, 103–120.

Khan, S. R. & Talbot, P. H. B. (1975). Monosporous sporangiola in *Mycotypha* and *Cunninghamella. Transactions of the British Mycological Society*, **65**, 29–39.

Kher, K., Greening, J. P., Hatton, J. P., Frazer, L. N. & Moore, D. (1992). Kinetics and mechanism of stem gravitropism in *Coprinus cinereus. Mycological Research*, **96**, 817–824.

Kibby, G. & Bingham, J. (2004). Fungal portraits. No. 20: *Phallus impudicus* var. *togatus. Field Mycology*, **5**, 111–112.

Kieslich, K. (1997). Biotransformations. In *Fungal Biotechnology*, ed. T. Anke. London: Chapman & Hall, pp. 297–399.

Kile, G. A. (1993). Plant diseases caused by species of *Ceratocystis sensu stricto* and *Chalara*. In *Ceratocystis and Ophiostoma: Taxonomy, Ecology and Pathology*, ed. M. J. Wingfield, K. A. Seifert & J. F. Webber. St. Paul: APS Press, pp. 173–83.

Killian, C. (1919). Sur la sexualité de l'ergot de Seigle, le *Claviceps purpurea* (Tulasne). *Bulletin de la Société Mycologique de France*, **35**, 182–197.

Kim, J. R., Yeon, S. H. & Ahny, J. (2002). Larvicidal activity against *Plutella xylostella* of cordycepin from the fruiting body of *Cordyceps militaris. Pest Management Science*, **58**, 713–717.

Kimbeng, C. A. (1999). Genetic basis of crown rust resistance in perennial ryegrass, breeding strategies, and genetic variation among pathogen populations: a review. *Australian Journal of Experimental Agriculture*, **39**, 361–378.

Kimbrough, J. W. (1994). Septal ultrastructure and ascomycete systematics. In *Ascomycete Systematics. Problems and Perspectives in the Nineties*, ed. D. L. Hawksworth. New York: Plenum Press, pp. 127–141.

King, D. S. (1977). Systematics of *Conidiobolus* (Entomophthorales) using numerical taxonomy. III. Descriptions of recognized species. *Canadian Journal of Botany*, **55**, 718–729.

King, J. E. & Coley-Smith, J. R. (1969). Production of volatile alkyl sulphides by microbial degradation of synthetic alliin and alliin-like compounds, in relation to germination of sclerotia of *Sclerotium cepivorum* Berk. *Annals of Applied Biology*, **64**, 303–314.

Kinloch, B. B. (2003). White pine blister rust in North America: past and prognosis. *Phytopathology*, **93**, 1044–1047.

Kinloch, B. B. & Dupper, G. E. (2002). Genetic specificity in the white pine-blister rust pathosystem. *Phytopathology*, **92**, 278–280.

Kinugawa, K. (1965). On the growth of *Dictyophora indusiata*. II. Relations between the change in osmotic value of expressed sap and the conversion of glycogen to reducing sugar in tissues during receptaculum elongation. *Botanical Magazine (Tokyo)*, **78**, 240–244.

Kirby, E. J. M. (1961). Host–parasite relations in the choke disease of grasses. *Transactions of the British Mycological Society*, **44**, 493–503.

Kirby, J. J. H., Webster, J. & Baker, J. H. (1990). A particle plating method for the analysis of fungal community composition and structure. *Mycological Research*, **94**, 621–626.

Kirk, P. M. (1977). Scanning electron microscopy of zygospore formation in *Choanephora circinans* (Mucorales). *Transactions of the British Mycological Society*, **68**, 429–434.

Kirk, P. M. (1984). A monograph of the Choanephoraceae. *Mycological Papers*, **152**, 1–61. Kew: Commonwealth Mycological Institute.

Kirk, P. M. (1993). Distribution of Zygomycetes – the tropical connection. In *Aspects of Tropical Mycology*, ed. S. Isaac, J. C. Frankland, R. Watling & A. J. S. Whalley. Cambridge: Cambridge University Press, pp. 91–102.

Kirk, P. M., Cannon, P. F., David, J. C. & Stalpers, J. A., eds. (2001). *Dictionary of the Fungi*, 9th edn. Wallingford, UK: CABI Publishing.

Kirk, T. K. & Cullen, D. (1998). Enzymology and molecular genetics of wood degradation by white-rot fungi. In *Environmentally Friendly Technologies for the Pulp and Paper Industry*, ed. R. A. Young & M. Akhtar. New York: Wiley, pp. 273–307.

Kislev, M. E. (1982). Stem rust of wheat 3300 years old found in Israel. *Science*, **216**, 993–994.

Kiss, L. (1997). Graminicolous powdery mildew fungi as new natural hosts of *Ampelomyces* parasites. *Canadian Journal of Botany*, **75**, 680–683.

Kitamoto, K., Yoshizawa, K., Ohsumi, Y. & Anraku, Y. (1988). Dynamic aspects of vacuolar and cytosolic amino acid pools of *Saccharomyces cerevisiae*. *Journal of Bacteriology*, **170**, 2683–2686.

Kitancharoen, N., Hatai, K., Ogihara, R. & Aye, D. N. N. (1995). A new record of *Achlya klebsiana* from snakehead, *Channa striatus*, with fungal infection in Myammar. *Mycoscience*, **36**, 235–238.

Klein, D. & Eveleigh, D. E. (1998). Ecology of *Trichoderma*. In *Trichoderma and Gliocladium. I. Basic Biology, Taxonomy and Genetics*, ed. C. P. Kubicek & G. E. Harman. London: Taylor & Francis, pp. 57–74.

Klein, S., Sherman, A. & Simchen, G. (1994). Regulation of meiosis and sporulation in *Saccharomyces cerevisiae*. In *The Mycota I: Growth, Differentiation and Sexuality*, ed. J. G. H. Wessels & F. Meinhardt. Berlin: Springer-Verlag, pp. 235–50.

Klich, M. A. (2002). *Identification of Common Aspergillus Species*. Utrecht: Centraalbureau voor Schimmelcultures.

Klich, M. A. & Cleveland, T. E. (2000). *Aspergillus* systematics and the molecular genetics of mycotoxin biosynthesis. In *Integration of Modern Taxonomic Methods for Penicillium and Aspergillus Classification*, ed. R. A. Samson & J. I. Pitt. Amsterdam: Harwood Academic Publishers, pp. 425–34.

Klionsky, D. J. (1997). Protein transport from the cytoplasm into the vacuole. *Journal of Membrane Biology*, **157**, 105–115.

Klionsky, D. J., Herman, P. K. & Emr, S. D. (1990). The fungal vacuole: composition, function, and biogenesis. *Microbiological Reviews*, **54**, 266–292.

Kluczewski, S. M. & Lucas, J. A. (1983). Host infection and oospore formation by *Peronospora parasitica* in agricultural and horticultural *Brassica* species. *Transactions of the British Mycological Society*, **81**, 591–596.

Kluyver, A. J. & van Niel, C. B. (1927). *Sporobolomyces*: ein Basidiomyzet? *Annales Mycologici*, **25**, 389–394.

Kniep, H. (1919). Untersuchungen über den Antherenbrand (*Ustilago violacea* Pers.). Ein Beitrag zum Sexualitätsproblem. *Zeitschrift für Botanik*, **11**, 275–284.

Ko, W. H. (1980). Hormonal regulation of sexual reproduction in *Phytophthora*. *Journal of General Microbiology*, **116**, 459–463.

Ko, W. H., Lee, C. J. & Su, H. J. (1986). Chemical regulation of mating type in *Phytophthora parasitica*. *Mycologia*, **78**, 134–136.

Kobayasi, Y. (1982). Keys to the taxa of the genus *Cordyceps* and *Torrubiella*. *Transactions of the Mycological Society of Japan*, **23**, 329–364.

Koch, E. & Slusarenko, A. (1990). *Arabidopsis* is susceptible to infection by a downy mildew fungus. *Plant Cell*, **2**, 437–445.

Köhler, E. (1956). Zur Kenntniss der Sexualität bei *Synchytrium*. *Berichte der Deutschen Botanischen Gesellschaft*, **69**, 121–127.

Kohlwein, S. D. (2000). The beauty of the yeast: live cell microscopy at the limits of optical resolution. *Microscopy Research and Technique*, **51**, 511–529.

Kohn, L. M. (1979). A monographic revision of the genus *Sclerotinia*. *Mycotaxon*, **9**, 365–444.

Kohn, L. M. (1992). Developing new characters for fungal systematics: an experimental approach for determining the rank of resolution. *Mycologia*, **84**, 139–163.

Kohn, L. M. & Grenville, D. J. (1989). Anatomy and histochemistry of stromatal anamorphs in the Sclerotiniaceae. *Canadian Journal of Botany*, **67**, 371–393.

Kole, A. P. (1965). Resting-spore germination in *Synchytrium endobioticum*. *Netherlands Journal of Plant Pathology*, **71**, 72–78.

Kollar, A. (1998). A simple method to forecast the ascospore discharge of *Venturia inaequalis*. *Zeitschrift für Pflanzenkrankheiten und Pflanzenschutz*, **105**, 489–495.

Kollár, R., Reinhold, B. B., Petráková, E., *et al.* (1997). Architecture of the yeast cell wall: β-(1,6)-glucan interconnects mannoprotein, β-(1,3)-glucan, and chitin. *Journal of Biological Chemistry*, **272**, 17 762–775.

Kolmer, J. A. (2005). Tracking wheat rust on a continental scale. *Current Opinion in Plant Biology*, **8**, 441–449.

Koltin, Y. & Day, P. R. (1975). Specificity of *Ustilago maydis* killer proteins. *Applied Microbiology*, **30**, 694–696.

Kombrink, E. & Somssich, I. E. (1997). Pathogenesis-related proteins and plant defense. In *The Mycota* VA: *Plant Relationships*, ed. G. C. Carroll & P. Tudzynski. Berlin: Springer-Verlag, pp. 107–28.

Konijn, T. M., van der Meene, J. G. C., Bonner, J. T. & Barkley, D. S. (1967). The acrasin activity of adenosine-3′-5′ cyclic phosphate. *Proceedings of the National Academy of Sciences of the USA*, **58**, 1152–1154.

Konishi, M. (1967). Growth-promoting effects of certain amino acids on the *Agaricus* fruit body. *Mushroom Science*, **6**, 121–134.

Kono, Y., Sekido, S., Yamaguchi, I., *et al.* (1991). Structures of two novel pyriculol-related compounds and identification of naturally produced epipyriculol from *Pyricularia oryzae*. *Agricultural and Biological Chemistry*, **55**, 2785–2791.

Kope, H. H. & Fortin, J. A. (1990). Germination and comparative morphology of basidiospores of *Pisolithus arhizus*. *Mycologia*, **82**, 350–357.

Kopecka, M., Fleet, G. H. & Phaff, H. J. (1995). Ultrastructure of the cell wall of *Schizosaccharomyces pombe* following treatment with various glucanases. *Journal of Structural Biology*, **114**, 140–152.

Korf, R. P. (1972). Synoptic key to genera of the Pezizales. *Mycologia*, **64**, 937–993.

Korhonen, K. & Hintikka, V. (1974). Cytological evidence for somatic diploidization in dikaryotic cells of *Armillaria mellea*. *Archiv für Mikrobiologie*, **95**, 187–192.

Korhonen, K. & Stenlid, J. (1998). Biology of *Heterobasidion annosum*. In *Heterobasidion annosum. Biology, Ecology, Impact and Control*, ed. S. Woodward, J. Stenlid, R. Karjalainan & A. Hüttermann. Wallingford, UK: CAB International, pp. 43–70.

Koske, R. E. & Duncan, I. W. (1974). Temperature effects on growth, sporulation and germination of some aquatic hyphomycetes. *Canadian Journal of Botany*, **52**, 1387–1391.

Kothe, E. (1996). Tetrapolar fungal mating types: sexes by the thousands. *FEMS Microbiology Reviews*, **18**, 65–87.

Kothe, E. (1999). Mating types and pheromone recognition in the homobasidiomycete *Schizophyllum commune*. *Fungal Genetics and Biology*, **27**, 146–152.

Kramer, C. L. (1961). Morphological development and nuclear behaviour in the genus *Taphrina*. *Mycologia*, **52**, 295–320.

Kramer, C. L. (1982). Production, release and dispersal of basidiospores. In *Decomposer Basidiomycetes, their Biology and Ecology*, ed. J. C. Frankland, J. N. Hedger & M. J. Swift. Cambridge: Cambridge University Press, pp. 33–49.

Kramer, C. L. (1987). The Taphrinales. *Studies in Mycology*, **30**, 151–166.

Krasnoff, S. B. & Gupta, S. (1992). Efrapeptin production by *Tolypocladium* fungi (Deuteromycotina, Hyphomycetes) – intraspecific and interspecific variation. *Journal of Chemical Ecology*, **18**, 1727–1741.

Krassilov, V. A. & Makulbekov, N. M. (2003). The first finding of Gasteromycetes in the Cretaceous of Mongolia. *Palaeontological Journal*, **37**, 439–442.

Kreger-van Rij, N. J. W. & Veenhuis, M. (1971). A comparative study of the cell wall structure of basidiomycetous and related yeasts. *Journal of General Microbiology*, **68**, 87–95.

Kreger-van Rij, N. J. W., Veenhuis, M. & Leemburg-van der Graaf, C. A. (1974). Ultrastructure of hyphae and ascospores in the genus *Eremascus* Eidam. *Antonie van Leeuwenhoek*, **40**, 533–542.

Krejzová, R. (1978). Taxonomy, morphology and surface structure of *Basidiobolus* sp. isolate. *Journal of Invertebrate Pathology*, **31**, 157–163.

Křen, V. & Cvak, L., eds. (1999). *Ergot. The Genus Claviceps*. Amsterdam: Harwood Academic Publishers.

Kronstad, J. W. & Staben, C. (1997). Mating-type in filamentous fungi. *Annual Review of Genetics*, **31**, 245–276.

Kronstad, J., de Maria, A. D., Funnell, D., *et al.* (1998). Signaling via cAMP in fungi: interconnections with mitogen-activated protein kinase pathways. *Archives of Microbiology*, **170**, 395–404.

Kroon, L. P. N. M., Bakker, F. T., van den Bosch, G. B. M., Bonants, P. J. M. & Flier, W. G. (2004). Phylogenetic analysis of *Phytophthora* species based on mitochondrial and nuclear DNA sequences. *Fungal Genetics and Biology*, **41**, 766–782.

Kropf, D. L. & Harold, F. M. (1982). Selective transport of nutrients via rhizoids of the water mold *Blastocladiella emersonii*. *Journal of Bacteriology*, **151**, 429–437.

Kropp, B. R. & Mueller, G. M. (1999). *Laccaria*. In *Ectomycorrhizal Fungi. Key Genera in Profile*, ed. J. W. G. Cairney & S. M. Chambers. Berlin: Springer-Verlag, pp. 65–88.

Krüger, D., Binder, M., Fischer, M. & Kreisel, H. (2001). The Lycoperdales. A molecular approach to the systematics of some gasteroid mushrooms. *Mycologia*, **93**, 947–957.

Kubicek, C. P., Eveleigh, D. E., Esterbauer, H., Steiner, W. & Kubicek-Pranz, E. M., eds. (1990). *Trichoderma reesii: Cellulases, Biodiversity, Genetics, Physiology and Applications*. Cambridge: Royal Society of Chemistry.

Kuck, K. H., Scheinpflug, H. & Pontzen, R. (1995). DMI fungicides. In *Modern Selective Fungicides. Properties, Applications, Mechanisms of Action*, 2nd edn, ed. H. Lyr. New York: Gustav Fischer Verlag, pp. 205–258.

Kucsera, J., Pfeiffer, I. & Ferenczy, L. (1998). Homothallic life cycle in the diploid red yeast *Xanthophyllomyces dendrorhous* (*Phaffia rhodozyma*). *Antonie van Leeuwenhoek*, **73**, 163–168.

Kucsera, J., Pfeiffer, I. & Takeo, K. (2000). Biology of the red yeast *Xanthophyllomyces dendrorhous* (*Phaffia rhodozyma*). *Mycoscience*, **41**, 195–199.

Kuehn, H. H. (1956). Observations on the Gymnoascaceae. III. Developmental morphology of *Gymnoascus reessii*, a new species of *Gymnoascus* and *Eidamella deflexa*. *Mycologia*, **48**, 805–820.

Kües, U. (2000). Life history and developmental processes in the basidiomycete *Coprinus cinereus*. *Microbiology and Molecular Biology Reviews*, **64**, 316–353.

Kües, U. & Casselton, L. A. (1992). Fungal mating-type genes – regulators of sexual development. *Mycological Research*, **96**, 993–1006.

Kuhlman, E. G. (1972). Variation in zygospore formation among species of *Mortierella*. *Mycologia*, **64**, 325–341.

Kuhls, K., Lieckfeldt, E., Samuels, G. J., *et al.* (1996). Molecular evidence that the asexual industrial fungus *Trichoderma reesei* is a clonal derivative of the ascomycete *Hypocrea jecorina*. *Proceedings of the National Academy of Sciences of the USA*, **93**, 7755–7760.

Kuldau, G. & Bacon, C. W. (2001). *Claviceps* and related fungi. In *Foodborne Disease Handbook* **3**: *Plant Toxicants*, 2nd edn, ed. H. Hui, R. A. Smith & D. G. Spoerke. New York: Marcel Dekker Inc., pp. 503–534.

Kuldau, G. A., Liu, J., White, F., Siegel, M. R. & Schardl, C. L. (1997). Molecular systematics of Clavicipitaceae supporting monophyly of genus *Epichloe* and form genus *Ephelis*. *Mycologia*, **89**, 431–441.

Kulka, M. & von Schmeling, B. (1995). Carboxin fungicides and related compounds. In *Modern Selective Fungicides. Properties, Applications, Mechanisms of Action*, ed. H. Lyr. Jena: Gustav Fischer Verlag, pp. 133–147.

Kulkarni, U. K. (1963). Initiation of the dikaryon in *Claviceps microcephala* (Wallr.) Tul. *Mycopathologia et Mycologia Applicata*, **21**, 19–22.

Kullnig-Gradinger, C. M., Szakags, G. & Kubicek, C. P. (2002). Phylogeny and evolution of the genus *Trichoderma*: a multigene approach. *Mycological Research*, **106**, 757–767.

Kumar, S. & Rzhetsky, A. (1996). Evolutionary relationships of eukaryotic kingdoms. *Journal of Molecular Evolution*, **42**, 183–193.

Kumar, J., Hückelhoven, R., Beckhove, U., Nagarajan, S. & Kogel, K.-H. (2001). A compromised Mlo pathway affects the response of barley to the necrotrophic fungus *Bipolaris sorokiniana* (teleomorph: *Cochliobolus sativus*) and its toxins. *Phytopathology*, **91**, 127–133.

Kumar, J., Schäfer, P., Hückelhoven, R., *et al.* (2002). *Bipolaris sorokiniana*, a cereal pathogen of global concern: cytological and molecular approaches towards better control. *Molecular Plant Pathology*, 3, 185–195.

Kunert, J. (2000). Physiology of keratinolytic fungi. In *Biology of Dermatophytes and Other Keratinolytic Fungi*, ed. R. K. S. Kushwara & J. Guarro. Bilbao: Revista Iberoamericana de Micología, pp. 77–85.

Kuninaga, S., Natsuaki, T., Takeuchi, T. & Yokosawa, R. (1997). Sequence variation of the rDNA ITS regions within and between anastomosis groups in *Rhizoctonia solani*. *Current Genetics*, 32, 237–243.

Kunoh, H. & Ishizaki, H. (1981). Cytological studies of early stages of powdery mildew in barley and wheat. VII. Reciprocal translocation of a fluorescent dye between barley coleoptile cells and conidia. *Physiological Plant Pathology*, 18, 207–211.

Kurahashi, H., Imai, Y. & Yamamoto, M. (2002). Tropomyosin is required for the cell fusion process during conjugation in fission yeast. *Genes to Cells*, 7, 375–384.

Kurtzman, C. P. (1995). Relationships among the genera *Ashbya*, *Eremothecium*, *Holleya* and *Nematospora* determined from rDNA sequence divergence. *Journal of Industrial Microbiology*, 14, 523–530.

Kurtzman, C. P. (1998). *Pichia* E. C. Hansen emend. Kurtzman. In *The Yeasts: A Taxonomic Study*, 4th edn, ed. C. P. Kurtzman & J. W. Fell. Amsterdam: Elsevier, pp. 273–352.

Kurtzman, C. P. & Fell, J. W., eds. (1998). *The Yeasts: A Taxonomic Study*, 4th edn. Amsterdam: Elsevier.

Kurtzman, C. P. & Robnett, C. J. (1998). Identification and phylogeny of ascomycetous yeasts from analysis of nuclear large subunit (26S) ribosomal DNA partial sequences. *Antonie van Leeuwenhoek*, 73, 331–371.

Kurtzman, C. P. & Robnett, C. J. (2003). Phylogenetic relationships among yeasts of the 'Saccharomyces complex' determined from multigene sequence analyses. *FEMS Yeast Research*, 3, 417–432.

Kurtzman, C. P. & Smith, M. T. (1998). *Saccharomycopsis* Schöning. In *The Yeasts: A Taxonomic Study*, 4th edn, ed. C. P. Kurtzman & J. W. Fell. Amsterdam: Elsevier, pp. 374–86.

Kurtzman, C. P. & Sugiyama, J. (2001). Ascomycetous yeasts and yeastlike taxa. In *The Mycota* **VIIA**: *Systematics and Evolution*, ed. D. J. McLaughlin, E. G. McLaughlin & P. A. Lemke. Berlin: Springer-Verlag, pp. 179–200.

Kusch, J., Meyer, A., Snyder, M. P. & Barral, Y. (2002). Microtubule capture by the cleavage apparatus is required for proper spindle positioning in yeast. *Genes and Development*, 16, 1627–1639.

Kushalappa, A. H. & Eskes, A. B. (1989). Advances in coffee rust research. *Annual Review of Phytopathology*, 27, 503–531.

Kwasna, H. & Kosiak, B. (2003). *Lewia avenicola* sp. nov., and its *Alternaria* anamorph from oat grain with a key to the species of *Lewia*. *Mycological Research*, 107, 371–376.

Kwok, O. C. H., Plattner, R., Weisleder, D. & Wicklow, D. T. (1992). A nematicidal toxin from *Pleurotus ostreatus* NRRL 3526. *Journal of Chemical Ecology*, 18, 127–136.

Kwon-Chung, K. J. (1973). Studies on *Emmonsiella capsulata*. I. Heterothallism and development of the ascocarp. *Mycologia*, 65, 109–121.

Kwon-Chung, K. J. (1998). *Filobasidiella* Kwon-Chung. In *The Yeasts: A Taxonomic Study*, 4th edn, ed. C. P. Kurtzman & J. W. Fell. Amsterdam: Elsevier, pp. 656–662.

Kwon-Chung, K. J. & Bennett, J. E. (1978). Distribution of α and a mating types of *Cryptococcus neoformans* among natural and clinical isolates. *American Journal of Epidemiology*, 108, 337–340.

Kwon-Chung, K. J. & Bennett, J. E. (1992). *Medical Mycology*. Malvern, USA: Lea and Febiger.

Kwon-Chung, K. J., Edman, J. C. & Wickes, B. L. (1992). Genetic association of mating types and virulence in *Cryptococcus neoformans*. *Infection and Immunity*, 60, 602–605.

Kybal, J. (1964). Changes in N and P content during growth of ergot sclerotia (*Claviceps purpurea*) due to nutrition supplied by rye. *Phytopathology*, 54, 244–245.

Labeyrie, E. S., Molloy, D. P. & Lichtwardt, R. W. (1996). An investigation of Harpellales (Trichomycetes) in New York State blackflies (Diptera: Simuliidae). *Journal of Invertebrate Pathology*, 68, 293–298.

Lacey, J. (1994). Aspergilli in feeds and seeds. In *The Genus Aspergillus. From Taxonomy and Genetics to Industrial Application*, ed. K. A. Powell, A. Renwick & J. F. Peberdy. New York: Plenum Press, pp. 73–92.

Lacey, J. (1996). Spore dispersal – its ecology and disease: the British contribution to fungal aerobiology. *Mycological Research*, 100, 641–660.

Lachance, M.-A., Pupovac-Velikonja, A., Natajan, S. & Schlag-Edler, B. (2000). Nutrition and phylogeny of predacious yeasts. *Canadian Journal of Microbiology*, 46, 495–505.

Lachke, S. A., Lockhart, S. R., Daniels, K. J. & Soll, D. R. (2003). Skin facilitates *Candida albicans* mating. *Infection and Immunity*, 71, 4970–4976.

Lafon, R. & Bulit, J. (1981). Downy mildew of the vine. In *The Downy Mildews*, ed. D. M. Spencer. London: Academic Press, pp. 601–614.

Laidlaw, W. M. R. (1985). A method for the detection of the resting sporangia of potato wart disease (*Synchytrium endobioticum*) in the soil of old outbreak sites. *Potato Research*, **28**, 223–232.

Laing, S. A. K. & Deacon, J. W. (1991). Video microscopical comparison of mycoparasitism by *Pythium oligandrum*, *P. nunn* and an unnamed *Pythium* species. *Mycological Research*, **95**, 469–479.

Lamb, B. C. (1996). Ascomycete genetics: the part played by ascus segregation phenomena in our understanding of the mechanisms of recombination. *Mycological Research*, **100**, 1025–1059.

Lampila, L. E., Wallen, S. E. & Bullermann, L. B. (1985). A review of factors affecting biosynthesis of carotenoids by the order Mucorales. *Mycopathologia*, **90**, 65–80.

Landvik, S., Egger, K. N. & Schumacher, T. (1997). Towards a subordinal classification of the Pezizales (Ascomycota): phylogenetic analyses of SSU rDNA sequences. *Nordic Journal of Botany*, **17**, 403–418.

Landvik, S., Schumacher, T. K., Eriksson, O. E. & Moss, S. T. (2003). Morphology and ultrastructure of *Neolecta* species. *Mycological Research*, **107**, 1021–1031.

Langan, T. A., Gautier, J., Lohka, M., *et al.* (1989). Mammalian growth-associated H1 histone kinase: a homolog of $cdc2^+/CDC28$ protein kinases controlling mitotic entry in yeast and frog cells. *Molecular and Cellular Biology*, **9**, 3860–3868.

Lange, L. (1987). *Synchytrium endobioticum*. In *Zoosporic Fungi in Teaching and Research*, ed. M. S. Fuller & A. Jaworski. Athens, USA: Southeastern Publishing Corporation, pp. 24–5.

Lange, L. & Insunza, V. (1977). Root-inhabiting *Olpidium* species: the *O. radicale* complex. *Transactions of the British Mycological Society*, **69**, 377–384.

Lange, L. & Olson, L. W. (1976a). The flagellar apparatus and striated rhizoplast of the zoospore of *Olpidium brassicae*. *Protoplasma*, **89**, 339–351.

Lange, L. & Olson, L. W. (1976b). The zoospore of *Olpidium brassicae*. *Protoplasma*, **90**, 33–45.

Lange, L. & Olson, L. W. (1978). The zoospore of *Synchytrium endobioticum*. *Canadian Journal of Botany*, **56**, 1229–1239.

Lange, L. & Olson, L. W. (1979). The uniflagellate Phycomycete zoospore. *Dansk Botanisk Arkiv*, **33**, 7–95.

Lange, L. & Olson, L. W. (1980). Transfer of the Physodermataceae from the Chytridiales to the Blastocladiales. *Transactions of the British Mycological Society*, **74**, 449–457.

Lange, L. & Olson, L. W. (1981). Development of the resting sporangia of *Synchytrium endobioticum*, the causal agent of potato wart disease. *Protoplasma*, **106**, 83–95.

Lange, L. & Olson, L. W. (1983). The fungal zoospore. Its structure and biological significance. In *Zoosporic Plant Pathogens. A Modern Perspective*, ed. S. T. Buczacki. London: Academic Press, pp. 1–42.

Lange, L., Sparrow, F. K. & Olson, L. W. (1987). *Physoderma* spp. In *Zoosporic Fungi in Teaching and Research*, ed. M. S. Fuller & A. Jaworski. Athens, USA: Southeastern Publishing Corporation, pp. 36–7.

Lange, M. (1952). Species concept in the genus *Coprinus*. *Dansk Botanisk Arkiv*, **14**, 1–140.

Lapeyrie, F. & Mendgen, K. (1993). Quantitative estimation of surface carbohydrates of ectomycorrhizal fungi in pure culture and during *Eucalyptus* root infection. *Mycological Research*, **97**, 603–609.

Large, E. C. (1940). *The Advance of the Fungi*. London: Jonathan Cape.

Larkin, R. P. & Fravel, D. R. (1999). Mechanisms of action and dose–response relationships governing biological control of fusarium wilt of tomato by nonpathogenic *Fusarium* spp. *Phytopathology*, **89**, 1152–1161.

Larsen, J., Mansfield-Giese, K. & Bødker, L. (2000). Quantification of *Aphanomyces euteiches* in pea roots using specific fatty acids. *Mycological Research*, **104**, 858–864.

Larsen, M. (2000). Prospects for controlling animal parasitic nematodes by predacious micro fungi. *Parasitology*, **120**, S121–S131.

Larsson, E. & Larsson, K.-H. (2003). Phylogenetic relationships of russuloid basidiomycetes with emphasis on aphyllophoralean taxa. *Mycologia*, **95**, 1037–1065.

Last, F. T., Mason, P. A., Wilson, J. & Deacon, J. W. (1983). Fine roots and sheathing mycorrhizas: their formation, function and dynamics. *Plant and Soil*, **71**, 9–21.

Last, F. T., Dighton, J. & Mason, P. A. (1987). Successions of sheathing mycorrhizal fungi. *Trends in Ecology and Evolution*, **2**, 157–161.

Latgé, J.-P. (1999). *Aspergillus fumigatus* and aspergillosis. *Clinical Microbiology Reviews*, **12**, 310–350.

Latgé, J.-P. (2001). The pathobiology of *Aspergillus fumigatus*. *Trends in Microbiology*, **9**, 382–389.

Latgé, J. P., Perry, D. F., Prévost, M. C. & Samson, R. (1989). Ultrastructural studies of primary spores of *Conidiobolus*, *Erynia*, and related Entomophthorales. *Canadian Journal of Botany*, **67**, 2576–2589.

Latunde-Dada, A. O. (2001). *Colletotrichum*: tales of forcible entry, stealth, transient confinement and breakout. *Molecular Plant Pathology*, **2**, 187–198.

Latunde-Dada, A. O., O'Connell, R. J., Nash, C., Pring, R. J., Lucas, J. A. & Bailey, J. A. (1996). Infection process and identity of the hemibiotrophic anthracnose fungus (*Colletotrichum destructivum*) from cowpea (*Vigna unguiculata*). *Mycological Research*, **100**, 1133–1141.

Lawrey, J. D. (1986). Biological role of lichen substances. *Bryologist*, **89**, 111–122.

Lawrey, J. D. & Diederich, P. (2003). Lichenicolous fungi: interactions, evolution, and biodiversity. *Bryologist*, **106**, 80–120.

Lawton, K., Weymann, K., Friedrich, L., Vernooij, B., Uknes, S. & Ryals, J. (1995). Systemic acquired resistance in *Arabidopsis* requires salicylic acid but not ethylene. *Molecular Plant–Microbe Interactions*, **8**, 863–870.

Lazéra, M. S., Pires, F. D. A., Camillo-Coura, L., *et al.* (1996). Natural habitat of *Cryptococcus neoformans* var. *neoformans* in decaying wood forming hollows in living trees. *Journal of Medical and Veterinary Mycology*, **34**, 127–131.

Leach, C. M. (1968). An action spectrum for light inhibition of the 'terminal phase' of photosporogenesis in the fungus *Stemphylium botryosum*. *Mycologia*, **60**, 532–546.

Leach, C. M. (1971). Regulation of perithecium development in *Pleospora herbarum* by light and temperature. *Transactions of the British Mycological Society*, **57**, 295–315.

Leach, C. M. (1976). An electrostatic theory to explain violent spore liberation by *Drechslera turcica* and other fungi. *Mycologia*, **68**, 63–86.

Leadbeater, G. & Mercer, C. (1957). Zygospores in *Piptocephalis*. *Transactions of the British Mycological Society*, **40**, 109–116.

Leake, J., Johnson, D., *et al.* (2004). Networks of power and influence: the role of mycorrhizal mycelium in controlling plant communities and agroecosystem functioning. *Canadian Journal of Botany*, **82**, 1016–1045.

Leander, C. A. & Porter, D. (2001). The Labyrinthulomycota is comprised of three distinct lineages. *Mycologia*, **93**, 459–464.

Leathers, T. D. (2003). Biotechnological production and applications of pullulan. *Applied Microbiology and Biotechnology*, **62**, 468–473.

Leberer, E., Thomas, D. Y. & Whiteway, M. (1997). Pheromone signalling and polarized morphogenesis in yeast. *Current Opinion in Genetics and Development*, **7**, 59–66.

Leclerc, M. C., Guillot, G. & Deville, M. (2000). Taxonomic and phylogenetic analysis of Saprolegniaceae (Oomycetes) inferred from LSU rDNA and ITS sequence comparisons. *Antonie van Leeuwenhoek*, **77**, 369–377.

Lee, M. G. & Nurse, P. (1987). Complementation used to clone a human homologue of the fission yeast cell cycle control gene cdc2. *Nature*, **327**, 31–35.

Lee, N., Bakkeren, G., Wong, K., Sherwood, J. E. & Kronstad, J. W. (1999). The mating-type and pathogenicity locus of the fungus *Ustilago hordei* spans a 500-kb region. *Proceedings of the National Academy of Sciences of the USA*, **96**, 15 026–15031.

Legard, D. E., Lee, T. Y. & Fry, W. E. (1995). Pathogenic specialization in *Phytophthora infestans*: aggressiveness on tomato. *Phytopathology*, **85**, 1356–1361.

Lehnen, L. P. & Powell, M. J. (1989). The role of kinetosome-associated organelles in the attachment of encysting secondary zoospores of *Saprolegnia ferax* to substrates. *Protoplasma*, **149**, 163–174.

Lei, J., Lapeyrie, F., Malajczuk, N. & Dexheimer, J. (1990). Infectivity of pine and eucalypt isolates of *Pisolithus tinctorius* (Pers.) Coker & Couch on roots of *Eucalyptus urophylla* S. T. Blake *in vitro*. II. Ultrastructural and biochemical changes at the early stage of mycorrhiza formation. *New Phytologist*, **116**, 115–122.

Leite, B. & Nicholson, R. L. (1992). Mycosporine-alanine: a self-inhibitor of germination from the conidial mucilage of *Colletotrichum graminicola*. *Experimental Mycology*, **16**, 76–86.

Leith, I. D. & Fowler, D. (1988). Urban distribution of *Rhytisma acerinum* (Pers.) Fries (tar spot) on sycamore. *New Phytologist*, **108**, 175–181.

LéJohn, H. B. (1972). Enzyme regulation, lysine pathways and cell wall structures as indicators of major lines of evolution in fungi. *Nature*, **231**, 164–168.

Lemke, A., Kiderlen, A. F. & Kayser, O. (2005). Amphotericin B. *Applied Microbiology and Biotechnology*, **68**, 151–162.

Lengeler, K. B., Fox, D. S., Fraser, J. A., *et al.* (2002). Mating-type locus of *Cryptococcus neoformans*: a step in the evolution of sex chromosomes. *Eukaryotic Cell*, **1**, 704–718.

LePage, B. A., Currah, R. S., Stockney, R. A. & Rothwell, G. W. (1997). Fossil ectomycorrhizae from the middle Eocene. *American Journal of Botany*, **84**, 410–412.

Leroux, P. & Gredt, M. (1997). Evolution of fungicide resistance in the cereal eyespot fungi *Tapesia yallundae* and *T. acuformis* in France. *Pesticide Science*, **51**, 321–327.

Leroux, P., Fritz, R., Debieu, D., *et al.* (2002). Mechanisms of resistance to fungicides in field strains of *Botrytis cinerea*. *Pest Management Science*, **58**, 876–888.

Letscher-Bru, V. & Herbrecht, R. (2003). Caspofungin: the first representative of a new antifungal class. *Journal of Antimicrobial Chemotherapy*, **51**, 513–521.

Leuchtmann, A. (1984). Über *Phaeosphaeria* Miyake und andere bitunicate Ascomyceten mit mehrfach querseptierten Ascosporen. *Sydowia*, **37**, 75–194.

Leuchtmann, A. (2003). Taxonomy and diversity of *Epichloe* endophytes. In *Clavicipitalean Fungi*, ed. J. F. White, C. W. Bacon, N. L. Hywel-Jones & J. W. Spatafora. New York: Marcel Dekker Inc., pp. 169–194.

Leuchtmann, A., Schardl, C. L. & Siegel, M. R. (1994). Sexual compatibility and taxonomy of a new species of *Epichloe* symbiotic with fine fescue grasses. *Mycologia*, **86**, 802–812.

Léveillé, J. H. (1851). Organisation et disposition méthodique des espèces qui composent le genre Erysiphé. *Annales des Sciences Naturelles. Botanique, Séries 3*, **15**, 109–179.

Lévesque, C. A. & de Cock, A. W. A. M. (2004). Molecular phylogeny and taxonomy of the genus *Pythium*. *Mycological Research*, **108**, 1363–1383.

Levetin, E. & Caroselli, N. E. (1976). A simplified medium for growth and sporulation of *Pilobolus* species. *Mycologia*, **68**, 1254–1258.

Levings, C. S., Rhoads, D. M. & Siedow, J. N. (1995). Molecular interactions of *Bipolaris maydis* T-toxin and maize. *Canadian Journal of Botany*, **73**, S483–S489.

Levisohn, I. (1927). Beitrag zur Entwicklungsgeschichte und Biologie von *Basidiobolus ranarum* Eidam. *Jahrbücher für Wissenschaftliche Botanik*, **66**, 513–555.

Lewin, B. (2000). *Genes VII*. New York: Oxford University Press.

Lewis, L. A. & Decaris, B. (1974). The induction of apothecial formation in *Ascobolus immersus* by a spermatization technique. *Transactions of the British Mycological Society*, **63**, 197–199.

Lewis, T. E., Nichols, P. D. & McMeekin, T. A. (1999). The biotechnological potential of thraustochytrids. *Marine Biotechnology*, **1**, 580–587.

Li, J., Heath, I. B. & Packer, L. (1993). The phylogenetic relationship of the anaerobic chytridiomycetous gut fungi (Neocallimasticaceae) and the Chytridiomycota. II. Cladistic analysis of structural data and description of the Neocallimasticales ord. nov. *Canadian Journal of Botany*, **71**, 393–407.

Li, L., Gerecke, E. E. & Zolan, M. E. (1999). Homolog pairing and meiotic progression in *Coprinus cinereus*. *Chromosoma*, **108**, 384–392.

Li, N., Erman, M., Pangborn, W., *et al.* (1999). Structure of *Ustilago maydis* killer toxin KP6 α-subunit. *Journal of Biological Chemistry*, **274**, 20 425–431.

Lichtwardt, R. W. (1967). Zygospores and spore appendages of *Harpella* (Trichomycetes) from larvae of Simuliidae. *Mycologia*, **59**, 482–491.

Lichtwardt, R. W. (1986). *The Trichomycetes. Fungal Associates of Arthropods*. Berlin: Springer-Verlag.

Lichtwardt, R. W. (1996). Trichomycetes and the arthropod gut. In *The Mycota **VI**: Human and Animal Relationships*, ed. D. J. Howard & J. D. Miller. Berlin: Springer-Verlag, pp. 315–330.

Lieckfeldt, E., Kullnig, C., Samuels, G. J. & Kubicek, C. P. (2000). Sexually competent, sucrose- and nitrate-assimilating strains of *Hypocrea jecorina* (*Trichoderma reesei*) from South American soils. *Mycologia*, **92**, 374–380.

Liew, E. C. Y., Aptroot, A. & Hyde, K. D. (2000). Phylogenetic significance of the pseudoparaphyses in Loculoascomycete taxonomy. *Molecular Phylogenetics and Evolution*, **16**, 392–402.

Lilley, J. H. & Roberts, R. J. (1997). Pathogenicity and culture studies comparing the *Aphanomyces* involved in epizootic ulcerative syndrome (EUS) with other similar fungi. *Journal of Fish Diseases*, **20**, 135–144.

Limpert, E., Godet, F. & Müller, K. (1999). Dispersal of cereal mildews across Europe. *Agricultural and Forest Meteorology*, **97**, 293–308.

Lincoff, G. & Mitchel, D. J. (1977). *Toxic and Hallucinogenic Mushroom Poisoning*. New York: Van Nostrand Reinhold.

Linder, D. H. (1929). A monograph of the helicosporous fungi imperfecti. *Annals of the Missouri Botanical Gardens*, **16**, 227–388.

Linderman, R. G. (1997). Vesicular–arbuscular mycorrhizal VAM) fungi. In *The Mycota **VB**: Plant Relationships*, ed. G. C. Carroll & P. Tudzynski. Berlin: Springer-Verlag, pp. 117–28.

Line, R. F. (2002). Stripe rust of wheat and barley in North America: a retrospective historical review. *Annual Review of Phytopathology*, **40**, 75–118.

Line, R. F. & Griffith, C. S. (2001). Research on the epidemiology of stem rust of wheat during the Cold War. In *Stem Rust of Wheat. From Ancient Enemy to Modern Foe*, ed. P. D. Peterson. St. Paul, USA: APS Press, pp. 83–118.

Lines, C. E. M., Ratcliffe, R. G., Rees, T. A. V. & Southon, T. E. (1989). A ^{13}C NMR study of photosynthate transport and metabolism in the lichen *Xanthoria calcicola* Oxner. *New Phytologist*, **111**, 447–456.

Lingappa, B. T. (1958a). Development and cytology of the evanescent prosori of *Synchytrium brownii* Karling. *American Journal of Botany*, **45**, 116–123.

Lingappa, B.T. (1958b). The cytology of development and germination of resting spores of *Synchytrium brownii*. *American Journal of Botany*, **45**, 613–620.

Lingle, W.L., Clay, R.P. & Porter, D. (1992). Ultrastructural analysis of basidiosporogenesis in *Panellus stypticus*. *Canadian Journal of Botany*, **70**, 2017–2027.

Lippincott, J. & Li, R. (1998). Sequential assembly of myosin II, an IQGAP-like protein, and filamentous actin to a ring structure involved in budding yeast cytokinesis. *Journal of Cell Biology*, **140**, 355–366.

List, P.H. & Freund, B. (1968). Geruchsstoffe der Stinkmorchel, *Phallus impudicus* L. *Planta Medica* (Supplement), 123–132.

Littlefield, L.J. & Heath, M.C. (1979). *Ultrastructure of Rust Fungi*. New York: Academic Press.

Liu, Y. (2003). Molecular mechanisms of entrainment in the *Neurospora* circadian clock. *Journal of Biological Rhythms*, **18**, 195–205.

Liu, Y.J. & Hall, B.D. (2004). Body plan evolution of ascomycetes, as inferred from an RNA polymerase II phylogeny. *Proceedings of the National Academy of Sciences of the USA*, **101**, 4507–4512.

Liu, Y.J., Whelen, S. & Hall, B.D. (1999). Phylogenetic relationships among ascomycetes: evidence from an RNA polymerase II subunit. *Molecular Biology and Evolution*, **16**, 1799–1808.

Liu, Z.-Y., Liang, Z.-Q., Whalley, A.J.S., Yao, Y.-J. & Liu, A.-Y. (2001). *Cordyceps brittlebankisoides*, a new pathogen of grubs and its anamorph *Metarhizium anisopliae* var *majus*. *Journal of Invertebrate Pathology*, **78**, 178–182.

Liu, Z.-Y., Liang, Z.-Q., Liu, A.-Y., Yao, Y.-J., Hyde, K.D. & Yu, Z.-N. (2002). Molecular evidence for teleomorph-anamorph connections in *Cordyceps* based on ITS-5.8S rDNA sequences. *Mycological Research*, **106**, 1100–1108.

Lizoň, P., Iturriaga, T. & Korf, R.P. (1998). A preliminary Discomycete flora of Macaronesia: Part 18, Leotiales. *Mycotaxon*, **67**, 73–83.

LoBuglio, K.F., Pitt, J.I. & Taylor, J.W. (1993). Phylogenetic analysis of two ribosomal DNA regions indicates multiple independent losses of a sexual *Talaromyces* state among asexual *Penicillium* species in subgenus *Biverticillium*. *Mycologia*, **85**, 592–604.

Lockhart, S.R., Daniels, K.J., Zhao, R., Wessels, D. & Soll, D.R. (2003). Cell biology of mating in *Candida albicans*. *Eukaryotic Cell*, **2**, 49–61.

Loeffler, R.S.T., Butters, J.A. & Hollomon, D.W. (1992). The sterol composition of powdery mildews. *Phytochemistry*, **31**, 1561–1563.

Lohmeyer, M. & Tudzynski, P. (1997). Claviceps alkaloids. In *Fungal Biotechnology*, ed. T. Anke. Weinheim, Germany: Chapman & Hall, pp. 173–85.

Long, P.E. & Jacobs, L. (1974). Aseptic fruiting of the cultivated mushroom *Agaricus bisporus*. *Transactions of the British Mycological Society*, **63**, 99–107.

Long, R.M., Singer, R.H., Meng, X., Gonzalez, I., Nasmyth, K. & Jansen, R.P. (1997). Mating type switching in yeast controlled by asymmetric localization of *ASH1* mRNA. *Science*, **277**, 383–387.

López-Franco, R., Howard, R.J. & Bracker, C.E. (1995). Satellite Spitzenkörper in growing hyphal tips. *Protoplasma*, **188**, 85–103.

Lorin, S., Dufour, E., Boulay, J., Begel, O., Marsy, S. & Sainsard-Chanet, A. (2001). Overexpression of the alternative oxidase restores senescence and fertility in a long-lived respiration-deficient mutant of *Podospora anserina*. *Molecular Microbiology*, **42**, 1259–1267.

Lösel, D.M. (1988). Fungal lipids. In *Microbial Lipids* **1**, ed. C. Ratledge & S.G. Wilkinson. London: Academic Press, pp. 699–806.

Lott, T.J., Holloway, B.P., Logan, D.A., Fundyga, R. & Arnold, J. (1999). Towards understanding the evolution of the human commensal yeast *Candida albicans*. *Microbiology*, **145**, 1137–1143.

Louis, C., Girard, M., Kuhl, G. & Lopez-Ferber, M. (1996). Persistence of *Botrytis cinerea* in its vector *Drosophila melanogaster*. *Phytopathology*, **86**, 934–939.

Lovett, J.S. (1975). Growth and differentiation of the water mold *Blastocladiella emersonii*. *Bacteriological Reviews*, **39**, 345–404.

Lowe, S.E., Griffith, G.G., Milne, A., Theodorou, M.K. & Trinci, A.P.J. (1987a). Life cycle and growth kinetics of an anaerobic rumen fungus. *Journal of General Microbiology*, **133**, 1815–1827.

Lowe, S.E., Theodorou, M.K. & Trinci, A.P.J. (1987b). Isolation of anaerobic fungi from saliva and faeces of sheep. *Journal of General Microbiology*, **133**, 1829–1834.

Lowy, B. (1952). The genus *Auricularia*. *Mycologia*, **44**, 656–692.

Lowry, D.S., Fisher, K.E. & Roberson, R.W. (1998). Establishment and maintenance of nuclear position during zoospore formation in *Allomyces macrogynus*. *Fungal Genetics and Biology*, **24**, 34–44.

Lowry, D.S., Fisher, K.E. & Roberson, R.W. (2004). Functional necessity of the cytoskeleton during cleavage membrane development and zoosporogenesis in *Allomyces macrogynus*. *Mycologia*, **96**, 211–218.

Lowry, R.J. & Sussman, A.S. (1958). Wall structure of ascospores of *Neurospora tetrasperma*. *American Journal of Botany*, **45**, 397–403.

Lowry, R.J. & Sussman, A.S. (1968). Ultrastructural changes during germination of ascospores of *Neurospora tetrasperma*. *Journal of General Microbiology*, **51**, 403–409.

Lu, B. C. (1964). Polyploidy in the basidiomycete *Cyathus stercoreus*. *American Journal of Botany*, **51**, 343–347.

Lu, G. (2003). Engineering *Sclerotinia sclerotiorum* resistance in oilseed crops. *African Journal of Biotechnology*, **2**, 509–516.

Lü, H. & McLaughlin, D. J. (1991). Ultrastructure of the septal pore apparatus and early septum initiation in *Auricularia auricula-judae*. *Mycologia*, **83**, 322–334.

Lu, S.-H. (1973). Effect of calcium on fruiting of *Cyathus stercoreus*. *Mycologia*, **65**, 329–334.

Lucarotti, C. J. (1981). Zoospore ultrastructure in *Nowakowskiella elegans* and *Cladochytrium replicatum* (Chytridiales). *Canadian Journal of Botany*, **59**, 137–148.

Lucarotti, C. J. (1987). *Cladochytrium replicatum*. In *Zoosporic Fungi in Teaching and Research*, ed. M. S. Fuller & A. Jaworski. Athens, USA: Southeastern Publishing Corporation, pp. 20–1.

Lucarotti, C. J. & Wilson, C. M. (1987). *Nowakowskiella elegans*. In *Zoosporic Fungi in Teaching and Research*, ed. M. S. Fuller & A. Jaworski. Athens, USA: Southeastern Publishing Corporation, pp. 22–3.

Lucas, J. A., Shattock, R. C., Shaw, D. S. & Cooke, L. R., eds. (1991). *Phytophthora*. Cambridge: Cambridge University Press.

Lucas, J. A., Dyer, P. S. & Murray, T. D. (2000). Pathogenicity, host-specificity, and population biology of *Tapesia* spp., causal agents of eyespot disease of cereals. *Advances in Botanical Research*, **33**, 225–258.

Ludwig-Müller, J. (1999). *Plasmodiophora brassicae*, the causal agent of clubroot disease: a review on molecular and biochemical events in pathogenesis. *Zeitschrift für Pflanzenkrankheiten und Pflanzenschutz*, **106**, 109–127.

Ludwig-Müller, J., Bendel, U., Thermann, P., *et al.* (1993). Concentrations of indole-3-acetic acid in plants of tolerant and susceptible varieties of Chinese cabbage infected with *Plasmodiophora brassicae* Woron. *New Phytologist*, **125**, 763–769.

Ludwig-Müller, J., Epstein, E. & Hilgenberg, W. (1996). Auxin-conjugate hydrolysis in Chinese cabbage: characterization of an amidohydrolase and its role during the clubroot disease. *Physiologia Plantarum*, **97**, 627–634.

Ludwig-Müller, J., Bennett, R. N., Kiddle, G., *et al.* (1999). The host range of *Plasmodiophora brassicae* and its relationship to endogenous glucosinolate content. *New Phytologist*, **141**, 443–458.

Lugones, L. G., Boscher, J. S., Scholtmeyer, K., De Vries, O. M. H. & Wessels, J. G. H. (1996). An abundant hydrophobin (ABH1) forms hydrophobic rodlet layers in *Agaricus bisporus* fruiting bodies. *Microbiology*, **142**, 1321–1329.

Lugones, L. G., Wösten, H. A. B., Birkenkamp, K. U., *et al.* (1999). Hydrophobins line air channels in fruiting bodies of *Schizophyllum commune* and *Agaricus bisporus*. *Mycological Research*, **103**, 635–640.

Lumbsch, H. T. & Lindemuth, R. (2001). Major lineages of *Dothideomycetes* (Ascomycota) inferred from SSU and LSU rDNA sequences. *Mycological Research*, **105**, 901–908.

Lumbsch, H. T., Schmitt, I., Döring, H. & Wedin, M. (2001). Molecular systematics supports the recognition of an additional order of Ascomycota: the Agyriales. *Mycological Research*, **105**, 16–23.

Lumbsch, H. T., Schmitt, I., Palice, Z., *et al.* (2004). Supraordinal phylogenetic relationships of Lecanoromycetes based on a Bayesian analysis of combined nuclear and mitochondrial sequences. *Molecular Phylogenetics and Evolution*, **31**, 822–832.

Lumsden, R. D. (1992). Mycoparasitism of soilborne plant pathogens. In *The Fungal Community. Its Organization and Role in the Ecosystem*, 2nd edn, ed. G. C. Carroll & D. T. Wicklow. New York: Marcel Dekker Inc., pp. 275–93.

Lundqvist, N. (1972). Nordic Sordariaceae *s. lat. Symbolae Botanicae Upsalienses*, **20**, 1–374.

Lundqvist, N., Mahoney, D. P., Bell, A. & Lorenzo, L. E. (1999). *Podospora austrohemisphaerica*, a new heterothallic ascomycete from dung. *Mycologia*, **91**, 405–415.

Lunney, C. Z. & Bland, C. E. (1976). An ultrastructural study of zoosporogenesis in *Pythium proliferum* de Bary. *Protoplasma*, **88**, 85–100.

Luttrell, E. S. (1951). Taxonomy of the Pyrenomycetes. *University of Missouri Studies*, **24**(3), 1–20.

Luttrell, E. S. (1955). The ascostromatic ascomycetes. *Mycologia*, **47**, 511–532.

Luttrell, E. S. (1963). Taxonomic criteria in *Helminthosporium*. *Mycologia*, **55**, 643–674.

Luttrell, E. S. (1965). Paraphysoids, pseudoparaphyses and apical paraphyses. *Transactions of the British Mycological Society*, **48**, 135–144.

Luttrell, E. S. (1973). Loculoascomycetes. In *The Fungi: An Advanced Treatise **IVA***, ed. G. C. Ainsworth, F. K. Sparrow & A. S. Sussman. New York: Academic Press, pp. 135–219.

Luttrell, E. S. (1977). The disease cycle and fungus–host relationships in dallisgrass ergot. *Phytopathology*, **67**, 1461–1468.

Luttrell, E. S. (1980). Host–parasite relationships and development of the ergot sclerotium in *Claviceps purpurea*. *Canadian Journal of Botany*, **58**, 942–958.

Luttrell, E. S. (1981). The pyrenomycete centrum. In *Ascomycete Systematics. The Luttrellian Concept*, ed. D. R. Reynolds. New York: Springer-Verlag, pp. 124–37.

Lutzoni, F., Pagel, M. & Reeb, B. (2001). Major fungal lineages are derived from lichen symbiotic ancestors. *Nature*, **411**, 937–940.

Ma, L.-J., Rogers, S. O., Catranis, C. M. & Starmer, W. T. (2000). Detection and characterization of ancient fungi entrapped in glacial ice. *Mycologia*, **92**, 286–295.

Macfarlane, I. (1952). Factors affecting the survival of *Plasmodiophora brassicae* Wor. in the soil and its assessment by a host test. *Annals of Applied Biology*, **39**, 239–256.

Macfarlane, I. (1968). Problems in the systematics of the Olpidiaceae. In *Marine Mykologie (Symposium über Niedere Pilze im Küstenbereich)*, ed. A. Gaertner. Bremerhaven: Institut für Meeresforschung, pp. 39–58.

Macfarlane, I. (1970). Germination of resting spores of *Plasmodiophora brassicae*. *Transactions of the British Mycological Society*, **55**, 97–112.

Macfarlane, T. D., Kuo, J. & Hilton, R. N. (1978). Structure of the giant sclerotium of *Polyporus mylittae*. *Transactions of the British Mycological Society*, **71**, 359–365.

MacHardy, W. E., Gadoury, D. M. & Gessler, C. (2001). Parasitic and biological fitness of *Venturia inaequalis*: relationship to disease management strategies. *Plant Disease*, **85**, 1036–1051.

Machlis, L. (1972). The coming of age of sex hormones in plants. *Mycologia*, **64**, 235–247.

Macko, V., Staples, R. C., Gershon, H. & Renwick, J. A. A. (1970). Self-inhibitor of bean rust uredospores: methyl 3,4-dimethoxycinnamate. *Science*, **170**, 539–540.

Maclean, D. J. (1982). Axenic culture and metabolism of rust fungi. In *The Rust Fungi*, ed. K. J. Scott & A. K. Chakravorty. London: Academic Press, pp. 37–84.

MacLeod, D. M., Müller-Kögler, E. & Wilding, N. (1976). *Entomophthora* species with *E. muscae*-like conidia. *Mycologia*, **68**, 1–29.

Madelin, M. F. (1984). Presidential address. Myxomycete data of ecological significance. *Transactions of the British Mycological Society*, **83**, 1–19.

Magee, B. B. & Magee, P. T. (2000). Induction of mating in *Candida albicans* by construction of MTLα and MTLa strains. *Science*, **289**, 310–313.

Magliani, W., Conti, S., Gerloni, M., Bertolotti, D. & Polonelli, L. (1997). Yeast killer systems. *Clinical Microbiology Reviews*, **10**, 369–400.

Magliani, W., Conti, S., Arseni, S., *et al.* (2005). Antibody-mediated protective immunity in fungal infections. *New Microbiologica*, **28**, 299–309.

Maheshwari, R. (1999). Microconidia of *Neurospora crassa*. *Fungal Genetics and Biology*, **26**, 1–18.

Maheshwari, R., Bharadwaj, G. & Bhat, M. K. (2000). Thermophilic fungi: their physiology and enzymes. *Microbiology and Molecular Biology Reviews*, **64**, 461–488.

Mahfoud, R., Maresca, M., Garmy, N. & Fantini, J. (2002). The mycotoxin patulin alters the barrier function of the intestinal epithelium: mechanism of action of the toxin and protective effects of glutathione. *Toxicology and Applied Pharmacology*, **181**, 209–218.

Mahr, F. A. & Bruzzese, E. (1998). The effect of *Phragmidium violaceum* (Shultz) Winter (Uredinales) on *Rubus fruticosus* L. agg. in south-eastern Victoria. *Plant Protection Quarterly*, **13**, 182–185.

Mai, S. H. (1976). Morphological studies in *Podospora anserina*. *American Journal of Botany*, **63**, 821–825.

Mai, S. H. (1977). Morphological studies in *Sordaria fimicola* and *Gelasinospora longispora*. *American Journal of Botany*, **64**, 489–495.

Maia, J. C. D. (1994). Hexosamine and cell wall biogenesis in the aquatic fungus *Blastocladiella emersonii*. *FASEB Journal*, **8**, 848–853.

Maia, L. C., Yano, A. M. & Kimbrough, J. W. (1996). Species of Ascomycota forming ectomycorrhizae. *Mycotaxon*, **57**, 371–390.

Maier, W., Begerow, D., Weiss, M. & Oberwinkler, F. (2003). Phylogeny of the rust fungi: an approach using nuclear large subunit ribosomal DNA sequences. *Canadian Journal of Botany*, **81**, 12–23.

Mains, E. B. (1957). Species of *Cordyceps* parasitic on *Elaphomyces*. *Bulletin of the Torrey Botanical Club*, **84**, 243–251.

Mainwaring, H. R. (1972). The fine structure of ascospore wall formation in *Sordaria fimicola*. *Archiv für Mikrobiologie*, **81**, 126–135.

Maitland, D. P. (1994). A parasitic fungus infecting yellow dung flies manipulates host perching behaviour. *Proceedings of the Royal Society of London*, **B 258**, 187–193.

Malcolmson, J. F. (1969). Races of *Phytophthora infestans* occurring in Great Britain. *Transactions of the British Mycological Society*, **53**, 417–423.

Malcolmson, J. F. (1970). Vegetative hybridity in *Phytophthora infestans*. *Nature*, **225**, 971–972.

Maleck, K., Neuenschwander, U., Cade, R. M., *et al.* (2002). Isolation and characterization of broad-spectrum disease-resistant *Arabidopsis* mutants. *Genetics*, **160**, 1661–1671.

Malik, M. M. S. (1974). Nuclear behaviour during teliospore germination in *Ustilago tritici* and *U. nuda*. *Pakistan Journal of Botany*, **6**, 59–63.

Mallett, K. J. & Harrison, I. M. (1988). The mating system of the fairy ring fungus *Marasmius oreades* and the genetic relationship of fairy rings. *Canadian Journal of Botany*, **66**, 1111–1116.

Malloch, D. W. (1987). The evolution of mycorrhizae. *Canadian Journal of Plant Pathology*, **9**, 398–402.

Malloch, D. & Blackwell, M. (1993). Dispersal biology of ophiostomatoid fungi. In *Ceratocystis and Ophiostoma: Taxonomy, Ecology and Pathology*, ed. M. J. Wingfield, K. A. Seifert & J. F. Webber. St. Paul, USA: APS Press, pp. 195–206.

Maloy, O. C. (1997). White pine blister rust control in North America: a case history. *Annual Review of Phytopathology*, **35**, 87–109.

Mamoun, M. & Olivier, J.-M. (1996). Receptivity of cloned hazels to artificial ectomycorrhizal infection by *Tuber melanosporum* and symbiotic competitors. *Mycorrhiza*, **6**, 15–19.

Manion, P. & Griffin, D. H. (1986). Sixty-five years of research on *Hypoxylon* canker of aspen. *Plant Disease*, **70**, 803–808.

Manners, D. J. & Meyer, M. T. (1977). The molecular structures of some glucans from the cell walls of *Schizosaccharomyces pombe*. *Carbohydrate Research*, **57**, 189–203.

Manners, J. M. (1989). The host–haustorium interface in powdery mildews. *Australian Journal of Plant Physiology*, **16**, 45–52.

Manocha, M. S. (1965). Fine structure of the *Agaricus* carpophore. *Canadian Journal of Botany*, **43**, 1329–1333.

Manocha, M. S. (1975). Host–parasite relations in a mycoparasite. III. Morphological and biochemical differences in the parasitic and axenic-culture spores of *Piptocephalis virginiana*. *Mycologia*, **67**, 383–391.

Manocha, M. S. (1981). Host specificity and mechanisms of resistance in a mycoparasitic system. *Physiological Plant Pathology*, **18**, 257–265.

Manocha, M. S. & Deven, J. M. (1975). Host–parasite relations in a mycoparasite. IV. A correlation between the levels of *gamma*-linolenic acid and parasitism in *Piptocephalis virginiana*. *Mycologia*, **67**, 1148–1157.

Manocha, M. S. & Golesorkhi, R. (1981). Host–parasite relations in a mycoparasite. 7. Light and scanning electron microscopy of interactions of *Piptocephalis virginiana* with host and non-host species. *Mycologia*, **73**, 976–987.

Manocha, M. S. & Lee, K. Y. (1971). Host–parasite relations in a mycoparasite. I. Fine structure of host, parasite and their interface. *Canadian Journal of Botany*, **49**, 1677–1681.

Manocha, M. S. & McCullough, C. M. (1985). Suppression of host cell wall synthesis at penetration sites in a compatible interaction with a mycoparasite. *Canadian Journal of Botany*, **63**, 967–973.

Manocha, M. S. & Zhonghua, Z. (1997). Immunocytochemical and cytochemical localization of chitinase and chitin in infected hosts of a biotrophic mycoparasite *Piptocephalis virginiana*. *Mycologia*, **89**, 185–194.

Manocha, M. S., Chen, Y. & Rao, N. (1990). Involvement of cell surface sugars in recognition, attachment and appressorium formation by a mycoparasite. *Canadian Journal of Microbiology*, **36**, 771–778.

Manocha, M. S., Xiong, D. & Govindasamy, V. (1997). Isolation and partial characterization of a complementary protein from the mycoparasite *Piptocephalis virginiana* that specifically binds to two glycoproteins at the host cell surface. *Canadian Journal of Microbiology*, **43**, 625–632.

Mantegani, S., Brambilla, E. & Varasi, M. (1999). Ergoline derivatives: receptor affinity and selectivity. *Farmaco*, **54**, 288–296.

Mantiri, F. R., Samuels, G. J., Rahe, J. E. & Honda, B. M. (2001). Phylogenetic relationships of *Neonectria* species having *Cylindrocarpon* anamorphs inferred from mitochondrial ribosomal DNA sequences. *Canadian Journal of Botany*, **79**, 334–340.

Maras, M., Saelens, X., Laroy, W., *et al.* (1997). *In vitro* conversion of the carbohydrate moiety of fungal glycoproteins to mammalian-type oligosaccharides. Evidence for *N*-acetylglucosaminyltransferase-I-accepting glycans from *Trichoderma reesei*. *European Journal of Biochemistry*, **249**, 701–707.

Marchant, R. & Wessels, J. G. H. (1974). An ultrastructural study of septal dissolution in *Schizophyllum commune*. *Archiv für Mikrobiologie*, **96**, 175–182.

Marek, L. E. (1984). Light affects *in vitro* development of gametangia and sporangia of *Monoblepharis macrandra* (Chytridiomycetes, Monoblepharidales). *Mycologia*, **76**, 420–425.

Margulis, L., Corliss, J. O., Melkonian, M. & Chapman, D. J., eds. (1990). *Handbook of the Protoctista*. Boston: Jones & Bartlett.

Markham, P. (1994). Occlusions of the septal pores of filamentous fungi. *Mycological Research*, **98**, 1089–1106.

Markham, P. & Collinge, A. J. (1987). Woronin bodies of filamentous fungi. *FEMS Microbiology Reviews*, **46**, 1–11.

Markovich, N. A. & Kononova, G. L. (2003). Lytic enzymes of *Trichoderma* and their role in plant defense from fungal diseases: a review. *Applied Biochemistry and Microbiology*, **39**, 341–351.

Marquina, D., Santos, A. & Peinado, J. M. (2002). Biology of killer yeasts. *International Microbiology*, **5**, 65–71.

Marshall, G. M. (1960). The incidence of certain seed-borne diseases in commercial seed-samples. IV. Bunt of wheat, *Tilletia caries* (DC.) Tul. V. Earcockles of wheat, *Anguina tritici* (Stein.) Filipjev. *Annals of Applied Biology*, **48**, 34–38.

Martin, E. (1940). The morphology and cytology of *Taphrina deformans*. *American Journal of Botany*, **27**, 743–751.

Martin, F., Díez, J., Dell, B. & Delaruelle, C. (2002). Phylogeography of the ectomycorrhizal *Pisolithus* species as inferred from nuclear ribosomal DNA ITS sequences. *New Phytologist*, **153**, 345–357.

Martin, G. W. (1945). The classification of the Tremellales. *Mycologia*, **37**, 527–542.

Martin, H. D., Ruck, C., Schmidt, M., *et al.* (1999). Chemistry of carotenoid oxidation and free radical reactions. *Pure and Applied Chemistry*, **71**, 2253–2262.

Martin, J. F., Gutiérrez, S. & Demain, A. L. (1997). Antibiotics: β-lactams. In *Fungal Biotechnology*, ed. T. Anke. Weinheim, Germany: Chapman & Hall, pp. 91–127.

Martin, M., Gay, J. L. & Jackson, G. V. H. (1976). Electron microscopic study of developing and mature cleistothecia of *Sphaerotheca mors-uvae*. *Transactions of the British Mycological Society*, **66**, 473–487.

Martín, M. P. (1996). *The Genus Rhizopogon in Europe*. Barcelona: Societat Catalana de Micologia.

Martinez, M., López-Ribot, J., Kirkpatrick, W. R., *et al.* (2002). Replacement of *Candida albicans* with *C. dubliniensis* in human immunodeficiency virus-treated patients with oropharyngeal candidiasis treated with fluconazole. *Journal of Clinical Microbiology*, **40**, 3135–3139.

Martínez-Espinoza, A. D., García-Pedrajas, M. D. & Gold, S. E. (2002). The Ustilaginales as plant pests and model systems. *Fungal Genetics and Biology*, **35**, 1–20.

Marx, D. H. (1977). Tree host range and world distribution of the ectomycorrhizal fungus *Pisolithus tinctorius*. *Canadian Journal of Microbiology*, **23**, 217–223.

Massicotte, H. B., Peterson, R. L. & Ashford, A. E. (1987). Ontogeny of *Eucalyptus pilularis*–*Pisolithus tinctorius* ectomycorrhiza. I. Light microscopy and scanning electron microscopy. *Canadian Journal of Botany*, **65**, 1927–1939.

Masterman, R., Ross, R., Mesce, K. & Spivak, M. (2001). Olfactory and behavioural response thresholds to odors of diseased brood differ between hygienic and non-hygienic honey bees (*Apis mellifera* L.). *Journal of Comparative Physiology A*, **187**, 441–452.

Masuch, G. (1993). *Biologie der Flechten.* Heidelberg and Wiesbaden: Quelle & Meyer.

Masuhara, G., Katsuya, K. & Yamaguchi, K. (1993). Potential for symbiosis of *Rhizoctonia solani* and binucleate *Rhizoctonia* with seeds of *Spiranthes sinensis* var. *amoena in vitro*. *Mycological Research*, **97**, 746–752.

Mata, J. & Nurse, P. (1998). Discovering the poles in yeast. *Trends in Cell Biology*, **8**, 163–167.

Mather, K. & Jinks, J. L. (1958). Cytoplasm in sexual reproduction. *Nature*, **182**, 1188–1190.

Mathew, K. T. (1961). Morphogenesis of mycelial strands in the cultivated mushroom, *Agaricus bisporus*. *Transactions of the British Mycological Society*, **44**, 285–290.

Mathre, D. E. (1996). Dwarf bunt: politics, identification, and biology. *Annual Review of Phytopathology*, **34**, 67–85.

Matossian, M. K. (1989). *Poisons of the Past: Molds, Epidemics and History.* New Haven: Yale University Press.

Matsumoto, C., Kageyama, K., Suga, H. & Hyakumachi, M. (1999). Phylogenetic relationships of *Pythium* species based on ITS and 5.8S sequences. *Mycoscience*, **40**, 321–331.

Mau, J. L., Beelman, R. B. & Ziegler, G. R. (1992). Effect of 10-oxo-*trans*-B decenoic acid on the growth of *Agaricus bisporus*. *Phytochemistry*, **31**, 4059–4064.

Mayer, A. M., Staples, R. C. & Gil-ad, N. L. (2001). Mechanisms of survival of necrotrophic fungal plant pathogens in hosts expressing the hypersensitive response. *Phytochemistry*, **58**, 33–41.

Mazars, C., Canivenc, E., Rossignol, M. & Auriol, P. (1991). Production of phomozin in sunflower following artificial inoculation with *Phomopsis helianthi*. *Plant Science*, **75**, 155–160.

Mazzola, M. (2002). Mechanisms of natural soil suppressiveness to soilborne diseases. *Antonie van Leeuwenhoek*, **81**, 557–564.

McAfee, B. J. & Fortin, J. A. (1986). Competitive interactions of ectomycorrhizal mycobionts under field conditions. *Canadian Journal of Botany*, **64**, 848–852.

McCabe, D. E., Humber, R. A. & Soper, R. S. (1984). Observation and interpretation of nuclear reductions during maturation and germination of entomophthoralean resting spores. *Mycologia*, **76**, 1104–1107.

McCartney, H. A., Schmechtel, D. & Lacey, M. E. (1993). Aerodynamic diameter of conidia of *Alternaria* species. *Plant Pathology*, **42**, 280–286.

McCreadie, J. W., Beard, C. E. & Adler, P. H. (2005). Context-dependent symbiosis between black flies (Diptera: Simuliidae) and trichomycete fungi (Harpellales: Legeriomycetaceae). *Oikos*, **108**, 362–370.

McDaniell, L. L. & Hindal, D. F. (1982). Spore swelling, germination, and germ tube formation in axenic culture among four species of *Piptocephalis*. *Mycologia*, **74**, 271–274.

McDowell, J. M., Cuzick, A., Can, C., *et al.* (2000). Downy mildew (*Peronospora parasitica*) resistance genes in *Arabidopsis* vary in functional requirements for *NDR1*, *EDS1*, *NPR1* and salicylic acid accumulation. *Plant Journal*, **22**, 523–529.

McGinnis, M. R. (1980). Recent taxonomic developments and changes in medical mycology. *Annual Review of Microbiology*, **34**, 109–135.

McGovern, P. E. (2003). *Ancient Wine. The Search for the Origins of Viticulture*. Princeton and Oxford: Princeton University Press.

McKeen, W. E., Mitchell, N. & Smith, R. (1967). The *Erysiphe cichoracearum* conidium. *Canadian Journal of Botany*, **45**, 1489–1496.

McKemy, J. M. & Morgan-Jones, G. (1991). Studies in the genus *Cladosporium sensu lato*. V. Concerning the type species, *Cladosporium herbarum*. *Mycotaxon*, **42**, 307–317.

McKendrick, S. L., Leake, J. R. & Read, D. J. (2000). Symbiotic germination and development of myco-heterotrophic plants in nature: transfer of carbon from ectomycorrhizal *Salix repens* and *Betula pendula* to the orchid *Corallorhiza trifida* through shared hyphal connections. *New Phytologist*, **145**, 539–548.

McLaughlin, D. J. (1973). Ultrastructure of sterigma growth and basidiospore formation in *Coprinus* and *Boletus*. *Canadian Journal of Botany*, **51**, 145–150.

McLaughlin, D. J. (1977). Basidiospore initiation and early development in *Coprinus cinereus*. *American Journal of Botany*, **64**, 1–16.

McLaughlin, D. J. (1982). Ultrastructure and cytochemistry of basidial and basidiospore development. In *Basidium and Basidiocarp: Evolution, Cytology, Function and Development*, ed. K. Wells & E. K. Wells. New York: Springer-Verlag, pp. 37–74.

McLaughlin, D. J., Frieders, E. M. & Lü, H. (1995). A microscopist's view of heterobasidiomycete phylogeny. *Studies in Mycology*, **38**, 91–109.

McLaughlin, D. J., McLaughlin, E. G. & Lemke, P. E., eds. (2001). *The Mycota* **VIIA** *and* **VIIB**. Berlin: Springer-Verlag.

McLaughlin, R. J., Wilson, C. L., Chalutz, D., *et al.* (1990). Characterization and reclassification of yeasts used for biological control of postharvest diseases of fruits and vegetables. *Applied and Environmental Microbiology*, **56**, 3583–3586.

McMeekin, D. (1960). The role of the oospores of *Peronospora parasitica* in downy mildew of crucifers. *Phytopathology*, **50**, 93–97.

McMillan, M. S., Schwartz, H. F. & Otto, K. L. (2003). Sexual stage development of *Uromyces appendiculatus* and its potential use for disease resistance screening of *Phaseolus vulgaris*. *Plant Disease*, **87**, 1133–1138.

McMorris, T. C. (1978). Sex hormones of the aquatic fungus *Achlya*. *Lipids*, **13**, 716–722.

McMorris, T. C., Seshadri, R., Weihe, G. R., Arsenault, G. P. & Barksdale, A. W. (1975). Structure of oogoniol-1, -2 and -3, steroidal sex hormones of the water mould *Achlya*. *Journal of the American Chemical Society*, **97**, 2544–3545.

Meehan, F. & Murphy, H. C. (1946). A new *Helminthosporium* blight of oats. *Science*, **104**, 413–414.

Mehta, B. J., Obraztsova, I. N. & Cerdá-Olmedo, E. (2003). Mutants and intersexual heterokaryons of *Blakeslea trispora* for production of β-carotene and lycopene. *Applied and Environmental Microbiology*, **69**, 4043–4048.

Meier, F. A., Scherrer, S. & Honegger, R. (2002). Faecal pellets of lichenivorous mites contain viable cells of the lichen-forming ascomycete *Xanthoria parietina* and its green algal photobiont, *Trebouxia arboricola*. *Biological Journal of the Linnean Society*, **76**, 259–268.

Meireles, M. C. A., Riet-Correa, F., Fischman, O., *et al.* (1993). Cutaneous pythiosis in horses from Brazil. *Mycoses*, **36**, 139–142.

Meistrich, M., Fork, R. L. & Matricon, J. (1970). Phototropism in *Phycomyces* as investigated by focused laser radiation. *Science*, **169**, 370–371.

Mendgen, K. (1997). The Uredinales. In *The Mycota* **VB**: *Plant Relationships*, ed. G. C. Carroll & P. Tudzynski. Berlin: Springer-Verlag, pp. 79–94.

Mendgen, K. & Deising, H. (1993). Infection structures of fungal plant pathogens – a cytological and physiological evaluation. *New Phytologist*, **124**, 193–213.

Mendgen, K. & Hahn, M. (2002). Plant infection and the establishment of fungal biotrophy. *Trends in Plant Science*, **7**, 352–356.

Mendoza, L., Villalobos, J., Calleja, C. E. & Solis, A. (1992). Evaluation of two vaccines for the treatment of pythiosis insidiosi in horses. *Mycopathologia*, **119**, 89–95.

Mendoza, L., Hernandez, F. & Ajello, L. (1993). Life cycle of the human and animal oomycete pathogen *Pythium insidiosum*. *Journal of Clinical Microbiology*, **31**, 2967–2973.

Menge, J. A. (1984). Inoculum production. In *VA Mycorrhiza*, ed. C. L. Powell & D. J. Bagyaraj. Boca Raton: CRC Press, pp. 187–203.

Menzies, J. G., Knox, R. E., Nielsen, J. & Thomas, P. L. (2003). Virulence of Canadian isolates of *Ustilago tritici*: 1964–1998, and the use of the geometric rule in understanding host differential complexity. *Canadian Journal of Plant Pathology*, **25**, 62–72.

Meredith, D. S. (1965). Violent spore release in *Helminthosporium turcicum*. *Phytopathology*, **55**, 1099–1102.

Meredith, D. S. (1966). Diurnal periodicity and violent liberation of conidia in *Epicoccum*. *Phytopathology*, **56**, 988.

Merkus, E. (1976). Ultrastructure of the ascospore wall in Pezizales (Ascomycetes). IV. Morchellaceae, Helvellaceae, Rhizinaceae, Thelebolaceae and Sarcoscyphaceae. General discussion. *Persoonia*, **9**, 1–38.

Merz, U. (1997). Microscopical observations of the primary zoospores of *Spongospora subterranea* f. sp. *subterranea*. *Plant Pathology*, **46**, 670–674.

Metcalf, D. A. & Wilson, C. R. (1999). Histology of *Sclerotium cepivorum* infection of onion roots and the spatial relationships of pectinases in the infection process. *Plant Pathology*, **48**, 445–452.

Metcalf, D. A. & Wilson, C. R. (2001). The process of antagonism of *Sclerotium cepivorum* in white rot affected onion roots by *Trichoderma koningii*. *Plant Pathology*, **50**, 249–257.

Metwalli, A. A. & Shearer, C. A. (1989). Aquatic hyphomycete communities in clear-cut and wooded areas of an Illinois stream. *Transactions of the Illinois Academy of Science*, **82**, 5–16.

Metzenberg, R. L. & Glass, N. L. (1990). Mating type and mating strategies in *Neurospora*. *Bioessays*, **12**, 53–59.

Meyer, R. J. & Fuller, M. S. (1985). Structure and development of hyphal septa in *Allomyces*. *American Journal of Botany*, **72**, 1458–1465.

Meyer, S. A., Payne, R. W. & Yarrow, D. (1998). *Candida* Berkhout. In *The Yeasts: A Taxonomic Study*, 4th edn, ed. C. P. Kurtzman & J. W. Fell. Amsterdam: Elsevier, pp. 454–573.

Meyer, W. & Gams, W. (2003). Delimitation of *Umbelopsis* (Mucorales, Umbelopsidaceae fam. nov.) based on ITS sequence and RFLP data. *Mycological Research*, **107**, 339–350.

Miadlikowska, J. & Lutzoni, F. (2004). Phylogenetic classification of peltigeralean fungi (Peltigerales, Ascomycota) based on ribosomal RNA small and large subunits. *American Journal of Botany*, **91**, 449–464.

Micali, C. O. & Smith, M. I. (2003). On the independence of barrage formation and heterokaryon incompatibility in *Neurospora crassa*. *Fungal Genetics and Biology*, **38**, 209–219.

Michaelides, J. & Kendrick, B. (1982). The bubble trap propagules of *Beverwijkella*, *Helicoon* and other aero-aquatic fungi. *Mycotaxon*, **14**, 247–260.

Michailides, T. J., Guo, L. Y. & Morgan, D. P. (1997). Factors affecting zygosporogenesis in *Mucor piriformis* and *Gilbertella persicaria*. *Mycologia*, **89**, 603–609.

Michelmore, R. W. & Ingram, D. S. (1980). Heterothallism in *Bremia lactucae*. *Transactions of the British Mycological Society*, **75**, 47–56.

Michelot, D. & Melendez-Howell, L. M. (2003). *Amanita muscaria*: chemistry, biology, toxicology and ethnomycology. *Mycological Research*, **107**, 131–146.

Michelot, D. & Tebbett, I. (1990). Poisoning by members of the genus *Cortinarius*. *Mycological Research*, **94**, 289–298.

Middleton, J. T. (1943). The taxonomy, host range and geographic distribution of the genus *Pythium*. *Memoirs of the Torrey Botanical Club*, **20**, 1–171.

Milburn, J. A. (1970). Cavitation and osmotic potentials of *Sordaria* ascospores. *New Phytologist*, **69**, 133–141.

Millay, M. A. & Taylor, T. N. (1978). Chytrid-like fossils of Pennsylvanian age. *Science*, **200**, 1147–1149.

Miller, C. E. & Dylewski, D. P. (1981). Syngamy and resting body development in *Chytriomyces hyalinus* (Chytridiales). *American Journal of Botany*, **68**, 342–349.

Miller, C. E. & Dylewski, D. P. (1987). *Chytriomyces hyalinus*. In *Zoosporic Fungi in Teaching and Research*, ed. M. S. Fuller & A. Jaworski. Athens, USA: Southeastern Publishing Corporation, pp. 12–13.

Miller, M. G. & Johnson, A. D. (2002). White–opaque switching in *Candida albicans* is controlled by mating-type locus homeodomain proteins and allows efficient mating. *Cell*, **110**, 293–302.

Miller, O. K. & Miller, H. H. (1988). *Gasteromycetes. Morphological and Development Features with Keys to the Orders, Families, and Genera*. Eureka, USA: Mad River Press.

Miller, O. K., Henkel, T. W., James, T. Y. & Miller, S. L. (2001). *Pseudotulostoma*, a remarkable new volvate genus in the Elaphomycetaceae from Guyana. *Mycological Research*, **105**, 1268–1272.

Miller, R. E. (1971). Evidence of sexuality in the cultivated mushroom. *Mycologia*, **63**, 630–634.

Miller, S. L. (1988). Early basidiospore formation in *Lactarius lignyotellus*. *Mycologia*, **80**, 99–107.

Mills, J. T. (1967). Spore dispersal and natural infection in the oat loose smut (*Ustilago avenae*). *Transactions of the British Mycological Society*, **50**, 403–412.

Mims, C. W. (1982). Ultrastructure of the haustorial apparatus of *Exobasidium camelliae*. *Mycologia*, **74**, 188–200.

Mims, C. W. & Nickerson, N. L. (1986). Ultrastructure of the host–pathogen relationship in red leaf disease of lowbush blueberry caused by the fungus *Exobasidium vaccinii*. *Canadian Journal of Botany*, **64**, 1338–1343.

Mims, C. W. & Seabury, F. (1989). Ultrastructure of tube formation and basidiospore development in *Ganoderma lucidum*. *Mycologia*, **81**, 754–764.

Mims, C. W., Richardson, E. A. & Roberson, R. W. (1987). Ultrastructure of basidium and basidiospore development in three species of the fungus *Exobasidium*. *Canadian Journal of Botany*, **65**, 1236–1244.

Mims, C. W., Richardson, E. A. & Timberlake, W. E. (1988). Ultrastructural analysis of conidiophore development in the fungus *Aspergillus nidulans* using freeze-substitution. *Protoplasma*, **144**, 132–141.

Minghetti, A. & Crespi-Perellino, N. (1999). The history of ergot. In *Ergot. The Genus Claviceps*, ed. V. Křen & L. Cvak. Amsterdam: Harwood Academic Publishers, pp. 1–24.

Minogue, K. P. & Fry, W. E. (1981). Effect of temperature, relative humidity, and rehydration rate on germination of dried sporangia of *Phytophthora infestans*. *Phytopathology*, **71**, 1181–1184.

Minter, D. W. (1984). New concepts in the interpretation of conidiogenesis in Deuteromycetes. *Microbiological Sciences*, **1**, 86–89.

Minter, D. W., Kirk, P. M. & Sutton, B. C. (1982). Holoblastic phialides. *Transactions of the British Mycological Society*, **79**, 75–93.

Minter, D. W., Kirk, P. M. & Sutton, B. C. (1983a). Thallic phialides. *Transactions of the British Mycological Society*, **80**, 39–66.

Minter, D. W., Sutton, B. C. & Brady, B. L. (1983b). What are phialides anyway? *Transactions of the British Mycological Society*, **81**, 109–120.

Minter, D. W., Cannon, P. F. & Peredo, H. L. (1987). South American species of *Cyttaria* (a remarkable and beautiful group of edible ascomycetes). *Mycologist*, **1**, 7–11.

Mirza, J. H. & Cain, R. F. (1969). Revision of the genus *Podospora*. *Canadian Journal of Botany*, **47**, 1999–2048.

Misra, J. K. (1998). Trichomycetes — fungi associated with arthropods: review and world literature. *Symbiosis*, **24**, 179–220.

Misra, J. K. & Lichtwardt, R. W. (2000). *Illustrated Genera of Trichomycetes. Fungal Symbionts of Insects and Other Arthropods*. Enfield, USA: Science Publishers Inc.

Mistry, A. (1977). Zygosporogenesis in *Blakeslea trispora* (Mucorales). *Microbios*, **20**, 73–79.

Mitani, S., Araki, S., Yamaguchi, T., *et al.* (2002). Biological properties of the novel fungicide cyazofamid against *Phytophthora infestans* on tomato and *Pseudoperonospora cubensis* on cucumber. *Pest Management Science*, **58**, 139–145.

Mitani, S., Sugimoto, K., Hayashi, H., *et al.* (2003). Effects of cyazofamid against *Plasmodiophora brassicae* Woronin on Chinese cabbage. *Pest Management Science*, **59**, 287–293.

Mitchell, R. T. & Deacon, J. W. (1986). Selective accumulation of zoospores of Chytridiomycetes and Oomycetes on cellulose and chitin. *Transactions of the British Mycological Society*, **86**, 219–223.

Mitchell, T. G. & Perfect, J. R. (1995). Cryptococcosis in the era of AIDS — 100 years after the discovery of *Cryptococcus neoformans*. *Clinical Microbiology Reviews*, **8**, 515–548.

Mithen, R. & Magrath, R. (1992). A contribution to the life history of *Plasmodiophora brassicae*: secondary plasmodia development in root galls of *Arabidopsis thaliana*. *Mycological Research*, **96**, 877–885.

Mitra, M. (1931). A new bunt of wheat in India. *Annals of Applied Biology*, **18**, 178–179.

Mix, A. J. (1949). A monograph of the genus *Taphrina*. *University of Kansas Science Bulletins*, **30**, 1–167.

Miyake, H., Takemaru, T. & Ishikawa, T. (1980). Sequential production of enzymes and basidiospore formation in fruiting bodies of *Coprinus macrorhizus*. *Archives of Microbiology*, **126**, 201–203.

Moake, M. M., Padilla-Zakour, O. I. & Worobo, R. W. (2005). Comprehensive review of patulin control methods in foods. *Comprehensive Reviews in Food Science and Food Safety*, **1**, 8–21.

Molina, A., Hunt, M. D. & Ryals, J. A. (1998). Impaired fungicide activity in plants blocked in disease resistance signal transduction. *Plant Cell*, **10**, 1903–1914.

Molina, M., Gil, C., Pla, J., Arroyo, J. & Nombrela, C. (2000). Protein localisation approaches for understanding yeast cell wall biogenesis. *Microscopy Research and Technique*, **51**, 601–612.

Molina, R. & Trappe, J. M. (1984). Mycorrhiza management in bareroot nurseries. In *Forest Nursery Manual: Production of Bareroot Seedlings*, ed. M. L. Duryea & T. D. Landis. Den Haag: Martinus Nijhoff and Dr. W. Junk Publishers, pp. 211–23.

Molina, R. & Trappe, J. M. (1994). Biology of the ectomycorrhizal genus, *Rhizopogon*. I. Host associations, host-specificity and pure culture syntheses. *New Phytologist*, **126**, 653–675.

Molina, R., Trappe, J.M., Grubisha, L.C. & Spatafora, J.W. (1999). *Rhizopogon*. In *Ectomycorrhizal Fungi: Key Genera in Profile*, ed. J.W.G. Cairney & S.M. Chambers. Berlin: Springer Verlag, pp. 129–161.

Moller, A.P. (1993). A fungus infecting domestic flies manipulates sexual behaviour of its host. *Behavioural Ecology and Sociobiology*, **33**, 403–407.

Mollicone, M.R.N. & Longcore, J.E. (1994). Zoospore ultrastructure of *Monoblepharis polymorpha*. *Mycologia*, **86**, 615–625.

Mollicone, M.R.N. & Longcore, J.E. (1999). Zoospore ultrastructure of *Gonapodya polymorpha*. *Mycologia*, **91**, 727–734.

Molnár, A., Sulyok, L. & Hornok, L. (1990). Parasexual recombination between vegetatively incompatible strains in *Fusarium oxysporum*. *Mycological Research*, **94**, 393–398.

Momany, M. & Hamer, J.E. (1997). Relationship of actin, microtubules, and crosswall synthesis during septation in *Aspergillus nidulans*. *Cell Motility and the Cytoskeleton*, **38**, 373–384.

Momany, M., Richardson, E.A., van Sickle, C. & Jedd, G. (2002). Mapping Woronin body position in *Aspergillus nidulans*. *Mycologia*, **94**, 260–266.

Monaco, L.C. (1977). Consequences of the introduction of coffee rust into Brazil. *Annals of the New York Academy of Sciences*, **287**, 57–71.

Moncalvo, J.-M., Lutzoni, F.M., Rehner, S.A., Johnson, J. & Vilgalys, R. (2000). Phylogenetic relationships of agaric fungi based on nuclear large subunit ribosomal DNA sequences. *Systematic Biology*, **49**, 278–305.

Moncalvo, J.-M., Vilgalys, R., Redhead, S.A., *et al.* (2002). One hundred and seventeen clades of euagarics. *Molecular Phylogenetics and Evolution*, **23**, 357–400.

Money, N.P. (1994). Osmotic adjustment and the role of turgor in mycelial fungi. In *The Mycota I: Growth, Differentiation and Sexuality*, ed. J.G.H. Wessels & F. Meinhardt. Berlin: Springer-Verlag, pp. 67–88.

Money, N.P. (1997). Wishful thinking of turgor revisited: the mechanisms of fungal growth. *Fungal Genetics and Biology*, **21**, 173–187.

Money, N.P. (1998). More g's than the space shuttle: ballistospore discharge. *Mycologia*, **94**, 547–558.

Money, N.P. & Harold, F.M. (1992). Extension growth of the water mold *Achlya*: interplay of turgor and wall strength. *Proceedings of the National Academy of Sciences of the USA*, **89**, 4245–4249.

Money, N.P. & Harold, F.M. (1993). Two water molds can grow without measurable turgor pressure. *Planta*, **190**, 426–430.

Money, N.P. & Hill, T.W. (1997). Correlation between endoglucanase secretion and cell wall strength in oomycete hyphae: implications for growth and morphogenesis. *Mycologia*, **89**, 777–785.

Money, N.P. & Webster, J. (1985). Water stress and sporangial emptying in *Achlya intricata*. *Botanical Journal of the Linnean Society*, **91**, 319–327.

Money, N.P. & Webster, J. (1988). Cell wall permeability and its relationship to spore release in *Achlya intricata*. *Experimental Mycology*, **12**, 169–179.

Money, N.P. & Webster, J. (1989). Mechanism of sporangial emptying in *Saprolegnia*. *Mycological Research*, **92**, 45–49.

Money, N.P., Webster, J. & Ennos, R. (1988). Dynamics of sporangial emptying in *Achlya intricata*. *Experimental Mycology*, **12**, 13–27.

Money, N.P., That, T.-C.C.-T., Frederick, B. & Henson, J.M. (1998). Melanin synthesis is associated with changes in hyphopodial turgor, permeability, and wall rigidity in *Gaeumannomyces graminis* var. *graminis*. *Fungal Genetics and Biology*, **24**, 240–251.

Monschau, N., Stahmann, K.-P., Pielken, P. & Sahm, H. (1997). *In vitro* synthesis of β-(1-3)-glucan with a membrane fraction of *Botrytis cinerea*. *Mycological Research*, **101**, 97–101.

Montemurro, N. & Visconti, A. (1992). *Alternaria* metabolites – chemical and biological data. In *Alternaria Biology, Plant Disease and Metabolites*, ed. J. Chelkowski & A. Visconti. Amsterdam: Elsevier, pp. 449–537.

Montgomery, G.W.G. & Gooday, G.W. (1985). Phospholipid–enzyme interactions of chitin synthase of *Coprinus cinereus*. *FEMS Microbiology Letters*, **27**, 29–33.

Monzer, J. (1995). Actin-filaments are involved in cellular graviperception of the basidiomycete *Flammulina velutipes*. *European Journal of Cell Biology*, **66**, 151–156.

Monzer, J. (1996). Cellular graviperception in the basidiomycete *Flammulina velutipes* – can the nuclei serve as fungal statoliths? *European Journal of Cell Biology*, **71**, 216–220.

Moore, D. (1994). Tissue formation. In *The Growing Fungus*, ed. N.A.R. Gow & G.M. Gadd. London: Chapman & Hall, pp. 423–465.

Moore, D. (1998). *Fungal Morphogenesis*. Cambridge: Cambridge University Press.

Moore, D. & Chiu, S.W. (2001). Fungal products as food. In *Bio-Exploitation of Filamentous Fungi*, ed. S.B. Pointing & K.D. Hyde. Hong Kong: Fungal Diversity Press, pp. 223–251.

Moore, D., Elhiti, M.M.Y. & Butler, R.D. (1979). Morphogenesis of the carpophore of *Coprinus cinereus*. *New Phytologist*, **83**, 695–722.

Moore, D., Hock, B., Greening, J.P., Kern, V.D., Frazer, L.N. & Monzer, J. (1996). Centenary Review. Gravimorphogenesis in agarics. *Mycological Research*, **100**, 257–273.

Moore, E.D. & Miller, C.E. (1973). Resting body formation by rhizoidal fusion in *Chytriomyces hyalinus*. *Mycologia*, **65**, 145–154.

Moore, E.J. (1962). The ontogeny of the sclerotia of *Pyronema domesticum*. *Mycologia*, **54**, 312–316.

Moore, E.J. (1963). The ontogeny of the apothecia of *Pyronema domesticum*. *American Journal of Botany*, **50**, 37–44.

Moore, E.J. & Korf, R.P. (1963). The genus *Pyronema*. *Bulletin of the Torrey Botanical Club*, **90**, 33–42.

Moore, N.Y., Pegg, K.G., Buddenhagen, I.W. & Bentley, S. (2001). Fusarium wilt of banana: a diverse clonal pathogen of a domesticated clonal host. In *Fusarium. Paul E. Nelson Memorial Symposium*, ed. B.A. Summerell, J.F. Leslie, D. Backhouse, W.L. Bryden & L.W. Burgess. St. Paul, USA: APS Press, pp. 212–224.

Moore, R.T. (1955). Index to the Helicosporae. *Mycologia*, **47**, 90–103.

Moore, R.T. (1985). The challenge of the dolipore/parenthesome septum. In *Developmental Biology of Higher Fungi*, ed. D. Moore, L.A. Casselton, D.A. Wood & J.C. Frankland. Cambridge: Cambridge University Press, pp. 175–212.

Moore, R.T. (1987). The genera of *Rhizoctonia*-like fungi: *Ascorhizoctonia*, *Ceratorhiza* gen. nov., *Epulorhiza* gen. nov., *Moniliopsis*, and *Rhizoctonia*. *Mycotaxon*, **29**, 91–99.

Moore, R.T. (1990). The genus *Lalaria* gen. nov.: Taphrinales anamorphosum. *Mycotaxon*, **38**, 315–330.

Moore, R.T. (1996). The dolipore/parenthesome in modern taxonomy. In *Rhizoctonia Species: Taxonomy, Molecular Biology, Ecology, Pathology and Disease Control*, ed. S. Sneh, S. Jabayi-Hare, S. Neate & G. Dijst. Dordrecht, Netherlands: Kluwer, pp. 13–35.

Moore, R.T. & Marchant, R. (1972). Ultrastructural characterization of the basidiomycete septum of *Polyporus biennis*. *Canadian Journal of Botany*, **50**, 2463–2469.

Moore-Landecker, E. (1975). Effect of cultural conditions on apothecial morphogenesis in *Pyronema domesticum*. *Canadian Journal of Botany*, **53**, 2759–2769.

Moore-Landecker, E. (1992). Physiology and biochemistry of ascocarp induction and development. *Mycological Research*, **96**, 705–716.

Morgan, W.M. (1983). Viability of *Bremia lactucae* oospores and stimulation of their germination by lettuce seedlings. *Transactions of the British Mycological Society*, **80**, 403–408.

Mori, Y., Sato, Y. & Takamatsu, S. (2000). Evolutionary analysis of the powdery mildew fungi using nucleotide sequences of the nuclear ribosomal DNA. *Mycologia*, **92**, 74–93.

Morte, A., Lovisolo, C. & Schubert, A. (2000). Effect of drought stress on growth and water relations of the mycorrhizal association *Helianthemum almeriense*–*Terfezia claveryi*. *Mycorrhiza*, **10**, 115–119.

Morton, C.O., Hirsch, P.R. & Kerry, B.R. (2004). Infection of plant-parasitic nematodes by nematophagous fungi – a review of the application of molecular biology to understand infection processes and to improve biological control. *Nematology*, **6**, 161–170.

Morton, D.J. (1961). Tryptan blue and boiling lactophenol for staining and clearing barley tissues infected with *Ustilago nuda*. *Phytopathology*, **51**, 27–29.

Morton, J.B. & Benny, G.L. (1990). Revised classification of arbuscular mycorrhizal fungi (Zygomycetes): a new order, Glomales, two new suborders, Glomineae and Gigasporineae, and two new families, Acaulosporaceae and Gigasporaceae, with an emendation of Glomaceae. *Mycotaxon*, **37**, 471–491.

Morton, J.B. & Bentivenga, S.P. (1994). Levels of diversity in endomycorrhizal fungi (Glomales, Zygomycetes) and their role in defining taxonomic and non-taxonomic groups. *Plant and Soil*, **159**, 47–59.

Morton, J.B., Bentivenga, S.P. & Wheeler, W.W. (1993). Germ plasm in the International Collection of Arbuscular and Vesicular–arbuscular Mycorrhizal Fungi (INVAM) and procedures for culture development, documentation and storage. *Mycotaxon*, **48**, 491–528.

Moseman, J.G. (1966). Genetics of powdery mildews. *Annual Review of Phytopathology*, **4**, 269–291.

Moseman, J.G. & Powers, H.R. (1957). Function and longevity of cleistothecia of *Erysiphe graminis* f. sp. *hordei*. *Phytopathology*, **47**, 53–56.

Moss, M.O. (1994). Biosynthesis of *Aspergillus* toxins – non-aflatoxins. In *The Genus Aspergillus. From Taxonomy and Genetics to Industrial Application*, ed. K.A. Powell, A. Renwick & J.F. Peberdy. New York: Plenum Press, pp. 29–50.

Moss, M.O. (2002). Mycotoxin review: 2. *Fusarium*. *Mycologist*, **16**, 158–161.

Moss, M.O. & Baker, T. (2002). Teaching techniques for mycology: 17. The phototropic response of *Phycomyces blakesleeanus*. *Mycologist*, **16**, 23–26.

Moss, M. O. & Long, M. T. (2002). Fate of patulin in the presence of the yeast *Saccharomyces cerevisiae*. *Food Additives and Contaminants*, **19**, 387–399.

Moss, S. T. (1975). Septal structure in the Trichomycetes with special reference to *Astreptonema gammari* (Eccrinales). *Transactions of the British Mycological Society*, **65**, 115–127.

Moss, S. T. (1986). Biology and phylogeny of the Labyrinthulales and Thraustochytriales. In *The Biology of Marine Fungi*, ed. S. T. Moss. Cambridge: Cambridge University Press, pp. 105–29.

Moss, S. T. & Descals, E. (1986). A previously undescribed stage in the life cycle of Harpellales (Trichomycetes). *Mycologia*, **78**, 213–222.

Moss, S. T. & Lichtwardt, R. W. (1976). Development of trichospores and their appendages in *Genistellospora homothallica* and other Harpellales and fine-structural evidence for the sporangial nature of trichospores. *Canadian Journal of Botany*, **54**, 2346–2364.

Moss, S. T. & Lichtwardt, R. W. (1977). Zygospores of the Harpellales: an ultrastructural study. *Canadian Journal of Botany*, **55**, 3099–3110.

Moss, S. T. & Young, T. W. K. (1978). Phyletic considerations of the Harpellales and Asellariales (Trichomycetes, Zygomycotina) and the Kickxellales (Zygomycetes, Zygomycotina). *Mycologia*, **70**, 944–963.

Mossebo, D. C. (1998). Étude de la polarité sexuelle chez *Dacrymyces stillatus* Nees:Fries (basidiomycète). *Cryptogamie Mycologie*, **19**, 285–300.

Mossebo, D.-C. & Amougou, A. (2001). Mise en évidence de la dissolution des cloisons dolipores préalable à la migration nucléaire chez *Dacrymyces stillatus* Nees:Fries (Basidiomycete). *Cryptogamie Mycologie*, **22**, 185–191.

Mostert, L., Crous, P. W. & Petrini, O. (2000). Endophytic fungi associated with shoots and leaves of *Vitis vinifera*, with specific reference to the *Phomopsis viticola* complex. *Sydowia*, **52**, 46–58.

Mostert, L., Crous, P. W., Kang, J.-C. & Phillips, A. J. L. (2001). Species of *Phomopsis* and a *Libertella* sp. occurring on grapevines with specific reference to South Africa: morphological, cultural, molecular and pathological characterization. *Mycologia*, **93**, 146–167.

Motta, J. (1967). A note on the mitotic apparatus in the rhizomorph meristem of *Armillaria mellea*. *Mycologia*, **59**, 370–375.

Mouchacca, J. (1997). Thermophilic fungi: biodiversity and taxonomic status. *Cryptogamie Mycologie*, **18**, 19–69.

Mouchacca, J. (2000). Thermophilic fungi and applied research: a synopsis of name changes and synonymies. *World Journal of Microbiology and Biotechnology*, **16**, 881–888.

Movahedi, S. & Heale, J. B. (1990). The roles of aspartic proteinase and endopectin lyase enzymes in the primary stages of infection and pathogenesis of various host tissues by different isolates of *Botrytis cinerea* Pers. ex Pers. *Physiological and Molecular Plant Pathology*, **36**, 303–324.

Mower, R. L. & Hancock, J. G. (1975a). Sugar composition of ergot honeydews. *Canadian Journal of Botany*, **53**, 2813–2825.

Mower, R. L. & Hancock, J. G. (1975b). Mechanism of honeydew formation by *Claviceps* species. *Canadian Journal of Botany*, **53**, 2826–2834.

Moxham, S. E. & Buczacki, S. T. (1983). Chemical composition of the resting spore wall of *Plasmodiophora brassicae*. *Transactions of the British Mycological Society*, **80**, 297–304.

Moyersoen, B., Beever, R. E. & Martin, F. (2003). Genetic diversity of *Pisolithus* in New Zealand indicates multiple long-distance dispersal from Australia. *New Phytologist*, **160**, 569–579.

Mtui, G. & Nakamura, Y. (2004). Lignin-degrading enzymes from mycelial cultures of basidiomycete fungi isolated in Tanzania. *Journal of Chemical Engineering of Japan*, **37**, 113–118.

Muehlstein, L. K., Porter, D. & Short, F. T. (1991). *Labyrinthula zosterae* sp. nov., the causative agent of wasting disease of eelgrass, *Zostera marina*. *Mycologia*, **83**, 180–191.

Mueller, G. M. & Pine, E. M. (1994). DNA data provide evidence on the evolutionary relationships between mushrooms and false truffles. *McIlvainea*, **11**, 61–74.

Mueller, W. C., Tessier, B. J. & Englander, L. (1986). Immunocytochemical detection of fungi in the roots of *Rhododendron*. *Canadian Journal of Botany*, **64**, 718–723.

Mugnier, J. & Mosse, B. (1987). Vesicular-arbuscular mycorrhizal infection in transformed root-inducing T-DNA roots grown axenically. *Phytopathology*, **77**, 1045–1050.

Mullens, B. A. & Rodriguez, J. L. (1985). Dynamics of *Entomophthora muscae* (Entomophthorales: Entomophthoraceae) conidial discharge from *Musca domestica* (Diptera: Muscidae) cadavers. *Environmental Entomology*, **14**, 317–322.

Muller, H. G. (1958). The constricting ring mechanism of two predacious Hyphomycetes. *Transactions of the British Mycological Society*, **41**, 341–364.

Müller, E. & Tomasevic, M. (1957). Kulturversuche mit einigen Arten der Gattung *Leptosphaeria* Ces. & de Not. *Phytopathologische Zeitschrift*, **29**, 287–294.

Müller, M. (1993). The hydrogenosome. *Journal of General Microbiology*, **139**, 2879–2889.

Müller, P. & Hilgenberg, W. (1986). Isomers of zeatin and zeatin riboside in clubroot tissue: evidence for *trans*-zeatin biosynthesis by *Plasmodiophora brassicae*. *Physiologia Plantarum*, **66**, 245–250.

Müller, W. H., Stalpers, J. A., van Aelst, A. C., van der Krift, T. P. & Boekhout, T. (1998a). Structural differences between the two types of basidiomycete septal pore caps. *Microbiology*, **144**, 1721–1730.

Müller, W. H., Stalpers, J. A., van Aelst, A. C., van der Krift, T. P. & Boekhout, T. (1998b). Field emission gun-scanning electron microscopy of septal pore caps of selected species in the *Rhizoctonia* s.l. complex. *Mycologia*, **90**, 170–179.

Müller, W. H., Koster, A. J., Humbel, B. M., *et al.* (2000). Automated electron tomography of the septal pore cap in *Rhizoctonia solani*. *Journal of Structural Biology*, **131**, 10–18.

Mullins, J. T. (1973). Lateral branch formation and cellulase production in the water molds. *Mycologia*, **65**, 1007–1014.

Mullins, J. T. & Raper, J. R. (1965). Heterothallism in biflagellate aquatic fungi: preliminary genetic analysis. *Science*, **150**, 1174–1175.

Multani, D. S., Meeley, R. B., Paterson, A. H., *et al.* (1998). Plant-pathogen microevolution: molecular basis for the origin of a fungal disease in maize. *Proceedings of the National Academy of Sciences of the USA*, **95**, 1686–1691.

Munn, A. L. (2000). The yeast endocytic membrane transport system. *Microscopy Research and Technique*, **51**, 547–562.

Munn, E. A., Orpin, C. G. & Hall, F. J. (1981) Ultrastructural studies of the free zoospore of the rumen phycomycete *Neocallimastix frontalis*. *Journal of General Microbiology*, **125**, 311–323.

Murphy, J. W. (1996). Cell-mediated immunity. In *The Mycota* **VI**: *Human and Animal Relationships*, ed. D. H. Howard & J. D. Miller. Berlin: Springer-Verlag, pp. 67–97.

Myers, R. B. & Cantino, E. C. (1974). The gamma particle. In *Monographs in Developmental Biology* 8, ed. A. Wolsky. New York: Karger, pp. 1–46.

Myllys, L., Källersjö, M. & Tehler, A. (1998). A comparison of SSU rDNA data and morphological data in Arthoniales (Euascomycetes) phylogeny. *Bryologist*, **101**, 70–85.

Myllys, L., Stenroos, S. & Thell, A. (2002). New genes for phylogenetic studies of lichenized fungi: glyceraldehyde-3-phosphate dehydrogenase and beta-tubulin genes. *Lichenologist*, **34**, 237–246.

Mylyk, O. M. (1976). Heteromorphism for heterokaryon incompatibility genes in natural populations of *Neurospora crassa*. *Genetics*, **83**, 275–284.

Nadeau, M. P., Dunphy, G. B. & Boisvert, J. L. (1995). Effects of physical factors on the development of secondary conidia of *Erynia conica* (Zygomycetes: Entomophthorales), a pathogen of adult black flies (Diptera: Simuliidae). *Experimental Mycology*, **19**, 324–329.

Nadeau, M. P., Dunphy, G. B. & Boisvert, J. L. (1996). Development of *Erynia conica* (Zygomycetes: Entomophthorales) on the cuticle of adult black flies *Simulium rostratum* and *Simulium decorum* (Diptera: Simuliidae). *Journal of Invertebrate Pathology*, **68**, 50–58.

Nagahama, T., Hamamoto, M., Nakase, T., Takami, H. & Horikoshi, K. (2001). Distribution and identification of red yeasts in deep-sea environments around the northwest Pacific Ocean. *Antonie van Leeuwenhoek*, **80**, 101–110.

Nagano, S. (2000). Modeling the model organism *Dictyostelium discoideum*. *Development, Growth and Differentiation*, **42**, 541–550.

Nagarajan, S. & Singh, D. V. (1990). Long-distance dispersal of rust pathogens. *Annual Review of Phytopathology*, **28**, 139–153.

Nagarajan, S., Aujla, S. S., Nanda, G. S., *et al.* (1997). Karnal bunt (*Tilletia indica*) of wheat – a review. *Review of Plant Pathology*, **76**, 1207–1214.

Nakagaki, T. (2001). Smart behavior of true slime mold in a labyrinth. *Researches in Microbiology*, **152**, 767–770.

Nakagaki, T., Yamada, H. & Hara, M. (2004). Smart network solutions in an amoeboid organism. *Biophysical Chemistry*, **107**, 1–5.

Nakamura, A. & Kohama, K. (1999). Calcium regulation of the actin–myosin interaction of *Physarum polycephalum*. *International Review of Cytology*, **191**, 53–98.

Nakase, T. (2000). Expanding world of ballistosporous yeasts: distribution in the phyllosphere, systematics and phylogeny. *Journal of General and Applied Microbiology*, **46**, 189–216.

Nakatsuji, N. & Bell, E. (1980). Control by calcium of the contractility of *Labyrinthula* slimeways and of the translocation of *Labyrinthula* cells. *Cell Motility*, **1**, 17–19.

Nannfeldt, J. A. (1932). Studien über die Morphologie und Systematik der nichtlichenisierten inoperculaten Discomyceten. *Nova Acta Regiae Societas Scientiarum Upsaliensis IV*, **8**(2), 1–368.

Nannfeldt, J.A. (1942). The Geoglossaceae of Sweden (with regard to surrounding countries). *Arkiv för Botanik*, **30A**(4), 1–67.

Nannfeldt, J.A. (1981). *Exobasidium*, a taxonomic reassessment applied to the European species. *Symbolae Botanicae Uppsaliensis*, **23**, 1–72.

Narisawa, K. & Hashiba, T. (1998). Development of resting spores on plants inoculated with a dikaryotic resting spore of *Plasmodiophora brassicae*. *Mycological Research*, **102**, 949–952.

Narisawa, K., Yamaoka, Y. & Katsuya, K. (1994). Mating type of isolates derived from the spermogonial state of *Puccinia coronata* var. *coronata*. *Mycoscience*, **35**, 131–135.

Narisawa, K., Kageyama, K. & Hashiba, T. (1996). Efficient root infection with single resting spores of *Plasmodiophora brassicae*. *Mycological Research*, **100**, 855–858.

Narisawa, K., Tokumasu, S. & Hashiba, T. (1998). Suppression of clubroot formation in Chinese cabbage by the root endophytic fungus, *Heteroconium chaetospira*. *Plant Pathology*, **47**, 206–210.

Nash, T.H., ed. (1996a). *Lichen Biology*. Cambridge: Cambridge University Press.

Nash, T.H. (1996b). Nitrogen, its metabolism and potential contribution to ecosystems. In *Lichen Biology*, ed. T.H. Nash. Cambridge: Cambridge University Press, pp. 121–135.

Nash, T.H. & Gries, C. (2002). Lichens as bioindicators of sulfur dioxide. *Symbiosis*, **33**, 1–21.

Nasmyth, K. (1983). Molecular analysis of a cell lineage. *Nature*, **302**, 670–676.

Nath, M.D., Sharma, S.L. & Kant, U. (2001). Growth of *Albugo candida* infected mustard callus in culture. *Mycopathologia*, **152**, 147–153.

Natvig, D.O. & Gleason, F.H. (1983). Oxygen uptake by obligately-fermentative aquatic fungi: absence of a cyanide-sensitive component. *Archives of Microbiology*, **134**, 5–8.

Navarre, D.A. & Wolpert, T.J. (1999). Victorin induction of an apoptotic/senescence-like response in oats. *Plant Cell*, **11**, 237–249.

Nawawi, A. (1985). Basidiomycetes with branched water-borne conidia. *Botanical Journal of the Linnean Society*, **91**, 51–60.

Nawawi, A. & Webster, J. (1982). *Sistotrema hamatum* sp. nov., the teleomorph of *Ingoldiella hamata*. *Transactions of the British Mycological Society*, **78**, 287–291.

Neergaard, P. (1945). *Danish species of Alternaria and Stemphylium*. Copenhagen: E. Munksgaard.

Neergaard, P. (1977). *Seed Pathology*, 2 volumes. London: Macmillan.

Neiman, A.M. (2005). Ascospore formation in the yeast *Saccharomyces cerevisiae*. *Microbiology and Molecular Biology Reviews*, **69**, 565–584.

Nelson, A.C. & Backus, M.P. (1968). Ascocarp development in two homothallic Neurosporas. *Mycologia*, **60**, 16–28.

Nelson, B.D. & Durán, R. (1984). Cytology and morphological development of basidia, dikaryons, and infective structures of *Urocystis agropyri* from wheat. *Phytopathology*, **74**, 299–304.

Nelson, J.C., Autrique, J.E., Fuentes-Dávila, G. & Sorrells, M.E. (1998). Chromosomal location of genes for resistance to Karnal bunt in wheat. *Crop Science*, **38**, 231–236.

Nelson, P.E., Toussoun, T.A. & Cook, R.J., eds. (1981). *Fusarium Diseases, Biology and Taxonomy*. London, PA: Pennsylvania State University Press.

Nelson, P.E., Toussoun, T.A. & Marasas, W.F.O. (1983). *Fusarium Species. An Illustrated Manual for Identification*. London, PA: Pennsylvania State University Press.

Nelson, R.T., Yangco, B.G., Strake, D.T. & Cochrane, B.J. (1990). Genetic studies in the genus *Basidiobolus*. 2. Phylogenetic relationships inferred from ribosomal DNA analysis. *Experimental Mycology*, **14**, 197–206.

Németh, É. (1999). Parasitic production of ergot alkaloids. In *Ergot. The Genus Claviceps*, ed. V. Křen & L. Cvak. Amsterdam: Harwood Academic Publishers, pp. 303–19.

Nes, W.D. (1990). Stereochemistry, sterol metamorphosis and evolution. In *The Biochemistry, Structure and Utilization of Plant Lipids*, ed. P.J. Quinn & J.L. Harwood. Portland: Portland Press, pp. 304–28.

Nes, W.D. & Stafford, A.E. (1984). Side chain structural requirements for sterol-induced regulation of *Phytophthora cactorum* physiology. *Lipids*, **9**, 596–612.

Nes, W.D., Patterson, G.W. & Bean, G.A. (1979). The effect of steroids and their solubilizing agents on mycelial growth of *Phytophthora cactorum*. *Lipids*, **14**, 458–462.

Newhook, F.J., Young, B.R., Allen, S.D. & Allen, R.N. (1981). Zoospore motility of *Phytophthora cinnamomi* in particulate substrates. *Phytopathologische Zeitschrift*, **101**, 202–209.

Newman, E.I. (1988). Mycorrhizal links between plants: their functioning and ecological significance. *Advances in Ecological Research*, **18**, 243–270.

Newsham, K. K., Low, M. N. R., McLeod, A. R., Greenslade, P. D. & Emmett, B. A. (1997). Ultraviolet-B radiation influences the abundance and distribution of phylloplane fungi on pedunculate oak (*Quercus robur*). *New Phytologist*, **136**, 287–297.

Niccoli, T., Arellano, M. & Nurse, P. (2003). Role of Tea1p, Tea3p and Pom1p in the determination of cell ends in *Schizosaccharomyces pombe*. *Yeast*, **20**, 1349–1358.

Nickerson, A. W. & Raper, K. B. (1973). Macrocysts in the life cycle of the Dictyosteliaceae. II. Germination of the macrocysts. *American Journal of Botany*, **60**, 247–254.

Niederhauser, J. S. (1991). *Phytophthora infestans*: the Mexican connection. In *Phytophthora*, ed. J. A. Lucas, R. C. Shattock, D. S. Shaw & L. R. Cooke. Cambridge: Cambridge University Press, pp. 25–45.

Nielsen, K. A., Nicholson, R. A., Carver, T. L. W., Kunoh, H. & Oliver, R. P. (2000). First touch: an immediate response to surface recognition in conidia of *Blumeria graminis*. *Physiological and Molecular Plant Pathology*, **56**, 63–70.

Nielsen, R. I. (1978). Sexual mutants of a heterothallic *Mucor* species, *Mucor pusillus*. *Experimental Mycology*, **2**, 193–197.

Nielsen, J. (1988). *Ustilago* spp., smuts. *Advances in Plant Pathology*, **6**, 483–490.

Niemelä, T. & Korhonen, K. (1998). Taxonomy of the genus *Heterobasidion*. In *Heterobasidion annosum. Biology, Ecology, Impact and Control*, ed. S. Woodward, J. Stenlid, R. Karjalainan & A. Hüttermann. Wallingford, UK: CAB International, pp. 27–33.

Nikoh, N. & Fukatsu, T. (2000). Interkingdom host jumping underground: phylogenetic analysis of entomopathogenic fungi of the genus *Cordyceps*. *Molecular Biology and Evolution*, **17**, 629–637.

Niks, R. E., Walther, U., Jaiser, H., *et al.* (2000). Resistance against barley leaf rust (*Puccinia hordei*) in West-European spring barley germplasm. *Agronomie*, **20**, 769–782.

Nishida, H. & Sugiyama, J. (1994). Archiascomycetes: detection of a major new lineage within the Ascomycota. *Mycoscience*, **35**, 361–366.

Noble, M. & Glynne, M. D. (1970). Wart disease of potatoes. *F. A. O. Plant Protection Bulletin*, **18**, 125–135.

Noble, R., Fermor, T. R., Lincoln, S., *et al.* (2003). Primordia initiation of mushroom (*Agaricus bisporus*) strains on axenic casing materials. *Mycologia*, **95**, 620–629.

Nordbring-Hertz, B. (1972). Scanning electron microscopy of the nematode-trapping organs in *Arthrobotrys oligospora*. *Physiologia Plantarum*, **26**, 279–284.

Nordbring-Hertz, B. (1977). Nematode-induced morphogenesis in the predacious fungus *Arthrobotrys oligospora*. *Nematologica*, **23**, 443–451.

Nordbring-Hertz, B. (1988). Ecology and recognition in the nematode–nematophagous fungus system. *Advances in Microbial Ecology*, **10**, 81–114.

Nordbring-Hertz, B. & Jansson, H.-B. (1984). Fungal development, predacity, and recognition of prey in nematode-destroying fungi. In *Current Perspectives in Microbial Ecology*, ed. M. J. Klug & C. A. Reddy. Washington: American Society for Microbiology, pp. 327–333.

Nordbring-Hertz, B., Jansson, H.-B., Persson, Y., Friman, E. & Dackman, C. (1995). Nematophagous fungi. VHS film C1851. Göttingen: Institut für den Wissenschaftlichen Film.

Normanly, J. (1997). Auxin metabolism. *Physiologia Plantarum*, **100**, 431–442.

Nosanchuk, J. D. (2005). Protective antibodies and endemic dimorphic fungi. *Current Molecular Medicine*, **5**, 435–442.

Nouhra, E. R., Horton, T. R., Cazares, E. & Castellano, M. (2005). Morphological and molecular characterization of selected *Ramaria* mycorrhizae. *Mycorrhiza*, **15**, 55–59.

Nout, M. J. R. & Aidoo, K. E. (2002). Asian fungal fermented food. In *The Mycota X: Industrial Applications*, ed. H. D. Osiewacz. Berlin: Springer-Verlag, pp. 23–47.

Nurse, P. (1990). Universal control mechanism regulating onset of M-phase. *Nature*, **344**, 503–508.

Nurse, P. (2002). Cyclin dependent kinases and cell cycle control (Nobel lecture). *Chembiochem*, **3**, 596–603.

Nuss, D. L. (1992). Biological control of chestnut blight: an example of virus-mediated attenuation of fungal pathogenesis. *Microbiological Reviews*, **56**, 561–576.

Nuss, D. L. (1996). Using hypoviruses to probe and perturb signal transduction processes underlying fungal pathogenesis. *Plant Cell*, **8**, 1845–1853.

Nuss, I., Jennings, D. H. & Veltkamp, C. J. (1991). Morphology of *Serpula lacrymans*. In *Serpula lacrymans: Fundamental Biology and Control Strategies*, ed. D. H. Jennings & A. F. Bravery. Chichester: Wiley, pp. 9–38.

Nutting, R., Rapoport, H. & Machlis, L. (1968). The structure of sirenin. *Journal of the American Chemical Society*, **90**, 6434–6438.

Nyhlén, L. & Unestam, T. (1980). Wound reactions and *Aphanomyces astaci* growth in crayfish cuticle. *Journal of Invertebrate Pathology*, **36**, 187–197.

Nylander, W. (1866). Les lichens du Jardin du Luxembourg. *Bulletin de la Société Botanique de France, Lettres Botaniques*, **13**, 364–372.

Obermayer, W. & Poelt, J. (1992). Contributions to the knowledge of the lichen flora of the Himalayas. III. On *Lecanora somervellii* Paulson (lichenized Ascomycotina, Lecanoraceae). *Lichenologist*, **24**, 111–117.

Oberwinkler, F. (1993). Genera in a monophyletic group: the Dacrymycetales. *Mycologia Helvetica*, **6**, 35–72.

O'Brien, E. & Dietrich, D. R. (2005). Ochratoxin A: the continuing enigma. *Critical Reviews in Toxicology*, **35**, 33–60.

O'Connell, R. J., Bailey, J. A. & Richmond, D. V. (1985). Cytology and physiology of infection of *Phaseolus vulgaris* infected by *Colletotrichum lindemuthianum*. *Physiological Plant Pathology*, **27**, 75–98.

O'Connell, R. J., Perfect, S., Hughes, B., *et al.* (2000). Dissecting the cell biology of *Colletotrichum* infection processes. In *Colletotrichum. Host Specificity, Pathology, and Host–Pathogen Interaction*, ed. D. Prusky, S. Freeman & M. B. Dickman. St. Paul, USA: APS Press, pp. 57–77.

Oda, T., Tanaka, C. & Tsuda, M. (2004). Molecular phylogeny and biogeography of the widely distributed *Amanita* species, *A. muscaria* and *A. pantherina*. *Mycological Research*, **108**, 885–896.

Odds, F. C. (1994). *Candida* species and virulence. *ASM News*, **60**, 313–318.

Odds, F. C. (2003). Coccidioidomycosis: flying conidia and severed heads. *Mycologist*, **17**, 37–40.

Odom, A., Muir, S., Lim, E., *et al.* (1997). Calcineurin is required for virulence of *Cryptococcus neoformans*. *EMBO Journal*, **16**, 2576–2589.

O'Donnell, K. (1996). Progress towards a phylogenetic classification of *Fusarium*. *Sydowia*, **48**, 57–70.

O'Donnell, K., Kistler, H. C., Cigelnik, E. & Ploetz, R. C. (1998). Multiple evolutionary origins of the fungus causing Panama disease of banana: concordant evidence from nuclear and mitochondrial gene genealogies. *Proceedings of the National Academy of Sciences of the USA*, **95**, 2044–2049.

O'Donnell, K. L. (1979). *Zygomycetes in Culture*. Athens, USA: Department of Botany, University of Georgia.

O'Donnell, K. L., Fields, W. G. & Hooper, G. R. (1974). Scanning ultrastructural ontogeny of cleistohymenial apothecia of the operculate Discomycete *Ascobolus furfuraceus*. *Canadian Journal of Botany*, **52**, 1653–1656.

O'Donnell, K. L., Hooper, G. R. & Fields, W. G. (1976). Zygosporogenesis in *Phycomyces blakesleeanus*. *Canadian Journal of Botany*, **54**, 2573–2586.

O'Donnell, K. L., Ellis, J. J., Hesseltine, C. W. & Hooper, G. R. (1977a). Zygosporogenesis in *Gilbertella persicaria*. *Canadian Journal of Botany*, **55**, 662–675.

O'Donnell, K. L., Ellis, J. J., Hesseltine, C. W. & Hooper, G. R. (1977b). Azygosporogenesis in *Mucor azygosporus*. *Canadian Journal of Botany*, **55**, 2712–2720.

O'Donnell, K. L., Ellis, J. J., Hesseltine, C. W. & Hooper, G. R. (1977c). Morphogenesis of azygospores induced in *Gilbertella persicaria* (+) by imperfect hybridization with *Rhizopus stolonifer*. *Canadian Journal of Botany*, **55**, 2721–2727.

O'Donnell, K. L., Flegler, S. L., Ellis, J. J. & Hesseltine, C. W. (1978a). The *Zygorhynchus* zygosporangium and zygospore. *Canadian Journal of Botany*, **56**, 1061–1073.

O'Donnell, K. L., Flegler, S. L. & Hooper, G. R. (1978b). Zygosporangium and zygospore formation in *Phycomyces nitens*. *Canadian Journal of Botany*, **56**, 91–100.

O'Donnell, K. L., Cigelnik, E., Weber, N. S. & Trappe, J. M. (1997). Phylogenetic relationships among ascomycetous truffles and the true and false morels inferred from 18S and 28S ribosomal DNA sequence analysis. *Mycologia*, **87**, 48–65.

O'Donnell, K. L., Cigelnik, E. & Benny, G. L. (1998). Phylogenetic relationships among the Harpellales and Kickxellales. *Mycologia*, **90**, 624–639.

O'Donnell, K. L., Lutzoni, F. M., Ward, T. J. & Benny, G. L. (2001). Evolutionary relationships among Mucoralean fungi (Zygomycota). Evidence for family polyphyly on a large scale. *Mycologia*, **93**, 286–296.

Oeser, B., Tenberge, K. B., Moore, S. M., *et al.* (2002). Pathogenic development of *Claviceps purpurea*. In *Molecular Biology of Fungal Development*, ed. H. D. Osiewacz. New York: Marcel Dekker Inc., pp. 419–455.

Ogawa, H. & Sugiyama, J. (2000). Evolutionary relationships of the cleistothecial genera with *Penicillium*, *Geosmithia*, *Merimbla* and *Sarophorum* anamorphs as inferred from 18S rDNA sequence divergence. In *Integration of Modern Taxonomic Methods for Penicillium and Aspergillus Classification*, ed. R. A. Samson & J. I. Pitt. Amsterdam: Harwood Academic Publishers, pp. 149–61.

Ogawa, H., Yoshimura, A. & Sugiyama, J. (1997). Phylogenetic origins of the anamorphic genus *Geosmithia* and the relationships of the cleistothecial genera: evidence from 18S, 5S and 28S sequence analyses. *Mycologia*, **89**, 756–771.

Ohtake, M., Yamamoto, H. & Uchiyama, T. (1999) Influences of metabolic inhibitors and hydrolytic enzymes on the adhesion of appressoria of *Pyricularia oryzae* to wax-coated cover-glasses. *Bioscience, Biotechnology, and Biochemistry*, **63**, 978–982.

Ojha, M. & Turian, G. (1971). Interspecific transformation and DNA characteristics in *Allomyces*. *Molecular and General Genetics*, **112**, 49–59.

Okada, G., Jacobs, K., Kirisits, T., *et al.* (2000). Epitypification of *Graphium penicillioides* Corda, with comments on the phylogeny and taxonomy of graphium-like synnematous fungi. *Studies in Mycology*, **45**, 169–188.

Okafor, J.I., Testrake, D., Mushinsky, H.R. & Yangco, B.G. (1984). A *Basidiobolus* sp. and its association with reptiles and amphibians in southern Florida. *Sabouraudia*, **22**, 47–51.

Okamoto, K. & Shaw, J.M. (2005). Mitochondrial morphology and dynamics in yeast and multicellular eukaryotes. *Annual Review of Genetics*, **39**, 503–536.

Olive, L.S. (1947). Notes on the Tremellales of Georgia. *Mycologia*, **39**, 90–108.

Olive, L.S. (1967). The Protostelida – a new order of the Mycetozoa. *Mycologia*, **59**, 1–29.

Olive, L.S. (1975). *The Mycetozoans*. New York: Academic Press.

Oliver, S.G. (1991). 'Classical' yeast biotechnology. In *Saccharomyces*, ed. M.F. Tuite & S.G. Oliver. London: Plenum Press, pp. 213–48.

Olivier, C., Berbee, M.L., Shoemaker, R.A. & Loria, R. (2000). Molecular phylogenetic support from ribosomal DNA sequences for origin of *Helminthosporium* from *Leptosphaeria*-like loculoascomycete ancestors. *Mycologia*, **92**, 736–746.

Olsen, J.H., Dragsted, L. & Autrup, H. (1988). Cancer risk and occupational exposure to aflatoxins in Denmark. *British Journal of Cancer*, **58**, 392–6

Olson, L.W. (1974). Meiosis in the aquatic Phycomycete *Allomyces macrogynus*. *Comptes Rendus des Travaux du Laboratoire Carlsberg*, **40**, 113–124.

Olson, L.W. (1984). *Allomyces* – a different fungus. *Opera Botanica*, **73**, 1–96.

Olson, L.W. & Fuller, M.S. (1968). Ultrastructural evidence for the biflagellate origin of the uniflagellate fungal zoospore. *Archiv für Mikrobiologie*, **62**, 237–250.

Olson, L.W. & Reichle, R. (1978). Synaptonemal complex formation and meiosis in the resting sporangium of *Blastocladiella emersonii*. *Protoplasma*, **97**, 261–273.

O'Neal, M.A. & Schoenenberger, K.R. (2003). A *Rhizocarpon geographicum* growth curve for the cascade range of Washington and northern Oregon, USA. *Quaternary Research*, **60**, 233–241.

Ono, Y. (2002). The diversity of nuclear cycle in microcyclic rust fungi (Uredinales) and its ecological and evolutionary implications. *Mycoscience*, **43**, 421–439.

Orlowski, M. (1991). *Mucor* dimorphism. *Microbiological Reviews*, **55**, 234–258.

Orlowski, M. (1995). Gene expression in *Mucor* dimorphism. *Canadian Journal of Botany*, **73**, S326–S334.

Ormrod, D.J., O'Reilly, H.J., van der Kamp, B.J. & Borno, C. (1984). Epidemiology, cultivar susceptibility, and chemical control of *Gymnosporangium fuscum* in British Columbia. *Canadian Journal of Plant Pathology*, **6**, 63–70.

Orpin, C.G. (1974). The rumen flagellate *Callimastix frontalis*: does sequestration occur? *Journal of General Microbiology*, **84**, 395–398.

Orpin, C.G. & Munn, E.A. (1986). *Neocallimastix patriciarum* sp. nov., a new member of the Neocallimasticaceae inhabiting the rumen of sheep. *Transactions of the British Mycological Society*, **86**, 178–181.

Ortega, J.K.E. (1990). Governing equations for plant cell growth. *Physiologia Plantarum*, **79**, 116–121.

Ortega, J.K.E., Lesh-Laurie, G.E., Espinosa, M.A., *et al.* (2003). Helical growth of stage-IVb sporangiophores of *Phycomyces blakesleeanus*: the relationship between rotation and elongation growth rates. *Planta*, **216**, 716–722.

Ortiz-García, S., Gernandt, D.S., Stone, J.K., *et al.* (2003). Phylogenetics of *Lophodermium* from pine. *Mycologia*, **95**, 846–859.

Orton, P.D. & Watling, R. (1979). Coprinaceae Part 1: *Coprinus*. *British Fungus Flora* **2**. Edinburgh: Royal Botanic Garden.

Osiewacz, H.D. (2002). Aging in fungi: role of mitochondria in *Podospora anserina*. *Mechanisms of Ageing and Development*, **123**, 755–764.

Osiewacz, H.D. & Kimpel, E. (1999). Mitochondrial–nuclear interactions and lifespan control in fungi. *Experimental Gerontology*, **34**, 901–909.

Oso, B. (1969). Electron microscopy of ascus development in *Ascobolus*. *Annals of Botany* N.S., **33**, 205–209.

Ota, Y., Fukuda, K. & Suzuki, K. (1998). The nonheterothallic life cycle of Japanese *Armillaria mellea*. *Mycologia*, **90**, 396–405.

Otani, H. & Kohmoto, K. (1992). Host-specific toxins of *Alternaria* species. In *Alternaria, Biology, Plant Diseases and Metabolites*, ed. J. Chelkowski & A. Visconti. Amsterdam: Elsevier, pp. 123–56.

Ott, S. (1987). Sexual reproduction and developmental adaptations in *Xanthoria parietina*. *Nordic Journal of Botany*, **7**, 219–228.

Ott, S. & Lumbsch, H.T. (2001). Morphology and phylogeny of ascomycete lichens. In *The Mycota* **IX**: *Fungal Associations*, ed. B. Hock. Berlin: Springer-Verlag, pp. 189–210.

Ott, S., Meier, T. & Jahns, H.M. (1995). Development, regeneration, and parasitic interactions between the lichens *Fulgensia bracteata* and *Toninia caeruleonigricans*. *Canadian Journal of Botany*, **73**, S595–S602.

Ouimette, D.G. & Coffey, M.D. (1990). Symplastic entry and phloem translocation of phosphonate. *Pesticide Biochemistry and Physiology*, **38**, 18–25.

Ower, R. (1982). Notes on the development of the morel ascocarp: *Morchella esculenta*. *Mycologia*, **74**, 142–144.

Oxenbøll, K. (1994). *Aspergillus* enzymes and industrial uses. In *The Genus Aspergillus. From Taxonomy and Genetics to Industrial Application*, ed. K.A. Powell, A. Renwick & J.F. Peberdy. New York: Plenum Press, pp. 147–154.

Oxford, A.E., Raistrick, H. & Simonart, P. (1939). Studies on the biochemistry of microorganisms. 60. Griseofulvin, a metabolic product of *Penicillium griseofulvum* Dierckx. *Biochemical Journal*, **33**, 240–248.

Pacioni, G. (1991). Effect of *Tuber* metabolites on the rhizosphere environment. *Mycological Research*, **95**, 1355–1358.

Pacioni, G., Bellina-Agostinone, C. & D'Antonio, M. (1990). Odour composition of the *Tuber melanosporum* complex. *Mycological Research*, **94**, 201–204.

Pady, S.M. & Johnston, C.O. (1955). The concentration of airborne rust spores in relation to epidemiology of wheat rusts in Kansas in 1954. *Plant Disease Reporter*, **39**, 463–466.

Page, R.M. (1959). Stimulation of asexual reproduction of *Pilobolus* by *Mucor plumbeus*. *American Journal of Botany*, **46**, 579–585.

Page, R.M. (1960). The effect of ammonia on growth and reproduction of *Pilobolus kleinii*. *Mycologia*, **52**, 480–489.

Page, R.M. (1964). Sporangium discharge in *Pilobolus*: a photographic study. *Science*, **146**, 925–927.

Page, R.M. & Curry, G.M. (1966). Studies on phototropism of young sporangiophores of *Pilobolus kleinii*. *Photochemistry and Photobiology*, **5**, 31–40.

Page, R.M. & Kennedy, D. (1964). Studies on the velocity of discharged sporangia of *Pilobolus kleinii*. *Mycologia*, **56**, 363–368.

Page, R.M. & Humber, R.A. (1973). Phototropism in *Conidiobolus coronatus*. *Mycologia*, **65**, 335–354.

Paktitis, S., Grant, B. & Lawrie, A. (1986). Surface changes in *Phytophthora palmivora* zoospores following induced differentiation. *Protoplasma*, **135**, 119–129.

Palecek, S.P., Parikh, A.S. & Kron, S.J. (2002). Sensing, signalling and integrating physical processes during *Saccharomyces cerevisiae* invasive and filamentous growth. *Microbiology*, **148**, 893–907.

Palfreyman, J.W. & White, N.A. (2003). Everything you wanted to know about the dry-rot fungus but were afraid to ask. *Microbiology Today*, **30**, 107–109.

Palfreyman, J.W., White, N.A., Buultjens, T.E.J. & Glancy, H. (1995). The impact of current research on the treatment of infestations by the dry rot fungus *Serpula lacrymans*. *International Biodeterioration and Biodegradation*, **35**, 369–395.

Palm, M.E. (1999). Mycology and world trade: a view from the front line. *Mycologia*, **91**, 1–12.

Palmer, C.-L. & Skinner, W. (2002). *Mycosphaerella graminicola*: latent infection, crop devastation and genomics. *Molecular Plant Pathology*, **3**, 63–70.

Panabières, F., Ponchet, M., Allasia, V., Cardin, L. & Ricci, P. (1997). Characterization of border species among Pythiaceae: several *Pythium* isolates produce elicitins, typical proteins of *Phytophthora* spp. *Mycological Research*, **101**, 1459–1468.

Pande, S., Siddique, K.H.M., Kishore, G.K., et al. (2005). *Ascochyta* blight of chickpea (*Cicer arietinum* L.): a review of biology, pathogenicity, and disease management. *Australian Journal of Agricultural Research*, **56**, 317–332.

Papavizas, G.C. (1985). *Trichoderma* and *Gliocladium*: biology, ecology and potential for biocontrol. *Annual Review of Phytopathology*, **23**, 23–54.

Papavizas, G.C. & Ayers, W.A. (1974). *Aphanomyces* species and their root rot diseases in pea and sugarbeet. *US Department of Agriculture Technical Bulletin*, **1485**, 1–158.

Papierok, B. & Hajek, A.E. (1997). Fungi: Entomophthorales. In *Manual of Techniques in Insect Pathology*, ed. L.A. Lacey. London: Academic Press, pp. 187–212.

Parbery, D.G. (1996a). Spermatial states of fungi are andromorphs. *Mycological Research*, **100**, 1400.

Parbery, D.G. (1996b). Trophism and the association of fungi with plants. *Biological Reviews*, **71**, 473–527.

Parguey-Leduc, A. & Janex-Favre, M.C. (1982). La paroi des asques chez les Pyrenomycètes. Étude ultrastructurale. I. Les asques bituniqués typiques. *Canadian Journal of Botany*, **60**, 1222–1230.

Parguey-Leduc, A. & Janex-Favre, M.C. (1984). La paroi des asques chez les Pyrenomycètes. Étude ultrastructurale. II. Les asques unituniqués. *Cryptogamie Mycologie*, **5**, 171–187.

Parguey-Leduc, A., Janex-Favre, M.C. & Montant, C. (1991). L'ascocarpe du *Tuber melanosporum* Vitt. (Truffe noire du Périgord. Discomycètes): structure de la glèbe. II. Les veines stériles. *Cryptogamie Mycologie*, **12**, 165–182.

Park, D. (1974) Aquatic hyphomycetes in non-aquatic habitats. *Transactions of the British Mycological Society*, **63**, 179–183.

Park, R. F., Burdon, J. J. & McIntosh, R. A. (1995). Studies on the origin, spread, and evolution of an important group of *Puccinia recondita* f. sp. *tritici* pathotypes in Australasia. *European Journal of Plant Pathology*, **101**, 613–622.

Park, R. F., Burdon, J. J. & Jahoor, A. (1999). Evidence for somatic hybridization in nature in *Puccinia recondita* f. sp. *tritici*, the leaf rust pathogen of wheat. *Mycological Research*, **103**, 715–723.

Parladé, J., Pera, J. & Alvarez, I. F. (1996). Inoculation of containerized *Pseudotsuga menziesii* and *Pinus pinaster* seedlings with spores of five species of ectomycorrhizal fungi. *Mycorrhiza*, **6**, 237–245.

Parry, D. W. (1990). *Plant Pathology in Agriculture*. Cambridge: Cambridge University Press.

Partida-Martinez, L. P. & Hertweck, C. (2005). Pathogenic fungus harbours endosymbiotic bacteria for toxin production. *Nature*, **437**, 884–888.

Pataky, J. K. & Chandler, M. A. (2003). Production of huitlacoche, *Ustilago maydis*: timing inoculation and controlling pollination. *Mycologia*, **95**, 1261–1270.

Patouillard, N. T. (1887). *Les Hyménomycètes d'Europe. Anatomie Générale et Classification des Champignons Supérieurs*. Paris: Librairie Paul Klincksieck.

Paul, D., Mukhopadhyay, R., Chatterjee, B. P. & Guha, A. K. (2002). Nutritional profile of food yeast *Kluyveromyces fragilis* biomass grown on whey. *Applied Biochemistry and Biotechnology*, **97**, 209–218.

Paulin-Mahady, A. E., Harrington, T. C. & McNew, D. (2002). Phylogenetic and taxonomic evaluation of *Chalara*, *Chalaropsis*, and *Thielaviopsis* anamorphs associated with *Ceratocystis*. *Mycologia*, **94**, 62–72.

Pažoutová, S., Olšovská, J., Linka, M., Kolínská, R. & Flieger, M. (2000). Chemoraces and habitat specialisation of *Claviceps purpurea* populations. *Applied and Environmental Microbiology*, **66**, 5419–5425.

Pazzagli, L., Cappugi, G., Manao, G., *et al.* (1999). Purification, characterization, and amino acid sequence of cerato-platanin, a new phytotoxic protein from *Ceratocystis fimbriata* f. sp. *platani*. *Journal of Biological Chemistry*, **274**, 24 959–24 964.

Pearce, D. A. (1998). PCR as a tool for the investigation of seed-borne disease. In *Applications of PCR in Mycology*, ed. P. D. Bridge, D. K. Arora, C. A. Reddy & R. P. Elander. Wallingford, UK: CAB International, pp. 309–24.

Peberdy, J. F. (1994). Protein secretion in filamentous fungi – trying to understand a highly productive black box. *Trends in Biotechnology*, **12**, 50–57.

Pedras, M. S. C. & Biesenthal, C. J. (1998). Production of the host-selective phytotoxin phomalide by isolates of *Leptosphaeria maculans* and its correlation with sirodesmin PL production. *Canadian Journal of Microbiology*, **44**, 547–553.

Pedras, M. S. C., Okanga, F. I., Zaharia, I. L. & Khan, A. Q. (2000). Phytoalexins from crucifers: synthesis, biosynthesis, and biotransformation. *Phytochemistry*, **53**, 161–176.

Pegg, G. F. (1985). Presidential address. Life in a black hole – the micro-environment of the vascular pathogen. *Transactions of the British Mycological Society*, **85**, 1–20.

Pegler, D. N. (1991). Taxonomy, identification and recognition of *Serpula lacrymans*. In *Serpula lacrymans. Fundamental Biology and Control Strategies*, ed. D. H. Jennings & A. F. Bravery. Chichester: John Wiley & Sons, pp. 1–7.

Pegler, D. N. (1996). Hyphal analysis of basidiomata. *Mycological Research*, **100**, 129–142.

Pegler, D. N. (2000). Taxonomy, nomenclature and description of *Armillaria*. In *Armillaria Root Rot: Biology and Control of Honey Fungus*, ed. R. T. V. Fox. Andover, UK: Intercept Ltd, pp. 81–93.

Pegler, D. N. (2001). Useful fungi of the world: mu-erh and silver ears. *Mycologist*, **15**, 19–20.

Pegler, D. N. & Young, T. W. K. (1971). Basidiospore morphology in the Agaricales. *Nova Hedwigia Beihefte*, **35**, 1–210.

Pegler, D. N. & Young, T. W. K. (1979). The gastroid Russulales. *Transactions of the British Mycological Society*, **72**, 353–388.

Pegler, D. N., Spooner, B. M. & Young, T. W. K. (1993). *British Truffles. A Revision of British Hypogeous Fungi*. Kew: Royal Botanic Gardens.

Pegler, D. N., Yao, Y.-J. & Li, Y. (1994). The Chinese caterpillar fungus. *Mycologist*, **8**, 3–5.

Pegler, D. N., Læssøe, T. & Spooner, B. M. (1995). *British Puffballs, Earthstars and Stinkhorns*. Kew: Royal Botanic Gardens.

Pei, M. H., Royle, D. J. & Hunter, T. (1993). Identity and host alternation of some willow rusts (*Melampsora* spp.) in England. *Mycological Research*, **97**, 845–851.

Pell, J. K., Eilenberg, J., Hajek, A. E. & Steinkraus, D. C. (2001). Biology and pest management potential of Entomophthorales. In *Fungi as Biocontrol Agents*, ed. T. M. Butt, C. Jackson & N. Magan. Wallingford, UK: CABI Publishing, pp. 73–152.

Pepin, R. (1980). Le comportement parasitaire de *Sclerotinia tuberosa* (Hedw.) Fuckel sur *Anemone nemorosa* L. Étude en microscopie photonique et électronique à balayage. *Mycopathologia*, **72**, 89–99.

Percudani, R., Trevisi, A., Zambonelli, A. & Ottonello, S. (1999). Molecular phylogeny of truffles (Pezizales: Terfeziaceae, Tuberaceae) derived from nuclear rDNA sequence analysis. *Molecular Genetics and Evolution*, **13**, 169–180.

Perea, S., López-Ribot, J. L., Kirkpatrick, W. R., *et al.* (2001). Prevalence of molecular mechanisms of resistance to azole antifungal agents in *Candida albicans* strains displaying high-level fluconazole resistance isolated from human immunodeficiency virus-infected patients. *Antimicrobial Agents and Chemotherapy*, **45**, 2676–2684.

Perez-Moreno, J. & Read, D. J. (2000). Mobilization and transfer of nutrients from litter to tree seedlings via the vegetative mycelium of ectomycorrhizal plants. *New Phytologist*, **145**, 301–309.

Perfect, J. R. (2004). Genetic requirements for virulence in *Cryptococcus neoformans*. In *The Mycota* **XII**: *Human Fungal Pathogens*, ed. J. E. Domer & G. S. Kobayashi. Berlin: Springer-Verlag, pp. 89–112.

Perfect, S. E., Green, J. R. & O'Connell, R. J. (2001). Surface characteristics of necrotrophic secondary hyphae produced by the bean anthracnose fungus, *Colletotrichum lindemuthianum*. *European Journal of Plant Pathology*, **107**, 813–819.

Perkins, D. D. (1992). *Neurospora*: the organism behind the molecular revolution. *Genetics*, **130**, 687–701.

Perkins, D. D. & Turner, B. C. (1988). *Neurospora* from natural populations – towards the population biology of a haploid eukaryote. *Experimental Mycology*, **12**, 91–131.

Perkins, D. D., Radford, A. & Sachs, M. S. (2000). *The Neurospora Compendium. Chromosomal Loci*. San Diego: Academic Press.

Perkins, F. O. (1972). The ultrastructure of holdfasts, 'rhizoids', and 'slime tracks' in thraustochytriaceous fungi and *Labyrinthula* spp. *Archiv für Mikrobiologie*, **84**, 95–118.

Persmark, L. & Nordbring-Hertz, B. (1997). Conidial trap formation of nematode-trapping fungi in soil and soil extracts. *FEMS Microbiology Ecology*, **22**, 313–324.

Persson, L., Bødker, L. & Larsson-Wikström, M. (1997). Prevalence and pathogenicity of foot and root rot pathogens of pea in Southern Scandinavia. *Plant Disease*, **81**, 171–174.

Petersen, J., Heitz, M. J. & Hagan, I. M. (1998). Conjugation in *S. pombe*: identification of a microtubule-organising centre, a requirement for microtubules and a role for Mad2. *Current Biology*, **8**, 963–966.

Petersen, R. H. (1973). Aphyllophorales II. The Clavariaceae and cantharelloid basidiomycetes.

In *The Fungi: An Advanced Treatise* **IVB**, ed. G. C. Ainsworth, F. K. Sparrow & A. S. Sussman. New York: Academic Press, pp. 351–368.

Petersen, R. H. & Cifuentes, J. (1994). Notes on mating systems of *Auriscalpium vulgare* and *A. villipes*. *Mycological Research*, **98**, 1427–1430.

Petersen, R. H. & Wu, Q. (1992). *Auriscalpium vulgare*: mating system and biological species. *Mycosystema*, **4**, 25–31.

Peterson, R. L. & Currah, R. S. (1990). Synthesis of mycorrhizae between protocorms of *Goodyera repens* (Orchidaceae) and *Ceratobasidium cereale*. *Canadian Journal of Botany*, **68**, 1117–1125.

Peterson, R. L., Bonfante, P., Faccio, A. & Uetake, Y. (1996). The interface between fungal hyphae and orchid protocorm cells. *Canadian Journal of Botany*, **74**, 1861–1870.

Peterson, R. L., Uetake, Y. & Zelmer, C. (1998). Fungal symbiosis with orchid protocorms. *Symbiosis*, **25**, 29–55.

Peterson, R. L., Massicotte, H. B. & Melville, L. H. (2004). *Mycorrhizas: Anatomy and Cell Biology*. Ottawa: NRC Press; Wallingford, UK: CABI Publishing.

Peterson, S. W. (2000a). Phylogenetic analysis of *Penicillium* species based on ITS and LSU-rDNA nucleotide sequences. In *Integration of Modern Taxonomic Methods for Penicillium and Aspergillus Classification*, ed. R. A. Samson & J. I. Pitt. Amsterdam: Harwood Academic Publishers, pp. 163–178.

Peterson, S. W. (2000b). Phylogenetic relationships in *Aspergillus* based on rDNA sequence analysis. In *Integration of Modern Taxonomic Methods for Penicillium and Aspergillus Classification*, ed. R. A. Samson & J. I. Pitt. Amsterdam: Harwood Academic Publishers, pp. 323–355.

Petrini, L. E. & Petrini, O. (1985). Xylariaceous fungi as endophytes. *Sydowia*, **38**, 216–234.

Petrini, O., Petrini, L. E. & Rodriguez, K. F. (1995). Xylariaceous endophytes: an exercise in biodiversity. *Fitopatalogia Brasiliera*, **20**, 531–539.

Peyton, G. A. & Bowen, C. C. (1963). The host–parasite interface of *Peronospora manshurica* on *Glycine max*. *American Journal of Botany*, **50**, 787–797.

Pfister, D. H. (1997). Castor, Pollux and life histories of fungi. *Mycologia*, **89**, 1–23.

Pfister, D. H. & Kimbrough, J. W. (2001). Discomycetes. In *The Mycota* **VIIA**: *Systematics and Evolution*, ed. D. J. McLaughlin, E. J. McLaughlin & P. A. Lemke. Berlin: Springer-Verlag, pp. 257–281.

Pfunder, M. & Roy, B.A. (2000). Pollinator-mediated interactions between a pathogenic fungus, *Uromyces pisi* (Pucciniaceae), and its host plant, *Euphorbia cyparissias* (Euphorbiaceae). *American Journal of Botany*, **87**, 48–55.

Pfunder, M., Schürch, S. & Roy, B.A. (2001). Sequence variation and geographic distribution of pseudo-flower-forming rust fungi (*Uromyces pisi s. lat.*) on *Euphorbia cyparissias*. *Mycological Research*, **105**, 57–66.

Pfyffer, G.E., Pfyffer, B.U. & Rast, D.M. (1986). The polyol pattern, chemotaxonomy, and phylogeny of the fungi. *Sydowia*, **39**, 160–201.

Phaff, H.J. (1990). Isolation of yeasts from natural sources. In *Isolation of Biotechnological Organisms from Nature*, ed. D.P. Labeda. New York: McGraw-Hill Publishing Co., pp. 53–79.

Phaff, H.J. & Starmer, W.T. (1987). Yeasts associated with plants, insects and soil. In *The Yeasts 1: Biology of Yeasts*, ed. A.H. Rose & J.S. Harrison. London: Academic Press, pp. 123–180.

Piepenbring, M., Bauer, R. & Oberwinkler, F. (1998a). Teliospores of smut fungi. General aspects of teliospore walls and sporogenesis. *Protoplasma*, **204**, 155–169.

Piepenbring, M., Bauer, R. & Oberwinkler, F. (1998b). Teliospores of smut fungi. Teliospore walls and the development of ornamentation studied by electron microscopy. *Protoplasma*, **204**, 170–201.

Piepenbring, M., Hagedorn, G. & Oberwinkler, F. (1998c). Spore liberation and dispersal in smut fungi. *Botanica Acta*, **111**, 444–460.

Piérard-Franchimond, C., Hermanns, J.F., Degreef, H. & Piérard, G.E. (2000). From axioms to new insights into dandruff. *Dermatology*, **200**, 93–98.

Pine, E.M., Hibbett, D.S. & Donoghue, M.J. (1999). Phylogenetic relationships of cantharelloid and clavarioid homobasidiomycetes based on mitochondrial and nuclear rDNA sequences. *Mycologia*, **91**, 944–963.

Pirozynski, K.A. & Dalpé, Y. (1989). Geological history of the Glomaceae with particular reference to mycorrhizal symbiosis. *Symbiosis*, **7**, 1–36.

Pirozynski, K.A. & Malloch, D.W. (1975). The origin of land plants: a matter of mycotrophism. *Biosystems*, **6**, 153–164.

Pitt, J.I. (1979). *The Genus Penicillium and its Teleomorphic States Eupenicillium and Talaromyces*. London: Academic Press.

Pitt, J.I. (2000). *A Laboratory Guide to Common Penicillium Species*, 3rd edn. North Ryde, Australia: Food Science Australia.

Pitt, J.I. (2002). Biology and ecology of toxigenic *Penicillium* species. *Advances in Experimental Medicine and Biology*, **504**, 29–41.

Pitt, J.I. & Hocking, A.D. (1985). *Fungi and Food Spoilage*. Sydney: Academic Press Australia.

Pitt, J.I., Samson, R.A. & Frisvad, J.C. (2000). List of accepted species and their synonyms in the family Trichocomaceae. In *Integration of Modern Taxonomic Methods for Penicillium and Aspergillus Classification*, ed. R.A. Samson & J.I. Pitt. Amsterdam: Harwood Academic Publishers, pp. 9–49.

Plattner, J.J. & Rapoport, H. (1971). The synthesis of *d*- and *l*-sirenin in their absolute configurations. *Journal of the American Chemical Society*, **90**, 1758–1761.

Pöder, R. (1983). Über Optimierungsstrategien des Basidiomycetenhymenophors. Morphologisch–phylogenetische Aspekte. *Sydowia*, **36**, 240–251.

Pöder, R. & Kirchmair, M. (1995). Gills and pores; the impact of geometrical constraints on form, size and number of basidia. *Documents Mycologiques*, **100**, 337–348.

Pöggeler, S., Risch, S., Kück, U. & Osiewacz, H.D. (1997). Mating-type genes from the homothallic fungus *Sordaria macrospora* are functionally expressed in a heterothallic ascomycete. *Genetics*, **147**, 567–580.

Poinar, G.O. & Thomas, G.M. (1982). An entomophthoralean fungus from Dominican amber. *Mycologia*, **74**, 332–334.

Poinar, G.O. & Thomas, G.M. (1984). A fossil entomogenous fungus from Dominican amber. *Experientia*, **40**, 578–579.

Polachek, I., Platt, Y. & Aronovitch, J. (1990). Catecholamines and virulence of *Cryptococcus neoformans*. *Infection and Immunity*, **58**, 2919–2922.

Polak, E., Hermann, R., Kües, U. & Aebi, M. (1997). Asexual sporulation in *Coprinus cinereus*: structure and development of oidiophores and oidia in an *Amut Bmut* homokaryon. *Fungal Genetics and Biology*, **22**, 112–126.

Põldmaa, K. (2000). Generic delimitation of the fungicolous Hypocreaceae. *Studies in Mycology*, **45**, 83–94.

Polley, R.W. & Clarkson, J.D.S. (1980). Take-all severity and yield in winter wheat: relationship established using a single plant assessment method. *Plant Pathology*, **29**, 110–116.

Poloczek, E. & Webster, J. (1994). Conidial traps in *Nematoctonus* (nematophagous basidiomycetes). *Nova Hedwigia*, **59**, 201–205.

Polonelli, L., Conti, S. & Gerloni, M. (1991). Antibiobodies: antibiotic-like antiidiotypic antibodies. *Journal of Medical and Veterinary Mycology*, **29**, 235–242.

Pommer, E.-H. (1995). Morpholine fungicides and related compounds. In *Modern Selective Fungicides. Properties, Applications, Mechanisms of Action*, 2nd edn, ed. H. Lyr. New York: Gustav Fischer Verlag, pp. 163–83.

Pommerville, J. & Fuller, M. S. (1976). The cytology of the gametes and fertilization of *Allomyces macrogynus*. *Archiv für Mikrobiologie*, **109**, 21–30.

Pommerville, J. & Olson, L. W. (1987). Evidence for male-produced pheromone in *Allomyces macrogynus*. *Experimental Mycology*, **11**, 245–248.

Pontecorvo, G. (1956). The parasexual cycle in fungi. *Annual Review of Microbiology*, **10**, 393–400.

Poon, H. & Day, A. W. (1974). 'Fimbriae' in the fungus *Ustilago violacea*. *Nature*, **250**, 648–649.

Poppe, J. (2000). Use of agricultural waste materials in the cultivation of mushrooms. In *Science and Cultivation of Edible Fungi*, ed. L. J. L. D. van Griensven. Rotterdam: A. Balkema, pp. 3–23.

Porter, C. A. & Jaworski, E. G. (1966). The synthesis of chitin by particulate preparations of *Allomyces macrogynus*. *Biochemistry*, **5**, 1149–1154.

Porter, D. (1972). Cell division in the marine slime mold, *Labyrinthula* sp., and the role of the bothrosome in extracellular membrane production. *Protoplasma*, **74**, 427–448.

Porter, D. (1990). Phylum Labyrinthulomycota. In *Handbook of Protoctista*, ed. L. Margulis, J. O. Corliss, M. Melkonian & D. J. Chapman. Boston: Jones & Bartlett, pp. 388–398.

Porter, D. & Lingle, W. L. (1992). Endolithic thraustochytrid marine fungi from planted shell fragments. *Mycologia*, **84**, 289–299.

Portman, R., Moseman, R. & Levetin, E. (1997). Ultrastructure of basidiospores in North American members of the genus *Calvatia*. *Mycotaxon*, **62**, 435–443.

Powell, K. A. & Rayner, A. D. M. (1983). Ultrastructure of the rhizomorph apex in *Armillaria bulbosa* in relation to mucilage production. *Transactions of the British Mycological Society*, **81**, 529–534.

Powell, M. J. (1983). Localization of antimonate-mediated precipitates of cations in zoospores of *Chytriomyces hyalinus*. *Experimental Mycology*, **7**, 266–277.

Powell, M. J. (1993). Looking at mycology with a Janus face: a glimpse at Chytridiomycetes active in the environment. *Mycologia*, **85**, 1–20.

Powell, M. J. & Roychoudhury, S. (1992). Ultrastructural organization of *Rhizophlyctis harderi* zoospores and redefinition of type 1 microbody–lipid globule complex. *Canadian Journal of Botany*, **70**, 750–761.

Powers, H. & Kuhlman, E. G. (1997). Rusts of hard pines: fusiform rust. In *Compendium of Conifer Diseases*, ed. E. M. Hansen & K. J. Lewis. St. Paul, USA: APS Press, pp. 27–9.

Pratt, J. E., Johansson, M. & Hüttermann, A. (1998). Chemical control of *Heterobasidion annosum*. In *Heterobasidion annosum. Biology, Ecology, Impact and Control*, ed. S. Woodward, J. Stenlid, R. Karjalainan & A. Hüttermann. Wallingford, UK: CAB International, pp. 259–82.

Preisig, O., Moleleki, N., Smit, W. A., Wingfield, D. B. & Wingfield, M. J. (2000). A novel RNA mycovirus in a hypovirulent isolate of the plant pathogen *Diaporthe ambigua*. *Journal of General Virology*, **81**, 3107–3114.

Premdas, P. D. & Kendrick, B. (1991). Colonization of autumn-shed leaves by four aero-aquatic fungi. *Mycologia*, **93**, 317–321.

Pring, R. J., Nash, C., Zakaria, M. & Bailey, J. A. (1995). Infection process and host range of *Colletotrichum capsici*. *Physiological and Molecular Plant Pathology*, **46**, 137–152.

Pringle, A., Patek, S. N., Fischer, M., Stolze, J. & Money, N. P. (2005). The captured launch of a ballistospore. *Mycologia*, **97**, 866–871.

Pringle, J. R., Jones, E. W. & Broach, J. R., eds. (1997). *The Molecular and Cellular Biology of the Yeast Saccharomyces*, **3**. Cold Spring Harbor and New York: Cold Spring Harbor Laboratory Press.

Prins, T. W., Tudzynski, P., von Tiedemann, A., *et al.* (2000). Infection strategies of *Botrytis cinerea* and related necrotrophic pathogens. In *Fungal Pathology*, ed. J. W. Kronstad. Dordrecht: Kluwer Academic Publishers, pp. 33–64.

Prusky, D., Freeman, S. & Dickman, M. B., eds. (2000). *Colletotrichum. Host Specificity, Pathology, and Host–Pathogen Interaction*. St. Paul, USA: APS Press.

Pruyne, D. & Bretscher, A. (2000a). Polarization of cell growth in yeast. I. Establishment and maintenance of polarity states. *Journal of Cell Science*, **113**, 365–375.

Pruyne, D. & Bretscher, A. (2000b). Polarization of cell growth in yeast. II. The role of the cortical actin cytoskeleton. *Journal of Cell Science*, **113**, 571–585.

Pruyne, D., Legesse-Miller, A., Gao, L., Dong, Y. & Bretscher, A. (2004). Mechanisms of polarized growth and organelle segregation in yeast. *Annual Review of Cell and Developmental Biology*, **20**, 559–591.

Pryor, B. M. & Gilbertson, R. L. (2000). Molecular phylogenetic relationships among *Alternaria* species and related fungi based upon analysis of nuclear ITS and mt SSU rDNA sequences. *Mycological Research*, **104**, 1312–1321.

Pukatzki, S., Kessin, R. H. & Mekalanos, J. J. (2002). The human pathogen *Pseudomonas aeruginosa* utilizes conserved virulence pathways to infect the social amoeba *Dictyostelium discoideum*. *Proceedings of the National Academy of Sciences of the USA*, **99**, 3159–3164.

Pukkila, P. J. & Casselton, L. A. (1991). Molecular genetics of the agaric *Coprinus cinereus*. In *More Gene Manipulations in Fungi*, ed. J. W. Bennett & L. L. Lasure. New York: Academic Press, pp. 126–150.

Punja, Z. K., Chittaranjan, S. & Gaye, M. M. (1992). Development of black root rot caused by *Chalara elegans* on fresh market carrots. *Canadian Journal of Plant Pathology*, **14**, 299–309.

Purdy, L. H. (1979). *Sclerotinia sclerotiorum*: history, diseases and symptomatology, host range, geographic distribution, and impact. *Phytopathology*, **69**, 875–880.

Purdy, L. H. & Schmidt, R. A. (1996). Status of cacao witches' broom: biology, epidemiology and management. *Annual Review of Phytopathology*, **34**, 573–594.

Purdy, L. H., Kendrick, E. L., Hoffmann, J. A. & Holton, C. S. (1963). Dwarf bunt of wheat. *Annual Review of Microbiology*, **17**, 199–222.

Purdy, L. H., Krupa, S. V. & Dean, J. L. (1985). Introduction of sugarcane rust into the Americas and its spread to Florida. *Plant Disease*, **69**, 689–693.

Purvis, O. W., Coppins, B. J., Hawksworth, D. L., James, P. W. & Moore, D. M., eds. (1992). *The Lichen Flora of Great Britain and Ireland*. London: Natural History Museum.

Raajmakers, J. M. & Weller, D. M. (1998). Natural plant protection by 2,4-diacetylphloroglucinol-producing *Pseudomonas* spp. in take-all decline soils. *Molecular Plant–Microbe Interactions*, **11**, 144–152.

Rabinovich, M. L., Bolobova, A. V. & Vasil'chenko, L. G. (2004). Fungal decomposition of natural aromatic structures and xenobiotics: a review. *Applied Biochemistry and Microbiology*, **40**, 1–17.

Radford, A., Stone, P. J. & Taleb, F. (1996). Cellulase and amylase complexes. In *The Mycota III: Biochemistry and Molecular Biology*, ed. R. Brambl & G. A. Marzluf. Berlin: Springer-Verlag, pp. 269–294.

Raghavendra Rao, N. N. & Pavgi, M. S. (1993). Life history of *Synchytrium* species parasitic on Cucurbitaceae. *Indian Phytopathology*, **46**, 36–43.

Raghukumar, S. (2002). Ecology of the marine protists, the Labyrinthulomycetes (Thraustochytrids and Labyrinthulids). *European Journal of Protistology*, **38**, 127–145.

Raguso, R. A. & Roy, B. A. (1998). 'Floral' scent production by *Puccinia* rust fungi that mimic flowers. *Molecular Ecology*, **7**, 1127–1136.

Rainieri, S., Zambonelli, C. & Kaneko, Y. (2003). *Saccharomyces sensu stricto*: systematics, genetic diversity and evolution. *Journal of Bioscience and Bioengineering*, **96**, 1–9.

Raju, N. B. (1992a). Genetic control of the sexual cycle in *Neurospora*. *Mycological Research*, **96**, 241–262.

Raju, N. B. (1992b). Functional heterothallism resulting from homokaryotic conidia and ascospores in *Neurospora tetrasperma*. *Mycological Research*, **96**, 103–116.

Raju, N. B. & Lu, B. C. (1970). Meiosis in *Coprinus*. III. Timing meiotic events in *C. lagopus* (*sensu* Buller). *Canadian Journal of Botany*, **48**, 2183–2186.

Raju, N. B. & Perkins, D. D. (1994). Diverse programs of ascus development in pseudohomothallic species of *Neurospora*, *Gelasinospora* and *Podospora*. *Developmental Genetics*, **15**, 104–118.

Rambold, G. & Triebel, D. (1992). *The Inter-Lecanoralean Associations. Bibliotheca Lichenologica*, **48**. Stuttgart: Gebr. Borntraeger.

Rambourg, A., Jackson, C. L. & Clermont, Y. (2001). Three dimensional configuration of the secretory pathway and segregation of secretion granules in the yeast *Saccharomyces cerevisiae*. *Journal of Cell Science*, **114**, 2231–2239.

Ramírez, C. (1982). *Manual and Atlas of the Penicillia*. New York: Elsevier Biomedical Press.

Ramos, S., Garcia Acha, I. & Peberdy, J. F. (1975). Wall structure and the budding process in *Pullularia pullulans*. *Transactions of the British Mycological Society*, **64**, 283–288.

Ramsay, L. M. & Gadd, G. M. (1997). Mutants of *Saccharomyces cerevisiae* defective in vacuolar function confirm a role for the vacuole in toxic metal ion detoxification. *FEMS Microbiology Letters*, **152**, 293–298.

Ramsbottom, J. (1938). Dry rot in ships. *Essex Naturalist*, **26**, 231–267.

Ramsbottom, J. (1953). *Mushrooms and Toadstools. A Study of the Activities of Fungi*. London: Collins.

Ramstedt, M. (1999). Rust disease on willows – virulence variation and resistance breeding strategies. *Forest Ecology and Management*, **121**, 101–111.

Rao, K. M. (1994). *Rice Blast Disease*. Delhi: Daya Publishing House.

Raper, C. A. (1985). Strategies for mushroom breeding. In *Developmental Biology of Higher Fungi*, ed. D. Moore, L. A. Casselton, D. A. Wood & J. C. Frankland. Cambridge: Cambridge University Press, pp. 513–528.

Raper, C. A., Raper, J. R. & Miller, R. E. (1972). Genetic analysis of the life cycle of *Agaricus bisporus*. *Mycologia*, **64**, 1088–1117.

Raper, J. R. (1939). Sexual hormones in *Achlya*. I. Indicative evidence of a hormonal coordinating mechanism. *American Journal of Botany*, **26**, 639–650.

Raper, J. R. (1950). Sexual hormones in *Achlya*. VII. The hormonal mechanism in homothallic species. *Botanical Gazette*, **112**, 1–24.

Raper, J. R. (1957). Hormones and sexuality in lower plants. *Symposia of the Society for Experimental Biology*, **11**, 143–165.

Raper, J. R. (1959). Sexual versatility and evolutionary processes in fungi. *Mycologia*, **51**, 107–124.

Raper, J. R. (1966). *Genetics of Sexuality in Higher Fungi*. New York: Ronald Press Co.

Raper, J. R. & Hoffman, R. M. (1974). *Schizophyllum commune*. In *Handbook of Genetics* **1**, ed. R. C. King. New York: Plenum Press, pp. 597–626.

Raper, K. B. (1984). *The Dictyostelids*. Princeton: Princeton University Press.

Raper, K. B. & Fennell, D. I. (1965). *The Genus Aspergillus*. Baltimore: Williams & Wilkins Company.

Rasmussen, H. N. (2002). Recent developments in the study of orchid mycorrhiza. *Plant and Soil*, **244**, 149–163.

Rast, D. & Hollenstein, G. O. (1977). Architecture of the *Agaricus bisporus* spore wall. *Canadian Journal of Botany*, **55**, 2251–2262.

Rast, D. M. & Pfyffer, G. E. (1989). Acyclic polyols and higher taxa of fungi. *Botanical Journal of the Linnean Society*, **99**, 39–57.

Ratledge, C. (1997). Microbial lipids. In *Biotechnology* **7**: *Products of Secondary Metabolism*, 2nd edn., ed. H. Kleinkauf & H. vonDöhren. Weinheim, Germany: VCH, pp. 133–197.

Ratner, D. I. & Kessin, R. H. (2000). Meeting report: *Dictyostelium* 2000: a conference on the cell and developmental biology of a social amoeba, Dundee, Scotland, July 30–August 4, 2000. *Protist*, **151**, 291–297.

Raudaskoski, M. (1972). Occurrence of microtubules in the hyphae of *Schizophyllum commune* during intercellular nuclear migration. *Archiv für Mikrobiologie*, **86**, 91–100.

Raudaskoski, M. & Vauras, R. (1982). Scanning electron microscope study of fruit body differentiation in *Schizophyllum commune*. *Transactions of the British Mycological Society*, **78**, 475–481.

Raudaskoski, M. & Viitanen, H. (1982). Effects of aeration and light on fruit-body induction in *Schizophyllum commune*. *Transactions of the British Mycological Society*, **78**, 89–96.

Rausch, T., Mattusch, P. & Hilgenberg, W. (1981). Influence of clubroot disease on the growth kinetics of Chinese cabbage. *Phytopathologische Zeitschrift*, **102**, 28–33.

Raybould, A. F., Gray, A. J. & Clarke, R. T. (1998). The long-term epidemic of *Claviceps purpurea* on *Spartina anglica* in Poole Harbour: pattern of infection, effects on seed production and the role of *Fusarium heterosporum*. *New Phytologist*, **138**, 495–507.

Rayner, A. D. M. (1991a). The challenge of the individualistic mycelium. *Mycologia*, **83**, 48–71.

Rayner, A. D. M. (1991b). The phytopathological significance of mycelial individualism. *Annual Review of Phytopathology*, **29**, 305–323.

Rayner, A. D. M. & Boddy, L. (1988). *Fungal Decomposition of Wood. Its Biology and Ecology*. Chichester, UK: John Wiley.

Rayner, A. D. M. & Todd, N. K. (1977). Intraspecific antagonism in natural populations of wood-decaying basidiomycetes. *Journal of General Microbiology*, **103**, 85–90.

Rayner, A. D. M. & Todd, N. K. (1979). Population and community structure and dynamics of fungi in decaying wood. *Advances in Botanical Research*, **7**, 333–420.

Rayner, A. D. M., Powell, K. A., Thompson, W. & Jennings, D. H. (1985). Morphogenesis of vegetative organs. In *Developmental Biology of Higher Fungi*, ed. D. Moore, L. A. Casselton, D. A. Wood & J. C. Frankland. Cambridge: Cambridge University Press, pp. 249–279.

Raziq, F. (2000). Biological and integrated control of *Armillaria* root rot. In *Armillaria Root Rot: Biology and Control of Honey Fungus*, ed. R. T. V. Fox. Andover, UK: Intercept Ltd, pp. 183–201.

Read, D. J. (1991). Mycorrhizas in ecosystems – nature's response to the 'law of the minimum'. In *Frontiers in Mycology*, ed. D. L. Hawksworth. Wallingford, UK: CAB International, pp. 101–130.

Read, D. J. (1996). The structure and function of the ericoid mycorrhizal root. *Annals of Botany* N.S., **77**, 365–374.

Read, N. D. & Beckett, A. (1985). The anatomy of the mature perithecium in *Sordaria humana* and its significance for fungal multicellular development. *Canadian Journal of Botany*, **63**, 281–296.

Read, N. D. & Beckett, A. (1996). Ascus and ascospore morphogenesis. *Mycological Research*, **100**, 1281–1314.

Read, S. J., Moss, S. T. & Jones, E. B. G. (1991). Attachment studies of aquatic hyphomycetes. *Philosophical Transactions of the Royal Society*, **334**, 449–457.

Read, S.J., Moss, S.T. & Jones, E.B.G. (1992a). Germination and development of attachment structures by conidia of aquatic hyphomycetes: light microscope studies. *Canadian Journal of Botany*, **70**, 831–837.

Read, S.J., Moss, S.T. & Jones, E.B.G. (1992b). Germination and development of attachment structures by conidia of aquatic hyphomycetes: a scanning electron microscopic study. *Canadian Journal of Botany*, **70**, 838–845.

Reavy, B., Arif, M., Kashiwazaki, S., Webster, K.D. & Barker, H. (1995). Immunity to potato mop-top virus in *Nicotiana benthamiana* plants expressing the coat protein gene is effective against fungal inoculation of the virus. *Molecular Plant–Microbe Interactions*, **8**, 286–291.

Reddy, M.S. & Kramer, C.L. (1975). A taxonomic zrevision of the Protomycetales. *Mycotaxon*, **3**, 1–50.

Reddy, P.V., Patel, R. & White, J.F. (1998). Phylogenetic and developmental evidence supporting reclassification of cruciferous pathogens *Phoma lingam* and *Phoma wasabiae* in *Plenodomus*. *Canadian Journal of Botany*, **76**, 1916–1922.

Redecker, D., Kodner, R. & Graham, L.E. (2000a). Glomalean fungi from the Ordovician. *Science*, **289**, 1920–1921.

Redecker, D., Morton, J.B. & Bruns, T.D. (2000b). Molecular phylogeny of the arbuscular mycorrhizal fungi *Glomus sinuosum* and *Sclerocystis coremioides*. *Mycologia*, **92**, 282–285.

Redhead, S.A. (1977). The genus *Neolecta* (Neolectaceae fam. nov., Lecanorales, Ascomycetes) in Canada. *Canadian Journal of Botany*, **55**, 301–306.

Redhead, S.A. (2001). Bully for *Coprinus* – a story of manure, minutiae and molecules. *McIlvainea*, **14**, 5–14.

Redhead, S.A., Vilgalys, R., Moncalvo, J.-M., Johnson, J. & Hopple, J.S. (2001). *Coprinus* Pers. and the disposition of *Coprinus* species *sensu lato*. *Taxon*, **50**, 203–241.

Redlin, S.C. (1991). *Discula destructiva* sp. nov., cause of dogwood anthacnose. *Mycologia*, **83**, 633–642.

Redman, R.S., Ranson, J.C. & Rodriguez, R.J. (1999a). Conversion of the pathogenic fungus *Colletotrichum magna* to a nonpathogenic, endophytic mutualist by gene disruption. *Molecular Plant–Microbe Interactions*, **12**, 969–975.

Redman, R.S., Freeman, S., Clifton, D.R., *et al.* (1999b). Biochemical analysis of plant protection afforded by a nonpathogenic endophytic mutant of *Colletotrichum magna*. *Plant Physiology*, **119**, 795–804.

Rees, B., Shepherd, V.A. & Ashford, A.E. (1994). Presence of a motile tubular vacuole system in

different phyla of fungi. *Mycological Research*, **98**, 985–992.

Reeves, F. (1967). The fine structure of ascospore formation in *Pyronema domesticum*. *Mycologia*, **59**, 1018–1033.

Reichle, R.E. & Fuller, M.S. (1967). The fine structure of *Blastocladiella emersonii* zoospores. *American Journal of Botany*, **54**, 81–92.

Reichle, R.E. & Lichtwardt, R.W. (1972). Fine structure of the Trichomycete, *Harpella melusinae*, from black-fly guts. *Archiv für Mikrobiologie*, **81**, 103–125.

Reid, D.A. (1974). A monograph of the British Dacrymycetales. *Transactions of the British Mycological Society*, **62**, 433–494.

Reijnders, A.F.M. (1963). *Problèmes du développement des carpophores des Agaricales et de quelques groupes voisins*. The Hague: Dr W. Junk.

Reijnders, A.F.M. (1986). Development of the primordium of the carpophore. In *The Agaricales in Modern Taxonomy*, 4th edn, ed. R. Singer. Königstein, Germany: Koeltz Scientific Books, pp. 20–29.

Reijnders, A.F.M. (2000). A morphogenetic analysis of the basic characters of the gasteromycetes and their relation to other basidiomycetes. *Mycological Research*, **104**, 900–910.

Reijnders, A.F.M. & Stalpers, J.A. (1992). The development of the hymenophoral trama in the Aphyllophorales and the Agaricales. *Studies in Mycology*, **34**, 1–109.

Reinhardt, M.O. (1892). Das Wachstum der Pilzhyphen. Ein Beitrag zur Kenntnis des Flächenwachstums vegetalischer Zellmembranen. *Jahrbücher für Wissenschaftliche Botanik*, **23**, 479–566.

Reischer, H.S. (1951). Growth of Saprolegniaceae in synthetic media. I. Inorganic nutrition. *Mycologia*, **43**, 142–155.

Renzel, S., Esselborn, S., Sauer, H.W. & Hildebrandt, A. (2000). Calcium and malate are sporulation-promoting factors of *Physarum polycephalum*. *Journal of Bacteriology*, **182**, 6900–6905.

Reshetnikov, S.V., Wasser, S.P., Nevo, E., Duckman, I. & Tsukor, K. (2000). Medicinal value of the genus *Tremella* Pers. (Heterobasidiomycetes). *International Journal of Medicinal Mushrooms*, **2**, 169–193.

Restrepo, A., McEwen, J.G. & Castañeda, E. (2001). The habitat of *Paracoccidioides brasiliensis*: how far from solving the riddle? *Medical Mycology*, **39**, 233–241.

Reynaga-Peña, C.G., Gierz, G. & Bartnicki-Garcia, S. (1997). Analysis of the role of the Spitzenkörper in fungal morphogenesis by computer simulation of apical branching in *Aspergillus niger*. *Proceedings of the National Academy of Sciences of the USA*, **94**, 9096–9101.

Reynolds, D.R. (1971). Wall structure of a bitunicate ascus. *Planta*, **98**, 244–257.

Reynolds, D.R. (1989). The bitunicate ascus paradigm. *Botanical Review*, **55**, 1–52.

Reynolds, D.R. & Taylor, J.W., eds. (1993). *The Fungal Holomorph: Mitotic, Meiotic and Pleomorphic Speciation in Fungal Systematics*. Wallingford, UK: CAB International.

Rghei, N.A., Castle, A.J. & Manocha, M.S. (1992). Involvement of fimbriae in fungal host–parasite interaction. *Physiological and Molecular Plant Pathology*, **41**, 139–148.

Ribeiro, O.K. (1983). Physiology of asexual sporulation and spore germination in *Phytophthora*. In *Phytophthora. Its Biology, Taxonomy, Ecology, and Pathology*, ed. D.C. Erwin, S. Bartnicki-Garcia & P.H. Tsao. St. Paul: APS Press, pp. 55–70.

Ribes, J.A., Vanover-Sams, C.L. & Baker, D.J. (2000). Zygomycetes in human disease. *Clinical Microbiology Reviews*, **13**, 236–301.

Rice, A.V. & Currah, R.S. (2005). *Oidiodendron*: a survey of the named species and related anamorphs of *Myxotrichum*. *Studies in Mycology*, **53**, 83–120.

Rice, S.L., Beuchat, L.R. & Worthington, R.E. (1977). Patulin production by *Byssochlamys* spp. in fruit juices. *Applied and Environmental Microbiology*, **34**, 791–796.

Richards, W.C. (1993). Cerato-ulmin: a unique wilt disease toxin of instrumental significance in the development of Dutch Elm disease. In *Dutch Elm Disease Research: Cellular and Molecular Approaches*, ed. M.B. Sticklen & J.L. Sherald. Berlin: Springer-Verlag, pp. 89–151.

Richardson, D.H.S. (1975). *The Vanishing Lichens. Their History, Biology and Importance*. Newton Abbot: David & Charles.

Richardson, D.H.S. (1988). Medicinal and other economic aspects of lichens. In *CRC Handbook of Lichenology III*, ed. M. Galun. Boca Raton: CRC Press, pp. 93–108.

Richardson, D.H.S. (1991). Lichens and man. In *Frontiers in Mycology*, ed. D.L. Hawksworth. Wallingford, UK: CAB International, pp. 187–210.

Richardson, D.H.S. (1999). War in the world of lichens: parasitism and symbiosis as exemplified by lichens and lichenicolous fungi. *Mycological Research*, **103**, 641–650.

Richardson, M.J. (1972). Coprophilous ascomycetes on different dung types. *Transactions of the British Mycological Society*, **58**, 37–48.

Richardson, M.J. (2001). Diversity and occurrence of coprophilous fungi. *Mycological Research*, **105**, 387–402.

Richardson, M.J. & Leadbeater, G. (1972). *Piptocephalis fimbriata* sp. nov., and observations on the occurrence of *Piptocephalis* and *Syncephalis*. *Transactions of the British Mycological Society*, **58**, 205–215.

Richardson, M.J. & Watling, R. (1997). *Keys to Fungi on Dung*. Stourbridge, UK: British Mycological Society.

Richter, S.G. & Barnard, J. (2002). The radiation resistance of ascospores and sclerotia of *Pyronema domesticum*. *Journal of Industrial Microbiology and Biotechnology*, **29**, 51–54.

Riehl, R.M., Toft, D.O., Meyer, M.D., Carlson, G.L. & McMorris, T.C. (1984). Detection of a pheromone-binding protein in the aquatic fungus *Achlya ambisexualis*. *Experimental Cell Research*, **153**, 544–549.

Riethmüller, A., Weiss, M. & Oberwinkler, F. (1999). Phylogenetic studies of Saprolegniomycetidae and related groups based on nuclear large subunit ribosomal DNA sequences. *Canadian Journal of Botany*, **77**, 1790–1800.

Riethmüller, A., Voglmayr, H., Göker, M., Weiss, M. & Oberwinkler, F. (2002). Phylogenetic relationships of the downy mildews (Peronosporales) and related groups based on nuclear large subunit ribosomal DNA sequences. *Mycologia*, **94**, 834–849.

Rifai, M.A. & Webster, J. (1966). Culture studies on *Hypocrea* and *Trichoderma*. III. *H. lactea* (= *H. citrina*) and *H. pulvinata*. *Transactions of the British Mycological Society*, **49**, 297–310.

Rikkinen, J. (2003). Ecological and evolutionary role of photobiont-mediated guilds in lichens. *Symbiosis*, **34**, 99–110.

Rikkinen, J. & Poinar, G. (2000). A new species of resinicolous *Chaenothecopsis* (*Mycocaliciaceae*, *Ascomycota*) from 20 million year old Bitterfeld amber, with remarks on the biology of resinicolous fungi. *Mycological Research*, **104**, 7–15.

Rikkinen, J., Oksanen, I. & Lohtander, K. (2002). Lichen guilds share related cyanobacterial symbionts. *Science*, **297**, 357.

Rinaldi, M.G. (1989). Zygomycosis. *Infectious Disease Clinics of North America*, **3**, 19–41.

Riousset, L., Riousset, G., Chevalier, G. & Bardet, M.C. (2001). *Truffes d'Europe et de la Chine*. Paris: INRA.

Riquelme, M., Reynaga-Peña, C.G., Gierz, G. & Bartnicki-García, S. (1998). What determines growth direction in fungal hyphae? *Fungal Genetics and Biology*, **24**, 101–109.

Rishbeth, J. (1952). Control of *Fomes annosus* Fr. *Forestry*, **25**, 41–50.

Rishbeth, J. (1961). Inoculation of pine stumps against *Fomes annosus*. *Nature*, **191**, 826–827.

Rishbeth, J. (1968). The growth rate of *Armillaria mellea*. *Transactions of the British Mycological Society*, **51**, 575–586.

Ristaino, J.B. (2002). Tracking historic migrations of the Irish potato famine pathogen, *Phytophthora infestans*. *Microbes and Infection*, **4**, 1369–1377.

Rivas, S. & Thomas, C.M. (2005). Molecular interactions between tomato and the leaf mold pathogen *Cladosporium fulvum*. *Annual Review of Phytopathology*, **43**, 395–436.

Robb, J. & Lee, B. (1986a). Developmental sequence of the attack apparatus of *Haptoglossa mirabilis*. *Protoplasma*, **135**, 102–111.

Robb, J. & Lee, B. (1986b). Ultrastructure of mature and fired gun cells of *Haptoglossa mirabilis*. *Canadian Journal of Botany*, **64**, 1935–1947.

Roberts, D.R., Mims, C.W. & Fuller, M.S. (1996). Ultrastructure of the ungerminated conidium of *Blumeria graminis* f. sp. *hordei*. *Canadian Journal of Botany*, **74**, 231–237.

Roberts, P. (1999). *Rhizoctonia-Forming Fungi. A Taxonomic Guide*. Kew: Royal Botanic Gardens.

Roberts, R.G., Reymond, S.T. & Andersen, B. (2000). RAPD fragment pattern analysis and morphological segregation of small-spored *Alternaria* species and species groups. *Mycological Research*, **104**, 151–160.

Robertson, N.F. (1965). Presidential address. The fungal hypha. *Transactions of the British Mycological Society*, **48**, 1–8.

Robertson, S.K., Bond, D.J. & Read, N.D. (1998). Homothallism and heterothallism in *Sordaria brevicollis*. *Mycological Research*, **102**, 1215–1223.

Robinow, C.F. (1963). Observations on cell growth, mitosis and division in the fungus *Basidiobolus ranarum*. *Journal of Cell Biology*, **17**, 123–152.

Roca, M.G., Davide, L.C. & Mendes-Costa, M.C. (2003). Cytogenetics of *Colletotrichum lindemuthianum* (*Glomerella cingulata* f. sp. *phaseoli*). *Fitopatologia Brasileira*, **28**, 367–373.

Rochon, D., Kakani, K., Robbins, M. & Reade, R. (2004). Molecular aspects of plant virus transmission by *Olpidium* and plasmodiophorid vectors. *Annual Review of Phytopathology*, **42**, 211–241.

Rodriguez, R.J. & Redman, R.S. (1997). Fungal life-styles and ecosystem dynamics: biological aspects of plant pathogens, plant endophytes and saprophytes. *Advances in Botanical Research*, **24**, 169–193.

Rodríguez-Peña, J.M., Cid, V.J., Arroyo, J. & Nombela, C. (2000). A novel family of cell wall related proteins regulated differently during the yeast life cycle. *Molecular and Cellular Biology*, **20**, 3245–3255.

Roelfs, A.P. (1984). Race specificity and methods of study. In *The Cereal Rusts* **1**: *Origins, Specificity, Structure, and Physiology*, ed. W.R. Bushnell & A.P. Roelfs. Orlando: Academic Press, pp. 131–64.

Roelfs, A.P. (1985). Wheat and rye stem rust. In *The Cereal Rusts* **2**: *Diseases, Distribution, Epidemiology, and Control*, ed. A.P. Roelfs & W.R. Bushnell. Orlando: Academic Press, pp. 3–37.

Roelfs, A.P. (1986). Development and impact of regional cereal rust epidemics. In *Plant Disease Epidemiology* **1**: *Population Dynamics and Management*, ed. K.J. Leonard & W.E. Fry. New York: McMillan, pp. 129–50.

Roelfs, A.P. & Bushnell, W.R., eds. (1985). *The Cereal Rusts* **2**: *Diseases, Distribution, Epidemiology, and Control*. Orlando: Academic Press.

Roelfs, A.P. & Martens, J.W. (1988). An international system of nomenclature for *Puccinia graminis* f. sp. *tritici*. *Phytopathology*, **78**, 526–533.

Roger, A.J., Smith, M.W., Doolittle, R.F. & Doolittle, W.F. (1996). Evidence for the Heterolobosea from phylogenetic analysis of genes encoding glyceraldehyde-3-phosphate dehydrogenase. *Journal of Eukaryotic Microbiology*, **43**, 475–485.

Rogers, H.J., Buck, K.W. & Brasier, C.M. (1986). Transmission of double-stranded RNA and a disease factor in *Ophiostoma ulmi*. *Plant Pathology*, **35**, 277–287.

Rogers, J.D. (1965). *Hypoxylon fuscum*. I. Cytology of the ascus. *Mycologia*, **57**, 789–803.

Rogers, J.D. (1975a). *Xylaria polymorpha*. II. Cytology of a form with typical robust stromata. *Canadian Journal of Botany*, **53**, 1736–1743.

Rogers, J.D. (1975b). *Hypoxylon serpens*: cytology and taxonomic considerations. *Canadian Journal of Botany*, **53**, 52–55.

Rogers, J.D. (1979). The Xylariaceae: systematic, biological and evolutionary aspects. *Mycologia*, **71**, 1–42.

Rogers, J.D. (2000). Thoughts and musings on tropical Xylariaceae. *Mycological Research*, **104**, 1412–1420.

Rogerson, C.T. (1970). The Hypocrealean fungi (Ascomycetes, Hypocreales). *Mycologia*, **62**, 865–910.

Rolinson, G.N. (1998). Forty years of β-lactam research. *Journal of Antimicrobial Chemotherapy*, **41**, 589–603.

Rollins, J.A. (2003). The *Sclerotinia sclerotiorum pac1* gene is required for sclerotial development and virulence. *Molecular Plant–Microbe Interactions*, **16**, 785–795.

Rollins, J.A. & Dickman, M.B. (2001). pH signalling in *Sclerotinia sclerotiorum*: identification of a *pacC/RIM1* homolog. *Applied and Environmental Microbiology*, **67**, 75–81.

Roncal, T., Cordobés, S., Sterner, O. & Ugalde, U. (2002). Conidiation in *Penicillium cyclopium* is induced by conidiogenone, an endogenous diterpene. *Eukaryotic Cell*, **1**, 823–829.

Roper, J. A. (1966). Mechanisms of inheritance. 3. The parasexual cycle. In *The Fungi. An Advanced Treatise* **II**: *The Fungal Organism*, ed. G. C. Ainsworth & A. S. Sussman. New York: Academic Press, pp. 589–617.

Rose, C. I. & Hawksworth, D. L. (1981). Lichen recolonization in London's cleaner air. *Nature*, **289**, 289–292.

Rosin, I. V. & Moore, D. (1985). Differentiation of the hymenium in *Coprinus cinereus*. *Transactions of the British Mycological Society*, **84**, 621–628.

Rosinski, M. A. (1961). Development of the ascus in *Ceratocystis ulmi*. *American Journal of Botany*, **48**, 285–293.

Ross, I. K. (1976). Nuclear migration rates in *Coprinus congregatus*: a new record? *Mycologia*, **68**, 418–422.

Rossignol, M. & Silar, P. (1996). Genes that control longevity in *Podospora anserina*. *Mechanisms of Ageing and Development*, **90**, 183–193.

Rossman, A. Y. (1983). The phragmosporous species of *Nectria* and related genera (*Calonectria, Ophionectria, Paranectria, Scoleconectria* and *Trichonectria*). *Mycological Papers*, **150**, 1–164.

Rossman, A. Y. (1996). Morphological and molecular perspectives on systematics of the Hypocreales. *Mycologia*, **88**, 1–19.

Rossman, A. Y. (2000). Towards monophyletic genera in the holomorphic Hypocreales. *Studies in Mycology*, **45**, 27–34.

Rossman, A. Y., Samuels, G. J., Rogerson, C. T. & Lowen, R. (1999). Genera of Bionectriaceae, Hypocreaceae and Nectriaceae (Hypocreales, Ascomycetes). *Studies in Mycology*, **42**, 1–248.

Rotem, J. (1994). *The Genus Alternaria. Biology, Epidemiology and Pathogenicity*. St Paul, USA: APS Press.

Rothman, J. E. & Orci, L. (1992). Molecular dissection of the secretory pathway. *Nature*, **355**, 409–415.

Rouxel, T. & Balesdent, M. H. (2005). The stem canker (blackleg) fungus, *Leptosphaeria maculans*, enters the genomic era. *Molecular Plant Pathology*, **6**, 225–241.

Roy, B. A. (1994). The effects of pathogen-induced pseudoflowers and buttercups on each other's insect visitation. *Ecology*, **75**, 352–358.

Roy, B. & Raguso, R. A. (1997). Olfactory versus visual cues in a floral mimicry system. *Oecologia*, **109**, 414–426.

Roy, B. A., Vogler, D. R., Bruns, T. D. & Szaro, T. M. (1998). Cryptic species in the *Puccinia monoica* complex. *Mycologia*, **90**, 846–853.

Rubiales, D. & Niks, R. E. (1996). Avoidance of rust infection by some genotypes of *Hordeum chilense* due to their relative inability to induce the formation of appressoria. *Physiological and Molecular Plant Pathology*, **49**, 89–101.

Rubner, A. (1996). Revision of the predacious hyphomycetes in the *Dactylella–Monacrosporium* complex. *Studies in Mycology*, **39**, 1–134.

Ruch, D. G. & Motta, J. J. (1987). Ultrastructure and cytochemistry of dormant basidiospores of *Psilocybe cubensis*. *Mycologia*, **79**, 387–398.

Ruch, D. G. & Nurtjahja, K. (1996). The fine structure and selected histochemistry of ungerminated basidiospores of *Agrocybe acericola*. *Canadian Journal of Botany*, **74**, 780–787.

Rugner, A., Rumbolz, J., Huber, B., *et al.* (2002). Formation of overwintering structures of *Uncinula necator* and colonization of grapevine under field conditions. *Plant Pathology*, **51**, 322–330.

Ruiz-Herrera, J. (1992). *Fungal Cell Wall: Structure, Synthesis and Assembly*. Boca Raton: CRC Press.

Ruiz-Herrera, J., León-Ramírez, C., Cabrera-Ponce, J. L., Martínez-Espinoza, A. D. & Herrera-Estrella, L. (1999). Completion of the sexual cycle and demonstration of genetic recombination in *Ustilago maydis* in vitro. *Molecular and General Genetics*, **262**, 468–472.

Rundel, P. W. (1969). Clinal variation in the production of usnic acid in *Cladonia subtenuis* along light gradients. *Bryologist*, **72**, 40–44.

Rupeš, I., Mao, W.-Z., Åström, H. & Raudaskoski, M. (1995). Effects of nocodazole and brefeldin A on microtubule cytoskeleton and membrane organization in the homobasidiomycete *Schizophyllum commune*. *Protoplasma*, **185**, 212–221.

Rush, C. M., Stein, J. M., Bowden, R. L., *et al.* (2005). Status of Karnal bunt of wheat in the United States 1996 to 2004. *Plant Disease*, **89**, 212–223.

Russell, I. & Stewart, G. G. (1995). Brewing. In *Biotechnology* **9**: *Enzymes, Biomass, Food and Feed*, ed. G. Reed & T. W. Nagodawithana. Weinheim, Germany: VCH, pp. 419–62.

Ryley, R., Bhuiyan, S., Herde, D. & Gordan, B. (2003). Efficacy, timing and method of application of fungicides for management of sorghum ergot caused by *Claviceps africana*. *Australasian Journal of Plant Pathology*, **32**, 329–338.

Saenz, G. S. & Taylor, J. W. (1999a). Phylogeny of the Erysiphales (powdery mildews) inferred from internal transcribed spacer ribosomal DNA sequences. *Canadian Journal of Botany*, **77**, 150–168.

Saenz, G. S. & Taylor, J. W. (1999b). Phylogenetic relationships of *Meliola* and *Meliolina* inferred from small subunit rRNA sequences. *Mycological Research*, **103**, 1049–1056.

Saikawa, M. & Morikawa, C. (1985). Electron microscopy on a nematode-trapping fungus, *Acaulopage pectospora*. *Canadian Journal of Botany*, **63**, 1386–1390.

Sakagami, Y., Yoshida, M., Isogai, A. & Suzuki, A. (1981). Peptidal sex hormones inducing conjugation tube formation in compatible mating-type cells of *Tremella mesenterica*. *Science*, **212**, 1525–1527.

Sakai, S., Kato, M. & Nagamasu, H. (2000). *Artocarpus* (Moraceae)-gall midge pollination mutualism mediated by a male-flower parasitic fungus. *American Journal of Botany*, **87**, 440–445.

Sakaki, H., Nochide, H., Komemushi, S. & Miki, W. (2002). Effect of active oxygen species on the productivity of torularhodin by *Rhodotorula glutinis* No. 21. *Journal of Bioscience and Bioengineering*, **93**, 338–340.

Salaman, R. N. (1949). *The History and Social Influence of the Potato*. Cambridge: Cambridge University Press.

Salamati, S., Zhan, J., Burdon, J. J. & McDonald, B. A. (2000). The genetic structure of field populations of *Rhynchosporium secalis* from three continents suggests moderate gene flow and regular recombination. *Phytopathology*, **90**, 901–908.

Salas, S. D., Bennett, J. E., Kwon-Chung, K. J., Perfect, J. R. & Williamson, P. R. (1996). Effect of the laccase gene, *CNLAC1*, on virulence of *Cryptococcus neoformans*. *Journal of Experimental Medicine*, **184**, 377–386.

Salmeron, J., Vernooij, B., Lawton, K., *et al.* (2002). Powdery mildew control through transgenic expression of antifungal proteins, resistance genes, and systemic acquired resistance. In *The Powdery Mildews. A Comprehensive Treatise*, ed. R. R. Bélanger, W. R. Bushnell, A. J. Dik & T. L. W. Carver. St. Paul, USA: APS Press, pp. 268–87.

Salvin, S. B. (1941). Comparative studies on the primary and secondary zoospores of the Saprolegniaceae. I. Influence of temperature. *Mycologia*, **53**, 592–600.

Samborski, D. J. (1985). Wheat leaf rust. In *The Cereal Rusts 2: Diseases, Distribution, Epidemiology, and Control*, ed. A. P. Roelfs & W. R. Bushnell. Orlando: Academic Press, pp. 39–59.

Samson, R. A. (2000). List of names of Trichocomaceae published between 1992 and 1999. In *Integration of Modern Taxonomic Methods for Penicillium and Aspergillus Classification*, ed. R. A. Samson & J. I. Pitt. Amsterdam: Harwood Academic Publishers, pp. 73–9.

Samson, R. A., Evans, H. G. & Latgé, J.-P. (1988). *Atlas of Entomopathogenic Fungi*. Berlin: Springer-Verlag.

Samson, R. A., Hoekstra, E. S., Frisvad, J. C. & Filtenborg, O. (2002). *Introduction to Food-and Airborne Fungi*, 6th edn. Utrecht: Centraalbureau voor Schimmelcultures.

Samson, R. A., Seifert, K. A., Kuijpers, A. F. A., Houbraken, J. A. M. P. & Frisvad, J. C. (2004). Phylogenetic analysis of *Penicillium* subgenus *Penicillium* using partial β-tubulin sequences. *Studies in Mycology*, **49**, 175–200.

Samuels, G. J. (1983). Ascomycetes of New Zealand. 6. *Atricordyceps harposporifera* gen. et sp. nov. and its *Harposporium* anamorph. *New Zealand Journal of Botany*, **21**, 171–176.

Samuels, G. J. (1996). *Trichoderma*: a review of biology and systematics of the genus. *Mycological Research*, **100**, 923–935.

Samuels, G. J. (2006). *Trichoderma*: systematics, the sexual state, and ecology. *Phytopathology*, **96**, 195–206.

Samuels, G. J. & Blackwell, M. (2001). Pyrenomycetes — fungi with perithecia. In *The Mycota* **VIIA**: *Systematics and Evolution*, ed. D. J. McLaughlin, E. G. McLaughlin & P. A. Lemke. Berlin: Springer-Verlag, pp. 221–55.

Samuels, G. J. & Brayford, D. (1990). Variation in *Nectria radicicola* and its anamorph, *Cylindrocarpon destructans*. *Mycological Research*, **94**, 433–442.

Samuels, G. J. & Lodge, D. J. (1996). Three species of *Hypocrea* with stipitate stromata and *Trichoderma* anamorphs. *Mycologia*, **88**, 302–315.

Samuels, G. J., Pardo-Schultheiss, R., Hebbar, K. P., *et al.* (2000). *Trichoderma stromaticum* sp. nov., a parasite of the cacao witches' broom. *Mycological Research*, **104**, 760–764.

Samuels, G. J., Nirenberg, H. I. & Seifert, K. A. (2001). Perithecial species of *Fusarium*. In *Fusarium. Paul E. Nelson Memorial Symposium*, ed. B. O. A. Summerell, J. F. Leslie, D. Backhouse, W. L. Bryden & L. W. Burgess. St. Paul, USA: APS Press, pp. 1–14.

Samuels, G. J., Dodd, S. L., Gams, W., Castlebury, L. A. & Petrini, O. (2002). *Trichoderma* species associated with the green mold epidemic of commercially grown *Agaricus bisporus*. *Mycologia*, **94**, 146–170.

Samuelson, D. A. (1978a). Asci of the Pezizales. I. The apical apparatus of iodine-positive species. *Canadian Journal of Botany*, **56**, 1860–1875.

Samuelson, D. A. (1978b). Asci of the Pezizales. VI. The apical apparatus of *Morchella esculenta*, *Helvella crispa*, and *Rhizina undulata*. *Canadian Journal of Botany*, **56**, 3069–3082.

Sánchez, C. (2004). Modern aspects of mushroom culture technology. *Applied Microbiology and Biotechnology*, **64**, 756–762.

Sanders, D. (1988). Fungi. In *Solute Transport in Plant Cells and Tissues*, ed. D. A. Baker & J. L. Hall. Harlow: Longman, pp. 106–65.

Sanders, P. F. & Webster, J. (1978). Survival of aquatic hyphomycetes in terrestrial situations. *Transactions of the British Mycological Society*, **71**, 231–237.

Sanders, P. F. & Webster, J. (1980). Sporulation responses of some aquatic hyphomycetes in flowing water. *Transactions of the British Mycological Society*, **74**, 601–605.

Sandmann, G. & Misawa, N. (2002). Fungal carotenoids. In *The Mycota* **X**: *Industrial Applications*, ed. H. D. Osiewacz. Berlin: Springer-Verlag, pp. 247–262.

Sangar, V. K. & Dugan, P. R. (1973). Chemical composition of the cell wall of *Smittium culisetae* (Trichomycetes). *Mycologia*, **65**, 421–431.

Sanglard, D. (2002). Clinical relevance of mechanisms of antifungal drug resistance in yeasts. *Enfermedades Infecciosas Microbiologia Clinica*, **20**, 462–470.

Sanglard, D. & Bille, J. (2002). Current understanding of the modes of action of and resistance to conventional and emerging antifungal agents for treatment of *Candida* infections. In *Candida and Candidiasis*, ed. R. A. Calderone. Washington: ASM Press, pp. 349–383.

Sanogo, S., Pomella, A., Hebbar, R. K., *et al.* (2002). Production and germination of conidia of *Trichoderma stromaticum*, a mycoparasite of *Crinipellis perniciosa* on cacao. *Phytopathology*, **92**, 1032–1037.

Sansome, E. (1963). Meiosis in *Pythium debaryanum* Hesse and its significance in the life history of the Biflagellatae. *Transactions of the British Mycological Society*, **46**, 63–72.

Sansome, E. & Sansome, F. W. (1974). Cytology and life-history of *Peronospora parasitica* on *Capsella bursa-pastoris* and of *Albugo candida* on *C. bursa-pastoris* and on *Lunaria annua*. *Transactions of the British Mycological Society*, **62**, 323–332.

Saupe, S. J. (2000). Molecular genetics of heterokaryon incompatibility in filamentous ascomycetes. *Microbiology and Molecular Biology Reviews*, **64**, 489–502.

Sauter, H., Steglich, W. & Anke, T. (1999). Strobilurins: evolution of a new class of active substances. *Angewandte Chemie International Edition*, **38**, 1328–1349.

Savile, D. B. O. (1968). Possible interrelationships between fungal groups. In *The Fungi, an Advanced Treatise* **III**: *The Fungal Population*, ed. G. C. Ainsworth & A. S. Sussman. New York: Academic Press, pp. 649–675.

Savile, D. B. O. & Hayhoe, H. N. (1978). The potential effect of drop size on efficiency of splash-cup and springboard dispersal devices. *Canadian Journal of Botany*, **56**, 127–128.

Sawin, K. E. & Nurse, P. (1998). Regulation of cell polarity by microtubules in fission yeast. *Journal of Cell Biology*, **142**, 457–471.

Sawin, K. E. & Snaith, H. A. (2004). Role of microtubules and tea1p in establishment and maintenance of fission yeast cell polarity. *Journal of Cell Science*, **117**, 689–700.

Scarborough, G. A. (1970). Sugar transport in *Neurospora crassa*. *Journal of Biological Chemistry*, **245**, 1694–1698.

Schaad, N. W., Frederick, R. D., Shaw, J., *et al.* (2003). Advances in molecular-based diagnostics in meeting crop biosecurity and phytosanitary issues. *Annual Review of Phytopathology*, **41**, 305–324.

Schardl, C. L. (1996). *Epichloe* species: fungal symbionts of grasses. *Annual Review of Phytopathology*, **34**, 109–130.

Schardl, C. L. & Clay, K. (1997). Evolution of mutualistic endophytes from plant pathogens. In *The Mycota* **VB**: *Plant Relationships*, ed. G. C. Carroll & P. Tudzynski. Berlin: Springer-Verlag, pp. 221–238.

Schardl, C. L. & Moon, C. D. (2003). Process of species evolution in *Epichloe/Neotyphodium* endophytes of grasses. In *Clavicipitalean Fungi*, ed. J. F. White, C. W. Bacon, N. L. Hywel-Jones & J. W. Spatafora. New York: Marcel Dekker Inc., pp. 273–310.

Scheffer, R. P. (1992). Ecological and evolutionary role of toxins from *Alternaria* species. In *Alternaria Biology, Plant Diseases and Metabolites*, ed. J. Chelkowski & A. Visconti. Amsterdam: Elsevier, pp. 101–122.

Schekman, R. (1992). Genetic and biochemical analysis of vesicular traffic in yeast. *Current Opinion in Cell Biology*, **4**, 587–592.

Schell, W. A. (2003). Dematiaceous hyphomycetes. In *Pathogenic Fungi in Humans and Animals*, ed. D. H. Howard. New York: Marcel Dekker Inc., pp. 565–636.

Scherrer, S., de Vries, O. M. H., Dudler, R., Wessels, J. G. H. & Honegger, R. (2000). Interfacial self-assembly of fungal hydrophobins of the lichen-forming ascomycetes *Xanthoria parietina* and *X. ectaneoides*. *Fungal Genetics and Biology*, **30**, 81–93.

Schieber, E. & Zentmyer, G. A. (1984). Coffee rust in the Western hemisphere. *Plant Disease*, **68**, 89–93.

Schiltz, P. (1981). Downy mildew of tobacco. In *The Downy Mildews*, ed. D. M. Spencer. London: Academic Press, pp. 577–599.

Schimek, C., Eibel, P., Grolig, F., *et al.* (1999). Gravitropism in *Phycomyces*: a role for sedimenting protein crystals and floating lipid globules. *Planta*, **210**, 132–142.

Schimek, C., Kleppe, K., Saleem, A.-R., *et al.* (2003). Sexual reactions in *Mortierellales* are mediated by the trisporic acid system. *Mycological Research*, **107**, 736–747.

Schipper, M. A. A. (1978). On certain species of *Mucor*, with a key to all accepted species. *Studies in Mycology*, **17**, 1–52.

Schipper, M. A. A. (1987). Mating ability and the species concept in the Zygomycetes. In *Evolutionary Biology of the Fungi*, ed. A. D. M. Rayner, C. M. Brasier & D. Moore. Cambridge: Cambridge University Press, pp. 261–269.

Schipper, M. A. A. & Stalpers, J. A. (1980). Various aspects of the mating system in Mucorales. *Persoonia*, **11**, 53–63.

Schippers, B. & van Eck, W. H. (1981). Formation and survival of chlamydospores in *Fusarium*. In *Fusarium: Diseases, Biology and Taxonomy*, ed. P. E. Nelson, T. A. Toussoun & R. J. Cook. London, PA: Pennsylvania State University Press, pp. 250–260.

Schmatz, D. M., Romancheck, M. A., Pittarelli, L. A., *et al.* (1990). Treatment of *Pneumocystis carinii* pneumonia with 1,3,-β-glucan synthesis inhibitors. *Proceedings of the National Academy of Sciences of the USA*, **87**, 5950–5954.

Schmidt, E. (1983). Spore germination of and carbohydrate colonization by *Morchella esculenta* at different soil temperatures. *Mycologia*, **75**, 870–875.

Schmidt, F. R. (2002). Beta-lactam antibiotics: aspects of manufacture and therapy. In *The Mycota X: Industrial Applications*, ed. H. D. Osiewacz. Berlin: Springer-Verlag, pp. 69–91.

Schmidt, O. & Moreth-Kebernik, U. (1991). Monokaryon pairing of the dry rot fungus *Serpula lacrymans*. *Mycological Research*, **95**, 1382–1386.

Schmidt, R. A. (2003). Fusiform rust of Southern pines: a major success for forest disease management. *Phytopathology*, **93**, 1048–1051.

Scholler, M., Hagedorn, G. & Rubner, A. (1999). A reevaluation of predatory orbiliaceous fungi. II. A new generic concept. *Sydowia*, **51**, 89–113.

Schoonbeek, H., del Sorbo, G. & de Waard, M. A. (2001). The ABC transporter BcatrB affects the sensitivity of *Botrytis cinerea* to the phytoalexin resveratol and the fungicide fenpiclonil. *Molecular Plant–Microbe Interactions*, **14**, 562–571.

Schreurs, W. J. A., Harold, R. L. & Harold, F. M. (1989). Chemotropism and branching as alternate responses to *Achlya bisexualis. Journal of General Microbiology*, **135**, 2519–2528.

Schroeder, W. A. & Johnson, E. A. (1993). Antioxidant role of carotenoids in *Phaffia rhodozyma. Journal of General Microbiology*, **139**, 907–912.

Schroeder, W. A. & Johnson, E. A. (1995). Carotenoids protect *Phaffia rhodozyma* against singlet oxygen damage. *Journal of Industrial Microbiology*, **14**, 502–507.

Schultz, M., Arendholz, W.-R. & Büdel, B. (2001). Origin and evolution of the lichenized ascomycete order Lichinales: monophyly and systematic relationships inferred from ascus, fruiting body and SSU rDNA evolution. *Plant Biology*, **3**, 116–123.

Schulz, T. R., Johnston, W. J., Goloh, C. T. & Maguire, J. D. (1993). Control of ergot in Kentucky bluegrass seeds using fungicides. *Plant Disease*, **77**, 685–687.

Schuman, G. L. (1991). *Plant Diseases: Their Biology and Social Impact.* St. Paul: APS Press.

Schüssler, A. & Kluge, M. (2001). *Geosiphon pyriforme*, an endocytosymbiosis between fungus and cyanobacteria, and its meaning as a model system for arbuscular mycorrhizal research. In *The Mycota IX: Fungal Associations*, ed. B. Hock. Berlin: Springer-Verlag, pp. 151–161.

Schüssler, A., Schwarzott, D. & Walker, C. (2001). A new fungal phylum, the *Glomeromycota*: phylogeny and evolution. *Mycological Research*, **105**, 1413–1421.

Schwendener, S. (1867). Über die wahre Natur der Flechtengonidien. *Verhandlungen der Schweizerischen Naturforschenden Gesellschaft*, **51**, 88–90.

Schwinn, F. J. & Staub, T. (1995). Phenylamides and other fungicides against Oomycetes. In *Modern Selective Fungicides. Properties, Applications, Mechanisms of Action*, 2nd edn, ed. H. Lyr. New York: Gustav Fischer Verlag, pp. 323–346.

Scorzetti, G., Fell, J. W., Fonseca, A. & Statzell-Tallman, A. (2002). Systematics of basidiomycetous yeasts: a comparison of large subunit D1/D2 and internal transcribed spacer rDNA regions. *FEMS Yeast Research*, **2**, 495–517.

Scott, W. W. (1961). A monograph of the genus *Aphanomyces. Virginia Agricultural Experiment Station Technical Bulletin*, **151**, 1–95.

Scrimshaw, N. S. & Murray, E. B. (1995). Nutritional value and safety of 'single cell protein'. In *Biotechnology 9: Enzymes, Biomass, Food and Feed*, ed. G. Reed & T. W. Nagodawithana. Weinheim, Germany: VCH, pp. 221–237.

Scudamore, K. A. (1994). *Aspergillus* toxins in food and animal feedingstuffs. In *The Genus Aspergillus. From Taxonomy and Genetics to Industrial Application*, ed.

K. A. Powell, A. Renwick & J. F. Peberdy. New York: Plenum Press, pp. 59–71.

Scutt, C. P., Li, Y., Robertson, S. E., Willis, M. E. & Gilmartin, P. M. (1997). Sex determination in dioecious *Silene latifolia*. Effects of the Y chromosome and the parasitic smut fungus (*Ustilago violacea*) on gene expression during flower development. *Plant Physiology*, **114**, 969–979.

Seaward, M. R. D. (1993). Lichens and sulphur dioxide air pollution: field studies. *Environmental Reviews*, **1**, 73–91.

Seaward, M. R. D. (1997). Urban deserts bloom: a lichen renaissance. *Bibliotheca Lichenologica*, **67**, 297–309.

Seaward, M. R. D. (2004). The use of lichens for environmental impact assessment. *Symbiosis*, **37**, 293–305.

Seaward, M. R. D. & Coppins, B. J. (2004). Lichens and hypertrophication. *Bibliotheca Lichenologica*, **88**, 561–572.

Seaward, M. R. D. & Edwards, H. G. M. (1997). Biological origin of major chemical disturbances on ecclesiastical architecture studied by Fourier transform Raman spectroscopy. *Journal of Raman Spectroscopy*, **28**, 691–696.

Seeger, M. & Payne, G. S. (1992). A role for clathrin in the sorting of vacuolar proteins in the Golgi complex of yeast. *EMBO Journal*, **11**, 2811–2818.

Seifert, K. A. (1983). Decay of wood by the Dacrymycetales. *Mycologia*, **75**, 1011–1018.

Seifert, K. A. (1985). A monograph of *Stilbella* and some allied hyphomycetes. *Studies in Mycology*, **27**, 1–224.

Seifert, K. A. (1993). Sapstain of commercial lumber by species of *Ophiostoma* and *Ceratocystis*. In *Ceratocystis and Ophiostoma. Taxonomy, Ecology and Pathogenicity*, ed. M. J. Wingfield, K. A. Seifert & J. F. Webber. St. Paul, USA: APS Press, pp. 141–151.

Seifert, K. A. & Gams, W. (2001). The taxonomy of anamorphic fungi. In *The Mycota* **VIIA**: *Systematics and Evolution*, ed. D. J. McLaughlin, E. G. McLaughlin & P. A. Lemke. Berlin: Springer-Verlag, pp. 307–347.

Seifert, K. A. & Samuels, G. J. (2000). How should we look at anamorphs? *Studies in Mycology*, **45**, 5–18.

Seifert, K. A., Wingfield, M. J. & Kendrick, W. B. (1993). A nomenclator for described species of *Ceratocystis, Ophiostoma, Ceratocystiopsis, Ceratostomella* and *Sphaeronaemella*. In *Ceratocystis and Ophiostoma. Taxonomy, Ecology and Pathogenicity*, ed. M. J. Wingfield, K. A. Seifert & J. F. Webber. St. Paul, USA: APS Press, pp. 269–287.

Seiler, S., Nargang, F. E., Steinberg, G. & Schliwa, M. (1997). Kinesin is essential for cell morphogenesis and polarized secretion in *Neurospora crassa*. *EMBO Journal*, **16**, 3025–3034.

Sela-Buurlage, M. B., Budai-Hadrian, O., Pan, Q., *et al.* (2001). Genome-wide dissection of *Fusarium* resistance in tomato reveals multiple complex loci. *Molecular Genetics and Genomics*, **265**, 1104–1111.

Sentandreu, R., Mormeneo, S. & Ruiz-Herrera, J. (1994). Biogenesis of the fungal cell wall. In *The Mycota* **I**: *Growth, Differentiation and Sexuality*, ed. J. G. H. Wessels & F. Meinhardt. Berlin: Springer-Verlag, pp. 111–124.

Seppelt, R. D. (1995). Phytogeography of continental Antarctic lichens. *Lichenologist*, **27**, 417–431.

Setliff, E. C. (1988). Hyphal deterioration in *Ganoderma applanatum*. *Mycologia*, **80**, 447–454.

Seviour, R. J., Willing, R. R. & Chilvers, G. A. (1973). Basidiocarps associated with ericoid mycorrhizas. *New Phytologist*, **72**, 381–385.

Seymour, R. L. (1970). The genus *Saprolegnia*. *Nova Hedwigia*, **19**, 1–124.

Seymour, R L. & Johnson, T. W. (1973). Saprolegniaceae: a keratinophilic *Aphanomyces* from soil. *Mycologia*, **65**, 1312–1318.

Shah, P. A., Aebi, M. & Tuor, T. (2000). Drying and storage procedures for formulated and unformulated mycelia of the aphid-pathogenic fungus *Erynia neoaphidis*. *Mycological Research*, **104**, 440–446.

Shamoun, S. F. (2000). Application of biological control to vegetation management in forestry. In *Proceedings of the X International Symposium on Biological Control of Weeds*, ed. N. R. Spencer. Bozeman, USA: Montana State University, pp. 87–96.

Shankar, M., Cowling, W. A. & Sweetingham, M. W. (1998). Histological observations of latent infection and tissue colonisation by *Diaporthe toxica* in resistant and susceptible narrow-leafed lupins. *Canadian Journal of Botany*, **76**, 1305–1316.

Shapira, R., Choi, G. H., Hillman, B. I. & Nuss, D. L. (1991). The contribution of defective RNAs to complexity of viral-encoded double-stranded RNA populations present in hypovirulent strains of the chestnut blight fungus, *Cryphonectria parasitica*. *EMBO Journal*, **10**, 741–746.

Sharland, P. & Rayner, A. D. M. (1986). Mycelial interactions in *Daldinia concentrica*. *Transactions of the British Mycological Society*, **86**, 643–649.

Sharma, R. & Cammack, R. H. (1976). Spore germination and taxonomy of *Synchytrium endobioticum* and *S. succisae*. *Transactions of the British Mycological Society*, **66**, 137–147.

Sharma, R., Rajak, R. C. & Pandey, A. K. (2002). Teaching techniques for mycology: 19. A micro-dilution drop-tail method for isolating Onygenalean ascomycetes from hair baits. *Mycologist*, **16**, 153–157.

Shattock, R.C. & Preece, T.F. (2000). Tranzschel revisited: modern studies of the relatedness of different rust fungi confirm his Law. *Mycologist*, **14**, 113–117.

Shattock, R.C., Tooley, P.W. & Fry, W.E. (1986a). Genetics of *Phytophthora infestans*: characterization of single oospore cultures from A1 isolates induced to self by intraspecific stimulation. *Phytopathology*, **76**, 407–410.

Shattock, R.C., Tooley, P.W. & Fry, W.E. (1986b). Genetics of *Phytophthora infestans*: determination of recombination, segregation and selfing by isozyme analysis. *Phytopathology*, **76**, 410–413.

Shaw, D.E. (1993). Honeybees collecting *Neurospora* spores from steamed *Pinus* logs in Queensland. *Mycologist*, **7**, 182–185.

Shaw, D.S. (1983). The cytogenetics and genetics of *Phytophthora*. In *Phytophthora. Its Biology, Taxonomy, Ecology, and Pathology*, ed. D.C. Erwin, S. Bartnicki-Garcia & P.H. Tsao. St. Paul: APS Press, pp. 81–94.

Shaw, G.C. & Kile, G.A., eds. (1991). *Armillaria Root Disease. Agricultural Handbook* **691**. Washington DC: USDA Forestry Service.

Shaw, J.D., Cummings, K.B., Huyer, G., Michaelis, S. & Wendland, B. (2001). Yeast as a model system for studying endocytosis. *Experimental Cell Research*, **271**, 1–9.

Shearer, C.A. (1992). The role of woody debris. In *The Ecology of Aquatic Hyphomycetes*, ed. F. Bärlocher. Berlin: Springer-Verlag, pp. 77–98.

Shelby, R.A. (1999). Toxicology of ergot alkaloids in agriculture. In *Ergot. The Genus Claviceps*, ed. V. Křen & L. Cvak. Amsterdam: Harwood Academic Publishers, pp. 46–51.

Sheu, Y.-J. & Snyder, M. (2001). Control of cell polarity and shape. In *The Mycota VIII: Biology of the Fungal Cell*, ed. R.J. Howard & N.A.R. Gow. Berlin: Springer-Verlag, pp. 19–53.

Shoemaker, R.A. (1955). Biology, cytology and taxonomy of *Cochliobolus sativus*. *Canadian Journal of Botany*, **33**, 562–576.

Shoemaker, R.A. & Brun, H. (2001). The teleomorph of the weakly aggressive segregate of *Leptosphaeria maculans*. *Canadian Journal of Botany*, **79**, 412–419.

Sholberg, P.L. & Gaudet, D.A. (1992). Grass as a source of inoculum for rot caused by *Coprinus psychromorbidus*. *Canadian Journal of Plant Pathology*, **14**, 221–226.

Shorrocks, B. & Charlesworth, P. (1982). A field study of the association between the stinkhorn *Phallus impudicus* Pers. and the British fungal-breeding *Drosophila*. *Biological Journal of the Linnean Society*, **17**, 307–318.

Shykoff, J.A. & Bucheli, E. (1995). Pollinator visitation patterns, floral rewards and the probability of transmission of *Microbotryum violaceum*, a venereal disease of plants. *Journal of Ecology*, **83**, 189–198.

Shykoff, J.A. & Kaltz, O. (1998). Phenotypic changes in host plants diseased by *Microbotryum violaceum*: parasite manipulation, side effects, and trade-offs. *International Journal of Plant Sciences*, **159**, 236–243.

Siddiqui, Z.A. & Mahmood, I. (1996). Biological control of plant parasitic nematodes by fungi: a review. *Bioresource Technology*, **58**, 229–239.

Sidhu, G. & Person, C. (1972). Genetic control of virulence in *Ustilago hordei*. III. Identification of genes for host resistance and demonstration of gene-for-gene relations. *Canadian Journal of Genetics and Cytology*, **14**, 209–213.

Sidney, E. (1846). *Blights of the Wheat and Their Remedies*. London: The Religious Tract Society.

Siegel, M.R. & Bush, L.P. (1997). Toxin production in grass/endophyte associations. In *The Mycota VA: Plant Relationships*, ed. G.C. Carroll & P. Tudzynski. Berlin: Springer-Verlag, pp. 185–207.

Siegel, M.R., Johnson, M.C., Varney, D.R., *et al.* (1984). A fungal endophyte in tall fescue: incidence and dissemination. *Phytopathology*, **74**, 932–937.

Sietsma, J.H. & Wessels, J.G.H. (1990). Occurrence of glucosaminoglycan in the cell wall of *Schizosaccharomyces pombe*. *Journal of General Microbiology*, **136**, 2261–2265.

Sietsma, J.H. & Wessels, J.G.H. (1994). Apical wall biogenesis. In *The Mycota I: Growth, Differentiation and Sexuality*, ed. J.G.H. Wessels & F. Meinhardt. Berlin: Springer-Verlag, pp. 125–141.

Sietsma, J.H., Rast, D. & Wessels, J.G.H. (1977). The effect of carbon dioxide on fruiting and on the degradation of a cell wall glucan in *Schizophyllum commune*. *Journal of General Microbiology*, **102**, 385–389.

Sigler, L. (2003). Ascomycetes: the Onygenaceae and other fungi from the order Onygenales. In *Pathogenic Fungi in Humans and Animals*, 2nd edn, ed. D.H. Howard. New York: Marcel Dekker, pp. 195–236.

Sigler, L., Flis, A.L. & Carmichael, J.W. (1998). The genus *Uncinocarpus* (Onygenaceae) and its synonym *Brunneospora*: new concepts, combinations and connections to anamorphs in *Chrysosporium*, and further evidence of its relationship with *Coccidioides immitis*. *Canadian Journal of Botany*, **76**, 1624–1636.

Silar, P. (2005). Peroxide accumulation and cell death in filamentous fungi induced by contact with a contestant. *Mycological Research*, **109**, 137–149.

Silar, P., Lalucque, H. & Vierney, C. (2001). Cell degeneration in the model system *Podospora anserina*. *Biogerontology*, **2**, 1–17.

Silliker, M. E., Monroe, J. A. & Jorden, M. A. (1997). Evaluation of the efficiency of sexual reproduction in restoring *Podospora anserina* mitochondrial DNA to wild-type. *Current Genetics*, **32**, 281–286.

Silva, M. C., Nicole, M., Guerra-Guimarães, L. & Rodrigues, C. J. (2002). Hypersensitive cell death and post-haustorial defence responses arrest the orange rust (*Hemileia vastatrix*) growth in resistant coffee leaves. *Physiological and Molecular Plant Pathology*, **60**, 169–183.

Silva-Hanlin, D. M. W. & Hanlin, R. T. (1999). Small subunit ribosomal RNA gene phylogeny of several loculoascomycetes and its taxonomic implications. *Mycological Research*, **103**, 153–160.

Silverman-Gavrila, L. B. & Lew, R. R. (2001). Regulation of the tip-high Ca^{2+} gradient in growing hyphae of the fungus *Neurospora crassa*. *European Journal of Cell Biology*, **80**, 379–390.

Silverman-Gavrila, L. B. & Lew, R. R. (2002). An IP_3-activated Ca^{2+} channel regulates fungal tip growth. *Journal of Cell Science*, **115**, 5013–5025.

Simmons, E. G. (1969). Perfect states of *Stemphylium*. *Mycologia*, **61**, 1–26.

Simmons, E. G. (1986). *Alternaria* themes and variations. *Mycotaxon*, **25**, 287–309.

Simmons, E. G. (1992). *Alternaria* taxonomy: current status, viewpoint, challenge. In *Alternaria Biology, Plant Diseases and Metabolites*, ed. J. Chelkowski & A. Visconti. Amsterdam: Elsevier, pp. 1–35.

Simmons, R. B. & Ahern, D. G. (1987). Cell wall ultrastructure and diazonium blue B reaction of *Sporopachydermia quercuum*, *Bullera tsugae*, and *Malassezia* spp. *Mycologia*, **79**, 38–43.

Simon, L., Bousquet, J., Lévesque, R. C. & Lalonde, M. (1993). Origin and diversification of endomycorrhizal fungi and coincidence with vascular land plants. *Nature*, **363**, 67–69.

Simons, M. D. (1985). Crown rust. In *The Cereal Rusts 2: Diseases, Distribution, Epidemiology, and Control*, ed. A. P. Roelfs & W. R. Bushnell. Orlando: Academic Press, pp. 131–172.

Simpson, R. M., van Hekezen, R., van Lune, F., *et al.* (2001). Extracellular enzymes of *Chondrostereum purpureum*, causal fungus of silverleaf disease. *New Zealand Plant Protection*, **54**, 202–208.

Sims, K. P., Watling, R. & Jeffries, P. (1995). A revised key to the genus *Scleroderma*. *Mycotaxon*, **56**, 403–420.

Singer, R., ed. (1986). *The Agaricales in Modern Taxonomy*, 4th edn. Königstein, Germany: Koeltz Scientific Books.

Singh, J., Bech-Andersen, J., Elborne, S. A., *et al.* (1993). The search for wild dry rot fungus (*Serpula lacrymans*) in the Himalayas. *Mycologist*, **7**, 124–130.

Singh, R. P., Huerta-Espino, J. & Roelfs, A. P. (2002). The wheat rusts. In *Bread Wheat. Improvement and Production*, ed. B. C. Curtis, S. Rajaram & H. Gómez Macpherson. Rome: Food and Agriculture Organization of the United Nations, pp. 227–249.

Sipiczki, M. & Bozsik, A. (2000). The use of morpho-mutants to investigate septum formation and cell separation in *Schizosaccharomyces pombe*. *Archives of Microbiology*, **174**, 386–392.

Sipman, H. J. M. & Aptroot, A. (2001). Where are the missing lichens? *Mycological Research*, **105**, 1433–1439.

Sivanesan, A. (1984). *The Bitunicate Ascomycetes and their Anamorphs*. Lehre, Germany: J. Cramer.

Sivanesan, A. (1987). Graminicolous species of *Bipolaris*, *Curvularia*, *Drechslera*, *Exserohilum* and their teleomorphs. *Mycological Papers*, **158**, 1–261.

Sivichai, S., Hywel-Jones, N. & Somrithipol, S. (2000). Lignicolous freshwater *Ascomycota* from Thailand: *Melanochaeta* and *Sporoschisma* anamorphs. *Mycological Research*, **104**, 478–485.

Sjamsuridzal, W., Tajiri, Y., Nishida, H., *et al.* (1997). Evolutionary relationships of members of the genera *Taphrina*, *Protomyces*, *Schizosaccharomyces*, and related taxa within the archiascomycetes: Integrated analysis of genotypic and phenotypic characters. *Mycoscience*, **38**, 267–280.

Sjollema, K. A., Dijksterhuis, J., Veenhuis, M. & Harder, W. (1993). An electron microscopical study of the infection of the nematode *Panagrellus redivivus* by the endoparasitic fungus *Verticillium balanoides*. *Mycological Research*, **97**, 479–484.

Skou, J. P. (1972). Ascosphaerales. *Friesia*, **10**, 1–24.

Skou, J. P. (1975). Two new species of *Ascosphaera* and notes on the conidial state of *Bettsia alvei*. *Friesia*, **11**, 62–74.

Skou, J. P. (1982). Ascosphaerales and their unique ascomata. *Mycotaxon*, **15**, 487–499.

Skou, J. P. (1988). More details in support of the class Ascosphaeromycetes. *Mycotaxon*, **31**, 191–198.

Skucas, G. P. (1967). Structure and composition of the resistant sporangial wall in the fungus *Allomyces*. *American Journal of Botany*, **54**, 1152–1158.

Skucas, G. P. (1968). Changes in wall and internal structure of *Allomyces* resistant sporangia during germination. *American Journal of Botany*, **55**, 291–295.

Slaninova, I., Kucsera, J. & Svoboda, A. (1999). Topology of microtubules and actin in the life cycle of *Xanthophyllomyces dendrorhous* (*Phaffia rhodozyma*). *Antonie van Leeuwenhoek*, **75**, 361–368.

Sláviková, E. & Vadkertiová, R. (2003). The diversity of yeasts in the agricultural soil. *Journal of Basic Microbiology*, **43**, 430–436.

Slayman, C. L. (1987). The plasma membrane ATPase of *Neurospora*: a proton-pumping electroenzyme. *Journal of Bioenergetics and Biomembranes*, **19**, 1–20.

Slayman, C. L. & Slayman, C. W. (1974). Depolarization of the plasma membrane of *Neurospora* during active transport of glucose: evidence for a proton-dependent cotransport system. *Proceedings of the National Academy of Sciences of the USA*, **71**, 1935–1939.

Slippers, B., Coutinho, T. A., Wingfield, B. D. & Wingfield, M. J. (2003). A review of the genus *Amylostereum* and its association with woodwasps. *South African Journal of Science*, **99**, 70–74.

Slutsky, B., Staebell, M., Anderson, J., *et al.* (1987). 'White–opaque transition': a second high-frequency switching system in *Candida albicans*. *Journal of Bacteriology*, **169**, 189–197.

Smalley, E. B. & Guries, R. P. (1993). Breeding elms for resistance to Dutch elm disease. *Annual Review of Phytopathology*, **31**, 325–352.

Smalley, E. B., Raffa, K. F., Proctor, R. H. & Klepzig, K. D. (1993). Tree responses to infection by species of *Ophiostoma* and *Ceratocystis*. In *Ceratocystis and Ophiostoma. Taxonomy, Ecology and Pathogenicity*, ed. M. J. Wingfield, K. A. Seifert & J. F. Webber. St. Paul, USA: APS Press, pp. 207–17.

Smart, C. D., Yuan, W., Foglia, R., *et al.* (2000). *Cryphonectria* hypovirus 3, a virus species in the family Hypoviridae with a single open reading frame. *Virology*, **265**, 66–73.

Smith, D. A. (1994). A local-oscillator theory of shuttle streaming in *Physarum polycephalum*. II. Phase control by cytoplasmic calcium. *Protoplasma*, **177**, 171–180.

Smith, I. M., Dunez, J., Lelliott, R. A., Phillips, D. H. & Archer, S. A., eds. (1988). *European Handbook of Plant Diseases*. Oxford: Blackwell Scientific Publications.

Smith, J. R. & Rubenstein, I. (1973). The development of 'senescence' in *Podospora anserina*. *Journal of General Microbiology*, **76**, 283–296.

Smith, M. F. & Callaghan, A. A. (1987). Quantitative survey of *Conidiobolus* and *Basidiobolus* in soils and litter. *Transactions of the British Mycological Society*, **89**, 179–185.

Smith, M. L., Bruhn, J. N. & Anderson, J. B. (1992). The fungus *Armillaria bulbosa* is among the largest and oldest living organisms. *Nature*, **356**, 428–431.

Smith, S. E. (1966). Physiology and ecology of orchid mycorrhizal fungi. *New Phytologist*, **65**, 488–499.

Smith, S. E. (1967). Carbohydrate translocation in orchid mycorrhizas. *New Phytologist*, **66**, 371–378.

Smith, S. E. & Read, D. J. (1997). *Mycorrhizal Symbiosis*, 2nd edn. London: Academic Press.

Smreciu, E. A. & Currah, R. S. (1989). Symbiotic germination of seeds of terrestrial orchids of North America and Europe. *Lindleyana*, **4**, 6–15.

Sneh, B., Burpee, L. & Ogoshi, A. (1991). *Identification of Rhizoctonia Species*. St. Paul: APS Press.

Sneh, B., Jabaji-Hare, S., Neate, S. M. & Dijst, G., eds. (1996). *Rhizoctonia Species: Taxonomy, Molecular Biology, Ecology, Pathology and Disease Control*. Dordrecht: Kluwer.

Snetselaar, K. M. & Mims, C. W. (1992). Sporidial fusion and infection of maize seedlings by the smut fungus *Ustilago maydis*. *Mycologia*, **84**, 193–203.

Snetselaar, K. M. & Mims, C. W. (1993). Infection of maize stigmas by *Ustilago maydis*: light and electron microscopy. *Phytopathology*, **83**, 843–850.

Snetselaar, K. M. & Mims, C. W. (1994). Light and electron microscopy of *Ustilago maydis* hyphae in maize. *Mycological Research*, **98**, 347–355.

Snetselaar, K. M., Bölker, M. & Kahmann, R. (1996). *Ustilago maydis* mating hyphae orient their growth toward pheromone sources. *Fungal Genetics and Biology*, **20**, 299–312.

Snider, P. J. (1959). Stages of development in rhizomorphic thalli of *Armillaria mellea*. *Mycologia*, **51**, 693–707.

Snider, P. J. (1965). Incompatibility and nuclear migration. In *Incompatibility in Fungi*, ed. K. Esser & J. R. Raper. Berlin: Springer-Verlag, pp. 52–70.

Snider, P. J. (1968). Nuclear movements in *Schizophyllum*. *Symposia of the Society for Experimental Biology*, **22**, 261–283.

Soares, M., Christen, P., Pandey, A. & Soccol, C. R. (2000). Fruity flavour production by *Ceratocystis fimbriata* grown on coffee husk in solid-state fermentation. *Process Biochemistry*, **35**, 857–861.

Söderhäll, K., Dick, M. W., Clark, G., Fürst, M. & Constantinescu, O. (1991). Isolation of *Saprolegnia parasitica* from the crayfish *Astacus leptodactylus*. *Aquaculture*, **92**, 121–125.

Soll, D. R. (2002). Molecular biology of switching in *Candida*. In *Fungal Pathogenesis. Principles and Clinical Applications*, ed. R. A. Calderone & R. L. Cihlar. New York: Marcel Dekker, pp. 161–82.

Soll, D. R. (2003). Mating-type locus homozygosis, phenotypic switching and mating: a unique sequence of dependencies in *Candida albicans*. *BioEssays*, **26**, 10–20.

Soll, D. R., Lockhart, S. R. & Zhao, R. (2003). Relationship between switching and mating in *Candida albicans*. *Eukaryotic Cell*, **2**, 390–397.

Solla, A. & Gil, L. (2002). Vessel diameter as a factor in resistance of *Ulmus minor* to *Ophiostoma novo-ulmi*. *Forest Pathology*, **32**, 123–134.

Solomon, J. M., Rupper, A., Cardelli, J. A. & Isberg, R. R. (2000). Intracellular growth of *Legionella pneumophila* in *Dictyostelium discoideum*, a system for genetic analysis of host–pathogen interactions. *Infection and Immunity*, **68**, 2939–2947.

Somers, E. & Horsfall, J. G. (1966). The water content of powdery mildew conidia. *Phytopathology*, **56**, 1031–1035.

Sonnenberg, A. S. M. (2000). Genetics and breeding of *Agaricus bisporus*. In *Science and Cultivation of Edible Fungi*, ed. L. J. L. D. van Griensven. Rotterdam: A. Balkema, pp. 25–39.

Sorrell, T. C., Chen, S. C. A., Ruma, P., *et al.* (1996). Concordance of clinical and environmental isolates of *Cryptococcus neoformans* var. *gattii* by random amplification of polymorphic DNA analysis and PCR fingerprinting. *Journal of Clinical Microbiology*, **34**, 1253–1260.

Soylu, S., Keshavarzi, M., Brown, I. & Mansfield, J. W. (2003). Ultrastructural characterisation of interactions between *Arabidopsis thaliana* and *Albugo candida*. *Physiological and Molecular Plant Pathology*, **63**, 201–211.

Sparrow, F. K. (1960). *Aquatic Phycomycetes*, 2nd edn. Ann Arbor: University of Michigan Press.

Sparrow, F. K. (1973). Chytridiomycetes. Hyphochytridiomycetes. In *The Fungi: An Advanced Treatise* **IVB**, ed. G. C. Ainsworth, F. K. Sparrow & A. S. Sussman. New York: Academic Press, pp. 85–110.

Spatafora, J. W. (1995). Ascomal evolution of filamentous ascomycetes: evidence from molecular data. *Canadian Journal of Botany*, **73**, S811–15.

Spatafora, J. W. & Blackwell, M. (1993). Molecular systematics of unitunicate perithecial ascomycetes. I. The Clavicipitales–Hypocreales connection. *Mycologia*, **85**, 912–922.

Spatafora, J. W. & Blackwell, M. (1994). The polyphyletic origins of ophiostomatoid fungi. *Mycological Research*, **98**, 1–9.

Speare, R. & Thomas, A. D. (1985). Kangaroos and wallabies as carriers of *Basidiobolus haptosporus*. *Australian Veterinary Journal*, **62**, 209–210.

Spellig, T., Bölker, M., Lottspeich, F., Frank, R. W. & Kahmann, R. (1994). Pheromones trigger filamentous growth in *Ustilago maydis*. *EMBO Journal*, **13**, 1620–1627.

Spencer, D. M., ed. (1978). *The Powdery Mildews*. London: Academic Press.

Spencer, D. M., ed. (1981). *The Downy Mildews*. London: Academic Press.

Spencer, J. F. T. & Spencer, D. M., eds. (1990). *Yeast Technology*. Berlin: Springer-Verlag.

Spencer, J. F. T., Ragout de Spencer, A. L. & Laluce, C. (2002). Non-conventional yeasts. *Applied Microbiology and Biotechnology*, **58**, 147–156.

Spencer, M. A., Vick, M. C. & Dick, M. W. (2002). Revision of *Aplanopsis*, *Pythiopsis* and 'subcentric' *Achlya* species (Saprolegniomycetidae) using 18S rDNA and morphological data. *Mycological Research*, **106**, 549–560.

Spencer-Phillips, P. T. N. (1997). Function of fungal haustoria in epiphytic and endophytic infections. *Advances in Botanical Research*, **24**, 309–333.

Spicher, G. & Brümmer, J.-M. (1995). Baked goods. In *Biotechnology* **9**: *Enzymes, Biomass, Food and Feed*, ed. G. Reed & T. W. Nagodawithana. Weinheim, Germany: VCH, pp. 241–319.

Spiegel, F. W. (1984). *Protostelium nocturnum*, a new, minute, ballistosporous protostelid. *Mycologia*, **76**, 443–447.

Spiegel, F. W. (1990). Phylum plasmodial slime molds, class Protostelida. In *Handbook of Protoctista*, ed. L. Margulis, J. O. Corliss, M. Melkonian & D. J. Chapman. Boston: Jones & Bartlett, pp. 484–497.

Spiegel, F. W. (1991). A proposed phylogeny of the flagellated protostelids. *BioSystems*, **25**, 113–120.

Spiegel, F. W., Olive, L. S. & Brown, R. M. (1979). Roles of actin during sporocarp culmination in the simple mycetozoan *Planoprotostelium aurantium*. *Proceedings of the National Academy of Sciences of the USA*, **76**, 2335–2339.

Spielman, L. J., Drenth, A., Davidse, L. C., *et al.* (1991). A second world-wide migration and population displacement of *Phytophthora infestans*? *Plant Pathology*, **40**, 422–430.

Spiers, A. G. & Hopcroft, D. H. (1994). Comparative studies of the poplar rusts *Melampsora medusae*, *M. larici-populina* and their interspecific hybrid *M. medusae-populina*. *Mycological Research*, **98**, 889–903.

Spiltoir, C. F. (1955). Life cycle of *Ascosphaera apis* (*Pericystis apis*). *American Journal of Botany*, **42**, 501–508.

Sprey, B. (1988). Cellular and extracellular localization of endocellulase in *Trichoderma reesei*. *FEMS Microbiology Letters*, **55**, 283–294.

Springer, M. L. & Yanofsky, C. (1989). A morphological and genetic analysis of conidiophore development in *Neurospora crassa*. *Genes and Development*, **3**, 559–571.

Sridhar, K. R. & Bärlocher, F. (1992a). Aquatic hyphomycetes in spruce roots. *Mycologia*, **84**, 580–584.

Sridhar, K. R. & Bärlocher, F. (1992b). Endophytic aquatic hyphomycetes of roots of spruce, birch and maple. *Mycological Research*, **96**, 305–308.

Srinivasan, M. C., Narasimhan, M. J. & Thirumalachar, M. J. (1964). Artificial culture of *Entomophthora muscae* and morphological aspects for differentiation of the genera *Entomophthora* and *Conidiobolus*. *Mycologia*, **56**, 683–691.

Srinivasan, S., Vargas, M. M. & Roberson, R. W. (1996). Functional, organizational, and biochemical analysis of actin in hyphal tip cells of *Allomyces macrogynus*. *Mycologia*, **88**, 57–70.

Staats, M., van Baarlen, P. & van Kan, J. A. L. (2005). Molecular phylogeny of the plant pathogenic genus *Botrytis* and the evolution of host specificity. *Molecular Biology and Evolution*, **22**, 333–346.

Staben, C. & Yanofsky, C. (1990). *Neurospora crassa a* mating-type region. *Proceedings of the National Academy of Sciences of the USA*, **87**, 4917–4921.

Stadler, M., Anke, H. & Sterner, O. (1993). Linoleic acid – the nematicidal principle of several nematophagous fungi and its production in trap-forming submerged cultures. *Archives of Microbiology*, **160**, 401–405.

Stahmann, K.-P. (1997). Vitamins, amino acids. In *Fungal Biotechnology*, ed. T. Anke. Weinheim, Germany: Chapman & Hall, pp. 81–90.

Stahmann, K.-P., Schimz, K.-L. & Sahm, H. (1993). Purification and characterization of four extracellular 1,3-β-glucanases of *Botrytis cinerea*. *Journal of General Microbiology*, **139**, 2833–2840.

Stahmann, K.-P., Revuelta, J. L. & Seulberger, H. (2000). Three biotechnological processes using *Ashbya gossypii*, *Candida famata*, or *Bacillus subtilis* compete with chemical riboflavin production. *Applied Microbiology and Biotechnology*, **53**, 509–516.

Staib, P., Wirsching, S., Strauss, A. & Morschhauser, J. (2001). Gene regulation and host adaptation mechanisms in *Candida albicans*. *International Journal of Medical Microbiology*, **291**, 183–188.

Stakman, E. C. & Christensen, C. M. (1946). Aerobiology in relation to plant disease. *Botanical Review*, **12**, 205–253.

Stakman, E. C. & Levine, M. N. (1922). The determination of biologic forms of *Puccinia graminis* on *Triticum* spp. *Minnesota University Agricultural Experiment Station Technical Bulletin*, **8**.

Stalpers, J. A. (1974). *Spiniger*, a new genus for the imperfect states of basidiomycetes. *Proceedings of the Koninklijke Nederlandse Akademie van Wetenschappen Series C*, **77**, 402–407.

Stamps, D. J., Waterhouse, G. M., Newhook, F. J. & Hall, G. S. (1990). Revised tabular key to the species of *Phytophthora*. *Mycological Papers*, **162**, 1–28.

Stanghellini, M. E. & Hancock, J. G. (1971). Radial extent of the bean spermosphere and its relation to the behaviour of *Pythium ultimum*. *Phytopathology*, **61**, 165–168.

Stanier, R. Y. (1942). The culture and nutrient requirements of a chytridiaceous fungus. *Journal of Bacteriology*, **43**, 499–520.

Stankis, M. M., Specht, C. A. & Giasson, L. (1990). Sexual incompatibility in *Schizophyllum commune*: from classical genetics to a molecular view. *Seminars in Developmental Biology*, **1**, 195–206.

Staples, R. C. (2000). Research on the rust fungi during the twentieth century. *Annual Review of Phytopathology*, **38**, 49–69.

Starks, P. T., Blackie, C. A. & Seeley, T. D. (2000). Fever in honeybee colonies. *Naturwissenschaften*, **87**, 229–231.

Starmer, W. T., Ganter, P. F., Aberdeen, V., Lachance, M.-A. & Phaff, H. J. (1987). The ecological role of killer yeasts in natural communities of yeasts. *Canadian Journal of Microbiology*, **33**, 783–796.

Statzell-Tallman, A. & Fell, J. W. (1998). *Sporidiobolus* Nyland. In *The Yeasts: A Taxonomic Study*, 4th edn, ed. C. P. Kurtzman & J. W. Fell. Amsterdam: Elsevier, pp. 693–699.

Staub, T. (1991). Fungicide resistance: practical experience with anti-resistance strategies and the role of integrated use. *Annual Review of Phytopathology*, **29**, 421–442.

Steinberg, G. (1998). Organelle transport and molecular motors in fungi. *Fungal Genetics and Biology*, **24**, 161–177.

Steinberg, G. (2000). The cellular roles of molecular motors in fungi. *Trends in Microbiology*, **8**, 162–168.

Steinberg, G. & Fuchs, U. (2004). The role of microtubules in cellular organization and endocytosis in the plant pathogen *Ustilago maydis*. *Journal of Microscopy*, **214**, 114–123.

Steinberg, G., Schliwa, M., Lehmler, C., et al. (1998). Kinesin from the plant pathogenic *Ustilago maydis* is involved in vacuole formation and cytoplasmic migration. *Journal of Cell Science*, **111**, 2235–2246.

Steinert, M. & Heuner, K. (2005). *Dictyostelium* as host model for pathogenesis. *Cellular Microbiology*, **7**, 307–314.

Stenroos, S. K. & DePriest, P. T. (1998). SSU rDNA phylogeny of cladoniiform lichens. *American Journal of Botany*, **85**, 1548–1559.

Stensrud, Ø., Hywel-Jones, N. L. & Schumacher, T. (2005). Towards a phylogenetic classification of *Cordyceps*: ITS nrDNA sequence data confirm divergent lineages and paraphyly. *Mycological Research*, **109**, 41–56.

Stensvand, A., Eikemo, H., Gadoury, D. M. & Seem, R. C. (2005). Use of a rainfall frequency threshold to adjust a degree-day model of ascospore maturity of *Venturia inaequalis*. *Plant Disease*, **89**, 198–202.

Stephenson, S. L. & Stempen, H. (1994). *Myxomycetes. A Handbook of Slime Molds*. Portland, Oregon: Timber Press.

Stergiopoulos, I., Gielkens, M. M., Goodall, S. D., Venema, K. & de Waard, M. A. (2002). Molecular cloning and characterisation of three new ATP-binding cassette transporter genes from the wheat pathogen *Mycosphaerella graminicola*. *Gene*, **289**, 141–149.

Sternberg, J. A., Geffken, D., Adams, J. B., *et al.* (2001). Famoxadone: the discovery and optimisation of a new agricultural fungicide. *Pest Management Science*, **57**, 143–152.

Stevens, R. B., ed. (1974). *Mycology Guidebook*. Seattle: University of Washington Press.

Stiles, C. M. & Glawe, D. A. (1989). Colonization of soybean roots by fungi isolated from cysts of *Heterodera glycines*. *Mycologia*, **81**, 797–799.

Stoll, M., Piepenbring, M., Begerow, D. & Oberwinkler, F. (2003). Molecular phylogeny of *Ustilago* and *Sporisorium* species (Basidiomycota, Ustilaginales) based on internal transcribed spacer (ITS) sequences. *Canadian Journal of Botany*, **81**, 976–984.

Straatsma, G. & Samson, R. A. (1993). Taxonomy of *Scytalidium thermophilum*, an important thermophilic fungus in mushroom compost. *Mycological Research*, **97**, 321–328.

Straatsma, G., Samson, R. A., Olijinsma, T. W., *et al.* (1994a). Ecology of thermophilic fungi in mushroom compost, with emphasis on *Scytalidium thermophilum* and growth stimulation of *Agaricus bisporus* mycelium. *Applied and Environmental Microbiology*, **60**, 454–458.

Straatsma, G., Olijinsma, T. W., Gerrits, J. P. G., *et al.* (1994b). Inoculation of *Scytalidium thermophilum* in button mushroom compost and its effect on yield. *Applied and Environmental Microbiology*, **60**, 3049–3054.

Straatsma, G., Samson, R. A., Olijinsma, T. W., *et al.* (1995). Bioconversion of cereal straw into mushroom compost. *Canadian Journal of Botany*, **73**, S1019–S1024.

Strickmann, E. & Chadefaud, M. (1961). Recherches sur les asques et sur les périthèces des *Nectria* et réflexions sur l'évolution des Ascomycètes. *Revue Générale de Botanique*, **68**, 725–770.

Stockdale, P. M. (1968). Sexual stimulation between *Arthroderma simii* Stockdale, Mackenzie and Austwick and related species. *Sabouraudia*, **6**, 176–181.

Stoev, S. D. (1998). The role of ochratoxin A as a possible cause of Balkan endemic nephropathy and its risk evaluation. *Veterinary and Human Toxicology*, **40**, 352–360.

Storck, R. & Morrill, R. C. (1977). Nuclei, nucleic acids and protein in sporangiospores of *Mucor bacilliformis*. *Mycologia*, **69**, 1031–1041.

Storsberg, J., Schulz, H. & Keller, E. R. J. (2003). Chemotaxonomic classification of some *Allium* wild species on the basis of their volatile sulphur compounds. *Journal of Applied Botany – Angewandte Botanik*, **77**, 160–162.

Strassmann, J. E., Zhu, Y. & Queller, D. C. (2000). Altruism and social cheating in the social amoeba *Dictyostelium discoideum*. *Nature*, **408**, 965–967.

Strathern, J. N. & Herskowitz, I. (1979). Asymmetry and directionality in production of new cell types during clonal growth: the switching pattern of homothallic yeast. *Cell*, **17**, 371–381.

Stringer, J. R. (1996). *Pneumocystis carinii*: What is it, exactly? *Clinical Microbiology Reviews*, **9**, 489–498.

Stringer, J. R. (2002). Pneumocystis. *International Journal of Medical Microbiology*, **292**, 391–404.

Stringer, J. R., Beard, C. B., Miller, R. F. & Wakefield, A. E. (2002). A new name (*Pneumocystis jiroveci*) for pneumocystis from humans. *Emerging Infectious Diseases*, **8**, 891–896.

Stringer, M. A., Dean, R. A., Sewall, T. C. & Timberlake, W. E. (1991). *Rodletness*, a new *Aspergillus* developmental mutant induced by directed gene inactivation. *Genes and Development*, **5**, 1161–1171.

Struck, C., Hahn, M. & Mendgen, K. (1996). Plasma membrane H^+-ATPase activity in spores, germ tubes, and haustoria of the rust fungus *Uromyces viciae-fabae*. *Fungal Genetics and Biology*, **20**, 30–35.

Strunz, G. M., Bethell, R., Dumas, M. T. & Boyonoski, N. (1997). On a new synthesis of sterpurene and the bioactivity of some related *Chondrostereum purpureum* sesquiterpene metabolites. *Canadian Journal of Chemistry*, **75**, 742–753.

Stubblefield, S. P. & Taylor, T. N. (1983). Studies of paleozoic fungi. I. The structure and organization of *Traquairia* (Ascomycota). *American Journal of Botany*, **70**, 387–399.

Stubblefield, S.P., Taylor, T.N., Miller, C.E. & Cole, G.T. (1983). Studies of carboniferous fungi. II. The structure and organization of *Myocarpon*, *Sporocarpon*, *Dubiocarpon* and *Coleocarpon* (Ascomycotina). *American Journal of Botany*, **70**, 1482–1498.

Stubbs, R.W. (1985). Stripe rust. In *The Cereal Rusts 2: Diseases, Distribution, Epidemiology, and Control*, ed. A.P. Roelfs & W.R. Bushnell. Orlando: Academic Press, pp. 61–101.

Stuteville, D.L. (1981). Downy mildew of forage legumes. In *The Downy Mildews*, ed. D.M. Spencer. London: Academic Press, pp. 355–66.

Suarit, R., Gopal, P.K. & Shepherd, M.G. (1988). Evidence for a glycosidic linkage between chitin and glucan in the cell wall of *Candida albicans*. *Journal of General Microbiology*, **134**, 1723–1730.

Suberkropp, K. & Klug, M.J. (1976). Fungi and bacteria associated with dead leaves during processing in a woodland stream. *Ecology*, **57**, 707–719.

Suberkropp, K. & Klug, M.J. (1980). The maceration of deciduous leaf litter by aquatic hyphomycetes. *Canadian Journal of Botany*, **58**, 1025–1031.

Subramanian, C.V. (1971). The phialide. In *Taxonomy of Fungi Imperfecti*, ed. B. Kendrick. Toronto: Toronto University Press, pp. 93–119.

Sugita, T., Suto, H., Unno, T., *et al.* (2001). Molecular analysis of *Malassezia* microflora on the skin of atopic dermatitis patients and healthy subjects. *Journal of Clinical Microbiology*, **39**, 3486–3490.

Sugiyama, J., ed. (1987). *Pleomorphic Fungi: The Diversity and its Taxonomic Implications*. Tokyo: Kodansha.

Sugiyama, J., Summerbell, R.C. & Mikawa, T. (2002). Molecular phylogeny of onygenalean fungi based on small subunit (SSU) and large subunit (LSU) ribosomal DNA sequences. *Studies in Mycology*, **47**, 5–23.

Summerbell, R.C. (2000). Form and function in the evolution of dermatophytes. In *Biology of Dermatophytes and Other Keratinolytic Fungi*, ed. R.K.S. Kushwara & J. Guarro. Bilbao: Revista Iberoamericana de Micología, pp. 30–43.

Summerbell, R. (2003). Ascomycetes: *Aspergillus*, *Fusarium*, *Sporothrix*, *Piedraia*, and their relatives. In *Pathogenic Fungi in Humans and Animals*, 2nd edn, ed. D.H. Howard. New York: Marcel Dekker, pp. 237–498.

Summerbell, R.C., Kane, J., Krajden, S. & Duke, E.E. (1993). Medically important *Sporothrix* species and related ophiostomatoid fungi. In *Ceratocystis and Ophiostoma. Taxonomy, Ecology and Pathogenicity*, ed. M.J. Wingfield, K.A. Seifert & J.F. Webber. St. Paul: APS Press, pp. 185–92.

Sun, N.C. & Bowen, C.C. (1972). Ultrastructural studies of nuclear division in *Basidiobolus ranarum* Eidam. *Caryologia*, **25**, 243–247.

Sundheim, L. (1982). Control of cucumber powdery mildew by the hyperparasite *Ampelomyces quisqualis* and fungicides. *Plant Pathology*, **31**, 209–214.

Sussman, A.S. (1968). Longevity and survivability of fungi. In *The Fungi. An Advanced Treatise* **III**: *The Fungal Population*, ed. G.C. Ainsworth & A.S. Sussman. New York Academic Press, pp. 447–86.

Sussman, A.S. & Halvorson, H.O. (1966). *Spores: Their Dormancy and Germination*. New York: Harper & Row.

Sutherland, M.L. & Brasier, C.M. (1995). Effect of d-factors on *in vitro* cerato-ulmin production by the Dutch-elm disease pathogen *Ophiostoma novo-ulmi*. *Mycological Research*, **99**, 1211–1217.

Sutter, R.P. (1987). Sexual development. In *Phycomyces*, ed. E. Cerdá-Olmedo & E.D. Lipson. Cold Spring Harbor, New York: Cold Spring Harbor Laboratory, pp. 317–36.

Sutton, B.C. (1980). *The Coelomycetes. Fungi Imperfecti with Pycnidia, Acervuli and Stromata*. Kew, UK: Commonwealth Mycological Institute.

Sutton, B.C. (1986). Presidential address. Improvisations on conidial themes. *Transactions of the British Mycological Society*, **86**, 1–38.

Sutton, B.C. (1992). The genus *Glomerella* and its anamorph *Colletotrichum*. In *Colletotrichum: Biology, Pathology and Control*, ed. J.A. Bailey & M.J. Jeger. Wallingford, UK: CAB International, pp. 1–26.

Sutton, D.K., MacHardy, W.E. & Lord, W.G. (2000). Effects of shredding or treating apple leaf litter with urea on ascospore dose of *Venturia inaequalis* and disease buildup. *Plant Disease*, **84**, 1319–1326.

Sutton, P.N., Henry, M.J. & Hall, J.L. (1999). Glucose, and not sucrose, is transported from wheat to wheat powdery mildew. *Planta*, **208**, 426–430.

Suyanto, Ohtsuki, T., Yazaki, S., Ui, S. & Mimura, A. (2003). Isolation of a novel thermophilic fungus, *Chaetomium* sp. nov. MS-017 and description of its palm-oil fiber-decomposing properties. *Applied Microbiology and Biotechnology*, **60**, 581–587.

Suzuki, M. & Nakase, T. (1999). A phylogenetic study of ubiquinone Q-8 species of the genera *Candida*, *Pichia*, and *Citeromyces* based on 18S ribosomal DNA sequence divergence. *Journal of General and Applied Microbiology*, **45**, 239–246.

Swann, E.C. & Taylor, J.W. (1993). Higher taxa of basidiomycetes: an 18S rRNA gene perspective. *Mycologia*, **85**, 923–936.

Swann, E. C. & Taylor, J. W. (1995). Phylogenetic perspectives on basidiomycete systematics: evidence from the 18S rRNA gene. *Canadian Journal of Botany*, **73**, S862–S868.

Swann, E. C., Frieders, E. M. & McLaughlin, D. J. (1999). *Microbotryum*, *Kriegeria* and the changing paradigm in basidiomycete classification. *Mycologia*, **91**, 51–66.

Swann, E. C., Frieders, E. M. & McLaughlin, D. J. (2001). Urediniomycetes. In *The Mycota* **VIIB**: *Systematics and Evolution*, ed. D. J. McLaughlin, E. G. McLaughlin & P. A. Lemke. Berlin: Springer-Verlag, pp. 37–56.

Swanson, J. A. & Taylor, D. L. (1982). Local and spatially coordinated movements in *Dictyostelium discoideum* amoebae during chemotaxis. *Cell*, **28**, 225–232.

Swedjemark, G. & Stenlid, J. (1993). Population dynamics of the root rot fungus *Heterobasidion annosum* following thinning of *Picea abies*. *Oikos*, **66**, 247–254.

Sykes, E. E. & Porter, D. (1980). Infection and development of the obligate parasite *Catenaria allomycis* on *Allomyces arbuscula*. *Mycologia*, **72**, 288–300.

Syrop, M. (1975). Leaf curl disease of almond caused by *Taphrina deformans* (Berk.) Tul. I. A light microscope study of the host/parasite relationship. *Protoplasma*, **85**, 39–56.

Syrop, M. J. & Beckett, A. (1972). The origin of ascospore-delimiting membranes in *Taphrina deformans*. *Archiv für Mikrobiologie*, **86**, 185–191.

Syrop, M. J. & Beckett, A. (1976). Leaf curl disease of almond caused by *Taphrina deformans* (Berk.) Tul. 3. Ultrastructural cytology of the pathogen. *Canadian Journal of Botany*, **54**, 293–305.

Szabo, S. P., O'Day, D. H. & Chagla, A. H. (1982). Cell fusion, nuclear fusion, and zygote differentiation during sexual development of *Dictyostelium discoideum*. *Developmental Biology*, **90**, 375–382.

Sziráki, I., Balázs, E. & Király, Z. (1975). Increased levels of cytokinin and indole-acetic acid in peach leaves infected with *Taphrina deformans*. *Physiological Plant Pathology*, **5**, 45–50.

Takamatsu, S., Hirata, T. & Sato, Y. (2000). A parasitic transition from trees to herbs occurred at least twice in tribe Cystotheceae (Erysiphaceae): evidence from nuclear ribosomal DNA. *Mycological Research*, **104**, 1304–1311.

Takamatsu, S., Niinomi, S., Cabrera de Álvarez, M. G., *et al.* (2005). *Caespithotheca* gen. nov., an ancestral genus in the *Erysiphales*. *Mycological Research*, **109**, 903–911.

Takeshige, K., Baba, M., Tsuboi, S., Noda, T. & Ohsumi, Y. (1992). Autophagy in yeast demonstrated with proteinase-deficient mutants and conditions for its induction. *Journal of Cell Biology*, **119**, 301–311.

Talbot, P. H. B. (1977). The *Sirex–Amylostereum–Pinus* association. *Annual Review of Phytopathology*, **15**, 41–54.

Tanabe, Y., O'Donnell, K., Saikawa, M. & Sugiyama, J. (2000). Molecular phylogeny of parasitic Zygomycota (Dimargaritales, Zoopagales) based on nuclear small subunit ribosomal DNA sequences. *Molecular Phylogenetics and Evolution*, **16**, 253–262.

Tanabe, Y., Saikawa, M., Watanabe, M. M. & Sugiyama, J. (2004). Molecular phylogeny of Zygomycota based on EF-1α and RPB1 sequences: limitations and utility of alternative markers to rDNA. *Molecular Phylogenetics and Evolution*, **30**, 438–449.

Tanabe, Y., Watanabe, M. M. & Sugiyama, J. (2005). Evolutionary relationships among basal fungi (Chytridiomycota and Zygomycota): insights from molecular phylogenetics. *Journal of General and Applied Microbiology*, **51**, 267–276.

Tanaka, K. (1970). Mitosis in the fungus *Basidiobolus ranarum* as revealed by electron microscopy. *Protoplasma*, **70**, 423–440.

Tanaka, K. & Hirata, A. (1982). Ascospore development in the fission yeasts *Schizosaccharomyces pombe* and *S. japonicus*. *Journal of Cell Science*, **56**, 263–279.

Tanaka, K. & Kanbe, T. (1986). Mitosis in the fission yeast *Schizosaccharomyces pombe* as revealed by freeze-substitution electron-microscopy. *Journal of Cell Science*, **80**, 253–268.

Tang, W., Yang, H. & Ryder, M. (2001). Research and application of *Trichoderma* spp. in biological control of plant pathogens. In *Bio-Exploitation of Filamentous Fungi*, ed. S. B. Pointing & K. B. Hyde. Hong Kong: Fungal Diversity Press, pp. 403–35.

Tange, Y. & Niwa, O. (1995). A selection system for diploid and against haploid cells in *Schizosaccharomyces pombe*. *Molecular and General Genetics*, **248**, 644–648.

Tapper, R. (1981). Direct measurement of translocation of carbohydrate in the lichen, *Cladonia convoluta*, by quantitative autoradiography. *New Phytologist*, **89**, 429–437.

Tate, K. G. & Wood, P. N. (2000). Potential ascospore production and resulting blossom blight by *Monilinia fructicola* in unsprayed peach trees. *New Zealand Journal of Crop and Horticultural Science*, **28**, 219–224.

Taylor, B. N., Harrer, T., Pscheidl, E., *et al.* (2003). Surveillance of nosocomial transmission of *Candida albicans* in an intensive care unit by DNA fingerprinting. *Journal of Hospital Infection*, **55**, 283–289.

Taylor, D. L. & Bruns, T. D. (1997). Independent specialized invasions of ectomycorrhizal mutualism by two nonphotosynthetic orchids. *Proceedings of the National Academy of Sciences of the USA*, **94**, 4510–4515.

Taylor, J. & Birdwell, D. O. (2000). A scanning electron microscopic study of the infection of water oak (*Quercus nigra*) by *Taphrina caerulescens*. *Mycologia*, **92**, 309–311.

Taylor, T. N., Remy, W. & Hass, H. (1992). Fungi from the Lower Devonian Rhynie chert: Chytridiomycetes. *American Journal of Botany*, **79**, 1233–1241.

Taylor, T. N., Remy, W., Hass, H. & Kerp, H. (1995). Fossil arbuscular mycorrhizae from the Early Devonian. *Mycologia*, **87**, 560–573.

Taylor, T. N., Hass, H. & Kerp, H. (1997). A cyanolichen from the Lower Devonian Rhynie chert. *American Journal of Botany*, **84**, 992–1004.

Taylor, T. N., Hass, H. & Kerp, H. (1999). The oldest fossil ascomycetes. *Nature*, **399**, 648.

Taylor, T. N., Hass, H., Kerp, H., Krings, M. & Hanlin, R. T. (2005). Perithecial ascomycetes from the 400 million year old Rhynie chert: an example of ancestral polymorphism. *Mycologia*, **97**, 269–285.

Tehler, A. (1996). Systematics, phylogeny and classification. In *Lichen Biology*, ed. T. H. Nash. Cambridge: Cambridge University Press, pp. 217–239.

Tehler, A., Little, D. P. & Farris, J. S. (2003). The full-length phylogenetic tree from 1551 ribosomal sequences of chitinous fungi. *Fungi. Mycological Research*, **107**, 901–916.

Teichert, U., Mechler, B., Müller, H. & Wolf, D. H. (1989). Lysosomal (vacuolar) proteinases of yeast are essential catalysts for protein degradation, differentiation, and cell survival. *Journal of Biological Chemistry*, **264**, 16 037–16 045.

Temmink, J. H. M. & Campbell, R. N. (1968). The ultrastructure of *Olpidium brassicae*. I. Formation of sporangia. *Canadian Journal of Botany*, **46**, 951–955.

Temmink, J. H. M. & Campbell, R. N. (1969a). The ultrastructure of *Olpidium brassicae*. II. Zoospores. *Canadian Journal of Botany*, **47**, 227–231.

Temmink, J. H. M. & Campbell, R. N. (1969b). The ultrastructure of *Olpidium brassicae*. III. Infection of host roots. *Canadian Journal of Botany*, **47**, 421–424.

Temmink, J. H. M., Campbell, R. N. & Smith, P. R. (1970). Specificity and site of *in vitro* acquisition of tobacco necrosis virus by zoospores of *Olpidium brassicae*. *Journal of General Virology*, **9**, 201–203.

Temple, B., Horgen, P. A., Bernier, L. & Hertz, W. E. (1997). Cerato-ulmin, a hydrophobin secreted by the causal agents of Dutch elm disease, is a parasitic fitness factor. *Fungal Genetics and Biology*, **22**, 39–53.

Tenberge, K. B. (1999). Biology and strategy of the ergot fungi. In *Ergot. The Genus Claviceps*, ed. V. Křen & L. Cvak. Amsterdam: Harwood Academic Publishers, pp. 25–56.

ten Have, R. & Teunissen, P. J. M. (2001). Oxidative mechanisms involved in lignin degradation by white-rot fungi. *Chemical Reviews*, **101**, 3397–3413.

Tenney, K., Hunt, I., Sweigard, J., *et al.* (2000). *hex-1*, a gene unique to filamentous fungi, encodes the major protein of the Woronin body and functions as a plug for septal pores. *Fungal Genetics and Biology*, **31**, 205–217.

Tereshina, V. M. & Feofilova, E. P. (1995). Two sporulation types in the Mucorous fungus *Blakeslea trispora* – morphological characteristics. *Microbiology*, **64**, 580–583.

Terhune, B. T. & Hoch, H. C. (1993). Substrate hydrophobicity and adhesion of *Uromyces* urediospores and germlings. *Experimental Mycology*, **17**, 241–252.

Terhune, B. T., Allen, E. A., Hoch, H. C., Wergin, W. P. & Erbe, E. F. (1991). Stomatal ontogeny and morphology in *Phaseolus vulgaris* in relation to infection structure initiation in *Uromyces appendiculatus*. *Canadian Journal of Botany*, **69**, 477–484.

Teunissen, H. A. S., Verkooijen, J., Cornelissen, B. J. C. & Haring, M. A. (2002). Genetic exchange of avirulence determinants and extensive karyotype rearrangements in parasexual recombinants of *Fusarium oxysporum*. *Molecular Genetics and Genomics*, **268**, 298–310.

Thaxter, R. (1888). The Entomophthoreae of the United States. *Memoirs of the Boston Society for Natural History*, **4**, 133–201.

Theodorou, M. K., Lowe, S. E. & Trinci, A. P. J. (1992). Anaerobic fungi and the rumen ecosystem. In *The Fungal Community. Its Organization and Role in the Ecosystem*, 2nd edn, ed. G. C. Carroll & D. T. Wicklow. New York: Marcel Dekker Inc., pp. 43–72.

Theodorou, M. K., Zhu, W.-Y., Rickers, A., *et al.* (1996). Biochemistry and ecology of anaerobic fungi. In *The Mycota VI: Human and Animal Relationships*, ed. D. H. Howard & J. D. Miller. Berlin: Springer-Verlag, pp. 265–295.

Thielke, C. (1982). Meiotic divisions in the basidium. In *Basidium and Basidiocarp: Evolution, Cytology, Function and Development*, ed. K. Wells & E. K. Wells. New York: Springer-Verlag, pp. 75–91.

Thiers, H. D. (1984). The secotioid syndrome. *Mycologia*, **76**, 1–8.

Thines, E., Eilbert, F., Sterner, O. & Anke, H. (1997). Signal transduction leading to appressorium formation in germinating conidia of *Magnaporthe grisea*: effects of second messengers diacylglycerols, ceramides and sphingomyelin. *FEMS Microbiology Letters*, **156**, 91–94.

Thines, E., Weber, R.W.S. & Talbot, N.J. (2000). MAP kinase and protein kinase A-dependent mobilization of triacylglycerol and glycogen during appressorium turgor generation by *Magnaporthe grisea*. *Plant Cell*, **12**, 1703–1718.

Thirumalachar, M.J. & Dickson, J.G. (1949). Chlamydospore germination, nuclear cycle and artificial culture of *Urocystis agropyri* on red top. *Phytopathology*, **39**, 333–339.

Thomas, D. des S. & McMorris, T.C. (1987). Allomonal functions of the steroid hormone, antheridiol, in water mold *Achlya*. *Journal of Chemical Ecology*, **13**, 1131–1137.

Thomas, P.L. (1984). Recombination of virulence genes following hybridization between isolates of *Ustilago nuda* infecting barley under natural conditions. *Canadian Journal of Plant Pathology*, **6**, 101–104.

Thomas, P.L. & Menzies, J.G. (1997). Cereal smuts in Manitoba and Saskatchewan, 1989–1995. *Canadian Journal of Plant Pathology*, **19**, 161–165.

Thomma, B.P.H.J. (2003). *Alternaria* spp.: from general saprophyte to specific parasite. *Molecular Plant Pathology*, **4**, 225–236.

Thompson, J.E., Fahnestock, S., Farrall, L., *et al.* (2000). The second naphthol reductase of fungal melanin biosynthesis in *Magnaporthe grisea*. *Journal of Biological Chemistry*, **275**, 34 867–34 872.

Thompson, W. & Rayner, A.D.M. (1982). Spatial structure in a population of *Tricholomopsis platyphylla* in a woodland site. *New Phytologist*, **92**, 103–114.

Thompson-Coffe, C. & Zickler, D. (1994). How the cytoskeleton recognizes and sorts nuclei of opposite mating-type during the sexual cycle in filamentous ascomycetes. *Developmental Biology*, **165**, 257–271.

Thomson, J., Melville, L.H. & Peterson, R.L. (1989). Interaction between the ectomycorrhizal fungus *Pisolithus tinctorius* and root hairs of *Picea mariana* (P). *American Journal of Botany*, **76**, 632–636.

Thorn, R.G., Moncalvo, J.M., Reddy, C.A. & Vilgalys, R. (2000). Phylogenetic analyses and the distribution of nematophagy support a monophyletic Pleurotaceae within the polyphyletic pleurotoid–lentinoid fungi. *Mycologia*, **92**, 241–252.

Thumm, M. (2000). Structure and function of the yeast vacuole and its role in autophagy. *Microscopy Research and Technique*, **51**, 563–572.

Tilston, E.L., Pitt, D. & Groenhof, A.C. (2002). Composted recycled organic matter suppresses soil-borne diseases of field crops. *New Phytologist*, **154**, 731–740.

Todd, N.K. & Rayner, A.D.M. (1980). Fungal individualism. *Science Progress*, **66**, 331–354.

Tomiyama, S., Sakuma, T., Ishizaka, N., *et al.* (1968). A new antifungal substance isolated from resistant potato tuber tissue infected by pathogens. *Phytopathology*, **58**, 115–116.

Tomlinson, J.A. (1958). Crook root of watercress. II. The control of the disease with zinc-fritted glass and the mechanism of its action. *Annals of Applied Biology*, **46**, 608–621.

Tommerup, I.C. & Broadbent, D. (1974). Nuclear fusion, meiosis and the origin of dikaryotic hyphae in *Armillaria mellea*. *Archiv für Mikrobiologie*, **103**, 279–282.

Tommerup, I.C. & Ingram, D.S. (1971). The life cycle of *Plasmodiophora brassicae* Woron. in *Brassica* tissue cultures and in intact roots. *New Phytologist*, **70**, 327–332.

Tommerup, I.C. & Sivasithamparam, K. (1990). Zygospores and asexual spores of *Gigaspora decipiens*, an arbuscular mycorrhizal fungus. *Mycological Research*, **94**, 897–900.

Ton, J., van Pelt, J.A., van Loon, L.C. & Pieterse, C.M. (2002). Differential effectiveness of salicylate-dependent and jasmonate/ethylene-dependent induced resistance in *Arabidopsis*. *Molecular Plant–Microbe Interactions*, **15**, 27–34.

Torralba, S., Raudaskoski, M. & Pedregosa, A.M. (1998). Effects of methyl benzimidazole-2-yl carbamate on microtubule and actin cytoskeleton in *Aspergillus nidulans*. *Protoplasma*, **202**, 54–64.

Torralba, S., Heath, I.B. & Ottensmeyer, F.P. (2001). Ca^{2+} shuttling in vesicles during tip growth in *Neurospora crassa*. *Fungal Genetics and Biology*, **33**, 181–193.

Torralba, S., Pisabarro, A.G. & Ramirez, L. (2004). Immunofluorescence microscopy of the microtubule cytoskeleton during conjugate division in the dikaryon of *Pleurotus ostreatus* N001. *Mycologia*, **96**, 41–51.

Tournas, V. (1994). Heat-resistant fungi of importance to the food and beverage industry. *Critical Reviews in Microbiology*, **20**, 243–263.

Townsend, B.B. (1954). Morphology and development of fungal rhizomorphs. *Transactions of the British Mycological Society*, **37**, 222–233.

Townsend, B.B. & Willetts, H.J. (1954). The development of sclerotia of certain fungi. *Transactions of the British Mycological Society*, **37**, 213–221.

Trail, F., Gaffoor, I. & Vogel, S. (2005). Ejection mechanisms and trajectory of the ascospores of *Gibberella zeae* (anamorph *Fusarium graminearum*). *Fungal Genetics and Biology*, **42**, 528–533.

Trappe, J.M. (1979). The orders, families, and genera of hypogeous Ascomycotina (truffles and their relatives). *Mycotaxon*, **9**, 297–340.

Trappe, J.M. & Maser, C. (1977). Ectomycorrhizal fungi: interactions of mushrooms and truffles with beasts and trees. In *Mushrooms and Man, an Interdisciplinary Approach to Mycology*, ed. T. Walters. Albany, Oregon: USDA Forestry Service, pp. 165–179.

Tredway, L.P., White, J.F., Gaut, B.S., *et al.* (1999). Phylogenetic relationships within and between *Epichloe* and *Neotyphodium* endophytes as estimated by AFLP markers and rDNA sequences. *Mycological Research*, **103**, 1593–1603.

Trembley, M.L., Ringli, C. & Honegger, R. (2002). Morphological and molecular analysis of early stages in the resynthesis of the lichen *Baeomyces rufus*. *Mycological Research*, **106**, 768–776.

Tribe, H.T. (1955). Studies on the physiology of parasitism. XIX. On the killing of plant cells by enzymes from *Botrytis cinerea* and *Bacterium aroideae*. *Annals of Botany* N.S., **19**, 351–368.

Tribe, H.T. (1957). On the parasitism of *Sclerotinia trifoliorum* by *Coniothyrium minitans*. *Transactions of the British Mycological Society*, **40**, 489–499.

Tribe, H.T. (1977). On 'Olpidium nematodeae Skvortzow'. *Transactions of the British Mycological Society*, **69**, 509–511.

Tribe, H.T. & Weber, R.W.S. (2002). A low-temperature fungus from cardboard, *Myxotrichum chartarum*. *Mycologist*, **16**, 3–5.

Trinci, A.P.J. (1991). 'Quorn' mycoprotein. *Mycologist*, **5**, 106–109.

Trinci, A.P.J., Davies, D.R., Gull, K., *et al.* (1994). Anaerobic fungi in herbivorous animals. *Mycological Research*, **98**, 129–152.

Trione, E.J. (1972). Isolation of *Tilletia caries* from infected wheat plants. *Phytopathology*, **62**, 1096–1097.

Tsai, H.-F., Liu, J.-S., Staben, C., *et al.* (1994). Evolutionary classification of fungal endophytes of tall fescue grass by hydridization of *Epichloe* species. *Proceedings of the National Academy of Sciences of the USA*, **91**, 2542–2546.

Tsao, P.H. (1983). Factors affecting isolation and quantitation of *Phytophthora* from soil. In *Phytophthora. Its Biology, Taxonomy, Ecology, and Pathology*, ed. D.C. Erwin, S. Bartnicki-Garcia & P.H. Tsao. St. Paul: APS Press, pp. 219–236.

Tsuneda, A. & Currah, R.S. (2004). Ascomatal morphogenesis in *Myxotrichum arcticum* supports the derivation of the Myxotrichaceae from a discomycetous ancestor. *Mycologia*, **96**, 627–635.

Tu, C.C., Kimbrough, J.W. & Aldrich, H.C. (1977). Cytology and ultrastructure of *Thanatephorus* and related taxa of the *Rhizoctonia* complex. *Canadian Journal of Botany*, **55**, 2419–2436.

Tu, J.C. (1985). Tolerance of white bean (*Phaseolus vulgaris*) to white mold (*Sclerotinia sclerotiorum*) associated with tolerance to oxalic acid. *Physiological Plant Pathology*, **26**, 111–117.

Tu, J.C. (1988). The role of white mold-infected white bean (*Phaseolus vulgaris* L.) seeds in the dissemination of *Sclerotinia sclerotiorum* (Lib.) de Bary. *Phytopathologische Zeitschrift*, **121**, 40–50.

Tucker, S.L. & Talbot, N.J. (2001). Surface attachment and pre-penetration stage development by plant pathogenic fungi. *Annual Review of Phytopathology*, **39**, 385–417.

Tudzynski, B. (1997). Fungal phytohormones in pathogenic and mutualistic associations. In *The Mycota VA: Plant Relationships*, ed. G.C. Carroll & P. Tudzynski. Berlin: Springer-Verlag, pp. 167–184.

Tudzynski, P. (1999). Genetics of *Claviceps purpurea*. In *Ergot. The Genus Claviceps*, ed. V. Křen & L. Cvak. Amsterdam: Harwood Academic Publishers, pp. 79–93.

Tudzynski, P., Correia, T. & Keller, U. (2001). Biotechnology and genetics of ergot alkaloids. *Applied Microbiology and Biotechnology*, **57**, 593–605.

Tulasne, L.R. & Tulasne, C.C. (1931). *Selecta Fungorum Carpologia I* (English translation by W.B. Grove). Oxford: Clarendon Press.

Tunlid, A., Jansson, H.-B. & Nordbring-Hertz, B. (1992). Fungal attachment to nematodes. *Mycological Research*, **96**, 401–412.

Tuno, N. (1998). Spore dispersal of *Dictyophora* fungi (Phallaceae) by flies. *Ecological Research*, **13**, 7–15.

Tuno, N. (1999). Insect feeding on spores of a bracket fungus, *Elfvingia applanata* (Pers.) Karst. (Ganodermataceae, Aphyllophorales). *Ecological Research*, **14**, 97–103.

Turner, B.C., Perkins, D.D. & Fairfield, A. (2001). *Neurospora* from natural populations: a global study. *Fungal Genetics and Biology*, **32**, 67–92.

Tuse, D. (1984). Single-cell protein: current status and future prospects. *Critical Reviews in Food Science and Nutrition*, **19**, 273–325.

Tyler, B.M. (2002). Molecular basis of recognition between *Phytophthora* pathogens and their hosts. *Annual Review of Phytopathology*, **40**, 137–167.

Tyrell, D. & MacLeod, D.M. (1975). *In vitro* germination of *Entomophthora aphidis* resting spores. *Canadian Journal of Botany*, **53**, 1188–1191.

Uchida, W., Matsunaga, S., Sugiyama, R., Kazama, Y. & Kawano, S. (2003). Morphological development of anthers induced by the dimorphic smut fungus *Microbotryum violaceum* in female flowers of the dioecious plant *Silene latifolia*. *Planta*, **218**, 240–248.

Uecker, F. A. (1976). Development and cytology of *Sordaria humana*. *Mycologia*, **68**, 30–46.

Uecker, F. A. (1988). *A World List of Phomopsis Names with Notes on Nomenclature, Morphology and Biology*. Berlin: J. Cramer.

Ueng, P. P., Subramaniam, K., Chen, W., *et al.* (1998). Intraspecific genetic variation of *Stagonospora avenae* and its differentiation from *S. nodorum*. *Mycological Research*, **102**, 607–614.

Uesugi, Y. (1998). Fungicide classes: chemistry, uses and mode of action. In *Fungicidal Activity. Chemical and Biological Approaches to Plant Protection*, ed. D. Hutson & J. Miyamoto. Chichester: John Wiley & Sons, pp. 23–56.

Uetake, Y., Kobayashi, K. & Ogoshi, A. (1992). Ultrastructural changes during the symbiotic development of *Spiranthes sinensis* (Orchidaceae) protocorms associated with binucleate *Rhizoctonia* anastomosis group C. *Mycological Research*, **96**, 199–209.

Ugadawa, S. (1984). Taxonomy of mycotoxin-producing *Chaetomium*. In *Toxigenic Fungi. Their Toxins and Health Hazards*, ed. H. Kurato & Y. Ueno. Tokyo: Elsevier, pp. 139–47.

Uhm, J. Y. & Fujii, H. (1983a). Ascospore dimorphism in *Sclerotinia trifoliorum* and cultural characters of strains from different-sized spores. *Phytopathology*, **73**, 565–569.

Uhm, J. Y. & Fujii, H. (1983b). Heterothallism and mating type mutation in *Sclerotinia trifoliorum*. *Phytopathology*, **73**, 569–572.

Ulken, A. & Sparrow, F. K. (1968). Estimation of chytrid propagules in Douglas Lake by the MPN-Pollen grain method. *Veröffentlichungen des Instituts für Meeresforschung Bremerhaven*, **11**, 83–88.

Ullrich, R. C. & Anderson, J. B. (1978). Sex and diploidy in *Armillaria mellea*. *Experimental Mycology*, **2**, 119–129.

Ullrich, R. C., Specht, C. A., Stankis, M. M., Giasson, L. & Novotny, C. P. (1991). Molecular biology of mating-type determination in *Schizophyllum commune*. In *Genetic Engineering, Principles and Methods* 13, ed. J. K. Setlow. New York: Plenum Press, pp. 279–306.

Ullstrup, A. J. (1972). The impacts of the southern corn leaf blight epidemics of 1970–71. *Annual Review of Phytopathology*, **10**, 37–50.

Umar, M. H. & van Griensven, L. J. L. D. (1997). Morphogenetic cell death in developing primordia of *Agaricus bisporus*. *Mycologia*, **89**, 274–277.

Untereiner, W. A. (2000). *Capronia* and its anamorphs: exploring the value of morphological and molecular characters in the systematics of the *Herpotrichiellaceae*. *Studies in Mycology*, **45**, 141–149.

Untereiner, W. A., Scott, J. A., Naveau, F. A., *et al.* (2004). The Ajellomycetaceae, a new family of vertebrate-associated Onygenales. *Mycologia*, **96**, 812–821.

Urashima, A. S., Igirashi, S. & Kato, H. (1993). Host range, mating type, and fertility of *Pyricularia grisea* from wheat in Brazil. *Plant Disease*, **77**, 1211–1261.

Urban, A., Neuner-Plattner, I., Krisai-Greilhuber, I. & Haselwandter, K. (2004). Molecular studies on terricolous microfungi reveal novel anamorphs of two *Tuber* species. *Mycological Research*, **108**, 749–758.

Urban, M., Bhargava, T. & Hamer, J. E. (1999). An ATP-driven efflux pump is a novel pathogenicity factor in rice blast disease. *EMBO Journal*, **18**, 512–521.

Urbanus, J. F., van den Ende, H. & Koch, B. (1978). Calcium oxalate crystals in the wall of *Mucor mucedo*. *Mycologia*, **70**, 829–842.

Urquhart, E. J. & Punja, Z. K. (2002). Hydrolytic enzymes and antifungal compounds produced by *Tilletiopsis* species, phyllosphere yeasts that are antagonists of powdery mildews. *Canadian Journal of Microbiology*, **48**, 219–229.

Valent, B. (1997). The rice blast fungus, *Magnaporthe grisea*. In *The Mycota VB: Plant Relationships*, ed. G. C. Carroll & P. Tudzynski. Berlin: Springer-Verlag, pp. 37–54.

van Brummelen, J. (1967). A world monograph of the genera *Ascobolus* and *Saccobolus* (Ascomycetes, Pezizales). *Persoonia Supplement* 1.

van Brummelen, J. (1981). The operculate ascus and allied forms. In *Ascomycete Systematics. The Luttrellian Concept*, ed. D. R. Reynolds. New York: Springer-Verlag, pp. 27–48.

van den Bogart, H. G., van den Ende, G., van Loon, P. C. & van Griensven, L. J. (1993). Mushroom worker's lung: serologic reactions to thermophilic actinomycetes present in the air of compost tunnels. *Mycopathologia*, **122**, 21–28.

van der Auwera, G., de Baere, R., van de Peer, Y., *et al.* (1995). The phylogeny of the Hyphochytriomycota as deduced from ribosomal RNA sequences of *Hyphochytrium catenoides*. *Molecular Biology and Evolution*, **12**, 671–678.

Vanderhaegen, B., Neven, H., Coghe, S., *et al.* (2003). Bioflavoring and beer refermentation. *Applied Microbiology and Biotechnology*, **62**, 140–150.

van der Plaats-Niterink, A. J. (1981). Monograph of the genus *Pythium*. *Studies in Mycology*, **21**, 1−242. Baarn: Centraalbureau voor Schimmelcultures.

van der Plank, J. E. (1963). *Plant Diseases: Epidemics and Control*. New York: Academic Press.

van der Walt, J. P. (1970). The perfect and imperfect states of *Sporobolomyces salmonicolor*. *Antonie van Leeuwenhoek*, **36**, 49−55.

van Donk, E. & Bruning, K. (1992). *Ecology of Aquatic Fungi In and On Algae*. Bristol: Biopress Ltd.

van Eijik, G. W. & Roeymans, H. J. (1982). Distribution of carotenoids and sterols in relation to the taxonomy of *Taphrina* and *Protomyces*. *Antonie van Leeuwenhoek*, **48**, 257−264.

van Griensven, L. J. L. D., ed. (1988). *The Cultivation of Mushrooms*. Rustington, UK: Darlington Mushroom Laboratories Ltd.

Vanittanakom, N., Cooper, C. R., Fisher, M. C. & Sirisanthana, T. (2006). *Penicillium marneffei* infection and recent advances in the epidemiology and molecular biology aspects. *Clinical Microbiology Reviews*, **19**, 95−110.

Vánky, K. (1987). *Illustrated Genera of Smut Fungi*. Stuttgart: Gustav Fischer.

Vánky, K. (1994). *European Smut Fungi*. Stuttgart: Gustav Fischer.

Vánky, K. (1998). The genus *Microbotryum* (smut fungi). *Mycotaxon*, **67**, 33−60.

Vánky, K. (1999). The new classificatory system for smut fungi, and two new genera. *Mycotaxon*, **70**, 35−49.

van Leeuwen, G. C. M., Baayen, R. P., Holb, I. J. & Jeger, M. J. (2002). Distinction of the Asiatic brown rot fungus *Monilia polystroma* sp. nov. from *M. fructigena*. *Mycological Research*, **106**, 444−451.

van Oorschot, C. A. N. (1985). Taxonomy of the *Dactylaria* complex. V. A review of *Arthrobotrys* and allied genera. *Studies in Mycology*, **26**, 61−96.

van Rinsum, J., Klis, F. M. & van den Ende, H. (1991). Cell wall glucomannoproteins of *Saccharomyces cerevisiae* mnn9. *Yeast*, **7**, 717−726.

Vanterpool, C. T. (1959). Oospore germination in *Albugo candida*. *Canadian Journal of Botany*, **37**, 169−172.

van Vuuren, H. J. J. & Jacobs, C. J. (1992). Killer yeasts in the wine industry: a review. *American Journal of Enology and Viticulture*, **43**, 119−128.

Varga, J., Rigo, K., Toth, B., Teren, J. & Kozakiewicz, Z. (2003). Evolutionary relationships among *Aspergillus* species producing economically important myco-toxins. *Food Technology and Biotechnology*, **41**, 29−36.

Varitchak, B. (1931). Contribution à l'étude du dével-oppement des Ascomycètes. *Botaniste*, **23**, 1−123.

Vasiliauskas, R., Lygis, V., Thor, M. & Stenlid, J. (2004). Impact of biological (Rotstop) and chemical (urea) treatments on fungal community structure in freshly cut *Picea abies* stumps. *Biological Control*, **31**, 405−413.

Vasiliauskas, R., Larsson, K. H., Larsson, K. H. & Stenlid, J. (2005). Persistence and long-term impact of Rotstop biological control agent on mycodiversity in *Picea abies* stumps. *Biological Control*, **32**, 295−304.

Vaughan-Martini, A. & Martini, A. (1998a). *Schizosaccharomyces* Lindner. In *The Yeasts. A Taxonomic Study*, 4th edn, ed. C. P. Kurtzman & J. W. Fell. Amsterdam: Elsevier, pp. 391−4.

Vaughan-Martini, A. & Martini, A. (1998b). *Saccharomyces* Meyen ex Reess. In *The Yeasts: A Taxonomic Study*, 4th edn, ed. C. P. Kurtzman & J. W. Fell. Amsterdam: Elsevier, pp. 358−71.

Veenhuis, M., Nordbring-Hertz, B. & Harder, W. (1985). An electron microscopical analysis of capture and initial stages of penetration of nematodes by *Arthrobotrys oligospora*. *Antonie van Leeuwenhoek*, **51**, 385−398.

Veenhuis, M., van Wijk, C., Wyss, U., Nordbring-Hertz, B. & Harder, W. (1989). Significance of electron dense microbodies in trap cells of the nematophagous fungus *Arthrobotrys oligospora*. *Antonie van Leeuwenhoek*, **56**, 251−261.

Vellinga, E. C. (2004). Genera in the family *Agaricaceae*: evidence from nrITS and nrLSU sequences. *Mycological Research*, **108**, 354−377.

Vellinga, E. C., De Kok, R. P. J. & Bruns, T. D. (2003). Phylogeny and taxonomy of *Macrolepiota* (Agaricaceae). *Mycologia*, **95**, 442−456.

Verkley, G. J. M. (1992). Ultrastructure of the apical apparatus of asci in *Ombrophila violacea*, *Neobulgaria pura* and *Bulgaria inquinans* (Leotiales). *Persoonia*, **15**, 3−22.

Verkley, G. J. M. (1993). Ultrastructure of the ascus apical apparatus in ten species of Sclerotiniaceae. *Mycological Research*, **97**, 179−194.

Verkley, G. J. M. (1994). Ultrastructure of the ascus apical apparatus in *Leotia lubrica* and some Geoglossaceae (Leotiales, Ascomycotina). *Persoonia*, **15**, 405−430.

Verkley, G. J. M. (1996). Ultrastructure of the ascus in the genera *Lachnum* and *Trichopeziza* (Hyaloscyphaceae, Ascomycotina). *Nova Hedwigia*, **63**, 215−228.

Versele, M. & Thorner, J. (2005). Some assembly required: yeast septins provide the instruction manual. *Trends in Cell Biology*, **15**, 414−424.

Verstrepen, K. J., Derdelinckx, G., Verachtert, H. & Delvaux, F. R. (2003). Yeast flocculation: what brewers should know. *Applied Microbiology and Biotechnology*, **61**, 197–205.

Viani, R. (2002). Effect of processing on ochratoxin A (OTA) content of coffee. *Advances in Experimental Medicine and Biology*, **504**, 189–193.

Vida, T. A. & Emr, S. D. (1995). A new vital stain for visualizing vacuolar membrane dynamics and endocytosis in yeast. *Journal of Cell Biology*, **128**, 779–792.

Vilgalys, R. & Cubeta, M. A. (1994). Molecular systematics and population biology of *Rhizoctonia*. *Annual Review of Phytopathology*, **32**, 135–155.

Vilgalys, R. & Gonzalez, D. (1990). Organization of ribosomal DNA in the basidiomycete *Thanatephorus praticola*. *Current Genetics*, **18**, 277–280.

Vinogradov, E., Petersen, B. O., Duus, J. Ø. & Wasser, S. (2004). The structure of the glucuronoxylomannan produced by culinary–medicinal yellow brain mushroom (*Tremella mesenterica* Ritz.:Fr., Heterobasidiomycetes) grown as one cell biomass in submerged culture. *Carbohydrate Research*, **339**, 1483–1489.

Vitkov, L., Krautgartner, W. D., Hannig, M., Weitgasser, R. & Stoiber, W. (2002). *Candida* attachment to oral epithelium. *Oral Microbiology Immunology*, **17**, 60–64.

Voegele, R. T. & Mendgen, K. (2003). Rust haustoria: nutrient uptake and beyond. *New Phytologist*, **159**, 93–100.

Vogel, H. J. (1964). Distribution of lysine pathways among fungi: evolutionary implications. *American Naturalist*, **98**, 435–446.

Vogel, S. (2005). Living in a physical world. III. Getting up to speed. *Journal of Bioscience*, **30**, 303–312.

Vogel, J. & Somerville, S. (2002). Powdery mildew of *Arabidopsis*: a model system for host–parasite interactions. In *The Powdery Mildews. A Comprehensive Treatise*, ed. R. R. Bélanger, W. R. Bushnell, A. J. Dik & T. L. W. Carver. St. Paul, USA: APS Press, pp. 161–168.

Vogler, D. R. & Bruns, T. D. (1998). Phylogenetic relationships among the pine stem rust fungi (*Cronartium* and *Peridermium* spp.). *Mycologia*, **90**, 244–257.

Voglmayr, H. (2000). Die aero-aquatischen Pilze des Sauwaldgebietes. *Beiträge zur Naturkunde Oberösterreichs*, **9**, 705–728.

Voglmayr, H., Bonner, L. M. & Dick, M. W. (1999). Taxonomy and oogonial ultrastructure of a new aero-aquatic peronosporomycete (*Medusoides* gen. nov.; Pythiogetonaceae fam. nov.). *Mycological Research*, **103**, 591–606.

Volk, T. J. & Leonard, T. J. (1989). Experimental studies on morel. I. Heterokaryon formation between monoascosporous strains of *Morchella*. *Mycologia*, **81**, 523–531.

Volk, T. J. & Leonard, T. J. (1990). Cytology of the life cycle of *Morchella*. *Mycological Research*, **94**, 399–406.

von Arx, J. A. (1950). Über die Ascusform von *Cladosporium herbarum*. *Sydowia*, **4**, 320–324.

von Arx, J. A. (1957). Die Arten der Gattung *Colletotrichum* Cda. *Phytopathologische Zeitschrift*, **29**, 413–468.

von Arx, J. A. (1986). The ascomycete genus *Gymnoascus*. *Persoonia*, **13**, 173–183.

von Arx, J. A., Guarro, J. & Figueras, M. J. (1986). The Ascomycete genus *Chaetomium*. *Nova Hedwigia, Beihefte*, **84**, 1–117.

von Röpenack, E., Parr, A. & Schulze-Lefert, P. (1998). Structural analyses and dynamics of soluble and cell wall-bound phenolics in a broad spectrum resistance to the powdery mildew fungus in barley. *Journal of Biological Chemistry*, **273**, 9013–9022.

Voos, J. R. (1969). Morphology and life cycle of a new chytrid with aerial sporangia. *American Journal of Botany*, **56**, 898–909.

Wagner, T. & Fischer, M. (2001). Natural groups and a revised system for the European poroid Hymenochaetales (Basidiomycota) supported by nLSU rDNA sequence data. *Mycological Research*, **105**, 773–782.

Wahl, I., Anikster, Y., Manisterski, J. & Segal, A. (1984). Evolution at the center of origin. In *The Cereal Rusts 1: Origins, Specificity, Structure, and Physiology*, ed. W. R. Bushnell & A. P. Roelfs. Orlando: Academic Press, pp. 39–77.

Waid, J. S. (1954). Occurrence of aquatic hyphomycetes upon the root surface of beech grown in woodland soil. *Transactions of the British Mycological Society*, **37**, 420–421.

Wakefield, A. E., Peters, S. E., Banerji, S., *et al.* (1993). *Pneumocystis carinii* shows DNA homology with the ustomycetous red yeast fungi. *Molecular Microbiology*, **6**, 1903–1911.

Walker, G. M. (1998). *Yeast Physiology and Biotechnology*. New York: John Wiley.

Walker, J. (2004). *Claviceps phalaridis* in Australia: biology, pathology and taxonomy with a description of the new genus *Cepsiclava* (Hypocreales, Clavicipitaceae). *Australasian Plant Pathology*, **33**, 211–239.

Walker, L. B. (1920). Development of *Cyathus fascicularis*, *C. striatus*, and *Crucibulum vulgare*. *Botanical Gazette*, **70**, 1–24.

Walker, R. F., West, D. C., McLaughlin, S. B. & Amundsen, C. C. (1989). Growth, xylem pressure potential, and nutrient absorption of loblolly pine on a reclaimed surface mine as affected by an induced *Pisolithus tinctorius* infection. *Forest Science*, **35**, 569–581.

Walkey, D. G. A. & Harvey, R. (1966a). Studies of the ballistics of ascospores. *New Phytologist*, **65**, 59–74.

Walkey, D. G. A. & Harvey, R. (1966b). Spore discharge rhythms in pyrenomycetes. I. A survey of the periodicity of spore discharge in pyrenomycetes. *Transactions of the British Mycological Society*, **49**, 583–592.

Wallace, D. R., MacLeod, C. R., Sullivan, C. R., Tyrell, D. & De Lyzer, A. J. (1976). Induction of resting spore germination of *Entomophthora aphidis* by long-day light conditions. *Canadian Journal of Botany*, **54**, 1410–1418.

Wallander, H. & Söderström, B. (1999). *Paxillus*. In *Ectomycorrhizal Fungi. Key Genera in Profile*, ed. J. W. G. Cairney & S. M. Chambers. Berlin: Springer-Verlag, pp. 231–52.

Walther, V., Rexer, K. & Kost, G. (2001). The ontogeny of the fruit bodies of *Mycena stylobates*. *Mycological Research*, **105**, 723–733.

Wanderlei-Silva, D., Neto, E. R. & Hanlin, R. (2003). Molecular systematics of the Phyllachorales (Ascomycota, Fungi) based on 18S ribosomal DNA sequences. *Brazilian Archives of Biology and Technology*, **46**, 315–322.

Wang, C., Xing, J., Chin, C.-K. & Peters, J. (2002). Fatty acids with certain structural characteristics are potent inhibitors of germination and inducers of cell death of powdery mildew spores. *Physiological and Molecular Plant Pathology*, **61**, 151–161.

Wang, Y. & Hall, I. R. (2004). Edible ectomycorrhizal mushrooms: challenges and achievements. *Canadian Journal of Botany*, **82**, 1063–1073.

Wang, Y., Aisen, P. & Casadevall, A. (1995). *Cryptococcus neoformans* melanin and virulence: mechanism of action. *Infection and Immunity*, **63**, 3131–3136.

Wang, Z., Binder, M., Dai, Y.-C. & Hibbett, D. S. (2004). Phylogenetic relationships of *Sparassis* inferred from nuclear and mitochondrial ribosomal DNA and RNA polymerase sequences. *Mycologia*, **96**, 1015–1029.

Warcup, J. H. & Talbot, P. H. B. (1962). Ecology and identity of mycelia isolated from soil. *Transactions of the British Mycological Society*, **45**, 495–518.

Warcup, J. H. & Talbot, P. H. B. (1966). Perfect states of some rhizoctonias. *Transactions of the British Mycological Society*, **49**, 427–435.

Ward, E. & Adams, M. J. (1998). Analysis of ribosomal DNA sequences of *Polymyxa* species and related fungi and the development of genus- and species-specific PCR primers. *Mycological Research*, **102**, 965–974.

Ward, H. M. (1882). Researches on the life-history of *Hemileia vastatrix*. *Journal of the Linnean Society*, **19**, 299–335.

Ward, H. M. (1888). A lily disease. *Annals of Botany*, **2**, 319–382.

Warren, A. C. & Colhoun, J. (1975). Viability of sporangia of *Phytophthora infestans* in relation to drying. *Transactions of the British Mycological Society*, **64**, 73–78.

Wasser, S. P. (2002). Medicinal mushrooms as a source of antitumor and immunomodulating polysaccharides. *Applied Microbiology and Biotechnology*, **60**, 258–274.

Wasson, R. G. (1968). *Soma: Divine Mushroom of Immortality*. New York: Harcourt Brace Jovanovich.

Wastie, R. L. (1991). Breeding for resistance. In *Phytophthora infestans, the Cause of Late Blight of Potato*, ed. D. S. Ingram & P. H. Williams. London: Academic Press, pp. 193–224.

Watanabe, T. (1994). Two new species of homothallic *Mucor* in Japan. *Mycologia*, **86**, 691–695.

Watanabe, T. (1997). Stimulation of perithecium and ascospore production in *Sordaria fimicola* by *Armillaria* and various fungal species. *Mycological Research*, **101**, 1190–1194.

Watanabe, T., Watanabe, Y., Fukatsu, T. & Kurane, R. (2001). *Mortierella tsukubaensis* sp. nov. from Japan, with a key to the homothallic species. *Mycological Research*, **105**, 506–509.

Waterhouse, G. M. (1963). Key to the species of *Phytophthora* de Bary. *Mycological Papers*, **92**, 1–22. Kew: Commonwealth Mycological Institute.

Waterhouse, G. M. (1967). Key to *Pythium* Pringsheim. *Mycological Papers*, **109**, 1–15. Kew: Commonwealth Mycological Institute.

Waterhouse, G. M. (1968). The genus *Pythium* Pringsheim. Diagnoses (or descriptions) and figures from the original papers. *Mycological Papers*, **110**, 1–71. Kew: Commonwealth Mycological Institute.

Waterhouse, G. M. (1970). The genus *Phytophthora* de Bary. Diagnoses (or descriptions) and figures from the original papers. *Mycological Papers*, **120**, 1–59. Kew: Commonwealth Mycological Institute.

Waterhouse, G. M. (1973). Plasmodiophoromycetes. In *The Fungi, An Advanced Treatise* **IVB**: *A Taxonomic Review with Keys*, ed. G. C. Ainsworth, F. K. Sparrow & A. S. Sussman. New York: Academic Press, pp. 75–82.

Waters, H., Butler, R.D. & Moore, D. (1975a). Structure of aerial and submerged sclerotia of *Coprinus lagopus*. *New Phytologist*, **74**, 199–205.

Waters, H., Butler, R.D. & Moore, D. (1975b). Morphogenesis of aerial sclerotia of *Coprinus lagopus*. *New Phytologist*, **74**, 207–213.

Waters, S.D. & Callaghan, A.A. (1999). Conidium germination in co-occurring *Conidiobolus* and *Basidiobolus* in relation to their ecology. *Mycological Research*, **103**, 1259–1269.

Watkinson, S.C. (1979). Growth of rhizomorphs, mycelial strands, coremia and sclerotia. In *Fungal Walls and Hyphal Growth*, ed. J.H. Burnett & A.P.J. Trinci. Cambridge: Cambridge University Press, pp. 93–113.

Watling, R. (1971). Polymorphism in *Psilocybe merdaria*. *New Phytologist*, **70**, 307–326.

Watling, R. & Harper, D.B. (1998). Chloromethane production by wood-rotting fungi and an estimate of the global flux to the atmosphere. *Mycological Research*, **102**, 769–787.

Watson, A.K., Gressel, J., Sharon, A. & Dinoor, A. (2000). *Colletotrichum* strains for weed control. In *Colletotrichum. Host Specificity, Pathology, and Host-Pathogen Interaction*, ed. D. Prusky, S. Freeman & M.B. Dickman. St. Paul, USA: APS Press, pp. 245–265.

Watson, I.A. & Luig, N.H. (1958). Somatic hybridization in *Puccinia graminis* var. *tritici*. *Proceedings of the Linnean Society of New South Wales*, **83**, 190–195.

Webb, J. & Theodorou, M.K. (1988). A rumen anaerobic fungus of the genus *Neocallimastix*: ultrastructure of the polyflagellate zoospore and young thallus. *BioSystems*, **21**, 393–401.

Webb, J. & Theodorou, M.K. (1991). *Neocallimastix hurleyensis* an anaerobic fungus from the bovine rumen. *Canadian Journal of Botany*, **69**, 1220–1224.

Webber, J.F. & Brasier, C.M. (1984). The transmission of Dutch elm disease: a study of the processes involved. In *Invertebrate–Microbial Interactions*, ed. J.M. Anderson, A.D.M. Rayner & D.W.H. Walton. Cambridge: Cambridge University Press, pp. 271–306.

Webber, J.F. & Gibbs, J.N. (1989). Insect dissemination of fungal pathogens of trees. In *Insect–Fungus Interactions*, ed. N. Wilding, N.M. Collins, P.M. Hammond & J.F. Webber. London: Academic Press, pp. 161–93.

Weber, R.W.S. (2002). Vacuoles and the fungal lifestyle. *Mycologist*, **16**, 10–20.

Weber, R.W.S. & Davoli, P. (2003). Teaching techniques for mycology: 20. Astaxanthin, a carotenoid of biotechnological importance from yeast and salmonid fish. *Mycologist*, **17**, 30–34.

Weber, R.W.S. & Pitt, D. (2001). Filamentous fungi – growth and physiology. In *Applied Mycology and Biotechnology 1: Agriculture and Food Production*, ed. G.G. Khachatourians & D.K. Arora. Amsterdam: Elsevier, pp. 13–54.

Weber, R.W.S. & Tribe, H.T. (2003). Oil as a substrate for *Mortierella* species. *Mycologist*, **17**, 134–139.

Weber, R.W.S. & Webster, J. (1998a). Teaching techniques for mycology: 5. *Basidiobolus ranarum*. *Mycologist*, **12**, 148–150.

Weber, R.W.S. & Webster, J. (1998b). Stimulation of growth and reproduction of *Sphaeronaemella fimicola* by other coprophilous fungi. *Mycological Research*, **102**, 1055–1061.

Weber, R.W.S. & Webster, J. (2000a). Teaching techniques for mycology: 9. *Olpidium* and *Rhizophlyctis*. *Mycologist*, **14**, 17–20.

Weber, R.W.S. & Webster, J. (2000b). Teaching techniques for mycology: 12. A demonstration of the teleomorph–anamorph connection using *Pleospora herbarum*. *Mycologist*, **14**, 171–173.

Weber, R.W.S. & Webster, J. (2001a). Teaching techniques for mycology: 13. Functioning of cleistothecia in *Phyllactinia guttata*. *Mycologist*, **15**, 26–30.

Weber, R.W.S. & Webster, J. (2001b). Teaching techniques for mycology: 14. Mycorrhizal infection of orchid seedlings in the laboratory. *Mycologist*, **15**, 55–59.

Weber, R.W.S. & Webster, J. (2003). Teaching techniques for mycology: 21. *Sclerotinia*, *Botrytis* and *Monilia* (Ascomycota, Leotiales, Sclerotiniaceae). *Mycologist*, **17**, 111–115.

Weber, R.W.S., Wakley, G.E. & Pitt, D. (1998). Histochemical and ultrastructural characterization of fungal mitochondria. *Mycologist*, **12**, 174–179.

Weber, R.W.S., Webster, J., Barnes, J.C. & Pitt, D. (1999). Teaching techniques for mycology: 7. Zoospore discharge by *Pythium* and *Phytophthora*. *Mycologist*, **13**, 117–119.

Weber, R.W.S., Wakley, G.E., Thines, E. & Talbot, N.J. (2001). The vacuole as central element of the lytic system and sink for lipid droplets in maturing appressoria of *Magnaporthe grisea*. *Protoplasma*, **216**, 101–112.

Webster, J. (1952). Spore projection in the Hyphomycete *Nigrospora sphaerica*. *New Phytologist*, **51**, 229–235.

Webster, J. (1964). Culture studies on *Hypocrea* and *Trichoderma* species. I. Comparison of the perfect and imperfect states of *Hypocrea gelatinosa, H. rufa* and *Hypocrea* sp. 1. *Transactions of the British Mycological Society*, **47**, 75–96.

Webster, J. (1966). Spore projection in *Epicoccum* and *Arthrinium*. *Transactions of the British Mycological Society*, **49**, 339–343.

Webster, J. (1980). *Introduction to Fungi*, 2nd edn. Cambridge: Cambridge University Press.

Webster, J. (1987). Convergent evolution and the functional significance of spore shape in aquatic and semi-aquatic fungi. In *Evolutionary Biology of the Fungi*, ed. A. D. M. Rayner, C. M. Brasier & D. Moore. Cambridge: Cambridge University Press, pp. 191–201.

Webster, J. (1992). Anamorph–teleomorph relationships. In *The Ecology of Aquatic Hyphomycetes*, ed. F. Bärlocher. Berlin: Springer-Verlag, pp. 99–117.

Webster, J., ed. (2006a). *Mycology Vol. 1*. Interactive DVD-ROM. Göttingen: IWF Wissen und Medien.

Webster, J., ed. (2006b). *Mycology Vol. 2*. Interactive DVD-ROM. Göttingen: IWF Wissen und Medien.

Webster, J. & Chien, C.-Y. (1990). Ballistospore discharge. *Transactions of the Mycological Society of Japan*, **31**, 301–305.

Webster, J. & Davey, R. A. (1984). Sigmoid conidial shape in aquatic fungi. *Transactions of the British Mycological Society*, **83**, 43–52.

Webster, J. & Dennis, C. (1967). The mechanism of sporangial discharge in *Pythium middletonii*. *New Phytologist*, **66**, 307–313.

Webster, J. & Descals, E. (1981). Morphology, distribution and ecology of conidial fungi in fresh-water habitats. In *Biology of Conidial Fungi 1*, ed. G. T. Cole & B. Kendrick. New York: Academic Press, pp. 295–355.

Webster, J. & Hard, T. (1998a). Alternation of generations in *Allomyces* (Blastocladiales). VHS Videotape 2656. Göttingen: Institut für den Wissenschaftlichen Film.

Webster, J. & Hard, T. (1998b). *Ballistospore Discharge in Basidiomycetes*. VHS Videotape C1993. Göttingen: Institut für den Wissenschaftlichen Film.

Webster, J. & Hard, T. (1999). *Pilobolus*, a specialized coprophilous fungus. VHS Videotape C1954. Göttingen: Institut für den Wissenschaftlichen Film.

Webster, J. & Towfik, F. H. (1972). Sporulation of aquatic hyphomycetes in relation to aeration. *Transactions of the British Mycological Society*, **59**, 353–364.

Webster, J. & Weber, R. W. S. (1997). Teaching techniques for mycology: 1. The bird's nest fungus, *Cyathus stercoreus*. *Mycologist*, **11**, 103–105.

Webster, J. & Weber, R. W. S. (1999). Teaching techniques for mycology: 8. Fruiting cultures of *Sphaerobolus stellatus*. *Mycologist*, **13**, 151–153.

Webster, J. & Weber, R. W. S. (2000). Rhizomorphs and perithecial stromata of *Podosordaria tulasnei*. *Mycologist*, **14**, 41–44.

Webster, J. & Weber, R. W. S. (2001). Teaching techniques for mycology: 15. Fertilization and apothecium development in *Pyronema domesticum* and *Ascobolus furfuraceus*. *Mycologist*, **15**, 126–131.

Webster, J. & Weber, R. W. S. (2004). Teaching techniques for mycology: 23. Eclosion of *Hypoxylon fragiforme* ascospores as a prelude to germination. *Mycologist*, **18**, 170–173.

Webster, J., Rifai, M. A. & El-Abyad, M. S. (1964). Culture observations on some discomycetes from burnt ground. *Transactions of the British Mycological Society*, **47**, 445–454.

Webster, J., Sanders, P. F. & Descals, E. (1978). Tetraradiate propagules in two species of *Entomophthora*. *Transactions of the British Mycological Society*, **70**, 472–479.

Webster, J., Davey, R. A., Duller, G. A. & Ingold, C. T. (1984a). Ballistospore discharge in *Itersonilia perplexans*. *Transactions of the British Mycological Society*, **82**, 13–29.

Webster, J., Davey, R. A. & Ingold, C. T. (1984b). Origin of the liquid in Buller's drop. *Transactions of the British Mycological Society*, **83**, 524–527.

Webster, J., Proctor, M. C. F., Davey, R. A. & Duller, G. A. (1988). Measurement of the electrical charge on some basidiospores and an assessment of two possible mechanisms of basidiospore propulsion. *Transactions of the British Mycological Society*, **91**, 193–203.

Webster, J., Davey, R. A. & Turner, J. C. R. (1989). Vapour as the source of water in Buller's drop. *Mycological Research*, **93**, 297–302.

Webster, J., Davey, R. A., Smirnoff, N., *et al.* (1995). Mannitol and hexoses are components of Buller's drop. *Mycological Research*, **99**, 833–838.

Webster, J., Pitt, D., Barnes, J. C. & Weber, R. W. S. (1999). Teaching techniques for mycology: 6. *Puccinia graminis*, cause of black stem rust. *Mycologist*, **13**, 15–18.

Webster, M. A. & Dixon, G. R. (1991). Boron, pH and inoculum concentration influencing colonization by *Plasmodiophora brassicae*. *Mycological Research*, **95**, 74–79.

Wedin, M., Wiklund, E., Crewe, A., *et al.* (2005). Phylogenetic relationships of *Lecanoromycetes* (Ascomycota) as revealed by analyses of mtSSU and nLSU rDNA sequence data. *Mycological Research*, **109**, 159–172.

Weeks, R. J., Padhye, A. A. & Ajello, L. (1985). *Histoplasma capsulatum* variety *farciminosum*: a new combination for *Histoplasma farciminosum*. *Mycologia*, **77**, 964–970.

Wehmeyer, L. E. (1926). A biologic and phylogenetic study of stromatic Sphaeriales. *American Journal of Botany*, **13**, 575–645.

Wehmeyer, L. E. (1955). Development of *Pleospora armeriae* of the *Pleospora herbarum* complex. *Mycologia*, **47**, 821–834.

Weiergang, I., Lyngs Jørgensen, H. J., Møller, I. M., Friis, P. & Smedegaard-Petersen, V. (2002). Correlation between sensitivity of barley to *Pyrenophora teres* toxins and susceptibility to the fungus. *Physiological and Molecular Plant Pathology*, **60**, 121–129.

Weijer, C. J. (2004). *Dictyostelium* morphogenesis. *Current Opinion in Genetics and Development*, **14**, 392–398.

Weiler, K. S. & Broach, J. R. (1992). Donor locus selection during *Saccharomyces cerevisiae* mating type interconversion responds to distal regulatory signals. *Genetics*, **132**, 929–942.

Weir, A. & Blackwell, M. (2001). Molecular data support the Laboulbeniales as a separate class of Ascomycota, the Laboulbeniomycetes. *Mycological Research*, **105**, 1182–1190.

Weiss, M. & Oberwinkler, F. (2001). Phylogenetic relationships in *Auriculariales* and related groups – hypotheses derived from nuclear ribosomal DNA sequences. *Mycological Research*, **105**, 403–415.

Weitzman, I. & Summerbell, R. C. (1995). The dermatophytes. *Clinical Microbiology Reviews*, **8**, 240–259.

Welch, A. M. & Bultmann, T. L. (1993). Natural release of *Epichloe typhina* ascospores and its temporal relationship to fly parasitism. *Mycologia*, **85**, 756–763.

Weller, D. M., Raajmakers, J. M., McSpadden Gardener, B. B. & Thomashow, L. S. (2002). Microbial populations responsible for specific soil suppressiveness to plant pathogens. *Annual Review of Phytopathology*, **40**, 309–348.

Wells, K. (1965). Ultrastructural features of developing and mature basidia of *Schizophyllum commune*. *Mycologia*, **57**, 236–261.

Wells, K. (1970). Light and electron microscopic studies of *Ascobolus stercorarius*. I. Nuclear divisions in the ascus. *Mycologia*, **62**, 761–790.

Wells, K. (1972). Light and electron microscopic studies of *Ascobolus stercorarius*. II. Ascus and ascospore ontogeny. *University of California Publications in Botany*, **62**, 1–93.

Wells, K. (1987). Comparative morphology, intracompatibility, and interincompatibility of several species of *Exidiopsis* (Exidiaceae). *Mycologia*, **79**, 274–288.

Wells, K. (1994). Jelly fungi, then and now! *Mycologia*, **86**, 18–48.

Wells, K. & Bandoni, R. J. (2001). Heterobasidiomycetes. In *The Mycota VIIB: Systematics and Evolution*, ed. D. J. McLaughlin, E. G. McLaughlin & P. A. Lemke. Berlin: Springer-Verlag, pp. 85–120.

Weresub, L. K. & LeClair, P. M. (1971). On *Papulaspora* and bulbilliferous basidiomycetes *Burgoa* and *Minimedusa*. *Canadian Journal of Botany*, **49**, 2203–2213.

Wessels, J. G. H. (1993a). Wall growth, protein excretion and morphogenesis in fungi. *New Phytologist*, **123**, 397–413.

Wessels, J. G. H. (1993b). Fruiting in the higher fungi. *Advances in Microbial Physiology*, **34**, 147–202.

Wessels, J. G. H. (1994). Development of fruit bodies in Homobasidiomycetes. In *The Mycota I: Growth, Differentiation and Sexuality*, ed. J. G. H. Wessels & F. Meinhardt. Berlin: Springer-Verlag, pp. 351–66.

Wessels, J. G. H. (1997). Hydrophobins: proteins that change the nature of the fungal surface. *Advances in Microbial Physiology*, **38**, 1–45.

Wessels, J. G. H. (2000). Hydrophobins, unique fungal proteins. *Mycologist*, **14**, 153–159.

Wessels, J. G. H. & Sietsma, J. H. (1981). Fungal cell walls: a survey. In *Encyclopedia of Plant Physiology, New Series 13B*, ed. W. Tanner & F. A. Loewus. New York: Springer Verlag, pp. 352–94.

Wessels, J. G. H., Kreger, D. R., Marchant, R., Regensburg, B. A. & de Vries, O. M. H. (1972). Chemical and morphological characterization of the hyphal wall surface of the basidiomycete *Schizophyllum commune*. *Biochimica et Biophysica Acta*, **273**, 346–358.

Wessels, J. G. H., Mol, P. C., Sietsma, J. H. & Vermeulen, C. A. (1990). Wall structure, wall growth and fungal cell morphogenesis. In *Biochemistry of Cell Walls and Membranes in Fungi*, ed. P. J. Kuhn, A. P. J. Trinci, M. J. Jung, M. W. Goosey & L. G. Copping. Berlin: Springer-Verlag, pp. 81–95.

West, J. S. (2000). Chemical control of *Armillaria*. In *Armillaria Root Rot: Biology and Control of Honey Fungus*, ed. R. T. V. Fox. Andover, UK: Intercept Ltd, pp. 173–82.

Western, J. H. & Cavett, J. J. (1959). The choke disease of cocksfoot (*Dactylis glomerata*) caused by *Epichloe typhina* (Fr.) Tul. *Transactions of the British Mycological Society*, **42**, 298–307.

Westfall, P. J. & Momany, M. (2002). *Aspergillus nidulans* septin AspB plays pre- and postmitotic roles in septum, branch, and conidiophore development. *Molecular Biology of the Cell*, **13**, 110–118.

Whalley, A. J. S. (1985). The Xylariaceae: some ecological considerations. *Sydowia*, **38**, 369–382.

Whalley, A. J. S. (1987). *Xylaria* inhabiting fallen fruits. *Agarica*, **8**, 68–72.

Whalley, A. J. S. (1996). Presidential address. The xylariaceous way of life. *Mycological Research*, **100**, 897–922.

Whalley, A. J. S. & Greenhalgh, G. N. (1973). Numerical taxonomy of *Hypoxylon*. II. A key to the identification of the British species of *Hypoxylon*. *Transactions of the British Mycological Society*, **61**, 455–459.

Whetzel, H. H. (1945). A synopsis of the genera and species of the Sclerotiniaceae, a family of stromatic inoperculate Discomycetes. *Mycologia*, **37**, 648–714.

Whetzel, H. H. (1946). The cypericolous and juncicolous species of *Sclerotinia*. *Farlowia*, **2**, 385–437.

Whisler, H. C. (1987). On the isolation and culture of water molds: the Blastocladiales and Monoblepharidales. In *Zoosporic Fungi in Teaching and Research*, ed. M. S. Fuller & A. Jaworski. Athens, USA: Southeastern Publishing Corporation, pp. 121–124.

Whisler, H. C. & Marek, L. E. (1987). *Monoblepharis macrandra*. In *Zoosporic Fungi in Teaching and Research*, ed. M. S. Fuller & A. Jaworski. Athens, USA: Southeastern Publishing Corporation, pp. 54–55.

Whisler, H. C., Zebold, S. L. & Shemanchuk, J. A. (1975). Life cycle of *Coelomomyces psophorae*. *Proceedings of the National Academy of Sciences of the USA*, **72**, 693–696.

White, J. A., Calvert, O. H. & Brown, M. F. (1973). Ultrastructure of the conidia of *Helminthosporium maydis*. *Canadian Journal of Botany*, **51**, 2006–2008.

White, J. F. (1988). Endophyte–host associations in forage grasses. XI. A proposal concerning origin and evolution. *Mycologia*, **80**, 442–446.

White, J. F. (1993). Endophyte–host associations in forage grasses. XIX. A systematic study of some sympatric species of *Epichloe* in England. *Mycologia*, **85**, 444–455.

White, J. F. (1997). Perithecial ontogeny in the fungal genus *Epichloe*: an examination of the clavicipitalean centrum. *American Journal of Botany*, **84**, 170–178.

White, J. F. & Bultman, T. L. (1987). Endophyte–host associations in forage grasses. VIII. Heterothallism in *Epichloë typhina*. *American Journal of Botany*, **74**, 1716–1721.

White, J. F., Morrow, A. C., Morgan-Jones, G. & Chambless, D. A. (1991). Endophyte–host associations in forage grasses. XIV. Primary stromata formation and seed transmission in *Epichloe typhina*: developmental and regulatory aspects. *Mycologia*, **83**, 72–81.

White, J. F., Martin, T. I. & Cabral, D. (1996). Endophyte associations in grasses. XXII. Conidia formation by *Acremonium* endophytes on the phylloplanes of *Agrostis hiemalis* and *Poa rigidifolia*. *Mycologia*, **88**, 174–178.

White, T. J., Bruns, T., Lee, S. & Taylor, J. (1990). Amplification and direct sequencing of fungal ribosomal RNA genes for phylogenetics. In *PCR Protocols: A Guide to Methods and Applications*, ed. M. A. Innis, D. H. Gelfand, J. J. Sninsky & T. J. White. New York: Academic Press, pp. 315–22.

Whiteside, W. C. (1957). Perithecial initials of *Chaetomium*. *Mycologia*, **49**, 420–425.

Whiteside, W. C. (1961). Morphological studies in the Chaetomiaceae. I. *Mycologia*, **53**, 512–523.

Whitney, K. D. & Arnott, H. J. (1986). Morphology and development of calcium oxalate deposits in *Gilbertella persicaria* (Mucorales). *Mycologia*, **78**, 42–51.

Whittaker, R. H. (1969). New concepts of kingdoms of organisms. *Science*, **163**, 150–160.

Wickes, B. L., Mayorga, M. E., Edman, U. & Edman, J. C. (1996). Dimorphism and haploid fruiting in *Cryptococcus neoformans*: association with the α-mating type. *Proceedings of the National Academy of Sciences of the USA*, **93**, 7327–7331.

Wicklow, D. T. (1979). Hair ornamentation and predator defence in *Chaetomium*. *Transactions of the British Mycological Society*, **72**, 107–110.

Wicklow, D. T., Langie, R., Crabtree, S. & Detroy, R. W. (1984). Degradation of lignocellulose in wheat straw *versus* hardwood by *Cyathus* and related species (Nidulariaceae). *Canadian Journal of Microbiology*, **30**, 632–636.

Wieland, T. (1996). Toxins and psychoactive compounds from mushrooms. In *The Mycota* **VI**: *Human and Animal Relationships*, ed. D. H. Howard & J. D. Miller. Berlin: Springer-Verlag, pp. 229–48.

Wieland, T. & Faulstich, H. (1991). Fifty years of amanitin. *Experientia*, **47**, 1186–1193.

Wilding, N. & Lauckner, F. B. (1974). *Entomophthora* infecting wheat bulb fly at Rothamsted, Hertfordshire. *Annals of Applied Biology*, **76**, 161–170.

Willetts, H. J. (1969). Structure of the outer surfaces of sclerotia of certain fungi. *Archiv für Mikrobiologie*, **69**, 48–53.

Willetts, H. J. (1971). The survival of fungal sclerotia under adverse environmental conditions. *Biological Reviews*, **46**, 387–407.

Willetts, H. J. (1972). The morphogenesis and possible evolutionary origins of fungal sclerotia. *Biological Reviews*, **47**, 515–536.

Willetts, H. J. & Bullock, S. (1992). Developmental biology of sclerotia. *Mycological Research*, **96**, 801–816.

Willetts, H. J. & Wong, J. A.-L. (1980). The biology of *Sclerotinia sclerotiorum*, *S. trifoliorum*, and *S. minor* with emphasis on specific nomenclature. *Botanical Review*, **46**, 101–165.

Williams, E. N. D., Todd, N. K. & Rayner, A. D. M. (1981). Spatial development of populations of *Coriolus versicolor*. *New Phytologist*, **89**, 307–319.

Williams, M. A. J., Beckett, A. & Read, N. D. (1985). Ultrastructural aspects of fruit body differentiation in *Flammulina velutipes*. In *Developmental Biology of Higher Fungi*, ed. D. Moore, L. A. Casselton, D. A. Wood & J. C. Frankland. Cambridge: Cambridge University Press, pp. 429–50.

Williams, P. G. (1984). Obligate parasitism and axenic culture. In *The Cereal Rusts* **1**: *Origins, Specificity, Structure, and Physiology*, ed. W. R. Bushnell & A. P. Roelfs. Orlando: Academic Press, pp. 399–430.

Williams, P. G., Scott, K. J. & Kuhl, J. L. (1966). Vegetative growth of *Puccinia graminis* f. sp. *tritici in vitro*. *Phytopathology*, **56**, 1418–1419.

Williams, R. H. & Fitt, B. D. L. (1999). Differentiating A and B groups of *Leptosphaeria maculans*, causal agent of stem canker (blackleg) of oilseed rape. *Plant Pathology*, **48**, 161–175.

Williams, R. H., Whipps, J. M. & Cooke, R. C. (1998). Role of soil mesofauna in dispersal of *Coniothyrium minitans*: mechanisms of transmission. *Soil Biology and Biochemistry*, **30**, 1937–1945.

Williams, R. J. (1984). Downy mildews of tropical cereals. *Advances in Plant Pathology*, **3**, 1–103.

Willocquet, L. & Clerjeau, M. (1998). An analysis of the effects of environmental factors on conidial dispersal of *Uncinula necator* (grape powdery mildew) in vineyards. *Plant Pathology*, **47**, 227–233.

Willoughby, L. G. (1962). The fruiting behaviour and nutrition of *Cladochytrium replicatum* Karling. *Annals of Botany* N.S., **26**, 13–36.

Willoughby, L. G. (1994). *Fungi and Fish Diseases*. Stirling: Pisces Press.

Willoughby, L. G. (1998a). *Saprolegnia polymorpha* sp. nov., a fungal parasite on Koi carp in the U.K. *Nova Hedwigia*, **66**, 507–511.

Willoughby, L. G. (1998b). A quantitative ecological study on the monocentric soil chytrid, *Rhizophlyctis rosea*, in Provence. *Mycological Research*, **102**, 1338–1342.

Willoughby, L. G. (2001). The activity of *Rhizophlyctis rosea* in soil: some deductions from laboratory observations. *Mycologist*, **15**, 113–117.

Willoughby, L. G. & Pickering, A. D. (1977). Viable Saprolegniaceae spores on the epidermis of the salmonid fish *Salmo trutta* and *Salvelinus alpinus*. *Transactions of the British Mycological Society*, **68**, 91–95.

Willoughby, L. G. & Roberts, R. J. (1992). Towards strategic use of fungicides against *Saprolegnia parasitica* in salmonid fish hatcheries. *Journal of Fish Diseases*, **15**, 1–13.

Wilson, C. M. (1952). Meiosis in *Allomyces*. *Bulletin of the Torrey Botanical Club*, **79**, 139–160.

Wilson, C. M. & Flanagan, P. W. (1968). The life cycle and cytology of *Brachyallomyces*. *Canadian Journal of Botany*, **46**, 1361–1367.

Wilson, I. M. (1952). The ascogenous hyphae of *Pyronema confluens*. *Annals of Botany* N.S., **16**, 321–339.

Wilson, J. P. & Hanna, W. W. (1998). Smut resistance and grain yield of pearl millet hybrids near isogenic at the *Tr* locus. *Crop Science*, **38**, 649–651.

Wilson, M. & Henderson, D. M. (1966). *British Rust Fungi*. Cambridge: Cambridge University Press.

Wilson, R. W. & Beneke, E. S. (1966). Basidiospore germination in *Calvatia gigantea*. *Mycologia*, **58**, 328–332.

Wingfield, B. D., Ericson, L., Szaro, T. & Burdon, J. J. (2004). Phylogenetic patterns in the Uredinales. *Australasian Plant Pathology*, **33**, 327–335.

Wingfield, M. J., Kendrick, W. B. & Schalk van Wyk, P. (1991). Analysis of conidium ontogeny in anamorphs of *Ophiostoma*: *Pesotum* and *Phialographium* are synonyms of *Graphium*. *Mycological Research*, **95**, 1328–1333.

Winka, K., Eriksson, O. E. & Bång, A. (1998). Molecular evidence for recognizing the Chaetothyriales. *Mycologia*, **90**, 822–830.

Winner, M., Giménez, A., Schmidt, H., *et al.* (2004). Unusual pulvinic acid dimers from the common fungi *Scleroderma citrinum* (common earthball) and *Chalciporus piperatus* (peppery bolete). *Angewandte Chemie International Edition*, **43**, 1883–1886.

Winterstein, D. (2001). *Cordyceps* (Kernkeulen), Gourmets in Pilzreich. *Der Tintling*, **6**, 24–30.

Win-Tin, & Dick, M. W. (1975). Cytology of Oomycetes. Evidence for meiosis and multiple chromosome associations in Saprolegniaceae and Pythiaceae, with an introduction to the cytotaxonomy of *Achlya* and *Pythium*. *Archives of Microbiology*, **105**, 283–293.

Wirth, V. (1995a). *Die Flechten Baden-Württembergs Teil 1*. Stuttgart: Eugen Ulmer.

Wirth, V. (1995b). *Die Flechten Baden-Württembergs Teil 2*. Stuttgart: Eugen Ulmer.

Witthuhn, R.C., Wingfield, B.D., Wingfield, M.J., Wolfaardt, M. & Harrington, T.C. (1998). Monophyly of the conifer species in the *Ceratocystis coerulescens* complex based on DNA sequence data. *Mycologia*, **90**, 96–101.

Woese, C.R. (1987). Bacterial evolution. *Microbiological Reviews*, **51**, 221–271.

Wolf, F.T. (1981). The biology of *Entomophthora*. *Nova Hedwigia*, **35**, 553–599.

Wolf, G. (1982). Physiology and biochemistry of spore germination. In *The Rust Fungi*, ed. K.J. Scott & A.K. Chakravorty. London: Academic Press, pp. 151–178.

Wolfe, L.M. & Rissler, L.J. (1999). Reproductive consequences of a gall-inducing fungal pathogen (*Exobasidium vaccinii*) on *Rhododendron calendulaceum* (Ericaceae). *Canadian Journal of Botany*, **77**, 1454–1459.

Wolpert, T.J., Dunkle, L.D. & Cuiffetti, L.M. (2002). Host-selective toxins and avirulence dominants: what's in a name? *Annual Review of Phytopathology*, **40**, 251–285.

Wong, G.J. (1993). Mating and fruiting studies of *Auricularia delicata* and *A. fuscosuccinea*. *Mycologia*, **85**, 187–194.

Wong, G.J. & Wells, K. (1987). Comparative morphology, compatibility, and interfertility of *Auricularia cornea*, *A. polytricha*, and *A. tenuis*. *Mycologia*, **79**, 847–856.

Wong, G.J., Wells, K. & Bandoni, R.J. (1985). Interfertility and comparative morphological studies of *Tremella mesenterica*. *Mycologia*, **77**, 36–49.

Wong, S., Fares, M.A., Zimmermann, W., Butler, G. & Wolfe, K.H. (2003). Evidence from comparative genomics for a complete sexual cycle in the 'asexual' pathogenic yeast *Candida glabrata*. *Genome Biology*, **4**, R10.

Wood, H.A. & Bozarth, R.F. (1973). Heterokaryon transfer of viruslike particles and a cytoplasmically inherited determinant in *Ustilago maydis*. *Phytopathology*, **63**, 1019–1021.

Wood, S.N. & Cooke, R.C. (1986). Effect of *Piptocephalis* species on growth and sporulation of *Pilaira anomala*. *Transactions of the British Mycological Society*, **86**, 672–674.

Wood, S.N. & Cooke, R.C. (1987). Nutritional competence of *Pilaira anomala* in relation to exploitation of faecal resource units. *Transactions of the British Mycological Society*, **88**, 247–255.

Woodham-Smith, C. (1962). *The Great Hunger. Ireland 1845–1849*. London: Hamish Hamilton.

Woods, J.P. (2002). *Histoplasma capsulatum* molecular genetics, pathogenesis, and responsiveness to its environment. *Fungal Genetics and Biology*, **35**, 81–97.

Woodward, R.C. (1927). Studies on *Podosphaera leucotricha* (Ell. & Ev.) Salm. I. The mode of perennation. *Transactions of the British Mycological Society*, **12**, 173–204.

Woronin, M. (1878). *Plasmodiophora brassicae*. Urheber der Kohlpflanzen-Hernie. *Jahrbücher für Wissenschaftliche Botanik*, **11**, 548–574.

Worrall, J.J., Anagnost, S.E. & Zabel, R.A. (1997). Comparison of wood decay among diverse lignicolous fungi. *Mycologia*, **89**, 199–219.

Wu, C.-G. & Kimbrough, J.W. (1992). Ultrastructural studies of ascosporogenesis in *Ascobolus immersus*. *Mycologia*, **84**, 459–466.

Wu, C.-G. & Kimbrough, J.W. (1993). Ultrastructure of ascospore ontogeny in *Aleuria*, *Octospora* and *Pulvinella* (Otideaceae, Pezizales). *International Journal of Plant Sciences*, **154**, 334–349.

Wu, C.-G. & Kimbrough, J.W. (2001). Ultrastructural studies of ascosporogenesis in *Ascobolus stictoideus* (Pezizales, Ascomycetes). *International Journal of Plant Sciences*, **162**, 71–102.

Wubah, D.A., Fuller, M.S. & Akin, D.E. (1991). Isolation of monocentric and polycentric fungi from the rumen and faeces of cows. *Canadian Journal of Botany*, **69**, 1232–1236.

Wyand, R.A. & Brown, J.K.M. (2003). Genetic and *forma specialis* diversity in *Blumeria graminis* of cereals and its implications for host–pathogen co-evolution. *Molecular Plant Pathology*, **4**, 187–198.

Wynn, A.R. & Epton, H.A.S. (1979). Parasitism of oospores of *Phytophthora erythroseptica* in soil. *Transactions of the British Mycological Society*, **73**, 255–259.

Xu, J., Kerrigan, R.W., Callac, P., Horgen, P.A. & Anderson, J.B. (1997). Genetic structure of natural populations of *Agaricus bisporus*, the commercial button mushroom. *Journal of Heredity*, **88**, 482–488.

Xu, J., Boyd, C.M., Livingston, E., *et al.* (1999). Species and genotypic diversities and similarities of pathogenic yeasts colonizing women. *Journal of Clinical Microbiology*, **37**, 3835–3843.

Xu, J., Vilgalys, R. & Mitchell, T.G. (2000). Multiple gene genealogies reveal recent dispersion and hybridization in the human pathogenic fungus *Cryptococcus neoformans*. *Molecular Ecology*, **9**, 1471–1481.

Xu, J.-R. (2000). MAP kinases in fungal pathogens. *Fungal Genetics and Biology*, **31**, 137–152.

Xu, J.-R. & Hamer, J.E. (1996). MAP kinase and cAMP signaling regulate infection structure formation and pathogenic growth in the rice blast fungus *Magnaporthe grisea*. *Genes and Development*, **10**, 2696–2706.

Xu, J.-T. & Mu, C. (1990). The relation between growth of *Gastrodia elata* protocorms and fungi. *Acta Botanica Sinica*, **32**, 26–31.

Xu, X. M. & Robinson, D. J. (2000). Epidemiology of brown rot (*Monilinia fructigena*) on apple: infection of fruits by conidia. *Plant Pathology*, **49**, 201–206.

Yamakazi, Y. & Ootaki, T. (1996). Roles of extracellular fibrils connecting progametangia in mating of *Phycomyces blakesleeanus*. *Mycological Research*, **100**, 984–988.

Yan, A. S. (1998). Contamination sources of *Pyronema domesticum* on Chinese cotton-based medical products. *Radiation Physics and Chemistry*, **52**, 7–9.

Yarrow, D. (1998). Methods for the isolation, maintenance and identification of yeasts. In *The Yeasts: A Taxonomic Study*, 4th edn, ed. C. P. Kurtzman & J. W. Fell. Amsterdam: Elsevier, pp. 77–100.

Yarwood, C. E. (1941). Diurnal cycle of ascus maturation of *Taphrina deformans*. *American Journal of Botany*, **28**, 355–357.

Ye, X. S., Lee, S.-L., Wolkow, T. D., *et al.* (1999). Interaction between developmental and cell cycle regulators is required for morphogenesis in *Aspergillus nidulans*. *EMBO Journal*, **18**, 6994–7001.

Yendol, W. G. & Pashke, J. D. (1965). Pathology of an *Entomophthora* infection in the Eastern subterranean termite *Reticulitermes flavipes* (Kollar). *Journal of Invertebrate Pathology*, **7**, 414–422.

Yokochi, T., Honda, D., Higashihara, T. & Nakahara, T. (1998). Optimization of docosahexaenoic acid production by *Schizochytrium limacinum* SR21. *Applied Microbiology and Biotechnology*, **49**, 72–76.

Yoon, K. S. & McLaughlin, D. J. (1979). Formation of the hilar appendix in basidiospores of *Boletus rubinellus*. *American Journal of Botany*, **66**, 870–873.

Yoon, K. S. & McLaughlin, D. J. (1984). Basidiosporogenesis in *Boletus rubinellus*. I. Sterigmal initiation and early spore development. *American Journal of Botany*, **71**, 80–90.

Yoon, K. S. & McLaughlin, D. J. (1986). Basidiosporogenesis in *Boletus rubinellus*. II. Late spore development. *Mycologia*, **78**, 185–197.

Yoon, S. I., Kim, S. Y., Lim, Y. W. & Jung, H. S. (2003). Phylogenetic evaluation of stereoid fungi. *Journal of Microbiology and Biotechnology*, **13**, 406–414.

Yoshikawa, M. & Masago, H. (1977). Effect of cyclic AMP in catabolite-repressed zoosporangial formation in *Phytophthora capsici*. *Canadian Journal of Botany*, **55**, 840–843.

Youatt, J. (1977). Chemical nature of *Allomyces* walls. *Transactions of the British Mycological Society*, **69**, 187–190.

Young, E. L. (1943). Studies on *Labyrinthula*. The etiological agent of the wasting disease of eelgrass. *American Journal of Botany*, **30**, 586–593.

Yu, K. W., Suh, H. J., Bae, S. H., *et al.* (2001). Chemical properties and physiological activities of stromata of *Cordyceps militaris*. *Journal of Microbiology and Biotechnology*, **11**, 266–274.

Yu, L., Lee, K. K., Ens, K., *et al.* (1994). Partial characterization of a *Candida albicans* fimbrial adhesin. *Infection and Immunity*, **62**, 2834–2842.

Yu, M. Q. & Ko, W. H. (1996). Mating-type segregation in *Choanephora cucurbitarum* following sexual reproduction. *Canadian Journal of Botany*, **74**, 919–923.

Yu, M. Q. & Ko, W. H. (1999). Azygospore formation following protoplast fusion in *Choanephora cucurbitarum*. *Mycological Research*, **103**, 684–688.

Yuan, G. F. & Yong, S. C. (1984). A new obligate azygosporic species of *Rhizopus*. *Mycotaxon*, **20**, 397–400.

Yuan, X., Xiao, S. & Taylor, T. N. (2005). Lichen-like symbiosis 600 million years ago. *Science*, **308**, 1017–1020.

Yuasa, K. & Hatai, K. (1996). Some biochemical characteristics of the genera *Saprolegnia*, *Achlya* and *Aphanomyces* isolated from fishes with fungal infection. *Mycoscience*, **37**, 477–479.

Yukawa, Y. & Tanaka, S. (1979). Scanning electron microscope observations on resting sporangia of *Plasmodiophora brassicae* in clubroot tissues after alcohol cracking. *Canadian Journal of Botany*, **57**, 2528–2532.

Yurlova, N. A., de Hoog, G. S. & Gerrits van den Ende, A. H. (1999). Taxonomy of *Aureobasidium* and allied genera. *Studies in Mycology*, **43**, 63–69.

Zadoks, J. C. & Bouwman, J. J. (1985). Epidemiology in Europe. In *The Cereal Rusts 2: Diseases, Distribution, Epidemiology, and Control*, ed. A. P. Roelfs & W. R. Bushnell. Orlando: Academic Press, pp. 329–369.

Zahari, P. & Shipton, W. A. (1988). Growth and sporulation responses of *Basidiobolus ranarum* to changes in environmental parameters. *Transactions of the British Mycological Society*, **91**, 141–148.

Zak, J. (1976). Pure culture synthesis of Pacific Madrone ectendomycorrhizae. *Mycologia*, **68**, 362–369.

Zambino, P. J. & Szabo, L. J. (1993). Phylogenetic relationships of selected cereal and grass rusts based on rDNA sequence analysis. *Mycologia*, **85**, 401–414.

Zambonelli, A., Rivetti, C., Percudani, R. & Ottonello, S. (2000). TuberKey: a DELTA-based tool for the description and interactive identification of truffles. *Mycotaxon*, **74**, 57–76.

Zeigler, R. S. (1998). Recombination in *Magnaporthe grisea*. *Annual Review of Phytopathology*, **36**, 249–275.

Zelmer, C. D. & Currah, R. S. (1995). Evidence for a fungal liaison between *Corallorhiza trifida* (Orchidaceae) and *Pinus contorta* (Pinaceae). *Canadian Journal of Botany*, **73**, 862–866.

Zelmer, C. D., Cuthbertson, L. & Currah, R. S. (1996). Fungi associated with terrestrial orchid mycorrhizas, seeds and protocorms. *Mycoscience*, **37**, 439–448.

Zemek, J., Marvanová, L., Kuniak, L. & Kadlecikova, B. (1985). Hydrolytic enzymes in aquatic hyphomycetes. *Folia Microbiologica*, **30**, 363–372.

Zentmyer, G. A. (1980). *Phytophthora cinnamomi* and the diseases it causes. Monograph No. 10. St. Paul, USA: APS Press.

Zeyen, R. J., Carver, T. L. W. & Lyngkjaer, M. F. (2002). Epidermal cell papillae. In *The Powdery Mildews. A Comprehensive Treatise*, ed. R. R. Bélanger, W. R. Bushnell, A. J. Dik & T. L. W. Carver. St. Paul, USA: APS Press, pp. 107–125.

Zhan, J., Mundt, C. C. & McDonald, B. A. (2001). Using restriction fragment length polymorphisms to assess temporal variation and estimate the number of ascospores that initiate epidemics in field populations of *Mycosphaerella graminicola*. *Phytopathology*, **91**, 1011–1017.

Zhang, G. & Berbee, M. L. (2001). *Pyrenophora* phylogenetics from ITS and glyceraldehyde-3-phosphate dehydrogenase gene sequences. *Mycologia*, **93**, 1048–1063.

Zhang, L., Baasiri, R. A. & van Alfen, N. K. (1998). Viral repression of fungal pheromone precursor gene expression. *Molecular and Cellular Biology*, **18**, 953–959.

Zhang, N. & Blackwell, M. (2001). Molecular phylogeny of dogwood anthracnose fungus (*Discula destructiva*) and the Diaporthales. *Mycologia*, **93**, 355–365.

Zhang, Q., Griffith, J. M. & Grant, B. R. (1992). Role of phosphatidic acid during differentiation of *Phytophthora palmivora* zoospores. *Journal of General Microbiology*, **138**, 451–459.

Zhang, W., Dick, W. A. & Hoitink, H. A. J. (1996). Compost-induced systemic acquired resistance in cucumber to *Pythium* root rot and anthracnose. *Phytopathology*, **86**, 1066–1070.

Zheng, R. Y. & Chen, G. Q. (2001). A monograph of *Cunninghamella*. *Mycotaxon*, **80**, 1–75.

Zhou, F. S., Zhang, Z. G., Gregersen, P. L., *et al.* (1998). Molecular characterization of the oxalate oxidase involved in the response of barley to the powdery mildew fungus. *Plant Physiology*, **117**, 33–41.

Zhou, X.-L., Stumpf, M. A., Hoch, H. C. & Kung, C. (1991). A mechanosensitive channel in whole cells and in membrane patches of the fungus *Uromyces*. *Science*, **253**, 1415–1417.

Zlotnik, H., Fernandez, M. P., Bowers, B. & Cabib, E. (1984). *Saccharomyces cerevisiae* mannoproteins form an external cell wall layer that determines wall porosity. *Journal of Bacteriology*, **159**, 1018–1026.

Zoberi, M. H. (1985). Propagule liberation in the Mucorales. *Botanical Journal of the Linnean Society*, **91**, 167–173.

Zugmaier, W., Bauer, R. & Oberwinkler, F. (1994). Mycoparasitism of some *Tremella* species. *Mycologia*, **86**, 49–56.

Zuppinger, C. & Roos, U.-P. (1997). Cell shape, motility and distribution of F-actin in amoebae of the mycetozoans *Protostelium mycophaga* and *Acrasis rosea*. A comparison with *Dictyostelium discoideum*. *European Journal of Protistology*, **33**, 396–408.

Zureik, M., Neukirch, C., Leynaert, B., *et al.* (2002). Sensitisation to airborne moulds and severity of asthma: cross sectional study from European Community respiratory health survey. *British Medical Journal*, **325**, 411–414.

Zycha, H. & Siepmann, R. & Linnemann, G. (1969). *Mucorales*. Lehre, Germany: J. Cramer.

Index

Page numbers with images are underlined, those with explanations of concepts are printed in bold.